COLLEGE PHYSICS

Paul A. Tipler

WORTH PUBLISHERS, INC.

For Becky and Ruth

COLLEGE PHYSICS

Paul A. Tipler

Copyright © 1987 by Worth Publishers, Inc.

All rights reserved.

Printed in the United States of America.

Library of Congress Catalog Card No. 86-062908

ISBN: 0-87901-268-4

First Printing, February 1987

Editors: Anne Vinnicombe and June Fox

Production and Makeup: Sarah Segal and Patricia Lawson

Design: Malcolm Grear Designers

Picture editor: David Hinchman

Composition: Progressive Typographers, Inc.

Printing and binding: W. A. Krueger Company

Cover: Sidewalk reflections (© *Pete Saloutos 1983/Stock Market.*)

Illustration credits appear on pp. 918–921, which constitute an extension of the copyright page.

Worth Publishers, Inc.

33 Irving Place

New York, NY 10003

The Greek Alphabet

Alpha	A	α	Iota	I	ι	Rho	P	ρ
Beta	B	β	Kappa	K	κ	Sigma	Σ	σ
Gamma	Γ	γ	Lambda	Λ	λ	Tau	T	τ
Delta	Δ	δ	Mu	M	μ	Upsilon	Y	υ
Epsilon	E	ϵ	Nu	N	ν	Phi	Φ	ϕ
Zeta	Z	ζ	Xi	Ξ	ξ	Chi	X	χ
Eta	H	η	Omicron	O	o	Psi	Ψ	ψ
Theta	Θ	θ	Pi	Π	π	Omega	Ω	ω

Abbreviations for Units

A	ampere	lb	pound
atm	atmosphere	m	metre
C	coulomb	MeV	megaelectronvolts
°C	degree Celsius	Mm	megametre (10^6 m)
cal	calorie	mi	mile
cm	centimetre	min	minute
dyn	dyne	mm	millimetre
eV	electronvolt	ms	millisecond
F	farad	N	newton
°F	degree Fahrenheit	nm	nanometre (10^{-9} m)
fm	femtometre, fermi (10^{-15} m)	pf	picofarad
		pm	picometre (10^{-12} m)
ft	foot	pt	pint
G	gauss	qt	quart
g	gram	rev	revolution
Gm	gigametre (10^9 m)	s	second
H	henry	T	tesla
h	hour	u	unified mass unit
Hz	hertz	V	volt
in	inch	W	watt
J	joule	Wb	weber
K	kelvin	y	year
keV	kiloelectronvolts	yd	yard
kg	kilogram	μC	microcoulomb
km	kilometre	μm	micrometre (10^{-6} m)
L	litre	μs	microsecond
		Ω	ohm

Preface

I have written this textbook for the two-semester algebra-based elementary physics course taken by students majoring in biology, environmental sciences, health sciences, premedicine, and other fields. Algebra and a small amount of trigonometry are used, but calculus is avoided. The usual topics in physics are covered in the traditional order that will fit most college courses: mechanics (Chapters 2 to 10), thermodynamics (Chapters 11 to 14), vibrations and waves including sound (Chapters 15 to 17), electricity and magnetism (Chapters 18 to 23), optics (Chapters 24 to 27), and modern physics (Chapters 28 to 33). Although the sequence of topics is standard, there are some features that are not often found in other books. A section on thermal energy and metabolic rate is included in Chapter 6 on work and energy to relate this subject to the familiar experience of students. Chapter 7 (Impulse, Momentum, and Center of Mass) includes a section on jet propulsion with a qualitative description of rocket motion. There is an entire chapter on gravity (Chapter 9), with a discussion of satellite motion and the problem of escaping the earth. Viscous flow is discussed in Chapter 10, and the equations for fluid flow, heat conduction, and electrical conduction are written in the same form so as to bring out their similarities. The chapter on the second law of thermodynamics (Chapter 14) relates entropy to the loss of available energy and to disorder and probability. The section on vibrations and waves (Chapters 15 to 17) follows thermodynamics, where it would naturally conclude the first semester. However, this material could easily be combined with the section on optics (Chapter 24 to 27) and taught either before or after the section on electricity and magnetism if desired.

In recent years there have been many textbooks with special sections or chapters that emphasize the application of physics to the life sciences. This is not one of them. I find that the physical principles are often lost in the maze of applications. I believe that these students are best served by a clear exposition of the principles of physics with many worked-out examples. I discuss applications of these principles to other fields of science or to everyday experience at the beginning of a section to motivate the learning of these principles or at the end of the section to illustrate their wide applicability.

In writing this text I have taken into account the important differences between the students taking this course and those taking the calculus-based course. The first is that most of these students will not take any more physics; in particular, they will not take a separate course in modern physics. For this reason, a significant part of the text is devoted to topics in modern physics. Secondly, these students have less need for, and very much less interest in, the detailed development of physics results. This does not mean that all derivations are omitted. It does mean, however, that more emphasis is placed on the plausibility of a result and its applications than on a mathematical derivation. Often a result is stated, its plausibility is discussed, and examples of applications are shown before a simple derivation is given. I have not hesitated to use special situations to "derive" an important result, or simply to state a plausible result without any derivation. Qualitative arguments are given wherever possible, and detailed mathematical analysis is avoided. Finally, many of these students are very anxious about using mathematics and solving problems. To help these students overcome their anxiety, problem solving is discussed with worked-out examples in Chapter 1, and again in Chapter 4 in conjunction with using Newton's laws. There is also a detailed review of the mathematics needed for the course in Appendix A.

One of my goals in teaching this course has been that students should enjoy their experience in learning physics. To make the text more fun, I have included chapter-opening quotations, many photographs (including movie stills), and cartoons. In addition, extensive use of lecture demonstrations is highly recommended both to illustrate the physics and to add excitement to the classroom. (To aid instructors, the *Instructor's Resource and Solutions Manual* includes chapter-by-chapter suggestions for classroom demonstrations.)

Within each chapter, there are questions at the ends of sections. Some are routine and can easily be answered from the material in the preceding section. Others are open-ended and can serve as a basis for classroom discussion. At the end of each chapter is a review section. This includes learning objectives, which list the information and skills that should have been learned from reading the chapter; a checklist of words and phrases that students are asked to "define, explain, or otherwise identify"; and a set of true-false questions. The page number where the definition of the term is given appears after each item in the checklist, and answers to all the true-false questions are given in the back of the book. Following the review section, there is an extensive set of exercises organized by the sections within the chapter. These exercises are not difficult, and each involves material in that section of the chapter only. The exercises are followed by a shorter set of problems. The problems tend to be more difficult than the exercises and require the student to assimilate material from the entire chapter. Answers to all the odd-numbered exercises and problems are given at the back of the book. (Detailed solutions are given to all the exercises and problems in the *Instructor's Resource and Solutions Manual.*) I suggest the assignment of many more exercises than problems; the exercises are designed to build the student's self-confidence, which can be easily destroyed by too many challenging problems. It would not be unreasonable to assign only exercises in the required homework with a few optional problems to challenge the better students.

Fifteen essays are presented for the enjoyment and enlightenment of students and instructors. Three are biographical (on Isaac Newton, Albert Einstein, and Benjamin Franklin), many are on applications (for example, xerography, lasers, and nerve impulse conduction), and others are on topics of current interest (for example, thermal pollution, energy resources, and the Voyager exploration of the solar system.) There are no exercises or problems associated with the essays.

For the sake of completeness and flexibility, the book contains more material than can be covered in two semesters. Some sections or chapters must therefore be omitted or skimmed over. One method for broadening a course (while sacrificing some of the depth) is to assign a chapter or section to be read without assigning the associated exercises. I would recommend this to those who are tempted to omit Section 15-5 on damped and driven oscillators because the important ideas of resonance should be presented to students as early as possible. The inclusion or omission of certain topics naturally depends on the judgment of the instructor and the particular needs of the students. For a two-semester course, I would suggest the omission of some of the following: Sections 5-5 (Drag Forces), 6-7 (Thermal Energy and Metabolic Rate), 7-5 (Jet Propulsion), 8-4 (Rolling Bodies), 8-5 (Motion of a Gyroscope), 9-5 (Escaping the Earth), 9-6 (Potential Energy, Total Energy, and Orbits), 10-5 (Surface Tension and Capillarity), 10-6 (Fluids in Motion and Bernoulli's Equation), 10-7 (Viscous Flow), 13-3 (The van der Waals Equation and Liquid-Vapor Isotherms), 13-4 (Humidity), 14-5 (The Heat Pump), 14-6 (Entropy and Disorder), 14-7 (Entropy and Probability), 17-5 (Harmonic Analysis and Synthesis), 18-5 (Gauss' Law), 18-6 (Electric Dipoles in Electric Fields), 20-6 (RC Circuits), 21-5 (Ampere's Law), 21-6 (Current Loops, Solenoids, and Magnets), 21-7 (Magnetism in Matter), 22-2 (Motional emf), 22-5 (LR Circuits and Magnetic Energy Density), 24-6 (Polarization), 27-7 (Diffraction Gratings), 29-4 (Compton Scattering), 31-4 (Molecular Bonding), 31-5 (Molecular Spectra), 32-3 (Nuclear Reactions), and Chapters 23 (Alternating Current Circuits), 26 (Optical Instruments), and 28 (Relativity).

Acknowledgements

Many people have contributed to this book. I would like to thank all those who have reviewed the various drafts and have offered many helpful suggestions. In particular, Roger Clapp (University of South Florida) not only reviewed the entire manuscript, but suggested many new examples and figures, and Gilbert H. Ward (Pennsylvania State University) reviewed the entire manuscript, suggested new exercises, and wrote the chapter introductions and marginal comments. Robin Macqueen (a graduate student at the University of British Columbia) contributed the suggestions for further readings for each chapter. John Hanneken (Memphis State University), Jeffry Mallow (Loyola University of Chicago), and Jack Prince (Bronx Community College) reviewed the entire second draft of the manuscript and contributed many useful suggestions. Many of the chapter-opening quotations were found by Donald Goldsmith who also contributed two of the essays. Stanley Shepherd (Pennsylvania State University) checked all of the exercises and problems and wrote the solutions, while Richard B. Minnix and D. Rae Carpenter, Jr. (both of Virginia Military Institute) prepared the chapter-by-chapter suggestions for classroom demonstrations for the *Instructor's Resource and Solutions Manual*. The student *Study Guide* was expertly written by Granvil C. Kyker (Rose-Hulman Institute of Technology) and all the exercises and problems in both the study guide and text were worked by Murray Kelley (a student at Rose-Hulman Institute of Technology). I am also grateful to the individuals listed below who reviewed various parts of the manuscript:

Naushad Ali, *University of Southern Illinois, Carbondale*

Robert P. Bauman, *University of Alabama, Birmingham*

James R. Benbrook, *University of Houston-University Park*

Joseph J. Boyle, *Miami-Dade Community College*

Ron Brown, *California Polytechnic State University, San Luis Obispo*

Fernando A. Díaz, *Universidad Metropolitana, Puerto Rico*

Lewis Ford, *Texas A&M University*

Philip J. Green, *Texas A&M University*

Robert Hart, *State University of New York, Binghamton*

Betty Howard, *University of British Columbia*

Gordon E. Jones, *Mississippi State University*

Floyd D. Lee, *Montana State University*

Lloyd Makarowitz, *State University of New York, Farmingdale*

A. Scott McRobbie, *State University of New York, Potsdam*

David S. Mills, *College of the Redwoods*

Richard A. Morrow, *University of Maine, Orono*

W. H. Potter, *University of California-Davis*

Dennis K. Ross, *Iowa State University*

John O. Thomson, *University of Tennessee, Knoxville*

I would like to thank the staff at Basis, Inc., in Berkeley, California, for their help with the word processing, and particularly Luap Relpit who did much of the typing. Finally, I would like to thank everyone at Worth Publishers for their help and encouragement — in particular, June Fox who developed and vastly improved my original manuscript and Anne Vinnicombe who worked tirelessly to turn this manuscript into a beautiful book. The considerable abilities and perseverance of these talented editors made this book possible.

Berkeley, California
February 1987

Paul Tipler

Contents in Brief

	To the Student	xiv
CHAPTER 1	Introduction	1

Part I Mechanics

CHAPTER 2	Motion in One Dimension	15
CHAPTER 3	Motion in Two and Three Dimensions	36
CHAPTER 4	Newton's Laws	58
CHAPTER 5	Applications of Newton's Laws	87
CHAPTER 6	Work and Energy	113
CHAPTER 7	Impulse, Momentum, and Center of Mass	147
CHAPTER 8	Rotation	173
CHAPTER 9	Gravity	198
CHAPTER 10	Solids and Fluids	221

Part II Thermodynamics

CHAPTER 11	Temperature	253
CHAPTER 12	Heat and the First Law of Thermodynamics	270
CHAPTER 13	Thermal Properties and Processes	292
CHAPTER 14	The Availability of Energy	324

Part III Vibrations and Waves

CHAPTER 15	Oscillations	352
CHAPTER 16	Mechanical Waves: Sound	373
CHAPTER 17	Interference, Diffraction, and Standing Waves	400

Part IV Electricity and Magnetism

CHAPTER 18	Electric Fields and Forces	423
CHAPTER 19	Electrostatics	453
CHAPTER 20	Electric Current and Circuits	488
CHAPTER 21	The Magnetic Field	529
CHAPTER 22	Magnetic Induction	562
CHAPTER 23	Alternating Current Circuits	583

Part V Optics

CHAPTER 24	Light	605
CHAPTER 25	Geometric Optics	639
CHAPTER 26	Optical Instruments	667
CHAPTER 27	Physical Optics: Interference and Diffraction	685

Part VI Modern Physics

CHAPTER 28	Relativity	711
CHAPTER 29	The Origins of Quantum Theory	746
CHAPTER 30	Electron Waves and Quantum Theory	771
CHAPTER 31	Atoms, Molecules, and Solids	789
CHAPTER 32	Nuclear Physics	833
CHAPTER 33	Elementary Particles	864

APPENDIX A	Review of Mathematics	882
APPENDIX B	SI Units	897
APPENDIX C	Numerical Data	898
APPENDIX D	Conversion Factors	901
APPENDIX E	Trigonometric Tables	902
APPENDIX F	Periodic Table of the Elements	903

Answers to Odd-Numbered Exercises and Problems	904
Illustration Credits	918
Index	922

Contents

To the Student xiv

CHAPTER 1   **Introduction**   1

1-1   Units   2

1-2   Conversion of Units   4

1-3   Dimensions of Physical Quantities   6

1-4   Scientific Notation   6

1-5   Significant Figures   8

1-6   Problem Solving   10

Summary, Suggestions for Further Reading, Review, Exercises, Problems   12

Part I Mechanics

CHAPTER 2   **Motion in One Dimension**   15

2-1   Speed, Displacement, and Velocity   17

2-2   Instantaneous Velocity   20

2-3   Acceleration   22

2-4   Motion with Constant Acceleration   24

Summary, Suggestions for Further Reading, Review, Exercises, Problems   29

CHAPTER 3   **Motion in Two and Three Dimensions**   36

3-1   The Displacement Vector   36

3-2   Addition of Vectors by Components   38

3-3   Velocity and Acceleration Vectors   41

3-4   Projectile Motion   44

3-5   Circular Motion   49

Summary, Suggestions for Further Reading, Review, Exercises, Problems   52

CHAPTER 4   **Newton's Laws**   58

4-1   Force and Mass   60

4-2   The Force Due to Gravity: Weight   63

4-3   Units of Force and Mass   64

4-4   Newton's Third Law   66

4-5   Springs, Strings, and Support Forces   68

4-6   Applications to Problem Solving: Constant Forces   69

Essay: Isaac Newton (1642–1727) *I. Bernard Cohen*   76

Summary, Suggestions for Further Reading, Review, Exercises, Problems   79

CHAPTER 5   **Applications of Newton's Laws**   87

5-1   Friction   88

5-2   Static Equilibrium of an Extended Body   92

5-3   Stability of Equilibrium and Balance   98

5-4   Circular Motion   100

5-5   Drag Forces   104

         Summary, Suggestions for
         Further Reading, Review,
         Exercises, Problems   106

CHAPTER 6   **Work and Energy**   **113**

6-1   Work and Kinetic Energy   114

6-2   Work Done by a Variable Force   119

6-3   Potential Energy   123

6-4   Conservative Forces and the
         Conservation of Energy   125

6-5   Simple Machines   132

6-6   Power   135

Essay: **Energy Resources**
         Laurent Hodges   136

6-7   Thermal Energy and Metabolic
         Rate   138

         Summary, Suggestions for
         Further Reading, Review,
         Exercises, Problems   139

CHAPTER 7   **Impulse, Momentum, and
            Center of Mass**   **147**

7-1   Impulse and Momentum   148

7-2   Conservation of Momentum   150

7-3   Center of Mass   153

7-4   Collisions   156

7-5   Jet Propulsion   165

         Summary, Suggestions for
         Further Reading, Review,
         Exercises, Problems   167

CHAPTER 8   **Rotation**   **173**

8-1   Angular Velocity   174

8-2   Torque and Moment of Inertia   177

8-3   Kinetic Energy and Angular
         Momentum   181

8-4   Rolling Bodies   186

8-5   Motion of a Gyroscope   188

         Summary, Suggestions for
         Further Reading, Review,
         Exercises, Problems   191

CHAPTER 9   **Gravity**   **198**

9-1   Kepler's Laws   199

9-2   Newton's Law of Gravity   202

9-3   The Cavendish Experiment   207

9-4   Gravitational and Inertial Mass   208

9-5   Escaping the Earth   209

9-6   Potential Energy, Total Energy,
         and Orbits   212

Essay: **Comets and Cosmic Archaeology**
         Donald Goldsmith   214

         Summary, Suggestions for
         Further Reading, Review,
         Exercises, Problems   217

CHAPTER 10 **Solids and Fluids**   **221**

10-1   Density   221

10-2   Stress and Strain   223

10-3   Pressure in a Fluid   227

10-4   Bouyancy and Archimedes'
         Principle   232

10-5   Surface Tension and Capillarity   236

10-6   Fluids in Motion and
         Bernoulli's Equation   239

10-7   Viscous Flow   244

         Summary, Suggestions for
         Further Reading, Review,
         Exercises, Problems   246

Part II Thermodynamics

CHAPTER 11 **Temperature**   **253**

11-1   The Celsius and Fahrenheit
         Temperature Scales   254

11-2   The Absolute Temperature Scale   256

11-3   The Ideal Gas Law   259

11-4   The Molecular Interpretation of
         Temperature   263

         Summary, Suggestions for
         Further Reading, Review,
         Exercises, Problems   266

CHAPTER 12 **Heat and the First Law of
            Thermodynamics**   **270**

12-1   Heat Capacity and Specific Heat   271

12-2   The First Law of
         Thermodynamics   275

12-3   Work and the *PV* Diagram for
         a Gas   279

12-4   Heat Capacities and the
         Equipartition Theorem   282

         Summary, Suggestions for
         Further Reading, Review,
         Exercises, Problems   287

CHAPTER 13 **Thermal Properties and
            Processes**   **292**

13-1   Thermal Expansion   292

13-2   Change of Phase and Latent Heat   296

13-3   The van der Waals Equation and Liquid-Vapor Isotherms   300

13-4   Humidity   302

13-5   The Transfer of Heat   304

Essay: Solar Energy
Laurent Hodges   314

Summary, Suggestions for Further Reading, Review, Exercises, Problems   318

CHAPTER 14   **The Availability of Energy**   **324**

14-1   Heat Engines and the Kelvin-Planck Statement of the Second Law of Thermodynamics   325

14-2   Refrigerators and the Clausius Statement of the Second Law of Thermodynamics   329

14-3   Equivalence of the Kelvin-Planck and Clausius Statements   330

14-4   The Carnot Engine   331

Essay: Power Plants and Thermal Pollution
Laurent Hodges   336

14-5   The Heat Pump   338

14-6   Entropy and Disorder   340

14-7   Entropy and Probability   344

Summary, Suggestions for Further Reading, Review, Exercises, Problems   346

Part III Vibrations and Waves

CHAPTER 15   **Oscillations**   **352**

15-1   Simple Harmonic Motion: A Mass on a Spring   353

15-2   Simple Harmonic Motion and Circular Motion   358

15-3   Energy in Simple Harmonic Motion   360

15-4   The Simple Pendulum   362

15-5   Damped and Driven Oscillations   364

Summary, Suggestions for Further Reading, Review, Exercises, Problems   367

CHAPTER 16   **Mechanical Waves: Sound**   **373**

16-1   Wave Pulses   374

16-2   Speed of Waves   378

16-3   Harmonic Waves   380

16-4   The Doppler Effect   384

16-5   Energy and Intensity   389

Essay: Sonic Booms
Laurent Hodges   393

Summary, Suggestions for Further Reading, Review, Exercises, Problems   395

CHAPTER 17   **Interference, Diffraction, and Standing Waves**   **400**

17-1   Interference of Waves from Two Point Sources   401

17-2   Diffraction   404

17-3   Beats   407

17-4   Standing Waves   409

17-5   Harmonic Analysis and Synthesis   416

Summary, Suggestions for Further Reading, Review, Exercises, Problems   419

Part IV Electricity and Magnetism

CHAPTER 18   **Electric Fields and Forces**   **423**

18-1   Electric Charge   424

18-2   Coulomb's Law   427

18-3   The Electric Field   431

18-4   Lines of Force   436

18-5   Gauss' Law   440

18-6   Electric Dipoles in Electric Fields   442

Essay: Benjamin Franklin (1706–1790)
I. Bernard Cohen   445

Summary, Suggestions for Further Reading, Review, Exercises, Problems   448

CHAPTER 19   **Electrostatics**   **453**

19-1   Electric Potential and Potential Difference   454

19-2   Electric Conductors   460

19-3   Equipotential Surfaces, Charge Sharing, and Dielectric Breakdown   464

19-4   Capacitance   469

19-5   Combinations of Capacitors   474

19-6   Electrical Energy Storage   478

Contents

Essay: Electrostatics and Xerography
         Richard Zallen   480

         Summary, Suggestions for
         Further Reading, Review,
         Exercises, Problems   482

CHAPTER 20   **Electric Current and Circuits**   **488**
20-1   Current and Motion of Charges   489
20-2   Ohm's Law and Resistance   490
Essay: Electrical Conduction in Nerve
         Cells
         Stephen Woods   496
20-3   Energy in Electric Circuits   500
20-4   Combinations of Resistors   505
20-5   Kirchhoff's Rules   510
20-6   *RC* Circuits   515
20-7   Ammeters, Voltmeters, and
         Ohmmeters   518

         Summary, Suggestions for
         Further Reading, Review,
         Exercises, Problems   521

CHAPTER 21   **The Magnetic Field**   **529**
21-1   Definition of the Magnetic Field   531
21-2   Torques on Magnets and
         Current Loops   535
21-3   Motion of a Point Charge in a
         Magnetic Field   539
21-4   Sources of the Magnetic Field   546
21-5   Ampere's Law   550
21-6   Current Loops, Solenoids, and
         Magnets   551
21-7   Magnetism in Matter   553

         Summary, Suggestions for
         Further Reading, Review,
         Exercises, Problems   556

CHAPTER 22   **Magnetic Induction**   **562**
22-1   Magnetic Flux and Faraday's
         Law   563
22-2   Motional emf   567
22-3   Eddy currents   569
22-4   Inductance   570
22-5   *LR* Circuits and Magnetic
         Energy Density   573
22-6   Generators and Motors   575

         Summary, Suggestions for
         Further Reading, Review,
         Exercises, Problems   578

CHAPTER 23   **Alternating Current Circuits**   **583**
23-1   Alternating Current in a Resistor   584
23-2   Alternating Current in
         Inductors and Capacitors   586
23-3   An *LCR* Circuit with a Generator   590
23-4   The Transformer   595
23-5   Rectification and Amplification   597

         Summary, Suggestions for
         Further Reading, Review,
         Exercises, Problems   600

Part V Optics

CHAPTER 24   **Light**   **605**
24-1   Waves or Particles   606
24-2   Electromagnetic Waves   609
24-3   The Speed of Light   613
24-4   Reflection   617
24-5   Refraction   619
24-6   Polarization   625
Essay: The New Age of Exploration:
         Voyagers to the Outer Solar
         System
         Donald Goldsmith   631

         Summary, Suggestions for
         Further Reading, Review,
         Exercises, Problems   634

CHAPTER 25   **Geometric Optics**   **639**
25-1   Plane Mirrors   640
25-2   Spherical Mirrors   642
25-3   Images Formed by Refraction   648
25-4   Thin Lenses   651
25-5   Aberrations   660

         Summary, Suggestions for
         Further Reading, Review,
         Exercises, Problems   661

CHAPTER 26   **Optical Instruments**   **667**
26-1   The Eye   668
26-2   The Simple Magnifier   672
26-3   The Camera   674
26-4   The Compound Microscope   677
26-5   The Telescope   678

         Summary, Suggestions for
         Further Reading, Review,
         Exercises, Problems   681

CHAPTER 27   Physical Optics: Interference
              and Diffraction   685

    27-1   Phase Difference and
              Coherence   686

    27-2   Interference in Thin Films   687

    27-3   The Michelson Interferometer   690

    27-4   The Two-Slit Interference
              Pattern   693

    27-5   Diffraction Pattern of a Single
              Slit   696

    27-6   Diffraction and Resolution   700

    27-7   Diffraction Gratings   703

              Summary, Suggestions for
              Further Reading, Review,
              Exercises, Problems   705

Part VI Modern Physics

CHAPTER 28   Relativity   711

    28-1   Newtonian Relativity   712

    28-2   The Michelson-Morley
              Experiment   714

    28-3   Einstein's Postulates and Their
              Consequences   717

    28-4   Clock Synchronization and
              Simultaneity   723

    Essay: Albert Einstein (1879–1955)
              Gerald Holton   728

    28-5   The Twin Paradox   730

    28-6   Relativistic Momentum   732

    28-7   Relativistic Energy   733

    28-8   General Relativity   737

              Summary, Suggestions for
              Further Reading, Review,
              Exercises, Problems   741

CHAPTER 29   The Origins of Quantum
              Theory   746

    29-1   Blackbody Radiation   748

    29-2   The Photoelectric Effect   750

    29-3   X-Rays   753

    29-4   Compton Scattering   756

    29-5   The Quantization of Atomic
              Energies: The Bohr Model   758

    Essay: X-rays and Medical Diagnosis
              John R. Cameron   763

              Summary, Suggestions for
              Further Reading, Review,
              Exercises, Problems   766

CHAPTER 30   Electron Waves and
              Quantum Theory   771

    30-1   The de Broglie Equations   772

    30-2   Electron Diffraction   775

    30-3   Wave-Particle Duality   777

    30-4   The Uncertainty Principle   778

    30-5   The Electron Wave Function   780

    30-6   A Particle in a Box   783

              Summary, Suggestions for
              Further Reading, Review,
              Exercises, Problems   785

CHAPTER 31   Atoms, Molecules, and Solids   789

    31-1   Quantum Theory of the
              Hydrogen Atom   791

    31-2   The Periodic Table   795

    31-3   Atomic Spectra   801

    31-4   Molecular Bonding   803

    31-5   Molecular Spectra   806

    31-6   Absorption, Scattering, and
              Stimulated Emission   810

    Essay: Lasers
              John R. Cameron   815

    31-7   Band Theory of Solids   817

    31-8   Semiconductor Junctions and
              Devices   820

    Essay: Electronics: Vacuum Tube to
              Solid State
              Larry C. Burton   823

              Summary, Suggestions for
              Further Reading, Review,
              Exercises, Problems   827

CHAPTER 32   Nuclear Physics   833

    32-1   Properties of Nuclei   834

    32-2   Radioactivity   840

    32-3   Nuclear Reactions   845

    32-4   Fission, Fusion, and Nuclear
              Reactors   848

    32-5   The Interaction of Particles
              with Matter   854

              Summary, Suggestions for
              Further Reading, Review,
              Exercises, Problems   859

CHAPTER **33** **Elementary Particles** **864**

33-1 Spin and Antiparticles 865

33-2 Hadrons and Leptons 867

33-3 The Conservation Laws 869

33-4 The Quark Model 871

33-5 Field Particles and Unified Theories 873

Essay: The Origin of the Universe
Donald Goldsmith 875

Summary, Suggestions for Further Reading, Review, Exercises, Problems 879

APPENDIX **A** **Review of Mathematics** **882**

APPENDIX **B** **SI Units** **897**

APPENDIX **C** **Numerical Data** **898**

APPENDIX **D** **Conversion Factors** **901**

APPENDIX **E** **Trigonometric Tables** **902**

APPENDIX **F** **Periodic Table of the Elements** **903**

Answers to Odd-Numbered Exercises and Problems **904**

Ilustration Credits **918**

Index **922**

To the Student

We have always been curious about the world around us. Since the beginnings of recorded thought, we have sought ways to impose order on the bewildering diversity of the events we observe. This search for order has taken a variety of forms: one is religion, another is art, and a third is science. Although the word *science* has its origins in a Latin verb meaning "to know," science has come to mean not merely knowledge but specifically knowledge of the natural world. Most importantly, science is a body of knowledge organized in a specific and rational way.

The roots of science are as deep as those of religion or art, but its traditions are much more modern. Only in the last few centuries have there been methods for studying nature systematically. They include techniques of observation, rules for reasoning and making predictions, the idea of planned experimentation, and ways for communicating experimental and theoretical results — all loosely referred to as the *scientific method*. An essential part of advances in our understanding of nature is the open communication of experimental results, theoretical calculations, speculations, and summaries of knowledge. A textbook is one form of open communication. An introductory textbook like this has two purposes. It is designed, first, to introduce newcomers to material that is already widely known in the scientific and technical community and that will form the basis of their more advanced studies. It also serves to acquaint students not majoring in science with information and a way of thinking that are having a cumulative effect upon our way of life.

Although we now think of science as being divided into several separate fields, this division occurred only in the last century or so. The separation of complex systems into smaller categories that can be more easily studied is one of the great successes of science in general. Biology, for example, is the study of living organisms. It can be further separated into zoology, botany, paleontology, macrobiology, microbiology, and so forth. Chemistry deals with the interaction of elements and compounds. Geology is the study of the earth. Astronomy is the study of the solar system, the stars and galaxies, and the universe as a whole. Physics deals with matter and energy, with the principles that govern the motion of particles and waves, with the interactions of particles, and with the properties of molecules, of atoms and atomic nuclei, and of larger scale systems such as gases, liquids, and solids.

Some consider physics the most fundamental science because it is the basis of biology, chemistry and all the other fields of science. Originally, all of these subjects were considered a single science called natural philosophy. Even today, with such a high degree of specialization, the boundaries between the various fields of science are not sharp, and we have many interdisciplinary fields such as biophysics, physical chemistry, biochemistry, and chemical physics.

In this book we shall study the usual subtopics of physics: mechanics, sound, light, heat, electricity, magnetism, atomic physics, and nuclear physics. These seemingly diverse subjects are unified by a few important laws and concepts, for example, Newton's laws of motion and the conservation of momentum and energy. The study of physics has wide application. Architects, nurses, physical therapists, engineers, doctors, musicians, and many others need to know about such subjects as heat transfer, fluid flow, sound waves, light, radioactivity, the balance of forces, and stresses in buildings or in bones. Physics also has wide applicability to questions in everyday life. Why does a car skid rounding a curve? In what sense are astronauts weightless? Why does sound travel around corners while light does not? Why does an oboe sound different from a flute playing the same note? Why do things appear larger through a magnifying glass? How does radioactive carbon dating work? Why is the sky blue? Why do metal objects feel colder than wood ones at the same temperature? These and countless other questions about our fascinating world can be answered from a basic knowledge of elementary physics.

A textbook is only one tool for learning physics. A good teacher, lecture demonstrations, films, and experimental work in the laboratory are also indispensible. Outside reading is highly recommended. While you are learning about the topics covered in your physics course, you should be broadening your familiarity with contemporary physics by reading widely among the many excellent popular and semipopular accounts of modern science such as those in *Scientific American.*

Like any other subject, physics has its own vocabulary that must be learned. Many common words such as *force, work,* and *momentum* have special meanings in physics. At the end of each chapter, there is a list of the new terms that have been introduced, allowing you to test yourself to see if you can identify them and explain their meaning.

An important part of learning physics is learning to solve problems. Most physics problems are stated in words but require the solving of mathematical equations. The only mathematics you will need to use for this text are algebra and trigonometry. A review of these subjects is given in Appendix A. Many students find the problem-solving aspect of physics a bit intimidating. Section 6 in Chapter 1 provides some hints and useful general techniques for working problems. More specific methods for solving problems using Newton's laws of motion are discussed in Section 6 of Chapter 4. A good way to gain practice in problem solving is to try to work through the examples in the text without referring to the solutions given. When you finish or get stuck, read through the worked example. Don't be discouraged if you have difficulty. Some of the examples introduce a new method of solution and some cover famous results and are an extension of the text. In these cases, you cannot expect to work the example without a great deal of thought and effort. At the end of each chapter are exercises and problems. (Chapters 1 and 33 have only exercises.) The exercises require material from just one section in the chapter and can usually be solved in a single step. If you have difficulty with an exercise, it is an indication that you should reread the corresponding section. The problems are much more challenging and often require assimilating material from throughout the chapter (and previous chapters) and applying it to a new situation. It is perfectly natural to have difficulty with many of the problems.

To supplement the exposition of basic principles, there are many essays throughout the book. These are for your pleasure and enlightenment and have no exercises or problems associated with them.

Your enjoyment of physics will be enhanced if you keep your eyes and ears open for examples in your everyday life of the physics you are learning. You should find many—the change in the sound of a passing car, the vibration of tree branches in the wind, the colored patterns in soap bubbles, the walking of water bugs on the surface of a pond, and so forth. See if you can explain the phenomena you observe using the principles of physics that you have learned. If you have questions, ask a fellow student or your instructor. One of the joys of learning physics is the greater awareness, understanding, and appreciation it brings of the world around us. I hope that this book will serve as your guide to that pleasure.

Berkeley, California
February 1987

Paul Tipler

CHAPTER ONE

Introduction

We all know of things that cannot be measured—the beauty of a flower, the sublimity of a Bach chorale, the love of a family. As certain as our knowledge of these things may be, however, we readily admit that it is not science. The ability not only to define but also to measure what is being considered is a requisite of science, and in physics, more than in any other field of knowledge, the precise definition of terms and the accurate measurement of quantities have led to great discoveries. In this course, numbers will be used constantly, but not advanced mathematics, contrary to what you may have heard. Students seldom find the mathematics used to be unfamiliar even though their mathematical skills may be a bit rusty.

In the first chapter, some preliminary chores—having to do mainly with establishing basic definitions—are taken care of, as is necessary at the beginning of any new endeavor, from taking up a new sport to learning to play a musical instrument. The fun and sense of accomplishment come later.

When you can measure what you are speaking about, and express it in numbers, you know something about it; but when you cannot measure it, when you cannot express it in numbers, your knowledge is of a meagre and unsatisfactory kind; it may be the beginning of knowledge, but you have scarcely, in your thoughts, advanced to the state of science.

WILLIAM THOMSON (LORD KELVIN)

The laws of physics express relationships among physical quantities such as length, time, force, energy, and temperature. The measurement of any specific physical quantity involves the comparison of it with some precisely defined unit value of the quantity. For example, to measure the distance between two points, we need a standard unit value of distance, such as a metrestick or a ruler. The statement that a certain distance is 25 metres means that it is 25 times the length of the unit metre; that is, a standard metrestick fits into the distance 25 times. It is important to include the unit "metre" along with the number "25" in expressing a distance because there are other units of distance, such as feet or miles, in common use. To say that a distance is "25" is meaningless. The magnitude of any physical quantity must include both a number and a unit.

We will begin our study of physics with a discussion of units and how to convert from one unit to another. We will then discuss the dimensions of physical quantities and how to express large or small numbers in scientific notation. We will conclude this introductory chapter with a discussion of problem solving.

1-1   Units

All physical quantities can be expressed in terms of a small number of fundamental units. For example, speed is expressed in terms of a unit of length and a unit of time, such as metres per second or miles per hour. Many of the concepts that we shall be studying, such as force, momentum, work, energy, and power, can be expressed in terms of three fundamental units of the quantities length, time, and mass. The choice of standard units for these basic quantities determines a system of units. In the system used universally in the scientific community, called the international system (SI),* the standard unit of length is the metre, the standard unit of time is the second, and the standard unit of mass is the kilogram.

The standard unit of length, the metre (abbreviated m), was originally indicated by two scratches on a bar made of a platinum-iridium alloy kept at the International Bureau of Weights and Measures in Sèvres, France. This length was chosen so that the distance between the equator and the North Pole along the meridian through Paris would be 10 million metres (see Figure 1-1). (After the construction of the standard metre bar, it was found that the actual distance differs from this figure by a few hundredths of a percent.) The standard metre is used to construct secondary standards that are used to calibrate measuring rods throughout the world. The standard metre is now defined in terms of the speed of light. The metre is the distance traveled by light in a vacuum during a time of $1/299{,}792{,}458$ second. (This makes the speed of light exactly $299{,}792{,}458$ m/s.) This new standard is the same length as the old, but it is much better because it is much more precise. There is also no need to keep a standard metre bar since the standard metre can be determined in any laboratory by measuring the speed of light.

The unit of time, the second (s), was originally defined in terms of the rotation of the earth as $\frac{1}{60} \times \frac{1}{60} \times \frac{1}{24}$ of the mean solar day. The second is now defined in terms of a characteristic frequency associated with the cesium atom. Every atom, when stimulated, emits light with wavelengths and frequencies characteristic of the particular element. There is a set of wavelengths and frequencies for each element, with a particular frequency and wavelength associated with each energy transition within the atom. As far as we know, these frequencies have remained the same since the beginning of time and will remain the same forever. The second is defined so that the frequency of the light of a certain energy transition in cesium is $9{,}192{,}631{,}770/\text{s}$.

The unit of mass, the kilogram (kg), which equals 1000 grams (g), is defined as the mass of a standard body, which is also kept at Sèvres. We shall discuss the concept of mass in detail in Chapter 4.

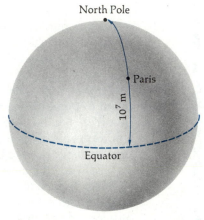

Figure 1-1
The SI standard of length, the metre, was originally chosen so that the distance from the equator to the North Pole along the meridian through Paris would be 10^7 m.

* SI stands for Système International. Complete definitions of all the SI units are given in Appendix B.

In our study of thermodynamics and electricity, we shall need three more fundamental physical units: the unit of temperature, the kelvin (K) (formerly called the degree Kelvin); the unit for the amount of a substance, the mole (mol); and the unit of current, the ampere (A). There is another fundamental unit, the candela (cd) for luminous intensity, which we shall have no occasion to use in this book. These seven fundamental units, the metre (m), the second (s), the kilogram (kg), the kelvin (K), the ampere (A), the mole (mol), and the candela (cd), constitute the international system of units (SI).

SI units

Cesium clock at the National Bureau of Standards.

Table 1-1

Prefixes for Powers of 10

Multiple	Prefix	Abbreviation
10^{18}	exa	E
10^{15}	peta	P
10^{12}	tera	T
10^{9}	giga	G
10^{6}	mega	M
10^{3}	kilo	k
10^{2}	hecto*	h
10^{1}	deka*	da
10^{-1}	deci*	d
10^{-2}	centi*	c
10^{-3}	milli	m
10^{-6}	micro	μ
10^{-9}	nano	n
10^{-12}	pico	p
10^{-15}	femto	f
10^{-18}	atto	a

* The prefixes hecto (h), deka (da), and deci (d) are not powers of 10^{3} or 10^{-3} and are rarely used. The other prefix which is not a power of 10^{3} or 10^{-3} is centi (c), now used only with the metre, as in 1 cm = 10^{-2} m.

The units of every physical quantity can be expressed in terms of these fundamental SI units. Some frequently used combinations are given special names. For example, the SI unit of force, $kg \cdot m/s^2$, is called a newton (N). Similarly, the SI unit of power, $kg \cdot m^2/s^3 = N \cdot m/s$, is called a watt (W).

Prefixes for the common multiples and submultiples of SI units are listed in Table 1-1. All these multiples are powers of 10. Such a system is called a *decimal system*. The decimal system based on the metre is called the *metric system*. These prefixes can be applied to any SI unit; for example, 0.001 second is 1 millisecond (ms), and 1,000,000 watts is 1 megawatt (MW).

Another decimal system still in use but gradually being replaced by the international system is the *cgs system*, based on the centimetre, gram, and second. The centimetre is now defined as 0.01 m and the gram as 0.001 kg. Originally, the gram was defined as the mass of one cubic centimetre of water. (The kilogram is then the mass of 1000 cubic centimetres or one litre of water). We shall generally work with SI units, giving the names of the corresponding cgs units and the appropriate conversion factors for any quantities you are likely to encounter in your studies.

In another system of units used in the United States, the *U.S. customary system*, a unit of force, the pound, is a fundamental unit. In this system, the unit of mass is defined in terms of the fundamental unit of force. (We shall discuss the relation between force and mass in detail in Chapter 4.) The pound (lb) is defined in terms of the gravitational attraction of the earth at a particular place for a standard body. The fundamental unit of length in this system is the foot (ft) and that of time is the second. The second is defined the same as in the international system. The foot is defined as exactly one-third of a yard (yd), which is now legally defined in terms of the metre:

$$1 \text{ yd} = 0.9144 \text{ m} \tag{1-1}$$

$$1 \text{ ft} = \tfrac{1}{3} \text{ yd} = 0.3048 \text{ m} \tag{1-2}$$

This makes the inch (in) exactly 2.54 cm. This system is not a decimal system. It is less convenient than the SI or other decimal systems because common multiples of a unit are not powers of 10. For example, 1 yd = 3 ft and 1 ft = 12 in. We shall see in Chapter 4 that mass is a better choice for a fundamental unit than is force because mass is an intrinsic property of an object independent of its location relative to the earth. The relations between the units of the U.S. customary system and SI units are given in Appendix D.

Roman-Greek tablet dating from A.D. 300–500 showing the length of the standard foot.

Questions

1. What are the advantages and disadvantages of using the length of a person's arm for a standard length?

2. A certain clock is consistently 10 percent fast compared with the standard cesium clock. A second clock varies in a random way by 1 percent. Which clock would make a more useful secondary standard for a laboratory? Why?

1-2   Conversion of Units

We have said that the magnitude of a physical quantity must include both a number and a unit. When such quantities are added, subtracted, multiplied, or divided in an algebraic equation, the unit can be treated like any other algebraic quantity. For example, suppose you wish to find the distance traveled in 3 hours (h) by a car moving at a constant rate of 80 kilometres per hour (km/h). The distance is the product of the speed, v, and the time, t:

$$x = vt = 80 \frac{\text{km}}{\text{h}} \times 3 \text{ h} = 240 \text{ km}$$

We cancel the unit of time, the hours, just as we would any algebraic quantity to obtain the distance in the proper unit of length, the kilometre. This method of treating units makes it easy to convert from one unit of distance to another.

Suppose we want to convert our answer of 240 km to miles (mi). We use the fact that

$$1 \text{ mi} = 1.61 \text{ km}$$

If we divide each side of this equation by 1.61 km, we obtain

$$\frac{1\ \text{mi}}{1.61\ \text{km}} = \frac{0.621\ \text{mi}}{1\ \text{km}} = 1$$

Since any quantity can be multiplied by 1 without changing its value, we can now change 240 km to miles by multiplying by the factor (1 mi)/(1.61 km):

$$240\ \text{km} = 240\ \cancel{\text{km}} \times \frac{1\ \text{mi}}{1.61\ \cancel{\text{km}}} = 149\ \text{mi}$$

We could, of course, obtain the same result by multiplying by the factor (0.621 mi)/(1 km). The factor (1 mi)/(1.61 km) or (0.621 mi)/(1 km) is called a *conversion factor*. All conversion factors have a value of 1 and are used to convert a quantity expressed in one unit of measure into its equivalent in another unit of measure. By writing out the units explicitly and then canceling them, we need not think about whether we should multiply by 1.61 or divide by 1.61 to change kilometres to miles because the units will tell us whether we have chosen the correct or incorrect factor.

Conversion factors

Example 1-1   What is the equivalent of 90 km/h in metres per second and in miles per hour?

We use the fact that 1000 m = 1 km, 60 s = 1 min, and 60 min = 1 h to convert to metres per second. The quantity 90 km/h is multiplied by a set of conversion factors each having the value 1 so that the value of the speed is not changed.

$$90\ \frac{\cancel{\text{km}}}{\cancel{\text{h}}} \times \frac{1000\ \text{m}}{1\ \cancel{\text{km}}} \times \frac{1\ \cancel{\text{h}}}{60\ \cancel{\text{min}}} \times \frac{1\ \cancel{\text{min}}}{60\ \text{s}} = 25\ \text{m/s}$$

To convert this speed into miles per hour, we use the fact that 1 mi = 1.61 km.

$$90\ \frac{\cancel{\text{km}}}{\text{h}} \times \frac{1\ \text{mi}}{1.61\ \cancel{\text{km}}} = 55.9\ \text{mi/h}$$

Converting feet to meters

Reprinted from *Never Eat Anything Bigger than Your Head and Other Drawings.*
New York: Workman Publishing Co. Copyright © 1976 by B. Kliban

1-3   Dimensions of Physical Quantities

The area of a surface is found by multiplying one length by another. For example, the area of a rectangle of sides 2 m and 3 m is $A = (2\ m)(3\ m) = 6\ m^2$. The units of this area are square metres. Because area is the product of two lengths, it is said to have the *dimensions* of length times length, or length squared, often written as L^2. The idea of dimensions is easily extended to other nongeometric quantities. For example, speed is said to have the dimensions length divided by time, or L/T. The dimensions of other quantities, such as force or energy, are written in terms of the fundamental quantities of length, time, and mass. Adding two physical quantities makes sense only if the quantities have the same dimensions. For example, we cannot add an area to a speed and obtain a meaningful sum. If we have an equation like

$$A = B + C$$

the quantities A, B, and C must all have the same dimensions. The addition of B and C also requires that these quantities be in the same units. For example, if B is an area of 500 in^2 and C is 4 ft^2, we must either convert B into square feet or C into square inches in order to find the sum of the two areas.

We can often find a mistake in a calculation by checking the dimensions or units of the quantities in the calculation. Suppose, for example, that we mistakenly use the formula $A = 2\pi r$ for the area of a circle. We can see immediately that this cannot be correct because the right side of the equation, $2\pi r$, has dimensions of length whereas area should have dimensions of length squared. For another example, consider a formula for distance x given by

$$x = vt + \tfrac{1}{2}at$$

where t is the time, v is the speed, and a is the acceleration, which (as we shall see) has dimensions of L/T^2. We can see by looking at the dimensions of each quantity that this formula cannot be correct. Since x has dimensions of length, each term on the right side of the equation must also have dimensions of length. The term vt has dimensions of length, but the dimensions of $\tfrac{1}{2}at$ are $(L/T^2)T = L/T$. Since this term does not have the correct dimensions, an error has been made somewhere in obtaining the formula. Dimensional consistency is a necessary condition for an equation to be correct. It is, of course, not sufficient. An equation can have the correct dimensions in each term and still not describe any physical situation.

1-4   Scientific Notation

Handling very large or very small numbers is simplified by using powers-of-10, or scientific, notation. In scientific notation, the number is written as a product of a number between 1 and 10 and a power of 10, such as 10^2 ($=100$) or 10^3 ($=1000$). For example, the number 12,000,000 is written 1.2×10^7; the distance from the earth to the sun, about 150,000,000,000 m, is written 1.5×10^{11} m. The number 7 in 10^7 is called the *exponent*. In this notation, 10^0 is defined to be 1. For numbers smaller than 1, the exponent is negative. For example, $0.1 = 10^{-1}$ and $0.0001 = 10^{-4}$. The diameter of a virus that is about 0.00000001 m is written 1×10^{-8} m.

In multiplication, exponents are added; in division, they are subtracted. These rules can be seen from some simple examples:

$$10^2 \times 10^3 = 100 \times 1000 = 100,000 = 10^{2+3} = 10^5$$

Similarly,

$$\frac{10^2}{10^3} = \frac{100}{1000} = \frac{1}{10} = 10^{2-3} = 10^{-1}$$

Example 1-2   Using scientific notation, compute (*a*) 120×6000 and (*b*) $3,000,000/0.00015$.

(*a*) $(1.20 \times 10^2)(6.00 \times 10^3) = (1.20)(6.00) \times 10^{2+3} = 7.20 \times 10^5$

(*b*) $(3.00 \times 10^6)/(1.50 \times 10^{-4}) = \dfrac{3.00}{1.50} \times 10^{6-(-4)} = 2.00 \times 10^{10}$

The two minus signs in (*b*) arise because -4 is subtracted from 6.

Example 1-3   A litre is the volume of a cube that is 10 cm by 10 cm by 10 cm. Find the volume of a litre in cubic centimetres and in cubic metres.

The volume of a cube of side L is L^3:

$$V = L^3 = 10 \text{ cm} \times 10 \text{ cm} \times 10 \text{ cm} = 10^3 \text{ cm}^3$$

To convert to m³, we use $1 \text{ cm} = 10^{-2}$ m:

$$10^3 \text{ cm}^3 = 10^3 \text{ cm}^3 \times \left(\frac{10^{-2} \text{ m}}{1 \text{ cm}}\right)^3$$

$$= 10^3 \text{ cm}^3 \times \left(\frac{10^{-6} \text{ m}^3}{1 \text{ cm}^3}\right) = 10^{-3} \text{ m}^3$$

Note that raising the conversion factor (which equals 1) to the third power does not change its value and enables us to cancel the units.

The addition or subtraction of two numbers written in scientific notation is slightly tricky when the exponents don't match. Consider, for example,

$$1.200 \times 10^2 + 8 \times 10^{-1} = 120.0 + 0.8 = 120.8$$

To find the sum without converting both numbers into ordinary decimal form, it is sufficient to rewrite either of the numbers so that its power of 10 is the same as that of the other. For example, we can find the sum by writing

$$1.200 \times 10^2 = 1200 \times 10^{-1}$$

and then adding

$$1200 \times 10^{-1} + 8 \times 10^{-1} = 1208 \times 10^{-1} = 120.8$$

If the exponents are very different, one of the numbers is much smaller than the other and can often be neglected in addition or subtraction. For example,

$$2 \times 10^6 + 9 \times 10^{-3} = 2{,}000{,}000 + 0.009$$

$$= 2{,}000{,}000.009$$

$$\approx 2 \times 10^6$$

where the symbol $\approx$ means "is approximately equal to."

1-5   Significant Figures

Many of the numbers in science are the result of measurements and are therefore known only within the limits of some experimental uncertainty. The magnitude of the uncertainty depends on the skill of the experimenter and on the apparatus used, and often can only be estimated. A rough indication of the uncertainty in a measurement is implied by the number of digits used. For example, if we say that a table is 2.50 m long, we are implying that its length is probably between 2.495 m and 2.505 m. That is, we know the length to about ± 0.005 m $= \pm 0.5$ cm. If we use a metrestick with millimetre markings and measure the table length carefully, we might estimate that we can measure the length to ± 0.5 mm rather than ± 0.5 cm. We would indicate this precision by using four digits, such as 2.503 m, to give the length. A reliably known digit (other than a zero used to locate the decimal point) is called a *significant figure*. The number 2.50 has three significant figures; 2.503 m has four. The number 0.00103 has three significant figures. (The first three zeroes are not significant figures but merely locate the decimal point.) In scientific notation, this number is 1.03×10^{-3}.

An error students commonly make, particularly since the advent of hand calculators, is to carry many more digits than are warranted. Suppose, for example, that you measure the area of a circular playing field by pacing off the radius and using the formula $A = \pi r^2$. If you estimate the radius to be 8 m by pacing and use a 10-digit calculator to compute the area, you obtain $\pi(8 \text{ m})^2 = 201.0619298 \text{ m}^2$. The digits after the decimal point are not only bulky to carry around, but they are also misleading about the accuracy with which you know the area. You would probably round the result off to 201 m². However, even this is misleading. If you found the radius by pacing, you might expect that your measurement was accurate to only about ± 0.5 m. That is, the radius could be as great as 8.5 m or as small as 7.5 m. If the radius is 8.5 m, the area is $\pi(8.5 \text{ m})^2 = 226.9800692 \text{ m}^2$, whereas if it is 7.5 m, the area is $\pi(7.5 \text{ m})^2 = 176.714587 \text{ m}^2$. A general rule for several numbers that have been multiplied or divided is

The number of significant figures in the result is no greater than the least number of significant figures in any of the numbers.

In this case, the radius is known to only one significant figure, so the area is also known to only one significant figure. It should be written as 2×10^2 m², which implies that the area is somewhere between 150 m² and 250 m².

In a textbook, it is cumbersome to write every number with the proper number of significant figures. Most of the examples and exercises will have data accurate to three (or sometimes four) significant figures, but occasionally we will say, for example, that a table top is 3 ft by 8 ft rather than taking the time and space to say that it is 3.00 ft by 8.00 ft. Any data you see in an example or exercise can be assumed to be known to three significant figures unless otherwise indicated.

In doing rough calculations or comparisons, we sometimes will round off a number to one significant figure or to the nearest power of 10, an expression that has no significant figures. A number rounded to the nearest power of 10 is called an *order of magnitude*. For example, the height of a small insect, say an ant, might be 8×10^{-4} m $\approx 10^{-3}$ m. We would say that the order of magnitude of the height of an ant is 10^{-3} m. Similarly, though the height of most people is about 2 m, we might round that off and say that the order of magnitude of the height of a person is 10^0 m. By this, we do not mean to imply that a typical person's height is really 1 m but that it is closer to 1 m than to 10 m or to $10^{-1} = 0.1$ m. We might say that a typical human being is 3 orders of magnitude taller than a typical ant, meaning that the ratio of their heights is about 1000 to 1.

Tables 1-2 to 1-4 give some typical order-of-magnitude values for some masses, time intervals, and lengths encountered in physics.

Table 1-2
Order of Magnitude of Some Masses

Mass	kg
Electron	10^{-30}
Proton	10^{-27}
Amino acid	10^{-25}
Hemoglobin	10^{-22}
Flu virus	10^{-19}
Giant amoeba	10^{-8}
Raindrop	10^{-6}
Ant	10^{-2}
Human being	10^2
Saturn 5 rocket	10^6
Pyramid	10^{10}
Earth	10^{24}
Sun	10^{30}
Milky Way galaxy	10^{41}
Universe	10^{52}

Table 1-3
Order of Magnitude of Some Time Intervals

Interval	s
Time for light to cross nucleus	10^{-23}
Period of visible-light radiation	10^{-15}
Period of microwaves	10^{-10}
Half-life of muon	10^{-6}
Period of highest audible sound	10^{-4}
Period of human heartbeat	10^0
Half-life of free neutron	10^3
Period of earth's rotation (day)	10^5
Period of revolution of earth (year)	10^7
Lifetime of a human	10^9
Half-life of plutonium 239	10^{12}
Lifetime of a mountain range	10^{15}
Age of earth	10^{17}
Age of universe	10^{18}

Table 1-4
Order of Magnitude of Some Lengths

Length	m
Radius of proton	10^{-15}
Radius of atom	10^{-10}
Radius of virus	10^{-7}
Radius of giant amoeba	10^{-4}
Radius of walnut	10^{-2}
Height of human being	10^0
Height of highest mountains	10^4
Radius of earth	10^7
Radius of sun	10^9
Earth–sun distance	10^{11}
Radius of solar system	10^{13}
Distance to nearest star	10^{16}
Radius of Milky Way galaxy	10^{21}
Radius of visible universe	10^{26}

1-6   Problem Solving

One of the best ways to find out if you understand a principle or law of physics is to apply it to a problem. Problem solving is the most difficult part of learning physics for many students. Fortunately, there are some standard techniques that apply to a wide variety of problems. We shall list some of these and illustrate their application to a common high school algebra problem that, though relatively easy, is of the type that can be mystifying to beginning students. Some of these techniques are

1. Try to guess the answer or the approximate answer without working the problem. Sometimes working a similar but easier problem will help you estimate the answer as well as lead you to the solution of the given problem.

2. List the information that is given in the form of mathematical equations or symbols. List the desired unknown quantity as a symbol and question mark (for example, if velocity is to be found, write $v = ?$).

3. Write an equation that relates the desired unknown quantity to the given information. (In more difficult problems it is sometimes hard to see how to do this step. It may then be useful just to list and calculate all the quantities that can be obtained from the given information. This will often be useful in finding the desired quantity.)

4. Solve the equation for the desired quantity and compare it with your estimate from step 1. If you could not guess or estimate the answer in step 1, try to check your answer in some other way to see if it is reasonable.

Example 1-4   Two house painters, June and Vicki, work together to paint a house. When working by herself, June could paint the house in 3 days whereas Vicki could paint it in 4 days. How long does it take when both work together?

We first note that the answer must be less than 3 days since June could do the job alone in this time and Vicki's help should certainly reduce the total time. (This rules out the popular but obviously incorrect answer of 7 days.) Next we note that if both painters worked at the same rate, the time would be half of that for either one working alone. Thus, if June worked at the slower rate of 4 days to paint the house alone, the time for both working together would be 2 days. Similarly, if Vicki could work at the faster rate of 3 days to paint the house alone, the time for both working would be 1.5 days. The answer must therefore be greater than 1.5 days and less than 2 days. A good guess would be 1.75 days. If this were a real problem, knowing the time within 0.25 day would certainly be good enough for any practical purposes.

For step 2, we note that the information given is the *rate* that each painter can work. These rates are

$$r_1 = 1 \text{ house per 3 days} = \tfrac{1}{3} \text{ house/day} \qquad \text{for June}$$

and $r_2 = \frac{1}{4}$ house/day   for Vicki

The quantity we wish to find is the time $t = ?$

   The relationship between the given rates and the unknown time is

$$N = rt$$

where N is the number of houses painted, r is the rate, and t is the time. The rate times the time for each painter gives the number of houses or the fraction of a house painted in that time. Since we want to get one house painted in some unknown time t, the correct algebraic equation is

$$r_1 t + r_2 t = (r_1 + r_2)t = 1 \text{ house}$$

Substituting the numbers into this equation, we obtain

$$(\tfrac{1}{3} \text{ house/day} + \tfrac{1}{4} \text{ house/day})t = 1 \text{ house}$$

Solving for t, we obtain

$$t = \tfrac{12}{7} \text{ days} = 1.71 \text{ days}$$

We note that our answer differs from our guess by only 0.04 day, which is less than $\frac{1}{2}$ hour for an 8-hour day.

Example 1-5   The universe is thought to have started with a "Big Bang" explosion. The galaxies that have moved the farthest are those with the greatest initial speeds, which are assumed to be approximately constant in time. If a galaxy 3×10^{21} km away is receding from us at 1.5×10^{11} km/y, calculate the age of the universe.

This problem is easier than the previous one, but it seems difficult because the numbers are very large and unfamiliar. Furthermore, our intuition tells us that finding the age of the universe should be a difficult problem. There is no simple way to guess the answer unless you happen to know already that the age of the universe is on the order of ten billion years (10^{10} y). The information given is

$$v = 1.5 \times 10^{11} \text{ km/y}$$

and

$$x = 3 \times 10^{21} \text{ km}$$

and the quantity desired is the time $t = ?$ These quantities are related by

$$x = vt$$

The desired time is then

$$t = \frac{x}{v} = \frac{3 \times 10^{21} \text{ km}}{1.5 \times 10^{11} \text{ km/y}} = 2 \times 10^{10} \text{ y}$$

   There are other problem-solving techniques that we shall add to this list for various kinds of problems. For example, in solving problems using Newton's laws, the first step is usually to draw a neat diagram including all the forces acting on each object in the problem. This and other techniques will be illustrated in later chapters.

Summary

1. Physical quantities, such as length, time, force, and energy, are expressed as a number plus a unit.

2. The fundamental units in the international system (SI) are the metre (m), the second (s), the kilogram (kg), the kelvin (K), the ampere (A), the mole (mol), and the candela (cd). The unit of every physical quantity can be expressed in terms of these fundamental units.

3. Units in equations are treated just like any other algebraic quantity.

4. Conversion factors, which are always equal to 1, provide a convenient method for converting from one kind of unit to another.

5. Very small and very large numbers are most easily handled using powers-of-10 (scientific) notation. When multiplying two numbers, the exponents are added; when dividing, the exponents are subtracted.

6. The number of significant figures in the result of a calculation should be no larger than the least number of significant figures in any of the numbers used.

7. Before starting to solve a problem, it is often useful to try to guess or estimate the answer. The solution of a problem should be compared with this estimate or checked against common sense.

Suggestions for Further Reading

Astin, Allen V.: "Standards of Measurement," *Scientific American,* June 1968, p. 50.

This article describes how the four "master measures" of the SI system—the metre, the kilogram, the second, and the kelvin—originated and how they are now defined.

McMahon, Thomas A., and John Tyler Bonner: "The Physics of Dimensions," Chapter 3 of *On Size and Life,* Scientific American Books, New York, 1983.

Submarines, airplanes, and stringed instruments are used as examples to illustrate the principles of dimensional analysis. This prepares the reader for later chapters in which the importance of scale to living things is considered.

Review

A. Objectives: After studying this chapter, you should:

1. Be able to define the units of length, time, and mass.

2. Know what is meant by SI units, U.S. customary units, and cgs units.

3. Know what conversion factors are and be able to use them to convert units from one system to another.

4. Know what is meant by the dimensions of a quantity.

B. Define, explain, or otherwise identify the following:

Units, pp. 1, 2
SI units, pp. 2, 3
Metric system, p. 3
U.S. customary system, p. 4
Conversion factor, p. 5
Dimensions, p. 6
Significant figure, p. 8

C. True or false: If a statement is true, explain why it is true. If it is false, give a counterexample, that is, a known example that contradicts the statement.

1. Two quantities to be added must have the same dimensions.

2. Two quantities to be multiplied must have the same dimensions.

3. All conversion factors have the value 1.

Exercises

Section 1-1   Units; Section 1-2   Conversion of Units; Section 1-3   Dimensions of Physical Quantities

1. Write the following using the prefixes listed in Table 1-1 and the abbreviations listed on the inside cover; for example, 10,000 metres = 10 km. (a) 1,000,000 watts; (b) 0.002 gram; (c) 3×10^{-6} metre; (d) 30,000 seconds.

2. Write the following without using prefixes: (a) 40 μW; (b) 4 ns; (c) 3 MW; (d) 25 km.

3. Write out the following (which are not SI units) without using any abbreviations; for example, 10^3 metres = 1 kilometre. (a) 10^{-12} boo; (b) 10^9 low; (c) 10^{-6} phone; (d) 10^{-18} boy; (e) 10^6 phone; (f) 10^{-9} goat; (g) 10^{12} bull.

4. In the following equations, the distance x is in metres, the time t in seconds, and the velocity v in metres per second. What are the SI units of the constants C_1 and C_2? (a) $x = C_1 + C_2 t$; (b) $x = C_1 t^2$; (c) $v^2 = 2C_1 x$; (d) $x = C_1 \cos C_2 t$; (e) $v = C_1 e^{-C_2 t}$. [Hint: The arguments of trigonometric functions and exponentials must be dimensionless. (The "argument" of $\cos \theta$ is θ and that of e^x is x.)]

5. What are the dimensions of the constants in Exercise 4?

6. In Exercise 4, if x is in feet, t is in seconds, and v is in feet per second, what are the units of the constants C_1 and C_2?

7. From the original definition of the metre in terms of the distance from the equator to the North Pole, find in metres (a) the circumference and (b) the radius of the earth. (c) Convert your answers to (a) and (b) from metres to miles.

8. Complete the following: (a) 100 km/h = _____ mi/h; (b) 60 cm = _____ in; (c) 100 yd = _____ m.

9. In the following, x is in metres, t is in seconds, v is in metres per second, and the acceleration a is in metres per second squared. Find the SI units of each of the following combinations: (a) v^2/xa; (b) $\sqrt{x/a}$; (c) $\frac{1}{2}at^2$.

10. Find the conversion factor to convert from miles per hour to kilometres per hour.

11. (a) Find the number of seconds in a year. (b) If one could count one dollar per second, how many years would it take to count 1 billion dollars (1 billion = 10^9)? (c) If one could count one molecule per second, how many years would it take to count the molecules in a mole? (The number of molecules in a mole is Avogadro's number, $N_A = 6 \times 10^{23}$.)

12. A cell membrane has a thickness of about 7 nm. How many cell membranes would it take to make a stack 1 in high?

Section 1-4   Scientific Notation; Section 1-5   Significant Figures

13. Express the following as decimal numbers without using powers-of-10 notation: (a) 3×10^4; (b) 6.2×10^{-3}; (c) 4×10^{-6}; (d) 2.17×10^{35}.

14. Write the following in scientific notation: (a) 100,000; (b) 0.0000000303; (c) 602,000,000,000,000,000,000,000; (d) $(5.14 \times 10^3) + (2.78 \times 10^2)$; (e) $(1.99 \times 10^2) + (9.99 \times 10^{-5})$.

15. Calculate the following, round off to the correct number of significant figures, and express your result in scientific notation: (a) $(1.14)(9.99 \times 10^4)$; (b) $(2.78 \times 10^{-8}) - (5.31 \times 10^{-9})$; (c) $12\pi/(4.56 \times 10^{-3})$; (d) $27.6 + (5.99 \times 10^2)$.

16. Calculate the following, round off to the correct number of significant figures, and express your result in scientific notation: (a) $(200.9)(569.3)$; (b) $(0.000000513)(62.3 \times 10^7)$; (c) $28,401 + (5.78 \times 10^4)$; (d) $63.25/(4.17 \times 10^{-3})$.

Section 1-6   Problem Solving

There are no exercises for this section.

Motion in One Dimension

The spirit of the time shall teach me speed.

WILLIAM SHAKESPEARE, *King John*

Now we come to the ideas with which the Renaissance genius Galileo Galilei launched modern science as we know it, with its constant interplay of theory and experiment. As a young professor at the University of Pisa, Galileo began studying how objects move, particularly what happens when they fall. Perhaps his most famous experiment involved dropping two balls, one of iron and one of wood, "from a high place"—the top of the Leaning Tower, according to legend—to demonstrate that they fell at the same rate, thus contradicting Aristotle, whose ideas had held sway for nearly 2000 years. Galileo completed his study of moving objects toward the end of his life while imprisoned by the Church in his country home for contradicting Church teachings about astronomy.

Over the last 400 years, many improvements have been made in the study of the movement of objects, most notably the introduction of algebra to Europe by Arab mathematicians. You will have the advantage of these improvements, but you will probably still experience some of the confusions and difficulties with which Galileo had to cope. Stick with it! A clear grasp of the ideas presented in this chapter and the next is essential for understanding Newton's discoveries discussed in Chapter 4, as well as the developments in modern physics covered in the final chapters of the book.

The description of the motion of objects is an important part of the description of the physical universe. One of the earliest scientific puzzles concerned the apparent motion of the sun across the sky and the seasonal motion of the planets and stars. A great triumph of newtonian mechanics was the discovery that the motion of the earth and other planets around the sun could be explained in terms of a force of attraction between the sun and the planets.

The motion of objects is intimately related to the important physical concepts of force, momentum, and energy. We all know, for example, that the motion of a car requires the expenditure of energy, usually energy stored in gasoline, and that when a person is jogging, chemical energy in the body is converted into heat energy. In this chapter and the next we shall be concerned with the *description* of motion (kinematics) without worrying about its causes. We shall consider the causes of motion in Chapter 4 when we study Newton's laws.

The description of the motion of objects was central to the development of science from Aristotle to Galileo. The laws of *how* things fall were developed long before Newton described *why* things fall. Galileo was one of a series of observers who tried to relate how far and how fast objects fall in various intervals of time. His empirical observations were the basis for the more sweeping theories of Newton relating motion to forces and, in particular, relating falling objects and the orbital motion of the planets to the gravitational attraction between two bodies.

To simplify our discussion of motion, we shall start with objects whose position can be described by locating one point. Such an object is called a *particle.* One tends to think of a particle as a very small object, but actually no size limit is implied by the word particle. Any object can be considered a particle if we are not interested in the size of the object or in its rotational motion. For example, it is sometimes convenient to consider the earth as a particle moving around the sun in a nearly circular path. (Certainly, when viewed from a distant planet or a distant galaxy, the earth looks like a point.) This is the case when we are interested only in the motion of the center of the earth and are ignoring the size of the earth and its rotation. In some astronomical problems, the entire solar system or even a whole galaxy is sometimes treated as a particle.

When we are interested in the internal motion or internal structure of an object, it can no longer be treated as a particle, but our study of particle motion will be useful even in these cases because any complex object can be treated as a collection or "system" of particles. Even such a small thing as an atomic nucleus turns out to be a rather complicated system of particles when its structure is examined in detail.

To describe the motion of a particle, we need the concepts of *displacement, velocity,* and *acceleration.* In the general motion of a particle in three dimensions, these quantities are vectors, which means that they have direction as well as magnitude. We will study vectors and the general motion of a particle in Chapter 3. In this chapter, we will confine our discussion to motion in one dimension, that is, motion along a straight line. For such restricted motion, there are only two possible directions, which we distinguish by designating one positive and the other negative. A simple example of one-dimensional motion is a car moving along a flat, straight, narrow road. We can choose any convenient point on the car for the location of the "particle."

Photo of the night sky showing a portion of the Milky Way galaxy. Sometimes even a galaxy can be considered to be a particle.

Question

1. Give several examples of the motion of large objects where treating the object as a particle is an adequate approximation. Give some other examples where it is not.

2-1   Speed, Displacement, and Velocity

We are all familiar with the concept of speed. We define the *average speed* of a particle as the ratio of the total distance traveled to the total time taken:

$$\left(\text{Average speed} = \frac{\text{total distance}}{\text{total time}}\right)$$

The SI units of average speed are metres per second (written m/s), and the U.S. customary units are feet per second (ft/s). A familiar everyday unit is kilometres per hour (km/h). If you drive 200 km in 5 h, your average speed is (200 km)/(5 h) = 40 km/h. Note that the average speed tells you nothing about the details of the trip. You may have driven at a steady rate of 40 km/h for the whole 5 h, or you may have driven faster part of the time and slower the rest of the time, or you may have stopped for an hour and then driven at varying rates during the other 4 h.

The concept of velocity is similar to that of speed but differs because it includes the *direction* of motion. To understand this concept, we must first understand the idea of displacement. Let us set up a coordinate system by choosing some reference point on a line as the origin O. To every other point on the line we assign a number x that indicates how far the point is from the origin. The value of x depends on the unit (ft, m, etc.) chosen to measure the distance. The sign of x depends on its position relative to the origin O. The usual convention we choose is that if x is to the right of the origin, it is positive; to the left, it is negative.

Suppose that our particle (a car in this case) is at position x_1 at some time t_1 and at point x_2 at time t_2. The change in the position of the particle, $x_2 - x_1$, is called the *displacement* of the particle (Figure 2-1). It is customary to use the Greek letter Δ (capital delta) to indicate the change in a quantity. Thus, the change in x is written Δx:

$$\left(\Delta x = x_2 - x_1 \hspace{3cm} \text{2-1} \quad \text{Displacement}\right)$$

The notation Δx (read "delta x") stands for a single quantity, the change in x. It is not a product of Δ and x any more than $\cos\theta$ is a product of cos and θ.

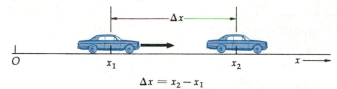

$$\Delta x = x_2 - x_1$$

Figure 2-1
When the car travels from point x_1 to point x_2, its displacement is $\Delta x = x_2 - x_1$.

Velocity is the rate at which the position changes. The *average velocity* of the particle is defined as the ratio of the displacement Δx to the time interval $\Delta t = t_2 - t_1$:

$$\left(\boxed{v_{\text{av}} = \frac{\Delta x}{\Delta t} = \frac{x_2 - x_1}{t_2 - t_1}} \hspace{2cm} \text{2-2} \quad \text{Average velocity}\right)$$

Note that the displacement and the average velocity may be either positive or negative, depending on whether x_2 is greater or less than x_1. A positive value indicates motion to the right, and a negative value indicates motion to the left.

Example 2-1 A snail is at $x_1 = 18$ mm at $t_1 = 2$ s and is later found at $x_2 = 14$ mm at $t_2 = 7$ s. Find the displacement and the average velocity of the snail for this time interval.

By the definition, the displacement of the snail is

$$\Delta x = x_2 - x_1 = 14 \text{ mm} - 18 \text{ mm} = -4 \text{ mm}$$

and the average velocity is

$$v_{av} = \frac{\Delta x}{\Delta t} = \frac{x_2 - x_1}{t_2 - t_1} = \frac{14 \text{ mm} - 18 \text{ mm}}{7 \text{ s} - 2 \text{ s}} = \frac{-4 \text{ mm}}{5 \text{ s}} = -0.8 \text{ mm/s}$$

The displacement and the average velocity are negative, indicating that the snail moved to the left, toward decreasing values of x.

Note that the unit millimetres per second is included as part of the answer for the average velocity in Example 2-1. Since there are many other possible choices for units of length (for example, feet, inches, miles, or light-years) and time (for example, hours, days, or years), it is essential to include the unit with the numerical answer. The statement "the average velocity of a particle is -3" is meaningless.

Example 2-2 How far does a car go in 5 min if its average velocity is 80 km/h during this period?

In this example, we are interested in the displacement during a time interval of 5 min. From Equation 2-2, the displacement Δx is given by

$$\Delta x = v_{av}\,\Delta t$$

Thus,

$$\Delta x = \frac{80 \text{ km}}{\text{h}} \times 5 \text{ min} = 400 \frac{\text{km} \cdot \text{min}}{\text{h}} \times \frac{1 \text{ h}}{60 \text{ min}} = 6.67 \text{ km}$$

We use the conversion factor 1 h $= 60$ min to change the unit of displacement from km·min/h to kilometres.

Example 2-3 A runner runs 100 m in 10 s and then turns around and jogs 50 m back toward the starting point in 30 s. What are her average speed and average velocity for the total trip?

The total distance traveled is 100 m $+$ 50 m $=$ 150 m, and the total time taken is 40 s. The average speed is therefore (150 m)/(40 s) $=$ 3.75 m/s. Note that this is not the average of her running and jogging speeds because she ran for 10 s but jogged for 30 s. To find the average *velocity* we first find the total displacement, which is 50 m (if x_1 is taken to be 0, then $x_2 = 50$ m). The average velocity is then

$$v_{av} = \frac{\Delta x}{\Delta t} = \frac{50 \text{ m}}{40 \text{ s}} = +1.25 \text{ m/s}$$

Again, this is *not* the average of her running velocity ($+10$ m/s) and her jogging velocity (-1.67 m/s) because the times are different.

It is often useful to interpret physical quantities graphically. Suppose we determine the position x of a particle at various times t. Figure 2-2 shows a graph of x versus t for some arbitrary motion along the x axis. The measured points have been connected with a smooth curve. How accurately this curve represents the motion of the particle depends upon the complexity of the motion. On the figure we have drawn a straight line between the initial position, labeled P_1, and the final position, labeled P_2. The displacement $\Delta x = x_2 - x_1$ and the time interval $\Delta t = t_2 - t_1$ for these points are indicated. The line between P_1 and P_2 is the hypotenuse of the triangle with sides Δt and Δx. The ratio $\Delta x / \Delta t$ is called the *slope* of this straight line. In geometric terms, the slope is a measure of the steepness of the straight line in the graph. For a given interval Δt, the steeper the line, the greater the value of $\Delta x / \Delta t$. Since the slope of this line is just the average velocity for the interval Δt, we have a geometric representation of the average velocity. It is the slope of the straight line connecting the points (x_1, t_1) and (x_2, t_2). Unless the velocity is constant, the average velocity will depend on the time interval chosen. For example, in Figure 2-2, if we chose a smaller time interval by choosing a time t_2' closer to t_1, the average velocity would be greater, as indicated by the greater steepness of the line connecting points P_1 and P_2'.

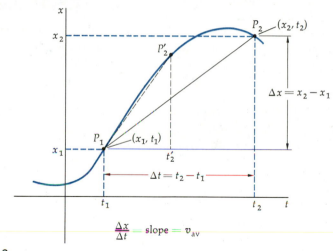

Figure 2-2
Graph of x versus t for a particle moving in one dimension. The initial and final points, P_1 and P_2, are connected by a straight line. The average velocity is the slope of this line, $\Delta x / \Delta t$, which depends on the time interval, as indicated by the fact that the line from P_1 to P_2' has a greater slope than the line from P_1 to P_2.

Questions

2. What sense, if any, does the following statement make? "The average velocity of the car at 9 A.M. was 60 km/h."

3. Is it possible for the average velocity for some time interval to be zero even though the average velocity for a shorter interval included in the first interval is not zero? Explain.

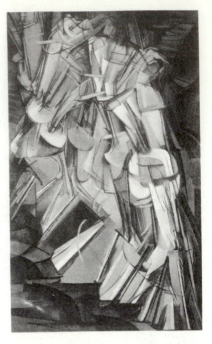

Motion as shown in the photograph of runners and in the painting "Nude Descending a Staircase, No. 2" by Marcel Duchamp. Can you tell from a single high-speed photograph that the runners are moving?

2-2   Instantaneous Velocity

To define the velocity of a particle at a single instant might seem impossible at first glance. At a specific time t_1, the particle is at a single point x_1. If it is at a single point, how can it be moving? On the other hand, if it is not moving, shouldn't it stay at the same point? This is an age-old paradox that can be resolved when we realize that, to observe motion and thus define it, we must look at the position of the object at more than one time. It is then possible to define the velocity at a specific instant by using what is called a "limiting" process.

Figure 2-3 is the same x-versus-t curve as Figure 2-2, showing a sequence of time intervals, Δt_1, Δt_2, Δt_3, and Δt_4, each smaller than the previous one. For each time interval Δt, the average velocity is the slope of the dashed line appropriate for that interval. The figure shows that, as the time intervals become smaller, the dashed lines get steeper, but they never incline more than the line tangent to the curve at point t_1. That is, as Δt decreases, the average velocity $\Delta x/\Delta t$ approaches a limiting value, which is the slope of the line tangent to the curve at point t_1. We define the slope of this tangent as the *instantaneous velocity* at the time t_1.

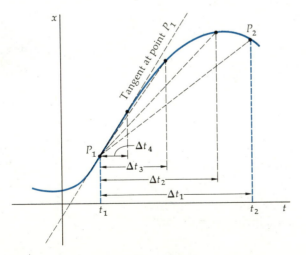

Figure 2-3
Graph of x versus t from Figure 2-2. As the time interval beginning at t_1 is decreased, the average velocity for the interval approaches the slope of the line tangent to the curve at time t_1. The instantaneous velocity is defined as the slope of this line.

The instantaneous velocity is the limit of the ratio Δx/Δt as Δt approaches zero.

$$v = \lim_{\Delta t \to 0} \frac{\Delta x}{\Delta t}$$

= slope of the line tangent to the *x*-versus-*t* curve

2-3   Instantaneous velocity

If you have studied calculus, you will recognize that the velocity is the *derivative* of *x* with respect to *t*. (It is usually written dx/dt.) In calculus, methods are developed for computing the velocity *v* from any function of time *t*. We need not worry about these methods. Instead, we need only remember the following.

The instantaneous velocity is the slope of the line tangent to the x-versus-t curve at a given point.

Since the slope may be positive (*x* is increasing) or negative (*x* is decreasing), the instantaneous velocity can be positive or negative (in one-dimensional motion). The magnitude of the instantaneous velocity is called the *instantaneous speed*.

Example 2-4   A particle has a position *x* that varies with time as shown in Figure 2-4. Find the instantaneous velocity at the time *t* = 2 s. When is the velocity greatest? When is it zero? Is it ever negative?

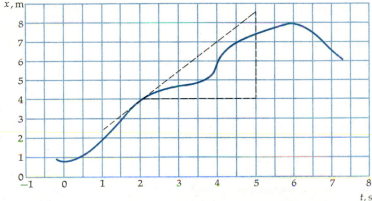

Figure 2-4
Graph of *x* versus *t* for Example 2-4. The instantaneous velocity at time *t* = 2 s can be found by measuring the slope of the line tangent to the curve at that time.

In the figure, we have sketched the line tangent to the curve at time *t* = 2 s. The slope of this line is measured from the figure to be (4.5 m)/(3 s) = 1.5 m/s. Thus *v* = 1.5 m/s at time *t* = 2 s. According to the figure, the slope is greatest at about *t* = 4 s. The velocity is zero at times *t* = 0 and *t* = 6 s, as indicated by the fact that the tangent lines at these times are horizontal and so have zero slopes. After *t* = 6 s, the curve has a negative slope, indicating that the velocity is negative. (The slope of the line tangent to a curve is often referred to merely as the *slope of the curve*.)

Example 2-5   A particle is at point *x* = 5 m at time *t* = 0 and moves with a constant velocity of 10 m/s. Sketch its position *x* as a function of time *t*.

A constant velocity means that the slope of the *x*-versus-*t* curve is constant.

The curve that has a constant slope is a straight line. In some time t, the distance covered by an object with constant speed v is $\Delta x = vt$. If the initial position of the object is x_0, its position after time t is given by

$$x = x_0 + vt \qquad\qquad 2\text{-}4$$

This is the general equation for x versus t for a particle moving with constant velocity. Substituting $x_0 = 5$ m and $v = 10$ m/s, we have

$$x = 5 \text{ m} + (10 \text{ m/s})t$$

This curve is sketched in Figure 2-5.

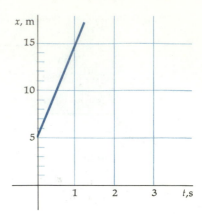

Figure 2-5
Graph of $x = 5$ m $+ (10$ m/s$)t$ for Example 2-5. When the graph of x versus t is a straight line, x is said to vary linearly with t.

It is important to distinguish carefully between average velocity and instantaneous velocity. By custom, the word velocity alone is assumed to mean instantaneous velocity.

Question

4. If the instantaneous velocity does not change from instant to instant, will the average velocities for different time intervals differ?

2-3   Acceleration

When the instantaneous velocity of a particle is changing with time, the particle is said to be accelerating. *Acceleration* is the *rate of change of the velocity* (just as *velocity* is the *rate of change of position*.)

The *average acceleration* for a particular time interval Δt is defined as the ratio $\Delta v/\Delta t$, where $\Delta v = v_2 - v_1$ is the change in instantaneous velocity for that time interval.

$$a_{av} = \frac{\Delta v}{\Delta t} \qquad\qquad 2\text{-}5 \qquad \text{Average acceleration}$$

Suppose, for example, that a car is traveling at 20 m/s at some time and that 2 seconds later it is traveling at 30 m/s in the same direction. The velocity thus changes by 10 m/s in 2 s, so the car's average acceleration is (10 m/s)/(2 s) = 5 m/s². We see that the dimensions of acceleration are length divided by time squared. Convenient units are metres per second per second (m/s²) or feet per second per second (ft/s²). For automobile accelerations, miles per hour per second $\left(\dfrac{\text{mi/h}}{\text{s}}\right)$ is often used. For example, a car that can accelerate from rest to 60 mi/h in 15 s has an acceleration of $4\,\dfrac{\text{mi/h}}{\text{s}}$.

Example 2-6   Convert the acceleration $4\,\dfrac{\text{mi/h}}{\text{s}} = 4\,\dfrac{\text{m}}{\text{h}\cdot\text{s}}$ to m/s².

We use the conversion factors 1 mi $= 1.61$ km $= 1610$ m and 1 h $= 60$ min $= 3600$ s.

$$4\,\frac{\text{mi/h}}{\text{s}} = 4\,\frac{\text{mi}}{\text{h}\cdot\text{s}} \times \frac{1610 \text{ m}}{1 \text{ mi}} \times \frac{1 \text{ h}}{3600 \text{ s}} = 1.79 \text{ m/s}^2$$

Instantaneous acceleration is the limit of $\Delta v / \Delta t$ as Δt approaches zero. If we plot the velocity versus time, the instantaneous acceleration is defined as the slope of the line tangent to the curve (Figure 2-6).

$$a = \lim_{\Delta t \to 0} \frac{\Delta v}{\Delta t} = \text{slope of the } v\text{-versus-}t \text{ curve} \qquad 2\text{-}6$$

Instantaneous acceleration

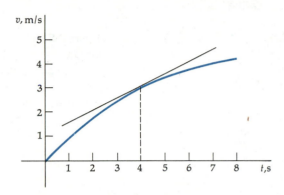

Figure 2-6
Graph of v versus t with a line drawn tangent to the curve at time $t = 4$ s. The instantaneous acceleration at some time t is the slope of the line tangent to the curve at that time.

Again, in calculus this limit is called the derivative of the velocity with respect to time and is written dv/dt. (Since v is the derivative of x with respect to time, a is the second derivative of x with respect to t.) We shall use the notation $\Delta v/\Delta t$ for acceleration and $\Delta x/\Delta t$ for velocity with the understanding that all the changes ($\Delta t, \Delta x, \Delta v,$) must be very small to obtain the instantaneous velocity or instantaneous acceleration. The acceleration of an object is important in physics because it is related to the forces acting on the object. We shall see in Chapter 4 that the acceleration of an object is directly proportional to the unbalanced force acting on it.

If the acceleration is zero, the velocity is constant, and the x-versus-t curve is a straight line with x given by Equation 2-4. On the other hand, if the acceleration is not zero, the velocity is not constant, and the x-versus-t curve is not a straight line. We can often get a qualitative idea of the acceleration from the x-versus-t curve, as discussed in the following example.

Example 2-7   After examining the x-versus-t curve in Figure 2-7, describe qualitatively the velocity and acceleration at the point $t = 5$ s.

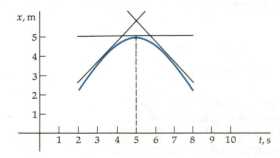

Figure 2-7
Graph of x versus t for Example 2-7. The slope of the line tangent to the curve at $t = 5$ s is zero, indicating that the instantaneous velocity is zero at this time. Just before this time, the slope is positive, and just after, it is negative. Since the slope is decreasing at $t = 5$ s, the acceleration at this time is negative.

At $t = 5$ s, the tangent line to the curve is horizontal with a zero slope. Thus, the instantaneous velocity at this point is zero. However, the fact that the velocity is zero does *not* imply that the acceleration must also be zero. The acceleration is the rate of change of the instantaneous velocity. To find the acceleration at some time, we must know how the velocity is changing at that time. For this example, we must know the velocity at other times in the neighborhood of $t = 5$ s to find the acceleration at $t = 5$ s. From the figure, we can see that the slope of the curve is positive just before $t = 5$ s and negative just after. Thus, the velocity changes from a positive value to zero to a negative value, so the acceleration at $t = 5$ s is negative.

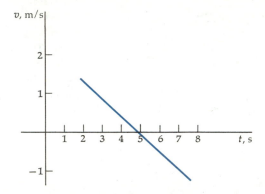

Figure 2-8
Graph of v versus t corresponding to the position curve in Figure 2-7. At $t = 5$ s, the velocity is zero and decreasing.

In Figure 2-8, we have sketched the v-versus-t curve corresponding to the position function given in Figure 2-7. This curve has a constant slope, indicating that the acceleration is constant. The slope is negative, so the acceleration is negative; that is, the velocity of the particle is decreasing with time. Note that at $t = 5$ s the velocity is zero; nevertheless, the slope of the v-versus-t curve is not zero. The slope is negative at that time, just as it is at all other times. We shall study motion with constant acceleration in the next section.

Questions

5. Give an example (sketch a graph of v versus t) of an object in motion for which the velocity is negative but the acceleration is positive.

6. Give an example of an object in motion for which both the acceleration and the velocity are negative.

7. Can an object be accelerating at an instant when its velocity is zero?

2-4   Motion with Constant Acceleration

The motion of a particle that has constant acceleration is important for several reasons. For one, this type of motion is common in nature. For example, near the surface of the earth all objects fall vertically with the constant acceleration of gravity if air resistance can be neglected and if there are no forces acting on the objects except the gravitational attraction of the earth. Even when the acceleration of a particle is not constant, we can sometimes learn much about the motion of the particle by using the con-

© 1987 Sidney Harris.

"It goes from zero to 60 in about 3 seconds."

stant-acceleration concepts developed in this section. For example, in a car crash the car and passengers undergo a large and variable acceleration as their velocity is reduced to zero. Though the acceleration is not constant, the forces exerted in such a crash can be estimated by assuming the acceleration to be constant and equal to its average value (see Example 2-12).

The acceleration of gravity is designated by g and has the approximate value

$$g = 9.81 \text{ m/s}^2 = 32.2 \text{ ft/s}^2 \qquad\qquad 2\text{-}7$$

(We often approximate this as 9.8 m/s^2 or 32 ft/s^2 or even 10 m/s^2 if no greater precision is required.)

A constant acceleration means that the slope of the v-versus-t curve is constant; that is, the v-versus-t curve is a straight line. In some time t, the velocity changes by at, where a is the acceleration. If the velocity is v_0 at time $t = 0$, its value at time t will be

$$v = v_0 + at \qquad\qquad 2\text{-}8 \qquad \text{Constant acceleration}(v \text{ versus } t)$$

If the particle starts at x_0 at time $t = 0$ and its position is x at time t, the displacement $\Delta x = x - x_0$ is given by

$$\Delta x = v_{av}t$$

(This is the same as Equation 2-2 with t replacing Δt because we have chosen the initial value of t to be zero.) For constant acceleration, the average velocity is the mean value of the initial and final velocities.*

$$v_{av} = \tfrac{1}{2}(v_0 + v) \qquad\qquad 2\text{-}9 \qquad \text{Constant acceleration, } v_{av}$$

Substituting $v = v_0 + at$ from Equation 2-8, we obtain

$$v_{av} = \tfrac{1}{2}(v_0 + v_0 + at) = v_0 + \tfrac{1}{2}at$$

We see that the average velocity equals the initial velocity plus half the increase in the velocity ($\tfrac{1}{2}at$), as shown in Figure 2-9. The displacement is then

$$\Delta x = (v_0 + \tfrac{1}{2}at)t = v_0t + \tfrac{1}{2}at^2 \qquad\qquad 2\text{-}10a$$

and the position function is

$$x = x_0 + v_0t + \tfrac{1}{2}at^2 \qquad\qquad 2\text{-}10b \qquad (\text{Constant acceleration, } x \text{ versus } t)$$

* This is not true if the acceleration is not constant.

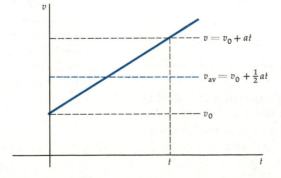

Figure 2-9
For constant acceleration, velocity varies linearly with time. The average velocity is then just the mean of the initial and final velocities. If v_0 is the initial velocity and $v_0 + at$ is the final velocity, the average velocity is $v_{av} = v_0 + \tfrac{1}{2}at$.

Example 2-8   A ball is thrown upward with an initial velocity of 30 m/s. If its acceleration is 10 m/s² downward, how long does the ball take to reach its highest point, and what is the distance to the highest point?

If we take the upward direction as positive, the initial velocity is positive and the acceleration is negative because it is downward. Then $v_0 = 30$ m/s and $a = -10$ m/s². As the ball moves upward (v is positive), the velocity decreases from its initial value until it is zero. When the velocity is zero, the ball is at its highest point. It then falls, and the velocity becomes negative, indicating that the ball is moving downward. We can find the time t it takes the ball to reach the top of its flight by using Equation 2-8 and setting $v = 0$.

$$v = v_0 + at$$

$$0 = 30 \text{ m/s} + (-10 \text{ m/s}^2)t$$

$$t = \frac{30 \text{ m/s}}{10 \text{ m/s}^2} = 3.0 \text{ s}$$

Note that the units work out correctly. Since the initial velocity is $+30$ m/s and the final velocity is zero, the average velocity for the upward motion is 15 m/s. The distance traveled is then

$$\Delta x = v_{av}t = (15 \text{ m/s})(3.0 \text{ s}) = 45 \text{ m}$$

We could have also found Δx from Equation 2-10a.

$$\Delta x = v_0 t + \tfrac{1}{2}at^2 = (30 \text{ m/s})(3.0 \text{ s}) + \tfrac{1}{2}(-10 \text{ m/s}^2)(3.0 \text{ s})^2$$

$$= +90 \text{ m} - 45 \text{ m} = 45 \text{ m}$$

Example 2-9   What is the total time the ball in Example 2-8 is in the air?

We could guess the answer to be 6 s since, by symmetry, if it takes 3.0 s to rise 45 m, it should take the same time to fall. This is correct. We can also find the time from Equation 2-10a by setting $\Delta x = 0$.

$$\Delta x = v_0 t + \tfrac{1}{2}at^2 = 0$$

Factoring, we obtain

$$t(v_0 + \tfrac{1}{2}at) = 0$$

The two solutions are $t = 0$, which corresponds to our initial condition, and

$$t = -\frac{2v_0}{a}$$

$$= -\frac{2(30 \text{ m/s})}{-10 \text{ m/s}^2} = 6 \text{ s}$$

Figures 2-10 and 2-11 show x-versus-t and v-versus-t curves for the ball in Examples 2-8 and 2-9. Note that at time $t = 3.0$ s, the velocity of the ball is zero but the slope of the v-versus-t curve is not. The slope of the v-versus-t curve has the value -10 m/s² at this time and at all other times because the acceleration is constant.

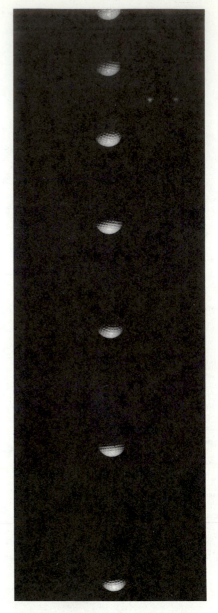

Multiflash photograph of a golf ball falling with constant acceleration. The position of the ball is recorded at 1/25-s intervals using a strobe light that flickers 25 times per second. The increase in spacing between successive positions indicates that the speed is increasing.

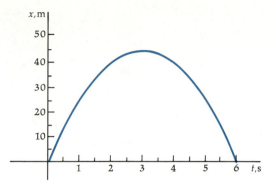

Figure 2-10
Graph of x versus t for the ball thrown into the air in Examples 2-8 and 2-9.

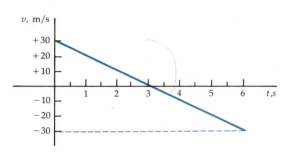

Figure 2-11
Graph of v versus t for the ball thrown into the air in Examples 2-8 and 2-9. The velocity decreases steadily from its initial value of 30 m/s to its final value of -30 m/s just before it hits the ground. At time $t = 3.0$ s, when the ball is at its highest point, the velocity is zero, but its rate of change is -10 m/s², the same as at any other time.

We sometimes want to find the final velocity of a particle after it moves through a given distance with constant acceleration. For example, we may want to know the velocity of a ball dropped from rest some height x. We could find it by first finding the time from $\Delta x = \frac{1}{2}at^2$ and then substituting the result into $v = at$ (for $v_0 = 0$). But if we are not interested in knowing the time of fall, it is convenient to eliminate the time from these two equations and develop a relationship between v, a, and Δx. We can find a general relationship for any initial velocity v_0 by solving Equation 2-8 for $t = (v - v_0)/a$ and substituting this result into Equation 2-10a.

$$\Delta x = v_0 t + \tfrac{1}{2}at = v_0 \left(\frac{v - v_0}{a} \right) + \tfrac{1}{2}a \left(\frac{v - v_0}{a} \right)^2$$

Multiplying each side by a and writing out the terms on the right side of the equation, we obtain

$$a\,\Delta x = v_0 v - v_0^2 + \tfrac{1}{2}v^2 - vv_0 - \tfrac{1}{2}v_0^2$$
$$= \tfrac{1}{2}v^2 - \tfrac{1}{2}v_0^2$$

or

$$v^2 = v_0^2 + 2a\,\Delta x \qquad\qquad \text{2-11} \quad \left(\textcolor{teal}{\text{Constant acceleration, } v \text{ versus } x}\right.$$

Example 2-10 A car traveling at 15 m/s (about 33 mi/h) brakes to a stop. If the acceleration is -5 m/s², how far does the car travel before stopping? This distance is called the *stopping distance*.

In this example, we choose the original direction of motion to be positive. The stopping distance will then also be positive, but the acceleration will be negative. When the speed is decreasing, as it is in this case, the (negative) acceleration is called *deceleration*. Setting $v = 0$ in Equation 2-11, we have

$$0 = (15 \text{ m/s})^2 + 2(-5 \text{ m/s}^2)\,\Delta x$$

$$\Delta x = \frac{-(15 \text{ m/s})^2}{2(-5 \text{ m/s}^2)} = 22.5 \text{ m}$$

The force that produces this deceleration is that of the friction between the car tires and the road. On wet pavement or gravel, the frictional force is smaller and the magnitude of the deceleration is even less than 5 m/s².

Example 2-11   What is the stopping distance under the same conditions as Example 2-10 if the car is initially traveling at 30 m/s?

From Equation 2-11, with $v = 0$ we see that the stopping distance is proportional to the square of the initial speed. If we double the speed, the stopping distance is increased by a factor of 4. The stopping distance at 30 m/s is therefore four times that at 15 m/s, or 4×22.5 m = 90 m. Note that this is a considerable distance.

Sometimes, even when the acceleration is not constant, valuable insight can be gained about the motion of an object by *assuming* that the constant-acceleration formulas still apply.

Example 2-12   A car traveling 100 km/h (about 62 mi/h) crashes into a concrete wall that does not move. How long does it take the car to come to rest, and what is its acceleration?

Test of car traveling at between 95 and 120 km/h crashing into a fixed object.

In this example, it is not accurate to treat the car as a particle because different parts of it have different accelerations. Moreover, the accelerations are not constant. Nevertheless, let us assume the constant acceleration of a particle. We need more information in order to find either the stopping time or the acceleration. The quantity missing is the stopping distance. We can estimate this from our practical knowledge. The center of the car certainly moves less than half the length of the car. A reasonable estimate for the stopping distance is probably between 0.5 and 1.0 m. Let us use 0.75 m as our estimate. We can then find the time it takes the car to stop from

$$\Delta x = v_{av} \, \Delta t$$

with

$$v_{av} = \tfrac{1}{2} v_0 = 50 \text{ km/h}$$

Converting this average velocity to metres per second, we have

$$v_{av} = 50\ \frac{km}{h} \left(\frac{1000\ m}{1\ km}\right)\left(\frac{1\ h}{3600\ s}\right) = 14\ m/s$$

(Since we are making estimates, two significant figures are sufficient). Then,

$$\Delta t = \frac{\Delta x}{v_{av}} = \frac{0.75\ m}{14\ m/s} = 0.054\ s$$

Since the car is brought from $v_0 = 100\ km/h = 28\ m/s$ to rest in this time, the acceleration is

$$a = \frac{\Delta v}{\Delta t} = \frac{0 - 28\ m/s}{0.054\ s} = -520\ m/s^2$$

To get a feeling for the magnitude of this acceleration, we note that it is about 52 times the acceleration of gravity, or "52 g" as the aerospace engineers would say.

Questions

8. Two boys standing on a bridge throw rocks straight down into the water below. They throw two rocks at the same time, but one hits the water before the other. How can this be if the rocks have the same acceleration?

9. A ball is thrown straight up. What is its velocity at the top of its flight? What is its acceleration at that point?

Summary

1. The average velocity is the ratio of the displacement Δx to the time interval Δt.

$$v_{av} = \frac{\Delta x}{\Delta t}$$

2. The instantaneous velocity v is the limit of this ratio as the time interval approaches zero. It is represented graphically as the slope of the x-versus-t curve. In one dimension, both average and instantaneous velocity can be either positive or negative. The magnitude of the instantaneous velocity is called the speed.

3. The average acceleration is the ratio of the change in velocity Δv to the time interval Δt.

$$a_{av} = \frac{\Delta v}{\Delta t}$$

The instantaneous acceleration is the limit of this ratio as the time interval approaches zero. The instantaneous acceleration is represented graphically as the slope of the v-versus-t curve.

4. In the special case of constant acceleration, the following formulas hold:

$$v = v_0 + at \qquad\qquad v_{av} = \tfrac{1}{2}(v_0 + v)$$

$$\Delta x = v_0 t + \tfrac{1}{2}at^2 \qquad\qquad v^2 = v_0^2 + 2a\,\Delta x$$

A particularly common example of motion with constant acceleration is the motion of an object near the surface of the earth on which the only force acting is that of gravity. In this case, the acceleration of the object is directed downward with a magnitude $9.81 \text{ m/s}^2 = 32.2 \text{ ft/s}^2$.

Another useful expression for constant-acceleration problems is

$$\Delta x = v_{av}t$$

This follows directly from the definition of average velocity and is therefore true in general, not just for constant acceleration.

Suggestions for Further Reading

Drake, Stillman: "Galileo's Discovery of the Law of Free Fall," *Scientific American*, May 1973, p. 84.

The author's reexamination of unpublished manuscripts clarifies the development of Galileo's ideas about motion.

Magie, W. F.: *A Source Book in Physics*, McGraw-Hill, New York and London, 1935.

This useful book contains short life histories and extracts from the important works of the great physicists from the time of Galileo to the year 1900.

Review

A. Objectives: After studying this chapter, you should:

1. Be able to define displacement, velocity, and acceleration.

2. Be able to distinguish between velocity and speed.

3. Be able to calculate the instantaneous velocity from a graph of position versus time.

4. Be able to state the important equations relating displacement, velocity, acceleration, and time that apply when acceleration is constant.

5. Be able to solve constant-acceleration problems.

B. Define, explain, or otherwise identify:

Particle, p. 16
Displacement, p. 17
Speed, p. 17
Average velocity, p. 17
Slope, p. 19

Instantaneous velocity, pp. 20, 21
Average acceleration, p. 22
Instantaneous acceleration, p. 23

C. True or false: If the statement is true, explain why it is true. If it is false, give a counterexample, that is, a known example that contradicts the statement.

1. The equation $\Delta x = v_0 t + \frac{1}{2}at^2$ is true for all motion in one dimension.

2. If the velocity is zero at a specific instant, the acceleration must also be zero at that instant.

3. If the acceleration of an object is zero, the object cannot be moving.

4. If the acceleration is zero, the x-versus-t curve is a straight line.

5. The equation $\Delta x = v_{av} \Delta t$ holds for all motion in one dimension.

Exercises

Section 2-1   Speed, Displacement, and Velocity

1. A jogger jogs 2 km in 10 min. What is his average speed in (*a*) metres per second and (*b*) miles per hour?

2. Find the average speed in metres per second and miles per hour for a runner who can run (*a*) 100 m in 10 s, (*b*) 1 mi in 4 min, and (*c*) 26 mi in 4 h.

3. As you drive into Kentucky, you notice that the mile marker on the interstate is 325. You drive straight through to mile marker 0 on the other side of the state in 6 h. You are then struck by a desire for fried chicken, so you turn around and drive back 25 mi for a snack. This return takes 30 min. (*a*) What was your average speed in miles per hour for the 350-mi trip? (*b*) What was your average velocity for the trip? (*c*) What was your average velocity for the last 25 mi?

4. A car travels in a straight line with average velocity of 96 km/h for 2.5 h and then with average velocity of 48 km/h for 1.5 h. (*a*) What is the total displacement for the 4-h trip? (*b*) What is the average velocity for the total trip?

5. Figure 2-12 shows the position of a particle versus time. Find the average velocity for the time intervals *a*, *b*, *c*, and *d* indicated.

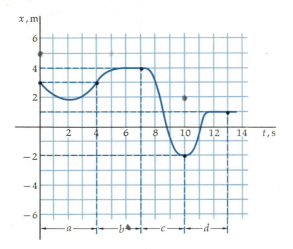

Figure 2-12
Graph of *x* versus *t* for Exercise 5.

6. A particle is at $x = +5$ m at $t = 0$, $x = -7$ m at $t = 6$ s, and $x = +2$ m at $t = 10$ s. Find the average velocity of the particle during the time intervals (*a*) 0 to 6 s, (*b*) 6 to 10 s, and (*c*) 0 to 10 s.

7. (*a*) How long does it take a supersonic jet flying at 2.4 times the speed of sound to fly across the Atlantic, which is about 5500 km wide? Take the speed of sound to be 350 m/s. (*b*) How long does it take a subsonic jet flying at 0.9 times the speed of sound to make the same trip? Assuming it takes 2 h at each end of the trip to get to or from the airport and to check or pick up your baggage, what is your average speed from your home to your final destination for (*c*) the supersonic jet and (*d*) the subsonic jet?

8. As you drive along the freeway, your speedometer quits working. You measure your speed by measuring the time it takes to travel between mile markers. (*a*) How many seconds should elapse between mile markers if your average speed is 55 mi/h? (*b*) What is your average speed in miles per hour if the time between mile markers is 45 s?

9. Light travels at a speed of 3×10^8 m/s. (*a*) How long does it take light to travel from the sun to earth, a distance of 1.5×10^{11} m? (*b*) How long does it take light to travel from the moon to the earth, a distance of 3.84×10^8 m? (*c*) A light-year is a unit of distance equal to that traveled by light in 1 year. Find the equivalent distance of 1 light-year in kilometres and in miles.

10. Continents move about 10 cm across the surface of the earth in 1 year. If North America was originally attached to Europe, how much time was needed for it to move 5500 km, thus forming the Atlantic Ocean? (Assume that Europe was stationary.)

Section 2-2 Instantaneous Velocity

11. For each of the four graphs of *x* versus *t* in Figure 2-13, indicate whether the velocity at time t_2 is greater than, less than, or equal to the velocity at time t_1. Also compare the speeds at these times.

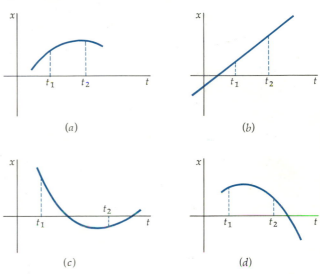

Figure 2-13
Graphs of *x* versus *t* for Exercise 11.

12. In the graph of *x* versus *t* in Figure 2-14, find (*a*) the average velocity between times $t = 0$ and $t = 2$ s and (*b*) the instantaneous velocity at time $t = 2$ s by measuring the slope of the tangent line indicated.

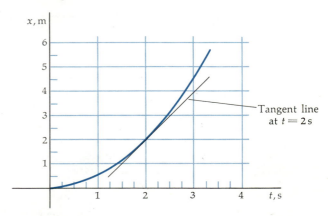

Figure 2-14
Graph of *x* versus *t* for Exercise 12 with a tangent line drawn at $t = 2$ s.

13. For the graph of x versus t in Figure 2-15, find the average velocity for the time intervals $\Delta t = t_2 - 0.75$ s when t_2 is 1.75, 1.5, 1.25, and 1.0 s. What is the instantaneous velocity at time $t = 0.75$ s?

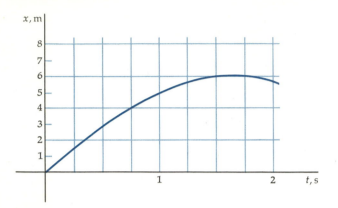

Figure 2-15
Graph of x versus t for Exercise 13.

14. For the graph of x versus t shown in Figure 2-16, find (*a*) the average velocity for the interval $t = 1$ s to $t = 5$ s. (*b*) Find the instantaneous velocity at $t = 4$ s. (*c*) At what time is the instantaneous velocity of the particle zero?

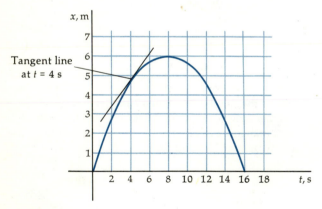

Figure 2-16
Graph of x versus t for Exercise 13 with a tangent line drawn at $t = 4$ s.

15. The position of a particle is given by the formula $x = (5 \text{ m/s}^2)t^2$. Make a table with columns t_2, Δt, Δx, and v_{av} when t_2 is 3, 2.5, 2.2, 2.1, 2.01, and 2.001 s. What is your estimate of the instantaneous velocity at time $t = 2$ s?

Section 2-3   Acceleration

16. A fast car can accelerate from 0 to 96 km/h in 5 s. What is the average acceleration during this interval? What is the ratio of this acceleration to the free-fall acceleration of gravity?

17. At $t = 5$ s, an object is traveling at 5 m/s. At $t = 8$ s, its velocity is -1 m/s. What is the average acceleration for this time interval?

18. A particle moves with velocity given by
$$v = (8 \text{ m/s}^2)t - 8 \text{ m/s}$$
(*a*) Find the average acceleration for the 1-s intervals beginning at $t = 3$ s and $t = 4$ s. (*b*) Sketch a graph of v versus t. What is the instantaneous acceleration at any time?

19. A car is traveling at 45 mi/h at time $t = 0$. It accelerates at a constant rate of $10\,\frac{\text{mi/h}}{\text{s}}$. (*a*) How fast is the car going at time $t = 1$ s? At $t = 2$ s? (*b*) Sketch a graph of its v versus t.

20. State whether the acceleration is positive, negative, or zero for each of the position functions shown in Figure 2-17.

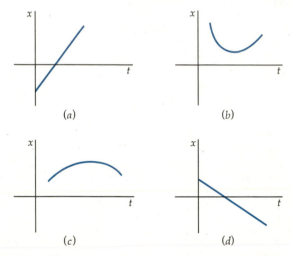

Figure 2-17
Graphs of x versus t for Exercise 20.

Section 2-4   Motion with Constant Acceleration

21. A car starts from rest and has a constant acceleration of 8 m/s². (*a*) How fast is it going after 10 s? (*b*) What is its average velocity for the time interval $t = 0$ to $t = 10$ s? (*c*) How far does it travel during this interval?

22. (*a*) How long does it take for a particle to travel 100 m if it begins from rest and accelerates at 10 m/s²? (*b*) What is its average velocity for the time interval from $t = 0$ to $t = 10$ s? (*c*) How far does it travel during this interval?

23. A world-land-speed car can decelerate at about 1 g. (That is, the magnitude of a is g.) (*a*) How long does it take the car to come to a stop from a record-setting run of 885 km/h? (*b*) How far does it travel while stopping?

24. A ball is dropped from a high cliff. (*a*) How far does it travel during the first second. (*b*) How fast is it moving after 1 s? (*c*) How fast is it moving after 2 s? (*d*) How far does it travel during the interval from $t = 1$ to $t = 2$ s?

25. What is the average acceleration of a rifle bullet that attains a speed of 600 m/s in a distance of 0.6 m?

26. A drag racer starts from rest and covers 0.35 mi at constant acceleration. Her average speed is 100 mi/h. (*a*) How long does it take her to cover the 0.35 mi? (*b*) What is her acceleration? (*c*) What is her final velocity?

27. A sailing spacecraft has been proposed that would be accelerated very slightly by the pressure of the sun's radiation. Suppose that such a craft started from rest and had an acceleration of 10^{-5} g. How many years would it take to get to Pluto, which is about 7.6×10^{12} m away?

28. A young person jumps from the ground with an initial upward speed of 3.5 m/s. How high would she rise (*a*) on earth and (*b*) on the moon, where $g_m = 1.6$ m/s²? (Neglect any effect of her space suit.)

29. In a sailing course, one is taught not to insist on the laws of the sea that say a motor ship must give way to a sailing vessel, at least not when approaching an oil tanker. If the oil tanker is moving at 10.3 m/s (about 20 knots) and decelerates at 0.01 m/s, find the distance required for it to stop.

30. A climber near the summit of Half Dome accidentally knocks loose a large rock. He sees it shatter at the bottom of the cliff 14 s later. (*a*) How far did the rock fall (assuming free-fall)? (*b*) How fast was the rock moving just before it hit bottom?

31. A modern jet fighter can achieve an acceleration of 1.1 g. Assuming constant acceleration during takeoff, how long a runway is needed for the jet to get from rest to its takeoff speed of 290 km/h?

32. A baseball is thrown straight up with initial speed of 30 m/s. (*a*) How high does it go (neglecting air resistance)? (*b*) How long is the ball in the air?

33. An electron in the electron gun of a TV tube undergoes an acceleration of 10^{14} m/s² while it travels 1 cm. It then travels 10 cm to the screen at constant speed. (*a*) What is its speed after traveling 1 cm? (*b*) How long does it take to travel the first centimetre? (*c*) How long does it take to travel the additional 10 cm to the screen?

34. An object with an initial velocity of $+6$ m/s has a constant acceleration of -2 m/s². (*a*) When will it be momentarily at rest? (*b*) How far does it travel before coming to rest momentarily? (*c*) When will it be back at its original position? (*d*) What will its velocity be then?

35. Blood flowing in an artery is found to have a speed 120 cm/s at one point and 90 cm/s at a point 40 cm further along the artery. (*a*) Assuming constant acceleration, what is the average velocity of the blood between the two points? (*b*) How long does it take to travel from one point to the other? (*c*) What is the acceleration?

36. A ball is thrown into the air with an initial speed that is great enough for the ball to take several seconds to reach its highest point. (*a*) What is the velocity of the ball 1 s before it reaches its highest point? (*b*) What is the velocity 1 s after it reaches its highest point? (*c*) What is the change in velocity during these two seconds? (*d*) Compute $\Delta v/\Delta t$ for this time interval.

37. The minimum distance for a controlled stop with no wheels locked for a certain car going 100 km/h is 52 m. (*a*) Find the acceleration, assuming it to be constant, and express your answer as a fraction of g. (*b*) How long does it take the car to stop? (*c*) What is the braking distance under the same conditions except that the initial speed is 125 km/h?

Problems

1. It has been found that galaxies are moving away from the earth at a speed that is proportional to their distance from the earth. This discovery is known as Hubble's law. The speed of a galaxy a distance r away is given by $v = Hr$, where H is Hubble's constant, which equals 1.58×10^{-18} s⁻¹. What is the speed of a galaxy (*a*) 5×10^{22} m from earth and (*b*) 2×10^{25} m from earth? (*c*) If each of these galaxies has traveled with constant speed, how long ago were they both located at the same place as the earth?

2. In the Blackhawk landslide in California, a mass of rock and mud fell 460 m down a mountain and then traveled 8 km across a level plain on a cushion of compressed air. Assume that the mud dropped with the free-fall acceleration of gravity, and then slid horizontally with constant deceleration. (*a*) How long did the mud take to drop the 460 m? (*b*) How fast was it traveling when it reached bottom? (*c*) How long did the mud take to slide the 8 km horizontally?

3. A ball is dropped from a height of 2 m and rebounds from the floor to a height of 1.5 m. (*a*) What is the velocity of the ball just before it hits the floor? (*b*) What is its velocity just as it leaves the floor? (*c*) If the ball is in contact with the floor for 0.02 s, what are the magnitude and the direction of the average acceleration of the ball while in contact with the floor?

4. A car traveling at constant speed of 20 m/s passes an intersection at time $t = 0$, and 54 s later a second car passes the same intersection traveling at 30 m/s in the same direction. (*a*) Sketch the position functions x versus t for each car. (*b*) Find when the second car overtakes the first. (*c*) How far have the cars traveled when this happens?

5. In this problem, we examine the benefit of wearing seat belts. Suppose you are in the front seat of a car traveling 90 km/h as it skids into a tree. The car comes to stop in a distance of 12 m as the front end crumples around the tree. (*a*) If you are wearing a seat belt, you stop with the same (negative) acceleration as the car. What is the magnitude of this acceleration? (*b*) If you are not wearing a seat belt, you continue forward at 90 km/h until you hit the dashboard. You are then brought to rest in a distance of about 15 cm as your body crumples against the dashboard. Find the magnitude of the acceleration of your body in this case.

6. Figure 2-18 shows the position of a car plotted as a function of time. At which of the times t_0 to t_7 is (*a*) the velocity negative, (*b*) the velocity positive, (*c*) the velocity zero, (*d*) the acceleration negative, (*e*) the acceleration positive, and (*f*) the acceleration zero?

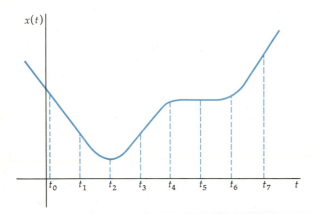

Figure 2-18
Graph of x versus t for the car in Problem 6.

7. A car making a 100-km journey covers the first 50 km at 40 km/h. How fast must it cover the second 50 km to average 50 km/h?

8. A glacier can surge as fast as 6×10^{-4} m/s. (*a*) What is this rate in metres per day? (*b*) What is the volume flow rate (m^3/day) for ice flowing into a valley if the glacier is 80 m high and 1200 m wide.

9. Hare and Tortoise begin a 10-km race at time $t = 0$. Hare runs at 4 m/s and quickly outdistances Tortoise, who runs at 1 m/s (about 10 times faster than a tortoise can actually run, but convenient for this problem). After running 5 min, Hare stops and falls asleep. His nap lasts 2 h and 15 min. He then awakens and runs again at 4 m/s. (*a*) Plot an x-versus-t curve for each on the same graph. (*b*) Who wins the race? By how much (distance and time)? (*c*) How long can Hare safely nap and still win the race?

10. The graphs in Figure 2-19 describe different motions of an object. For each graph, indicate (*a*) at which times the acceleration of the object is positive, negative, or zero. (*b*) At which times is the acceleration constant? (*c*) At which times is the instantaneous velocity zero?

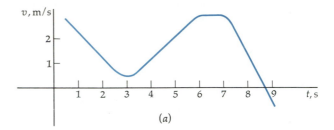

(*a*)

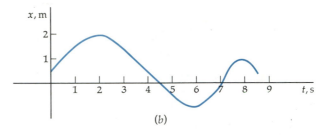

(*b*)

Figure 2-19
Graphs of (*a*) v versus t and (*b*) x versus t for Problem 10. These graphs represent two different and unrelated motions.

11. A car is traveling 80 km/h in a school zone. A police car starts from rest just as the speeder passes it and accelerates at a constant rate of 8 km/h·s. (*a*) Make a sketch of x versus t for both cars. (*b*) When does the police car catch the speeding car? (*c*) How fast is the police car traveling when it finally catches the speeder?

12. The police car in Problem 11 chases a speeder traveling 125 km/h. The maximum speed of the police car is 190 km/h. The police car travels from rest with a constant acceleration of 8 km/h·s until its speed reaches 190 km/h and then moves at constant speed. (*a*) When does it catch the speeder if it starts just as the speeder passes? (*b*) How far has each car traveled? (*c*) Sketch the x-versus-t graph curve for each car.

13. When the police car of Problem 12, traveling at 190 km/h, is 100 m behind the speeder, traveling at 125 km/h, the speeder sees the police car and slams on his brakes, locking the wheels. (*a*) Assuming that each car can brake at 6 m/s² and that the driver of the police car brakes as soon as she sees the brake lights of the speeder (that is, with no reaction time), show that the cars collide. (*b*) What is the speed of the police car relative to the speeder when they collide? How long after the brakes are applied do they collide? (c) The time interval between the police officer's seeing the brake lights and hitting her brakes is called her reaction time T. Estimate or measure your reaction time and, assuming you were the police officer, discuss how it affects this problem. (See Problem 14.)

14. You can check human reaction times in the following simple way. Have a friend rest his or her hand on the edge of a table with fingers over the edge and the separated thumb and finger near the 0 cm mark of a ruler that you hold at its top. Then drop the ruler and have your friend catch it. (a) If your friend grabs the ruler at the 7 cm mark, what is his or her reaction time? (b) If your friend's reaction time is 0.2 s, where will he or she catch the ruler? (c) If the reaction time is 0.2 s, estimate the velocity of a nerve impulse, assuming a reasonable distance from hand to brain. (d) Would you replace the ruler with a dollar bill and allow your friend to keep the bill if he or she catches it?

15. A passenger is running at his maximum speed of 8 m/s to catch a train. When he is a distance d from the nearest entry to the train, the train starts from rest and moves away from the passenger with a constant acceleration of 0.4 m/s^2. (a) If d = 30 m and the passenger keeps running, will he be able to jump onto the train? (b) Sketch the position function x versus t for the train choosing x = 0 at t = 0. On the same graph, sketch the position function for the passenger for various values of separation distances d, including d = 30 m and the critical value d_c, which is such that he just catches the train. (c) For the critical separation distance d_c, what is the speed of the train when the passenger catches it? (d) What is its average speed for the time t = 0 until he catches it? (e) What is the value of d_c?

16. A flower pot falls from the ledge of an apartment building. A person in an apartment below just happens to have a stop watch handy and notices that it takes 0.2 s for the pot to fall past his window, which is 4 m high. How far above the top of the window is the ledge from which the pot fell?

CHAPTER THREE

Motion in Two and Three Dimensions

I shot an arrow into the air,
It fell to earth, I knew not where.

HENRY WADSWORTH LONGFELLOW

In this chapter, you will build on the work you have done with motion in one dimension and acceleration. You will learn, as Galileo did, what causes the beautiful parabolic arcs of fountains and fireworks. You will also learn the vector ideas needed to understand how to sail against the wind, how to head an airplane if you want to go east when there's a northerly cross-wind, and why satellites don't fall out of the sky. All this calls for a little trigonometry, but only as much as can be written out on half a notebook page.

We now extend our description of the motion of a particle to the more general cases of motion in two and three dimensions. In these cases, displacement, velocity, and acceleration are *vectors*, quantities that have direction in space as well as magnitude. In future chapters, we shall encounter many other vector quantities, such as force, momentum, and the electric field. Quantities that have magnitude only and no associated direction, such as distance, speed, mass, or temperature, are called *scalars*.

In this chapter, we shall investigate the properties of vectors in general, and we shall study displacement, velocity, and acceleration in particular. Many of the interesting features of motion in more than one dimension can be seen with two-dimensional motion. Since this motion is easily illustrated on paper or a blackboard, most of our examples will be limited to two dimensions. Two important special types of motion in a plane, projectile motion and circular motion, will be discussed in detail.

3-1   The Displacement Vector

If you asked somebody where the post office is and they said it is 10 blocks away, you would most likely ask in which direction before you set off to find it. Whether it is 10 blocks east, 10 blocks north, or 6 blocks west and 8 blocks south (and therefore 14 blocks walking distance but 10 blocks "as the crow flies") makes quite a difference. The quantity that gives the straight-line distance and the direction from one point in space to another is called the *displacement vector*. A vector is represented by an arrow showing the direction of the vector with a length proportional to the magnitude of the vector.

Figure 3-1 shows a curved path in the *xy* plane of a particle that moves from point P_1 to a second point P_2 and then to a third point P_3. The displacement from point P_1 to point P_2 is represented by the arrow **A**. Note that the displacement **A** does not depend on the path taken by the particle as it moves from P_1 to P_2 but only on the endpoints P_1 and P_2. A second displacement from P_2 to P_3 is indicated by the arrow **B**. The resultant displacement from P_1 to P_3 is represented by the arrow **C**. The resultant displacement vector **C** is the sum of the two successive displacements, **A** and **B**.

$$\mathbf{C} = \mathbf{A} + \mathbf{B} \qquad\qquad\qquad 3\text{-}1$$

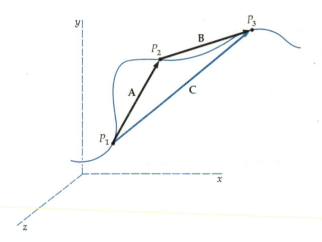

Figure 3-1
Addition of vectors. The displacement **C** is equivalent to the successive displacements **A** and **B**; that is, $\mathbf{C} = \mathbf{A} + \mathbf{B}$.

It is important to distinguish vector quantities from scalar quantities. We denote vector quantities with boldface type, as in **A**, **B**, and **C**. (In handwriting, we indicate a vector by drawing an arrow over the symbol.) The *magnitude* of a vector **A** is written |**A**| or simply *A*. The magnitude of a vector ordinarily has physical units. For example, a displacement vector has a magnitude that can be expressed in feet, metres, or any other measure of distance. Note that the sum of the magnitudes of **A** and **B** does not equal the magnitude of **C** unless **A** and **B** are in the same direction. That is, $\mathbf{C} = \mathbf{A} + \mathbf{B}$ does *not* imply that $C = A + B$.

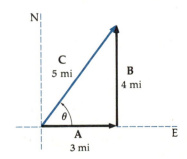

Example 3-1   A man walks 3 mi east and then 4 mi north. What is his resultant displacement?

The two displacements and the resultant displacement are shown in Figure 3-2. Since these three vectors form a right triangle, we can find the magnitude of the resultant displacement from the pythagorean theorem.

$$C^2 = A^2 + B^2 = (3 \text{ mi})^2 + (4 \text{ mi})^2 = 25 \text{ mi}^2$$

$$C = 5 \text{ mi}$$

Figure 3-2
Displacement vectors for Example 3-1. The magnitude of the resultant displacement vector **C** can be found from the pythagorean theorem.

To describe the resultant displacement, we need to give the direction as well as the magnitude. If θ is the angle between the east axis and the resultant displacement, we have from Figure 3-2

$$\tan \theta = \frac{4 \text{ mi}}{3 \text{ mi}} = 1.33$$

We can find θ using trigonometric tables or a calculator that has trigonometric functions.

$$\theta = 53.1°$$

The resultant displacement is therefore 5 mi at 53.1° north of east.

Example 3-2   A man walks 3 mi east and then 4 mi at 60° north of east. What is his resultant displacement?

The vector displacements for this example are shown in Figure 3-3. In this case, the triangle formed by the three vectors is not a right triangle, so we cannot use the pythagorean theorem to find the resultant displacement. In the next section, we shall learn how to find the resultant vector for a case like this using vector components. For now, we find the resultant vector graphically by drawing each of the displacements to scale and measuring the resultant displacement with a ruler. For example, if we draw the first displacement vector, **A**, 3 cm long and the second one, **B**, 4 cm long, we find the resultant vector, **C**, to be about 6 cm long. Thus, the magnitude of the resultant displacement is 6 mi. The angle made between the resultant displacement and the east direction is about 35° as measured with a protractor.

Questions

1. Can the displacement of a particle have a magnitude that is less than the distance traveled by the particle along its path? Can its magnitude be more than the distance traveled? Explain.

2. Give an example in which the distance traveled is a significant amount yet the corresponding displacement is zero.

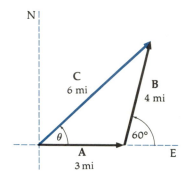

Figure 3-3
Displacement vectors for Example 3-2. Since **A** and **B** are not perpendicular, the pythagorean theorem cannot be used to find the magnitude of **C**. Instead, C can be found graphically.

3-2   Addition of Vectors by Components

In the previous section, we saw that to add two vectors graphically we simply place the tail of one vector at the tip of the other and draw the resultant vector, as was done in Examples 3-1 and 3-2. Figure 3-4 shows an alternate method called the *parallelogram method*. From this figure, we see that it makes no difference in which order we add two vectors.

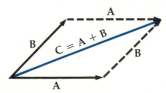

Figure 3-4
Parallelogram method of graphically adding two vectors. The resultant **C** = **A** + **B** is along the diagonal of the parallelogram formed when **A** and **B** are placed tail to tail. We can see from this figure that the order of addition makes no difference; that is, **A** + **B** is the same as **B** + **A**.

We can subtract vector **B** from vector **A**, as shown in Figure 3-5*a*, by simply adding $-\mathbf{B}$, which has the same magnitude as **B** but points in the opposite direction, to **A**. The result is $\mathbf{C} = \mathbf{A} + (-\mathbf{B}) = \mathbf{A} - \mathbf{B}$. Another method of subtraction, illustrated in Figure 3-5*b*, is to draw the two vectors **A** and **B** tail to tail and then note that the vector $\mathbf{C} = \mathbf{A} - \mathbf{B}$ is that vector that must be added to **B** to obtain the resultant vector **A**.

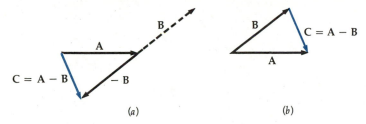

        (*a*)   (*b*)

Figure 3-5
Subtraction of vectors. (*a*) Here $\mathbf{C} = \mathbf{A} - \mathbf{B}$ is found by adding $-\mathbf{B}$ to **A**. (*b*) An alternate method of finding $\mathbf{A} - \mathbf{B}$ is to find the vector **C** that when added to **B** gives the vector **A**.

It is often more convenient to add or subtract vectors analytically rather than graphically. We can do this by first breaking down the vectors into their components. In Figure 3-6, we see that a vector **A** can be thought of as the resultant vector of two vectors, $\mathbf{A}_x$ parallel to the *x* axis and $\mathbf{A}_y$ parallel to the *y* axis.

$$\mathbf{A} = \mathbf{A}_x + \mathbf{A}_y \qquad\qquad 3\text{-}2$$

$\mathbf{A}_x$ and $\mathbf{A}_y$ are called the component vectors of **A**. The quantities A_x and A_y are called the *components* of the vector **A**. They are the "projections" of **A** on the *x* and *y* axes. They can be positive or negative. For example, if **A** points in the negative *x* direction, A_x is negative. If θ is the angle between the vector **A** and the *x* axis, we see from Figure 3-6 that

$$\tan \theta = A_y / A_x \qquad\qquad 3\text{-}3$$

$$\sin \theta = A_y / A \qquad\qquad 3\text{-}4$$

$$\cos \theta = A_x / A \qquad\qquad 3\text{-}5$$

where A is the magnitude of **A**. We can therefore determine the components of **A** analytically from the magnitude A and the angle θ by

$$A_x = A \cos \theta \qquad\qquad 3\text{-}6a$$

and

$$A_y = A \sin \theta \qquad\qquad 3\text{-}6b$$

Conversely, if we know the components A_x and A_y, we can find the angle θ from Equation 3-3 and the magnitude A from the pythagorean theorem.

$$A = \sqrt{A_x^2 + A_y^2} \qquad\qquad 3\text{-}7$$

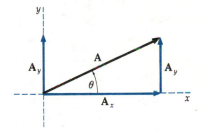

Figure 3-6
A vector **A** can be thought of as the sum of the component vectors $\mathbf{A}_x$ and $\mathbf{A}_y$.

Components

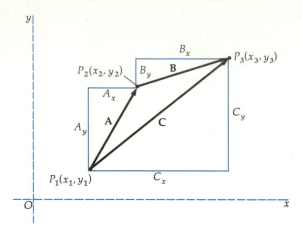

Figure 3-7
The x and y components of the vectors **A**, **B**, and **C** = **A** + **B**. We see that $C_x = A_x + B_x$ and $C_y = A_y + B_y$.

We now use the vector components to add vectors analytically. Figure 3-7 shows vectors **A** and **B**, whose sum equals **C**. The components of each vector are shown in the figure. We can see from the figure that the x component of **C** equals the sum of the x components of **A** and **B** and that the y component of **C** equals the sum of the y components of **A** and **B**.

$$C_x = A_x + B_x \qquad\qquad\qquad 3\text{-}8a$$

and

$$C_y = A_y + B_y \qquad\qquad\qquad 3\text{-}8b$$

Example 3-3   Work Example 3-2 using vector components.

In this example, a man first walked east 3 mi. If **A** represents his displacement and we take the x axis in the easterly direction, the components of **A** are

$$A_x = 3 \text{ mi}$$

and

$$A_y = 0$$

The next displacement was of magnitude 4 miles at 60° north of east. The vector **B** representing this displacement has components

$$B_x = (4 \text{ mi}) \cos 60° = (4 \text{ mi})(0.5) = 2 \text{ mi}$$

and

$$B_y = (4 \text{ mi}) \sin 60° = (4 \text{ mi})(0.866) = 3.46 \text{ mi}$$

The components of the resultant displacement are thus

$$C_x = A_x + B_x = 3 \text{ mi} + 2 \text{ mi} = 5 \text{ mi}$$

and

$$C_y = A_y + B_y = 0 + 3.46 \text{ mi} = 3.46 \text{ mi}$$

We obtain the magnitude of the resultant displacement **C** from the pythagorean theorem.

$$C^2 = C_x^2 + C_y^2 = (5 \text{ mi})^2 + (3.46 \text{ mi})^2 = 37 \text{ mi}^2$$

So,

$$C = \sqrt{37} \text{ mi} = 6.1 \text{ mi}$$

The angle between **C** and the x axis is found from

$$\tan \theta = C_y/C_x = 3.46/5 = 0.692$$

Thus, the angle θ is 34.7°. This agrees with the results of Example 3-2 within the accuracy of our measurement in that example.

Questions

3. Can two vectors of unequal magnitude be added to give zero?

4. How would you subtract two vectors using the component method?

3-3   Velocity and Acceleration Vectors

Suppose that you are driving a car at 50 km/h as indicated by the speedo-meter reading and are headed south as indicated by a compass. The speedo-meter reading, the speed, is the magnitude of the velocity whereas the compass reading gives the direction of the velocity. The instantaneous ve-locity vector is a vector that points in the direction of motion and has magnitude equal to the speed of the car. It equals the rate of change of the displacement vector.

Figure 3-8 shows a particle moving along some curve in space. At some time t_1, it is at point P_1, indicated by the position vector $\mathbf{r}_1$ drawn from the origin to P_1. At some later time t_2, it is at point P_2, indicated by the position vector $\mathbf{r}_2$. The displacement is the change in the position vector

$$\Delta \mathbf{r} = \mathbf{r}_2 - \mathbf{r}_1 \qquad \text{3-9} \qquad \text{Displacement vector}$$

(This is analogous to our one-dimensional definition in Chapter 2 where the displacement is the change in the position coordinate x.) The new position vector is the sum of the original position vector $\mathbf{r}_1$ and $\Delta \mathbf{r}$ as shown. The ratio of the displacement vector to the time interval $\Delta t = t_2 - t_1$ is the average velocity.

$$\mathbf{v}_{av} = \frac{\Delta \mathbf{r}}{\Delta t} \qquad \text{3-10} \qquad \text{Average velocity vector}$$

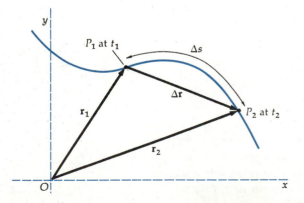

Figure 3-8
A particle moving along some arbitrary curve in space with position vectors $\mathbf{r}_1$ and $\mathbf{r}_2$ at two different times t_1 and t_2. The displacement vector $\Delta \mathbf{r}$ is then the difference in the position vectors $\Delta \mathbf{r} = \mathbf{r}_2 - \mathbf{r}_1$.

We note from Figure 3-8 that the magnitude of the displacement vector is *not* equal to the distance traveled Δs as measured along the curve. It is, in fact, less than this distance (unless the particle traveled in a straight line between points P_1 and P_2). However, if we consider smaller and smaller time intervals, as indicated in Figure 3-9, the magnitude of the displacement approaches the distance traveled by the particle along the curve and the direction of $\Delta \mathbf{r}$ approaches the direction of the line tangent to the curve at point P_1. We define the instantaneous-velocity vector as the limiting value of the average velocity as the time interval Δt approaches zero. The direction of the instantaneous velocity is along the line tangent to the curve traveled by the particle in space. The magnitude of the instantaneous velocity is the instantaneous speed.

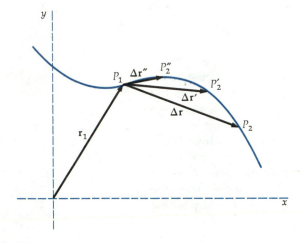

Figure 3-9
As smaller and smaller time intervals are considered, the magnitude of the displacement vector approaches the distance traveled along the curve. The direction of the displacement vector approaches the direction of the line tangent to the curve at point P_1.

Example 3-4 A sailboat has initial coordinates $(x_1, y_1) = (100 \text{ m}, \ 200 \text{ m})$. After 2.00 min, it has the coordinates $(x_2, y_2) = (120 \text{ m}, \ 210 \text{ m})$. What are the components, magnitude, and direction of its average velocity for this 2.00-min interval?

$$v_{x,\text{av}} = \frac{x_2 - x_1}{\Delta t} = \frac{120 - 100 \text{ m}}{2.00 \text{ min}} = 10.0 \text{ m/min}$$

$$v_{y,\text{av}} = \frac{y_2 - y_1}{\Delta t} = \frac{210 - 200 \text{ m}}{2.00 \text{ min}} = 5.0 \text{ m/min}$$

$$v_{\text{av}} = \sqrt{(v_{x,\text{av}})^2 + (v_{y,\text{av}})^2}$$
$$= \sqrt{10.0^2 + 5.0^2}$$
$$= \sqrt{125} = 11.2 \text{ m/min}$$

$$\tan \theta = \frac{v_{y,\text{av}}}{v_{x,\text{av}}} = \frac{5.0 \text{ m/min}}{10.0 \text{ m/min}} = 0.500$$
$$\theta = 26.6°$$

The velocity of an object is, of course, measured relative to some coordinate system, but this coordinate system may be moving relative to a second coordinate system. For example, suppose that you swim with velocity $\mathbf{v}_{yw}$

relative to still water but you are swimming across a river that has a velocity $\mathbf{v}_{ws}$ relative to the shore. Your velocity relative to the shore, $\mathbf{v}_{ys}$, is the sum of these two velocities

$$\mathbf{v}_{ys} = \mathbf{v}_{yw} + \mathbf{v}_{ws} \qquad\qquad 3\text{-}11$$

The addition of relative velocities is done in the same way as the addition of displacements, either graphically by placing the tail of one velocity vector at the tip of the other or analytically using vector components.

Example 3-5   A river flows from west to east with a speed of 3 m/s. A boy swims north across the river with a speed of 2 m/s relative to the water. What is the velocity of the boy relative to the shore?

Figure 3-10 shows the velocity vectors for this problem. The velocity of the boy relative to the shore is the vector sum of the velocity of the boy relative to the water, $\mathbf{v}_{bw}$, and the velocity of the water relative to the shore, $\mathbf{v}_{ws}$, shown in the figure. The magnitude of this velocity is

$$v = \sqrt{v_{ws}^2 + v_{bw}^2} = \sqrt{2^2 + 3^2} = \sqrt{13} = 3.61 \text{ m/s}$$

The direction is at an angle θ to the shore where $\tan \theta = v_{bw}/v_{ws} = \frac{3}{2}$. Thus, $\theta = 56.3°$.

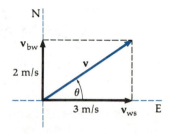

Figure 3-10
Velocity vectors for Example 3-5.

The average acceleration vector is defined as the ratio of the change in the instantaneous velocity vector $\Delta \mathbf{v}$ to the time interval Δt.

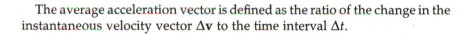

$$\mathbf{a}_{av} = \frac{\Delta \mathbf{v}}{\Delta t} \qquad\qquad 3\text{-}12 \qquad \text{Average acceleration vector}$$

The instantaneous acceleration vector is the limit of this ratio as the time interval approaches zero.

It is particularly important to note that the velocity vector may be changing in magnitude, direction, or both. If the velocity vector is changing in any way, the particle is accelerating. We are perhaps most familiar with acceleration in which the velocity changes in magnitude; that is, the speed changes. However,

A particle can be traveling with constant speed and still be accelerating if the direction of the velocity vector is changing.

A particularly important example of this is circular motion, which is discussed in Section 3-5. This acceleration is just as real as when the speed is changing. We shall see in the next chapter that a force is necessary to produce an acceleration of a particle. The force required to produce a given acceleration (say, 1 m/s² downward) on a given particle is the same whether this acceleration is associated with a change in the magnitude of the velocity vector, a change in its direction, or both.

Questions

5. For an arbitrary motion of a given particle, does the direction of the velocity vector have any particular relation to the direction of the position vector?

6. Give examples in which the directions of the velocity and position vectors are (*a*) opposite, (*b*) the same, and (*c*) mutually perpendicular.

7. How is it possible for a particle moving at constant speed to be accelerating? Can a particle with constant velocity be accelerating at the same time?

8. Is it possible for a particle to round a curve without accelerating?

3-4   Projectile Motion

An interesting application of motion in two dimensions is that of a projectile, which is a body that is launched into the air and is allowed to move freely. The general motion of a projectile is complicated by air resistance, the rotation of the earth, and the variation in the acceleration of gravity. For simplicity, we shall neglect these complications. The projectile then has a constant acceleration directed vertically downward with magnitude $g = 9.81 \text{ m/s}^2 = 32.2 \text{ ft/s}^2$. If we take the y axis as vertical with the positive direction upward and the x axis as horizontal with the positive direction in the direction of the original horizontal component of the projectile's velocity, we have for the acceleration

$$a_y = -g$$
$$a_x = 0$$

3-13

Since there is no horizontal acceleration, the horizontal component of the velocity is constant. On the other hand, the vertical motion is simply motion with constant acceleration, identical to that studied in Chapter 2.

Illuminated fountains, St. Louis, Missouri. The jets follow parabolic paths like those followed by projectiles.

Suppose we launch the projectile from the origin with an initial speed v_0 at angle θ with the horizontal axis (Figure 3-11). The initial velocity then has the components

$$v_{0x} = v_0 \cos \theta \quad \text{and} \quad v_{0y} = v_0 \sin \theta \qquad \text{3-14}$$

Since there is no horizontal acceleration, the x component of the velocity is constant:

$$v_x = v_{0x} \qquad \text{3-15}$$

The y component varies with time according to

$$v_y = v_{0y} - gt \qquad \text{3-16}$$

(This is the same as Equation 2-8 with $a = -g$.) The components of the displacement of the projectile are

$$\Delta x = v_{0x}t$$
$$\Delta y = v_{0y}t - \tfrac{1}{2}gt^2 \qquad \text{3-17}$$

(Compare this equation for y with Equation 2-10.)

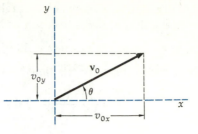

Figure 3-11
The components of the initial velocity of a projectile are $v_{0x} = v_0 \cos \theta_0$ and $v_{0y} = v_0 \sin \theta_0$, where θ_0 is the angle made by v_0 and the horizontal x axis.

Equations for projectile motion

Example 3-6   A ball is thrown into the air with an initial velocity of 50 m/s at 37° to the horizontal. Find the total time the ball is in the air and the total horizontal distance it travels using the approximation $g = 10$ m/s^2.

The components of the initial velocity vector are

$$v_{0x} = (50 \text{ m/s}) \cos 37° = 40 \text{ m/s}$$

$$v_{0y} = (50 \text{ m/s}) \sin 37° = 30 \text{ m/s}$$

Figure 3-12 shows the height y versus t for this example. This curve is identical to the one in Figure 2-8 for Examples 2-8 and 2-9 because the vertical acceleration and the velocities are the same for these two examples. Since the ball moves 40 m horizontally during each second, we can interpret this curve as a sketch of y versus x if we change the horizontal axis from a time scale to a distance scale by multiplying the time values by 40 m/s.

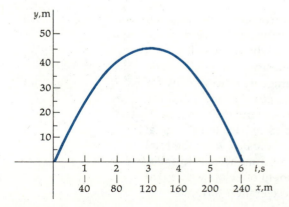

Figure 3-12
Graph of y versus t and y versus x for the ball in Example 3-6. The time scale can be converted into a horizontal distance scale by multiplying each time by 40 m/s because x is related to t by $x = (40 \text{ m/s})t$.

The curve y versus x is a parabola. The total time the ball is in the air is twice the time it takes to reach its highest point. We find this time from

$$v_y = v_{0y} - gt = 30 \text{ m/s} - (10 \text{ m/s}^2)t$$

and solving for the time t when v_y is zero. The result is

$$t = \frac{30 \text{ m/s}}{10 \text{ m/s}^2} = 3.0 \text{ s}$$

This is, of course, the same result found in Example 2-8. The total time the ball is in the air is then 6 s. Since the ball moves horizontally with the constant velocity of 40 m/s, the horizontal distance traveled is 40 m/s × 6 s = 240 m. This distance is called the *range* of a projectile.

In the preceding example, we found the range of the projectile by finding the total time the projectile was in the air from the vertical component of the velocity and then using this time and the horizontal component of the velocity to find the horizontal distance traveled. This method works whether or not the projectile lands at the same elevation as that from which it was projected.

Example 3-7   The ball in Example 3-6 is projected with the same initial velocity but from a cliff that is 55 m above the plane below (Figure 3-13). Where does the ball land?

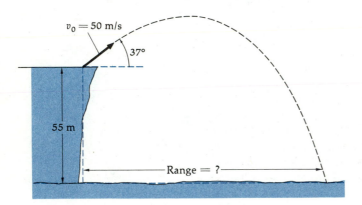

In this case, the time for the ball to reach its maximum height is again 3 s, but the time of fall is longer because it falls a greater distance. We find the distance of this fall by first finding the maximum height of the ball as was done in Example 2-7. The vertical component of velocity is initially 30 m/s, and at the ball's greatest height, it is zero. The average upward velocity for the 3 s it takes to reach its maximum height is therefore 15 m/s, and the maximum height is $y = (15 \text{ m/s})(3 \text{ s}) = 45 \text{ m}$, as was found in Example 2-7. After reaching this maximum height, the ball falls 45 m back to its original level of projection plus an additional 55 m to the plane below, so the total distance of fall is 45 m + 55 m = 100 m. The time for the ball to fall a distance of 100 m from "rest" (the vertical component of velocity is zero at the top of the path) is found from the constant acceleration formula

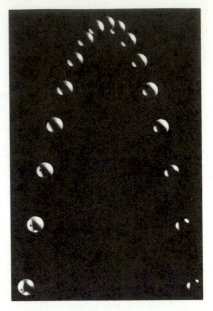

Multiflash photograph of a ball thrown into the air. The position of the ball is recorded at approximately 0.43-s intervals.

Figure 3-13
A ball thrown from a cliff in Example 3-7. The range is found by first finding the total time the ball is in the air and then multiplying by the x component of velocity.

$$\Delta y = -\tfrac{1}{2}gt^2 = -\tfrac{1}{2}(10 \text{ m/s}^2)t^2 = -100 \text{ m}$$

or

$$t = \sqrt{20 \text{ s}^2} = 4.5 \text{ s}$$

The total time the ball is in the air is thus $3 \text{ s} + 4.5 \text{ s} = 7.5 \text{ s}$, and the horizontal distance covered in this time is

$$x = v_{0x}t = (40 \text{ m/s})(7.5 \text{ s}) = 300 \text{ m}$$

We could have used Equation 3-17 directly to determine the time without considering the problem in two parts. Since the ball lands 55 m below its starting point, we set $\Delta y = -55$ m. The time is then found from

$$\Delta y = v_{0y}t - \tfrac{1}{2}gt^2$$

$$-55 \text{ m} = (30 \text{ m/s})t - \tfrac{1}{2}(10 \text{ m/s}^2)t^2$$

There are two solutions for this quadratic equation, $t = -1.5$ s and $t = +7.5$ s. The negative time corresponds to the time of 1.5 s it would take for the ball to reach its starting point at $y = 0$ if it were projected from $y = -55$ m with an initial speed such that it would be moving at 30 m/s when it reached $y = 0$.

For the special case in which the initial and final elevations are equal, we can derive a general formula for the range of a projectile in terms of its initial speed and the angle of projection. The time for the projectile to reach its maximum height is found by setting the vertical component of its velocity equal to zero.

$$v_y = v_{0y} - gt = 0$$

or

$$t = \frac{v_{0y}}{g}$$

The range R is then the distance traveled in twice this time.

$$R = v_{0x} \times \frac{2v_{0y}}{g} = \frac{2v_{0x}v_{0y}}{g}$$

In terms of the initial speed v_0 and angle θ_0, the range is

$$R = \frac{2(v_0 \cos \theta)(v_0 \sin \theta)}{g} = \frac{2v_0^2 \sin \theta \cos \theta}{g}$$

This can be further simplified by using the trigonometric identity for the sine of twice an angle

$$\sin 2\theta = 2 \sin \theta \cos \theta$$

We have then

$$R = \frac{v_0^2 \sin 2\theta}{g} \qquad \text{3-18 Range of a projectile}$$

Since the maximum value of $\sin 2\theta_0$ is 1 when $2\theta_0 = 90°$ or $\theta_0 = 45°$, the range is a maximum v_0^2/g when $\theta_0 = 45°$.

Equation 3-18 would be useful if we wanted to find the range for many projectile problems where the initial and final elevations were the same. Note, however, that the formula could not be used in Example 3-7 because the elevations there were not the same. More importantly, Equation 3-18 is useful because we can learn something about the dependence of the range on the initial angle of projection. For example, we just saw that the range is maximum when the angle is 45°.

We note that the horizontal distance traveled is the product of the initial horizontal velocity component v_{0x} and the time the projectile is in the air, which is in turn proportional to v_{0y}. The maximum range occurs when these components are equal. For that case, the angle of projection is 45°. In some practical applications, other considerations are important. For example, in the shot put, the ball is projected from an initial height of about 1 m rather than from the ground level. This initial height has the effect of increasing the time the ball is in the air. The range is maximum when v_{0x} is somewhat greater than v_{0y}; that is, at an angle somewhat smaller than 45°. Studies of the form of the best shot putters show that maximum range occurs with an initial angle of about 42°. With artillery shells, air resistance must be taken into account to predict the range accurately. As expected, air resistance reduces the range for a given angle of projection. It also decreases the optimum angle of projection slightly. An interesting effect caused by the rotation of the earth is important for long-range ballistic missiles. The motion of the missile does not take place entirely in a plane. Instead, there is a slight drift to the right in the northern hemisphere and to the left in the southern hemisphere. It is caused by the *Coriolis effect,* which arises because the surface of the earth is accelerating as a consequence of the earth's rotation.

According to our analysis of projectile motion, an object dropped from a height h above the ground will hit the ground in the same time as one projected horizontally from the same height. In each case, the distance the object *falls* is given by $y = \frac{1}{2}gt^2$ (measuring y downward from the original height). This remarkable fact can easily be demonstrated. In fact, it was first commented upon during the Renaissance by Galileo Galilei (1564–1642), the first person to give the modern, quantitative description of projectile motion we have been discussing. Galileo used this observation to illustrate the validity of treating the horizontal and vertical components of a projectile's motion as independent motion.

Galileo Galilei (1564–1642).

Example 3-8 A park ranger with a tranquilizer dart gun wishes to shoot a monkey hanging from a branch of a tree. The ranger aims right at the monkey, not realizing that the dart will follow a parabolic path and thus fall below the monkey. The monkey, however, sees the dart leave the gun and lets go of the branch, expecting to avoid the dart. Show that the monkey will be hit regardless of the initial velocity of the dart so long as the velocity is great enough for the dart to travel the horizontal distance to the tree before hitting the ground.

This problem is often demonstrated using a target suspended by an electromagnet. When the dart leaves the gun, the circuit is broken and the target falls. Let the horizontal distance to the tree be x and the original height of the monkey be h, as shown in Figure 3-14. The dart is projected at an angle

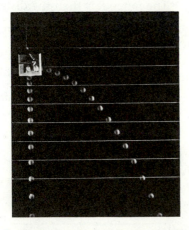

Comparison of a ball dropped with one projected horizontally. As can be seen, the vertical position is independent of the horizontal motion.

given by $\tan \theta = h/x$. If there were no gravity, the dart would reach the height h in the time t it takes to travel the horizontal distance x:

$$y = v_{0y}t = h$$

$$t = x/v_{0x}$$

However, because of gravity, the dart has a vertical acceleration downward. In time $t = x/v_{0x}$, the dart reaches a height given by

$$y = v_{0y}t - \tfrac{1}{2}gt^2 = h - \tfrac{1}{2}gt^2$$

This is lower than h by $\tfrac{1}{2}gt^2$, which is exactly the amount the monkey falls in the same time. In the usual lecture demonstration, the initial velocity of the dart is varied so that for a large v_0 the target is hit very near its original height and for a small v_0 it is hit just before it reaches the floor.

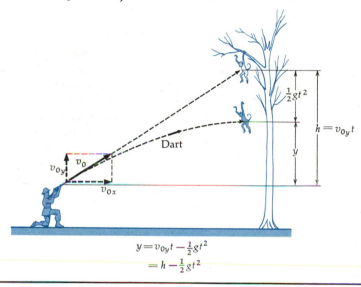

$$y = v_{0y}t - \tfrac{1}{2}gt^2$$
$$= h - \tfrac{1}{2}gt^2$$

Figure 3-14
The monkey and dart from Example 3-8. The height of the dart at any time is $y_2 = v_{0y}t - \tfrac{1}{2}gt^2$. The height of the monkey at any time is $h - \tfrac{1}{2}gt^2$, which is the same as that of the dart because $h = v_{0y}t$. The dart will therefore always hit the monkey if the monkey drops the instant the dart is fired.

Questions

9. What is the acceleration of a projectile at the top of its flight?

10. Can the velocity of an object change direction while its acceleration is constant in both magnitude and direction? If so, give an example.

3-5   Circular Motion

Circular motion is common in nature and in our everyday experience. The earth revolves in a nearly circular orbit around the sun; the moon, around the earth. Wheels rotate in circles, cars travel on circular arcs as they round corners, and so forth. In this section, we consider a particle with constant speed moving in a circle. At first, we might be tempted to say that since the speed is constant the particle is not accelerating. We have, however, defined acceleration as the rate of change of the velocity vector, and when a particle moves in a circle, its velocity vector is continually changing direction.

Newton was one of the first to recognize the importance of circular motion. He showed that when a particle moves with a constant speed v in a circle of radius r, it has an acceleration with a magnitude of v^2/r directed toward the center of the circle. This acceleration is called *centripetal acceleration*. Newton also proposed that an unbalanced force is needed to produce acceleration, so for a particle to move in a circle at constant speed, an unbalanced force pushing or pulling it toward the center of the circle is needed. In the case of the earth revolving around the sun, the unbalanced force is the force of gravitational attraction of the sun on the earth. For a car rounding a curve, the force is provided by the friction between the tires and the road. If the frictional force is not great enough, the car will not make the curve but will move off at a tangent to the curve.

Figure 3-15 shows a satellite moving in a circular orbit around the earth.

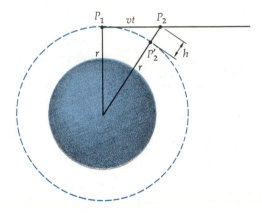

Figure 3-15
Satellite moving with speed v in a circular orbit of radius r about the earth. If the satellite did not accelerate toward the earth, it would move in a straight line from point P_1 to point P_2 in time t. Because of its acceleration, it falls a distance h in this time. For small values of t, $h = \frac{1}{2}(v^2/r)t^2 = \frac{1}{2}at^2$.

Why doesn't the satellite fall toward the earth? The answer is *not* that there is no force of gravity acting on the satellite. At 200 km above the surface of the earth, the gravitational force on the satellite is about 94 percent of what it would be were the satellite at the earth's surface. The satellite *does* "fall" toward the earth, but because of its tangential velocity, it continually misses the earth. To see this, consider Figure 3-15. If the satellite were not accelerating, it would move from point P_1 to P_2 in some time t. Instead, it arrives at point P_2' on its circular orbit. Thus in a sense, the satellite "falls" the distance h shown. If we take the time t to be very small, the points P_2 and P_2' are approximately on a radial line, as shown in the figure, and we can use the approximation that h is much smaller than orbit radius r. (From the figure we can see that the smaller we make the time t, the shorter the distance vt and the smaller the distance h will be for any given orbit radius r.) We can then calculate h from the right triangle with the sides vt, r, and $r + h$. Since $r + h$ is the hypotenuse of the right triangle, the pythagorean theorem gives

$$(r + h)^2 = (vt)^2 + r^2$$

$$r^2 + 2hr + h^2 = v^2t^2 + r^2$$

or

$$h(2r + h) = v^2t^2$$

For very short times t, h will be much less than r, so we can neglect h as compared with $2r$ for the term in parentheses. Thus,

$$2rh \approx v^2 t^2 \qquad \text{or} \qquad h \approx \frac{1}{2}\left(\frac{v^2}{r}\right)t^2$$

Comparing this with the constant-acceleration expression $h = \frac{1}{2}at^2$, we see that the magnitude of the acceleration of the satellite is

$$a = \frac{v^2}{r} \hspace{4cm} \text{3-19} \quad \text{Centripetal acceleration}$$

and the direction is inward toward the center of the circle. This result can be shown to hold in general for any particle in circular motion with a constant speed, such as a stone on a string or a car rounding a curve.

Example 3-9   A tether ball moves in a horizontal circle with a radius of 2 m. It makes one revolution in 3 s. Find its acceleration.

To find the acceleration, we must first find the speed. Since the ball moves a distance $2\pi r = 2\pi(2 \text{ m}) = 12.6 \text{ m}$ in 3 s, its speed is $(12.6 \text{ m})/(3 \text{ s}) = 4.2$ m/s. The magnitude of its acceleration is then

$$\frac{v^2}{r} = \frac{(4.2 \text{ m/s})^2}{2 \text{ m}} = 8.82 \text{ m/s}$$

This is about 0.9 g. The direction of the acceleration is inward toward the center of the circle.

Example 3-10   A satellite moves at constant speed in a circular orbit about the center of the earth and near the surface of the earth. If its acceleration is 9.81 m/s^2, what is its speed and how long does it take to make one complete revolution?

This acceleration is the same as that for any body falling freely near the surface of the earth. We take the radius of the earth, about 6370 km, to be the approximate radius of the orbit. (For actual satellites put into orbit a few hundred kilometres above the earth's surface, the radius of the orbit will be slightly greater, of course, and the acceleration will be slightly less than 9.81 m/s^2 because of the decrease in the gravitational force with distance from the center of the earth.) The speed of the satellite can be found from Equation 3-19:

$$v^2 = rg = (6370 \text{ km})(9.81 \text{ m/s}^2)$$

$$v = 7.91 \text{ km/s}$$

The time for the satellite to make one complete revolution is called the period T. Since it travels a distance of $2\pi r$ in this time T at a speed of v, the period is

$$T = \frac{2\pi r}{v} = \frac{2\pi(6370 \text{ km})}{7.91 \text{ km/s}} = 5060 \text{ s} = 84.3 \text{ min}$$

Figure 3-16 is a drawing from Newton's *System of the World* that illustrates the connection between projectile motion and satellite motion. Newton's description reads as follows:*

> That by means of centripetal forces the planets may be retained in certain . . . orbits, we may easily understand, if we consider the motions of projectiles; for a stone that is projected is by the pressure of its own weight forced out of the rectilinear path, which by the initial projection alone it should have pursued, and made to describe a curved line in the air; and through that crooked way is at last brought down to the ground; and the greater the velocity is with which it is projected, the farther it goes before it falls to the earth. We may therefore suppose the velocity to be so increased, that it would describe an arc of 1, 2, 5, 10, 100, 1000 miles before it arrived at the earth, till at last, exceeding the limits of the earth, it should pass into space without touching it.

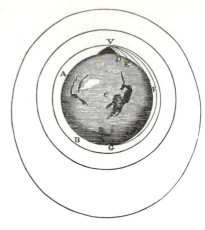

Figure 3-16
A drawing from Newton's *System of the World,* published in 1729, illustrating the connection between projectile motion and satellite motion.

Example 3-11   A car rounds a curve of radius 100 m at a speed of 90 km/h. What is the magnitude of its centripetal acceleration?

To obtain the acceleration in metres per second squared, we need to change the speed to metres per second. Using the result of Example 1-1 that 90 km/h = 25 m/s, we have $v = 25$ m/s. The acceleration is then

$$a = \frac{v^2}{r} = \frac{(25 \text{ m/s})^2}{100 \text{ m}} = 6.25 \text{ m/s}^2$$

If a particle moves in a circle with speed that is varying, there is a component of acceleration that is tangent to the circle as well as the centripetal acceleration inward. The tangential component of the acceleration is simply the rate of change of the speed, whereas the radially inward component has the magnitude v^2/r. For any general motion of a particle along a curve, we can treat a portion of the curve as an arc of a circle.

* From Sir Isaac Newton, *System of the World,* trans. Andrew Motte, 1729, University of California Press, Berkeley, 1960. Reprinted by permission of the Regents of the University of California.

Summary

1. Quantities that have magnitude and direction, such as displacement, velocity, and acceleration, are vectors.

2. Vectors can be added graphically by placing the tail of one vector at the tip of the other and drawing the resultant vector from the tail of the first to the tip of the second. Subtracting a vector **B** is the same as adding −**B**, where −**B** is a vector with magnitude equal to **B** but in the opposite direction.

3. Vectors can be added analytically by first finding the components of the vectors given by

$$A_x = A \cos \theta$$
$$A_y = A \sin \theta$$

The x component of the resultant vector is the sum of the x components of the individual vectors and the y component is the sum of the y components of the individual vectors.

4. The displacement vector points from the initial position to the final position. The velocity vector is the rate of change of the position vector. Its magnitude is the speed, and it points in the direction of motion, tangent to the curve along which the particle is traveling.

5. In projectile motion, the horizontal and vertical motions are independent. The vertical motion is the same as the motion in one dimension with the constant acceleration of gravity g downward. The horizontal motion is a constant velocity equal to the horizontal component of the original velocity. The total distance traveled by the projectile, called the range, is found by first finding the total time the projectile is in the air and then multiplying this time by the constant horizontal velocity. For the special case in which the initial and final elevations are the same, the range is maximum when the angle of projection is 45°.

6. When an object moves in a circle with constant speed, it is accelerating because its velocity is changing in direction. The acceleration is called centripetal acceleration because it points toward the center of the circle. The magnitude of the centripetal acceleration is v^2/r, where v is the speed and r is the radius of the circle.

Suggestions for Further Reading

Drake, Stillman, and James MacLachan: "Galileo's Discovery of the Parabolic Trajectory," *Scientific American*, March 1975, p. 102.

Galileo knew that a falling body with a horizontal component of velocity describes a parabola 30 years before he published the fact, according to unpublished manuscripts. (See also Chapter 2 references.)

Review

A. After studying this chapter, you should:

1. Be able to add and subtract vectors graphically.

2. Be able to determine the components of vectors and use them to add and subtract vectors.

3. Know that in projectile motion the horizontal and vertical motions are independent and be able to use this fact in working projectile problems.

4. Know that when a particle moves in a circle with constant speed it has centripetal acceleration of magnitude v^2/r directed toward the center of the circle.

B. Define, explain, or otherwise identify:

Vector, p. 36
Scalar, p. 36
Displacement vector, pp. 37, 41
Component of a vector, p. 39
Velocity vector, p. 41
Acceleration vector, p. 43
Range, pp. 46, 47
Centripetal acceleration, pp. 50, 51

C. True or false: If the statement is true, explain why it is true. If it is false, give a counterexample; that is, a known example that contradicts the statement.

1. The instantaneous velocity vector is always in the direction of motion.

2. The instantaneous acceleration vector is always in the direction of motion.

3. If the speed is constant, the acceleration must be zero.

4. If the acceleration is zero, the speed must be constant.

5. The magnitude of the sum of two vectors must be greater than the magnitude of either vector.

6. It is impossible to go around a curve with no acceleration.

7. The time required for a bullet that is fired horizontally to reach the ground is the same as if the bullet were dropped from rest from the same height.

Exercises

Section 3-1   The Displacement Vector

1. A bear walks northeast for 10 m and then east for 10 m. (*a*) Show each displacement graphically and (*b*) find the resultant displacement vector.

2. (*a*) A man walks along a circular arc from the position $x = 5$ m and $y = 0$ to a final position $x = 0$ and $y = 5$ m. What is his displacement? (*b*) A second man walks from the same initial position along the *x* axis to the origin and then along the *y* axis until he reaches $y = 5$ m and $x = 0$. What is his displacement?

3. A hiker sets off at 8 A.M. on level terrain. At 9 A.M. she is 2 km due east of her starting point. At 10 A.M. she is 1 km northwest of where she was at 9 A.M. At 11 A.M. she is 3 km due north of where she was at 10 A.M. (*a*) Make a drawing showing these successive displacements as vectors, with the tail of each placed at the tip of the previous one. What are the magnitudes and directions of these displacements? (Specify the direction of the vectors by giving their angle with the eastward direction.) (*b*) What are the north and east components of these displacements? (*c*) How far is the hiker from her starting point at 11 A.M.? In what direction? (*d*) Add the three displacement vectors by drawing them to scale. Do these successive straight lines represent the actual path the hiker followed? (*e*) Is the distance she walked the sum of the lengths of the three displacement vectors?

4. The displacement vectors **A** and **B** shown in Figure 3-17 both have magnitude 2 m. Find graphically (*a*) **A** + **B**, (*b*) **A** − **B**, and (*c*) **B** − **A**

Figure 3-17
Exercise 4.

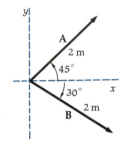

Section 3-2   Addition of Vectors by Components

5. What are the components of the displacement vector for part (*a*) of Exercise 2?

6. Find the *x* and *y* components of the vectors that lie in the *xy* plane, have the magnitude *A*, and make an angle θ with the *x* axis, as shown in Figure 3-18, for the following values of *A* and θ: (*a*) $A = 10$ m, $\theta = 30°$; (*b*) $A = 5$ m, $\theta = 45°$; (*c*) $A = 7$ km, $\theta = 60°$; (*d*) $A = 5$ km, $\theta = 90°$; (*e*) $A = 15$ km/s, $\theta = 150°$; and (*f*) $A = 10$ m/s, $\theta = 240°$.

Figure 3-18
Exercise 6.

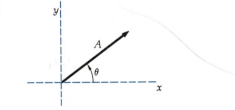

7. Find the magnitude and direction of the vectors that have the following components: (*a*) $A_x = 5$ m, $A_y = 3$ m; (*b*) $B_x = 10$ m/s, $B_y = -7$ m/s; and (*c*) $C_x = -2$ m, $C_y = -3$ m.

8. A plane is inclined at an angle of 30° with the horizontal. Choose the *x* axis parallel to the plane pointing down the slope and the *y* axis perpendicular to the plane pointing away from the plane. Find the *x* and *y* components of the accelera-

tion of gravity, which has the magnitude 9.8 m/s² and points vertically down.

9. Find the magnitude and direction of **A**, **B**, and **A** + **B** for vectors **A** and **B** that have the components: (a) $A_x = -4$ m, $A_y = -7$ m, $B_x = 3$ m, $B_y = -2$ m; and (b) $A_x = 1$ m, $A_y = -4$ m, $B_x = 2$ m, $B_y = 6$ m.

10. Describe the following vectors by giving the x and y components: (a) a velocity of 10 m/s at an angle of elevation of 60°; (b) a vector **A** of magnitude $A = 5$ m and $\theta = 225°$; and (c) a displacement from the origin to the point $x = 14$ m, $y = -6$ m.

11. For the displacement vectors **A** and **B** in Figure 3-19, find (a) their x and y components, (b) the x and y components, the magnitude, and the direction of the sum of **A** + **B**, and (c) the x and y components, the magnitude, and the direction of the difference **A** − **B**.

Figure 3-19
Exercise 11.

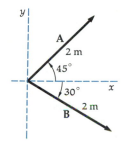

Section 3-3 Velocity and Acceleration Vectors

12. A particle's position coordinates (x, y) are (2 m, 3 m) at $t = 0$; (6 m, 7 m) at $t = 2$ s; and (13 m, 14 m) at $t = 5$ s. (a) Find v_{av} from $t = 0$ to $t = 2$ s. (b) Find v_{av} from $t = 0$ to $t = 5$ s.

13. A stationary radar operator determines that a ship is 10 km south of him. An hour later the same ship is 20 km southeast of him. If the ship moved at constant speed always in the same direction, what was its velocity during this time?

14. A ball is thrown directly upward. Consider the 2-s time interval $\Delta t = t_2 - t_1$, where t_1 is 1 s before the ball reaches its highest point. Find (a) the change in speed, (b) the change in velocity, and (c) the average acceleration for this time interval.

15. Figure 3-20 shows the path of an automobile, which is made up of segments of straight lines and arcs of circles. The automobile starts from rest at point A. After it reaches point B, it travels at constant speed until it reaches point E. It comes to rest at point F. (a) At the middle of each segment (AB, BC, CD, DE, and EF) what is the direction of the velocity vector? (b) At which of these points does the automobile have an acceleration? In those cases, what is the direction of the acceleration? (c) How do the magnitudes of the acceleration compare for segments BC and DE?

Figure 3-20
Exercise 15.

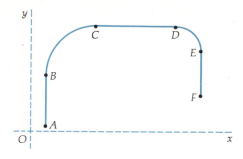

16. A swimmer heads directly across a river swimming at 1.6 m/s relative to still water. She arrives at a point 40 m downstream from the point directly across the river, which is 80 m wide. (a) What is the speed of the river current? (b) What is her swimming speed relative to the shore? (c) What direction should she head so as to arrive at a point directly opposite her starting point?

17. A plane flies at a speed of 250 km/h relative to still air. There is a wind blowing at 80 km/h in the northeast direction (that is, at 45° to the east of north). (a) What direction should the plane head so as to fly due north? (b) What is the speed of the plane relative to the ground?

18. Initially, a particle is moving due west with a speed of 40 m/s, and 5 s later it is moving north with a speed of 30 m/s. (a) Indicate on a sketch the initial velocity vector, the final velocity vector, and the change in velocity Δv. (b) Find the magnitude and direction of Δv. (c) What are the magnitude and direction of a_{av} for this interval?

Section 3-4 Projectile Motion

19. A bullet is fired horizontally with an initial velocity of 245 m/s. The gun is 1.5 m above the ground. (a) How long is the bullet in the air? (b) How far does the bullet travel before striking the ground?

20. A supersonic transport is flying horizontally at an altitude of 20 km with speed of 2500 km/h when an engine falls off. (a) How long does it take the engine to hit the ground? (b) How far horizontally is the engine from where it fell off when it hits the ground? (c) How far is the engine from the aircraft (assuming that the aircraft continues to fly as if nothing had happened) when the engine hits the ground? Neglect air resistance.

21. A cannon is elevated at an angle of 45°. It fires a ball with a speed of 300 m/s. (a) What height does the ball reach? (b) How long is the ball in the air? (c) What is the horizontal range?

22. A pitcher throws a fastball at 140 km/h toward home plate, which is 18.4 m away. Neglecting air resistance (not a good idea if you are the batter), find how far the ball drops because of gravity by the time it reaches home plate.

23. A projectile is fired with initial velocity of 50 m/s at 60° to the horizontal. (*a*) At the highest point, what is the velocity? (*b*) What is its acceleration?

24. A ball is thrown at 30° to the horizontal with an initial speed of 30 m/s. A second ball is thrown with the same initial speed but at 60° to the horizontal. (*a*) Find the time that each ball is in the air. (*b*) Find the distance traveled by each ball before it reaches its original height. (*c*) Sketch the paths of the two balls on the same diagram.

Section 3-5 Circular Motion

25. An object travels with a constant speed v in a circular path of radius r. (*a*) If v is doubled, how is the acceleration a affected? (*b*) If r is doubled, how is a affected?

26. A particle travels in a circular path of radius 5 m at a constant speed of 15 m/s. What is the magnitude of its acceleration?

27. A boy whirls a ball on a string in a horizontal circle of radius 1.5 m. (*a*) What must the speed of the ball be if its acceleration toward the center of the circle is to have the same magnitude as the acceleration of gravity? (*b*) How many revolutions per minute does the ball make at this speed.

28. An airplane pilot pulls out of a dive by following an arc of a circle whose radius is 300 m. At the bottom of the circle, where her speed is 180 km/h, what are the direction and magnitude of her acceleration?

29. A car rounds a curve of radius 40 m. The frictional force between the road and the tires is such that the maximum possible inward acceleration is 0.6 g. Find the maximum speed of the car in kilometres per hour and miles per hour if the car's centripetal acceleration is 0.6 g.

30. The moon moves around the earth in a circle of radius 3.84×10^8 m. The time for one complete revolution (the period) is 27.3 days. (*a*) Find the speed of the moon in metres per second. (*b*) Find the centripetal acceleration of the moon in metres per second squared.

Problems

1. A circular path has a radius of 10 m. An *xy* coordinate system is established so that the center of the circle is on the *y* axis at $y = 10$ m and the circle passes through the origin. A man starts at the origin and walks around the path at a steady speed, returning to the origin exactly 1 min after he started. (*a*) Find the magnitude and direction of his displacement from the origin 15, 30, 45, and 60 s after he starts. (*b*) Find the magnitude and direction of his displacement for each of the four successive 15-s intervals of his walk. (*c*) How is his displacement for the first 15 s related to that for the second 15 s? (*d*) How is his displacement for the second 15-s interval related to that for the last 15-s interval?

2. Draw any three nonparallel vectors **A**, **B**, and **C** that lie in a plane and show that the sum $(\mathbf{A} + \mathbf{B}) + \mathbf{C}$ equals $(\mathbf{A} + \mathbf{C}) + \mathbf{B}$.

3. At $t = 0$, a particle located at the origin has a speed of 40 m/s at $\theta = 45°$. At $t = 3$ s the particle is at $x = 100$ m, $y = 80$ m with speed of 30 m/s at $\theta = 50°$. Calculate (*a*) the average velocity and (*b*) the average acceleration of the particle during this interval.

4. Sketch the trajectory of a projectile and indicate the velocity and acceleration vectors at several points. (*a*) What is the magnitude of the acceleration at the top of the path? (*b*) Does the magnitude or direction of the acceleration change from point to point?

5. A particle travels counterclockwise with constant speed in a circular path of radius 5 m around the origin. It begins at $t = 0$ at $x = 5$ m, $y = 0$ and takes 100 s to make a complete revolution. (*a*) What is the speed of the particle? (*b*) Give the

magnitude and direction of the position vector **r** at the times $t = 50$ s, $t = 25$ s, $t = 10$ s, and $t = 0$. (*c*) Find the magnitude and indicate graphically the direction of $\mathbf{v}_{av}$ for each of the following time intervals: $t = 0$ to $t = 50$ s; $t = 0$ to $t = 25$ s; and $t = 0$ to $t = 10$ s. (*d*) How does $\mathbf{v}_{av}$ for the interval $t = 0$ to $t = 10$ s compare with the instantaneous velocity at $t = 0$?

6. A projectile is launched with a speed v_0 at an angle θ_0 with the horizontal. Find the expression for the maximum height it reaches above its starting point in terms of v_0, θ_0, and g.

7. A projectile is fired into the air from the top of a 200-m cliff above a valley (Figure 3-21). Its initial velocity is 60 m/s at 60° to the horizontal. Neglecting air resistance, where does the projectile land?

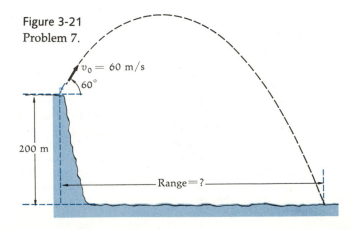

Figure 3-21
Problem 7.

$v_0 = 60$ m/s

60°

200 m

Range = ?

8. An acceleration of 31 g was withstood for 5 s by R. F. Gray in 1959. How many revolutions per minute does a centrifuge have to make to produce an acceleration of 31 g on somebody positioned in the arm at a radius of 5 m?

9. A stone thrown horizontally from the top of a tower hits the ground at a distance 18 m from the base of the tower. (*a*) Find the speed at which the stone was thrown if the tower is 24 m high. (*b*) Find the speed of the stone just before it hits the ground.

10. A particle moves in a circle of radius 4 cm. It takes 8 s to make a complete round-trip. Draw the path of the particle to scale and indicate the positions at 1-s intervals. Draw the displacement vectors for these 1-s intervals. These vectors also are the average velocity vectors for these intervals. (*a*) Find graphically the change in the average velocity $\Delta \mathbf{v}_{av}$ for two consecutive 1-s intervals. (*b*) Compare $\Delta \mathbf{v}_{av}/\Delta t$ measured in this way with the instantaneous acceleration computed from $a_r = v^2/r$.

11. A baseball is struck by a bat, and 3 s later it is caught 30 m away. (*a*) If the baseball was 1 m above the ground when it was struck and caught, what was the greatest height it reached above the ground? (*b*) What were its horizontal and vertical components of velocity when it was struck? (*c*) What was its speed when it was caught? (*d*) At what angle with the horizontal did it leave the bat? (Neglect air resistance.)

12. A freight train is moving at a constant speed of 10 m/s. A man standing on a flatcar throws a ball into the air and catches it as it falls. Relative to the flatcar, the initial velocity of the ball is 15 m/s straight up. (*a*) What are the magnitude and direction of the initial velocity of the ball as seen by a second man standing next to the track? (*b*) How long is the ball in the air according to the man on the train? According to the man on the ground? (Take $g = 10$ m/s².) (*c*) What horizontal distance has the ball traveled by the time it is caught according to the man on the train? According to the man on the ground? (*d*) What is the minimum speed of the ball during its flight according to the man on the train? According to the man on the ground? (*e*) What is the acceleration of the ball according to the man on the train? According to the man on the ground?

13. A baseball is thrown toward a player with an initial speed of 20 m/s at 45° with the horizontal. At the moment the ball is thrown, the player is 50 m from the thrower. At what speed and in what direction must he run to catch the ball at the same height at which it was released?

14. A large boulder rests on a cliff 400 m above a small village in such a position that, if it should roll off, it would

leave with a speed of 50 m/s (Figure 3-22). There is a pond with a diameter of 200 m and its edge 100 m from the base of the cliff, as shown in the figure. The village houses are at the other edge of the pond. (*a*) A physics student says that the boulder will land in the pond. Is she right? (*b*) How fast will the boulder be going when it hits? (*c*) What will the horizontal component of its velocity be when it hits? (*d*) How long will the boulder be in the air?

Figure 3-22
Problem 14.

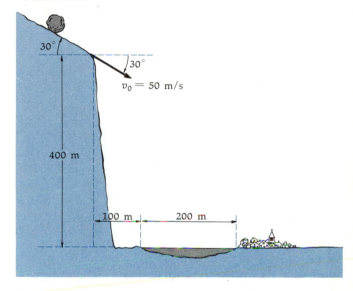

15. A trail bike comes to a ditch. A ramp with an angle of 10° has been built so that the bike can jump the ditch. If the bike needs to jump a horizontal distance of 7 m to clear the ditch, how fast must it be traveling when it leaves the ramp?

16. A gun shoots bullets that leave the muzzle at 250 m/s. If the bullet is to hit a target 100 m away at the level of the muzzle, the gun must be aimed at a point above the target. How far above the target is this point? (Neglect air resistance.)

17. A baseball just clears a 3-m wall 120 m from home plate. If it leaves the bat at 45° and 1.2 m above the ground, what must its initial velocity be (again making the unrealistic assumption that air resistance can be ignored)?

18. A hockey puck struck at ice level just clears the top of a glass wall 2.80 m high. The flight time to this point was 0.650 s and the horizontal distance was 12.0 m. Find (*a*) the initial speed of the puck and (*b*) the maximum height it will reach.

Newton's Laws

Your study of motion up to this point has been entirely about how motion occurs; that is, how bodies gain speed and cover distance when they have uniform acceleration. But we have not yet discussed the causes of accelerations nor why bodies fall at the same rate regardless of how heavy they are. Sir Isaac Newton, who was born in 1642, the year Galileo died, undertook the task of explaining the causes of motion. As a student at Cambridge, where he was later a mathematics professor, Newton learned of the work of Galileo and Kepler on the motions of the planets. He wanted to figure out why the planets move in ellipses at speeds dependent on their distance from the sun and even why the solar system stays together at all. As an initial step, he conceived of the idea of the *mass* of a body—its resistance to changes in velocity—as an important property of the body. He went on to develop his law of universal gravitation to explain what holds the solar system together.

Newton's law of universal gravitation will be discussed in detail in Chapter 9. In this chapter, we take up the three laws of motion with which he began his great masterpiece *Philosophiae Naturalis Principia Mathematica*. You will find that an amazing variety of phenomena can be explained by Newton's laws of motion and that the world around you will seem quite different once you have grasped their significance.

Isaac Newton (1642–1727).

Classical, or newtonian, mechanics is a theory of motion based on the ideas of mass and force and the laws connecting these physical concepts to the kinematic quantities—displacement, velocity, and acceleration—discussed in the preceding chapters. All phenomena in classical mechanics can be described using just three simple laws called Newton's laws of motion. We will begin by stating Newton's laws, and then we will illustrate their applications with rather simple problems involving constant forces. In the next chapter, we will discuss some more general applications.

It is interesting to read Newton's version of the laws of motion:*

> Law I. Every body continues in its state of rest, or in uniform motion in a right line unless it is compelled to change that state by forces impressed upon it.
>
> Law II. The change of motion is proportional to the motive force impressed; and is made in the direction of the right line in which that force is impressed.
>
> Law III. To every action there is always opposed an equal reaction; or, the mutual actions of two bodies upon each other are always directed to contrary parts.
>
> Corollary I. A body, acted on by two forces simultaneously, will describe the diagonal of a parallelogram in the same time as it would describe the sides by those forces separately.

A modern version of these laws is

Law 1. A body continues in its initial state of rest or motion with uniform velocity unless acted on by an unbalanced external force.

Newton's laws of motion

Law 2. The acceleration of a body is inversely proportional to its mass and directly proportional to the resultant external force acting on it.

$$a = \frac{\Sigma \mathbf{F}}{m}$$

or

$$\Sigma \mathbf{F} = \mathbf{F}_{net} = m\mathbf{a} \qquad\qquad 4\text{-}1$$

Law 3. Forces always occur in pairs. If body A exerts a force on body B, an equal but opposite force is exerted by body B on body A.

Corollary. Forces add as vectors.

The Greek capital letter sigma (Σ) is used to indicate a summation. $\Sigma \mathbf{F}$ therefore means the sum of all the forces acting on the body. This sum is called the resultant force acting on the body. It is also sometimes called the *net force* or the *unbalanced force*.

Newton's laws relate the acceleration of an object to its mass and the forces acting on it. We have intuitive ideas about the words "force" and "mass." We think of a force as being a push or a pull, like that exerted by our muscles, and we visualize a massive body as something large or heavy. These intuitive notions are all right for everyday conversation but not for the application of Newton's laws to problems in physics or even for a precise statement of the laws. To understand Newton's laws fully and to be able to apply them, we must define these words carefully. This we do by outlining methods for their measurement. We shall then find that Newton's second law follows directly from the definitions of force and mass.

* Sir Isaac Newton, *Philosophiae Naturalis Principia Mathematica*, 1686, trans. Andrew Motte, 1729, University of California Press, Berkeley, 1960. Reprinted by permission of the Regents of the University of California.

4-1   Force and Mass

Consider a body, such as a block of wood or metal, resting on a horizontal surface, for example, a table. We observe that, if the body is at rest (relative to the table), it remains at rest unless we push on it or pull on it. If we project the body along the table, it slides along for a ways, but eventually the speed decreases and it comes to rest. We attribute the decrease in speed to the *frictional force* exerted on the body by the table. If we polish the surfaces of the table and body, the body slides farther and its decrease in velocity in a given time is smaller. If we support the body by a thin cushion of air (this is possible with an air table or with a glider on an air track), the body will glide for a considerable time and distance with almost no perceptible change in its velocity. We can extrapolate from this to the idea of an *ideal frictionless* surface that in no way impedes the movement of a body, and we can state that on such a surface the velocity of a body will not change. We have thus *defined* a situation in which there are *no* (horizontal) forces acting on the body. If there are no forces acting on the body, the velocity of the body remains constant. This is Newton's first law, the *law of inertia*.   Law of inertia

If the velocity of a body is not constant, we can conclude that a net force is acting on the body. Our next problem, then, is to develop a quantitative measure of force. We do this by *defining the magnitude and direction of a given force in terms of the acceleration it produces on a particular object, which we shall call our standard body.* The international standard body is a cylinder of platinum alloy carefully preserved at the International Bureau of Weights and Measures at Sèvres, France. The mass of the standard body is 1 kg, the SI unit of mass. (We shall discuss the definition and measurement of mass shortly.) The force required to produce an acceleration of 1 m/s² on the   Force of one newton defined
standard body is defined as 1 newton (N). Similarly, the force that produces an acceleration of 2 m/s² is defined as 2 N. Thus, we define force in terms of the acceleration it produces on a standard body.

A convenient agent for exerting forces on bodies is a spring. It takes a push or pull to compress or extend a spring from its natural length; the greater the push or pull, the greater the compression or extension. Consider a particular spring attached to our standard body on a frictionless horizontal surface, as shown in Figure 4-1. If the spring is extended from its natural length, the body accelerates. By noting the extension needed to produce a particular acceleration of our standard body and using our definition of force, we can calibrate our spring in units of force. A plot of force versus extension for a typical spring is shown in Figure 4-2. Using the same standard body, we can calibrate other springs in a similar way. We can then do experiments to see how several forces combine.

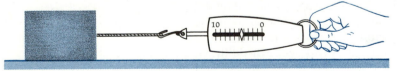

Figure 4-1
A horizontal force is applied to a body by the extended spring. The spring can be calibrated by noting the extension produced by a given force as measured by the acceleration produced.

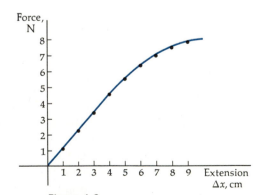

Figure 4-2
Calibration curve for a spring scale like that in Figure 4-1.

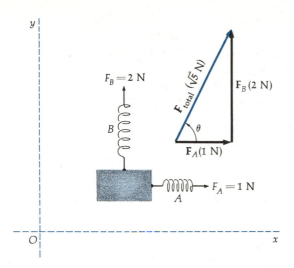

Figure 4-3
Forces are vectors. Forces of 1 N in the x direction and 2 N in the y direction combine as shown to give a resultant force of $\sqrt{5}$ N at an angle θ to the x axis given by $\tan\theta = 2$.

Consider two calibrated springs attached to our standard body and extended so that they exert forces in different directions, as shown in Figure 4-3. We observe experimentally that the forces add as vectors; that is, the resultant acceleration is found by adding the vector accelerations each force would produce if it were acting alone. For example, if spring A exerts a force of 1 N along the x axis and spring B exerts a force of 2 N along the y axis, as in Figure 4-3, the acceleration observed has a magnitude $a = \sqrt{1^2 + 2^2}\ \text{m/s}^2 = \sqrt{5}\ \text{m/s}^2$ and makes an angle θ with the x axis given by $\tan\theta = 2$, or $\theta = 63.4°$.

Forces add as vectors

We can use our calibrated springs to measure other types of forces. For example, the force exerted by the gravitational attraction of the earth for an object is called its *weight*. If we hang our standard body by one of our calibrated springs, as in Figure 4-4, we find that the spring must exert a force of 9.81 N to balance the downward gravitational force exerted by the earth. Our standard body thus weighs 9.81 N at this point near the surface of the earth.

We are now ready to define mass, an intrinsic property of a body that measures its resistance to acceleration. We shall do this by investigating the effects of a given force on different bodies. If we use one of our calibrated springs to produce a given force on a second body, we find that the acceleration produced for a given spring extension is not necessarily the same as it was for our standard body. If our second body is "more massive" (according to our everyday use of this term), the acceleration produced by a force of 1 N is observed to be less than 1 m/s^2. If the second body is less massive, the acceleration produced is greater. For example, if we connect two identical bodies, the acceleration produced by a given force is exactly half that produced by the same force acting on just one of the bodies. We define the ratio of the mass of one body to that of another body to be the *inverse* ratio of the accelerations produced in those two bodies by the *same force*. If a given force produces an acceleration a_1 when it acts on body 1 and an acceleration a_2 on body 2, the ratio of the masses of the two bodies is defined as

$$\frac{m_2}{m_1} = \frac{a_1}{a_2}$$

4-2   **Mass defined**

Figure 4-4
Forces on a body suspended from a spring. The upward force due to the spring balances the downward force due to the gravitational attraction of the earth, that is, to the weight of the body.

Spring force

Weight

Thus, we can find the ratio of the masses of any two bodies by applying the same force to each and comparing their accelerations. The ratio of the accelerations a_1/a_2 produced by the same force on two bodies is independent of the magnitude or direction of the force. It is also independent of the kind of force used, that is, whether the force is due to springs, the pull of gravity, electric or magnetic attraction or repulsion, or whatever. We also find that if mass m_2 is found to be twice the mass m_1 by direct comparison and if a third mass m_3 is found to be four times the mass m_1, m_3 will be twice the mass m_2 when the two are compared directly. We can therefore set up a mass scale by choosing one particular body to be a standard body and assigning it a mass of 1 unit. (The 1-kg mass of the standard body was originally intended to be equal to the mass of 1000 cm^3 = 1 litre of water, but that relationship proved to be inexact by a very small amount.) The standard body can be used to produce secondary standards by direct comparison, and the mass of any other body can then be found by comparing the acceleration produced on it by a given force with that produced on one of the secondary standards.

Example 4-1   A given force produces an acceleration of 5 m/s^2 on the standard body. When the same force is applied to a second body, its acceleration is 15 m/s^2. What is the mass of the second body and what is the magnitude of the force?

Since the acceleration of the second body is 3 times that of the standard body under the influence of the same force, the mass of the second body is $\frac{1}{3}$ that of the standard body, or 0.33 kg. The magnitude of the force is $F = (1 \text{ kg})(5 \text{ m/s}^2) = 5$ N.

Note that we have defined the concepts of force and mass so that Newton's second law, $\Sigma\mathbf{F} = m\mathbf{a}$, follows directly from the definitions. These definitions agree with our intuitive notions of the meaning of force and mass. They are useful because they allow us to describe a wide variety of physical phenomena using just a few, relatively simple force laws. For example, with the addition of Newton's law of gravitational attraction between two bodies, we can calculate and explain the motion of the moon, the orbits of all the planets around the sun, the orbits of artificial satellites, the variation in the acceleration of gravity g with altitude and with latitude, the variation in the acceleration of gravity due to the presence of mineral deposits, the paths of ballistic missiles, and many other phenomena.

Questions

1. If a body has no acceleration, can you conclude that no forces are acting on it?

2. If only a single force acts on a body, must it accelerate? Can it ever have zero velocity?

3. Is there an unbalanced force acting when (a) a body moves at constant speed in a circle, (b) a body moving in a straight line slows down, (c) a body moves at constant speed in a straight line?

In truth, not all of the laws of motion were original with Newton. While trying to find the speed of a ball rolling down an incline, Galileo constructed a device that caused the ball to roll off the incline and onto a level track. He found that the more he smoothed the track — and to this end he even lined it with kid leather — the farther the ball went. He concluded that if there were no friction, the ball would keep on rolling forever. So Newton's first law was actually discovered by Galileo. Such are the vagaries of fame.

4. Is it possible for an object to round a curve without any force acting on it?

5. If a single known force acts on a body, can you tell *in which direction* the body will *move* from this information alone?

6. If several forces of different magnitudes and directions are applied to a body initially at rest, how can you predict the direction in which the body will move?

7. Can you judge the mass of an object by its size? If *A* is twice as big as *B*, does that mean that $m_A = 2m_B$?

8. Can the mass of a body be negative?

9. Mass is sometimes said to be the measure of the quantity of matter in a body. How does this definition compare with the definition discussed in this section?

4-2   The Force Due to Gravity: Weight

The force most common in our everyday experience is the force of the gravitational attraction of the earth for an object. This force is called the *weight* of the object. If we drop an object near the surface of the earth and neglect air resistance so that the only force acting on the object is the force due to gravity, the object accelerates toward the earth with acceleration of 9.81 m/s². (This situation is called *free-fall*.) At any given point in space, this acceleration is found to be the same for all objects independent of their mass. Since the acceleration of an object is the resultant force divided by the object's mass, we can conclude that the force due to gravity on an object is proportional to the mass of the object:

$$\mathbf{F}_g = \mathbf{w} = m\mathbf{g} \qquad\qquad 4\text{-}3$$

The vector **g** in Equation 4-3 is called the *gravitational field* of the earth. It is the force per unit mass exerted by the earth on any object, and it is equal to the free-fall acceleration experienced by an object when the only force acting on it is the gravitational force of the earth. Near the surface of the earth, *g* has the value

$$g = 9.81 \text{ N/kg} = 9.81 \text{ m/s}^2$$

Gravitational field **g**

Careful measurements of *g* at various places show that it does not have the same value everywhere. *The force of attraction of the earth for an object varies with location.* Thus weight, unlike mass, is not an intrinsic property of a body; that is, it is not a property of the body itself as is mass. In particular, at points above the surface of the earth, the force due to gravity varies inversely with the square of the distance of the object from the center of the earth. Thus, a body weighs less at very high altitudes than it does at sea level. The gravitational field also varies slightly with latitude because the earth is not exactly spherical but is slightly flattened at the poles.

Near the surface of the moon, the gravitational attraction of the moon is much stronger than that of the earth. The force exerted on a body by the moon is usually called the weight of the body when it is near the moon. Note again that the mass of a body is the same whether it is on the earth, on the moon, or somewhere else in space. Mass is a property of the body itself, whereas weight depends on the nature and distance of other objects that exert gravitational forces on the body.

Since, at any particular location, the weight of a body is proportional to its mass, we can conveniently compare the mass of one body with that of another by comparing their weights as long as we determine the weights at the same location.

Our sensation of our own weight usually comes from other forces that balance it. For example, sitting on a chair, we feel the force exerted by the chair that balances our weight and thereby prevents us from falling to the floor. When we stand on a spring scale, our feet feel the force exerted on us by the scale. The scale is calibrated to read the force it must exert (by the compression of its springs) to balance our weight. The force that balances our weight is called our *apparent weight*. It is the apparent weight that is given by a spring scale. If there is no force to balance your weight, as in free-fall, your apparent weight is zero. This condition, called *weightlessness,* is experienced by astronauts in orbiting satellites. Consider a satellite in a circular orbit near the surface of the earth with a centripetal acceleration v^2/r, where r is the orbit radius and v is the speed. The only force acting on the satellite is its weight. Thus, it is in free-fall with the acceleration of gravity. An astronaut in the satellite is also in free-fall. The only force on the astronaut is his or her weight, which produces the acceleration $g = v^2/r$. Since there is no force balancing the force of gravity, the astronaut's apparent weight is zero.

Apparent weight

Weightlessness

Weightlessness in a space capsule. The astronauts are in free-fall, accelerating toward the earth with the acceleration of gravity.

Questions

10. From our definitions of mass and weight, would it be conceivable to use the same units for both?

11. Suppose an object were sent far out in space, away from galaxies, stars, or other bodies. How would its mass change? Its weight?

12. How would an astronaut in a condition of weightlessness be aware of his or her mass?

13. Under what circumstances would your apparent weight be greater than your true weight?

4-3   Units of Force and Mass

Like the units of time (the second) and length (the metre), the kilogram is a fundamental unit in SI. The unit of force, the newton, and the units for other quantities, such as momentum and energy, that we shall study are derived from these three units. Since 1 N produces an acceleration of 1 m/s² when acting on 1 kg, we have from $F = ma$

$$1\,\text{N} = 1\,\text{kg}\cdot\text{m/s}^2 \qquad\qquad 4\text{-}4$$

Although we generally use SI units in this book, we need to know about two other systems: the cgs system, the metric system based on the centimetre, gram, and second, which is closely related to the international system and is used by many scientists, and the U.S. customary system, based on the foot, the second, and a force unit (the pound), which is still used today in the United States.

The unit of mass in the cgs system, the gram (g), is now defined to be exactly one one-thousandth of the mass of the standard kilogram:

$$1 \text{ g} = 10^{-3} \text{ kg} \qquad\qquad 4\text{-}5$$

The gram was originally chosen to be the mass of one cubic centimetre of water at standard pressure and temperature. The unit of force in the cgs system, called the *dyne (dyn)*, is the force that, when applied to a one-gram mass, produces an acceleration of one centimetre per second squared:

$$1 \text{ dyn} = 1 \text{ g} \cdot \text{cm/s}^2 \qquad\qquad 4\text{-}6$$

Because the units are so small, the cgs system is less convenient than the international system for practical work. For example, the mass of a penny is about 3 g. Since the free-fall acceleration of gravity is 981 cm/s², the weight of a penny in the cgs system is about

$$w = mg = (3 \text{ g})(981 \text{ cm/s}^2) = 2.94 \times 10^3 \text{ dyn}$$

The dyne is a very small unit of force. The relation between the dyne and the newton is

$$1 \text{ dyn} = \frac{1 \text{ g} \cdot \text{cm}}{\text{s}^2} \times \frac{1 \text{ kg}}{10^3 \text{ g}} \times \frac{1 \text{ m}}{10^2 \text{ cm}} = 10^{-5} \text{ kg} \cdot \text{m/s}^2$$

or

$$1 \text{ dyn} = 10^{-5} \text{ N} \qquad\qquad 4\text{-}7$$

The U.S. customary system differs from both the international and the cgs systems in that a unit of force rather than a unit of mass has been chosen as a fundamental unit. The pound (lb) was originally defined as the weight of a **Pound** particular standard body at a particular location. It is now defined as 4.448222 N. (This is the weight of a body of mass 0.45359237 kg at a point where g has the value 9.80665 m/s² = 32.1740 ft/s².) Rounding to three places, we have

$$1 \text{ lb} \approx 4.45 \text{ N}$$

Since 1 kg weighs 9.81 N, its weight in pounds is

$$9.81 \text{ N} \times \frac{1 \text{ lb}}{4.45 \text{ N}} = 2.20 \text{ lb}$$

The unit of mass in the U.S. customary system is defined as the mass that will be given an acceleration of one foot pound per second squared when a force of one pound is applied to it. This unit, called a *slug*, is the mass of a body that weighs 32.2 pounds. We shall not need to use this unit. Instead, when working problems in this system, we shall substitute w/g for the mass m, where w is the weight in pounds and g the acceleration of gravity in feet per second squared.

The standard kilogram, kept at the International Bureau of Weights and Measures, Sèvres, France.

-2   The net force acting on a 10.0-lb body is 3.00 lb. What is its acceleration?

The acceleration is the force divided by the mass:

$$a = \frac{F}{m} = \frac{F}{w/g} = \frac{3.00 \text{ lb}}{(10.0 \text{ lb})/(32.2 \text{ ft/s}^2)} = 9.66 \text{ ft/s}^2$$

Although the weight of an object varies from place to place because of changes in g, this variation is too small to be noticed in most practical applications. Thus, in our everyday experience, the weight of a body appears to be as much a constant characteristic of the body as its mass.

Questions

14. What is your weight in newtons?

15. What is your mass in kilograms?

16. What would your weight be in pounds on the moon, where objects fall freely with acceleration of about 5.33 ft/s²?

4-4   Newton's Third Law

Newton's third law, the *law of interaction*, describes an important property of forces, that they always occur in pairs. For example, if a force is exerted on some body A, there must be an external agent, another body B, exerting the force. The third law states that body A exerts an equal but opposite force on body B. For example, the earth exerts a gravitational force $\mathbf{F}_g$ on a projectile, causing it to accelerate toward the earth. According to the third law, the projectile in turn exerts a force on the earth equal in magnitude and opposite in direction. Thus, the projectile exerts a force $\mathbf{F}'_g$ on the earth toward the projectile. If this were the only force acting on the earth, the earth would accelerate toward the projectile. (Since there are many other forces acting on the earth and the earth has a very large mass, the acceleration it experiences due to this force is negligible and unobserved.)

In discussions of Newton's third law, the words "action" and "reaction" are frequently used. If the force exerted on body A is called the *action* of B upon A, then the force A exerts back on body B is called the *reaction* of A upon B. It does not matter which force in such a pair is called the action and which the reaction. The important points are that forces always occur in action–reaction pairs and that the reaction force is equal in magnitude and opposite in direction to the action force.

Note that the action and reaction forces can never balance each other because they act on *different objects*. This is illustrated in Figure 4-5, which shows two action–reaction pairs for a book resting on a table. The force acting downward on the book is the weight $\mathbf{w}$ due to the attraction of the earth. An equal and opposite force $\mathbf{w}'$ is exerted by the book *on the earth*.

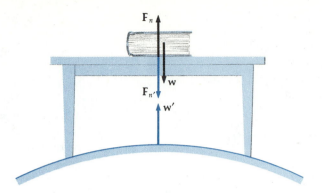

Figure 4-5
Action–reaction forces. The weight **w** is the force exerted on the book by the earth. The equal and opposite reaction force **w'** is the force exerted on the earth by the book. Similarly, the table exerts a force $\mathbf{F}_n$ on the book, and the book exerts an equal and opposite force $\mathbf{F}'_n$ on the table. Action–reaction forces are exerted on different objects and therefore cannot balance.

These forces are an action–reaction pair. If they were the only forces acting, the book would accelerate downward because it would have only a single force acting on it. However, the table in contact with the book exerts an upward force $\mathbf{F}_n$ on it. This force balances the weight of the book. The book also exerts a force $\mathbf{F}'_n$ downward on the table. The forces $\mathbf{F}_n$ and $\mathbf{F}'_n$ are also an action–reaction pair.

Example 4-3   A horse refuses to pull a cart. The horse reasons, "according to Newton's third law, whatever force I exert on the cart, the cart will exert an equal and opposite force on me, so the resultant force will be zero and I will have no chance of accelerating the cart." What is wrong with this reasoning?

Figure 4-6 is a sketch of a horse pulling a cart. Since we are interested in the motion of the cart, we have circled it and indicated the forces acting on it. The force exerted by the horse is labeled **T**. Other forces on the cart are its

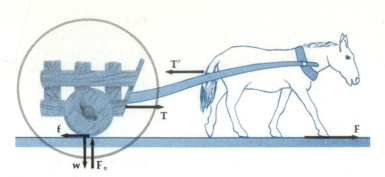

Figure 4-6
Horse pulling a cart. The cart will accelerate to the right if the force **T** exerted on it by the horse is greater than the friction force **f** exerted on it by the ground. The force **T'** is equal and opposite to **T**. Because it is exerted on the *horse,* it has no effect on the motion of the cart.

weight **w**, the vertical support force of the ground $\mathbf{F}_n$, and the horizontal force exerted by the ground labeled **f** (for friction). The vertical forces **w** and $\mathbf{F}_n$ balance. (We know this because we know the cart does not accelerate vertically.) The horizontal forces are **T** to the right and **f** to the left. The cart will accelerate if **T** is greater than **f**. Note that the reaction force to **T**, which we call **T'**, is exerted on the *horse,* not on the cart. It has no effect on the motion of the cart. It does affect the motion of the horse, however. If the horse is to accelerate to the right, there must be a force **F** (to the right) exerted by the ground on the horse's feet that is greater than **T'**. This example illustrates the importance of a simple diagram in solving mechanics problems. Had the horse drawn a simple diagram, he would have seen that he need only push back hard against the ground so that the ground would push him forward.

4-5   Springs, Strings, and Support Forces

Most forces are exerted by one body in contact with another. (Exceptions are the force of gravity and the electric and magnetic forces.) A spring made by winding a stiff wire into a helix is a familiar device, which we discussed briefly in Section 4-1. The force exerted by the spring when it is compressed or extended is the result of complicated intermolecular forces in the spring, but an empirical description of the macroscopic behavior of the spring is sufficient for most applications. If the spring is compressed or extended and is then released, it returns to its original, or natural, length, provided that the displacement is not too great. There is a limit to such displacements beyond which the spring does not return to its original length but remains permanently deformed. If we allow only displacements below this limit, we can calibrate the extension or compression Δx in terms of the force needed to produce the extension or compression, as outlined in Section 4-1. It has been found experimentally that, for small Δx, the force exerted by the spring is approximately proportional to Δx. This relationship, known as *Hooke's law,* can be written

$$F_x = -k(x - x_0) = -k\,\Delta x \qquad\qquad \text{4-8}$$

Hooke's law

where the constant k is called the *force constant* of the spring. The distance x is the coordinate of the free end of the spring or of any object attached to that end of the spring. The constant x_0 is the value of this coordinate when the spring is unstretched, that is, in its equilibrium position. There is a negative sign in Equation 4-8 because, if the string is stretched (Δx is positive), the force F_x is negative, whereas if the string is compressed (Δx is negative), F_x is positive. Such a force is called a *restoring force because it tends to restore the spring to its initial configuration.*

The force exerted by a spring is similar to that exerted by one atom on another in a molecule or in a solid. It is often useful to visualize the atoms in a molecule or solid as being connected by springs. For example, if we slightly increase the separation of the atoms in the molecule and then release them, we would expect the atoms to oscillate back and forth as if they were two masses connected by a spring.

If we pull on a flexible string, the string stretches slightly and pulls back with an equal but opposite force (unless the string breaks). We can think of a string as a spring with such a large force constant that the extension of the string is negligible. If the string is flexible, however, we cannot exert a force of compression on it. When we push on a string, it merely flexes or bends.

When two bodies are in contact with each other, they exert forces on each other due to the interaction of the molecules in one body with those of the other. Consider a book resting on a horizontal table. The weight of the book pulls the book downward, pressing it against the table. Because the molecules in the table have a great resistance to compression, the table exerts a force upward on the book perpendicular, or normal, to the surface with no noticeable compression of the table. (Careful measurement would show that the supporting surface always bends slightly in response to a load.) Similarly, the book exerts a downward force on the table. Note that this normal force exerted by one surface on the other can vary over a wide range of values. For example, unless the book is so heavy that the table breaks, the table will exert an upward support force on the book exactly equal to the weight of the book, no matter how large or small the weight may be. Of course, if you press down on the book, the table will exert a support force greater than the weight of the book to prevent it from accelerating downward.

Under certain circumstances, bodies in contact will exert forces on each other parallel to the surfaces in contact. The parallel component of the contact force exerted by one body on another is called a *frictional force.* We shall study friction in the next chapter.

Robert Hooke was apparently a bit suspicious of Isaac Newton, his fellow-member of the Royal Society. Indeed, he developed the fear that Newton's friends might try to give Newton credit for Hooke's law. So he wrote the law out in Latin—ut tensio, sic vis, "as the extension, so the force"—scrambled the letters in an anagram, and sent it to a friend in a letter for safekeeping.

4-6   Applications to Problem Solving: Constant Forces

Newton's laws are used to determine the acceleration of a particle from a knowledge of all the forces acting on it or to determine the forces acting on a particle given its acceleration. In this section, we shall illustrate the application of Newton's laws to problem solving by considering some simple examples of motion under constant forces. Careful study of these simple examples will make you aware of the content of newtonian mechanics and how it can be applied. Practical problems are generally more complex than these examples, but the methods of solving them are natural extensions of those illustrated here. Some more general examples of the applications of Newton's laws will be discussed in Chapter 5.

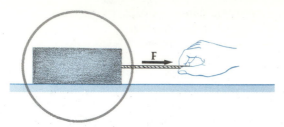

Figure 4-7
A box on a frictionless horizontal
surface with a horizontal force
exerted on it through a string. The
first step in solving the problem is to
isolate the system to be analyzed. In
this case, the circle isolates the box
from its surroundings.

Consider a box resting on a frictionless horizontal table and pulled with a force **F** applied to a light string, as shown in Figure 4-7. To find the motion of the box, we need to find the resultant force acting on it. The first step is to choose the object whose acceleration is to be determined and upon which the forces to be considered act. In the figure, a circle has been drawn around the box to help us isolate it mentally from its surroundings. There are two classes of forces that can act on a body:

1. *Contact forces,* which are forces exerted by objects such as springs, strings, or surfaces that are in direct contact with the body.

2. *Action-at-a-distance forces,* which are forces exerted by objects not in direct contact with the body. The force of gravity (the weight of a body) and electric and magnetic forces are familiar examples of action-at-a-distance forces. To determine whether such forces are acting, we must know whether there is any body nearby that can exert significant gravitational force or there are any electric charges nearby that can exert electric or magnetic forces.

Two classes of forces

Three significant external forces act on the box in this example. They are indicated in Figure 4-8. Such a diagram is called a *free-body diagram.* The three forces are

Free-body diagram

1. The weight of the box, **w**. This is an action-at-a-distance force.

2. The contact force exerted by the table, $\mathbf{F}_n$. Since we are assuming that the table is frictionless, the contact force is perpendicular to the table. Such a force is called a *normal force.* (The word "normal" means perpendicular.)

3. The contact force exerted by the spring, **T**, called the *tension* in the string. Assuming the mass of the string to be negligible compared with that of the box, this force equals the force applied to the string by the hand: $\mathbf{T} = \mathbf{F}$. The string merely acts as a device for transmitting the force exerted by the hand to the box. In general, a light string connecting two points has a tension that has constant magnitude and acts along the string. (This also holds for a string that passes over a frictionless peg or a pulley of negligible mass as long as there are no tangential forces on the string between the two points being considered.)

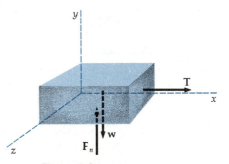

Figure 4-8
A free-body diagram for the box in
Figure 4-7. The three significant
forces acting on the box are the
force **w** exerted by the earth, the
normal force $\mathbf{F}_n$ exerted by the
table, and the force **T** exerted by
the string.

A convenient coordinate system is also indicated in Figure 4-8. Note that the normal force $\mathbf{F}_n$ and the weight **w** are drawn with equal magnitude. We know these forces have equal magnitude because the box does not accelerate vertically. Since the resultant force is in the x direction and has magnitude T, Newton's second law gives $T = ma_x$. Thus $a_x = T/m = F/m$.

Even in this simple example, both kinds of applications of Newton's laws are used: the horizontal acceleration is found in terms of the given force **F**, and the vertical force F_n exerted by the table is found from the fact that the box remains on the table and thus $a_y = 0$. Conditions on the motion of an object such as the fact that the box remains on the table are called *constraints*.

Constraints

According to Newton's third law, forces always act in pairs. In Figure 4-8 we have shown only those forces exerted on the box. Figure 4-9 shows the reaction forces to the forces shown in Figure 4-8. These are the gravitational force **w′** exerted *by the block on the earth*, the force $\mathbf{F}'_n$ exerted *by the box on the table*, and the force **T′** exerted *by the box on the string*. Since these forces are not exerted *on* the box, they have nothing to do with its motion. They are therefore omitted in the application of Newton's second law to the motion of the box.

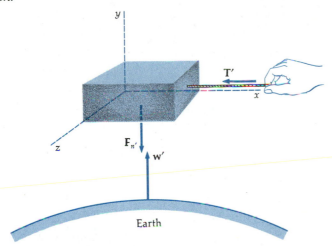

Figure 4-9
The reaction forces corresponding to the three forces shown in Figure 4-8. Note that these forces *do not* act on the box: **T′** acts on the string, $\mathbf{F}'_n$ acts on the table, and **w′** acts on the earth.

This simple example illustrates a general method of attack for problems using Newton's laws, which consists of the following steps:

1. Draw a neat diagram.   *free body force diagram*

General method for problem solving

2. Isolate the body (particle) of interest and draw a free-body diagram, indicating each external force acting on the body. If there is more than one body of interest in the problem, draw a separate diagram for each.

3. Choose a convenient coordinate system for each body and apply Newton's law $\Sigma \mathbf{F} = m\mathbf{a}$ in component form.

4. Solve the resulting equations for the unknowns using whatever additional information is available. The unknowns may include the masses, the components of the accelerations, or the components of some of the forces.

5. Finally, inspect the results carefully, checking whether they correspond to reasonable expectations. Predictions of the solution are particularly helpful when variables are assigned extreme values. In this way you can check your work for errors.

We shall now give a variety of examples.

Example 4-4   Find the acceleration of a block of mass m that slides down a frictionless fixed surface inclined at an angle θ to the horizontal.

There are only two forces acting on the block, the weight **w** and the normal force $\mathbf{F}_n$ exerted by the incline (see Figure 4-10). (For real surfaces, there would be a force of friction parallel to the incline, but here we are assuming an ideal frictionless surface.) Since the two forces are not along the same line, they cannot add to zero and the block must therefore accelerate. The acceleration is along the incline, which is another example of a constraint. For this problem, it is convenient to choose a coordinate frame with one axis parallel to the incline and the other perpendicular to it, as shown in Figure 4-10. Then, the acceleration has only one component, a_x. For this choice, $\mathbf{F}_n$ is in the y direction, and the weight **w** has the components

$$w_x = w \sin \theta = mg \sin \theta$$
$$w_y = -w \cos \theta = -mg \cos \theta$$

4-9

where m is the mass and g is the acceleration due to gravity (see Figure 4-11).

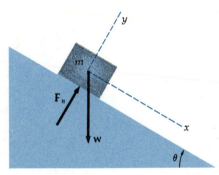

Figure 4-10
Forces acting on a block of mass m on a frictionless incline. It is convenient to choose the x axis parallel to the incline.

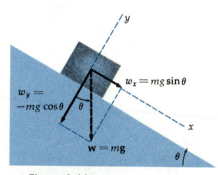

Figure 4-11
The weight of the block can be replaced by its components, $mg \sin \theta$ parallel to the incline and $mg \cos \theta$ perpendicular to the incline. The component $mg \cos \theta$ is balanced by the normal force (not shown).

The resultant force in the y direction is $F_n - mg \cos \theta$. From Newton's second law and the fact that $a_y = 0$,

$$\Sigma F_y = ma_y = F_n - mg \cos \theta = 0$$

and thus

$$F_n = mg \cos \theta$$

4-10

Similarly, for the x components,

$$\Sigma F_x = ma_x = mg \sin \theta$$
$$a_x = g \sin \theta$$

4-11

The acceleration down the incline is constant and equal to $g \sin \theta$. It is useful to check our results for the extreme values of inclination, $\theta = 0$ and $\theta = 90°$. At $\theta = 0$, the surface is horizontal. The weight has only a y component, which is balanced by the normal force F_n. The acceleration is of course zero; $a_x = g \sin 0° = 0$. At the opposite extreme, $\theta = 90°$, the incline is vertical.

The weight has only an x component along the incline, and the normal force is zero; $F_n = mg \cos 90° = 0$. The acceleration is $a_x = g \sin 90° = g$ since the block is in free-fall.

When friction is neglected, the body slides down the incline with acceleration $g \sin \theta$.

Example 4-5   A picture weighing 8 newtons is supported by two wires of tension $\mathbf{T}_1$ and $\mathbf{T}_2$, as shown in Figure 4-12a. Find the tension in the wires.

This is a problem in static equilibrium. Since the picture does not accelerate, the net force acting on it must be zero. The three forces acting on the picture, its weight $m\mathbf{g}$, the tension $\mathbf{T}_1$, and the tension $\mathbf{T}_2$, must therefore sum to zero. Since the weight has only a vertical component mg downward, the horizontal components of the tensions $\mathbf{T}_1$ and $\mathbf{T}_2$ must balance each other, and the vertical components of the tensions must balance the weight:

$$\Sigma F_x = T_{1x} + T_{2x} = T_1 \cos 30° - T_2 \cos 60° = 0$$

$$\Sigma F_y = T_{1y} + T_{2y} = T_1 \sin 30° + T_2 \sin 60° - mg = 0$$

Using $\cos 30° = \sqrt{3}/2 = \sin 60°$ and $\sin 30° = \frac{1}{2} = \cos 60°$, we have

$$T_1\left(\frac{\sqrt{3}}{2}\right) = \frac{T_2}{2}$$

or

$$T_2 = \sqrt{3}T_1$$

and

$$T_1\left(\frac{1}{2}\right) + T_2\left(\frac{\sqrt{3}}{2}\right) = mg$$

Multiplying by 2 and substituting $\sqrt{3}T_1$ for T_2, we obtain

$$T_1 + (\sqrt{3}T_1)(\sqrt{3}) = 2\,mg$$

Thus,

$$T_1 = \tfrac{1}{2}mg = 4 \text{ N}$$

$$T_2 = \sqrt{3}T_1 = \frac{\sqrt{3}}{2}\,mg = 6.93 \text{ N}$$

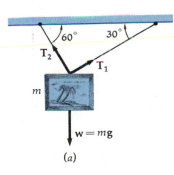

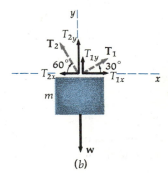

Figure 4-12
(a) Picture supported by two wires in Example 4-5. (b) Choice of coordinate system and resolution of the forces into their x and y components.

Example 4-6 A block hangs by a (massless) string that passes over a friction-less peg and is connected to another block on a frictionless table. Find the acceleration of each block and the tension in the string.

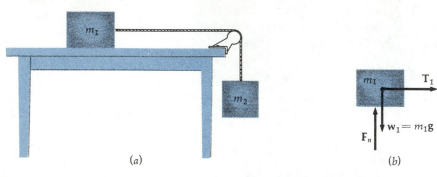

(a)   (b)   (c)

Figure 4-13 shows the important elements of this problem. The string tensions T_1 and T_2 have equal magnitudes because the string is assumed to be massless and there are no tangential forces acting on it (the peg is friction-less). For block 1 on the table, the vertical forces F_n and w_1 have equal magnitudes because of the constraint that the vertical acceleration is zero for m_1. Newton's second law applied to the horizontal component gives

$$T = m_1 a_1 \qquad\qquad 4\text{-}12$$

Figure 4-13
(a) The two blocks of Example 4-6.
(b) Free-body diagram for m_1.
(c) Free-body diagram for m_2.

where a_1 is the acceleration of m_1 along the horizontal surface. If we take the downward direction to be positive for the acceleration a_2 of block 2, the equation of motion for m_2 is

$$m_2 g - T = m_2 a_2 \qquad\qquad 4\text{-}13$$

We can simplify these equations by noting that, if the connecting string does not stretch, the accelerations a_1 and a_2 are both positive and equal in magnitude (but not in direction). Let us call this magnitude a. We then have

$$T = m_1 a \qquad\qquad 4\text{-}14$$

$$m_2 g - T = m_2 a \qquad\qquad 4\text{-}15$$

We solve these two equations for the two unknowns T and a by first eliminating one of the unknowns. For example, we can eliminate T by substituting $m_1 a$ from Equation 4-14 for T in Equation 4-15. We then obtain

$$m_2 g - m_1 a = m_2 a$$

or

$$a = \frac{m_2}{m_1 + m_2} g \qquad\qquad 4\text{-}16$$

This expression for a can then be substituted into Equation 4-14 to find T. We obtain

$$T = \frac{m_1 m_2}{m_1 + m_2} g \qquad\qquad 4\text{-}17$$

Note that the result for the magnitude a is the same as that for a mass $m = m_1 + m_2$ acted on by a force $m_2 g$.

Example 4-7   A man of mass m stands on a scale fastened to the floor of an elevator, as illustrated in Figure 4-14. What does the scale read when the elevator is accelerating (*a*) up and (*b*) down?

(*a*) Since the man is at rest relative to the elevator, he is also accelerating up. The forces on the man are F_n upward, exerted by the scale platform on which he stands, and w downward, the force of gravity. The resultant force is $F_n - w$ upward in the direction of the acceleration a. Newton's second law gives

$$F_n - w = ma$$

or

$$F_n = w + ma = mg + ma \qquad\qquad 4\text{-}18$$

The force $\mathbf{F}'_n$ exerted by the man on the scale determines the reading on the scale, which is his apparent weight. Since $\mathbf{F}'_n$ and $\mathbf{F}_n$ are an action–reaction pair, they are equal in magnitude. Thus, when the elevator accelerates up, the apparent weight of the man is greater than his true weight by the amount ma.

(*b*) For the case of the elevator accelerating downward, let us call the magnitude of the acceleration a'. In this case the resultant force must be downward, implying that the weight is greater than F_n. Again Newton's second law gives

$$w - F_n = ma'$$

or

$$F_n = w - ma' = mg - ma' \qquad\qquad 4\text{-}19$$

Again, the scale reading, or apparent weight, equals F_n. In this case, the apparent weight is less than mg. If $a' = g$, as it would if the elevator were in free-fall, the man would be apparently weightless. What if the acceleration of the elevator is greater than g? (For this to happen, something in addition to gravity would have to pull down on the elevator.) Assuming that the surface of the scale is not sticky, the scale cannot exert a force down on the man. Since the downward force on the man cannot be greater than w, the scale will soon leave the man. The man will have the acceleration g, which is less than that of the elevator, so he will eventually hit the ceiling of the elevator. Then, if the ceiling is strong enough, it can provide the force downward necessary to give him the acceleration a'.

Questions

17. A picture is supported by two wires as in Example 4-5. Do you expect the tension to be greater or less in the wire that is more nearly vertical?

18. A weight is hung on a wire that is originally horizontal. Can the wire remain horizontal? Explain.

19. What effect does the velocity of the elevator have on the apparent weight of the man in Example 4-7?

Figure 4-14
Man on a scale in an accelerating elevator. The scale indicates his *apparent weight* F_n, which is greater than mg when the acceleration is upward and less than mg when the acceleration is downward.

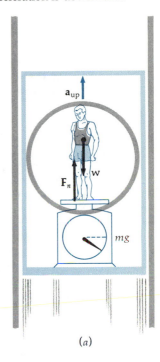

(*a*)

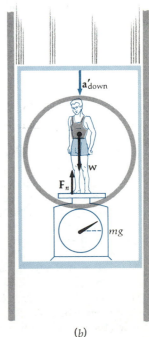

(*b*)

Isaac Newton (1642 – 1727)

I. Bernard Cohen
Harvard University

When Isaac Newton was once asked how he had made his great discoveries, he replied, "By always thinking unto them." He is also reported to have said, "I keep the subject constantly before me and wait till the first dawnings open little by little into the full light." This ability to concentrate is a particular quality of Newton's genius, and it fits in well with his character and personality. For he was a solitary man, without close and intimate friends or confidants. He never married, and he spent his early boyhood deprived of father (who died before young Isaac was born on Christmas day in 1642) and of mother (who remarried within two years and left Isaac to be reared by an aged grandmother).

A lonesome man, Newton developed "unusual powers of continuous concentrated introspection." In these words, his biographer Lord Keynes, the economist, epitomized Newton's "power of holding continuously in his mind a purely mental problem . . . for hours and days and weeks until it surrendered to him its secret." And then, in keeping with this character of inwardness, he was satisfied to keep his discoveries to himself, neither rushing into print, as his contemporary fellow scientists were wont to do, nor even communicating to his associates what he had accomplished. Accordingly, it has been said that every discovery of Newton's had two phases; Newton made the discovery, and then others had to find out that he had done so.

In 1684, Edmund Halley (the astronomer after whom Halley's comet is named) went from London to Cambridge, where Newton was a professor, to ask him about a fundamental problem that had baffled the major scientists of the Royal Society (the world's oldest existing scientific organization). What "curve would be described by the Planets," Halley asked Newton, "supposing the force of attraction towards the Sun to be reciprocal to the square of their distance from it?" Newton "replied immediately that it would be an Ellipsis," that is, an ellipse. Halley, "struck with joy and amazement asked him how he knew it. Why, saith he, I have calculated it." In these final four words, Newton revealed to Halley that he had solved the major scientific problem of the century: to find the law of force that holds the solar system together and regulates the motion of its component planets and their satellites. Encouraged and stimulated by Halley, Newton went on to show that this is a universal force, one that acts between any two bodies anywhere in the universe and that varies directly as a product of the two masses and inversely as the square of the distance between them.

The seeds of Newton's great achievements in science go back to a period of some 18 months after his graduation from college (1665 – 1667), when fear of the plague caused the

university to be shut. Newton returned to the family farm in Woolsthorpe, Lincolnshire, where he had been born. Here he worked out his most fundamental contributions to mathematics, the methods of the differential and integral calculus (an innovation for which he must share the credit with the German philosopher and mathematician Leibniz, an independent codiscoverer). During this same time, he found the inverse-square law of gravitational force and tested it by a rough calculation of the moon's motion. Newton knew that the moon is about 60 times as far from the center of the earth as an apple. Accordingly, it follows from the inverse-square law that as the moon "falls" toward the earth (that is, as it constantly "falls" away from straight-line inertial motion along a tangent), the force or acceleration should be just $1/60^2$, or $1/3600$, of the acceleration of free-fall of an apple. The calculations, Newton said, agreed "pretty nearly." In his own words, "I thereby compared the force requisite to keep the Moon in her orb with the force of gravity at the surface of the Earth."

A decade or so later, in 1679 and 1680, Newton learned from Robert Hooke how to analyze curved motion by considering it to be composed of two separate components: a uniform inertial motion along the tangent to the curve and a linear accelerated motion toward the center. Newton gave his centrally directed component the name "vis centripeta" or "centripetal force," by which name we have known it ever since. It was by using this new form of analysis that Newton solved the problem of planetary motion. Using a radical new measure of force of his own invention, Newton proceeded in two stages. First he showed that Kepler's law of areas (Chapter 9) is both a necessary and sufficient condition that a force be directed toward the center with respect to which the equal areas are computed. Then, he proved that if the curve (or orbit) is an ellipse, this centrally directed force varies inversely as the square of the distance. Only later did Newton prove the converse, that if the force varies inversely as the square of the distance, the orbit will be one of the conic sections—an ellipse, but possibly a parabola or hyperbola depending on the conditions.

Later, he used his mathematical prowess to prove the fundamental theorem that a uniform sphere will attract gravitationally as if all its mass were concentrated at its geometric center.

Other discoveries made by Newton during his period of retreat during the plague years were in the field of optics. Newton is celebrated not only for his work in pure mathematics and for having founded the science of celestial mechanics; he is also known as an experimental scientist, primarily for his discoveries relating to light and color and that

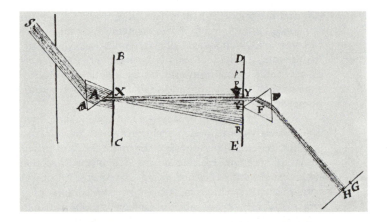

Figure 1
Manuscript drawing in Newton's papers, showing the "crucial experiment." Sunlight enters a darkened room at the left-hand side through a small hole in the shutter *S*, producing a spectrum on passing through the first prism *A*. A hole at *Y* in the board *DE* permits light of a single color to pass through the second prism *F*, with the result of a deviated beam of light, without any further alteration in color.

class of effects known today as *interference phenomena*. He bought a triangular glass prism, he tells us, to try "the celebrated phenomenon of colours."

Newton's experiments on color showed that "white" light, or sunlight, is a mixture of light of all the different colors. He demonstrated that the prism separates out the different colors because each color has its own index of refraction, or degree of being bent or deviated from its original path by the prism. Hence when a beam of white light enters a prism, each color is refracted by the prism to a different degree, causing the light to spread out in a spectrum in which the different colors appear in a continuous succession from deep red to blue or violet. The violet light he found to be deviated the most; the red light, the least. In order to prove that this is so, he devised what he called a "crucial experiment" (see Figure 1). He allowed a narrow beam of sunlight to enter a darkened room through a small hole in a window shutter. This beam then passed through a prism and produced a spectrum. Using an opaque board with a small hole in it, Newton could separate out of this spectrum a beam of light of a single color — red, orange, green, blue — and allow this monochromatic light to pass through a second prism. If prisms produce a spectrum by "altering" the incident light in some way, this effect should be apparent in the way the second prism would alter or affect an incident beam of monochromatic light. When monochromatic light entered the second prism, it emerged with no change of color; the second prism had produced only a further deviation, or bending, of the path of the light.

These investigations not only led Newton to an understanding of why objects appear to have the colors they display in terms of the colors of the light they absorb, transmit, or reflect, but they also showed him that telescopes with simple or single lenses have definite limitations. Since a biconvex lens can be thought of as a pair of prisms base to base (see Figure 2), we can see at once that there will be a different focus for each color — an effect called *chromatic aberration* — and so the image will be poorer. Newton accordingly proceeded to invent a wholly new kind of telescope, one which reflected light rather than refracted it to form the image thus avoiding chromatic aberration.

A most notable set of investigations centered on interference patterns of light produced in thin plates or films, as in the familiar colored patterns seen on films of oil. Newton studied these alternating fringe rings of lightness and darkness that are produced by the thin layer of air between a flat glass surface and the convex side of a plano-convex lens, a phenomenon known today as Newton's rings (see Figure 3).

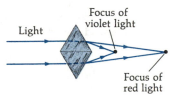

Figure 2
Two prisms, one on top of the other, act like a lens that is thick at the middle and thin at the edges. Each of the component colors of light passing through these prisms will be bent, or refracted, a different amount, thus coming to a focus at a different point.

Essay: Isaac Newton (continued)

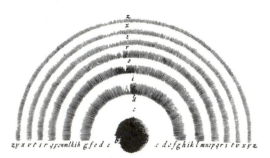

zyxvtsrqponmlkih gfed c   c defghiklmnopqrs tv xyz

Figure 3
A drawing from Newton's *Optiks* showing the
phenomenon of Newton's rings.

His measurements were so precise that they were used a
hundred years later by Thomas Young to compute the wavelength of light.

Newton's achievements during his early years are remarkable. We can only contemplate in amazement how, in so
short a time, a young man barely out of college should have
made such fundamental contributions to pure mathematics,
to theoretical celestial mechanics and dynamics, and to experimental physics. As Newton himself later said, "All this
was in the two plague years of 1665 and 1666, for in those
days I was in the prime of my age for invention, and minded
mathematics and philosophy more than at any time since."
(The word "philosophy" as used at that time meant "natural
philosophy," or science.)

The remainder of Newton's scientific life was devoted to
elaboration of the fundamental discoveries we have described. He became Lucasian Professor of Mathematics at
Cambridge, and in 1672 his paper on light and color, together with a description of the newly invented telescope,
was read to the Royal Society. After some aspects of his work
had been misunderstood and criticized, he decided that he
would publish no more. His mathematical innovations were
distributed only in manuscripts until later in his life when
some were printed. Then, in 1684, after Halley's visit—and
with Halley's urging—he wrote up his discoveries in dynamics and celestial mechanics in the *Philosophiae Naturalis
Principia Mathematica* (London, 1687), which may be translated as *The Mathematical Principles of Natural Philosophy*.
Here he used the law of universal gravitation to deduce the
shape of the earth (flattened at the poles and bulging at the
equator), to show how the acceleration of freely falling
bodies varies with terrestrial latitude, and to explain how the
tides in the oceans are caused by the gravitational pull of the

sun and moon on the water. He also introduced the three
famous laws of motion that bear his name and that were the
"axioms" on which his system of dynamics was constructed.
In this work, too, he set forth for the first time a clear concept
of mass and momentum and set up a series of beautiful
experiments to show that mass (as measured by inertia) is
proportional to weight.

In the 1690s, Newton became bored or disenchanted with
the cloistered life of a university professor. He moved to
London, where he became director of the Mint, a post he
held until his death. He became president of the Royal Society and ruled British science with a firm control. During the
London years, he brought out two revised editions of his
Principia (1713, 1726) and published his book on optics
(1704 and later editions) and various tracts on mathematics.
He died in 1727 and was buried with state honors in West-
minster Abbey.

From early manhood, Newton devoted only a small fraction of his creative intellectual life to orthodox scientific pur-
suits: pure and applied mathematics, astronomy and celestial
mechanics, dynamics, experimental physics, and optics.
During all these years he was an ardent student of theology,
reading and taking innumerable notes and writing tracts and
even whole books on religious subjects. Some of these dealt
with fundamental questions of interpretation of theological
doctrines, others with the meaning of the prophetic books of
the Bible, and yet others with the problem of unraveling
Church history. He also developed a wholly new system of
world chronology, in part based on astronomy, that proved
to be of little value. And his chief subject of study was
alchemy—reading extensively, copying out whole sections
of books, and making experiments, all to what end we do not
know.

Through his readings of mystical philosophers, theolo-
gians, and speculative alchemists, Newton may have gained
a vision of a unified system of knowledge that would em-
brace both physical and divine science, revealing both the
laws of the created world and the plan of its Creator. Various
hints of this aspect of Newton's thought appear in his writ-
ings, for example, "And thus to discourse about God from
phenomena does belong to experimental philosophy." It is
this vision of a knowledge that he never attained that may
have caused him to undervalue his monumental scientific
achievements. For shortly before his death he said, "I do not
know what I may appear to the world; but to myself I seem to
have been only like a boy, playing on the sea-shore, and
diverting myself in now and then finding a smoother pebble
or a prettier shell than ordinary, whilst the great ocean of
truth lay all undiscovered before me."

Summary

1. The fundamental relations of classical mechanics are contained in Newton's laws of motion:

   Law 1. A body continues in its initial state of rest or motion with uniform velocity unless acted on by an unbalanced or resultant force.

   Law 2. The acceleration of a body is inversely proportional to its mass and directly proportional to the resultant external force acting on it

$$\mathbf{a} = \frac{\Sigma \mathbf{F}}{m} \quad \text{or} \quad \Sigma \mathbf{F} = m\mathbf{a}$$

   Law 3. Forces always occur in pairs. If body A exerts a force on body B, an equal but opposite force is exerted by body B on body A.

2. Force is defined in terms of the acceleration it produces on a given body. A force of 1 newton (N) is the force that produces an acceleration of $1 \ m/s^2$ on a standard body with a mass of 1 kilogram (kg).

3. Mass is an intrinsic property of a body that measures its resistance to acceleration. The mass of one body can be compared to that of another by applying an equal force to each body and measuring their accelerations. The ratio of the masses of the bodies is then the inverse ratio of the accelerations of the bodies produced by the same force: $m_1/m_2 = a_2/a_1$. The mass of a body does not depend on the location of the body.

4. The weight of a body is the force of gravitational attraction between the body and the earth. It is proportional to its mass m and to the gravitational field $\mathbf{g}$ (which also equals the free-fall acceleration of gravity)

$$\mathbf{w} = m\mathbf{g}$$

Weight is not an intrinsic property of a body. It does depend on the location of the body because $\mathbf{g}$ does.

5. The general method of attack for solving a problem using Newton's laws includes the following steps:

   1. Draw a neat diagram.

   2. Isolate the body (particle) of interest and draw a free-body diagram, indicating each external force acting on the body. Draw a separate diagram for each body of interest.

   3. Choose a convenient coordinate system for each body and apply Newton's second law in component form.

   4. Solve the resulting equations for the unknowns using whatever additional information is available.

   5. Check your results to see if they are reasonable. Examine your solutions carefully when the variables are assigned extreme values.

Suggestions for Further Reading

Cohen, I. Bernard: "Isaac Newton," *Scientific American*, December 1955, p. 73.

This is a short biographical sketch of Newton.

Feld, Michael S., Ronald E. McNair, and Stephen R. Wilk: "The Physics of Karate," *Scientific American*, April 1979, p. 150.

An analysis of the forces that a karate expert can apply with bare hands to blocks of wood or concrete.

Walker, Jearl: "The Amateur Scientist: In Judo and Aikido Application of the Physics of Forces Makes the Weak Equal to the Strong," *Scientific American*, July 1980, p. 150.

Many diagrams help to make this an entertaining and instructive exposition of physical principles employed in two martial arts.

Review

A. Objectives: After studying this chapter, you should:

1. Be able to discuss the definitions of force and mass and to state Newton's laws of motion.

2. Be able to distinguish between mass and weight.

3. Be familiar with the following units and know how they are defined: kilogram, newton, dyne, and pound.

4. Be able to distinguish between action–reaction pairs of forces that act on different bodies and balancing forces that act on the same body.

5. Be able to apply Newton's laws in a systematic way to the solution of a variety of mechanics problems.

B. Define, explain, or otherwise identify:

Force, p. 60	Gram, p. 65
Kilogram, p. 60	Dyne, p. 65
Mass, p. 61	Pound, p. 65
Weight, p. 63	Free-body diagram, p. 70
Apparent weight, p. 64	Normal force, p. 70
Newton, p. 64	Constraint, p. 71

C. Write out the steps involved in the general method of attack for solving problems using Newton's laws of motion.

D. True or false: If the statement is true, explain why. If it is false, give a counterexample.

1. If there are no forces acting on a body, the body will not accelerate.

2. If a body is not accelerating, there must be no forces acting on it.

3. The motion of a body is always in the direction of the resultant force.

4. Action–reaction forces never act on the same body.

5. The mass of a body depends on its location.

6. Action equals reaction only if the bodies are not accelerating.

Exercises

Section 4-1   Force and Mass

1. A force of 15 N is applied to a box of mass m. The box moves in a straight line with its speed increasing by 10 m/s every 2 s. Find the mass of the box.

2. A 3-kg box is pulled in a straight line along a level, frictionless surface by a constant force F_0. Its speed increases by 5 km/h in 10 s. (*a*) Find F_0 in newtons. (*b*) When a second constant force is applied in the same direction in addition to the first force, the speed increases by 15 km/s in a 10-s interval. Find the magnitude of the second force.

3. An object experiences an acceleration of 4 m/s² when a certain force F_0 acts on it. (*a*) What is the acceleration when the force is doubled? (*b*) A second object experiences an acceleration of 8 m/s² under the influence of force F_0. What is the ratio of the masses of the two objects? (*c*) If the two objects are tied together, what acceleration will the force F_0 produce?

4. Figure 4-15 shows the path taken by a car. It consists of straight lines and arcs of circles. The car starts from rest at point A, and its speed increases until it reaches point B. It then proceeds at constant speed until it reaches point E. From point E on it slows down, coming to rest at point F. What is the direction of the resultant force, if there is any, on the car at the midpoint of each section of the path?

Figure 4-15
Path taken by automobile in Exercise 4.

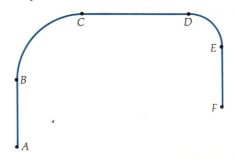

5. A certain force applied to a particle of mass m_1 gives it an acceleration of 20 m/s². The same force applied to a particle of mass m_2 gives it an acceleration of 30 m/s². If the two particles are tied together and the same force is applied to the combination, find the acceleration.

6. A constant force $\mathbf{F}_0$ causes an acceleration of 5 m/s² when it acts on an object of mass m. Find the acceleration of the same mass when it is acted on by the forces shown in Figure 4-16a and b.

Figure 4-16
Forces acting on object for Exercise 6.

(a)

Frictionless surface

(b)

7. A single force of 10 N acts on a particle of mass m. The particle starts from rest and travels in a straight line for a distance of 18 m in 6 s. Find m.

8. Find the average force exerted on a 5.00-g bullet as it acquires a speed of 600 m/s in a rifle barrel of length 65 cm. (The reaction force exerted by the bullet on the rifle is transmitted to the shoulder of the person shooting and is called the "kick".)

9. In order to drag a 100-kg log along the ground at constant velocity, you have to pull on it with a force of 300 N (horizontally). (a) What is the resistive force exerted by the ground on the log? (b) What force must you exert if you want to give the log an acceleration of 2 m/s²?

Section 4-2   The Force Due to Gravity: Weight;
Section 4-3   Units of Force and Mass

10. Find the weight of a 50-kg girl in (a) newtons and (b) pounds.

11. Find the mass of a 175-lb man in (a) kilograms and (b) grams.

12. A newspaper reports that 6000 lb of grass was found in a ship docked at Los Angeles. What is its mass in kilograms?

13. Find the weight of a 50-g object in (a) dynes and (b) newtons.

Section 4-4   Newton's Third Law

14. A 2-kg body hangs at rest from a string attached to the ceiling. (a) Draw a diagram showing the forces acting on the body and indicate each reaction force. (b) Do the same for the forces acting on the string.

15. In a tug of war, two teams pull on a rope, each trying to pull the other over a line midway between their original positions. According to Newton's third law, the forces exerted by each team on the other (via the rope) are equal and opposite. Show with a force diagram how one team can win.

16. A box slides down a rough inclined plane. ("Rough" means that there is friction between the plane and the box.) Draw a diagram showing the forces acting on the box. For each force in your diagram, indicate the reaction force.

17. A hand pushes two bodies on a frictionless horizontal surface, as shown in Figure 4-17. The masses of the bodies are 2 and 1 kg. The hand exerts a force of 5 N on the 2-kg body. (a) What is the acceleration of the system? (*Hint:* Since the two bodies move together, you may consider them as a single body of mass 3 kg.) (b) What is the acceleration of the 1-kg body? Find the resultant force acting on this body. What is the origin of the force exerted on this body? (c) Show all the forces acting on the 2-kg body. What is the resultant force acting on this body?

Figure 4-17
Exercise 17.

Section 4-5   Springs, Strings, and Support Forces

18. A spring has a force constant $k = 200$ N/m. A 5-kg object is suspended motionless from the spring. Find (*a*) the numerical values of all the forces acting on the object and (*b*) the extension of the spring from its equilibrium position.

19. A 100-kg block is pulled with constant acceleration along a frictionless table by a cable that stretches 0.3 cm. The block starts from rest and moves 4 m in 4 s. Assuming that the cable obeys Hooke's law, find (*a*) its force constant and (*b*) how far the cable stretches if the block is suspended vertically from the cable at rest.

20. A body of mass m is attached to two springs along a line, as shown in Figure 4-18. Each spring is stretched from its equilibrium position. The force constants of the springs are k_1 and k_2. (*a*) Find the ratio of the amounts the springs stretch. (*b*) Show that, if the body is displaced a small distance x from equilibrium, the net restoring force is the same as if the body were attached to a single spring with force constant of $k = k_1 + k_2$.

Figure 4-18
Exercise 20.

21. A 2-kg box rests on a frictionless incline of angle 30° supported by a spring (see Figure 4-19). The spring stretches 3 cm. (*a*) Find the force constant of the spring. (*b*) If the box is pulled down the incline 5 cm from its equilibrium position and is then released, what will its initial acceleration be?

Figure 4-19
Exercise 21.

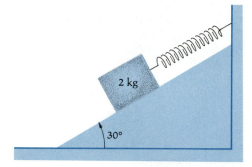

Section 4-6   Applications to Problem Solving: Constant Forces

22. In Figure 4-20, the bodies are attached to spring balances calibrated in newtons. Give the readings of the balances in each case, assuming the strings to be massless and the incline to be frictionless.

Figure 4-20
Exercise 22.

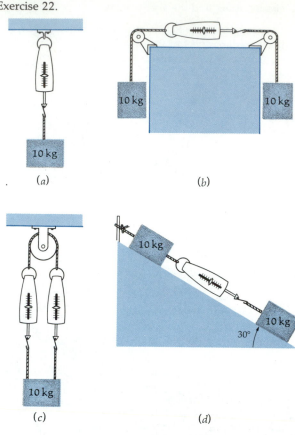

(*a*)   (*b*)

(*c*)   (*d*)

23. A body is held in position by a cable along a smooth incline (see Figure 4-21). (*a*) If $\theta = 60°$ and $m = 50$ kg, find the tension in the cable and the normal force exerted by the incline. (*b*) Find the tension as a function of θ and m, and check your result for $\theta = 0°$ and $\theta = 90°$.

Figure 4-21
Body held on frictionless incline for Exercise 23.

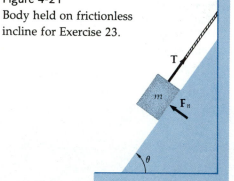

24. A 2-kg picture is hung by two wires of equal length that make an angle θ with the horizontal, as shown in Figure 4-22.

Figure 4-22
Exercise 24.

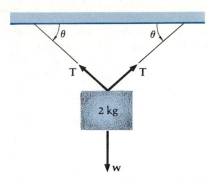

(a) If $\theta = 30°$, find the tension in the wires. (b) Find the tension for general values of θ and the weight **w** of the picture. For what angle θ is the tension the least? The greatest?

25. A 5-kg object is pulled along a frictionless horizontal surface by a horizontal force of 10 N. (a) If the object is at rest at $t = 0$, how fast is it moving after 3 s? (b) How far does it travel from $t = 0$ to $t = 3$ s? (c) Indicate on a diagram all the forces acting on the object.

26. A 10-kg object is subjected to two forces F_1 and F_2, as shown in Figure 4-23. (a) Find the acceleration **a** of the object. (b) A third force F_3 is applied so that the object is in static equilibrium. Find F_3.

Figure 4-23
Exercise 26.

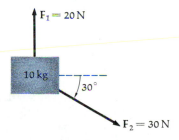

27. A vertical force **T** is exerted on a 5-kg body near the surface of the earth, as shown in Figure 4-24. Find the acceleration of the body if (a) **T** = 5 N, (b) **T** = 10 N, and (c) **T** = 100 N. (Take $g = 10$ m/s².)

Figure 4-24
Exercise 27.

28. A 1-tonne (1000-kg) load is being moved upward by a crane. Find the tension in the cable that supports the load as (a) the load is accelerated upward at 2 m/s², (b) the load is lifted at constant speed, and (c) the load moves upward but its speed decreases by 2 m/s each second.

29. A man holding a 12-kg body on a cord made to withstand 160 N steps into an elevator. When the elevator starts up, the cord breaks. What was the minimum acceleration of the elevator?

30. A 2-kg body hangs from a spring balance (calibrated in newtons) attached to the ceiling of an elevator (Figure 4-25). What does the balance read (a) when the elevator is moving upward with a constant velocity of 30 m/s, (b) when the elevator is moving downward with a constant velocity of 30 m/s, and (c) when the elevator is accelerating upward at 10 m/s²? (d) From time $t = 0$ to $t = 2$ s, the elevator moves upward at 10 m/s. Its velocity is then reduced uniformly to zero in the next 2 s so that it is at rest at $t = 4$ s. Describe the reading on the balance during the time $t = 0$ to $t = 4$ s.

Figure 4-25
Exercise 30.

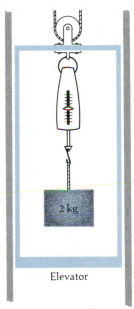

31. A 60-kg girl weighs herself by standing on a scale in an elevator. What does the scale read when (a) the elevator is descending at a constant rate of 10 m/s, (b) the elevator is accelerating downward at 2 m/s², and (c) the elevator is ascending at 10 m/s but its speed is decreasing by 2 m/s in each second. (Take $g = 10$ m/s².)

32. Two 5-kg bodies are connected by a light string, as shown in Figure 4-26. The table is frictionless, and the string passes over a frictionless peg. Find the acceleration of the masses and the tension in the string.

Figure 4-26
Exercise 32.

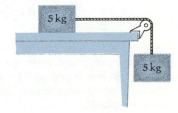

33. A 2.8-kg box slides along a frictionless table. It is attached to a 0.2-kg box by a light string, as shown in Figure 4-27. Find the time it takes for the 0.2-kg box to fall 2 m to the floor if the system starts from rest.

Figure 4-27
Exercise 33.

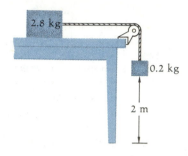

34. A 2-kg body rests on a 4-kg body that rests on a frictionless horizontal surface. A force **F** of magnitude 3 N is exerted on the 4-kg body, as shown in Figure 4-28. (*a*) If the surface between the bodies is frictionless, find the acceleration of each body. (*b*) If the surface between the bodies is rough enough so that the 2-kg body does not slide relative to the 4-kg body, find the acceleration of the two bodies. (*c*) Find the resultant force acting on each body in (*b*).

Figure 4-28
Exercise 34.

Problems

1. Two college students on vacation visit a 20,000-tonne cruise ship (1 tonne = 1000 kg). Seeing that the ship is standing next to the pier with its hawsers hanging loose, they decide to see if they can push it 1 m. They push with a combined force of 350 N (80 lb). Find (*a*) the acceleration of the ship, (*b*) the time for it to move 1 m, and (*c*) the final velocity of the ship.

2. A mother holds her 10-kg baby in her lap while riding in a car moving at 11 m/s (about 25 mi/h). The car collides with one in front and stops in 1 m. Assuming the mother has her own seat belt on, with what force must she hold the baby to keep it in her grasp, assuming the baby also moves 1 m during the collision? Give your answer in both newtons and pounds.

3. The acceleration versus spring length observed when a 0.5-kg mass is pulled along a frictionless table by a single spring is

L, cm	4	5	6	7	8	9	10	11	13	14
a, m/s²	0	2.0	3.8	5.6	7.4	9.2	11.2	12.8	14.6	14.6

(*a*) Make a plot of the force exerted by the spring versus the spring length. (*b*) If the spring is extended to 12.5 cm, what force does it exert? (*c*) By how much is the spring extended when the mass is suspended from it at rest near sea level, where $g = 9.81$ N/kg?

4. Your car is stuck in a mudhole. You are alone, but you have a long, strong rope. Having studied physics, you tie the rope tautly to a tree and pull on it sideways, as shown in Figure 4-29. (*a*) Find the force exerted by the rope on the car when the angle θ is 3° and you are pulling with a force of

400 N but the car does not move. (*b*) How strong must the rope be if it takes a force of 600 N at an angle of $\theta = 3°$ to move the car?

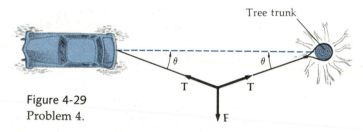

Figure 4-29
Problem 4.

5. A car traveling 90 km/h comes to a sudden stop to avoid an accident. Fortunately, the driver is wearing a seat belt. Using reasonable values for the mass of the driver and the time it takes the car to come to a stop, estimate the force (assumed to be constant) exerted on the driver by the seat belt.

6. Two bodies of mass m_1 and m_2 rest on a horizontal frictionless table, as shown in Figure 4-30. A force **F** is applied as shown. (*a*) If $m_1 = 2$ kg, $m_2 = 4$ kg, and $F = 3$ N, find the acceleration of the bodies and the contact force F_c exerted by one body on the other. (*b*) Find the contact force for general masses of the bodies.

Figure 4-30
Problem 6.

7. A 100-kg box is pulled along a frictionless surface by a force **F** so that its acceleration is 6 m/s² (see Figure 4-31). A 20-kg box slides along the top of the 100-kg box and has an acceleration of 4 m/s². (It thus slides back relative to the 100-kg box.) (*a*) What is the frictional force exerted by the

100-kg box on the 20-kg box? (b) What is the resultant force on the 100-kg box? What is the force **F**? (c) After the 20-kg box falls off of the 100-kg box, what is the acceleration of the 100-kg box?

Figure 4-31
Problem 7.

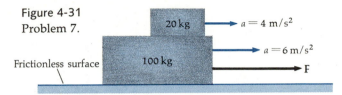

8. A 2-kg body rests on a smooth wedge that has an inclination of 60° (Figure 4-32) and an acceleration a to the right such that the body remains stationary relative to the wedge. (a) Find a. (b) What would happen if the wedge were given a greater acceleration?

Figure 4-32
Problem 8.

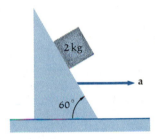

9. Two climbers on an icy (frictionless) slope and tied together on a 30-m rope are in the predicament indicated in Figure 4-33. At time $t = 0$ the speed of each is zero, but the top climber Paul (mass 52 kg) has taken one step too many and his friend Jay (mass 74 kg) has dropped his pick. (a) Find the tension in the rope as Paul falls and his speed just before he hits the ground. (b) If Paul unties his rope after hitting the ground, find Jay's speed as he hits the ground.

Figure 4-33
Problem 9.

10. A simple pendulum can be used as an accelerometer. A small body is suspended from a string attached to a fixed point in the accelerating object, for example, to the ceiling of a passenger car. When there is an acceleration, the body will deflect and the string will make some angle with the vertical. (a) How is the direction in which the suspended body deflects related to the direction of the acceleration? (b) Show

that the acceleration a is related to the angle θ the string makes by $a = g \tan \theta$. (c) Suppose the accelerometer is attached to the ceiling of a car that brakes to rest from 50 km/h in a distance of 60 m. What angle will the accelerometer make? Will the body swing forward or backward?

11. A man stands on a scale in an elevator that has an upward acceleration a. The scale reads 960 N. When the man picks up a 20-kg box, the scale reads 1200 N. Find the mass of the man, his weight, and the acceleration a.

12. Two bodies are connected by a light string, as shown in Figure 4-34. The incline and peg are frictionless. Find the acceleration of the bodies and the tension in the string (a) for $\theta = 30°$ and $m_1 = m_2 = 5$ kg and (b) for general values of θ, m_1, and m_2.

Figure 4-34
Problem 12.

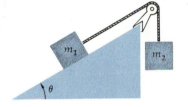

13. A 65-kg girl weighs herself by standing on a scale mounted on a skateboard that rolls down an incline, as shown in Figure 4-35. Assume no friction so that the force exerted by the incline on the skateboard is perpendicular to the incline. What is the reading on the scale if $\theta = 30°$.

Figure 4-35
Problem 13.

14. It occasionally happens that one steps off an unexpected step or curb and lands on a straight leg with the knee joint locked so that one's downward motion is stopped in the distance that the leg and spine can contract—perhaps 2.0 cm. If a step is 15 cm high and a man's mass is 80 kg, find his velocity at the instant of impact and the average acceleration and average force exerted by the ground on him to stop him in a distance of 2.0 cm.

15. A 60-kg house painter stands on a 15-kg aluminum platform. A rope attached to the platform and passing over an overhead pulley allows the painter to raise herself and the platform (see Figure 4-36). (*a*) To get started, she accelerates herself and the platform at a rate of 0.8 m/s². With what force must she pull on the rope? (*b*) After 1 s, she pulls so that she and the platform go up at a constant speed of 1 m/s. What force must she exert on the rope?

Figure 4-37
Problem 16.

Figure 4-36
Problem 15.

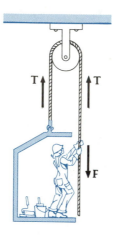

16. The apparatus shown in Figure 4-37, called an *Atwood's machine,* is used to measure the acceleration of gravity *g* by measuring the acceleration of the bodies. Assuming the string to be massless and the peg to be frictionless, show that the magnitude of the acceleration of either body and the tension in the string are

$$a = \frac{m_1 - m_2}{m_1 + m_2} g \quad \text{and} \quad T = \frac{2 m_1 m_2}{m_1 + m_2} g$$

17. A pulley is attached to a beam 12.0 m above the ground. A construction worker of mass 90 kg pulls an empty barrel up on a rope over the pulley and ties the rope at ground level. He climbs up and fills the barrel with spare bricks, giving the loaded barrel a total mass of 180 kg. The worker then climbs down and unties the rope but holds onto it. (*a*) Find his speed as his hard hat hits the beam, assuming that he hangs onto the rope long enough to rise vertically 12.0 m. (*b*) How does his speed when he hits the beam compare with his speed when he hits the ground after he lets go of the rope upon hitting his head?

18. A student has to escape from his girlfriend's dormitory window that is 15.0 m above the ground. He has a heavy rope 20 m long, but it will break when the tension exceeds 360 N and he weighs 600 N. The student will be injured if he hits the ground with speed greater than 10 m/s. (*a*) Show that he cannot safely slide down the rope. (*b*) Find a strategy using the rope that will permit the student to reach the ground safely.

CHAPTER FIVE

Applications of Newton's Laws

The origin of science is in the desire to know causes; and the origin of all false science and impostures is in the desire to accept false causes rather than none; or, which is the same thing, in the unwillingness to acknowledge our ignorance.

WILLIAM HAZLITT

Now that you know what Newton's laws of motion have to say, you are ready for the systematic expansion of these ideas to explain innumerable properties of the world you live in. One thinks first of the engineering applications of scientific principles, but Newton's laws have more personal applications; indeed, understanding them can be a matter of life or death. Why, for instance, should you not hold a baby in your lap while riding in a car? What happens if the car comes to a sudden stop? By the end of this chapter, you should be able to back up your intuitive answers to these questions with calculations of the forces involved.

You will also find that you better understand the way things work and why things are built the way they are. You will be able to find, for example, the answers to such questions as: Are there really such things as "*g*-forces"? How can a space craft stop or turn with nothing to push against, whereas a car on ice is almost helpless? Why can you go upside down on a roller coaster? Are you in increasing danger of having a ladder slip out from under you as you go higher on it? Is there any limit to the speed at which humans can travel on the surface of the earth? Though not stated explicitly, the answers to all these questions are in this chapter.

In Chapter 4, we discussed Newton's laws of motion and their application to the motion of a particle under the influence of constant forces. We now will consider some more general applications. The methods of problem solving discussed in Section 4-6 have wide applicability to many problems in mechanics and should be reread and learned well. In particular, you should develop the habit of approaching a problem by first drawing a picture and then indicating the important forces acting on each object in a free-body diagram.

5-1   Friction

If there is a large box resting on the floor and you push on it with a small horizontal force, it is possible that the box won't move at all. The reason is that the floor is exerting a horizontal force, called the force of *static friction* f_s, that balances the force **F** you are exerting (Figure 5-1). This frictional force is due to the bonding between the molecules of the box and the floor at those places where the surfaces are in very close contact. It is somewhat like a support force in that it can vary from zero to some maximum force $f_{s,\text{max}}$, depending on how hard you push. If you push hard enough, the box will slide across the floor. When the box is sliding, molecular bonds are continually being made and broken and small pieces of the surfaces are being broken off. The result is a force of *sliding friction* or *kinetic friction* f_k that opposes the motion. To keep the box sliding with constant velocity, you must exert a force equal and opposite to this force of kinetic friction.

We shall consider static friction first. We might expect the maximum force of static friction to be proportional to the area of contact between the two surfaces. However, it has been shown experimentally that, to a good approximation, this force is independent of the area and is simply proportional to the normal force exerted by one surface on the other. Figure 5-2 shows that the actual microscopic area of contact where the molecules of the box and the floor can bond together is just a small fraction of the total macroscopic area. The following is a possible model of static friction that is consistent with our intuition and with empirical findings. The maximum force of friction *is* proportional to the microscopic area of contact, as expected, but the microscopic area is proportional to the total macroscopic area A and to the normal force per unit area F_n/A exerted between the surfaces. The product of A and F_n/A is thus independent of the total macroscopic area A.

Consider, for example, a 1-kg box with a side area of 60 cm² and an end area of 20 cm². When the box is on its side on the floor, a small fraction of the total 60 cm² is actually in microscopic contact. When the box is placed on end, the fraction of the total area actually in microscopic contact is increased by a factor of three because the force per unit area is three times as great. However, since the area of the end is one-third that of the side, the actual microscopic area of contact is unchanged. The maximum force of static friction $f_{s,\text{max}}$ is thus proportional to the normal force between the surfaces

$$f_{s,\text{max}} = \mu_s F_n \qquad\qquad 5\text{-}1$$

where μ_s is called the *coefficient of static friction*. It depends on the nature of the surfaces of the box and floor. If we exert a smaller horizontal force on the box, the frictional force will just balance the horizontal force. In general we can write

$$f_s \le \mu_s F_n \qquad\qquad 5\text{-}2$$

Kinetic friction, like static friction, is a complicated phenomenon that even today is not completely understood. The *coefficient of kinetic friction μ_k* is defined as the ratio of the magnitude of the frictional force f_k to the magnitude of the normal force F_n. Thus,

$$f_k = \mu_k F_n \qquad\qquad 5\text{-}3$$

Figure 5-1
When you try to push a large box across the floor, friction opposes the motion. The floor exerts a force of static friction f_s that balances the applied force **F** unless the applied force is greater than the maximum possible static frictional force.

Static friction

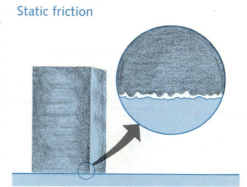

Figure 5-2
The microscopic area of contact between the box and the floor is only a small fraction of the macroscopic area of the box. This fraction is proportional to the normal force exerted between the surfaces.

Kinetic friction

Cross-country skiers use a special ski wax to maximize static friction (so that they can slide up an incline) and to minimize kinetic friction.

Experimentally, it has been found that

1. μ_k is less than μ_s.

2. μ_k depends on the relative speed of the surfaces, but for speeds in the range from about 1 cm/s to several metres per second, μ_k is approximately constant.

3. μ_k (like μ_s) depends on the nature of the surfaces but is independent of the macroscopic area of contact.

We shall neglect any variation in μ_k with speed and assume that it is a constant that depends only on the nature of the surfaces.

We can measure μ_s for two surfaces simply by placing a block on a plane surface and inclining the plane until the block begins to slide. Let θ_c be the critical angle at which the sliding starts. For angles of inclination less than this, the block is in static equilibrium under the influence of its weight $m\mathbf{g}$, the normal force $\mathbf{F}_n$, and the force of static friction $\mathbf{f}_s$ (see Figure 5-3). Choosing the x axis parallel to the plane and the y axis perpendicular to the plane, we have

$$\Sigma F_y = F_n - mg \cos \theta = 0$$

and

$$\Sigma F_x = mg \sin \theta - f_s = 0$$

Eliminating the weight mg from these two equations gives

$$f_s = mg \sin \theta = \frac{F_n}{\cos \theta} \sin \theta = F_n \tan \theta$$

At the critical angle θ_c, the force of static friction equals its maximum value, and we can replace f_s by $\mu_s F_n$. Then

$$\mu_s = \tan \theta_c \qquad\qquad 5\text{-}4$$

The coefficient of static friction equals the tangent of the angle of inclination at which the block just begins to slide.

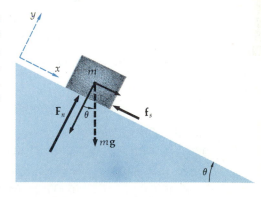

Figure 5-3
Forces on a block on a rough inclined plane. At angles less than the critical angle θ_c, the frictional force balances the component $mg \sin \theta$ down the incline. At greater angles, the block slides down the incline. The critical angle is related to the coefficient of friction by $\tan \theta = \mu_s$.

Example 5-1   A man pulls his children on a sled. The total mass of the sled plus children is 60 kg. The coefficients of static and sliding friction are $\mu_s = 0.2$ and $\mu_k = 0.15$. The sled is pulled by a rope that makes an angle of 40° with the horizontal, as shown in Figure 5-4. Find the frictional force and the acceleration of the children and sled if the tension in the rope is (*a*) 100 N and (*b*) 160 N. (Take $g = 10$ N/kg.)

Figure 5-4
Man pulling children on sled for Example 5-1.

(*a*) The vertical and horizontal components of the tension are

$$T_y = T \sin 40° = (100 \text{ N})(0.64) = 64 \text{ N}$$

and

$$T_x = T \cos 40° = (100 \text{ N})(0.77) = 77 \text{ N}$$

Since there is no vertical acceleration, the net vertical force must be zero. Then $\Sigma F_y = F_n + T_y - mg = 0$. The normal force is therefore

$$F_n = mg - T_y = (60 \text{ kg})(10 \text{ N/kg}) - 64 \text{ N}$$
$$= 600 \text{ N} - 64 \text{ N} = 536 \text{ N}$$

The maximum possible static frictional force is

$$f_{s,\text{max}} = \mu_s F_n = (0.2)(536 \text{ N}) = 107 \text{ N}$$

Since the applied horizontal force T_x is 77 N, which is less than the maximum force of static friction, the sled remains at rest. The frictional force is therefore 77 N to the left to balance the applied horizontal force to the right.

There are two important points to note about this example: (1) the normal force is *not* equal to the weight of the children and sled because the vertical component of the tension helps lift the sled off the ground, and (2) the force of static friction is *not* equal to $\mu_s F_n$. It is less than this maximum possible limiting value.

(b) When the tension is increased to 160 N, its vertical and horizontal components are

$$T_y = (160 \text{ N})(\sin 40°) = 102 \text{ N}$$

and

$$T_x = (160 \text{ N})(\cos 40°) = 123 \text{ N}$$

The normal force is then

$$F_n = mg - T_y = 600 \text{ N} - 102 \text{ N} = 498 \text{ N}$$

and the maximum possible force of static friction is

$$f_{s,max} = \mu_s F_n = (0.2)(498 \text{ N}) = 100 \text{ N}$$

Since this maximum static frictional force is less than the applied horizontal force, the sled will slide. The frictional force on the sled will thus be due to kinetic friction and will have the value

$$f_k = \mu_k F_n = (0.15)(498 \text{ N}) = 75 \text{ N}$$

The resultant force in the horizontal direction is thus

$$\Sigma F_x = T_x - f_k = 123 \text{ N} - 75 \text{ N} = 48 \text{ N}$$

and the acceleration is

$$a_x = \Sigma F_x/m = (48 \text{ N})/(60 \text{ kg}) = 0.80 \text{ m/s}^2$$

Figure 5-5 shows the forces on a car that is just starting up from rest. The weight of the car is balanced by the normal force exerted on the tires. (In general these normal forces will not be equal.) To start the car moving, the engine delivers a torque to the drive shaft that makes the front wheels rotate. (We shall discuss torque in the next section. For a rear-wheel-drive car, the torque is delivered to the rear wheels.) If the road were perfectly frictionless, the wheels would merely spin, with the surface of the tire in contact with the road moving backward. If there is friction and the torque delivered by the engine is not too great, the tire will not slip on the road because of static friction. This force is in the forward direction and provides the acceleration needed for the car to start moving forward. If the wheel rolls without slipping, the tire surface in contact with the road is at rest relative to the road and the friction is static friction. (The tire surface in contact with the road is moving backward relative to the axle, but the axle is moving forward as the car moves forward.) If the engine torque is too great, the wheels spin; that is, the tire surface slips backward relative to the road. Then the force that accelerates the car is kinetic friction, which is less than static friction. We can see from this that if we are stuck on ice or snow, our chances of getting

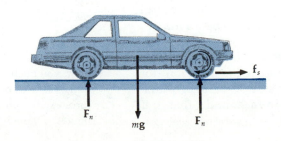

Figure 5-5
Forces on a car. The normal forces F_n are not generally equal on the front and rear tires. As the engine turns the wheel, the force of static friction prevents the tire from slipping on the road, causing the car to roll forward. If the wheel spins, the smaller force of kinetic friction pushes the car forward.

free are better if we use a light touch on the accelerator pedal of the car so that the wheels do not slip.

Questions

1. The friction between two surfaces can be reduced by polishing both, but if the surfaces are polished until they are extremely smooth and flat, friction increases again. Explain.

2. Various objects lie on the floor of a truck. If the truck accelerates, what force acts on the objects to cause them to accelerate?

3. Any object resting on the floor of a truck will slip if the acceleration of the truck is too great. How does the critical acceleration at which a small box will slip compare with that at which a much heavier object will slip?

5-2   Static Equilibrium of an Extended Body

If a particle is in static equilibrium—that is, if the particle is at rest and remains at rest—the resultant force acting on it must be zero. We used this condition in the previous chapter in solving static equilibrium problems such as those that dealt with hanging pictures. When we consider the equilibrium of an extended body, such as a stick, the condition that the resultant force be zero is still necessary, but it is not sufficient because the body may rotate even with no unbalanced force acting on it. For example, consider a uniform stick resting on a horizontal table (assumed to be frictionless), as shown in Figure 5-6a. The stick is subjected to two forces of equal magnitude and opposite direction, but the forces are not exerted along the same line. Although the center of the stick will remain at rest, the forces F_1 and F_2 will cause the stick to rotate in the clockwise direction (as viewed from above the table). The stick is clearly not in static equilibrium. On the other hand, if the forces are applied along the same line, as shown in Figure 5-6b, they will not cause rotation and the stick will be in equilibrium. For extended bodies, the point of application of a force, as well as the magnitude and direction of the force, is important.

Suppose you wish to push open a heavy swinging door that is hinged along the right side (see Figure 5-7). No amount of force applied at the hinged side will cause the door to swing open. It is intuitively clear that a given force will have the most effect if it is applied at the left side of the door as far from the hinges as possible. The quantity that measures the effectiveness of a force for producing rotation is called the *torque* τ. A torque may be either clockwise or counterclockwise, depending on the sense of the rotation it tends to produce. For the case of a force **F** applied perpendicularly to the hinged door, the torque is the product of the force and the distance L from the point of application to the hinged side of the door. If you apply a force F_1 at a distance L_1 from the hinged side to try to swing the door open, the torque you apply is F_1L_1 in the clockwise direction as viewed from above. This torque could be balanced by one applied in the opposite direction, for example, by somebody else pushing on the other side of the door with a force F_2 at a distance L_2. If the torques are equal in magnitude, that is, if $F_1L_1 = F_2L_2$, the door will not rotate.

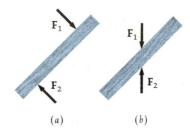

Figure 5-6
(*a*) The two forces F_1 and F_2 are equal and opposite, but the stick is not in static equilibrium because these forces tend to cause it to rotate clockwise. (*b*) Here the two forces have the same line of action, so they have no tendency to cause the stick to rotate.

Figure 5-7
To push open a heavy swinging door, you push as far from the hinges as possible. No amount of force exerted on the door at the hinges will push it open.

Figure 5-8 shows a stick pivoted at a point O with a force $\mathbf{F}$ applied at a distance r from the pivot and at an angle θ to the radius vector $\mathbf{r}$. The force $\mathbf{F}$ can be resolved into components $F_\perp$ perpendicular to $\mathbf{r}$ and $F_\parallel$ parallel to $\mathbf{r}$. Clearly the parallel component does not tend to produce any rotation of the stick about the pivot. For this general case, we define the torque about the pivot O to be the product of $F_\perp$ and r.

$$\tau = F_\perp r = (F \sin \theta) r \qquad\qquad 5\text{-}5 \qquad \text{Torque defined}$$

In the figure, we have indicated the line of action of the force with a dashed line. The perpendicular distance from the point O to the line of action of the force is $L = r \sin \theta$. This distance L is called the *lever arm* of the force. From Equation 5-5 we see that the torque about O can also be written

$$\tau = F(r \sin \theta) = FL \qquad\qquad 5\text{-}6$$

Figure 5-8
Force $\mathbf{F}$ exerted on a stick pivoted at point O. The rotational effect of this force is measured by the torque τ defined as $\tau = F_\perp r = (F \sin \theta) r = FL$.

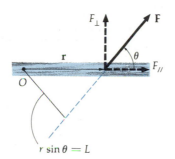

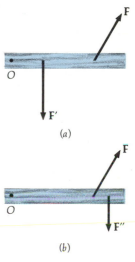

If $\mathbf{F}$ is the only force that produces a torque about O, the stick will rotate about point O. Point O remains fixed because of the force exerted by the pivot. We can prevent the rotation of the stick by applying a second force $\mathbf{F}'$ such that it produces a torque of equal magnitude but opposite sense, as indicated in Figure 5-9a. Note that $\mathbf{F}'$ is not necessarily equal to $\mathbf{F}$. In Figure 5-9a, $\mathbf{F}'$ is considerably greater in magnitude, but its lever arm is smaller so that $F'L' = FL$. Rotation could also be prevented by applying a smaller force $\mathbf{F}''$ with a greater lever arm, as in Figure 5-9b.

Figure 5-9
The torque produced about O by force $\mathbf{F}$ can be balanced by an equal torque produced by a larger force $\mathbf{F}'$ at a smaller distance from O, as in (a), or by a smaller force $\mathbf{F}''$ at a larger distance from O, as in (b).

The torque exerted by a wrench on a nut is proportional to the force and the lever arm. Charlie could exert a greater torque with the same force if he held the wrenches nearer their ends.

We can summarize the preceding discussion by stating the second condition for static equilibrium:

For a body to be in static equilibrium, the net clockwise torque about any point must equal the net counterclockwise torque about that point.

If we call counterclockwise torques positive and clockwise torques negative, this condition is equivalent to saying that the algebraic sum of the torques about any point must be zero.

We now have two conditions for the static equilibrium of an extended body:

1. The resultant external force must be zero:

$$\Sigma \mathbf{F} = 0 \qquad\qquad 5\text{-}7a$$

2. The resultant external torque about any point must be zero:

$$\Sigma \tau = 0 \qquad\qquad 5\text{-}7b$$

Conditions for equilibrium

Note that we have stated our second condition for equilibrium in terms of torques about *any point.* If a body is not rotating about one particular point, it is not rotating about *any* point. This fact is often useful in solving problems because we can choose the point for computing torques at our convenience.

Example 5-2   A 3-m board of negligible weight rests with its ends on scales, as shown in Figure 5-10. A small, 60-N weight rests on the board 2.5 m from the left end and 0.5 m from the right end, as also shown. Find the readings on the scales.

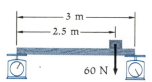

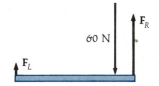

Figure 5-10
Board of negligible weight on scales for Example 5-2.

Figure 5-11
Free-body diagram for the board in Example 5-2.

An extended body in equilibrium. What keeps Buster Keaton from rotating?

Figure 5-11 shows the free-body diagram for the board. The force $\mathbf{F}_L$ is that exerted by the scale at the left end of the board. Since the board exerts an equal but opposite force on the scale, the magnitude of $\mathbf{F}_L$ is the reading on the left scale. Similarly, the magnitude of $\mathbf{F}_R$ is the reading on the right scale. From our first condition of equilibrium (the resultant force must be zero), we know that

$$F_L + F_R = 60 \text{ N}$$

We get a second relation between F_L and F_R by considering the torques. If we consider the point at the weight to be the "pivot" point, we have two torques, a clockwise torque of magnitude $F_L(2.5 \text{ m})$ and a counterclockwise torque of magnitude $F_R(0.5\ \text{m})$. Equating these torques gives

$$0.5F_R = 2.5F_L$$

or

$$F_R = 5F_L$$

Substituting this in the previous equation, we obtain

$$F_L + 5F_L = 60\ \text{N}$$

or

$$F_L = 10\ \text{N}$$

and

$$F_R = 60\ \text{N} - F_L = 50\ \text{N}$$

The scale readings are thus 10 N for the left scale and 50 N for the right scale. The scale on the right supports the greater weight, as expected.

Although there is nothing incorrect with the above analysis, there is an easier way to solve the problem without having to solve two equations for two unknowns. If we compute the torques about a point on the line of action of one of the unknown forces, that force will not enter into the equation because its lever arm will be zero. We first consider the torques about the left scale. The weight produces a clockwise torque of magnitude $(60\ \text{N})(2.5\ \text{m}) = 150\ \text{N·m}$, and F_R produces a counterclockwise torque $F_R(3\ \text{m})$. Setting their magnitudes equal, we get

$$F_R(3\ \text{m}) = 150\ \text{N·m}$$

or

$$F_R = 50\ \text{N}$$

We can then find F_L immediately from $F_L = 60\ \text{N} - 50\ \text{N} = 10\ \text{N}$ or by considering the torques about the right scale. We then have $F_L(3\ \text{m}) = (60\ \text{N})(0.5\ \text{m}) = 30\ \text{N·m}$ or $F_L = 10\ \text{N}$. Whenever there are two ways of solving a problem, it is a good idea to use one method to check the results of the other.

In general, if we have several unknown forces, we can reduce the work involved in solving the problem by computing the torques about a point on the line of action of one of the unknown forces so that that force does not enter into the equation.

We neglected the weight of the board in Example 5-2 to simplify the problem. In general, of course, extended bodies do have weight. The force of gravity acts on each part of an extended body, as is indicated in Figure 5-12 where we have divided up a body into many smaller bodies. If we make the divisions small enough, we can consider the smaller bodies to be particles. The force of attraction of the earth on each of these particles is indicated by the vectors $\mathbf{w}_i$. There is one point about which these parallel weight vectors produce zero torque. This point is called the *center of gravity*. We can represent the force of attraction of the earth for any body as a single vector $\mathbf{W}$, the weight of the body acting at the center of gravity. For a uniform body, such as a uniform sphere or stick, the center of gravity is at the body's geometric center. We can often locate the center of gravity of an object such as a stick by balancing it on a pivot. The pivot point at which the stick will balance is the center of gravity of the stick.

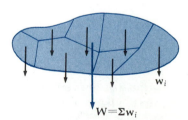

Figure 5-12
The weights of all the particles of a body can be replaced by the total weight $\mathbf{W}$ of the body acting at the center of gravity, which is the point about which the resultant torque due to the forces $\mathbf{w}_i$ is zero.

If we suspend a body from a pivot (not at the center of gravity) so that the body is free to rotate about the pivot under the action of gravity, the body will hang in equilibrium with the center of gravity directly below the pivot. (At any other position the attraction of the earth will produce a torque about the pivot.) We can use this property to find the center of gravity of a plane figure, as illustrated in Figure 5-13. Here we first suspend the body from point A and draw a vertical line on the body from the pivot point when the body is in equilibrium. Since the center of gravity lies directly below the pivot point, it must be somewhere on this line. We then suspend the body from another point, point B, and again draw a vertical line on the body. The intersection of the two lines is the center of gravity. We can check our measurement by suspending the body from a third point, point C, as shown in the figure.

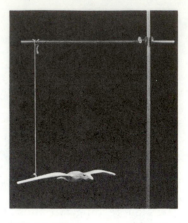

This bird looks symmetrical, but evidently its mass distribution is not. Where is the center of gravity?

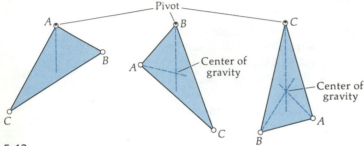

Figure 5-13
The center of gravity of an irregularly shaped object lies directly below any point about which the object is pivoted. This property can be used to locate the center of gravity.

Example 5-3   A 60-N weight is held in the hand with the forearm at 90° to the upper arm, as shown in Figure 5-14. The biceps muscle exerts a force $\mathbf{F}_m$ that is 3.4 cm from the pivot point O at the elbow joint. Neglecting the weight of the arm and hand, what is the magnitude of $\mathbf{F}_m$ if the distance from the weight to the pivot point is 30 cm.

The counterclockwise torque about O exerted by the muscle must balance the clockwise torque exerted by the weight. Setting these torques equal, we have

$$F_m(3.4 \text{ cm}) = (60 \text{ N})(30 \text{ cm}) \quad \text{or} \quad F_m = \frac{(60 \text{ N})(30 \text{ cm})}{3.4 \text{ cm}} = 529 \text{ N}$$

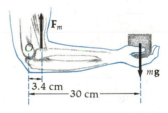

Figure 5-14
Hand holding weight for Example 5-3.

The force exerted by the muscle is much greater than the 60-N weight because of the very short lever arm of the muscle about the pivot point.

In the figure we have isolated the forearm, hand, and weight. Since this system is not translating (as well as not rotating), the net vertical *force* on the system must be zero. We have calculated the upward force of the muscle to be 529 N to ensure rotational equilibrium about point O. Since the downward force of the weight is only 60 N, there must be an additional downward force of $529 - 60 \text{ N} = 469 \text{ N}$. The line of action of this force must pass through O; otherwise there would be an additional torque about O. This force is exerted by the upper arm at the elbow joint.

Example 5-4 A uniform 5-m ladder weighs 12 N and leans against a friction-less vertical wall (see Figure 5-15). The foot of the ladder is 3 m from the wall. What is the minimum coefficient of friction necessary between the ladder and the floor if the ladder is not to slip?

The forces acting on the ladder are the force due to gravity **w** acting downward at the center of gravity, the force $\mathbf{F}_1$ exerted horizontally by the wall (since the wall is frictionless, it exerts only a normal force), and the force exerted by the floor, which consists of a normal force $\mathbf{F}_n$ and a horizontal force of friction **f**. From the first condition of equilibrium we have

$$F_n = w = 12 \text{ N}$$

and

$$F_1 = f$$

Since we know neither f nor F_1, we must use the second condition of equilibrium and compute the torques about some convenient point. We choose the point of contact between the ladder and the floor because both $\mathbf{F}_n$ and **f** act at this point and will therefore not appear in our torque equation. The torque exerted by the force of gravity about this point is clockwise with a magnitude 12 N times the lever arm 1.5 m. The torque exerted by $\mathbf{F}_1$ about the point of contact of the ladder and the floor is counterclockwise with a magnitude F_1 times the lever arm 4 m. The second condition of equilibrium thus gives

$$F_1(4 \text{ m}) - (12 \text{ N})(1.5 \text{ m}) = 0$$

$$F_1 = 4.5 \text{ N}$$

This equals the magnitude of the frictional force. Since the frictional force **f** is related to the normal force $\mathbf{F}_n$ by

$$f \leq \mu_s F_n$$

we have

$$\mu_s \geq \frac{f}{F_n} = \frac{4.5 \text{ N}}{12 \text{ N}} = 0.375$$

where μ_s is the coefficient of static friction.

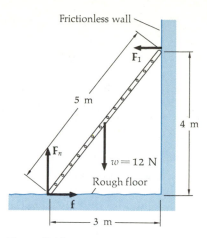

Figure 5-15
Ladder on rough floor leaning against a frictionless wall for Example 5-4.

Example 5-5 A sign of mass 20 kg hangs from the end of a rod of length 2 m and mass 4 kg. A wire is attached to the end of the rod and to a point 1 m above point O (see Figure 5-16). Find the tension in the wire and the force exerted by the wall on the rod at point O.

Figure 5-17 shows the forces acting on the rod. Using $g = 9.8$ N/kg, the weight of the sign is 196 N and that of the rod is 39.2 N. The force exerted by the wall has components F_x and F_y. The tension **T** has been resolved into its x and y components. Since we do not know the force exerted by the wall, we choose point O for computing torques. The weight of the sign and that of the rod produce clockwise torques about O, and tension in the wire produces a counterclockwise torque about O. Setting these torques equal, we have

$$T_y(2 \text{ m}) = (196 \text{ N})(2 \text{ m}) + (39.2 \text{ N})(1 \text{ m}) = 432 \text{ N} \cdot \text{m}$$

$$T_y = 216 \text{ N}$$

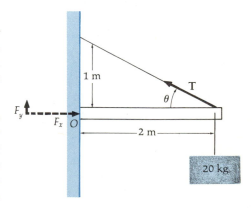

Figure 5-16
Sign hanging from rod for Example 5-5.

The x component of the tension can be found from

$$\frac{T_y}{-T_x} = \tan \theta = \frac{1}{2}$$

$$T_x = -2T_y = -432 \text{ N}$$

(T_x is negative because it points to the left.) The magnitude of the tension is

$$T = \sqrt{T_x^2 + T_y^2} = \sqrt{(216 \text{ N})^2 + (-432 \text{ N})^2} = 490 \text{ N}$$

The force exerted by the wall on the rod at O is found from the first condition of equilibrium. The horizontal component F_x must equal 432 N to balance the horizontal component of the tension. Setting the upward forces equal to the downward forces, we have

$$F_y + T_y = 196 \text{ N} + 39.2 \text{ N} = 235 \text{ N}$$

$$F_y = 235 \text{ N} - T_y = 235 \text{ N} - 216 \text{ N} = 19 \text{ N}$$

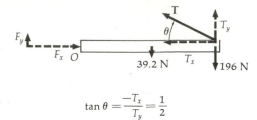

$$\tan \theta = \frac{-T_x}{T_y} = \frac{1}{2}$$

Figure 5-17
Free-body diagram for the rod in Figure 5-16. The tension in the wire is resolved into its x and y components. The unknown force exerted by the wall on the rod has the components F_x and F_y.

Questions

4. Must there be any matter at the location of the center of gravity of a body?

5. Can a ladder in equilibrium stand on a frictionless horizontal floor leaning against a rough vertical wall? Why or why not?

5-3   Stability of Equilibrium and Balance

The equilibrium of a body can be classified into three categories: stable, unstable, or neutral. Stable equilibrium occurs when the resulting torques or forces that arise from a small displacement of the body urge the body back toward its equilibrium position. Stable equilibrium is illustrated in Figures 5-18 and 5-19. When the box in Figure 5-18 is rotated slightly about one end, the resulting torque about the pivot point restores the box to its original position. Similarly, if the marble resting at the bottom of the hemispherical bowl in Figure 5-19 is displaced slightly, the resultant force urges it back toward its initial position.

Figure 5-18
An example of stable equilibrium. When the box is rotated slightly, the torque exerted by the weight about the pivot point tends to restore the box to its original position.

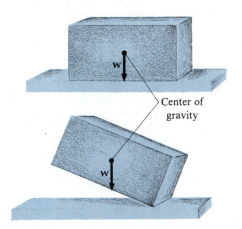

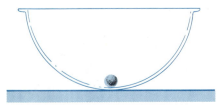
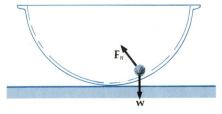

Figure 5-19
Another example of stable equilibrium. If the marble is displaced slightly in any direction, there is an unbalanced force that pushes the marble back toward its original position.

Unstable equilibrium is illustrated in Figures 5-20 and 5-21. A slight rotation of the narrow stick in Figure 5-20 causes it to fall over because the resulting torque due to its weight urges it away from its original position. Similarly, if the marble on top of the overturned bowl in Figure 5-21 is displaced slightly, the resultant force urges the marble away from its original position.

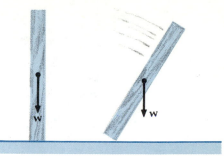

Figure 5-20
An example of unstable equilibrium. If the stick is rotated slightly, the torque exerted by the weight about the pivot point tends to rotate the stick away from its original position.

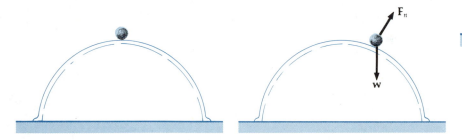

Figure 5-21
Another example of unstable equilibrium. If the marble is displaced slightly, there is an unbalanced force that pushes the marble farther from its original position.

The cylinder resting on a horizontal surface in Figure 5-22 illustrates neutral equilibrium. If the cylinder is rotated slightly, there is no torque or force urging it either back toward its original position or away from it. In summary, when a system is disturbed slightly from its equilibrium position, the equilibrium is stable if the system returns to its original position, unstable if it moves farther away, and neutral if there are no torques or forces urging it in either direction.

Since a "slight disturbance" is a relative term, stability is also relative. One example of equilibrium may be more or less stable than another. Figure 5-23 shows a stick balanced on end that is not as narrow as the one in Figure 5-20. Here, if the disturbance is very small (Figure 5-23b), the stick will move back toward its original position, but if the disturbance is great enough so that the center of gravity no longer lies over the base of support (Figure 5-23c), the stick will fall. We can improve the stability of a system by either lowering the

Figure 5-22
An example of neutral equilibrium. If the cylinder is rotated slightly, it is again in equilibrium. There are no torques that tend to rotate the cylinder either toward or away from its original position.

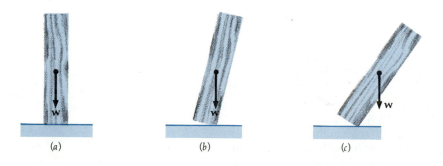

(a)   (b)   (c)

Figure 5-23
Stability of equilibrium is relative. If the stick in (a) is rotated slightly, as in (b), it returns to its original equilibrium position as long as the center of gravity lies over the base of support. If the rotation is too great, as in (c), the center of gravity is no longer over the base of support and the stick falls.

center of gravity or widening the base of support. Figure 5-24 shows a nonuniform stick that is loaded so that its center of gravity is near one end. If the stick is stood on its heavy end so that the center of gravity is low (Figure 5-24a), it is much more stable than if it is stood on the other end so that the

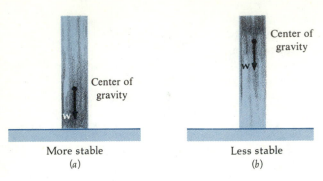

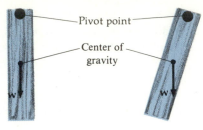

Figure 5-24
A nonuniform stick. When the stick rests on its heavy end with its center of gravity low, as in (a), the equilibrium is more stable than when the center of gravity is high, as in (b).

Figure 5-25
When a stick is pivoted so that its center of gravity is below the pivot point, the equilibrium is stable no matter how far the stick is displaced from its equilibrium position.

center of gravity is high (Figure 5-24b). In Figure 5-25, the center of gravity lies below the point of support of the system. This system is stable for any displacement because the resulting torque always rotates the system back toward its equilibrium position.

A human standing or walking upright has difficulty maintaining balance because the center of gravity must be kept over the base of support, the feet, which is relatively small. An animal walking on four legs has a much easier time because the base of support is larger and the center of gravity is lower.

Questions

6. How can a hiker with a heavy backpack stand so as to maintain balance?

7. How does a lineman in football improve his balance?

8. Why is it so much more difficult to stand on one foot than on two?

Is this equilibrium stable?

5-4   Circular Motion

The application of Newton's laws to circular motion is straightforward, but it seems to cause much confusion for students. When a particle moves with speed v in a circle of radius r, it has an acceleration of magnitude v^2/r directed toward the center of the circle whether or not the speed is changing. If the speed is changing, there is also a component of acceleration tangential to the circle and equal to the rate of change of the speed.

The centripetal acceleration is related to the change in the direction of the velocity of the particle, as discussed in Chapter 3. As with any acceleration, there must be a resultant force in the direction of the acceleration to produce it. This resultant force is called the *centripetal force*. It is important to understand that the centripetal force is not a new kind of force that we have not yet studied. It is merely a name for the resultant inward force that must be present to provide the centripetal acceleration needed for circular motion.

Centripetal force

The centripetal force may be due to a string, a spring, or a contact force such as a normal force or friction; it may be an action-at-a-distance type of force such as a gravitational force; or it may be a combination of any of these familiar forces. Whenever an object moves in a circle, there must be an unbalanced force to provide its centripetal acceleration just as there must be an unbalanced force to provide linear acceleration. We shall illustrate with some examples.

It is often convenient to describe the motion of a particle moving in a circle with constant speed in terms of the time required for one complete revolution T, called the *period*. If the radius of the circle is r, the particle travels a distance $2\pi r$ during one period, so its speed is related to the radius and the period by

Period of circular motion

$$v = \frac{2\pi r}{T} \qquad\qquad 5\text{-}8$$

Example 5-6   A tetherball of mass m is suspended from a rope of length L and travels at constant speed v in a horizontal circle of radius r. The rope makes an angle θ given by $\sin \theta = r/L$, as shown in Figure 5-26. Find the tension in the rope and the speed of the ball.

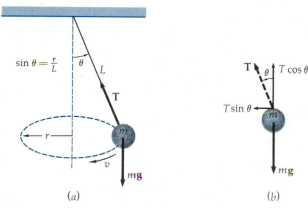

(a)   (b)

Figure 5-26
(a) Tetherball for Example 5-6.
(b) Free-body diagram for the tetherball. (See also Exercise 16 and Problem 9.)

The two forces acting on the ball are therefore due to the earth's attraction $m\mathbf{g}$, which acts vertically downward, and the tension $\mathbf{T}$, which acts along the rope. In this problem, we know that the acceleration is horizontal, toward the center of the circle, and of magnitude v^2/r. Thus, the vertical component of the tension must balance the weight $m\mathbf{g}$. The horizontal component of the tension is the resultant centripetal force. The vertical and horizontal components of $\Sigma\mathbf{F} = m\mathbf{a}$ therefore give

$$T \cos \theta - mg = 0$$

$$T \sin \theta = ma = \frac{mv^2}{r}$$

The tension is found directly from the first equation since θ is given. We can find the speed v in terms of the known quantities r and θ by dividing one equation by the other to eliminate T. We thereby obtain

$$\tan \theta = \frac{v^2}{rg} \qquad \text{or} \qquad v = \sqrt{rg \tan \theta}$$

The forces acting on this child who is moving in a circle are similar to those acting on the tetherball of Example 5-6.

Example 5-7   A pail of water is whirled in a vertical circle of radius r. If its speed is v_t at the top of the circle, find the force exerted on the water by the pail at the top of the circle. Find also the minimum value of v_t for the water to remain in the pail.

The forces on the water at the top of the circle are shown in Figure 5-27. They are the force of gravity $m\mathbf{g}$ and the force $\mathbf{F}_p$ exerted by the pail. Both of these forces act downward. The acceleration, which is toward the center of the circle, is also downward at this point. Newton's second law gives

$$F_p + mg = m\frac{v_t^2}{r}$$

The force exerted by the pail is therefore $F_p = m(v_t^2/r) - mg$. Note that both the force of gravity and the contact normal force exerted by the pail contribute to the necessary centripetal force.

If we increase the speed of the pail, the bottom of the pail will exert a larger force on the water to keep it moving in a circle. If we decrease the speed, F_p will decrease. Since the pail cannot exert an upward force on the water, the minimum speed the water can have at the top of the circle occurs when $F_p = 0$. Then,

$$mg = m\frac{v_{t,min}^2}{r}$$

or

$$v_{t,min} = \sqrt{rg} \qquad\qquad 5\text{-}9$$

When the water is moving at this minimum speed, its acceleration at the top of the path is g, the free-fall acceleration of gravity, and the only force acting on the water is the gravitational attraction of the earth, the weight mg.

If you wish to try this, you might want to know how fast you need to move the pail so you won't get wet. If your arm is 70 cm long and there is another 30 cm from your hand to the water, 1 m is a reasonable value for r. The minimum speed at the top is then

$$v = \sqrt{rg} = \sqrt{(1\text{ m})(9.8\text{ m/s}^2)} = 3.13\text{ m/s}$$

If you rotate the pail at constant speed, the maximum period of revolution is found from Equations 5-8 and 5-9:

$$T = \frac{2\pi r}{v} = \frac{(2)(3.14)(1\text{ m})}{3.13\text{ m/s}} \approx 2\text{ s}$$

Example 5-8   The largest blades on a modern power-plant turbine are about 1 m long and rotate at about 60 revolutions per second. Calculate the force on a piece of the blade of mass 100 g at the end of the blade.

Since the blade makes 60 rev/s, the period for 1 rev is $1/60 = 0.0167$ s. We find the speed of the tip of the blade from Equation 5-8:

$$v = \frac{2\pi r}{T} = \frac{(2)(3.14)(1\text{ m})}{(0.0167\text{ s})} = 377\text{ m/s}$$

The force needed to keep a piece of mass 100 g $= 0.1$ kg moving in a circle of radius 1 m at this speed is then

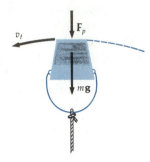

Figure 5-27
Pail of water whirled in a vertical circle. At the top, the forces on the water are its weight $m\mathbf{g}$ and the normal force exerted by the pail bottom $\mathbf{F}_p$. Both of these forces act downward at this point toward the center of the circle.

Motorcycle moving in a vertical circle. This is not a good time to have engine failure.

$$F = \frac{mv^2}{r} = \frac{(0.1 \text{ kg})(377 \text{ m/s})^2}{(1 \text{ m})} = 14,200 \text{ N}$$

This large force (about 3200 pounds) on the tip of the blade is provided by the piece of blade next to it, which is in turn rotating and requires a large force to keep it moving in a circle. The total force needed to hold the blade to the shaft is over 100,000 N. Rotating machinery involves enormous forces that place great stress on the materials used for the blades and on the engineers who design the machinery.

Example 5-9   A car travels on a horizontal road in a circle of radius 30 m. If the coefficient of static friction is $\mu_s = 0.6$, how fast can the car travel without slipping?

Figure 5-28 shows the free-body diagram for the car. The normal force $\mathbf{F}_n$ balances the downward force due to gravity $m\mathbf{g}$. The only horizontal force is due to friction. Its maximum value is $f_{s,max} = \mu_s F_n = \mu_s mg$. In this case, the frictional force is the centripetal force. The maximum speed of the car v_{max} occurs when the frictional force equals its maximum value. Newton's second law then gives

$$f_{s,max} = \mu_s mg = m \frac{v_{max}^2}{r}$$

or

$$v_{max} = \sqrt{\mu_s gr} = \sqrt{(0.6)(9.81 \text{ m/s}^2)(30 \text{ m})} = 13.3 \text{ m/s}$$

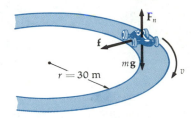

Figure 5-28
Car traveling in a horizontal circle for Example 5-9. The normal force $\mathbf{F}_n$ balances the weight of the car $m\mathbf{g}$ and the frictional force $\mathbf{f}$ provides the centripetal acceleration.

This speed is approximately equal to 47.8 km/h = 29.7 mi/h. If the car travels at a speed greater than 13.3 m/s, the force of static friction will not be great enough to provide the acceleration needed for the car to travel in a circle. The car will slide out away from the center of the circle; that is, it will tend to travel in a straight line.

If the road is not horizontal but is banked, the normal force of the road will have a component inward toward the center of the circle, which will contribute to the centripetal force. The banking angle can be chosen in such a way that, for a given speed, no friction is needed for the car to make the curve.

Example 5-10   A curve of radius 30 m is banked at an angle θ, as shown in Figure 5-29. Find θ so that a car can round the curve at 40 km/h even if the road is frictionless.

In this example, the normal force of the road acting on the car has a component inward toward the center of the circle that provides the centripetal force. From the figure we see that the normal force $\mathbf{F}_n$ makes an angle θ with the vertical. The vertical component of this force, $F_n \cos \theta$, must balance the weight of the car:

$$F_n \cos \theta = mg$$

The horizontal component of the normal force, $F_n \sin \theta$, provides the centripetal force:

$$F_n \sin \theta = m \frac{v^2}{r}$$

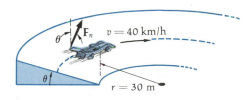

Figure 5-29
Car on banked curve for Example 5-10. Here the normal force has a component $F_n \sin \theta$ toward the center of the circle. If the speed of the car is small enough, this component can provide centripetal acceleration even when there is no friction.

This race track is banked so that the normal force exerted by the track on the car has a horizontal component toward the center of the circle to provide the centripetal force.

If we divide this equation by the previous one, we eliminate m and F_n and obtain an equation relating θ to the speed v and the radius r:

$$\tan \theta = \frac{v^2}{rg}$$

Substituting $v = 40$ km/h $= 11.1$ m/s, $r = 30$ m, and $g = 9.81$ m/s^2, we obtain

$$\tan \theta = \frac{(11.1 \text{ m/s})^2}{(30 \text{ m})(9.81 \text{ m/s}^2)} = 0.419$$

Thus, $\theta = 22.7°$.

Questions

9. Explain why the following statement is incorrect: In the circular motion of a ball on a string, the ball is in equilibrium because the outward force of the ball due to its motion is balanced by the tension in the string.

10. Discuss the following statement: The centripetal force arises because of the circular motion of a body.

5-5   Drag Forces

When an object moves through a fluid, such as air or water, the fluid exerts a retarding or drag force that tends to reduce the speed of the object. This drag force depends on the shape of the object, on the properties of the fluid, and on the speed of the object relative to the fluid. Like the force of friction, this drag force is very complicated. In general, as the speed of the object increases, the drag force increases. For small speeds, the drag force is approximately proportional to the speed of the object; for higher speeds, it is more nearly proportional to the square of the speed.

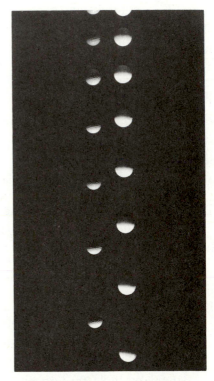

A golf ball (left) and a styrofoam ball falling in air. The air resistance is negligible for the heavier golf ball which falls with essentially constant acceleration. The styrofoam ball initially accelerates, but then the drag force becomes equal to the weight and the ball falls with constant speed (called the terminal speed) as indicated by the equal spacing at the bottom.

Consider a person falling from rest under the influence of the force of gravity, which we assume to be constant, and a retarding force of magnitude bv^n, where b and n are constants. We then have a downward constant force mg and an upward force bv^n (see Figure 5-30). If we take the downward direction to be positive, we obtain from Newton's second law

$$F = mg - bv^n = ma \qquad \text{5-10}$$

At $t = 0$, when the object is dropped, the speed is zero, so the retarding force is zero and the acceleration is g downward. As the speed of the object increases, the drag force increases and the acceleration is less than g. Eventually the speed is great enough for the drag force bv^n to equal the force of gravity mg, and the acceleration is zero. The object then continues moving at a constant speed v_t, called its *terminal speed*. Setting the acceleration a equal to zero in Equation 5-10, we obtain

$$bv_t^n = mg$$

Solving for the terminal speed, we obtain

$$v_t = \left(\frac{mg}{b}\right)^{1/n} \qquad \text{5-11}$$

The larger the constant b, the smaller the terminal speed. The constant b depends on the shape of the object. A parachute is designed so that b will be large and the terminal speed will be small. On the other hand, cars are designed so that b will be small to reduce the effect of wind resistance.

For a sky diver with parachute closed, the terminal speed is about 200 km/h. When the parachute is opened, the drag force is greater than the force of gravity and the sky diver has an upward acceleration as he or she falls; that is, the speed downward decreases, decreasing the drag force, until a new terminal speed (about 20 km/h) is reached.

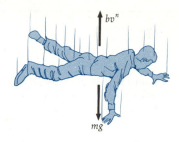

Figure 5-30
Free-body diagram showing forces on a body falling with air resistance. The downward force is the weight mg. The upward force is the drag force due to air resistance bv^n, which depends on the speed v of the body. As the speed increases, the drag force increases until it equals the weight, after which the body falls with constant speed called the terminal speed.

When the drag forces equal their weight, the sky divers fall with terminal speed, which is about 200 km/h without parachutes.

Questions

11. How does a sky diver vary the value of b?

12. How would you expect the value of b for air resistance to depend on the density of air?

Summary

1. When two bodies are in contact, they can exert frictional forces on each other. The frictional force is parallel to the surface at the points of contact. When the surfaces are at rest relative to each other, the friction is static friction, which can vary anywhere from 0 to its maximum value, $\mu_s F_n$, where F_n is the normal force of contact and μ_s is the coefficient of static friction. When the surfaces are moving relative to each other, the friction is kinetic or sliding friction and has the value $\mu_k F_n$, where μ_k is the coefficient of kinetic friction. The coefficient of kinetic friction is slightly less than the coefficient of static friction.

2. The torque exerted by a force about some pivot point is the product of the force and the lever arm, which is the distance from the line of action of the force to the pivot. The conditions for static equilibrium of an extended body are

   1. $\Sigma \mathbf{F} = 0$

   2. $\Sigma \tau = 0$

3. The equilibrium of an extended body can be stable, neutral, or unstable. A body resting on some surface will be in equilibrium or balanced if its center of gravity lies over its base of support. The stability or balance of a body can be improved by lowering the center of gravity or by increasing the size of the base of support.

4. When a body moves in a circle, it has centripetal acceleration directed toward the center of the circle and must therefore have a resultant force acting in that direction. This force is called the centripetal force. It may be due to a string, a spring, or a contact force such as a normal force or friction; or it may be an action-at-a-distance force such as gravity; or it may be a combination of these forces.

5. When a body moves through a fluid, such as air or a liquid, the body experiences a drag force that opposes the motion. The drag force increases with increasing speed. If the body is dropped from rest, its speed increases until the drag force equals the force of gravity, after which time it moves with constant speed called its terminal speed. The terminal speed depends on the shape of the body and on the medium through which it is falling.

Suggestions for Further Reading

Armstrong, H. L.: "How Dry Friction Really Behaves," *American Journal of Physics,* vol. 53, no. 9, 1985, p. 910.

This one-page article discusses the dependence of friction on velocity and points out that the change from static to kinetic friction is not instantaneous but takes place gradually without discontinuity.

Brancazio, Peter J.: "Trajectory of a Fly Ball," *The Physics Teacher,* January 1985, p. 20.

This article shows why drag forces cannot be neglected in any realistic analysis of the motion of a baseball.

Gross, A. C., C. R. Kyle, and D. J. Malewicki: "The Aerodynamics of Human-Powered Land Vehicles," *Scientific American*, December 1983, p. 142.

Drag forces play an important role in determining the maximum velocity a bicycle can attain. By applying the principles of aerodynamics, a bicycle that can travel 60 miles per hour on level ground can be constructed.

Mark, Robert: "The Structural Analysis of Gothic Cathedrals," *Scientific American*, November 1972, p. 90.

The technique of optical stress analysis of scale models aids in an analysis of the static and wind-loaded behaviors of these cathedrals.

Steinman, David B.: "Bridges," *Scientific American*, November 1954, p. 60.

This article presents a history of bridge building, an empirical discipline until the eighteenth century, when physical principles began to be studied and applied to it. The aerodynamics of the Tacoma Narrows Bridge and the reasons for its collapse are also considered.

Review

A. Objectives: After studying this chapter, you should:

1. Know that the maximum static friction force and the kinetic friction force are proportional to the normal force between the surfaces involved.

2. Be able to apply Newton's laws to problems involving frictional forces.

3. Be able to compute the torque exerted by a force about any point.

4. Be able to state the two conditions for static equilibrium and apply them to problems.

5. Be able to discuss stable, unstable, and neutral equilibrium.

6. Be able to apply Newton's laws to circular-motion problems.

7. Be able to describe qualitatively motion with a velocity-dependent retarding force.

B. Define, explain, or otherwise identify:

Friction force, p. 88
Coefficient of friction, p. 88
Torque, p. 92
Center of gravity, p. 95
Stable equilibrium, p. 98
Unstable equilibrium, p. 99
Centripetal force, p. 100
Terminal speed, p. 105

C. True or false: If the statement is true, explain why. If it is false, give a counterexample.

1. The force of static friction always equals μF_n.

2. $\Sigma \mathbf{F} = 0$ is sufficient for static equilibrium.

3. $\Sigma \mathbf{F} = 0$ is necessary for static equilibrium.

4. In static equilibrium, the resultant torque about any point is zero.

5. The center of gravity is always at the geometric center of a body.

6. The terminal speed of an object depends on its shape.

Exercises

Section 5-1 Friction

1. The coefficient of static friction between a box and an inclined plane is 0.4. What is the greatest angle of incline for which the box can rest on the plane without slipping?

2. In a lecture demonstration, a physics teacher stands on a board whose angle of inclination can be varied. (*a*) When he stands in stocking feet on a smooth portion of the board, he begins to slip when the angle is 20°. Find the coefficient of friction between the board and his stocking feet. (*b*) When he stands in his running shoes on a rough part of the board, he

doesn't begin to slip until the angle is 40°. What is the coefficient of friction between his shoes and that part of the board?

3. The coefficients of friction between the tires of a car and the road are $\mu_s = 0.6$ and $\mu_k = 0.5$. (*a*) If the resultant force on the car is the force of static friction exerted by the road, what is the maximum acceleration of the car? (*b*) What is the least distance in which the car can stop if it is initially traveling at 30 m/s and the wheels do not slip? (*c*) What is the stopping distance if the wheels skid so that stopping force is provided by kinetic friction.

4. A car accelerates along a flat road. If the car is to accelerate from 0 to 99 km/h in 12 s at constant acceleration, what is the minimum coefficient of friction needed between the road and tires? (Assume that the wheels do not spin.)

5. A chair is sliding across a polished floor with an initial speed of 3 m/s. It comes to rest after sliding 2 m. What is the coefficient of kinetic friction between the floor and chair?

6. The coefficient of static friction between the floor of a truck and a box resting on it is 0.30. The truck is traveling at 80 km/h. What is the least distance in which the truck can stop if the box is not to slide?

7. A 50-kg box must be moved across a level floor. The coefficient of static friction between the box and the floor is 0.6. One method is to push down on the box at an angle θ with the horizontal. Another method is to pull up on the box at an angle θ with the horizontal. (*a*) Explain why one method is better than the other. (*b*) Calculate the force necessary to move the box by each method if $\theta = 30°$ and compare these results with those for $\theta = 0°$.

Section 5-2   Static Equilibrium of an Extended Body

8. Each of the objects shown in Figure 5-31 is suspended from the ceiling by a thread attached to the point marked ✕ on the object. Describe the orientation of each suspended object with a diagram.

Figure 5-31
Exercise 8.

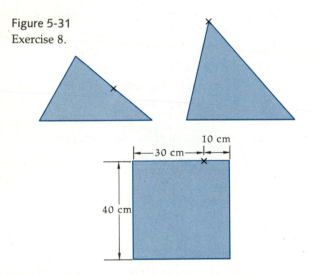

9. A seesaw consists of a 4-m board pivoted at the center. A 28-kg child sits on one end of the board. Where should a 40-kg child sit to balance the seesaw?

10. The height of the center of gravity of a man standing erect is determined by weighing the man as he lies on a board of negligible weight supported by two scales, as shown in Figure 5-32. If the man's height is 6 ft 2 in and the left scale reads 100 lb and the right scale reads 90 pounds, where is his center of gravity relative to his feet?

Figure 5-32
Man on a board on two scales to determine the location of his center of gravity for Exercise 10.

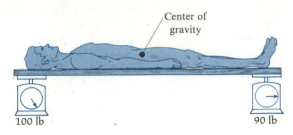

11. In Figure 5-33, Marsha is about to do a push-up. Her center of gravity lies directly above point P on the floor, which is 0.9 m from her feet and 0.6 m from her hands. If her mass is 54 kg, what is the force exerted by the floor on her hands?

Figure 5-33
Marsha doing a push-up for Exercise 11.

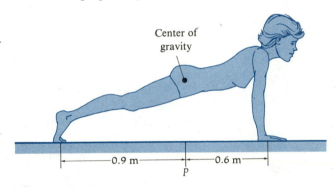

12. John and Marsha are carrying a 60-kg weight on a 4-m board, as shown in Figure 5-34. The mass of the board is 10 kg. Since John spends most of his time reading cookbooks, whereas Marsha regularly does push-ups, they place the weight 2.5 m from John and 1.5 m from Marsha. Find the force in newtons exerted by each to carry the weight.

Figure 5-34
Exercise 12.

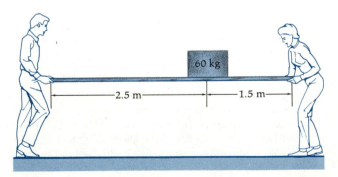

13. Marsha wishes to measure the strength of her biceps muscle by exerting a force on a test strap, as shown in Figure 5-35. The strap is 28 cm from the pivot point at the elbow and her biceps muscle is attached at a point 5 cm from the pivot point. If the scale reads 18 N when she exerts her maximum force, what force is exerted by the biceps muscle?

Figure 5-35
Exercise 13.

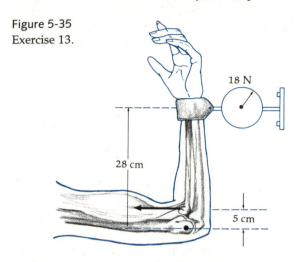

14. A 90-N board 12 m long rests on two supports, each 1 m from the end of the board. A 360-N block is placed on the board 3 m from one end, as shown in Figure 5-36. Find the force exerted by each support on the board.

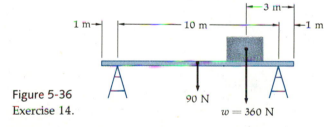

Figure 5-36
Exercise 14.

Section 5-3   Stability of Equilibrium and Balance

15. A box 2 by 1 m of uniform mass is placed on end on a rough hinged plank, as shown in Figure 5-37. The plank is inclined at an angle θ, which is slowly increased. The coefficient of friction is great enough to prevent the box from sliding before it tips over. Find the greatest angle that can be applied without tipping the box over.

Figure 5-37
Exercise 15.

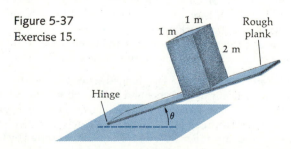

Section 5-4   Circular Motion

16. A 2-kg stone attached to a string is whirled in a horizontal circle of radius 40 cm (like the tetherball in Figure 5-26). The string makes an angle of 30° with the vertical. Find the tension in the string and the speed of the stone.

17. A model airplane of mass 500 g flies in a horizontal circle of radius 6 m attached to a horizontal string. (The weight of the plane is balanced by the upward "lift" force of the air on the wings of the plane.) The plane makes 1 rev every 4 s. (a) What is the speed of the plane? (b) What is the acceleration of the plane? (c) What is the tension in the string?

18. A penny is placed on a record that is gradually accelerated to 78 rev/min. The penny stays on the record if it is placed within 8 cm of the center but slides off if it is placed farther from the center. What is the coefficient of friction between the penny and the record?

19. A car rounds an unbanked curve with a radius of curvature 40 m. The coefficient of friction between the tires and the road is 0.6. What is the maximum speed the car can travel without slipping?

20. A race car rounds a curve of radius 120 m at a speed of 200 km/h. At what angle should the curve be banked?

21. A pilot comes out of a vertical dive in a circular arc such that her upward acceleration is 9 g. (a) If the mass of the pilot is 50 kg, what is the magnitude of the force exerted on her by the airplane seat at the bottom of the arc? (b) If the speed of the plane is 320 km/h, what is the radius of the circular arc?

22. In a carnival ride, the passenger sits on a seat in a compartment that rotates in a vertical circle of radius 5 m (see Figure 5-38). Find the minimum frequency of rotation if the seat belt exerts no force on the passenger at the top of the ride.

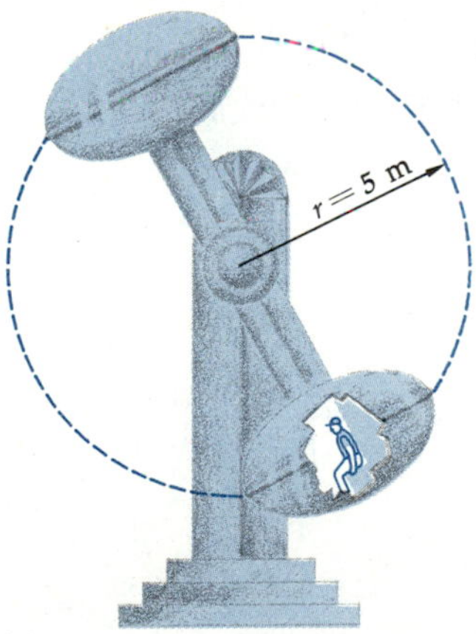

Figure 5-38
Exercise 22.

Section 5-5   Drag Forces

23. A sky diver of mass 60 kg can slow herself to a constant speed of 90 km/h by adjusting her form. (a) What is the magnitude of the upward drag force on the sky diver? (b) If the drag force is equal to bv^2, what is the value of b?

Problems

1. Figure 5-39 shows a mobile consisting of four weights hanging on three rods of negligible mass. Find each of the unknown weights if the mobile is to be balanced. (*Hint:* Find weight w_1 first.)

Figure 5-39
Problem 1.

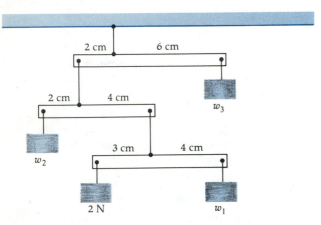

2. A crutch is pressed against the sidewalk with a force $\mathbf{F}_c$ along its own direction, as shown in Figure 5-40. This force is balanced by a normal force $\mathbf{F}_n$ and a frictional force f_s, as shown in the figure. Show that when the force of friction is at its maximum value, the coefficient of friction is related to the angle θ by $\mu_s = \tan \theta$. Explain how this result applies to the forces on your foot when you are not using a crutch. Why is it advantageous to take short steps when walking on ice?

Figure 5-40
Crutch pressing against sidewalk for Problem 2.

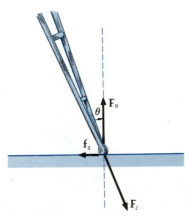

3. The coefficient of friction between box A and the cart in Figure 5-41 is 0.6. The box has a mass of 2 kg. (*a*) Find the

minimum acceleration a of the cart and box if the box is not to fall. (*b*) What is the magnitude of the frictional force in this case? (*c*) If the acceleration is greater than this minimum, will the frictional force be greater than in part (*b*)? Explain. (*d*) Show that, in general, a box of any mass will not fall if the acceleration is $a \geq g/\mu_s$, where μ_s is the coefficient of static friction.

Figure 5-41
Problem 3.

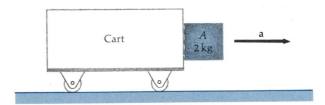

4. An 80-N weight is supported by a cable attached to a strut hinged at point A (see Figure 5-42). The strut is supported by a second cable under tension $\mathbf{T}_2$, as shown in the figure. The mass of the strut is negligible. (*a*) What are the three forces acting on the strut? (*b*) Show that the vertical component of the tension $\mathbf{T}_2$ must equal 80 N. (*c*) Find the force exerted on the strut by the hinge.

Figure 5-42
Problem 4.

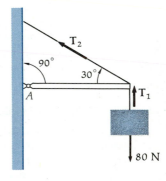

5. A platform scale calibrated in newtons is placed on the bed of a truck driven at a constant speed of 14 m/s. A box weighing 500 N is placed on the scale. Find the reading on the scale if (*a*) the truck passes over the crest of a hill with a radius of curvature 100 m and (*b*) the truck passes through the bottom of a dip with a radius of curvature 80 m.

6. A 3-m board of mass 5 kg is hinged at one end. A force $\mathbf{F}$ is applied vertically at the other end to lift a 60-kg box, which rests on the board 80 cm from the hinge, as shown in Figure 5-43. (*a*) Find the magnitude of the force $\mathbf{F}$ needed to hold the board stationary at $\theta = 30°$. (*b*) Find the force exerted by the hinge at this angle. (c) Find the force **F** and the force exerted by the hinge if **F** is exerted perpendicular to the board when $\theta = 30°$.

Figure 5-43
Problem 6.

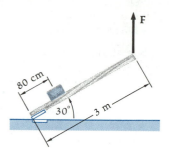

7. Find the force exerted by the hinge *A* on the strut for the arrangement shown in Figure 5-44 if (*a*) the strut is weightless and (*b*) the strut weighs 20 N.

Figure 5-44
Problem 7.

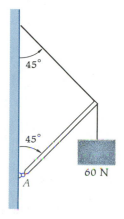

8. A block of mass m_1 is attached to a cord of length L_1, which is fixed at one end. The mass moves in a horizontal circle supported by a frictionless table. A second block of mass m_2 is attached to the first by a cord of length L_2. It also moves in a circle, as shown in Figure 5-45. If the period of the motion is *T*, find the tension in each cord.

Figure 5-45
Problem 8.

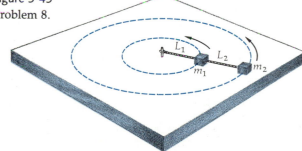

9. Show that, for the tetherball of Figure 5-26, the angle made by the string with the vertical is given by $\cos \theta = gT^2/4\pi^2L$, where *L* is the length of the string and *T* is the period.

10. A horizontal plank 8.0 m long is used by pirates to make their victims walk the plank. A victim of mass 63 kg is walking the plank while a pirate of mass 105 kg is standing on the shipboard end of the plank to prevent it from tipping. Find the maximum distance the plank can overhang the water as the victim walks to the end if (*a*) the mass of the plank is negligible and (*b*) the mass of the plank is 25 kg.

11. In an amusement-park ride, riders stand against the wall of a spinning cylinder and are held up by friction as the floor falls away. If the radius of the cylinder is 4 m, find the minimum frequency in revolutions per minute necessary when the coefficient of friction between a rider and the wall is 0.4.

12. In Figure 5-46, the mass $m_2 = 10$ kg slides on a frictionless table. The coefficients of static and kinetic friction between m_2 and $m_1 = 5$ kg are $\mu_s = 0.6$ and $\mu_k = 0.4$. (*a*) What is the maximum acceleration of m_1? (*b*) What is the maximum value of m_3 if m_1 moves with m_2 without slipping? (*c*) If $m_3 = 30$ kg, find the acceleration of each body and the tension in the string.

Figure 5-46
Problem 12.

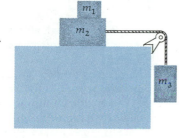

13. Two painters are working from a board 5.0 m long suspended from the top of a building by two ropes attached to the ends of the plank. Either rope will break when the tension exceeds 1 kilonewton (kN). Painter *A*, mass 80 kg, is working at a distance of 1.0 m from one end. Find the range of positions available to painter *B* if his mass is 60 kg and the plank's is 20 kg. (Use $g \approx 10$ m/s².)

14. A 4-kg block rests on a 30° incline attached to a cord that passes over a smooth peg and is attached to a second block of mass *m*, as shown in Figure 5-47. The coefficient of static friction between the block and the incline is 0.4. (*a*) Find the range of possible values of *m* such that the system will be in static equilibrium. (*b*) If $m = 1$ kg, the system will be in static equilibrium. What is the frictional force on the 4-kg block in this case?

Figure 5-47
Problem 14.

15. A 2-kg block sits on a 4-kg block resting on a frictionless table (see Figure 5-48). The coefficients of friction between the blocks are $\mu_s = 0.3$ and $\mu_k = 0.2$. (a) What is the maximum F that can be applied if the 2-kg block is not to slide on the 4-kg block? (b) If F is twice this maximum, find the acceleration of each block. (c) If F is half this maximum, find the acceleration of each block and the friction force acting on each block.

Figure 5-48
Problem 15.

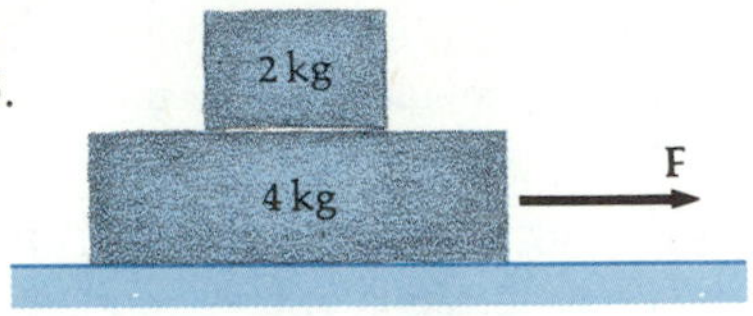

2 kg

4 kg

F

16. A block is on an incline whose angle can be varied. The angle is gradually increased from $0°$. At $30°$, the block starts to slide down the incline. It slides 3 m in 2 s. Calculate the coefficients of static and kinetic friction between the block and incline.

17. (a) A parachute creates enough air resistance to hold an 80-kg parachutist to a constant downward speed of 6.0 m/s after it has been open for a while. Assuming that the force of the air resistance is given by $f = bv^2$, calculate b for this case. (b) A sky diver drops in free-fall until his speed is 60 m/s before opening his chute. If the chute opens instantaneously, calculate the initial upward force exerted by the chute on the sky diver moving at 60 m/s. Explain why it is important that the chute take a few seconds to open.

18. A 400-N box rests on a horizontal table. The coefficient of static friction is 0.6. The box is pulled by a massless rope with a force **F** at an angle θ, as shown in Figure 5-49. The minimum value of the force needed to move the box depends on θ. Describe qualitatively how you would expect this force to depend on θ. Compute the force for the angles $\theta = 0, 10, 20, 30, 40, 50,$ and $60°$, and make a plot of F versus θ. From your plot, at what angle is it most efficient to apply the force to move the box?

Figure 5-49
Problem 18.

F

θ

m

19. A uniform door, 2.0 m high by 0.8 m wide and mass 18 kg, is hung from two hinges that are 20 cm from the top and 20 cm from the bottom, respectively. If each hinge supports half the weight of the door, find the magnitude and direction of the two horizontal components of the forces exerted by the hinges on the door.

20. A 900-N boy sits on top of a ladder of negligible weight that rests on a frictionless floor. There is a cross brace halfway up the ladder (see Figure 5-50). The angle at the apex is $\theta = 30°$. (a) What is the force exerted by the floor on each leg of the ladder? (b) Find the tension in the cross brace. (c) If the cross brace is moved down toward the bottom of the ladder (with the same angle θ), will its tension be greater or less?

Figure 5-50
Problem 20.

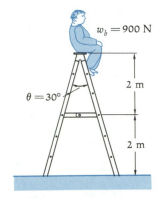

$w_b = 900$ N

$\theta = 30°$

2 m

2 m

21. A wheel of mass m and radius r rests on a horizontal surface against a step of height h ($h < r$). The wheel is to be raised over the step by a horizontal force F applied to the axle of the wheel. Find the force F necessary to raise the wheel over the step.

22. A light ladder rests on a rough floor and leans against a frictionless vertical wall at an angle θ with the horizontal. Show that if L is the length of the ladder, a person can climb no further than $\mu_s L \tan \theta$ before the ladder slips, where μ_s is the coefficient of friction between the floor and the ladder and it is assumed that the weight of the ladder is negligible compared with that of the climber.

CHAPTER SIX

Work and Energy

Energy is Eternal Delight.

WILLIAM BLAKE

In this chapter, we begin our investigations of energy, which may be the most important quantity in science. The full scope and importance of energy was not appreciated until the 1840s when Julius Mayer in Germany and James Joule in England independently discovered that heat is simply another form of energy. Today, thanks to the genius of Einstein, we know that matter itself can be viewed as a form of energy.

As you work on this chapter, you will find that some problems that have previously required several calculations can be solved in one step using the concept of conservation of energy. Other problems that can't be solved at all with the methods you have learned so far can be solved just as easily.

The concepts of work and energy are among the most important concepts in physics and science, and they play an important role in our everyday life. In science, work is given a precise definition that differs from our everyday usage. Work is done only when a force moves through a distance and there is a component of the force along the line of motion. Thus, when you exert a force to pull a sled across the snow, you do work, but no work is done as you sit at your desk and study or if you merely hold a heavy object at rest. Closely associated with the concept of work is the concept of energy, which may be thought of as the capacity to do work. Whenever work is done, energy is transferred from one system to another. For example, when you pull a sled, the work you do goes partly into the energy of motion of the sled, called kinetic energy, and partly into heat energy if there is friction between the sled and the snow. At the same time, as you pull the sled, the internal chemical energy in your body is decreased. The net result is the transfer of some of the internal chemical energy of your body to the external kinetic energy of the sled plus heat energy. One of the most important principles in science is the law of conservation of energy: the total energy of a system and its surroundings does not change. When the energy of a system decreases,

there is always a corresponding increase in the energy of its surroundings or of another system.

There are many forms of energy. Kinetic energy is associated with the motion of a body. Potential energy is associated with the position of the body relative to other bodies, such as the earth. Internal energy is associated with the motion of the molecules within the body and is closely connected with the temperature of the body.

In this chapter, we will explore the concepts of work, kinetic energy, and potential energy and show how to use the law of conservation of energy to solve various problems. Later, when we study thermodynamics, we will look at heat energy, which is energy transferred due to differences in temperature, and will explore the internal molecular energy of systems in more depth.

6-1   Work and Kinetic Energy

When a force moves through a distance and there is a component of the force parallel to the displacement, the force does work. For simplicity, we first consider the special case of constant forces and motion in one dimension. We define the work done by such a force as the product of the component of the force along the line of motion and the displacement. If the force $\mathbf{F}$ makes an angle θ with the line of the displacement Δx, as shown in Figure 6-1, the work done is

$$W = F \cos \theta \, \Delta x = F_x \, \Delta x \qquad\qquad 6\text{-}1$$

Work by a constant force

The work is positive if Δx and F_x are in the same direction and negative if they are in opposite directions. The dimensions of work are those of force times distance. The SI unit of work and energy is the joule (J), which equals the product of a newton and a metre.

$$1 \, \text{J} = 1 \, \text{N} \cdot \text{m} \qquad\qquad 6\text{-}2$$

In the U.S. customary system, the unit of work is the foot-pound. The relation between these units is easily found using the relations between pounds and newtons and between metres and feet. The result is

$$1 \, \text{J} = 0.738 \, \text{ft} \cdot \text{lb} \qquad\qquad 6\text{-}3$$

Figure 6-2 shows a boy pulling a sled across the snow by exerting a force $\mathbf{F}$ at an angle θ to the horizontal. For this discussion, we will assume that friction can be neglected. The other forces acting on the sled are its weight $m\mathbf{g}$ and a vertical support force $\mathbf{F}_n$. Since the sled does not accelerate vertically, the vertical upward force $F \sin \theta + F_n$ must equal the weight mg. Then the horizontal force $F \cos \theta$ is the resultant force. We call the work done by the resultant force the net work W_{net}. When there are several forces that do work, the net work can be found by first summing the forces to find the resultant force or by computing the work done by each force and then summing. There is an important relation between the net work and the initial and final speeds of the body. Newton's second law gives

$$F_x = ma_x$$

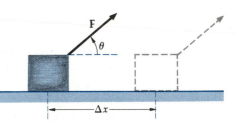

Figure 6-1
When a force $\mathbf{F}$ acts through a distance Δx, the work done is $F \cos \theta \, \Delta x = F_x \, \Delta x$.

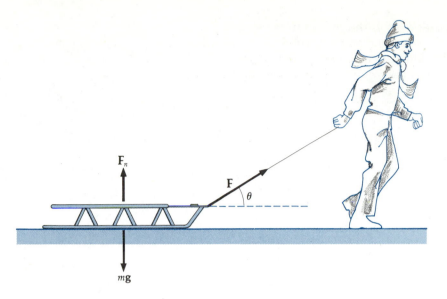

Figure 6-2
Boy pulling sled. The work done by the boy as he pulls the sled a distance Δx is $F \cos \theta\, \Delta x$.

For a constant force, the acceleration is constant, and we can relate the distance moved to the initial and final speed by the constant-acceleration formula (Equation 2-11). If the initial speed is v_1 and the final speed v_2, we have

$$v_2^2 = v_1^2 + 2a\, \Delta x \tag{6-4}$$

Then

$$W_{net} = F_x\, \Delta x = ma_x\, \Delta x$$

Substituting $\frac{1}{2}(v_2^2 - v_1^2)$ for $a\, \Delta x$, we obtain

$$W_{net} = \tfrac{1}{2}mv_2^2 - \tfrac{1}{2}mv_1^2 \tag{6-5}$$

The quantity $\frac{1}{2}mv^2$ is called the *kinetic energy* E_k of the body. It is a scalar quantity that depends on the body's mass and speed:

$$E_k = \tfrac{1}{2}mv^2 \tag{6-6} \qquad \text{Kinetic energy}$$

The work done by the rope equals the force times the distance the body is raised.

The quantity on the right side of Equation 6-5 is the change in the kinetic energy of the body, that is, the kinetic energy $\frac{1}{2}mv_2^2$ at the end of the interval minus the kinetic energy $\frac{1}{2}mv_1^2$ at the beginning of the interval. The net work done by the resultant force is therefore equal to the change in the kinetic energy of the body:

$$W_{net} = \Delta E_k = \tfrac{1}{2}mv_2^2 - \tfrac{1}{2}mv_1^2 \qquad\qquad 6\text{-}7 \qquad \text{Work-energy theorem}$$

This result is known as the work-energy theorem.

The net work done by the resultant force acting on a body equals the change in the kinetic energy of the body.

The work-energy theorem holds whether the resultant force is constant or not, as will be shown in the next section. As already mentioned, the work done by the resultant force is also the net work done by all the forces acting on the body, which can be found by summing the work done by each force.

Example 6-1   If the mass of the sled in Figure 6-2 is 5 kg and the boy exerts a force of 8 N at 30°, find the work done by the boy and the final speed of the sled after it moves 3 m assuming that it starts from rest.

The forces acting on the sled are shown in the figure. The vertical forces are the force due to gravity,

$$mg = (5 \text{ kg})(9.81 \text{ N/kg}) = 49 \text{ N}$$

the upward component of the force of the rope,

$$F_y = 8 \text{ N sin } 30° = 4 \text{ N}$$

and the vertical support force of the ground, which equals $mg - F_y$ because there is no vertical acceleration,

$$F_n = mg - F_y = 49 \text{ N} - 4 \text{ N} = 45 \text{ N}$$

The only horizontal force is

$$F_x = 8 \text{ N cos } 30° = 6.93 \text{ N}$$

The work done by the force **F** is the product of the component in the direction of motion (6.93 N) and the distance traveled (3 m):

$$W = F \cos \theta \, x = (8 \text{ N})(0.866)(3 \text{ m}) = 20.8 \text{ J}$$

Since the horizontal component of the force **F** is the resultant force, the work done by this force equals the change in the kinetic energy of the sled. If the sled starts from rest, its kinetic energy after traveling 3 m is therefore 20.8 J. Its speed can be found from

$$\tfrac{1}{2}mv^2 = E_k = 20.8 \text{ J}$$

$$v = \sqrt{\frac{2E_k}{m}} = \sqrt{\frac{2(20.8 \text{ J})}{5 \text{ kg}}} = 2.88 \text{ m/s}$$

Example 6-2   A car of mass 1200 kg moving at 25 m/s brakes and locks its wheels, skidding to a stop. If the coefficient of kinetic friction is 0.3, find the work done by friction and the stopping distance.

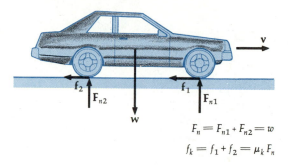

$$F_n = F_{n1} + F_{n2} = w$$

$$f_k = f_1 + f_2 = \mu_k F_n$$

Figure 6-3
Forces on a car skidding to a stop. The total frictional force $f = f_1 + f_2$ equals $\mu_k F_n$, where $F_n = F_{n1} + F_{n2} = w$ is the total normal force exerted by the road on the car.

The forces acting on the car are shown in Figure 6-3. The normal forces of the road balance the weight of the car. The force of friction is therefore $f_k = \mu_k F_n = \mu_k mg$. Since the force of friction is opposite the direction of motion, it does negative work. The initial kinetic energy of the car is

$$E_k = \tfrac{1}{2}mv^2$$

$$= \tfrac{1}{2}(1200 \text{ kg})(25 \text{ m/s})^2 = 3.75 \times 10^5 \text{ kg} \cdot \text{m}^2/\text{s}^2 = 3.75 \times 10^5 \text{ J}$$

Since the car comes to rest, its final kinetic energy is zero. The change in kinetic energy of the car is thus

$$\Delta E_k = E_{kf} - E_{ki}$$

$$= 0 - 3.75 \times 10^5 \text{ J} = -3.75 \times 10^5 \text{ J}$$

If Δx is the stopping distance, the work done by friction is

$$W = F_x \, \Delta x = -\mu_k mg \, \Delta x$$

$$= -(0.3)(1200 \text{ kg})(9.81 \text{ N/kg}) \, \Delta x = -(3530 \text{ N}) \, \Delta x$$

The force of friction is the resultant force acting on the car. Setting the work done by friction equal to the change in kinetic energy, we have

$$W = -(3530 \text{ N}) \, \Delta x = -3.75 \times 10^5 \text{ J}$$

or

$$\Delta x = \frac{3.75 \times 10^5 \text{ J}}{3530 \text{ N}} = 106 \text{ m}$$

We could have found the final speed in Example 6-1 and the stopping distance in Example 6-2 by first finding the acceleration and then using the constant-acceleration formulas. The work-energy theorem provides an alternative method to Newton's laws of motion for solving problems in mechanics. When the forces are not constant, the work-energy theorem is often much easier to use than are Newton's laws.

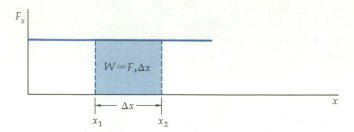

Figure 6-4
The work done by a constant force is interpreted graphically as the area under the F_x-versus-x curve.

In Figure 6-4, work by a constant force is interpreted graphically as the area under the force-versus-position curve. We note that for work to be done, the force must act through a distance. Consider a person holding a weight a distance h off the floor, as shown in Figure 6-5. In everyday usage, we might say that it takes work to do this, but by our scientific definition, no work is done by a force acting on a stationary object. We could eliminate the effort of holding the weight by merely tying the rope to some object, and the weight would be supported with no help from us.

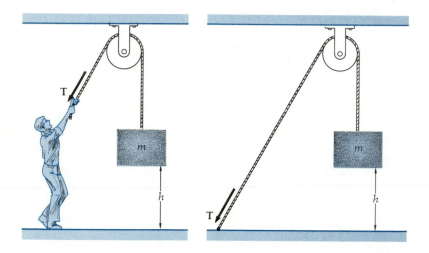

Figure 6-5
No work is done by the man holding the weight at a fixed position. The same task could be accomplished by tying the rope to a fixed point.

If a force is to do work, it must move through a distance and energy must be supplied from somewhere. For example, if we wish to lift the weight in Figure 6-5 to a greater height, we must pull the rope through a distance, doing work with our muscles. Some of the internal chemical energy of our body is used, and we must eventually replenish this energy by taking in nourishment. If we attach the rope to an electric motor to raise the weight, we expend electric energy to perform this work.

Although there is no *external* work done when we merely hold a stationary weight, we do become tired from this activity because *internal work* is being done in our body. While we hold the weight, nerve impulses inside our body continually trigger contractions of muscle fibers. The muscle fibers contract and relax, exerting a force through a small distance and thus doing work inside our body. In this process, some of the internal chemical energy in our body is converted into heat.

Even though no external work is done holding a stationary weight, internal work is done by the repeated contraction and relaxation of the muscles.

Questions

1. A heavy box is to be moved from the top of one table to the top of another of the same height on the other side of the room. Is work required to do this?

2. To get out of bed in the morning, do you have to do work?

6-2   Work Done by a Variable Force

Many forces are not constant but vary with position. For example, when you stretch a spring, the force exerted by the spring is proportional to the amount the spring is stretched. Similarly, the gravitational force exerted by the earth on a rocket ship varies inversely with the square of the distance from the rocket ship to the center of the earth. We can use the graphical interpretation of work as the area under the force-versus-position curve to extend our definition of work to cases in which the force varies with position.

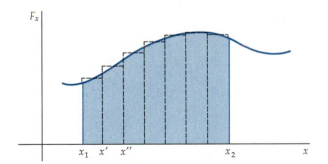

Figure 6-6
A variable force can be approximated by a series of constant forces over small intervals. The work done by each constant force equals the rectangular area indicated. The total work done by the variable force from x_1 to x_2 is the total area under the curve for this interval.

Figure 6-6 shows a general force as a function of position x. In the figure, we have divided the interval from x_1 to x_2 into a set of smaller subintervals Δx_i. If each subinterval is small enough, we can approximate the varying force by a series of constant forces, as shown in the figure. For each subinterval, the work done by the constant force is the area of the rectangle shown. The sum of the rectangular areas is the sum of the work done by the series of constant forces that approximates the varying force. As can be seen from the figure, this area is approximately equal to the area under the curve.

We can therefore define the work done by a varying force as the area under the F-versus-x curve.

We will now show that the net work done by the resultant force equals the change in kinetic energy even when the resultant force varies with position; that is, we will show that Equation 6-7 holds in general. Consider Figure 6-6. For each rectangular area the force is constant, so for each subinterval the work done by the constant force equals the change in the kinetic energy over that subinterval. When we sum the changes in kinetic energy for all the subintervals, we obtain the total or net change in kinetic energy for the entire interval from x_1 to x_2. The total area under the curve therefore equals the sum of the changes in kinetic energy for each subinterval, which equals the net change in kinetic energy for the complete interval, $\frac{1}{2}mv_2^2 - \frac{1}{2}mv_1^2$. This is the same as Equation 6-7.

Example 6-3   A 4-kg block resting on a frictionless table (see Figure 6-7a) is attached to a horizontal spring that obeys Hooke's law and exerts a force $F_x = -kx$, where x is measured from the equilibrium length of the spring and the force constant is $k = 400$ N/m. The spring is compressed to $x_1 = -5$ cm. Find (a) the work done by the spring as the block moves from $x_1 = -5$ cm to its equilibrium position $x_2 = 0$ and (b) the speed of the block at $x_2 = 0$.

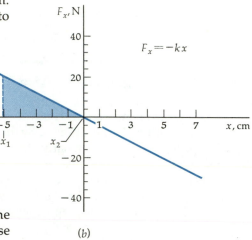

(b)

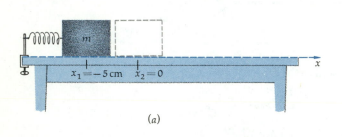

(a)

(a) Figure 6-7b is a plot of the force versus distance. The work done as the block moves from x_1 to x_2 equals the area under the curve between these limits, as indicated by the shaded area in the figure. This area is one-half the base times the height of the triangle. The base is 5 cm = 0.05 m, and the height is the value of the force at x_1, which is

$$F_x = -(400 \text{ N/m})(-0.05 \text{ m}) = +20 \text{ N}$$

The work done is thus

$$W = \tfrac{1}{2}(0.05 \text{ m})(20 \text{ N}) = 0.500 \text{ N·m} = 0.500 \text{ J}$$

The work is positive because the force is acting in the direction of motion. This is indicated in the figure by the fact that the area is above the x axis.

(b) As in Example 6-1, we can find the speed of the block from the fact that its kinetic energy is 0.500 J. The speed is then

$$v = \sqrt{2E_k/m} = \sqrt{2(0.500 \text{ J})/4 \text{ kg}} = 0.50 \text{ m/s}$$

Note that we could *not* have found this by first finding the acceleration and then using the constant-acceleration expressions. Since the force varies with position, the acceleration varies also and is therefore not constant over time.

Figure 6-7
(a) Block attached to a spring for Example 6-3. The spring is compressed from equilibrium and released. (b) Plot of F_x versus x for the block attached to the spring. The work done by the spring as the block moves from x_1 to x_2 is the shaded area indicated.

Example 6-4   Find the speed of the block in Example 6-3 when the spring is at its equilibrium position if the coefficient of kinetic friction between the table and block is 0.200.

In this case, the work done by the spring is not the net work done on the block because the frictional force also does (negative) work. Since the frictional force is constant, the work done by it is just the force times the distance. This work is negative because the force $F_x = -\mu_k F_n = -\mu_k mg$ is acting opposite to the direction of motion. The work done by the frictional force is

$$W_f = -\mu_x mg \Delta x$$

$$= -(0.200)(4.00 \text{ kg})(9.81 \text{ N/kg})(0.05 \text{ m})$$

$$= -0.392 \text{ J}$$

The net work done on the block is then

$$W_{net} = 0.500 \text{ J} - 0.392 \text{ J} = 0.108 \text{ J}$$

The net work equals the change in kinetic energy, which is $\frac{1}{2}mv^2$:

$$W_{net} = 0.108 \text{ J} = \tfrac{1}{2}mv^2 = \tfrac{1}{2}(4.00)v^2$$

$$v^2 = 0.054 \text{ m}^2/\text{s}^2$$

$$v = 0.232 \text{ m/s} = 23.2 \text{ cm/s}$$

Example 6-5   A skier of mass m skis down a frictionless hill that has a constant angle of inclination θ, as shown in Figure 6-8. The skier begins at rest from a height h. Find the work done by all the forces and the speed of the skier at the bottom of the hill.

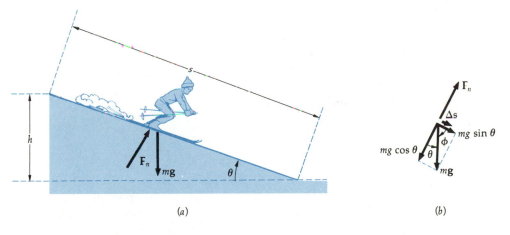

(a)   (b)

Figure 6-8
(a) Skier skiing down a hill of constant slope for Example 6-5. (b) Free-body diagram for the skier. The resultant force is $mg \sin \theta$, which is the component of the weight in the direction of the displacement $\Delta \mathbf{s}$.

The forces acting on the skier are the force of gravity $m\mathbf{g}$ and the contact force $\mathbf{F}_n$ exerted by the snow, as indicated in the figure. Since the force $\mathbf{F}_n$ is perpendicular to the hill and to the motion of the skier, it does no work on the skier. The only force that does work is the weight $m\mathbf{g}$. The angle between this force and the displacement is $\phi = 90° - \theta$ (see Figure 6-8b). The component of the weight in the direction of motion is therefore $mg \cos \phi = mg \cos (90° - \theta) = mg \sin \theta$. When the skier moves a distance Δs down the hill, the earth does work ($mg \sin \theta$) Δs. Since the force exerted by the earth is constant, the total work done when the skier moves a distance s down the hill is merely ($mg \sin \theta$)s. We see from Figure 6-8 that the total distance s measured along the incline of the hill is related to the initial height h by $\sin \theta = h/s$ or $h = s \sin \theta$, so the work done by the earth is

$$W = (mg \sin \theta)s = mgh$$

Since this is the total work done by all the forces, the work-energy theorem gives

$$W_{net} = mgh = \Delta E_k = \tfrac{1}{2}mv^2 - 0$$

The speed at the bottom of the incline is thus given by

$$v = \sqrt{2gh}$$

This result is the same as if the skier had dropped the total distance h in free-fall. The work done by the earth on the skier, mgh, is independent of the angle of the hill. If the angle θ were increased, the skier would travel a smaller distance s to drop the same vertical distance h, but the component of mg parallel to the motion, $mg \sin \theta$, would be greater, making the work done, ($mg \sin \theta$)$s = mgh$, the same.

The results of Example 6-5 can be generalized. Consider a skier skiing down a hill of arbitrary shape. Figure 6-9 shows a small displacement Δs parallel to the hill. The work done by the earth during this displacement is ($mg \cos \phi$)Δs, where ϕ is the angle between the displacement and the downward force of gravity. The quantity $\Delta s \cos \phi$ is just Δh, the vertical distance dropped. As the skier skis down the hill, the angle ϕ varies, but for each displacement Δs, the downward component of the displacement parallel to the weight is $\Delta s \cos \phi = \Delta h$ and the work done by the earth is $mg \Delta s \cos \phi = mg \Delta h$. Thus, the total work done by the earth is mgh, where h is the total vertical distance descended. If the hill is frictionless, the weight is the only force that does work. In this case, the speed of the skier after descending a vertical distance h is obtained from

$$W_{net} = mgh = \Delta E_k = \tfrac{1}{2}mv^2 - 0$$

assuming the initial speed is zero. The speed at the bottom of the hill is given by

$$v = \sqrt{2gh}$$

The work done by the earth on the skier is mgh, where h is the total vertical distance dropped. If there were no friction, the speed of the skier at the bottom of the hill would be $\sqrt{2gh}$, the same as if he fell through a distance h.

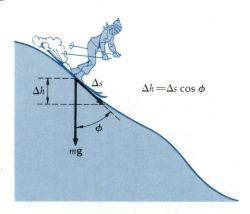

$\Delta h = \Delta s \cos \phi$

Figure 6-9
Skier skiing down a hill of arbitrary shape. The work done by the earth when the displacement is Δs is ($mg \cos \phi$) $\Delta s = mg \Delta h$, where Δh is the vertical component of the displacement.

If the hill is not frictionless, the frictional force will do work (which will be negative because the frictional force will be acting in the direction opposite to the motion). The work done by the frictional force depends on the shape of the hill, the coefficient of friction, and the total distance traveled.

Questions

3. A block is pulled up an inclined plane a certain distance. It is then pulled down the same distance. How does the work done by friction as the block moves up compare with that done by friction as the block moves down?

4. How does the work done by gravity as the block in Question 3 moves up compare with the work done by gravity as it moves down?

5. Is it possible to exert a force that does work on a body without increasing its kinetic energy? If so, give an example.

6. How does the kinetic energy of a car change when its speed is doubled?

7. A body moves in a circle at constant speed. Does the force that accounts for its acceleration do work on it? Explain.

6-3   Potential Energy

Potential energy is energy possessed by a body because of its position. A rock poised at the top of a cliff has gravitational potential energy. If given a slight push, the rock comes tumbling down, gaining speed and therefore kinetic energy. As the rock falls, its potential energy is converted into kinetic energy.

Consider a skier going up a mountain on a ski lift. If the lift moves at constant speed and the angle of inclination is θ, the force exerted by the lift in the direction up the incline is $mg \sin \theta$. This force does work, but the work does not result in an increase in the kinetic energy of the skier because the earth exerts an equal and opposite force so that the resultant force is zero. If s is the total distance the skier moves up the incline, the work done by the lift is $(mg \sin \theta)s = mgh$, where h is the net vertical elevation gained. At the same time, the earth does work $-mgh$ on the skier, so the net work done on the skier is zero and there is no change in his or her kinetic energy. Instead, the work done by the lift is stored as the potential energy mgh of the skier. This potential energy is converted into kinetic energy when the skier skis down the incline. (If there is friction between the skis and the snow, some of the original potential energy is converted into heat.)

If we throw a ball vertically into the air with some initial speed v_0, it will travel upward until its kinetic energy is zero. Its kinetic energy decreases because the gravitational attraction of the earth, the weight $m\mathbf{g}$, is the only force acting on the ball, and since this force is opposite the direction of motion, it does negative work. As the ball moves upward, its decrease in kinetic energy is matched by an equal increase in its potential energy. At the top of its flight, the potential energy mgh of the ball equals its initial kinetic energy $\frac{1}{2}mv^2$. Then, as the ball falls back, it loses potential energy and gains kinetic energy. At any point in its motion, the sum of the ball's kinetic energy

This boulder, located in the Arches National Park in Utah, has gravitational potential energy relative to ground level.

and its potential energy is constant. This is an example of the law of *conservation of energy*.

The gravitational potential energy U_g of a body of mass m at some height h is given by

$$U_g = mgh \qquad\qquad 6\text{-}8$$

Gravitational potential energy

The work done by the earth *decreases* the gravitational potential energy.

$$W = -\Delta U \qquad\qquad 6\text{-}9$$

Note that the value of the gravitational potential energy U_g depends on where we measure h from, which is arbitrary; that is, it is only the *change* in potential energy that is important. For a skiing problem, we would naturally choose h to be zero at the bottom of the hill so that the potential energy of the skier is zero at that point. But we could just as well choose h to be zero at the top of the hill, in which case the potential energy would be zero there and would become negative as the skier moves down the hill. In your laboratory, you would probably choose h to be zero at the floor, but choosing h to be zero at the top of your work table would be just as valid.

There are other kinds of potential energy besides gravitational potential energy. In Examples 6-3 and 6-4, the work done to compress the spring was stored in the spring as potential energy. When the spring was released, its potential energy was converted into kinetic energy of the block. We can find the potential-energy function for a spring by computing the work it takes to stretch or compress the spring. Figure 6-10 shows a mass m attached to a spring and resting on a frictionless table. When the spring is at its equilibrium length, there is no net force acting on the mass. If we wish to stretch the spring, we must exert a force equal and opposite to that exerted by the spring. Since the spring exerts a force $-kx$ when it is stretched a distance x, we must exert a force $F_x = +kx$. To stretch the spring from $x = 0$ to some value x_1, we must do the work indicated by the shaded area of Figure 6-11. This work is $\frac{1}{2}(x_1)(kx_1) = \frac{1}{2}kx_1^2$. This work is stored as potential energy in the spring. Thus, for a general extension or compression x of a spring, the potential energy of the spring is

$$U_s = \tfrac{1}{2}kx^2 \qquad\qquad 6\text{-}10$$

Potential energy of a spring

This climber does work in increasing his gravitational potential energy.

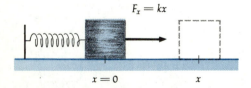

$F_x = kx$

$x = 0$ x

Figure 6-10
Mass attached to a spring. To stretch the spring, a force $F_x = +kx$ must be applied.

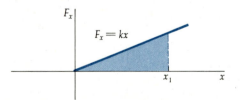

F_x

$F_x = kx$

x_1 x

Figure 6-11
F_x versus x for the force F_x needed to stretch the spring in Figure 6-10. The work done in stretching the spring from $x = 0$ to x_1 is the shaded area indicated.

If we release the spring, the spring will do work decreasing its potential energy and increasing the kinetic energy of the mass.

Although we have talked about the potential energy of an object at some height above the surface of the earth or of an object attached to a spring as if potential energy were associated with a single object, this description is a simplification. More generally, the concept of potential energy applies to a *system of particles.* Consider, for example, two objects that exert gravitational forces of attraction on each other. If we apply an external force to each object equal but opposite to the force of attraction between them and pull them apart with no acceleration, we do work on the system. The work done by our external applied forces increases the potential energy of the two-object system. When the objects are released so that they can move under the influence of their mutual attraction, the work done by the forces of attraction between them decreases the potential energy of the system and increases its kinetic energy. If one of the objects is much larger than the other, the motion of the larger object is negligible, which is the case when one of the objects is an ordinary ball and the other is the earth. We separate this pair of objects by "lifting" the ball, and we release them by "dropping" the ball. It is common practice, then, to think of the potential energy of this system as being associated with just the smaller object.

A slingshot is an example of elastic potential energy. The work done in stretching the rubber is stored as potential energy and then converted into kinetic energy of the missile.

Questions

8. How can a man who can exert a maximum force of 400 N raise an 800-N box from the ground to a shelf at a height 2 m above the ground? Could a boy who can exert only 200 N raise the box to the shelf? How much work would each have to do?

9. When you climb a mountain, is the work done on you by gravity different if you take a short, steep trail instead of a long, gentle trail? If not, why would you find one trail easier than the other?

10. Estimate the amount by which your gravitational potential energy changes when you walk up a flight of stairs.

6-4   Conservative Forces and the Conservation of Energy

We saw in the previous section that if a ski lift carries a skier up a mountain slope, the work done by the lift against gravity is stored as the gravitational potential energy of the skier and can be recovered. There are many situations in which the work done is not readily recovered, however. If we push a chair across a rough floor, the work done against friction is not recovered. When we stop pushing and release the chair, it slides to a stop. The work done does not reappear as kinetic energy of the chair. The work is not lost completely, though; it appears as thermal energy in the surroundings. As we will see when we study thermodynamics, this thermal energy is not easily reconverted into mechanical energy.

The force of friction is called a *nonconservative force*, or sometimes a *dissipative force*, because the work done by friction dissipates mechanical energy into thermal energy. Another type of nonconservative force is the one involved in large deformations of a body. For example, if a spring is stretched beyond its elastic limit, it becomes permanently deformed, and the work done in the stretching is not recovered when the spring is released. Again, the work done in deforming the spring is dissipated into thermal energy. The spring becomes warmer. You may have observed that when you bend a coat hanger back and forth until it breaks, the coat hanger becomes warm. The work you do bending the coat hanger is not stored as potential energy but is instead dissipated into heat energy.

Nonconservative forces

Forces such as gravity and that of a spring that do work (or have work done against them) that is recoverable are called *conservative forces*. An important property of a conservative force is that the work it does on an object is independent of the path taken by the object and depends only on the object's initial and final positions. We saw an example of this when we found that the work done by the force of gravity on a skier skiing along a curve of any shape is just mgh, where h is the total vertical distance dropped. The work done depends only on the initial and final elevations of the skier. We can use this property to give a general definition of the potential-energy function U associated with a conservative force. U is *defined* such that the work done *by* a conservative force equals the *decrease* in the potential-energy function:

Conservative forces

$$W = -\Delta U \qquad\qquad 6\text{-}11$$

The work done by nonconservative forces generally depends on things other than the initial and final positions of the object (or objects in a two-particle system). For example, the work done by frictional forces or by drag forces may depend on the speed of the object, on the total distance traveled, or on the particular path through space taken by the object as it moves from one position to another. A potential-energy function cannot be defined for a nonconservative force.

We recall from Section 6-1 that, in general, the net work done by all the forces acting on an object equals the change in the kinetic energy of the object. If a conservative force is the only force that does work, the work done also equals the decrease in potential energy. Thus,

$$W = -\Delta U = +\Delta E_k$$

Then

$$\Delta E_k + \Delta U = \Delta(E_k + U) = 0 \qquad\qquad 6\text{-}12$$

or

$$E_k + U = \text{constant} \qquad\qquad 6\text{-}13$$

The sum of the kinetic and potential energies of a system is called the *total mechanical energy* of that system. When only conservative forces do work, the total mechanical energy of a system remains constant because the increase in kinetic energy equals the corresponding decrease in potential energy. This is known as the *law of conservation of (mechanical) energy* and is the origin of the expression *conservative* force. If E_{k1} and U_1 are the initial kinetic and potential energies of a particle and E_{k2} and U_2 are its kinetic and

Conservation of mechanical energy

potential energies at some later time, the law of conservation of energy gives

$$E_{k1} + U_1 = E_{k2} + U_2 \qquad\qquad 6\text{-}14$$

For the example of a skier skiing down a frictionless slope, if his or her speed is v_1 at height h_1 and v_2 at h_2, conservation of energy gives

$$\tfrac{1}{2}mv_1^2 + mgh_1 = \tfrac{1}{2}mv_2^2 + mgh_2 \qquad\qquad 6\text{-}15$$

The principle of conservation of energy is one of the most important in physics. In the macroscopic world, nonconservative forces such as friction are nearly always present. Because of this, the importance of energy and its conservation was not realized until the nineteenth century. Then it was discovered that the disappearance of *macroscopic mechanical energy* is always accompanied by the appearance of thermal or internal energy, usually indicated by an increase in the temperature of the body and its surroundings. We now know that, on the microscopic scale, this internal energy consists of the kinetic and potential energy of molecular motion. When the concept of energy is generalized to include this internal energy, the total energy of an object plus its surroundings does not change even when nonconservative forces such as friction are present.

In this chapter, we will consider only macroscopic mechanical energy. When both conservative and nonconservative forces do work, the work done by the nonconservative forces equals the change in the total mechanical energy of the system. This is a generalization of the work-energy theorem of Equation 6-7. If we call W_{nc} the work done by nonconservative forces, we have

$$W_{nc} = \Delta(E_k + U) = \Delta E_{\text{total}} \qquad\qquad 6\text{-}16$$

We will now give some examples illustrating the application of the work-energy theorem and the conservation of mechanical energy to problem solving.

Internal chemical energy is converted into kinetic energy when the pole vaulter begins to run. Some of the energy is then converted into elastic potential energy of deformation of the pole and eventually into gravitational potential energy which in turn is converted into kinetic energy as he drops. The mechanical energy is eventually lost to heat when he reaches the ground.

Generalized work-energy theorem

Example 6-6   A skier skis down a frictionless hill of height h and of arbitrary shape (see Figure 6-12). Find her speed at the bottom of the hill.

We have already solved this problem in Section 6-2. We repeat it here to illustrate the use of the law of conservation of energy. The only force that does work is the force of gravity, which is conservative. The normal force exerted by the snow on the skis acts to constrain the skier to follow a certain

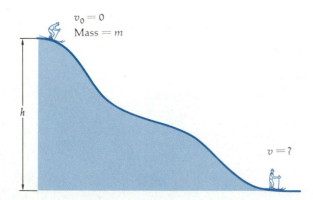

$v_0 = 0$
Mass $= m$

h

$v = ?$

Figure 6-12
Skier on a frictionless hill for Example 6-6. The speed at the bottom can be found from the conservation of energy without any knowledge of the shape of the hill.

path, but it does no work. Taking the potential energy to be zero at the bottom of the hill and assuming the initial speed at the top of the hill to be zero, we obtain from the conservation of energy

$$(E_k + U)_{top} = (E_k + U)_{bottom}$$

$$0 + mgh = \tfrac{1}{2}mv^2 + 0$$

or

$$v = \sqrt{2gh}$$

Note that we could not have solved this problem using Newton's second law because we cannot compute the acceleration at each point without knowing the shape of the curve. This example demonstrates the great power of the principle of conservation of energy. Even though we do not know the acceleration at any point, we can find the final speed quite easily.

Example 6-7   A pendulum consists of a bob of mass m attached to a string of length L. The bob is pulled aside so that the string makes an angle θ_0 with the vertical and is then released from rest (see Figure 6-13). Find (*a*) the speed v and (*b*) the tension T in the string at the bottom of the swing.

(*a*) The two forces acting on the bob (neglecting air resistance) are the force of gravity $m\mathbf{g}$, which is conservative, and the tension $\mathbf{T}$, which is perpendicular to the motion and therefore does no work. The mechanical energy of the bob is thus conserved in this problem.

Let us choose the gravitational potential energy to be zero at the bottom of the swing. At its original position, a height h above the bottom, the bob has potential energy mgh and no kinetic energy because it is at rest. As it swings down, the potential energy is converted into kinetic energy. At the bottom of the swing, the conservation of energy gives us

$$\tfrac{1}{2}mv^2 = mgh$$

We now must find h in terms of the initial angle θ_0. According to Figure 6-13, the distance h is related to θ_0 and the length of the pendulum L by

$$h = L - L \cos \theta_0 = L(1 - \cos \theta_0)$$

Thus, the speed at the bottom is

$$v = \sqrt{2gL(1 - \cos \ \theta_0)}$$

(*b*) To find the tension at the bottom, we use Newton's second law. (This is one of the things that makes physics challenging. We must apply everything we learned in previous chapters as well as the material in this chapter.) At the bottom of the swing, the forces on the bob are the weight $m\mathbf{g}$ acting downward and the tension $\mathbf{T}$ acting upward. Since the bob is moving in a circle of radius L with speed v, it has a centripetal acceleration of v^2/L. This acceleration is directed toward the center of the circle, which is upward at this point. Newton's second law thus gives

$$T - mg = \frac{mv^2}{L} = 2mg(1 - \cos \theta_0)$$

$$T = mg + 2mg(1 - \cos \theta_0)$$

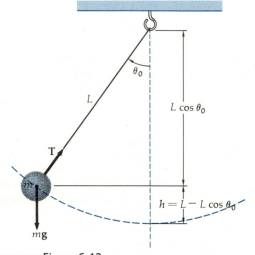

Figure 6-13
Simple pendulum for Example 6-7. The tension is perpendicular to the motion and does no work. The speed of the bob at the bottom is found from the conservation of energy, $\tfrac{1}{2}mv^2 = mgh$, where the initial height h above the bottom is related to the initial angle by $h = L - L \cos \theta_0$.

Multiflash photograph of a simple pendulum. As the bob descends, gravitational potential energy is converted into kinetic energy and the speed increases, as indicated by the increased spacing of the recorded positions. The speed decreases as the bob moves up, and the kinetic energy is changed into potential energy.

We note that if the bob is released from $\theta_0 = 90°$, the tension at the bottom is three times the force of gravity mg.

The speed of the bob at the bottom could also be found using Newton's laws, but the solution is difficult and requires calculus because the acceleration tangential to the curve varies with the angle θ and therefore with time so the constant acceleration formulas do not apply.

Example 6-8   A 2-kg block is pushed against a spring of force constant 500 N/m, compressing it 20 cm. It is then released, and the spring projects the block along a frictionless horizontal surface and then up a frictionless incline of angle 45°, as shown in Figure 6-14. (*a*) What is the speed of the block when it leaves the spring? (*b*) How far up the incline does the block travel?

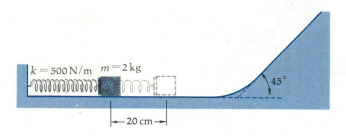

Figure 6-14
Example 6-8.

(*a*) After the spring is released, the only forces that do work are the forces exerted by the spring and the force of gravity. Since both these forces are conservative, the total mechanical energy of the block–spring–earth system is conserved. In this case, the total mechanical energy consists of the kinetic energy of the block $\frac{1}{2}mv^2$, the potential energy of the spring $\frac{1}{2}kx^2$, and the gravitational potential energy mgh.

When the block leaves the spring, its kinetic energy equals the original potential energy of the spring.

$$E_k = \tfrac{1}{2}mv^2 = \tfrac{1}{2}kx^2$$

$$= \tfrac{1}{2}(500 \text{ N/m})(0.2 \text{ m})^2 = 10 \text{ J}$$

The speed is then

$$v = \sqrt{\frac{2E_k}{m}} = \sqrt{\frac{2(10 \text{ J})}{2 \text{ kg}}} = 3.16 \text{ m/s}$$

(b) The block slides up the incline until its original kinetic energy is completely converted into gravitational potential energy. The height it reaches is given by

$$mgh = \tfrac{1}{2}mv^2 = 10 \text{ J} \qquad \text{or} \qquad h = \frac{10 \text{ J}}{(2 \text{ kg})(9.81 \text{ m/s}^2)} = 0.51 \text{ m}$$

The distance up the incline is found from $\cos 45° = h/s = 0.707$, giving $s = 0.721$ m.

Example 6-9   A child of mass 20 kg slides down a rough slide inclined at 30° (see Figure 6-15). The coefficient of friction between the child and the slide is 0.2. If the child starts from rest at the top of the slide a height 4 m above the bottom, how fast is she traveling when she reaches the bottom?

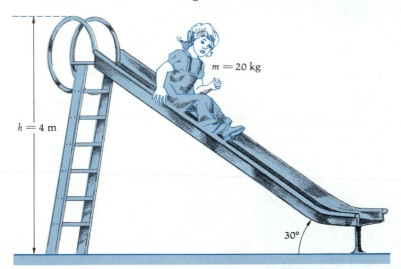

m = 20 kg

h = 4 m

30°

Figure 6-15
Child on slide for Example 6-9.

This problem could be readily solved by applying Newton's second law to the child and then finding v from the constant-acceleration formulas. We use the work-energy theorem instead to illustrate this alternative method of analyzing a problem. The forces acting on the child are gravity, friction, and the normal force exerted by the slide. Since the normal force is perpendicular to the motion, it does no work. The initial mechanical energy of the child is her potential energy, which is

$$U_1 = mgh_1 = (20 \text{ kg})(9.81 \text{ N/kg})(4 \text{ m}) = 785 \text{ J}$$

The final energy when she reaches the bottom is her kinetic energy $\tfrac{1}{2}mv^2$. The work done by the nonconservative force of friction is the force μF_n times the distance traveled s, where $F_n = mg \cos 30°$ is the normal force exerted by the slide. Since the original height h is 4 m and $\sin 30° = h/s = 0.5$, the distance s is 8 m. The work done by friction is thus

$$W_{nc} = -fs = -(\mu mg \cos 30°)(s)$$

$$= -(0.2)(20 \text{ kg})(9.81 \text{ N/kg})(0.866)(8 \text{ m}) = -272 \text{ J}$$

The work done by friction is negative because the force is opposite the direction of motion. According to the work-energy theorem, the work done by friction, -272 J, equals the change in the total mechanical energy.

$$W_{nc} = \Delta(E_k + U) = (E_{k2} + mgh_2) - (E_{k1} + mgh_1)$$

$$= (\tfrac{1}{2}mv^2 + 0) - (0 + 785 \text{ J}) = -272 \text{ J}$$

or

$$\tfrac{1}{2}mv^2 = 785 \text{ J} - 272 \text{ J} = 513 \text{ J}$$

$$v = \sqrt{\frac{2E_k}{m}} = \sqrt{\frac{2(513 \text{ J})}{20 \text{ kg}}} = 7.16 \text{ m/s}$$

The total mechanical energy is thus decreased from its original value of 785 J to its final value $785 \text{ J} - 272 \text{ J} = 513 \text{ J}$, which equals the kinetic energy $\tfrac{1}{2}mv^2$ at the bottom.

Example 6-10   Two bodies of mass m_1 and m_2 are attached to a light string that passes over a light, frictionless pulley, as shown in Figure 6-16. Find the speed of either body when the heavier one falls a distance h.

This device, called *Atwood's machine,* was developed in the eighteenth century to measure the acceleration of gravity g. As we shall see, if the values of m_1 and m_2 are not too different, the acceleration of either body is a small fraction of g. This acceleration could be easily measured with the rather crude timing devices available in the eighteenth century, whereas a direct measurement of g was difficult if not impossible.

If the friction in the pulley bearings is negligible and the pulley mass is so small that the work done to speed it up is negligible, the tension in the string is uniform. Then, the net work done by the tension is zero because when the lighter body moves upward in the direction of **T**, the heavier body moves downward the same distance in the opposite direction. The only force that does net work on the two-body system is the force of gravity. Therefore, the total mechanical energy of the system is conserved.

Let us assume that m_2 is greater than m_1 and choose the potential energy of the system to be zero at the initial position of the masses. Since they are at rest at this position, the total energy of the system is zero. Calling this initial energy E_i, we have

$$E_i = 0$$

Let v be the speed of m_1 after it has moved up a distance h. Then its kinetic energy is $\tfrac{1}{2}m_1v^2$, and its potential energy is m_1gh. Since the connecting string does not stretch, m_2 must move down the same distance h and acquire the same speed v. Its kinetic energy is $\tfrac{1}{2}m_2v^2$, and its potential energy is $-m_2gh$. The total energy at this time E_f is thus

$$E_f = \tfrac{1}{2}m_1v^2 + m_1gh + \tfrac{1}{2}m_2v^2 - m_2gh$$

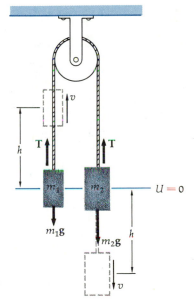

Figure 6-16
Atwood's machine for Example 6-10.

Setting the final total energy equal to the initial total energy, we obtain

$$\tfrac{1}{2}m_1v^2 + m_1gh + \tfrac{1}{2}m_2v^2 - m_2gh = 0$$

or

$$\tfrac{1}{2}(m_1 + m_2)v^2 = (m_2 - m_1)gh$$

Note that the left side of this equation is the kinetic energy gained, and the right side is the net potential energy lost by the two masses. Solving for v^2 gives

$$v^2 = \frac{2(m_2 - m_1)}{m_1 + m_2}gh \qquad\qquad 6\text{-}17$$

which relates the speed of either mass to the distance it moves. Comparing this with the constant-acceleration equation $v^2 = 2ah$, we see that the acceleration is given by

$$a = \frac{(m_2 - m_1)}{m_1 + m_2}g \qquad\qquad 6\text{-}18$$

This problem can also be solved by applying Newton's second law to each of the bodies and eliminating the tension T from the two equations obtained. In the laboratory experiment, the masses are chosen to be nearly equal so that the acceleration a is small and can be easily measured. Then the unknown value of g is obtained from $g = a(m_1 + m_2)/(m_2 - m_1)$.

The waterfall in this 1961 lithograph by the Swiss artist M. C. Escher violates the law of the conservation of energy. When the water falls, part of its potential energy is converted into the kinetic energy of the waterwheel. How then does the water get back to the top of the waterfall?

6-5   Simple Machines

A simple machine is a device for converting a small input force into a large output force by causing the input force to move through a much larger distance than the output force. This type of device is often used to lift a heavy weight when the force available is much smaller than the weight. Examples of simple machines are levers, inclined planes, and pulley systems. The *mechanical advantage* of a simple machine is defined as the ratio of the output force to the input force

$$MA = \frac{\text{force out}}{\text{force in}} \qquad\qquad 6\text{-}19 \qquad \text{Mechanical advantage}$$

We will first consider an inclined plane. Suppose we have a large weight mg that we wish to lift to a height h, but the force available F_{in} is much smaller than mg. We can lift the weight by putting it on rollers to reduce friction and then pushing or pulling it up an inclined plane (see Figure 6-17). If the angle of the incline is θ, the force we need to exert to push the weight up the incline is $mg \sin \theta$, which balances the component of the gravitational force acting down the incline. If S is the length of the incline, the work we must do to push the weight through this distance is $W = F_{\text{in}}S = (mg \sin \theta)S = mgh$. Here we are neglecting friction so that the input work FS equals the output work mgh, which is stored as potential energy. The mechanical advantage of the inclined plane is

$$MA = \frac{mg}{F_{\text{in}}} = \frac{mg}{mg \sin \theta} = \frac{1}{\sin \theta}$$

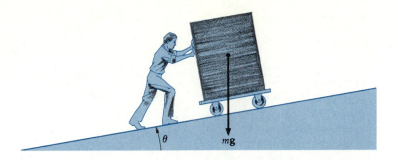

Figure 6-17
A heavy weight *mg* can be lifted by a small force by putting the weight on rollers to reduce friction and then pushing it up an incline. The force needed is then only *mg* sin θ.

We can make the mechanical advantage of the incline large by making the angle of the incline small so that sin θ is small. For example, if the weight is 1000 N (about 445 pounds), we can lift this weight with a small force of 100 N (44.5 pounds) if the mechanical advantage is 10, which occurs when sin $\theta = 0.1$ or $\theta = 5.7°$.

It is important to realize that a simple machine does not multiply work. In the best case, when there is no friction or other source of mechanical energy loss, the work out equals the work in. In actual practice, there is always some dissipative force present, and so the work out is always somewhat less than the work in. In the case of the inclined plane, if there is friction, the force needed to push the weight up the incline is greater than *mg* sin θ and the mechanical advantage is therefore less than $1/\sin \theta$. We can define an *ideal simple machine* as one in which there are no energy losses so that the work in equals the work out. If F_{in} is the input force, S_{in} is the distance through which it moves, F_{out} is the output force, and S_{out} is the distance though which it moves, we have for an ideal simple machine:

$$F_{in}S_{in} = F_{out}S_{out}$$

or

$$MA_{ideal} = \frac{F_{out}}{F_{in}} = \frac{S_{in}}{S_{out}} \qquad \text{6-20}$$

Ideal mechanical advantage

Figure 6-18 shows another simple machine, the lever. The lever is pivoted so that the lever arm of the applied or input force is larger than that of the output force. Let L_{in} and L_{out} be these lever arms. If we wish to lift a weight without acceleration, the input torque must equal the output torque:

$$F_{in}L_{in} = F_{out}L_{out}$$

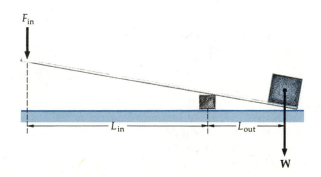

Figure 6-18
A lever can also be used to lift a heavy weight. Here the force needed is found from $F_{in}L_{in} = WL_{out}$.

The ideal mechanical advantage of an ideal lever is thus

$$MA_{ideal} = \frac{F_{out}}{F_{in}} = \frac{L_{in}}{L_{out}}$$

We note that this result is the same as Equation 6-20 since the distances moved by the input and output forces are proportional to the respective lever arms.

Figure 6-19 shows still another example of a simple machine, a pulley. A rope passes under a movable pulley to which a weight mg is attached. The rope then passes over a fixed pulley, which does not affect the mechanical advantage of the machine but merely changes the direction of the necessary applied force for convenience. We note that when the rope is pulled through a distance S_{in}, the movable pulley and therefore the weight moves up through half this distance. The ideal mechanical advantage of this machine is thus 2. More complicated pulley systems can be constructed using several pulleys. To find the ideal mechanical advantage of such a system, one needs only to figure out the ratio of the distance through which the applied force moves to the distance moved by the weight.

"There goes Archimedes with his confounded lever again."

Figure 6-19
A pulley with two lengths of rope supporting the weight has a mechanical advantage of 2 because the applied force F_{in} moves twice as far as the weight.

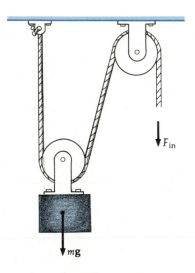

Example 6-11 Find the ideal mechanical advantage of the wheel and axle, another type of simple machine, shown in Figure 6-20. The radius of the axle or inner wheel is a and that of the outer wheel is b.

When the wheel turns though one revolution, the force **F** moves through a distance equal to the circumference of the outer wheel, $S_{in} = 2\pi b$. At the same time, the weight moves up through a distance equal to the circumference of the inner wheel, $S_{out} = 2\pi a$. The ratio of these distances equals the ideal mechanical advantage:

$$MA_{ideal} = \frac{2\pi b}{2\pi a} = \frac{b}{a}$$

The ideal mechanical advantage is thus the ratio of the radii of the wheels b/a. Note that this is the ratio of the weight to the input force when the torques produced by each about the center of the wheel are equal.

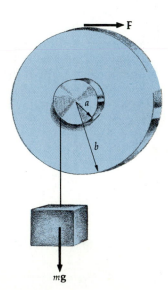

Figure 6-20
A small force F applied to the rim of the wheel can lift a large weight mg hung from the axle.

6-6   Power

When you say that someone is powerful, you probably are referring to that person's strength (or if it is in a political context, to that person's influence). In science, *power* is the *rate* at which work is done. A powerful car is one that can accelerate to a high speed quickly; that is, a powerful engine can increase the kinetic energy of the car in a short time. If W is the work done by some force in time t, the power input of the force is

$$P = \frac{W}{t}$$

6-21   Power defined

Suppose that a force $\mathbf{F}$ acts on a body that moves a distance Δs in some time interval Δt. The work done by the force is $F_s \, \Delta s$ where F_s is the component of the force in the direction of motion. The power input is the rate at which this work is done

$$P = \frac{F_s \, \Delta s}{\Delta t} = F_s \frac{\Delta s}{\Delta t} = F_s v$$

6-22

where $v = \Delta s/\Delta t$ is the speed of the body.

The SI unit of power, one joule per second, is called a watt (W).

$$1 \text{ watt} = \frac{1 \text{ joule}}{1 \text{ second}}$$

6-23   Watt defined

What you buy from a "power company" is not power but energy. You are usually charged by the kilowatt-hour (kW·h). A kilowatt-hour of energy is

$$1 \text{ kW·h} = (10^3 \text{ W}) \times (3600 \text{ s})$$

$$= 3.6 \times 10^6 \text{ W·s} = 3.6 \text{ MJ}$$

6-24

In the U.S. customary system, the unit for energy is the foot-pound, and the unit for power is the foot-pound per second. A common multiple of this unit is called a horsepower (hp):

$$1 \text{ hp} = 550 \text{ ft·lb/s} = 746 \text{ W}$$

Example 6-12   A small motor is used to power a lift that raises a load of bricks weighing 800 N to a height of 10 m in 20 s. What is the minimum power motor needed?

Assuming that the bricks are lifted without acceleration, the upward force equals the force of gravity 800 N. The speed of the bricks is 10 m/20 s = 0.5 m/s. Since the velocity and applied force are in the same direction, the power input of the force is

$$P = Fv$$

$$= (800 \text{ N})(0.5 \text{ m/s}) = 400 \text{ N·m/s}$$

$$= 400 \text{ J/s} = 400 \text{ W}$$

If there are no energy losses, to frictional forces, for example, the motor must have a power output of at least 400 W or approximately $\frac{1}{2}$ horsepower.

Energy Resources

Laurent Hodges
Iowa State University

One of the most notable characteristics of modern society is its extreme reliance on energy resources. Coal is used to extract metals from their ores, petroleum fuels are used for ground and air transportation, natural gas or fuel oil is used for heating buildings, and electricity is used for industrial machinery and home appliances.

Primary energy resources, that is, resources from the environment that are the ultimate sources of our energy, include the fossil fuels—coal, oil, and gas—and three types of electricity—hydroelectricity, which is generated by falling water; electricity from nuclear power plants, which ultimately comes from the energy of fission of ^{235}U; and electricity generated by geothermal power, which is the energy of the steam and hot water that originate beneath the earth. These are really only the so-called *commercial* primary energy resources. Other primary resources such as wood, peat, and cow or camel dung that are absent from international trade account for perhaps only 3 percent of world energy consumption, although they may represent up to about half the energy consumption in certain developing countries.

The energy we actually use is often in a modified form, referred to as a *secondary energy resource.* For example, about 80 percent of the electricity used in the United States is generated by the combustion of fossil fuels; in this case the fossil fuel is the primary energy resource and the electricity is the secondary resource.

Figure 1 shows the consumption of commercial energy resources by the United States and the world in the years 1950 and 1980. During that period, coal declined in relative importance while oil and gas grew in popularity because of their greater convenience to transport and use and their lesser negative impact on the environment. Nuclear power was first used during that period but still makes only a small contribution. In the early 1950s, the United States was producing approximately as much energy as it consumed, but by 1980, it was importing about 15 percent of the energy it consumed, and most of these imports were from developing countries.

The fossil fuels that constitute the bulk of the energy resources consumed are finite and nonrenewable, and so cannot meet the world's energy needs forever. Although the recoverable reserves of these fuels can be estimated only imperfectly and future energy demands are uncertain, it seems clear that oil and gas can continue to be a major resource for only a few more decades at most and coal only a few centuries. Nuclear energy, in the form of ^{238}U or ^{232}Th used in breeder reactors or deuterium or other light isotopes used in controlled thermonuclear fusion reactors (should they prove feasible), could supply projected energy needs for

much longer—thousands or even millions of years, assuming a leveling off of the growth of energy consumption.

In addition, there is increasing concern over the environmental costs of energy production and use: land devastation by the strip mining of coal, high acid concentrations in the streams of coal mining areas, particulate matter and sulfur oxide air pollution from coal combustion, water pollution from oil-field operations, air pollution from petroleum refining, sulfur oxide pollution from petroleum combustion, carbon monoxide and hydrocarbon air pollution from the use of gasoline in motor vehicles, nitrogen oxide air pollution from any high-temperature combustion process, radiation releases from uranium mining and from nuclear power plants, and so forth.

More extensive use of renewable energy resources may be necessary if the world is to have abundant and environmen-

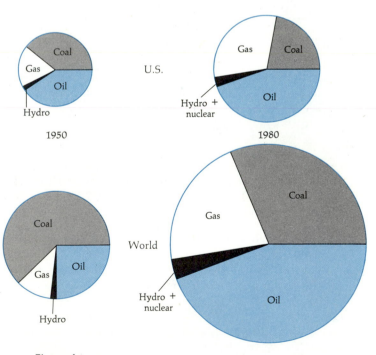

Figure 1

Commercial energy consumption in the United States and the world in 1950 and 1980.

tally tolerable energy. It is also especially advantageous if we are to conserve fossil fuels for their other valuable uses, such as the manufacture of chemicals and drugs. Renewable energy resources which are of negligible importance today that may make significant contributions to the energy supply in the future are solar energy, wind power, organic wastes, and tidal energy. Solar energy is especially abundant, though its low power density may make it necessary to collect solar radiation over large land areas.

Figure 2 shows the per capita energy consumption in 1980 in different parts of the world. As can be seen, U.S. per capita consumption is about 2 times that of the other developed countries (Canada, Europe, Australia, New Zealand, Japan, Israel, and Soviet Russia) and 25 to 30 times that of the developing countries. It is often stated that the United States cannot cut its energy consumption significantly without returning to some sort of "dark age." Actually, a drastic 50 percent cut would only put U.S. per capita energy consumption at the level of that in the other developed countries, where the standard of living is certainly decent and civilized. A less drastic but supposedly intolerable cut of 30 percent would return us to consumption levels of the early 1960s, which hardly qualifies as a dark age.

The increase in U.S. per capita consumption in the 1970s about equaled the total per capita consumption in developing countries. It would have been more equitable for the consumption increases to come in the developing countries, especially since they are providing most of the world's energy production increases.

Figure 3 shows a breakdown of energy consumption in the United States in the early 1980s by end uses. About 40 percent of U.S. energy consumption is by the industrial sector; 25 percent is by ground, air, and water transportation; 20 percent by residences; and 15 percent by commercial establishments (stores, offices, restaurants, etc.). All the small electrical appliances in use—clocks, radios, televisions, power tools, hair dryers—account for much less than 1 percent of total energy consumption in the United States.

Much of the energy used today is wasted. Energy conservation has become an important policy to implement. If we can use energy more efficiently, if we can devise new products with lower energy requirements (think of pocket calculators!), and if we can change our lifestyles to reduce energy consumption, we can slow the rate of depletion of our nonrenewable energy resources and simultaneously decrease the adverse environmental effects of energy production and use. Fortunately, improved efficiency in industry, homes, and transportation has led to a rather stable energy consumption level in the United States beginning in the 1970s.

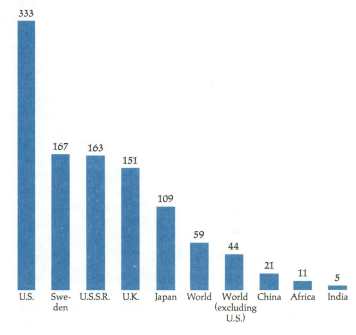

Figure 2

Per capita energy consumption (in 10^9 joules) in 1980 in different parts of the world.

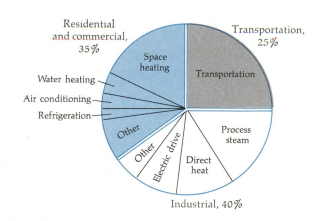

Figure 3

Approximate energy use in the United States as of the early 1980s.

6-7   Thermal Energy and Metabolic Rate

We have mentioned that the work done by nonconservative forces such as friction appears as heat or thermal energy. Before it was understood that heat was just another form of energy — energy that flows from one body to another because of a difference in temperature — separate units were defined for heat. The calorie (cal) was originally defined as the amount of heat needed to raise the temperature of one gram of water one Celsius degree. Similarly, the British thermal unit (Btu) was defined as the amount of heat needed to raise the temperature of one pound of water one Fahrenheit degree. In the 1840s, James Joule (1818–1889), after whom the SI unit of energy is named, showed that the appearance or disappearance of a given quantity of mechanical energy is always accompanied by the disappearance or appearance of an equivalent quantity of heat. Joule did a series of careful measurements of the amount of mechanical energy needed to raise one gram of water one Celsius degree and thereby determined the equivalence of the unit of heat and the unit of mechanical energy. Today the calorie and the Btu are defined in terms of the joule

$$1 \text{ cal} = 4.184 \text{ J} \qquad\qquad 6\text{-}25$$

$$1 \text{ Btu} = 1054 \text{ J} = 252 \text{ cal} \qquad\qquad 6\text{-}26$$

Calorie and Btu defined

We will discuss these experiments in more detail in Chapter 12. We mention them here only to introduce the calorie and the Btu and relate these familiar units to the joule, which is probably less familiar to you.

The calorie referred to by nutritionists in labeling the energy content of foods is actually 1000 calories or 1 kilocalorie (kcal). Our bodies obtain energy for heat and the performance of work from the chemical energy in the foods we eat. The rate at which we convert this chemical energy into heat and work is called our *metabolic rate*. The metabolic rate depends on such factors as one's weight, activity, efficiency, and so forth. A typical metabolic rate for a 70-kg person (about 150 lb) at rest is about 120 watts (W). We can use Equation 6-25 to convert from watts to kilocalories per hour or kilocalories per day:

$$1 \text{ W} = \frac{1 \text{ J}}{1 \text{ s}} \times \frac{1 \text{ kcal}}{4184 \text{ J}} \times \frac{3600 \text{ s}}{1 \text{ h}} = \frac{0.861 \text{ kcal}}{h}$$

If we multiply this result by 24 h/day, we obtain

$$1 \text{ W} = 20.65 \text{ kcal/day} \qquad\qquad 6\text{-}27$$

Thus, a metabolic rate of 120 W is equivalent to $120 \times 20.65 = 2478$ kcal/day, or about 2500 kcal/day. If one's average metabolic rate is 120 W, a food intake of 2500 kcal is required to maintain an even weight. To gain or lose weight, one needs to take in more or less food energy than one expends. Approximately 3500 kcal of food energy is equivalent to a change of 1 lb in body weight. However, as many people are aware, gaining or losing body weight is a tricky business. For example, studies have shown that when a person goes on a very low calorie diet to lose weight, the body automatically lowers its metabolic rate to compensate for the reduction in food energy intake. Then, not only is the weight loss slower than expected, but when food intake is increased, the lower metabolic rate results in a rapid weight gain.

The metabolic rate of a person engaged in some activity is usually measured by measuring the oxygen consumed in a given time interval (normally a few minutes). This is done by collecting the air exhaled and measuring the oxygen remaining. The oxygen consumed reacts with carbohydrates, fats, and proteins in the body, releasing about 20 kJ (4.8 kcal) of energy per litre (L) of oxygen consumed ($1 \text{ L} = 10^3 \text{ cm}^3 = 10^{-3} \text{ m}^3$).

Example 6-13   A 90-kg man hikes up a 1000 m mountain. If his body is 20 percent efficient in converting food energy into mechanical energy, how many kilocalories does he burn up? How many pounds of body weight does he lose, assuming 3500 kcal is equivalent to 1 lb?

The work done by the man's body in lifting his weight to a height h is

$$W = mgh = (90 \text{ kg})(9.81 \text{ N/kg})(1000 \text{ m})$$

$$= 8.83 \times 10^5 \text{ J}$$

We convert this to kilcalories using the relation 4184 J = 1 kcal.

$$8.83 \times 10^5 \text{ J} = 8.83 \times 10^5 \text{ J} \left(\frac{1 \text{ kcal}}{4184 \text{ J}} \right) = 211 \text{ kcal}$$

Since the man's body is only 20 percent efficient, he burns five times this amount of energy to produce the necessary work; that is, for every 5 kcal burned, only 1 kcal is converted into mechanical work. The energy burned is therefore

$$5 \times 211 \text{ kcal} = 1.06 \times 10^3 \approx 1000 \text{ kcal}$$

The amount of body weight lost in this exercise is

$$1000 \text{ kcal} \times \frac{1 \text{ lb}}{3500 \text{ kcal}} = 0.28 \text{ lb}$$

Example 6-14   A woman jogging on a treadmill consumes oxygen at a rate of 1.5 L/min. What is her metabolic rate while jogging?

Using 1 L oxygen = 20 kJ of energy, we have

$$\left(\frac{1.5 \text{ L}}{1 \text{ min}} \right) \left(\frac{20 \text{ kJ}}{1 \text{ L}} \right) \left(\frac{1 \text{ min}}{60 \text{ s}} \right) = 0.5 \text{ kJ/s} = 500 \text{ J/s} = 500 \text{ W}$$

Her metabolic rate is thus 500 W.

Summary

1. The work done by a constant force is the product of the component of the force in the direction of motion and the displacement, which can be represented graphically as the area under the F_x-versus-x curve.

2. The SI unit of work and energy is the joule, which equals the product of a newton and a metre.

3. Work done by a variable force equals the area under the force-versus-displacement curve

4. Kinetic energy is the energy associated with the motion of an object and is related to its mass and speed by

$$E_k = \tfrac{1}{2}mv^2$$

5. The work done by the resultant force equals the change in the kinetic energy of the object.

6. Potential energy is the energy associated with the position of an object. The gravitational potential energy of an object of mass m at a height h above some reference point is

$$U_g = mgh$$

The absolute value of potential energy is unimportant because the choice of the reference point for measuring h is arbitrary. Only changes in U are important.

The potential energy of a spring of force constant k when stretched or compressed a distance x from equilibrium is given by

$$U_s = \tfrac{1}{2}kx^2$$

7. If the work done on an object by a force depends only on the initial and final positions of the object, the force is said to be conservative. The force of gravity is an example of a conservative force. Associated with each conservative force is a potential-energy function. The work done on an object by a conservative force equals the *decrease* in the potential energy of the object.

8. If only conservative forces do work on an object, the sum of the kinetic and potential energies of the object remains constant. This is known as the law of conservation of mechanical energy.

9. Nonconservative forces such as friction dissipate mechanical energy into thermal energy. When both conservative and nonconservative forces do work on an object, the work done by the nonconservative force equals the change in the total mechanical energy of the object. This is called the generalized work-energy theorem,

$$W_{nc} = \Delta E_{total}$$

where

$$E_{total} = \tfrac{1}{2}mv^2 + U$$

10. A simple machine is a device for converting a small input force to a large output force. The mechanical advantage of a simple machine is the ratio of the output force to the input force. For an ideal machine (no dissipative forces), the work in equals the work out and the mechanical advantage also equals the ratio of the input distance to the output distance. In practice, there are always dissipative forces such as friction, and the work out is always somewhat less than the work in. Examples of simple machines are inclined planes, levers, pulley systems, and the wheel and axel.

11. Power is the rate of doing work

$$P = \frac{W}{t} = F_s v$$

where F_s is the component of the force in the direction of motion and v is the speed. The SI unit of power is the watt (W), which equals one joule per second. A common energy unit is the kilowatt-hour (kW·h), which equals 3.6 megajoules (MJ).

12. Units of energy often used in describing thermal processes are the calorie (cal), which equals 4.184 joules, and the kilocalorie (kcal), which equals 4184 joules.

13. The rate at which our bodies convert chemical energy derived from food into mechanical and thermal energy is our metabolic rate. A typical metabolic rate is about 120 W, which is equivalent to about 2500 kcal/day.

Suggestions for Further Reading

Dyson, Freeman J.: "Energy in the Universe," *Scientific American*, September 1971, p. 50.

This article introduces a Scientific American *special issue on energy by examining the different forms in which energy manifests itself and how it is converted from one form to another.*

Gosz, James R., Richard T. Holmes, Gene E. Likens, and F. Herbert Bermann: "The Flow of Energy in a Forest Ecosystem," *Scientific American*, March 1978, p. 92.

The various ways in which energy from the sun is transferred among living things in a forest, and how efficiently they use it are examined.

Heinrich, Bernd: "The Energetics of the Bumblebee," *Scientific American*, April 1973, p. 96.

Kingsolver, Joel G.: "Butterfly Engineering," *Scientific American*, August 1985, p. 106.

These two articles describe how bumblebees and butterflies metabolize nectar and absorb thermal energy from the sun to achieve body temperatures high enough to enable them to fly.

Walker, Jearl: "The Amateur Scientist: Fly Casting Illuminates the Physics of Fishing," *Scientific American*, July 1985, p. 122.

The conservation of kinetic energy helps to explain how it is possible to cast a fly so far; a consideration of drag makes the model more realistic; and the forces involved in fighting a fish once it is on the line are analyzed.

Review

A. Objectives: After studying this chapter, you should:

1. Know the definitions of work, kinetic energy, potential energy, and power.

2. Be able to distinguish between conservative and nonconservative forces and know the criterion for a force to be conservative.

3. Be able to state the law of conservation of energy and use it in solving problems.

4. Be able to state the work-energy theorem and use it in solving problems.

5. Be able to find the mechanical advantage of a simple machine.

B. Define, explain, or otherwise identify:

Work, p. 114
Joule, p. 114
Foot-pound, p. 114
Kinetic energy, p. 115
Potential energy, pp. 123–124
Nonconservative force, p. 126

Conservative force, p. 126
Total mechanical energy, p. 126
Work-energy theorem, p. 127
Mechanical advantage, p. 132
Power, p. 135
Watt, p. 135
Kilowatt-hour, p. 135
Horsepower, p. 135
Calorie, p. 138
Metabolic rate, p. 138

C. True or false: If the statement is true, explain why. If it is false, give a counterexample.

1. Only the resultant force acting on an object can do work.

2. No work is done on a particle that remains at rest.

3. Work is the area under the force-versus-time curve.

4. A force that is always perpendicular to the velocity of a particle does no work on the particle.

5. A kilowatt-hour is a unit of power.

6. Only conservative forces can do work.

7. If only conservative forces act, the kinetic energy of a particle does not change.

8. The work done by a conservative force decreases the potential energy associated with that force.

Exercises

Section 6-1   Work and Kinetic Energy

1. A 5-kg body is raised at constant velocity a distance of 10 m by a force F_0. Find the work done on the body (a) by the force F_0 and (b) by the earth. (c) What is the net work done by all the forces acting?

2. A 10-kg box rests on a horizontal table. The coefficient of friction between the box and table is 0.4. A force F_x pulls the box at constant velocity a distance of 5 m. Find the work done (a) by F_x and (b) by friction.

3. The box described in Exercise 2 is pulled by a force F_x so that the box has an acceleration of 2 m/s². (a) Find the work done by F_x while the box is pulled a distance of 5 m. (b) What is the work done by the frictional force in this case ($\mu_k = 0.4$)?

4. (a) From the conversion factors for newtons to pounds and metres to feet, derive the relation between the joule and the foot-pound expressed in Equation 6-3. (b) The cgs unit of work is the erg, which is defined as a dyne-centimetre. Find the number of ergs in a joule.

5. A 10-g bullet has a speed of 1.2 km/s. (a) What is its kinetic energy in joules? (b) If the speed is halved, what is its kinetic energy?

6. A 0.145-kg baseball moves with speed 40 m/s. What is its kinetic energy in joules?

7. A 60-kg jogger runs at a steady nine-minute-per-mile pace. What is her kinetic energy?

8. A 5-kg sled, initially at rest on a level snow-covered road, is pulled by a constant force of 10 N for a distance of 6 m. The force makes an angle of 20° with the horizontal, and friction between the sled and the snow is negligible. Find (a) the final kinetic energy and (b) the final speed of the sled.

9. A 5-kg mass is raised a distance of 4 m by a vertical force of 80 N. Find (a) the work done by the force and the work done by the earth and (b) the final kinetic energy of the mass if it was originally at rest.

10. An 8-kg sled is initially at rest on a horizontal road. The coefficient of kinetic friction between the sled and the road is 0.4. The sled is pulled a distance of 3 m by a horizontally applied force of 50 N. (a) Find the work done by the applied force. (b) Find the work done by friction. (c) Find the change in the kinetic energy of the sled. (d) Find the speed of the sled after it has traveled 3 m.

11. A 4-kg object is lifted by a rope exerting a force T so that it accelerates upward at 6 m/s². The object is lifted 3 m. Find (a) the tension T, (b) the work done by T, (c) the work done by the earth, and (d) the final kinetic energy of the object after being lifted 3 m if it was originally at rest.

12. A 2000-kg car moving with initial speed of 25 m/s is stopped in 60 m by a constant frictional force. (a) What is the kinetic energy of the car? (b) How much work is done by the frictional force in stopping the car? (c) What is the coefficient of friction?

13. A block of mass 6 kg slides down a rough incline starting from rest. The coefficient of kinetic friction is 0.2, and the angle of incline is 60°. (a) List all the forces acting on the block, and find the work done by each force when the block slides 2 m (measured along the incline). (b) What is the net work done on the block? (c) What is the velocity of the block after it has slid 2 m?

14. A 100-kg cart is lifted up a 1-m step by rolling the cart up an incline formed when a plank of length L is laid from the lower level to the top of the step. (Assume the rolling is equivalent to sliding without friction.) (a) Find the force parallel to the incline needed to push the cart up without acceleration for values of L of 3, 4, and 5 m. (b) Calculate directly from Equation 6-1 the work needed to push the cart up the incline for each of these values of L. (c) Since the work found in (b) is the same for each value of L, what advantage, if any, is there in choosing one length over another?

Section 6-2   Work Done by a Variable Force

15. A 2-kg particle is moving with speed 3 m/s when it is at $x = 0$. It is subjected to a single force F_x, which varies with position x as shown in Figure 6-21. (a) What is the kinetic energy of the particle when it is at $x = 0$? (b) How much work is done by the force as the particle moves from $x = 0$ to $x = 4$ m? (c) What is the speed of the particle when it is at $x = 4$ m?

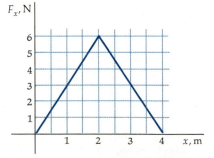

Figure 6-21
Exercise 15.

16. A 4-kg particle is initially at rest at $x = 0$. It is subjected to a single force F_x, which varies with position x as shown in Figure 6-22. (a) Find the work done by the force as the particle moves from $x = 0$ to $x = 3$ m and (b) as it moves from

$x = 3$ to $x = 6$ m. Find the kinetic energy of the particle when it is at (c) $x = 3$ m and (d) $x = 6$ m.

Figure 6-22
Exercise 16.

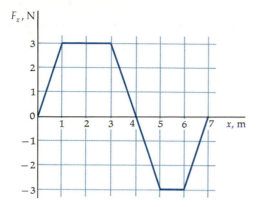

Section 6-3   Potential Energy; Section 6-4
Conservative Forces and the Conservation of Energy

17. A man of mass 80 kg climbs up a stairway of height 5 m. (a) What is the increase in the potential energy of the man? (b) What is the least amount of work the man must do to climb the stairway? (c) Explain why the work done by the man may be more than this minimum amount.

18. State which of the following forces are conservative and which are nonconservative: (a) the frictional force exerted on a sliding box, (b) the force exerted by a spring obeying Hooke's law, (c) the force of gravity, (d) the wind resistance on a moving car, (e) the force exerted by a boy on a ball when he throws it, and (f) the force exerted by a floor on a dropped egg.

19. A 2-kg book is held at a height of 20 m above the ground and is then released at $t = 0$. (a) What is the initial potential energy of the book relative to the ground? (b) From Newton's laws, find the distance the book falls in 1 s and its speed at $t = 1$ s. (c) Find the potential energy, the kinetic energy, and the total mechanical energy of the book at $t = 1$ s. (d) Find the kinetic energy and the speed of the book just before it hits the ground.

20. A 2-kg box slides along a long frictionless incline of angle $30°$. It starts from rest at time $t = 0$ at the top of the incline, which is 20 m above the ground. (a) What is the initial potential energy of the box relative to the ground? (b) From Newton's laws, find the distance the body travels in 1 s and its speed at $t = 1$ s. (c) Find the potential energy, the kinetic energy, and the total mechanical energy of the box at $t = 1$ s. (d) Find the kinetic energy and the speed of the box just as it reaches the bottom of the incline.

21. How high must a body be lifted to gain an amount of potential energy equal to the kinetic energy it has when moving at a speed of 20 m/s?

22. A 0.5-kg ball is thrown straight up with an initial speed of 10 m/s. (a) Find the initial kinetic energy of the ball and its

total mechanical energy, assuming that the initial potential energy is zero. (b) Find the kinetic energy, potential energy, and total mechanical energy of the ball at the top of its flight. (c) Use the law of conservation of total mechanical energy to find the maximum height of the ball. (Take $g = 10$ m/s² for convenience.) (d) Compare this method of finding the maximum height of the ball with using the constant-acceleration formulas from Chapter 2. Can you find the time needed for the ball to reach its maximum height from the conservation of mechanical energy?

23. A man places a 2-kg block against a horizontal spring with a force constant $k = 300$ N/m and compresses it 9 cm. (a) Find the work done by the man and the work done by the spring. (b) The spring is released from its compression of 9 cm and returns to its original position. Find the work done by the spring and the kinetic energy of the block at this point (assuming that only the spring does work on the block).

24. A spring with a force constant $k = 10^4$ N/m obeys Hooke's law. How far must it be stretched for its potential energy to be (a) 100 J and (b) 50 J?

25. A 2-kg block slides down a frictionless curved ramp from rest at a height of 3 m (see Figure 6-23). It then slides 9 m on a rough horizontal surface before coming to rest. (a) What is the speed of the block at the bottom of the ramp? (b) How much work is done by friction on the block? (c) What is the coefficient of friction between the block and the horizontal surface?

Figure 6-23
Exercise 25.

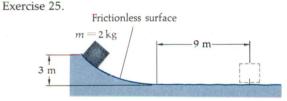

26. A 3-kg object is released from rest at a height of 5 m on a curved frictionless ramp. At the foot of the ramp is a spring of force constant $k = 400$ N/m (see Figure 6-24). The object slides down the ramp and onto the spring, compressing it a distance x before coming momentarily to rest. (a) Find x. (b) What happens to the object after it comes to rest?

Figure 6-24
Exercise 26.

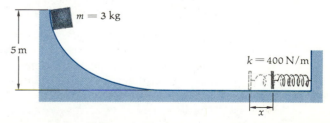

27. For the Atwood's machine shown in Figure 6-25, if the system starts from rest when the lower string is cut, find the speed of the objects when they are at the same height.

28. A particle slides along the frictionless track shown in Figure 6-26. Initially it is at point P, headed downhill with speed v. Describe the motion in as much detail as you can if (*a*) $v = 7$ m/s and (*b*) $v = 12$ m/s. (*c*) What is the least speed needed by the mass to get past point Q?

Figure 6-26
Exercise 28.

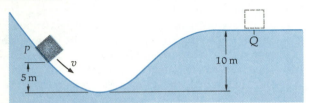

29. Find the speed of the 2-kg mass in Figure 6-27 after it has fallen a distance of 2 m from rest, assuming no friction.

Figure 6-27
Exercise 29.

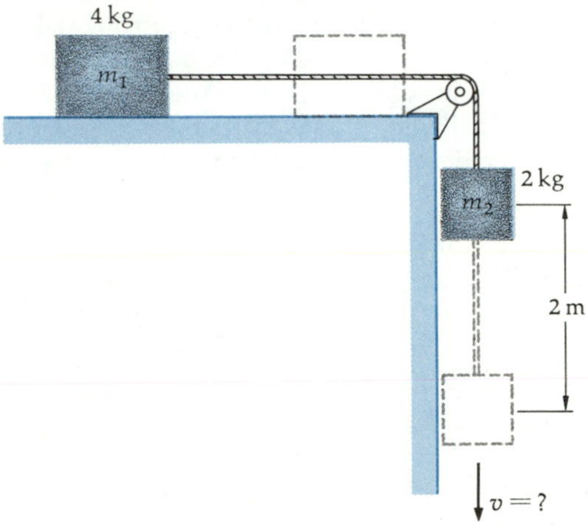

Figure 6-25
Exercise 27.

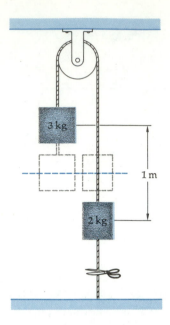

Section 6-5   Simple Machines

30. A large box on rollers weighing 500 N is to be lifted to a loading dock, which is a height of 3 m above the street. The maximum force available is 180 N. (*a*) What is the greatest angle of an incline that can be used to push the box up to the dock? (*b*) How long an incline is needed?

31. (*a*) What is the mechanical advantage of the nutcracker shown in Figure 6-28? (*b*) If a force of 50 N is needed to crack a tough nut, how much force must be applied to the handle?

32. When the handle on an automobile jack moves 0.6 m, the automobile moves upward 1.2 cm. (*a*) What is the (ideal) mechanical advantage of the jack? (*b*) If the automobile weighs 10,000 N, what force must be applied to the jack handle?

Section 6-6   Power

33. A 4-kg mass is lifted by a force equal to the weight of the mass. The mass moves upward at a constant velocity of 2 m/s. (*a*) What is the power input of the force? (*b*) How much work is done by the force in 2 s?

34. A constant horizontal force $F = 3$ N drags a box along a rough horizontal surface at a constant speed v. The force does work at the rate of 5 W. (*a*) What is the speed v? (*b*) How much work is done by F in 3 s?

35. A constant force of 4 N acts at an angle of 30° above the horizontal on a box of mass 2 kg resting on a rough horizontal table. The box is dragged with a constant speed of 50 cm/s. (*a*) Find the normal force exerted by the table on the box and the coefficient of friction. (*b*) What is the power input of the applied force? (*c*) How much work is done by the frictional force in 3 s?

36. An engine is rated at 400 horsepower. How much work is done when it operates at full power for 1 min? Express your answer in foot-pounds and in kilowatt-hours.

Figure 6-28
Exercise 31.

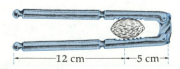

Section 6-7   Thermal Energy and Metabolic Rate

37. A 70-kg girl climbs 800 m on a hike. (*a*) If the efficiency of her muscles in converting internal chemical energy into work is 25 percent, how much energy does she use in the climb? (*b*) How much weight (in pounds) does she lose because of this exercise, assuming 1 lb lost for each 3500 kcal of energy expended?

38. An 80-kg man runs up some stairs of height 12 m in 14 s. (*a*) Calculate his power output in watts. (*b*) If his efficiency is 22 percent, what is the least his metabolic rate can be during this exercise?

39. A runner uses oxygen at 3 litres per minute. What is his metabolic rate?

40. The total resistance force acting on a bicyclist is 21 N. (*a*) If she travels at 5 m/s, what power is required? (*b*) If her efficiency is 26 percent, how many kilocalories does she burn up in 1 h of riding at this pace?

41. A 3.0-kg hawk flies to a height of 510 m above ground. (*a*) What is the hawk's gravitational potential energy at this height relative to the ground? (*b*) How many kilocalories of food energy are needed, assuming 100 percent efficiency? (*c*) If at this altitude, the hawk folded its wings and plummeted down, at what speed would it hit the ground, neglecting air resistance?

Problems

1. A 3-kg body is moving with speed 1.50 m/s in the x direction. When it passes the origin, it is acted on by a single force F_x, which varies with x as shown in Figure 6-29. (*a*) Find the work done by the force from $x = 0$ to $x = 2$ m. (*b*) What is the kinetic energy of the body at $x = 2$ m? (*c*) What is the speed of the body at $x = 2$ m? (*d*) Find the work done on the body from $x = 0$ to $x = 4$ m. (*e*) What is the speed of the body at $x = 4$ m?

Figure 6-29
Problem 1.

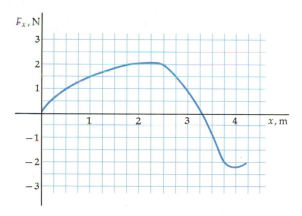

2. A 5-kg block is held against a spring of force constant 20 N/cm, compressing it 3 cm. The block is released and the spring expands, pushing the block along a rough horizontal surface. The coefficient of friction between the surface and the block is 0.2. (*a*) Find the work done on the block by the spring as it extends from its compressed position to its equilibrium position. (*b*) Find the work done on the block by friction while it moves the 3 cm to the equilibrium position of the spring. (*c*) What is the speed of the block when the spring reaches its equilibrium position? (*d*) If the block is not at-

tached to the spring, how far will it slide along the rough surface before coming to rest?

3. A cyclist is climbing a hill inclined at 6° to the horizontal at speed 2.8 m/s. If the frictional resistance force to the motion is 0.8 percent of the weight, and the mass of the cyclist plus bike is 85 kg, find the power delivered by the cyclist.

4. A T-bar tow is required to pull a maximum of 80 skiers up a 600-m slope inclined 15° to the horizontal at a speed of 2.5 m/s. The coefficient of friction is 0.06. Find the motor power required if the mass of the average skier is 75 kg.

5. A car of mass 1500 kg travels at 22 m/s up a hill 2.0 km long and rising 120 m. If the frictional effects are negligible, find (*a*) the work done on the car and (*b*) the average power delivered by the car motor.

6. A ski jumper of mass 70 kg starts from rest at A (see Figure 6-30). His speed is 30.0 m/s at B and 23.0 m/s at C, and the distance from B to C is 30 m. (*a*) How much work is done by friction on the skier as he moves from B to C? (*b*) Find the greatest height attained by the jumper above the level of C.

Figure 6-30
Problem 6.

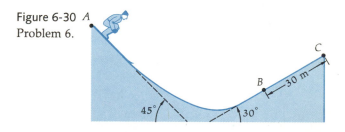

7. A train with a total mass of 300 Mg rises 707 m over a travel distance of 62 km at an average speed of 15.0 km/h. If the frictional force is 0.8 percent of the weight, (*a*) find the kinetic energy of the train, (b) the total change in potential energy, (c) the work done against friction, and (d) the power input of the engines of the train.

8. A child is swinging from a suspended rope 4.0 m long that will break if the tension becomes twice the weight of the child. (a) What is the greatest angle θ_0 the rope can make with the vertical during the swing if the rope is not to break? (b) What is the speed of the child when the rope breaks if the angle found in (a) is slightly exceeded?

9. An elastic string 25 cm long obeys Hooke's law, $F_x = -kx$, where x is the extension from equilibrium. When a 0.15-kg object is suspended from the string, the string stretches 5 cm. If the object is attached to the end of the string and is dropped from the point of support of the string, find the distance it falls before first coming to rest.

10. A small bead of mass m slides without friction along the loop-the-loop track shown in Figure 6-31. The bead starts from point P a distance h above the bottom of the loop. (a) What is the kinetic energy of the bead when it reaches the top of the loop? (b) What is its acceleration at the top of the loop assuming that it stays on the track? (c) What is the least value of h if the bead is to reach the top of the loop without leaving the track?

Figure 6-31
Problem 10.

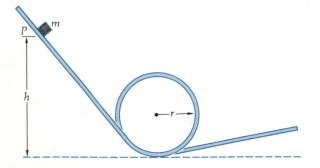

11. A particle of mass m moves in a horizontal circle of radius r on a rough table. It is attached to a string fixed at the center of the circle. The speed of the particle is initially v_0. After completing one full trip around the circle, the speed of the particle is $\frac{1}{2}v_0$. (a) Find the work done by friction during that 1 rev in terms of m, v_0, and r. (b) What is the coefficient of sliding friction? (c) How many more revolutions will the particle make before coming to rest?

12. A 2-kg box is projected with initial speed of 3 m/s up a rough plane inclined at 60° to the horizontal. The coefficient of kinetic friction is 0.3. (a) List all the forces acting on the box, and find the work done by each as the box slides up the plane. (b) How far does the box slide along the plane before it stops momentarily? (c) Find the work done by each force acting on the box as it slides back down the plane. (d) Find the speed of the box when it reaches its initial position.

13. A skier starts from rest at height h above the center of a rounded hummock of radius 4.0 m, as shown in Figure 6-32. There is negligible friction. Find the maximum value of h for which the skier will remain in contact with the snow at the peak of the hummock.

Figure 6-32
Problem 13.

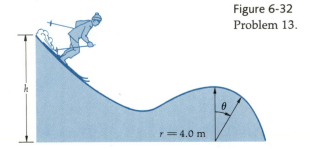

14. When you walk along a horizontal surface, you lift part of your weight a small amount with each step and therefore do work against the force of gravity. (a) Estimate the work done with each step and the total work you do in lifting yourself when you walk 1 km along a horizontal surface. (b) If you walk 1 km in 20 min, what is your power output?

15. A 0.015-kg ball is shot from a spring gun whose spring has a force constant of 600 N/m. The spring can be compressed 5 cm. (a) How high can the ball be shot if the gun is aimed vertically? (b) What is the greatest possible horizontal range for the ball fired from this compression?

16. In one model of running, it is assumed that the energy expended goes into accelerating and decelerating the legs. If the mass of the leg is m and the running speed is v, the energy to accelerate the leg from rest to v is $\frac{1}{2}mv^2$, and the same energy is needed to decelerate the leg back to rest for the next stride. Thus, mv^2 is required for each stride. Assume that the mass of a man's leg is 10 kg and he runs at 3 m/s with 1 m between one footfall and the next. In each second, then, he must provide $3 \times mv^2$ energy to his legs. Calculate the man's rate of energy expenditure using this model, assuming his muscles are 25 percent efficient.

CHAPTER SEVEN

Impulse, Momentum, and Center of Mass

Sliding headfirst is the safest way to get to the next base, I think. And the fastest. You don't lose your momentum.

PETE ROSE

Although Isaac Newton missed discovering the importance of energy, he knew the quantity dealt with in this chapter—momentum—very well. In fact, his second law states not that force is equal to mass times acceleration but that force is equal to the rate of change of momentum.

A slightly simpler quantity than energy, momentum explains the behavior of tennis balls and molecules and the astonishing trick by which an interplanetary space probe can use a near encounter with one planet to gain the energy to visit another even further away. Moreover, it is by combining momentum equations with energy equations in the interpretation of experimental data that atomic and nuclear physicists have made most of the famous discoveries about the nature of the atom.

When two objects collide such as a car and a tree or a baseball and a bat, they exert very large forces on each other for a very brief time. The force exerted by a bat on a baseball may be several thousand times the weight of the ball, but this enormous force is exerted for only a millisecond or so. Such forces are called *impulsive forces*. We will study these forces in this chapter and will introduce an important new quantity, *momentum*, which is the product of the mass of a particle and its velocity. Momentum is important because, as we will see, if there is no resultant external force acting on a system of particles, the total momentum of the system remains constant. It follows from this that the total momentum of the universe remains constant over time. Like energy, then, momentum is a conserved quantity in an isolated system. Momentum conservation is very useful in analyzing collisions between billiard balls, automobiles, or subatomic particles in a nuclear reaction; the motion of a jet plane or a rocket; and the recoil of a rifle. Momentum conservation is also the basis for the operation of the balistocardiograph, a device for monitoring the pumping action of the heart by measuring the recoil of the patient's body.

In this chapter, we will also study an important concept for the description of the motion of a complex system of particles—the center of mass. The center of mass is important because its acceleration equals the total external force acting on the system divided by its total mass. (In nearly all the situations we will be interested in, the center of mass of a system coincides with its center of gravity, which we studied in Chapter 5.) The motion of any system of particles, no matter how complex, can be described in terms of the motion of its center of mass (which may be thought of as the bulk motion of the system) plus the motion of the individual particles in the system relative to the center of mass.

7-1   Impulse and Momentum

Figure 7-1 shows a graph of a typical force exerted by one body on another in a collision. The force rises sharply to a very large value as the bodies come together and then decreases to zero. For most purposes, we can replace this variable force with the average force F_{av}, which is constant over the time interval of the collision Δt, as shown in the figure. According to Newton's second law, the average force acting on a body equals the mass of the body times the average acceleration:

$$\mathbf{F}_{av} = m\,\frac{\Delta \mathbf{v}}{\Delta t}$$

The *impulse* of a force is defined as the product of the average force and the time interval Δt:

$$\text{Impulse} = \mathbf{F}_{av}\,\Delta t = m\,\Delta \mathbf{v} = \Delta(m\mathbf{v}) \qquad 7\text{-}1$$

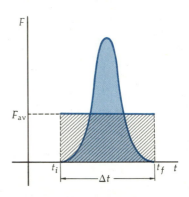

Figure 7-1
Force exerted by one body on another during a collision. The force becomes very large, but it occurs only during a very short time interval. The area under the F_{av}-versus-t curve is the same as the area under the F-versus-t curve and therefore gives the same impulse.

Impulse defined

(There is no commonly used symbol for impulse.) The product of the mass of a body times its velocity is called the *momentum* of the body, $\mathbf{p}$:

$$\mathbf{p} = m\mathbf{v} \qquad 7\text{-}2$$

Momentum defined

Momentum is a vector quantity that can be thought of as measuring the difficulty of bringing the body to rest. For example, a heavy truck has more momentum than a light car traveling at the same speed. It therefore takes a given force a greater time to stop the truck than to stop the car, or it takes a greater force to stop the truck than it does the car in a given time. The units of momentum are $kg \cdot m/s = N \cdot s$. From Equation 7-1, we can see that the impulse of a force equals the change in momentum:

$$\mathbf{F}_{av}\,\Delta t = \Delta \mathbf{p} \qquad 7\text{-}3$$

Example 7-1   A large truck of mass 30,000 kg moves at 20 m/s (about 45 mi/h). What is the magnitude of its momentum and how great a force is needed to stop the truck in 3 s?

The magnitude of the truck's momentum is $p = mv = (30{,}000$ kg$) \times 20$ m/s$) = 6 \times 10^5$ kg·m/s. To stop the truck, its momentum must be changed from this value to zero. The force required to do this in 3 s is

$$F = \frac{\Delta p}{\Delta t} = \frac{6 \times 10^5 \text{ kg·ms}}{3 \text{ s}} = 2 \times 10^5 \text{ N}$$

This force is about 45,000 lb or about 22.5 tons.

The hammer delivers a large impulsive force to drive the nail into the board.

Example 7-2   Assuming that the force that stops the truck in Example 7-1 is constant during the 3 s, how far does the truck travel in that time?

Since the acceleration is constant, the initial velocity is 20 m/s, and the final velocity is 0, the average velocity is 10 m/s during this time. The distance traveled is the average velocity times the time, or $x = v_{av}t = (10$ m/s$)(3$ s$) = 30$ m.

We can often estimate the force exerted by one body on another by finding the time of the collision from a reasonable estimate of the distance traveled. Consider, for example, an egg rolling off a table 1 m high and splattering on the floor. We can find the speed of the egg just before it hits the floor from the constant-acceleration formula, $v^2 = 2gy$, where g is the acceleration of gravity and y is the distance the egg falls. For $y = 1$ m, we obtain

$$v^2 = 2gy = 2(10 \text{ m/s}^2)(1 \text{ m}) = 20 \text{ m}^2/\text{s}^2$$

where we have used the approximation $g = 10$ m/s^2 for simplicity. The speed is then

$$v = \sqrt{20 \text{ m}^2/\text{s}^2} \approx 4.5 \text{ m/s}$$

A reasonable estimate for the distance the egg moves from the time of initial contact with the floor until it stops is about 2 cm. (This is about half the smaller diameter of a typical egg.) Taking $\frac{1}{2}$ (4.5 m/s) $= 2.2$ m/s for the average speed, the time of the collision is

$$\Delta t = \frac{\Delta y}{v_{av}} = \frac{0.02 \text{ m}}{2.2 \text{ m/s}} = 0.009 \text{ s}$$

For an egg of mass 50 g, the momentum just before it strikes the floor is

$$p = mv = (0.05 \text{ kg})(4.5 \text{ m/s}) = 0.22 \text{ kg·m/s}$$

Since the final momentum is zero, this is the magnitude of the momentum change. The average force is then

$$F_{av} = \frac{\Delta(mv)}{\Delta t} = \frac{(0.05 \text{ kg})(4.5 \text{ m/s})}{0.009 \text{ s}} = 25 \text{ N}$$

This is about 50 times the weight of the egg.

Example 7-3   Estimate the force exerted by the seat belt on a 80-kg car driver when his car, originally moving at 25 m/s (about 56 mi/h), crashes into a fixed object.

We will assume that the car travels about 1 m during the collision as the front end of the car crumples against the fixed object. This is also the distance traveled by the driver during the collision if he is wearing a seat belt. Since the average speed of the car during the collision is half its initial speed, or 12.5 m/s, the time of the collision is

$$\Delta t = \frac{1 \text{ m}}{12.5 \text{ m/s}} = 0.08 \text{ s}$$

The average acceleration is then

$$a_{av} = \frac{\Delta v}{\Delta t} = \frac{25 \text{ m/s}}{0.08 \text{ s}} = 312 \text{ m/s}^2$$

This acceleration is about 32 g, that is, about 32 times the free-fall acceleration of gravity. The average force on the driver is then

$$F_{av} = m a_{av} = (80 \text{ kg})(312 \text{ m/s}^2) = 25,000 \text{ N}$$

This force is great enough to break the driver's ribs and cause other chest injuries, but he may survive the crash. If he were not wearing a seat belt, he would continue moving at 25 m/s until he hit the dashboard or windshield. His stopping distance would then be considerably less than 1 m, which would result in a correspondingly greater acceleration and force.

Questions

1. Why can friction and the force of gravity usually be neglected in collision problems?

2. Explain how a safety net can save the life of a circus performer.

3. How might you estimate the collision time of a baseball and bat.

4. Why might a wine glass survive a fall onto a carpet but not onto a concrete floor?

7-2   Conservation of Momentum

The concept of momentum is important because, in many situations, the total momentum of a system of particles is conserved, that is, it remains constant. Consider a collision of two bodies. Let $\mathbf{p}_1 = m_1\mathbf{v}_1$ be the momentum of one body and $\mathbf{p}_2 = m_2\mathbf{v}_2$ be that of the other body. Newton's second law for each body is

$$\mathbf{F}_{1,av} = m_1 \frac{\Delta \mathbf{v}_1}{\Delta t} = \frac{\Delta \mathbf{p}_1}{\Delta t} \quad \text{and} \quad \mathbf{F}_{2,av} = m_2 \frac{\Delta \mathbf{v}_2}{\Delta t} = \frac{\Delta \mathbf{p}_2}{\Delta t}$$

Since the force exerted on body 1 is equal and opposite that exerted on body 2, we can add these equations and eliminate the forces. We then obtain

$$0 = \frac{\Delta \mathbf{p}_1}{\Delta t} + \frac{\Delta \mathbf{p}_2}{\Delta t}$$

$$\Delta \mathbf{p}_1 + \Delta \mathbf{p}_2 = \Delta(\mathbf{p}_1 + \mathbf{p}_2) = 0$$

or

$$\mathbf{p}_1 + \mathbf{p}_2 = \text{constant} \qquad \qquad 7\text{-}4$$

The sum $p_1 + p_2$ is the total momentum P of the system.

$$P = p_1 + p_2 = \text{constant} \qquad\qquad 7\text{-}5$$

If P_i is the total momentum of the system before the collision and P_f is that after the collision, Equation 7-5 says that

$$P_f = P_i \qquad\qquad 7\text{-}6$$

This result is known as the *law of conservation of momentum*. It holds for any system of particles as long as there is no resultant *external* force acting on the system. When the forces acting on the particles in such a system are added, the result is zero because for each force acting on one particle there is an equal and opposite force acting on another particle in the system.

If the resultant external force acting on a system of particles is zero, the total momentum of the system remains constant.

Conservation of momentum

It is important to remember that momentum is a *vector* quantity. In a collision, the momentum vectors of the two bodies are often oppositely directed. In some cases, the magnitudes of the momenta of the two bodies are equal, so the total momentum of the system before the collision is zero. Then, according to Equation 7-6, the total momentum of the system must be zero after the collision also.

Momentum conservation also holds (approximately) in collisions even if there are external forces acting because, as we saw in the previous section, the forces exerted by one body on another in a collision are very large compared with the usual external forces such as gravity or friction. The deformation of the golf ball shown in Figure 7-2 illustrates how large the internal force in a typical collision can be. Because the time of a typical collision is so small (around 0.01 s or less), the impulses delivered by such external forces as gravity or friction are nearly always negligible compared with those of the very large internal forces of the collision.

The law of conservation of momentum is one of the most important laws in physics. It applies to such everyday events as collisions between automobiles as well as to the more esoteric collisions of nuclear particles produced in an accelerator. It is also very useful in analyzing the complicated phenomena of rocket propulsion. The thrust of a rocket is essentially due to the force exerted on the rocket by the spent fuel exhausted out the back of the rocket. The momentum lost by the exhaust gases is balanced by the momentum gained by the rocket.

The law of conservation of momentum is more generally applicable than the law of conservation of mechanical energy. Since the internal forces exerted by one particle in a system on another are often not conservative, they can change the total mechanical energy of the system. However, since these internal forces always occur as action–reaction pairs, they cannot change the total momentum of the system.

Figure 7-2
Golf ball during collision with club. Note the deformation of the ball, which indicates that the force exerted by the club is very large.

Example 7-4   A man of mass 70 kg and a boy of mass 35 kg are standing together on a smooth ice surface for which friction is negligible. If the two push each other apart, and the man moves away with speed 0.3 m/s relative to the ice, how far apart are they after 5 s (see Figure 7-3)?

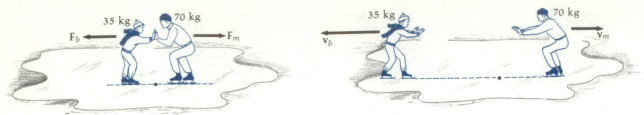

We take the man and boy together as the system. The force of gravity on each is balanced by a corresponding normal force of the ice. Since there is no friction, the resultant force on the system is zero. The force exerted by the man on the boy is equal and opposite to that exerted by the boy on the man. The total momentum of the system is zero since they are standing at rest initially. Therefore, after they push each other, they must have equal and opposite momenta. Since the motion takes place in one dimension, we can take care of the vector nature of the momentum by choosing the momentum to the right to be positive and that to the left to be negative. Assuming the man to be moving to the right, his momentum is

$$p_m = m_m v_m = (70 \text{ kg})(0.3 \text{ m/s}) = 21 \text{ kg} \cdot \text{m/s}$$

The momentum of the boy is

$$p_b = m_b v_b = (35 \text{ kg}) v_b$$

Setting the total momentum equal to zero, we obtain

$$p_m + p_b = 21 \text{ kg} \cdot \text{m/s} + (35 \text{ kg})(v_b) = 0$$

$$v_b = -\frac{21 \text{ kg} \cdot \text{m/s}}{35 \text{ kg}} = -0.6 \text{ m/s}$$

Since the man has twice the mass of the boy and moves in one direction with speed 0.3 m/s, the boy moves in the opposite direction with speed 0.6 m/s. After 5 s the man has moved 1.5 m and the boy 3 m, and so they are 4.5 m apart. Note that the mechanical energy of this system is not conserved. The force exerted by each on the other is not conservative. In this case, the mechanical energy is increased since the kinetic energy was initially zero and the potential energy did not change. The additional energy comes from a decrease in the internal chemical energy of the man and boy.

Figure 7-3
The resultant external force on the man–boy system is zero, so the total momentum, which is originally zero, remains zero. Since the mass of the man is twice that of the boy, the speed of the boy is twice that of the man.

Example 7-5   A bullet of mass 10 g moves horizontally with speed 400 m/s and embeds itself in a block of mass 390 g initially at rest on a frictionless table. What is the final velocity of the bullet and block (see Figure 7-4)?

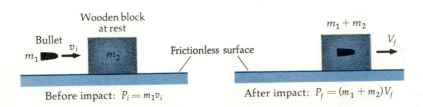

Figure 7-4
Bullet striking block in Example 7-5. Since there is no net external force on the bullet–block system, the momentum of the system is conserved.

Since there are no horizontal forces on the bullet–block system, the horizontal component of the momentum of the system is conserved. The total initial horizontal momentum P_i before the bullet strikes the block is just that of the bullet:

$$P_i = m_1 v_i = (10 \text{ g})(400 \text{ m/s}) = 4000 \text{ g} \cdot \text{m/s} = 4 \text{ kg} \cdot \text{m/s}$$

Afterward, the bullet and block move together with a common velocity V_f. The total final momentum P_f is

$$P_f = (m_1 + m_2)V_f = (0.4 \text{ kg})V_f$$

Since the total momentum is conserved, the final momentum equals the initial momentum:

$$(0.4 \text{ kg})V_f = 4 \text{ kg} \cdot \text{m/s}$$
$$V_f = 10 \text{ m/s}$$

A simple computation of the initial and final energies shows that, again, mechanical energy is not conserved. The original kinetic energy is

$$E_{ki} = \tfrac{1}{2}m_1 v_i^2 = \tfrac{1}{2}(0.01 \text{ kg})(400 \text{ m/s})^2 = 800 \text{ J}$$

and the final kinetic energy is

$$E_{kf} = \tfrac{1}{2}(m_1 + m_2)V_f^2 = \tfrac{1}{2}(0.04 \text{ kg})(10 \text{ m/s})^2 = 20 \text{ J}$$

In this case, most of the original mechanical energy (780 J out of 800 J) is lost (converted to heat energy) because of the large nonconservative forces between the bullet and the block that deform the bodies. A bullet embedding itself in a block is an example of an *inelastic collision*. We will study collisions in more detail in Section 7-4.

Questions

5. A pendulum bob swings back and forth. Is the momentum of the bob conserved? Explain why or why not.

6. How is the recoil of a rifle or cannon related to momentum conservation?

7. A man is stranded in the middle of an ice rink that is perfectly frictionless. How can he get to the edge?

8. A boy jumps from a boat to a dock. Why does he have to jump with more energy than if he were jumping the same distance from one dock to another?

7-3   Center of Mass

Figure 7-5 shows a multiflash photograph of a baton thrown into the air. Although the motion of the baton is quite complicated, one point on the baton, the *center of mass* (which coincides with the center of gravity of the baton), follows a simple parabolic path. This is the same path that would be followed by a small ball or stone thrown with the same initial velocity. In this section, we will show that, in general, the center of mass of an object or a system of particles moves as though it were a particle with a mass equal to the total mass of the system under the influence of the net external force acting on the system. For most systems that are small enough so that the

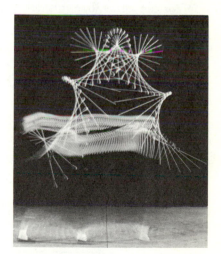

Figure 7-5
Multiflash photograph of a baton thrown into the air. The center of mass follows the same simple parabolic path it would if it were a single point particle.

gravitational field **g** is the same throughout the system, the center of mass coincides with the center of gravity, which we defined in Chapter 5 as the point at which the total weight of a body can be considered to be concentrated. (These two points would not coincide for a very large system, such as a mountain, for which the gravitational field **g** varies over the system.)

Let us first consider a system of point masses aligned along the x axis as shown in Figure 7-6. The resultant torque about some arbitrary point O on the axis is the sum of the torques of all the masses, $\Sigma m_i g x_i$. It also equals MgX_{cg} where M is the total mass and X_{cg} is the distance from point O to the center of gravity. Then

$$MgX_{cg} = m_1 g x_1 + m_2 g x_2 + \cdots$$

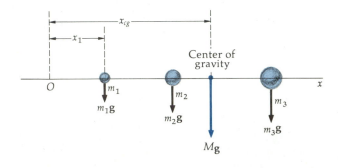

Center of
gravity

Figure 7-6
A line of particles on the x axis. The center of gravity is that point at which the total mass M could be placed so as to produce a torque about any arbitrary point O that is equal to the sum of the torques about the same point produced by the particles in the system. When **g** is the same at each mass point, the center of gravity coincides with the center of mass.

If g has the same value at each particle, we can divide each term by g and eliminate it. The position of the center of mass X_{cm} then coincides with that of the center of gravity X_{cg}. The location of the center of mass X_{cm} is given by

$$MX_{cm} = m_1 x_1 + m_2 x_2 + \cdots \qquad \text{7-7}$$

We can generalize this definition of the location of the center of mass by replacing x with the position vector **r** for each particle

$$M\mathbf{R}_{cm} = m_1 \mathbf{r}_1 + m_2 \mathbf{r}_2 + \cdots \qquad \text{7-8}$$ Center of mass

Suppose that in some time interval Δt the ith particle has displacement $\Delta \mathbf{r}_i$. Then

$$\frac{M \Delta \mathbf{R}_{cm}}{\Delta t} = \frac{m_1 \Delta \mathbf{r}_1}{\Delta t} + \frac{m_2 \Delta \mathbf{r}_2}{\Delta t} + \cdots$$

If we take the time interval Δt to be very small, we have

$$M\mathbf{V}_{cm} = m_1 \mathbf{v}_1 + m_2 \mathbf{v}_2 + \cdots$$ Total momentum of a system

The right side of this equation is the total momentum of the system of particles. The total momentum of a system of particles is thus the product of the total mass M and the velocity of the center of mass $\mathbf{V}_{cm}$.

$$\mathbf{P} = M\mathbf{V}_{cm} = m_1 \mathbf{v}_1 + m_2 \mathbf{v}_2 + \cdots \qquad \text{7-9}$$

If we compute the change in velocity $\Delta \mathbf{v}_i$ for each particle for a small time interval Δt and divide by the time interval, we obtain from Equation 7-9

$$M\mathbf{A}_{cm} = m_1 \mathbf{a}_1 + m_2 \mathbf{a}_2 + \cdots \qquad \text{7-10}$$

According to Newton's second law, the resultant force acting on the ith particle equals $m_i\mathbf{a}_i$. Equation 7-10 thus states that the total mass M times the acceleration of the center of mass equals the sum of all the forces acting on all the particles in the system. The forces acting on a particle can be separated into two types: the forces internal to the system, that is, those forces exerted by other particles in the system; and forces, such as gravity, that are external to the system. When we sum all the forces acting on the system, the internal forces sum to zero because for each internal force acting on one particle there is an equal and opposite force acting on another particle in the system. The sum of all the forces thus equals the resultant external force acting on the system. Replacing the right side of Equation 7-10 with the resultant external force acting on the system and rearranging, we have

$$\mathbf{F}_{\text{net,ext}} = M\mathbf{A}_{cm} \qquad\qquad 7\text{-}11$$

Motion of the center of mass

Equation 7-11 shows that the acceleration of the center of mass equals the resultant external force divided by the total mass. This is important because it shows us how to describe the motion of one point, the center of mass, for any system of particles. The center of mass behaves just like a single particle that is subject only to the external forces. The simple parabolic path followed by the center of mass of the baton in Figure 7-5 is an example of this. Equation 7-11 is the justification for our earlier treatment of large objects as point particles.

Example 7-6   A projectile of mass $2m$ explodes into two equal pieces, each of mass m, at the top of its flight. One of the pieces drops straight down from rest after the explosion while the other moves off horizontally so that the two pieces land simultaneously. Where does the second piece land?

Considering the projectile to be the system (whether it is in one piece or two), the only external force acting on it (until the masses hit the ground) is that due to gravity. The forces exerted in the explosion are internal forces, which do not affect the motion of the center of mass. After the explosion, the center of mass traces out the rest of the parabola just as if there had been no explosion. Since the center of mass is halfway between the two equal masses and we know that one mass drops straight down, the other must land at an equal distance from the center of mass, as shown in Figure 7-7.

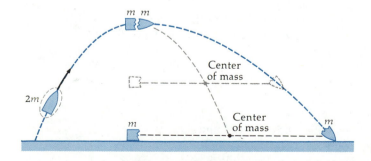

Figure 7-7
A projectile exploding into two equal fragments at the top of its flight in such a way that the fragments land simultaneously. The internal forces of the explosion have no effect on the motion of the center of mass.

Example 7-7   A cylinder of mass M rests on a rough piece of paper on a table. A force f is applied to the cylinder by pulling the paper to the right, as shown in Figure 7-8. Which way does the cylinder move?

This is a somewhat puzzling situation. We know that the cylinder will roll backward (to the left) on the paper. Yet the only force acting on it is the frictional force exerted by the paper, and this force is to the right. The center of mass of a uniform cylinder is at its geometric center (that is, on the axis, halfway between the faces). Applying Equation 7-11 to the cylinder as our system, we have

$$\mathbf{f} = M\mathbf{a}_{cm}$$

The acceleration of the center of mass of the cylinder is in the direction of the force $\mathbf{f}$ and has the magnitude f/M. Since $\mathbf{f}$ is in the direction the paper is pulled, $\mathbf{a}_{cm}$ is to the right. What happens is that the cylinder rolls backward on the paper because the paper accelerates to the right with an acceleration greater than that of the cylinder.

You can easily do this experiment with a can resting on a piece of paper on a table. The forward acceleration of the can can be demonstrated by marking the initial position of the can on the *table* with a piece of chalk. The can accelerates backward relative to the paper, but forward relative to the table. Note that Equation 7-11 does not describe the rotation of the cylinder but only the motion of its center of mass.

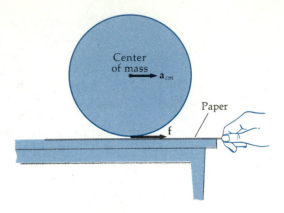

Figure 7-8
Cylinder rolling on a piece of paper pulled to the right. Since the resultant external force on the cylinder is the frictional force exerted by the paper to the right, the center of mass of the cylinder must accelerate to the right. The cylinder rolls backward relative to the paper because the acceleration of the paper is greater than that of the cylinder.

Questions

9. In Example 7-4, where the man and the boy are at the center of a smooth ice surface and are pushing each other, what happens to the center of mass of the man–boy system?

10. A rocket ship is initially at rest (relative to the distant stars) in empty space when its engines are turned on. What happens to the center of mass of the rocket and its original contents? How can the rocket move?

7-4   Collisions

We saw in Example 7-5 that, in a collision, mechanical energy is sometimes lost to heat energy. When this happens, the collision is said to be *inelastic*. If the total mechanical energy is conserved in a collison, the collision is *elastic*. An example of a collision that is approximately elastic is one between two billiard balls. If the cue ball is struck so that it has no spin, it stops when it strikes another ball, and the struck ball moves off with a speed approximately the same as the original speed of the cue ball.

Figure 7-9*a* shows a 2-kg body moving to the right at 6 m/s and a 4-kg body moving to the left at 3 m/s. What happens when they collide? Since we are considering a one-dimensional collision only, we can take care of the vector nature of momentum by choosing the momentum to the right to be

Elastic collision

positive and that to the left to be negative. The momentum of the 2-kg body is

$$p_1 = (2 \text{ kg})(6 \text{ m/s}) = 12 \text{ kg} \cdot \text{m/s}$$

and that of the 4-kg body is

$$p_2 = (4 \text{ kg})(-3 \text{ m/s}) = -12 \text{ kg} \cdot \text{m/s}$$

The total initial mechanical energy is

$$E_k = \tfrac{1}{2}(2 \text{ kg})(6 \text{ m/s})^2 + \tfrac{1}{2}(4 \text{ kg})(3 \text{ m/s})^2$$

$$= 36 \text{ J} + 18 \text{ J} = 54 \text{ J}$$

The bodies have equal and opposite momenta, so the total momentum of the two-body system is zero. Since there is no resultant external force acting on this two-body system, the total momentum after the collision must also be zero.

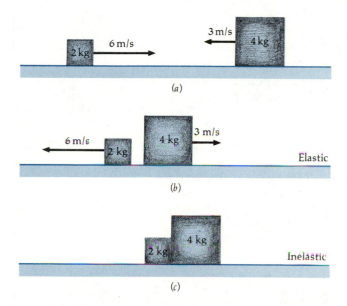

(a)

(b)

(c)

Figure 7-9
(a) A 2-kg body moving to the right at 6 m/s collides with a 4-kg body moving to the left at 3 m/s. (b) If the collision is elastic, the bodies are merely turned around and recede with their original speeds. (c) If the collision is perfectly inelastic, the bodies stick together. Since the original momentum of the system was zero the bodies are at rest after the collision.

It is not hard to see that there are many possible final results that conserve momentum. One is that each body turns around and leaves with the same speed that it had when approaching. That is, the 2-kg body moves to the left with velocity -6 m/s and the 4-kg body moves to the right with velocity $+3$ m/s (see Figure 7-9b). If this happens, the total mechanical energy after the collision equals that before the collision since each body still has the same energy that it had initially. This is the case of an elastic collision with both momentum and energy conserved.

The other extreme is if the bodies stick together and come to rest with each having zero momentum and zero mechanical energy (see Figure 7-9c). When the two bodies stick together after the collision, the collision is called *perfectly inelastic*. In a perfectly inelastic collision, the loss in mechanical energy is the maximum amount that is consistent with the conservation of momentum. In this instance, all of the original mechanical energy is lost to heat energy, but this occurs only if the total momentum of the system is zero.

Perfectly inelastic collision

In a more general perfectly inelastic collision, where the initial total momentum is not zero, the bodies stick together and move with a velocity such that the final momentum equals the initial momentum. Only part of the initial mechanical energy is lost to heat in that case.

Another possibility for the motion of the system after the collision is for the 2-kg body to move to the left with velocity -2 m/s and the 4-kg body to move to the right with velocity $+1$ m/s. In this case, some but not all of the mechanical energy is lost to heat. Any motion with the 4-kg body moving to the right and the 2-kg body moving to the left with twice that speed is consistent with the law of conservation of momentum. We could even have a gain in mechanical energy if the blocks were equipped with explosive caps that detonated on impact. In Example 7-4, where the boy and man pushed each other on the ice, we had a gain in mechanical energy at the expense of a loss in internal chemical energy.

The collisions illustrated in Figure 7-9 are very special cases because the total momentum of the two-body system is zero. Since the total momentum of a system equals the total mass times the velocity of the center of mass, if the total momentum is zero, the velocity of the center of mass must be zero. The reference frame in which the center of mass is at rest is called the *center of mass reference frame*. The analysis of collisions viewed in the center of mass reference frame is particularly simple. If a collision is perfectly inelastic, the objects are at rest after the collision. If a collision is elastic, the objects are merely turned around by the collision, and each object leaves with the same speed it had before the collision. We will now look at perfectly inelastic and elastic collisions in a more general reference frame in which the total momentum is not zero.

Figure 7-10 shows a 2-kg body moving at 9 m/s that collides with a 4-kg body that is initially at rest. (This is how the collision of Figure 7-9 would look to us if we were moving to the left at 3 m/s relative to the coordinate system of Figure 7-9, that is, if we were riding on the 4-kg block.) We will first consider a perfectly inelastic collision and then an elastic collision.

This collision will probably not be elastic.

Center of mass reference frame

Figure 7-10
Collision of a 2-kg body moving to the right at 9 m/s with a stationary 4-kg body. This is the collision of Figure 7-9a viewed in a coordinate system moving to the left at 3 m/s.

Inelastic Collision

If the collision is perfectly inelastic, the two bodies stick together and move with a common velocity after the collision. The total momentum of the two-body system before the collision is just that of the 2-kg body since the 4-kg body is initially at rest:

$$P_i = m_1 v_1 = (2 \text{ kg})(9 \text{ m/s}) = 18 \text{ kg} \cdot \text{m/s}$$

If V is the common velocity after the collision, the total momentum is

$$P_f = (m_1 + m_2)V = (2 \text{ kg} + 4 \text{ kg})V = (6 \text{ kg})V$$

Comparing this equation with Equation 7-9, we note that the velocity V is the velocity of the center of mass of the system. Since there are no net external forces acting on this system, the velocity of the center of mass does not change. And since the two bodies move together after the collision, they must move with the velocity of the center of mass. Setting the final momentum equal to the initial momentum, we have

$$P_f = P_i$$

$$(6 \text{ kg})V = 18 \text{ kg} \cdot \text{m/s}$$

$$V = 3 \text{ m/s}$$

Thus, the two bodies move at 3 m/s after the collision, as shown in Figure 7-11.

We will now calculate the mechanical energy lost to heat in this collision. The initial kinetic energy is

$$E_{ki} = \tfrac{1}{2}(2 \text{ kg})(9 \text{ m/s})^2 = 81 \text{ J}$$

and the final kinetic energy is

$$E_{kf} = \tfrac{1}{2}(6 \text{ kg})(3 \text{ m/s})^2 = 27 \text{ J}$$

The mechanical energy lost is thus 81 J − 27 J = 54 J. This is the same amount of mechanical energy loss we found for the same collision viewed in the center of mass reference frame. That is, if we ride along with the center of mass, moving at 3 m/s to the right, the 2-kg body is moving at 6 m/s to the right relative to us, and the 4-kg body is moving at 3 m/s to the left, just as in Figure 7-9a. The total initial energy computed for that situation was 54 J. After a perfectly inelastic collision in that reference frame, the two bodies are at rest with all of the initial mechanical energy converted into heat energy.

Elastic Collision

If the collision of Figure 7-10 is elastic, the two bodies move with different velocities after the collision. Let v_{1f} be the velocity of the 2-kg body after the collision and v_{2f} be that of the 4-kg body. Conservation of momentum gives

$$m_1 v_{1f} + m_2 v_{2f} = m_1 v_1 + m_2 v_2$$

$$(2 \text{ kg})v_{1f} + (4 \text{ kg})v_{2f} = (2 \text{ kg})(9 \text{ m/s}) + (4 \text{ kg})(0) = 18 \text{ kg} \cdot \text{m/s}$$

Dividing each term by 2 kg, we obtain

$$v_{1f} + 2v_{2f} = 9 \text{ m/s} \qquad\qquad 7\text{-}12$$

We can obtain a second equation for the unknown final velocities from the conservation of mechanical energy. The initial mechanical energy is 81 J, as computed before. The final energy is

$$\tfrac{1}{2}(2 \text{ kg})v_{1f}^2 + \tfrac{1}{2}(4 \text{ kg})v_{2f}^2$$

Setting this equal to the initial energy, we obtain

$$v_{1f}^2 + 2v_{2f}^2 = 81 \text{ J/kg} \qquad\qquad 7\text{-}13$$

Equations 7-12 and 7-13 can be solved simultaneously for the two unknown velocities. We will drop the units in these equations to simplify the writing.

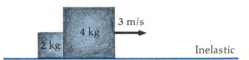

Figure 7-11
If the collision in Figure 7-10 is perfectly inelastic, the bodies stick together. They move with the velocity of the center of mass, which is 3 m/s found from conservation of momentum.

An elastic collision of two equal cannonballs from Johann Marcus Marci, *De Proportione Motus*, 1639.

Substituting $v_{1f} = (9 - 2v_{2f})$ from Equation 7-12 into Equation 7-13, we obtain

$$(9 - 2v_{2f})^2 + 2v_{2f}^2 = 81$$

$$81 - 36v_{2f} + 4v_{2f}^2 + 2v_{2f}^2 = 81$$

$$6v_{2f}^2 - 36v_{2f} = 0$$

Dividing by 6 and factoring, we obtain

$$v_{2f}(v_{2f} - 6) = 0$$

The two solutions are $v_{2f} = 0$ and $v_{2f} = 6$ m/s. If $v_{2f} = 0$, Equation 7-12 gives $v_{1f} = 9$ m/s. This corresponds to no collision at all, which would, of course, conserve both momentum and energy. For $v_{2f} = 6$ m/s, we obtain from Equation 7-12

$$v_{1f} = 9 \text{ m/s} - 2v_{2f} = 9 \text{ m/s} - 12 \text{ m/s} = -3 \text{ m/s}$$

It is straightforward to calculate the final kinetic energy and show that it equals the initial kinetic energy of 81 J. After the elastic collision, the 4-kg body moves to the right at 6 m/s and the 2-kg body moves to the left at 3 m/s (see Figure 7-12). The relative velocity of recession is 9 m/s; that is, if we ride on one body, the other one recedes after the collision at 9 m/s. This equals the relative velocity of approach before the collision. This is a general, and very useful, result for elastic collisions.

In an elastic collision, the relative speed of recession after the collision equals the relative speed of approach before the collision.

$v_{2f} - v_{1f} = -(v_{2i} - v_{1i})$	elastic collision	7-14

The proof of this result is outlined in Problem 22. It is usually much easier to use this result (rather than the equation for energy conservation) along with the conservation of momentum to find the final velocities in an elastic collision. We will illustrate this in Example 7-9.

We have seen that, for an elastic collision, the relative velocity of recession equals that of approach, whereas for a perfectly inelastic collision, the relative velocity of recession is zero. In general, a collision is somewhere between perfectly inelastic, where the bodies stick together, and perfectly elastic, where the mechanical energy is conserved. The *coefficient of restitution e* is defined as the ratio of the relative velocity of recession and the relative velocity of approach. The relative velocity of recession is then

$v_{2f} - v_{1f} = -e(v_{2i} - v_{1i})$	7-15

For a perfectly elastic collision, $e = 1$; for a perfectly inelastic collision, $e = 0$.

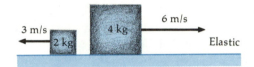

Figure 7-12
If the collision in Figure 7-10 is elastic, the bodies move off with velocities such that both momentum and kinetic energy is conserved. In an elastic collision, the relative velocity of recession after the collision, 9 m/s in this case, equals the relative velocity of approach before the collision.

Example 7-8   A 10-g bullet is fired with speed 300 m/s into the bob of a pendulum, which has a mass of 1.99 kg (see Figure 7-13). How high does the pendulum bob (plus the bullet) swing after the collision?

This device is called a *ballistic pendulum.* It is used to measure the initial speed of the bullet by measuring the height h and the masses of the bullet and block. This problem needs to be solved in two steps. The collision itself

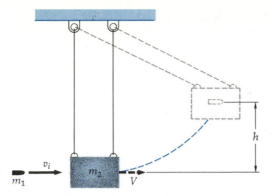

Figure 7-13
Ballistic pendulum. The height h is related to the speed of the bullet–block system v after the collision by conservation of energy. During the collision, energy is not conserved. The speed V can be found by conservation of momentum of the system.

is inelastic since the bullet is embedded in the block. Kinetic energy is therefore *not* conserved, but momentum is. Since the duration of the collision is very small, we can neglect any motion of the block during the collision. The conservation of momentum equation allows us to calculate the speed of the block–bullet combination after the collision from the initial speed of the bullet. Once the bullet stops relative to the block, the only forces acting on the system are gravity and the constraint forces of the strings. Thus, after the collision, mechanical energy is conserved as the block–bullet system swings to its maximum height.

The momentum of the bullet before the collision is

$$p = m_1 v_i = (0.01 \text{ kg})(300 \text{ m/s}) = 3.00 \text{ kg} \cdot \text{m/s}$$

The mass of the bullet–block system is $0.01 \text{ kg} + 1.99 \text{ kg} = 2.00 \text{ kg}$. If the speed after the collision is V, the conservation of momentum gives

$$(2.00 \text{ kg})V = p = 3.00 \text{ kg} \cdot \text{m/s}$$

or

$$V = 1.50 \text{ m/s}$$

After the collision, the original kinetic energy of the bullet–block system is converted into gravitational potential energy as the block swings upward. The maximum height of the swing is therefore obtained from

$$\tfrac{1}{2}MV^2 = Mgh$$

where M is the mass of the bullet–block system and v is the speed just after the collision. Solving for h, we obtain

$$h = \frac{V^2}{2g} = \frac{(1.5 \text{ m/s})^2}{2 \times 9.81 \text{ m/s}^2} = 0.115 \text{ m} = 11.5 \text{ cm}$$

Example 7-9   A 3-kg body moving to the right at 6 m/s makes an elastic collision with a 6-kg body initially at rest (see Figure 7-14). Find the final velocities of each body after the collision.

Figure 7-14
A 3-kg block moving at 6 m/s toward a stationary 6-kg block for Example 7-9.

Let v_{1f} and v_{2f} be the final velocities of the 3-kg and 6-kg bodies. The initial momentum of this system is just that of the 3-kg body before the collision, which is

$$P_i = (3 \text{ kg})(6 \text{ m/s}) = 18 \text{ kg} \cdot \text{m/s}$$

Conservation of momentum gives us one equation relating the final velocities:

$$P_f = 18 \text{ kg} \cdot \text{m/s} = (3 \text{ kg})(v_{1f}) + (6 \text{ kg})(v_{2f})$$

Dividing by 3 kg, we obtain

$$v_{1f} + 2v_{2f} = 6 \text{ m/s} \qquad\qquad 7\text{-}16$$

A second relation between the two final velocities could be obtained from conservation of energy, but it is easier to use the result that the relative velocity of recession after the collision equals the relative velocity of approach before the collision. The relative velocity of approach before the collision is 6 m/s. After the collision, the relative velocity of recession is $v_{2f} - v_{1f}$. Thus,

$$v_{2f} - v_{1f} = 6 \text{ m/s} \qquad\qquad 7\text{-}17$$

We can eliminate v_{1f} by adding these two equations. The result is

$$3v_{2f} = 12 \text{ m/s}$$

or

$$v_{2f} = 4 \text{ m/s}$$

We can now find v_{1f} from either Equation 7-16 or Equation 7-17. Using Equation 7-16, we obtain

$$v_{1f} = 6 \text{ m/s} - 2v_{2f} = 6 \text{ m/s} - 2(4 \text{ m/s}) = -2 \text{ m/s}$$

The minus sign means that the 3-kg body moves to the left after the collision. It is a good idea to check our work by computing the initial and final kinetic energies of the bodies. The initial kinetic energy is

$$E_{ki} = \tfrac{1}{2}(3 \text{ kg})(6 \text{ m/s})^2 = 54 \text{ J}$$

and the final kinetic energy is

$$E_{kf} = \tfrac{1}{2}(3 \text{ kg})(-2 \text{ m/s})^2 + \tfrac{1}{2}(6 \text{ kg})(4 \text{ m/s})^2 = 6 \text{ J} + 48 \text{ J} = 54 \text{ J}$$

Example 7-10   The coefficient of restitution for steel on steel is measured by dropping a steel ball onto a steel plate rigidly attached to the earth. If the ball is dropped from height h_i and rebounds to height h_f, what is the coefficient of restitution?

The speed of the ball just before it strikes the plate is found from conservation of energy:

$$\tfrac{1}{2}mv_i^2 = mgh_i$$

or

$$v_i = \sqrt{2gh_i}$$

Similarly, for the ball to reach height h_f, its speed just after the collision must be

$$v_f = \sqrt{2gh_f}$$

Since the plate is attached to the earth, we can neglect its rebound velocity. The relative speeds in this case are thus just v_i and v_f.

$$v_f = ev_i$$
$$\sqrt{2gh_f} = e\sqrt{2gh_i}$$
$$e = \sqrt{h_f/h_i}$$

For example, if the final height is 80 percent of the original height, $h_f/h_i = 0.8$ and $e = \sqrt{0.8} = 0.9$.

For collisions in three dimensions, the law of conservation of linear momentum is a vector equation. Perfectly inelastic collisions present no special difficulty. The initial momentum vector is found for each body before the collision, and the total momentum **P** is found by vector addition. After the collision, the bodies stick together and move with velocity **V** given by

$$\mathbf{V} = \frac{\mathbf{P}}{(m_1 + m_2)}$$

where m_1 and m_2 are the masses of the bodies. Elastic collisions are more complicated and will not be treated here. An interesting special case occurs when the two bodies are of equal mass and one is initially at rest. After an elastic collision, the two bodies move off with velocity vectors perpendicular to each other, as shown in Figures 7-15 and 7-16.

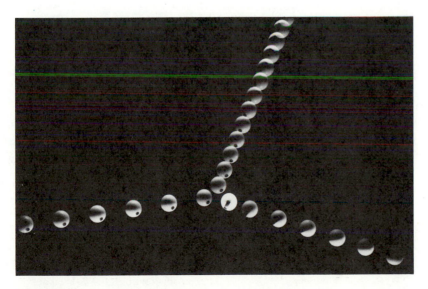

Figure 7-15
Multiflash photograph of an elastic off-center collision of two balls of equal mass. The dotted ball comes from the left and strikes the striped ball, which is initially at rest. The final velocities of the two balls are perpendicular to each other.

Figure 7-16
Proton–proton collision in a liquid-hydrogen bubble chamber. The incident proton enters from the left and interacts with a stationary proton in the chamber. The two protons move off at right angles after the collision. (The slight curvature of the tracks is due to a magnetic field.)

Example 7-11   A small 1.2-Mg car traveling east at 60 km/h collides at an intersection with a 3-Mg truck traveling north at 40 km/h (see Figure 7-17). The car and truck stick together. Find the velocity of the wreckage just after the collision.

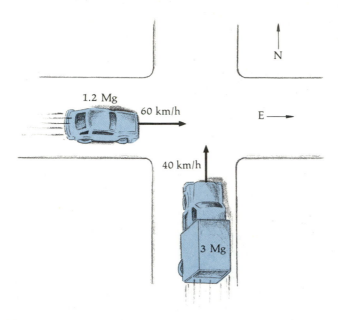

Figure 7-17
Perfectly inelastic collision of a car and a truck. After the collision, the wreckage moves off with a velocity whose magnitude and direction are determined by conservation of momentum.

The initial momentum of the car is

$$p_c = (1.2 \text{ Mg})(60 \text{ km/h}) = 72 \text{ Mg} \cdot \text{km/h} \qquad \text{east}$$

and the momentum of the truck is

$$p_t = (3 \text{ Mg})(40 \text{ km/h}) = 120 \text{ Mg} \cdot \text{km/h} \qquad \text{north}$$

These momentum vectors are shown in Figure 7-18. We find the magnitude of the total momentum P from the pythagorean theorem.

$$P^2 = (72 \text{ Mg} \cdot \text{km/h})^2 + (120 \text{ Mg} \cdot \text{km/h})^2$$

$$= 1.96 \times 10^4 \ (\text{Mg} \cdot \text{km/h})^2$$

$$P = \sqrt{1.96 \times 10^4} \text{ Mg} \cdot \text{km/h}$$

$$= 140 \text{ Mg} \cdot \text{km/h}$$

Since the total mass of the wreckage is 4.2 Mg, the magnitude of the final velocity is

$$V = \frac{P}{M} = \frac{140 \text{ Mg} \cdot \text{km/h}}{4.2 \text{ Mg}} = 33.3 \text{ km/h}$$

The direction of the final velocity is the same as that of the momentum vector. The angle θ in the figure is found from

$$\tan \theta = 120/72 = 1.67$$

$$\theta = 59°$$

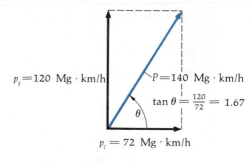

Figure 7-18
Momentum vector diagram for the collision of Figure 7-17. The original momentum, which equals the final momentum, is determined by vector addition of the individual momenta.

7-5   Jet Propulsion

Jet propulsion is an interesting application of Newton's third law and the law of conservation of momentum. Squids and octopuses propel themselves by squirting water from their bodies with great force. The water expelled exerts an equal and opposite force on the squid or octopus, propelling it forward. A rocket gets its thrust by burning fuel and exhausting the gases created out the rear of the rocket. The rocket exerts a force on the exhaust gas, and by Newton's third law, the gas exerts an equal and opposite force on the rocket, propelling it forward.

This octopus moves by jet propulsion. It takes in water and then suddenly expells it propelling itself forward by recoil.

Initially, when the rocket is at rest, the momentum of the exhaust gases in the rearward direction is balanced by the momentum gained by the rocket in the forward direction. When the rocket is moving at high speed, the momentum lost by the gases being expelled equals the momentum gained by the rocket (see Figure 7-19). If the rocket is traveling in free space, there are no external forces acting on it (neglecting the gravitational attraction of the sun or nearby planets or moons). Then the center of mass of the rocket and the exhaust gases remains at rest or moves with its initial constant velocity. We are usually not interested in the total system of the rocket plus the exhaust gas. Instead, we normally want to describe the motion of the rocket alone. This description is complicated because the mass of the rocket changes continuously as it burns fuel and expels the exhaust gases. We can get an idea of some of the features of rocket motion by considering a simple problem about the recoil of a gun.

Suppose a gun of mass M_g fires a bullet of mass m_b with a speed v_b. If the gun is not held but is resting in free space (or on a frictionless table), it recoils with speed V_g, which can be obtained by using conservation of momentum.

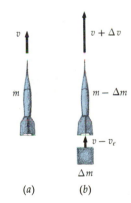

Figure 7-19
Rocket motion. (*a*) Rocket of mass m moving with speed v has momentum mv. (*b*) Gas of mass Δm is exhausted out the rear of the rocket with speed v_e relative to the rocket. The momentum lost by the gas equals the momentum gained by the rocket.

The momenta of the bullet and gun are equal in magnitude and opposite in direction. Thus,

$$M_g V_g = m_b v_b$$

or

$$V_g = \frac{m_b}{M_g} v_b \tag{7-18}$$

Now, suppose that we have a rocket of mass 30 kg, which includes 10 kg of fuel. It ejects the fuel at a uniform rate as high-speed gas moving at speed $v_e = 300$ m/s. While this is happening, the mass of the rocket decreases from 30 kg to 20 kg. If we use Equation 7-18 for the recoil of the gun, with m_b replaced by the mass of the fuel (10 kg), v_b replaced by the exhaust speed (300 m/s), and M_g replaced by the average mass of the rocket (25 kg), we obtain for the final speed of the rocket:

$$v_r = \frac{10 \text{ kg}}{25 \text{ kg}} \, 300 \text{ m/s} = 120 \text{ m/s}$$

There are two reasons why this approximation is inaccurate. First, the mass of the rocket changes continuously. We should think of the change in the speed of the rocket and the expulsion of the gas as the summation of a series of small bursts, like the firing of a machine gun. For each burst, the mass of the rocket is slightly less than it was for the previous one. Adding these up is not the same as using the average mass of the rocket for a single burst. The second inaccuracy comes from the fact that the exhaust speed of the gas v_e is measured relative to the rocket, whose speed is continuously changing, whereas the speed of the bullet v_b was measured relative to the fixed earth.

The correct expression for the final speed of the rocket, assuming it starts from rest, is

$$v_r = v_e \ln \frac{m_i}{m_f} \tag{7-19}$$

Rocket equation

where m_i is the initial mass of the rocket including the fuel, m_f is the final mass of the rocket, v_e is the exhaust speed of the gas relative to the rocket, and ln stands for the natural logarithm (to the base $e = 2.718$ . . .). Substituting the numbers for our example into this expression, we obtain

$$v_r = (300 \text{ m/s}) \ln \frac{30 \text{ kg}}{20 \text{ kg}} = 122 \text{ m/s}$$

The mass of a rocket without any fuel is called the *payload*. We can see from Equation 7-19 that, to make the final speed of the rocket as great as possible, we need to make the exhaust speed v_e as large as possible and the initial mass of the rocket mostly that of the fuel. However, the slowly varying nature of the logarithmic function severely restricts the final speeds obtainable. For example, if the fuel is 90 percent of the initial mass and the payload is 10 percent, the ratio m_f/m_i is 10, and the final rocket speed will be

$$v_r = v_e \ln 10 = 2.3 v_e$$

If we reduce the payload and increase the fuel so that the fuel is 99 percent of the initial mass, m_f/m_i will be 100. But the (natural) logarithm of 100 is just two times that of 10. The final rocket speed will then be just

$$v_r = v_e \ln 100 = 4.6 v_e$$

There was a misconception prevalent before rockets in space were commonplace that a rocket needs air to push against. This is, of course, not true. The rocket pushes against its own exhaust gases, which push back against the rocket. Prior to World War II, much of the research on the principles of jet propulsion was done by Robert Goddard, physics professor at Clark College in Worchester, Massachusetts. A quotation from an editorial in *The New York Times* in 1921 illustrates a common attitude toward his work at that time.

> That Professor Goddard with his "chair" at Clark College and the countenance of the Smithsonian Institution does not know the relation between action and reaction, and the need to have something better than a vacuum against which to react—to say that would be absurd. Of course, he only seems to lack the knowledge ladled out daily in high schools.*

Fortunately for rocket research, Professor Goddard had a better understanding of Newton's third law than did the editorial writer for the *Times*.

* Quoted from Nancy P. Gamarra, *Erroneous Predictions and Negative Comments Concerning Exploration, Territorial Expansion, Scientific and Technological Development*, Library of Congress Legislative Reference Service, Washington, 1967, p. 33.

Robert Goddard single-handedly did most of the good scientific work on rocket principles up to the beginning of World War II. While struggling to obtain support for his work on liquid-fueled rockets at Clark College, Goddard was forced to vehemently deny that he was interested in space travel. He was, of course, but too many of the people he was trying to get money from would have thought that made him a crackpot. Finally, with the help of Charles Lindberg among others, he managed to get one grant—$30,000 from the Guggenheim foundation.

Summary

1. The impulse of a force is defined as the product of the average force and the time interval over which the force acts.

2. The momentum of a body is defined as the mass times its velocity. The change in the momentum of a body equals the impulse of the force exerted on it.

3. If the resultant external force on a system is zero, the total momentum of the system is conserved.

4. The center of mass of a system of particles is defined by

$$M\mathbf{R}_{cm} = m_1\mathbf{r}_1 + m_2\mathbf{r}_2 + \cdots$$

where M is the total mass of the system and $\mathbf{R}_{cm}$ is the position vector from some origin to the center of mass. In a uniform gravitational field, the center of mass coincides with the center of gravity.

5. The total mass of a system times the velocity of the center of mass equals the total momentum of the system:

$$\mathbf{P} = M\mathbf{V}_{cm} = m_1\mathbf{v}_1 + m_2\mathbf{v}_2 + \cdots$$

6. The center of mass moves like a particle with a mass equal to the total mass of the system under the influence of the resultant external force acting on the system.

$$\mathbf{F}_{net,ext} = M\mathbf{A}_{cm}$$

7. In a perfectly inelastic collision, mechanical energy is converted into heat energy. After the collision, the bodies stick together and move with the velocity of the center of mass.

8. In an elastic collision, the total mechanical energy as well as the total momentum is conserved. The relative velocity of recession of the bodies after the collision equals the relative velocity of approach before the collision.

9. The coefficient of restitution e is defined as the ratio of the relative velocity of recession to the relative velocity of approach. For an elastic collision, $e = 1$; for a perfectly inelastic collision, $e = 0$.

10. A rocket gets its thrust by burning fuel and exhausting the resulting gases. The force exerted by the exhaust gases on the rocket propels the rocket forward.

Suggestions for Further Reading

Walker, Jearl: "The Amateur Scientist: Success in Racquetball Is Enhanced by Knowing the Physics of the Collision of Ball with Wall," *Scientific American*, September 1984, p. 215.

The laws of conservation of momentum and energy are applied to the collisions of a racquetball.

Review

A. Objectives: After studying this chapter, you should:

1. Be able to state the definitions of impulse and momentum.

2. Be able to estimate the magnitude of the large forces that occur in collisions.

3. Be able to explain why the forces exerted on a automobile driver in a collision are so much smaller when the driver wears a seat belt than when he or she does not.

4. Be able to state the law of conservation of momentum and use this law to solve collision problems.

5. Be able to describe the motion of the center of mass of a system of particles.

B. Define, explain, or otherwise identify:

Impulse, p. 148
Momentum, p. 148
Law of conservation of momentum, p. 151
Center of mass, pp. 153–154
Elastic collision, p. 156

Perfectly inelastic collision, p. 157
Center of mass reference frame, p. 158
Coefficient of restitution, p. 160

C. True or false: If the statement is true, explain why. If it is false, give a counterexample.

1. The momentum of a heavy object is greater than that of a light object moving at the same speed.

2. In a perfectly inelastic collision, all the kinetic energy of the bodies is lost.

3. The momentum of a system may be conserved even when mechanical energy is not.

4. Mechanical energy is conserved in an elastic collision.

5. In an elastic collision, the relative speed of recession after the collision equals the relative speed of approach before the collision.

6. The velocity of the center of mass of a system equals the total momentum of the system divided by its total mass.

Exercises

Section 7-1 Impulse and Momentum

1. Find the magnitude of the momentum of (*a*) a 20-g bullet traveling at 300 m/s, (*b*) an 80-kg jogger running at 3.3 m/s, and (*c*) a 2-Mg car traveling at 80 km/h.

2. A football player running at 3.4 m/s has momentum 400 kg·m/s. What is his mass?

3. A soccer ball of mass 0.43 kg leaves the foot of the kicker with an initial speed of 25 m/s. (*a*) What is the impulse delivered to the ball by the kicker? (*b*) If the foot of the kicker is in contact with the ball for 0.008 s, what is the average force exerted by the foot on the ball?

4. A 0.6-kg body initially moving at 15 m/s is stopped by a constant force of 50 N, which lasts for a short time Δ*t*. (*a*) What is the impulse of this force? (*b*) Find the time interval Δ*t*.

5. Find the impulse necessary to stop a 1800-kg car traveling at 70 km/h.

6. A 0.4-kg ball is dropped from a height of 2 m above the floor. It rebounds to a height of 1.5 m. (*a*) Find the impulse exerted by the floor on the ball. (*b*) If the ball was in contact with the floor for 1 ms, find the average force exerted by the floor on the ball during that time.

7. A 0.3-kg brick is dropped from a height of 8 m. It hits the ground and comes to rest. (*a*) What is the impulse exerted by the ground on the brick. (*b*) If it takes 1.3 ms for the brick to come to rest, what is the average force exerted by the ground during this time?

8. When hit, the velocity of an 0.15 kg baseball changes from +20 m/s to −20 m/s. (*a*) What is the magnitude of the impulse delivered by the bat to the ball? (*b*) If the baseball is in contact with the bat for 0.0013 s, find the average force exerted on the ball.

Section 7-2 Conservation of Momentum

9. A cart of mass 15 kg is rolling along a level floor at a speed of 6 m/s. A 5-kg block dropped from rest lands in the cart. (*a*) Find the momentum of the cart before the block was dropped? (*b*) What is the momentum of the cart and block after the block is dropped? (*c*) What is the speed of the cart and block?

10. A 70,000-kg railroad car is traveling at 1.2 m/s. It collides with a 45,000-kg car traveling on the same tracks in the same direction at 0.6 m/s. The two cars couple together. (*a*) Find the total momentum of the system before and after the collision. (*b*) Find the speed of the cars after the collision.

11. An open-topped freight car of mass 15 Mg is rolling along a track at 6 m/s. Rain is falling vertically downward into the car. After the car has collected 3 Mg of water, what is its speed?

12. A 20-g bullet is fired horizontally with speed 250 m/s from a 1.5-kg gun. If the gun were held loosely in the hand, what would its recoil speed be?

Section 7-3 Center of Mass

13. A 24-kg child is 20 m from an 86-kg man. Where is the center of mass of this system?

14. A 5-kg body is at *x* = 10 cm and a 10-kg body is at *x* = 25 cm. Where is the center of mass of this two-body system?

15. Find the center of mass for the system shown in Figure 7-20.

Figure 7-20
Exercise 15.

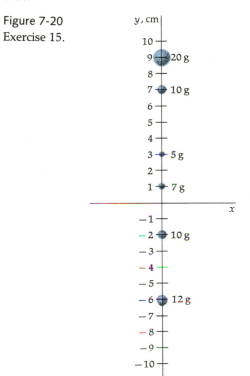

16. A 1.5-Mg car is moving to the west with speed 20 m/s and a 3.0-Mg car is moving to the east with speed 16 m/s. (*a*) Find the total momentum of this system. (*b*) Find the velocity of the center of mass of this system.

Section 7-4 Collisions

17. A 2-Mg car is moving at 30 m/s to the right. It is chasing another 2-Mg car moving at 10 m/s, also to the right. (*a*) If the two cars collide and stick together, what is their speed just after the collision? (*b*) How much mechanical energy is lost in this collision?

18. An 85-kg running back moving at 7 m/s makes an inelastic collision with a 105-kg linebacker who is initially at rest. What is the speed of the players just after the collision?

19. A 3-kg fish is swimming at 1.5 m/s to the right. It swallows a $\frac{1}{4}$-kg fish swimming toward it at 4 m/s (see Figure 7-21). What is the velocity of the larger fish immediately after lunch?

Figure 7-21
Big fish eating little fish for Exercise 19.

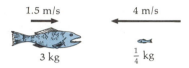

20. A 4-kg block is moving to the right at 3 m/s. It collides with a 6-kg block moving to the left at 2 m/s. (a) What is the total momentum of the two-block system? Find the final velocities of each block (b) if the collision is perfectly inelastic and (c) if the collision is elastic.

21. A 70-kg baseball player jumps vertically into the air to catch a 140-g baseball traveling horizontally at 40 m/s. If the velocity of the baseball player is 15 cm/s upward just before he catches the ball, what is his velocity just afterward.

22. A 12-g bullet is fired with velocity of 320 m/s into a pendulum bob that has a mass of 990 g. How high does the pendulum bob (plus the bullet) swing after the collision?

23. A 25-Mg railway car stands on a hill with its brakes set. The brakes are released, and the car rolls down to the bottom of the hill 9 m below its original position (see Figure 7-22). It collides with a 12-Mg car resting (with its brakes off) at the bottom of the track. The two cars couple together and roll up the track to a height h. Find h.

Figure 7-22
Exercise 25.

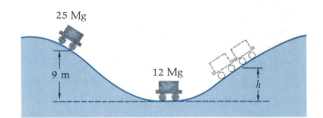

24. A 1.5-Mg car traveling north at 70 km/h collides at an intersection with a 2-Mg car traveling west at 55 km/h. The two cars stick together. (a) What is the total momentum of the system before the collision? (b) Find the magnitude and direction of the velocity of the combined wreckage just after the collision.

25. A ball bounces off the floor to 70 percent of its original height. What is the coefficient of restitution of the ball–floor system?

Section 7-5   Jet Propulsion

There are no exercises for this section.

Problems

1. A small car of mass 800 kg is parked behind a small truck of mass 1600 kg on a level road. The brakes of both the car and the truck are off so that they are both free to roll with negligible friction. A man sitting on the tailgate of the truck exerts a constant force on the car with his feet, as shown in Figure 7-23. The car is observed to accelerate at 1.2 m/s². (a) What is the acceleration of the truck? (b) What is the magnitude of the force exerted on either the truck or the car?

2. A 16-g bullet is fired into the bob of a ballistic pendulum of mass 1.6 kg. When the bob is at its maximum height, the strings make an angle of 60° with the vertical. The length of the pendulum is 2.3 m. Find the speed of the bullet.

3. A 3-kg body moving at 4 m/s makes an elastic collision with a stationary body of mass 2 kg. Use conservation of momentum and the fact that the relative velocity of recession equals the relative velocity of approach to find the velocity of each body after the collision. Check your answer by computing the initial and final kinetic energies of each body.

Figure 7-23
Small car parked behind small truck for Problem 1.

4. A projectile is launched at 20 m/s at an angle of 30° with the horizontal. In the course of its flight, it explodes, breaking into two parts, one of which has twice the mass of the other. The two fragments land simultaneously. The lighter fragment lands 20 m from the launch point in the direction the projection was fired. Where does the other fragment land?

5. A boy on a stationary skateboard (combined mass of 40 kg) throws a ball of mass 0.4 kg forward. Find the initial recoil speed of the boy and skateboard if the speed of the ball is such that it would rise to a height of 8 m if thrown vertically upward.

6. A man of mass 70 kg is riding on a small cart of mass 20 kg that is rolling along a level floor at speed 2 m/s. He jumps off the back of the cart so that his speed relative to the ground is 0.8 m/s in the direction opposite the motion of the cart. (*a*) What is the velocity of the center of mass of the man–cart system after he jumps? (*b*) What is the velocity of the cart after the man jumps? (*c*) What is the velocity of the center of mass of the system after the man hits the ground and comes to rest? (*d*) What force is responsible for the change in the velocity of the center of mass?

7. The payload of a rocket is 5 percent of its total mass, the rest being fuel. If the rocket starts from rest, what is its final velocity if the exhaust velocity is $v_e = 5$ km/s? (*Hint:* Use Equation 7-19.)

8. A rocket exhausts gases with velocity $v_e = 6$ km/s. Find the ratio of the fuel to the payload if the rocket starts from rest and attains a final velocity of (*a*) 12 km/s and (*b*) 30 km/s.

9. A 2000-kg car traveling at 90 km/h crashes into a concrete wall that does not give at all. Estimate the time interval of the collision, assuming that the center of the car travels about halfway to the wall with constant deceleration. (Use any reasonable length for the car.) Estimate the average force exerted by the wall on the car.

10. A 6-kg projectile is launched at $\theta = 30°$ with an initial speed 40 m/s. At the top of its flight, it explodes into two parts of mass 2 and 4 kg. The fragments move horizontally just after the explosion, and the 2-kg piece lands back at the initial launch point. (*a*) Where does the 4-kg piece land? (*b*) Compute the kinetic energy of the projectile just before the explosion and the kinetic energy of the fragments just after the explosion, and find the energy of the explosion.

11. An 80-kg man 2 m tall steps off a ledge, falls 6 m, and makes an inelastic collision with the ground. (*a*) How fast is he traveling just before he hits? (*b*) What is the impulse exerted on the man by the ground? (*c*) Estimate the time of collision, assuming that the force on the man is constant and that he travels 1 m during the collision. (*d*) Using your results for (*b*) and (*c*), find the average force exerted by the ground on the man.

12. A 3-kg bomb slides along a frictionless horizontal plane in the *x* direction at 6 m/s. It explodes into two pieces, one of mass 2 kg and the other of mass 1 kg. The 1-kg piece moves in the horizontal plane in the *y* direction at 4 m/s after the explosion. (*a*) Find the velocity of the 2-kg piece after the explosion. (*b*) What is the velocity of the center of mass after the explosion?

13. A neutron of mass m moves with speed v and collides head-on with a carbon atom of mass $12m$, initially at rest. Assuming the collision to be perfectly elastic, find (*a*) the speed of the neutron and that of the carbon atom after the collision and (*b*) the fraction of the initial kinetic energy of the neutron that is transferred to the carbon atom. The elastic collision is an important mechanism in slowing down neutrons released in fission in a reactor so that they can be captured and produce another fission reaction.

14. A 2-kg body moving at 3 m/s to the right collides with a 3-kg body moving at 2 m/s to the left. The coefficient of restitution is 0.4. Find the velocity of each body after the collision.

15. A 1-kg body moving at 5 m/s makes an elastic collision with a stationary 1-g body. Find the velocity of each body after the collision. This problem illustrates a general result that when a very heavy body strikes a very light body initially at rest, the velocity of the heavy body is hardly changed whereas that of the light body is approximately twice the initial velocity of the heavy body.

16. A hammer of mass 0.6 kg is used to drive nails of mass 25 g into wood. When the hammer has an impact speed of 4.0 m/s, the nail penetrates 1.5 cm at a single blow. Find (*a*) the common speed of the hammer and nail immediately after impact, assuming the collision to be perfectly inelastic; (*b*) the time the nail is in motion, assuming that the initial speed is acquired in a negligible time after which there is uniform deceleration; and (*c*) the average resisting force of the wood against the nail as it penetrates.

17. A ballistic pendulum consists of a pumpkin of mass 1.4 kg suspended on a 1.2-m string. A bullet of mass 12 g has initial speed of 350 m/s and passes through the pumpkin. After the collision, the pendulum swings up to a maximum angle of 20°. Find the final speed of the bullet after the collision and the energy lost in the collision.

18. You throw a 200-g ball to a height of 20 m. Use a reasonable value for the distance the ball moves while it is in your hand to calculate the average force exerted by your hand and also the time the ball is in your hand while you are throwing it. Is it reasonable to neglect the weight of the ball while it is being thrown?

19. A 2-kg body moves at 6 m/s and collides with a 4-kg body initially at rest. After the collision, the 2-kg body is observed to be moving backward at 1 m/s. (*a*) Find the velocity of the 4-kg body after the collision. (*b*) Find the energy lost in the collision. (c) What is the coefficient of restitution for this collision?

20. A 50-kg boy gets on his 12-kg wagon on level ground with two 5-kg bricks. He throws the bricks horizontally off the back of the wagon one at a time, each with a speed of 4 m/s relative to himself. How fast does he go after throwing the second brick? How fast would he go if he threw both bricks at the same time at 4 m/s relative to himself?

21. A 300-g glob of putty is thrown at a 12-kg block initially at rest on a level floor. The putty hits the block and sticks to it. The block and putty slide 16 cm along the floor. If the coefficient of friction is 0.3, what was the original speed of the putty?

22. In this problem, you are to derive Equation 7-14 for an elastic collision, which states that the relative velocity of recession equals the relative velocity of approach. Let the masses of the two bodies be m_1 and m_2, their initial velocities be v_{1i} and v_{2i}, and their final velocities be v_{1f} and v_{2f}. (a) Write equations for the conservation of mechanical energy and the conservation of momentum. (b) Show that the equation for the conservation of energy can be rearranged to obtain

$$m_2(v_{2f}^2 - v_{2i}^2) = m_1(v_{1i}^2 - v_{1f}^2)$$

(c) Factor the differences in the squares in this equation to obtain

$$m_2(v_{2f} - v_{2i})(v_{2f} + v_{2i}) = m_1(v_{1i} - v_{1f})(v_{1i} + v_{1f})$$

(d) Show that the conservation of momentum equation can be written

$$m_2(v_{2f} - v_{2i}) = m_1(v_{1i} - v_{1f})$$

(e) Divide the equation obtained in (d) into that obtained in (c) and rearrange to get the desired result:

$$v_{2f} - v_{1f} = -(v_{2i} - v_{1i})$$

Rotation

In 1851, when Jean Foucault hung a pendulum from the dome of the Pantheon in Paris, proving that the earth turned on its axis was still considered to be a worthwhile pursuit. As the pendulum swung, it traced out a vertical plane that appeared to rotate gradually around the room during the course of a day. In fact, though, the plane of the pendulum was constant. The earth moved around it. Today, a replica of Foucault's pendulum is one of the most popular exhibits at the National Museum of Science in Washington, as it is in virtually every museum that has one.

The next year, Foucault built a gyroscope. Spinning with a high rotational speed, the gyroscope appeared to turn over as the day progressed. But, like the pendulum, the gyroscope actually remained in a fixed orientation in space while the earth moved. Today, this property of the gyroscope is exploited in modern versions of the device that provide a constant heading to guide ships and airplanes, missiles and space probes.

The topic of this chapter, rotation, is one of the two fundamental types of motion. The other is translational motion, which you have already studied. You will find that your work so far has paved the way for your study of rotation, for many of the translational concepts have rotational analogs. By the end of the chapter, you will finally understand many things that you have probably wondered about, like how ice skaters slow or speed up their spins or why, when you tip up a bicycle, the wheel starts to turn.

There are many examples of rotational motion in the world around us. Merry-go-rounds rotate. High divers rotate as they somersault and ice skaters as they pirouette. The earth rotates about its axis. Balls and cylinders rotate as they roll. Knives are sharpened with rotating stone disks.

Star tracks in a time exposure of the night sky.

In this chapter, we will learn how to describe rotational motion. We will first define the kinematic quantities of angular displacement, angular velocity, and angular acceleration, quantities that are the rotational analogs of displacement, velocity, and acceleration, which we use to describe linear motion. We will see that the rotation of a rigid body with constant angular acceleration can be described by equations that are the rotational analogs of the equations we have already dealt with for linear motion with constant acceleration. We will then study the dynamics of rotational motion and introduce the rotational analog of Newton's second law of motion. Again, we will find that the dynamic quantities we encountered in our study of linear motion—force, mass, and momentum—have corresponding rotational analogs—torque, moment of inertia, and angular momentum, respectively.

8-1   Angular Velocity and Angular Acceleration

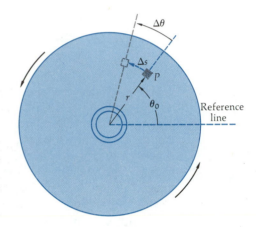

Figure 8-1
A disk rotating about a fixed axis through its center. The distance Δs moved by a particle P on the disk depends on r, but the angular displacement $\Delta \theta$ is the same for all points on the disk.

Figure 8-1 shows a disk that is free to rotate about its axis of symmetry. When the disk rotates, different parts of it move at different speeds. For example, a point near the rim moves faster than a point near the axis. It therefore makes no sense to speak of the speed or velocity of the disk as it rotates. However, in a given time, each point on the disk moves through the same angle. For example, when a point near the rim rotates through $360° = 2\pi$ radians, so does any other point. The angle through which each point on the disk rotates is thus a characteristic of the disk as a whole, as is the rate at which the angle changes. To develop the concepts of angular velocity and angular acceleration, we consider the disk to be made up of many small point particles. As the disk turns, the distance between any two particles remains fixed. Such a system is called a *rigid body*.

Let us focus our attention on a typical particle P on the disk at a distance r from the center. During a small time interval Δt, the particle moves along the arc of a circle a distance Δs given by

$$\Delta s = v \, \Delta t \qquad\qquad 8\text{-}1$$

where v is the speed of the particle. During this time, the radial line to the particle sweeps out an angle $\Delta \theta$ relative to a reference line fixed in space. The measure of this angle in radians is the arc length Δs divided by r:

$$\Delta \theta = \frac{\Delta s}{r} \qquad\qquad 8\text{-}2 \qquad \text{Angular displacement}$$

(If you are unfamiliar with the radian measure of angles, you should read the trigonometry review section in Appendix A.) Although the distance Δs varies from particle to particle, the angle $\Delta \theta$ swept out in a given time is the same for all the particles of the disk. The term $\Delta \theta$ is called the *angular displacement* of the disk. When the disk makes one complete revolution, the arc length Δs is $2\pi r$ and the angular displacement is

$$\Delta \theta = \frac{2\pi r}{r} = 2\pi \text{ rad} = 360°$$

The rate of change of the angle with respect to time, $\Delta\theta/\Delta t$, is the same for all the particles of the disk. It is called the *angular velocity* ω of the disk:

$$\omega = \frac{\Delta\theta}{\Delta t}$$

8-3   Angular velocity

The angular velocity ω is positive or negative, depending on whether θ is increasing (counterclockwise rotation) or decreasing (clockwise rotation). The units of angular velocity are radians per second. Since a radian is a dimensionless unit of angle, the dimensions of angular velocity are those of reciprocal seconds (s^{-1}). The magnitude of the angular velocity is called the *angular speed*. Although the angular motion of a disk is often described using other units, such as revolutions per minute, it is important to remember that many of the equations we will use involving the angular velocity ω are valid only when the angles are expressed in radians and angular velocity is expressed in radians per second. To convert from revolutions per second or revolutions per minute, we use

$$2\pi \text{ rad} = 1 \text{ rev}$$

The rate of change of angular velocity with respect to time is called the *angular acceleration* α.

$$\alpha = \frac{\Delta\omega}{\Delta t}$$

8-4   Angular acceleration

The units of angular acceleration are radians per second per second (rad/s^2).

We can relate the linear speed of a particle on the disk to the angular speed of the disk using Equations 8-2 and 8-3.

$$v = \frac{\Delta s}{\Delta t} = \frac{r\,\Delta\theta}{\Delta t}$$

or

$$v = r\omega$$

8-5   Tangential velocity

Similarly, the tangential acceleration a_t of a particle on the disk is related to the angular acceleration of the disk:

$$a_t = \frac{\Delta v}{\Delta t} = r\frac{\Delta\omega}{\Delta t}$$

Thus,

$$a_t = r\alpha$$

8-6   Tangential acceleration

Each particle of the disk also has a *radial* linear acceleration, which is the centripetal acceleration a_c that points inward along the radial line and has the magnitude

$$a_c = \frac{v^2}{r} = r\omega^2$$

8-7   Centripetal acceleration

The three quantities angular displacement θ, angular velocity ω, and angular acceleration α are analogous to linear displacement x, linear velocity v, and linear acceleration a in one-dimensional motion (Chapter 2). Because of the similarity of the definitions of rotational and linear quantities, much of what we learned in Chapter 2 will be useful in treating problems dealing with the rotation of a rigid body. For example, the equations for constant angular acceleration are the same as Equations 2-8 and 2-10 with θ replacing x, ω replacing v, and α replacing a. Thus,

$$\omega = \omega_0 + \alpha t \qquad\qquad 8\text{-}8$$

Constant-angular-acceleration equations

is the rotational analog of

$$v = v_0 + at$$

and

$$\theta = \theta_0 + \omega_0 t + \tfrac{1}{2}\alpha_0 t^2 \qquad\qquad 8\text{-}9$$

where ω_0 and θ_0 are the initial values of angular velocity and angular position, respectively, is the rotational analog of

$$x = x_0 + v_0 t + \tfrac{1}{2}at^2$$

As with the constant-linear-acceleration formulas, we can eliminate time from these equations to obtain an equation relating angular displacement, angular velocity, and angular acceleration,

$$\omega^2 = \omega_0^2 + 2\alpha(\theta - \theta_0) \qquad\qquad 8\text{-}10$$

which is the rotational analog of

$$v^2 = v_0^2 + 2a(x - x_0)$$

Example 8-1   A disk rotates with constant angular acceleration $\alpha = 2 \text{ rad/s}^2$. If the disk starts from rest, how many revolutions does it make in 10 s?

This is analogous to the linear problem of finding the distance traveled by a particle in a given time if it starts from rest with constant acceleration.
   The number of revolutions is related to the angular displacement by the fact that each revolution is an angular displacement of 2π rad. Thus we need to find the angular displacement $\theta - \theta_0$ in radians for a time of 10 s and multiply it by the conversion factor $(1 \text{ rev})/(2\pi \text{ rad})$.
   Equation 8-9 relates the angular displacement to the time. We are given $\omega_0 = 0$ (the disk starts from rest). Thus,

$$\theta - \theta_0 = \tfrac{1}{2}\alpha t^2 = \tfrac{1}{2}(2 \text{ rad/s}^2)(10 \text{ s})^2 = 100 \text{ rad}$$

The number of revolutions is therefore

$$100 \text{ rad} \times \frac{1 \text{ rev}}{2\pi \text{ rad}} = 15.9 \text{ rev}$$

Example 8-2   Find the angular speed of the disk in Example 8-1 after 10 s.
   Using Equation 8-8, we have

$$\omega = \omega_0 + at = 0 + (2 \text{ rad/s}^2)(10 \text{ s}) = 20 \text{ rad/s}$$

As a check of this result and that of the previous example, we can also find the angular speed from Equation 8-10,

$$\omega^2 = 2\alpha(\theta - \theta_0) = 2(2 \text{ rad/s}^2)(100 \text{ rad})$$

$$= 400 \text{ rad}^2/\text{s}^2$$

or

$$\omega = \sqrt{400 \ \text{rad}^2/\text{s}^2} = 20 \ \text{rad/s}$$

This bowl is shaped by rotating it on a potter's wheel.

Question

1. Two points are on a disk turning at constant angular velocity, one point on the rim and the other halfway between the rim and the axis. Which point moves the greater distance in a given time? Which turns through the greater angle? Which has the greater speed? The greater angular velocity? The greater tangential acceleration? The greater angular acceleration? The greater centripetal acceleration?

8-2   Torque and Moment of Inertia

In our discussion of the static equilibrium of a rigid body in Chapter 5, we found that a necessary condition for a body not to rotate is that the resultant torque about any point must be zero. Although this is a *necessary* condition, it is not a *sufficient* condition for a body to be static. If a body is rotating about an axis, it will continue to rotate with constant angular velocity if there is no resultant external torque acting on it. (This is similar to the situation for linear motion: if a body is moving with some velocity, it will continue to move with constant velocity if there is no resultant external *force* acting on it.) We will now show that if there is a resultant external torque acting about the point of rotation of a rigid body, the angular velocity of the body does not remain constant but changes with an angular acceleration that is proportional to the external torque.

We first consider a simple system consisting of a particle of mass m attached to a thin rod of negligible mass that is pivoted at a point O a distance r from the particle. The particle is then constrained to move in a circle of radius r. Suppose that there is a force $\mathbf{F}$ acting on the particle perpendicular to the rod and thus tangential to the circle, as shown in Figure 8-2. For the particle, Newton's second law gives

$$F = ma = mr\alpha$$

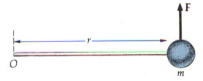

Figure 8-2
Force $\mathbf{F}$ exerted on a point mass m constrained to move in a circle of radius r.

where we have used the relation $a = r\alpha$ between the tangential acceleration of the mass and the angular acceleration. If we multiply each side by r, we obtain

$$rF = mr^2\alpha \qquad\qquad 8\text{-}11$$

The left side of Equation 8-11 is the torque, $\tau = rF$, exerted by the force F Torque
about the pivot O. We thus have

$$\tau = mr^2\alpha \qquad\qquad 8\text{-}12$$

If we have a group of particles or an extended body, such as a disk, we can apply Equation 8-12 to each part of the body and sum over the entire body. Applying this equation to the ith particle of mass m_i, we have

$$\tau_i = m_i r_i^2 \alpha$$

If we now sum over all the particles in the body, we obtain

$$\Sigma\tau_i = \Sigma m_i r_i^2 \alpha \qquad\qquad 8\text{-}13$$

The quantity $\Sigma\tau_i$ is the resultant torque acting on the body, which we shall denote by τ_{net}. For a rigid body, the angular acceleration is the same for all particles and can therefore be taken out of the sum. The quantity $\Sigma m_i r_i^2$ is a property of the body and the axis of rotation called the *moment of inertia I.*

$$\boxed{I = \Sigma m_i r_i^2} \qquad\qquad 8\text{-}14 \qquad \text{Moment of inertia}$$

The moment of inertia is a measure of the resistance of a body to changes in its rotational motion. It depends on the mass distribution relative to the axis of rotation of the body. It is a property of the body (and the axis of rotation) just as the mass m is a property of the body that measures its resistance to changes in translational motion. For systems consisting of a small number of discrete particles, we can compute the moment of inertia about a given axis directly from its definition (Equation 8-14). In the more common case of a continuous body, such as a wheel, the moment of inertia about a given axis is computed using calculus. Table 8-1 lists the moments of inertia for various uniform bodies.

In terms of the moment of inertia, Equation 8-13 becomes

$$\boxed{\tau_{net} = I\alpha} \qquad\qquad 8\text{-}15$$

Equation 8-15 is the rotational analog of Newton's second law, $F_{net} = ma$, for linear motion. The resultant torque is analogous to the resultant force, the moment of inertia is analogous to the mass, and the angular acceleration is analogous to the linear acceleration.

Example 8-3   Four particles of mass m are connected by massless rods to form a rectangle of sides $2a$ and $2b$ as shown in Figure 8-3. The system rotates about an axis in the plane of the figure through the center. Find the moment of inertia about this axis.

From the figure, we see that the distance from each particle to the axis of rotation is a. The moment of inertia of each particle about this axis is therefore ma^2, and since there are four particles, the total moment of inertia of the body is $I = 4ma^2$. The distance b plays no role at all because it is not related to the distance from any mass to the axis of rotation.

Example 8-4   Find the moment of inertia of the system of Example 8-3 for rotation about an axis parallel to the first axis but passing through two of the masses as shown in Figure 8-4.

For this rotation, two of the masses are at a distance $2a$ from the axis of rotation and two are on the axis (and therefore at zero distance). The moment of inertia is thus

$$I = \Sigma m_i r_i^2$$

$$= m(0)^2 + m(0)^2 + m(2a)^2 + m(2a)^2 = 8ma^2$$

This example illustrates the fact that the moment of inertia depends on the axis of rotation. We note that the moment of inertia is larger about this axis than about the one parallel to it through the center of mass.

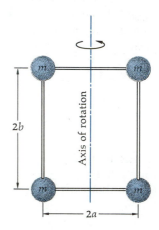

Figure 8-3
Four particles of equal mass connected by massless rods and rotating about an axis in the plane of the particles and through the center of mass for Example 8-3.

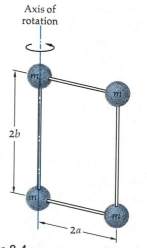

Figure 8-4
Same system as in Figure 8-3 except that the axis of rotation passes through two of the particles for Example 8-4.

Table 8-1

Moments of Inertia of Uniform Bodies of Various Shapes

Cylindrical shell about axis $I = MR^2$	Thin rod about perpendicular line through center $I = \frac{1}{12}ML^2$
Solid cylinder about axis $I = \frac{1}{2}MR^2$	Thin rod about perpendicular line through one end $I = \frac{1}{3}ML^2$
Hollow cylinder about axis $I = \frac{1}{2}M(R_1^2 + R_2^2)$	Thin spherical shell about diameter $I = \frac{2}{3}MR^2$
Cylindrical shell about diameter through center $I = \frac{1}{2}MR^2 + \frac{1}{12}ML^2$	Solid sphere about diameter $I = \frac{2}{5}MR^2$
Solid cylinder about diameter through center $I = \frac{1}{4}MR^2 + \frac{1}{12}ML^2$	Solid rectangular parallelepiped about axis through center perpendicular to face $I = \frac{1}{12}M(a^2 + b^2)$

Example 8-5   A string is wound around the rim of a uniform disk pivoted to rotate without friction about a fixed axis through its center (see Figure 8-5). The mass of the disk is 3 kg, and its radius is 25 cm. The string is pulled with a force of 10 N. If the disk is initially at rest, what is its angular velocity after 5 s?

From Table 8-1, the moment of inertia of a uniform disk about its axis is

$$I = \frac{1}{2}MR^2$$

$$= \frac{1}{2}(3 \text{ kg})(0.25 \text{ m})^2 = 9.38 \times 10^{-2} \text{ kg} \cdot \text{m}^2$$

Since the direction of the string as it leaves the rim of the disk is always tangent to the disk, the lever arm of the force it exerts is just R. The applied torque is thus

$$\tau = TR = (10 \text{ N})(0.25 \text{ m}) = 2.5 \text{ N} \cdot \text{m}$$

Figure 8-5
String wrapped around a disk for Example 8-5.

To find the angular velocity, we first find the angular acceleration from Newton's second law for rotational motion (Equation 8-15)

$$\alpha = \frac{\tau_{net}}{I} = \frac{2.5 \text{ N} \cdot \text{m}}{0.0938 \text{ kg} \cdot \text{m}^2} = 26.7 \text{ rad/s}^2$$

Since α is constant, we find ω from Equation 8-8, setting $\omega_0 = 0$:

$$\omega = \omega_0 + \alpha t = 0 + (26.7 \text{ rad/s}^2)(5 \text{ s}) = 133 \text{ rad/s}$$

Example 8-6   A body of mass m is tied to a light string wound around a wheel of moment of inertia I and radius R (see Figure 8-6). The wheel bearing is frictionless, and the string does not slip on the rim. Find the tension in the string and the acceleration of the body.

The only force exerting a torque on the wheel is the tension T in the string. It has a lever arm R. Hence,

$$TR = I\alpha \qquad\qquad \text{8-16}$$

Two forces act on the suspended body, the upward tension T and the downward force of gravity mg. Taking the downward direction to be positive, we have from Newton's second law

$$mg - T = ma \qquad\qquad \text{8-17}$$

There are three unknowns, T, a, and α, in these two equations. The string provides a constraint by which we can relate a and α. As the wheel turns through angle θ, an amount of string of length $s = R\theta$ unwraps and the body falls a distance $y = R\theta$. The velocity of the body is then $v = R\omega$ and its acceleration is

$$a = R\alpha \qquad\qquad \text{8-18}$$

Substituting a/R for α in Equation 8-16 gives

$$TR = \frac{Ia}{R} \qquad\qquad \text{8-19}$$

or

$$a = \frac{TR^2}{I}$$

Substituting this result for a into Equation 8-17 gives

$$mg - T = \frac{mTR^2}{I}$$

or

$$T\left(1 + \frac{mR^2}{I}\right) = mg$$

$$T = \frac{mI}{I + mR^2} g$$

We can use this value for T in Equation 8-19 to find a:

$$a = \frac{mR^2}{I + mR^2} g$$

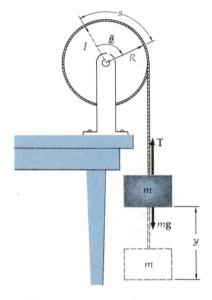

Figure 8-6
Body tied to a string that is wrapped around a rotating wheel for Example 8-6. When the wheel turns through angle θ, a length of string $R\theta$ unwraps and the body drops a distance $y = R\theta$.

Questions

2. Can an object rotate if there is no torque acting on it?

3. Can a given rigid body have more than one moment of inertia?

4. Does an applied resultant torque always increase the angular speed of a body?

5. If the angular velocity of a body is zero at some instant, does this mean that the resultant torque on the body must be zero?

8-3   Kinetic Energy and Angular Momentum

When a body rotates about an axis, it has kinetic energy associated with the rotational motion. This is not surprising since it would take work (a force acting over a distance) to cause the body to rotate. The kinetic energy of a rotating system is just the sum of the kinetic energies of the individual particles in the system. Let us consider again the rotating disk. If v_i is the speed of the ith particle of mass m_i, the total kinetic energy of the disk is

$$E_k = \Sigma \tfrac{1}{2} m_i v_i^2$$

$$= \Sigma \tfrac{1}{2} m_i (r_i \omega)^2$$

$$= \tfrac{1}{2} (\Sigma m_i r_i^2) \omega^2$$

or

$$E_k = \tfrac{1}{2} I \omega^2 \qquad\qquad \text{8-20}$$

Rotational kinetic energy

Note the similarity of this expression and that for linear motion

$$E_k = \tfrac{1}{2} m v^2$$

In rotational motion, the angular velocity ω plays the role of the velocity v, and the moment of inertia I plays the role of the mass m.

Example 8-7   Find the kinetic energy of the disk of Example 8-5 when it is rotating at 133 rad/s.

In Example 8-5, we computed the moment of inertia of the disk to be 9.38×10^{-2} kg·m². Its kinetic energy is thus

$$E_k = \tfrac{1}{2} I \omega^2$$

$$= \tfrac{1}{2}(9.38 \times 10^{-2} \text{ kg·m}^2)(133 \text{ rad/s})^2$$

$$= 830 \text{ kg·m}^2/\text{s}^2$$

$$= 830 \text{ J}$$

Note that the dimensionless unit, the radian, can be simply omitted from the final units. Since a kg·m/s² is a newton, and a newton-meter is a joule, the unit kg·m²/s² is equal to a joule.

Figure 8-7 shows a force **F** applied tangentially to a rotating disk. When the disk turns through an angle $\Delta\theta$, the force moves through a distance $\Delta s = R\,\Delta\theta$. The work done by the force is thus

$$\Delta W = F\,\Delta s = FR\,\Delta\theta = \tau\,\Delta\theta \qquad \text{8-21}$$

where we have used the fact that FR is the torque τ. The rate of doing work $\Delta W/\Delta t$ is the power input of the torque:

$$P = \tau\frac{\Delta\theta}{\Delta t} = \tau\omega \qquad \text{8-22}$$

Power

Equation 8-22 is the rotational analog of the linear equation $P = Fv$.

The *angular momentum L* of a body is defined as the product of the moment of inertia and the angular velocity:

$$L = I\omega \qquad \text{8-23}$$

Angular momentum

The generalized form of Newton's second law for rotation is

$$\tau_{\text{net}} = \frac{\Delta L}{\Delta t} = \frac{\Delta(I\omega)}{\Delta t} \qquad \text{8-24}$$

The rate of change of the angular momentum of a system equals the net torque acting on the system.

Generalization of Newton's second law for rotation

For a rigid body, the moment of inertia does not change, and Equation 8-24 is the same as Equation 8-15.

$$\tau_{\text{net}} = \frac{\Delta(I\omega)}{\Delta t} = I\frac{\Delta\omega}{\Delta t} = I\alpha$$

However, for a general system of particles, the moment of inertia need not remain constant. Equation 8-24 holds in general whether or not the moment of inertia is constant.

When there is no external torque acting on a system, we have from Equation 8-24

$$\frac{\Delta L}{\Delta t} = 0$$

or

$$L = \text{constant} \qquad \text{8-25}$$

This is analogous to the linear case that the momentum is constant when the net external force is zero:

$$F_{\text{net}} = \frac{\Delta P}{\Delta t} = 0$$

$$P = \text{constant}$$

Equation 8-25 is a statement of the *law of conservation of angular momentum.*

If the resultant external torque acting on a system is zero, the total angular momentum of the system is constant.

Conservation of angular momentum

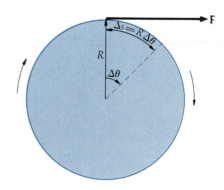

Figure 8-7
When the wheel rotates through $\Delta\theta$, the force moves through $\Delta s = R\,\Delta\theta$. The work done is therefore $F\,\Delta s = FR\,\Delta\theta = \tau\,\Delta\theta$.

If a system is isolated from its surroundings, there can be no external forces or torques. We therefore have a third conservation law for isolated systems. Energy, linear momentum, and angular momentum are all conserved. The law of conservation of angular momentum is a fundamental law of nature. Even on a microscopic scale in atomic and nuclear physics, where newtonian mechanics does not hold, the angular momentum of an isolated system is constant over time.

There are many examples of the conservation of angular momentum in everyday life. Consider, for example, an ice skater spinning on her toes (see Figure 8-8). Because the torque exerted by the ice is small, the angular momentum of the skater is approximately constant. When she draws in her arms to her body, the moment of inertia of her body about a vertical axis through her body is reduced. Since the angular momentum, $L = I\omega$, must remain constant, as I decreases her angular velocity ω increases; that is, she spins at a faster rate.

Figure 8-8
Because the torque exerted by the ice is small, the angular momentum of the skater is approximately constant. When she reduces her moment of inertia by drawing in her arms, her angular velocity increases.

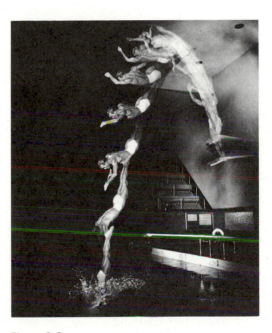

Figure 8-9
Multiflash photograph of a diver. His initial angular momentum is provided by the torque exerted by the diving board. He can increase his angular velocity by pulling in his arms and legs, thus reducing his moment of inertia.

Figure 8-9 is a multiflash photograph of a diver. The diver's center of mass moves in a parabolic path after the diver leaves the board. The angular momentum is provided by the initial torque due to the force of the board. (This force does not pass through the center of mass if the diver leans forward as he jumps.) If the diver wants to somersault in the air, he draws in his arms and legs, decreasing his moment of inertia to increase his angular velocity.

Table 8-2

Comparison of Linear Motion and Rotational Motion

Linear motion		Rotational motion	
Displacement	Δx	Angular displacement	$\Delta\theta$
Velocity	$v = \dfrac{\Delta x}{\Delta t}$	Angular velocity	$\omega = \dfrac{\Delta\theta}{\Delta t}$
Acceleration	$a = \dfrac{\Delta v}{\Delta t}$	Angular acceleration	$\alpha = \dfrac{\Delta\omega}{\Delta t}$
Constant-acceleration equations	$v = v_0 + at$ $\Delta x = v_{av}\,\Delta t$ $v_{av} = \frac{1}{2}(v_0 + v)$ $x = x_0 + v_0 t + \frac{1}{2}at^2$ $v^2 = v_0^2 + 2a\,\Delta x$	Constant-angular-acceleration equations	$\omega = \omega_0 + \alpha t$ $\Delta\theta = \omega_{av}\,\Delta t$ $\omega_{av} = \frac{1}{2}(\omega_0 + \omega)$ $\theta = \theta_0 + \omega_0 t + \frac{1}{2}\alpha t^2$ $\omega^2 = \omega_0^2 + 2\alpha\,\Delta\theta$
Mass	m	Moment of inertia	I
Momentum	$\mathbf{p} = m\mathbf{v}$	Angular momentum	$L = I\omega$
Force	$\mathbf{F}$	Torque	τ
Power	$P = Fv$	Power	$P = \tau\omega$
Newton's second law	$\Sigma\mathbf{F} = \dfrac{\Delta\mathbf{p}}{\Delta t} = m\mathbf{a}$ if $\mathbf{F}_{net} = 0$, $\mathbf{P} = $ constant	Newton's second law	$\Sigma\tau = \dfrac{\Delta L}{\Delta t} = I\alpha$ if $\tau_{net} = 0$, $L = $ constant

Table 8-2 lists some of the rotational equations we have developed in this chapter beside the analogous equations for linear motion.

Example 8-8   A merry-go-round of radius 2 m and of moment of inertia 500 kg·m² is rotating without friction. It makes 1 rev every 5 s. A child of mass 25 kg sitting at the center crawls out to the rim. Find the new angular speed of the merry-go-round.

Since there is no external torque on the child–merry-go-round system, the angular momentum of the system remains constant. Originally, the child is at the center and thus has no appreciable moment of inertia about the axis and therefore no angular momentum. When he is at the rim, he has angular momentum $I_c\omega_f$, where $I_c = mr^2$ is the moment of inertia of the child relative to the axis of the merry-go-round and ω_f is the final angular velocity of the child and the merry-go-round. Since the mass of the child is 25 kg and r is 2 m, the moment of inertia of the child when he is at the rim is

$$I_c = (25 \text{ kg})(2 \text{ m})^2 = 100 \text{ kg·m}^2$$

Writing I_m for the moment of inertia of the merry-go-round and ω_i for its initial angular velocity, we have from conservation of angular momentum

$$I_m\omega_i = I_m\omega_f + I_c\omega_f = (I_m + I_c)\omega_f$$

$$\omega_f = \frac{I_m}{I_m + I_c}\,\omega_i = \frac{500 \text{ kg·m}^2}{(500 + 100) \text{ kg·m}^2}\,\omega_i = \tfrac{5}{6}\omega_i$$

The rotation of this nonrigid body, the Whirlpool Galaxy, is indicated by its spiral shape.

Since the merry-go-round originally makes 1 rev every 5 s, its initial angular velocity is $\frac{1}{5}$ rev/s or $0.4\,\pi$ rad/s. The final angular velocity is then

$$\omega_f = \tfrac{5}{6}(\tfrac{1}{5}\text{ rev/s}) = \tfrac{1}{6}\text{ rev/s}$$

Thus, after the child reaches the rim, the merry-go-round rotates at 1 rev every 6 s.

When the child is at the center of the merry-go-round, he is at rest. As he crawls out, he begins moving in a circle. The force that accelerates the child is the force of friction exerted by the merry-go-round. This force is tangential to the circle and thus produces a torque that increases the angular momentum of the child. The child exerts an equal and opposite frictional force on the merry-go-round. The torque associated with this force decreases the angular momentum of the merry-go-round.

Example 8-9   A disk with moment of inertia I_1 is rotating with angular velocity ω_i about a frictionless shaft. It drops onto another disk with moment of inertia I_2, which is initially at rest on the same shaft (see Figure 8-10). Because of surface friction, the two disks eventually attain a common angular velocity, ω_f. Find this final common angular velocity.

Each disk exerts a torque on the other, but there is no torque external to the two-disk system. The angular momentum of the first disk about its center of mass is

$$L_i = L_1 \omega_i$$

When both disks are rotating together, the total angular momentum is

$$L_f = I_1 \omega_f + I_2 \omega_f = (I_1 + I_2)\omega_f$$

Setting the final angular momentum equal to the initial angular momentum, we obtain

$$(I_1 + I_2)\omega_f = I_1 \omega_f$$

The final angular velocity is therefore

$$\omega_f = \frac{I_1}{I_1 + I_2}\,\omega_i$$

This interaction of the disks is analogous to an inelastic collision between two masses in one dimension. Energy is not conserved in this collision. We can see this by writing the energy in terms of the angular momentum. The initial kinetic energy is

$$E_{ki} = \tfrac{1}{2}I_1 \omega_i^2 = \frac{(I_1 \omega_i)^2}{2I_1} = \frac{L_i^2}{2I_1}$$

The final energy is

$$\tfrac{1}{2}(I_1 + I_2)\omega_f^2 = \frac{[(I_1 + I_2)\omega_f]^2}{2(I_1 + I_2)} = \frac{L_f^2}{2(I_1 + I_2)}$$

Since $L_f = L_i$, the final kinetic energy is less than the initial kinetic energy by the factor $I_1/(I_1 + I_2)$.

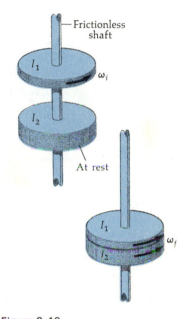

Figure 8-10
Inelastic rotational collision of Example 8-9. Since the only torques acting are internal to the system, angular momentum is conserved.

Question

6. A woman sits on a spinning piano stool with her arms folded. If she extends her arms, what happens to her angular velocity?

8-4   Rolling Bodies

Figure 8-11 shows a ball of radius R rolling along a plane surface. If the ball rolls without slipping, the rotational and translational motions are simply related. As the ball turns through an angle ϕ, as shown in the figure, the point of contact between the ball and plane moves a distance s related to ϕ by

$$s = R\phi \qquad\qquad 8\text{-}26$$

Since the center of the ball lies directly over the point of contact, the center of mass of the ball also moves the distance s. The velocity of the center of mass and the angular velocity of rotation are therefore related by

$$v = \frac{\Delta s}{\Delta t} = R\,\frac{\Delta \phi}{\Delta t}$$

or

$$v_{cm} = R\omega \qquad\qquad 8\text{-}27$$

Equation 8-27 (or the equivalent Equation 8-26) is called the *rolling condition*. It holds whenever a ball or cylinder rolls without slipping. We can also write the rolling condition in terms of the linear and angular accelerations:

$$\frac{\Delta v}{\Delta t} = R\,\frac{\Delta \omega}{\Delta t}$$

or

$$a_{cm} = R\alpha \qquad\qquad 8\text{-}28$$

When a ball rotates with angular speed ω, a point on the edge of the ball has speed $R\omega$ relative to the center of the ball. Since the center of the ball is moving with speed $R\omega$ relative to the surface, and the edge in contact is moving backward with this same speed relative to the center of the ball, the edge of the ball in contact with the surface is instantaneously at rest relative to the surface (see Figure 8-12). If there is a frictional force exerted by the surface on the ball, it is static friction and no energy is dissipated.

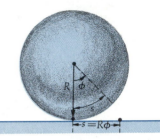

Figure 8-11
As a ball or cylinder rolls without slipping, the center of mass moves a distance $s = R\phi$ as the ball rotates through angle ϕ.

Rolling condition

Figure 8-12
(*a*) Translation without rotation. The top and bottom of the ball move with the same velocity. (*b*) Rotation without translation. The top of the ball moves to the right with speed $v = R\omega$ relative to the center, which is at rest. The bottom moves to the left with speed $v = R\omega$ relative to the center. (*c*) Rolling without slipping is a combination of translation and rotation. If the center moves with speed v, the top moves with speed $2v$ and the bottom of the ball is momentarily at rest.

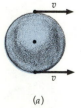

(*a*)
Translation
without rotation

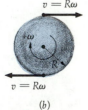

(*b*)
Rotation
without translation

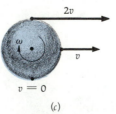

(*c*)
Translation
with rotation

We can apply these ideas to several interesting problems. We first consider what happens when a bowling ball is thrown so that it initially has no rotation. As it slides along the floor, there is a force of sliding friction **f** opposing the motion (see Figure 8-13). This frictional force causes the ball to rotate, and it also reduces the linear velocity of the center of mass of the ball. The angular velocity increases and the linear velocity decreases until the rolling condition $v = R\omega$ is met. Then the ball rolls without slipping, and there is no sliding friction between the surfaces.

We next consider the problem of where to strike the cue ball in pool or billiards. If the cue ball is struck at the center with a horizontal force F from the cue stick, the ball will initially have no rotation, similar to the bowling ball we just considered. If it is struck below the center, it will start moving with backspin. The force of sliding friction will reduce the backspin and will eventually produce forward spin until the rolling condition is met. Where, then, should the cue ball be struck so that it will begin to roll without slipping? We can find this by requiring that the linear acceleration of the ball and the angular acceleration produced by the cue stick obey the rolling condition of Equation 8-28. Let F be the force exerted by the cue stick at a height x above the center of the ball (see Figure 8-14). Since the cue stick imparts a very large impulsive force for a very short time, we can neglect friction during this time. Newton's second law for linear motion gives

$$F = ma = mR\alpha \qquad\qquad 8\text{-}29$$

where we have substituted in the rolling condition $a = R\alpha$. Newton's second law for rotation gives

$$\tau = Fx = I\alpha \qquad\qquad 8\text{-}30$$

Dividing Equation 8-30 by Equation 8-29 to eliminate F and α gives

$$x = \frac{I}{mR}$$

For a ball, the moment of inertia (Table 8-1) is $I = \frac{2}{5}mR^2$. Thus,

$$x = \frac{2}{5}R$$

Figure 8-13
A bowling ball thrown with no initial rotation. The frictional force **f** exerted by the floor reduces the speed of the center of mass and increases the angular speed until the rolling condition $v = R\omega$ is reached.

Figure 8-14
Cue stick hitting cue ball a distance x above the center. If x is chosen properly, the cue ball will begin rolling without slipping.

If the ball is struck at $\frac{2}{5}R$ above the center, it will begin to roll without slipping. If it is struck at a point higher than this, it will have top spin. Then there will be a force of friction in the direction of motion (see Figure 8-15) that increases the linear velocity of the ball and decreases its rotational velocity until the rolling condition is met.

The final problem we consider is that of balls and cylinders rolling down an incline. If we place a ball and a cylinder at the top of an incline and release them at the same time, which will reach the bottom first? Figure 8-16 shows a ball rolling down an incline. If the ball is to roll without slipping, its

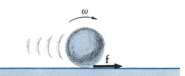

Figure 8-15
Cue ball with top spin. The force of friction increases the speed of the center of mass and reduces the angular speed until the rolling condition is met.

angular acceleration and linear acceleration must be related by Equation 8-28. The angular acceleration is the result of the force of static friction exerted by the surface on the ball. This force provides a torque about the center of mass of the ball that causes its angular speed to increase as its linear speed increases to maintain rolling.

We can find the speed of a ball or cylinder at the bottom of the incline from energy considerations. Whichever has the greater speed at the bottom must have the greater average speed so that it reaches the bottom first. We consider the ball first. At the top of the incline, the energy of the ball is mgh where h is the height of the incline. At the bottom, the ball has both rotational and translational kinetic energy. Conservation of energy gives

$$\tfrac{1}{2}mv^2 + \tfrac{1}{2}I\omega^2 = mgh$$

We can use the rolling condition, $v = R\omega$, to eliminate either v or ω. Eliminating ω, we obtain

$$\tfrac{1}{2}mv^2 \tfrac{1}{2}I(v/R)^2 = mgh$$

Solving for v^2, we obtain

$$v^2 = \frac{2mgh}{m + I/R^2} \qquad\qquad 8\text{-}31$$

For a ball, the moment of inertia (Table 8-1) is $\tfrac{2}{5}mR^2$. Substituting this into Equation 8-31, we obtain

$$v^2 = \frac{2mgh}{m + 2m/5} = \tfrac{5}{7}(2gh) \qquad \text{ball} \qquad 8\text{-}32$$

This is less than the result $v^2 = 2gh$ that is obtained for a body sliding down an incline without friction. Equation 8-31 also applies to a cylinder, which has moment of inertia $I = \tfrac{1}{2}mR^2$. Substituting this for I in Equation 8-31, we obtain

$$v^2 = \frac{2mgh}{m + m/2} = \tfrac{2}{3}(2gh) \qquad \text{cylinder} \qquad 8\text{-}33$$

Since $\tfrac{5}{7} = 0.71$ is greater than $\tfrac{2}{3} = 0.67$, the speed of the ball is greater and it will reach the bottom first if the two are released together. Note that Equations 8-32 and 8-33 do not contain the radius of the ball or cylinder. Any ball will beat any cylinder, assuming they each have uniform mass distributions.

Question

7. A wheel rolls along a level surface at speed v without slipping. Relative to the surface, what is the speed of the top of the wheel?

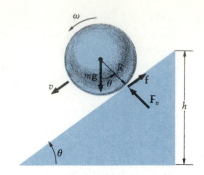

Figure 8-16
Ball rolling down incline. There is a frictional force acting up the incline that exerts a torque about the center of mass of the ball, which increases its angular velocity and maintains the rolling condition as the ball accelerates down the incline.

8-5   Motion of a Gyroscope

For general rotational motion, the quantities torque, angular velocity, angular acceleration, and angular momentum are vector quantities. We did not have to worry about the vector nature of these quantities in our treatment so far because we considered only rotation about a fixed axis (or, in the case of rolling bodies, about an axis that moved parallel to itself). Our results were therefore analogous to linear motion in one dimension, where we could use a + or − sign to indicate the direction of the velocity or acceleration. When the axis of rotation is not fixed in space, the motion is very complicated. In this section, we will discuss the motion of a gyroscope, a device consisting of a disk or wheel with a large moment of inertia that is usually mounted so that its axis is free to turn in any direction.

A large demonstration gyroscope can be made from a bicycle wheel by loading the rim (to increase its moment of inertia). If you have access to such a device (most physics department stockrooms have one), you should play with it to experience its somewhat surprising properties. For example, if the gyroscope is set spinning with its axis horizontal, and you push horizontally on one side of the axle, the axle tends to move up or down rather than in the direction it is pushed. To understand this behavior, we need to take into account the vector nature of the rotational quantities we have been using.

Figure 8-17 shows a wheel rotating in the plane of the paper. We wish to assign a direction to the angular velocity ω. We cannot use any direction in the plane of the paper because all directions in this plane are equivalent. We could say the wheel is rotating clockwise or counterclockwise, but this depends on how we look at the wheel. Furthermore, we couldn't use these terms to describe a change in the direction of the axis, for example, if the axis turned so that it was no longer perpendicular to the paper. The direction we choose to describe the rotation of the wheel is the direction of the axis of rotation, which is perpendicular to the wheel. To completely specify the rotation, we must also specify the sense of rotation, which we do by using the right-hand rule illustrated in Figure 8-18. To apply this rule, we place the fingers of our right hand on the axis or the rim of the wheel pointing in the direction of the motion of the rim, and the thumb points in the direction of ω. Since α is the rate of change of ω, its direction follows from that of ω. The direction of torque is that of the angular acceleration it would produce if it were acting alone. For example, the torque τ produced by the force **F** in Figure 8-19 is into the paper. The angular momentum **L** is the product of the moment of inertia I and the angular velocity ω, so its direction is that of ω.*

* In our discussion, we have somewhat oversimplified the concept of angular momentum. In advanced treatments, it is defined more generally. When an object rotates about an axis that is not an axis of symmetry, the angular momentum is not necessarily equal to $I\omega$, nor is it parallel to the direction of ω. We need not be concerned with such complications here.

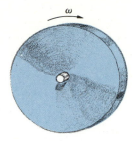

Figure 8-17
Wheel rotating about an axis through its center and perpendicular to its plane. All directions in the plane of the wheel are equivalent. The angular velocity can be described by a vector along the axis of the wheel perpendicular to the plane of the wheel.

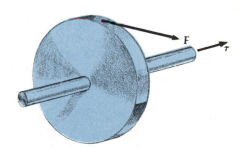

Figure 8-19
The torque τ exerted by the force **F** shown is along the axis of rotation and in the direction of the angular acceleration that would result if **F** were acting alone.

Figure 8-18
The right-hand rule for determining the direction of the angular velocity ω.

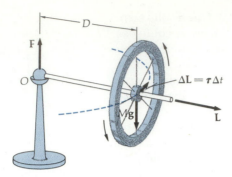

Figure 8-20

Figure 8-20
Gyroscope. The force of gravity Mg produces a torque about the pivot into the paper, which causes a *change* in the angular momentum in that direction. If the wheel is initially spinning with **L** along its axle, the change is perpendicular to **L**, and the axle moves in the direction of the torque. This motion is called *precession*.

A gyroscope mounted so that it is free to turn in any direction. When the gyroscope is set spinning, its axis maintains a fixed direction in space no matter how the base is moved.

We are now ready to consider the motion of a gyroscope in which the axis of rotation changes direction. Figure 8-20 shows such a system consisting of a bicycle wheel free to turn on an axle. The axle is pivoted at a point a distance D from the center of the wheel, but it is free to turn in any direction. We will try to obtain a qualitative understanding of the complicated motion of such a system by using the general form of Newton's second law for rotation in vector form:

$$\tau_{net} = \frac{\Delta \mathbf{L}}{\Delta t}$$

All that we really need to remember is that the *change* in angular momentum of the wheel must be in the direction of the net torque acting on the wheel.

When the axle is held horizontal and is then released, if the wheel is not spinning, it simply falls. The torque about point O is MgD in the direction into the page. As the wheel falls, it has angular velocity ω in the direction of the torque. This angular velocity (and angular momentum) is due to motion of the wheel itself; that is, it is associated with the motion of the center of mass of the wheel. (Since the center of mass accelerates downward, the upward force F exerted by the support at O is evidently less than Mg.)

We now consider what happens when the wheel is initially set spinning and the axle is held horizontal and is then released. The wheel now has a large angular momentum along its axle (to the right in the figure) due to its spin. If the wheel were to fall as before, the axle would point downward, giving a large component of angular momentum in the downward direction. But there is no torque in the downward direction. In Figure 8-20, we have indicated a large component of angular momentum along the axis of the wheel. We have also indicated a *change* in angular momentum, $\Delta \mathbf{L}$, in the direction of the torque, which is into the paper. What actually happens when the axle of the spinning wheel is held horizontal and then released is that the axle moves in a horizontal plane in the direction into the paper. The wheel must move this way when it is spinning so that the change in angular momentum is in the direction of the net torque. This motion, which is often quite surprising when first encountered, is called *precession*. If the angular momentum due to the spin of the wheel is very large, the precession will be very slow; that is, the angular velocity of the precession associated with the axle of the wheel moving in a horizontal plane will be very small.

Precession

If you perform this experiment or demonstration, you will note a small up and down oscillation of the axis. This motion is called *nutation*. It can be

Nutation

eliminated if the axle is given a slight horizontal push when it is released. We can understand nutation qualitatively from a consideration of the directions of the angular momentum of the wheel and the torques exerted on it. Even when the precession is slow, there is a small amount of angular momentum associated with the bulk motion of the center of mass of the wheel. The direction of this small angular momentum is upward. But if the axle is simply released with no push, there is no component of torque in the upward or downward direction. Thus, unless the wheel is given an initial push (and thus an initial torque in the upward direction), the axle of the wheel will dip down slightly to provide a small component of angular momentum in the downward direction associated with the spin to cancel out the small upward angular momentum associated with the motion of the wheel itself. A detailed analysis of the motion of the gyroscope shows that if the wheel is not given an initial push, the axle will initially dip down, then overshoot, and then oscillate up and down.

Summary

1. The angular velocity of a body is defined as the rate of change of the angle between a line fixed on the body and one fixed in space:

$$\omega = \frac{\Delta\theta}{\Delta t}$$

The angular acceleration is defined as the rate of change of the angular velocity:

$$\alpha = \frac{\Delta\omega}{\Delta t}$$

2. The linear speed of a particle a distance r from the axis of rotation is related to the angular velocity:

$$v = r\omega$$

Similarly, the linear tangential acceleration of such a particle is related to the angular acceleration of the body by

$$a_t = r\alpha$$

3. Equations for the rotation of a rigid body with constant angular acceleration are

$$\omega = \omega_0 + \alpha t$$
$$\theta = \theta_0 + \omega_0 t + \tfrac{1}{2}\alpha_0 t^2$$

and

$$\omega^2 = \omega_0^2 + 2\alpha_0(\theta - \theta_0)$$

These equations are analogous to those for motion with constant acceleration in one dimension.

4. The moment of inertia of a system of particles is defined as

$$I = \Sigma m_i r_i^2$$

The moment of inertia for rotational motion is analogous to the mass in translational motion.

5. The kinetic energy of a rotating body is given by

$$E_k = \tfrac{1}{2}I\omega^2$$

and the power input of a torque is

$$P = \tau\omega$$

6. The angular momentum of a system of particles is

$$L = I\omega$$

7. The generalized form of Newton's second law for rotational motion is

$$\tau_{\text{net}} = \frac{\Delta(I\omega)}{\Delta t}$$

For a rigid body, the moment of inertia is constant, and this becomes

$$\tau_{\text{net}} = I\frac{\Delta\omega}{\Delta t} = I\alpha$$

8. In an isolated system, the net torque is zero and the angular momentum of the system is conserved.

9. When a ball or cylinder of radius R rolls without slipping, the velocity of the center of mass of the body is related to the angular velocity by the rolling condition

$$v = R\omega$$

The acceleration is similarly related to the angular acceleration by

$$a = R\alpha$$

10. The complicated motion of a gyroscope can be understood in terms of the directional properties of angular velocity, angular acceleration, and torque. When a system such as a bicycle wheel has a large initial angular momentum, and the net torque acting on it is perpendicular to the angular momentum, the system moves such that the change in the angular momentum is in the direction of the torque. This motion is called precession.

Suggestions for Further Reading

Frohlich, Cliff: "The Physics of Somersaulting and Twisting," *Scientific American*, March 1980, p. 154.

Though they may appear to, the midair rotations of divers and cats do not violate the law of conservation of angular momentum. With the aid of excellent illustrations, this article explains why.

Jones, David E. H.: "The Stability of the Bicycle," *Physics Today*, vol. 23, no. 4, 1970.

An amusing account of the author's attempt to experimentally test the idea, among others, that the gyroscopic effect of the spinning front wheel of a bicycle is responsible for its stability.

Laws, Kenneth: "The Physics of Dance," *Physics Today*, February 1985, p. 25.

Walker, Jearl: "The Amateur Scientist: The Essence of Ballet Maneuvers Is Physics," *Scientific American*, June 1982, p. 146.

Both articles present analyses of various dance movements from a physical point of view, showing that dance may be appreciated on more than just a purely aesthetic level. Various jumps and turns are discussed, as well as the dancer's problem of maintaining balance.

Walker, Jearl: "The Amateur Scientist: The Physics of Spinning Tops, Including Some Far-out Ones," *Scientific American*, March 1981, p. 182.

Precession, nutation, and other more unusual aspects of top mechanics are explained in a nonmathematical fashion.

Walker, Jearl: "The Amateur Scientist: Thinking About Physics While Scared to Death (on a Falling Roller Coaster)," *Scientific American,* October 1983, p. 162.

Momentum and its contribution to a thrilling roller-coaster ride, the interconversion of potential and kinetic energy, and the rotational dynamics of several amusement-park rides are discussed.

Review

A. Objectives: After studying this chapter, you should:

1. Be able to state the definitions of angular velocity, angular acceleration, and moment of inertia.

2. Be able to state the equations for rotation with constant angular acceleration and use these equations to work problems.

3. Be able to state the analog of Newton's second law for rotational motion and apply it to solve problems.

4. Be able to apply the law of conservation of angular momentum to problems in which the moment of inertia changes.

5. Be able to state the condition for rolling without slipping and apply it to problems involving translation and rotation.

6. Be able to describe qualitatively the motion of a gyroscope.

B. Define, explain, or otherwise identify:

Angular velocity, p. 175
Angular acceleration, p. 175

Moment of inertia, p. 178
Angular momentum, p. 182
Rolling condition, p. 186

C. True or false: If the statement is true, explain why. If it is false, give a counterexample.

1. Angular velocity and linear velocity have the same dimensions.

2. All parts of a rotating wheel have the same angular velocity.

3. All parts of a rotating wheel have the same angular acceleration.

4. The moment of inertia of a body depends on the location of the axis of rotation.

5. The moment of inertia of a body depends on the angular velocity of the body.

6. If the resultant torque on a body is zero, the angular momentum must be zero.

7. If the resultant torque on a rotating system is zero, the angular velocity of the system cannot change.

Exercises

Section 8-1   Angular Velocity and Angular Acceleration

1. A particle moves in a circle of radius 80 m with a constant speed of 15 m/s. (*a*) What is the angular velocity of the particle in radians per second? (*b*) How many revolutions does the particle make in 12 s?

2. Find a formula to convert radians per second into revolutions per minute.

3. Convert the following to radians: (*a*) 30°, (*b*) 60°, (*c*) 180°, and (*d*) 2 rev.

4. A Ferris wheel of radius 12 m rotates once in 25 s. (*a*) What is its angular velocity in radians per second? (*b*) What is the linear speed of a passenger? (*c*) What is the centripetal acceleration of a passenger?

5. A 30-cm diameter record rotates at 33.3 rev/min. (*a*) What is its angular velocity in radians per second? (*b*) Find the speed of a point on the rim of the record.

6. A record takes 3 s to start up from rest to 78 rev/min. How many revolutions does it make in this time?

7. A wheel starts from rest and has a constant angular acceleration of 2.5 rad/s². (*a*) What is its angular velocity after 6 s? (*b*) How fast is a point 0.4 m from the axis of rotation moving after 6 s? (*c*) Through what angle has the wheel turned after 6 s?

8. A cyclist starts from rest and pedals such that the wheels of his bike have a constant angular acceleration. After 10 s, the wheels have made 5 rev. (*a*) What is the angular acceleration of the wheels? (*b*) What is the angular velocity of the wheels after 10 s? (*c*) If the radius of the wheel is 36 cm, how far has the cyclist traveled in 10 s?

9. A turntable rotating at 33.3 rev/min is shut off. It brakes with constant angular acceleration and comes to rest in 1.5 min. (*a*) Find the angular acceleration. (*b*) What is the average angular velocity of the turntable? (*c*) How many revolutions does it make before stopping.

10. A disk starts from rest and rotates with a constant angular acceleration of 6 rad/s². What is its angular velocity after it has made 20 rev?

11. When a turntable is turned on, it rotates through $10°$ in the first $\frac{1}{4}$ s. Assuming constant angular acceleration, how long will it be before it is rotating at 33.3 rev/min?

12. An ultracentrifuge rotates at 70,000 rev/min. (*a*) What is its angular velocity in radians per second? (*b*) What is the centripetal acceleration of a particle at 5 cm from the axis of rotation?

Section 8-2   Torque and Moment of Inertia

13. The objects in Figure 8-21 are connected by a very light rod whose moment of inertia may be neglected. (*a*) Find the moment of inertia of the system about the *y* axis. (*b*) What torque is needed to bring the system to rotating at 2 rad/s in 3 s?

Figure 8-21
Exercises 13 and 21.

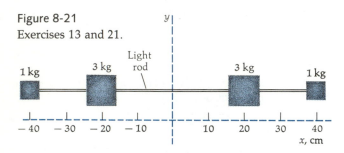

14. A penny has a mass of 3 g and a diameter of 2 cm. What is its moment of inertia about its diameter, assuming its thickness to be negligible?

15. Find the moment of inertia about the *y* axis of the system in Figure 8-22, assuming the connecting rods to have negligible mass.

Figure 8-22
Exercise 15.

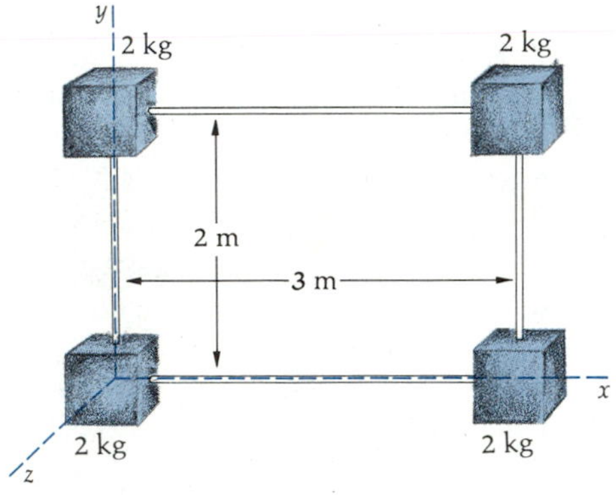

16. A tennis ball has a mass of 57 g and a diameter of 7 cm. Find the moment of inertia about its diameter.

17. A uniform disk of radius 0.14 m and mass 6 kg is pivoted so that it rotates freely about its axis. A string wrapped around the disk is pulled with a force of 25 N (see Figure 8-23). (*a*) What is the torque exerted on the disk? (*b*) What is the angular acceleration of the disk? (*c*) If the disk starts from rest, what is its angular velocity after 5 s?

Figure 8-23
Exercise 17.

18. To start a playground merry-go-round rotating, a rope is wrapped around it and pulled. A force of 200 N is exerted on the rope for 10 s. During this time the merry-go-round, which has a radius of 1.6 m, makes one complete rotation. (*a*) Find the angular acceleration of the merry-go-round, assuming it to be constant. (*b*) What torque is exerted by the rope on the merry-go-round? (*c*) What is the moment of inertia of the merry-go-round?

19. A 4-kg cylinder of radius 12 cm is free to rotate about its axis. A rope is wrapped around it and pulled with a force of 18 N. The cylinder is initially at rest. Find (*a*) the torque exerted by the rope, (*b*) the moment of inertia of the cylinder, and (*c*) the angular acceleration of the cylinder. (*d*) Find the angular velocity of the cylinder after 5 s.

20. A disk-shaped grindstone of mass 2 kg and radius 7 cm is spinning at 700 rev/min. After the power is shut off, a man continues to sharpen his ax by holding it against the grindstone for 10 s until the grindstone stops rotating. (*a*) Find the angular acceleration of the grindstone, assuming it to be constant. (*b*) What is the torque exerted by the ax on the grindstone? (Assume no other frictional torques.)

Section 8-3   Kinetic Energy and Angular Momentum

21. The system shown in Figure 8-21 is rotating at 2 rad/s. (*a*) Find the linear speed and kinetic energy of each body. (*b*) Find the total kinetic energy of the system from (*a*) and also from $E_k = \frac{1}{2}I\omega^2$.

22. Find the initial kinetic energy of the grindstone of Exercise 20.

23. A solid ball of mass 1.2 kg and diameter 16 cm is rotating about its diameter at 30 rev/min. Find (*a*) the kinetic energy of the ball and (*b*) its angular momentum.

24. Assuming the earth to be a homogeneous sphere of mass 6.0×10^{24} kg and radius 6.4×10^6 m, find its kinetic energy and angular momentum due to its rotation of 1 rev/day about its diameter.

25. Find the power input of the torque when the cylinder of Exercise 19 is rotating at 2 rev/s.

26. An engine develops 400 N·m of torque at 3200 rev/min. Find the power developed by the engine.

27. A 15-g coin of diameter 1.5 cm is spinning about its vertical diameter at 10 rev/s. Find (*a*) its kinetic energy and (*b*) its angular momentum.

28. Explain why a helicopter with just one main rotor has a second smaller rotor mounted on a horizontal axis at the rear (see Figure 8-24). Describe the resultant motion of the helicopter if this rotor fails during flight.

Figure 8-24
Helicopter with two rotors for Exercise 28.

29. A man stands at the center of a circular platform holding his arms extended horizontally with a 4-kg block in each hand. The platform is rotating freely at 0.5 rev/s. The moment of inertia of the platform plus the man is 1.6 kg·m², which is assumed to be constant. The blocks are 90 cm from the axis of rotation. The man now pulls the blocks in toward his body until they are only 15 cm from the axis of rotation. Find his new angular velocity.

30. The merry-go-round of Example 8-8 (radius 2 m and moment of inertia 500 kg·m²) is rotating without friction at 1 rev every 4 s with the 25-kg child sitting at the edge. The child now crawls to the center of the merry-go-round. (a) Find the new angular velocity of the merry-go-round. (b) Calculate the initial and final kinetic energies of the child and merry-go-round. Where did the additional energy come from?

31. A disk is rotating freely at 1800 rev/min about a vertical axis through its center. A second disk mounted on the same shaft above the first is initially at rest. The moment of inertia of the second disk is twice that of the first. The second disk is dropped onto the first one, and the two eventually rotate together with a common angular velocity. Find the new angular velocity.

Section 8-4   Rolling Bodies

32. A homogeneous cylinder of radius 16 cm and mass 400 kg is rolling without slipping along a horizontal floor at 6 m/s. How much work is needed to stop the disk?

33. Work Exercise 32 for a uniform sphere of the same mass, radius, and speed.

34. Work Exercises 32 for a hoop of the same mass, radius, and speed.

35. Find the percentage of the total kinetic energy associated with rotation and with translation for rolling without slipping if the object is (a) a uniform sphere, (b) a uniform cylinder, and (c) a hoop.

36. A hoop of radius 0.50 m and mass 0.8 kg is rolling without slipping at a speed of 20 m/s toward an incline of slope 30°. How far up the incline will the hoop roll?

37. A ball rolls without slipping along a horizontal plane. Show that the frictional force on the ball must be zero. (*Hint:* Consider a possible direction for a frictional force and what effect such a force would have on the ball's velocity and its angular velocity.)

38. A spinning ball is set onto a horizontal plane. Draw a diagram showing the forces acting on the ball and describe its translational and rotational motion.

Section 8-5   Motion of a Gyroscope

39. The angular momentum of the propeller of a small airplane points forward. (a) As the plane takes off, the nose lifts up and the airplane tends to veer to one side. Which side and why? (b) If the plane is flying horizontally and suddenly turns to the right, does the nose of the plane tend to move up or down? Why?

Problems

1. A wheel mounted on a horizontal axle has a radial line painted on it that is used to determine its angular position relative to the vertical. The wheel has a constant angular acceleration. At time $t = 0$, the line points straight up. At $t = 1$ s, the line is again vertical, having turned through 1 rev. At $t = 2$ s, the line points straight down, having turned through 1.5 rev more since $t = 1$ s. (a) What is the angular acceleration of the wheel? (b) What was its angular velocity at $t = 0$?

2. A hoop starts from height h and rolls without slipping down an incline. Find its speed at the bottom. How does its time to roll down compare with that of a cylinder and a uniform sphere?

3. A flywheel is a uniform disk of mass 100 kg and radius 0.3 m. It rotates with angular velocity of 1200 rev/min. (a) A constant tangential force is applied. How much work must be done to stop the wheel? (b) If the wheel is brought to rest in 2 min, what torque does the force produce? (*c*) What is the magnitude of the force? (*d*) How many revolutions does the wheel make in these 2 min?

4. A 1.2-Mg car is being unloaded by a winch, as shown in Figure 8-25. Suddenly, the winch gearbox shaft breaks, and the car falls from rest. The moment of inertia of the winch drum is 310 kg·m² and its radius is 0.8 m. Find the speed of the car as it hits the water. (*Hint:* Use conservation of energy.)

Figure 8-25
Problem 4.

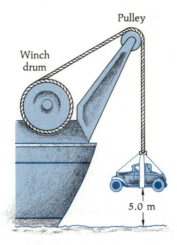

Pulley

Winch drum

5.0 m

5. The system of Figure 8-26 is released from rest. The 30-kg body is 2 m above the floor. The pulley is a uniform disk with a radius of 10 cm and a mass of 5 kg. Find the speed of the 30-kg body just before it hits the floor.

Figure 8-26
Problem 5.

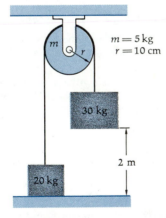

$m = 5$ kg
$r = 10$ cm

30 kg

2 m

20 kg

6. A circular platform is mounted on a vertical frictionless axle. Its radius is $r = 3$ m, and its moment of inertia is $I = 200$ kg·m². It is initially at rest. A 80-kg man stands on the edge of the platform and begins to walk along the edge at speed 1.0 m/s *relative to the ground*. (*a*) What is the angular velocity of the platform? (*b*) When the man has walked once around the platform so that he is at his original position on it, what is his angular displacement relative to the ground?

7. A metrestick is pivoted at one end so it can swing freely in a vertical plane. It is released from rest in a horizontal position. What is the angular velocity of the stick when it is vertical? (*Hint:* Use conservation of energy.)

8. A wheel mounted on an axis that is not frictionless is at rest initially. A constant external torque of 50 N·m is applied to the wheel for 20 s. At the end of the 20 s, the wheel has an angular velocity of 600 rev/min. The external torque is then removed and the wheel comes to rest after 120 s more. (*a*) What is the moment of inertia of the wheel? (*b*) What is the frictional torque, assuming it is constant?

9. A uniform sphere of mass *M* and radius *R* is free to rotate about a horizontal axis through its center. A string is wrapped around the sphere and is attached to a body of mass *m*, as shown in Figure 8-27. Find the acceleration of the body and the tension *T* in the string.

Figure 8-27
Problem 9.

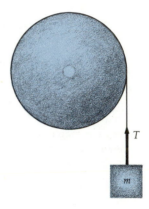

T

m

10. A 120-kg uniform disk of radius 0.5 m is placed flat on some smooth ice. Two skaters wind ropes around the disk in the same sense (see Figure 8-28). At time *t* = 0, the skaters pull on their ropes and skate away, exerting constant forces of 45 N and 65 N. Describe the motion of the disk, giving the acceleration, velocity, and position of the center of mass as functions of time and the angular acceleration and angular velocity as functions of time.

Figure 8-28
Problem 10.

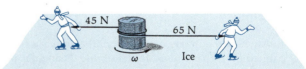

45 N

65 N

Ice

ω

11. A uniform ball of radius *r* rolls without slipping along the loop-the-loop track in Figure 8-29. It starts at rest at height *h*. If the ball is not to leave the track at the top of the loop, what is the least value *h* can have (in terms of the radius *R* of the loop)? (*Hint:* The speed of the ball at the top of the track can be found from conservation of energy. If the ball is to stay on the track at the top, its speed must be great enough so that its centripetal acceleration is at least g.)

Figure 8-29
Problem 11.

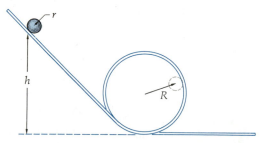

12. A uniform cylinder of mass M and radius R has a string wrapped around it. The string is held fixed, and the cylinder falls vertically, as shown in Figure 8-30. (*a*) Show that the acceleration of the cylinder is downward with magnitude $a = 2g/3$. (*b*) Find the tension T in the string.

Figure 8-30
Problem 12.

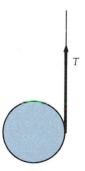

13. In Problem 12, replace the cylinder with a uniform sphere of mass M and radius R with the string wound around it. Find both the acceleration of the sphere and the tension in the string.

14. A typical car engine delivers about 2 MJ of mechanical energy per kilometre on the average. A car is designed to use the energy stored in a large flywheel in a vacuum container. If the mass of the flywheel is not to exceed 100 kg and its angular velocity is not to exceed 400 rev/s, find the smallest radius of the flywheel (assumed to be a uniform cylinder) such that the car can travel 300 km without recharging the flywheel.

15. For a car powered by a flywheel, discuss problems that would arise if there were just a single flywheel with angular momentum **L**. For example, what would happen if **L** points horizontally and the car attempts to turn to the left or to the right, or if **L** is vertical and the car travels over a hill or through a valley.

CHAPTER NINE

Gravity

In this chapter, we come at last to Isaac Newton's crowning achievement—his law of universal gravitation. As you will see, this proposal was an amazing intellectual leap. Indeed, the world had to wait for 112 years before the strength of the gravitational force Newton hypothesized was first measured. And to this day, truly accurate measurements of this force are difficult to obtain and much about it is still unknown.

Nevertheless, Newton's success had a profound impact on his age and our own. It gave the world of learning such confidence in the power of science and logic that it ushered in a new intellectual era, the Age of Reason. From that time on, every aspect of the world and of human life has been considered an appropriate subject for scientific scrutiny and independent thought.

"But the Solar System!" I protested. "What the deuce is it to me?" [Sherlock Holmes] interrupted impatiently: "You say that we go round the sun. If we went around the moon it would not make a pennyworth of difference to me. . . ."

ARTHUR CONAN DOYLE

All the different forces observed in nature can be explained in terms of four basic interactions that occur between elementary particles: the gravitational force, the electromagnetic force, the strong nuclear force (also called the hadronic force), and the weak nuclear force. Although of negligible importance in the interactions of elementary particles, gravity is of primary importance in the interactions of large objects. It is gravity that binds us to the earth and that holds the earth and the other planets in place in the solar system. The gravitational force plays an important role in the evolution of stars and in the behavior of galaxies. In a sense, it is gravity that holds the universe together.

In this chapter, we will study the force of gravity in some detail. We will begin by stating Kepler's empirical laws of planetary motion, and we will then discuss how these laws are related to Newton's law of gravity.

"Castle in the Pyrenees" by René Magritte.

198

9-1   Kepler's Laws

The night sky, with its myriad of stars and shining planets, has fascinated humankind since the beginning of time. The apparent movements of the stars and planets relative to the earth have been observed and charted by astronomers over the centuries.

   The first model of the universe, published by Ptolemy in about A.D. 140, had the earth at the center, with the sun and stars moving around the earth in simple circles and the planets moving around the earth in more complex paths consisting of small circles, called epicycles, superimposed on larger circles (see Figure 9-1). This somewhat complicated model was in basic agreement with the naked-eye observations of that time and was universally accepted for over 14 centuries. It was replaced by the simpler but more controversial model of Copernicus in 1543, which proposed that the sun and other stars were fixed and that the planets, including the earth, revolved in circles around the sun. (This model was controversial because it did not place the earth at the center of the universe, contrary to the religious doctrine of the day.)

Johannes Kepler (1571–1630).

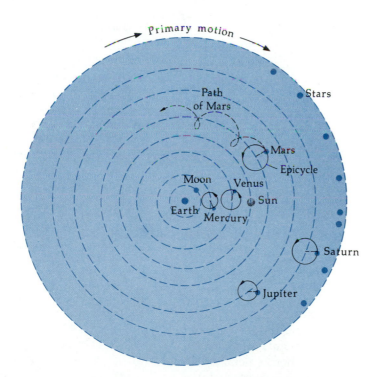

Figure 9-1
Ptolemy's model of the universe. The primary motion of the planets and stars is clockwise with a period of 1 day. To account for the motion of the planets relative to the stars, the planets are given a smaller counterclockwise angular velocity. This model is called a geocentric model because it places the earth at the center of the universe.

   Toward the end of the sixteenth century, the Danish astronomer Tycho Brahe made observations of the motions of the planets that were considerably more accurate than previous ones had been. Using Brahe's data, Johannes Kepler discovered, after much trial and error, that the actual paths of the planets about the sun were ellipses. He also showed that the planets did not move with constant speed, but moved faster when close to the sun than when farther away. Finally, Kepler was able to give a precise mathematical relation between the period of a planet and its average distance from the sun. Kepler stated his results in three empirical laws of planetary motion, which formed the basis for Newton's discovery of the universal law of gravitation.

Kepler's three laws are

> *Law 1. All planets move in elliptical orbits with the sun at one focus.*   Kepler's laws
>
> *Law 2. A line joining any planet to the sun sweeps out equal areas in equal times.*
>
> *Law 3. The square of the period of any planet is proportional to the cube of the planet's mean distance from the sun.*

Figure 9-2 shows an ellipse. The points labeled F are called the focal points. The distance a is called the semimajor axis, and b is called the semiminor axis. You can draw an ellipse by taking a piece of string, fixing each end at a focal point (for example, with thumb tacks), and using it to guide your pencil, as is shown in the figure. As the focal points are moved

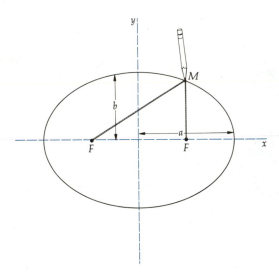

Figure 9-2
Ellipse with two focal points F. Such a figure can be drawn by fastening a string at the focal points and using it to guide a pencil (shown at point M) to trace out the ellipse. The distance a is called the semimajor axis and b the semiminor axis of the ellipse. If the focal points coincide, a and b are equal and the ellipse is a circle.

closer together, the ellipse begins to resemble a circle. A circle is a special case of an ellipse in which the two focal points coincide. Figure 9-3 shows an elliptical path of a planet with the sun at one focus. The point P, at which the planet is closest to the sun, is called the perihelion, whereas point A, the farthest distance, is called the aphelion. The earth's orbit is nearly circular, with the distance to the sun at the perihelion being 91.5 million miles and that at the aphelion being 94.5 million miles. The semimajor axis equals half the sum of these distances, which is 93 million miles for the earth's orbit. This turns out to be the mean distance from the earth to the sun during its orbit.

Figure 9-4 illustrates Kepler's second law, the law of equal areas. A planet moves faster when it is closer to the sun than when it is farther away. The law of equal areas is related to the conservation of angular momentum, as we will see in the next section.

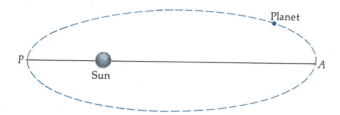

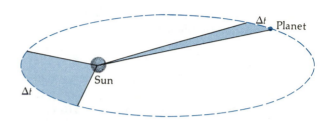

Figure 9-3
Elliptical path of a planet with the sun at one focal point. The point P, where the planet is closest to the sun, is called the perihelion; point A, where it is farthest, is called the aphelion. The average distance of the planet to the sun is the semimajor axis.

Figure 9-4
When a planet is close to the sun, it moves faster than when it is farther away such that the areas swept out in a given time are equal.

Kepler's third law relates the period of any planet to its mean distance from the sun, which turns out to be the semimajor axis of the elliptical path. In algebraic form, if r is the mean distance between a planet and the sun, and T is the period of revolution, Kepler's third law states that

$$T^2 = Cr^3 \qquad \text{9-1}$$ Kepler's third law

where the constant C has the same value for all the planets. We will show in the next section that (for the special case of a circular orbit) this law is a simple consequence of the fact that the force exerted by the sun on a planet varies inversely with the square of the distance from the sun to the planet.

Example 9-1 The mean distance of Jupiter from the sun is 5.20 times that of the earth. What is the period of Jupiter?

From Kepler's third law, the square of the period is proportional to the cube of the mean distance from the sun. Taking the square root of both sides of Equation 9-1, we have

$$T = \sqrt{C}\, r^{3/2}$$

If T_E and r_E are the period and the mean distance for the earth and T_J and r_J are the period and the mean distance for Jupiter, we have

$$\frac{T_J}{T_E} = \left(\frac{r_J}{r_E}\right)^{3/2} = (5.20)^{3/2} = 11.9$$

Since the period of the earth is 1 y, that of Jupiter is 11.9 y.

9-2   Newton's Law of Gravity

Although Kepler's laws were an important step in the understanding of the motion of the planets, they were still merely empirical rules obtained from the astronomical observations of Brahe. It remained for Newton to take the giant step forward and attribute the acceleration of a planet in its orbit to a force exerted by the sun on the planet that varied inversely with the square of the distance between the sun and the planet. It had been proposed by others besides Newton that such a force existed, but Newton was able to prove that a force that varied inversely with the square of the separation distance would result in the elliptical orbits observed by Kepler. He then made the bold assumption that such a force existed between any two bodies in the universe. (Before Newton, it was not even generally accepted that the laws of physics observed on earth were applicable to the heavenly bodies.)

Newton's law of gravity postulates that every body exerts an attractive force on every other body that is proportional to the masses of the two bodies and inversely proportional to the square of the distance separating them. The magnitude of the gravitational force exerted by a particle of mass m_1 on another particle of mass m_2 a distance r away is thus given by

$$F = G \frac{m_1 m_2}{r^2}$$ 9-2   Newton's law of gravity

where the constant G, the *universal gravitational constant*, has the value

$$G = 6.67 \times 10^{-11} \text{ N} \cdot \text{m}^2/\text{kg}^2$$ 9-3   Universal gravitational constant

Newton published his theory of gravitation in 1686, but it was not until more than a century later that an accurate experimental determination of G was made by Cavendish, as will be discussed in Section 9-3.

We can use the known value of G to compute the gravitational attraction between two ordinary objects.

Example 9-2   Find the force of attraction between two balls, each of mass 1 kg, when their centers are 10 cm apart.

We can treat each ball as if its total mass were at a point at its center. The magnitude of the force on either ball exerted by the other is then

$$F = \frac{(6.67 \times 10^{-11} \text{ N} \cdot \text{m}^2/\text{kg}^2)(1 \text{ kg})(1 \text{ kg})}{(0.1 \text{ m})^2}$$

$$= 6.67 \times 10^{-9} \text{ N}$$

This example demonstrates that the gravitational force exerted by one object of ordinary size on another such object is extremely small. For example, the weight of a 1-g object is 9.81×10^{-3} N, more than a million times the force computed in Example 9-2. We can usually neglect the gravitational force between objects compared with the other forces acting on them. The gravitational attraction can be noticed only if at least one of the objects is extremely massive, as with two mountains or with an ordinary body and the earth, or if great care is taken to eliminate the other forces on the objects, as Cavendish did in determining G.

Newton showed that, in general, when an object (such as a planet or comet) moves about a force center (such as the sun) to which it is attracted by a force that varies as $1/r^2$, the path of the object is an ellipse, a parabola, or a hyperbola. The parabolic and hyperbolic paths apply to objects (if there are any) that make one pass by the sun and never return. Such orbits are not closed orbits. The only possible closed orbits in an inverse square force field are ellipses. (A circle is a special case of an ellipse.) Thus, Kepler's first law is a direct consequence of Newton's law of gravity.

Kepler's second law, the law of equal areas, follows from the fact that the force exerted by the sun on a planet is directed toward the sun. Such a force is called a *central force*. Since the force on a planet is along the line from the planet to the sun, it exerts no torque about the sun. We know from our study of angular momentum that if the net torque on a body is zero, the angular momentum of the body is conserved. Figure 9-5 shows a planet moving in an elliptical orbit around the sun. The angular momentum of the planet is

$$L = I\omega = mr^2\omega \qquad\qquad 9\text{-}4$$

where m is the mass of the planet and ω is its angular velocity. The component of the linear velocity v_p perpendicular to the radius $\mathbf{r}$ is related to the angular velocity by

$$v_p = r\omega \qquad\qquad 9\text{-}5$$

The angular momentum can thus be written

$$L = mr^2\omega = mrv_p \qquad\qquad 9\text{-}6$$

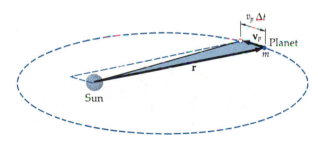

In Figure 9-5, we have indicated a parallelogram whose area is twice that swept out by the planet in some time interval Δt. The area of the parallelogram is $rv_p\,\Delta t$. The area swept out by the planet in time Δt is thus

$$\Delta A = \tfrac{1}{2}rv_p\,\Delta t \qquad\qquad 9\text{-}7$$

Then,

$$\frac{\Delta A}{\Delta t} = \tfrac{1}{2}rv_p$$

Comparing this with Equation 9-6, we see that

$$\frac{\Delta A}{\Delta t} = \frac{L}{2m} \qquad\qquad 9\text{-}8$$

Thus, Kepler's observation that a planet sweeps out equal areas in equal times implies that the angular momentum L of the planet must remain constant as the planet moves about the sun. This, in turn, implies that the force exerted on the planet is directed toward the sun.

© 1987 Sidney Harris.

"You can't go on like this. Why don't you see a physicist."

Figure 9-5
The area swept out in time Δt is half the area of the parallelogram shown. This area is $\tfrac{1}{2}rv_p\,\Delta t$, which is proportional to the angular momentum of the planet.

We will now show that Newton's law of gravity implies Kepler's third law for the special case of a circular orbit. We consider a planet moving about the sun with speed v in a circular orbit of radius r. Since the planet is moving in a circle, we know it has centripetal acceleration v^2/r. This acceleration is provided by the force of attraction between the sun and the planet, which is given by Newton's law of gravitation:

$$F = \frac{GM_s m_p}{r^2} = m_p a = m_p \frac{v^2}{r} \qquad\qquad 9\text{-}9$$

where M_s is the mass of the sun and m_p is that of the planet. Solving for v^2, we obtain

$$v^2 = \frac{GM_s}{r} \qquad\qquad 9\text{-}10$$

We now relate the speed v of the planet to its period T. Since the planet moves a distance $2\pi r$ in time T, we have

$$2\pi r = vT$$

or

$$v = \frac{2\pi r}{T} \qquad\qquad 9\text{-}11$$

Substituting this expression for v in Equation 9-10, we obtain

$$v^2 = \frac{4\pi^2 r^2}{T^2} = \frac{GM_s}{r}$$

or

$$T^2 = \left(\frac{4\pi^2}{GM_s}\right) r^3 \qquad\qquad 9\text{-}12 \qquad \text{Kepler's third law}$$

Equation 9-12 is Kepler's third law. For a general elliptical path, the radius r is replaced by the mean distance between the planet and the sun, which equals the semimajor axis of the ellipse. This equation also applies to the orbits of the moons of any planet if we replace the mass of the sun M_s with the mass of the planet. For example, it applies to the earth's moon and to all artificial satellites orbiting the earth if the mass M_s is replaced with the earth's mass M_E. We note that since G is known, we can determine the mass of a planet by measuring the period T and the radius r of a moon orbiting it.

Example 9-3   Mars has a satellite with a period of 460 min and mean orbit radius of 9.4 Mm. What is the mass of Mars?

Replacing M_s in Equation 9-9 with the mass of Mars M and using $r = 9.4 \times 10^6$ m, $T = 460(60)$ s, and $G = 6.67 \times 10^{-11}$ N·m²/kg², we obtain

$$M = \frac{4\pi^2 r^3}{GT^2} = \frac{4\pi^2 (9.4 \times 10^6)^3}{(6.67 \times 10^{-11})[460(60)]^2} = 6.45 \times 10^{23} \text{ kg}$$

As a further check on the validity of the inverse square nature of the gravitational force, Newton compared the acceleration of the moon in its orbit with the acceleration of objects near the surface of the earth (such as the legendary apple). He made the bold assumption that the force that causes the moon to circle the earth has the same origin as the force that causes objects near the earth's surface to fall toward the earth, namely the gravitational attraction due to the earth. He first assumed that the earth and the moon could be treated as point particles with their total masses concentrated at their centers. Since the distance to the moon is about 60 times the radius of the earth, the acceleration of objects near the surface of the earth ($g = 9.81$ m/s^2) should be $(60)^2 = 3600$ times the acceleration of the moon. The moon's centripetal acceleration can be calculated from its known distance from the center of the earth and its period:

$$a_m = \frac{v^2}{r} = \frac{(2\pi r/T)^2}{r} = \frac{4\pi^2 r}{T^2}$$

Using $r = 3.84 \times 10^8$ m and $T = 27.3$ d, the acceleration of the moon is $a_m = 2.72 \times 10^{-3}$ m/s^2. Comparing this result with g, we have

$$\frac{g}{a_m} = \frac{9.81 \text{ m/s}^2}{2.72 \times 10^{-3} \text{ m/s}^2} = 3607 \approx (60)^2$$

The calculations agreed "pretty nearly," Newton said. "I thereby compared the force requisite to keep the Moon in her orb with the force of gravity at the surface of the Earth, and found them answer pretty nearly."

   The assumption that the earth and moon can be treated as point particles in the calculation of the force on the moon is reasonable because the moon is far from the earth compared with the radius of either the earth or the moon, but it is certainly questionable when applied to finding the force exerted by the earth on a body near its surface. After considerable effort, Newton was able to prove that the force exerted by any spherically symmetrical body on or outside its surface is the same as if all the mass of the body were concentrated at its center. The proof involves integral calculus, which Newton developed for this problem. Using $r = R_E$, the radius of the earth, in Equation 9-2, we can find the gravitational force acting on a body of mass m near the surface of the earth. This force is directed toward the center of the earth and has the magnitude

$$F = \frac{GM_E m}{R_E^2}$$

Earthrise as seen from the moon.

where M_E is the mass of the earth. The acceleration of a body in free-fall near the earth's surface is then

$$g = \frac{F}{m} = \frac{GM_E}{R_E^2} \qquad\qquad\qquad 9\text{-}13$$

Since $g = 9.81$ m/s^2 is easily measured and the radius of the earth is known, Equation 9-10 can be used to determine either the constant G or the mass of the earth M_E if one of these quantities is known. Newton estimated the value of G from an estimate of the mass of the earth. When Cavendish determined G some 100 years later by measuring the force between two small spheres of known mass and separation, he called his experiment "weighing the earth."

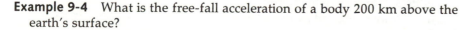

Example 9-4   What is the free-fall acceleration of a body 200 km above the earth's surface?

The acceleration due to gravity at a distance r from the center of the earth is given by F/m, where F is the gravitational force given by Equation 9-2 with $m_1 = M_E$ and $m_2 = m$.

$$g(r) = \frac{F}{m} = \frac{GM_E}{r^2}$$

From Equation 9-13, we have $GM_E/R_E^2 = 9.81$ m/s². The acceleration at distance r is then

$$g(r) = (9.81 \text{ m/s}^2)\frac{R_E^2}{r^2}$$

Using $R_E = 6.37$ Mm $= 6370$ km and $r = R_E + 200$ km $= 6570$ km, we obtain for the acceleration due to gravity

$$g = (9.81 \text{ m/s}^2)\left(\frac{6370 \text{ km}}{6570 \text{ km}}\right)^2 = 9.22 \text{ m/s}^2$$

Example 9-5   A satellite orbits the earth in a circular orbit. Find its period (a) if the satellite is just above the surface of the earth (assume it is high enough for air resistance to be neglected) and (b) if the satellite is at an altitude of 300 km.

(a) We can apply Equation 9-12 (Kepler's third law) to satellites orbiting the earth if we replace the mass of the sun M_s with the mass of the earth M_E. We then have

$$T^2 = \frac{4\pi^2}{GM_E} r^3$$

Kepler's third law for satellites

It is convenient to replace GM_E with gR_E^2 from Equation 9-13. We then have

$$T^2 = \frac{4\pi^2}{gR_E^2} r^3$$

If the satellite is just above the surface of the earth, $r = R_E$ and

$$T^2 = \frac{4\pi^2}{gR_E^2} R_E^3 = \frac{4\pi^2 R_E}{g}$$

Then,

$$T = 2\pi \sqrt{\frac{R_E}{g}} = 2\pi \sqrt{\frac{6.37 \times 10^6 \text{ m}}{9.81 \text{ m/s}^2}} = 5.06 \times 10^3 \text{ s} = 84.4 \text{ min}$$

(b) At an altitude of 300 km above the earth's surface, $r = 6370$ km $+$ 300 km $= 6670$ km $= 6.67$ Mm. Since T is proportional to $r^{3/2}$, we can find T at this distance from

$$T = (84.4 \text{ min})\left(\frac{r}{R_E}\right)^{3/2} = (84.4 \text{ min})\left(\frac{6.67 \text{ Mm}}{6.37 \text{ Mm}}\right)^{3/2} = 90.4 \text{ min}$$

Questions

1. Why don't you feel the gravitational attraction of a large building when you walk near it?

2. An astronaut orbiting the earth in a satellite 300 km above the surface of the earth feels weightless. Why? Is the force of gravity exerted by the earth on the astronaut negligible at this height?

9-3   The Cavendish Experiment

The first measurement of the gravitational constant G was made by Henry Cavendish in 1798. Figure 9-6 shows a schematic drawing of the apparatus he used to measure the gravitational force between two small bodies. The two small bodies of mass m_2 are at the ends of a light rod suspended by a fine fiber. A torque is required to turn the two masses through the angle θ from their equilibrium position because the fiber must be twisted. Careful measurement shows that the torque required to turn the fiber through a given angle is proportional to the angle. The constant of proportionality can be determined, and the fiber and the suspended masses can be used to measure very small torques. This arrangement, called a *torsion balance*, was invented in the eighteenth century by John Michell. The French physicist Charles Augustin de Coulomb used a similar device in 1785 to determine the electrical force between charged particles, which is now referred to as Coulomb's law (Chapter 18). Cavendish used a refined and especially sensitive torsion balance in his determination of G.

In the Cavendish experiment, two large masses m_1 are placed near the small masses m_2, as shown in Figure 9-6a. The apparatus is allowed to come to equilibrium in this position. Since the apparatus is so sensitive and the gravitational force is so small, this takes hours. Instead of measuring the deflection angle directly, the positions of the large masses are reversed as shown by the dashed lines in Figure 9-6b. If the balance is allowed to come to equilibrium again, it will turn through the angle 2θ in response to the reversal of the torque. From the measurement of the angle and the torsion constant, the force between the masses m_1 and m_2 can be determined. When the masses and their separations are known, G can be calculated. Cavendish obtained a value for G within about 1 percent of the presently accepted value as given by Equation 9-3.

The very small magnitude of G means that the gravitational force exerted by one object of ordinary size on another such object is extremely small. For example, the attractive force between two objects each of mass 1 kg separated by 1 m is only 6.67×10^{-11} N. Such forces can be observed only if extreme care is taken to balance all the other forces acting on the objects, as must be done in the Cavendish experiment to measure G.

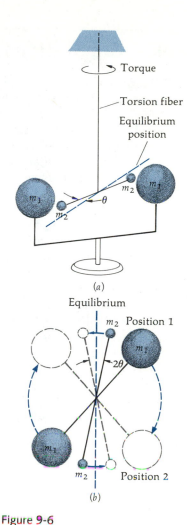

(a)

(b)

Figure 9-6
Schematic drawing of the Cavendish apparatus for determining G. (a) Because of the gravitational attraction of the large masses m_1 for the nearby small masses m_2, the fiber is turned through a very small angle θ from its equilibrium position. (b) As seen from above, the large masses are reversed so that they are at the same distance from the equilibrium position of the balance but on the other side. The fiber then turns through the angle 2θ. Measurement of this angle and of the torsion constant of the fiber makes it possible to determine the force exerted by m_1 on m_2, which in turn allows G to be determined.

Original drawing of the Cavendish apparatus.

9-4   Gravitational and Inertial Mass

The property of a body responsible for the gravitational force it exerts on another body is called its *gravitational mass*. On the other hand, the property of a body that measures its resistance to acceleration is called its *inertial mass*. We have used the same symbol m for these two properties because, experimentally, the gravitational and inertial masses of a body are equal. The fact that the gravitational force exerted by a body is proportional to its inertial mass is a characteristic unique to the force of gravity among all the forces we know, and it is a matter of considerable interest. One consequence is that all objects near the earth fall with the same acceleration if air resistance is neglected. This fact has seemed surprising to all since ancient times. The well-known story of how Galileo demonstrated it by dropping objects from the Leaning Tower of Pisa is just one example of the excitement this discovery aroused in the sixteenth century.

Pen-and-ink drawing of Galileo's legendary experiment from the Leaning Tower of Pisa.

We could easily imagine that the gravitational and inertial masses of a body were not the same. Suppose we write m_G for the gravitational mass and m for the inertial mass. The force exerted by the earth on an object near its surface would then be

$$F = \frac{GM_E m_G}{R_E^2} = m_G g \qquad\qquad 9\text{-}14$$

where $g = M_E G / R_E^2$ and M_E is the gravitational mass of the earth. The free-fall acceleration of the body near the earth's surface would then be

$$a = \frac{F}{m} = \frac{m_G}{m} g \qquad\qquad 9\text{-}15$$

It might be reasonable to expect that the ratio m_G / m would depend on such things as the chemical composition of the body, its temperature, or some other physical characteristics of the body. The free-fall acceleration would then be different for different objects.

The experimental fact, however, is that a is the same for all bodies. This means that for every body, m_G / m has the same ratio. Since this is the case, we need not maintain the distinction between m_G and m and can put $m_G = m$. (This amounts to choosing the constant of proportionality equal to 1, which in turn determines the magnitude and units of G in the law of gravity.) We must keep in mind, however, that this is an experimental conclusion limited by the accuracy of the experiment. As an experimental law, the statement that m_G is equal to m is known as the *principle of equivalence*. It is the foundation of Einstein's general theory of relativity, published in 1916. Experiments testing the equivalence principle were carried out by Simon Stevin in the 1580s. (He also discovered the law of vector addition of forces.) Galileo publicized this law widely, and his contemporaries made considerable improvements on the experimental accuracy with which the law was established.

The most precise early comparisons of m_G and m were made by Newton. Performing experiments using simple pendulums rather than falling bodies, Newton was able to establish the equivalence between gravitational and

That's funny—when Galileo did it, it worked perfectly.

© 1987 Sidney Harris.

inertial mass to an accuracy of about 1 part in 1000. Experiments comparing gravitational and inertial mass have improved steadily over the years. Their equivalence is now established to about 1 part in 10^{12}. Thus, the principle of equivalence is one of the best established of all the physical laws.

9-5   Escaping the Earth

In the last three decades, the idea of escaping from the earth's gravitational field has changed from fantasy to reality. Space probes have been sent out to the far reaches of the solar system. Some of these probes are expected to orbit the sun, whereas others will leave the solar system and drift on into outer space. In this section, we will look at the problem of escaping from the gravitational field of the earth (or of the sun). We will see that there is a minimum initial speed, called the *escape speed,* that is required for a body to escape from the earth, and we will calculate this escape speed using Newton's law of gravitation.

Near the surface of the earth, the force of attraction between the earth and some body of mass m is a constant mg, which is independent of the height of the body above the earth's surface. The force per unit mass g is called the *gravitational field.* If we project a body upward with an initial speed v in such a uniform gravitational field, the body will rise to a maximum height h, which we can easily calculate from conservation of energy. (We will neglect air resistance, which is important for practical calculations but does not affect the essential ideas of this discussion.) Taking the gravitational potential energy to be zero at the earth's surface, the initial potential energy is zero and the initial kinetic energy is $\frac{1}{2}mv^2$. At the maximum height h, the kinetic energy is zero and the potential energy is mgh. Conservation of energy gives

$$\frac{1}{2}mv^2 = mgh \qquad\qquad 9\text{-}16$$

or

$$v^2 = 2gh \qquad\qquad 9\text{-}17$$

According to Equation 9-17, if we increase the initial speed of the object, the maximum height attained by the object increases. For any initial speed, no matter how great, there is some maximum height h. That is, there is no initial speed v great enough to escape from a *uniform* gravitational field.

However, we know from Newton's law of gravity that the gravitational field of the earth is not uniform but decreases as $1/r^2$, where r is the distance from the center of the earth. If we project an object upward with a very great initial speed so that the object moves a distance comparable to the radius of the earth, we must take the decrease in the gravitational force on the object into account to correctly calculate the maximum height attained. We will again use conservation of energy, but we need to find the correct expression for the potential energy of an object in a gravitational field that varies as $1/r^2$.

July 16, 1969 blast-off of Saturn V rocket carrying Apollo II astronauts on man's first voyage to the moon.

Figure 9-7 shows the magnitude of the gravitational force exerted by the earth on an object of mass m as a function of the distance r from the center of the earth. We wish to find the work needed to lift the object from its initial position on the surface of the earth, $r_i = R_E$, to some final position r, a distance $h = r - R_E$. This work will equal the increase in the potential energy of the object. The gravitational force on the object varies from its initial maximum value $GM_E m / R_E^2$ to its final minimum value $GM_E m / r^2$. To find the work done, we must find the average force F_{av} and multiply it by the distance $r - R_E$. The correct expression for the average force, which can be derived using calculus, is

$$F_{av} = \frac{GM_E m}{r R_E} \qquad\qquad 9\text{-}18$$

The work needed to lift the object up from its initial position on the surface of the earth, a distance R_E from the center, to its final position at distance r is this average force times the distance:

$$W = F_{av} h = \frac{GM_E m}{r R_E}(r - R_E)$$

$$= \frac{GM_E m}{R_E} - \frac{GM_E m}{r}$$

If we choose the potential energy of the object to be zero when it is on the surface of the earth at distance R_E, this work equals the potential energy of the object at distance r:

$$U = \frac{GM_E m}{R_E} - \frac{GM_E m}{r} \qquad\qquad 9\text{-}19 \qquad \text{Gravitational potential energy}$$

where $U = 0$ at the earth's surface. Figure 9-8 shows this potential energy as a function of r. The dashed line is the potential-energy function, $mgh = mg(r - R_E)$, for a constant gravitational force. We note that the actual potential-energy function for the varying gravitational force does not increase indefinitely as r increases. As r becomes larger and larger, the second term in Equation 9-19 gets smaller and smaller, and the potential energy approaches a maximum value U_{max} given by

$$U_{max} = \frac{GM_E m}{R_E} \qquad\qquad 9\text{-}20$$

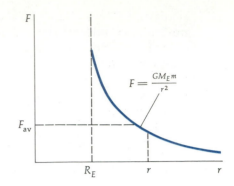

Figure 9-7
Magnitude of the gravitational force $GM_E m / r^2$ versus r. The work done in lifting a mass m from the surface of the earth, $r_i = R_E$, to a distance r is the average force F_{av} shown times the distance $r - R_E$.

Figure 9-8
The gravitational potential energy $U = GM_E m / R_E - GM_E m / r$ versus r. At large values of r, this potential-energy function approaches $U_{max} = GM_E m / R_E$. The colored dashed line shows the potential energy, $mgh = mg(r - R_E)$, for a constant force mg.

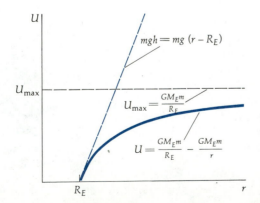

Example 9-6   A projectile is projected straight up from the surface of the earth with an initial speed of $v_1 = 8$ km/s. Find the maximum height it reaches, neglecting air resistance.

We use conservation of energy. Initially, at the surface of the earth, the potential energy of the projectile is zero and its kinetic energy is $\frac{1}{2}mv_1^2$. When it is at height $h = r - R_E$, its potential energy is given by Equation 9-19. We wish to find the height such that its final speed $v_2 = 0$. Conservation of energy gives

$$\tfrac{1}{2}mv_1^2 + U_1 = \tfrac{1}{2}mv_2^2 + U_2$$

$$\tfrac{1}{2}mv_1^2 + 0 = 0 + \left(\frac{GM_E m}{R_E} - \frac{GM_E m}{r} \right)$$

Simplifying this equation, we obtain

$$v_1^2 = \frac{2GM_E}{R_E} - \frac{2GM_E}{r} = \frac{2GM_E}{R_E}\left(1 - \frac{R_E}{r} \right)$$

We can simplify this further by noting that $g = GM_E / R_E^2$ from Equation 9-13, so $GM_E / R_E = gR_E$. Then,

$$v_1^2 = 2gR_E \left(1 - \frac{R_E}{r} \right)$$

It is sufficient to solve for R_E / r.

$$1 - \frac{R_E}{r} = \frac{v_1^2}{2gR_E}$$

$$\frac{R_E}{r} = 1 - \frac{v_1^2}{2gR_E}$$

$$= 1 - \frac{(8000 \text{ m/s})^2}{2(9.81 \text{ m/s}^2)(6.37 \times 10^6 \text{ m})} = 1 - 0.512$$

$$= 0.488$$

Then

$$r = \frac{R_E}{0.488} = 2.05 R_E$$

The maximum height reached is therefore $h = r - R_E = 1.05 R_E$.

We are now ready to find the initial speed it is necessary to give an object so that it can escape from the earth's gravitational field. We note that if we project an object upward from the earth with some initial kinetic energy, the kinetic energy decreases and the potential energy increases as the object rises. But the potential energy cannot increase by more than the amount U_{max}. Therefore, this is the most that the kinetic energy can decrease. If the initial kinetic energy is greater than U_{max}, the object will still have some kinetic energy when r is very great (or even when r is infinite). Thus, the object will escape from the earth if the initial kinetic energy is greater than

U_{max} as given by Equation 9-20. The critical speed v_e associated with this energy is the escape speed, which is found from

$$\tfrac{1}{2}mv_e^2 = U_{max} = \frac{GM_E m}{R_E} \qquad\qquad\text{9-21}$$

or

$$v_e = \sqrt{\frac{2GM_E}{R_E}} \qquad\qquad\text{9-22}$$

Again, we can put this into a slightly more useful form by replacing GM_E with gR_E^2 from Equation 9-13. We then have

$$v_e = \sqrt{2gR_E} \qquad\qquad\text{9-23}\qquad\text{Escape speed}$$

Using $g = 9.81$ m/s^2 and $R_E = 6.37$ Mm, we obtain

$$v_e = \sqrt{2(9.81 \text{ m/s}^2)(6.37 \times 10^6 \text{ m})} = 11.2 \text{ km/s}$$

This is about 6.95 mi/s or about 25,000 mi/h.

The magnitude of the escape speed of a planet or moon relative to the thermal speeds of gas molecules determines the kind of atmosphere the planet or moon may have. In equilibrium, the average kinetic energy of gas molecules, $\tfrac{1}{2}mv^2$, is proportional to the absolute temperature T (see Chapter 11). The average speed of gas molecules, then, depends on the temperature and varies inversely with the mass of the molecule. On earth, the speeds of oxygen and nitrogen molecules are much lower than the escape speed, so these gases can exist in our atmosphere. For the lighter molecules of hydrogen and helium, however, a considerable fraction of the molecules have speeds greater than the escape speed at the earth's surface. Hydrogen and helium are thus not found in our atmosphere.

The escape speed at the surface of the moon can be calculated from Equation 9-23 with the acceleration of gravity on the moon and the radius of the moon replacing g and R_E. The escape speed for the moon is 2.3 km/s, considerably smaller than that for the earth and too small for any atmosphere to exist.

Questions

3. What is the effect of air resistance on the escape speed near the earth's surface?

4. Would it be possible, in principle, for the earth to escape from the solar system?

9-6   Potential Energy, Total Energy, and Orbits

The gravitational potential-energy function given by Equation 9-19 was found by calculating the work needed to lift an object up from the surface of the earth. We chose the potential energy to be zero at the surface of the earth. Since only *changes* in potential energy are important, we could have chosen the potential energy to be zero at any arbitrary position. Although it seems natural to choose the potential energy to be zero at the earth's surface when we are working problems with objects near the earth's surface, there are many situations in which this is not at all the most convenient choice. For example, when considering the potential energy associated with a planet and the sun, there is no reason to want the potential energy to be zero at the surface of the sun. In fact, it is nearly always more convenient to chose the gravitational potential energy of a two-body system to be zero when the separation of the two bodies is infinite. From Equation 9-22, we see that we can change the choice of zero for the potential energy so that it is zero when r is infinite by subtracting the constant $GM_E m/R_E$ from the right side of the equation. The gravitational potential energy of an object of mass m a distance r from the center of the earth is then just

$$U = -\frac{GM_E m}{r} \qquad\qquad 9\text{-}24 \qquad \text{Gravitational potential energy}$$

where $U = 0$ at $r = \infty$. (We assume that r is greater than the radius of the earth, that is, the object is not inside the earth.) This choice of zero for the potential energy seems to have the disadvantage of making the potential energy always negative, but this is not a real disadvantage. It is like defining the potential energy to be zero at the ceiling of a room rather than at the floor when you are doing a problem with an object in the room. The condition for escaping the gravitational field of the earth can now be simply stated. The potential energy of an object of mass m at the surface of the earth is now not zero but $-GM_E m/R_E$. If the initial kinetic energy is great enough so that the total energy is zero or greater than zero, the object will not return to the earth.

With this choice of zero for the gravitational potential energy, we can discuss the question of the kind of orbit taken by an object that moves in the gravitational field of the sun. The potential energy of an object of mass m at a distance r from the sun is

$$U = -\frac{GM_s m}{r} \qquad\qquad 9\text{-}25$$

where M_s is the mass of the sun. The kinetic energy of the object is, of course, $\frac{1}{2}mv^2$. If the total energy, kinetic plus potential, is less than zero, the orbit will be an ellipse (or a circle) and the object will be bound to the sun; that is, it cannot escape from the gravitational field of the sun. On the other hand, if the total energy is positive, the orbit will be a hyperbola, and the object will make one swing around the sun and then leave, never to return again. If the total energy is exactly zero, the orbit will be a parabola, and again the object will escape. That is, when the total energy is zero or positive, the object is not bound to the gravitational field of the sun.

Given the assumption that potential energy is zero when the separation between two bodies is infinite, the bound state of the moon and all of our artificial satellites to the earth is indicated by the fact that the total energy—kinetic plus potential—of these bodies is negative. This is similar to the bound state of orbital electrons in an atom. Thus, giving an object enough energy to escape the earth is somewhat like ionizing an atom.

Question

5. An object (such as a newly discovered comet) enters the solar system and makes a pass around the sun. How can we tell if the object will return in many years or if it will never return?

Comets and Cosmic Archaeology

Donald Goldsmith*
President, Interstellar Media

Comets, the oldest and least altered parts of our solar system, are lumps of ices and dust that have orbited the sun for over 4.6 billion years, ever since the sun and its planets began to form. Although the details of the formation of the solar system remain the subject of intense debate, the broad outlines of the process are generally agreed upon by astronomers. Picture, then, a conglomeration of interstellar gas and dust, made up primarily of hydrogen and helium—as are most objects in the universe—with a sprinkling of the other elements. If this cloud happens to become somewhat denser than usual, perhaps as the result of a shock wave from a nearby exploding star, it will begin to contract as a result of its own gravitational force. Each piece of the cloud will attract every other piece, and the strength of these attractions will increase as the cloud grows smaller; "self-gravitational" contraction is an escalating process in which the initial steps take much more time than the final ones.

Now, assume that the cloud is rotating. (All clouds of interstellar matter appear to have at least some gentle rotation.) As the cloud contracts, it will rotate more and more rapidly because of the conservation of angular momentum discussed in Chapter 8. This rotation makes it more difficult for the cloud to contract in directions *perpendicular* to its axis of rotation than in directions *parallel* to its axis. The cloud therefore becomes progressively flatter and more pancake-shaped as it contracts. Before long—only a few tens of millions of years—it will be a flat disk, rotating perhaps once every hundred thousand years, with a high concentration of matter toward its center.

Within the rotating disk, individual clumps of matter are forming both from self-gravitation and from the fact that their electromagnetic forces sometimes hold atoms and molecules together when they collide. The clumps of matter vary in size from a few millimetres up to a few kilometres. The cloud will not have the density of matter required to form clumps larger than this until it contracts to a much smaller size and thereby raises its density.

Comets are these original clumps, the largest no more than a few kilometres across, that were left behind when the "proto–solar system" contracted to the size of the planets' present orbits around the sun. Thus, they have orbits that are typically tens of thousands of times larger than the earth's orbit, reflecting their origin at the time the rotating disk that became the solar system was much larger than when comet-like chunks of matter agglomerated to form the planets.

* Donald Goldsmith's most recent book, *Nemesis: The Death-Star and Other Theories of Mass Extinction,* received the 1986 AIP Science Writing Award for the best popular book on physics or astronomy by a scientist.

Each comet is a "frozen snowball," to use Fred Whipple's model and phrase, a mountain-sized lump of various ices frozen around primordial dust grains and pebbles. They number not in the millions or billions but in the trillions, though the sum of their masses does not equal the mass of the planet Jupiter.

The totality of the comets orbiting the sun forms the *Oort cloud,* first hypothesized by the Dutch astronomer Jan Oort. Despite the fact that trillions of comets in the Oort cloud slowly and majestically orbit the sun, collisions are rare. The Oort cloud fills a great deal more space than that of the "inner solar system," the realm of the planets. Indeed, the outer edge of the Oort cloud almost rubs shoulders with the Oort clouds (no doubt named after great alien astronomers) of the sun's closest neighbors.

If this were the entire story of comets, we would never have seen one, for even our best telescopes cannot begin to detect the weakly reflected sunlight of a comet at their typical distances of 10,000 to 40,000 times that between the earth and the sun. But at rare intervals, something—perhaps a passing star, perhaps a close encounter with another comet—will make one of the comets in the Oort cloud deviate from the orbit along which it has quietly moved for billions of years. Some of these deviations send comets completely out of the Oort cloud to become interstellar wanderers, but others send comets toward the inner solar system. Two main possibilities await an in-bound comet. It may pass once around the sun and then head into interstellar space forever, or it may be "captured" by the sun in a much smaller orbit than it had originally. This capture occurs as the result of the gravitational forces of the *planets,* particularly that of Jupiter, the largest and most massive planet. All of the familiar comets, including the most famous, Halley's comet, have been captured in this manner.

Close approach to the sun does more to a comet than expose it to orbital capture—it makes the comet visible to us. When a comet approaching the sun comes within a distance a few times that between the earth and the sun, the sun's heat begins to vaporize some of the cometary ices. The gas and dust released from the comet spread out to form a fuzzy envelope, the *coma,* around the actual comet, called the *nucleus.* The nucleus and coma together make up the *head* of the comet that we see. As the comet approaches still closer to the sun, it develops a *tail* of gas and dust that can extend for millions of kilometres from the nucleus. Radiation pressure from sunlight and particles expelled from the sun in the solar wind push on the atoms and molecules in the tail, causing the tail always to point *away* from the sun. Hence, a comet approaches the sun with its tail "dragging," but it recedes

Halley's comet as photographed on
March 7, 1986

from the sun, on its return to the frozen outback of the solar system, with its tail preceding its head. Since each close passage to the sun vaporizes part of a comet, any comet that undergoes such passages cannot survive for more than a few million years.

Modern astrophysicists have used spectroscopic analysis — the study of light at various wavelengths — to unlock the mystery of the composition of comets. Each time that a comet approaches the sun and the earth, we obtain information about what it consists of by studying how the comet reflects and absorbs sunlight. This technique, developed to its greatest extent thus far for the return of Halley's comet in 1986, has revealed that comets indeed consist largely of the molecules that form easily from the most abundant elements in the universe. These molecules include H_2O, CO, CO_2, NH, and NH_2, the same sorts of molecules that we think existed near the earth's surface soon after it formed. The dust grains and pebbles, which cannot be studied spectroscopically, are presumed to be either silicon-oxygen (quartz) compounds or carbon-rich (graphite) molecular chains or possibly a combination of the two. Sunlight reflected from the tail gives comets their spectacular, awe-inspiring appearance, even though the matter in the gauzy tail is so rarefied that all of it could easily be packed into a suitcase with room to spare for a change of clothes.

Halley's comet owes its name to Edmond Halley, a man of many abilities — scientist, sea captain, and eventually Astronomer Royal — who persuaded Isaac Newton to publish his key work, the *Principia Mathematica*. Halley noted that, among all the reports of comets throughout history, an interval of approximately 77 years stood out. He hypothesized that this interval implied a single comet, moving in a highly elongated orbit around the sun, that would return to the earth's vicinity in 1758. When the comet did return (after Halley's death), although somewhat behind schedule as the

result of perturbations in its orbit caused by the gravitational forces of Jupiter and Saturn, it was given Halley's name.

We now know that Halley's comet has an orbit that takes it out past the orbit of Neptune to 33 times the earth's distance from the sun and in to a closest solar approach of half that distance. In 1986, a bevy of spacecraft flew by the comet for the first ever close-up studies. These studies revealed, among a host of other discoveries, that the comet is a pickle-shaped object $15 \times 8 \times 8$ kilometres in size and that its surface is remarkably dark, apparently because it is coated with carbon-rich compounds. The composition of Halley's comet, which is assumed to be similar to that of other comets, lends credence to the hypothesis that comet-like objects may have played a key role in depositing "volatiles" — compounds that vaporize easily — on the surfaces of the earth and the other inner planets during the formation of the solar system.

Comet Showers and the Death of the Dinosaurs

Comets may be more than the oldest relics of the solar system and lovely, rare objects for us to admire. Some scientists have recently hypothesized that, for one reason or another, comets are diverted from the Oort cloud in massive numbers every 26 million years or so. As a result, a "comet shower" occurs all through the inner solar system soon after such a diversion. Although the bulk of these comets would never intersect the earth's orbit, some would be likely to pass close to the earth, and some of these, aided by the earth's gravitation, would strike our planet.

What would be the result of a collision between the earth and a comet ten kilometres wide? The collision would produce a giant crater, something like the eroded craters that dot the earth's crust, and it would lift a vast cloud of dust and grit into the atmosphere. During the months required for the dust and grit to settle back to the surface, this cloud would interfere with sunlight's penetration of the atmosphere. The result would be a "nuclear winter" effect—months of semi-darkness during which a new, harsh climate would exist on the Earth's surface, especially in regions far from the oceans. (Because ocean waters store heat, coastal areas would be buffered against the bitter cold that would sweep over the land.)

Sixty-five million years ago, the dinosaurs—lords of the earth for more than a hundred million years—vanished in a geological instant. Furthermore, dozens of other species vanished at the same time. This "mass extinction," one of several found in the geological record, differs from the extinctions of individual species that occur continually, even without an assist from humanity. Recent evidence has shown that layers of rock laid down about that time contain far more than the normal amount of iridium, an element that is rare near the earth's surface because most of it apparently "migrated" toward the earth's core when the earth was young and relatively plastic. The excess iridium can be explained as the result of the impact of an object 10 to 20 kilometres in diameter—a comet would do nicely—that contained the amount of iridium typical of the sun and other stars and, we believe, of all unaltered pieces of the solar system. A tempting conclusion fits all these data together: The dinosaurs died 65 million years ago when a comet hit the earth. This collision raised a shroud of dust that for some reason left the dinosaurs at a competitive disadvantage compared to our mammalian ancestors, small ratlike and shrewlike creatures who had scampered at the feet of the giant reptiles for tens of millions of years. Freed from their overlords, mammals radiated into a thousand ecological niches, evolving into a wide variety of creatures that, for the last few million years, has included the hominids from which we have descended.

This "impact theory" of the mass extinction that killed the dinosaurs has been sharply criticized, not the least because some evidence suggests that the dinosaurs died out over hundreds of thousands of years, not in just the few months or years that would be expected if the effects of a comet's impact left them incapable of survival. Undaunted, the theory's proponents have gone on to hypothesize a far more startling extension of the impact theory: Impacts recur cyclically, so mass extinctions occur about every 26 million years. This cyclical-extinction theory puts a new wrinkle in the model of evolution on earth, since it hypothesizes that cosmic influences (whatever produces the "comet showers") have a strong effect on the evolution of life on earth.

Arguments for and against the cyclical theory have centered on the statistical analysis of the data relating both to mass extinctions and to the ages of impact craters on the earth's surface. Do these data reveal a periodic cycle of 26

million years? The proponents insist that they do, but their case is as yet unproven. However, *if* the cycle is ever verified, an immediate question will arise: Since no terrestrial phenomenon has a cycle as long as the 26 million years claimed for the mass extinctions, what *extraterrestrial* cause can explain the cyclical impacts?

Theories to answer this question are not lacking. In fact, there are at least three. All of them explain the cyclical occurrence of mass extinctions and impact craters as the result of gravitational forces that perturb the comets in the Oort cloud at regularly spaced time intervals. One school of thought holds that an undiscovered "Planet X," farther from the sun than even Pluto, the farthest known planet, perturbs the comets in the Oort cloud at these regular intervals. Another school claims that the perturbations of the comets arise from interactions between the Oort cloud and the disklike configuration of matter in the Milky Way Galaxy. As the sun orbits the galactic center, taking 240 million years for each orbit, it bobs up and down through the midplane of the galaxy where the galaxy's mass is most strongly concentrated. Each passage through the midplane could cause perturbations of the comets in the Oort cloud. Since these passages occur at intervals of about 30 million years, the time scales approximately match.

Finally, a few bold scientists have suggested that the sun has a hitherto undiscovered companion star, a "brown dwarf" of low luminosity, which they call "Nemesis." They hypothesize that this solar companion moves in a huge, elongated orbit, well beyond the Oort cloud, with a period of 26 million years. When Nemesis approaches relatively close to the sun, they argue, it perturbs the comets in the Oort cloud. If the Nemesis theory is correct, then this closest star to the sun must be extremely faint since it has never been detected. Still, it would play a very important role in explaining how life on earth evolved to produce humans rather than, say, intelligent dinosaurs.

Today, we are fairly sure of the orbits of comets, their ages, their composition, and their role in the history of the solar system. By the time Halley's comet next returns to our vicinity in 2061, we should be able to mount a true expedition to "land" on the comet. (A comet has such a small gravitational force that you could launch yourself into space from it with a good running start.) This will enable us to do what scientists have long dreamed of—examine an archaeological relic 4.6 billion years old to learn more about how the solar system began. Meanwhile, the luck of the cosmic draw has given us the chance to see and admire the comets that we do, and scientific investigation has already made comets yield some of the clues they hold as to how the solar system formed and how life on earth evolved.

Summary

1. Kepler's three laws are

> Law 1. All planets move in elliptical orbits with the sun at one focus.

> Law 2. A line joining any planet to the sun sweeps out equal areas in equal times.

> Law 3. The square of the period of any planet is proportional to the cube of the planet's mean distance from the sun.

Kepler's laws can be derived from Newton's law of gravity. The first and third laws result from the fact that the force exerted by the sun on the planets varies inversely as the square of the separation distance. The second law follows from the fact that the force exerted by the sun on a planet is along the line joining them, and thus the angular momentum of the planet is conserved. Kepler's laws also hold for any body orbiting another in a $1/r^2$ field, for example, a satellite orbiting a planet.

2. Newton's law of gravity postulates that every body exerts an attractive force on every other body that is proportional to the masses of the two bodies and is inversely proportional to the square of the distance separating them. The magnitude of the gravitational force exerted by a particle of mass m_1 on another particle of mass m_2 a distance r away is given by

$$F = G\frac{m_1 m_2}{r^2}$$

where G, the universal gravitational constant, has the value

$$G = 6.67 \times 10^{-11} \text{ N} \cdot \text{m}^2/\text{kg}^2$$

3. The gravitational constant G was determined by Cavendish in 1798 by a laboratory measurement of the gravitational force of attraction between two bodies.

4. The fact that gravitational mass and inertial mass are equal is known as the principle of equivalence. It is because of this that all bodies have the same free-fall acceleration near the surface of the earth. This equivalence has been established experimentally to a very high degree of precision.

5. The gravitational potential energy of a body of mass m at a distance r from the center of the earth does not increase indefinitely as r increases. Instead, it approaches a maximum value $U_{max} = GM_E m/R_E = mgR_E$ greater than the potential energy at the earth's surface. If a body is given an initial kinetic energy at the earth's surface greater than or equal to this maximum potential energy, the body will escape from the earth. The speed needed for escape is approximately 11.2 km/s.

6. It is convenient and customary to choose the gravitational potential energy of two bodies to be zero when the separation of the bodies is infinite. The potential energy of a mass m a distance r from another mass M is then

$$U = -\frac{GMm}{r}$$

With this choice, the orbits in a gravitational field can be simply classified. If the total energy of an orbiting body is less than zero, the body is bound and the orbit is an ellipse. If the total energy is greater than or equal to zero, the body is not bound and the orbit is a parabola (for zero energy) or a hyperbola (for positive energy).

Suggestions for Further Reading

Cohen, I. Bernard: "Newton's Discovery of Gravity," *Scientific American*, March 1981, p. 166.

An historical article in which the development of Newton's ideas leading to his formulation of the universal law of gravitation is put into the context of the ideas of other natural philosophers of the time.

Whipple, Fred L., and J. Allen Hynek: "Observations of Satellite I," *Scientific American*, December 1957, p. 37.

An account of the American effort to determine precisely the orbit of the first artificial earth satellite, launched by the U.S.S.R. in 1957.

Review

A. Objectives: After studying this chapter, you should:

1. Be able to state Kepler's three empirical laws of planetary motion.

2. Be able to derive Kepler's third law of planetary motion: the square of the period is proportional to the cube of the radius for circular orbits.

3. Be able to discuss gravitational and inertial mass.

4. Be able to give the value for the speed needed to escape from the earth.

5. Be able to sketch gravitational potential energy as a function of separation distance and discuss the calculation of escape speed.

6. Be able to discuss the relationship between the total energy of an orbiting body and the type of orbit.

B. Define, explain, or otherwise identify:

Kepler's laws, p. 200
Ellipse, p. 200
Gravitational constant, p. 202
Gravitational mass, p. 208
Inertial mass, p. 208
Principle of equivalence, p. 208
Escape speed, pp. 211–212

C. True or false: If the statement is true, explain why. If it is false, give a counterexample.

1. Kepler's law of equal areas implies that gravity varies inversely with the square of the distance.

2. The planet closest to the sun on the average has the shortest period of revolution about the sun.

3. The force that causes an apple to fall has the same origin as the force that causes the moon to move in a circle about the earth.

Exercises

Section 9-1   Kepler's Laws

1. Halley's comet has a period of about 76 y. What is its mean distance from the sun? Give your answer in astronomical units, AU, where

$$1 \text{ AU} = 1.50 \times 10^{11} \text{ m}$$

is the mean distance between the earth and the sun.

2. Suppose a small planet were discovered with a period of 5 y. What would be its mean distance from the sun?

3. The mean distance of Saturn from the sun is 9.54 AU. What is the period of Saturn? [See Exercise 1 for definition of astronomical units (AU).]

4. The comet Kohoutek has a period estimated to be at least 10^6 y. What is its mean distance from the sun?

5. The radius of the earth's orbit is 1.50×10^{11} m and that of Uranus is 2.87×10^{12} m. What is the period of Uranus?

6. Mars has a period of 1.88 y. What is its mean distance from the sun?

7. The mean distance of Pluto from the sun is 39.5 AU. Find the period of Pluto. [See Exercise 1 for definition of astronomical units (AU).]

Section 9-2   Newton's Law of Gravity

8. Calculate the mass of the earth using the known values of G, g, and R_E.

9. A body is dropped from a height of 6.37 Mm above the surface of the earth. What is its initial acceleration?

10. At what distance above the surface of the earth is the acceleration of gravity half its value at sea level?

11. Find the gravitational force that attracts a 65-kg boy to a 50-kg girl when they are 0.5 m apart. (Assume that they are point masses.)

12. One of Jupiter's moons, Io, has a mean orbit radius of 422 Mm and a period of 1.53×10^5 s. (a) Find the mean radius of Jupiter's moon Callisto, whose period is 1.44×10^6 s. (b) Use the known value of G to compute the mass of Jupiter.

13. Uranus has a moon, Umbriel, whose mean orbit radius is 267 Mm and whose period is 3.58×10^5 s. (a) Find the mass of Uranus. (b) Find the period of Uranus's moon Oberon, whose mean orbit radius is 586 Mm.

14. The mass of Saturn is 5.69×10^{26} kg. (a) Find the period of its moon Mimas, whose mean orbit radius is 186 Mm. (b) Find the mean orbit radius of its moon Titan, whose period is 1.38×10^6 s.

15. Find the radius of a circular orbit of a satellite that orbits the earth with a period of 1 d. (If such a satellite is above the equator and moves in the same direction as the rotation of the earth, it appears stationary relative to the earth.)

16. Suppose you land on a planet of another solar system that has the same mass per unit volume as the earth but has 10 times its radius. What would you weigh on this planet compared with what you weigh on earth?

Section 9-3   The Cavendish Experiment

17. Calculate the mass of the earth from the values of the period of the moon (27.3 d), the mean orbit radius of the moon (0.384 Gm), and the known value of G.

18. The masses in a Cavendish apparatus are $m_1 = 12$ kg and $m_2 = 15$ g, and the separation of their centers is 5 cm. (a) What is the force of attraction between these two masses? (b) If the rod separating the two small masses is 18 cm long, what torque must be exerted by the suspension to balance the torque exerted by gravity?

Section 9-4   Gravitational and Inertial Mass

There are no exercises for this section.

Section 9-5   Escaping the Earth

19. The planet Saturn has a mass 95.2 times that of the earth and a radius 9.47 times that of the earth. Find the escape speed for objects near the surface of Saturn.

20. Find the escape speed for a rocket leaving the moon. The acceleration of gravity on the moon is 0.166 times that on earth, and the moon's radius is $0.273R_E$.

Section 9-6   Potential Energy, Total Energy, and Orbits

21. (a) Find the potential energy (relative to zero at infinity) of a 100-kg mass at the surface of the earth. (Take 6.37 Mm for the earth's radius.) (b) Find the potential energy of the same mass at a height above the earth's surface equal to the earth's radius. What would be the escape speed for a body projected from this height?

Problems

1. The earth orbits the sun in a nearly circular orbit of radius 1.50×10^{11} m. Its period is 1 y. Use these data to calculate the mass of the sun.

2. A particle is projected from the surface of the earth with a speed equal to twice the escape speed. When it is very far from the earth, what is its speed?

3. A space probe is to be sent from the earth so that it has a speed of 50 km/s when it is very far from the earth. What speed is needed for the probe at the surface of the earth?

4. Show that Equation 9-19 can be written

$$U = mgR_E\left(1 - \frac{R_E}{r}\right)$$

where g is the acceleration of gravity at the earth's surface.

5. An object is dropped from rest from a height of 4×10^6 m above the surface of the earth. If there were no air resistance, what would its speed be when it strikes the earth?

6. Two planets of equal mass orbit a much more massive star (see Figure 9-9). Planet m_1 moves in a circular orbit of radius

100 Gm with a period of 2 y. Planet m_2 moves in an elliptical orbit with its closest distance $r_1 = 100$ Gm and its farthest distance $r_2 = 180$ Gm, as shown in the figure. (a) Using the fact that the mean radius of an elliptical orbit is the length of the semimajor axis, find the period of m_2's orbit. (b) What is the mass of the star? (c) Which planet has the greater speed at point P? Which has the greater total energy? (d) How does the speed of m_2 at point P compare with its speed at point A?

Figure 9-9
Problem 6.

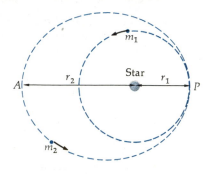

7. (a) Calculate the energy in joules necessary to send a 1-kg mass away from the earth with the escape speed. (b) Convert this energy to kilowatt-hours. (c) If energy can be obtained at 10 cents per kilowatt-hour, what is the minimum cost of giving an 80-kg astronaut enough energy to escape the earth's gravitational field?

8. Show that the potential energy of a particle of mass m at a height h above the earth's surface can be written

$$U = mgR_E[1 - (1 + h/R_E)^{-1}]$$

When x is much smaller than 1, we can use the approximation $(1 + x)^{-1} \approx 1 - x$. Use this with $x = h/R_E$ to show that when h is much smaller than the radius of the earth, the gravitational potential energy is given by

$$U = mgR_E[1 - (1 + h/R_E)^{-1}] \approx mgh$$

CHAPTER TEN

Solids and Fluids

We now leave cosmology to look again at our immediate environment—the solids and fluids that make up our physical world. Some of the characteristics of these states of matter that we will discuss may surprise you, though you have probably taken them for granted for years. For example, some solids like rubber are elastic, and we associate that with their being soft. But steel is hard yet more highly elastic than rubber, which is why it is used in springs. So what does "highly elastic" really mean? Water, a liquid, slides out of the way when we enter a pool yet somehow supports nearly our whole weight. How does it do it? Air, a gas, doesn't support much of our weight, yet airplanes stay aloft. Why? You will find out as you study this chapter.

God in the beginning formed matter in solid, massy, hard, impenetrable, movable particles, of such sizes and figures, and with such other properties, and in such proportion to space, as most conduced to the end for which he formed them.

ISAAC NEWTON

The states of matter in bulk can conveniently be divided into solids and fluids. Solids tend to be rigid and to maintain their shape, whereas fluids do not maintain their shape but flow. Fluids include both liquids, which flow under gravity until they occupy the lowest possible regions of their containers, and gases, which expand to fill their containers regardless of their shapes. The distinction between solids and liquids is not sharp. Glasses, and even rocks under great pressure, tend to flow slightly over long periods of time.

In this chapter, we will look at some of the mechanical properties of solids and fluids at rest and in motion.

10-1   Density

An important property of a substance is the ratio of its mass to its volume, which is called its *density*.

$$\text{Density} = \frac{\text{mass}}{\text{volume}}$$

The greek letter ρ (rho) is usually used to denote density:

$$\rho = \frac{m}{V}$$
  10-1   Density

Since the mass unit, the gram, was originally chosen to be the mass of one cubic centimetre of water, the density of water in cgs units is 1 g/cm^3. Converting to SI units of kilograms per cubic metre, we obtain for the density of water

$$\rho = \frac{1 \text{ g}}{\text{cm}^3} \times \frac{1 \text{ kg}}{10^3 \text{ g}} \times \left(\frac{100 \text{ cm}}{\text{m}}\right)^3 = 10^3 \text{ kg/m}^3$$
  10-2

The density of water varies with temperature. Equation 10-2 gives its maximum value, which occurs at $4°C$.

A convenient unit of volume is the litre (L):

$$1 \text{ L} = 10^3 \text{ cm}^3 = 10^{-3} \text{ m}^3$$

In terms of this unit, the density of water is 1.00 kg/L. When an object's density is greater than that of water, it will sink in water. When its density is less, it will float. In fact, we will show in Section 10-4 that, for floating objects, the fraction of the volume of the object that is submerged in any liquid equals the ratio of the density of that object to the density of the liquid. For example, ice has a density of approximately 0.92 g/cm^3 and floats in water with about 92 percent of its volume submerged. Since the density of water is 1 in cgs units, they are slightly more convenient than SI units. The ratio of the density of a substance to that of water is called the *specific gravity*   Specific gravity
of the substance. The specific gravity is a dimensionless number that equals the magnitude of the density when expressed in grams per cubic centimetre. For example, the specific gravity of ice is 0.92. The specific gravities of objects that sink in water range from 1 to about 22 (for the densest element, osmium). Table 10-1 lists the densities of some common materials.

Although most solids and liquids expand slightly when heated and contract slightly when subjected to an increase in external pressure, these changes in volume are relatively small, so we can say that the densities of most solids and liquids are approximately independent of temperature and pressure. On the other hand, the density of a gas depends strongly on the pressure and temperature, so the pressure and temperature must be specified when giving the densities of gases. In Table 10-1, the densities are given at *standard conditions* (atmospheric pressure at sea level and a temperature of $0°C$). Note that the densities of gases are considerably less than those of liquids or solids. For example, the density of water is about 800 times that of air under standard conditions.

Most solids are more dense than their corresponding liquids. The reverse relationship between ice and water is thus somewhat unusual (though it is not unique). If this were not the case, though, ice would form at the bottom of lakes rather than the top during the winter—bad news for the fish, to say the least.

In the U.S. customary system of units, *weight density*, which is defined as the ratio of the weight of an object to its volume, is often used. Weight density is the product of the density and the acceleration of gravity ρg:

$$\rho g = \frac{w}{V} = \frac{mg}{V}$$
  10-3   Weight density

The weight density of water is

$$\rho_w g = 62.4 \text{ lb/ft}^3$$
  10-4

The weight density of any other material can be found by multiplying the specific gravity by 62.4 lb/ft^3.

Example 10-1   A lead brick is $5 \times 10 \times 20$ cm. How much does it weigh?

From Table 10-1 the density of lead is 11.3×10^3 kg/m³. The volume of the brick is

$$V = (5 \text{ cm})(10 \text{ cm})(20 \text{ cm}) = 1000 \text{ cm}^3 = 10^{-3} \text{ m}^3$$

Its mass is then

$$m = (11.3 \times 10^3 \text{ kg/m}^3)(10^{-3} \text{ m}^3) = 11.3 \text{ kg}$$

and its weight is

$$w = mg = (11.3 \text{ kg})(9.81 \text{ N/kg}) = 111 \text{ N} \approx 25 \text{ lb}$$

Example 10-2   A 200-mL flask is filled with water at 4°C. When the flask is heated to 80°C, 6 g of water spill out. What is the density of water at 80°C?

Since the density of water at 4°C is 1 g/cm³ and 200 mL = 200 cm³, the mass of the water originally in the flask is

$$m = \rho V = (1 \text{ g/cm}^3)(200 \text{ cm}^3) = 200 \text{ g}$$

Since 6 g spill out of the flask at 80°C, the volume of water at this temperature must be 206 cm³. The density of water at 80°C is thus

$$\rho = \frac{m}{V} = \frac{200 \text{ g}}{206 \text{ cm}^3} = 0.97 \text{ g/cm}^3$$

Question

1. What is the approximate specific gravity of your body?

Table 10-1

Densities of Selected Substances*

Substance	Density, kg/m³
Aluminum	2.70×10^3
Bone	$1.7 - 2.0 \times 10^3$
Brick	$1.4 - 2.2 \times 10^3$
Cement	$2.7 - 3.0 \times 10^3$
Copper	8.96×10^3
Earth (average)	5.52×10^3
Glass (common)	$2.4 - 2.8 \times 10^3$
Gold	1.93×10^3
Ice	0.92×10^3
Iron	7.96×10^3
Lead	11.3×10^3
Wood (oak)	$0.6 - 0.9 \times 10^3$
Alcohol (ethanol)	0.806×10^3
Gasoline	0.68×10^3
Mercury	13.6×10^3
Seawater	1.025×10^3
Water	1.00×10^3
Air	1.293
Helium	0.1786
Hydrogen	0.08994
Steam (100°C)	0.6

* $t = 0°C$ and $P = 1$ atm unless otherwise indicated.

10-2   Stress and Strain

If a solid object is in equilibrium but is subjected to forces that tend to stretch, shear, or compress it, the shape of the object changes. If the object returns to its original shape when the forces are removed, it is said to be elastic. Most objects are elastic for forces up to a certain limit called the *elastic limit.* If the forces are too great, so that the elastic limit is exceeded, the object does not return to its original shape but is permanently deformed.

Figure 10-1a shows a solid bar subjected to a force **F** to the right and an equal but opposite force to the left. In Figure 10-1b, we concentrate on a small element of the bar of length L. Since this element is in equilibrium, the forces exerted on it by adjacent elements to the right must equal those exerted by adjacent elements to the left. If the element is not too near the end of the bar, these forces will be distributed uniformly over the cross-sectional area of the bar. The ratio of the force to the cross-sectional area of the bar is called the *tensile stress:*

$$\text{Stress} = \frac{F}{A} \qquad\qquad 10\text{-}5$$

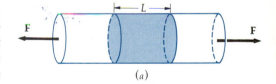

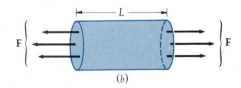

Figure 10-1
(a) Solid bar subjected to a stretching force **F**. (b) A section of the bar. The force per unit area is the stress S.

The forces exerted on the bar tend to stretch the bar. The fractional change in length of the bar $\Delta L/L$ is called the *strain*:

$$\text{Strain} = \frac{\Delta L}{L}$$

10-6   Strain

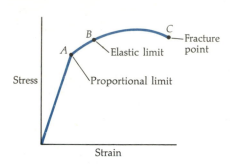

Stress

Strain

Figure 10-2

Stress versus strain. Up to point A, the strain is proportional to the stress. Beyond the elastic limit at point B, the bar will not return to its original length when the stress is removed. At point C, the bar fractures.

Figure 10-2 shows a graph of the strain versus the stress for a typical solid bar. The graph is linear up to point A. This is known as *Hooke's law*. (It is the same behavior as that of a coiled spring for small stretchings, but a coiled spring is more complicated because the stretching force is a combination of tensile forces and shearing forces, which will be discussed shortly.) Point B in Figure 10-2 is the elastic limit of the material. If the bar is stretched beyond this point, it does not return to its original length but is permanently deformed. If an even greater stress is applied, the material eventually breaks, as indicated by point C. The ratio of the stress to the strain in the linear region of the graph is a constant called *Young's modulus Y*:

$$Y = \frac{\text{stress}}{\text{strain}} = \frac{F/A}{\Delta L/L}$$

10-7   Young's modulus

The units of Young's modulus are newtons per square metre (or pounds per square inch). Approximate values of Young's modulus for various materials are listed in Table 10-2.

Table 10-2

Young's Modulus Y and Strengths of Various Materials*

Material	Y, GN/m²†	Tensile strength, MN/m²	Compressive strength, MN/m²
Aluminum	70	90	—
Bone			
Tensile	16	200	—
Compressive	9	—	270
Brass	90	370	—
Concrete	23	2	17
Copper	110	230	—
Iron (wrought)	190	390	—
Lead	16	12	—
Steel	200	520	520

* These values are representative. Actual values for particular samples may differ.
† $1\ \text{GN} = 10^3\ \text{MN} = 10^9\ \text{N}$.

Example 10-3   A certain man's biceps muscle has a maximum cross-sectional area of 12 cm^2 = 1.2 × 10^{-3} m^2. What is the stress in the muscle if it exerts a force of 300 N?

From the definition of tensile stress, we have

$$\text{Stress} = \frac{F}{A} = \frac{300 \text{ N}}{1.2 \times 10^{-3} \text{ m}^2} = 2.5 \times 10^5 \text{ N/m}^2$$

The maximum stress that can be exerted is approximately the same for all muscles. Greater forces can be exerted by muscles with greater cross-sectional areas.

Example 10-4   A 500-kg load is hung from a 3-m steel wire with a cross-sectional area of 0.15 cm^2. By how much does the wire stretch?

The weight of a 500-kg mass is

$$mg = (500 \text{ kg})(9.81 \text{ N/kg}) = 4.90 \times 10^3 \text{ N}$$

From Table 10-2, we find Young's modulus for steel to be 200 GN/m^2 = 2 × 10^{11} N/m^2. The stress S of the wire is

$$S = \frac{4.9 \times 10^3 \text{ N}}{0.15 \text{ cm}^2} = 3.27 \times 10^4 \text{ N/cm}^2$$

$$= 3.27 \times 10^8 \text{ N/m}^2$$

The strain is therefore

$$\frac{\Delta L}{L} = \frac{S}{Y} = \frac{3.27 \times 10^8 \text{ N/m}^2}{2.0 \times 10^{11} \text{ N/m}^2} = 1.63 \times 10^{-3}$$

Since the wire is 300 cm long, the amount it stretches is

$$\Delta L = (1.63 \times 10^{-3})L = (1.63 \times 10^{-3})(300 \text{ cm})$$

$$= 0.49 \text{ cm}$$

If an object is subjected to forces that tend to compress it rather than stretch it, the stress is called *compressive stress*. For many (but not all) materials, Young's modulus for compressive stress is the same as that for tensile stress if ΔL in Equation 10-7 is taken to be the decrease in the length. (Bone is an important exception that has a different Young's modulus for compression and extension.)

If the tensile or compressive stress is too great, the object breaks. The stress at which breakage occurs is called the tensile strength or, in the case of compression, the compressive strength. Approximate values of the tensile and compressive strengths for various materials are also listed in Table 10-2.

In Figure 10-3, a force is applied tangentially to the top of a book. Such a force is called a *shear force*. The ratio of the shear force F_s to the area is called the *shear stress*:

$$\text{Shear stress} = \frac{F_s}{A}$$

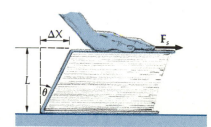

Figure 10-3
Application of a horizontal force to a book causes a shearing stress. The ratio $\Delta X/L = \tan \theta$ is the shear strain.

A shear stress tends to deform the book, as shown in Figure 10-3. The ratio $\Delta X/L$ is called the *shear strain:*

$$\text{Shear strain} = \frac{\Delta X}{L} = \tan \theta$$

where θ is the shear angle, as shown in the figure. The ratio of the shear strain to the shear stress is called the *shear modulus* M_s:

$$M_s = \frac{\text{shear stress}}{\text{shear strain}} = \frac{F_s/A}{\Delta X/L} = \frac{F_s/A}{\tan \theta} \qquad \text{10-8}$$

When a body is submerged in a fluid such as water, the increase in pressure (force per unit area) due to the fluid pressing against the body tends to compress the body. The ratio of the increase in pressure, ΔP, to the fractional decrease in volume, $-\Delta V/V$, is called the *bulk modulus* B:

$$B = -\frac{\Delta P}{\Delta V/V} \qquad \text{10-9}$$

The minus sign in Equation 10-9 is introduced to make B positive since all materials decrease in volume when subjected to an increase in external pressure. The inverse of the bulk modulus is called the *compressibility* k:

$$k = \frac{1}{B} = -\frac{\Delta V/V}{\Delta P} \qquad \text{10-10}$$

The concepts of bulk modulus and compressibility can be applied to liquids and gases as well as to solids. Solids and liquids are relatively incompressible; that is, they have small values of compressibility and large values of the bulk modulus, and these values are relatively independent of temperature and pressure. Gases, on the other hand, are easily compressed, and the values of B and k depend strongly on the pressure and temperature. Approximate values of the shear modulus and the bulk modulus for various materials are listed in Table 10-3.

The karate chop delivered by physicist Ronald McNair produces compressive stress at the top and tensile stress at the bottom of the patio blocks. Since concrete is weaker under tension than under compression, it breaks first at the bottom.

Table 10-3

Approximate Values of the Shear Modulus M_s and the Bulk Modulus B of Various Materials

Material	M_s, GN/m²	B, GN/m²
Aluminum	30	70
Brass	36	61
Copper	42	140
Iron	70	100
Lead	5.6	7.7
Mercury		27
Steel	84	160
Tungsten	150	200
Water		2.0

10-3   Pressure in a Fluid

Fluids differ from solids in that they are unable to support a shear stress. Thus, they can deform to fill any shape. We can define the pressure in a fluid as follows. Consider a small surface of area A, such as a small card submerged in the fluid. The fluid on one side of the card exerts a force F on the card that is balanced by an equal but opposite force exerted by the fluid on the other side of the card. If the card is very small so that we can neglect the difference in the depth of the fluid over its surface, the force F is independent of the orientation of the card. The *pressure P* is defined as the ratio of the magnitude of the force to the area of the card:

$$P = \frac{F}{A} \qquad\qquad 10\text{-}11 \qquad \textcolor{blue}{\textbf{Pressure defined}}$$

The SI unit of pressure is the newton per square metre (N/m²), which is called the pascal (Pa):

$$1\ \text{Pa} = 1\ \text{N/m}^2 \qquad\qquad 10\text{-}12$$

In the U.S. customary system, pressure is usually given in pounds per square inch (lb/in²). Another common unit is the atmosphere (atm), which is approximately air pressure at sea level. One atmosphere is now defined to be 101.325 kilopascal (kPa), which is approximately 14.70 lb/in²:

$$1\ \text{atm} = 101.325\ \text{kPa} = 14.70\ \text{lb/in}^2 \qquad\qquad 10\text{-}13$$

Other units of pressure in common use will be discussed later.

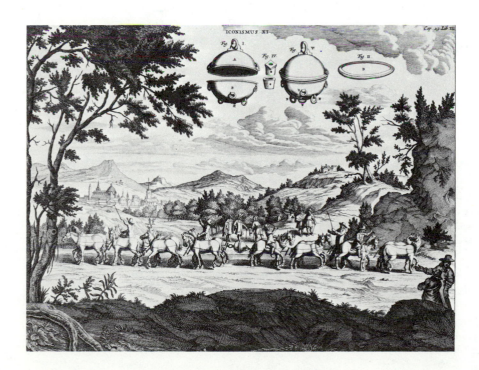

Otto von Guericke's demonstration of the large forces due to atmospheric pressure. The forces exerted by the horses were not great enough to separate the two hollow hemispheres, which had been placed together and evacuated.

As any scuba diver knows, the pressure in a lake or the ocean increases as we go to greater and greater depths. Similarly, the atmospheric pressure decreases as we go to greater and greater altitudes (which is why aircraft cabins must be pressurized). For a liquid such as water whose density is constant throughout, the pressure increases linearly with depth. We can see this by considering a column of liquid of height h and cross-sectional area A, as shown in Figure 10-4. The pressure at the bottom of the column must be greater than the pressure at the top of the column to support the weight of the column. The mass of this liquid column is

$$m = \rho V = \rho A h$$

and its weight is

$$w = mg = \rho A h g$$

If P_0 is the pressure at the top of the column and P is the pressure at the bottom, the net upward force exerted by this pressure difference is $PA - P_0 A$. Setting this net upward force equal to the weight of the column, we obtain

$$PA - P_0 A = \rho A h g$$

or

$$P = P_0 + \rho g h \qquad\qquad 10\text{-}14$$

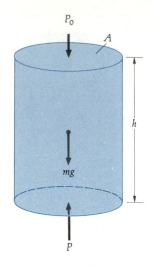

Figure 10-4
Column of water of height h and cross-sectional area A. The pressure P at the bottom must be greater than the pressure P_0 at the top to balance the weight of the water.

Example 10-5   Find the pressure at a depth of 10 m below the surface of a lake if the pressure at the surface is 1 atm.

Using Equation 10-14 with $P_0 = 1$ atm $= 101$ kPa, $\rho = 10^3$ kg/m³, and $g = 9.81$ N/kg, we have

$$P = 101 \text{ kPa} + (10^3 \text{ kg/m}^3)(9.81 \text{ N/kg})(10 \text{ m})$$

$$= 101 \text{ kPa} + 98.1 \text{ kPa}$$

$$= 199 \text{ kPa} = 1.97 \text{ atm}$$

The pressure at a depth of 10 m is nearly twice that at the surface.

The result that the pressure at a depth h is greater than that at the top by the amount $\rho g h$ holds for a liquid in any container, independent of the shape of the container. The pressure is the same at all points at the same depth. If we increase P_0, say, by inserting a piston at the top surface and pressing down on it, the increase in pressure is the same throughout the liquid. This is known as *Pascal's principle*, after Blaise Pascal (1623–1662):

Pressure applied to an enclosed liquid is transmitted undiminished to every point in the fluid and to the walls of the container.

Pascal's principle

A common application of Pascal's principle is the hydraulic lift shown in Figure 10-5. When a force F_1 is applied to the smaller piston, the pressure in

© 1987 Sidney Harris.

"Of course your head hurts. The pressure down here is 20 million newtons per square meter."

the liquid increases by F_1/A_1. The upward force exerted by the liquid on the larger piston of the lift is this increase in pressure times the area A_2. Calling this force F_2, we have

$$F_2 = \frac{F_1}{A_1} A_2 = \frac{A_2}{A_1} F_1 \qquad\qquad 10\text{-}15$$

If A_2 is much greater than A_1, a small force F_1 can be used to exert a much larger force F_2 to lift a weight placed on the larger piston.

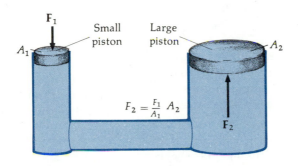

Figure 10-5
Hydraulic lift. A small force F_1 on the small piston produces a change in pressure, which is transmitted by the liquid to the large piston. Since the area of the large piston is much greater than that of the small piston, the force F_2 is much greater than F_1.

Example 10-6   The large piston in a hydraulic lift has a radius of 20 cm. What force must be applied to the small piston of radius 2 cm to raise a car of mass 1500 kg?

The weight of the car is $mg = (1500 \text{ kg})(9.81 \text{ N/kg}) = 1.47 \times 10^4$ N. The force that must be applied is then

$$F_1 = \frac{A_1}{A_2} mg = \frac{\pi r_1^2}{\pi r_2^2} mg$$

$$= \frac{(2 \text{ cm})^2}{(20 \text{ cm})^2} (1.47 \times 10^4 \text{ N}) = 147 \text{ N} \approx 33 \text{ lb}$$

Figure 10-6 shows water in a container made up of parts of different shapes. At first glance, it might seem that the pressure in the largest part of the container would be greater and that the water would therefore be forced into the smallest part of the container to a greater height. This does not happen, however. The pressure depends only on the depth of the water and not on the shape of the container, so at the same height, the pressure is the same in all parts of the container, as can be shown experimentally. This is known as the hydrostatic paradox. Although the water in the largest part of the container weighs more than that in the smaller parts, some of the weight is supported by the normal force exerted by the sides of the largest part of the container, which in this case has a component upward. In fact, the shaded portion of the water is completely supported by the sides of the container.

Figure 10-6
The hydrostatic paradox. The water level is the same regardless of the shape of the vessel. The shaded portion of the water is supported by the sides of the container.

We can use the result that the pressure difference is proportional to the depth of the fluid to measure unknown pressures. Figure 10-7 shows the simplest pressure gauge, the open-tube manometer. The top of the tube is open to the atmosphere at pressure P_{at}. The other end of the tube is at pressure P, which is to be measured. The difference $P - P_{at}$ is equal to $\rho g h$, where ρ is the density of the liquid in the tube. The difference between the "absolute" pressure P and atmospheric pressure P_{at} is called the *gauge pressure*. The pressure you measure in your automobile tire is gauge pressure. When the tire is absolutely flat, the gauge pressure is zero, but the absolute pressure is atmospheric pressure. The absolute pressure is obtained from the gauge pressure by adding atmospheric pressure to it:

$$P = P_{gauge} + P_{at} \qquad\qquad 10\text{-}16$$

Figure 10-8 shows a mercury barometer used to measure atmospheric pressure. The top end of the tube has been closed off and evacuated so that the pressure there is zero. The other end is open to the atmosphere at pressure P_{at}. The pressure P_{at} is given by $P_{at} = \rho g h$, where ρ is the density of mercury.

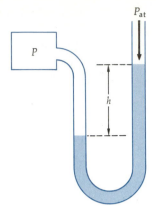

Figure 10-7
Open-tube manometer for measuring an unknown pressure P. The difference $P - P_{at}$ equals $\rho g h$.

Example 10-7   At $0°C$ the density of mercury is $13.595 \times 10^3 \, kg/m^3$. What is the height of a mercury column if the pressure is $1 \, atm = 101.325 \, kPa$?

We have

$$h = \frac{P}{\rho g} = \frac{1.01325 \times 10^5 \, N/m^2}{(13.595 \times 10^3 \, kg/m^3)(9.81 \, N/kg)}$$

$$= 0.7597 \, m \approx 760 \, mm$$

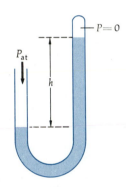

Figure 10-8
U-tube barometer for measuring atmospheric pressure P_{at}.

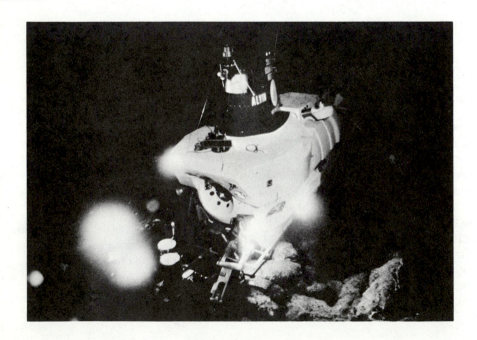

The Deep Submergence Vehicle ALVIN can dive to depths of more than 3 km. It makes more than 100 dives per year, mostly for research purposes.

In practice, pressure is often measured in millimetres of mercury (commonly called torr, after the Italian physicist Torricelli) and inches or feet of water (written inH$_2$O or ftH$_2$O). They are related as follows:

$$1 \text{ atm} = 760 \text{ mmHg} = 760 \text{ torr} = 29.9 \text{ inHg}$$

$$= 33.9 \text{ ftH}_2\text{O} = 101.325 \text{ kPa} \qquad \qquad \text{10-17}$$

$$1 \text{ mmHg} = 1 \text{ torr} = 1.316 \times 10^{-3} \text{ atm} = 133.3 \text{ Pa} \qquad \text{10-18}$$

Other units commonly used on weather maps are the bar and the millibar, which are defined as

$$1 \text{ bar} = 10^3 \text{ millibar} = 100 \text{ kPa} \qquad \qquad \text{10-19}$$

A pressure of 1 bar is just slightly less than 1 atm.

Example 10-8   The (gauge) pressure in the aorta varies from a maximum of about 120 torr (systolic blood pressure) to a minimum of about 80 torr (diastolic blood pressure). Convert the average blood pressure of 100 torr to pascals and pounds per square inch.

We can use the conversion factors implied in Equation 10-17. For example, we have that 760 torr = 101.325 kPa. Thus,

$$P = 100 \text{ torr} \left(\frac{101.325 \text{ kPa}}{760 \text{ torr}} \right) = 13.3 \text{ kPa}$$

and

$$P = 100 \text{ torr} \left(\frac{14.7 \text{ lb/in}^2}{760 \text{ torr}} \right) = 1.93 \text{ lb/in}^2$$

The relation between pressure and altitude for a gas, such as air, is more complicated because the density of a gas is not constant but depends on pressure. In fact, to a good approximation, the density of a gas is proportional to the pressure. The pressure in a column of air decreases with height as you go up from the surface of the earth, like the pressure in a water column decreases as you go up from the bottom, but unlike the case with water pressure, the decrease in air pressure is not linear with distance. Instead, the air pressure decreases by a constant fraction for a given increase in height, as shown in Figure 10-9. This type of decrease is called an *exponential decrease*. At a height of about 5.5 km (18,000 ft), air pressure is half its value at sea level. If we go up another 5.5 km, to an altitude of 11 km (a typical altitude for airliners), the pressure is again halved so that it is one-fourth its value at sea level, and so on. Since the density of air is proportional to the pressure, the density of air also decreases with altitude. Thus, for example, there is less oxygen available on a mountain than at normal elevations, making exercise difficult. At very high altitudes, such as in a jet plane, oxygen must be supplied to the pilot and passengers.

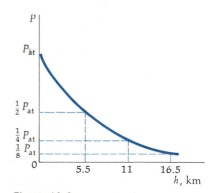

Figure 10-9

Variation in pressure with height above the earth's surface. For each increase in height of 5.5 km, the pressure decreases by 50 percent. This is an example of exponential decrease.

10-4   Buoyancy and Archimedes' Principle

If a heavy object submerged in water is "weighed" by suspending it from a scale (see Figure 10-10*a*), the scale reads less than when the object is weighed in air. Evidently, the water exerts an upward force that partially balances the force of gravity. This force is even more evident when we submerge a piece of cork: it accelerates up toward the surface, where it floats, partially submerged. The submerged cork experiences an upward force from the water greater than the forces of gravity.

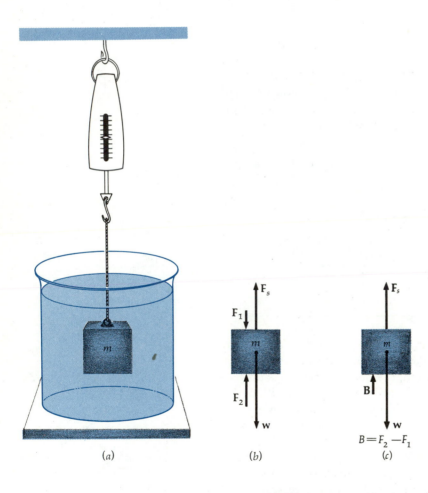

(a)   (b)   (c)

Figure 10-10
(*a*) Weighing a body submerged in a fluid. (*b*) Free-body diagram for the object showing the weight **w**, the force of the spring $\mathbf{F}_s$, and the forces $\mathbf{F}_1$ and $\mathbf{F}_2$ exerted by the surrounding fluid. (*c*) The buoyant force $B = F_2 - F_1$ is the net force exerted by the fluid. It is upward because the pressure at the bottom of the body is greater than that at the top.

The force exerted by a fluid on a body submerged in it, called the *buoyant force*, depends on the density of the fluid and the volume of the body but not on the composition or shape of the body. The force is equal in magnitude to the weight of the fluid displaced by the body. This result is known as *Archimedes' principle*:

A body wholly or partially submerged in a fluid is buoyed up by a force equal to the weight of the displaced fluid.

Archimedes' principle

Archimedes (287 – 212 B.C.) had been given the task of determining whether a crown made for King Hieron II was pure gold or whether it contained some cheaper metal, such as silver. The problem was to determine the density of the irregularly shaped crown without destroying it. As the story goes, Archimedes came upon the solution while bathing, and immediately rushed naked through the streets of Syracuse shouting "Eureka" ("I have found it"). This flash of insight preceded Newton's laws, from which Archimedes' principle can be derived, by some 1900 years. What Archimedes found was a simple and accurate way to determine the specific gravity of the crown, which he could then compare with the specific gravity of gold. The specific gravity of an object is the weight of the object divided by the weight of an equal volume of water. And the weight of an equal volume of water equals the buoyant force on the object, which equals the loss of weight of the object when it is weighed while submerged in water. Thus, the specific gravity of the crown equals the weight of the crown divided by the loss of weight when the crown is submerged in water. It can be determined by weighing it in air and then again when submerged in water.

Archimedes (287 – 212 B.C.)

Example 10-9   The specific gravity of gold is 19.3. If a crown made of pure gold weighs 8 N in air, what should its weight be when submerged in water?

The specific gravity of an object is

$$\text{Specific gravity} = \frac{\text{weight of object in air}}{\text{weight of equal volume of water}}$$

$$= \frac{\text{weight of object in air}}{\text{buoyant force on object when submerged}}$$

or

$$\text{Specific gravity} = \frac{\text{weight of object in air}}{\text{weight loss when submerged in water}} \qquad 10\text{-}20$$

The weight loss in water is thus

$$\text{Weight loss} = \frac{\text{weight in air}}{\text{specific gravity}} = \frac{8 \text{ N}}{19.3} = 0.052 \text{ N}$$

The crown should weigh 8 N − 0.052 N = 7.95 N.

We can derive Archimedes' principle from Newton's laws by considering the forces acting on a portion of a fluid and noting that in static equilibrium the resultant force must be zero. Figure 10-10b shows the vertical forces acting on an object being weighed while submerged. These are the force of gravity $\mathbf{w}$ down, the force of the spring balance $\mathbf{F}_s$ acting up, a force $\mathbf{F}_1$ acting down because of the fluid pressing on the top surface of the object, and a force $\mathbf{F}_2$ acting up because of the fluid pressing on the bottom surface of the object. Since the spring balance reads a force less than the weight, the force $\mathbf{F}_2$ must be greater in magnitude than the force $\mathbf{F}_1$. The difference in magnitude of these two forces is the buoyant force $B = F_2 - F_1$. The buoyant force occurs because the pressure of the fluid at the bottom of the object is greater than that at the top.

In Figure 10-11, we have eliminated the spring and have replaced the submerged object by an equal volume of fluid (indicated by the dashed lines). The buoyant force $B = F_2 - F_1$ acting on this volume of fluid is the same as that acting on our original object since the fluid surrounding that space is the same. Since this volume of fluid is in equilibrium, the resulting force acting on it must be zero. The upward buoyant force thus equals the downward weight of the fluid in this volume:

$$B = w_f \qquad\qquad 10\text{-}21$$

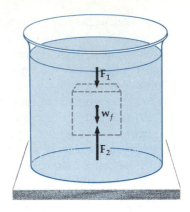

Figure 10-11

Figure 10-10 with the submerged body replaced by an equal volume of fluid. The forces $\mathbf{F}_1$ and $\mathbf{F}_2$ due to the pressure of the fluid are the same as in Figure 10-10. The buoyant force is thus equal to the weight of the displaced fluid.

We note that this result does not depend on the shape of the submerged object. If we consider any irregularly shaped portion of fluid, there must be a buoyant force acting on it due to the surrounding fluid equal to the weight of that portion. We have thus derived Archimedes' principle.

We can see from this that a body will float in fluid if the density of the body is less than that of the fluid. Let ρ_f be the density of the fluid. A volume V of the fluid then has mass $\rho_f V$ and weight

$$w_f = \rho_f g V = B$$

The weight of the object can be written

$$w_o = \rho g V$$

where ρ is the density of the object. If the density of the object is greater than that of the fluid, the weight will be greater than the buoyant force and the object will sink unless supported. If ρ is less than ρ_f, the buoyant force will be greater than the weight and the object will accelerate up to the top of the fluid unless held down. It will float in equilibrium with a fraction of its volume submerged so that the weight of the displaced fluid equals the weight of the object.

Example 10-10   A block of an unknown material weighs 3 N in air and 1.89 N when submerged in water, as shown in Figure 10-10a. What is the material? What correction must be made for the buoyancy of air when the block is weighed in air? (The density of air is approximately 1 kg/m³.)

We determine the material from which the block is constructed by finding its density. According to Equation 10-20, the specific gravity of an object equals the weight of the object in air divided by the loss in weight when the object is submerged in water. Since for this example the loss in weight is $3\text{ N} - 1.89\text{ N} = 1.11\text{ N}$, we have

$$\text{Specific gravity} = \frac{3\text{ N}}{1.11\text{ N}} = 2.70$$

The density of this material is thus 2.70 times that of water or 2.70×10^3 kg/m³. Comparing this with the densities listed in Table 10-1, we see that this block must be made of aluminum.

In general, when an object is weighed in a fluid, as in Figure 10-10a, the measured weight is less than the true weight because of the buoyant force exerted by the fluid. Figure 10-10c shows the forces acting on such an object. The upward forces on the block are the buoyant force B and that of the spring scale F_s. The only downward force is the weight. Since the block is in static equilibrium, the net vertical force on it must be zero. Then,

$$F_s + B = w$$

The force exerted by the spring scale F_s equals the measured weight of the object. We can write the weight and buoyant force in terms of the volume of the object V:

$$B = \rho_f g V$$

and

$$w = \rho g V$$

where ρ_f is the density of the fluid and ρ is that of the object. Then,

$$F_s = w - B = \rho g V - \rho_f g V = \rho g V \left(1 - \frac{\rho_f}{\rho}\right)$$

or

$$F_s = w \left(1 - \frac{\rho_f}{\rho}\right) \qquad\qquad 10\text{-}22$$

The measured weight F_s is less than the true weight w by the factor $(1 - \rho_f/\rho)$. For aluminum in air, we have

$$\frac{\rho_a}{\rho} = \frac{1 \text{ kg/m}^3}{2.7 \times 10^3 \text{ kg/m}^3} = 3.7 \times 10^{-4}$$

and

$$1 - \frac{\rho_a}{\rho} = 1 - 0.00037$$

This differs from 1 by only 0.037 percent, showing that we can usually neglect the buoyant force of air.

The buoyant force of air is not always negligible. Here it is approximately equal to the weight of the balloons plus the child so that they float in the air.

Example 10-11   A cork has a density of 200 kg/m³. Find the fraction of the volume of the cork that is submerged when the cork floats in water.

Let V be the volume of the cork and V' be the volume that is submerged when the cork floats on the water. The weight of the cork is $\rho g V$, and the buoyant force is $\rho_w g V'$. Since the cork is in equilibrium, the buoyant force equals the weight:

$$B = w$$

$$\rho_w g V' = \rho g V$$

The fraction of the cork submerged is then

$$\frac{V'}{V} = \frac{\rho}{\rho_w} \qquad\qquad 10\text{-}23$$

For this example,

$$\frac{V'}{V} = \frac{200 \text{ kg/m}^3}{1000 \text{ kg/m}^3} = \frac{1}{5}$$

Thus, one-fifth of the cork is submerged.

For any object floating in a fluid, Equation 10-23 gives the fraction of the object that is submerged if we replace ρ_w by ρ_f, the density of the fluid. Since the density of ice is 920 kg/m³ and that of seawater is 1025 kg/m³, the fraction of an iceberg that is submerged in seawater is

$$\frac{V'}{V} = \frac{\rho}{\rho_f} = \frac{920 \text{ kg/m}^3}{1025 \text{ kg/m}^3} = 0.90$$

The great danger of icebergs to ships is related to the fact that only about 10 percent of an iceberg is visible above the water.

Questions

2. How can you measure your average density?

3. Why can you see "only the tip of the iceberg"?

4. Smoke usually rises from a smokestack, but it may sink on a very humid day. What can be concluded about the relative densities of humid and dry air?

5. Why is it easier to float in saltwater than in freshwater?

6. Fish can adjust their volume by varying the amount of oxygen and nitrogen gas (obtained from the blood) in a thin-walled sac under their spinal column called a swim bladder. Explain how this helps them swim.

This man's density is very nearly equal to that of water.

10-5   Surface Tension and Capillarity

A needle can be made to "float" on a water surface if it is placed there carefully. The forces that support the needle are not buoyant forces; they are due to *surface tension*. In the interior of a liquid, a molecule is surrounded on all sides by other molecules, but at the surface, there are no molecules above the surface molecules. If a surface molecule is raised slightly, the molecular bonds between it and the adjacent molecules are stretched, and there is a restoring force that pulls the molecule back toward the surface. Similarly, when a needle is placed carefully on the surface, the surface molecules are depressed slightly and the adjacent molecules exert an upward restoring force on them, supporting the needle. Thus, the surface of a liquid is rather like a stretched elastic membrane. The force necessary to break the surface can be measured by lifting a thin wire off the surface, as shown in Figure 10-12. This force is found to be proportional to the length of the surface broken, which is twice the length of the wire since there is a surface film on both sides of the wire. If the wire has a mass m and length L, the force F needed to lift it off the surface is

$$F = \gamma 2L + mg \qquad \text{10-24}$$

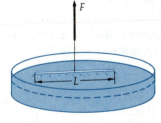

Figure 10-12
A wire of length L being lifted off the surface of a liquid. Surface tension exerts a force on the wire toward the surface.

Coefficient of surface tension

where γ is the *coefficient of surface tension*. The value of γ for water is about 0.073 N/m. It is because of surface tension that small droplets of a liquid tend to be spherical. As the drop is formed, surface tension pulls the molecules together, minimizing the surface area and making the drop spherical. This is how BB's and other small spheres are made.

Because of surface tension, the water breaks up into drops of nearly spherical shape.

The attractive forces between a molecule in a liquid and other like molecules in the liquid are called *cohesive forces*. The force between a molecule of the liquid and another substance, such as the wall of a thin glass tube, is called an *adhesive force*. When the adhesive forces are large relative to the cohesive forces, as in the case of water and a glass surface, the liquid is said to wet the surface of the other substance. In this case, the surface of a column of liquid in a tube is concave upward, as shown in Figure 10-13*a*. In the figure, the force **A**, which is directed toward the wall, is the adhesive force acting on a molecule near the surface. The cohesive force **C** points downward and toward the left because there are no molecules above or to the right of this molecule. The resultant force **R** is perpendicular to the surface. (Otherwise the molecule would slide along the surface.) The contact angle θ_c between the wall and the tangent to the surface measures the relative magnitudes of the cohesive force and the adhesive force. For a liquid that wets the surface, the contact angle is less than 90°, as in Figure 10-13*a*. When the adhesive forces are small relative to the cohesive forces, as is the case for mercury and glass, the liquid does not wet the surface of the other substance, and the surface of the liquid is convex, as shown in Figure 10-13*b*. In this case, the contact angle is greater than 90°.

Cohesive and adhesive forces are difficult to calculate theoretically, but the angles of contact θ_c in Figure 10-13 can be measured. For water and glass, this angle is approximately 0°. For mercury and glass, the contact angle is about 140°.

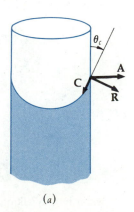

(a)

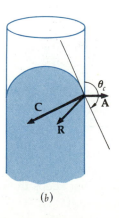

(b)

Figure 10-13
Forces exerted on a molecule of a liquid near the surface. The force **R** is the resultant of the adhesive force **A** exerted by the glass and the cohesive force **C** exerted by the other molecules in the liquid. (*a*) When the adhesive force is much greater than the cohesive force, the surface of the liquid is concave and the contact angle θ_c is less than 90°. (*b*) When the cohesive force is much greater than the adhesive force, the surface of the liquid is convex, and the contact angle θ_c is greater than 90°.

When the surface of a liquid is concave upward, the surface tension at the wall of the tube has a component upward, as shown in Figure 10-14. The liquid will rise in the tube until the net upward force on it due to the surface tension is balanced by the weight of the liquid. This rise is called *capillary action*, or just *capillarity*, and the tube is called a capillary tube. (The smallest blood vessels are also called capillaries.)

In Figure 10-14, the liquid has risen to a height h in a thin capillary tube of radius r. The tube is open to atmospheric pressure at the top. The force holding the liquid up is the vertical component of the surface tension, $F \cos \theta_c$. Since the length of the contact surface is $2\pi r$, this vertical force is $\gamma 2\pi r \cos \theta_c$. If the slight curvature of the surface is neglected, the volume of the liquid in the tube is $\pi r^2 h$. Setting the net upward force equal to the weight, we get

$$\gamma 2\pi r \cos \theta_c = \rho(\pi r^2 h)g$$

or

$$h = \frac{2\gamma \cos \theta_c}{\rho r g} \qquad \text{10-25}$$

Capillarity is responsible for the rise of a liquid in blotting paper and for the rise of lamp oil in the wick. The most important application of capillarity is in holding water in the soil in the small spaces between soil particles. If it were not for capillarity, all the rain water would trickle down to the water table, leaving the upper soil dry. Farming could then only be done in swampy areas, as is the case for rice farming.

Capillarity

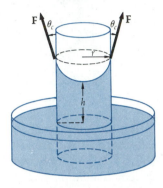

Figure 10-14
Liquid rising in a capillary tube. The upward force due to surface tension supports the weight of the column of liquid.

Example 10-12   How high will water rise in a tube of radius 0.1 mm if the contact angle is zero?

Using $\gamma = 0.073$ N/m for water, we have from Equation 10-25

$$h = \frac{2(0.073 \text{ N/m}) \cos 0°}{(1000 \text{ kg/m}^3)(0.0001 \text{ m})(9.81 \text{ N/kg})} = 0.149 \text{ m} = 14.9 \text{ cm}$$

Example 10-13   The coefficient of surface tension for mercury is 0.465 N/m and $\theta_c = 140°$. A glass capillary tube of radius 3 mm is placed in a bowl of mercury. What is the height of the mercury in the tube relative to that in the bowl?

From Equation 10-25, we obtain

$$h = \frac{2(0.465 \text{ N/m})(\cos 140°)}{(13.6 \times 10^3 \text{ kg/m}^3)(3 \times 10^{-3} \text{ m})(9.81 \text{ N/kg})}$$

$$= -1.78 \times 10^{-3} \text{ m} = -1.78 \text{ mm}$$

The mercury in the capillary tube is depressed 1.78 mm below the surface of the mercury in the bowl.

Question

7. A water bug walks on the surface of a lake. What keeps the bug from sinking?

A water bug "walking on" water.

10-6 Fluids in Motion and Bernoulli's Equation

The general flow of a fluid can be very complicated. Consider, for example, the rising smoke from a burning cigarette (see Figure 10-15). At first the smoke rises in a regular stream, but soon turbulence sets in and the smoke swirls irregularly. Turbulent flow is very difficult to treat even qualitatively. We will therefore be concerned only with nonturbulent, steady-state flow, for example, the smooth rising of the smoke.

We will first consider a fluid flowing with no dissipation of energy. Such a fluid is called nonviscous. (We will treat viscosity in the next section.) We will also assume that the fluid is incompressible, which is a good approximation for most liquids. In an incompressible fluid, the density is constant throughout the fluid. We will use the work-energy theorem to derive an important equation known as Bernoulli's equation, which is useful in understanding many diverse phenomena from the lift of airplane wings to the curving of the path of a baseball's flight.

Figure 10-16 shows a fluid flowing in a tube of varying cross-sectional area. The shading on the left indicates the volume of fluid flowing in at point 1 in some time Δt. If the speed of the fluid at this point is v_1 and the cross-sectional area of the tube is A_1, the volume flowing into the tube in time Δt is

$$\Delta V = A_1 v_1 \, \Delta t$$

Since we are assuming the fluid to be incompressible, an equal volume of fluid must flow out of the tube at point 2, as indicated by the shading on the right. If the speed of the fluid at this point is v_2 and the cross-sectional area is A_2, the volume is $\Delta V = A_2 v_2 \, \Delta t$. Since these volumes are equal, we have

$$A_1 v_1 \, \Delta t = A_2 \, v_2 \, \Delta t$$

$$A_1 v_1 = A_2 v_2 \qquad\qquad 10\text{-}26$$

Equation 10-26 is called the *continuity equation*. The quantity Av is called the *volume flow rate* I_v. In the steady flow of an incompressible fluid, the volume flow rate is the same at any point in the fluid.

$$I_v = Av = \text{constant} \qquad\qquad 10\text{-}27$$

Example 10-14 Blood flows in an aorta of radius 1.00 cm at 30 cm/s. What is the volume flow rate?

From Equation 10-27,

$$I_v = Av = \pi(0.01 \text{ m})^2(0.30 \text{ m/s}) = 9.42 \times 10^{-5} \text{ m}^3/\text{s}$$

It is customary to give the pumping rate of the heart in litres per minute. Using $1 \text{ L} = 10^{-3} \text{ m}^3$ and $1 \text{ min} = 60 \text{ s}$, we have

$$I_v = (9.42 \times 10^{-5} \text{ m}^3/\text{s}) \times \frac{1 \text{ L}}{10^{-3} \text{ m}^3} \times \frac{60 \text{ s}}{1 \text{ min}} = 5.65 \text{ L/min}$$

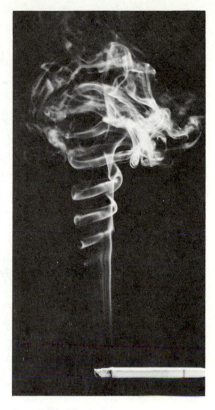

Figure 10-15
Smoke rising from a cigarette. The simple, streamlined flow quickly becomes turbulent.

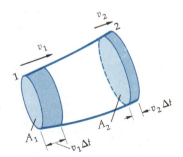

Figure 10-16
Incompressible fluid flowing in a tube of varying cross section. The shaded volumes are equal.

Example 10-15   Blood flows in a large artery of radius 0.3 cm at 10 cm/s into a region where the radius has been reduced to 0.2 cm because of thickening of the walls (arteriosclerosis). What will be the speed of the blood in the narrower region?

If v_1 and v_2 are the initial and final speeds and A_1 and A_2 are the initial and final areas, Equation 10-26 gives

$$v_2 = \frac{A_1}{A_2} v_1 = \frac{\pi(0.3 \text{ cm})^2}{\pi(0.2 \text{ cm})^2} (10 \text{ cm/s}) = 22.5 \text{ cm/s}$$

We now consider a fluid flowing in a tube that varies in elevation as well as in cross-sectional area, as shown in Figure 10-17a. We apply the work-energy theorem to the fluid initially contained between points 1 and 2. After some time Δt, this fluid will have moved along the tube and will be contained in the region between points 1′ and 2′ in Figure 10-17b. The only change between Figures 10-17a and b is for the heavily shaded portions of the fluid mass. Let $\Delta m = \rho \Delta V$ be the mass of this fluid. The net effect on the fluid in time Δt is that the mass of fluid Δm is lifted from height y_1 to height y_2 and its speed is increased from v_1 to v_2. The change in potential energy of this mass is

$$\Delta U = \Delta m \, g y_2 - \Delta m \, g y_1 = \rho \, \Delta V \, g(y_2 - y_1)$$

and the change in kinetic energy is

$$\Delta E_k = \tfrac{1}{2}(\Delta m)v_2^2 - \tfrac{1}{2}(\Delta m)v_1^2 = \tfrac{1}{2}\rho \, \Delta V(v_2^2 - v_1^2)$$

The fluid following it in the pipe (to the left) exerts a force on it of magnitude $F_1 = P_1 A_1$, where P_1 is the pressure at point 1. This force does work:

$$W_1 = F_1 \, \Delta x_1 = P_1 A_1 \, \Delta x_1 = P_1 \, \Delta V$$

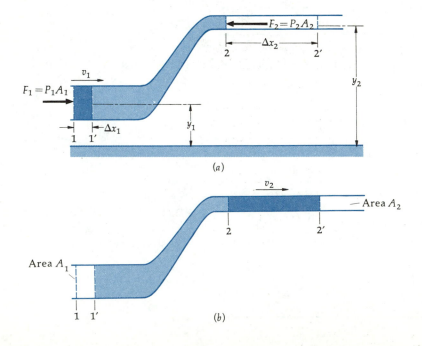

(a)

(b)

Figure 10-17
Fluid moving in a pipe. The net work done by the forces $P_1 A_1$ and $P_2 A_2$ has the effect of raising the part of the fluid indicated by the heavy shading from height y_1 to y_2 and changing its speed from v_1 to v_2.

At the same time, the fluid preceding that under consideration (to the right) exerts a force $F_2 = P_2 A_2$ to the left in the figure. This force does negative work (since it opposes the motion):

$$W_2 = -F_2\,\Delta x_2 = -P_2 A_2\,\Delta x_2 = -P_2\,\Delta V$$

The net work done by these forces is

$$W_{net} = P_1\,\Delta V - P_2\,\Delta V = (P_1 - P_2)\,\Delta V$$

Thus, the work-energy theorem gives

$$W_{net} = \Delta U + \Delta E_k$$

or

$$(P_1 - P_2)\,\Delta V = \rho\,\Delta V\, g(y_2 - y_1) + \tfrac{1}{2}\rho\,\Delta V\,(v_2^2 - v_1^2)$$

If we divide by ΔV, we obtain

$$P_1 - P_2 = \tfrac{1}{2}\rho v_2^2 - \tfrac{1}{2}\rho v_1^2 + \rho g y_2 - \rho g y_1$$

When we collect all the quantities with subscript 2 on one side and all those with subscript 1 on the other, this equation becomes

$$P_1 + \rho g y_1 + \tfrac{1}{2}\rho v_1^2 = P_2 + \rho g y_2 + \tfrac{1}{2}\rho v_2^2 \qquad \text{10-28}$$

This result can be restated as

$$P + \rho g y + \tfrac{1}{2}\rho v^2 = \text{constant} \qquad \text{10-29}$$ Bernoulli's equation

meaning that this combination of quantities has the same value at any point along the pipe. Equation 10-29 is known as *Bernoulli's equation* for the steady, nonviscous flow of an incompressible fluid. To some extent, we can apply Bernoulli's equation to compressible fluids like gases.

A special application of Bernoulli's equation occurs for a fluid at rest. Then $v_1 = v_2 = 0$, and we obtain

$$P_1 - P_2 = \rho g(y_2 - y_1) = \rho g h$$

where $h = y_2 - y_1$ is the difference in height between points 2 and 1. This is the same as Equation 10-14. We will now give some examples using Bernoulli's equation in nonstatic situations.

Example 10-16   A large tank of water has a small hole a distance h below the surface. Find the speed of the water as it flows out the hole.

We apply Bernoulli's equation to points a and b in Figure 10-18. Since the diameter of the hole is assumed to be much smaller than the diameter of the tank, we can neglect the velocity of the water at the top (point a). We then have

$$P_a + \rho g y_a = P_b + \tfrac{1}{2}\rho v_b^2 + \rho g y_b$$

Since both points a and b are open to the atmosphere, the pressures P_a and P_b are both equal to atmospheric pressure. Then

$$v_b^2 = 2g(y_a - y_b) = 2gh$$

and

$$v_b = \sqrt{2gh}$$

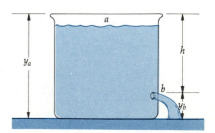

Figure 10-18
Water tank with small hole near the bottom. The speed of the water emerging from the hole is the same as if the water dropped a distance $h = y_a - y_b$, a result known as Torricelli's law.

The water emerges with a speed equal to that it would attain if it dropped in free-fall a distance h. This is known as *Torricelli's law*.

Torricelli's law

Example 10-17   Water flows through a horizontal pipe that has a constricted section, as shown in Figure 10-19. Show that the pressure is *reduced* in the constriction.

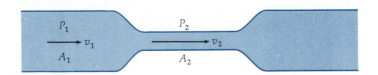

Figure 10-19
Constriction in a pipe carrying a moving fluid. The pressure is smaller in the narrow section of the pipe where the fluid is moving faster.

Since both sections of the pipe are at the same elevation, we take $y_1 = y_2$ in Equation 10-28. Then Bernoulli's equation is

$$P + \tfrac{1}{2}\rho v^2 = \text{constant} \qquad\qquad 10\text{-}30$$

When the fluid moves into the constriction, the area A gets smaller, so the speed v must get larger (since Av is constant). But according to Equation 10-30, as the speed gets larger, the pressure must get smaller if $P + \tfrac{1}{2}\rho v^2$ is to remain constant.

Example 10-18   Water moves through a pipe at 4 m/s under a pressure of 200 kPa. The pipe narrows to half its original diameter. What is the pressure in the narrow part of the pipe?

Since the area of the pipe is proportional to the square of the diameter, the area of the narrow part of the pipe is one-fourth that of the original area. Then, from the continuity equation, $Av = \text{constant}$, the speed in the narrow part of the pipe must be four times that in the wide part of the pipe, or 16 m/s. Equation 10-30 then gives

$$P_1 + \tfrac{1}{2}\rho v_1^2 = P_2 + \tfrac{1}{2}\rho v_2^2$$

$$200 \text{ kPa} + \tfrac{1}{2}(1000 \text{ kg/m}^3)(4 \text{ m/s})^2 = P_2 + \tfrac{1}{2}(1000 \text{ kg/m}^3)(16 \text{ m/s})^2$$

$$200 \text{ kPa} + 8000 \text{ Pa} = P_2 + 128{,}000 \text{ Pa}$$

$$P_2 = 200 \text{ kPa} + 8 \text{ kPa} - 128 \text{ kPa} = 80 \text{ kPa}$$

Note that when we put ρ and v in their proper SI units, the units of $\tfrac{1}{2}\rho v^2$ are pascals. We can check this by writing out the units of ρv^2:

$$\frac{\text{kg}}{\text{m}^3} \times \frac{\text{m}^2}{\text{s}^2} = \frac{\text{kg}\cdot\text{m/s}^2}{\text{m}^2} = \frac{\text{N}}{\text{m}^2} = \text{Pa}$$

where we have used $1 \text{ kg}\cdot\text{m/s}^2 = 1 \text{ N}$.

Equation 10-30 is a general and important result that applies to many situations in which we can ignore changes in height.

When the speed of a fluid increases, the pressure drops.

Venturi effect

This result is often referred to as the *Venturi effect*.

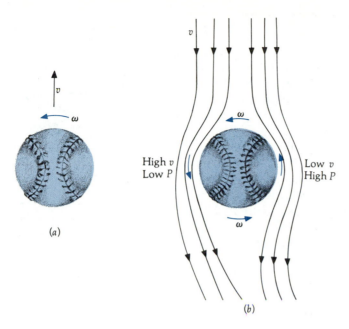

(a)

(b)

Figure 10-20
(a) Top view of baseball thrown with counterclockwise spin, as would be the case when a right-handed pitcher throws a curve. (b) In the frame of reference of the ball, the ball is stationary (but spinning) and the air rushes past it. Because of its rough cover, the spinning ball drags air around it, making the air speed greater on the left side and less on the right. The pressure is less on the left side, where the air speed is greater, so the ball curves to the left.

The Venturi effect can be used qualitatively to understand the lift of an airplane wing and the curve of the path of a baseball. An airplane wing is designed so that the air moves faster over the top of the wing than it does under the bottom of the wing, thus making the air pressure less on top of the wing than underneath. Figure 10-20a shows a top view of a curve ball thrown by a right-handed pitcher. As the baseball spins, it tends to drag the air around with it. Figure 10-20b is drawn from the point of view of the ball with the air rushing past it. The air movement caused by the drag of the spinning ball adds to the velocity of the air rushing by on the left of the ball and subtracts from it on the right. Thus, the speed of the air is greater on the left side of the ball than on the right, and so according to Equation 10-30, the pressure on the left is less than the pressure on the right. The path of the ball therefore curves to the left.

Figure 10-21 shows an atomizer. When the bulb is squeezed, air is forced through a nozzle, reducing the pressure below atmospheric pressure. The liquid in the jar is thus pumped through the connecting tube and emerges as a fine spray. A similar effect occurs in the carburetor of a gasoline engine.

Although Bernoulli's equation is very useful for qualitative descriptions of many features of fluid flow, it is often grossly inaccurate when compared with quantitative experiments. Gases like air are, of course, hardly incompressible, and liquids like water have viscosity, which invalidates the assumption of conservation of mechanical energy. In addition, it is often difficult to maintain steady-state, streamlined flow with no turbulence.

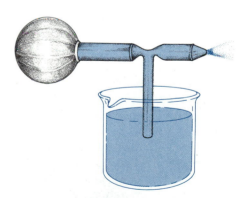

Figure 10-21
Atomizer. The air is forced through the narrow constriction by squeezing the bulb at the left. Because the pressure in the constriction is below atmospheric pressure, the liquid in the jar below is forced into the airstream, resulting in a fine spray of droplets.

Questions

8. In Figure 10-17, the water entering the narrow part of the pipe is accelerated to a greater speed. What forces act on the water to produce this acceleration?

9. In a department store, a beach ball is supported by the airstream from a hose connected to the exhaust of a vacuum cleaner. Does the air blow under or over the ball to support it? Why?

10-7   Viscous Flow

According to Bernoulli's equation, if a fluid flows steadily through a long, narrow, horizontal pipe of constant cross section, the pressure will be constant along the pipe. In practice, we observe a pressure drop as we move along the direction of the flow. Looking at this in another way, a pressure difference is required to push a fluid through a horizontal pipe. This pressure difference is due to the viscosity of the fluid. The pipe exerts a resistive drag on the fluid next to it, and the fluid layers exert viscous drag forces on adjacent layers. The fluid velocity is also not constant across the diameter of the pipe. It is greatest near the center of the pipe and least near the edge where the fluid is in contact with the walls of the pipe (see Figure 10-22). Let P_1 be the pressure at point 1 and P_2 that at point 2, a distance L downstream from point 1. The pressure drop $\Delta P = P_1 - P_2$ is proportional to the flow rate. The proportionality constant is called the resistance R.

$$\Delta P = P_1 - P_2 = I_v R \qquad\qquad \text{10-31}$$

where $I_v = vA$ is the volume flow rate. The resistance to flow R depends on the length of the pipe L, the radius r, and the viscosity of the fluid.

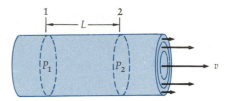

Figure 10-22
When a viscous fluid flows through a pipe, the speed is greater at the center of the pipe. Near the pipe walls, the fluid tends to remain stationary.

Example 10-19   As blood flows from the aorta through the arteries, arterioles, capillaries, venules, and veins to the left atrium, the (gauge) pressure drops from about 100 torr to zero. If the flow rate is 0.08 L/s, find the total resistance of the circulatory system.

In Example 10-8 we found that 100 torr = 13.3 kPa = 1.33×10^4 N/m². Using 1 L = 1000 cm³ = 10^{-3} m³, we have from Equation 10-31,

$$R = \frac{\Delta P}{I_v} = \frac{1.33 \times 10^4 \text{ N/m}^2}{8 \times 10^{-5} \text{ m}^3/\text{s}}$$

$$= 1.66 \times 10^8 \text{ N} \cdot \text{s/m}^5$$

The coefficient of viscosity η of a fluid is defined as follows. In Figure 10-23, a fluid is confined between two parallel plates, each of area A, separated by a distance z. The upper plate is pulled at a constant speed v by a force **F** while the bottom plate is held at rest. A force is needed to pull the upper plate because the fluid next to the plate exerts a viscous drag force opposing the motion. The speed of the fluid between the plates is essentially v near the upper plate and zero near the lower plate, and it varies linearly with the separation between the plates. The magnitude of the force **F** is found to be proportional to v and A and inversely proportional to the plate separation z. The proportionality constant η is the coefficient of viscosity

$$F = \eta \frac{vA}{z} \qquad\qquad \text{10-32}$$

The SI unit of viscosity is the N·s/m² = Pa·s. An older cgs unit in common use is the dyn·s/cm², called a *poise* after the French physicist Poiseuille.

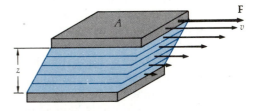

Figure 10-23
Two plates of equal area with a viscous fluid between them. When the upper plate is moved relative to the lower one, each layer of the fluid exerts a drag force on adjacent layers. The force needed to pull the upper plate is proportional to the area A and inversely proportional to the separation z.

These units are related by

$$1 \text{ Pa} \cdot \text{s} = 10 \text{ poises} \qquad \qquad 10\text{-}33$$

Table 10-4 lists the coefficients of viscosity for various fluids at various temperatures. Generally, the viscosity of a fluid increases as the temperature decreases. Thus, in cold climates, for example, a thinner grade of oil is used to lubricate automobile engines in the winter than in the summer.

In terms of the coefficient of viscosity, the resistance to fluid flow R in Equation 10-31 for steady flow through a circular tube of radius r can be shown to be

$$R = \frac{8\eta L}{\pi r^4} \qquad \qquad 10\text{-}34$$

Equations 10-31 and 10-34 can be combined to give the pressure change in a circular tube:

$$\Delta P = \frac{8\eta L}{\pi r^4} I_v \qquad \qquad 10\text{-}35 \qquad \text{Poiseuille's law}$$

Equation 10-35 is known as *Poiseuille's law*. Note the inverse r^4 dependence of the resistance. If the radius of the tube is halved, the pressure drop for a given flow rate is increased by a factor of 16, or a pressure 16 times as great is needed to pump the fluid through the tube at the original flow rate. For example, if the diameter of the blood vessels or arteries is reduced for some reason, either the blood flow is greatly reduced or the heart must work much harder to maintain the flow. For water flowing through a long garden hose, the pressure drop is fixed. It equals the difference in the pressure at the water source and atmospheric pressure at the open end. The flow rate is then proportional to the fourth power of the radius. If the radius is halved, the flow rate drops by a factor of 16.

Poiseuille's law applies only to the laminar (nonturbulent) flow of a fluid whose viscosity is constant, independent of the velocity of the fluid. Blood is a complicated fluid consisting of solid particles of various shapes suspended in a liquid. The red blood cells, for example, are disc-shaped objects that are randomly oriented at low velocities but tend to orient themselves at high velocities to facilitate the flow. Thus, the viscosity of blood decreases as the flow velocity increases, and Poiseuille's law is not strictly valid. Nevertheless, Poiseuille's law is a good approximation that is very useful in obtaining a qualitative understanding of blood flow.

When the flow velocity of a fluid becomes sufficiently great, laminar flow breaks down and turbulence sets in. The critical velocity above which the flow through a tube is turbulent depends on the density and viscosity of the fluid and on the radius of the tube. The flow of a fluid can be characterized by a dimensionless number called the *Reynolds number* N_R defined by

$$N_R = \frac{2r\rho v}{\eta} \qquad \qquad 10\text{-}36 \qquad \text{Reynolds number}$$

where v is the average velocity of the fluid. Experiments have shown that the flow will be laminar if the Reynolds number is less than about 2000 and turbulent if it is greater than 3000. Between these values, the flow is unstable and may change from one type to the other.

Table 10-4

Coefficients of Viscosity of Various Fluids

Fluid	t, °C	η, mPa·s
Water	0	1.8
	20	1.00
	60	0.65
Blood (whole)	37	4.0
Engine oil (SAE 10)	30	200
Glycerin	0	10,000
	20	1,410
	60	81
Air	20	0.018

Example 10-20 Calculate the Reynolds number for blood flowing at 30 cm/s through an aorta of radius 1.0 cm. Assume the viscosity of the blood is 4 mPa·s.

Since the Reynolds number is dimensionless, we can use any set of units as long as we are consistent. We shall put each quantity in Equation 10-36 in SI units. Then,

$$N_R = \frac{2r\rho v}{\eta} = \frac{2(0.01 \text{ m})(1000 \text{ kg/m}^3)(0.3 \text{ m/s})}{4 \times 10^{-3} \text{ Pa·s}} = 1500$$

Since the Reynolds number is less than 2000, this flow will be laminar rather than turbulent.

Summary

1. The density of a substance is the ratio of its mass to its volume:

$$\text{Density} = \frac{\text{mass}}{\text{volume}}$$

$$\rho = \frac{m}{V}$$

The specific gravity is the ratio of its density to that of water. An object sinks or floats in a given fluid depending on whether its density is greater than or less than that of the fluid. The densities of most solids and liquids are approximately independent of temperature and pressure, whereas those of gases depend strongly on these quantities. Objects that sink in water have specific gravities that range from 1 to about 22. Weight density is the density times g. The weight density of water is 62.4 lb/ft³.

2. Tensile stress is the force per unit area applied to a body:

$$\text{Stress} = \frac{F}{A}$$

Strain is the fractional change in the length of the body:

$$\text{Strain} = \frac{\Delta L}{L}$$

Young's modulus is the ratio of stress to strain:

$$Y = \frac{\text{stress}}{\text{strain}} = \frac{\Delta F/A}{\Delta L/L}$$

The (negative) ratio of the change in pressure to the fractional change in volume of an object is called its bulk modulus:

$$B = -\frac{\Delta P}{\Delta V/V}$$

The inverse of this ratio is the compressibility.

3. The SI unit of pressure is the pascal (Pa), which is a newton per square metre:

$$1 \text{ Pa} = 1 \text{ N/m}^2$$

Many other units, such as the atmosphere, bar, torr, pound per square inch, or millimetres of mercury, are often used. These are related by

$$1 \text{ atm} = 101.325 \text{ kPa} = 760 \text{ mmHg} = 760 \text{ torr}$$
$$= 29.9 \text{ inHg} = 33.9 \text{ ftH}_2\text{O} = 14.71 \text{ lb/in}^2$$

Gauge pressure is the difference between the absolute pressure and atmospheric pressure.

4. Pascal's principle states: Pressure applied to an enclosed liquid is transmitted undiminished to every point in the fluid and to the walls of the container.

5. In a liquid such as water, the pressure increases linearly with depth:

$$P = P_0 + \rho g h$$

In a gas such as air, pressure decreases exponentially with altitude.

6. Archimedes' principle states: A body wholly or partially submerged in a fluid is buoyed up by a force equal to the weight of the displaced fluid.

7. Objects can "float" on the surface of a less dense fluid because of surface tension, which is the result of molecular forces at the surface of the fluid. These molecular forces are also responsible for the rise of a liquid in a thin tube, which is known as capillary action.

8. In the steady-state flow of an incompressible fluid, the volume flow rate is the same throughout the fluid:

$$I_v = Av = \text{constant}$$

9. Bernoulli's equation

$$P + \rho g y + \tfrac{1}{2}\rho v^2 = \text{constant}$$

applies to steady-state, nonviscous flow without turbulence in which mechanical energy is conserved. For many situations in which we can ignore changes in height, we have the general and important result that when the speed of the fluid increases, the pressure drops. This result, known as the Venturi effect, can be applied to explain partially the lift on an airplane wing and the curve of the path of a baseball's flight.

10. In viscous flow through a tube, the pressure drop is proportional to the volume flow rate and to the resistance, which is in turn inversely proportional to the fourth power of the radius of the tube. This is Poiseuille's law:

$$\Delta P = I_v R = \frac{8\eta L}{\pi r^4} I_v$$

Suggestions for Further Reading

Hazen, David C., and Rudolf F. Lehnert: "Low-Speed Flight," *Scientific American*, April 1956, p. 46.

How an airplane wing generates lift and how the viscosity of air helps to determine the minimum possible flying speed of an airplane.

Walker, Jearl: "The Amateur Scientist: Looking into the Ways of Water Striders, the Insects that Walk (and Run) on Water," *Scientific American*, November 1983, p. 188.

How surface tension enables water striders to stay dry and how they are able to propel themselves.

Walker, Jearl: "The Amateur Scientist: What Forces Shape the Behavior of Water as a Drop Meanders Down a Windowpane?" *Scientific American,* September 1985, p. 138.

The effects of surface tension on both resting drops and small streams on glass or plexiglass are investigated.

Review

A. Objectives: After studying this chapter, you should:

1. Be able to state the definitions of density, specific gravity, weight density, Young's modulus, and bulk modulus.

2. Be able to explain the buoyancy of boats, hot-air balloons, and the like using Archimedes' principle.

3. Be able to work problems involving buoyant forces on submerged or floating objects.

4. Be able to explain the rise of a liquid in a capillary tube in terms of the surface tension of the liquid.

5. Be able to explain the lift of an airplane wing and the curving of the path of a baseball's flight using Bernoulli's equation.

B. Define, explain, or otherwise identify:

Density, p. 221
Weight density, p. 222
Stress, p. 223
Strain, p. 224
Young's modulus, p. 224

Shear modulus, p. 226
Bulk modulus, p. 226
Pascal's principle, p. 228
Gauge pressure, p. 230
Archimedes' principle, p. 232
Surface tension, p. 236
Capillarity, p. 238
Bernoulli's equation, p. 241
Poiseuille's law, p. 245

C. True or false: If the statement is true, explain why. If it is false, give a counterexample.

1. The buoyant force on a submerged object depends on the shape of the object.

2. Young's modulus has the same dimensions as pressure.

3. If the density of an object is greater than that of water, it cannot float on the surface of the water.

4. In viscous flow, the pressure drop along a pipe is proportional to the flow rate.

Exercises

Section 10-1   Density

1. A copper cylinder is 6 cm long and has a radius of 2 cm. Find its mass.

2. Find the mass of a lead sphere of radius 2 cm.

3. A solid metal cube 8 cm on a side has a mass of 4.08 kg. (*a*) What is the density of the cube? (*b*) If the cube is made from a single element, what is the element?

4. Find the mass of air in a room 4 m by 5 m by 4 m.

5. A small flask used for measuring densities of liquids (called a pyconmeter) has a mass of 22.71 g. When it is filled with water, the total mass is 153.38 g, and when it is filled with milk, the total mass is 157.67 g. Find the density of milk.

6. A solid oak door is 200 cm high, 75 cm wide, and 4 cm thick. How much does it weigh?

7. A 60-mL flask is filled with mercury at 0°C. When the temperature rises to 80°C, 1.47 g of mercury spills out of the

flask. Assuming the volume of the flask is constant, find the density of mercury at 80°C if its density at 0°C is 13,645 kg/m³.

Section 10-2   Stress and Strain

8. A wire 1.5 m long has a cross-sectional area of 2.4 mm². It is hung vertically and stretches 0.32 mm when a 10-kg block is attached to it. Find (*a*) the stress, (*b*) the strain, and (*c*) Young's modulus for the wire.

9. A 50-kg ball is suspended from a steel wire of length 5 m and radius 2 mm. By how much does the wire stretch?

10. Copper has a breaking stress of about 3×10^8 N/m³. (*a*) What is the maximum load that can be hung from a copper wire of diameter 0.42 mm? (*b*) If half this maximum load is hung from the copper wire, by what percentage of its length will it stretch?

11. What pressure is required to reduce the volume of 1 kg of water from 1.00 L to 0.99 L?

12. Seawater has a bulk modulus of 23×10^8 N/m². Find the density of seawater at a depth where the pressure is 800 atm if the density at the surface is 1024 kg/m³.

13. As a runner's foot touches the ground, the shearing force acting on an 8-mm thick sole is as shown in Figure 10-24. If the force of 25 N is distributed over an area of 15 cm², find the angle of shear θ shown, given that the shear modulus of the sole is 1.9×10^5 N/m².

Figure 10-24
Shear forces on a shoe sole for
Exercise 13.

25 N ←   → 25 N

θ

Section 10-3   Pressure in a Fluid

14. Barometer readings are commonly given in inches of mercury. Find the pressure in inches of mercury equal to 101 kPa.

15. The pressure at the surface of a lake is atmospheric pressure $P_{at} = 101$ kPa. (a) At what depth is the pressure twice atmospheric pressure? (b) If the pressure at the top of a deep pool of mercury is P_{at}, at what depth is the pressure $2P_{at}$?

16. Find (a) the absolute pressure and (b) the gauge pressure at the bottom of a diving pool of depth 5.0 m.

17. The top of a card table is 80 cm by 80 cm. What is the force exerted on it by the atmosphere? Why doesn't the table collapse?

18. When a woman in high heels takes a step, she momentarily places her entire weight on one heel of her shoe, which has a radius of 0.4 cm. If her mass is 56 kg, what is the pressure exerted on the floor by her heel?

19. A hydraulic lift is used to raise an automobile of mass 1500 kg. The radius of the shaft of the lift is 8 cm and that of the piston is 1 cm. How much force must be applied to the piston to raise the automobile?

20. Blood flows into the aorta through a circular opening of radius 0.9 cm. If the blood pressure is 120 torr, how much force must be exerted by the heart?

21. A car misses a turn and sinks into a shallow lake to a depth of 8 m. If the area of the car door is 0.5 m², what is the net force exerted on the door, assuming the inside of the car is at atmospheric pressure? What should the occupant do to get the door open?

22. A 1500-kg car rests on four tires, each of which is inflated to a gauge pressure of 200 kPa. What is the area of contact of each tire with the road, assuming the four tires support the weight equally?

23. Blood plasma flows from a bag through a tube into a patient's vein, where the blood pressure is 12 mmHg. The specific gravity of blood plasma at 37°C is 1.03. What is the minimum elevation of the bag so that the pressure of the plasma as it flows into the vein is at least 12 mmHg?

24. In the seventeenth century, Pascal performed the experiment shown in Figure 10-25. A wine barrel was filled with water and then connected to a long tube. Water was added to the tube until the barrel burst. (a) If the radius of the lid was 20 cm and the height of the water in the tube was 12 m, calculate the force exerted on the lid. (b) If the tube had an inner radius of 3 mm, what mass of water in the tube caused the pressure that burst the barrel?

Figure 10-25
Pascal's experiment for Exercise 24.
The wine barrel is filled with water,
and water is then poured into the
long tube until the barrel bursts.

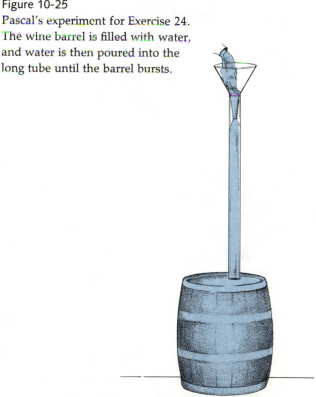

25. Many people believe that, if they float the top of a flexible snorkel tube out of the water (see Figure 10-26), they will be able to breathe with it while walking underwater. But the water pressure opposes the expansion of the chest and the inflation of the lungs. Suppose you can just breathe while lying on the floor with a 225-N (50-lb) weight on your chest.

How far below the surface of the water could your chest be for you to breathe, assuming your chest has a frontal area of 0.09 m²?

Figure 10-26
Walking underwater with a floating snorkel. The water pressure on your chest makes breathing difficult.

26. During floods, when the ground becomes saturated with water, pressure develops in the ground similar to that in water. This pressure forces water through the joints in concrete-block cellar walls. If this happens quickly enough to fill up the cellar with water, there may be no further damage. Otherwise, the upward pressure on the floor may act to float the house like a ship. What upward force would be applied to lift a small house with a 10 m by 10 m basement floor if the floor is 2 m below the surface of the water?

Section 10-4   Buoyancy and Archimedes' Principle

27. A 500-g piece of copper (specific gravity 9.0) is submerged in water and suspended from a spring scale (see Figure 10-27). What force does the spring scale read?

28. When a 60-N stone is attached to a spring scale and submerged in water, the scale reads 40 N. Find the specific gravity of the stone.

29. A block of an unknown material weighs 5 N in air and 4.55 N when submerged in water. (*a*) What is the density of the material? (*b*) Of what material is the block made?

30. A 5-kg iron block is attached to a spring scale and submerged in a fluid of unknown density. The spring scale reads 6.16 N. What is the density of the fluid?

Figure 10-27
Exercise 27.

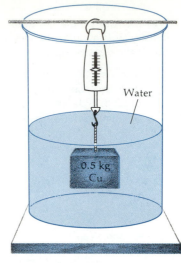

Water

0.5 kg Cu

31. Under standard conditions, the density of air is 1.29 kg/m³ and that of helium is 0.178 kg/m³. A helium balloon lifts a basket and cargo of total weight 2000 N. What must be the volume of the balloon?

Section 10-5   Surface Tension and Capillarity

32. A wire 12.0 cm long and 0.8 mm in diameter is pulled off a water surface with its length parallel to the surface. What force in addition to the weight of the wire is needed?

33. Figure 10-28 shows a small sphere supported on the surface of water by surface tension. The upward force due to surface tension is $2\pi r\gamma \cos\theta_c$. An insect of mass 0.002 g is supported on the surface of a lake by its six legs each of which has an approximately spherical base of radius 0.02 mm. (*a*) Using

$$\tfrac{1}{6}w = \tfrac{1}{6}mg = 2\pi r\gamma \cos\theta_c$$

where w is the weight of the insect, and the factor $\tfrac{1}{6}$ is used because the six legs support the insect equally, calculate the angle θ_c. (*b*) If the mass of the insect increases, with no increase in r, the angle θ_c decreases until the critical value $\theta_c = 0$ is reached. If the mass increases further, the insect cannot be supported by surface tension. Find the critical mass of the insect that can be supported by legs of this size.

Figure 10-28
Exercise 33 and Problem 13.

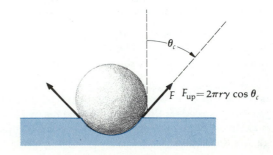

θ_c

$F \quad F_{up} = 2\pi r\gamma \cos\theta_c$

34. When a capillary tube with diameter 0.8 mm is dipped into methanol, the methanol rises to a height of 15.0 mm. If the angle of contact is zero, find the surface tension of methanol (specific gravity 0.79).

35. The tiny tubes called xylem that carry nutrients upward in a plant have a radius of about 0.01 mm. How high will water rise by capillary action in such a tube, assuming the contact angle to be zero?

Section 10-6   Fluids in Motion and Bernoulli's Equation

36. Blood flows in an aorta of radius 9 mm at 30 cm/s. (*a*) Calculate the volume flow rate in litres per minute. (*b*) Although the cross-sectional area of a capillary is much smaller than that of the aorta, there are many capillaries, so the total cross-sectional area is much larger. If all the blood from the aorta flows into the capillaries and the flow rate through the capillaries is 1.0 mm/s, calculate the total cross-sectional area of the capillaries.

37. Water flows through a 3-cm diameter hose at 0.65 m/s. The diameter of the nozzle is 0.30 cm. (*a*) At what speed does the water pass through the nozzle? (*b*) If the pump at one end and the nozzle at the other end are at the same height, and the pressure at the nozzle is atmospheric, what is the pressure at the pump?

38. Water is flowing at 3 m/s in a horizontal pipe under a pressure of 200 kPa. The pipe narrows to half its original diameter. (*a*) What is the speed of flow in the narrow section? (*b*) What is the pressure in the narrow section? (*c*) How do the volume flow rates in the two sections compare?

39. A large tank of water is tapped a distance *h* below the water's surface by a small pipe, as shown in Figure 10-29. Find the distance *x* reached by the water flowing out the pipe.

Figure 10-29
Exercise 39.

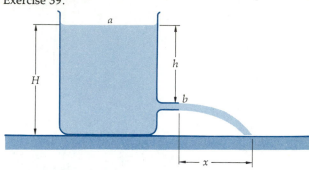

40. The pressure in a section of horizontal pipe of 2-cm diameter is 142 kPa. Water flows through the pipe at 2.80 L/s. What should the diameter of a constricted section of the pipe be for the pressure to be 101 kPa there?

41. During very high winds, the atmospheric pressure inside a house may blow off the roof because of the reduced pressure outside. Calculate the force on a square roof 15 m on a side if the wind speed over the roof is 30 m/s.

Section 10-7   Viscous Flow

42. A horizontal tube with inside diameter 1.2 mm and length 25 cm has water flowing through it at 0.30 mL/s. Find the pressure difference required to drive this flow if the viscosity of water is 1.00 mPa·s.

43. Find the diameter of a tube that would give double the flow rate for the same pressure difference as in Exercise 42.

44. Blood takes about 1.0 s to pass through a 1-mm long capillary in the human circulatory system. If the diameter of the capillary is 7 μm and the pressure drop is 2.60 kPa, find the viscosity of the blood.

Problems

1. A manometer is a U-tube that contains a liquid and is designed to measure small pressure differences between its two arms. If an oil manometer ($\rho = 900$ kg/m³) can be read to ±0.05 mm, what is the smallest pressure change that can be detected?

2. A 3-m by 3-m raft is 10 cm thick and is made of wood with specific gravity 0.6. How many 70-kg people can stand on the raft without getting their feet wet when the water is calm?

3. An object has neutral buoyancy when its density equals that of the liquid in which it is submerged so that it neither

floats nor sinks. If the average density of an 85-kg diver is 0.96 kg/L, what mass of lead should be added to give him neutral buoyancy?

4. When you weigh yourself in air, the scale reading is less than the force exerted on you by gravity because of the buoyant force of air. Estimate the correction to the scale reading you should make to get your true weight.

5. A beaker of mass 1 kg contains 2 kg of water and rests on a scale. A 2-kg block of aluminum (specific gravity 2.70) suspended from a spring scale is submerged in water as shown in Figure 10-30. Find the readings of both scales.

Figure 10-30
Problem 5.

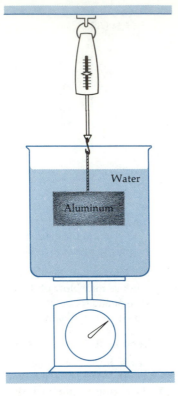

6. A force F is applied to a long wire of length L and cross-sectional area A. Show that if the wire is considered to be a spring, the force constant k is given by $k = AY/L$, where Y is Young's modulus.

7. The demolition of a building is to be accomplished by swinging a 400-kg steel ball on the end of a 30-m steel wire of diameter 5.0 mm hanging from a tall crane. The ball is swung through an arc from side to side, the wire making an angle of 50° with the vertical at the top of the swing. Find the amount by which the wire is stretched at the bottom of the swing.

8. A firefighter holds a hose with a bend in it, as shown in Figure 10-31. Water is expelled from the hose in a stream of radius 1.5 cm at a speed of 30 m/s. (a) What is the mass of water that emerges from the hose in 1 s? (b) What is the horizontal momentum of the water? (c) Before reaching the bend, the water has momentum upward, whereas afterward, its momentum is horizontal. Draw a vector diagram of the initial and final momentum vectors and find the change in the momentum of the water at the bend in 1 s. From this, find the force exerted on the water by the hose.

Figure 10-31
Problem 8.

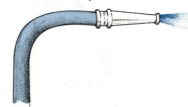

9. A fountain designed to spray a column of water 12 m into the air has a 1-cm diameter nozzle at ground level. The water pump is 3 m below the ground. The pipe to the fountain has a diameter of 2 cm. Find the necessary pump pressure (neglecting the viscosity of the water).

10. A ship sails from seawater (specific gravity 1.03) into freshwater and therefore sinks slightly. When its load of 600 Mg is removed, it returns to its original level. Assuming the sides of the ship are vertical at the water line, find the mass of the ship before it was unloaded.

11. The hydrometer shown in Figure 10-32 is a device for measuring the density of liquids. The bulb contains lead shot, and the density can be read directly from the liquid level on the stem when the hydrometer is calibrated. The bulb's volume is 20 mL, the stem is 15 cm long with diameter 5.00 mm, and the mass of the glass is 6.0 g. (a) What mass of lead shot must be added so that the least density of a liquid that can be measured is 0.9 kg/L? (b) What is the maximum density of a liquid that can then be measured?

Figure 10-32
Problem 11.

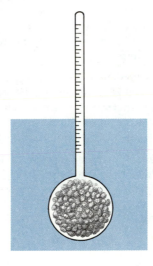

12. A hollow can with a small hole of radius 0.1 mm is pushed under water. At what depth will the water start to flow into the can through the hole if the surface tension of water is 0.73 N/m?

13. A small sphere rests on the surface of a liquid as shown in Figure 10-28. (a) Show that the angle θ_c is related to the radius r and density ρ of the sphere and the surface tension γ of the liquid by

$$\cos \theta_c = \tfrac{2}{3} \frac{r^2 \rho g}{\gamma}$$

(b) Find the radius of the largest copper sphere that can rest on water without sinking.

CHAPTER ELEVEN

Temperature

*Our temperatures differ in
capacity of heat, or,
we boil at different degrees.*

RALPH WALDO EMERSON

In our continuing examination of the material world, we now
begin our investigation of a new field of physics —
thermodynamics. To start off, we will concentrate on temperature, which may at first strike you as a humdrum subject.
However, it is of absorbing interest to a potter trying to recapture
a shimmering glaze, an astronomer correctly classifying a star
between the red giants and the white dwarfs, a metallurgist
tempering a piece of steel, and parents with a sick child. For us,
temperature is the basic measuring tool that we will need in our
study of thermodynamics.

Thermodynamics is the study of temperature, heat, and energy exchange. It
has practical applications in all branches of science and engineering as well
as in all aspects of our daily lives, from dealing with the weather to cooking.

In this chapter, we will study temperature. Familiar to us all as the measure of hotness or coldness, temperature is, more precisely, a measure of the
average internal molecular kinetic energy of an object or substance. Defining and determining temperature is much more subtle than you might
expect, however. For example, as we will see shortly, it is quite difficult to
define temperature so that different thermometers will agree with each
other in the measurement of the temperature of a substance. In this chapter,
we will see that the properties of gases at low densities allow us to define an
ideal-gas temperature scale and to construct gas thermometers that do agree
in the measurement of a given temperature independent of what gas is used.
In later chapters, we will see that the second law of thermodynamics allows
the definition of an absolute temperature scale that is independent of the
properties of any substance and can be chosen to coincide with the ideal-gas
scale in the range for which that scale can be used.

11-1   The Celsius and Fahrenheit Temperature Scales

When an object is heated or cooled, some of its physical properties change. Most solids and liquids expand when heated. A gas, if it is allowed to, will also expand when heated, or, if its volume is kept constant, its pressure will rise. If an electrical conductor is heated, its electrical resistance changes. A physical property that varies with temperature, such as the length of a metal rod, the volume of a liquid, the volume or pressure of a gas, or the resistance of a conductor, is called a *thermometric property*. Any one of these properties can be used to establish a temperature scale and construct a thermometer.

Figure 11-1 shows a common mercury thermometer consisting of a glass bulb and tube containing a fixed amount of mercury. When the mercury is heated by placing the thermometer in contact with a warmer body, the mercury expands more than the glass, and the length of the mercury column increases. Temperatures are measured by comparing the end of the mercury column with markings on the glass tube. These markings are constructed as follows. The thermometer is first placed in ice and water in equilibrium at a pressure of 1 atm. If the thermometer was originally in a room that was warmer than the ice water, the length of the mercury column will decrease, but eventually it will cease to change. The thermometer is now in *thermal equilibrium* with the ice water. Two systems in thermal contact are in thermal equilibrium when their thermometric properties cease to change. The position of the mercury column is marked on the glass tube. This is the *ice-point temperature* (also called the normal freezing point of water). The thermometer is now placed in boiling water at a pressure of 1 atm, and the length of the mercury column increases until the thermometer is in thermal equilibrium with the boiling water. The new position of the column is marked as the *steam-point temperature* (also called the normal boiling point of water). The *Celsius* temperature scale* is constructed by defining the ice-point temperature as 0°C and the steam-point temperature as 100°C. The space on the glass tube between the ice-point mark and the steam-point mark is then divided into 100 equal intervals or degrees, and the degree markings are extended below the ice point and above the steam point. The temperature of another system can now be measured by placing the mercury thermometer in contact with it, waiting for thermal equilibrium to be established, and noting the position of the mercury column. For example, the normal temperature of the human body measured in this way is about 37°C.

The *Fahrenheit temperature scale* is constructed by defining the ice-point temperature as 32°F and the steam-point temperature as 212°F. Fahrenheit wanted all measurable temperatures to be positive. Originally, he chose 0°F for the coldest temperature he could obtain with a mixture of ice and salt water and 96°F (a convenient number with many factors for subdivision) for the temperature of the human body. He then modified his scale slightly

Figure 11-1
A simple mercury thermometer. When the thermometer is in contact with a warm body, the mercury expands and the column rises.

Thermal equilibrium

---

* This temperature scale was formerly called the centigrade scale.

to make the ice-point and steam-point temperatures whole numbers. This resulted in the temperature of the human body being between 98° and 99°F.

Since the Fahrenheit scale is in common use in the United States and the Celsius scale is used in scientific work and throughout the rest of the world, we often need to convert temperatures between these two scales. We note that there are 100 Celsius degrees and 180 Fahrenheit degrees between the ice point and the steam point. A temperature change of 1 Fahrenheit degree is therefore smaller than a change of 1 Celsius degree. A temperature change of 1 Celsius degree (written 1 C° to distinguish it from a temperature of 1°C) equals a change of $\frac{9}{5}$ Fahrenheit degrees. To convert a temperature given on one scale to that of the other scale, we must also take into account the fact that the zero points of the two scales are shifted. The general relation between a Fahrenheit temperature t_F and a Celsius temperature t_C is

$$t_C = \tfrac{5}{9}(t_F - 32 \text{ F}°) \hspace{4cm} \text{11-1}$$

To convert from a Celsius temperature to a Fahrenheit temperature, we merely solve Equation 11-1 for t_F:

$$t_F = \tfrac{9}{5}t_C + 32 \text{ F}° \hspace{4cm} \text{11-2}$$

We can understand the application of these formulas better by considering some examples.

Example 11-1   Find the temperature on the Celsius scale equivalent to 41°F.

From Equation 11-1, we obtain

$$t_C = \tfrac{5}{9}(41 - 32) = \tfrac{5}{9}(9) = 5°C$$

We can understand this formula as follows. A temperature of 41°F is 9 Fahrenheit degrees above the ice point ($41 - 32 = 9$). But 9 Fahrenheit degrees above the ice point is equivalent to 5 Celsius degrees above the ice point. Since the ice point on the Celsius scale is 0°C, a temperature of 41°F corresponds to 5°C.

Example 11-2   Find the temperature on the Fahrenheit scale equivalent to −10°C.

The temperature −10°C is 10 C° below the ice point. This is equivalent to 18 F° below the ice point on the Fahrenheit scale. Since the Fahrenheit ice point is 32°F, −10°C is equivalent to 14°F. This result can be obtained directly from Equation 11-2 by substituting $t_C = -10°C$.

Other thermometric properties can be used to set up thermometers and construct temperature scales. Figure 11-2 shows a bimetallic strip consisting of two different metals bonded together. Because of the difference in the rates of thermal expansion of the two metals, when the strip is heated or

Galileo's thermoscope used the expansion of air in a bulb to measure temperature. It was somewhat unusual in that the reading went down as the environment got warmer. Since the thermoscope responds to changes in both temperature and pressure, it does a rather poor job of measuring either.

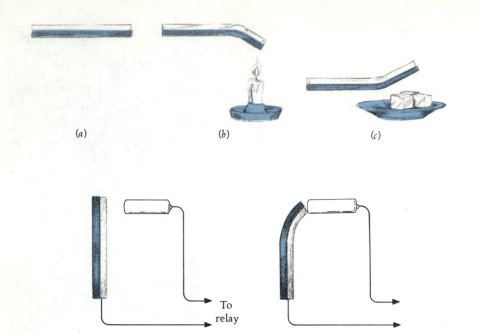

(a)　　　　　　　　(b)　　　　　　　　(c)

Figure 11-2
(a) A bimetallic strip. (b) When heated, the two metals expand at different rates, causing the strip to bend. (c) When cooled, the two metals contract at different rates, and so the strip bends in the opposite direction.

To relay

Figure 11-3
A bimetallic strip can be used in a thermostat to break or make an electrical contact.

cooled, it bends to accommodate the difference in expansion. One application of such a strip is in the thermostat used to regulate a furnace. The bending of the strip can be used to make or break an electrical contact (see Figure 11-3), thus turning the furnace on or off. Figure 11-4 shows a thermometer consisting of a coil of a bimetallic strip with a pointer to indicate temperature. When the thermometer is heated, the coil bends and the pointer moves to the right. The thermometer is calibrated in the same way as our mercury thermometer was, that is, by determining the ice point and the steam point and dividing the resulting interval into 100 Celsius degrees (or 180 Fahrenheit degrees). Other thermometers can be similarly calibrated.

Questions

1. Which is greater, an increase in temperature of 1 C° or one of 1 F°?

2. One body is at −2°C and another is at +20°F. Which body is colder?

3. Distinguish between 1°C and 1 C°.

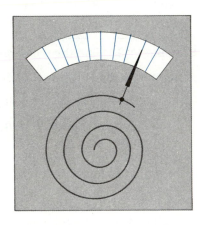

Figure 11-4
A thermometer using a bimetallic strip.

11-2　The Absolute Temperature Scale

Although thermometers and temperature scales based on the changes in the volume of mercury, the length of a metal rod, or the electrical resistance of a wire are useful in everyday life, they are not sufficiently accurate for scientific work. The reason is that two thermometers based on two different thermometric properties do not agree with each other when they are used to measure the temperature of some body. For example, if we measure the normal boiling point of sulfur (about 444°C), we get different numbers using different thermometers. (All thermometers agree, of course, at the ice point and steam point of water, where they have been calibrated to agree.) The reason for the disagreement is that none of these thermometric properties vary linearly with temperature when the temperature scale is defined using a different property. For example, the change in the length of a metal rod for a temperature change of 20 C° is not actually twice that for a temperature change of 10 C° if the temperature changes are measured with a mercury thermometer.

There is, however, one group of thermometers for which measured temperatures are in very close agreement at all temperatures. These are gas thermometers. In one version, the constant-volume gas thermometer, the volume of the gas is kept constant, and the pressure of the gas is used as the thermometric property. In another version, the constant-pressure gas thermometer, the pressure is kept constant, and the volume is used to measure the temperature. At sufficiently low pressures, the pressure of a gas at constant volume (or the volume at constant pressure) is found to vary linearly with temperature no matter what kind of gas is used in the thermometer.

Figure 11-5 shows a constant-volume gas thermometer. The volume is kept constant by raising or lowering tube B_3. The pressure of the gas is read by reading the height h of the mercury column. The ice-point pressure and the steam-point pressure are measured in the usual way, and the resulting interval is divided into 100 equal degrees (for the Celsius scale). If we measure the temperature of a given state of some system (the boiling point of sulfur, for example), we get close agreement no matter what kind of gas is used as long as we use a very small amount of gas in the thermometer.

Figure 11-6 shows the results of measurements of the boiling point of sulfur using constant-volume gas thermometers filled with various gases. The pressure of the steam point P_s for each thermometer is varied by varying the amount of gas in the thermometer. We can see from this figure that when we reduce the amount of gas in the thermometers, the agreement improves. In the limit, as the amount of gas in the thermometer goes to zero, all gas thermometers give exactly the same value for the temperature.

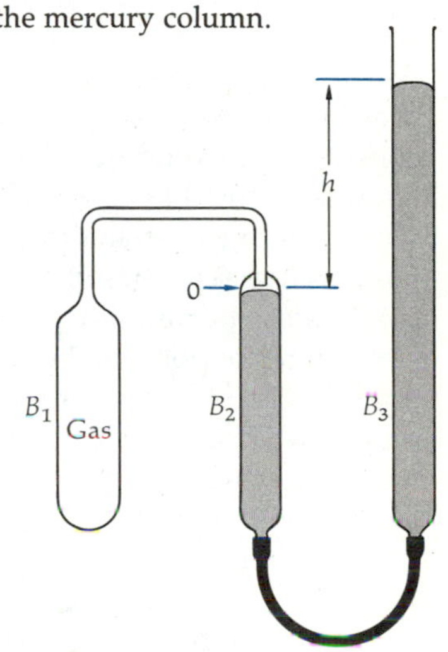

Figure 11-5
A constant-volume gas thermometer. The volume is kept constant by raising or lowering tube B_3 until the mercury in tube B_2 is at the zero mark. The temperature is proportional to the pressure of the gas as determined by the height h of the mercury column.

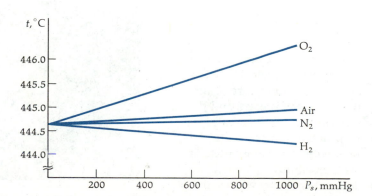

Figure 11-6
Temperature of the boiling point of sulfur measured with constant-volume gas thermometers filled with various gases. The pressure of the steam point of water P_s is varied by varying the amount of gas in the thermometer. As the amount of gas in the thermometer is reduced, the measured temperatures approach the value 444.60°C for all thermometers.

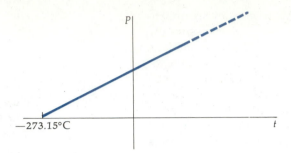

Figure 11-7
Plot of pressure versus temperature
for a constant-volume gas
thermometer. When extrapolated to
zero pressure, the graph intersects
the temperature axis at the value
273.15°C.

Figure 11-7 shows a plot of temperature versus pressure measured by a constant-volume gas thermometer containing a very small amount of gas. As the pressure approaches zero, the temperature approaches −273.15°C. This limit is the same no matter what kind of gas is used, but there is often considerable experimental uncertainty in the measurement of this number because of the difficulty in duplicating the ice-point and steam-point states with high precision in different laboratories. Because of these difficulties, a temperature scale based on a single fixed point was adopted in 1954 by the International Committee on Weights and Measures.

A reference point that is much more easily reproduced than either the ice point or the steam point is the *triple point of water*, the single temperature and pressure at which water, water vapor, and ice coexist in equilibrium. If we put ice, water, and water vapor in a sealed container with no air, the system will eventually reach an equilibrium temperature and pressure at which no ice will melt or evaporate, no water will freeze or evaporate, and no vapor will condense or freeze. This equilibrium temperature and pressure is called the triple point. The triple-point pressure is 4.58 mmHg, and the triple-point temperature is 0.01°C. The *absolute temperature scale* is defined so that the temperature of the triple point is 273.16 kelvins (K). (The kelvin, formerly called the degree Kelvin, is the same size as the Celsius degree; that is, a temperature change of 1 K is the same as a temperature change of 1 C°).

Triple point

The temperature of any state of a system measured in a constant-volume gas thermometer is defined to be proportional to the pressure of the gas in that thermometer. If P_3 is the pressure of the triple point of water for a given thermometer, the *absolute temperature* of a system is given by

$$T = \frac{273.16 \text{ K}}{P_3} P \qquad\qquad 11\text{-}3$$

where P is the pressure of the gas in the thermometer when the thermometer is in thermal equilibrium with the system. The temperature scale defined by Equation 11-3 has the advantage that the measured temperature of any state of a system is the same no matter what kind of gas is used, and the reference state is easily reproduced throughout the world. This scale depends on the properties of gases but not on the properties of any particular gas. The lowest temperature that can be measured with a gas thermometer is about 1 K, using helium for the gas. Below this temperature, helium liquefies. (All other gases liquefy at higher temperatures.) Later, we will see that the second law of thermodynamics can be used to define the absolute

Lord Kelvin (1824–1907)

temperature scale independent of the properties of any substance and with no limitation on the range of temperatures that can be measured. Temperatures as low as a millionth of a kelvin have been measured. The absolute scale so defined is identical to that defined by Equation 11-3 in the range that gas thermometers can be used.

The difference between the absolute scale (also called the Kelvin scale) and the Celsius scale lies only in the choice of zero temperature. To convert from degrees Celsius to kelvins, we merely add 273.15. (From now on, we will usually round off the temperature of absolute zero to $-273°C$.)

$$T = t_c + 273 \text{ K} \qquad\qquad 11\text{-}4$$

To convert from kelvins to degrees Fahrenheit, it is usually easiest to convert to degrees Celsius first by subtracting 273 and then use Equation 11-2. It is important to remember that, since the Celsius degree is the same size as the kelvin, temperature *differences* are the same on the two scales. That is, a temperature difference of 10 K is the same as a temperature difference of 10 C°.

Although the Celsius and Fahrenheit scales are convenient for everyday use, the absolute scale is much more convenient for scientific purposes, partly because many formulas are more simply expressed when the absolute scale is used, and partly because the absolute temperature can be given a more fundamental interpretation.

Oliver is shivering in spite of the fact that the temperature is over 270 K.

Example 11-3   What is the Kelvin temperature corresponding to 70°F?

From Equation 11-1, the Celsius temperature equal to 70°F is

$$t_c = \tfrac{5}{9}(70 - 32) = \tfrac{5}{9}(38) = 21°C$$

The Kelvin temperature is then obtained from Equation 11-4:

$$T = t_c + 273 = 21 + 273 = 294 \text{ K}$$

Questions

4. Why might the Celsius and Fahrenheit scales be more convenient for ordinary nonscientific purposes?

5. The temperature at the surface of the sun is said to be about 6000 degrees. Do you think that this is degrees Celsius or kelvins, or does it matter?

11-3   The Ideal-Gas Law

If we compress a gas while keeping the temperature constant, we find that the pressure increases as the volume decreases, whereas if we cause a gas to expand, its pressure decreases as the volume increases. To a good approximation, the pressure of a gas varies inversely with its volume. This implies that, at constant temperature, the product of the pressure and volume of a gas is constant. This experimental result is known as *Boyle's law:*

$$PV = \text{constant} \qquad \text{(constant temperature)}$$

Boyle's law

This law holds approximately for all gases at low pressures. But according to Equation 11-3, the absolute temperature of a gas at low pressures is proportional to the pressure at constant volume. The absolute temperature is also proportional to the volume of a gas if the pressure is kept constant. Thus, at low pressures, the product PV is approximately proportional to the temperature T:

$$PV = CT$$

where C is a constant of proportionality that depends on the amount of gas.

The pressure exerted by a gas on the walls of its container is the result of collisions of the gas molecules with the walls. If we double the number of gas molecules in a given container, we double the number of collisions that take place in some time interval and, consequently, we double the pressure. We therefore expect the constant C in the preceding equation to be proportional to the number of molecules of the gas in the container:

$$C = kN$$

where N is the number of molecules of the gas and k is a constant. We then have

$$PV = NkT \qquad\qquad\qquad 11\text{-}5$$

The constant k is called *Boltzmann's constant*. It has been found experimentally to have the same value for any kind or amount of gas. Its value in SI units is

$$k = 1.38 \times 10^{-23} \text{ J/K} \qquad\qquad 11\text{-}6 \qquad \text{Boltzmann's constant}$$

It is often convenient to write the amount of gas in terms of the number of moles. A *mole* (mol) of any substance is the amount of that substance that contains Avogadro's number of atoms or molecules. *Avogadro's number N_A* is defined as the number of carbon atoms in 12 g of ^{12}C. The value of Avogadro's number is

$$N_A = 6.022 \times 10^{23} \text{ molecules/mol} \qquad 11\text{-}7 \qquad \text{Avogadro's number}$$

The mass of 1 mol is called the *molar mass M*. (The term "molecular weight" is also sometimes used.) The molar mass of ^{12}C is, by definition, 12 g or 12×10^{-3} kg. The molar masses of the elements are given in the periodic table in Appendix F. The molar mass of a molecule, such as CO_2, is the sum of the molar masses of the elements in the molecule. Since the molar mass of oxygen is 16 g/mol (actually, it is 15.999 g/mol), the molar mass of O_2 is 32 g/mol and that of CO_2 is $12 + 32 = 44$ g/mol. If we have n mol of a substance, the number of molecules is

Molar mass

$$N = nN_A \qquad\qquad\qquad 11\text{-}8$$

Equation 11-5 is then

$$PV = nN_A kT = nRT \qquad\qquad 11\text{-}9$$

where

$$R = kN_A \qquad\qquad\qquad 11\text{-}10$$

is called the *universal gas constant*. Its value, which is the same for all gases, is

$$R = 8.314 \text{ J/mol} \cdot \text{K} = 0.08206 \text{ L} \cdot \text{atm/mol} \cdot \text{K} \qquad 11\text{-}11 \qquad \text{Universal gas constant}$$

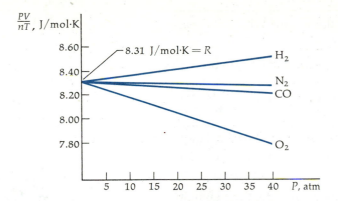

Figure 11-8
Plot of PV/nT versus P for real gases. As the density of the gas is reduced, thereby reducing the pressure, the ratio PV/nT approaches the same value, 8.31 J/mol·K, for all gases. This value is the universal gas constant R. The ideal-gas equation, $pV = nRT$, is a good approximation for all real gases at low pressures, that is, up to a few atmospheres.

Figure 11-8 shows plots of PV/nT versus pressure P for several gases. We can see that, for real gases, PV/nT is very nearly constant over a rather large range of pressures. (Even for oxygen, which has the largest variation in this graph, the variation is only about 1 percent for 0 to 5 atm.) We define an *ideal gas* as one for which PV/nT is constant for all pressures. We can see from Figure 11-8 that N_2 approximates this ideal. For an ideal gas, the pressure, volume, and temperature are related by

$$PV = nRT \qquad\qquad\qquad 11\text{-}12$$ Ideal-gas law

Equation 11-12, which relates P, V, and T for a given amount of gas, is called an *equation of state*. This equation of state applies to an ideal gas, a concept that is an extrapolation of the behavior of real gases at low pressures to ideal behavior. As can be seen from Figure 11-8, at reasonably low pressures, many gases differ little from an ideal gas. Thus Equation 11-12 is quite useful in describing the properties of real gases. At higher gas densities and pressures, corrections must be made to Equation 11-12 if it is to apply to real gases. One set of these corrections will be discussed in Chapter 13.

Example 11-4   What volume is occupied by 1 mol of gas at standard temperature of 0°C and normal atmospheric pressure of 1 atm?

The absolute temperature corresponding to 0°C is 273 K. From the ideal-gas law (Equation 11-12), we have

$$V = \frac{nRT}{P} = \frac{(1 \text{ mol})(0.0821 \text{ L·atm/mol·K})(273 \text{ K})}{1 \text{ atm}} = 22.4 \text{ L}$$

Example 11-5   How many molecules are there in one cubic centimetre of gas under standard conditions?

Here $P = 1$ atm, $V = 1$ cm^3 = 10^{-3} L, and $T = 273$ K. Thus,

$$n = \frac{PV}{RT} = \frac{(1 \text{ atm})(10^{-3} \text{ L})}{(0.0821 \text{ L·atm/mol·K})(273 \text{ K})} = 4.46 \times 10^{-5} \text{ mol}$$

The number of molecules in this many moles is

$$N = nN_A = (4.46 \times 10^{-5} \text{ mol})(6.02 \times 10^{23} \text{ molecules/mol})$$

$$= 2.68 \times 10^{19} \text{ molecules}$$

Example 11-6   The molar mass of hydrogen is 1.008 g/mol. What is the mass of one hydrogen atom?

Since there are N_A hydrogen atoms in 1 mol, the mass of one atom is

$$m = \frac{1.008 \text{ g/mol}}{6.022 \times 10^{23} \text{ atoms/mol}} = 1.67 \times 10^{-24} \text{ g/atom}$$

Example 11-7   A gas has a volume of 2 L, a temperature of 30°C, and a pressure of 1 atm. It is heated to 60°C and compressed to a volume of 1.5 L. Find its new pressure.

For a fixed amount of gas, we see from Equation 11-12 that the quantity PV/T is constant. If we use the subscript 1 for initial values and 2 for final values, we have

$$\frac{P_2 V_2}{T_2} = \frac{P_1 V_1}{T_1} \qquad\qquad\qquad 11\text{-}13$$

or

$$P_2 = \frac{T_2 V_1}{T_1 V_2} P_1$$

Since we are dealing with ratios, we can express the pressure and volume in any units, but we must remember that the ideal-gas equation holds only if the temperatures are absolute temperatures. Using $T_1 = 273 + 30 = 303$ K and $T_2 = 273 + 60 = 333$ K, we have

$$P_2 = \frac{(333 \text{ K})(2 \text{ L})}{(303 \text{ K})(1.5 \text{ L})} (1 \text{ atm}) = 1.47 \text{ atm}$$

Example 11-8   In a 55-L container, 100 g of CO_2 is at a pressure of 1 atm. (a) Find the temperature. (b) If the volume is increased to 80 L and the temperature is kept constant, what is the new pressure?

(a) We can find the temperature from the ideal-gas equation (Equation 11-12) if we first find the number of moles. Since the molar mass of CO_2 is 44 g/mol, the number of moles is

$$n = \frac{m}{M} = \frac{100 \text{ g}}{44 \text{ g/mol}} = 2.27 \text{ mol}$$

The absolute temperature is then

$$T = \frac{PV}{nR} = \frac{(1 \text{ atm})(55 \text{ L})}{(2.27 \text{ mol})(0.0821 \text{ L} \cdot \text{atm/mol} \cdot \text{K})} = 295 \text{ K}$$

(b) Using Equation 11-13 with $T_2 = T_1$, we have

$$P_2 V_2 = P_1 V_1$$

$$P_2 = \frac{V_1}{V_2} P_1 = \frac{55 \text{ L}}{80 \text{ L}} (1 \text{ atm}) = 0.688 \text{ atm}$$

11-4   The Molecular Interpretation of Temperature

In this section, we will show that the absolute temperature of a gas is a measure of the average kinetic energy of the gas molecules. We will do this by using a simple model to calculate the pressure exerted by the gas on the walls of its container. In the microscopic view, the pressure is the result of collisions between the gas molecules and the walls. We can calculate this pressure by calculating the rate of change of momentum of the gas molecules due to the collisions with the walls. By Newton's second law, this rate of change equals the force exerted by the walls on the gas molecules, and by Newton's third law, this force equals the force exerted by the molecules on the walls. The force per unit area equals the pressure.

We first assume that we have a rectangular container of volume V containing N molecules of mass m moving with speed v. We wish to calculate the force exerted by these molecules on the right wall perpendicular to the x axis. The horizontal component of the momentum of a molecule before it hits the wall is $+mv_x$, and after the molecule makes an elastic collision with the wall, its momentum is $-mv_x$. The magnitude of the change in momentum during the collision with the wall is thus $2mv_x$. The total change in the momentum of all the molecules in some time Δt is $2mv_x$ times the number of molecules that hit the wall during this interval. We will now show that the number of molecules that hit the wall in any time interval is proportional to v_x, so the total change in momentum is proportional to v_x^2. Before we do this, let us use some unrealistic numbers to understand why the number of molecules that hit the wall in some time interval is proportional to v_x. If v_x were $+1$ mm/s, the number that hit the right wall in 1 s would be those molecules a distance of 1 mm from the wall that are moving to the right. (Those farther away would not reach the wall in 1 s.) But if v_x were $+2$ mm/s, all those molecules within 2 mm moving to the right would hit the wall in 1 s; that is, twice as many would hit in the given time.

Figure 11-9 shows gas molecules in a rectangular container. The number of molecules that hit the right wall of area A in the time interval Δt is the number that are within the distance $v_x \Delta t$ and are moving to the right. This number is the number per unit volume N/V times the volume $v_x \Delta t\, A$ times one-half because, on the average, half the molecules will be moving to the right and half to the left. The total change in the momentum of the gas molecules in the time interval Δt is this number times $2mv_x$, the change in momentum per molecule.

$$\Delta \text{ momentum} = \frac{1}{2}\frac{N}{V}\,(v_x \Delta t\, A)\, 2mv_x$$

The force exerted by the wall on the molecules and by the molecules on the wall is this change in momentum divided by the time interval Δt. The pressure is this force divided by the area. Thus, dividing the change in momentum by the time Δt and by the area A, we obtain

$$P = \frac{N}{V} mv_x^2 \qquad\qquad 11\text{-}14$$

and

$$PV = Nmv_x^2 \qquad\qquad 11\text{-}15$$

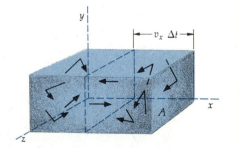

Figure 11-9
Gas molecules in a rectangular container. In a time interval Δt, the molecules at a distance $v_x \Delta t$ will hit the right wall if they are moving to the right. The number of molecules in this distance is proportional to v_x and to the number of molecules per unit volume.

Of course, all the molecules in the container do not have the same speed. To take this into account, we merely replace v_x^2 with the average value $(v_x^2)_{av}$. Replacing v_x^2 by its average value and writing Equation 11-15 in terms of the energy associated with motion along the x axis, $\frac{1}{2}mv_x^2$, we have

$$PV = 2N(\tfrac{1}{2}mv_x^2)_{av} \qquad \text{11-16}$$

Comparing this with Equation 11-5, we have

$$PV = NkT = 2N(\tfrac{1}{2}mv_x^2)_{av}$$

or

$$(\tfrac{1}{2}mv_x^2)_{av} = \tfrac{1}{2}kT \qquad \text{11-17}$$

The average kinetic energy associated with translational motion in the x direction is thus $\frac{1}{2}kT$. Since x was an arbitrarily selected direction, we can write similar equations for $(\frac{1}{2}mv_y^2)_{av}$ and $(\frac{1}{2}mv_z^2)_{av}$:

$$(\tfrac{1}{2}mv_y^2)_{av} = \tfrac{1}{2}kT \qquad \text{11-18}$$

$$(\tfrac{1}{2}mv_z^2)_{av} = \tfrac{1}{2}kT \qquad \text{11-19}$$

The result that the energy per molecule associated with motion in each of the three directions is $\frac{1}{2}kT$ is a special case of a more general result known as the *equipartition theorem*, which will be discussed in the next chapter. Adding Equations 11-17, 11-18, and 11-19 and writing $v^2 = v_x^2 + v_y^2 + v_z^2$, we have

$$(\tfrac{1}{2}mv^2)_{av} = \tfrac{3}{2}kT \qquad \text{11-20}$$

Average molecular kinetic energy

The absolute temperature is thus a measure of the average translational kinetic energy of the molecules. We include the word translational because a molecule may have other kinds of kinetic energy, such as rotational or vibrational kinetic energy, for example. Only the translational kinetic energy is involved in the calculation of the pressure exerted on the walls of the container. However, as we will see later, because the molecules make collisions with each other, there are exchanges among the different kinds of molecular energy (translational, rotational, and vibrational), and the total energy is "equally partitioned" among the various forms. The absolute temperature is thus a measure of the average molecular energy of any type.

   The total translational kinetic energy of n mol of a gas containing N molecules is

$$E_k = N(\tfrac{1}{2}mv^2)_{av} = \tfrac{3}{2}NkT = \tfrac{3}{2}nRT \qquad \text{11-21}$$

The translational kinetic energy is thus $\frac{3}{2}kT$ per molecule and $\frac{3}{2}RT$ per mole.

   We can use these results to estimate the order of magnitude of the kinetic energies and speeds of the molecules in a gas. The average kinetic energy of translation of a molecule is $\frac{3}{2}kT$. At a temperature $T = 300$ K $= 27°$C $= 81°$F, $\frac{3}{2}kT$ has the value 6.21×10^{-21} J. The square root of $(v^2)_{av}$ is the *root mean square* (rms) speed. By Equation 11-20, the average value of v^2 is

$$(v^2)_{av} = \frac{3kT}{m} = \frac{3N_A kT}{N_A m} = \frac{3RT}{M}$$

where $M = N_A m$ is the mass of 1 mol, the molar mass. Thus, the rms speed is

$$v_{rms} = [(v^2)_{av}]^{1/2} = \left(\frac{3kT}{m}\right)^{1/2}$$

$$= \left(\frac{3RT}{M}\right)^{1/2}$$

11-22

Example 11-9   The molar mass of oxygen gas (O_2) is 32 g/mol and that of hydrogen gas (H_2) is 2 g/mol. (*a*) Calculate the rms speed of an oxygen molecule when the temperature is 300 K. (*b*) Calculate the rms speed of a hydrogen molecule.

(*a*) To make the units come out right, we must put the molar mass in units of kilograms per mole in Equation 11-22. We then have

$$v_{rms} = \sqrt{\frac{3(8.31 \text{ J/mol} \cdot \text{K})(300 \text{ K})}{32 \times 10^{-3} \text{ kg/mol}}} = 483 \text{ m/s}$$

(*b*) Since the molar mass of hydrogen is one sixteenth that of oxygen and v_{rms} is proportional to $1/\sqrt{M}$, the rms speed of hydrogen is four times that of oxygen or about 1.93 km/s.

The rms speed of oxygen molecules is slightly greater but of the same order of magnitude as the speed of sound in air, which is about 347 m/s at 300 K. In Chapter 16, we will see that the speed of sound in air is given by $v_{sound} = (1.4RT/M)^{1/2}$, where M is the molar mass of air, which is about 29×10^{-3} kg/mol.

The rms speed of hydrogen molecules is about one-sixth the escape speed at the surface of the earth (11.2 km/s, as discussed in Section 9-5). The reason that hydrogen escapes from the gravitational field of the earth is that a considerable fraction of the molecules of a gas in equilibrium at some temperature have speeds greater than the rms speed. Figure 11-10 shows the distribution of speeds of gas molecules at some temperature T. In this figure, the fraction f of the total number of molecules that have speed v (in some narrow range of speeds) is plotted versus v. The speed v_m at which the distribution is maximum is the most probable speed. The mean speed (labeled $\bar{v}$) and the rms speed are somewhat greater than the most probable speed. We can see from this figure that there is a considerable fraction of molecules that have speeds much greater than the mean speed or the rms speed. This accounts for the fact that, even when the mean speed of the molecules of a particular gas is only about one-sixth the escape speed for a planet, enough of the molecules have speeds greater than the escape speed so that the gas cannot exist in the atmosphere of that planet.

Questions

6. By what factor must the absolute temperature of a gas be increased to double the rms speed of its molecules?

7. How does the average translational kinetic energy of a molecule in a gas change if the pressure is doubled while the volume is kept constant? If the volume is doubled while the pressure is kept constant?

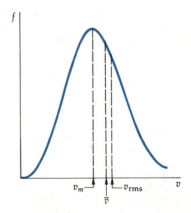

Figure 11-10
Distribution of molecular speeds in a gas. The fraction f of the number of molecules having a particular speed in a narrow range of speeds is plotted versus speed v. The mean speed $\bar{v}$ and the rms speed v_{rms} are both slightly greater than the most probable speed v_m.

8. Why would you not expect all the molecules in a gas to have the same speed?

9. Two different gases have the same temperature. What can you say about the rms speeds of the molecules? What can you say about the average kinetic energies of the molecules?

10. If a gas is heated at constant volume, explain in terms of molecular motion why its pressure on the walls of its container increases.

11. If the volume of a gas is reduced at constant temperature, explain in terms of molecular motion why its pressure on the walls of its container rises.

Summary

1. A temperature scale is constructed by choosing some thermometric property and defining the temperature to vary linearly with that property using two fixed points, such as the ice point and the steam point of water, or defining the temperature to be proportional to that property using one fixed point, such as the triple point of water.

2. In general, different thermometers do not agree with each other on the measurement of a temperature except at the fixed points. Gas thermometers have the unique property of agreeing with each other in the measurement of any temperature as long as only a small amount of gas is used so that the gas density in the thermometer is very low.

3. Temperatures on the Fahrenheit and Celsius scales are related by

$$t_C = \tfrac{5}{9}(t_F - 32 \text{ F}°)$$

or

$$t_F = \tfrac{9}{5}t_C + 32 \text{ F}°$$

where t_C is the temperature in degrees Celsius and t_F is the temperature in degrees Fahrenheit. The absolute or Kelvin temperature scale is related to the Celsius scale by

$$T = t_C + 273 \text{ K}$$

where T is the absolute temperature in kelvins. The size of the kelvin is the same as the Celsius degree, so temperature differences in kelvins or degrees Celsius are the same. The Fahrenheit degree is smaller than the Celsius degree. A temperature difference of 9 F° equals a temperature difference of 5 C° or of 5 K.

4. At low pressures, all gases obey the ideal-gas law:

$$PV = nRT$$

where R is the universal gas constant, which is related to Avogadro's number N_A and Boltzmann's constant k by

$$R = kN_A$$

A form of the ideal-gas law that is useful for solving problems involving a fixed amount of gas is

$$\frac{P_1 V_1}{T_1} = \frac{P_2 V_2}{T_2}$$

5. The absolute temperature T is a measure of the average molecular energy. For an ideal gas, the average kinetic energy of translation of the molecules is

$$(\tfrac{1}{2}mv^2)_{av} = \tfrac{3}{2}kT$$

The total kinetic energy of n mol of a gas is given by

$$E_k = \tfrac{3}{2}nRT$$

6. The rms speed of a molecule in a gas is related to the absolute temperature by

$$v_{rms} = \sqrt{3RT/M}$$

where M is the molar mass.

Suggestions for Further Reading

Hall, Marie Boas: "Robert Boyle," *Scientific American*, August 1967, p. 96.

This English investigator of the seventeenth century performed careful experiments leading to his discovery of the inverse relationship between pressure and temperature in a gas, and he obtained numerous results in chemistry, guided by his views on the "mechanical" nature of matter.

Romer, Robert H.: "Temperature Scales: Celsius, Fahrenheit, Kelvin, Reamur, and Romer," *The Physics Teacher*, vol. 20, no. 7, 1982, p. 450.

A compact and well-researched history of the development of temperature scales.

Review

A. Objectives: After studying this chapter, you should:

1. Be able to discuss how a mercury thermometer is calibrated using two fixed points for either the Celsius or the Fahrenheit temperature scales.

2. Be able to convert temperatures from Celsius to Fahrenheit and from Fahrenheit to Celsius.

3. Be able to convert temperatures given on either the Celsius or the Fahrenheit scales to kelvins.

4. Be able to work problems using the ideal gas equation, $PV = nRT$.

5. Be able to discuss the molecular interpretation of temperature.

B. Define, explain, or otherwise identify:

Ice point, p. 254
Steam point, p. 254
Celsius temperature scale, p. 254
Fahrenheit temperature scale, p. 254–255
Absolute temperature scale, p. 258
Kelvin, p. 258
Boltzmann's constant, p. 260
Universal gas constant, p. 260
Ideal-gas law, p. 261

C. True or false: If the statement is true, explain why. If it is false, give a counterexample.

1. The Fahrenheit and Celsius temperature scales differ only in the choice of the zero temperature.

2. The kelvin is the same size as the Celsius degree.

3. If the pressure of a gas increases, the temperature must increase.

4. The absolute temperature of a gas is a measure of the average translational kinetic energy of the gas molecules.

Exercises

Section 11-1   The Celsius and Fahrenheit Temperature Scales; Section 11-2   The Absolute Temperature Scale

1. A certain ski wax is rated for use between -12 and $-7°C$. What is this temperature range on the Fahrenheit scale?

2. The length of the column of a mercury thermometer is 4.0 cm when the thermometer is immersed in ice water and 24.0 cm when it is immersed in boiling water. (*a*) What should the length be at a room temperature of $22.0°C$. (*b*) The mercury column is 25.4 cm long when the thermometer is placed in a boiling chemical solution. What is the temperature of the solution?

3. The boiling point of O_2 is $-182.86°C$. Express this temperature in (*a*) kelvins and (*b*) degrees Fahrenheit.

4. Find the Fahrenheit temperature corresponding to 0 K.

5. Find the Celsius temperature corresponding to the normal temperature of the human body of $98.6°F$.

6. The boiling point of tungsten is $5900°C$. Find this temperature in (*a*) kelvins and (*b*) degrees Fahrenheit.

7. The highest and lowest temperatures ever recorded in the United States are $134°F$ (in California in 1913) and $-80°F$ (in Alaska in 1971). Express these temperatures on the Celsius scale.

8. In September 1933 in Portugal, the temperature rose to $70°C$ for 2 min. What is this temperature in degrees Fahrenheit?

9. The temperature of the interior of the sun is about 10^7 K. What is this temperature (*a*) on the Celsius scale and (*b*) on the Fahrenheit scale?

10. A constant-volume gas thermometer reads 50 torr pressure at the triple point of water. (*a*) What will the pressure be when the thermometer measures a temperature of 300 K? (*b*) What temperature corresponds to a pressure of 678 torr?

11. A constant-volume gas thermometer has a pressure of 30 torr when it reads a temperature of 373 K. (*a*) What is its triple-point pressure? (*b*) What temperature corresponds to a pressure of 0.175 torr?

Section 11-3   The Ideal-Gas Law

12. A gas is kept at constant pressure. If the temperature is changed from 50 to $100°C$, by what factor does its volume change?

13. A pressure as low as 1×10^{-8} torr can be achieved with an oil-diffusion pump. How many molecules are there in 1 cm^3 at this pressure if the temperature is 300 K?

14. A 10-L vessel contains gas at $0°C$ and a pressure of 4 atm. (*a*) How many moles of gas are there in the vessel? (*b*) How many molecules?

15. (*a*) What is the molecular mass of H_2O? (*b*) How many moles are there in 1 L of water?

16. How many moles of lead ($M = 207.2$ g/mol) are there in a lead brick 5 cm by 10 cm by 20 cm?

17. One mole of a gas is at a pressure of 1 atm and a temperature of 300 K. (*a*) What is the volume of this gas? (*b*) If the gas expands at constant temperature to twice its original volume, what is the new pressure. (*c*) On a graph, plot P versus V for this *isothermal expansion*.

18. A room has dimensions 6 m by 5 m by 3 m. (*a*) If the pressure is 1 atm and the temperature is 300 K, find the number of moles of air in the room. (*b*) If the temperature rises by 5 K and the pressure remains constant, how many moles of air leave the room?

19. (*a*) If 2 mol of gas occupies a volume of 30 L at a pressure of 1 atm, what is the temperature of the gas? (*b*) The container is fitted with a piston so that the volume can change. The gas is heated at constant pressure and expands to a volume of 40 L. What is the temperature in kelvins? In degrees Celsius? (*c*) The volume is now fixed at 40 L and the gas is heated at constant volume until its temperature is 350 K. What is its pressure?

20. A container fitted with a piston holds 1.5 mol of gas at initial pressure and temperature of 2 atm and 300 K. (*a*) What is the initial volume of the gas? (*b*) The gas is allowed to expand at constant temperature until the pressure is 1 atm. What is the new volume? (*c*) The gas is compressed and heated at the same time until it is back to its original volume, at which time the pressure is 2.5 atm. What is its temperature now?

21. A cubic metal box of sides 20 cm contains air at a pressure of 1 atm and a temperature of 300 K. It is sealed so that the volume is constant and is heated to a temperature of 400 K. Find the net force on each wall of the box.

Section 11-4   The Molecular Interpretation of Temperature

22. Find v_{rms} for an argon atom ($M = 40$ g/mol) when the temperature is 300 K.

23. Find the rms speed and average kinetic energy of a hydrogen atom in a gas at a temperature of 10^7 K. (At this temperature, which is of the order of magnitude of the temperatures in the interior of stars, the hydrogen atom is ionized and consists merely of a single proton.)

24. Show that $\sqrt{3RT/M}$ has the correct units of metres per second if R is in joules per mole-kelvin, T is in kelvins, and M is in kilograms per mole.

25. Find v_{rms} for an oxygen molecule in a gas at temperature $-30°C$.

26. Find the total translational kinetic energy of 1 L of oxygen gas held at a temperature of $0°C$ and a pressure of 1 atm.

27. At what temperature will the rms speed of an H_2 molecule equal 332 m/s?

Problems

1. The mass of Avogadro's number of ^{12}C atoms is 12 g. Each ^{12}C atom contains six protons and six neutrons of approximately equal mass. The mass of Avogadro's number of protons or neutrons is thus approximately 1 g, and the mass of a proton or neutron is approximately $1/N_A$ g. Calculate $1/N_A$ and compare it with the mass of the proton given in Appendix C.

2. Show that the ideal-gas law can be written

$$P = \rho \frac{RT}{M} \qquad\qquad 11\text{-}23$$

where ρ is the density of the gas and M is its molar mass.

3. Use Equation 11-23 to calculate the density of (a) N_2 at a pressure of 1 atm and a temperature of $20°C$ and (b) air ($M = 29.0\,g/mol$) at $P = 1\,atm$ and $T = 273\,K$. Compare this result with the value given in Table 10-1.

4. (a) Find the temperature that has the same numerical value on the Fahrenheit and Celsius scales. (b) Show that the conversion formulas can be written

$$t_C = \tfrac{5}{9}(t_F + 40\ F°) - 40\ C°$$

and

$$t_F = \tfrac{9}{5}(t_C + 40\ C°) - 40\ F°$$

5. An automobile tire is pumped to a gauge pressure of 200 kPa when it is at an air temperature of $20°C$. After the car has been driven at high speed, the tire temperature has increased to $50°C$. (a) Assuming that the volume of the tire has not changed, find the new gauge pressure of the air in the tire, assuming the air to be an ideal gas. (b) Calculate the gauge pressure if the tire expands so that the volume increases by 10 percent.

6. A cylindrical container with radius 2.5 cm and height 20 cm is open at the top. It contains air at 1 atm pressure. A tightly fitting piston of mass 1.2 kg is inserted and gradually lowered until the increased pressure in the container supports the weight of the piston. (a) What is the force exerted on the top of the piston due to atmospheric pressure? (b) What is the force that must be exerted by the gas in the container below the piston to keep it in equilibrium? What is the pres-

sure in the container? (c) Assuming that the temperature of the gas in the container remains constant, what is the height of the equilibrium position of the piston?

7. (a) Show that the rms speed of gas molecules is given by

$$v_{rms} = \sqrt{3P/\rho}$$

where ρ is the density of the gas and P is the pressure. (b) Find the rms speed of gas molecules if the density of the gas is 3.5 g/L and the pressure is 300 kPa.

8. Gas is confined in a steel cylinder at $20°C$ at a pressure of 5 atm. (a) If the cylinder is surrounded by boiling water and allowed to come to thermal equilibrium, what will the new pressure be? (b) If some of the gas is then allowed to escape until the pressure is again 5 atm (and the temperature is still $100°C$), what fraction of the original gas (by weight) will escape? (c) If the temperature of the remaining gas in the cylinder is returned to $20°C$, what will its final pressure be?

9. (a) If 2 mol of hydrogen gas ($M = 2\ g/mol$) is at atmospheric pressure and room temperature ($20°C$), what is the average translational kinetic energy of a hydrogen molecule in this gas? (b) What is v_{rms} for the hydrogen molecules? (c) What is the total kinetic energy of translation of all the molecules in this amount of hydrogen gas?

10. At the surface of the sun, the temperature is about 6000 K, and all materials are gaseous. From data given by the spectrum of light, it is known that most elements are present on the sun. (a) What is the average kinetic energy of translation of an atom at the surface of the sun? (b) What is the range of v_{rms} at the surface of the sun if the atoms present range from hydrogen ($M = 1\ g/mol$) to uranium ($M = 238\ g/mol$)?

11. The escape speed on Mars is 5.0 km/s, and the surface temperature is typically $0°C$. Calculate the rms speed for H_2, O_2, and CO_2 at this temperature, and comment on the possibility of these gases being in the atmosphere of Mars. (Use the criterion that all these molecules will have escaped by this time if the rms speed is greater than one-sixth of the escape speed.)

12. Repeat Problem 11 for Jupiter, whose escape speed is 60 km/s and whose temperature is typically $-150°C$.

CHAPTER TWELVE

Heat and the First Law of Thermodynamics

If you can't stand the heat
get out of the kitchen.

HARRY TRUMAN

Human beings have understood the importance of heat for thousands of years. In fact, from the Bronze and Iron Ages at the beginnings of history all the way up to the present, much of the advance of civilization has been closely tied to the improved use of heat sources. The exploitation of coal reserves in Europe and Great Britain, as much as such inventions as the steam engine, brought on the Industrial Revolution. Later, the discovery and use of oil and the invention of the gasoline and diesel engines initiated a new revolution, chiefly in transportation. Recently, the nuclear reactor, also a heat source, has provided the compact, long-lasting power sources needed for our first, tentative explorations of space.

In spite of the tremendous impact heat has always had on human life, its nature remained shrouded in mystery until the middle of the last century. In this and the next two chapters, you will see this mystery unveiled. We will learn that heat is simply a form of energy, thermal energy.

Heat is energy that is transferred from one object to another because of a difference in temperature. In the seventeenth century, Galileo, Newton, and other scientists generally supported the theory of the ancient Greek atomists who considered heat to be a manifestation of molecular motion. In the next century, methods were developed for making quantitative measurements of the amount of heat that leaves or enters a body, and it was found that when two bodies are in thermal contact, the amount of heat that leaves one body equals the amount that enters the other. This discovery led to the development of an apparently successful theory of heat as a conserved material substance — an invisible fluid called *caloric* — that was neither created nor destroyed but merely flowed out of one body and into another. The caloric theory of heat served quite well in the description of heat transfer, but it was eventually discarded when it was observed that caloric could apparently be generated endlessly by friction with no corresponding disappearance of caloric somewhere else. In other words, the principle of the conservation of caloric, which had been the experimental foundation of this theory of heat, proved to be false.

The first clear experimental observations showing that caloric could not be conserved were made at the end of the eighteenth century by Benjamin Thompson, an American who emigrated to Europe, became the director of the Bavarian arsenal, and was given the title Count Rumford. Thompson supervised the boring of cannons for Bavaria. Because of the heat generated by the boring tool, water was used for cooling. The water had to be replaced continually because it boiled away during the boring. According to the caloric theory, as the metal from the bore was cut into small chips, its ability to retain caloric was decreased. It therefore released caloric to the water, heating the water and causing it to boil. Thompson noticed, however, that even when the drill was too dull to cut the metal, the water still boiled away as long as the drill was turned. Apparently, caloric was produced merely by friction and could be produced endlessly. Thompson suggested that heat is not a substance that is conserved but is some form of motion communicated from the bore to the water. He showed, in fact, that the heat produced was approximately proportional to the work done by the boring tool.

The caloric theory of heat continued to be the leading theory for some 40 years after Thompson's work, but it was gradually weakened as more and more examples of the nonconservation of heat were observed. The modern mechanical theory of heat did not emerge until the 1840s. In this view, heat is another form of energy, exchangeable at a fixed rate with the various forms of mechanical energy we have already discussed. The most varied and precise experiments demonstrating this were performed, starting in the late 1830s, by James Joule (1818–1889), after whom the SI unit of energy is named. Joule showed that the appearance or disappearance of a given quantity of heat is always accompanied by the disappearance or appearance of an equivalent quantity of mechanical energy. The experiments of Joule and others showed that neither heat nor mechanical energy is conserved independently but that the mechanical energy lost always equals the heat produced. What *is* conserved is the total mechanical energy plus heat energy.

12-1   Heat Capacity and Specific Heat

When heat energy is added to a substance, the temperature of the substance usually rises. (An exception occurs during a change of phase, such as when water freezes or evaporates. We will consider phase changes in Chapter 13.) The amount of heat energy Q needed to raise the temperature of a substance is proportional to the temperature change and to the mass of the substance:

$$Q = C \, \Delta T = mc \, \Delta T \qquad\qquad 12\text{-}1 \qquad \text{Heat capacity}$$

where C is the *heat capacity* of the substance, which is defined as the heat energy needed to raise the temperature of a substance by one degree. The *specific heat c* is the heat capacity per unit mass:

$$C = mc \qquad\qquad 12\text{-}2 \qquad \text{Specific heat}$$

The historical unit of heat energy, the calorie, was originally defined as the amount of heat energy needed to raise the temperature of one gram of water one Celsius degree (or one kelvin since both the Celsius degree and the kelvin measure the same temperature change). The kilocalorie, then, is the amount of heat energy needed to raise the temperature of one kilogram of water by one Celsius degree. (The "calorie" used in measuring the energy equivalent of foods is actually the kilocalorie.) Since we now recognize that heat energy is just another form of energy, we do not need any special units for heat that differ from other energy units. Thus, the calorie is now defined in terms of the SI unit of energy, the joule:

$$1 \text{ cal} = 4.184 \text{ J} \qquad\qquad 12\text{-}3 \qquad \text{Calorie defined}$$

The U.S. customary unit of heat is the Btu (for British thermal unit), which was originally defined as the amount of energy needed to raise the temperature of one pound of water one Fahrenheit degree. The Btu is related to the calorie and to the joule by

$$1 \text{ Btu} = 252 \text{ cal} = 1.054 \text{ kJ} \qquad\qquad 12\text{-}4 \qquad \text{Btu defined}$$

It should be evident from the original definitions of the calorie and the Btu that the specific heat of water is

$$c_{\text{water}} = 1 \text{ cal/g} \cdot \text{C}° = 1 \text{ kcal/kg} \cdot \text{C}°$$
$$= 1 \text{ kcal/kg} \cdot \text{K} = 1 \text{ Btu/lb} \cdot \text{F}° \qquad\qquad 12\text{-}5$$
$$= 4.184 \text{ kJ/kg} \cdot \text{K}$$

Careful measurement shows that the specific heat of water varies slightly with temperature. It has its minimum value at about 30°C and varies by only about one percent over the entire temperature range from 0 to 100°C. We will neglect this small variation and take the specific heat of water to be 1 kcal/kg·K = 4.18 kJ/kg·K.

The heat capacity per mole is called the *molar heat capacity* C_m. The molar heat capacity equals the specific heat (heat capacity per mass) times the mass per mole M, the molar mass:

$$C_m = Mc \qquad\qquad 12\text{-}6 \qquad \text{Molar heat capacity}$$

The heat capacity of n mol of a substance is then

$$C = nC_m \qquad\qquad 12\text{-}7$$

Specific heats and molar heat capacities have been measured for many substances. Table 12-1 lists the specific heats and molar heat capacities of some solids and liquids.

Example 12-1   How much heat is needed to raise the temperature of 3 kg of copper by 20 C°?

From Table 12-1, the specific heat of copper is 0.386 kJ/kg·K. Equation 12-1 gives

$$Q = mc \, \Delta T = (3 \text{ kg})(0.386 \text{ kJ/kg} \cdot \text{K})(20 \text{ K}) = 23.2 \text{ kJ}$$

Note that we used the fact that 20 C° = 20 K. Alternatively, we could have expressed the specific heat as 0.386 kJ/kg·C° and written the temperature change as 20 C°.

Table 12-1

Specific Heat and Molar Heat Capacity for Various Solids and Liquids at 20°C

Substance	c, kJ/kg·K	c, kcal/kg·K or Btu/lb·F°	C_m, J/mol·K
Aluminum	0.900	0.215	24.3
Bismuth	0.123	0.0294	25.7
Copper	0.386	0.0923	24.5
Gold	0.126	0.0301	25.6
Ice ($-10°C$)	2.05	0.49	36.9
Lead	0.128	0.0305	26.4
Silver	0.233	0.0558	24.9
Tungsten	0.134	0.0321	24.8
Zinc	0.387	0.0925	25.2
Alcohol (ethyl)	2.4	0.58	111
Mercury	0.140	0.033	28.3
Water	4.18	1.00	75.2

We see from Table 12-1 that the specific heat of water is considerably larger than that of other substances. For example, it is more than 10 times the specific heat of copper. Because of its very large heat capacity, water is an excellent coolant. Also, a large body of water, such as a lake or an ocean, tends to moderate the variations of temperature near it because it can absorb or release large quantities of heat energy with only a very small change in temperature.

Note that the molar heat capacities listed in Table 12-1 are about the same for all the metals. To a good approximation, they are all about $3R$, where R is the universal gas constant.

$$C_m \approx 3R = 24.9 \text{ J/mol·K} \qquad \text{12-8}$$ Dulong-Petit law

This result is known as the *Dulong-Petit law*. We will discuss the significance of this result in Section 12-4.

Since the specific heat of water is practically constant over a wide range of temperatures, the specific heat of an object can be conveniently measured by heating the object to some easily measured temperature, placing it in a water bath of known mass and temperature, and measuring the final equilibrium temperature. If the whole system is isolated from its surroundings, the heat leaving the body equals the heat entering the water. This procedure is called *calorimetry*. Let m be the mass of the body, c its specific heat, and T_b its initial temperature. If T_f is the final temperature of the body in the water bath, the heat leaving the body is

$$Q_{out} = mc(T_b - T_f)$$

Similarly, if T_w is the initial temperature of the water and T_f its final temperature (the final temperature of the body and the water are the same since they eventually come to equilibrium), the heat absorbed by the water is

$$Q_{in} = m_w c_w (T_f - T_w)$$

where m_w is the mass of the water and $c_w = 4.18$ kJ/kg·K is its specific heat.

Note that in these equations we have chosen to write the temperature differences so that both the heat in and the heat out are positive quantities. (We know that if we place a heated body in cooler water, the final temperature T_f will be greater than the initial temperature of the water bath and less than the initial temperature of the body.) Since these amounts of heat are equal, the specific heat c of the body can be calculated by setting the heat out equal to the heat in:

$$mc(T_b - T_f) = m_w c_w (T_f - T_w) \qquad\qquad 12\text{-}9$$

Since only temperature differences occur in Equation 12-9 and since the kelvin and the Celsius degree are the same size, the temperatures can be measured on either the Celsius or the Kelvin scale without affecting the result.

Example 12-2   Aluminum shot of mass 100 g is heated to 100°C and is then placed in 500 g of water initially at 18.3°C. The final equilibrium temperature of the mixture is 21.7°C. What is the specific heat of aluminum?

Since the temperature change of the water is $21.7°C - 18.3°C = 3.4\ C° = 3.4$ K, the heat absorbed by the water is

$$Q_w = m_w c_w\, \Delta T_w = (0.5\ \text{kg})(4.18\ \text{kJ/kg}\cdot\text{K})(3.4\ \text{K}) = 7.11\ \text{kJ}$$

The temperature change of the aluminum is $100°C - 21.7°C = 78.3\ C° = 78.3$ K, and the heat given off by the aluminum is

$$Q_{Al} = mc\, \Delta T_{Al} = (0.1\ \text{kg})(c)(78.3\ \text{K})$$

Setting $Q_{Al} = Q_w$ and solving for c gives

$$c = \frac{7.11\ \text{kJ}}{(0.1\ \text{kg})(78.3\ \text{K})} = 0.908\ \text{kJ/kg}\cdot\text{K}$$

Note that the specific heat of aluminum is considerably less than that of water.

The amount of heat needed to raise the temperature of a body depends on whether the substance is allowed to expand while its temperature is rising. If it is allowed to expand, it will do work on its surroundings (on the air, if nothing else), and more heat will be needed to do the work as well as raise the temperature. It is very difficult to prevent a solid or liquid from expanding when heated (though the expansion is usually very small), and the pressure on the substance is usually constant (atmospheric pressure) during the heating. The specific heats listed in Table 12-1 for solids and liquids are the specific heats at constant pressure c_p. For gases, the pressure may be constant during heating or cooling, as when air is heated in a house, or, as is common in the laboratory, the gas may be heated in a closed container at constant volume. When a gas is heated at constant volume, no work is done by the gas, and the specific heat c_v is less than that at constant pressure c_p. The difference is large for gases, but it is small enough to be ignored in most cases for solids and liquids. We will calculate the difference in the specific heat at constant volume and that at constant pressure for an ideal gas in Section 12.4.

Questions

1. A potato is wrapped in aluminum foil and baked in an oven. After it is removed from the oven and the foil is removed, the foil cools much faster than the potato? Why?

2. Body *A* has twice the mass and twice the specific heat of body *B*. If they are supplied with equal amounts of heat, how do their temperature changes compare?

12-2   The First Law of Thermodynamics

The first law of thermodynamics is a statement of the conservation of energy. It is the result of many experiments relating the work done on a system, the heat added to or subtracted from the system, and the internal energy of the system. From the definition of the calorie, we know that it takes one calorie of heat to raise the temperature of one gram of water by one Celsius degree. But we can also raise the temperature of water or of any other system by doing work on it without adding any heat.

Figure 12-1 is a schematic illustration of Joule's most famous experiment, in which he determined the amount of work required to produce a given amount of heat; that is, the amount of work needed to raise the temperature of one gram of water by one Celsius degree. Once the experimental equivalence of heat and energy was established, Joule's experiment could be described as determining the size of the calorie in the usual energy units or as measuring the specific heat of water. The water in Figure 12-1 is enclosed by

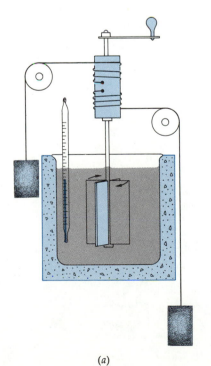

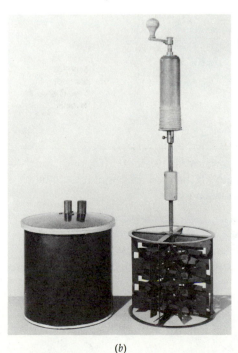

(a)   (b)

Figure 12-1
(a) Schematic diagram of Joule's experiment to determine the amount of work required to produce a given temperature increase in water. The work done is determined by measuring the loss in the potential energy of the weights. (b) Closeup view of Joule's apparatus showing the paddle wheel and its container.

insulating walls so that the temperature of the system cannot be affected by heat entering or leaving. The weights, falling at constant speed, turn a paddle wheel, which does work against the water. Assuming negligible energy loss through friction in the bearings and so forth, the work done by the paddle wheel against the water equals the loss of mechanical energy of the falling weights. This loss is easily measured by determining the distance through which the weights fall. The results of Joule's experiment and of many others after his is that it takes about 4.18 units of mechanical work (joules) to raise the temperature of one gram of water by one Celsius degree. The result that 4.18 J of mechanical energy is equivalent to 1 cal of heat energy was known as the *mechanical equivalence of heat*. Historically, it was customary to express heat energy in calories and then use the mechanical equivalence of heat to convert to the standard units of mechanical energy. Today, all forms of energy are usually expressed in joules.

There are other ways of doing work on a system. For example, we could merely drop an insulated container of water from some height h, letting the system make an inelastic collision with the ground and then measuring the increase in the temperature of the system. Or we could do mechanical work to generate electricity and then use the electricity to heat the water (see Figure 12-2). In all such experiments, the same amount of work is required to produce a given temperature change in a given system. By conservation of energy, the work done must go into an increase in the *internal energy U* of the system. The change in the internal energy of the system is usually measured by the change in the temperature of the system.

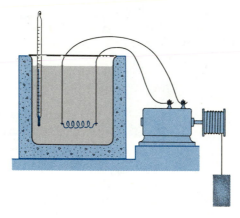

Figure 12-2
Another method of doing work on a thermally insulated container of water. Electrical work is done on the system by the generator, which is driven by the falling weight.

Example 12-3   A thermally insulated container of water is dropped from a large height h and collides inelastically with the ground. What must h be so that the temperature of the water increases by 1 C°?

To increase the temperature by 1 C°, we need to increase the internal energy of the water by 4.18 J for each gram of water or 4.18 kJ for each kilogram. If the mass of the water is m, the increase in internal energy is $m(4.18 \text{ kJ/kg})$. Setting this equal to the loss in potential energy, mgh, we have

$$mgh = m(4.18 \text{ kJ/kg})$$

$$h = \frac{4.18 \text{ kJ/kg}}{9.81 \text{ N/kg}} = 0.426 \text{ km} = 426 \text{ m}$$

We note that this result is independent of the mass of the water because the loss in potential energy and the energy needed to raise the temperature by 1 C° are both proportional to the mass. We note also that this is a rather large distance. This result illustrates one of the difficulties in Joule's experiment. A large amount of work must be done to produce a measurable change in the temperature of a system of water.

Let us now consider doing Joule's experiment, but let us replace the insulating walls of our water system with conducting walls. We find that the work needed to produce a given temperature change in the system depends on how much heat is added to or subtracted from the system by conduction through the walls. However, if we measure the heat added or subtracted and the work done on the system, we find that the sum of the work done on the system and the heat added to the system is always the same for a given temperature change.

The sum of the heat added to a system and the work done on the system equals the change in the internal energy of the system.

First law of thermodynamics

This is a statement of the first law of thermodynamics. It is customary to write W for the work done *by* the system on its surroundings. Then $-W$ is the work done *on* the system. For example, if a gas expands against a piston and does work on the surroundings, W is positive; if work is done to compress the gas, W is negative. The heat Q is taken to be positive if it is put *into* the system and negative if it is taken *out* of the system (see Figure 12-3).

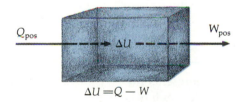

$$\Delta U = Q - W$$

Figure 12-3
Sign conventions for quantities in the first law of thermodynamics. Heat Q is positive when it goes into the system, whereas work W is positive when it is done by the system.

Using these conventions and writing U for the internal energy of a system, the first law of thermodynamics is

$$Q + (-W) = \Delta U$$

or, as it is usually written:

$$Q = \Delta U + W \qquad\qquad \text{12-10}$$

The heat added to a system equals the change in the internal energy of the system plus the work done by the system.

First law of thermodynamics

Again, we stress that the first law of thermodynamics is merely a statement of conservation of energy. The heat energy put into a system is accounted for by the work done by the system or by the increase in the internal energy of the system or by some combination of both.

When heat is added to a system while the volume of the system is kept constant and no other work is done on the system, the heat goes entirely into increasing the internal energy of the system, and the increase in the temperature of the system is given by

$$Q = \Delta U = mc_v\,\Delta T \qquad\qquad\qquad 12\text{-}11$$

Example 12-4 A system consists of 3 kg of water at 80°C. Stirring the system with a paddle wheel does 25 kJ of work on it while 15 kcal of heat is removed. What is the change in the internal energy of the system? What is the final temperature of the system?

We calculate the change in the internal energy of the system from the first law of thermodynamics as expressed in Equation 12-10. We first change the units of the heat removed from calories to joules so that both the work and heat are expressed in the same units. Since 1 kcal = 4.18 kJ, 15 kcal = 15 × 4.18 kJ = 62.7 kJ. We thus have 25 kJ of work done on the system and 62.7 kJ of heat removed. The heat added *to* the system is thus − 62.7 kJ, and the work done *on* the system is W = − 25 kJ. Equation 12-10 gives

$$Q = \Delta U + W$$

$$-62.7 \text{ kJ} = \Delta U + (-25 \text{ kJ})$$

$$\Delta U = -62.7 \text{ kJ} + 25 \text{ kJ} = -37.7 \text{ kJ}$$

The net change in the internal energy is negative because more energy is removed from the system as heat than is added to the system by doing work on it. Since 4.18 kJ will change the temperature of 1 kg of water 1 C°, the change in temperature is

$$\Delta T = \frac{-37.7 \text{ J}}{(4.18 \text{ J/kg}\cdot\text{C}°)(3 \text{ kg})} = -2.98 \text{ C}°$$

The final temperature is thus 77.02°C.

It is important to understand that U in Equation 12-10 represents the internal energy *of* the system, but Q is the heat put *into* the system and W is the work done *by* the system. Internal energy is a property of the system that depends on the state of the system. It is therefore called a *state function*. Other state functions are pressure P, volume V, and temperature T. For a gas, the state of the system is determined by any two properties, such as P and V. Suppose that the gas is in some initial state P_1, V_1. If we compress the gas or let it expand, add or remove heat from it, and do work on it or let it do work, the gas moves through a sequence of states, that is, it has different values of the state functions P, V, T, and U. If the gas is now returned to its original state P_1, V_1, the temperature T and the internal energy U must equal their original values. On the other hand, the heat Q and work W are not functions of the state of a system. It is correct to say that a system has a large amount of internal energy, but it is not correct to say that a system has a large amount of heat or a large amount of work. In Example 12-4, for instance, it is not correct to say that the system contains 62.7 kJ less heat because we removed this much heat from the system. Nor can we say that the system contains 25 kJ more work because we did that much work on the system. What we *can* say is that after doing that much work on the system and removing that much heat, the internal energy of the system is 37.7 kJ less than it was originally. To increase the internal energy of a system by 2 J, we may add 2 J of heat, or we may add no heat but instead do 2 J of work on the system, or we may do 3 J of work on the system and remove 1 J of heat. Heat, then, is not something that is contained in a system. Instead, it is energy that flows from one system to another because of a difference in temperature.

Questions

3. The experiment of Joule discussed in this section involved the conversion of mechanical energy into internal energy. Can you give examples in which the internal energy of a system is converted into mechanical energy?

4. Can a system absorb heat with no change in its internal energy?

5. The temperature of 2 L of water is to be raised from 20 to 25°C. What is the minimum amount of heat needed to do this? What is the minimum amount of work that must be done to do this? Explain.

12-3   Work and the *PV* Diagram for a Gas

In many engines, work is done by a gas expanding against a movable piston. For example, in a steam engine, water is heated in a boiler to produce steam that then expands and drives a piston that does work. In an automobile engine, gasoline is ignited, causing an explosion that results in rapidly expanding gas that drives a piston. In this section, we will consider an ideal situation in which the expansion is slow enough so that the work done can be easily calculated.

Figure 12-4 shows a gas confined in a container with a tightly fitting movable piston, which we will assume to be frictionless. When the piston moves, the volume of the gas changes, and either the temperature or the pressure or both must change since these three variables are related by an equation of state, such as $PV = nRT$ for an ideal gas. Let us begin with a gas at a fairly high pressure and let it expand slowly so that the piston does not accelerate as it moves. The force F exerted by the gas on the piston is PA, where A is the area of the piston and P is the gas pressure, which generally changes as the gas expands. (The pressure could be kept constant by heating the gas as it expands.) Since we are assuming that the piston does not accelerate, there must be an external force pushing against the piston also equal to PA. The piston does work on the agent providing this external force. If the piston moves a small distance Δx, the work done *by the gas* is

$$W = F \Delta x = PA \Delta x = P \Delta V \qquad\qquad 12\text{-}12$$

where $\Delta V = A \Delta x$ is the increase in the volume of the gas. Equation 12-12 also holds for a compression, in which case ΔV is negative, indicating that the gas does negative work. Equation 12-12 holds only for small values of ΔV, where any pressure changes during the expansion can be neglected. In

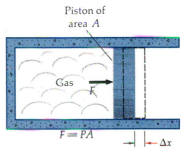

Piston of area A

Figure 12-4
Gas confined in a thermally insulated cylinder with a movable piston. When the piston moves a distance Δx, the volume changes by $\Delta V = A \Delta x$. The work done by the gas is $PA \Delta x = P \Delta V$, where P is the pressure.

Work done by gas

order to calculate the work done by the gas for an expansion where the volume changes appreciably, we would need to know how the pressure varies during the expansion.

In the previous section, we mentioned that the state of a gas is determined by two variables, such as P and V. It is customary and convenient to represent these states on a diagram of P versus V. Each point on such a diagram represents a state of the gas with pressure P and volume V. Since P and V are related to the absolute temperature T by an equation of state ($PV = nRT$ for an ideal gas), there is also a temperature T associated with each point. (There can be more than one point associated with the same temperature, as we will see.) There is also a value of U associated with each point on the PV diagram. Figure 12-5 shows a PV diagram with a horizontal line representing a series of states corresponding to the same value of P. This line represents an expansion at constant pressure. For a volume change of ΔV, the work done is $P\,\Delta V$, which equals the area under the curve, as indicated in the figure. In general, the work done by the gas (or on the gas) is represented by the area under the P-versus-V curve.

$$W = \text{area under the } P\text{-versus-}V \text{ curve} \qquad 12\text{-}13$$

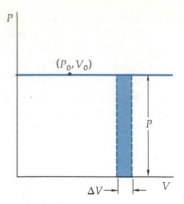

Figure 12-5
A particular state P_0, V_0 is represented by a point on the PV diagram. The horizontal line represents a series of states of constant pressure P_0. The work done by a gas as it expands is represented by the shaded area $P_0\,\Delta V$.

Example 12-5   Three litres of an ideal gas is at a pressure of 2 atm. It is heated so that it expands at constant pressure until its volume is 5 L. Find the work done by the gas.

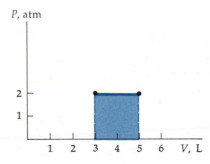

Figure 12-6
Expansion of a gas for Example 12-5.

This expansion is shown in Figure 12-6. Since the pressure is constant, the work done is

$$W = P(V_2 - V_1) = (2\text{ atm})(5\text{ L} - 3\text{ L}) = 4\text{ L}\cdot\text{atm}$$

Since pressures are often given in atmospheres and volumes in litres, it is convenient to have a conversion factor between litre-atmospheres and joules:

$$1\text{ L}\cdot\text{atm} = (10^{-3}\text{ m}^3)(101.3 \times 10^3\text{ N/m}^2) = 101.3\text{ J} \qquad 12\text{-}14$$

Figure 12-7 shows three different possible paths in the PV diagram for a gas that is originally in state P_1, V_1 and is finally in state P_2, V_2. We assume that the gas is ideal, and we have chosen the original and final states to have the same temperature so that $P_1V_1 = P_2V_2 = nRT$. For an ideal gas, the internal energy depends only on the temperature (this will be discussed

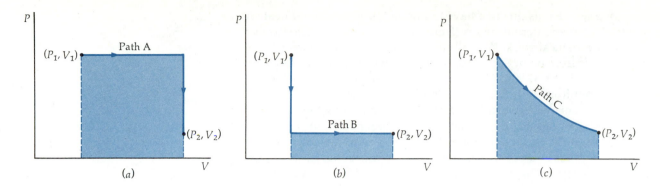

Figure 12-7
Three paths in the *PV* diagram connecting an initial state P_1, V_1 and a final state P_2, V_2. The work done along each path is indicated by the shaded area.

more fully in the next section), and so the internal energy of the initial and final states is the same. In Figure 12-7*a*, the gas is heated at constant pressure until its volume is V_2, after which it is cooled at constant volume until its pressure is P_2. The work done along this path is $P_1 (V_2 - V_1)$ for the horizontal part of the path and zero along the constant-volume part. This work is indicated by the shaded area under the curve. In Figure 12-7*b*, the gas is first cooled at constant volume until its pressure is P_2, after which it is heated at constant pressure until its volume is V_2. The work done for this path is $P_2(V_2 - V_1)$, which is much less than that along the first path as is indicated by the shaded region in the figure. In Figure 12-7*c*, both the pressure and volume vary for each part of the path. This path is one of constant temperature, $PV =$ constant. It is called an *isothermal expansion*. The work is again indicated by the shaded area. We could determine the amount of this work by measuring the area indicated. It can also be calculated using calculus. The result is

$$W_{\text{isothermal}} = nRT \ln \frac{V_2}{V_1}$$

12-15   Work done in isothermal expansion

where the symbol ln stands for the natural logarithm, the logarithm to the base *e*.

It should be clear from Figure 12-7 that the work done is different for each different process illustrated. Since the change in internal energy must be the same for each of the paths, which begin and end at the same states, the net amount of heat added must be different for each of the processes. In this case, the final internal energy equals the initial internal energy, so the net amount of heat added for each process equals the net work done by the gas. This discussion illustrates the fact that both the work done and the heat added (or subtracted) depend on just how the system moves from one state to another even though the change in internal energy (zero in this case) does not.

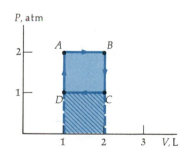

Figure 12-8
Cycle for Example 12-6. The work done as the gas expands from *A* to *B* minus that done on the gas as it is compressed from *C* to *D* is the net work done by the gas for the cycle and is represented by the enclosed area.

Example 12-6   An ideal gas undergoes the cyclic process illustrated in Figure 12-8. It begins at volume 1 L and pressure 2 atm and expands at constant pressure until the volume is doubled to 2 L, after which it is cooled at constant volume until its pressure is 1 atm. It is then compressed at constant pressure until its volume is again 1 L and is then heated at constant volume until it is back in its original state. Find the total work done by the gas for this cycle and the total heat added.

Cyclic processes such as this one have important applications for heat engines, which convert internal energy (from steam, gasoline, oil, and the like) into useful mechanical work. We will study heat engines in Chapter 14. As the gas expands from point A to point B at constant pressure, the work done is $W_{AB} = P(V_B - V_A) = (2\text{ atm})(2\text{ L} - 1\text{ L}) = 2\text{ L}\cdot\text{atm}$. This work is the area under the curve from A to B, as indicated by the shading in Figure 12-8. Using the conversion between litre-atmospheres and joules of Equation 12-14, this work equals 202.6 J. As the gas cools from point B to point C, the volume is constant so no work is done. As the gas is compressed at constant pressure from point C to point D, it does negative work; that is, work must be done *on* the gas. The work done *by* the gas is $W_{CD} = P(V_D - V_C) = (1\text{ atm})(1\text{ L} - 2\text{ L}) = -1\text{ L}\cdot\text{atm} = -101.3\text{ J}$. The magnitude of this work is the area under the curve CD in Figure 12-8, indicated by the hatching. As the gas is heated at constant volume back to its original state A, no work is done (because V is constant). The net work done by the gas for the entire cycle is therefore $202.6 - 101.3\text{ J} = 101.3\text{ J}$. The net work done by the gas during the complete cycle is represented in the figure by the area that is shaded but not hatched. Since the gas is back in its original state, all the state properties, P, V, T, and U, are at their original values. There is thus no *net* change in internal energy. However, 101.3 J of work has been done by the gas, so by the first law of thermodynamics, 101.3 J of heat must have been put into the gas during the cycle.

12-4   Heat Capacities and the Equipartition Theorem

The determination of the heat capacity of an object provides information about its internal energy, which in turn provides information about its molecular structure. As we have mentioned, for solids and liquids it matters little whether the heat is added at constant pressure or at constant volume because the work done in expansion at constant pressure is usually negligible. However, for gases we must distinguish between the heat capacity at constant volume C_v and that at constant pressure C_p because at constant pressure gases readily expand and do work.

When heat is added to a gas at constant volume, no work is done by or on the gas, so the heat added equals the increase in the internal energy of the gas. Writing Q_v for the heat added at constant volume, we have

$$Q_v = C_v \Delta T$$

By the first law of thermodynamics,

$$Q_v = \Delta U + W = \Delta U$$

since $W = 0$. Thus,

$$\Delta U = C_v \Delta T \qquad\qquad 12\text{-}16$$

If we add heat at constant pressure, the gas will expand and do work on its surroundings. Only a part of the heat added will go into an increase in the internal energy of the gas. Since the absolute temperature of the gas T is a measure of its internal energy, we expect that more heat must be added at constant pressure than at constant volume to achieve the same increase in temperature. In other words, we expect the heat capacity at constant pressure to be greater than that at constant volume. We will now show that, for an ideal gas, C_p is greater than C_v by the amount nR where n is the number of moles and R is the gas constant.

If we denote the heat added at constant pressure by Q_p, we have from the definition of C_p

$$Q_p = C_p \, \Delta T$$

From the first law of thermodynamics,

$$Q_p = \Delta U + W = \Delta U + P \, \Delta V$$

Thus,

$$C_p \, \Delta T = \Delta U + P \, \Delta V$$

But $\Delta U = C_v \, \Delta T$ from Equation 12-16. Therefore,

$$C_p \, \Delta T = C_v \, \Delta T + P \, \Delta V \qquad\qquad 12\text{-}17$$

The pressure, volume, and temperature for an ideal gas are related by

$$PV = nRT$$

When the gas expands at constant pressure, its change in volume is related to its change in temperature by

$$P \, \Delta V = nR \, \Delta T \qquad \text{(constant pressure)}$$

Substituting this into Equation 12-17, we obtain

$$C_p \, \Delta T = C_v \, \Delta T + nR \, \Delta T$$

Thus,

$$C_p = C_v + nR \qquad\qquad 12\text{-}18$$

Table 12-2 (p. 284) lists the measured molar heat capacities C_{mp} and C_{mv} for several gases. We note from this table that the ideal-gas prediction that $C_{mp} - C_{mv} = R$ holds quite well for all gases. We also note that C_{mv} is approximately $1.5R$ for all monatomic gases, $2.5R$ for all diatomic gases, and greater than $2.5R$ for gases consisting of more complex molecules. We can understand these results by considering the molecular model of a gas discussed in Section 11-4. In that section, we showed that the total translational kinetic energy of n mol of a gas is $E_k = \frac{3}{2}nRT$ (Equation 11-21). Thus, if the internal energy of a gas consists of translational kinetic energy only, we have

$$U = \tfrac{3}{2}nRT \qquad\qquad 12\text{-}19$$

A change in temperature ΔT is then related to a change in internal energy ΔU by

$$\Delta U = \tfrac{3}{2}nR \, \Delta T \qquad\qquad 12\text{-}20$$

Table 12-2

Molar Heat Capacities in J/mol·K of Various Gases at 25°C*

Gas	C_{mp}	C_{mv}	$\dfrac{C_{mv}}{R}$	$C_{mp} - C_{mv}$	$\dfrac{C_{mp} - C_{mv}}{R}$
Monatomic					
He	20.79	12.52	1.51	8.27	0.99
Ne	20.79	12.68	1.52	8.11	0.98
Ar	20.79	12.45	1.50	8.34	1.00
Kr	20.79	12.45	1.50	8.34	1.00
Xe	20.79	12.52	1.51	8.27	0.99
Diatomic					
N_2	29.12	20.80	2.50	8.32	1.00
H_2	28.82	20.44	2.46	8.38	1.01
O_2	29.37	20.98	2.52	8.39	1.01
CO	29.04	20.74	2.49	8.30	1.00
Polyatomic					
CO_2	36.62	28.17	3.39	8.45	1.02
N_2O	36.90	28.39	3.41	8.51	1.02
H_2S	36.12	27.36	3.29	8.76	1.05

* The values of C_{mp} are measured values taken from *The American Institute of Physics Handbook,* 3d ed., McGraw-Hill, New York, 1972. The values of C_{mv} are calculated from those of C_{mp}.

The heat capacity at constant volume C_v is related to the change in internal energy by Equation 12-16.

$$C_v = \frac{\Delta U}{\Delta T} = \tfrac{3}{2}nR \qquad\qquad\qquad 12\text{-}21$$

The heat capacity at constant pressure is then

$$C_p = C_v + nR = \tfrac{5}{2}nR \qquad\qquad\qquad 12\text{-}22$$

We note that these predictions agree well with the experimental results in Table 12-2 for monatomic gases, but for other gases, the heat capacities in the table are greater than those predicted by Equations 12-21 and 12-22. The internal energy for a gas consisting of diatomic or more complicated molecules is evidently greater than $\tfrac{3}{2}nRT$. The reason is that such molecules can have other types of energy, such as energy of rotation or vibration, in addition to the kinetic energy of translation.

In Section 11-4, we showed that the average kinetic energy associated with translation motion in the x direction is $\tfrac{1}{2}kT$ per molecule, where k is Boltzmann's constant (Equation 11-17), or equivalently, $\tfrac{1}{2}RT$ per mole. We argued that since x was an arbitrary direction, equivalent relations must also hold for the average kinetic energy of translation associated with motion in the y and z directions. This sharing of energy equally among the three translational energy terms is a special case of the *equipartition theorem*. Each coordinate, velocity component, angular velocity component, and so forth that appears in the expression for the energy of a molecule* is called a *degree of freedom*.

The equipartition theorem states:

In equilibrium, there is associated with each degree of freedom an average energy of $\frac{1}{2}kT$ per molecule or $\frac{1}{2}RT$ per mole.

Degree of freedom

Equipartition theorem

According to Table 12-2, nitrogen, oxygen, hydrogen, and carbon monoxide all have molar heat capacities at constant volume of about $\frac{5}{2}R$. The internal energy of n mol of any of these gases must therefore be

$$U = \tfrac{5}{2}nRT$$

indicating that for each of these gases there are five degrees of freedom. Clausius speculated (about 1880) that these gases must be diatomic gases that can rotate about two axes as well as translate, which gives them two additional degrees of freedom. Figure 12-9 shows a rigid-dumbbell model of a diatomic molecule. The two degrees of freedom in addition to the three for translation are now known to be associated with rotation about each of the two axes, x' and y', perpendicular to the line joining the atoms. If the molecule rotates about the line joining the atoms, there should be an additional degree of freedom. Similarly, we would not expect a real diatomic molecule to be rigid but, instead, that the two atoms would vibrate along the line joining them. We would then have two more degrees of freedom, corresponding to the kinetic energy and potential energy of vibration. According to the measured values of the molar heat capacities in Table 12-2, diatomic gases apparently do not rotate about the line joining the atoms, nor do they vibrate. The equipartition theorem gives no explanation for this nor for the fact that monatomic atoms apparently do not rotate about any of the three possible perpendicular axes in space. Furthermore, the equipartition theorem predicts constant values for the heat capacities of gases, whereas careful measurement shows that these quantities are somewhat dependent on temperature. The equipartition theorem fails because classical mechanics itself breaks down when applied to atomic and molecular systems and must be replaced by quantum mechanics.

Despite these failures, the equipartition theorem is very useful in giving us insight into molecular structure. For example, from the equipartition theorem and a simple model of a solid element, we can understand the Dulong-Petit law for the heat capacities of solids. (We need not distinguish between C_v and C_p here because, for a solid, they are almost equal.) Figure 12-10 shows a simple model of a solid consisting of a regular array of atoms, each having a fixed equilibrium position and being connected by springs to each of its neighbors. Each atom can oscillate in the x, y, and z directions. We will discuss oscillations in more detail in Chapter 15; for now, all we need to

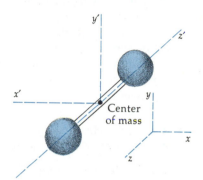

Figure 12-9
Rigid-dumbbell model of a diatomic gas molecule. Assuming no rotation about the z' axis, there are five degrees of freedom, three associated with translation in the x, y, and z directions and two associated with rotation about the x' and y' axes.

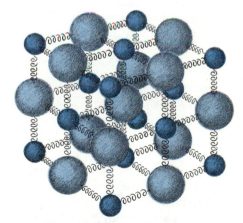

Figure 12-10
Model of a solid consisting of atoms connected to each other by springs. The internal energy of the solid then consists of the kinetic and potential energies of vibration.

* Strictly speaking, the coordinate must appear as a squared term in the expression for the energy of a molecule. Typical degrees of freedom are associated with the kinetic energy of translation, rotation, and vibration and the potential energy of vibration.

know is that the energy of vibration consists of two terms for each dimension, one for kinetic energy and one for potential energy. Thus, for oscillation in three dimensions, there are six degrees of freedom and the equipartition theorem predicts that the average energy is $6 \times \frac{1}{2}kT$ per atom and $6 \times \frac{1}{2}RT$ per mole. The internal energy of n mol of a solid is thus

$$U = 6 \times \tfrac{1}{2}nRT = 3nRT \qquad\qquad 12\text{-}23$$

and the molar heat capacity is $3R$, which is the Dulong-Petit law. Again, this result holds for many solid elements (but not for all) at room temperature, and again the reason for disagreement between prediction and experimental results is the breakdown of classical physics in the atomic realm. It is interesting to note that the successes of the equipartition theorem in explaining the measured heat capacities of gases and solids led to the first real understanding of molecular structure in the nineteenth century, whereas its failures played an important role in the development of quantum mechanics in the twentieth century.

Example 12-7   The molecular mass of copper is 63.5 g/mol. Use the Dulong-Petit law to calculate the specific heat of copper.

According to the Dulong-Petit law, the molar heat capacity of any solid element is $3R = 3(8.31 \text{ J/mol} \cdot \text{K}) = 24.9 \text{ J/mol} \cdot \text{K}$. The specific heat of copper should then be

$$c = \frac{(24.9 \text{ J/mol} \cdot \text{K})}{(63.5 \text{ g/mol})} = 0.392 \text{ J/g} \cdot \text{K} = 0.392 \text{ kJ/kg} \cdot \text{K}$$

This is fairly close to the measured value of $0.386 \text{ kJ/kg} \cdot \text{K}$.

Example 12-8   One mole of oxygen gas is heated from room temperature of 20°C and pressure of 1 atm to a temperature of 100°C. How much heat must be supplied if the volume is kept constant during the heating? How much heat must be supplied if the pressure is constant? (Assume that oxygen is an ideal gas.)

The heat capacity of oxygen at constant volume is

$$C_v = \tfrac{5}{2}nR = 20.8 \text{ J/K}$$

using $n = 1$ mol and $R = 8.31 \text{ J/K}$. The heat added to raise the temperature by 80 C° = 80 K is then

$$Q_v = C_v \, \Delta T = (20.8 \text{ J/K})(80 \text{ K}) = 1.66 \text{ kJ}$$

No work is done when the volume is constant, so the internal energy of the gas must increase by 1.66 kJ. Since the internal energy depends only on the temperature, this is the increase in internal energy when the gas temperature changes from 20 to 100°C by any process.

If we keep the pressure constant, the heat that must be added to raise the temperature by 80 K is

$$Q_p = C_p \, \Delta T = \tfrac{7}{2}(8.31 \text{ J/K})(80 \text{ K}) = 2.33 \text{ kJ}$$

where we have used $C_p = C_v + \tfrac{1}{2}R = \tfrac{7}{2}R$ for 1 mol. Since the increase in internal energy is again 1.66 kJ, the work done must equal

2.33 kJ − 1.66 kJ = 0.67 kJ. We can check this result by calculating directly the work done when the heat is added at constant pressure. Since the volume of 1 mol of any gas is 22.4 L at the standard conditions of 1 atm and $0°C = 273$ K, and since the volume is proportional to the absolute temperature if the pressure is constant, the initial volume V_1 at $20°C = 293$ K is

$$V_1 = (22.4 \text{ L}) \frac{293 \text{ K}}{273 \text{ K}} = 24.0 \text{ L}$$

When the temperature is raised by 80 K to 373 K at constant pressure, the volume increases to V_2, which is given by

$$V_2 = (22.4 \text{ L}) \frac{373 \text{ K}}{273 \text{ K}} = 30.6 \text{ L}$$

The work done is thus

$$W = P \, \Delta V = (1 \text{ atm})(30.6 \text{ L} - 24.0 \text{ L}) = 6.6 \text{ L} \cdot \text{atm}$$

Converting this to joules, we obtain

$$W = (6.6 \text{ L} \cdot \text{atm}) \frac{101.3 \text{ J}}{1 \text{ L} \cdot \text{atm}} = 0.67 \text{ kJ}$$

which agrees with our result from the first law of thermodynamics.

Summary

1. Heat capacity is the energy needed to raise the temperature of an object by one degree. The heat capacity of an object can be measured by heating it to a known temperature, placing it in an insulated water bath, and measuring the final common temperature of the system.

2. The calorie, originally defined as the heat necessary to raise the temperature of one gram of water by one Celsius degree, is now defined to be 4.184 joules:

1 cal = 4.184 J

3. The first law of thermodynamics is a statement of the conservation of energy: The heat added to a system equals the change in the internal energy of the system plus the work done by the system:

$Q = \Delta U + W$

The internal energy of a system is a property of the state of the system as are pressure, volume, and temperature, but heat and work are not.

4. When a system expands slightly, the work done by the system is given by

$W = P \, \Delta V$

The work for a finite expansion can be calculated if P is known as a function of V for the expansion. The work done by a gas is interpreted graphically as the area under the P-versus-V curve.

5. The heat capacity at constant volume is related to the change in internal energy by

$$C_v = \frac{\Delta U}{\Delta T}$$

For an ideal gas, the heat capacity at constant pressure is greater than that at constant volume by the amount nR:

$$C_p = C_v + nR$$

The heat capacity at constant pressure is greater because at constant pressure the gas expands and does work, so it takes a greater amount of heat to achieve the same temperature change.

6. The heat capacity at constant volume for monatomic gases is

$$C_v = \tfrac{3}{2}nR$$

For diatomic gases it is

$$C_v = \tfrac{5}{2}nR$$

7. The equipartition theorem states: In equilibrium, there is associated with each degree of freedom an average energy of $\tfrac{1}{2}kT$ per molecule or $\tfrac{1}{2}RT$ per mole. Monatomic gases apparently have three degrees of freedom associated with the kinetic energy of translation in three dimensions. Diatomic gases have two additional degrees of freedom associated with rotation.

8. The molar heat capacity of most solid elements is $3R$, a result known as the Dulong-Petit law. This result can also be understood from the equipartition theorem by assuming a model in which each atom in the solid can vibrate in three dimensions and therefore has a total of six degrees of freedom, three associated with the kinetic energy and three with the potential energy of vibration.

Suggestions for Further Reading

"Heat," *Scientific American* (Special Issue), September 1954.

This issue discusses many aspects of heat and includes articles on early humans' use of fire, the temperature range in which life is possible, the temperature of flames, high-temperature resistant materials, chemistry at high temperatures, and the ultrahigh temperatures in the interior of stars.

Review

A. Objectives: After studying this chapter, you should:

  1. Be able to work calorimetry problems.

  2. Be able to state the first law of thermodynamics and use it in solving problems.

  3. Be able to state the Dulong-Petit law and use it to estimate the heat capacity of a given solid or to calculate the molar mass of a solid from its specific heat.

  4. Be able to state the equipartition theorem and use it to relate the molar heat capacity of a gas to a mechanical model of gas molecules.

B. Define, explain, or otherwise identify:

  Heat capacity, p. 271
  Specific heat, p. 271
  Calorie, p. 272
  Btu, p. 272
  Molar heat capacity, p. 272

Internal energy, p. 276
First law of thermodynamics, p. 277
PV diagram, p. 280
Equipartition theorem, p. 285

C. True or false: If the statement is true, explain why. If it is false, give a counterexample.

1. The heat capacity of a body is the amount of heat it can store at a given temperature.

2. When a system goes from state 1 to state 2, the amount of heat added is the same for all processes.

3. When a system goes from state 1 to state 2, the work done on the system is the same for all processes.

4. When a system goes from state 1 to state 2, the change in internal energy is the same for all processes.

5. For any material that expands when heated, C_p is greater than C_v.

Exercises

Section 12-1   Heat Capacity and Specific Heat

1. How many joules of heat energy are required to raise the temperature of 20 kg of water from 10 to 20°C, assuming no work is done?

2. Find the number of Btu needed to raise the temperature of 1 gal of water from 32 to 212°F (1 gal contains 8 pints and 1 pint of water weighs 1 lb).

3. How much heat is needed to raise the temperature of 2 kg of copper from 20 to 100°C, assuming no work is done?

4. A 3-kg block of lead is at 180°C. It cools to 20°C, giving off heat *Q*. Find *Q*.

5. A solar home contains 100 Mg of concrete (specific heat 1.00 kJ/kg·K). How much heat is given off by the concrete when it cools from 25 to 20°C?

6. How much heat is needed to heat a lead brick 5 cm by 10 cm by 20 cm from 20 to 25°C?

7. Lead shot of mass 200 g is heated to 90°C and placed in 500 g of water initially at 20°C. Neglecting the heat capacity of the container and any heat loss from the system, find the final temperature of the lead and water.

8. Repeat Exercise 7 with copper shot instead of lead shot.

9. The specific heat of a certain metal is determined by measuring the temperature change that occurs when a heated piece of the metal is placed in an insulated container made of the same material and containing water. The piece of metal has a mass of 100 g and is initially at 100°C. The container has a mass of 200 g and contains 500 g of water at an initial temperature of 17.3°C. The final temperature of the system is 22.7°C. What is the specific heat of the metal?

Section 12-2   The First Law of Thermodynamics

10. A body of water is heated at constant pressure from 20 to 40°C. Explain why it is incorrect to say that the water at 40°C contains more heat than it did at 20°C. Is it correct to say that the water has more internal energy at 40°C than at 20°C?

11. If 20 J of work is done on a system and 80 cal of heat is added, what is the change in the internal energy of the system?

12. If 400 kcal is added to a gas that expands and does 800 kJ of work, what is the change in the internal energy of the gas?

13. In a Joule-type experiment, the paddle wheel is turned by a 4-kg weight that falls at constant speed through a distance of 1.5 m. If the system contains 0.6 kg of water, by how much is the temperature of the water increased?

14. (*a*) How much work is required to change the temperature of 1 kg of water from 20 to 25°C, assuming no heat is lost or gained. (*b*) In an actual experiment, the work done to produce this temperature change is 22.5 kJ. How much heat escaped to the surroundings?

15. In a lecture demonstration, a box of lead shot is thrown vertically into the air to a height of 4 m and is allowed to fall to the floor. The original temperature of the lead is 20°C. Five such throws are made, and the temperature of the shot is then measured. What result do you expect if there are no heat losses during the demonstration?

16. A lead bullet moving at 200 m/s is stopped in a block of wood. Assuming that all the energy change goes into heating the bullet, find the final temperature of the bullet if its initial temperature is 20°C.

17. The water at Niagara Falls drops 50 m. (*a*) If the change in potential energy goes into the internal energy of the water, compute the increase in its temperature. (*b*) Do the same for Yosemite Falls, where the water drops 740 m. (These temperature rises are not observed because the water cools by evaporation as it falls.)

Section 12-3   Work and the *PV* Diagram for a Gas

18. A gas initially at pressure 4 atm and volume 1.5 L expands at constant pressure until its volume is 4.5 L. Find the work done by the gas in joules.

19. A gas is initially at a pressure of 300 kPa and a volume of 15 L. It is compressed at constant pressure to a volume of 12.5 L. How much work is done *by* the gas?

20. An ideal gas is originally at pressure 4 atm and volume 1 L. It expands isothermally until its pressure is 1 atm and its volume is 4 L. Calculate the work done using Equation 12-15. (*Hint:* The value of the natural log of a quantity can be found in tables or from a calculator. The value of nRT for this gas can be found from $PV = nRT$.)

21. An ideal gas is originally at a pressure of 100 kPa and a volume of 20 L. It is compressed isothermally until its volume is 10 L and its pressure is 200 kPa. Calculate the work needed to perform this compression (see Exercise 20).

Section 12-4   Heat Capacities and the Equipartition Theorem

22. The specific heat of steam ($M = 18.0$ g/mol) is measured at constant pressure to be 2.50 kJ/kg·K. Assuming steam to be an ideal gas, what is the specific heat capacity at constant volume?

23. The specific heat of air ($M = 29.0$ g/mol) at 0°C is listed in a handbook as having the value 1.00 J/g·K measured at constant pressure. Assuming air to be an ideal gas, what is the specific heat at 0°C at constant volume?

24. The heat capacity at constant volume of a certain amount of a monatomic gas is 49.8 J/K. (*a*) Find the number of moles of the gas. (*b*) What is the internal energy of this gas at $T = 300$ K? (*c*) What is the heat capacity at constant pressure?

25. For a certain gas, the heat capacity at constant pressure is greater than that at constant volume by 29.1 J/K. (*a*) How many moles of gas are there? (*b*) If the gas is monatomic, what are C_v and C_p? (*c*) If the gas is diatomic, what are C_v and C_p?

26. The Dulong-Petit law was originally used to determine the molar mass of a substance from its measured heat capacity. Given that the specific heat of a certain solid is measured to be 0.447 kJ/kg·K, find the molar mass of the substance. What element is this?

27. The specific heat of a certain solid element is 0.13311 kJ/kg. Find the molar mass of the solid, assuming that it obeys the Dulong-Petit law, and state what element it is.

Problems

1. A 500-watt microwave oven is used to heat 250 mL of water to make coffee. How long does it take to heat the water from 20 to 90°C?

2. A 200-g insulated aluminum can contains 50 g of water at 20°C. Aluminum shot of mass 300 g is heated to 100°C and placed in the can. Using the value of the specific heat of aluminum given in Table 12-1, find the final temperature of the system, assuming that no heat is lost to the surroundings.

3. For the isothermal expansion of Exercise 20, make a careful sketch of P versus V. Calculate the work done by counting squares to get the area under the P-versus-V curve and compare this result with that obtained by calculation from Equation 12-15.

In Problems 4 to 7, 1 mol of an ideal gas is originally in state $P_1 = 3$ atm, $V_1 = 1$ L, and $U_1 = 456$ J. Its final state is $P_2 = 2$ atm, $V_2 = 3$ L, and $U_2 = 912$ J. All processes occur slowly enough so they can be represented on a PV diagram.

4. The gas expands at constant pressure to a volume of 3 L. It then cools at constant volume until its pressure is 2 atm. (*a*) Indicate this process on a PV diagram and calculate the work done by the gas. (*b*) Find the heat added during the process.

5. The gas is first cooled at constant volume until its pressure is 2 atm. It then expands at constant pressure until its volume is 3 L. (*a*) Indicate this process on a PV diagram and calculate the work done by the gas. (*b*) Find the heat added during this process.

6. The gas expands isothermally until its volume is 3 L and its pressure is 1 atm. It is then heated at constant volume until its pressure is 2 atm. (*a*) Indicate this process on a PV diagram and calculate the work done by the gas. (*b*) Find the heat added during this process.

7. The gas expands and heat is added so that the gas follows a straight-line path on the PV diagram from its initial state to its final state. (*a*) Indicate this process on a PV diagram and calculate the work done by the gas. (*b*) Find the heat added for this process.

8. One mole of a monatomic ideal gas is at 273 K and 1 atm. Find the initial and final internal energies and the work done by the gas when 500 J of heat is added (*a*) at constant pressure and (*b*) at constant volume.

9. For each of the four parts of Example 12-6, calculate the heat input or output and the internal energy change, assuming that the gas is an ideal monatomic gas.

10. One mole of an ideal monatomic gas is heated at constant volume from 300 to 600 K. (*a*) Find the heat added, the work done, and the change in internal energy. (*b*) Find these same quantities if the gas is heated from 300 to 600 K at constant pressure.

11. One mole of an ideal gas is in an initial state of $P = 2$ atm and $V = 10$ L indicated by point A on the PV diagram of Figure 12-11. It expands at constant pressure to point B, where its volume is 30 L, and is then cooled at constant volume until its pressure is 1 atm at point C. It is then compressed at constant pressure to its original volume at point D, and finally, it is heated at constant volume until it is back at its original state. (*a*) Find the temperature of each state, A, B, C, and D. (*b*) Assuming the gas to be monatomic, find the heat added along each path of the cycle. (*c*) Calculate the work done along each path. (*d*) Find the internal energy of each state, A, B, C, and D. (*e*) What is the net work done by the gas for the complete cycle? What is the net amount of heat added during the complete cycle?

Figure 12-11
Cycle for Problem 11.

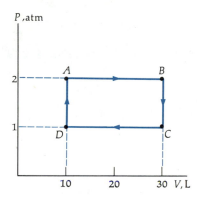

CHAPTER THIRTEEN

Thermal Properties and Processes

This chapter deals with the flow of heat and the things that it can do. You will be familiar with many of these things already, but some of them may surprise you. Why, for instance, do we boil eggs rather than bake them in the oven? Why is the climate near the ocean more moderate than it is further inland? How does a pressure cooker cook potatoes in a quarter of the normal time? Why is steam so much more dangerous to the skin than equally hot air? How does a thermos bottle keep soup hot and iced tea cold? Why does a metal surface at room temperature feel cold? Why does a goose down jacket keep you so warm? Most of these questions will be answered directly; the rest you will be able to figure out for yourself as you read this chapter.

Heat and cold are nature's two hands by which she chiefly worketh.

FRANCIS BACON

When a body is heated, various changes take place. The temperature of the body may rise, accompanied by an expansion or contraction of the body; or the body may liquefy or vaporize with no change in temperature. In this chapter, we will examine thermal processes such as the expansion of metals when they are heated, the freezing of water into ice, cooling by evaporation, the direct change of dry ice into vapor, the formation of dew on a cool evening, the conduction of heat through windowpanes, and the radiation of heat from a hot body. We will learn to calculate such interesting things as the relative humidity from the measurement of the dew point, the effectiveness of thermal insulation, and the temperature of a star from its radiation spectrum.

Warm air currents rising from fingers in still air are shown in this photograph which uses the schlieren technique that is sensitive to small changes in the density of air. The temperature difference is estimated to be 10 F°.

13-1 Thermal Expansion

When the temperature of a body increases, it usually expands. Thermal expansion is often used to measure temperature changes, as with a mercury thermometer. To understand thermal expansion, we first consider a long rod of length L at temperature T. When the temperature changes by ΔT, we find experimentally that the change in length ΔL is proportional to ΔT and to the original length L.

$$\Delta L = \alpha L \, \Delta T \qquad \text{13-1}$$

where α is called the *coefficient of linear expansion*. The coefficient of linear expansion of a solid or liquid usually does not vary much with pressure, but it may vary with temperature, so Equation 13-1 should be used to calculate ΔL only for small temperature changes for which the variation in α can be neglected. We can see from Equation 13-1 that the coefficient of thermal expansion is the ratio of the fractional change in length to the change in temperature:

$$\alpha = \frac{\Delta L / L}{\Delta T} \qquad \text{13-2}$$

Its units are reciprocal Celsius degrees $(1/C°)$ or reciprocal kelvins $(1/K)$.

The coefficient of volume expansion β, sometimes called the *expansivity*, is similarly defined as the ratio of the fractional change in volume to the change in temperature:

$$\beta = \frac{\Delta V / V}{\Delta T} \qquad \text{13-3}$$

For a given material, the coefficient of volume expansion is three times the coefficient of linear expansion (see Problem 4).

$$\beta = 3\alpha \qquad \text{13-4}$$

The increase in size of any part of a body for a given temperature change is proportional to the original size of that part of the body. If we increase the temperature of a steel ruler, for example, the effect will be similar to that of a (very slight) photographic enlargement. Lines that were previously equally spaced will still be equally spaced, but the spaces will be slightly larger and the width of the ruler will be slightly greater. Values of α and β for various substances are given in Table 13-1.

Table 13-1

Approximate Values of Thermal-Expansion Coefficients

Material	α, K^{-1}	Material	β, K^{-1}
Aluminum	24×10^{-6}	Acetone	1.5×10^{-3}
Brass	19×10^{-6}	Air	3.67×10^{-3}
Carbon		Alcohol	1.1×10^{-3}
Diamond	1.2×10^{-6}	Mercury	0.18×10^{-3}
Graphite	7.9×10^{-6}	Water (20°C)	0.207×10^{-3}
Copper	17×10^{-6}		
Glass			
Ordinary	9×10^{-6}		
Pyrex	3.2×10^{-6}		
Ice	51×10^{-6}		
Invar	1×10^{-6}		
Steel	11×10^{-6}		

These railway tracks were not designed to accommodate thermal expansion.

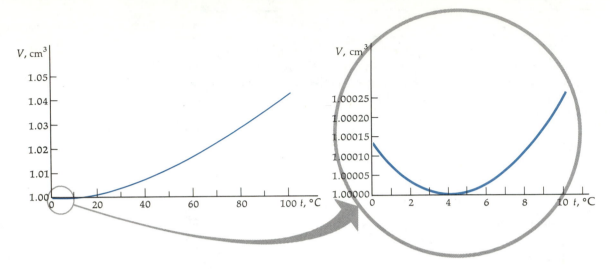

Figure 13-1
Volume of 1 g of water at atmospheric pressure versus temperature. The minimum volume corresponding to the maximum density occurs at 4°C.

Though most materials expand when heated, the behavior of water at temperatures between 0 and 4°C is an important exception. Figure 13-1 shows the volume occupied by 1 g of water as a function of temperature. The volume is minimum and the density is therefore maximum at 4°C. Thus, when water is heated at temperatures below 4°C, it contracts rather than expands. This property has important consequences for the ecology of lakes. At temperatures above 4°C, the water in a lake becomes denser as it cools and sinks to the bottom. But at temperatures below 4°C, water becomes less dense as it cools, so it stays at the surface. Ice therefore forms first on the surface of a lake, and being less dense than water, it remains there and acts as a thermal insulator for the water below. If water contracted when it freezes, as most substances do, the ice would sink and new water would be exposed at the surface and would freeze. Lakes would then fill with ice from the bottom up and would be much more likely to become completely frozen in the winter, killing the fish and other aquatic life.

The expansivity of a gas is not even approximately temperature independent. It is, in fact, simply related to the absolute temperature. Consider an ideal gas that obeys the equation of state (Equation 11-12):

$$PV = nRT$$

If the change in volume is ΔV for a temperature change ΔT at constant pressure, we have

$$P\,\Delta V = nR\,\Delta T$$

The coefficient of volume expansion is then

$$\beta = \frac{\Delta V/V}{\Delta T} = \frac{(nR\,\Delta T/PV)}{\Delta T}$$

$$= \frac{nR}{PV} = \frac{1}{T}$$

13-5

Thus, the expansivity is just the reciprocal of the absolute temperature.

Once ice covers the surface of a lake, it insulates the water below from the cold air above. Then, during the winter, the water at the bottom actually gets warmer, heated by the ground beneath the lake, which has a temperature of about 10°C. In the spring, when the ice on top melts, the less dense water at the bottom rises, carrying nutrients to the surface. At the same time, the more dense surface water sinks, carrying oxygen to the depths.

Example 13-1   A steel bridge is 1000 m long. By how much does it expand when the temperature rises from 0 to 30°C?

From Table 13-1, the coefficient of linear expansion is 11×10^{-6} K^{-1} for steel. The change in length for a 30-K rise in temperature is then

$$\Delta L = \alpha L \, \Delta T = (11 \times 10^{-6} \text{ K}^{-1})(1000 \text{ m})(30 \text{ K})$$

$$= 0.33 \text{ m} = 33 \text{ cm}$$

To relieve the stresses that would occur if this expansion were not allowed, expansion joints must be included in the bridge. We can calculate the stress involved if the bridge were not allowed to expand by using Equation 10-7 for Young's modulus, $Y = \text{stress/strain} = (\Delta F/A)/(\Delta L/L)$. If the bridge could not expand, the stress would be

$$\frac{\Delta F}{A} = Y \frac{\Delta L}{L}$$

$$= 2 \times 10^{11} \text{ N/m}^2 \left(\frac{0.33 \text{ m}}{1000 \text{ m}} \right) = 6.6 \times 10^7 \text{ N/m}^2$$

where we have used the value of Young's modulus from Table 10-2. This stress is about one third of the compression breaking stress for steel. A steel bridge under a compression stress of this magnitude would certainly buckle and become permanently deformed.

An expansion joint allows the bridge to expand and contract without causing damage.

Example 13-2   A 1-L glass flask is filled to the brim with alcohol at 10°C. If the temperature warms to 30°C, how much alcohol spills out of the flask?

Using $\Delta T = 20 \text{ C}° = 20 \text{ K}$ and the expansion coefficients from Table 13-1, we can obtain the change in the volume of the glass

$$\Delta V = \beta V \, \Delta T = (9 \times 10^{-6} \text{ K}^{-1})(1 \text{ L})(20 \text{ K})$$

$$= 1.8 \times 10^{-4} \text{ L} = 0.18 \text{ mL}$$

and the change in the volume of the alcohol

$$\Delta V = \beta V \, \Delta T = (1.1 \times 10^{-3} \text{ K}^{-1})(1 \text{ L})(20 \text{ K})$$

$$= 2.2 \times 10^{-2} = 22.0 \text{ mL}$$

The amount of alcohol that spills out is thus 22.0 mL − 0.18 mL ≈ 21.8 mL.

Example 13-3   A copper bar is heated to 300°C and is then clamped rigidly between two fixed points so that it can neither expand nor contract. If the breaking stress of copper is 230 MN/m², at what temperature will the bar break as it cools?

In this example, the change in length ΔL that would occur if the bar contracted as it cooled must be offset by an equal stretching due to tensile stress in the bar. From the definition of Young's modulus (Equation 10-7), we can find the stretching ΔL caused by a tensile stress F/A:

$$\Delta L = L \frac{F/A}{Y}$$

Setting this equal to the change in length that would occur if the bar could contract, we have

$$\Delta L = L\alpha\,\Delta T = L\,\frac{F/A}{Y}$$

Solving for ΔT and using $Y = 110$ GN/m² from Table 10-2, $\alpha = 17 \times 10^{-6}$ K^{-1} from Table 13-1, and $F/A = 230$ MN/m² for the breaking stress, we obtain

$$\Delta T = \frac{F/A}{\alpha Y}$$

$$= \frac{230 \times 10^6 \text{ N/m}^2}{(17 \times 10^{-6}\text{ K}^{-1})(110 \times 10^9 \text{ N/m}^2)}$$

$$= 123 \text{ K} = 123 \text{ C}°$$

Since the original temperature is 300°C, the temperature at which the bar will break is

$$t = 300°\text{C} - 123 \text{ C}° = 177°\text{C}$$

Questions

1. If a metal sheet with a hole in it expands, does the hole get larger or smaller?

2. If mercury and glass had the same coefficient of expansion, could a glass-mercury thermometer be built?

13-2   Change of Phase and Latent Heat

When heat is supplied to a body at constant pressure, the result is usually an increase in the temperature of the body. Sometimes, though, a body can absorb large amounts of heat without any change in temperature. This happens when the physical condition, or *phase,* of the material is changing from one form to another. Examples of phase changes are *fusion,* the change of a liquid to a solid (such as the freezing of water into ice); *vaporization,* the change of a liquid to a vapor or gas (such as in the evaporation of water); and *sublimation,* the change of a solid directly into a gas (such as the vaporization of moth balls or of solid carbon dioxide, sometimes called dry ice). There are other phase changes in which a solid changes from one crystalline form to another.

These phenomena can be understood in terms of molecular theory. An increase in the temperature of a body reflects an increase in the kinetic energy of motion of its molecules. When a material changes from liquid to gaseous form, its molecules, which were originally held together by their natural attraction for one another, are moved far apart from each other. This requires that work be done against the attractive forces; that is, energy must be supplied to the molecules to separate them. This energy goes into an

Ice cream undergoes a phase change on a warm day.

increase in the potential energy of the molecules rather than an increase in their kinetic energy. Therefore, the temperature of the body, which is a measure of the average molecular kinetic energy, does not change.

For a pure substance, a change in phase at a given pressure occurs only at a particular temperature. For example, pure water at atmospheric pressure changes from solid to liquid at 0°C and from liquid to gas at 100°C. The first temperature is called the normal melting point and the second the normal boiling point. These points were those chosen by Celsius for his temperature scale based on two fixed points.

A specific quantity of heat energy is required to change the phase of a given amount of a substance. The required heat is proportional to the mass of the substance. The heat required to melt a body of mass m is written

$$Q = mL_f \qquad\qquad\qquad 13\text{-}6$$

where L_f is called the *latent heat of fusion* of the substance. For the melting of ice to water at atmospheric pressure, the latent heat of fusion is 333.5 kJ/kg = 79.7 kcal/kg. If the phase change is from liquid to gas, the heat required is written

> Latent heat of fusion

$$Q = mL_v \qquad\qquad\qquad 13\text{-}7$$

where L_v is the *latent heat of vaporization*. For water at atmospheric pressure, the latent heat of vaporization is 2.26 MJ/kg = 540 kcal/kg. Table 13-2 gives the normal melting and boiling points and the latent heats of fusion and vaporization at 1 atm for various substances.

> Latent heat of vaporization

Table 13-2

Normal Melting Point (MP), Latent Heat of Fusion L_f, Normal Boiling Point (BP), and Latent Heat of Vaporization L_v for Various Substances at 1 atm

Substance	MP, K	L_f, kJ/kg	BP, K	L_v, kJ/kg
Alcohol, ethyl	159	109	351	879
Bromine	266	67.4	332	369
Carbon dioxide	—	—	194.6*	573*
Copper	1356	205	2839	4726
Gold	1336	62.8	3081	1701
Helium	1.76	2.1	4.2	21
Lead	600	24.7	2023	858
Mercury	234	11.3	630	296
Nitrogen	63	25.7	77.35	199
Oxygen	54.4	13.8	90.2	213
Silver	1234	105	2436	2323
Sulfur	388	38.5	717.75	287
Water	273.15	333.5	373.15	2257
Zinc	692	102	1184	1768

* These values are for sublimation. Carbon dioxide does not have a liquid state at 1 atm.

Many substances have several different solid or liquid phases. For example, eight different forms or phases of ice are known, each with properties different from the others. The transitions between these forms occur at definite temperatures and pressures, just like the transitions between the solid and liquid phases. Helium has two liquid phases. Figure 13-2 shows the specific heat of helium as a function of temperature in the range of 1.2 to 3 K. The specific heat has a sharp discontinuity at the temperature of 2.17 K. Because the shape of this curve resembles the Greek letter lambda (λ), this transition temperature is called the *lambda point*. The liquid at temperatures above the lambda point is called helium I, and that at temperatures below the lambda point is called helium II. These two fluids are very different. For example, liquid helium II has zero viscosity and is a perfect heat conductor. Because of these remarkable properties, it is called a superfluid. On the other hand, the viscosity and thermal properties of liquid helium I are similar to those of an ordinary liquid.

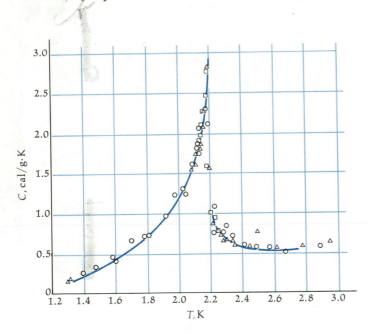

Figure 13-2
Specific heat of liquid helium versus temperature between 1.2 and 3 K. Because of the resemblance of this curve to the Greek letter λ, this phase-transition point is called the lambda point.

Example 13-4   If 1 kg of ice at $-20°C$ is heated at atmospheric pressure until all the ice has been changed into steam, how much heat is required?

Assuming the heat capacity of ice to be constant and equal to 2.05 kJ/kg·K (Table 12-1), we find that the heat energy needed to raise the temperature of the ice from -20 to $0°C$ is

$$Q_1 = mc\,\Delta T = (1\text{ kg})(2.05\text{ kJ/kg·K})(20\text{ K}) = 41\text{ kJ}$$

The heat needed to melt 1 kg of ice is

$$Q_2 = mL_f = 334\text{ kJ}$$

The heat needed to raise the temperature of the resulting 1 kg of water from 0 to 100°C is

$$Q_3 = mc\,\Delta T = (1\text{ kg})(4.18\text{ kJ/kg·K})(100\text{ K}) = 418\text{ kJ}$$

where we have neglected any variation in the heat capacity of water over this range of temperatures. Finally, the heat needed to vaporize 1 kg of water at 100°C is

$$Q_4 = mL_v = 2.26 \text{ MJ} = 2260 \text{ kJ}$$

The total amount of heat required is

$$Q = Q_1 + Q_2 + Q_3 + Q_4 = 3.05 \text{ MJ}$$

Note that most of this heat is needed to change the phase of the water, not to raise its temperature.

This tea kettle illustrates a familiar phase change of water.

Figure 13-3 shows a graph of the temperature versus time for Example 13-4, assuming the heat is added at a constant rate of 1 kJ/s. The temperature of the ice increases at a steady rate until it reaches 0°C, the normal melting point. The temperature then remains constant at 0°C as the ice melts. Note that it takes longer to melt the ice than it does to raise its temperature from −20°C to 0°C. As calculated in the example, 334 kJ are needed to melt the ice, whereas only 41 kJ are needed to raise its temperature by 20 C°. So at 1 kJ/s, it takes only 41 s to raise the temperature of the ice 20 C°, but it takes 334 s to melt the ice. When all the ice is melted into water at 0°C, the temperature of the water begins to rise. Since the specific heat of water is about twice that of ice, the temperature of the water increases at a slower rate than did the temperature of the ice. As calculated, it takes 418 kJ to raise the temperature by 100 C°, so at 1 kJ/s it takes 418 s for the water to reach the boiling point. When the temperature reaches 100°C, the normal boiling point, the water vaporizes (boils), and the temperature remains constant until all of the water has turned to vapor (steam). Since 2260 kJ are needed to vaporize 1 kg of water, it takes considerably longer to vaporize the water than it does to melt the ice or to raise the temperature of the water by 100 C°. When all of the water has vaporized, the temperature again rises as heat is added.

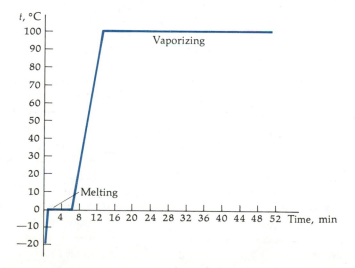

Figure 13-3
Plot of temperature versus time when heat at a constant rate of 1 kJ/s is added to 1 kg of ice originally at −20°C. The temperature first rises to 0°C, at which point the ice melts at constant temperature. When the ice has melted, the temperature of the water rises to 100°C and then the water vaporizes at constant temperature. If heat continues to be added after all the water has turned to steam, the temperature of the steam will again rise.

13-3   The van der Waals Equation and Liquid-Vapor Isotherms

Although most gases behave as ideal gases at ordinary pressures, this ideal behavior breaks down at temperatures low enough to result in a high density, as when the gas is about to condense into a liquid. An equation of state called the *van der Waals equation* describes the behavior of many real gases over a wide range of pressures more accurately than does the ideal-gas equation of state ($PV = nRT$). For one mole, the van der Waals equation is written

$$\left(P + \frac{a}{V^2}\right)(V - b) = RT \qquad \text{13-8} \qquad \text{van der Waals equation}$$

The constant b in this equation arises because the gas molecules are not points but have a finite size; therefore, the volume available for the molecules to move about in is reduced. The magnitude of b is the molecular volume of one mole of gas molecules. The term a/V^2 accounts for the reduced pressure due to the attraction of the molecules for each other when the density of the gas is high enough for the molecules to be relatively close together. Both of these terms are negligible when the volume V is large, that is, at low densities. Thus, at low densities the van der Waals equation approaches the ideal-gas law, whereas at high densities it is a much better description of the behavior of real gases.

Figure 13-4 shows the P-versus-V isothermal curves for various temperatures for a real gas. For temperatures above some *critical temperature* T_c on the diagram, these curves are described quite accurately by the van der Waals equation and can be used to determine the constants a and b. (Since the constant b is the actual volume of one mole of gas molecules, the determination of its value can be used to estimate the size of a molecule.) At temperatures below T_c, the van der Waals equation describes the curves outside the shaded region in Figure 13-4 but not those inside the shaded region.

Suppose we have a gas at a temperature below T_c that is initially at a low pressure and large volume (point A in the figure), and we begin to compress it isothermally. At first the pressure rises, but when we reach the dashed line on the curve (point B), the pressure ceases to rise and the gas begins to liquefy at constant pressure. Along the horizontal line DB in the figure, the gas and liquid are in equilibrium. As we continue to compress the gas, more and more gas liquefies until we have only liquid (point D on the curve). Then, if we try to compress further, the pressure rises sharply because a liquid is nearly incompressible. The constant pressure for which the gas and liquid are in equilibrium at a given temperature is called the *vapor pressure*. When this equilibrium exists and the liquid is heated slightly (or the pressure is reduced slightly), the liquid boils.

It is clear from the figure that the vapor pressure of a gas depends on temperature. If we had started compressing the gas at a lower temperature, A' in the figure, the vapor pressure would be lower, as indicated by the constant-pressure horizontal line at a lower pressure. The temperature for which the vapor pressure of a substance equals one atmosphere is the normal boiling point of that substance (see Table 13-2). Table 13-3 gives the

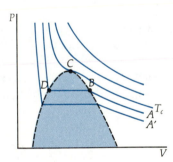

Figure 13-4
Isotherms on the PV diagram for a real substance. For temperatures above the critical temperature T_c, the substance remains a gas at all pressures and is described by the van der Waals equation. The pressure for the horizontal curves in the shaded region is the vapor pressure for which the vapor and the liquid are in equilibrium. To the left of the shaded region for temperatures below the critical temperature, the substance is a liquid and is nearly incompressible.

vapor pressure of water as a function of temperature. The vapor pressure of water is 1 atm at a temperature of 373 K = 100°C, the normal boiling point of water. If we are at a high altitude, such as on the top of a mountain, the pressure is less than 1 atm. Water then boils at a lower temperature than 373 K. At temperatures greater than the critical temperature T_c, the gas does not liquefy at any pressure. The critical temperature of water vapor is 647 K = 374°C. The point at which the critical isotherm intersects the dashed curve in Figure 13-4 (point C) is called the *critical point*.

Consider putting a liquid such as water in an evacuated container that is sealed so that its volume is constant. At first, some of the water will evaporate, and the water-vapor molecules will fill the previously empty space in the container. When there are water-vapor molecules in the container, some of them will condense again into liquid water. At first, the rate of evaporation will be greater than the rate of condensation, and the density of vapor molecules will increase. Then, as the number of water-vapor molecules in the container increases, the condensation rate will become greater and greater until it equals the rate of evaporation and equilibrium is established. The pressure of the water vapor in equilibrium will be the vapor pressure for that temperature. If we now heat the container to a greater temperature, more liquid will evaporate and a new equilibrium will occur at a higher vapor pressure.

Figure 13-5 is a plot of *pressure* versus *temperature* called a *phase diagram*. The portion of this diagram between points O and C shows the vapor pressure versus the temperature for a constant volume. As we continue to heat the container, the density of the liquid decreases and the density of the vapor increases. At point C in Figure 13-5, these densities are equal. This point is the critical point, whose temperature is labeled T_c on Figure 13-4. (Critical temperatures for various substances are listed in Table 13-4, p. 302.) At this point and above it there is no distinction between the liquid and the gas.*

If we now cool our container, the vapor condenses into a liquid, and we move along curve OC of Figure 13-5 until we reach point O, where the liquid begins to solidify. At this one point, the vapor, liquid, and solid phases of a substance can coexist in equilibrium. This occurs at a unique temperature and pressure called the *triple point*. For water, the triple-point temperature is 273.16 K = 0.01°C and the triple-point pressure is 4.58 mmHg.

At temperatures and pressures below the triple point, the liquid cannot exist. Curve OA on the phase diagram of Figure 13-5 is the locus of pressures and temperatures for which the solid and vapor coexist in equilibrium. The direct change from a solid to a vapor is called *sublimation*. If you put ice cubes in the freezer compartment of a refrigerator (especially a self-defrosting refrigerator), they eventually disappear due to sublimation. Since atmospheric pressure is well above the triple-point pressure of water, equilibrium is never established between the ice and water vapor. The triple-point pressure and temperature of carbon dioxide (CO_2) are 216.55 K and 3880 mmHg. Liquid CO_2 can therefore exist only at pressures above 3880 mmHg = 5.1 atm. At ordinary atmospheric pressures, liquid carbon

---

* The word *vapor* is often used if the temperature is below the critical temperature, and the word *gas* is used if the temperature is above the critical temperature, though there is really no need for such a distinction.

Table 13-3

Vapor Pressure of Water versus Temperature

t, °C	P, mmHg	P, kPa
0	4.581	0.611
10	9.209	1.23
15	12.653	1.69
20	17.535	2.34
30	31.827	4.24
40	55.335	7.38
50	92.55	12.3
60	149	19.9
70	233.8	31.2
80	355	47.4
90	526	70.1
100	760	101.3
110	1074	143.3
120	1489	198.5
130	2026	270.1

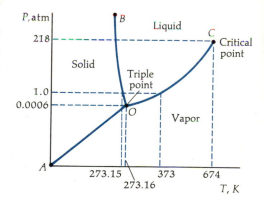

Figure 13-5
Phase diagram for water. The pressure and temperature scales are not linear but have been compressed so as to show the interesting points. Curve OC is the curve of vapor pressure versus temperature. Curve OB is the melting curve and curve OA is the sublimation curve.

dioxide cannot exist at any temperature. Solid carbon dioxide sublimates directly into gaseous CO_2 without going through the liquid phase, hence the name "dry ice."

Curve OB in Figure 13-5 is the melting curve, which separates the liquid and solid phases. In this figure, the OB curve is drawn for a substance such as water for which the melting temperature decreases as the pressure increases. For many other substances, the melting temperature increases as the pressure increases, and the OB curve for such a substance slopes upward to the right from the triple point.

Boiling is an especially rapid form of vaporization in which bubbles form inside the liquid. For this to happen, the vapor in the forming bubble must push the liquid aside to make room and to rise to the surface of the liquid against the applied pressure (usually air pressure). For the bubbles to form, the vapor pressure must be at least equal to the applied pressure. If the applied pressure is increased, as in a pressure cooker, the temperature of the liquid must be raised further before boiling can occur.

Vaporization in general, and boiling in particular, are cooling processes for the liquid left behind. If water is caused to boil by evacuating the air above it, with no heat added, the energy needed is taken from the liquid left behind, cooling it. In fact, it will cool down to the point that ice forms on the top of the boiling water. On the other hand, if the water is boiled in the usual way, with heat input, the cooling effect keeps the temperature of the liquid constant at the boiling point.

Questions

3. At high altitudes, as in the mountains, it takes longer to cook things in boiling water than at sea level. Why?

4. What is the advantage of a pressure cooker?

5. Why is helium so difficult to liquefy?

Table 13-4

Critical Temperatures T_c for Various Substances

Substance	T_c, K
Argon	150.8
Carbon dioxide	304.2
Chlorine	417.2
Helium	5.3
Hydrogen	33.3
Neon	44.4
Nitric oxide	180.2
Oxygen	154.8
Sulfur dioxide	430.9
Water	647.4

13-4   Humidity

Air is made up of nitrogen (about 78 percent), oxygen (about 21 percent), and small amounts of other gases such as argon, carbon dioxide, and water vapor. The pressure exerted by air molecules is the sum of the partial pressures exerted by each of the various gases that make up air. It is found that the partial pressure of any particular gas—nitrogen, oxygen, water vapor, or whatever—in a given volume is the same as if the gas occupied that volume alone. That is, the presence of other gases does not alter the partial pressure of any given gas. This result is known as Dalton's law.

If more water vapor is added to a given volume of air, the partial pressure of the water vapor is increased. When this partial pressure equals the vapor pressure for that temperature, the air is said to be saturated. The water vapor then begins to condense into liquid water if the temperature is above the melting point or into ice crystals (snow or frost) if the temperature is below the melting point. The ratio of the partial pressure of water vapor to the vapor pressure for that temperature is called the *relative humidity,* which is usually expressed as a percentage:

$$\text{Relative humidity} = \frac{\text{partial pressure}}{\text{vapor pressure}} \times 100\%$$ 13-9   Relative humidity

The relative humidity can be increased either by increasing the amount of water vapor in the air at a given temperature or by lowering the temperature and thereby lowering the vapor pressure. The temperature at which the air becomes saturated with water vapor—that is, when the relative humidity equals 100 percent—is called the *dew point.* When the surface of the earth   Dew point cools below the dew point at night (due to radiation, which will be discussed in the next section), dew forms if the dew point is above 0°C or frost forms if it is below 0°C.

Dew drops form on this spider web when the dew point is above 0°C.

Frost forms on a window when the dew point is below 0°C.

Example 13-5   On a humid 20°C day, the dew point is measured by cooling a metal container until moisture forms on its surface. This happens when the temperature of the can is 15°C. What is the relative humidity?

At the dew point of 15°C, the partial pressure of the water vapor in the air equals the vapor pressure, which according to Table 13-3 is 1.69 kPa. This, then, is the original partial pressure of the water vapor at the original temperature of 20°C. Since the vapor pressure at 20°C is 2.34 kPa (Table 13-3), the relative humidity by Equation 13-9 is

$$\text{Relative humidity} = \frac{1.69 \text{ kPa}}{2.34 \text{ kPa}} \times 100\% = 72.2\%$$

13-5   The Transfer of Heat

Thermal energy is transferred from one place to another by three main processes: conduction, convection, and radiation. In heat conduction, thermal energy is transferred by the interactions of molecules, though there is no transport of the molecules themselves. For example, if one end of a solid bar is heated, the atoms in the heated end vibrate with greater energy than do those at the cooler end. Because of the interaction of the more energetic atoms with their neighbors, this energy is transported along the bar. If the solid is a metal, the transport of thermal energy is helped by free electrons, which move throughout the metal, receiving and giving off thermal energy as they collide with the atoms. In a gas, heat is conducted by direct collisions between the gas molecules. Molecules in the warmer part of the gas have higher energies than the average. They lose some of this energy in collisions with molecules from the cooler part of the gas that have lower average energies. These molecules thus gain energy.

In convection, heat is transferred by direct mass transport. For example, if air near the floor is warmed, it expands and rises because of its lower density. Thermal energy in this warm air is thus transported from the floor to the ceiling along with the mass of warm air.

In heat radiation, energy is emitted and absorbed by bodies in the form of electromagnetic radiation. This radiation moves through space with the speed of light. In fact, thermal radiation, light waves, radio waves, television waves, and x-rays are all electromagnetic radiation that differ from one another only in their wavelengths and frequencies. All bodies emit and absorb electromagnetic radiation. When a body is in thermal equilibrium with its surroundings, it emits and absorbs energy at the same rate. If it is warmed to a higher temperature than its surroundings, it radiates away more energy than it absorbs, thus cooling down as its surroundings warm.

Conduction and Convection

Figure 13-6 shows a solid bar of cross-sectional area A. If we keep one end of the bar at a high temperature (for example, in a steam bath) and the other end at a lower temperature (for example, in an ice bath), thermal energy is continually conducted down the bar from the hot end to the cold end. In the steady state, the temperature varies uniformly (if the bar is uniform) from the high-temperature end to the low-temperature end. The rate of change of the temperature along the bar $\Delta T / \Delta x$ is called the *temperature gradient.* Let us consider a small portion of the bar, a slab of thickness Δx, and let ΔT be the temperature difference across the slab. If Q is the amount of heat energy conducted through the slab in some time t, the rate of conduction of thermal

Figure 13-6
(*a*) A conducting bar between two heat reservoirs. (*b*) A section of the bar of length Δx. The rate at which heat is conducted is proportional to the area A of the section and to the temperature difference ΔT and is inversely proportional to the length Δx of the section.

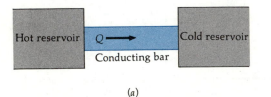

(*a*)

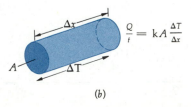

$$\frac{Q}{t} = kA\frac{\Delta T}{\Delta x}$$

(*b*)

energy Q/t is called the heat current I. Experimentally, it is found that the heat current is proportional to the temperature gradient and to the area A:

$$I = \frac{Q}{t} = kA\frac{\Delta T}{\Delta x}$$

13-10   Thermal conduction

where the proportionality constant k is called the *coefficient of thermal conductivity*. It depends on the composition of the bar. In SI units, the heat current is in watts (joules per second) and the thermal conductivity has the units of watts per metre-kelvin, though in some tables the energy may be given in calories or kilocalories and the thickness in centimetres. In practical calculations in the United States for finding the heat conducted through the walls of a room, for example, the heat current is usually expressed in Btu per hour, the area in square feet, the thickness in inches, and the temperature difference in Fahrenheit degrees. The thermal conductivity in this system is given in $Btu \cdot in/h \cdot ft^2 \cdot F°$. Table 13-5 lists values of thermal conductivity for various materials in both sets of units.

Table 13-5
Thermal Conductivities k for Various Materials

Material	k, $W/m \cdot K$	k, $Btu \cdot in/h \cdot ft^2 \cdot F°$
Air (27°C)	0.026	0.18
Ice	0.592	4.11
Water (27°C)	0.609	4.22
Aluminum	237	1644
Copper	401	2780
Gold	318	2200
Iron	80.4	558
Lead	353	2450
Silver	429	2980
Steel	46	319
Oak	0.15	1.02
Maple	0.16	1.1
White pine	0.11	0.78
Brick	0.4–0.9	3–6
Concrete	0.19–1.3	6–9
Cork board	0.04	0.3
Glass	0.7–0.9	5–6
Glass wool	0.042	0.29
Masonite	0.048	0.33
Plaster	0.3–0.7	2–5
Rock wool	0.039	0.27

If we solve Equation 13-10 for the temperature difference, we obtain

$$\Delta T = \frac{\Delta x}{kA} I \qquad\qquad 13\text{-}11$$

The quantity $\Delta x/kA$ is called the *thermal resistance R*.

$$R = \frac{\Delta x}{kA} \qquad\qquad 13\text{-}12 \quad \text{Thermal resistance}$$

Equation 13-11 is then written

$$\Delta T = IR \qquad\qquad 13\text{-}13$$

Equation 13-13 is of the same form as Equation 10-31 for the viscous flow of a fluid through a pipe, except that I now stands for the flow of heat energy rather than the flow of a fluid, and the pressure difference ΔP is replaced by the temperature difference ΔT. (When we study electricity, we will encounter a similar equation known as Ohm's law for the flow of electric charge. In that case, ΔT in Equation 13-13 is replaced by the voltage difference, I becomes the electric current, and R the electric resistance.)

In many practical problems, we are interested in the flow of heat through two or more conductors (or insulators) in series. For example, we may wish to know the effect of adding insulating material of a certain thickness and thermal conductivity to the space between two layers of plasterboard. Figure 13-7 shows two slabs of the same area but of different materials and different thicknesses. Let T_1 be the temperature on the warm side, T_2 be that at the interface between the slabs, and T_3 be that on the cool side. Under conditions of steady-state heat flow, the thermal current must be the same through each slab. This follows from energy conservation; the energy going in must equal that coming out in the steady state. If R_1 and R_2 are the thermal resistances of the two slabs, we have from Equation 13-13 for each slab

$$T_1 - T_2 = IR_1 \quad \text{and} \quad T_2 - T_3 = IR_2$$

Adding gives

$$\Delta T = T_1 - T_3 = (R_1 + R_2)I = R_{eq}I \qquad\qquad 13\text{-}14$$

where the equivalent resistance is

$$R_{eq} = R_1 + R_2 \qquad\qquad 13\text{-}15 \quad \text{Resistors in series}$$

For thermal resistors in series, the equivalent resistance is the sum of the individual resistances. This result can be applied to any number of resistances in series. It is the same result we will obtain later for electric resistances in series.

In the building industry, the thermal resistance in U.S. customary units for a square foot of material is called the R *factor*, which we will designate by R_f. The R factor is merely the thickness divided by the thermal conductivity:

$$R_f = \frac{\Delta x}{k} = RA \qquad\qquad 13\text{-}16 \quad R \text{ factor}$$

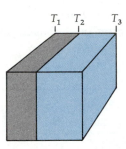

Figure 13-7
Two conducting slabs of different materials in series. The thermal resistance of conductors in series is the sum of the individual thermal resistances. The heat current is the same through each slab.

Table 13-6 lists R factors for several materials. To find the rate of heat conduction per unit area in Btu/h·ft² through several slabs of materials, we divide the temperature difference in Fahrenheit degrees by the sum of the R factors:

$$\frac{I}{A} = \frac{\Delta T}{R_{f,eq}}$$

13-17

Table 13-6

R Factors $\Delta x/k$ and $1/k$ for Various Building Materials

Material	Thickness, in	R_f, h·ft²·F°/Btu	R_f per in, h·ft²·F°/Btu·in
Building board			
Gypsum or plasterboard	0.375	0.32	
Plywood (Douglas fir)	0.5	0.62	
Plywood or wood panels	0.75	0.93	
Particle board, medium density			1.06
Finish flooring materials			
Carpet and fibrous pad			2.08
Tile		0.5	
Wood, hardwood finish	0.75	0.68	
Insulating materials			
Blanket and batt	~3–3.5	11	
Mineral fiber	~6–7	22	
Board and slabs			
Expanded polystyrene, extruded		5	
Molded beads		3.6	
Loose fill			
Cellulosic insulation			3.1–3.7
Sawdust or shavings			2.2
Vermiculite, exfoliated			2.13
Roof insulation			2.8
Roofing			
Asphalt roll roofing		0.15	
Asphalt shingles		0.44	
Windows			
Single-pane		0.9	
Double-pane		1.8	
Triple-pane		2.7	

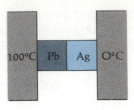

Example 13-6   Two metal cubes, one lead (Pb) and one silver (Ag), with 2-cm edges are arranged as shown in Figure 13-8. Find the total thermal current through the cubes and the temperature at the interface.

We first find the thermal resistance of each cube using Equation 13-12 with values for the conductivity k from Table 13-5. For the lead cube, we have

$$R_{Pb} = \frac{\Delta x}{kA} = \frac{0.02 \text{ m}}{(353 \text{ W/m} \cdot \text{K})(0.02 \text{ m})^2} = 0.142 \text{ K/W}$$

Figure 13-8
Two thermally conducting cubes in series for Example 13-6.

Similarly, the resistance of the silver cube is

$$R_{Ag} = \frac{0.02 \text{ m}}{(429 \text{ W/m} \cdot \text{K})(0.02 \text{ m})^2} = 0.117 \text{ K/W}$$

The total thermal resistance for the two cubes in series is the sum of R_{Pb} and R_{Ag}:

$$R_{eq} = R_{Pb} + R_{Ag} = 0.142 + 0.117 = 0.259 \text{ W/K}$$

The thermal current is then given by

$$I = \frac{\Delta T}{R_{eq}} = \frac{100 \text{ K}}{0.259 \text{ W/K}} = 386 \text{ W}$$

We find the temperature at the interface by noting that the temperature drop between the reservoir at 100°C and the interface must equal the current I times the thermal resistance of the lead cube. Calling this temperature t_{if}, we have

$$100°\text{C} - t_{if} = IR_{Pb} = (386 \text{ W})(0.142 \text{ K/W})$$
$$= 54.8 \text{ K} = 54.8 \text{ C}°$$

The temperature at the interface is thus

$$t_{if} = 100°\text{C} - 54.8 \text{ C}° = 45.2°\text{C}$$

Snow patterns on the roof of a house indicate the conduction—or lack of conduction—of heat through the roof. Note the snow over the unheated garage.

Example 13-7   A 60- by 20-ft roof is made of 1-in pine board with asphalt shingles. If the overlap in the shingles is neglected, how much heat is conducted through the roof when the inside temperature is 70°F and that outside is 40°F? By what factor is the heat loss reduced if 2 in of roof insulation is added?

Since the thermal conductivity of pine board from Table 13-5 is 0.78 in U.S. customary units, the R factor for a board of 1-in thickness is

$$R_f = \frac{\Delta x}{k} = \frac{1 \text{ in}}{0.78 \text{ Btu} \cdot \text{in}/\text{h} \cdot \text{ft}^2 \cdot \text{F}°} = 1.28 \text{ h} \cdot \text{ft}^2 \cdot \text{F}°/\text{Btu}$$

From Table 13-6, the R factor for asphalt shingles is 0.44 h · ft² · F°/Btu. The R factor for the combination is therefore $1.28 + 0.44 = 1.72$ h · ft² · F°/Btu. For a temperature difference of 30 F°, the heat conducted per square foot is

$$\frac{I}{A} = \frac{\Delta T}{R_{f,\text{eq}}} = \frac{30 \text{ F}°}{1.72 \text{ h} \cdot \text{ft}^2 \cdot \text{F}°/\text{Btu}} = 17.4 \text{ Btu}/\text{h} \cdot \text{ft}^2$$

The rate of conduction through the area of 60 by 20 ft = 1200 ft² is

$$I = (17.4 \text{ Btu}/\text{h} \cdot \text{ft}^2)(1200 \text{ ft}^2) = 21{,}000 \text{ Btu}/\text{h}$$

From Table 13-6, the R factor for roof insulation is 2.8 per inch, so for 2 in the R factor is 5.6. Adding the insulation thus increases the R factor from 1.72 to $1.72 + 5.6 = 7.32$. Since $(7.32)/(1.72) = 4.26$, the rate of heat conducted through the roof is decreased by a factor of 4.26. For the temperature difference given, the rate of heat conduction with the insulation will be

$$I = \frac{21{,}000 \text{ Btu}/\text{h}}{4.26} = 4900 \text{ Btu}/\text{h}$$

To calculate the amount of heat leaving a room by conduction in a given time, we need to know how much leaves through the walls, the windows, the floor, and the ceiling. Such a problem involves what are called "parallel paths" for heat flow. The walls, windows, floor, and ceiling all represent independent paths for heat loss to the outside. The temperature difference is the same for each path, but the heat current is different. The total heat current is the sum of the heat currents through each of the independent or parallel paths:

$$I_\text{tot} = I_1 + I_2 + \cdots = \frac{\Delta T}{R_1} + \frac{\Delta T}{R_2} + \cdots$$

or

$$I_\text{tot} = \frac{\Delta T}{R_\text{eq}} \qquad\qquad\qquad \textbf{13-18}$$

where the equivalent thermal resistance in this case is given by

$$\frac{1}{R_\text{eq}} = \frac{1}{R_1} + \frac{1}{R_2} + \cdots \qquad\qquad \textbf{13-19} \qquad \text{\color{blue}Resistors in parallel}$$

(We will encounter this equation again when we study electrical conduction through parallel resistors.)

The thermal conductivity of air is very small compared with that of solid materials, so air is a very good insulator. However, the efficiency of a large air gap—as between a storm window and the inside window—is greatly reduced because of convection. As soon as there is a temperature difference between different parts of the airspace, convection currents act quickly to equalize the temperature, and the effective conductivity is greatly increased. Air gaps of about 1 to 2 cm are optimal. Wider air gaps actually reduce the thermal resistance of a double-pane window because of convection. It is possible to use the insulating properties of air if it can be trapped in small pockets that are separated from each other so that convection cannot take place. For example, goose down is a good thermal insulator because it fluffs up, trapping air so that it cannot circulate and transport heat by convection. Another example is Styrofoam, a cellular material with tiny pockets of air separated by poorly conducting cell walls that prevent convection. The thermal conductivity of such a material is essentially the same as that of air.

If you touch the inside glass surface of a window when it is cold outside, you will observe that the surface is considerably colder than the inside air. The thermal resistance of a window is due mainly to thin insulating air films that adhere to either side of the glass surface. The thickness of the glass has little effect on the overall thermal resistance. An air film typically adds an R factor of about 0.45, so the R factor of a window with N glass layers is approximately 0.9N (because of the two sides of each layer). Under windy conditions, the outside air film may be greatly decreased, leading to a smaller R factor for the window.

It is possible to write an equation for the thermal energy transported by convection and to define a coefficient of convection, but the analyses of practical problems are very difficult and will not be discussed here. The heat transferred from a body to its surroundings is approximately proportional to the area of the body and to the difference in temperature between the body and its surroundings.

Radiation

The third mechanism for heat transfer is radiation in the form of electromagnetic waves. The rate at which a body radiates thermal energy is proportional to the area of the body and to the fourth power of its absolute temperature. This result, found empirically by Josef Stefan in 1879, is written

$$I = e\sigma A T^4 \qquad\qquad \text{13-20}$$

Stefan-Boltzmann law

where I is the power radiated in watts, A is the area, e is a fraction between 0 and 1 called the *emissivity* of the body, and σ is a universal constant called *Stefan's constant*, which has the value

$$\sigma = 5.6703 \times 10^{-8} \text{ W/m}^2 \cdot \text{K}^4 \qquad\qquad \text{13-21}$$

Equation 13-20 was derived theoretically by Ludwig Boltzmann about five years later and is now called the *Stefan-Boltzmann law*.

When radiation falls on an opaque body, part of the radiation is reflected and part is absorbed. Light-colored bodies reflect most of the visible radiation, whereas dark bodies absorb most of it. The radiation absorbed is proportional to the area of the body and to the fourth power of the temperature of the surroundings. For the absorption of radiation, we can write

$$I_a = a\sigma A T^4 \qquad\qquad\qquad 13\text{-}22$$

where a is the *coefficient of absorption*, which like e is a fraction between 0 and 1.

Consider a hot body placed in an environment that is at a lower temperature. The body emits more radiation than it absorbs, so it cools down while the surroundings absorb radiation from the body and warm up. Eventually, the body and the surroundings come to the same temperature and are in thermal equilibrium. Then the body will absorb radiation at the same rate as it emits it. The coefficient of absorption a must therefore be equal to the emissivity e. For the net power radiated by a body at temperature T in an environment at temperature T_0, we can write

$$I_{\text{net}} = e\sigma A(T^4 - T_0^4) \qquad\qquad 13\text{-}23$$

A body that absorbs all the radiation incident upon it has an emissivity equal to 1 and is called a *blackbody*. A blackbody is also an ideal radiator. Materials like black velvet or lampblack come close to being ideal blackbodies, but the best practical realization of an ideal blackbody is a small hole leading into a cavity (see Figure 13-9), for example, a keyhole in a closet door. Radiation incident on the hole has little chance of being reflected back out of the hole before it is absorbed by the walls of the cavity. The concept of an ideal blackbody is important because the characteristics of the radiation emitted by such a body can be calculated theoretically.

Blackbody

Figure 13-9
Cavity approximating an ideal blackbody. Radiation entering the cavity has little chance of leaving the cavity before it is completely absorbed.

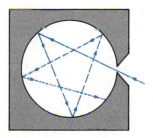

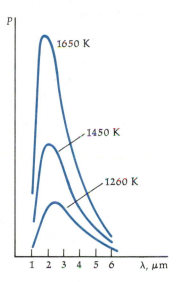

Figure 13-10
Radiated power versus wavelength for radiation from a blackbody. The wavelength of the maximum power varies inversely with the absolute temperature of the blackbody.

At ordinary temperatures (below about 600°C), the thermal radiation emitted by a body is not visible. Most of it is concentrated in wavelengths much longer than those of visible light. (When we study light, we will see that visible light is electromagnetic radiation with wavelengths between about 400 and 700 nm.) As the body is heated, the quantity of energy emitted increases (Equation 13-20), and the energy radiated extends to shorter and shorter wavelengths. Between about 600 and 700°C, there is enough energy in the visible spectrum for the body to glow a dull red; at higher temperatures, it becomes bright red or even "white hot." Figure 13-10 shows the power radiated by a blackbody as a function of wavelength for several different temperatures. The wavelength at which the power is a maximum varies inversely with the temperature, a result known as *Wien's displacement law.* This wavelength has been found to be

$$\lambda_m = \frac{2.898 \text{ mm} \cdot \text{K}}{T} \qquad\qquad 13\text{-}24 \qquad \text{Wien's displacement law}$$

Wien's displacement law is used to determine the temperature of stars from analyses of their radiation.

The calculation of the shape of the spectral-distribution curves shown in Figure 13-10 played an important role in the history of physics. It was the discrepancy between the theoretical calculations of these curves using classical thermodynamics and the experimental measurements of them that led to the first ideas of the quantization of energy by Max Planck in 1897. We will discuss this matter further in Chapter 29.

Example 13-8   The surface temperature of the sun is about 6000 K. If the sun is assumed to be a blackbody radiator, at what wavelength λ_m would its spectrum peak?

From Wien's displacement law (Equation 13-24), we have

$$\lambda_m = \frac{2.898 \text{ mm} \cdot \text{K}}{6000 \text{ K}} = 483 \times 10^{-9} \text{ m} = 483 \text{ nm}$$

This wavelength is in the middle of the visible spectrum. The blackbody radiation spectrum describes the sun's radiation fairly well, so the sun is indeed a good example of a blackbody.

Example 13-9   Calculate λ_m for a blackbody at room temperature, $T = 300$ K.

From Equation 13-24, we have

$$\lambda_m = \frac{2.898 \text{ mm} \cdot \text{K}}{300 \text{ K}} = 9.66 \times 10^{-6} \text{ m} = 9660 \text{ nm}$$

This is an infrared wavelength, far longer than those visible to the eye. This obscures the fact that surfaces not black to our eyes may act as blackbodies for infrared radiation and absorption. For example, it has been found experimentally that the skin of all human races is black to infrared radiation; hence, its emissivity is 1.00 for its own radiation process.

Example 13-10   Calculate the net loss in radiated energy of a naked person in a room at 20°C, assuming the person to be a blackbody with a surface area of 1.4 m² and a surface temperature of 33°C = 306 K. (The surface temperature of the human body is slightly less than the internal temperature, 37°C, because of the thermal resistance of the skin.)

Using Equation 13-23, we have

$$I_{net} = (1)(5.67 \times 10^{-8} \text{ W/m}^2 \cdot \text{K}^4)(1.4 \text{ m}^2)[(306)^4 - (293)^4]$$

$$= 111 \text{ W}$$

This is a large energy loss. It is approximately equal to the basic metabolism rate of 120 watts we calculated in Section 6-7 assuming a food intake of 2500 kilocalories per day. We protect ourselves from such a great energy loss by wearing clothing, which, because of its low thermal conductivity, has a much lower outside temperature and therefore a much lower rate of heat radiation.

Thermograph of a boy and his dog. The bright regions are those of high temperature. Note the dog's cold nose.

In practice, all three mechanisms of heat transfer often occur simultaneously, though one may be more effective than the others. An ordinary space heater with a metal heating element, for example, heats partly by direct radiation and partly by convection. Because the emissivity of metals is fairly small, the main mechanism is convection. The heater warms the nearby air, which then rises and is replaced by cooler air. Often a fan is included in the heater to speed the convection process. (This is called forced convection.) On the other hand, a space heater with a quartz filament heats mainly by radiation because the filament has a large emissivity. Coffee in a ceramic coffee cup cools by radiation as well as conduction (and evaporation) because the cup has a relatively large emissivity, whereas if a metal cup is used, the cooling is mainly by conduction (and evaporation).

Questions

6. In a cool room, a metal or marble table top feels much colder to the touch than a wood surface does even though they are at the same temperature. Why?

7. Which heat-transfer mechanisms are most important in the warming effect of a fire in a fireplace? In the transfer of energy from the sun to the earth?

8. Do you think that the fact that the radiant energy from the sun is concentrated in the visible range of wavelengths is an accident or not? If not, what is the reason?

Solar Energy

Laurent Hodges
Iowa State University

Until the late nineteenth century, we in the United States depended on renewable solar energy in its many forms as our sole energy resource. Sunlight and wood provided warmth, beasts of burden provided mechanical power, wind enabled ships to transport passengers and freight, and wind and water power provided mechanical energy for such purposes as grinding grain. The United States was predominantly a solar-energy economy until the 1880s, when coal displaced wood as the major energy resource. By that time, Great Britain and other European countries had already been relying on coal for some time. (Coal and the other fossil fuels are nonrenewable energy resources, although their ultimate source was also solar energy.)

Forms of Solar Energy

"Solar energy" includes not only direct sunlight but also several indirect forms, such as photosynthetic fuels and the energy from water power and wind power (see Table 1). These are regarded as solar energy because their energy ultimately derives from the sun: the energy of wood and other plant materials is solar energy fixed by photosynthesis; wind power is solar energy because it is the heating of land, air, and water by solar radiation that produces winds; and water power is possible because sunlight drives the hydrological cycle whereby water evaporates, rains on high ground, and returns to the oceans by gravity.

Advantages and Disadvantages of Solar Energy

Solar energy was neglected as an energy resource for many years. Today there is renewed interest in it because of its many advantages. It is available to at least some extent everywhere in the world, unlike the fossil and nuclear fuels. Solar energy itself costs nothing and is thus exempt from rising energy prices. It can be used in many different ways, to provide heating, cooling, lighting, mechanical power, electricity, and transportation. Few environmental problems are created by its use.

Solar energy also has some disadvantages. It is not highly concentrated, although enough solar energy for some important purposes can be collected by using a modest area of land or a rooftop or wall. It is intermittent, its flow being interrupted by nights and cloudy days, but there are some

Table 1

Forms of Solar Energy

Form	Explanation	Use

Solar radiation and solar thermal energy	From direct sunlight	Heating buildings and water; process heat for industry and agriculture
		Cooling buildings
		Generating electricity:
		   Photovoltaic cells
		   Power towers
		   Ocean thermal-energy conversion

Photosynthesis	Solar energy converted to chemical energy of plants and fossil fuels	Solid fuels (wood, coal) Liquid fuels (alcohols) Gaseous fuels (methane)
Water power	Sunlight drives hydrological cycle (water evaporates, rains on high ground, and returns to ocean by gravity)	Generating electricity (hydroelectricity) Mechanical power (water mills)
Wind power	Heating of land, air, and water by solar radiation produces winds	Generating electricity (wind generators) Mechanical power (windmills) Sailing vessels

good and often inexpensive ways of storing it for these periods. Many solar-energy systems require a large capital investment, but the amortization costs are often more than offset by the savings in energy costs.

Despite these disadvantages, it is likely that in the next 50 years, solar energy in its various forms will become an important energy resource in the United States and much of the world—a world that can no longer afford to waste nonrenewable energy resources.

Uses of Solar Energy

Some of the present and possible uses of solar energy are listed in Table 1. They are currently the subject of considerable research and development by both government and industry. As will be discussed shortly, some of these uses are clearly economical today, others appear to be marginally economical, and still others need considerable development before they are competitive.

Hydroelectricity has been a significant source of electricity for many decades and is the cheapest type of electricity available today. Unfortunately, its potential is limited by the availability of suitable sites for and the environmental disruption caused by hydroelectric reservoirs.

The next most economical type of solar energy appears to be the solar space heating of small buildings, particularly passive solar space heating. Solar heating involves the use of (1) collectors to collect solar radiation, (2) thermal storage to store the excess energy for use at night or on cloudy days, and (3) a distribution system to move solar energy from the collector to storage or living space and from storage to living space.

When the distribution system requires external energy inputs (such as electricity for fans or pumps), the solar-energy system is called an *active system;* when the distribution is by natural means (conduction, radiation, and natural convection), the system is called a *passive system.* Passive systems generally use large south-facing windows as the collector and large amounts of concrete, masonry, or water for thermal storage. As will be discussed later, a well-insulated passive solar home in the northern states can have extremely low heating bills. Passive solar space-heating systems add very little to the cost of a home and are virtually maintenance-free; in the northern states, they are the preferred solar space-heating system for new construction and can be successfully used in many existing buildings.

Active systems have also been used for homes, but they tend to be more expensive and require occasional maintenance. They are most economical for water heating, especially if the only alternative is electric heating, and are sometimes economical for the retrofit of existing homes for space heating. They are also becoming feasible for industrial- and agricultural-process heat applications.

Since space heating and water heating are the two largest sources of home energy consumption nationally, the fact that solar-energy systems are economical for these purposes means that solar energy can become a very important energy resource in the near future.

Wind-generated electricity is marginally economic today. Although a wind electric system for a residence costs several thousand dollars, it may be economical in locations where electricity is very expensive. In most places the high capital costs make wind-generated electricity more expensive than utility electricity; however, it may well prove to be cheaper over the lifetime of the system if electricity continues to increase in price and the wind system is reliable. Current research and development includes small wind generators for individual use and large generators for incorporation into electric utility grids.

Photovoltaic or solar cells are semiconductor devices usually (but not always) made of silicon. They produce electricity directly from solar radiation, acting much like a battery when the sun is shining on them. Although their price has dropped substantially in recent years, they are still far too expensive to be economical for home use. Some scientists are optimistic that efficient solar cells will be so cheap by the end of the century that homeowners can install them on or near their homes to meet all or part of their electricity requirements. It appears that the most economical method of solar air conditioning in the future will be the use of ordinary electric air conditioners operating from photovoltaic cells. New buildings should be designed so that photovoltaic cells can easily be added when they have become economical. Large photovoltaic power plants in space may someday supply electric power to earth; the energy would be transported to earth as microwave energy.

There are several ways in which solar energy might be used for transportation, which accounts for about 25 percent of United States energy consumption. Solar electricity from photovoltaic cells or other solar electricity might be used with electric vehicles when they become more common. Solar electricity can also be used to generate hydrogen (by the electrolysis of water), which is a possible future fuel for transportation. Biological materials can be converted into alcohol fuels, which are already being mixed, up to 10 percent, with gasoline to form gasohol. None of these methods is currently economical.

Essay: Solar Energy (continued)

Solar Radiation

The solar radiation that lands on the earth resembles the spectral distribution of a blackbody at about 6000 K, the temperature of the sun's surface. About 47 percent of this solar energy is in the visible spectrum, 45 percent is in the infrared, and 8 percent is in the ultraviolet.

The earth is 150×10^6 km from the sun, so it intercepts only a small part—about 0.5 billionths—of the sun's total radiation. The average energy flux at this distance on a surface perpendicular to the sun's rays is about 1353 W/m². Since the earth-to-sun distance varies by about six percent over the course of the year, from perihelion in early January to aphelion in early July, this "solar constant" varies from about 1308 to 1398 W/m².

These values apply to the top of the earth's atmosphere. The atmosphere reduces the flux considerably, typically to 1 kW/m² on a clear day and a few watts per square metre on an overcast day. When the sun's rays intercept a surface at an angle, the flux onto the surface is further reduced for geometrical reasons.

The total solar radiation reaching the top of the atmosphere is about 1.7×10^{17} W, only about half of which actually reaches the ground. Distributed over the whole globe, this amounts to about 170 W/m² averaged over the whole year, day and night. Thus, one square metre of the earth's surface receives about 4 kW·h daily or 1460 kW·h annually. For the land area of the United States, the average is a little higher, about 4.7 kW·h/m² daily or a total of over 10^{16} kW·h annually for the entire country; this is about 600 times greater than the total energy consumption in the United States.

The total amount of solar radiation falling on a horizontal surface is much greater in summer, when the days are longer and the sun's altitude is very high at midday, than it is in winter. In the Midwest, for example, a horizontal surface might receive 30 MJ/m², or a little more, on a sunny day in June but rarely over 10 MJ/m² on sunny days in December. The situation is different for a vertical south-facing surface, which receives far more in winter (when the sun is lower in the sky) than in summer. A vertical south-facing solar collector thus is ideal for collecting solar energy for winter space heating. A south-facing surface tilted up at 60° or so above the horizontal has considerably less variation over the course of a year and would thus be well suited for collecting solar energy on a year-round basis, as for water heating. These results are illustrated in Figure 1, which shows the daily incident solar energy per unit area for surfaces of various inclinations. Figure 2 shows a rather large array of solar cells for collecting solar radiation.

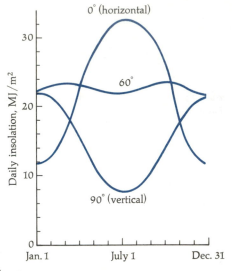

Figure 1
Plot of incident solar energy per unit area for surfaces of various inclinations.

Figure 2
This array of solar cells at Barstow, California, produces 10 MW of thermal electric power.

Passive Solar Heating: A Case Study

The Hodges residence (Figure 3) in Ames, Iowa, has a passive solar system. It consists of about 40 m² of vertical south-facing double-pane glass serving as solar collector and over 100 Mg of concrete floors and walls serving as thermal storage.

The heat transfers in the house are very simple. Solar radiation passes through the glass into the house. Much of it is directly absorbed by the surface of the concrete and then conducted into it. Part of the solar radiation heats the interior air, which comes in contact with the concrete and also transfers heat into it. When the air begins to cool down, the concrete radiates heat back into the air. This occurs naturally, without any controls or mechanical system. Since the radiant-heating system is the Rolls-Royce of heating systems, this natural radiation makes for an extremely comfortable interior. All parts of the house are within a few feet of an exposed radiant concrete surface. Natural convection in the interior of the house tends to keep the upper level a few degrees warmer, which is desirable since the upper level is the living area and the bedrooms are on the lower level.

On a sunny day the 40 m² of south-facing glass may collect 14 MJ/m², or a total of 560 MJ. The heat loss of the house is about 300 W/K, so that the heat loss would total about 500 MJ on a 0°C day and 1000 MJ on a −20°C day. Internal heat (from lights, cooking, appliances, people, etc.) contributes about 100 MJ per day, reducing the needed heat to about 400 MJ or 900 MJ, respectively. Since the thermal storage has a heat capacity of about 100 MJ/K, a change of 5 C° (9 F°) in its temperature corresponds to an exchange of 500 MJ in or out of storage. On a sunny day averaging 0°C, the 560 MJ collected would provide an excess of 160 MJ, which would be absorbed by a 1.6 C° (2.9 F°) temperature rise in the storage.

These figures show that the house is very slow to change in temperature, since its time constant (ratio of heat capacity to heat loss) is about 4 days. Whenever the indoor temperature drops too low because of cloudiness or cold weather, auxiliary heat is needed. Much of the winter the interior temperature is in the range of 20 to 25°C, and the furnace is often not used at all for several consecutive days.

In its first seven winters, the Hodges residence required about 120 MJ of heat (less than $1000 worth of natural gas). The family uses more energy for water heating than for space heating.

Figure 3
The Hodges residence.

Summary

1. The coefficient of linear expansion is the ratio of the fractional change in length to the change in temperature:

$$\alpha = \frac{\Delta L/L}{\Delta T}$$

2. The coefficient of volume expansion, called the expansivity, is the ratio of the fractional change in volume to the temperature change. The expansivity is three times the coefficient of linear expansion.

$$\beta = \frac{\Delta V/V}{\Delta T} = 3\alpha$$

The expansivity of an ideal gas equals the reciprocal of the absolute temperature:

$$\beta = \frac{1}{T}$$

3. The heat needed to melt a body is proportional to the mass of the body and to the latent heat of fusion L_f:

$$Q = mL_f$$

The heat needed to vaporize a liquid is proportional to the mass of the liquid and to the latent heat of vaporization L_v:

$$Q = mL_v$$

Both melting and evaporation occur at a constant temperature. For water, $L_f = 333.5$ kJ/kg and $L_v = 2.26$ MJ/kg. The heat required to melt one gram of ice or to vaporize one gram of water are both large compared with the heat required to raise the temperature of one gram of water by one degree.

4. The van der Waals equation of state describes the behavior of real gases over a wide range of temperatures and pressures, taking into account the space occupied by the gas molecules themselves and the attraction of the molecules for one another. The van der Waals equation for one mole is

$$\left(P + \frac{a}{V^2}\right)(V - b) = RT$$

5. Vapor pressure is the pressure of a vapor in equilibrium with its liquid at a given temperature. If the external pressure equals the vapor pressure, the liquid boils at that temperature.

6. The direct change of a solid into a vapor is called sublimation.

7. The triple point is the temperature and pressure at which the gas, liquid, and solid phases of a substance can coexist. The triple-point temperature and pressure for water are 273.16 K (0.01°C) and 4.58 mmHg (0.006 atm). At temperatures and pressures below the triple point, the liquid phase of a substance cannot exist. The triple-point temperature and pressure of carbon dioxide are 216.55 K (-56.6°C) and 3880 mmHg (5.1 atm), so liquid carbon dioxide does not exist under ordinary conditions.

8. Relative humidity is the ratio of the partial pressure of water vapor in air to its vapor pressure at that temperature. For a given amount of water vapor, the temperature at which the partial pressure equals the vapor pressure is called the dew point. At the dew point temperature, the water vapor begins to condense into liquid water or into ice crystals depending on whether the dew point temperature is above or below 0°C.

9. The three mechanisms of heat transfer are conduction, convection, and radiation.

10. The rate of conduction of heat is given by

$$I = \frac{Q}{t} = kA\frac{\Delta T}{\Delta x}$$

where I is the heat current and k is the coefficient of heat conduction. This equation can be written

$$\Delta T = IR$$

where R is the thermal resistance

$$R = \frac{\Delta x}{kA}$$

The thermal resistance for a unit area of a slab of material is called the R factor, R_f.

$$R_f = RA = \frac{\Delta x}{k}$$

The equivalent thermal resistance of a series of thermal resistances equals the sum of the individual resistances.

11. The net thermal power radiated by a body at temperature T in an environment at temperature T_0 is given by

$$I_{net} = e\sigma A(T^4 - T_0^4)$$

Materials that are good heat absorbers are good heat radiators. A blackbody is a body that absorbs all the radiation incident upon it. It is also a perfect radiator.

12. The spectrum of electromagnetic energy radiated by a blackbody has a maximum at a wavelength λ_m, which varies inversely with the absolute temperature of the body. This is known as Wien's displacement law:

$$\lambda_m = \frac{2.898 \text{ mm} \cdot \text{K}}{T}$$

Suggestions for Further Reading

Allen, Philip B.: "Conduction of Heat," *The Physics Teacher*, vol. 21, no. 9, 1983, p. 582.

This is an advanced article on our present understanding of heat conduction by solids and gases.

Velarde, Manuel G., and Christiane Normand: "Convection," *Scientific American*, July 1980, p. 92.

Interesting illustrations and a discussion of Lord Rayleigh's theory of convection make this article useful, though the modern theories are quite mathematical.

Wilson, Mitchell: "Count Rumford," *Scientific American*, October 1960, p. 158.

The terms "specific heat" and "latent heat" originated during the time of the predominance of the caloric theory of heat, which was dealt heavy blows by the experiments of this highly unusual man.

Review

A. Objectives: After studying this chapter, you should:

1. Be able to calculate the linear expansion and volume expansion of a substance given the change in its temperature.

2. Be able to work calorimetry problems that include latent heats of fusion and of vaporization.

3. Be able to determine the relative humidity at a given temperature from a knowledge of the vapor pressure and the dew point.

4. Be able to calculate the rate of heat conduction for various thermal resistors in series or in parallel.

5. Be able to state the Stefan-Boltzmann law of radiation and use it to calculate the power radiated by an object at a given temperature.

6. Be able to state the Wien displacement law and use it to relate the absolute temperature to the wavelength at which the power radiated by a blackbody is maximum.

B. Define, explain, or otherwise identify:

Coefficient of linear expansion, p. 293
Expansivity, p. 293
Phase transition, p. 296
Latent heat of fusion, p. 297
Latent heat of vaporization, p. 297
Vapor pressure, p. 300

Critical temperature, pp. 300–301
Triple point, p. 301
Sublimation, p. 301
Relative humidity, pp. 302–303
Dew point, p. 303
Thermal resistance, p. 306
R factor, p. 306
Stefan-Boltzmann law, p. 310
Emissivity, p. 310
Blackbody, p. 311
Wien's displacement law, p. 311

C. True or false: If the statement is true, explain why. If it is false, give a counterexample.

1. All materials expand when heated.

2. During a phase change, the temperature remains constant.

3. The temperature at which water boils depends on the pressure.

4. The vapor pressure of a gas depends on the temperature.

5. The melting temperature of a substance depends on the pressure.

6. The emissivity and the coefficient of absorption of any body must be equal.

Exercises

Section 13-1   Thermal Expansion

1. A steel ruler has a length of 30 cm at 20°C. What is its length at 100°C?

2. Use Table 13-1 to find the coefficient of volume expansion for steel.

3. A 100-m long bridge is built of steel. If it is built as a single, continuous structure, how much will its length change from the coldest winter days (−30°C) to the hottest summer days (40°C)?

4. The experimental value of β for N_2 gas at 0°C and 1 atm is 0.003673 K^{-1}. Compare this with the theoretical value $\beta = 1/T$ for an ideal gas.

5. A car has a 60-L steel gas tank filled to the top with gasoline whose temperature is 10°C and whose coefficient of volume expansion is $\beta = 0.900 \times 10^{-3} \ K^{-1}$. Taking the expansion of the steel tank into account, find how much gasoline spills out when the car is parked in the sun and its temperature rises to 25°C.

Section 13-2   Change of Phase and Latent Heat

6. How much heat is needed to melt a 20-g piece of lead that is originally at 300 K?

7. How much heat is needed to melt a 20-g piece of copper that is originally at 300 K?

8. A 200-g piece of ice at 0°C is put into 500 g of water at 20°C. The system is in a container of negligible heat capacity that is insulated from its surroundings. (a) What is the final equilibrium temperature of the system? (b) How much of the ice melts?

9. A 50-g piece of ice at 0°C is placed in 500 g of water at 20°C as in Exercise 8. What is the final temperature of the system assuming no heat loss to the surroundings?

10. How much heat is removed when 100 g of steam at 150°C is cooled and frozen into 100 g of ice at 0°C? (Take the specific heat of steam to be 2.01 kJ/kg·K.)

11. A 50-g piece of aluminum (specific heat 0.90 kJ/kg·K) at 20°C is placed in a large container of liquid nitrogen at its normal boiling point of 77 K. How much nitrogen is vaporized in cooling the aluminum to 77 K?

12. In a container of negligible heat capacity, 300 g of steam at 100°C is added to 300 g of ice at 0°C. (*a*) What is the final temperature of the system? (*b*) How much steam remains?

13. A lead bullet initially at 30°C just melts upon striking a target. Assuming that all of the initial kinetic energy of the bullet goes into internal energy of the bullet to raise its temperature and melt it, calculate its initial speed.

14. An arctic explorer decides to conserve his fuel by using his hand-cranked radio generator to heat water for his morning cup of coffee. He winds a piece of wire into a heating coil and connects it to his generator. He then immerses it in 250 g of snow at −30°C in a Styrofoam cup. Neglecting the heat absorbed by the cup or lost in other ways, how long does he have to crank the generator to raise the snow to the boiling point if he works at 40 W (which is about the limit for steady work with the arms)? (Take the specific heat of snow to be 2.05 kJ/kg·K.)

Section 13-3   The van der Waals Equation and Liquid-Vapor Isotherms

15. Which gases in Table 13-4 cannot be liquefied by applying pressure at 20°C?

Section 13-4   Humidity

16. On a certain day the temperature is 30°C and the relative humidity is 80 percent. What is the partial pressure of the water vapor in the air?

17. If the temperature is 20°C and the dew point is 10°C, what is the relative humidity?

18. If the partial pressure of water vapor is 3.00 kPa when the temperature is 30°C, what is the relative humidity?

Section 13-5   The Transfer of Heat

19. A copper bar 2-m long has a circular cross section of radius 1 cm. One end is kept at 100°C and the other at 0°C, and the surface is insulated so that there is negligible heat loss through the surface. Find (*a*) the thermal resistance of the bar, (*b*) the thermal current I, (*c*) the temperature gradient $\Delta T/\Delta x$, and (*d*) the temperature 25 cm from the hot end.

20. A slab of insulation 20 by 30 ft has an R factor of 11. How much heat (in Btu per hour) is conducted through the slab if the temperature on one side is 68°F and that on the other side is 30°F?

21. Two metal cubes, one copper (Cu) and one aluminum (Al), with 3-cm edges are arranged as shown in Figure 13-11. Find (*a*) the thermal resistance of each cube, (*b*) the thermal resistance of the two-cube system, (*c*) the thermal

current I, and (*d*) the temperature at the interface of the two cubes.

Figure 13-11
Two metal cubes in series for Exercises 21 and 23.

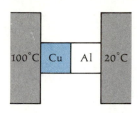

22. A house has a window area of 300 ft² of single-pane glass. Calculate the rate of heat loss by conduction in Btu per hour through the windows if the inside temperature is 68°F and the outside temperature is −10°F. (*Hint:* Use Table 13-6). Why is the thickness of the glass unimportant for this calculation?

23. Find the temperature at the interface of the two cubes in Exercise 21 if the cubes are interchanged.

24. The same metal cubes of Exercise 21 are arranged as shown in Figure 13-12. Find (*a*) the thermal current carried by each cube from one reservoir to the other, (*b*) the total thermal current, and (*c*) the equivalent thermal resistance of the two-cube system.

Figure 13-12
Two metal cubes in parallel for Exercise 24.

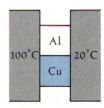

25. The thermal conductivity of an insulating material is measured by constructing a cubical box of side 0.5 m and thickness 2 cm out of the material. When a 135-W heater is placed inside the box, the steady-state inside temperature is greater than the outside temperature by 60°C. Calculate the thermal conductivity of the material.

26. If the absolute temperature of a blackbody is doubled, by what factor does its total radiated power increase?

27. Calculate λ_m for a human blackbody radiator, assuming the skin-surface temperature to be 33°C. (Radiation of this wavelength is called *infrared radiation*. It lies just outside the visible spectrum toward the long-wavelength side and can be detected by infrared detectors.)

28. Calculate the temperature of a blackbody that radiates energy with a peak in the spectrum at (*a*) $\lambda_m = 3$ cm (microwave region) and (*b*) $\lambda_m = 3$ m (FM radio waves).

Problems

1. A mercury thermometer is made from an ordinary glass tube of inside diameter 0.60 mm. The distance between the ice point and the steam point is to be 20.0 cm. Find the volume of mercury needed in the bulb and tube.

2. Steam at 100°C is passed into a flask containing 100 g of ice at -10°C. Neglecting any heat loss from the flask, determine the mass of water at 100°C that will finally be produced in the flask.

3. A 1-kW electric heater has heating wires that are "red hot" at a temperature of 900°C. Assuming that 100 percent of the heat output is due to radiation and that the wires act as blackbody radiators, find the effective area of the radiating surface. Take the temperature of the surroundings to be 20°C.

4. In this problem you are to derive the result that $\beta = 3\alpha$ for small increases in temperature. Suppose you have a cube of side L and volume $V = L^3$ that is heated so that its temperature change is ΔT and each side expands by ΔL. Compute

$$V + \Delta V = (L + \Delta L)^3 = L^3(1 + \Delta L/L)^3$$

and show that, if $\Delta L/L$ is small,

$$\Delta V \approx V(3\,\Delta L/L) = 3\alpha V\,\Delta T$$

5. If 500 g of molten lead at 327°C is poured into a cavity in a large block of ice at 0°C, how much ice melts?

6. The surface temperature of the filament of an incandescent lamp is 1300°C. If the input electric power is doubled, what will the temperature become? (Treat the filament as a blackbody, and neglect the temperature of the surroundings.)

7. A steel tube has an outside diameter of 3.000 cm at room temperature (20°C). A brass tube has an inside diameter of 2.997 cm at the same temperature. To what temperature must the ends of the tubes be heated if the steel tube is to be inserted into the brass tube?

8. One way to construct a device with two points whose separation remains the same in spite of temperature changes is to use rods with different expansion coefficients (see Figure 13-13). The two rods are bolted together at one end. Show that the distance L will not change with temperature if the lengths L_A and L_B are chosen so that $L_A/L_B = \alpha_B/\alpha_A$. If material B is steel, material A is brass, and $L_A = 250$ cm at 0°C, what is the value of L?

Figure 13-13
Problem 8.

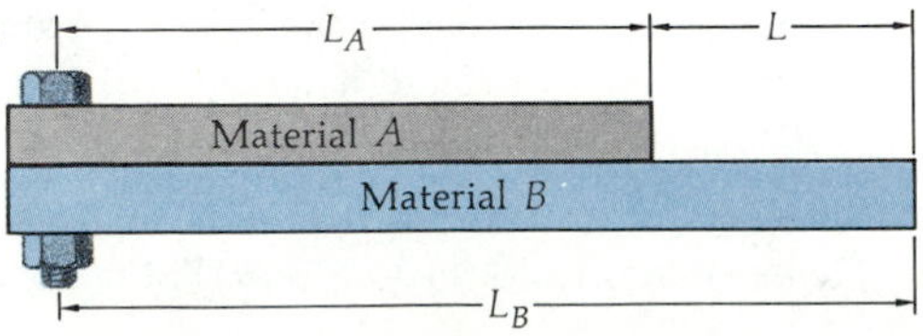

9. A spherical balloon filled with helium in a house at 20°C has a diameter of 36 cm. Find the diameter when the balloon is taken outside at a temperature of -10°C, assuming that the balloon has negligible elastic forces so that the internal pressure is always 1.0 atm.

10. A piece of ice is dropped from a height h. Find the minimum value of h such that the ice melts when it makes an inelastic collision with the ground. Is it reasonable to neglect the variation in the acceleration of gravity in doing this problem? What effect would air resistance have on your answer?

11. Steam at 100°C is added to a 400-g block of ice at 0°C until the ice melts and the water temperature is 60°C. What is the mass of the water produced at that temperature?

12. Use the values in Table 13-3 to draw a graph of the vapor pressure of water versus temperature. From your graph find (a) the temperature at which water boils on a mountain where the atmospheric pressure is 70 kPa, (b) the temperature at which water boils when the pressure is 2 atm, and (c) the pressure at which the temperature of boiling water is 115°C.

13. Use your graph from Problem 12 to find (a) the relative humidity when the temperature is 25°C and the dew point is 5°C, (b) the dew point when the temperature is 20°C and the relative humidity is 85 percent, and (c) the temperature at which the relative humidity is 60 percent when the dew point is 10°C.

14. An insulated cylinder with a movable piston to maintain constant pressure initially contains 100 g of ice at -10°C. Heat is supplied to the cylinder at a constant rate by a 100-W heater. Draw a graph showing the temperature of the cylinder contents as a function of time, starting at time $t = 0$ when the temperature is -10°C and ending when the temperature is 110°C. (Use 2.0 kJ/kg·K for the specific heat of ice and steam and neglect the heat capacity of the cylinder.)

15. A 200-g aluminum calorimeter can contains 500 g of water at 20°C. A 100-g piece of ice at -20°C is placed in the can. (a) Find the final temperature of the system, assuming no heat losses. (Take 2.0 kJ/kg·K for the specific heat of ice.) (b) A second 200-g piece of ice at -10°C is added. How much ice remains in the system after it reaches equilibrium? (c) Would your answer to part (b) be different if both pieces were added at the same time?

16. A saucepan 15 cm in diameter has boiling water at 100°C resting on the upper surface of the bottom and a 2000°C flame playing on its lower surface. If the pan bottom is made of copper 3.00 mm thick, how long will it take to boil away 0.8 L of water after it reaches the boiling point? (The answer obtained is ridiculously small because the bottom of the pan will not reach 2000°C. The heat from the flame will pass through the pan much too fast to heat the bottom to 2000°C. See Problem 19.)

17. The *solar constant* is the power per unit area received from the sun at the earth on an area perpendicular to the sun's rays. Its value at the upper atmosphere of the earth is about 1.35 kW/m². Calculate the effective surface temperature of the sun, assuming it to be a blackbody. (The radius of the sun is 696 Mm.)

18. Using the solar constant from Problem 17, (a) calculate the total radiant power per unit area emerging from the sun's surface. (b) Calculate the power radiated per unit mass of the sun ($M_s = 1.99 \times 10^{30}$ kg). Compare this with the power per unit mass generated by a human of mass 70 kg who generates 2500 kcal per day.

19. A copper-bottomed saucepan containing 0.8 L of boiling water boils dry in 10 min. Assuming that all the heat flows through the flat copper bottom of diameter 15 cm and thickness 3.00 mm, calculate the temperature of the outside of the copper bottom while some water is still present.

20. A room is maintained at 0°C by brine at −16°C circulating through copper pipes with a wall thickness of 1.5 mm. By what factor is the heat transfer reduced when the pipes are coated with a 5-mm layer of ice?

21. A picnic cooler made of Styrofoam is a simple bucket with a tightly fitting lid of the same material. It has an effective area of 1800 cm² and an average thickness of 2.0 cm. The thermal conductivity is 0.05 W/m·C°. The cooler is filled with 3.0 kg of water and 2.0 kg of ice, all at 0°C. The temperature of the outside surface is at 15°C and remains at that value. (a) What is the heat current into the interior? (b) How much time will be needed to melt all of the ice?

22. A 0.2-mm-diameter steel wire stretched between two rigid supports 30 cm apart is under a tension of 36 N at 25°C. If the breaking stress is 1.2 GN/m², find the temperature at which the wire will snap as it cools.

23. A piece of copper of mass 100 g is heated in a furnace at temperature t. The copper is then inserted into a copper calorimeter of mass 150 g containing 200 g of water initially at 16°C. The final temperature after equilibrium is established is 38°C. When the calorimeter and contents are weighed, it is found that 1.2 g of water has evaporated. What was the temperature t of the furnace?

24. Three metal cubes with edges of 3 cm, one of lead (Pb), one of copper (Cu), and one of aluminum (Al), are arranged as shown in Figure 13-14. The heat reservoirs are maintained

at the temperatures indicated. After a steady state is reached, find (a) the temperatures t_1 and t_2 at the interfaces and (b) the temperatures t_1 and t_2 if the lead and copper blocks are interchanged.

Figure 13-14
Problem 24.

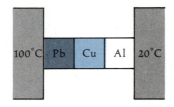

25. Liquid helium is stored at its boiling point (4.2 K) in a spherical can. The can is separated by a vacuum space from a surrounding shield maintained at the temperature of liquid nitrogen (77 K). If the can is 30 cm in diameter and is blackened on the outside so that it acts as a blackbody, how much helium boils away per hour?

26. For the phase diagram for water given in Figure 13-15, state what changes (if any) occur for each line segment, *AB*, *BC*, *CD*, and *DE*, in (a) volume (small increase, large increase, and so forth) and (b) phase (give the new phase). (c) For what type of substance would *OH* be replaced by *OG*? (d) What is the significance of point *F*?

Figure 13-15
Phase diagram for Problem 26.

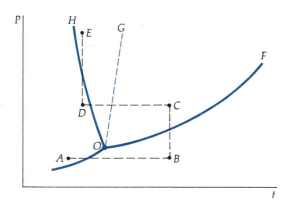

27. A steel bar of radius 2.2 cm and length 60 cm is jammed horizontally perpendicular to two stationary vertical concrete walls at a temperature of 20°C. The bar is heated with a blow torch to 60°C. Find the force exerted by the bar on each wall.

CHAPTER FOURTEEN

The Availability of Energy

The motive power of heat is independent of the agents employed to realize it. The efficiency is fixed solely by the temperatures of the bodies between which, in the last resort, the transfer of heat is effected.

SADI CARNOT

As we saw in the last chapter, the first law of thermodynamics is really quite simple once you are familiar with the terminology. It says merely that work, heat flow, and changes in internal energy are all interchangeable. But now we come to the second law of thermodynamics, which is another matter altogether. In various guises, it pops up all over the fields of science and engineering— even in communications and computer science. In fact, many people agree with the English novelist C. P. Snow that one mark of a well educated person, no matter what their field, is a familiarity with the second law of thermodynamics. It explains why it is difficult to tap the inexhaustible thermal energy of the tropic seas, why those spool-shaped cooling towers are needed at power stations even though they complicate weather patterns and threaten the water supplies with thermal pollution, and why the most modern power stations are only about 40 percent efficient. Understanding the second law requires some hard thinking, but be assured, your efforts will be repaid.

There has been much debate in recent years about the energy crisis. Since 1974, the price of oil has fluctuated wildly, increasing more than tenfold, then dropping by more than 50 percent, and so on. This has played a major role in the economy of the United States, which consumes more oil per person than any other country. We are continually told to conserve energy. Yet, according to the first law of thermodynamics, energy is always conserved. In a closed system, the total amount of energy cannot change. What then does it mean to conserve energy if the total amount of energy in the universe does not change no matter what we do? In this chapter, we will see that although the first law of thermodynamics is correct, that is, energy *is* conserved, it does not tell the whole story. Some forms of energy are more useful than others. The possibility or impossibility of putting energy to use is the subject of the second law of thermodynamics. In simple words, it is easy to convert mechanical work completely into heat or into the internal energy of a system with no other changes, but it is impossible to remove heat or internal energy from a system and convert it completely into mechanical work with no other changes. This experimental fact is known as the Kelvin-Planck statement of the second law of thermodynamics. There is thus a lack of symmetry in the roles played by heat and work that is not evident from the first law. This lack of symmetry is related to the fact that some processes are irreversible.

A common example in which mechanical energy is converted into internal energy is the work done against friction. Consider, for example, a block with some initial kinetic energy sliding along a rough table. In this process, mechanical energy, the initial kinetic energy of the block, is converted into the internal energy of the block and the table. The reverse process never occurs. The internal energy of the block and table is never spontaneously converted into the kinetic energy of the block, sending it sliding along the table while the table and the block cool. Yet, such an amazing occurrence would not violate the conservation of energy or any other physical laws we have encountered so far. It does, however, violate the second law of thermodynamics, as we will discuss shortly. There are many other irreversible processes, seemingly quite different from one another, but all related to the second law. Suppose, for example, we pour a layer of black sand into a jar and cover it with a layer of white sand. If we now shake the jar, the sand becomes mixed. We cannot "unshake" the jar; that is, we cannot reverse the shaking until the black sand is again on the bottom and the white on top. A third example of an irreversible process is heat conduction. If we place a hot body in contact with a cold body, heat will flow from the hot body to the cold body until they are at the same temperature. However, the reverse does not occur. Two bodies in contact at the same temperature remain at the same temperature; heat does not flow from one to the other and make one colder and the other warmer. The second law of thermodynamics summarizes the fact that processes of this type do not occur. There are many different ways of stating the second law. We will study several and show them to be equivalent. But before giving precise statements of the second law, we will briefly consider heat engines and refrigerators. It was the study of the efficiency of heat engines that gave rise to the first clear statements of the second law of thermodynamics.

14-1   Heat Engines and the Kelvin-Planck Statement of the Second Law of Thermodynamics

The first practical heat engine was the steam engine, invented in the seventeenth century for pumping water out of coal mines. Today, the primary use of steam engines is for generating electrical energy. In a typical steam engine, water is heated under pressure (normally several hundred atmospheres) until it vaporizes into steam at a high temperature (normally around 500°C). The steam expands and does work against the rotating blades of a turbine. The steam exits at a much lower temperature and is further cooled until it condenses. It is then pumped back into the boiler to be heated again.

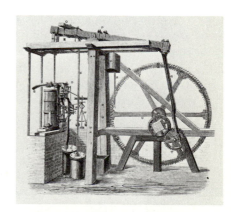

Early steam engine designed by James Watt in the mid-1700s.

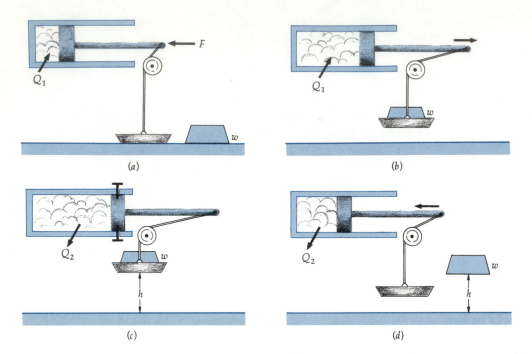

(a)   (b)

(c)   (d)

Figure 14-1
A simple engine that does work in a cycle. (*a*) A gas enclosed in a
cylinder is heated while the piston is held fixed. The pressure
increases from P_1 to P_2, as shown in the *PV* diagram of Figure 14-2.
(*b*) The force holding the piston is removed, and an extra weight w
is added so that the pressure is constant as the piston moves.
(*c*) The piston is held fixed while the gas is cooled back to its original
pressure. (*d*) The weight is removed and the gas is compressed at
constant pressure back to its original state. The net result of the
cycle is that work has been done in lifting the weight to some
height h.

Figure 14-1 illustrates how a simple heat engine performs work in a cyclic
process. A cylinder containing gas at pressure P_1 is fitted with a piston. In
step (*a*), the piston is held fixed by a force F as the gas is heated to a new
pressure P_2. The increase in pressure is shown in the *PV* diagram in Figure
14-2. In step (*b*), the force is removed and a weight w is added, so as the gas is
heated, it expands at constant pressure, lifting the weight to some height h.
In step (*c*), the piston is again held fixed as heat is removed, reducing the
pressure back to P_1. In step (*d*), the weight is removed, so the piston is again
in equilibrium. The gas is compressed at constant pressure, and heat is
exhausted until the gas is back to its original volume. The net result is that an
amount of heat Q_1 is put into the system, some heat Q_2 is exhausted, and
work $W = mgh = Q_1 - Q_2$ is performed.

Figure 14-3 shows a schematic diagram of a practical engine, the internal
combustion engine used in most automobiles. With the exhaust valve
closed, a mixture of gasoline vapor and air enters the combustion chamber

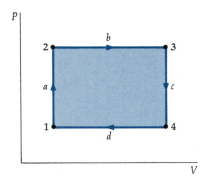

Figure 14-2
PV diagram for the simple engine in
Figure 14-1. Work is done by the
gas during step (*b*) and on the gas in
step (*c*). The net work done by the
gas is the enclosed area in the
diagram.

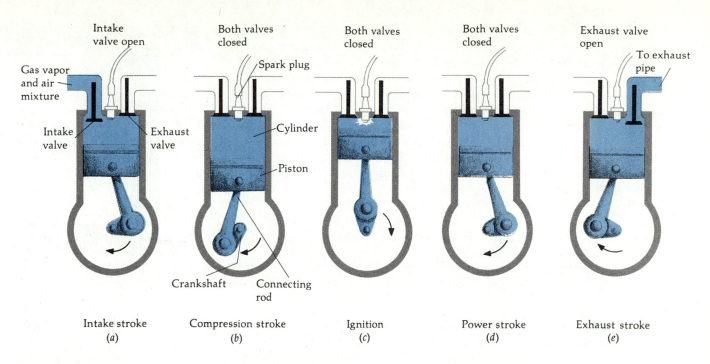

Figure 14-3
Internal combustion engine. In (a) a mixture of air and gas vapor enters the piston chamber as the piston moves down. The piston then moves up in (b), compressing the gas for ignition in (c). The hot gases expand, moving the piston down in (d), the power stroke. In (e) the piston moves up again to exhaust the burned gases. The cycle then repeats.

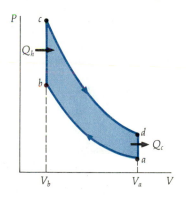

Figure 14-4
Otto cycle representing the internal combustion engine. The air and gas vapor enter at a and are adiabatically compressed to b, where they are heated (by ignition from the spark plug) at constant volume to c. The power stroke is represented by the adiabatic expansion from c to d. The cooling at constant volume from d to a represents the exhausting of the burned gases and the intake of fresh air and gas vapor.

as the piston moves down during the intake stroke. The mixture is then compressed (compression stroke), after which it is ignited by a spark from the spark plug. The hot gases then expand against the piston, doing work during the power stroke. The gases are then exhausted through the exhaust valve and the cycle repeats. Figure 14-4 shows an *Otto cycle*, which is an idealized model of the processes of the internal combustion engine. The gas–air mixture enters the cylinder at point a. The compression ab takes place with no heat exchange. A process with no heat exchange is called an *adiabatic process*. The combustion of the gases is represented by heat input Q_h, which raises the temperature at constant volume from b to c. The power stroke is an adiabatic expansion from c to d. Heat Q_c leaves the system during exhaust from d to a. Since an equivalent amount of gas then enters at a, we can treat the process as if the same gas were used over. The net work done during the cycle is indicated by the shaded area.

The basic features of any heat engine are that a substance or system called the *working substance* (water in the case of the steam engine, air and gasoline vapor for an internal combustion engine) absorbs a quantity of heat Q_h at a high temperature T_h, does work W, and rejects heat Q_c at a lower temperature T_c. The working substance then returns to its original state. The engine is therefore a cyclic device. The purpose of a heat engine is to convert as much of the heat input Q_h into work as possible. Figure 14-5 shows a schematic representation of a basic heat engine. The heat input is represented as coming from a "heat reservoir" at temperature T_h, and the exhaust goes into a "heat reservoir" at a lower temperature T_c. (A heat reservoir is an idealized body or system that has a very large heat capacity so that it can absorb or give off heat with no appreciable change in temperature. In practice, the surrounding atmosphere or a lake often acts as a heat reservoir.) Since the initial and final states of the engine and working substance are the same, the final internal energy must equal the initial internal energy. Then, according to the first law of thermodynamics, the work done equals the net heat absorbed:

$$W = Q_h - Q_c \qquad\qquad 14\text{-}1$$

The *efficiency* ϵ of the engine is defined as the ratio of the work done to the heat absorbed from the hot reservoir:

$$\epsilon = \frac{W}{Q_h} = \frac{Q_h - Q_c}{Q_h} = 1 - \frac{Q_c}{Q_h} \qquad\qquad 14\text{-}2$$

Since the heat Q_h is usually produced by burning coal, oil, or some other kind of fuel that must be paid for, one tries to design a heat engine with the greatest possible efficiency. Efficiencies of steam engines are typically about 40 percent; those of internal combustion engines may be 50 percent. We can see from Equation 14-2 that we want to eject as small a fraction of the heat absorbed as possible. For perfect efficiency ($\epsilon = 1 = 100$ percent), $Q_c = 0$ and no heat is rejected in a cycle. In that case, all the heat absorbed from the first reservoir would be converted into work, and none would be rejected or lost to the second reservoir.

Although the efficiency of heat engines has been greatly increased since the early steam engines, it is impossible to make a heat engine that is 100 percent efficient, that is, one that would reject no heat to a reservoir at a lower temperature. This experimental result is the *Kelvin-Planck statement of the second law of thermodynamics*.

It is impossible for an engine working in a cycle to produce no other effect than that of extracting heat from a reservoir and performing an equivalent amount of work.

Second law (Kelvin-Planck statement)

It is important to include the word "cycle" in this statement of the second law. It is possible to convert heat completely into work in a noncyclic process. An ideal gas undergoing an isothermal expansion does just this. But after the expansion, the gas is not in its original state. In order to bring the gas back to its original state, work must be done on the gas and some heat will be exhausted. Essentially, the second law tells us that if we want to extract energy from a reservoir to do work, we must have available a colder reservoir in which to exhaust a part of the energy. If this were not true, we could design a heat engine for a ship that would extract energy from the ocean (a convenient heat reservoir with an enormous supply of energy) and use it to power the ship. Unfortunately, this enormous reservoir of energy is not available for such use because of the lack of a colder reservoir for the exhaust.

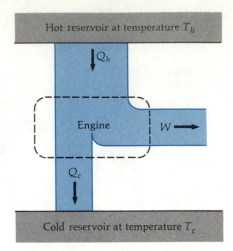

Figure 14-5
Schematic representation of a heat engine that removes heat energy Q_h from a hot reservoir at temperature T_h, does work W, and rejects heat Q_c to a cold reservoir at temperature T_c.

Example 14-1   An engine absorbs 200 J of heat from a hot reservoir, does work, and exhausts 160 J to a cold reservoir. What is its efficiency?

By the first law, the work done is

$$W = 200 \text{ J} - 160 \text{ J} = 40 \text{ J}$$

The efficiency is thus

$$\epsilon = \frac{40 \text{ J}}{200 \text{ J}} = 0.20 = 20\%$$

Questions

1. Where does the energy come from in an internal combustion engine? In a steam engine?

2. How does friction in an engine affect its efficiency?

14-2   Refrigerators and the Clausius Statement of the Second Law of Thermodynamics

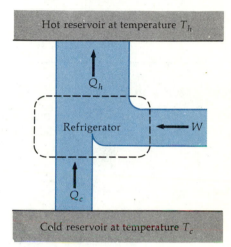

Figure 14-6
Schematic representation of a refrigerator that removes heat energy Q_c from a cold reservoir and rejects heat Q_h to a hot reservoir using work W.

A refrigerator is like a heat engine run backward. Work is put into the refrigerator to extract heat from a cold reservoir and transfer it to a hot reservoir. Figure 14-6 is a schematic representation of such a device. It is desirable to remove as much heat Q_c as possible while doing as little work W as possible. It has been found from experience that some work must always be done. This result is the *Clausius statement of the second law of thermodynamics.*

It is impossible for a refrigerator working in a cycle to produce no other effect than the transfer of heat from a colder body to a hotter body.

Second law (Clausius statement)

If the Clausius statement were not true, it would be possible, in principle, to cool our homes in the summer with a refrigerator that pumps heat to the outside without using any electricity or other energy.

A measure of the performance of a refrigerator is the ratio Q_c/W, called the *coefficient of performance* COP:

$$\text{COP} = \frac{Q_c}{W}$$

14-3   Coefficient of performance

The greater the coefficient of performance, the better the refrigerator. Typical refrigerators have coefficients of performance of about 5 or 6. In terms of this ratio, the Clausius statement of the second law is that the coefficient of performance of a refrigerator cannot be infinite.

Example 14-2   A refrigerator has a coefficient of performance of 5.5. Ice cubes are to be made from 1 L of water at 10°C. How much work is needed?

The mass of 1 L of water is 1 kg. The heat energy that must be removed to reduce the temperature by 10 K is

$$Q_1 = mc\,\Delta T = (1\text{ kg})(4.18\text{ kJ/kg}\cdot\text{K})(10\text{ K}) = 41.8\text{ kJ}$$

The heat of fusion of water is 333.5 kJ/kg, so 333.5 kJ must be removed to freeze the water into ice cubes. The total amount of heat that must be removed is $333.5 + 41.8 = 375.3 \approx 375$ kJ. From Equation 14-3, the work needed is

$$W = \frac{Q_c}{\text{COP}} = \frac{375\text{ kJ}}{5.5} = 68.2\text{ kJ}$$

Question

3. To cool his kitchen, a man leaves his refrigerator door open. Explain why this does not have the desired effect and, in fact, heats the kitchen.

Rudolf Julius Emanuel Clausius (1822–1888).

14-3   Equivalence of the Kelvin-Planck and Clausius Statements

The Kelvin-Planck and Clausius statements are both statements of the second law of thermodynamics. (There are many other ways to state the second law.) Although they seem quite different, they are actually equivalent. That is, it can be proved that the two statements must be either both true or both false. Thus, the truth of one statement implies the truth of the other. We will illustrate this using a numerical example that shows that if the Clausius statement is false, the Kelvin-Planck statement must also be false.

Figure 14-7a shows an ordinary engine that removes 100 J of energy from a hot reservoir, does 40 J of work, and exhausts 60 J into a cold reservoir. The efficiency of this engine is 40 percent. We now assume that the Clausius statement is false; that is, we assume that we can remove heat from the cold reservoir and transfer it to the hot reservoir with no other effects. In Figure 14-7b, we have added a "perfect refrigerator" that removes 60 J from the cold reservoir and transfers it to the hot reservoir with no work required. The net effect of the ordinary engine and the perfect refrigerator working together is to remove 40 J from the hot reservoir and do 40 J of work with no energy exhausted (see Figure 14-7c). This violates the Kelvin-Planck statement of the second law. We thus see that if the Clausius statement is false, the Kelvin-Planck statement is false. That is, if we can find a perfect refrigerator that transfers heat from a cold reservoir to a hot one without requiring any work, we can use this refrigerator in conjunction with an ordinary engine to construct a perfect engine that removes heat from a reservoir and converts it completely into work.

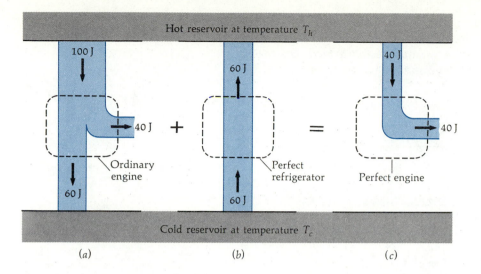

Figure 14-7
(*a*) An ordinary heat engine that removes 100 J from a hot reservoir, does 40 J of work, and rejects 60 J to the cold reservoir. (*b*) A perfect refrigerator that violates the Clausius statement of the second law by removing 60 J from the cold reservoir and transferring it to the hot reservoir with no other effect. (*c*) The two engines in (*a*) and (*b*) working together make a perfect heat engine that violates the Kelvin-Planck statement of the second law by removing 40 J of heat from the hot reservoir and converting it completely into work with no other effect.

It can similarly be shown that if a perfect engine exists, we can use it in conjunction with an ordinary refrigerator to construct a perfect refrigerator. This is left as an exercise for the students (see Exercise 11). Thus, if the Kelvin-Planck statement is false, the Clausius statement is false. It then follows that if either statement is true, the other must also be true. The two statements are therefore equivalent.

14-4 The Carnot Engine

We have seen that, according to the second law of thermodynamics, it is impossible for an engine working between two heat reservoirs to be 100 percent efficient. We now ask, "What is the maximum possible efficiency for such an engine?" This question was answered in 1824 by a young French engineer, Sadi Carnot, before the first law of thermodynamics was established. Carnot found that all *reversible* engines working between two heat reservoirs have the same efficiency and that no other engine could have a greater efficiency than that of a reversible engine. This result is known as the *Carnot theorem:*

No engine working between two given heat reservoirs can be more efficient than a reversible engine working between those reservoirs.

Let us look at what makes a process reversible or irreversible. We have already mentioned that the conversion of mechanical energy into heat by friction is irreversible. The reverse process of converting heat into mechanical energy with no other effects violates the Kelvin-Planck statement of the second law. Similarly, the conduction of heat from a hot body to a cold body is irreversible. The reverse process, conduction of heat from a cold body to a hot body with no other effect, violates the Clausius statement of the second

Nicolas Léonard Sadi Carnot (1796–1832) at age seventeen, in a drawing by L. L. Boilly.

law. A third type of irreversibility occurs when a system passes through nonequilibrium states, such as when there is turbulence in a gas or an explosion. For a process to be reversible, we must be able to move the system back through the same equilibrium states in the reverse order.

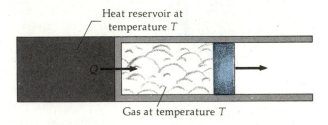

Heat reservoir at
temperature T

Q

Gas at temperature T

Figure 14-8
An isothermal expansion of a gas. As the piston slowly moves out, heat is conducted into the gas from the reservoir, keeping the temperature of the gas constant. If the expansion is slow enough, the pressure can adjust so that the gas is always essentially in an equilibrium state. The equilibrium states traversed by the gas can be plotted on a PV diagram, as shown in Figure 14-9.

Figure 14-8 illustrates a reversible process, a very slow isothermal expansion of a gas. In this process, the gas is at some temperature T in a container in thermal contact with a heat reservoir at the same temperature. As the gas expands, the volume increases and the pressure decreases, and heat enters the gas from the reservoir. We can think of the expansion as a series of tiny steps. For each step, the gas comes quickly to equilibrium at its new pressure and volume. We can therefore plot the states of the gas on a PV diagram, as shown in Figure 14-9. If we compress the gas slowly, the gas moves through the same states in the reverse direction on the PV diagram. Work is then done on the gas, and heat is transferred into the heat reservoir. If the expansion or compression were done rapidly, turbulence would result and we could not plot the states on the PV diagram. In turbulent motion, there is no one pressure associated with the entire system.

We now illustrate the Carnot theorem with a numerical example. Figure 14-10a shows a reversible engine that removes 100 J from a hot reservoir, does 40 J of work, and exhausts 60 J to the cold reservoir. Its efficiency is 40 percent. In Figure 14-10b, the engine is reversed. Here, 40 J of work are done to remove 60 J from the cold reservoir and to exhaust 100 J to the hot reservoir. (If the engine were not reversible, more than 40 J of work would have to be done to remove 60 J from the cold reservoir.) We now assume that there is another engine working between the same two reservoirs that is more than 40 percent efficient and show that, if this is so, the second law of thermodynamics is violated. Figure 14-10c shows an engine with an efficiency of 45 percent. The net effect of running this engine through one cycle and the reversible engine backward through one cycle is to remove 5 J of heat from the lower reservoir and change it completely into work, violating the Kelvin-Planck statement of the second law.

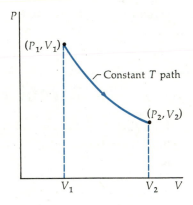

P

(P_1, V_1)

Constant T path

(P_2, V_2)

V_1 V_2 V

Figure 14-9
PV diagram for the isothermal expansion of Figure 14-8. When the gas is compressed isothermally, it traverses the same path in the diagram, going through the same states as during the expansion. The process is thus reversible.

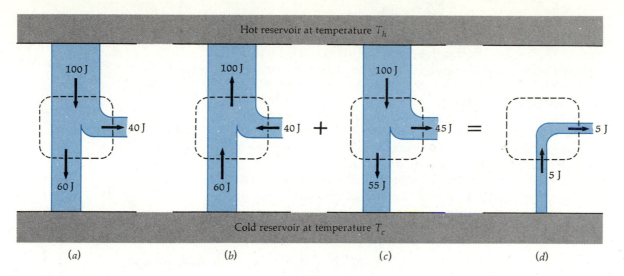

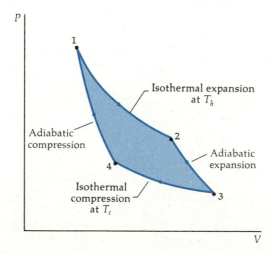

Figure 14-10
Illustration of the Carnot theorem. (a) A reversible heat engine with
40 percent efficiency. (b) The same engine run backward as a
refrigerator. (c) An assumed heat engine working between the same
two reservoirs with 45 percent efficiency, which is greater than that
of the reversible engine in (a). (d) The net effect of the engine in (c)
and the refrigerator in (b) is a perfect engine that removes 5 J from
the cold reservoir and converts it completely into work with no
other effect, violating the Kelvin-Planck statement of the second law.

If no other engine can have a greater efficiency than a reversible engine, it
follows that all reversible engines working between the same two reservoirs
have the same efficiency. This efficiency is independent of the working
substance of the engine; thus, it can depend only on the temperatures of the
reservoirs. A reversible engine is called a *Carnot engine*. Figure 14-11 shows
a *Carnot cycle* for an ideal gas used as the working substance for the Carnot
engine. The cycle starts at point 1 with an isothermal absorption of heat

Carnot cycle

Figure 14-11
A Carnot cycle for an ideal gas.
Heat is absorbed during the
isothermal expansion from 1 to 2 at
temperature T_h. The gas expands
adiabatically from 2 to 3, until its
temperature is reduced to T_c. It is
then compressed isothermally from
3 to 4, rejecting heat, after which it
is compressed adiabatically until its
temperature is again T_h. All
processes are reversible.

from the hot reservoir at temperature T_h. Since the heat is absorbed isothermally, the process could be reversed without violating the second law of thermodynamics. Work is done during this expansion to point 2. From point 2 to point 3, the gas expands adiabatically, that is, with no heat exchange. If this expansion is done slowly, it will be reversible. More work is done, and the temperature of the gas is reduced to T_c. The third part of the cycle, from points 3 to 4, is an isothermal compression at temperature T_c. During this part of the cycle, work is done *on* the gas. The last part of the cycle, from points 4 to 1, is an adiabatic compression back to the original state. Again, work is done on the gas during this compression. The net work done during the cycle is represented by the area enclosed by the curve in the figure. The work done by or on the gas and the heat added or subtracted can be calculated using calculus for each part of the cycle and can be used to calculate the efficiency of this cycle. The result is

$$\epsilon_c = 1 - \frac{T_c}{T_h} \qquad \qquad \text{14-4} \qquad \text{Carnot efficiency}$$

Comparing this with Equation 14-2 for the general efficiency of any engine, we see that for the Carnot engine, the ratio of the heat rejected to that absorbed is

$$\frac{Q_c}{Q_h} = \frac{T_c}{T_h} \qquad \qquad \text{14-5}$$

The efficiency given by Equation 14-4 is called the *Carnot efficiency*. It represents the greatest possible efficiency for an engine working between temperatures T_h and T_c. No engine can have an efficiency greater than the Carnot efficiency given by Equation 14-4. The existence of an engine with a greater efficiency than this would violate the second law of thermodynamics.

Example 14-3   A steam engine works between a hot reservoir at $100°C = 373$ K and a cold reservoir at $0°C = 273$ K. What is the maximum possible efficiency of such an engine?

From Equation 14-4, the Carnot efficiency is

$$\epsilon_C = 1 - \frac{T_c}{T_h} = 1 - \frac{273 \text{ K}}{373 \text{ K}} = 0.268 = 26.8\%$$

Even though this efficiency seems to be quite low, it is the greatest efficiency possible for any engine working between these temperatures. Real engines will have lower efficiencies because of friction, heat conduction, or other irreversible processes.

We can see that it is very useful to know the Carnot efficiency. Suppose, for example, we had a real engine that removed 100 J from a hot reservoir at 373 K, did 25 J of work, and exhausted 75 J to a cold reservoir at 273 K. Its efficiency would be only 25 percent, but this would be a very good engine. If we wanted to improve this engine by reducing friction, eliminating various irreversibilities, and so forth, the best we could do would be to obtain an efficiency of 26.8 percent as calculated in the preceding example. That is, the most work we could possibly get using an ideal engine would be 26.8 J for every 100 J removed from the hot reservoir.

The ratio of the efficiency of an engine to the Carnot efficiency is called the *second-law efficiency* ϵ_{SL}.

$$\epsilon_{SL} = \frac{\text{actual efficiency}}{\text{Carnot efficiency}} \qquad\qquad 14\text{-}6 \qquad \text{Second-law efficiency}$$

For a real engine that has an efficiency of 25 percent when working between 373 K and 273 K, the second-law efficiency is

$$\epsilon_{SL} = \frac{25\%}{26.8\%} = 0.93 = 93\%$$

Example 14-4   An engine removes 200 J from a heat reservoir at 373 K, does 48 J of work, and exhausts 152 J to a cold reservoir at 273 K. How much work per cycle is "lost" due to the irreversibilities of this engine?

The efficiency of this engine is $\epsilon = 48/200 = 24$ percent. As we just saw, this is a very good engine. Its second-law efficiency is $24/26.8 = 90$ percent. A Carnot engine working between these reservoirs and removing 200 J from the upper reservoir could do $0.268 \times 200 = 53.6$ J of work per cycle. Since the real engine does 48 J of work per cycle, it could do only 5.6 J more work if it were an ideal engine. Thus only 5.6 J of work is lost per cycle by this engine. This energy is, of course, not lost completely since the total energy is conserved. It is simply exhausted into the cold reservoir.

Example 14-5   If 200 J of heat are conducted from a heat reservoir at 373 K to one at 273 K, how much work capability is lost in this process?

We have seen in the previous examples that a Carnot engine working between these two reservoirs could do 53.6 J of work if it extracted 200 J from the 373-K reservoir and exhausted to a 273-K reservoir. Thus if 200 J is conducted directly from the hot reservoir to the cold reservoir with no work done, 53.6 J of this energy is wasted; that is, it has been "lost" in the sense that it could have been converted into useful work.

Questions

4. Why do power-plant designers try to increase the temperature of the steam fed to engines as much as possible?

5. Some think that electrical heat for the home is more efficient because all the electrical energy goes into heat in the home whereas, with a gas or oil furnace, some of the heat goes out the chimney. However, electrical energy is usually generated in a steam power plant, which is only about 40 percent efficient. In light of this fact, discuss the relative merits of electrical heat versus oil or gas heat for buildings.

Power Plants and Thermal Pollution

Laurent Hodges
Iowa State University

Scientists' interest in heat engines is focused primarily on the work obtained from a given heat input. Society, however, is also interested in the waste heat output, especially when it leads to undesirable effects, or *thermal pollution.*

The largest heat engines in modern society—those from which thermal pollution can be the most damaging—are the steam engines used in electric power plants to drive an alternating-current generator and produce electricity. About 85 percent of the electricity in the United States is produced by such steam engines, the other 15 percent being generated by water power in hydroelectric plants.

The operation of a steam electric power plant is explained in Figure 1. The *heat source* is usually either the combustion of a fossil fuel (coal, oil, or natural gas) or the fission of uranium 235; occasionally, geothermal power or the combustion of solid wastes is used. The *boiler* heats water, converting it into high-temperature, high-pressure steam. A nuclear power plant has a reactor core in which water is heated into steam (boiling-water reactor, or BWR) or into pressurized water (pressurized-water reactor, or PWR); in the PWR, heat is then transferred to a steam cycle in a heat exchanger. Steam expands and cools as it passes through the *turbine,* converting heat energy into the mechanical energy of a rotating shaft. A *generator* converts the mechanical energy into electric energy (alternating current). The *condenser* cools the steam exhausted from the turbine and reduces its pressure, thereby increasing the efficiency of the power plant.

Typical temperatures and efficiencies for steam electric plants are shown in Table 1. The average efficiency for all power plants in the United States is about 34 percent. The best fossil-fuel plants have about 40 percent efficiency, but many smaller or older plants have efficiencies of 25 percent or less. The efficiencies of BWR and PWR nuclear plants are about 34 percent, but those of some experimental nuclear plants (cooled by gas or liquid sodium) are closer to 40 percent.

Existing power plants must dispose of waste heat amounting to twice the electrical energy generated. A popular method, *once-through cooling,* involves passing water from a river or lake through the condenser once and then returning it at a warmer temperature. This works well only if adequate water is present. Many plants of 1000 MW_e (e = electric power) are being built today, sometimes several at one location. Even on a large river, the resulting temperature rise may be significant: a 3000 MW_e–9000 MW_t (t = thermal power) installation using a flow of 120 m^3/s (the average yearly minimum flow of the Missouri River at its mouth) would cause a 12 C° rise by its 6000 MW of waste heat.

Although the temperature changes produced in a river by once-through cooling may not be as large as natural daily or seasonal fluctuations, possible adverse ecological effects include greater growth of bacteria, pathogens, or undesirable blue-green algae or physiological and behavioral changes in aquatic life (such as fish). As an example, large-mouth bass

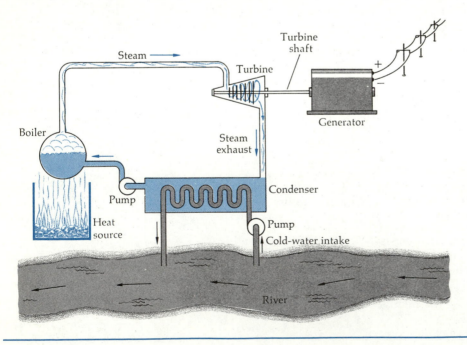

Figure 1
Schematic diagram of major components of a steam electric power plant.

Table 1

Typical Higher and Lower Temperatures and Efficiencies for Steam Electric Power Plants

	Temperature, °C		Efficiency, %	
	High	Low	Carnot	Actual
Large fossil-fuel plant	380	40	52	40
Boiling-water reactor	285	40	44	34
Pressurized-water reactor	315	40	47	34

acclimated to 30°C show a 50 percent mortality rate within 72 hours if the water is warmed to 34°C.

Warmer water also contains less of the dissolved oxygen necessary for aquatic life and for the decomposition of organic wastes; for example, the dissolved oxygen at saturation in water is 11.3 mg/L at 10°C and 6.6 mg/L at 40°C. Warmer water is also less viscous, permitting faster deposition of the sediment load, which can affect aquatic food supplies.

Most of the cooling water used in the United States is used by electric power plants. The rapid growth of electric energy consumption, averaging 7 percent annually for many decades, has led to greater use of alternatives to once-through cooling. One of these is the cooling pond, really a private lake, but the land requirements for such a pond are often too expensive to allow this alternative to be considered.

The most common alternative, the cooling tower, transfers the waste heat to the atmosphere (see Figure 2). Usually, these are evaporative towers, which use the heat to evaporate water. Evaporative cooling towers are either large hyperbolic types, with a natural draft, or mechanical-draft (fan) types. Nonevaporative cooling towers, which transfer heat to the air by heat exchangers, are expensive but may be needed in moist climates. At ordinary temperatures, the evaporation of water requires 2.26 MJ/kg, so the water requirements are much less than in once-through cooling, although the water is consumed in the process. The evaporated water can cause fog, increased precipitation, and (in certain climates) the icing of nearby roads. Furthermore, the cooling tower and its plume may be blots on the landscape, so evaporative towers are not a completely satisfactory answer to the problem of thermal pollution.

Figure 2
Hyperbolic natural-draft evaporative cooling towers at the Fort Martin Power Station, Fairmont, West Virginia.

14-5   The Heat Pump

About 35 percent of our energy use is for the space heating of our homes and office buildings. Much of this heating is done by the on-site combustion of fossil fuels (oil, coal, or natural gas). Although this is relatively efficient, it is a major source of pollution in our cities. The alternative of electric heat for homes seems, at first glance, to be efficient and nonpolluting, but this is far from the truth. Most of our electricity is generated in power plants by the combustion of fossil fuels. (A small fraction is generated from waterfalls or from nuclear reactors.) This leads to the same kind of air pollution as the combustion of fossil fuels in the home. Also, the heat exhausted by a power plant leads to thermal pollution of our lakes and rivers and the atmosphere. In addition, the Carnot efficiency of a power plant is only about 50 percent and the actual efficiency is around 40 percent. An efficient alternative to direct combustion as a method of heating homes and office buildings in moderate climates is the heat pump discussed in this section.

A *heat pump* is essentially a refrigerator that is used to pump thermal energy from a cold reservoir (for example, the cold air outside a house) to a hot reservoir (for example, the hot air inside the house). If work W is done to remove heat Q_c from the cold reservoir and reject heat $Q_h = W + Q_c$ to the hot reservoir, the coefficient of performance is

$$\text{COP} = \frac{Q_c}{W} \qquad\qquad 14\text{-}7$$

Using $W = Q_h - Q_c$, this can be written

$$\text{COP} = \frac{Q_c}{Q_h - Q_c}$$

$$= \frac{Q_c/Q_h}{1 - Q_c/Q_h} \qquad\qquad 14\text{-}8$$

The maximum coefficient of performance occurs for a reversible or Carnot heat pump. Then Q_c and Q_s are related by Equation 14-5

$$\frac{Q_c}{Q_h} = \frac{T_c}{T_h}$$

Substituting $Q_c/Q_h = T_c/T_h$ into Equation 14-8, we can obtain the maximum coefficient of performance

$$\text{COP}_{\text{max}} = \frac{T_c/T_h}{1 - T_c/T_h}$$

$$= \frac{T_c}{T_h - T_c}$$

or

$$\text{COP}_{\text{max}} = \frac{T_c}{\Delta T} \qquad\qquad 14\text{-}9 \qquad \text{Carnot coefficient of performance}$$

where ΔT is the difference in temperature between the cold and hot reservoirs. Real heat pumps and refrigerators have COPs less than the Carnot

COP because of friction, heat conduction, or other irreversible processes. The second-law efficiency of a heat pump or refrigerator is the ratio of the actual coefficient of performance to the Carnot coefficient of performance.

$$\epsilon_{SL} = \frac{COP_{actual}}{COP_{max}}$$

14-10   Second-law efficiency

We are usually interested in the work that must be done to exhaust a given amount of heat Q_h into the hot reservoir. (For heating a house, the hot reservoir may be the hot-air supply for the heating blower in the house.) Using $Q_h = Q_c + W$, we can write Equation 14-7

$$COP = \frac{Q_c}{W} = \frac{Q_h - W}{W}$$

$$W(COP) = Q_h - W$$

or

$$Q_h = W(1 + COP) \qquad\qquad 14\text{-}11$$

Example 14-6   A heat pump with 75 percent second-law efficiency is used to pump heat from the outside air at $0°C$ to the hot-air blower unit in a house at $40°C$. The pump is run by electric energy. How much energy is required to pump 1 kJ of heat into the house?

From Equation 14-8, the Carnot coefficient of performance for a heat pump operating at $T_c = 273$ K and $\Delta T = 40$ K is

$$COP_{max} = \frac{T_c}{\Delta T}$$

$$= \frac{273 \text{ K}}{40 \text{ K}} = 6.83$$

The actual coefficient of performance is 75 percent of this, or

$$COP_{actual} = 75\% \; COP_{max}$$

$$= (0.75)(6.83) = 5.12$$

We can calculate the work needed from Equation 14-11 with $Q_h = 1$ kJ,

$$W = \frac{Q_h}{1 + COP}$$

$$= \frac{1 \text{ kJ}}{(1 + 5.12)} = 0.163 \text{ kJ}$$

Thus, only 0.163 kJ of work is needed to pump 1 kJ of heat into the hot-air blower in the house.

We see from this example and from Equation 14-11 that the heat pump effectively multiplies the energy needed to run the pump by $(1 + COP)$. In the example, with $COP = 5.12$, if we use 1 kJ to run the engine, we can exhaust 6.12 kJ of heat into the house.

14-6   Entropy and Disorder

We have seen that the second law of thermodynamics is related to the fact that some processes go in one direction only. Essentially, mechanical energy can be converted into internal thermal energy with 100 percent efficiency, but internal thermal energy cannot be converted completely into mechanical energy with no other effects. There are many other irreversible processes that are related to the second law, but they are not easily described by the Kelvin-Planck or Clausius statements. Consider, for example, a glass falling off a table and shattering when it hits the floor. Clearly, this process is irreversible. The mechanical potential energy of the glass in its original position on the table goes into the breaking of the molecular bonds that held the glass together. Neither this process nor the mixing of black and white sand, also clearly irreversible, are easily related to the two statements of the second law that we have been discussing. In all irreversible processes, the system plus its surroundings move toward a less ordered state. We will illustrate this with a simple example.

Consider a box containing a gas at atmospheric pressure and room temperature moving with velocity **V** along a frictionless table and colliding inelastically with a fixed wall (see Figure 14-12). Let us assume that all the original kinetic energy of the box goes into the internal thermal energy of the gas molecules. Before the collision, the velocity of each gas molecule consists of two parts: a large random velocity associated with the internal energy of the gas, as measured by its temperature T, and the velocity **V** associated with the motion of the center of mass. The kinetic energy of the motion of the center of mass is ordered energy. It could be used to perform work, for example, by attaching a string to the box and using it to lift a weight. However, the inelastic collision, which is clearly irreversible, converts this ordered energy into internal thermal energy, which is not ordered.

Figure 14-12
(*a*) A box of gas molecules moving to the right with speed V. The velocity of a molecule is the vector sum of the velocity of the center of mass **V** and the random velocity associated with the temperature of the gas. The energy associated with the motion of the center of mass is ordered energy. It could be used to do work, for example, to lift a weight. (*b*) After the box has collided inelastically with the wall, the molecules have the same total energy as before, but now the energy is completely disordered and is associated with the new temperature, which is higher than that before the collision.

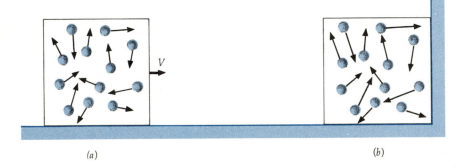

(*a*)   (*b*)

After the collision, the gas molecules have the same kinetic energy as before, but this energy is now entirely associated with random motion and the temperature of the gas is greater.

There is a thermodynamic function called *entropy S* that measures the "disorder" of a system. Like the pressure P, volume V, temperature T, and internal energy U, entropy is a function of the state of a system. As with internal energy, it is the *change* in entropy that is important. The change in

Drawing by Levin; © 1985
The New Yorker Magazine, Inc.

entropy ΔS of a system when it goes from one state to another is defined as

$$\Delta S = \frac{\Delta Q_{rev}}{T}$$

14-12   Entropy

where ΔQ_{rev} is the heat that must be added to the system *in a reversible process* to bring it from its initial state to its final state. If heat is subtracted from the system, ΔQ_{rev} is negative and the entropy change of the system is negative. The entropy of a given system can increase or decrease, but the entropy of the "universe" ΔS_u, that is, the entropy of the system plus its surroundings, never decreases. This is in fact a statement of the second law of thermodynamics that is equivalent to the Kelvin-Planck or Clausius statements.

The entropy of the universe never decreases.

Second law in terms of entropy

In a reversible process, the entropy change of the universe (or of any isolated system) is zero; in an irreversible process, the entropy of an isolated system increases.

Equation 14-12 does not imply that reversible heat transfer must take place for the entropy of a system to change. Indeed, there are many examples in which the entropy of a system changes with no heat transfer whatsoever. Equation 14-12 gives a method of calculating the entropy difference between two states of a system. When a system moves from one state to another, the change in entropy, like the change in internal energy, is the same no matter what the process. Both entropy and internal energy are "state functions." Changes in them depend only on the initial and final states of the system. However, to *calculate* the change in entropy, we must find a reversible process connecting the initial and final states, find the heat added or subtracted during this reversible process, and use Equation 14-12. We will illustrate this somewhat subtle point by calculating the entropy change in various simple examples.

© 1987 Sidney Harris.

We first consider a simple example of a block of mass m falling from a height h and making an inelastic collision with the ground. Let the block, ground, and atmosphere all be at temperature T, which is not significantly changed by this process. Clearly, this is an irreversible process. If we consider the block, ground, and atmosphere as our isolated system, there is no heat conducted into or out of the system. (The heat flow from the block into the ground and atmosphere as a result of the collision is internal to our large system.) The state of the ground plus atmosphere is changed because its internal energy has been increased by the amount mgh. This is the same as if we added heat $\Delta Q = mgh$ to the system at constant temperature T. To calculate the change in entropy of the system, we thus consider a reversible process in which heat $Q_{rev} = mgh$ is added at a constant temperature T. According to Equation 14-12, the change in entropy is then

$$\Delta S = \frac{\Delta Q}{T} = \frac{mgh}{T}$$

This is the entropy change of the universe ΔS_u for the block of mass m falling to the ground. In this process, energy is "wasted." Initially, there was a block of mass m at a height h. The potential energy mgh could have been used to do useful work, such as lifting a weight. After the inelastic collision, this energy is no longer available. It is now internal, disordered energy of the surroundings. The energy "lost" equals $T \Delta S_u$. This is a general result.

In an irreversible process, energy equal to the entropy change of the universe times the temperature of the coldest available reservoir becomes unavailable for doing work.

For simplicity, we will call the energy that becomes unavailable for doing work the "work lost":

$$W_{lost} = T \Delta S_u \qquad\qquad 14\text{-}13$$

We next consider heat conduction. Let heat Q be conducted from a hot reservoir at temperature T_h to a cold reservoir at T_c. The state of a heat reservoir is determined only by its temperature and its internal energy. The change in entropy of a heat reservoir when heat is conducted out of it or into it is the same whether the heat exchange is reversible or not. If heat Q is put into a reservoir at temperature T, the entropy of the reservoir increases by Q/T. If heat is removed, the entropy of the reservoir decreases. In this case, the hot reservoir loses heat, so its entropy change is

$$\Delta S_h = -\frac{Q}{T_h}$$

The cold reservoir absorbs heat, so its entropy change is

$$\Delta S_c = +\frac{Q}{T_c}$$

The net entropy change of the universe is

$$\Delta S_u = \frac{Q}{T_c} - \frac{Q}{T_h}$$

According to our general statement that the work lost equals the entropy change times the temperature of the coldest reservoir available, the work lost in this process is

$$W_{\text{lost}} = T_c\, \Delta S_u = Q\left(1 - \frac{T_c}{T_h}\right)$$

We can see that this is just the work that could have been done by a Carnot engine running between these reservoirs, removing heat Q from the upper reservoir and doing work $\epsilon_c Q$, where $\epsilon_c = 1 - T_c/T_h$.

Example 14-7   Each cycle, a Carnot engine removes 100 J of energy from a heat reservoir at 400 K, does work, and exhausts heat to a reservoir at 300 K. Compute the entropy change of each reservoir for each cycle and show that the entropy change of the universe is zero for this reversible process.

The efficiency of a Carnot engine working between these two reservoirs is $\epsilon_c = 1 - T_c/T_h = 1 - (300/400) = 0.25$. The engine thus does 25 J work each cycle and exhausts 75 J to the cold reservoir. Since the hot reservoir loses heat, its entropy change is negative:

$$\Delta S_{400} = -\frac{100\text{ J}}{400\text{ K}} = -0.250\text{ J/K}$$

The cold reservoir gains heat, so its entropy increases:

$$\Delta S_{300} = \frac{75\text{ J}}{300\text{ K}} = +0.250\text{ J/K}$$

Since the entropy change of the engine working in a cycle is zero and the entropy increase of the cold reservoir equals the entropy decrease of the hot reservoir, the net entropy change of the universe is zero for this process.

Example 14-8   A real heat engine with 60 percent second-law efficiency removes 100 J each cycle from the heat reservoir at 400 K as in Example 14-7 and exhausts heat to a reservoir at 300 K. Compute the entropy change of each reservoir for each cycle.

The efficiency of this engine is

$$\epsilon = 0.6\epsilon_c = (0.6)(0.25) = 0.15$$

The engine thus does 15 J of work each cycle and exhausts 85 J to the cold reservoir. The entropy change of the hot reservoir is the same as in Example 14-7:

$$\Delta S_{400} = -\frac{100\text{ J}}{400\text{ K}} = -0.250\text{ J/K}$$

The entropy change of the cold reservoir is

$$\Delta S_{300} = \frac{85\text{ J}}{300\text{ K}} = +0.283\text{ J/K}$$

In this case, the entropy increase of the cold reservoir is greater than the decrease for the hot reservoir. The entropy change of the universe for one cycle is $0.283 - 0.250 = 0.033$ J/K.

14-7   Entropy and Probability

In the previous section, we stated that entropy is a measure of disorder of a system and that, though the entropy of a given *system* can decrease during an irreversible process, the net change in the entropy of the universe (the system plus its surroundings) is always positive. The universe always moves toward a state of less order. In this section, we will consider a simple example to show that the entropy of a state is related to *probability*. Essentially, a state of high order has a low probability, whereas a state of low order has a high probability. In an irreversible process, the universe moves from a state of low probability to one of high probability.

The process we will consider to illustrate this is a free adiabatic expansion of an ideal gas. In Figure 14-13, we have a gas in a compartment of volume V_1 and an adjacent compartment of equal volume that is evacuated. The compartments are separated by a stopcock, and the whole system is surrounded by a rigid, nonconducting wall. When the stopcock is opened, the gas rushes into the evacuated compartment and eventually comes to equilibrium, filling both compartments. There is no work done and no heat added, so the final energy of the gas is the same as the initial energy. Its final temperature is also the same as its initial temperature. The state of the gas has changed merely by having its volume doubled and its pressure halved. Since this process is clearly irreversible, we expect that the entropy of the universe has increased.

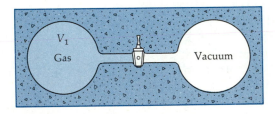

We can calculate the entropy change of the gas if we can find a *reversible* process that takes the gas from the same initial state to the same final state. A convenient process is an isothermal expansion during which the gas absorbs heat Q from a heat reservoir at the temperature T of the gas while the gas expands against a piston and does work $W = Q$. During this reversible isothermal expansion, the entropy of the heat reservoir decreases by Q/T and that of the gas increases by an equal amount Q/T. As we have mentioned, the work done by the gas equals the heat absorbed and can be calculated using calculus. The result is Equation 12-15 in Section 12-3:

$$W = Q = nRT \ln \frac{V_2}{V_1} \qquad\qquad 14\text{-}14$$

Since $V_2 = 2V_1$, the entropy change of the gas is

$$\Delta S = \frac{Q}{T} = nR \ln 2$$

Figure 14-13
A free expansion of a gas. When the stopcock is opened, the gas rushes into the compartment on the right and eventually comes to equilibrium, occupying both compartments. There is no work done, and no heat transferred to or from the gas, so the final energy is the same as the initial energy. This is an irreversible process. Left to itself, the gas will not move back into the left compartment and leave the right compartment empty.

In the free expansion, the entropy change of the gas is also $nR \ln 2$, but there is no heat reservoir, and no work is done. Thus, the entropy change of the universe in the free expansion is $nR \ln 2$. This process wastes energy in the sense that the gas could have been used to absorb heat Q from a reservoir and change it completely into work W. After the free expansion, this capability is no longer there because the gas is now at a greater volume and lower pressure. Thus, in this irreversible process, energy becomes unavailable for doing work and the entropy of the universe increases.

Why is this process irreversible? Why can't the gas compress by itself back into its original volume? Since there is no energy change involved, such a process does not violate the first law of thermodynamics. We will see that the reason the gas does not compress itself is merely that such a compression is extremely improbable. Let us first assume that there are only 10 molecules in the gas and that, initially, they occupy the entire volume. If we assume that each molecule moves about the total volume randomly, there will be an equal chance of a given molecule being in the left or right half of the volume. The chance that any 1 molecule will be in the left half at any given time will be $1/2$. The chance that any 2 particular molecules will both be in the left half will be $(1/2) \times (1/2) = 1/4$. (This is the same as the chance that a coin flipped twice will come up heads twice.) The chance that 3 particular molecules will be in the left side will be $(1/2) \times (1/2) \times (1/2) = (1/2)^3 = 1/8$. The chance that all 10 will be in the left half is $(1/2)^{10} = 1/1024$. That is, there is 1 chance in 1024, or about one chance in a thousand, that all 10 molecules will be in the left compartment at any given time. Though this probability is small, we would not be too surprised to see all 10 molecules in one side. For example, if we made measurements 1024 times, we would expect that, on the average, during one of the measurements, all 10 of the molecules would be in the left half of the volume. This would seemingly violate the second law of thermodynamics. If we started with all 10 molecules randomly distributed and we found them all in the left half of the volume, the entropy of the universe would have decreased. (Note, however, that the decrease would be extremely small because the number of moles n corresponding to 10 molecules is very small. For 6×10^{23} molecules, $n = 1$, so for 10 molecules, $n = 1.67 \times 10^{-23}$.) Of course, we have other difficulties applying thermodynamics to a system of just 10 molecules. For example, our concepts of temperature and pressure apply only to systems with a very large number of molecules. However, this example illustrates the connection between irreversible processes and probability. If we have 6×10^{23} molecules rather than just 10, we can see that, if they start in random positions, the chance that all will wind up in the left half of the volume is extremely small, so small that it is essentially zero.

Some people are dismayed to learn that the second law of thermodynamics seems to say only that certain processes (such as the spontaneous conduction of heat from a cold body to a hot body) are improbable, not impossible. In particular, it is dismaying to learn that if we apply the second law to a system with a very small number of molecules, there is a reasonable chance of its being violated. However, thermodynamics itself is applicable only to systems with a very large number of molecules, that is, to macroscopic systems. When applied to macroscopic systems, the probability of a decrease in entropy of the universe is so extremely small that the distinction between improbable and impossible becomes blurred.

Summary

1. A heat engine removes heat Q_h from a hot reservoir, does work W, and exhausts heat Q_c to a cold reservoir. The efficiency of a heat engine is the ratio of the work done to the total heat absorbed:

$$\epsilon = \frac{W}{Q_h} = 1 - \frac{Q_c}{Q_h}$$

2. A refrigerator uses work W to remove heat Q_c from a cold reservoir and exhausts heat $Q_h = Q_c + W$ to a warmer reservoir. The coefficient of performance of a refrigerator is the ratio of the heat removed to the work that must be added:

$$\text{COP} = \frac{Q_c}{W}$$

3. According to the Kelvin-Planck statement of the second law of thermodynamics, it is impossible for a heat engine working in a cycle to remove heat from a reservoir and convert it completely into work with no other effect. According to the Clausius statement of the second law, it is impossible to transfer heat from a cold reservoir to a hot reservoir with no other effect. These two statements of the second law of thermodynamics are equivalent.

4. The Carnot theorem states that no engine working between two heat reservoirs can be more efficient than a reversible engine. A reversible engine is called a Carnot engine. The efficiency of a Carnot engine is

$$\epsilon_C = 1 - \frac{T_c}{T_h}$$

For a Carnot engine, the ratio of the heat exhausted to the cold reservoir to the heat removed from the hot reservoir is

$$\frac{Q_c}{Q_h} = \frac{T_c}{T_h}$$

The coefficient of performance of a Carnot refrigerator or heat pump is

$$\text{COP}_{\max} = \frac{T_c}{\Delta T}$$

where $\Delta T = T_h - T_c$. The second-law efficiency of a heat engine is the ratio of the actual efficiency to the Carnot efficiency:

$$\epsilon_{\text{SL}} = \frac{\epsilon}{\epsilon_C}$$

The second-law efficiency of a refrigerator or heat pump is the ratio of the actual coefficient of performance to the Carnot coefficient of performance:

$$\epsilon_{\text{SL}} = \frac{\text{COP}}{\text{COP}_{\max}}$$

5. A heat pump is a device that removes energy from a cold reservoir, such as the air outside a house, and exhausts heat into a hot reservoir, such as the air inside a house. The work W needed to exhaust heat Q_h into a hot reservoir is related to the coefficient of performance by

$$W = \frac{Q_h}{\text{COP} + 1}$$

6. During an irreversible process, the universe moves toward a state of less order. A measure of disorder and of the reversibility of a process is the entropy function. The change in entropy of a system is given by

$$\Delta S = \frac{\Delta Q_{\text{rev}}}{T}$$

where Q_{rev} is the heat added in a reversible process connecting the initial and final states of the system.

7. The entropy of a system can increase or decrease, but the entropy of the universe or of any other isolated system never decreases. For a reversible process, the entropy of the universe remains constant; for an irreversible process, it increases. The statement that the entropy of the universe never decreases is a statement of the second law of thermodynamics, equivalent to the Kelvin-Planck and Clausius statements.

8. During an irreversible process, the entropy of the universe S_u increases and a certain amount of energy

$$W_{\text{lost}} = T\,\Delta S_u$$

becomes unavailable for doing work.

9. Entropy is related to probability. A highly ordered system is one of low probability and low entropy. An isolated system moves toward a state of low order, high probability, and high entropy.

Suggestions for Further Reading

Angrist, Stanley W.: "Perpetual Motion Machines," *Scientific American*, January 1968, p. 114.

An amusing account of some of these machines, all of which would violate either the first or second law of thermodynamics.

Dyson, Freeman J.: "What Is Heat," *Scientific American*, September 1954, p. 58.

A discussion of the relationship of energy and entropy to heat, and a short history of our evolving concepts of heat.

Ehrenberg, W.: "Maxwell's Demon," *Scientific American*, November 1967, p. 103.

Maxwell's thought experiment involving a violation of the second law of thermodynamics by a small, intelligent being has troubled physicists for over 100 years.

Tribus, Myron, and Edward C. McIrvine: "Energy and Information," *Scientific American*, September 1971, p. 179.

The concept of entropy can also be applied to the transmission of information, forming the basis of modern information theory.

Wilson, S. S.: "Sadi Carnot," *Scientific American*, August 1981, p. 134.

A brief biography of this French investigator, who was concerned mainly with the practical improvement of steam engines and other heat engines.

Review

A. Objectives: After studying this chapter, you should:

1. Be able to give the definition of the efficiency of a heat engine and the coefficient of performance of a refrigerator.

2. Be able to give both the Kelvin-Planck and Clausius statements of the second law of thermodynamics and illustrate their equivalence with a numerical example.

3. Be able to state the Carnot theorem and illustrate it with a numerical example.

4. Be able to give the expression for the Carnot efficiency of a heat engine.

5. Be able to describe how a heat pump works.

6. Be able to discuss the concept of entropy, including its interpretation in terms of the unavailability of energy.

B. Define, explain, or otherwise identify:

Heat engine, pp. 326–328
Efficiency of a heat engine, p. 328
Kelvin-Planck statement of the second law, p. 328
Clausius statement of the second law, p. 329
Coefficient of performance, p. 329
Carnot theorem, p. 331
Carnot engine, p. 333
Carnot efficiency, p. 334

Second-law efficiency, p. 335
Heat pump, p. 338
Entropy, p. 340
Unavailable energy, p. 342

C. True or false: If the statement is true, explain why. If it is false, give a counterexample.

1. Work can never be converted completely into heat.

2. Heat can never be converted completely into work.

3. All heat engines have the same efficiency.

4. It is impossible to transfer a given quantity of heat from a cold reservoir to a hot reservoir.

5. The coefficient of performance of a refrigerator cannot be greater than 1.

6. All Carnot engines are reversible.

7. The entropy of a system can never decrease.

8. The entropy of the universe can never decrease.

Exercises

Section 14-1   Heat Engines and the Kelvin-Planck Statement of the Second Law of Thermodynamics

1. An engine with 20 percent efficiency does 100 J of work in each cycle. (*a*) How much heat is absorbed in each cycle? (*b*) How much heat is rejected in each cycle?

2. An engine removes 230 J from a heat reservoir and does 38 J of work in each cycle. (*a*) What is its efficiency? (*b*) How much heat does it exhaust in each cycle?

3. An engine with 35 percent efficiency exhausts 45 J of energy each cycle. (*a*) How much heat does it absorb during each cycle? (*b*) How much work does it do in each cycle?

4. An engine absorbs 100 J and rejects 60 J in each cycle. (*a*) What is its efficiency? (*b*) If each cycle takes 0.5 s, find the power output of this engine in watts.

5. An engine with an output of 200 W has an efficiency of 30 percent. It works at 10 cycles/s. (*a*) How much work is done in each cycle? (*b*) How much heat is absorbed and how much is rejected in each cycle?

Section 14-2   Refrigerators and the Clausius Statement of the Second Law of Thermodynamics

6. A refrigerator absorbs 15 kJ from a cold reservoir and rejects 20 kJ to a hot reservoir. (*a*) How much work must be added? (*b*) What is the coefficient of performance of this refrigerator?

7. A refrigerator has a coefficient of performance of 5.0. It removes 25 kJ of heat from a cold reservoir in each cycle. (*a*) How much work must be done? (*b*) How much heat is rejected to the hot reservoir?

8. A refrigerator requires 80 J of work per cycle to remove 300 J from a cold reservoir. (*a*) What is its coefficient of performance. (*b*) How much heat is rejected to the hot reservoir?

9. A refrigerator absorbs 5 kJ from a cold reservoir and rejects 8 kJ. (*a*) What is its coefficient of performance? (*b*) The refrigerator is reversible and is run backward as a heat engine between the same two reservoirs. What is its efficiency?

Section 14-3   Equivalence of the Kelvin-Planck and Clausius Statements

10. A certain engine running at 30 percent efficiency draws 200 J of heat from a hot reservoir. Assume that the Clausius statement is false, and show how this engine combined with a perfect refrigerator can violate the Kelvin-Planck statement of the second law.

11. A certain refrigerator takes in 500 J of heat from a cold reservoir and rejects 800 J to a hot reservoir. Assume that the Kelvin-Planck statement is false, and show how a perfect engine working with this refrigerator can violate the Clausius statement.

Section 14-4   The Carnot Engine

12. A heat engine works between a hot reservoir at 500°C and a cold reservoir at 150°C. What is its maximum possible efficiency?

13. A Carnot engine removes 200 kJ in each cycle from a hot reservoir at 300°C. It does 150 kJ of work in each cycle. What is the temperature of the cold reservoir?

14. A Carnot engine works between reservoirs at 600 and 300°C. It absorbs 80 J of heat from the hot reservoir during each cycle. (a) How much heat is rejected to the cold reservoir during each cycle? (b) How much work is done in each cycle? (c) What is the efficiency of this engine?

15. A Carnot engine removes 200 J of heat from a hot reservoir and rejects 60 J to the atmosphere at 20°C. (a) How much work is done? (b) What is the efficiency of this engine? (c) What is the temperature of the hot reservoir?

16. An engine is designed to exhaust heat to the atmosphere at 20°C. What is the least possible temperature of the hot reservoir if the engine is to have an efficiency of 20 percent?

17. A Carnot engine works between two heat reservoirs at temperatures of 300 and 200 K. (a) What is its efficiency? (b) If it absorbs 120 J from the hot reservoir during each cycle, how much work does it do? (c) What is the coefficient of performance of this engine when it works backward as a refrigerator between these same two reservoirs?

18. An engine draws heat from a reservoir at 100°C. If the engine is to be 40 percent efficient, what is the greatest possible temperature of the cold reservoir?

19. In each cycle, an engine removes 150 J from a reservoir at 100°C and rejects 125 J to a reservoir at 20°C. (a) What is the efficiency of this engine? (b) What is its second-law efficiency?

20. An engine has a second-law efficiency of 85 percent. In each cycle, it removes 200 kJ of heat from a hot reservoir at 500 K and exhausts heat to a reservoir at 200 K. (a) What is the efficiency of this engine? (b) How much work is done in each cycle? (c) How much heat is exhausted in each cycle?

21. An engine removes 250 J from a reservoir at 300 K and exhausts 200 J to a reservoir at 200 K. (a) What is its efficiency? (b) What is its second-law efficiency? (c) How much more work could be done if the engine were reversible?

22. An engine has a second-law efficiency of 75 percent. In each cycle, it removes 300 J from a reservoir at 200°C and rejects 100 J to a cold reservoir. (a) What is the efficiency of this engine? (b) What would be the Carnot efficiency of an engine working between the same two reservoirs? (c) What is the temperature of the cold reservoir? (d) How much more work could be done by a Carnot engine working between the same reservoirs if it removed 300 J in each cycle?

23. A steam power plant has a second-law efficiency of 58 percent. It takes in steam at 245°C and exhausts it at 37°C. (a) What is the actual efficiency? (b) How much heat is exhausted when it does 1000 J of work?

Section 14-5   The Heat Pump

24. A heat pump delivers 20 kW to heat a house. The outside temperature is −10°C and the inside temperature of the hot-air blower is 40°C. (a) What is the coefficient of performance of a Carnot heat pump operating between these temperatures? (b) What is the minimum power engine needed to run the heat pump? (c) If the second-law efficiency of the heat pump is 60 percent, what power engine is needed?

25. Rework Exercise 24 if the outside temperature is −20°C.

26. A refrigerator is rated at 370 W. (a) What is the maximum amount of heat it can remove in 1 s if the inside is at 0°C and it exhausts into a room at 20°C? (b) If the second-law efficiency of the refrigerator is 70 percent, how much heat can it remove in 1 s?

27. Rework Exercise 26 if the room is 35°C.

Section 14-6   Entropy and Disorder

28. A 5-kg block is dropped from rest at a height of 6 m above the ground. It hits the ground and comes to rest. The block, ground, and atmosphere are all at 300 K. What is the entropy change of the universe for this process?

29. If 500 J of heat is conducted from a reservoir at 400 K to one at 300 K, find the entropy change of (a) the hot reservoir, (b) the cold reservoir, and (c) the universe. (d) How much work could have been done with the 500 J if it were removed by a Carnot engine working between these two reservoirs?

30. An engine working in a cycle removes 200 J from a hot reservoir at 600 K, does 80 J of work, and exhausts 120 J to a cold reservoir at 300 K. Find the entropy change of (a) the hot reservoir, (b) the cold reservoir, and (c) the universe. (d) How much work could have been done by a Carnot engine removing 200 J from the hot reservoir and exhausting heat to the same cold reservoir. (e) Show that the difference between your answer to (d) and the 80-J work done by the original engine is $T \, \Delta S_u$, where $T = 300$ K.

31. Calculate the entropy change of 1 kg of water at 100°C when it changes to steam under standard conditions.

32. A 200-kg block of ice at 0°C is placed in a large lake. The temperature of the lake is just slightly higher than 0°C, and the ice melts. (*a*) What is the entropy change of the ice? (*b*) What is the entropy change of the lake? (*c*) What is the entropy change of the universe?

33. Calculate the rate at which the entropy of the universe increases because of the transfer of 2000 kcal per day from a human body at 37°C to the atmosphere at 20°C.

Section 14-7   Entropy and Probability

There are no exercises for this section.

Problems

1. A 1500-kg car traveling at 100 km/h crashes into a concrete wall. If the temperature of the air is 20°C, calculate the entropy change of the universe.

2. (*a*) Show that if the Clausius statement of the second law were not true, the entropy of the universe could decrease. (*b*) Show that if the Kelvin-Planck statement were not true, the entropy of the universe could decrease.

3. An engine operates with a working substance that is 1 mol of an ideal gas with $C_v = \frac{3}{2}R$ and $C_p = \frac{5}{2}R$. The cycle begins at $P_1 = 1$ atm and $V_1 = 24.6$ L. The gas is heated at constant volume to $P_2 = 2$ atm. It then expands at constant pressure until $V_2 = 49.2$ L. During these two steps, heat is absorbed. The gas is then cooled at constant volume until its pressure is again 1 atm. It is then compressed at constant pressure back to its original state. During the last two steps, heat is rejected. (*a*) Indicate this cycle on a *PV* diagram. Find (*b*) the work done, (*c*) the heat absorbed, and (*d*) the heat rejected during each cycle. (*e*) Find the efficiency of this engine.

4. Show that the coefficient of performance of a Carnot refrigerator working between two reservoirs at temperatures T_h and T_c is related to the efficiency of a Carnot engine by

$$\text{COP}_{\text{max}} = \frac{T_c}{\epsilon_C T_h}$$

5. A Carnot engine extracts 4 kJ from a hot reservoir at 120°C to blow up a balloon, increasing its volume by 4 L at a constant pressure of 1 atm. (*a*) How much heat is exhausted to the cold reservoir? (*b*) What is the temperature of the cold reservoir?

6. A reversible engine working between reservoirs at temperatures T_h and T_c has an efficiency of 30 percent. Working as a heat engine it rejects 140 J of heat to the cold reservoir. A second engine working between the same two reservoirs also rejects 140 J to the cold reservoir. Show that if the second engine has an efficiency of greater than 30 percent, the two engines can work together to violate the Kelvin-Planck statement of the second law.

7. A reversible engine working between reservoirs at temperatures T_h and T_c has an efficiency of 20 percent. Working

as a heat engine, it does 100 J of work in each cycle. A second engine working between the same two reservoirs also does 100 J of work in each cycle. Show that if the efficiency of the second engine is greater than 20 percent, the two engines can work together to violate the Clausius statement of the second law.

8. One mole of an ideal gas first undergoes an adiabatic free expansion from $V_1 = 12.3$ L and $T_1 = 300$ K to $V_2 = 24.6$ L and $T_2 = 300$ K. It is then compressed isothermally and reversibly back to its original state. (*a*) What is the entropy change of the gas for the complete cycle? (*b*) What is the entropy change of the universe for the complete cycle?

9. A refrigerator has a temperature of −5°C in its ice-cube compartment. It exhausts heat into a room at 25°C. (*a*) Calculate the minimum energy input to the refrigerator needed to make ice cubes of 1 kg of water initially at 20°C. (*b*) Calculate the energy needed if the second-law efficiency of the refrigerator is 70 percent. (Assume the compartment remains at −5°C, and calculate only the energy needed to freeze the water at 0°C.)

10. A Carnot engine works between two heat reservoirs as a refrigerator. It removes 100 J of heat from the cold reservoir and rejects 150 J to the hot reservoir during each cycle. (*a*) What is its coefficient of performance? (*b*) What is the efficiency of the Carnot engine when it is working as a heat engine between the same two reservoirs? (*c*) Show that no other engine working as a refrigerator between the same two reservoirs can have a coefficient of performance greater than that found in (*a*).

11. A Carnot engine works between two heat reservoirs at temperatures $T_h = 300$ K and $T_c = 200$ K. (*a*) What is its efficiency? (*b*) If it absorbs 100 J from the hot reservoir during each cycle, how much work does it do? How much heat does it reject during each cycle? (*c*) What is the coefficient of performance of this engine when it works as a refrigerator between these two reservoirs?

12. A heat engine works in a cycle between reservoirs at 400 and 200 K. It absorbs 1000 J of heat from the hot reservoir and does 200 J of work in each cycle. (*a*) What is its efficiency? (b) Find the entropy change of the engine, each reservoir, and the universe for each cycle. (c) How much work could be done by a Carnot engine working between these same reservoirs and absorbing 1000 J of heat in each cycle? (d) Show that the difference in the work done by a Carnot engine and the work done by the original engine is $T_c \Delta S_u$, where ΔS_u is that calculated in part (b).

13. Two moles of an ideal gas at $T = 400$ K expand isothermally and reversibly from an initial volume of 40 L to a final volume of 80 L. (a) Find the entropy change of the gas for this process. (b) Find the entropy change of the universe for this process. (c) This same gas is at 400 K and is originally in one-half of an 80-L container that is shielded from its surroundings. The partition is broken and the gas undergoes a free adiabatic expansion from 40 L to 80 L and eventually comes to equilibrium. Find the entropy change of the gas for this process. (d) Find the entropy change of the universe for this process.

14. A steam engine takes in superheated steam at 270°C and discharges condensed steam from its cylinder at 50°C. Its second-law efficiency is 75 percent. (a) If the useful power output of the engine is 200 kW, how much heat does the engine discharge in 1 h? (b) If a Carnot engine worked between these reservoirs with a power output of 200 kW, how much heat would it discharge in 1 h?

15. An engine using 1 mol of an ideal gas initially at $V_1 = 24.6$ L and $T = 400$ K performs a cycle consisting of four steps: (1) isothermal expansion at $T = 400$ K to twice its volume, (2) cooling at constant volume to $T = 300$ K, (3) isothermal compression to its original volume, and (4) heating at constant volume to its original temperature of $T = 400$ K. Assume that $C_v = 21$ J/K. Sketch the cycle on a PV diagram and calculate its efficiency.

CHAPTER FIFTEEN

Oscillations

With this chapter, we return to the field of dynamics, specifically to the study of vibrations and waves. Both of these terms are familiar. *Vibration* may call to mind the twang of a guitar string, the blur of a hummingbird's wings, or the nearly silent vibration of the balance wheel of a fine watch. Similarly, *wave* may make us think of the rolling surface of the ocean or ripples on a pond. Familiar as these things are, their significance pales in comparison to the fact that most of our contact with the world outside ourselves — everything we see and hear — comes to us via the vibrations and waves responsible for sight and sound. We will begin our study of vibrations and waves with the basics, oscillations.

In the case of a cock putting its head into an empty utensil of glass where it crowed so that the utensil thereby broke, the whole cost shall be payable.

THE TALMUD

There are many familiar examples of oscillations: Small boats bob up and down, clock pendulums swing back and forth, and the strings and reeds of musical instruments vibrate. Other, less familiar examples are the oscillations of air molecules in a sound wave and the oscillations of electrical currents in radios or television sets. The most recognizable characteristic of oscillatory motion is that the motion repeats itself. The time required for each repetition is called the *period*.

Oscillation occurs when a system is disturbed from a position of stable equilibrium. The three kinds of equilibrium, stable, neutral, and unstable, were discussed in Section 5-3. Figure 15-1a, which shows an object resting at the bottom of a bowl, illustrates stable equilibrium. If the object is displaced slightly, the resultant force urges the object toward its equilibrium position, and the object oscillates about this position. Unstable equilibrium is illustrated in Figure 15-1b by the object resting on top of an inverted bowl. If the object is displaced slightly from its position of unstable equilibrium, it tends to move farther away. In Figure 15-1c, the object is in a position of neutral equilibrium. When it is displaced from that position, there is no tendency for it to move either toward or away from equilibrium. Oscillations do not occur near positions of unstable or neutral equilibrium.

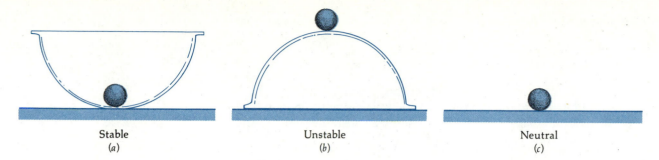

Stable
(a)

Unstable
(b)

Neutral
(c)

Wave motion is closely related to oscillatory motion. Sound waves, for example, can be produced by a vibrating string (such as a violin string), a vibrating oboe reed, a vibrating drum head, or the vibrations of your vocal cords when you speak. In each case, the vibrating system sets up oscillations in the nearby air molecules, and these oscillations are propagated through the air (or some other medium, such as water or a solid).

In this chapter, we will study some of the fundamentals of oscillating systems. In Chapter 16, we will study mechanical waves—sound waves and waves on strings.

Figure 15-1
(a) Stable equilibrium. If the object is displaced slightly from its equilibrium position at the bottom of the bowl, it oscillates about its equilibrium position. (b) Unstable equilibrium. If the object is displaced slightly, it moves away from its equilibrium position. (c) Neutral equilibrium. If the object is displaced, it has no tendency to move either toward or away from its equilibrium position.

15-1   Simple Harmonic Motion: A Mass on a Spring

A common and very important kind of oscillatory motion is *simple harmonic motion*. Simple harmonic motion occurs when the restoring force is proportional to the displacement from equilibrium. This condition is nearly always met, at least approximately, when the displacement from equilibrium is small.

A typical system that exhibits simple harmonic motion is an object attached to a spring, as illustrated in Figure 15-2. In equilibrium, the spring exerts no force on the object. When the object is displaced an amount x from its equilibrium position, the spring exerts a force $-kx$, as given by Hooke's law (Section 4-5).

$$F = -kx \qquad\qquad 15\text{-}1$$

The minus sign in Hooke's law arises because the force is opposite the direction of the displacement. If we choose x to be positive for displacements to the right, the force is negative (to the left) when x is positive and positive (to the right) when x is negative. Combining Equation 15-1 with Newton's second law, we have

$$F = -kx = ma$$

or

$$a = -\left(\frac{k}{m}\right)x \qquad\qquad 15\text{-}2$$

The acceleration is proportional to the displacement and is oppositely directed. This is a general characteristic of simple harmonic motion and can, in

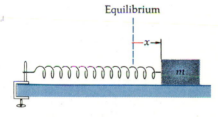

Equilibrium

Figure 15-2
An object on a spring resting on a frictionless table. The displacement x is measured from the equilibrium position.

fact, be used to identify systems that will exhibit simple harmonic motion:

Whenever the acceleration is proportional to the displacement and is oppositely directed, simple harmonic motion occurs.

Condition for simple harmonic motion

If we displace the object from equilibrium and release it, it oscillates back and forth about its equilibrium position with simple harmonic motion. The time for the object to make a complete oscillation is called the *period T*. The reciprocal of the period is the *frequency f*.

$$f = \frac{1}{T} \qquad\qquad 15\text{-}3$$

Frequency and period

The unit of frequency is the reciprocal second (s^{-1}), which is called a hertz (Hz). For example, if the time for one complete oscillation is 2 s, the frequency is 0.5 Hz. The maximum displacement from equilibrium is called the *amplitude A*.

The displacement *x* as a function of time *t* can be obtained experimentally. Let us attach a marking pen to an object and place it over a strip of paper that can be moved perpendicular to the direction of oscillation, as shown in Figure 15-3. We then displace the object a distance *A* and pull the paper to the left with constant speed as we release the object. The pen traces out the sinusoidal curve shown in the figure.

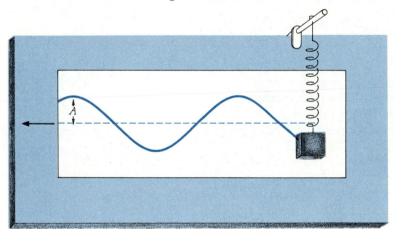

Figure 15-3
As the paper is moved with constant speed to the left, the marker attached to the object on the spring traces out the displacement *x* as a function of time *t*.

The equation for this curve is

$$x = A \cos 2\pi ft \qquad\qquad 15\text{-}4$$

where *f* is the frequency. The corresponding equations for the velocity and acceleration of the object are

$$v = -(2\pi f)A \sin 2\pi ft \qquad\qquad 15\text{-}5$$

Displacement, velocity, and acceleration for simple harmonic motion

and

$$a = -(2\pi f)^2 A \cos 2\pi ft \qquad\qquad 15\text{-}6$$

We will derive these equations in Section 15-2. Since the displacement is given by $x = A \cos 2\pi ft$, Equation 15-6 can be written

$$a = -(2\pi f)^2 x \qquad\qquad 15\text{-}7$$

Acceleration is proportional to displacement

Equation 15-7 is a general result relating the acceleration, displacement, and frequency of simple harmonic motion. The variations of displacement, velocity, and acceleration with time are plotted in Figure 15-4. Initially, at time $t = 0$, the velocity is zero. It then becomes negative as the object moves back up toward its equilibrium position. The acceleration is negative at $t = 0$ and is given by $a = -(2\pi f)^2 A$. The object is at its equilibrium position $x = 0$ when $\cos 2\pi ft = 0$. The acceleration at these times is also zero, and the speed has its maximum value of $(2\pi f)A$ because, when $\cos 2\pi ft$ is zero, $\sin 2\pi ft$ is either $+1$ or -1.

Comparing Equation 15-7 with $a = -(k/m)x$ from Equation 15-2, we see that the frequency of oscillation is related to the spring constant k and the mass m by

$$(2\pi f)^2 = k/m$$

$$f = \frac{1}{2\pi} \sqrt{\frac{k}{m}} \qquad\qquad 15\text{-}8$$

The period is then given by

$$T = \frac{1}{f} = 2\pi \sqrt{\frac{m}{k}} \qquad\qquad 15\text{-}9$$

Frequency and period for mass on spring

From this result, we can see that when k is large, as in the case of a stiff spring, the frequency is large. Similarly, if the mass is large, the frequency is small.

Example 15-1   A 2-kg object stretches a spring 10 cm when it hangs vertically in equilibrium. The object is then attached to the same spring resting on a frictionless table and fixed at one end, as shown in Figure 15-2. The object is held a distance 5 cm from the equilibrium position and released. Find the amplitude A, period T, and frequency f.

The force constant of the spring is determined from the first measurement. In equilibrium, the downward force mg must equal the upward force kx_0 where x_0 is the amount the spring is stretched from its natural length. Using $g = 9.81 \text{ m/s}^2$, $m = 2 \text{ kg}$, and $x_0 = 10 \text{ cm} = 0.1 \text{ m}$, we have

$$k = \frac{mg}{x_0} = \frac{(2 \text{ kg})(9.81 \text{ m/s}^2)}{0.1 \text{ m}} = 196 \text{ N/m}$$

In the horizontal position, the spring is initially stretched 5 cm from equilibrium and released. The maximum displacement is 5 cm, so the amplitude A is 5 cm. The period is related to the force constant k and the mass m by Equation 15-9:

$$T = 2\pi \sqrt{\frac{m}{k}} = 2\pi \sqrt{\frac{2 \text{ kg}}{196 \text{ N/m}}} = 0.63 \text{ s}$$

The frequency is the reciprocal of the period

$$f = 1/T = 1.58 \text{ s}^{-1} = 1.58 \text{ Hz}$$

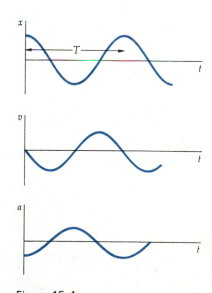

Figure 15-4
Plots of the displacement x, the velocity v, and the acceleration a as functions of time t for simple harmonic motion such as that of a mass on a spring.

Example 15-2   What is the maximum speed of the object on the spring in Example 15-1, and when does this maximum speed first occur?

The velocity of the object at any time is given by Equation 15-5. The maximum value of the velocity occurs when $\sin 2\pi ft = 1$ (or -1). The magnitude of the velocity at this time is

$$v_{max} = 2\pi fA = (2\pi)(1.58 \text{ s}^{-1})(5 \text{ cm}) = 49.6 \text{ cm/s}$$

This maximum speed first occurs after one-fourth of a period, when the object first reaches the equilibrium position. This is $(0.63 \text{ s})/4 = 0.16$ s after the object is released. At this time, $\sin 2\pi ft = +1$, and the velocity is negative, indicating that the object is moving to the left.

Example 15-3   A second spring, identical to that in Example 15-1, is attached to a second object, also of mass 2 kg. This spring is stretched a distance 10 cm from equilibrium and released at the same time as the first, which is stretched to 5 cm. Which object reaches the equilibrium position first?

Figure 15-5 shows the initial positions of the objects, and Figure 15-6 shows plots of the position functions for the two objects. Since both springs have the same force constant and the masses are equal, the frequencies and periods of the motions are equal. Only the amplitudes differ. But according to Equation 15-9, the period depends only on k and m and not on the amplitude. *Thus, both objects reach the equilibrium position at the same time.* The second object has twice as far to go to reach the equilibrium position as the first object, but it also begins with twice the initial acceleration.

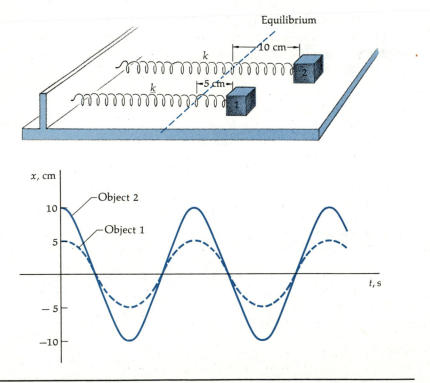

Figure 15-5
Two objects on identical springs released simultaneously. They reach their equilibrium positions at the same time because the period depends on the mass and the force constant, which are the same, and not on the amplitude.

Figure 15-6
Plots of displacement versus time for the objects in Figure 15-5.

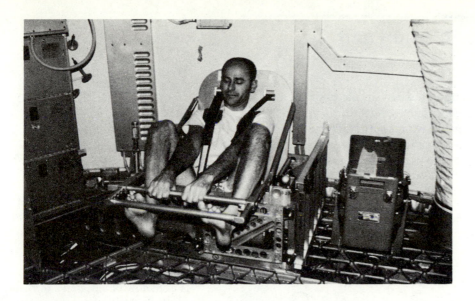

Astronaut Alan L. Bean measuring his body mass during the second Skylab mission. The total mass of the astronaut plus the apparatus is related to the period of vibration by Equation 15-9.

This last example illustrates a very important general property of simple harmonic motion.

In simple harmonic motion, the frequency and period are independent of the amplitude.

The fact that the frequency in simple harmonic motion is independent of the amplitude has important consequences in many fields. In music for example, it means that the frequency of a note struck on the piano does not depend on how loud the note is played and does not change as the amplitude of the note decreases. For many musical instruments, there is a slight dependence of the frequency on amplitude (the pitch of an oboe reed, for example, depends slightly on how hard it is blown because the vibration is not exactly simple harmonic), but a skilled musician can adjust for this. If changes in amplitude had a large effect on frequency, musical instruments would be unplayable.

Questions

1. Give some additional examples of oscillatory motion.

2. How far does a particle oscillating with amplitude A move in one full period? What is its displacement after one full period?

3. If you know that the speed of an oscillator of amplitude A is zero at certain times, can you say exactly what its displacement is at those times?

4. What is the magnitude of the acceleration of an oscillator of amplitude A and frequency f when its speed is maximum? When its displacement is maximum?

15-2 Simple Harmonic Motion and Circular Motion

There is a simple but important relation between simple harmonic motion and circular motion with constant speed. Figure 15-7 shows a peg on the rim of a rotating turntable and an object hanging on a spring. The shadow of the peg and that of the object are projected on a screen. If the period of rotation is adjusted so that it is equal to that of the oscillating object and the amplitude of the spring system equals the radius of the circle, the shadows move together:

The projection on a straight line of a particle moving with uniform circular motion is simple harmonic motion.

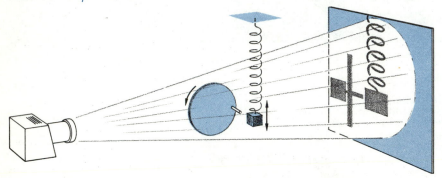

Figure 15-7
The shadow of a peg on a turntable is projected on a screen along with the shadow of an object on a spring. When the period of rotation of the turntable equals the period of oscillation of the object on the spring, the shadows move together.

We can use this result to obtain Equations 15-4, 15-5, 15-6, and 15-7 for the position, velocity, and acceleration versus time in simple harmonic motion.

Figure 15-8 shows a particle moving in a circle of radius A with constant speed v_0. Its angular velocity ω (see Chapter 8) is also constant and is related to its linear velocity by

$$v_0 = A\omega \qquad \text{15-10}$$

Since the particle travels a distance $2\pi A$ during one revolution, the period and frequency of the circular motion are found from

$$v_0 T = 2\pi A$$

$$T = \frac{2\pi A}{v_0} = \frac{2\pi}{\omega} \qquad \text{15-11}$$

and

$$f = \frac{1}{T} = \frac{v_0}{2\pi A} = \frac{\omega}{2\pi} \qquad \text{15-12}$$

The angular velocity of the particle, which equals $2\pi f$, is also called the angular frequency. If the particle begins on the x axis at time $t = 0$, its angular displacement at a later time is given by

$$\theta = \omega t = 2\pi ft$$

From the figure, we can see that the x component of the position of the particle is given by

$$x = A \cos \theta = A \cos 2\pi ft \qquad \text{15-13}$$

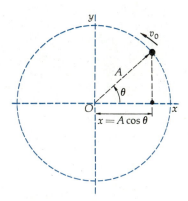

Figure 15-8
A particle moving in a circle with constant speed. The angle θ increases with time, $\theta = 2\pi ft$, where f is the frequency of the circular motion, which is related to the speed v_0 and radius A by $f = v_0/(2\pi A)$. The x component of the motion is simple harmonic motion.

This is the same as Equation 15-4 for simple harmonic motion. Figure 15-9 shows the velocity and acceleration vectors of the particle moving in a circle. The x component of the velocity is

$$v_x = -v_0 \sin 2\pi ft$$

Using $v_0 = \omega A = 2\pi fA$, we obtain

$$v_x = -(2\pi f)A \sin 2\pi ft \qquad\qquad 15\text{-}14$$

which is the same as Equation 15-5. The acceleration of the particle is the centripetal acceleration directed toward the center of the circle with a magnitude

$$a = \frac{v_0^2}{A} = \frac{(2\pi fA)^2}{A} = A(2\pi f)^2$$

The x component of the acceleration is

$$a_x = -A(2\pi f)^2 \cos 2\pi ft \qquad\qquad 15\text{-}15$$

or

$$a_x = -(2\pi f)^2 x \qquad\qquad 15\text{-}16$$

which are the same as Equations 15-6 and 15-7, respectively.

Example 15-4   A particle on the rim of a record moves in a circle of radius 15 cm, making 33.3 revolutions per minute. (a) Write equations for the x components of the velocity and the acceleration of the particle, assuming it is at $x = A = 15$ cm at $t = 0$. (b) What are the maximum magnitudes of v_x and a_x, and when do they first occur?

(a) The frequency of this motion is

$$f = (33.3 \text{ rev/min})(1 \text{ min}/60 \text{ s}) = 0.555/\text{s} = 0.555 \text{ Hz}$$

Note that the dimensionless unit "rev" can be dropped (1 rev/s = 1/s = 1 Hz). The period is

$$T = \frac{1}{f} = \frac{1}{0.555/\text{s}} = 1.80 \text{ s}$$

The x component of the velocity of the particle is given by Equation 15-14. It is slightly more convenient to write this equation in terms of the angular frequency

$$\omega = 2\pi f = (2)(3.14)(0.555/\text{s}) = 3.49 \text{ rad/s}$$

(The dimensionless unit "rad" can also be dropped whenever it is convenient.) Equation 15-14 then gives

$$v_x = -(2\pi f)A \sin 2\pi ft = -\omega A \sin \omega t$$

$$= -(3.49/\text{s})(0.15 \text{ m}) \sin \omega t = (0.523 \text{ m/s}) \sin \omega t$$

The x component of the acceleration is given by Equation 15-15. Writing ω for $2\pi f$, we have

$$a_x = -A\omega^2 \cos \omega t = -(0.15 \text{ m})(3.49/\text{s})^2 \cos \omega t$$

$$= -(1.83 \text{ m/s}^2) \cos \omega t$$

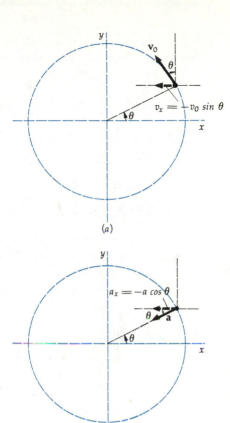

(a)

(b)

Figure 15-9
(a) Velocity vector for a particle moving in a circle with constant speed. The x component is $v_x = -v \sin \theta = -(2\pi f)A \sin 2\pi ft$.
(b) Acceleration vector for a particle moving in a circle with constant speed. The x component is $a_x = -a \cos \theta = -(v_0^2/A) \cos \theta = -(2\pi f)^2 A \sin 2\pi ft$.

(b) The acceleration a_x has its maximum magnitude of 1.83 m/s² when $\cos \omega t = \pm 1$, which first occurs at $t = 0$. At this time, the acceleration is negative, meaning it is to the left. At $t = 0$, $\sin \omega t = 0$ and $v_x = 0$. The velocity v_x has its maximum magnitude when $\sin \omega t = \pm 1$. This first occurs at $t = T/4$, that is, after one fourth of the period when the particle is at the top of the circle. Then $v_x = -0.523$ m/s. Since the period is 1.80 s, the particle is at the top of the circle at $t = (1.80 \text{ s})/4 = 0.45$ s.

15-3   Energy in Simple Harmonic Motion

When an object oscillates on a spring, it has kinetic and potential energy, both of which vary with time. The sum of these, which is the total energy, is constant. The potential energy for a spring of force constant k stretched a distance x from equilibrium is Equation 6-10:

$$U = \tfrac{1}{2}kx^2 \tag{15-17}$$

The kinetic energy is

$$E_k = \tfrac{1}{2}mv^2 \tag{15-18}$$

The total energy is the sum of the potential and kinetic energies:

$$E_{\text{tot}} = U + E_k = \tfrac{1}{2}kx^2 + \tfrac{1}{2}mv^2 \tag{15-19}$$

When the displacement is maximum, $x = A$, the velocity is zero, and the total energy is

$$E_{\text{tot}} = \tfrac{1}{2}kA^2 \tag{15-20}$$

This is an important general property of simple harmonic motion.

The total energy of an object oscillating with simple harmonic motion is proportional to the square of the amplitude.

    If we begin with the object at its maximum displacement, the total energy is originally just potential energy. The object then moves toward equilibrium, with its potential energy decreasing and its kinetic energy increasing. At the equilibrium point, its speed is maximum, the potential energy is zero, and the total energy is kinetic. As it moves beyond its equilibrium point, the kinetic energy decreases and the potential energy increases until it is at its maximum displacement in the other direction. At this time, it stops; its kinetic energy is zero and its potential energy is again maximum. At all times, the sum of the potential and kinetic energies is constant.

    We can demonstrate that the total energy is constant for an object oscillating on a spring by substituting the expressions of Equations 15-4 and 15-5 for x and v into Equations 15-17 and 15-18 for the potential and kinetic energies:

$$U = \tfrac{1}{2}k(A \cos 2\pi ft)^2 \tag{15-21}$$

and

$$E_k = \tfrac{1}{2}m[(2\pi fA) \sin 2\pi ft]^2$$

If we write the frequency f in terms of the force constant k from Equation 15-8, we have

$$(2\pi f)^2 = k/m$$

The kinetic energy can then be written

$$E_k = \tfrac{1}{2}kA^2 \sin^2 2\pi ft \qquad\qquad 15\text{-}22$$

Adding the potential and kinetic energies to get the total energy, we obtain

$$E_{tot} = \tfrac{1}{2}kA^2 \cos^2 2\pi ft + \tfrac{1}{2}kA^2 \sin^2 2\pi ft$$
$$= \tfrac{1}{2}kA^2 (\cos^2 2\pi ft + \sin^2 2\pi ft)$$
$$= \tfrac{1}{2}kA^2$$

where we have used the trigonometric identity $\cos^2\theta + \sin^2\theta = 1$.

Example 15-5   A 3-kg object attached to a spring oscillates with amplitude 4 cm and period 2 s. What is the total energy? What is the maximum speed?

The total energy is $\tfrac{1}{2}kA^2$, where k is the force constant of the spring, which is related to the period by

$$T = 2\pi\sqrt{m/k}$$

The force constant for this spring is thus

$$k = \frac{(2\pi)^2 m}{T^2}$$

$$= \frac{(4\pi^2)(3\text{ kg})}{4\text{ s}^2} = 29.6\text{ N/m}$$

The total energy is thus

$$E_{tot} = \tfrac{1}{2}kA^2$$
$$= \tfrac{1}{2}(29.6\text{ N/m})(0.04\text{ m})^2 = 2.37 \times 10^{-2}\text{ J}$$

We use this to find the maximum speed. When the speed is maximum, the potential energy is zero and the total energy is kinetic energy.

$$E_{tot} = \tfrac{1}{2}mv_{max}^2 = 2.37 \times 10^{-2}\text{ J}$$

The maximum speed is therefore

$$v_{max} = \sqrt{\frac{2E_{tot}}{m}}$$

$$= \sqrt{\frac{2(2.37 \times 10^{-2}\text{ J})}{2\text{ kg}}} = 0.126\text{ m/s}$$

We could have found the maximum speed from Equation 15-5, as we did in Example 15-2. Using $f = 1/T = 0.5$ s, we have

$$v_{max} = 2\pi fA = 2(3.14)(0.5\text{ s}^{-1})(4\text{ cm})$$
$$= 12.6\text{ cm/s} = 0.126\text{ m/s}$$

15-4   The Simple Pendulum

A familiar example of oscillatory motion is that of a pendulum. The motion of a pendulum is simple harmonic motion only if the amplitude of the motion is small. Figure 15-10 shows a simple pendulum consisting of a string of length L and a bob of mass m. The forces acting on the bob are its weight and the tension in the string. When the string makes an angle θ with the vertical, the weight has components $mg \cos \theta$ along the string and $mg \sin \theta$ perpendicular to the string. If the angle θ is small, the component perpendicular to the string is approximately in the negative x direction. Newton's law for motion along the x axis is then

$$F_x = ma_x = -mg \sin \theta \qquad \text{15-23}$$

From the figure, we can see that $\sin \theta = x/L$. The acceleration, from Equation 15-23, is then

$$a_x = -g \sin \theta = -g \frac{x}{L} = -\left(\frac{g}{L}\right)x \qquad \text{15-24}$$

We see that for small angles, the acceleration is proportional to the displacement and oppositely directed. The motion is therefore simple harmonic. Comparing this with Equation 15-7

$$a = -(2\pi f)^2 x$$

we have

$$(2\pi f)^2 = \frac{g}{L}$$

or

$$f = \frac{1}{2\pi}\sqrt{\frac{g}{L}} \qquad \text{15-25}$$

The period of the motion is then

$$T = \frac{1}{f} = 2\pi \sqrt{\frac{L}{g}} \qquad \text{15-26}$$

This is consistent with experimental observation that the greater the length of a pendulum, the greater the period. Note that the period does not depend on the mass. This is because the restoring force is proportional to the mass. The acceleration, $a = F/m$, is therefore independent of the mass. Note also that the frequency and period are independent of the amplitude of vibration, a general feature of simple harmonic motion.

Example 15-6   Find the period of a pendulum of length 1 m.

From Equation 15-26, we have

$$T = 2\pi \sqrt{\frac{1 \text{ m}}{9.81 \text{ m/s}^2}} = 2.01 \text{ s}$$

The validity of this result is easily demonstrated experimentally.

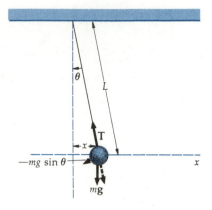

Figure 15-10
A simple pendulum. The forces on the bob are its weight $m\mathbf{g}$ and the tension $\mathbf{T}$. The force perpendicular to the string is $mg \sin \theta = mg(x/L)$. For small displacements, the motion of the pendulum is simple harmonic motion.

Frequency and period of a pendulum

Galileo discovered the fact that the period of a pendulum depends only on its length and not on either its mass or its amplitude. He went on to suggest that doctors use small pendulums for timing a patient's pulse by varying the length of the pendulum until its period was the same as the pulse. When Christian Huygens heard about Galileo's work, he used the new discoveries about pendulums to build a clock.

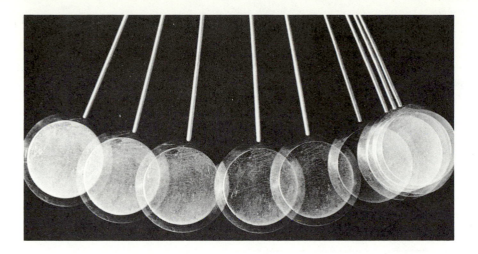

Motion of a simple pendulum. The bob is shown at equal time intervals. It moves faster at the bottom, as shown by the greater spacing of the images.

The acceleration of gravity can be easily measured using a simple pendulum. You need only measure the length L with a metrestick and the period T by determining the time for one oscillation. (One usually measures the time for n vibrations and then divides by n to reduce the error in the time measurement.) The acceleration of gravity is then determined by solving Equation 15-26 for g:

$$g = \frac{4\pi^2 L}{T^2}$$ 15-27

When the amplitude of vibration is not small, the motion of a pendulum is periodic, but it is not simple harmonic. In particular, the period of vibration does have a slight dependence on the amplitude. This dependence is usually written in terms of the *angular amplitude* θ_0, the maximum angle made by the string with the vertical. For amplitudes that are not necessarily small, the period can be shown to be given by

$$T = T_0 \left(1 + \frac{1}{2^2} \sin^2 \frac{1}{2}\theta_0 + \frac{1}{2^2}\frac{3}{4^2} \sin^4 \frac{1}{2}\theta_0 + \cdots \right)$$ 15-28

Period of a pendulum for a large amplitude

where $T_0 = 2\pi\sqrt{L/g}$ is the period for very small amplitudes.

Example 15-7   A simple pendulum clock is calibrated to keep accurate time at an angular amplitude of $\theta_0 = 10°$. When the amplitude has decreased so that it is very small, how much time does the clock gain in one day?

From Equation 15-28, the original period is approximately

$$T \approx T_0 (1 + \tfrac{1}{4} \sin^2 5°)$$

since $\sin^4 \frac{1}{2}\theta_0$ in Equation 15-28 is much smaller than $\sin^2 \frac{1}{2}\theta_0$. When the amplitude is very small, the period is T_0. Since this is less than T, the frequency is greater and the clock gains time. The difference in these two periods is

$$T - T_0 = \frac{T_0}{4} \sin^2 5° = \frac{T_0}{4} (0.0872)^2 \approx 2 \times 10^{-3}\, T_0$$

The percentage change in the period is

$$\frac{T - T_0}{T_0} \times 100\% = 0.2\%$$

This is a very small percentage change. However, it leads to an inaccuracy in time keeping that is intolerable by today's time-keeping standards. The number of minutes in one day is

$$\frac{24 \text{ h}}{1 \text{ d}} \times \frac{60 \text{ min}}{1 \text{ h}} = 1440 \text{ min/d}$$

A gain of 0.2 percent, then, leads to an accumulated error in one day of 2×10^{-3} (1440 min) = 2.88 min. The clock thus gains nearly three minutes each day. For this reason, pendulum clocks are designed to maintain a constant amplitude.

15-5 Damped and Driven Oscillations

In all real oscillatory motion, mechanical energy is dissipated because of some kind of frictional force. Left to itself, a spring or a pendulum eventually stops oscillating. When the mechanical energy of oscillatory motion decreases with time, the motion is said to be *damped*. If the frictional or damping forces are small, the motion is nearly periodic, except that the amplitude decreases slowly with time, as illustrated in Figure 15-11. Since the energy of an oscillator is proportional to the square of the amplitude, if the amplitude decreases slowly with time, the energy does also. For the case of the small damping shown in Figure 15-11, both the amplitude and the energy of the oscillations decrease by a constant percentage in a given time interval. For example, the energy of a slightly damped pendulum may decrease by 10 percent every minute. Then after 1 min, the energy would be 90 percent of the original energy, after 2 min it would be 90 percent of 90 percent or 81 percent of the original energy, and so on. This type of decrease is called *exponential decrease*. (We encountered exponential decrease in Section 10-3, when we studied the decrease in air pressure with altitude.)

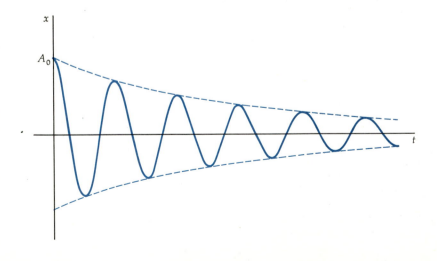

Figure 15-11
Displacement versus time for a slightly damped oscillator. The motion is approximately simple harmonic motion with an amplitude that decreases slowly with time.

The damping of a slightly damped oscillator is usually described by a dimensionless quantity Q called the *quality factor* or *Q factor*. If E is the total energy and $|\Delta E|$ is the energy loss in one period, the Q factor is defined as

$$Q = 2\pi \frac{E}{|\Delta E|}$$ 15-29   Q factor

The Q factor is thus a measure of the fractional energy loss per cycle.

$$\frac{|\Delta E|}{E} = \frac{2\pi}{Q}$$ 15-30

The higher the value of Q, the smaller the damping.

Example 15-8   A simple pendulum loses 1 percent of its energy during each oscillation. What is its Q factor?

Since the energy loss per period is 1 percent,

$$\frac{|\Delta E|}{E} = \frac{1}{100}$$

The Q factor is then

$$Q = 2\pi \frac{E}{|\Delta E|} = 2\pi(100) = 628$$

In Figure 15-12, the oscillation of the body is damped because of the motion of the plunger in the liquid. The rate of energy loss can be varied by changing the size of the plunger or the viscosity of the liquid. If the damping is gradually increased, it eventually reaches a critical value such that no oscillation occurs. When displaced from equilibrium, the object merely returns slowly to its equilibrium position. The oscillator is then said to be *critically damped*. If the damping is increased further, the system is said to be overdamped. For overdamping, the object returns even more slowly to its equilibrium position after being displaced. This is illustrated in Figure 15-13, which shows the displacement versus time for a critically damped system and an overdamped system. Compare these curves with Figure 15-11 for the case of a slightly damped oscillator.

A practical example of a heavily damped oscillating system is a car with shock absorbers. Without shock absorbers, the car would vibrate many times after hitting a bump. Shock absorbers damp the oscillations so that the car oscillates only a few times or not at all. You can test the damping on your car by pushing down the front or back of the car and releasing it. If the car moves back up to equilibrium with no oscillation, the system is critically damped or overdamped. Usually, you will note one or two oscillations, indicating that the damping is just under the critical value. Near critical damping is often desirable for a system that you want to return to equilibrium quickly with little or no oscillation after being displaced.

Driven Oscillations and Resonance

We have seen that in damped oscillations, energy is continually dissipated and the amplitude of the oscillations decreases. To keep a damped system going, energy must be put into the system. When this is done, the oscillator

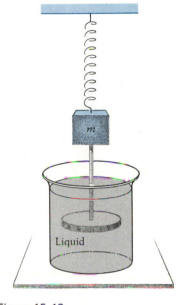

Figure 15-12
Physical arrangement of a damped oscillator. The motion is damped by the plunger immersed in the liquid.

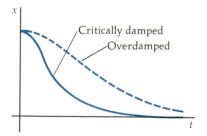

Figure 15-13
Displacement versus time for a critically damped and an overdamped oscillator.

is said to be driven or forced. When you sit on a swing and keep it going by "pumping," that is, by moving your body and legs, you are driving the oscillator. If you put energy into the system at a greater rate than it is being dissipated by damping, the energy increases with time, as indicated by the increase in the amplitude. If you put energy in at the same rate as it is dissipated, the amplitude remains constant over time.

Figure 15-14 shows an object on a spring being driven by moving the point of support up and down. Similarly, a simple pendulum can be driven by moving the support back and forth. If the point of support of an object on a spring or a simple pendulum is moved back and forth with simple harmonic motion of small amplitude and with frequency f, the system will begin oscillating. At first, the motion is complicated, but eventually a steady state is reached in which the system oscillates with a constant amplitude and therefore constant energy. In the steady state, the energy put into the system by the driving force during one cycle equals the energy dissipated per cycle because of the damping. The frequency of oscillation in the steady state is the frequency f of the driver. If the driving frequency is equal to or nearly equal to the natural frequency of the system f_0, the system will oscillate with a large amplitude. (The natural frequency of a system is the frequency of the undriven oscillator with no damping, $(1/2\pi)\sqrt{k/m}$ for a spring and $(1/2\pi)\sqrt{g/L}$ for a pendulum.) This phenomenon is called *resonance*. When the driving frequency equals the natural frequency of the oscillator, the energy absorbed by the oscillator is maximum. The natural frequency of the system is thus called the *resonance frequency* of the system.

Figure 15-15 shows the power absorbed by a driven oscillator as a function of the driving frequency for two different values of damping. These curves are called *resonance curves*. When the damping is small, as indicated by a large Q value, the oscillator absorbs a large amount of energy from the driving force at resonance, and the resonance is sharp; that is, the resonance curve is narrow, indicating that the energy is large only near the resonance frequency. When the damping is large (small Q), the resonance curve is broad. The width of the resonance curve Δf is indicated in the figure. For the case of relatively small damping, the ratio of the resonance frequency f_0 to the width of the resonance curve equals the Q factor.

$$Q = \frac{f_0}{\Delta f}$$ 15-31

The Q factor is thus a measure of the sharpness of the resonance.

You can compare the relative Q values of good crystal glasses by tapping them and listening to how long they ring. The longer the glass rings, the less the damping and the higher the Q value. A glass of high Q can be broken by an intense sound wave of a frequency equal to or very nearly equal to that of the glass. This is often done in a physics lecture demonstration using an audio oscillator and an amplifier. There are many stories of singers breaking a glass with a loud high note. None of these stories has ever been documented. The problem a singer has trying to break a glass is that it is nearly impossible for the singer to concentrate enough energy in a narrow frequency range.

When we study waves and the vibrations of piano or violin strings, drum heads, and the like, we will see that such systems have a whole series of

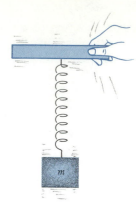

Figure 15-14
An object on a vertical spring can be driven by moving the point of support up and down.

Resonance frequency

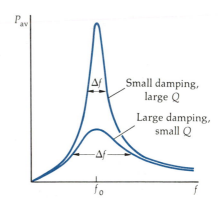

Figure 15-15
Power absorbed by a driven oscillator versus driving frequency for different values of damping. Resonance occurs when the frequency of the driving force equals the natural frequency of the system. The resonance is sharp when the damping is small.

A child on a swing. The child puts energy into the swing by pumping at a frequency equal to the natural frequency of the swing.

natural frequencies rather than just a single natural frequency. Such systems can resonate at any one of their natural frequencies.

There are many familiar examples of resonance. When you sit on a swing and pump, the driving force is not simple harmonic motion. However, it is periodic, and you learn intuitively to pump with the same frequency as the natural frequency of the swing. When soldiers march across a small bridge, they often break step because it is possible that the frequency of their step might be near one of the resonance frequencies of the bridge and set the bridge into resonance vibration. Most machines have rotating parts that are often not in perfect balance and so vibrate. (Observe a washing machine on the spin cycle.) If such a machine is supported by a structure that can vibrate, the structure becomes a driven oscillatory system that is set in motion by the machine. If the structure resonates with the applied vibration, considerable damage can occur, or it may produce unwanted sound throughout the building. Great engineering efforts are put into balancing rotary parts and isolating such machines from their building supports or into damping the vibrations.

Question

5. Give examples of other common situations that can be described as oscillators driven by periodic driving forces.

Summary

1. In simple harmonic motion, the acceleration is proportional to the displacement and is oppositely directed. In general, the acceleration is related to the displacement by

$$a_x = -(2\pi f)^2 x$$

where f is the frequency of oscillation.

2. The position function x for simple harmonic motion of amplitude A and frequency f is given by

$$x = A \cos 2\pi f t$$

The velocity of the particle is given by

$$v = -(2\pi f)A \sin 2\pi f t$$

3. The period of oscillation is the reciprocal of the frequency

$$T = \frac{1}{f}$$

The period and frequency in simple harmonic motion are independent of the amplitude. For the motion of an object of mass m on a spring of force constant k, the period is given by

$$T = 2\pi \sqrt{\frac{m}{k}}$$

The period of the motion of a simple pendulum of length L is

$$T = 2\pi \sqrt{\frac{L}{g}}$$

4. When a particle moves in a circle with constant speed, the x component of its position varies with simple harmonic motion.

5. The total energy in simple harmonic motion is proportional to the square of the amplitude. For a mass on a spring of force constant k, it is given by

$$E = \tfrac{1}{2}kA^2$$

6. In the oscillation of real systems, the motion is damped because of frictional forces or other forces that dissipate energy. If the damping is greater than some critical value, the system does not oscillate but merely returns to its equilibrium position if disturbed. The motion of a slightly damped system is nearly simple harmonic with an amplitude that decreases exponentially with time. The energy of such a system also decreases exponentially with time. For a slightly damped oscillator, the damping is measured by the Q factor, which is related to the fractional energy loss per period by

$$\frac{|\Delta E|}{E} = \frac{2\pi}{Q}$$

7. When a slightly damped system is driven by an external force that varies sinusoidally with time, the system oscillates with a frequency equal to the driving frequency and an amplitude that depends on the driving frequency. If the driving frequency is equal to or nearly equal to the natural frequency of the system, the system oscillates with a large amplitude. This is called resonance. The Q factor is a measure of the sharpness of the resonance. Systems with low damping and therefore high Q factors show a sharply peaked resonance curve. For such systems, the ratio of the resonance frequency f_0 to the width of the resonance curve Δf equals the Q factor:

$$Q = \frac{f_0}{\Delta f}$$

Suggestions for Further Reading

Walker, Jearl: "The Amateur Scientist: Strange Things Happen when Two Pendulums Interact through a Variety of Interconnections," *Scientific American*, October 1985, p. 176.

Various arrangements of pendulums, as well as the Wilberforce pendulum, are used to illustrate the general features of a system of coupled oscillators.

Review

A. Objectives: After studying this chapter, you should:

1. Be able to describe the general characteristics of simple harmonic motion.

2. Be able to describe the relationship between simple harmonic motion and circular motion.

3. Be able to work problems with springs and pendulums.

4. Be able to give definitions of the frequency, period, angular frequency, and amplitude of simple harmonic motion.

5. Be able to describe the motion of a damped oscillator.

6. Be able to discuss the motion of a forced oscillator and sketch a typical resonance curve.

B. Define, explain, or otherwise identify:

Simple harmonic motion, pp. 353–354
Period, p. 354
Frequency, p. 354
Amplitude, p. 354
Damped oscillations, p. 364
Q factor, p. 365
Critical damping, p. 365
Resonance, p. 366

C. True or false: If the statement is true, explain why. If it is false, give a counterexample.

1. In simple harmonic motion, the period is proportional to the square of the amplitude.

2. In simple harmonic motion, the frequency does not depend on the amplitude.

3. In simple harmonic motion, the total energy is proportional to the square of the amplitude.

4. The motion of a simple pendulum is simple harmonic for any initial angular displacement.

5. The motion of a simple pendulum is periodic for any initial angular displacement.

6. If the acceleration of a particle is proportional to the displacement and is oppositely directed, the motion is simple harmonic.

7. The energy of a damped (unforced) oscillator decreases exponentially with time.

8. Resonance occurs when the driving frequency equals the natural frequency.

9. If the Q value is high, the resonance is sharp.

Exercises

Section 15-1   Simple Harmonic Motion: A Mass on a Spring

1. A 2-kg object is attached to a spring of force constant $k = 5$ kN/m. The spring is stretched 10 cm from equilibrium and released. Find (*a*) the frequency, (*b*) the period, and (*c*) the amplitude of the motion. (*d*) What is the maximum speed? (*e*) What is the maximum acceleration? (*f*) When does the object first reach its equilibrium position? What is its acceleration at this time?

2. Answer the questions in Exercise 1 for a 3-kg object attached to a spring of force constant $k = 600$ N/m if the spring is initially stretched 8 cm from equilibrium.

3. When a 2-kg object is hung vertically from a spring, the spring stretches 3 cm. (*a*) What is the force constant of the spring? (*b*) What is the frequency of oscillation of the object on the spring? (*c*) What is the period?

4. A 3-kg object is attached to a spring and oscillates with an amplitude $A = 10$ cm and a frequency $f = 2$ Hz. (a) What is the force constant of the spring? (b) What is the period of the motion? (c) What is the maximum speed of the object? (d) What is its maximum acceleration?

5. If the object in Exercise 4 is suspended from the spring vertically, by how much does the spring stretch from its natural length when the object is in equilibrium?

6. An object on a spring oscillates with a period of 4 s. If the object is suspended from the spring vertically, by how much does the spring stretch from its natural length when the object is in equilibrium?

7. An 80-kg person steps into a car of mass 2400 kg and causes it to sink 2.50 cm on its springs. If the shock absorbers are ineffective, with what frequency will the car and passenger vibrate on the springs?

8. A spring whose force constant is 300 N/m is hanging with no load attached. A 1.5-kg object is attached to it and released from rest. (a) How far will the object move downward before it stops and starts up again? (b) With what frequency will it oscillate?

9. A 5-kg object oscillates on a spring with amplitude 4 cm. Its maximum acceleration is 24 m/s². Find (a) the force constant k, (b) the frequency, and (c) the period of the motion.

10. An object oscillates with amplitude 6 cm on a spring of force constant 2 kN/m. Its maximum speed is 2.20 m/s. Find (a) the mass of the object, (b) the frequency of the motion, and (c) the period of the motion.

11. The position of a particle is given by

$x = (5 \text{ cm}) \cos 4\pi t$

where t is in seconds. What is (a) the frequency, (b) the period, and (c) the amplitude of the motion? (d) What is the first time after $t = 0$ that the particle is in its equilibrium position? What direction is it moving at that time?

12. A particle of mass m begins at rest from $x = +25$ cm and oscillates about its equilibrium position at $x = 0$ with a period of 1.5 s. Write equations for (a) the position x versus time t, (b) the velocity v versus t, and (c) the acceleration a versus t.

13. (a) What is the maximum speed of the particle in Exercise 11? (b) What is its maximum acceleration?

14. Find (a) the maximum speed and (b) the maximum acceleration of the particle in Exercise 12. (c) What is the first time that the particle is at $x = 0$ and moving to the right?

Section 15-2   Simple Harmonic Motion and Circular Motion

15. A particle moves in a circle of radius 40 cm with a constant speed of 80 cm/s. Find (a) the frequency and (b) the

period of the motion. (c) Write an equation for the x component of the position of the particle as a function of time, assuming the particle is on the x axis at time $t = 0$.

16. For the particle in Exercise 15, write equations for the x components of the velocity and the acceleration of the particle.

17. A particle moves in a circle of radius 15 cm, making 1 rev every 3 s. (a) What is the speed of the particle? (b) What is its angular velocity ω? (c) Write an equation for the x component of the position of the particle as a function of time, assuming the particle is on the x axis at time $t = 0$.

Section 15-3   Energy in Simple Harmonic Motion

18. Find the total energy of the mass oscillating on a spring of Exercise 1.

19. Find the total energy of the mass oscillating on a spring of Exercise 2.

20. A 1.5-kg object oscillates with simple harmonic motion on a spring of force constant $k = 500$ N/m. Its maximum speed is 70 cm/s. (a) What is the total energy? (b) What is the amplitude of the oscillation?

21. A 3-kg object oscillates on a spring of force constant 2 kN/m with total energy 0.9 J. (a) What is the amplitude of the motion? (b) What is the maximum speed?

22. Make a sketch of the potential energy versus time for a particle undergoing simple harmonic motion. On the same sketch, show the kinetic energy as a function of time (see Equations 15-21 and 15-22).

23. An object oscillates on a spring with amplitude 4.5 cm. Its total energy is 1.4 J. What is the force constant of the spring?

Section 15-4   The Simple Pendulum

24. Find the length of a simple pendulum if the period is 5 s at a point where $g = 9.81$ m/s².

25. Find the period of the pendulum in Exercise 24 on the moon, where the acceleration of gravity is one-sixth that on earth.

26. If the period of a pendulum 70 cm long is 1.68 s, what is the value of g at the location of the pendulum?

27. Find the length of a pendulum whose period is 1 s at a point where $g = 9.81$ m/s².

28. A pendulum set up in the stairwell of a 10-story building consists of a heavy weight suspended on a 34.0-m wire. If $g = 9.81$ m/s², what is the period of oscillation?

29. Find the length of a simple pendulum if the period is 0.5 s at a point where $g = 9.81$ m/s².

30. Two simple pendulums, each of length of 10 m, are side by side so that their bobs just touch, as shown in Figure 15-16. The bob on the right is displaced 15 cm to the right, measured along the arc, and the one on the left is displaced 2 cm to the left. (Both of these angular displacements are very small, so the corrections to the period given by Equation 15-28 can be neglected.) The bobs are released from rest simultaneously. Where do they meet?

Figure 15-16
Two pendulums for
Exercise 30.

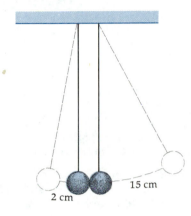

15 cm

2 cm

Section 15-5   Damped and Driven Oscillations

31. Find the resonance frequency for each of the three systems shown in Figure 15-17.

Figure 15-17
Exercise 31.

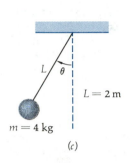

$k = 400$ N/m $k = 800$ N/m

10 kg   5 kg

(a)   (b)

L / θ

$L = 2$ m

$m = 4$ kg

(c)

32. An oscillator has a Q factor of 200. By what percentage does its energy decrease during one period?

33. A damped oscillator loses 2 percent of its energy each cycle. (a) What is its Q factor? (b) If its resonance frequency is 300 Hz, what is the width Δf of the resonance curve when the oscillator is driven?

Problems

1. The period of an oscillating particle is 8 s. At $t = 0$, the particle is at $x = A = 10$ cm. (a) Sketch x as a function of t. (b) Find the distance traveled in the first second, the second second, the third second, and the fourth second after $t = 0$.

2. A 4-kg object is attached to a spring of force constant 2 kN/m. It is displaced 6 cm from equilibrium and released. (a) What is the total energy? (b) Write an equation for the potential energy U as a function of time, and sketch U versus t on a graph on which the vertical axis is energy in joules and the horizontal axis is time in seconds. (c) On the same graph, sketch the kinetic energy as a function of time.

3. The bow of a destroyer in heavy seas undergoes simple harmonic vertical pitching motion with a period of 8.0 s and an amplitude of 2.0 m. (a) What is its maximum velocity? (b) What is its maximum acceleration? (c) An 80-kg sailor is standing on a scale in his cabin. What are the maximum and minimum readings on the scale in newtons?

4. A system oscillates with simple harmonic motion with amplitude 10 cm and period 4 s. At $t = 0$, the displacement is 10 cm. (a) Find the displacement x at the following times: 0.5 s, 1.0 s, 1.5 s, and 2.0 s. (b) Find the velocity v at times $t = 0$, 0.5 s, 1.0 s, 1.5 s, and 2.0 s.

5. For the object in Problem 4, write equations for (a) the displacement x as a function of time t, (b) the velocity v as a function of time t, and (c) the acceleration a as a function of time t. Sketch each of these functions on a graph using the same time scale for each function.

6. A 3-kg object oscillates on a spring with amplitude 8 cm. Its maximum acceleration is 3.50 m/s². Find the total energy.

7. An object of mass m is supported by a vertical spring with force constant 1800 N/m. When pulled down 2.5 cm from equilibrium and released, it oscillates at 5.5 Hz. (a) Find m. (b) Find the amount the spring is stretched from its natural length when the object is in equilibrium. (*c*) Write expressions for the displacement *x*, velocity *v*, and acceleration *a* as functions of time *t*.

8. Military specifications often call for electronic assemblies to be subjected to accelerations of 10 $g = 98$ m/s^2. To attain such accelerations, manufacturers use a shaking table that can vibrate the device at various specified frequencies and amplitudes. If the device is given a vibration of amplitude 1.5 cm, what should its frequency be?

9. A child on a swing oscillates with a period of 3 s. The mass of the child and the swing is 35 kg. The swing is pushed by a patient father so that the amplitude remains constant. At the bottom of the swing, the speed of the child and the swing is 2.0 m/s. (*a*) What is the total energy of the child and the swing? (*b*) If $Q = 20$, how much energy is dissipated by the swing in each oscillation? (*c*) What is the power input of the father?

10. It has been stated that the vibrating earth has a resonance period of 54 min and a Q of about 400 and that, after a large earthquake, the earth "rings" (continues to vibrate) for about 2 months. (*a*) Find the pecentage of energy of vibration lost to damping forces per cycle. (*b*) Show that after *n* periods, the energy is $E_n = (0.984)^n E_0$ where E_0 is the original energy. (*c*) If the original energy of vibration of an earthquake is E_0, what is it after 2 days?

11. A block of wood slides on a frictionless horizontal surface. It is attached to a spring and oscillates with a period of 3 s. A second block rests on top of the first one. The coefficient of static friction between the two blocks is 0.25. (*a*) If the amplitude of oscillation is 1 cm, will the block on top slip? (*b*) What is the greatest amplitude of oscillation for which the top block will not slip?

12. Show that for the situations in Figure 15-18*a* and *b*, the object oscillates with a frequency $f = (1/2\pi) \sqrt{k_{eff}/m}$, where k_{eff} is given by (*a*) $k_{eff} = k_1 + k_2$ and (*b*) $1/k_{eff} = 1/k_1 + 1/k_2$. (*Hint:* Find the resultant force *F* on the object for a small displacement *x* and write $F = -k_{eff}x$. Note that in (*b*) the springs stretch by different amounts, the sum of which is *x*.)

Figure 15-18
Problem 12.

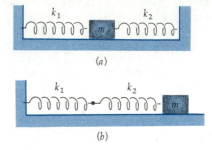

(*a*)

(*b*)

13. A pendulum clock that has run down to a very small amplitude gains 5 min each day. What is the proper (angular) amplitude for the pendulum to keep correct time?

14. The iron pendulum of an eighteenth-century grandfather clock has a period of 2.000000 s, averaged over a day when the average temperature is 21°C. (*a*) For a day when the average temperature is 15°C, find the new period. (*b*) Will the clock be fast or slow on that day? (*c*) By how many seconds will it be fast or slow? (Take $\alpha = 12 \times 10^{-6}$ K^{-1} for the coefficient of linear expansion for iron.)

15. A 20-kg ball is hung on a steel wire of 1.00-mm diameter and 3.00-m length. It is set into vertical vibration and is found to have a frequency of 8.14 Hz. (*a*) Find the force constant of the wire. (*b*) Show that Young's modulus for the wire is related to the force constant by $Y = kL/A$. (*c*) Find Young's modulus for this wire.

CHAPTER SIXTEEN

Mechanical Waves: Sound

Music, when soft voices die,
Vibrates in the memory. . . .

PERCY SHELLEY

A pebble is tossed into a lake, and circular waves spread out, becoming less and less distinct with distance. The waves from a second pebble pass right through the first, neither set being deflected by the other. The up and down vibrations of the water where the pebbles went in have propagated in all directions; they have escaped their source. This is how waves differ from vibrations; they are extended in space and may have consequences far away and long after the vibrations which caused them have died away.

Wave motion can be thought of as the transport of energy and momentum from one point in space to another without the transport of matter. In mechanical waves (waves in a string or sound waves in air, for example), the energy and momentum are transported by means of a disturbance in the medium that is propagated because of the elastic properties of the medium. On the other hand, in electromagnetic waves (such as light, radio waves, television waves, or x-rays), the energy and momentum are carried by electric and magnetic fields that can propagate through a vacuum. (These electric and magnetic fields, which we will define in later chapters, are caused by the oscillation of electric charges in atoms, molecules, or perhaps a transmitting antenna. Television waves, radio waves, light, radar, microwaves, and x-rays are all examples of electromagnetic waves that differ only in the frequency of vibration.)

Although the variety of wave phenomena observed in nature is immense, many features are common to all kinds of waves, and others are shared by a wide range of wave phenomena. In this chapter and the next, we will study waves in strings and sound waves. Many of the ideas and results discussed in these two chapters will be applied later when we study light and other electromagnetic waves.

16-1   Wave Pulses

When a string stretched under tension is given a flip, as illustrated in Figure 16-1, the shape of the string changes in time in a regular way. The bump produced by the flip travels down the string as a *wave pulse*. The disturbance in the medium in this case is the change in the shape of the string from its equilibrium shape, which is that of a taut string. The wave pulse travels down the string at a definite speed that depends on the tension in the string and on its mass per unit length. In general,

The speed of propagation of a wave depends on the properties of the medium.

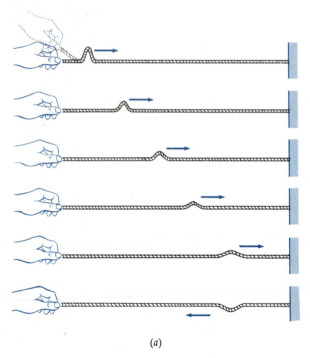

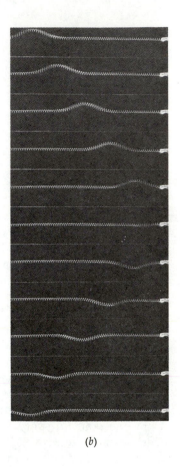

(a)

(b)

Figure 16-1
(a) and (b) Wave pulse moving to the right on a stretched string. When the pulse arrives at the rigid support, it is reflected and inverted.

The fate of the pulse at the other end of the string depends on how the string is fastened there. If it is tied to a rigid support as shown here, the pulse will be reflected and will return inverted. If the string is tied to another string of different mass density, part of the pulse will be transmitted and part will be reflected. If the second string is heavier than the first, the wave speed in that string will be smaller and the reflected part of the pulse will be inverted (see Figure 16-2a and b). If the second string is lighter, the wave speed will be greater and the reflected pulse will not be inverted (see Figure 16-2c and d). In either case, the transmitted pulse is not inverted.

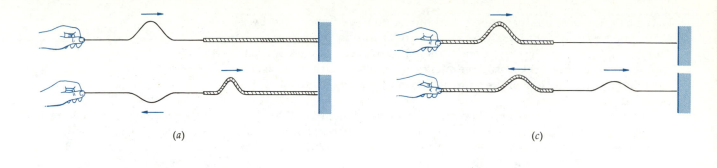

(a)   (c)

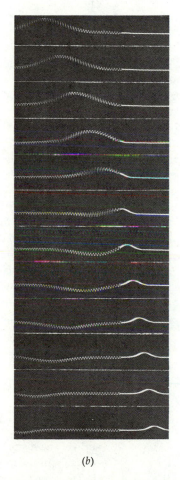

(b)

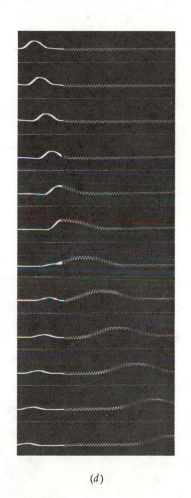

(d)

Figure 16-2
(a) and (b) A pulse traveling along a light string attached
to a heavier string in which the wave speed is smaller. The
reflected pulse is inverted. (c) and (d) A pulse traveling
along a heavy string attached to a light string in which the
wave speed is greater. In this case, the reflected pulse is
not inverted.

We can demonstrate that energy and momentum are transported by a wave pulse by hanging a weight on a string under tension (see Figure 16-3) and giving the string a flip at one end. When the pulse arrives at the weight, the weight is momentarily lifted. The energy and momentum introduced by the hand flipping the end of the string is transmitted along the string and is received by the weight.

It is important to understand that it is not the mass elements of the string that are transported in wave motion but rather the *disturbance in the shape* of the string caused by flipping one end. The mass elements of the string, in fact, move in a direction perpendicular to the string and thus perpendicular to the direction of motion of the pulse (see Figure 16-4). A wave in which the disturbance is perpendicular to the direction of propagation is called a *transverse wave*. (Other examples of transverse waves are electromagnetic waves, including light waves, which we will study later.)

Sound waves in air are not transverse waves. In sound waves, a disturbance in the pressure and density of air is set up by the vibration of a body (say, a tuning fork or violin string), and the disturbance is propagated through the air by the collisions of air molecules. The vibration of the air molecules is in the direction of the propagation of the wave. Once again, it is only the disturbance that is propagated; the air molecules merely vibrate back and forth about their equilibrium positions. A wave in which the disturbance is parallel to the direction of propagation is called a *longitudinal wave*. Sound consists of longitudinal waves. A longitudinal pulse in a spring analogous to a sound pulse can be produced by suddenly compressing the spring (see Figure 16-5a). (Figure 16-5b shows transverse waves produced in the same spring.) Similarly, if a long rod is tapped at one end, a longitudinal wave pulse travels down the rod.

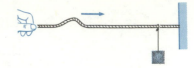

Figure 16-3
Waves transmit both energy and momentum, as indicated here by the upward motion of the weight when the wave pulse arrives.

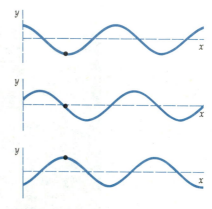

Figure 16-4
Three successive drawings of a wave traveling to the right. An element on the string moves up and down in simple harmonic motion.

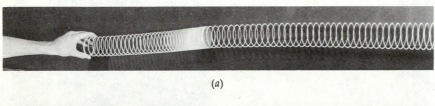

(a)

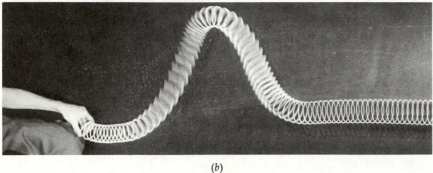

(b)

Figure 16-5
(a) Longitudinal wave pulse in a spring. The disturbance is in the direction of the motion of the wave. (b) Transverse wave pulse in the same spring. The disturbance is perpendicular to the direction of the motion of the wave.

Water waves are neither completely transverse nor completely longitudinal, but a combination of the two. Figure 16-6 shows the motion of a water particle in a water wave. The water particle moves in an elliptical path with both transverse and longitudinal components.

Figure 16-7*a* shows two pulses in a string moving in opposite directions. The shape of the string when they meet can be found by adding the displacements produced by each pulse separately, as shown in the figure. The property of wave motion that the resultant wave is the sum of two or more individual waves is called the *principle of superposition*. In the special case where two pulses are identical except that one is inverted relative to the other (see Figure 16-7*b*), there will be one time when the pulses exactly overlap and add to zero. At this time, the string is horizontal but it is not at rest. A short time later, the pulses emerge, each continuing in its original direction. The addition of separate waves to produce the resultant wave is called *interference*. Interference is a characteristic property of wave motion. There is no analogous situation in particle motion. Two particles never overlap or add together in this way. Thus, interference is unique to wave motion.

Principle of superposition

Interference

When one of the two interfering pulses is inverted relative to the other, as in Figure 16-7*b*, the resultant pulse is smaller than the larger pulse and may be smaller than either pulse. The pulses tend to cancel each other when they overlap. This is called *destructive interference*. (Complete cancellation occurs only if the magnitudes of the pulses are identical, as in this figure.) On the other hand, if the displacements of the two pulses are in the same direction, as in Figure 16-7*a*, the resultant pulse when they overlap is greater than either pulse by itself. This is called *constructive interference.*

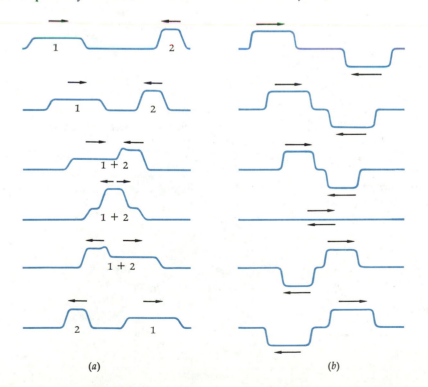

(a)   (b)

Figure 16-7
Two wave pulses moving in opposite directions on a string. The shape of the string when the pulses meet is found by adding the displacements of the separate pulses. (*a*) When the displacements of the pulses are in the same direction, they add. This kind of superposition is called constructive interference. (*b*) When the displacements of the pulses are in opposite directions, the algebraic addition of the displacements amounts to a subtraction of the magnitudes. This is called destructive interference.

Questions

1. Give examples of wave pulses in nature in addition to those mentioned in the text. In each case, explain what kind of disturbance from equilibrium occurs.

2. When two waves moving in opposite directions interfere, does either impede the progress of the other?

16-2   Speed of Waves

A general property of waves is that their speed depends on the properties of the medium but is independent of the motion of the source relative to the medium. For example, the speed of a sound wave produced by a train whistle depends only on the properties of air and not on the motion of the train. Similarly, the speed of a wave in a string depends only on the properties of the string.

Formulas for the speed of a wave in a string or a sound wave can be derived from Newton's laws, but the derivations are somewhat difficult, so we will merely state the results and give some examples of their applications. If we send wave pulses down a long rope, we can easily demonstrate that the speed of propagation of the wave pulses increases as we increase the tension in the rope. If we have two ropes, a light rope and a heavy rope under the same tension, wave pulses will propagate more slowly in the heavy rope. The speed of propagation v of waves in a string or rope is related to the tension F and the mass per unit length μ by

$$v = \sqrt{\frac{F}{\mu}}$$

16-1   Speed of waves in a string

(We use F for tension here rather than T to avoid confusion with the use of T for temperature in the expression for the speed of sound waves.) The dependence of the speed on F/μ is not surprising when we realize that it is essentially the tension that accelerates the mass elements in a rope to provide the wave pulse.

Example 16-1   The tension in a string is provided by hanging a mass of 3 kg at one end, as shown in Figure 16-8. The mass density of the string is 0.02 kg/m. What is the speed of the waves in the string?

The tension in the string is

$$F = mg = (3 \text{ kg})(9.81 \text{ N/kg}) = 29.4 \text{ N}$$

The speed is therefore

$$v = \sqrt{\frac{F}{\mu}} = \sqrt{\frac{29.4 \text{ N}}{0.02 \text{ kg/m}}} = 38.3 \text{ m/s}$$

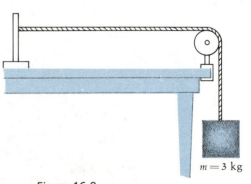

$m = 3$ kg

Figure 16-8
Example 16-1. The tension in the string is provided by the weight of the block.

For ordinary sound waves in which the pressure variation is not too great, the speed of sound is given by

$$v = \sqrt{\frac{B}{\rho_0}} \qquad\qquad 16\text{-}2$$

where ρ_0 is the equilibrium density of the medium and B is the bulk modulus (Section 10-2). Comparing Equations 16-1 and 16-2, we can see that the wave speed for both waves in strings and sound waves depends on (1) an elastic property of the medium (the tension F for string waves and the bulk modulus B for sound waves) and (2) an inertial property of the medium (the linear mass density or the volume mass density). Both these results follow from Newton's laws.

For sound waves in air, the bulk modulus can be shown to be

$$B = \gamma P \qquad\qquad 16\text{-}3$$

where P is the pressure and γ is the ratio of the heat capacity at constant pressure to that at constant volume. (For air, γ has the value 1.4.) The pressure P can be related to the mass density of a gas, which is given by

$$\rho_0 = \frac{m}{V} = \frac{nM}{V}$$

where M is the molar mass. Using the ideal gas law

$$PV = nRT$$

the pressure is

$$P = \frac{n}{V} RT \qquad\qquad 16\text{-}4$$

Substituting ρ_0/M for n/V, we obtain

$$P = \frac{\rho_0 RT}{M} \qquad\qquad 16\text{-}5$$

Equation 16-3 for the bulk modulus then gives

$$B = \gamma P = \frac{\gamma \rho_0 RT}{M}$$

Substituting this result into Equation 16-2 we obtain the speed of sound in an ideal gas:

$$v = \sqrt{\frac{\gamma RT}{M}} \qquad\qquad 16\text{-}6 \qquad \text{Speed of sound in air}$$

Example 16-2   Calculate the speed of sound in air at 0°C and at 20°C.

The temperature T in Equation 16-6 is the absolute temperature, which is related to the Celsius temperature by

$$T = t_C + 273$$

The molar mass of air is 29.0 g/mol = 29.0 × 10⁻³ kg/mol, and the value of the universal gas constant is $R = 8.31$ J/mol·K. The speed of sound at $0°C$ is therefore

$$v = \sqrt{\frac{(1.4)(8.31 \text{ J/mol·K})(273 \text{ K})}{29.0 \times 10^{-3} \text{ kg/mol}}} = 331 \text{ m/s}$$

To find the speed at $20°C = 293$ K, we note that the speed is proportional to the square root of the absolute temperature. Its value at 293 K, v_{293}, is related to that at 273 K, v_{273}, by

$$\frac{v_{293}}{v_{273}} = \frac{\sqrt{293}}{\sqrt{273}}$$

or

$$v = \sqrt{\frac{293}{273}} (331 \text{ m/s}) = 343 \text{ m/s}$$

Questions

3. Two strings are stretched between the same two posts. One weighs twice as much as the other. How should their tensions be adjusted so that waves travel along both at the same speed?

4. Do you expect the speed of sound waves in helium gas to be greater or less than that in air? Why?

5. Although the density of most solids is more than 1000 times that of air, the speed of sound in a solid is usually greater than it is in air. Why?

16-3 Harmonic Waves

If, instead of flipping the end of a string, we move the end up and down in simple harmonic motion (as though it were tied to a vibrating tuning fork), a sinusoidal wave train propagates down the string. Such a wave is called a *harmonic wave*. The shape of the string at some instant of time is that of a sine function, as shown in Figure 16-9. (The question of whether this is a sine function or a cosine function depends merely on where the origin is chosen on the axis.) Such a figure can be obtained by taking a snapshot of the string.

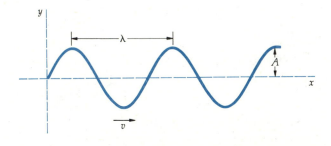

Figure 16-9
Harmonic wave at some instant in time. λ is the wavelength and A is the amplitude. For transverse waves on a string, this figure can be obtained by taking a snapshot of the string.

The distance between two successive wave crests is called the *wavelength* Wavelength
λ. The wavelength is the distance in space within which the wave repeats
itself. As the wave propagates down the string, each point on the string
moves up and down in simple harmonic motion with the frequency f of the
tuning fork or whatever agent is driving the end of the string. There is a
simple relation between the frequency f, the wavelength λ, and the speed v
of the harmonic wave. During the time of one period, $T = 1/f$, the wave
moves a distance of one wavelength, so the speed is given by

$$v = \frac{\lambda}{T} = f\lambda \qquad\qquad 16\text{-}7$$

Equation 16-7 applies to all types of waves, transverse or longitudinal. It
arises simply from the definitions of wavelength and frequency. Since the
speed of wave propagation is determined by the tension and mass density of
the string (Equation 16-1), the wavelength is determined by the frequency
of the source through $\lambda = v/f$. The greater the frequency, the smaller the
wavelength.

Harmonic sound waves can be generated by a source, such as a tuning
fork or a whistle, that vibrates with simple harmonic motion or by a speaker
driven by an audio oscillator. The vibrating source sets up compressions and
rarefactions in the surrounding medium (such as air). These compressions
and rarefactions also vary harmonically. As we have mentioned, the air
molecules in a sound wave oscillate back and forth along the direction of
motion of the wave. These displacements lead to variations in the density of
the air and in the air pressure (see Figure 16-10). The graph of density (or
pressure) versus position at some instant shown in Figure 16-11 is similar to
the "snapshot" of a transverse wave in Figure 16-9. As the wave moves in
time, the displacement of air molecules, the density, and the pressure at one
point all vary sinusoidally with a frequency f that is related to the wavelength and wave speed by Equation 16-7. Again, since the speed of waves is
determined by the properties of the medium, the wavelength of a sound
wave varies inversely as the frequency, $\lambda = v/f$.

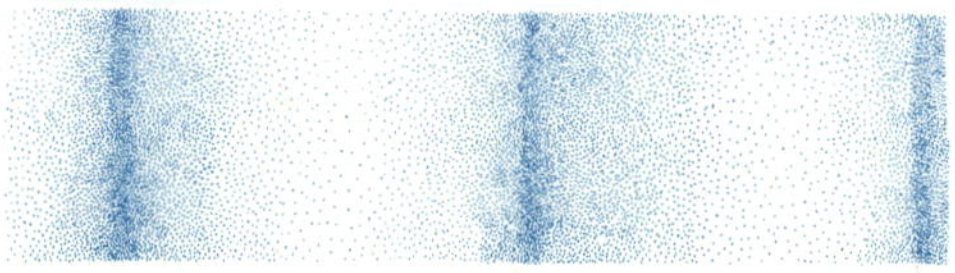

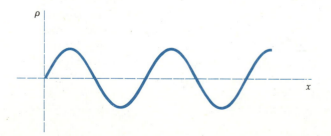

Figure 16-10
Variations in the density of air for a
harmonic sound wave. The pressure
is also greatest at points where the
density is greatest.

Figure 16-11
Graph of density ρ versus position
for a harmonic sound wave. A
graph of pressure versus position
would look the same. This figure is
similar to Figure 16-9 for a
harmonic transverse wave.

So far, we have discussed one-dimensional waves, that is, waves that propagate in a straight line. Figure 16-12 shows circular waves on the surface of water in a ripple tank generated by a point source moving up and down with simple harmonic motion. The wavelength here is the distance between successive wave crests, which are concentric circles. These circles are called *wavefronts*. If we have a point source of sound or light, the waves are emitted in three dimensions. They move out in all directions, and the wavefronts are concentric spherical surfaces. The motion of any set of wavefronts can be indicated by *rays*, which are directed lines perpendicular to the wavefronts (see Figure 16-13).

Wavefront

Figure 16-12
Circular wavefronts diverging from a point source in a ripple tank.

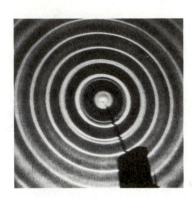

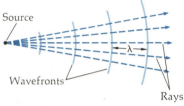

Figure 16-13
The motion of wavefronts can be represented by rays drawn perpendicular to the wavefronts. For a point source, the rays are radial lines diverging from the source.

For spherical waves, the rays are radial lines. At a great distance from a point source, a small part of the wavefront can be approximated by a plane, and the rays are approximately parallel lines. Such a wave is called a *plane wave* (see Figure 16-14). The two-dimensional analog of a plane wave is a line wave, which is a small part of a circular wavefront at a great distance from the source. Line waves can also be produced in a ripple tank by a line source, as illustrated in Figure 16-15.

Figure 16-14
Plane waves. At great distances from a point source, the wavefronts are approximately parallel planes, and the rays are parallel lines perpendicular to the planes.

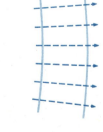

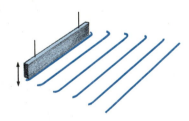

Figure 16-15
A two-dimensional analog of a plane wave can be generated in a ripple tank with a flat board that oscillates up and down in the water and produces wavefronts that are straight lines.

Example 16-3   The human ear can hear sound of frequencies in the range from about 20 Hz to 20,000 Hz. (Many people have rather limited hearing above about 15,000 Hz.) If the speed of sound in air is 340 m/s, what are the wavelengths corresponding to these extreme frequencies?

Taking 340 m/s for the speed of sound in air at ordinary temperatures, the wavelength corresponding to the lowest audible frequency is

$$\lambda = \frac{v}{f} = \frac{340 \text{ m/s}}{20 \text{ Hz}} = 17 \text{ m}$$

and that corresponding to the highest audible frequency is

$$\lambda = \frac{v}{f} = \frac{340 \text{ m/s}}{20 \text{ kHz}} = 1.7 \text{ cm}$$

Sound waves with frequencies above 20,000 Hz are called *ultrasonic waves*. Bats navigate by emitting ultrasonic waves and using their echos to detect objects. Ships use ultrasonic waves called sonar (from *sound naviga-tion* and *ranging*) to detect submarines and other submerged objects in much the same way. Ultrasonic waves are used in medicine for diagnostic purposes. For example, a "sonagram," a picture constructed from ultrasonic waves, is sometimes taken during pregnancy to check the size and position of the fetus or to aid in detecting certain abnormalities. Ultrasonic waves have the advantage that, because of their very small wavelengths, narrow beams can be sent out and reflected from small objects. In the next chapter, we will see that, in general, waves do not travel strictly in straight lines but bend around corners, a phenomenon called diffraction. Because diffraction is most pronounced for waves of long wavelength, ultrasonic waves of high frequencies and short wavelengths are needed to locate small objects. In general, waves cannot be used to locate objects that are smaller than the wavelength.

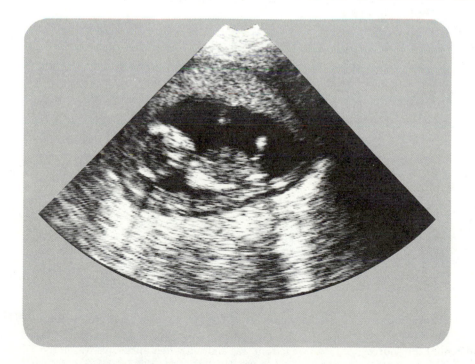

Sonogram of a pregnant woman showing four-month-old fetus in the womb.

16-4   The Doppler Effect

When a wave source and a receiver are moving relative to each other, the frequency observed by the receiver is not the same as that emitted by the source. When they are moving toward each other, the observed frequency is greater than the source frequency; when they are moving away from each other, the observed frequency is less than the source frequency. This is called the *doppler effect*. A familiar example is the change in the pitch of a car horn as the car approaches or recedes.

The change in frequency of a sound wave is slightly different depending on whether the source or the receiver is moving relative to the medium. When the source moves, the wavelength changes, and the new frequency f' is found by first finding the new wavelength and then computing $f' = v/\lambda'$. On the other hand, when the source is stationary and the receiver moves, the frequency is different simply because the receiver moves past more or fewer waves in a given time. We will first consider the case of a moving source. Figure 16-16a shows waves in a ripple tank produced by a source moving to the right with a speed less than the wave speed. We can see that the waves in front of the source are compressed so that the wavefronts are closer together than they would be for a stationary source, whereas behind the source, they are farther apart. We can calculate the wavelength in front of the source λ_f' and that behind the source λ_b' as follows. Let the frequency of the source be f_0. In some time t, the source produces a number of waves N given by $N = f_0 t$. The first wavefront moves a distance vt while the source moves a distance $u_s t$, where u_s is the speed of the source (see Figure 16-16b).

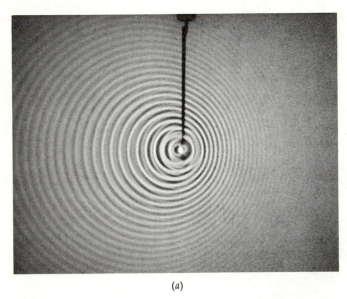

(a)

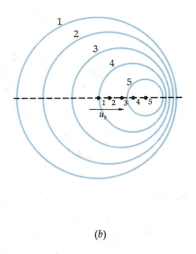

(b)

Figure 16-16
(a) Waves in a ripple tank produced by a point source moving to the right with speed less than the wave speed. The wavefronts are closer together in front of the source and farther apart behind the source than they would be for a stationary source. (b) Successive wavefronts emitted by a point source moving with speed u_s. Each numbered wavefront was emitted when the source was at the correspondingly numbered position.

Since these N waves are contained in the distance $(v - u_s)t$, the wavelength in front of the source is obtained by dividing this distance by N:

$$\lambda'_f = \frac{(v - u_s)t}{N} = \frac{(v - u_s)t}{f_0 t}$$

or

$$\lambda'_f = \frac{v - u_s}{f_0} \qquad\qquad 16\text{-}8$$

The frequency in front of the moving source is thus

$$f'_f = \frac{v}{\lambda'_f} = \frac{v}{(v - u_s)} f_0$$

or

$$f'_f = \left(\frac{1}{1 - u_s/v}\right) f_0 \qquad\qquad 16\text{-}9$$

Behind the source, the N waves are contained in the distance $(v + u_s)t$, so the wavelength behind the source is

$$\lambda'_b = \frac{(v + u_s)t}{f_0 t} = \frac{v + u_s}{f_0} \qquad\qquad 16\text{-}10$$

The frequency behind the source is thus

$$f'_b = \frac{v}{\lambda'_b} = \frac{v}{v + u_s}$$

or

$$f'_b = \left(\frac{1}{1 + u_s/v}\right) f_0 \qquad\qquad 16\text{-}11$$

When the source is at rest and the receiver is moving relative to the medium, there is no change in the wavelength, but the frequency with which waves pass the receiver is increased when the receiver moves toward the source and decreased when the receiver moves away from the source. The number of waves passing a stationary receiver in time t is the number in the distance vt, which is vt/λ_0. When the receiver moves toward the source with speed u_r, it passes an additional number of waves given by $u_r t/\lambda_0$ (see Figure 16-17). The total number of waves passing the receiver in time t is then

$$N = \frac{vt + u_r t}{\lambda_0} = \frac{v + u_r}{\lambda_0} t$$

The frequency observed is this number divided by the time interval:

$$f' = \frac{N}{t} = \frac{v + u_r}{\lambda_0} = \frac{v}{\lambda_0}\left(1 + \frac{u_r}{v}\right)$$

or

$$f' = f_0\left(1 + \frac{u_r}{v}\right) \qquad\qquad 16\text{-}12$$

"I love hearing that lonesome wail of the train whistle as the frequency of the wave changes due to the doppler effect."

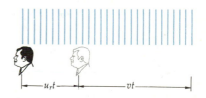

Figure 16-17
The number of waves passing a stationary receiver in time t is the number in the distance vt, where v is the wave speed. If the receiver moves toward the source with speed u_r, he or she passes the additional number of waves in the distance $u_r t$. The frequency of waves received is therefore increased.

If the receiver moves away from the source with speed u_r, similar reasoning leads to the observed frequency

$$f' = f_0 \left(1 - \frac{u_r}{v}\right) \qquad\qquad \text{16-13}$$

When both the source and the receiver are moving relative to the medium, the results expressed in Equations 16-9 and 16-11 and Equations 16-12 and 16-13 can be combined:

$$f' = f_0 \frac{(1 \pm u_r/v)}{(1 \mp u_s/v)} \qquad\qquad \text{16-14}$$

The correct choice of the plus (+) or minus (−) sign is most easily obtained by remembering that when either the source or the receiver is approaching the other, the frequency increases, whereas when they are receding, the frequency decreases. Thus, for example, if the source and receiver are approaching each other, the + sign is used in the numerator and/or the − sign in the denominator.

If the medium is moving (for example, if there is a wind blowing), the wave speed v is replaced by $v' = v \pm u_w$, where u_w is the speed of the medium.

Example 16-4   The frequency of a car horn is 400 Hz. What frequency is observed if the car moves toward a stationary receiver with speed $u_s = 34$ m/s (about 76 mi/h)? Take the speed of sound in air to be 340 m/s. (The car speed in this example is rather high, but it was chosen because it is 10 percent of the speed of sound in air.)

According to Equation 16-8, the wavelength in front of the car is

$$\lambda_f' = \frac{v - u_s}{f_0}$$

$$= \frac{340 - 34 \text{ m/s}}{400 \text{ s}^{-1}} = \frac{306}{400} \text{ m} = 0.765 \text{ m}$$

The frequency observed is thus

$$f' = \frac{v}{\lambda_f'}$$

$$= \frac{340 \text{ m/s}}{0.765 \text{ m}} = 444 \text{ Hz}$$

We note that the observed frequency is about 10 percent higher (actually 11 percent) than the original frequency of the car horn.

Example 16-5   The horn of a stationary car has a frequency of 400 Hz. What frequency is observed by a receiver moving toward the car at 34 m/s?

For a moving receiver, the wavelength does not change; the receiver merely passes more waves in a given time. The observed frequency is (Equation 16-12):

$$f' = f_0 \left(1 + \frac{u_r}{v}\right) = (400 \text{ Hz}) \left(1 + \frac{34}{340}\right) = (400 \text{ Hz})(1.10) = 440 \text{ Hz}$$

In this case, the increase in the observed frequency is exactly 10 percent.

A familiar example of the use of the doppler effect is police radar used to measure the speed of a car. The waves are emitted by the police radar transmitter and strike the moving car. The car acts as both a moving receiver and a moving source as the waves reflect off of it and back to the police receiver. Another application involves the famous red shift in the light from distant galaxies. Because the galaxies are moving away from us, the light they emit is shifted toward the longer, red wavelengths. The speed of the galaxies relative to us is determined by measuring this shift.

In Examples 16-4 and 16-5, we saw that the doppler shift in frequency depends on whether the source or the receiver moves relative to the medium. In the first example, the source moved with speed 34 m/s relative to still air, and the frequency shifted from 400 to 444 Hz. In the second example, the receiver moved with speed 34 m/s and the frequency shifted from 400 to 440 Hz. These shifts are approximately, but not exactly, equal. When the receiver moves relative to the medium, we can see from Equation 16-12 that the fractional shift in frequency, $\Delta f/f_0$, is just the ratio of the speed of the receiver to the wave speed, u_r/v. In our example, this ratio was 10 percent and the frequency increased by 10 percent. However, when the source moves, the calculation (Equation 16-9) is slightly more complicated, and the fractional frequency shift is not exactly u_s/v_0. In our two examples, the frequency shifted by 10 percent in one case and 11 percent in the other, a difference of 1 percent, which is 10 percent of 10 percent or $(u/v)^2 = 1/100 = 1$ percent, where u is the speed of either the source or the receiver. This is a general result:

The difference in the frequency shift between the source moving with speed u and the receiver moving with speed u is of the order of $(u/v)^2$.

In most practical situations, this difference can be neglected because u is usually much smaller than v. The fractional shift in frequency is then

$$\frac{\Delta f}{f_0} = \frac{u}{v} \qquad\qquad 16\text{-}15$$

where u is the speed of either the source or the receiver. However, the difference is real and is of theoretical importance because it shows that the two situations are really different. Not only is the *relative* motion of the source and receiver important but so are the "absolute" motions of both relative to the medium. Thus, if we can measure the doppler shift in frequency to order $(u/v)^2$, we can tell whether it is the source or the receiver that is moving relative to the medium. If you move relative to still air, you feel air rushing past you. In your reference frame, there is a wind. For sound

Absolute motion

waves in air, therefore, we can tell whether the source or the receiver is moving by noting which feels the wind in its own reference frame.

However, a problem arises with regard to light and other electromagnetic waves that propagate through a vacuum. Our equations for the doppler effect in sound seem to imply that we could detect absolute motion relative to a vacuum if we could measure the doppler shift accurately enough. This contradicts the principle of relativity enunciated by Albert Einstein in 1905, which states that it is impossible to detect absolute motion. It turns out that for light and other electromagnetic waves, our equations for the doppler effect are only approximately correct. When the proper relativistic corrections are made, the doppler shift for electromagnetic waves is the same for a moving source as for a moving receiver; that is, it depends only on the relative motion of the source and receiver.

In our derivations of the doppler-shift expressions, we have assumed that the speed u of the source or receiver is less than the wave speed v. If a source moves with a speed greater than the wave speed, there will be no waves in front of the source. The waves behind the source pile up on top of one another to form a shock wave that, for example, is heard as a sonic boom when it arrives at the receiver or is seen as a boat's wake. The shock wave is confined to a cone that narrows as u increases. We can calculate the angle of this cone. Figure 16-18a shows a source originally at point P_1 moving to the right with speed u. After some time t, the wave emitted from point P_1 will have traveled a distance vt. The source will have traveled a distance ut and will be at point P_2. The line from this new position of the source to the wavefront emitted when the source was at P_1 makes an angle θ with the path of the source given by

$$\sin \theta = \frac{vt}{ut} = \frac{v}{u} \qquad\qquad 16\text{-}16$$

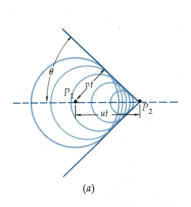

(a)

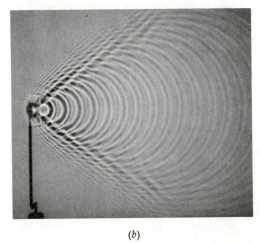

(b)

Figure 16-18
(a) A source moving from P_1 to P_2 with speed u greater than the wave speed v. The envelope of the wavefronts forms a cone with the source at the apex. The angle θ of this cone is given by $\sin \theta = v/u$.
(b) Waves in a ripple tank produced by a source moving with a speed greater than the wave speed.

Question

6. If the source and receiver are at rest relative to each other but the wave medium is moving relative to them, will there be any doppler shift in frequency?

16-5   Energy and Intensity

An important property of waves is that they transport energy and momentum. The transport of energy in a wave is described in terms of the *wave intensity*. Intensity is the average power per unit area that is incident perpendicular to the direction of propagation. The units of intensity are watts per square metre.

$$I = \frac{P_{av}}{A}$$

16-17   Intensity defined

If a point source emits power P uniformly in all directions, the intensity will decrease with the square of the distance from the source. We can see this from the fact that, at a distance r from the source, the power will be uniformly spread over a spherical surface of area $4\pi r^2$. In this case, the intensity will be

$$I = \frac{P}{4\pi r^2}$$

16-18

There is a simple relation between the intensity of a wave and the energy per unit volume in the medium carrying the wave. Consider the spherical wave that has just reached the radius r_1 in Figure 16-19. The volume inside the radius r_1 contains energy because its particles are oscillating with simple harmonic motion. The region outside r_1 contains no energy because the wave has not yet reached that region. After a short time Δt, the wave moves out past r_1 a short distance $\Delta r = v\,\Delta t$. The total energy in the medium is increased by the energy in the spherical shell of thickness $v\,\Delta t$. The additional energy in the spherical shell can be written

$$\Delta E = \eta A v\,\Delta t$$

where η is the average energy per unit volume in the shell, A is the surface area of the shell, and $Av\,\Delta t$ is the volume of the shell. The rate of increase of energy in the medium is the power passing into the shell. The source of this energy is, of course, at the center of the sphere from which the wave is radiating. Thus, the average incident power is

$$P_{av} = \frac{\Delta E}{\Delta t} = \eta A v$$

The intensity of the wave at a point is found by dividing the incident power by the area A.

$$I = \frac{P_{av}}{A} = \eta v$$

16-19

The intensity equals the product of the wave speed v and the average energy per unit volume.

This result is applicable to all waves.

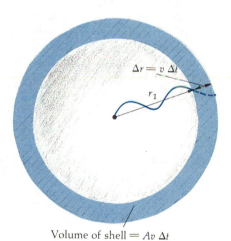

Volume of shell $= Av\,\Delta t$

Figure 16-19
A spherical wave that has just reached the distance r_1 at this time. All the energy is contained in the sphere of this radius. In some time Δt, the wave moves out a distance $\Delta r = v\,\Delta t$ and occupies the additional volume $Av\,\Delta t$ of the spherical shell, where A is the surface area of the shell. The energy transmitted past r_1 is $\eta A v\,\Delta t$, where η is the energy per unit volume.

The energy of a wave is the energy of the simple harmonic motion of the elements of the medium. Since the energy of simple harmonic motion is proportional to the square of the amplitude (Chapter 15), the intensity of a wave is proportional to the square of the amplitude of the vibrations. In a sound wave, air molecules vibrate in simple harmonic motion along the direction of the wave propagation. The intensity of a sound wave is proportional to the square of the amplitude of these vibrations. As we have mentioned, a sound wave can also be thought of as a pressure wave. The oscillation of the air molecules causes variations in the density of the air and therefore in the air pressure. The intensity of a sound wave is also proportional to the square of the amplitude of the pressure variations.

The human ear can accommodate a rather large range of sound-wave intensities, from about 10^{-12} W/m^2 (which is usually taken to be the threshold of hearing) to about 1 W/m^2 (which produces a sensation of pain in most people). The pressure variations corresponding to these extreme intensities are about 3×10^{-5} Pa for the hearing threshold and 30 Pa for the pain threshold. (Recall that a pascal is a newton per square metre.) These small pressure variations are superimposed on the normal constant atmospheric pressure of about 101 kPa.

Because of the enormous range of intensities to which the ear is sensitive and because the psychological sensation of loudness varies with intensity not directly but more nearly logarithmically, a logarithmic scale is used to describe the *intensity level* of a sound wave. The *intensity level β*, measured in *decibels* (dB), is defined as

$$\beta = 10 \log \frac{I}{I_0} \qquad\qquad\qquad \text{16-20} \quad \text{Intensity level}$$

where I is the intensity corresponding to the level β and I_0 is a reference level, which we will take to be the threshold of hearing:

$$I_0 = 10^{-12} \text{ W/m}^2 \qquad\qquad\qquad \text{16-21}$$

On this scale, the threshold of hearing is

$$\beta = 10 \log \frac{I_0}{I_0}$$

$$= 0 \text{ dB}$$

Similarly, the pain threshold is

$$\beta = 10 \log \frac{1}{10^{-12}}$$

$$= 10 \log 10^{12} = 120 \text{ dB}$$

Thus, the range of sound intensities from 10^{-12} W/m^2 to 1 W/m^2 corresponds to a range of intensity levels from 0 dB to 120 dB. Table 16-1 lists the intensity levels of some common sounds.

Table 16-1

Intensity and Intensity Level of Some Common Sounds, $I_0 = 10^{-12}$ W/m²

Source	I/I_0	dB	Description
	10^0	0	Hearing threshold
Normal breathing	10^1	10	Barely audible
Rustling leaves	10^2	20	
Soft whisper (at 5 m)	10^3	30	Very quiet
Library	10^4	40	
Quiet office	10^5	50	Quiet
Normal conversation (at 1 m)	10^6	60	
Busy traffic	10^7	70	
Noisy office with machines; average factory	10^8	80	
Heavy truck (at 15 m); Niagara Falls	10^9	90	Constant exposure endangers hearing
Old subway train	10^{10}	100	
Construction noise (at 3 m)	10^{11}	110	
Rock concert with amplifiers (at 2 m); jet takeoff (at 60 m)	10^{12}	120	Pain threshold
Pneumatic riveter; machine gun	10^{13}	130	
Jet takeoff (nearby)	10^{15}	150	
Large rocket engine (nearby)	10^{18}	180	

Example 16-6   A dog barking delivers about 1 mW of power. If this power is uniformly distributed in all directions, what is the sound intensity level at a distance of 5 m? What would the intensity level be if two dogs barking at the same time each delivered 1 mW of power?

The intensity at a distance of 5 m is the power divided by the area (Equation 16-18):

$$I = \frac{P}{4\pi r^2} = \frac{10^{-3} \text{ W}}{4\pi(25 \text{ m}^2)} = 3.18 \times 10^{-6} \text{ W/m}^2$$

The intensity level at this distance is then

$$\beta = 10 \log \frac{I}{I_0} = 10 \log \left(\frac{3.18 \times 10^{-6}}{10^{-12}} \right) = 10 \log (3.18 \times 10^6)$$

$$= 10(\log 3.18 + \log 10^6) = 10(0.50 + 6) = 65 \text{ dB}$$

If there are two dogs barking at the same time, the intensity will be twice as great, or

$$I = 2(3.18 \times 10^{-6} \text{ W/m}^2) = 6.36 \times 10^{-6} \text{ W/m}^2$$

Then I/I_0 will be

$$\frac{I}{I_0} = 6.36 \times 10^6$$

and the intensity level will be

$$\beta = 10 \log (6.36 \times 10^6) = 10(0.80 + 6) = 68.0 \text{ dB}$$

We can see from this example that if the intensity is doubled, the intensity level increases by 3 dB.

Example 16-7 A sound absorber attenuates the sound level by 30 dB. By what factor is the intensity decreased?

From Table 16-1, we note that for each drop in intensity of a factor of 10, the decibel level drops by 10 dB. Thus, a drop of 30 dB corresponds to a decrease in intensity by a factor of $10^3 = 1000$.

The sensation of loudness depends on the frequency as well as the intensity of the sound. Figure 16-20 is a plot of intensity level versus frequency for sounds of equal loudness to the human ear. (In this figure, the frequency is plotted on a logarithmic scale to allow the display of the wide range of frequencies from 20 Hz to 10 kHz.) The lowest curve corresponds to the threshold of hearing for someone with a very good ear. We can see from this curve that the threshold is 0 dB at 1 kHz, but it is about 50 dB at 60 Hz. About 1 percent of the population have a hearing threshold this low. The second lowest curve is a more typical hearing threshold curve; it applies to about 50 percent of the population. The upper curve represents the pain threshold. Note that it does not vary with frequency as much as the lower-level curves do. We note from this figure that the human ear is most sensitive at about 4 kHz for all intensity levels.

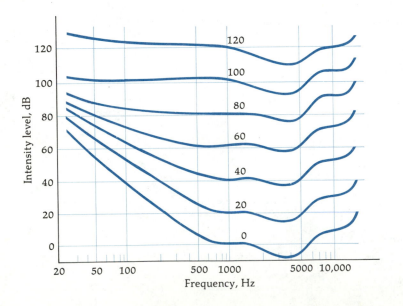

Figure 16-20
Intensity level versus frequency for sounds of equal loudness. The lowest curve is below the threshold for hearing of all but about 1 percent of the population. The second lowest curve is approximately the hearing threshold for about 50 percent of the population.

Sonic Booms

Laurent Hodges
Iowa State University

One of the most common examples of a shock wave resulting from a source traveling faster than the wave-propagation speed is the sonic boom generated by an airplane whose speed exceeds the speed of sound. A common misconception is that a sonic boom is produced when the airplane first accelerates past the speed of sound ("crashes through the sound barrier"). In fact, the shock wave exists during the whole time that the airplane is traveling supersonically, and a boom is heard every time the shock wave sweeps over a person with good hearing.

The airplane actually has two shock waves, associated with its front and back (Figure 1). These bow and tail shock waves have nearly the shape of cones (*Mach cones*) with their apexes at the aircraft's bow and tail, respectively. (Some deviation from a perfect cone shape results from variations in sound speeds in different parts of the air.) The cones have a half angle θ given by $\sin \theta = v/u_a$, where v is the speed of sound and u_a is the speed of the airplane.

The air pressure in the region of overpressure between the two shock waves differs from normal atmospheric pressure, the bow and tail waves being associated with pressures higher and lower than normal, respectively. This pressure deviation produces the sensation of sound in an observer. As the bow wave sweeps over the observer, the pressure increases by the *overpressure* ΔP in a rise time τ. It then decreases to approximately ΔP below normal atmospheric pressure before suddenly returning to normal at the tail wave. A plot of the time dependence of the pressure at a point has the shape of a slanted N and is therefore called the *N signature* (Figure 2).

The major pressure changes experienced at the ear occur as the bow and tail shock waves reach the observer, each producing an explosive sound. This explains the double crack of a sonic boom. A supersonic bullet also has bow and tail shock waves, but its two booms are so close together in time that only one crack is heard (see Figure 3).

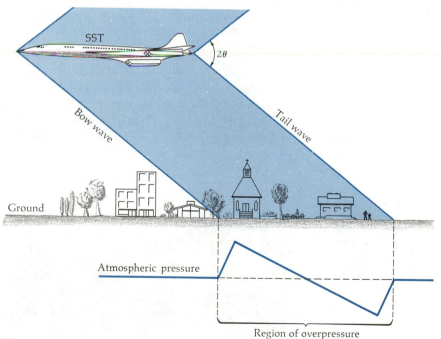

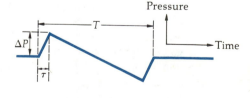

Figure 1
Cross-sectional representation of a supersonic transport in flight, showing bow and tail shock waves, Mach cone angle 2θ, and a plot of the pressure at ground level. The deviations from normal atmospheric pressure occur in the region of overpressure between the two shock waves.

Figure 2
A plot of the N signature, or time dependence, of the pressure at a given point where a sonic boom occurs. Shown are the overpressure ΔP, the rise time τ, and the time duration T.

Essay: Sonic Booms (continued)

The startling nature of a sonic boom and the resulting annoyance are a function of the rise time, the total duration of the N signature, and the overpressure. The rise time is of the order of 3 ms, and the time duration varies from about 0.1 s for a supersonic fighter to about 0.4 s for the British–French Concorde supersonic transport (SST). A typical overpressure for an SST 20 km overhead would be 100 N/m² = 100 Pa. The overpressure is less at places on either side of the plane's path, but booms can typically be heard over a zone 80 km wide.

It is not possible to assign a decibel reading to a sonic boom because of its short duration and its highly nonsinusoidal form. A sinusoidal sound wave with an rms pressure of 100 Pa would correspond to 134 dB, but the overpressure is not the only relevant parameter. Shorter rise times, for example, are associated with "louder" booms. U.S. Air Force studies have shown that an overpressure of 100 Pa corresponds to 110 to 120 on the perceived-noise decibel scale. (This modification of the decibel scale incorporates the psychological effects of the frequency spectrum of aircraft noises.)

There have been several systematic studies of the effects of sonic-boom exposures, including some carried out by the U.S. Air Force, which has tried to make the sonic boom psychologically acceptable by referring to it as the "sound of freedom." Besides startling people and animals, sonic booms can rattle dishes, shatter glass, and even lead to structural damage to some buildings. One study has indicated that approximately $600 in damage claims would result from every million person-booms,* and another researcher has estimated that each SST might generate 1000 claims totaling $500,000 annually.†

Some people are not bothered much by sonic booms. Many others will presumably learn to tolerate them if supersonic flight becomes commonplace, but even they will suffer the adverse physiological effects that result from any loud noise. Perhaps 25 to 50 percent of the United States population will not be able to adapt to sonic booms, and frequent supersonic flights will doubtlessly encounter stiff public opposition.

* Karl D. Kryter, "Sonic Booms from Supersonic Transport," *Science*, vol. 163, p. 359, 1969.

† William A. Shurcliff, *S/S/T and Sonic Boom Handbook*, Ballantine Books, New York, 1970. For further information see Harvey H. Hubbard, "Sonic Booms," *Physics Today*, vol. 21, p. 31, February 1968; and Herbert A. Wilson, Jr., "Sonic Boom," *Scientific American*, vol. 206, p. 36, January 1962.

Figure 3
The shock waves in air produced by a supersonic .22-caliber bullet.

Soviet Russia and a British–French consortium built SSTs, but only the state-owned airlines of the three producing countries ever purchased the aircraft (see Figure 4). The SST program ended in the United States in 1971 after Congress cut off federal funding, although there are occasional attempts to revive the project. Many countries have instituted restrictions, for example, no supersonic flights over land, that may well make the SST highly uneconomical. The SST's high fuel consumption has also worked against it.

Figure 4
SST Concorde taking off from John F. Kennedy airport in New York.

Summary

1. Wave motion is the propagation of a disturbance in a medium. In transverse waves, such as waves in a string, the disturbance is perpendicular to the direction of propagation. In longitudinal waves, such as sound waves, the disturbance is along the direction of propagation. Both energy and momentum are transported by a wave.

2. The speed of a wave depends on the density and elastic properties of the medium. It is independent of the motion of the wave source. The speed of waves in a string is related to the tension F and the mass per unit length μ by

$$v = \sqrt{\frac{F}{\mu}}$$

The speed of sound waves in air is related to the absolute temperature T by

$$v = \sqrt{\frac{\gamma RT}{M}}$$

where R is the gas constant, M is the molar mass, and γ is the ratio of the heat capacity at constant pressure to that at constant volume. For air, $\gamma = 1.4$.

3. When two waves meet at the same place, the disturbances add algebraically, resulting in either constructive or destructive interference.

4. In harmonic waves, the disturbance varies sinusoidally in time and space. The speed of a harmonic wave equals the frequency times the wavelength:

$$v = f\lambda$$

The human ear is sensitive to sound waves of frequencies in the range of about 20 Hz to 20 kHz.

5. When the source and the receiver are in relative motion, the observed frequency is increased if they move toward each other and decreased if they move away from each other. This is known as the doppler effect. The observed frequency f' is related to the source frequency f_0 by

$$f' = f_0 \frac{(1 \pm u_r/v)}{(1 \mp u_s/v)}$$

When the relative speed u of the source and receiver is much less than the wave speed v, the doppler shift is approximately the same whether the source or the receiver moves and is given by

$$\frac{\Delta f}{f_0} = \frac{u}{v}$$

6. The intensity of a wave is the power divided by the area. The intensity from a point source varies inversely as the square of the distance from the source:

$$I = \frac{P}{4\pi r^2}$$

The intensity of a harmonic wave is proportional to the square of the amplitude of the wave.

7. Sound intensities are measured on a logarithmic scale. The sound intensity level β in decibels (dB) is related to the intensity I by

$$\beta = 10 \log \frac{I}{I_0}$$

where $I_0 = 10^{-12}$ W/m^2, which is approximately the threshold of hearing. On this scale, the threshold of hearing is 0 dB and the pain threshold is 120 dB.

Suggestions for Further Reading

Boore, David M.: "The Motion of the Ground in Earthquakes," *Scientific American*, December 1977, p. 68.

The sudden slippage of one continental plate relative to its neighbor along a fault line excites mechanical waves in the earth. The study of this phenomenon may lead to improved building designs for areas at high risk for earthquakes.

Devey, Gilbert B., and Peter N. T. Wells: "Ultrasound in Medical Diagnosis," *Scientific American*, May 1978, p. 98.

The physics and technology of imaging using ultrahigh-frequency sound waves and the interpretation of the medical images produced are discussed.

Fletcher, Neville H., and Suzanne Thwaites: "The Physics of Organ Pipes," *Scientific American*, January 1983, p. 94.

Modern measurements show that the competing theories of nineteenth-century physicists Helmholtz and Rayleigh explaining the sound production of organ pipes are both correct but apply to different pressure ranges.

Jeans, Sir James: *Science and Music.* The University Press, Cambridge, England, 1953 (paperback, Dover, New York, 1968).

The basic features of sound production by vibrating strings and columns of air as well as harmony are well treated for the general reader in this book, originally published in 1937.

The Physics of Music, readings from *Scientific American* with an introduction by Carleen M. Hutchins, Scientific American, Inc., distributed by W. H. Freeman and Co., New York, 1978.

This collection of articles includes analyses of the acoustics of the singing voice, the piano, woodwinds, brasses, violins, and the bowed string as well as architectural acoustics.

Pierce, John R.: *The Science of Musical Sound,* Scientific American Books, Inc., distributed by W. H. Freeman and Co., New York, 1983.

This excellent book discusses the physical and mathematical aspects of sound waves that underlie our experience of music as well as the quantifiable aspects of sound perception.

Quate, Calvin F.: "The Acoustic Microscope," *Scientific American*, October 1979, p. 62.

The development of an acoustic microscope employing sound with a frequency 1000 times higher than that used in medical imaging technology is described. The images produced are comparable in quality to light micrographs.

Review

A. Objectives: After studying this chapter, you should:

1. Be able to describe the common features of all types of wave motion.

2. Be able to state the quantities on which the speed of a mechanical wave depends.

3. Be able to state the relationship among the speed, wavelength, and frequency of a wave.

4. Be able to explain why the pitch of a car horn goes down as the car passes and calculate the doppler frequency shift for various other examples of moving sources and receivers.

B. Define, explain, or otherwise identify:

Transverse wave, p. 376	Doppler effect, p. 384
Longitudinal wave, p. 376	Intensity, p. 389
Superposition, p. 377	Intensity level, p. 390
Harmonic wave, p. 380	Decibel, p. 390
Wavelength, p. 381	

C. True or false: If the statement is true, explain why. If it is false, give a counterexample.

1. Wave pulses in strings are transverse waves.

2. Sound waves in air are transverse waves.

3. When a wave pulse is reflected, it is always inverted.

4. The speed of sound at 20°C is twice that at 5°C.

5. A 60-dB sound has twice the intensity of a 30-dB sound.

6. The doppler shift of sound waves depends only on the relative motion of the source and the receiver.

Exercises

Section 16-1 Wave Pulses; Section 16-2 Speed of Waves

1. A steel wire 7 m long has a mass of 100 g and is under a tension of 900 N. What is the speed of transverse waves in this wire?

2. The pulse in Figure 16-21 is moving to the right at 2 cm/s. Sketch the shape of the string at times $t = 1$ s, 2 s, and 4 s.

Figure 16-21
Wave pulse for Exercise 2.

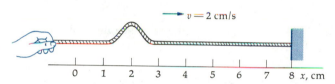

3. A steel piano wire is 0.7 m long and has a mass of 5 g. It is stretched with a tension of 500 N. (a) What is the speed of transverse waves in the wire? (b) To reduce the wave speed by a factor of 2 without changing the tension, copper wire is wrapped around the steel wire. What mass of copper wire is needed?

4. Show that the units of $\sqrt{F/\mu}$ are metres per second if F is in newtons and μ is in kilograms per metre.

5. Transverse waves travel at 231 m/s in a wire of mass 15 g and length 2 m. (a) What is the tension in the wire? (b) If the tension is doubled, what is the speed of transverse waves in the wire?

6. Transverse waves travel at 150 m/s in a wire of length 80 cm that is under a tension of 500 N. What is the mass of the wire?

7. A common lecture demonstration of wave pulses uses a piece of rubber tubing tied at one end to a fixed post and passed over a pulley to a suspended weight at the other end. The tubing is 10 m long and has a mass of 0.7 kg, and the suspended weight is 110 N. If the tubing is given a transverse blow at one end, how long will it take the resulting pulse to reach the other end?

8. (a) What are the SI units of the bulk modulus? (b) Show that $\sqrt{B/\rho}$ has units of metres per second if B has the correct SI units and ρ is in kilograms per cubic metre.

9. Find the speed of sound in air at (a) 40°C and (b) −20°C.

10. Find the speed of sound in air at $T = 300$ K in kilometres per hour.

11. Calculate the speed of sound at $T = 300$ K in helium, for which $M = 4$ g/mol and $\gamma = 1.67$.

12. Calculate the speed of sound at $T = 300$ K in (a) neon, for which $M = 20.2$ g/mol; and (b) argon, for which $M = 39.9$ g/mol. For both these gases, $\gamma = 1.67$.

13. Aluminum has a bulk modulus of 7.0×10^{10} N/m² and a density of 2.7×10^3 kg/m³. Calculate the speed of sound in aluminum.

14. The bulk modulus of water is 2.0×10^9 N/m². Find the speed of sound in water.

15. The speed of sound in mercury is 1410 m/s. The specific gravity of mercury is 13.6. Find the bulk modulus of mercury.

Section 16-3 Harmonic Waves

16. Middle C on the musical scale has a frequency of 262 Hz. (a) What is the wavelength of this note in air? (b) The frequency of the C one octave above middle C is twice that of middle C. Find the wavelength of this note.

17. A tuning fork has a frequency of 440 Hz. (Because orchestras normally tune to this frequency, it is sometimes called "concert A.") (a) What is the wavelength in air for waves of this frequency? (b) What is the wavelength in water, where the speed of sound is 1.44 km/s?

18. Equation 16-7 applies to any type of wave, including light waves, which travel at 3×10^8 m/s. The range of wavelengths of light to which the eye is sensitive is about 4×10^{-7} to 7×10^{-7} m. What are the corresponding frequencies of these light waves?

19. Find the frequency of microwaves of wavelength 3 cm. (Microwaves are electromagnetic waves; like all electromagnetic waves, they travel at 3×10^8 m/s.)

20. A wave in a string has a frequency of 250 Hz and an amplitude of 1 cm. It travels with a speed of 10 m/s. (a) Find its wavelength. (b) Make a sketch to scale of the shape of the string at some instant. (c) Sketch the same wave at a time 10 ms later.

21. A bat emits ultrasonic waves with wavelength 3.3 mm at 0°C. What is the frequency corresponding to this wavelength?

22. One end of a string 6 m long is moved up and down with simple harmonic motion of frequency 60 Hz. The waves reach the other end of the string in 0.5 s. Find the wavelength of the waves in the string.

Section 16-4   The Doppler Effect

23. This exercise is a doppler-effect analogy. A conveyor belt moves to the right with speed $v = 300$ m/min. A very fast piemaker puts pies on the belt at a rate of 20 per minute, and they are received at the other end by a pie eater. (a) If the piemaker is stationary, find the spacing λ between the pies and the frequency f with which they are received by the stationary pie eater. (b) The piemaker now walks with speed 30 m/min toward the pie eater while continuing to put pies on the belt at 20 per minute. Find the spacing of the pies and the frequency with which they are received by the stationary pie eater. (c) Repeat your calculations for a stationary piemaker and a pie eater who moves toward the piemaker at 30 m/min.

In Exercises 24 to 27, a source emits sound waves of frequency 200 Hz, which move through still air at 340 m/s.

24. The source moves with speed 80 m/s relative to still air toward a stationary listener. (a) Find the wavelength of the sound between the source and listener. (b) Find the frequency heard by the listener.

25. The source moves with speed 80 m/s away from a stationary listener. Find (a) the wavelength of the sound waves between the source and listener and (b) the frequency heard by the listener.

26. The listener moves at 80 m/s relative to still air toward a stationary source. (a) What is the wavelength of the sound between the source and listener? (b) What is the frequency heard by the listener?

27. The listener moves at 80 m/s away from a stationary source. Find the frequency heard by the listener.

28. A whistle of frequency 500 Hz moves in a circle of radius 1.2 m, making 3 revolutions per second. What are the maximum and minimum frequencies heard by a stationary listener?

29. A car traveling at 100 km/h blows its horn, which has a frequency of 400 Hz. Find the frequency heard by a stationary listener when (a) the car is approaching and (b) the car is receding.

Section 16-5   Energy and Intensity

30. A sound source emits sound uniformly in all directions with power 2.0 W. (a) Find the sound intensity at a distance of 5 m from the source. (b) Find the intensity level in decibels at this distance.

31. Find the intensity level in decibels for a sound wave of intensity (a) 10^{-10} W/m² and (b) 10^{-2} W/m².

32. Find the intensity in W/m² corresponding to an intensity level of (a) $\beta = 10$ dB and (b) $\beta = 3$ dB.

33. What fraction of the acoustic power of a noise would have to be eliminated to lower its sound intensity level from 90 dB to 70 dB?

34. A tape recorder lists the signal to noise ratio as 50 dB, meaning that when music is being played back, the intensity level of the music is 50 dB greater than that of noise (tape hiss and so forth). By what factor is the sound intensity of the music greater than that of the noise?

35. A quiet listening level for music is about 60 dB. When sound of this level impinges on an ear, the sound power entering the ear is given by the product of the intensity and the effective receiving area. Assuming the effective receiving area of the ear canal is 0.80 cm², find the typical sound power received by the ear when listening to quiet music.

Problems

1. A common rule of thumb for finding the distance to a lightning flash is to begin counting when the flash is observed and stop when the thunder clap is heard. The number of seconds counted is then divided by 3 to get the distance in kilometres. Why is this procedure justified? What is the speed of sound in kilometres per second? How accurate is this procedure? Is a correction for the time it takes for the light to reach you important? (The speed of light is about 3×10^8 m/s.)

2. A method for measuring the speed of sound using an ordinary watch is to stand some distance from a large flat wall and clap your hands rhythmically in such a way that the echo from the wall is heard halfway between each two claps. Show that the speed of sound is then given by $v = 4LN$, where L is the distance to the wall and N is the number of claps per second. What is a reasonable value for L for this experiment to be feasible? (If you have access to a flat wall outdoors somewhere, try this method and compare your result with the standard value.)

3. A tuning fork attached to a stretched wire generates transverse harmonic waves. The vibration of the fork is perpendicular to the wire. Its frequency is 440 Hz, and its amplitude of oscillation is 0.5 mm. The wire has a linear mass density of 0.01 kg/m and is under a tension of 1.2 kN. (*a*) Find the period and frequency of the waves in the wire. (*b*) What is the speed of the waves? (*c*) What is the wavelength?

4. An automobile has an acoustic noise output of 0.01 W. If the sound is radiated uniformly in all directions, (*a*) what is the intensity at a distance of 30 m? (*b*) What is the sound intensity level in decibels at this distance? (*c*) At what distance is the sound intensity level 40 dB?

5. A car moves with speed 17 m/s toward a stationary wall. Its horn emits 200-Hz sound waves, which move at 340 m/s. (*a*) Find the wavelength of the sound in front of the car and the frequency with which the waves strike the wall. (*b*) Since the waves reflect off the wall, the wall acts as a source of sound waves at the frequency found in part (*a*). What frequency does the driver of the car hear reflected from the wall?

6. A man drops a stone from a high bridge and hears it strike the water below exactly 4 s later. (*a*) Estimate the distance to the water on the assumption that the travel time for the sound to reach the man is negligible. (*b*) Improve your estimate by using your result in part (*a*) for this distance to estimate the time for sound to travel this distance. Then calculate the distance the stone falls in 4 s minus this time.

7. A stationary destroyer is equipped with sonar that sends out pulses of sound at 40 MHz. The sound travels at 1.54 km/s in seawater. Directly below the destroyer is a submarine that is moving downward at 8 m/s. (*a*) Find the frequency with which the sound waves hit the submarine. (*b*) The submarine acts as a moving source of the frequency found in (*a*). Find the frequency of the reflected waves detected by the destroyer.

8. The sound intensity from a loudspeaker at a rock concert is 10^{-2} W/m² at 20 m from the speaker. (*a*) What is the intensity level at this distance? (*b*) Assuming that the intensity decreases as $1/r^2$ from the speaker, at what distance will the intensity level be at the pain threshold of 120 dB?

9. Everyone at a party is speaking equally loudly. When only one person is talking, the sound level is 72 dB. What is the sound level when all 40 people are talking?

10. In this problem, we derive a convenient formula for the speed of sound in air at temperature t in Celsius degrees. We begin by writing the temperature in Equation 16-6 as $T = T_0 + t$, where $T_0 = 273$ K. (*a*) Show that Equation 16-6 can be written

$$v = \sqrt{\frac{\gamma R T_0}{M}}\left(1 + \frac{t}{T_0}\right)^{1/2}$$

$$= 331 \text{ m/s} \left(1 + \frac{t}{T_0}\right)^{1/2}$$

(*b*) Use the binomial expansion $(1 + x)^n \approx 1 + nx$ to write this result as

$$v \approx 331 \text{ m/s} \left(1 + \frac{t}{2T_0}\right)$$

(*c*) Show that this expression can be further simplified to

$$v \approx (331 + 0.606t) \text{ m/s}$$

where t is in Celsius degrees.

11. The noise level in an empty examination hall is 40 dB. When 100 students are writing an exam, the sound of heavy breathing and pens traveling rapidly over paper causes the noise level to rise to 60 dB (not counting groans). Assuming that each student contributes an equal amount of noise power, find the noise level to the nearest decibel after 50 students have left.

CHAPTER SEVENTEEN

Interference, Diffraction, and Standing Waves

We shall see in this chapter how sounds quarrel, fight, and when they are of equal strength, destroy one another, and give place to silence.

ROBERT BALL,
Wonders of Acoustics (1867)

We've already mentioned that light and sound are both waves, so why can we hear around corners but not see around corners? The answer to this question lies in one of the fundamental properties of waves, diffraction, which you will learn about in this chapter. You will also learn about another fundamental property of waves, interference, and how interference of two identical waves traveling in opposite directions can create waves which stand still. Standing waves are particularly interesting. Musical instruments operate by creating standing waves within them. By the time you finish the chapter, you'll know why violins are small and bass viols are large, why a clarinet sounds different from an oboe, and how music synthesizers can mimic the sounds of other instruments.

In Section 16-1, we noted that when two wave pulses in a string meet, they add or subtract depending on whether one of the pulses is inverted relative to the other. In this chapter, we will look at the interference of two or more harmonic waves. We will first look at the interference of waves of the same frequency from two sources. Then, we will discuss the bending of waves around corners, a phenomenon known as *diffraction*. Next, we will examine the phenomenon of beats, which are due to the interference of two waves of slightly different frequencies. We will then study standing waves. Finally, we will consider the analysis of a complex musical tone in terms of its component harmonic waves and the inverse problem of the synthesis of harmonic waves to produce a complex tone.

17-1   Interference of Waves from Two Point Sources

The interference of harmonic waves depends on the *phase* difference between the waves. If we have a single harmonic wave, the displacement y_1 at some fixed point varies with time as

$$y_1 = A \cos 2\pi ft \qquad\qquad 17\text{-}1$$

where A is the amplitude and f is the frequency. This is the same as Equation 15-4 for the displacement in simple harmonic motion. For this equation, we have chosen a zero time such that $y_1 = A$, the maximum displacement, when $t = 0$. This choice is convenient, but it is not necessary. If, for example, we prefer to choose a zero time such that the displacement is zero at $t = 0$, the equation for the displacement would be $y = A \sin (2\pi ft) = A \cos (2\pi ft - 90°)$.

If we have a second harmonic wave at the same point in space, and we have already chosen the zero time, the general equation for the displacement y_2 can be written

$$y_2 = A \cos (2\pi ft + \delta) \qquad\qquad 17\text{-}2$$

where δ is called the *phase constant*. The argument of the cosine function ($2\pi ft + \delta$ in this case) is called the phase of the wave. The two waves described by Equations 17-1 and 17-2 are said to differ in phase by δ. Figure 17-1 shows a plot of y_1 and y_2 as functions of time. If the phase difference δ is zero, the waves are in phase, and the two waves will interfere constructively. If δ is 180°, the two waves are said to be 180° out of phase. For $\delta = 180°$, Equation 17-2 can be written

Phase difference

$$y_2 = A \cos (2\pi ft + 180°) = -A \cos 2\pi ft \qquad\qquad 17\text{-}3$$

If we now add y_1 to y_2, we get 0; the two waves thus interfere destructively.

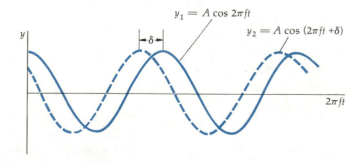

$y_1 = A \cos 2\pi ft$

$y_2 = A \cos (2\pi ft + \delta)$

$2\pi ft$

Figure 17-1

Displacement versus $2\pi ft$ for two harmonic waves of the same frequency and wavelength but differing in phase by δ.

A common cause of a phase difference between two waves is a path difference from the source to the point of interference. We will see that a path difference of half a wavelength leads to a phase difference of 180°.

Figure 17-2a shows the wave pattern produced by two point sources, separated by a small distance, oscillating in phase, each producing circular waves of wavelength λ. We can construct a similar pattern with a compass by drawing circular arcs to represent the wave crests from each source at some particular time (see Figure 17-2b). At the points where the crests from each source overlap, the waves add constructively. At these points, the paths for the waves from the two sources either are equal in length or differ by an integral number of wavelengths. The dashed lines indicate points that are equidistant from the sources or where the path difference is one wavelength, two wavelengths, or three wavelengths.

Figure 17-2
(a) Interference of two point sources oscillating in phase in a ripple tank. (b) Geometric construction of interference pattern in (a). The waves add constructively at the points of intersection. These points occur whenever the difference in path length from the two sources, Δr, is zero or an integral number of wavelengths.

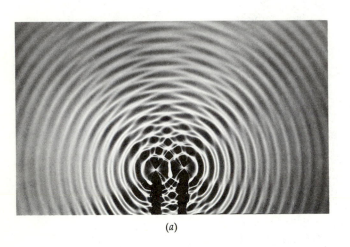

(a)

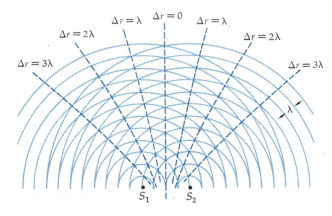

Figure 17-3 shows two harmonic waves of the same wavelength from two sources one wavelength apart along a line joining them. The sources are emitting waves in phase with each other; for example, when a positive crest leaves one source, a positive crest leaves the other source at the same time. To the right of the sources along this line, the path difference is one wavelength. We can see from this figure that a path difference of one wavelength or of any integral number of wavelengths is equivalent to no path difference at all. The two waves fall on top of one another with their maxima occurring at the same time and their minima also occuring at the same time. If the waves have equal amplitudes, as is the case here, the amplitude of the resulting wave will be twice the amplitude of each separate wave. Since the

(q)

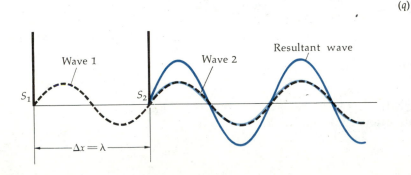

Figure 17-3
Constructive inteference. When waves from two sources oscillating in phase have a path difference Δx of one wavelength (or any integral number of wavelengths), the waves are in phase and add. The resultant amplitude is the sum of the amplitudes of the individual waves.

energy of a wave is proportional to the square of its amplitude, the energy of the resultant wave will be four times that of either wave by itself.

Between each set of interference maxima in Figure 17-1 there are interference minima where the path difference is an odd number of half wavelengths. Two harmonic waves with a path difference of one-half wavelength or an odd number of half wavelengths are shown in Figure 17-4. In this figure, the maximum of one wave falls at the minimum of the other and vice versa. If the waves are of equal amplitude, as they are here, the sum of the waves is zero. A path difference of one-half wavelength (or any odd integral number of half wavelengths) leads to a phase difference of 180° and the waves cancel. The lines in Figure 17-2b along which the waves cancel completely are called *nodes* or nodal lines.

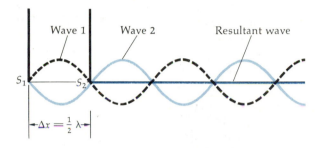

Figure 17-4
Destructive interference. When waves from two sources oscillating in phase have a path difference Δx of one-half wavelength (or any odd integral number of half wavelengths), they are out of phase and the amplitude of the resultant wave is the difference of the individual amplitudes.

Figure 17-5 shows the intensity of the resultant wave from two sources as a function of path difference. At points where the interference is constructive, the intensity is four times that due to either source because the amplitude of the resultant wave is twice that of either wave alone. At points of destructive interference, the intensity is zero. The average intensity, shown by the dashed line in the figure, is twice the intensity from either source alone, a result required by conservation of energy.

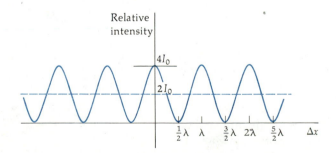

Figure 17-5
Plot of relative intensity versus path difference for two sources in phase. When the path difference Δx is an integral number of wavelengths, the intensity is $4 I_0$, where I_0 is the intensity from each source individually. When the path difference is an odd number of half wavelengths, the intensity is zero. The dashed line indicates the average intensity, which is twice that from either source.

The interference of two sound sources can be demonstrated by driving two separated speakers with the same amplifier (to ensure that they are always in phase) fed by an audio-signal generator. Moving about the room, one can detect by ear the positions of constructive and destructive interference. (In this demonstration, the sound intensity will not be quite zero at the points of destructive interference of the direct sound waves because of sound reflections from the walls or objects in the room.)

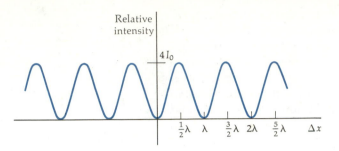

Relative intensity

$4I_0$

$\frac{1}{2}\lambda$ λ $\frac{3}{2}\lambda$ 2λ $\frac{5}{2}\lambda$ Δx

Figure 17-6
Plot of relative intensity versus path difference for two sources 180° out of phase. The pattern is the same as that in Figure 17-5 except that it is shifted by one-half wavelength.

Two sources need not be in phase to produce an interference pattern. Figure 17-6 shows the intensity of the resultant wave from two sources that are 180° out of phase as a function of path difference. (Two sound sources that are in phase can be made to be out of phase by 180° by merely switching the leads to one of the speakers.) The pattern is the same as that in Figure 17-5 except that the maximum and minimum are interchanged. Points equidistant from the sources or those for which the distance differs by an integral number of wavelengths are nodes because the waves are 180° out of phase and cancel (see Figure 17-7). At points where the path difference is an odd number of half wavelengths, the waves are now in phase because the 180° phase difference of the sources is offset by the 180° phase difference due to the path difference. It should be evident that similar interference patterns will be produced no matter what the phase difference between the sources is as long as the phase difference is constant over time.

Two sources that are in phase or have a constant phase difference are called *coherent sources.* Coherent sources of water waves in a ripple tank are easy to produce by driving both sources with the same motor. Similarly, coherent sound sources can be obtained by driving two speakers with the same audio-signal generator and amplifier. There are many examples of sound and light sources that are not coherent, that is, sources whose phase difference is not constant over time but varies randomly. For example, two speakers driven by different amplifiers or any two separate light sources, such as two candles, are incoherent. For such sources, the interference at a particular point varies rapidly back and forth from constructive to destructive, so no interference pattern is observed.

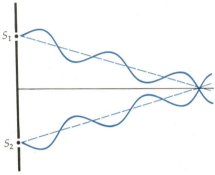

Figure 17-7
At a point equidistant from two sources that are out of phase, the waves cancel.

17-2 Diffraction

When a portion of a wavefront is cut off by an obstruction, the propagation of the wave is complicated. The portion of the wavefront that is not obstructed does not simply propagate in the direction of straight rays as might be expected. The situation that does occur is illustrated in Figure 17-8, which shows plane waves in a ripple tank meeting a barrier with a small opening. The waves to the right of the barrier are not confined to the narrow angle of the rays from the source that can pass through the opening; instead they are circular waves, just as if there were a point source at the opening. We can understand this by noting that the motion of the water at the opening due to the incoming waves is no different from the motion that would be produced by a point source at the opening (assuming the opening to be very small).

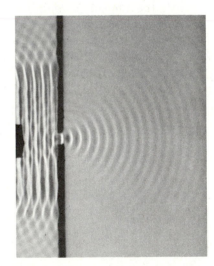

Figure 17-8
Plane waves in a ripple tank meeting a barrier with a small opening. The waves to the right of the barrier are circular waves concentric about the opening just as if there were a point source at the opening.

The propagation of a wave is thus quite different from the propagation of a stream of particles. In Figure 17-9a, the arrows indicate a stream of particles that hit a barrier with a small opening. Those that get through the opening are confined to a small angle. In Figure 17-9b, the arrows are rays that describe the propagation of a wave toward the barrier. The rays appear to bend around the edges of the opening. This bending of the rays, which always occurs to some extent when part of the wavefront is limited, is called *diffraction.**

In Figure 17-8, the opening in the barrier was much smaller than the wavelength and could be considered to be a point. Figure 17-10a shows plane waves hitting a barrier with an opening much larger than the wavelength. The waves beyond the barrier are similar to plane waves in the region far from the edges of the opening. (Near the edges, the wavefront is distorted, and the waves appear to bend slightly.) If the waves are limited by a barrier with an opening that is neither much smaller nor much larger than the wavelength but of the same order, the situation is more complicated. The transmitted waves do not approximate either plane waves or circular waves (see Figure 17-10b).

Although the detailed calculation of the wave pattern of an obstructed wave is much too complicated to be considered here, we can describe the propagation of the wave using a geometric method discovered by Christian

---

* Diffraction should not be confused with *refraction,* which is the bending of a wave when it encounters a surface separating two media in which the wave speed differs. For example, a light beam is refracted when it enters glass from air.

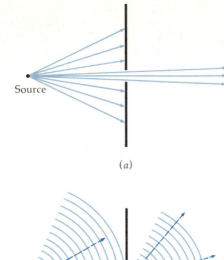

(a)

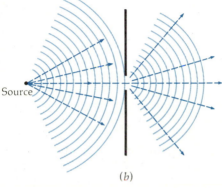

(b)

Figure 17-9
Comparison of the transmission through a narrow opening in a barrier of (a) a beam of particles and (b) a wave. In (a) the transmitted particles are confined to a narrow angle. In (b) the opening acts as a point source of circular waves that are radiated to the right through a much wider angle than are the particles in (a).

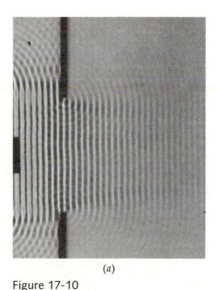

(a)

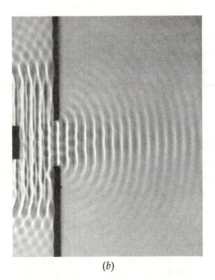

(b)

Figure 17-10
Plane waves in a ripple tank meeting a barrier with an opening. (a) When the opening is very large compared with the wavelength, the effect of the barrier is noticeable only near the edges. (b) Here the width of the opening is five times as large as the wavelength. When the opening is neither small nor large compared with the wavelength, the wave pattern is much more difficult to analyze.

Huygens in about 1678. Huygens' method considers each point on a given wavefront to be a point source of waves. The new wavefront at some later time is then the interference pattern produced by these sources. This turns out to be the surface enveloping all the little spherical wavelets emitted by these point sources. Figure 17-11 shows Huygens' construction for the propagation of a plane wave and a spherical wave. (Of course if each point on a wavefront were really a point source, there would be waves in the backward direction as well. Huygens ignored these back waves, but in a refinement of his method by Kirchhoff in the nineteenth century, the intensity of the wavelets was shown to be zero in the backward direction.) When the wavefront is limited by an aperture or obstacle, the resulting wave pattern can be found by calculating the interference pattern of the line of point sources located on the unobstructed portion of the wavefront. We will use this idea when we study optics in Chapter 27 to calculate the diffraction pattern of light passing through a single slit. Here, we merely state the important general result known as the *ray approximation:*

If the aperture or obstacle is large compared with the wavelength, the bending of the wavefront is not noticeable and the wave propagates in straight lines or rays, much as a beam of particles does.

Because the wavelength of audible sound ranges from a few centimetres to several metres (Example 16-3) and is often large compared with apertures and obstacles (doors or windows, for example), the bending of sound waves around corners is an everyday phenomenon. On the other hand, the wavelengths of visible light range from about 4×10^{-7} to 7×10^{-7} m. Because these wavelengths are so small compared with the size of ordinary objects and apertures, diffraction of light is negligible and so usually is not observed. That is, you can hear but not see around corners. It was for this reason that Newton and others thought that light was a beam of particles rather than a wave motion.

Diffraction effects place a limitation on the use of waves to locate small objects or to identify the details of a small object. Essentially, no detail on a scale smaller than the wavelength used can be observed. For example, if audible sound waves are used to detect or locate some object, the precision of the location cannot be greater than the wavelength used. Since the smallest wavelength of audible sound is about 2 cm (corresponding to a frequency of 17 kHz), one cannot locate an object to better than ± 2 cm using audible sound. In addition, no appreciable reflection of waves occurs unless the object is at least as large as the wavelength. Bats emit ultrasonic waves and detect their reflection to locate prey, such as moths. They can hear frequencies up to about 120,000 Hz, which corresponds to a wavelength of (340 m/s)/(120 kHz) = 2.8 mm. The shortest wavelength of visible light is about 4×10^{-7} m, which is much smaller than most objects but is about 4000 times larger than the diameter of a typical atom. Thus, atoms can never be seen with visible light.

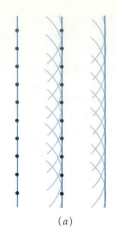

(a)

(b)

Figure 17-11
Huygens' construction for the propagation of (a) plane waves to the right and (b) outgoing spherical or circular waves.

Question

1. You may have noticed that when you hear music coming from another room around a corner, there is a loss in the high frequencies compared with the rest of the sound. Can you explain this?

17-3 Beats

The interference of two waves of nearly equal frequency produces the interesting phenomenon known as *beats*. A familiar example is the beats produced by the sound waves from two tuning forks or two guitar strings of nearly equal but not identical frequencies. What is heard is a tone whose intensity varies alternately between loud and soft. The frequency of this variation in intensity is called the *beat frequency*.

Figure 17-12a shows a plot of two harmonic waves of nearly equal frequency. We have chosen the time $t = 0$ to be when the waves are in phase and interfere constructively. Because their frequencies differ, the waves gradually become out of phase, and at time t_1 they are 180° out of phase and interfere destructively. (Complete cancellation will, of course, occur only if the amplitudes of the two waves are equal.) At an equal time interval later, the two waves will again be in phase and interfere constructively (time t_2 in the figure). The resultant interference pattern is shown in Figure 17-12b. The frequency of the maxima in the envelope in that figure (the dashed line) is the beat frequency. The closer the two waves are in frequency, the longer it takes them to move out of phase and back in phase, and so the closer the waves are in frequency, the smaller the beat frequency. In fact, the beat frequency turns out to be just equal to the *difference* between the two frequencies:

$$f_{beat} = \Delta f \qquad\qquad \text{17-4 Beat frequency}$$

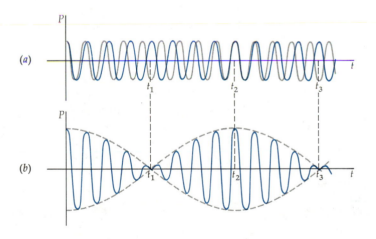

Figure 17-12
Beats. (a) Two waves of different frequencies that are in phase at $t = 0$ are 180° out of phase at some later time t_1. At a still later time t_2, they are back in phase. (b) The resultant of the two waves shown in (a). The frequency of the rapid oscillation is about the same as that of the original waves, but the amplitude is modulated as indicated by the dashed envelope. The amplitude is maximum at times 0 and t_2 and is zero at times t_1 and t_3.

For example, if we strike two tuning forks of frequency 241 Hz and 243 Hz, respectively, we will hear a pulsating tone that has maximum intensity 2 times per second; that is, the beat frequency is 2 Hz. (The frequency of this tone is actually the mean frequency 242 Hz.) The phenomenon of beats is often used to compare an unknown frequency with a known frequency, as when tuning a piano string with a tuning fork. The ear can detect up to about 15 to 20 beats per second. Above this frequency, the fluctuations in loudness are too rapid to be heard. Beats are also used to detect small frequency changes, such as those produced in a radar beam reflected from a moving car. The shift in frequency of the reflected beam is due to the doppler effect, which depends on the speed of the car. The speed can be determined by measuring the beats produced when the reflected radar beam is combined with the original radar beam.

An interesting phenomenon somewhat related to beats is the moiré pattern produced by two sets of parallel lines of slightly different spacing (see Figure 17-13). Moiré patterns can often be observed, for example, when one looks through two parallel fences along the road or through two window screens.

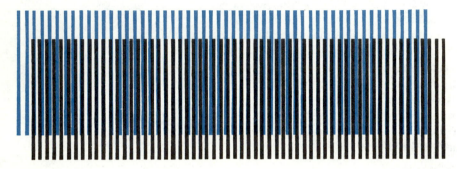

Figure 17-13
Moiré pattern showing the beats produced by two sets of parallel lines when the spacing of one set differs slightly from that of the other.

Example 17-1   When a 440-Hz tuning fork is struck simultaneously with the playing of the note A on a guitar, 3 beats per second are heard. When the guitar string is tightened slightly to increase its frequency, the beat frequency increases to 6 beats per second. What is the frequency of the guitar string after it has been tightened?

Since 3 beats per second were heard, the original frequency of the guitar string was either 443 Hz or 437 Hz. If it were 437 Hz, increasing the frequency by slightly tightening the string would decrease the beat frequency because the frequency would now be closer to that of the tuning fork. Since the beat frequency increased to 6 beats per second, the original frequency was 443 Hz and the new frequency is 446 Hz. To tune the string to 440 Hz, the string should be loosened slightly to lower its frequency.

Questions

2. When musical notes are sounded together to make chords, beats are not heard. Why not?

3. About how accurately do you think you can tune a piano wire to a tuning fork?

17-4   Standing Waves

When waves are confined in space, as they are with a piano string or an organ pipe, there are reflections at both ends, so there are waves traveling in both directions. These waves combine according to the general law of wave interference. For a given string or pipe, there are certain frequencies for which this interference results in a stationary vibration pattern called a *standing wave*. The study of standing waves has many applications in the field of music and in nearly all areas of science and technology. We will first consider standing waves in strings, which are easily demonstrated and visualized. We will then consider standing sound waves.

String Fixed at Both Ends

If we fix one end of a long rope and move the other end up and down with simple harmonic motion of small amplitude, we find that standing-wave patterns such as those shown in Figure 17-14 are produced at certain frequencies. The frequencies that produce such patterns are called the resonance frequencies of the rope (or string) system. The lowest resonance frequency is called the fundamental frequency f_1. It produces the standing-wave pattern shown in Figure 17-14a, which is called the *fundamental mode of vibration* or the *first harmonic*. The second lowest frequency f_2 produces the pattern shown in Figure 17-14b. It is found to be twice the fundamental frequency. This mode of vibration is called the *second harmonic*. The third lowest frequency f_3 is three times the fundamental frequency and produces the *third-harmonic* pattern shown in Figure 17-14c. We note from Figure 17-14 that there are certain points on the rope that do not move. For example, the midpoint in Figure 17-14b is at rest. Such points are called *nodes*. Midway between each pair of nodes is a point of maximum amplitude of vibration called an *antinode*. The fixed end of the rope is, of course, a node,

Standing waves are produced on a violin string by bowing the string. The length of the string is varied by clamping one end with the finger.

Fundamental

Nodes and antinodes

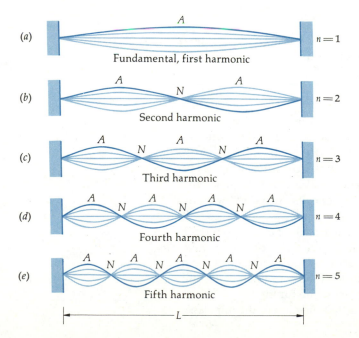

(a) Fundamental, first harmonic $n=1$

(b) Second harmonic $n=2$

(c) Third harmonic $n=3$

(d) Fourth harmonic $n=4$

(e) Fifth harmonic $n=5$

L

Figure 17-14
Standing waves in a string that is fixed at both ends. The points labeled A are antinodes and those labeled N are nodes. In general, the nth harmonic has n antinodes.

and the end that is vibrated is approximately a node because the amplitude of vibration of that end is much smaller than the amplitude at the antinodes. We note that the fundamental or first harmonic has one antinode, the second harmonic has two antinodes, and so on.

We can easily relate the resonance frequencies to the wave speed in the rope and the length of the rope. We can see from Figure 17-14 that the length of the rope L equals half the wavelength for the first harmonic, whereas it equals two halves of the wavelength for the second harmonic, three halves for the third harmonic, and so forth. In general, for the nth harmonic we have

$$L = n\left(\frac{\lambda_n}{2}\right) \qquad n = 1, 2, 3, 4, \ldots \qquad \text{17-5}$$

Resonance condition for a string fixed at both ends

This result is known as the *resonance condition*. We can find the frequency from the fact that the wave speed v equals the frequency times the wavelength:

$$f_n = \frac{v}{\lambda_n} = \frac{v}{(2L/n)}$$

or

$$f_n = n\frac{v}{2L} = nf_1 \qquad n = 1, 2, 3, 4, \ldots \qquad \text{17-6}$$

Resonance frequencies

where

$$f_1 = \frac{v}{2L}$$

Since the wave speed v is $\sqrt{F/\mu}$, where F is the tension and μ is the mass per unit length,

$$f_1 = \frac{v}{2L} = \frac{1}{2L}\sqrt{\frac{F}{\mu}} \qquad \text{17-7}$$

The easiest way to remember the resonance frequencies given by Equation 17-6 is to sketch Figure 17-14 to remind yourself of the resonance condition $L = n\lambda/2$ and then use $f = v/\lambda$.

We can understand the production of standing waves in terms of resonance. Let us consider a string attached at one end to a vibrating body, such as a tuning fork, and fixed at the other end. The first wave sent out by the vibrating tuning fork travels down the string a distance L and is reflected and inverted at the fixed end. It then travels back to the tuning fork, where it is again reflected. It is again inverted and starts back down the string. If the time for the wave to travel the complete distance of $2L$ is exactly equal to the period of the vibrating tuning fork, the twice reflected wave will exactly overlap the second wave produced by the fork, and the two waves will interfere constructively; that is, the two waves will add together to produce a wave of twice the amplitude of either original wave. The combined wave will then travel down the string and back, where it will add to the third wave produced by the fork, and so on. The amplitude of the combined wave will continue to grow as the string absorbs energy from the tuning fork. Various damping effects, such as the loss of energy due to reflection or the imperfect flexibility of the string, put a limit on the maximum amplitude that can be reached. This maximum amplitude is much larger than that of the tuning fork. Thus the tuning fork is in *resonance* with the string. Resonance will also occur if the time for the first wave to travel the distance $2L$ is twice the period of the fork or any integer n times the period. Since the time for a wave to travel the distance $2L$ is $2L/v$, where v is the wave speed, we can write the resonance condition as

$$\frac{2L}{v} = nT = \frac{n}{f}$$

or

$$f = n\frac{v}{2L}$$

where $T = 1/f$ is the period. This is the same result we found by fitting an integral number of half wavelengths into the distance L.

The frequencies given by Equation 17-6 are called the resonance frequencies or *natural frequencies* of the string. This resonance phenomenon is analogous to the resonance of a simple harmonic oscillator with a harmonic driving force that we discussed in Section 15-5. If the frequency of the driving force equals the natural frequency of a simple harmonic oscillator, the oscillator absorbs the maximum amount of energy from the driving force. Note, however, that a vibrating string has not just one natural frequency, but a series of natural frequencies that are integral multiples of the fundamental frequency. This series is called a *harmonic series*. The first frequency f_1 is the first harmonic, the second $f_2 = 2f_1$ is the second harmonic, and so on. In the terminology often used in music, the second harmonic is called the first *overtone*, the third harmonic the second overtone, and so forth.

Harmonic series

Gusting winds excited standing waves in the Tacoma Narrows suspension bridge, leading to its collapse on

November 7, 1940, only 4 months after it had been opened to traffic.

String Fixed at One End

Standing waves can also be produced in a string with one end free instead of both ends fixed. A nearly free end can be obtained by tying the string to a very long, light thread, as shown in Figure 17-15.

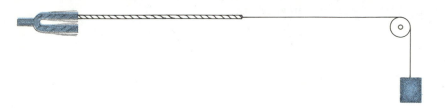

Figure 17-15
An approximation of a string fixed at one end and free at the other end can be produced by connecting the "free" end of the string to a very long, light thread. Since the amplitude of the tuning fork is very small, the end attached to the fork is approximately fixed.

The standing-wave patterns that result are shown in Figure 17-16. In this case, the free end is an antinode. In the fundamental mode of vibration, the length of the string L equals $\lambda/4$. In the next highest mode, $L = 3\lambda/4$. The standing-wave condition can thus be written

$$n\frac{\lambda}{4} = L \qquad n = 1, 3, 5, 7, \ldots \qquad\qquad 17\text{-}8$$

The frequencies are thus given by

$$f_n = n\frac{v}{4L} = nf_1 \qquad n = 1, 3, 5, 7, \ldots \qquad\qquad 17\text{-}9$$

where

$$f_1 = \frac{v}{4L} \qquad\qquad 17\text{-}10$$

is the fundamental frequency. The natural frequencies of this system occur in the ratios $1:3:5:7$. The even harmonics are missing.

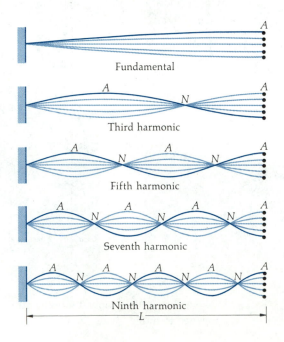

Fundamental

Third harmonic

Fifth harmonic

Seventh harmonic

Ninth harmonic

L

Figure 17-16
Standing waves in a string fixed at only one end. The free end is an antinode.

Example 17-2   A string is stretched between two fixed supports 1 m apart, and the tension is adjusted until the fundamental frequency of the string is 440 Hz. What is the speed of transverse waves in the string?

From the condition for resonance (Equation 17-5), the wavelength for the first harmonic is $\lambda = 2L = 2$ m. Hence, the wave speed is

$$v = \lambda f = (2 \text{ m})(440 \text{ Hz}) = 880 \text{ m/s}$$

Standing Sound Waves

Much of what we have learned about standing waves in strings can be applied to standing sound waves. Figure 17-17 shows a tube of air of length L closed at the right end and fitted with a movable piston at the left end. Because the air cannot vibrate past the closed end, this point must be a node for the displacement of the air molecules. If the vibration of the piston at the left end is of small amplitude, that end is also approximately a displacement node. (This approximation is similar to that for the vibrating string.) The standing-wave condition for this system is the same as for the string fixed at both ends, and all the same equations apply. The distance $2L$ must contain an integral number of wavelengths:

$$n\lambda = 2L$$

The allowed wavelengths are those that can be fitted into the length of the tube with displacement nodes at each end. As we have mentioned, sound waves can be thought of as pressure waves; that is, the longitudinal vibrations of the air molecules back and forth cause variations in the air pressure. These pressure variations are sinusoidal if the displacement is sinusoidal, as is the case for a harmonic wave. However, the pressure and displacement variations in a sound wave are 90° out of phase. In a standing sound wave, the displacement nodes are pressure antinodes and vice versa.

There are no musical instruments based on tubes closed at both ends. However, there are electrical devices (antennas and cavity resonators) that are analogous to this case.

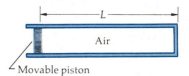

Figure 17-17
Air confined to a tube closed at both ends. There is a displacement node at each end if the amplitude of the oscillation of the piston is small. The condition for standing sound waves is the same as that for standing waves in a string fixed at both ends.

Example 17-3   If the speed of sound is 340 m/s, what are the allowed frequencies and wavelengths for standing sound waves in a tube 1 m long that is closed at both ends?

The fundamental frequency is the lowest allowed frequency, and it corresponds to the longest wavelength $\lambda_1 = 2L = 2$ m. Thus, the fundamental frequency is $f_1 = v/\lambda_1 = (340 \text{ m/s})/(2 \text{ m}) = 170$ Hz. The frequencies of the other harmonics are $f_2 = 2f_1 = 340$ Hz, $f_3 = 3f_1 = 3(170 \text{ Hz}) = 510$ Hz, and $f_n = nf_1 = 170n$ Hz. The wavelengths are $\lambda_2 = 2L/2 = 1$ m, $\lambda_3 = 2L/3 = 0.67$ m, and $\lambda_n = (2L/n) = (2 \text{ m})/n$.

If the end of the tube at the right in Figure 17-17 is not closed but is open to the atmosphere, the open end is a displacement antinode. (It is also a pressure node since the pressure is fixed at atmospheric pressure.) The conditions for the resonance frequencies and wavelengths for this system are the same as those for a string with one end fixed and one end free. The wavelength of the fundamental mode is four times the length of the tube, and only odd harmonics are present. For the tube of length 1 m in Example 17-3, the wavelength of the fundamental is $4L = 4$ m and its frequency is $f_1 = 340/4 = 85$ Hz. The other allowed frequencies are $f_3 = 3f_1 = 255$ Hz; $f_5 = 5f_1 = 425$ Hz, and so on.

The result that the open end of the tube is a displacement antinode (and a pressure node) is based on the assumption that the sound wave in the tube is a one-dimensional wave, which is approximately true if the diameter of the tube is very small compared with the wavelength of the sound wave. In practice, the displacement antinode and pressure node are slightly beyond the open end of the tube. The effective length of the tube is thus somewhat longer than its true length. If L is the length of the tube and ΔL is the correction, the effective length of the tube is

$$L_{\text{eff}} = L + \Delta L$$

The end correction ΔL is of the order of the radius of the tube. Note that the distance between two successive nodes or antinodes is still $\frac{1}{2}\lambda$ even though the distance from the open end of the tube to the first displacement node is slightly less than $\frac{1}{4}\lambda$ because of the end correction.

Figure 17-18 shows an apparatus for measuring the speed of sound in air. A narrow vertical tube is partially filled with water. Since the water level can be adjusted, the length of the air column L can be adjusted. A tuning fork of frequency f vibrates at the open end of the tube and sends sound waves down the tube. They are reflected at the adjustable bottom, the water level. When the water level is such that the standing-wave condition is met, reasonance occurs and the increased energy of the sound wave can be detected by the ear. In this experiment, the wavelength is $\lambda_0 = v/f_0$, where v is the speed of sound in air. The wavelength is fixed since v is determined by the air temperature and f_0 is determined by the tuning fork. If ΔL is the end correction term for the tube, the tube will resonate if its length is L_1, which is given by

$$L_1 + \Delta L = \tfrac{1}{4}\lambda_0$$

It will also resonate at lengths $L_3, L_5, \ldots, L_n$, which are given by

$$L_3 + \Delta L = \tfrac{3}{4}\lambda_0$$

$$L_5 + \Delta L = \tfrac{5}{4}\lambda_0$$

$$L_n + \Delta L = \frac{n}{4}\lambda_0$$

The wavelength, and thus the speed of sound, can be measured by measuring the distance between two consecutive resonance lengths, for example,

$$L_3 - L_1 = \tfrac{1}{2}\lambda_0 = \frac{v}{2f_0}$$

An organ pipe is a familiar example of the use of standing waves in air columns. In the flue-type organ pipe (see Figure 17-19), a stream of air is directed against the sharp edge of an opening (point A in the figure). The complicated swirling motion of the air near the edge sets up vibrations in the air column above. The resonance frequencies of the pipe depend on the length of the pipe and on whether the other end is closed or open.

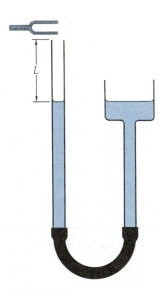

Figure 17-18
Apparatus for determining the speed of sound in air. Sound waves of the frequency of the tuning fork are excited in the tube on the left, whose length L can be adjusted by adjusting the level of the water. Resonance occurs when the effective length of the tube equals $\tfrac{1}{4}\lambda_0, \tfrac{3}{4}\lambda_0, \tfrac{5}{4}\lambda_0, \ldots$, where λ_0 is the wavelength of the sound.

In a closed organ pipe, there is a displacement node at the closed end and an antinode near the opening (point A in the figure). The resonance frequencies for such a pipe are therefore those for a tube open at one end and closed at the other. The wavelength of the fundamental is approximately four times the length of the pipe, and only odd harmonics are present. The clarinet family of musical instruments are essentially open–closed cylinders with odd harmonics for the lowest notes of the lowest register. Figure 17-20 shows the standing-wave patterns in a pipe open at one end and closed at the other.

In an open organ pipe, there is a displacement antinode (and a pressure node) near both ends of the pipe. The resonance frequencies for a pipe open at both ends are the same as for one closed at both ends except that there is an end correction at each end. The wavelength of the fundamental is two times the effective length of the pipe, and all harmonics are present. Flutes and recorders behave as open–open pipes with all the harmonics present. The standing-wave patterns for a pipe open at both ends are shown in Figure 17-21.

Most musical instruments are much more complicated than simple cylindrical tubes. The conical tube, which forms the basis for the oboe, bassoon, English horn, and saxophone, has a complete harmonic series with its fundamental wavelength equal to twice the length of the cone. Brass instruments are combinations of cones and cylinders. The fact that they have

Figure 17-19
Flue-type organ pipe. A stream of air is blown against the edge A, causing a swirling motion of the air near A that excites standing waves in the pipe. There is a pressure node near point A, which is open to the atmosphere. The resonance frequencies of the pipe depend on the length of the pipe and on whether the other end is open or closed.

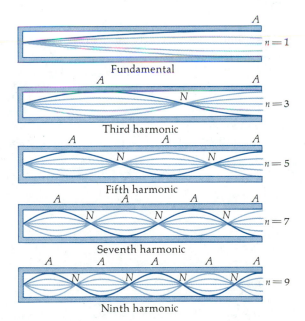

Figure 17-20
Standing-wave patterns in a pipe that is open at one end and closed at the other. These patterns are for the displacement, which has a node at the closed end and an antinode at the open end. As is the case for a string fixed at one end, only the odd harmonics are present.

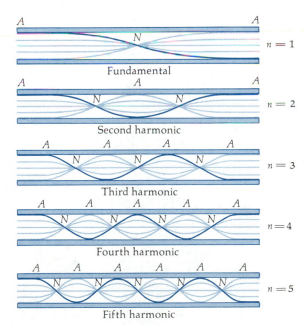

Figure 17-21
Standing-wave patterns in a pipe that is open at both ends. The displacement has an antinode at each end. As is the case for a string fixed at both ends, all harmonics are present.

nearly harmonic series is a triumph of educated trial and error rather than one of mathematical calculation. The analysis of these instruments is extremely complex.

Questions

4. When a guitar string is tightened, its frequency increases. Explain why.

5. How do the resonance frequencies of an organ pipe change when the air temperature increases?

17-5   Harmonic Analysis and Synthesis

When an oboe and a violin in an orchestra play the same note, for example, concert A, they sound quite different. Both tones have the same "pitch," a psychological sensation as to the highness or lowness of the note, which is strongly correlated with its frequency—the higher the frequency, the higher the pitch. However, the notes differ in what is called *tone quality*. The principal reason for the difference in tone quality is that, although both the violin and the oboe are producing vibrations at the fundamental frequency of 440 Hz, each instrument is also producing harmonics whose relative intensities depend on the instrument and how it is played. If each instrument produced only the fundamental frequency of 440 Hz, they would sound the same.

Figure 17-22 shows plots of pressure variations versus time for a tuning fork, a clarinet, and a cornet, each playing the same note. These patterns are called *waveforms*. The waveform for the tuning fork is very nearly a pure

Pipe organ at the Christian Science Center in Boston. The fundamental frequency of each pipe is determined by its length and whether the pipe is open or closed.

Figure 17-22
Waveforms of (*a*) a tuning fork, (*b*) a clarinet, and (*c*) a cornet, each at a frequency of 440 Hz and at approximately the same intensity.

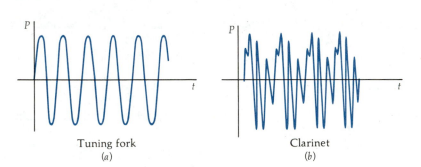

| Tuning fork | Clarinet | Cornet |
| (*a*) | (*b*) | (*c*) |

sine wave, but those for the clarinet and cornet are clearly not simple sine waves. These and other such waveforms can be analyzed in terms of what harmonics are present. Such analysis is called *harmonic analysis*. (It is also sometimes called *Fourier analysis* after the French mathematician Fourier who developed the mathematics for analyzing periodic functions.) Figure 17-23 shows a plot of the relative intensity of the harmonics that correspond to the waveforms in Figure 17-22. The waveform of the tuning fork contains only the fundamental frequency. That of the clarinet contains the fundamental plus large amounts of the third, fifth, and seventh harmonics in

Harmonic analysis

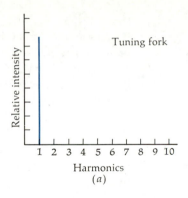

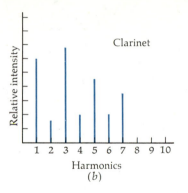

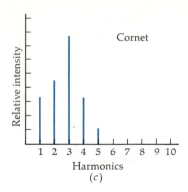

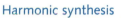

addition to lesser amounts of the second, fourth, and sixth harmonics. For the cornet, there is far more energy in the third harmonic than in the fundamental.

Although the harmonic content of steady tones is important in identifying the instrument, other factors are also important, especially the way the tone begins (the attack), the presence of vibrato or tremolo (variations in pitch and loudness), and the rate of buildup of the harmonics.

The inverse of harmonic analysis is *harmonic synthesis*, the construction of an arbitrary periodic waveform from its harmonic components. Figure 17-24 shows a square wave that can be produced by an electronic square-wave generator and its approximate synthesis using only three harmonics.

Figure 17-23
Relative intensities of the harmonics in the waveforms shown in Figure 17-22 for the tuning fork, clarinet, and cornet.

Harmonic synthesis

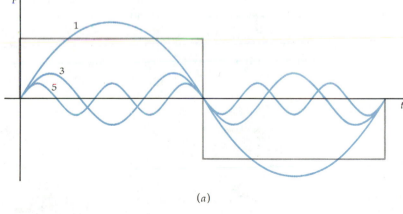

(a)

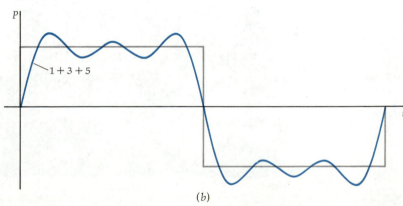

(b)

Figure 17-24
(a) A square wave and the first three odd harmonics used to synthesize the wave. (b) Synthesis of the square wave using the first three odd harmonics.

Relative
amplitude

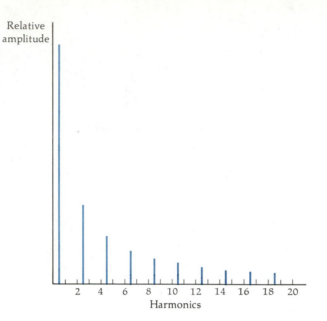

Harmonics

Figure 17-25
Relative amplitudes of the
harmonics needed to synthesize the
square wave shown in Figure
17-24. The more harmonics that are
used, the better the approximation
of the square wave.

The relative intensities of the harmonics are shown in Figure 17-25. The
more harmonics used, the better the approximation of the actual waveform.

An electronic music synthesizer produces a series of harmonics whose
relative amplitudes can be adjusted to produce the desired waveform. In
addition, each note can be shaped by adjusting such parameters as the
attack, the duration, the amount of vibrato or tremolo, the decay (the rate at
which the sound decreases from it peak level), and the release (the rate at
which the sound drops off at the end). Sophisticated modern synthesizers
can produce tones that sound like any instrument in the orchestra.

(a) Synthesizer designed by Heinrich
von Helmholtz and constructed by
Rudolph Koenig. Each electrically
driven tuning fork has a resonator
which can be mechanically opened
or closed to vary the intensity of
that harmonic. (b) In the more
modern Moog synthesizer, shown
here with Robert Moog, the
harmonics are produced and mixed
electronically.

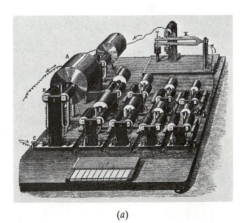

(a)

(b)

Summary

1. Two wave sources that are in phase or have a phase difference that is constant over time are coherent.

2. Waves from two coherent sources interfere because of the path difference from the sources. If the sources are in phase, the interference will be constructive when the path difference is zero or a whole number of wavelengths and destructive when the path difference is an odd number of half wavelengths.

3. Diffraction is the bending of a wave around an obstacle or an edge of an aperture. It occurs whenever the wavefront is limited by an obstacle or aperture. When the wavelength is much smaller than the size of obstacles or apertures, diffraction is negligible and the wave propagates in straight lines like a beam of particles. This is known as the ray approximation. Because of diffraction, waves can be used to locate an object only to within a wavelength or so.

4. Beats are the result of the interference of two waves of slightly different frequency. The beat frequency equals the difference in the frequencies of the two sources.

$$f_{beat} = \Delta f$$

5. When waves are confined in space — for example, waves in a piano wire or sound waves in an organ pipe — standing waves occur. For a string fixed at both ends or a pipe that is either open or closed at both ends, the standing-wave condition can be found by fitting waves into the string or pipe with a node at each end. The result is that an integral number of half wavelengths must fit into the length of the string or tube. The standing wave condition is thus

$$n\frac{\lambda}{2} = L \qquad n = 1, 2, 3, 4, \ldots$$

The allowed waves form a harmonic series, the frequencies of which are given by

$$f_n = nf_1$$

where $f_1 = v/2L$ is the lowest frequency, called the fundamental. If the string has one end fixed and one end free or the pipe is closed at one end and open at the other end, there is a node at one end and an antinode at the other. The standing-wave condition in this case is

$$n\frac{\lambda}{4} = L \qquad n = 1, 3, 5, 7, \ldots$$

Only the odd harmonics are present. Their frequencies are given by

$$f_n = nf_1$$

where $f_1 = v/4L$.

6. Sounds of different tone quality contain different mixtures of harmonics. The analysis of a particular tone in terms of its harmonic content is called harmonic analysis. Harmonic synthesis is the construction of a tone by adding the proper mixture of harmonics.

Suggestions for Further Reading

See the Suggestions for Further Reading for Chapter 16, particularly *The Science of Musical Sound*, which contains discussions of beats, standing waves, and harmonic analysis.

Review

A. Objectives: After studying this chapter, you should:

1. Be able to describe the interference pattern produced by two wave sources.

2. Be able to discuss the conditions under which the bending of waves around corners is significant.

3. Be able to describe how a piano tuner uses beats to tune a piano.

4. Be able to sketch the standing-wave patterns for vibrating strings and vibrating air columns in organ pipes and from them obtain the possible frequencies for standing waves.

5. Be able to explain why the tone quality is different for different musical instruments.

B. Define, explain, or otherwise identify:

Constructive interference, p. 402
Destructive interference, p. 403
Coherent sources, p. 404
Diffraction, p. 405
Beats, p. 407
Node, p. 409
Antinode, p. 409
Harmonic series, p. 411
Overtone, p. 411

Harmonic analysis, p. 416
Harmonic synthesis, p. 417

C. True or false: If the statement is true, explain why. If it is false, give a counterexample.

1. The waves from two sources in phase interfere constructively everywhere in space.

2. Two wave sources that are out of phase by 180° are incoherent.

3. Interference patterns are observed only for coherent sources.

4. Diffraction occurs whenever the wavefront is limited.

5. Diffraction occurs only for transverse waves.

6. The beat frequency between two sound waves of nearly equal frequencies equals the difference in the frequencies of the individual sound waves.

7. The frequency of the fifth harmonic is five times that of the fundamental mode.

8. In a pipe open at one end and closed at the other, the even harmonics are not excited.

9. When a violin string is bowed, it vibrates with a single frequency equal to its fundamental frequency.

Exercises

Use $v = 340 \ m/s$ for the speed of sound in air in all exercises unless an exercise specifies otherwise.

Section 17-1   Interference of Waves from Two Point Sources

1. Two speakers (coherent sound sources) are separated by a distance of 6 m. A listener sits directly in front of one speaker a distance 8 m from it so that the two speakers and the listener form a right triangle. Find the two lowest frequencies for which the listener observes (*a*) destructive interference and (*b*) constructive interference.

2. With a compass, draw circular arcs representing wave crests for each of two point sources a distance $d = 6$ cm apart using 1 cm for the wavelength (see Figure 17-2*b*). Connect the intersections corresponding to points of constant path difference and label the path difference for each line.

3. Two speakers are driven in phase by an audio-signal generator and amplifier. A listener is stationed 2 m from one speaker and 2.4 m from the other. Find the two lowest frequencies for which the listener observes (*a*) destructive interference and (*b*) constructive interference.

4. Two violinists are standing a few feet apart and playing the same notes. Are there places in the room where certain notes are not heard because of destructive interference? Explain.

5. Two loudspeakers separated by some distance emit sound waves of the same frequency. At some point P, the intensity due to each speaker is I_0. The path distance from P to one of the speakers is $\frac{1}{2}\lambda$ greater than that from P to the other speaker. What is the intensity at P if (a) the speakers are coherent and in phase, (b) the speakers are coherent and out of phase by $180°$, and (c) the speakers are incoherent?

6. Answer the questions of Exercise 5 for a point P' for which the distance to the far speaker is 1λ greater than the distance to the near speaker.

Section 17-2 Diffraction

7. If the wavelength is much larger than the diameter of a loudspeaker, the speaker radiates in all directions, much like a point source. On the other hand, if the wavelength is much smaller, the sound travels in approximately a straight line in front of the speaker. Find the frequency of sound for which the wavelength is (a) 10 times and (b) one-tenth the diameter of a 30-cm speaker. (c) Do the same for a 6-cm diameter speaker.

8. Discuss the relationship between diffraction and interference. Can diffraction occur without interference? Can interference occur without diffraction?

Section 17-3 Beats

9. Two tuning forks have frequencies 256 and 260 Hz. What is the beat frequency if both forks vibrate at the same time?

10. When a violin string is played (without fingering) simultaneously with a tuning fork of frequency 440 Hz, beats are heard at a rate of 3 per second. When the tension in the string is increased slightly, the beat frequency increases. What was the initial frequency of the violin string?

11. Two tuning forks are struck simultaneously, and 4 beats per second are heard. The frequency of one fork is 500 Hz. (a) What are the possible values of the frequency of the other fork? (b) A piece of wax is placed on the fork of unknown frequency to lower its frequency, and 6 beats per second are heard. What is the frequency of this fork without the wax?

Section 17-4 Standing Waves

12. A string fixed at both ends is 3 m long. It resonates in its second harmonic at a frequency of 60 Hz. What is the speed of transverse waves in the string?

13. A string 3 m long fixed at both ends is vibrating in its third harmonic. The speed of waves in the string is 50 m/s. (a) What are the wavelength and frequency of this wave? (b) What are the wavelength and frequency of the fundamental?

14. A 5-g steel wire 1.2 m long is under tension of 968 N. (a) What is the speed of transverse waves in this wire? (b) Find the wavelength and frequency of the fundamental. (c) Find the frequency of the second and third harmonics.

15. A rope 4 m long is fixed at one end and is attached to a light string at the other end so that it is free to move there. The speed of waves in the rope is 20 m/s. Find the frequency of (a) the fundamental, (b) the first overtone, and (c) the second overtone.

16. Calculate the fundamental frequency for a 10-m organ pipe that is (a) open at both ends and (b) closed at one end.

17. The normal range of hearing is about 20 Hz to 20 kHz. What is the greatest length of an organ pipe that would have its fundamental mode in this range if (a) it is closed at one end and (b) it is open at both ends?

18. The shortest pipes used in organs are about 7.5 cm long. (a) What is the fundamental frequency of a pipe this long that is open at both ends? (b) For such a pipe, what is the highest harmonic that is within audible range?

Section 17-5 Harmonic Analysis and Synthesis

There are no exercises for this section.

Problems

Use $v = 340$ m/s for the speed of sound in air in all problems unless a problem specifies otherwise.

1. The G string on a violin is 30 cm long. When played without fingering, it vibrates at 196 Hz. The next higher notes on the scale are A (220 Hz), B (247 Hz), C (262 Hz), and D (294 Hz). How far from the end of the string must a finger be placed to play these notes?

2. A string with a mass density of 4×10^{-3} kg/m is under tension of 360 N and is fixed at both ends. One of its resonance frequencies is 375 Hz. The next higher resonance frequency is 450 Hz. (a) What is the fundamental frequency of this string? (b) Which harmonics are the ones given? (c) What is the length of the string?

3. Two loudspeakers radiate in phase at 170 Hz. An observer sits 8 m from one speaker and 11 m from the other. The intensity level at the observer from either speaker acting alone is 60 dB. (a) Find the intensity level at the observer when both speakers are on together. (b) Find the intensity level at the observer when both speakers are on together but one has its leads reversed so that the speakers are $180°$ out of phase. (c) Find the intensity level at the observer if the speakers are incoherent.

4. Three successive resonance frequencies in an organ pipe are 1310, 1834, and 2358 Hz. (a) Is the pipe closed at one end or open at both ends? (b) What is the length of the pipe?

5. A string fastened at both ends has successive resonances

with wavelengths of 0.54 m for the nth harmonic and 0.48 m for the $(n + 1)$th harmonic. (a) Which harmonics are these? (b) What is the length of the string?

6. The use of a vernier scale is related to beats. It is easiest to learn how such a scale works by constructing one. Along the edge of a card (such as a 3 by 5 index card) make a set of nine marks equally spaced by 0.9 cm apart, and number the marks (call the edge 0). Your scale can now be used to interpolate between centimetre marks on a scale that is not divided into millimetres. Place your card along a scale ruled in centimetres so that the edge is somewhere between the 2- and 3-cm marks. Then note which mark on your card is aligned with a centimetre mark on the scale. If, for example, the fourth mark on your scale is aligned with a centimetre mark, the edge of the card is at 2.4 cm. Explain how this works and relate it to beats. If the card is now moved 2 mm to the right, which mark on the card is aligned with a centimetre mark on the scale?

7. A radar trap radiates microwaves with a frequency of 2.00 GHz. When the waves are reflected from a moving car, the beat frequency is 293 beats per second. Find the speed of the car. (Review Section 16-4 on the doppler effect. You may use Equation 16-15 for a moving source here.)

8. In a lecture demonstration of standing waves, a string is attached to a tuning fork which vibrates at 60 Hz and sets up transverse waves of that frequency in the string. The other end of the string passes over a pulley and the tension is varied by attaching weights to that end. The string has approximate nodes at the tuning fork and at the pulley. If the string is 2.54 m long (from the tuning fork to the pulley) and it has a mass of 20 g, what weights must be attached for the string to oscillate in its fundamental mode? Find the tensions for vibrations in the first three overtones.

9. With a tuning fork of frequency 500 Hz held above the tube in Figure 17-18, resonances are found when the water level is at distances of 16, 50.5, 85, and 119.5 cm. Use these data to calculate the speed of sound in air. How far outside the open end of the tube is the pressure node?

10. The tuning fork in Figure 17-18 is replaced by a wire of mass 1 g and length 50 cm stretched with a tension of 440 N. It is stroked with a violin bow so that it oscillates with its fundamental frequency. The water level in the tube is lowered until resonance is first obtained at 18.9 cm below the top of the tube. Use these data to determine the speed of sound in air. Why is this method not very accurate?

11. The tuning fork in Figure 17-18 is replaced by a 40-cm long wire of mass 0.01 kg vibrating in its second harmonic. Resonance in the tube is observed when the water level is 1 m below the top of the tube. Assuming the speed of sound in air to be 340 m/s, find (a) the frequency of oscillation of the air column in the tube, (b) the speed of waves along the wire, and (c) the tension in the wire.

12. A tuning fork of frequency f can be "stopped" by looking at it in a dark room with a strobe light of the same frequency. If the strobe frequency is slightly different from f, the fork appears to vibrate in slow motion. Discuss how this phenomenon is related to beats.

CHAPTER EIGHTEEN

Electric Fields and Forces

With this chapter we enter a new realm of physics, that of electric charges and their effects. This begins our investigation of the second of the four known forces in the world, the electromagnetic force. (The remaining two are strong and weak nuclear forces, which you'll learn about in Chapter 33.) The electromagnetic force is responsible for all of the forces we have dealt with so far, except gravitation. Contact forces, friction, surface tension, the forces that hold solids and liquids together, the force that causes atoms to form molecules—all these are electromagnetic in nature.

We start our study of the electromagnetic force with three chapters devoted to various aspects of electricity. In this one, you will learn the basics about electric charges and electric fields.

The field concept, one of the most mysterious in physics, has gradually forced itself on the attention of scientists trying to interpret the interactions among different parts of matter. . . . Physicists became convinced that, if a body placed in an empty region of space is acted upon by a force, it is because that region of space is not really structureless but has a power of action of its own.

LOUIS DE BROGLIE

Electricity is in such common use today that we don't give it much thought. Yet just a century ago, there were no electric lights, electric heaters, electric motors, radios, or television sets. Although these devices are all products of the twentieth century, the study of electricity has a long and interesting history. We will therefore begin our study with a short discussion of the historical development of the concept of electric charge. We will then study the force exerted by one electric charge on another and introduce the concept of the electric field. In Section 18-4, we will show how the electric field can be described by drawing lines of force to indicate the magnitude and direction of the field. We will then define electric flux and discuss Gauss' law, which relates the net electric flux through a closed surface to the net charge inside that surface. Finally, we will study the behavior of electric dipoles in electric fields.

18-1 Electric Charge

Observations of electrical attraction can be traced back to the ancient Greeks.* The Greek philosopher Thales of Miletus (640–546 B.C.) observed that when amber is rubbed, it attracts small objects such as straw or feathers. (This attraction was often confused with the magnetic attraction of lodestone for iron.) Electrical phenomena were little understood, however, until the sixteenth century, when the English physician William Gilbert (1540–1603) studied electrical and magnetic phenomena systematically. Gilbert showed that many substances besides amber acquire an attractive property when rubbed. He was one of the first to understand clearly the distinction between this attraction and the magnetic attraction, and he introduced the terms *electric force, electric attraction,* and *magnetic pole.* (The word "electric" comes from the Greek *elektron,* meaning amber, and the word "magnetic" comes from *Magnesia,* a area in Thessaly, part of ancient Greece, where magnetic iron ore was found.) Gilbert is perhaps best known for his discovery that the behavior of a compass needle results from the fact that the earth itself is a large magnet with poles at the north and the south. The orientation of a compass needle near the surface of the earth is merely an example of the general phenomenon of the attraction or repulsion of the poles of two magnets. (Gilbert apparently failed to observe electrical repulsion.)

Around 1729 Stephen Gray, an Englishman, discovered that electrical attraction and repulsion can be passed from one body to another if the bodies are connected by certain substances, particularly metals. This discovery was of great importance, since previously experimenters could electrify an object only by rubbing it. The discovery of electric conduction also implied that electricity has an existence of its own and is not merely a property somehow brought out in a body by rubbing.

* Much of this discussion follows Sir Edmund Whittaker's excellent *A History of Theories of Aether and Electricity,* Nelson, London, 1953; Torchbook edition, Harper, New York, 1960.

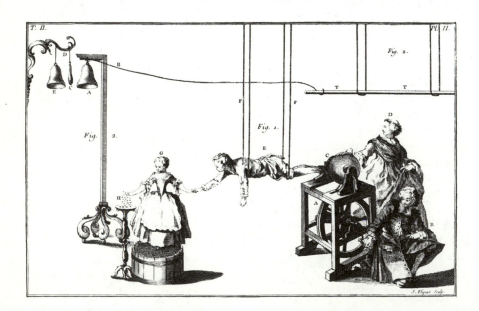

A 1745 woodcut showing electrical effects being transmitted through two persons before attracting feathers or bits of paper at the table on the left.

The existence of two kinds of electricity was suggested by Charles François Du Fay (1698–1739), who described an experiment in which a gold foil was attracted by a piece of glass rod that had previously been rubbed. When the foil was touched by the glass, it acquired the "electric virtue" and then repelled the glass.

> It is certain that bodies which have become electric by contact are repelled by those which have rendered them electric; but are they repelled likewise by other electrical bodies of all kinds?*

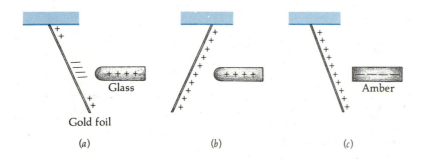

(a)   (b)   (c)

Du Fay answered this question by noting that the gold foil, which was repelled by the glass rod, was attracted by an electrified piece of amber or resin. If the gold foil was electrified by touching it to an electrified piece of amber, however, it was repelled by the amber but attracted by the electrified glass (see Figure 18-1). Du Fay gave the two kinds of electricity the names *vitreous* and *resinous* and postulated the existence of two fluids that become separated by friction.†

In 1747, the great American statesman and scientist Benjamin Franklin proposed a one-fluid model of electricity, which is described in the following experiment. If person *A* standing on wax (to provide insulation from the ground) rubs a glass tube with a piece of silk cloth, and another person *B*, also standing on wax, touches the glass tube, both *A* and *B* become electrified. They can each give a spark to person *C*, who is standing on the ground. However, if *A* and *B* touch each other before touching *C*, the electricity of *A* and *B* is neutralized. Franklin proposed that every body has a "normal" amount of electricity. When a body is rubbed against another, some of the electricity is transferred from one body to the other; thus, one has an excess and the other has an equal deficiency. The excess and deficiency can be described with plus and minus signs: one body is plus and the other minus. An important feature of Franklin's model is the implication of the conservation of electricity, now known as the *law of the conservation of charge*. The electric charge is not created by the rubbing; it is merely transferred. Since Franklin chose to call Du Fay's vitreous kind of electricity positive, Du Fay's resinous electricity was merely the lack of vitreous electricity and was therefore negative. A glass rod when rubbed acquires an excess of electricity and is positive, whereas an amber rod loses electricity when rubbed and becomes negative (see Figure 18-2).

* *Memoir de l'Académie* (1733), pp. 43 and 44, as quoted by Whittaker.

† We know now that the transfer of charge from one substance to another is not associated with friction but with the close contact of the substances achieved by rubbing them together.

Figure 18-1
(a) An uncharged gold foil is attracted to a charged glass rod. (b) After the foil has been touched by the rod, it is repelled by the rod. (c) The foil, charged by touching the glass rod, is attracted to a charged amber rod. In (a) the uncharged gold foil is attracted because the positively charged glass rod attracts mobile electrons in the foil, making the region near the rod negative and leaving further away regions positive. The neutral foil would also be attracted by the negatively charged amber rod because it would repel electrons in the foil from the region nearest the rod.

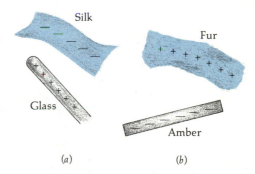

(a)   (b)

Figure 18-2
(a) When a glass rod is rubbed with silk, the glass becomes positively charged and the silk becomes negatively charged. (b) When an amber (or rubber) rod is rubbed with fur, the amber becomes negatively charged and the fur becomes positively charged.

Franklin's choice was unfortunate because we now know that it is electrons that are transferred in the rubbing process, and according to Franklin's convention, the electrons have a negative charge. Thus, when we say that a (positive) charge is transferred from body A to body B, it is really electrons that are transferred from body B to body A. When glass is rubbed with silk, electrons are transferred from the glass to the silk, leaving the glass positive and the silk negative; when amber is rubbed with fur, electrons are transferred from the fur to the amber.

In the course of his experiments, Franklin noticed that small cork balls inside a metal cup seemed to be completely unaffected by the electricity of the cup. He asked his friend Joseph Priestley (1733–1804) to check this fact, and Priestley began experiments that showed that there is no electricity on the inside surface of a hollow metal vessel (except near the opening). From this result, Priestley made the remarkable deduction that the force between two charges varies as the inverse square of the distance between them, just like the gravitational force between two masses. We will discuss this result further in Section 18-4.

The inverse-square law for electricity was confirmed by the experiments of Charles Coulomb (1736–1806), for which he used a torsion balance of his own invention. Coulomb's experimental apparatus was essentially the same as that described for the Cavendish experiment (Chapter 9) with the masses replaced by small charged balls. For the magnitudes of charges easily transferred by rubbing, the gravitational attraction of the balls is negligible compared with the electric attraction or repulsion. Coulomb also used his torsion balance to confirm that the force between two magnetic poles varies as the inverse square of the distance.

The nineteenth century saw rapid growth in the understanding of electricity and magnetism, culminating with the great experiments of Michael Faraday (1791–1867) and others and the mathematical theory of electromagnetism of James Clerk Maxwell (1831–1879).

The experiments conducted around the beginning of the twentieth century on the atomic nature of charge should be mentioned in this brief historical discussion of electrical charge. In 1897, the English physicist J. J. Thomson showed that all materials contain particles that seemed to be identical in that they had the same ratio of mass to electric charge. We now know that these particles, subsequently called electrons, are a fundamental part of the makeup of all atoms. In 1909, the American physicist Robert Millikan discovered that electric charge always occurs in integral amounts of a fundamental unit; that is, electric charge is *quantized*. Any charge can be written $q = Ne$, where N is an integer and e is the magnitude of the fundamental charge unit.*

The electron has a charge of $-e$, and the proton $+e$. The electron and the proton are very different particles. For example, the mass of the proton is nearly 2000 times that of the electron. Yet the magnitude of the charge of the electron is exactly equal to that of the proton. Most elementary particles have either no charge (for example, the neutron), a charge of $+e$, or a charge of $-e$. Particles made up of combinations of elementary particles may have

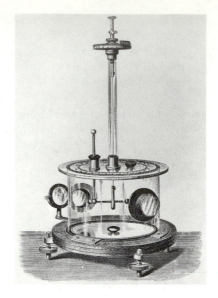

Coulomb's torsion balance.

Charge quantization

---

* In the quark model of elementary particles, protons, neutrons, and some other elementary particles are thought to be made up of particles called quarks, which carry charges of $\pm \frac{1}{3}e$ or $\pm \frac{2}{3}e$. Such particles apparently cannot be observed individually, but only in combinations that result in a net charge of $\pm Ne$ or 0 charge. We will discuss this model further in Chapter 33.

charges of $2e$, $3e$, and so forth. For example, the alpha particle, which is the nucleus of the helium atom, consists of two protons and two neutrons bound tightly together and has a charge of $+2e$. An atom with Z protons in its nucleus (Z is the *atomic number*) has Z electrons outside the nucleus and is electrically neutral. When one or more of the electrons are removed, the atom becomes a positively charged ion. If the atom takes on additional electrons, it becomes a negatively charged ion.

In this brief historical survey, we have mentioned three important properties of electric charge:

1. Charge is conserved.

2. Charge is quantized.

3. The force between two point charges varies as the inverse square of the distance between the charges.

Questions

1. After you stroke your cat, do you think the cat is charged positively or negatively?

2. Why might it be more convenient to choose electrons to be positive and protons to be negative?

3. How might the world be different if the charge of the proton were slightly greater in magnitude than that of the electron?

18-2   Coulomb's Law

The experiments of Coulomb and others on the forces exerted by one point charge on another are summarized in *Coulomb's law:*

The force exerted by one point charge on another is along the line joining the charges. It is repulsive if the charges have the same sign and attractive if the charges have opposite signs (see Figure 18-3). The force varies inversely as the square of the distance separating the charges and is proportional to the product of the charges.

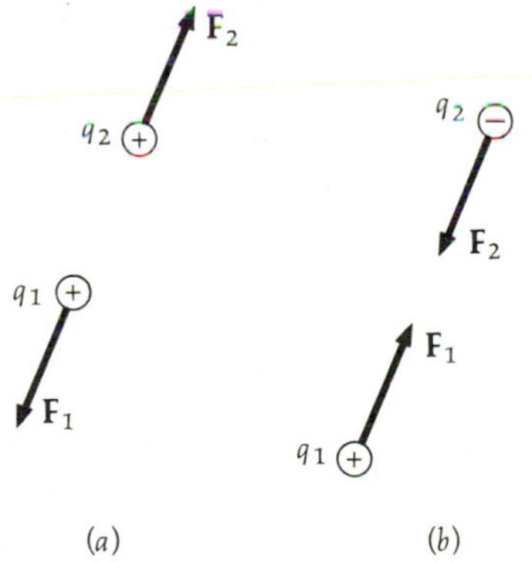

The electric force exerted by one charge on another, like the gravitational force exerted by one mass on another, is an example of "action at a distance." That is, there is no visible agent between the particles that transmits the force. Coulomb's law can be stated more simply with an algebraic expression. If q_1 and q_2 are two point charges separated by a distance r, the magnitude of the force exerted by one charge on the other is

$$F = k \frac{q_1 q_2}{r^2}$$

18-1   Coulomb's law

where k is a constant known as the Coulomb constant. The SI unit of charge

Figure 18-3
(a) Like charges repel, whereas (b) unlike charges attract.

is the coulomb (C), which is defined in terms of the unit of electrical current, the ampere. (The ampere is defined in terms of a magnetic-force measurement, which we will discuss in Chapter 21. It is the practical unit of current used in everyday electrical work.) The coulomb is the amount of charge that follows past a point in a wire in one second when the current is one ampere. The value of the Coulomb constant k is

$$k = 8.99 \times 10^9 \text{ N} \cdot \text{m}^2/\text{C}^2 \qquad\qquad 18\text{-}2$$

In most calculations, it is convenient to use the approximation

$$k = 9 \times 10^9 \text{ N} \cdot \text{m}^2/\text{C}^2 \qquad\qquad 18\text{-}3 \qquad \text{The Coulomb constant}$$

If both charges have the same sign, that is, if they are both positive or both negative, the force is repulsive. If the two charges have opposite signs, the force is attractive. Note the similarity between Coulomb's law and Newton's law of gravitation (Equation 9-2). Both are inverse-square laws. The gravitational force between two particles is proportional to the masses of the particles and is always attractive, whereas the electrical force is proportional to the charges of the particles and may be attractive or repulsive. The fundamental unit of electric charge e is related to the coulomb by

$$e = 1.60 \times 10^{-19} \text{ C} \qquad\qquad 18\text{-}4 \qquad \text{Fundamental unit of charge}$$

Typical laboratory charges of the order of 10 nC (1 nC = 10^{-9} C) to 0.1 μC (1 μC = 10^{-6} C) can be produced by putting certain objects in intimate contact, often by simply rubbing their surfaces together. These charges involve the transfer of many electrons and produce relatively large forces.

Example 18-1   Two point charges of 0.05 μC each are separated by 10 cm. Find the force exerted by one charge on the other and the number of fundamental units of charge in each.

From Coulomb's law, the magnitude of the force is

$$F = \frac{(9 \times 10^9 \text{ N} \cdot \text{m}^2/\text{C}^2)(0.05 \times 10^{-6} \text{ C})(0.05 \times 10^{-6} \text{ C})}{(10^{-1} \text{ m})^2}$$

$$= 2.25 \times 10^{-3} \text{ N}$$

The number of electrons that must be transferred to produce a charge of 0.05 μC is found from

$$q = Ne$$

$$N = \frac{q}{e} = \frac{0.05 \times 10^{-6} \text{ C}}{1.6 \times 10^{-19} \text{ C}} = 3.31 \times 10^{11}$$

A charge of this size does not reveal that electric charge is quantized. A million electrons could be added to or subtracted from this charge without being detected by ordinary instruments.

Charles Augustin de Coulomb (1736–1806).

Example 18-2   Find the force exerted by a charge equal to Avogadro's number of electrons on a similar charge a distance of 12,700 km away. (Avogadro's number of electrons is the number in 1 g of hydrogen. The separation distance equals the diameter of the earth.)

Using $N_A = 6.02 \times 10^{23}$ for Avogadro's number, we have

$$F = \frac{k(N_A e)^2}{r^2} = \frac{(9 \times 10^9 \text{ N} \cdot \text{m}^2/\text{C}^2)[(6.02 \times 10^{23})(1.60 \times 10^{-19} \text{ C})]^2}{(1.27 \times 10^7 \text{ m})^2}$$

$$= 5.18 \times 10^5 \text{ N}$$

This example illustrates the enormous strength of the electric force. In U.S. customary units, this force is about 58 tons. In the hydrogen atom, the positive charge of the proton exactly balances the negative charge of the electron.

Since both the electrical force and the gravitational force between two particles vary inversely with the square of the separation, the ratio of these forces is independent of the separation. We can therefore compare the relative strengths of these forces for elementary particles such as two protons, two electrons, or an electron and a proton.

Example 18-3   Compute the ratio of the electric force to the gravitational force exerted by a proton on an electron.

Since the proton charge is $+e$ and the electron charge $-e$, the electric force is attractive and has the magnitude

$$F_e = \frac{ke^2}{r^2}$$

The gravitational force is also attractive and is given by Newton's law of gravitation:

$$F_g = \frac{Gm_e m_p}{r^2}$$

where m_e is the mass of the electron and m_p is that of the proton. The ratio of these forces is independent of the separation distance r.

$$\frac{F_e}{F_g} = \frac{ke^2}{Gm_e m_p}$$

With $k = 9 \times 10^9$ N·m^2/C^2, $e = 1.6 \times 10^{19}$ C, $G = 6.67 \times 10^{-11}$ N·m^2/kg^2, $m_e = 9.11 \times 10^{-31}$ kg, and $m_p = 1.67 \times 10^{-27}$ kg, we obtain

$$\frac{F_e}{F_g} = \frac{(9 \times 10^9 \text{ N} \cdot \text{m}^2/\text{C}^2)(1.6 \times 10^{19} \text{ C})^2}{(6.67 \times 10^{-11} \text{ N} \cdot \text{m}^2/\text{kg}^2)(9.11 \times 10^{-31} \text{ kg})(1.67 \times 10^{-27} \text{ kg})}$$

$$\approx 2.3 \times 10^{39}$$

We can see from this example that the gravitational force between two elementary particles is so much smaller than the electric force between them (assuming they are charged) that it can always be neglected in describing their interaction. It is only because large masses such as the earth contain almost exactly equal numbers of positive and negative charges that the gravitational force is important. If the positive and negative charges in bodies did not almost exactly cancel each other, the electric forces between them would be much greater than the gravitational forces.

In a system of charges, each charge exerts the force given by Equation 18-1 on every other charge. The resultant force on any charge is the vector sum of the individual forces exerted on that charge by all the other charges in the system.

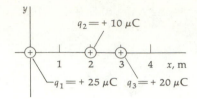

Figure 18-4
Point charges on the x axis for Example 18-4.

Example 18-4   Three positive point charges lie on the x axis; $q_1 = +25$ nC is at the origin, $q_2 = +10$ nC is at $x = 2$ m, and $q_3 = +20$ nC is at $x = 3$ m (see Figure 18-4). Find the resultant force on q_3.

The force on q_3 due to q_2, which is 1 m away, is in the positive x direction and has the magnitude

$$F_{23} = k \frac{q_2 q_3}{r_{23}^2} = \frac{(9 \times 10^9 \ \text{N} \cdot \text{m}^2/\text{C}^2)(10 \times 10^{-9} \ \text{C})(20 \times 10^{-9} \ \text{C})}{(1 \ \text{m})^2}$$

$$= 1.8 \times 10^{-6} \ \text{N}$$

The force on q_3 due to q_1, which is 3 m away, is also in the positive x direction. Its magnitude is

$$F_{13} = k \frac{q_1 q_3}{r_{13}^2} = \frac{(9 \times 10^9 \ \text{N} \cdot \text{m}^2/\text{C}^2)(25 \times 10^{-9} \ \text{C})(20 \times 10^{-9} \ \text{C})}{(3 \ \text{m})^2}$$

$$= 5.0 \times 10^{-7} \ \text{N}$$

Since both forces are in the positive x direction, the resultant force is also in that direction and is just the sum of the magnitudes:

$$F = F_{23} + F_{13} = (1.8 \times 10^{-6} \ \text{N}) + (0.5 \times 10^{-6} \ \text{N})$$

$$= 2.3 \times 10^{-6} \ \text{N}$$

Note that the charge q_2, which is between q_1 and q_3, has no effect on the force F_{13} exerted by q_1 on q_3, just as the charge q_1 has no effect on the force exerted by q_2 on q_3. There is, of course, a repulsive force between charges q_1 and q_2, but this force has no effect on the forces acting on q_3.

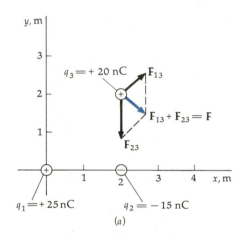

Example 18-5   Charge $q_1 = +25$ nC is at the origin, charge $q_2 = -15$ nC is on the x axis at $x = 2$ m, and charge $q_3 = +20$ nC is at the point $x = 2$ m, $y = 2$ m, as shown in Figure 18-5a. find the resultant force on q_3.

Since q_2 and q_3 have opposite signs, the force exerted by q_2 on q_3 is attractive in the negative y direction, as shown in the figure. Its magnitude is

$$F_{23} = \frac{(9 \times 10^9 \ \text{N} \cdot \text{m}^2/\text{C}^2)(15 \times 10^{-9} \ \text{C})(20 \times 10^{-9} \ \text{C})}{(2 \ \text{m})^2}$$

$$= 6.75 \times 10^{-7} \ \text{N}$$

The distance between q_1 and q_3 is $2\sqrt{2}$ m. The force exerted by q_1 on q_3 is directed along the line from q_1 to q_3 and has the magnitude

$$F_{13} = \frac{(9 \times 10^9 \ \text{N} \cdot \text{m}^2/\text{C}^2)(25 \times 10^{-9} \ \text{C})(20 \times 10^{-9} \ \text{C})}{(2\sqrt{2} \ \text{m})^2}$$

$$= 5.62 \times 10^{-7} \ \text{N}$$

The resultant force is the vector sum of these two forces. Since $\mathbf{F}_{13}$ makes an

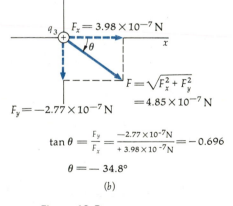

Figure 18-5
(a) Force diagram for Example 18-5. The resultant force on charge q_3 is the vector sum of the forces $\mathbf{F}_{13}$ due to q_1 and $\mathbf{F}_{23}$ due to q_2. (b) Diagram showing the resultant force in (a) and its x and y components.

angle of 45° with the x and y axes, its x and y components are equal to each other and to $F_{13}/\sqrt{2}$.

$$F_{13x} = F_{13y} = \frac{5.62 \times 10^{-7}\text{ N}}{\sqrt{2}} = 3.98 \times 10^{-7}\text{ N}$$

The x and y components of the resultant force are therefore

$$F_x = F_{23x} + F_{13x} = (0\text{ N}) + (3.98 \times 10^{-7}\text{ N})$$

$$= 3.98 \times 10^{-7}\text{ N}$$

$$F_y = F_{23y} + F_{13y} = (-6.75 \times 10^{-7}\text{ N}) + (3.98 \times 10^{-7}\text{ N})$$

$$= -2.77 \times 10^{-7}\text{ N}$$

The magnitude of the resultant force is

$$F = \sqrt{F_x^2 + F_y^2}$$

$$= \sqrt{(3.98 \times 10^{-7})^2 + (2.77 \times 10^{-7})^2} = 4.85 \times 10^{-7}\text{ N}$$

The resultant force points to the right and downward, as indicated in Figure 18-5b, making an angle θ with the x axis given by

$$\tan \theta = \frac{F_y}{F_x} = \frac{-2.77}{3.98} = -0.696$$

$$\theta = -34.8°$$

Question

4. If the sign convention for charge were changed so that the electron charge were positive and the proton charge were negative, would Coulomb's law be written the same or differently?

18-3   The Electric Field

In Chapter 4, we defined the gravitational field of the earth as the force exerted by the earth on a mass m divided by the mass m. The gravitational field of the earth **g** describes the property of the space around the earth such that, when a mass m is placed at some point, the force exerted by the earth is m**g**. The electric field **E** is defined in a similar way. The concept of a field, though somewhat abstract, is very useful, particularly in describing electrical interactions.

Figure 18-6 shows a set of point charges, q_1, q_2, and q_3, arbitrarily arranged in space. If we place a small "test" charge q_0 at some point near this system of charges, there will be a force exerted on this charge due to the other charges. This force is the vector sum of the forces exerted on q_0 by each of the other charges in the system. Since each of these forces is proportional to q_0, the resultant force will be proportional to q_0. We define the *electric field* **E** as the resultant force on the test charge divided by the charge q_0:

$$E = \frac{F}{q_0} \qquad\qquad 18\text{-}5$$

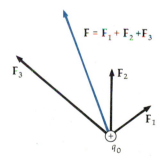

Figure 18-6
A small test charge q_0 in the vicinity of a system of charges, q_1, q_2, and q_3, experiences a force **F** that is proportional to q_0. The ratio **F**/q_0 is the electric field at that point.

The SI unit of the electric field is the newton per coulomb (N/C). Table 18-1 lists some values of electric-field magnitudes found in nature.

Table 18-1

Some Electric Fields in Nature

	E, N/C
In household wires	10^{-2}
In radio waves	10^{-1}
In the atmosphere	10^2
In sunlight	10^3
Under a thundercloud	10^4
In a lightning bolt	10^4
In an x-ray tube	10^6
At the electron in a hydrogen atom	6×10^{11}
At the surface of a uranium nucleus	2×10^{21}

We note that the electric field is a vector. It can be found by computing the electric field due to each charge in the system separately from Equation 18-5 and then adding these vectors to obtain the resultant electric field vector due to all the charges.

The test charge q_0 is a special charge. It must be small enough so that the charges in the system are not disturbed by its presence. (If, for example, some of the charges in the system are on conductors, the presence of a large test charge would alter their positions.) We usually think of the test charge as being positive, but its sign makes no difference; if it were negative, the force on it would be in the direction opposite to $\mathbf{E}$, but the ratio $\mathbf{F}/q_0$ would not change.

We can think of the electric field as a condition in space set up by the system of point charges. This condition is described by the vector $\mathbf{E}$. By moving the test charge q_0 from point to point we can find the electric field vector $\mathbf{E}$ at any point (except one occupied by a charge q). The electric field $\mathbf{E}$ is thus a vector function of position. The force exerted on a test charge q_0 at any point is related to the electric field at that point by

$$\mathbf{F} = q_0\mathbf{E} \qquad\qquad 18\text{-}6$$

Example 18-6   When a 5-nC test charge is placed at a point, it experiences a force of 2×10^{-4} N in the x direction. What is the electric field at that point?

Since the force on the positive test charge is in the x direction, the electric field vector is also in the x direction. From the definition, the electric field is

$$E_x = \frac{2 \times 10^{-4}\ \text{N}}{5 \times 10^{-9}\ \text{C}} = 4 \times 10^4\ \text{N/C}$$

Example 18-7 What is the force on an electron placed at the point in Example 18-6, where the electric field is $E_x = 4 \times 10^4$ N/C?

Since the charge of the electron is $-e = -1.6 \times 10^{-19}$ C, the force is

$$F_x = qE_x = (-1.6 \times 10^{-19} \text{ C})(4 \times 10^4 \text{ N/C}) = -6.4 \times 10^{-15} \text{ N}$$

The negative sign indicates that the force is to the left, in the negative x direction.

The electric field is more than a calculation device. This concept enables us to avoid the problem of action at a distance. We can think of the force exerted on charge q_0 at point P as being exerted *by the field at point P* rather than by the charges, which are some distance away. Of course, the field at point P is produced by the other charges, but the field is not produced instantaneously. Instead, the electric field propagates with the speed of light. (The gravitational field does also.) For example, if one of the charges in a system is moved suddenly, the effect some distance away is not instantaneous. There is a time delay due to the propagation time for the field. Because the speed of propagation is so great (3×10^8 m/s), this time delay is usually very small for ordinary distances. In this chapter, we will be concerned only with static electric fields.

The electric field due to a single point charge q at the origin is easily calculated from Coulomb's law. If we place a small, positive test charge q_0 at some point a distance r away, the force on it is in the radial direction and has the magnitude

$$F_r = \frac{kqq_0}{r^2}$$

The magnitude of the electric field at that point due to the charge q is thus

$$E_r = \frac{kq}{r^2} \qquad \text{18-7} \quad \text{Electric field of a point charge}$$

Example 18-8 A positive charge $q_1 = +8$ nC is at the origin, and a second positive charge $q_2 = +12$ nC is on the x axis at $x = 4$ m (see Figure 18-7). Find the electric field at point P_1 on the x axis at $x = 7$ m and at point P_2 on the x axis at $x = 3$ m.

The point $x = 7$ m is to the right of both charges. The electric field due to each charge is in the positive x direction. The resultant electric field thus has only an x component. The electric field at $x = 7$ m due to the 8-nC charge at the origin, a distance $r_1 = 7$ m away, is

$$E_{1x} = \frac{kq_1}{r_1^2} = \frac{(9 \times 10^9 \text{ N} \cdot \text{m}^2/\text{C}^2)(8 \times 10^{-9} \text{ C})}{(7 \text{ m})^2} = 1.47 \text{ N/C}$$

The 12-nC charge at $x = 4$ m is $r_2 = 7$ m $- 4$ m $= 3$ m away from the point $x = 7$ m. The electric field due to this charge is

$$E_{2x} = \frac{kq_2}{r_2^2} = \frac{(9 \times 10^9 \text{ N} \cdot \text{m}^2/\text{C}^2)(12 \times 10^{-9} \text{ C})}{(3 \text{ m})^2} = 12.00 \text{ N/C}$$

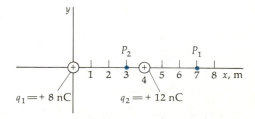

Figure 18-7
Two point charges on the x axis for Example 18-8. The electric field is to the right at point P_1 and to the left at point P_2.

The resultant field at $x = 7$ m is the sum of the fields due to the individual charges.

$$E_x = E_{1x} + E_{2x} = 1.47 \text{ N/C} + 12.00 \text{ N/C} = 13.47 \text{ N/C}$$

The point $x = 3$ m is between the charges. A positive test charge placed at this point would experience a repulsive force to the right due to the 8-nC charge and a repulsive force to the left due to the 12-nC charge. The electric field due to the 8-nC charge, which is 3 m away, is

$$E_{1x} = \frac{kq_1}{r_1^2} = \frac{(9 \times 10^9 \text{ N} \cdot \text{m}^2/\text{C}^2)(8 \times 10^{-9} \text{ C})}{(3 \text{ m})^2} = 8.00 \text{ N/C}$$

The electric field due to the 12-nC charge 1 m to the right of this point is to the left. E_{2x} is negative and given by

$$E_{2x} = -\frac{kq_2}{r_2^2} = -\frac{(9 \times 10^9 \text{ N} \cdot \text{m}^2/\text{C}^2)(12 \times 10^{-9} \text{ C})}{(1 \text{ m})^2} = -108 \text{ N/C}$$

The resultant electric field at this point between the charges is

$$E_x = E_{1x} + E_{2x} = (+8.00 \text{ N/C}) + (-108 \text{ N/C}) = -100 \text{ N/C}$$

The electric field at this point is in the negative x direction because the contribution to the field due to the 12-nC charge, which is 1 m away, is larger than that due to the 8-nC charge, which is 3 m away. As we move toward the 8-nC charge at the origin, the magnitude of E_{1x} increases and that of E_{2x} decreases. Clearly, there is one point between the charges where the electric field due to the two charges is zero. At this point, a positive test charge would experience no resultant force because the repulsive force to the right due to the 8-nC charge would just balance the repulsive force to the left due to the 12-nC charge. If we continue to move closer to the 8-nC charge at the origin, the electric field E_x becomes positive, indicating that the field is in the positive x direction.

Example 18-9 Find the electric field at a point P_3 on the y axis at $y = 3$ m for the charges in Example 18-8.

The fields due to each charge at this point on the y axis are shown in Figure 18-8a. The field $\mathbf{E}_1$ due to the $+8$-nC charge is in the $+y$ direction and has a magnitude

$$E_1 = \frac{kq_1}{r_1^2} = \frac{(9 \times 10^9 \text{ N} \cdot \text{m}^2/\text{C}^2)(8 \times 10^{-9} \text{ C})}{(3 \text{ m})^2} = 8.00 \text{ N/C}$$

The field $\mathbf{E}_2$ due to the $+12$-nC charge is in the direction of the line from that charge to point P_3. The distance from the $+12$-nC charge to point P_3 is found from the pythagorean theorem

$$r_2^2 = (3 \text{ m})^2 + (4 \text{ m})^2 = 25 \text{ m}^2$$

or

$$r_2 = 5 \text{ m}$$

The magnitude of $\mathbf{E}_2$ is

$$E_2 = \frac{(9 \times 10^9 \text{ N} \cdot \text{m}^2/\text{C}^2)(12 \times 10^{-9} \text{ C})}{(5 \text{ m})^2} = 4.32 \text{ N/C}$$

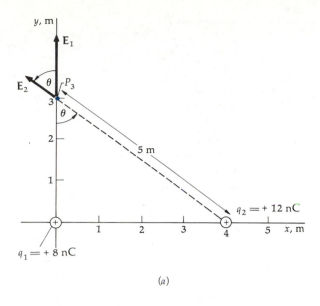

(a)

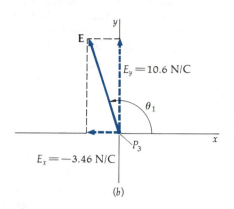

(b)

Figure 18-8
Example 18-9. (a) On the y axis, the electric field $\mathbf{E}_1$ due to charge q_1 is along the y axis, and the field $\mathbf{E}_2$ due to charge q_2 makes an angle θ with the y axis. The resultant field is the vector sum $\mathbf{E} = \mathbf{E}_1 + \mathbf{E}_2$. (b) The resultant field and its x and y components.

The contribution to the electric field due to the 12-nC charge has a component in the positive y direction equal to $E_2 \cos \theta$ and a component in the negative x direction equal to $-E_2 \sin \theta$. From the triangle in Figure 18-8a, we can see that $\cos \theta = 3/5 = 0.6$ and $\sin \theta = 4/5 = 0.8$. The x and y components of $\mathbf{E}_2$ are thus

$$E_{2x} = -E_2 \sin \theta = -(4.32 \text{ N/C})(0.8) = -3.46 \text{ N/C}$$

and

$$E_{2y} = E_2 \cos \theta = (4.32 \text{ N/C})(0.6) = 2.59 \text{ N/C}$$

We can obtain the x and y components of the resultant electric field from

$$E_x = E_{1x} + E_{2x} = 0 + (-3.46 \text{ N/C}) = -3.46 \text{ N/C}$$

and

$$E_y = E_{1y} + E_{2y} = 8.00 \text{ N/C} + 2.59 \text{ N/C} = 10.6 \text{ N/C}$$

The magnitude of the resultant electric field is

$$E = \sqrt{E_x^2 + E_y^2} = \sqrt{(-3.46 \text{ N/C})^2 + (10.6 \text{ N/C})^2} = 11.2 \text{ N/C}$$

The resultant electric field $\mathbf{E}$ makes an angle θ_1 with the x axis (see Figure 18-8b) given by

$$\tan \theta_1 = \frac{E_y}{E_x} = \frac{10.6 \text{ N/C}}{-3.46 \text{ N/C}} = -3.06$$

$$\theta_1 = 108°$$

Question

5. When a hand calculator is used to find the angle whose tangent is -3.06 for Example 18-9, the calculator gives $-72°$, which is the same as $288°$. Why is the correct answer $108°$?

18-4   Lines of Force

It is convenient to picture the electric field by drawing *lines of force* to indicate the direction of the field at any point. The field vector **E** is tangent to the line at each point and indicates the direction of the electric force experienced by a positive test charge placed at that point. Lines of force are also called *electric field lines*. At any point near a positive charge, a positive test charge is repelled by the charge; the field lines therefore diverge from a point occupied by a positive charge. Similarly, the electric field near a negative point charge points inward toward the charge, so the lines of force point inward and converge at the point occupied by a negative charge.

Figure 18-9 shows the lines of force, or electric field lines, of a single positive point charge. As we move away from the charge, the electric field becomes weaker and the lines are further apart. There is, in fact, a connection between the spacing of the lines and the strength of the electric field. Consider a spherical surface of radius r with its center at the charge. We are interested in the number of lines per unit area of the sphere, which we will call the density of the lines. If we make r larger, the area of the sphere increases but the same number of lines pass through it. The number of lines per unit area thus decreases as r increases. The area of the sphere is given by $A = 4\pi r^2$. The number of lines per unit area of the sphere thus decreases inversely with the square of the distance from the point charge. But the electric field strength, $E = kq/r^2$, also decreases inversely with the square of this distance. Thus, if we adopt the convention of drawing a fixed number of lines from a point charge, the number being proportional to the charge q, and if we draw the lines symmetrically about the point charge, the field strength is indicated by the density of the lines. The more closely spaced the lines, the stronger the electric field.

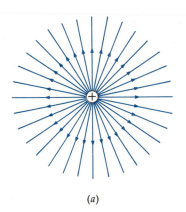

(a)

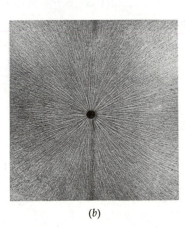

(b)

Figure 18-9
(*a*) Electric field lines of a single charge. If the charge were negative, the arrows would be reversed. (*b*) Bits of thread suspended in oil near an electric charge. The electric field of the charged object in the center induces opposite charges on the ends of each bit of thread, causing the threads to align themselves parallel to the field.

Figure 18-10 shows the lines of force for two equal positive point charges separated by a distance a. We can sketch this pattern without calculating the field at each point. We use the fact that the contribution to the field due to each of the charges varies as $1/r^2$, where r is the distance from that charge. So at a point near one of the charges, the field is approximately due to that charge alone because the other charge is relatively far away. If we are very

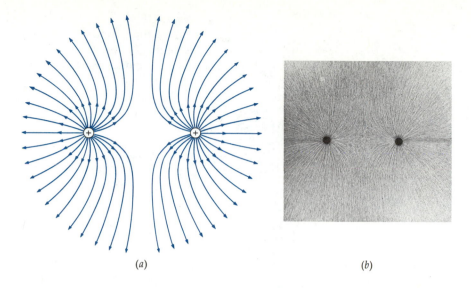

(a)   (b)

Figure 18-10
(a) Electric field lines due to two positive point charges. The arrows would be reversed if both charges were negative. (b) The same electric field lines shown by bits of thread in oil.

near one of the charges, we can ignore the contribution due to the other, far away charge. Thus, on a sphere of very small radius about one of the charges, the field lines are radial and equally spaced. Since the charges are equal, we draw an equal number of lines from each charge. On the other hand, at very large distances from the charges, the field is approximately the same as that due to a point charge of magnitude $2q$. For example, if the two charges were one millimetre apart and you were looking at a point 100 kilometres from the charges, they would look like a single charge. So on a sphere of radius r that is much greater than a, the lines are approximately equally spaced. (In our two-dimensional drawings, these spheres are indicated by circles.) We can see by merely looking at the figure that the electric field in the space between the charges is weak because there are few lines in this region compared with the region just to the right or left of the charges, where the lines are more closely spaced. This information can, of course, also be obtained by direct calculation of the field at points in these regions.

We can apply the above reasoning to drawing the lines of force for any system of point charges. Very near each charge, the field lines are equally spaced and leave or enter the charge depending on the sign of the charge. Very far from all the charges, the detailed structure of the system cannot be important, and the field lines are the same as those of a single point charge equal to the net charge of the system. For future reference, we will now summarize our rules for drawing lines of electric force.

1. The number of lines leaving a positive charge or entering a negative charge is proportional to the charge.

2. The lines are drawn symmetrically leaving or entering a point charge.

3. The lines begin on positive charges and end on negative charges.

4. The density of lines (the number per unit area perpendicular to the lines) is proportional to the magnitude of the field.

5. No two field lines can cross.

Rules for drawing lines of force

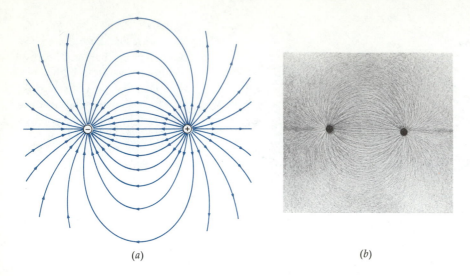

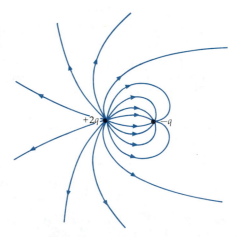

Figure 18-11
(a) Electric field lines for two equal
charges of opposite sign. (b) The
same field lines shown by bits of
thread in oil.

(a)   (b)

Rule 5 follows from the fact that **E** has a unique direction at any point in space (except at a point occupied by a point charge). If two lines crossed, two directions would be indicated for **E** at the point of intersection.

Figure 18-11 shows the lines of force due to two equal but opposite charges. Such a charge system is called an *electric dipole*. Very near the positive charge, the lines are radially outward. Very near the negative charge, the lines are radially inward. Since the charges have equal magnitude, the number of lines that begin at the positive charge equals the number that end at the negative charge. In this case, the field is strongest in the region between the charges, as indicated by the high density of field lines.

Figure 18-12 shows the lines of force for a negative charge $-q$ a distance a from a positive charge $+2q$. Since the positive charge has twice the magnitude of the negative charge, twice as many lines leave the positive charge as enter the negative charge. Half the lines beginning on the positive charge leave the system. The other half leaving the positive charge $+2q$ enter the negative charge $-q$. On a sphere of radius r, where r is much larger than the separation of the charges, the lines leaving the system are approximately symmetrically spaced and point radially outward, the same as the lines from a single positive point charge $+q$. Thus, at great distances from the charges, the system looks like a single charge $+q$. Note that this would be true no matter how many charges there were as long as the system has one more positive charge than negative charges. At a great distance from a charge system, it is only the net or unbalanced charge that is important.

The convention of indicating the electric field strength by the lines of force works because the electric field varies inversely as the square of the distance from a point charge. Since the gravitational field of a point mass also varies inversely as the square of the distance, the concept of lines of force is also useful in picturing the gravitational field. Near a point mass, the gravitational field lines converge toward the mass just like the electric field lines converge toward a negative charge. There are no points in space where gravitational field lines diverge like the electric field lines do near a positive charge. The gravitational force is always attractive, whereas the electric force can be either repulsive or attractive, depending on whether the charges have like or unlike signs.

Figure 18-12
Electric field lines for a point charge
$+2q$ and a second point charge $-q$.
At great distances from the charges,
the lines are the same as for a single
charge $+q$.

As an illustration of the usefulness of the concept of lines of force, we now consider the electric field due to a spherically symmetric shell of charge of radius R and negligible thickness. Let the total charge on the shell be Q. (Such a charge distribution can be realized in practice by placing the charge on a spherical conductor. As we will see in the next section, the charge will distribute itself uniformly over the surface of such a conductor.) We choose the origin to be at the center of the shell. We first consider the lines outside the shell. Because of the symmetry of the spherical shell, the only possible direction for the electric field lines is along radii, either toward the origin or away from it. Let us assume that the charge is positive. At great distances compared with the shell radius R, the field must look like that of a point charge Q. Thus, far from the shell, the lines are radially outward and equally spaced.

We now consider the region of space inside the shell. Again, because of the spherical symmetry, the lines inside must also be radial, either inward or outward. If they were inward, they would converge at the center of the spherical shell. But electric field lines only converge to a point occupied by a negative charge, and there is no negative charge at the center of the shell. Similarly, if the field lines were outward, they would diverge from the center of the shell. But electric field lines diverge only from a point occupied by a positive charge. We are therefore forced to conclude that there can be no electric field lines inside the shell. The lines begin at the positive charge on the shell. We have just discovered a remarkable result. *A spherically symmetric shell of charge produces no electric field inside it.* The lines of force for this system are shown in Figure 18-13. Outside the shell, the lines are exactly the same as those for a point charge. There are no lines inside the shell.

From this discussion of the electric field lines, we can write an algebraic expression for the electric field produced by a spherically symmetric shell of charge. The field has only a radial component E_r given by

$$E_r = \frac{kQ}{r^2} \qquad r > R \text{ (outside the shell)} \qquad\qquad 18\text{-}8a$$

$$E_r = 0 \qquad r < R \text{ (inside the shell)} \qquad\qquad 18\text{-}8b$$

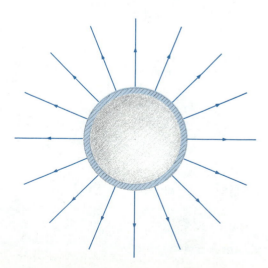

Figure 18-13
Electric field lines for a charge distribution on a spherical shell. Outside the shell, the lines are the same as those from a point charge at the center of the shell. Inside the shell, there are no field lines, indicating that the electric field there is zero.

If we place a test charge q_0 anywhere inside the shell, it will experience no electric attraction to or repulsion from the shell. This is easily understood for a test charge at the center of the shell because of the symmetry, but it is also true at every other point inside the shell. This remarkable result is a consequence of the inverse-square nature of the electric force. It can be derived directly from Coulomb's law using integral calculus, but the derivation is quite difficult. It was Priestley's observation that there was no electric field inside a hollow spherical conductor that led him to conclude that the force between electric charges decreases as the square of their separation.

18-5   Gauss' Law

The qualitative description of an electric field using lines of force is related to a mathematical equation known as Gauss' law, which relates the electric field on a closed surface to the net charge within the surface. Figure 18-14 shows a surface of arbitrary shape enclosing a point charge Q. The product of the component of the electric field perpendicular to the surface E_n and an element of area ΔA is called the *electric flux* $\Delta\phi$ through the area:

$$\Delta\phi = E_n\,\Delta A \qquad\qquad\text{18-9}$$

The flux is positive if E_n points outward from the surface and negative if it points inward. Since the number of field lines per unit area is proportional to the electric field, the flux through an area is proportional to the number of field lines that pass outward through the area. We can compute the total or net flux ϕ_{net} through the closed surface by computing $E_n\,\Delta A$ for each element and then summing over the entire surface.

$$\phi_{net} = \Sigma E_n\,\Delta A \qquad\qquad\text{18-10}$$

Since the flux through each area element is proportional to the number of field lines passing through that element, the net flux through the entire surface is proportional to the total number of field lines through the surface, which is in turn proportional to the charge Q inside the surface:

$$\phi_{net} = CQ \qquad\qquad\text{18-11}$$

where C is some constant. We note that the number of lines passing through a surface enclosing a charge is independent of the shape or location of the surface as long as the charge is inside the surface. We can use this fact to find the constant C in Equation 18-11. We consider a spherical surface of radius R with its center at the charge Q (see Figure 18-15). The electric field everywhere on this surface is perpendicular to the surface and has the magnitude

$$E_n = \frac{kQ}{R^2}$$

We can calculate the net flux through this spherical surface from Equation 18-11:

$$\phi_{net} = \Sigma E_n\,\Delta A = E_n\,\Sigma\,\Delta A$$

where we have taken E_n out of the sum because it is constant everywhere on

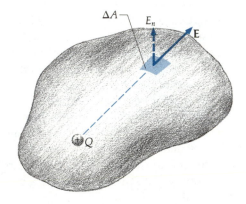

Figure 18-14
A surface of arbitrary shape around a point charge. The electric flux through the area element ΔA is the product of the perpendicular component of the field E_n and the area ΔA.

Figure 18-15
A spherical surface enclosing the point charge Q. The same number of electric field lines pass through this surface as pass through the surface in Figure 18-14 or through any surface that encloses Q. The flux ϕ_{net} is easily calculated for a spherical surface. It equals E_n times the surface area $4\pi R^2$.

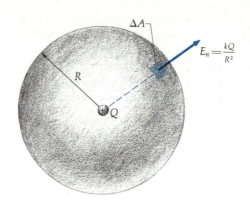

the surface. The sum of ΔA over the surface is just the total area of the surface, which equals $4\pi R^2$. Using this and substituting kQ/R^2 for E_n, we obtain

$$\phi_{net} = \frac{kQ}{R^2} \, 4\pi R^2 = 4\pi kQ$$

The constant C in Equation 18-11 is thus $4\pi k$.

In Figure 18-16, we have drawn a surface of arbitrary shape enclosing the charge distribution of Figure 18-12, consisting of a positive charge $+2q$ and a negative charge $-q$. Consider those lines that begin on the positive charge and end on the negative charge. Some of them do not pass through the surface at all. Those that do pass through the surface, go out of the surface at one point and back in through the surface at another point. At points where the lines go into the surface, the electric field E_n points into the surface, and the flux through the area element around that point is negative. If we count the lines passing into the surface as negative, the net number of lines passing out of the surface is proportional to the net charge inside the surface. For this figure, the net number is the same as for a single positive charge $+q$. The net flux out of the surface is thus also proportional to the net charge inside the surface. It is equal to $4\pi k$ times this net charge. This is *Gauss' law*:

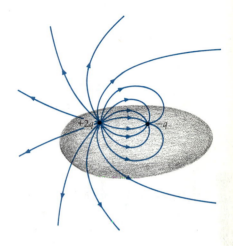

Figure 18-16
A surface of arbitrary shape enclosing the charges $+2q$ and $-q$. The field lines that end on $-q$ either do not pass through the surface or they exit once and enter once. The net number that exit is the same as for a single charge equal to the net charge within the surface.

The net electric flux through a closed surface equals $4\pi k$ times the net charge inside the surface.

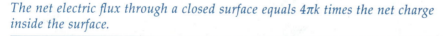

$$\phi_{net} = \Sigma E_n \, \Delta A = 4\pi k Q_{inside} \qquad\qquad 18\text{-}12 \qquad \text{Gauss' law}$$

Gauss' law is somewhat more general than Coulomb's law in that it holds for moving as well as for stationary charges. It can be used to calculate the electric field for some special cases of charge distributions that have a high degree of symmetry, for example, spherically symmetric charge distributions, such as the spherical shell discussed in the previous section, or cylindrically symmetric charge distributions, as illustrated in the following example.

Example 18-10 A very long straight wire contains a uniform positive charge with a linear charge density of λ coulombs per metre. Find the electric field at a distance r from the wire at a point far from either end, assuming that r is much smaller than the length of the wire.

In Figure 18-17, we have drawn a cylindrical surface of length L and radius r around the wire. By symmetry, at points far from the ends of the wire, the electric field lines radiate out from the wire uniformly. Thus, on the cylindrical surface, the electric field is perpendicular to the surface and has the same value everywhere on the surface. The electric flux is then just the product of the electric field E_n and the area of the cylinder, which is $2\pi rL$.

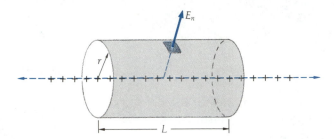

Figure 18-17
A cylindrical surface enclosing a long charged wire for Example 18-10. The flux ϕ_{net} through the surface is E_n times the surface area $2\pi rL$.

The net charge inside this surface is the charge per unit length λ times the length L. Gauss's law then gives

$$\phi_{net} = E_n 2\pi rL = Q_{inside} = \lambda L$$

$$E_n = \frac{\lambda}{2\pi r}$$

18-13

This result can also be obtained from Coulomb's law, but the calculation is difficult and requires the use of calculus.

18-6   Electric Dipoles in Electric Fields

Although atoms and molecules are electrically neutral, they are affected by electric fields because they contain positive and negative charges. We can think of an atom as a very small positively charged nucleus surrounded by a negatively charged electron cloud. Since the radius of the nucleus is about 100,000 times smaller than that of the electron cloud, we can consider it to be a point charge. In most atoms and molecules, the electron cloud is spherically symmetric, so its "center of charge" is at the center of the atom or molecule coinciding with that of the positive nucleus. Such an atom or molecule is said to be *nonpolar*. However, in the presence of an external electric field, the center of negative charge does not coincide with the center of positive charge. The electric field exerts a force on the positively charged nucleus in the direction of the field and a force on the negatively charged electron cloud in the opposite direction. The positive and negative charges therefore separate until the attractive force they exert on each other balances

Nonpolar molecules

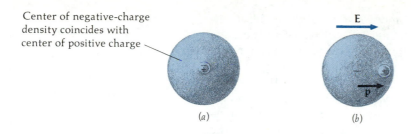

Center of negative-charge
density coincides with
center of positive charge

(a)

(b)

Figure 18-18
Schematic diagram of a nonpolar
molecule. (*a*) In the absence of an
external electric field, the center of
the negative charge coincides with
that of the positive charge. (*b*) In
the presence of an external electric
field, the centers of positive and
negative charge are displaced,
producing a dipole moment **p** in the
direction of the external field **E**.

the forces on them due to the external electric field (see Figure 18-18). Such a
charge distribution behaves like an electric dipole, which is two equal and
opposite point charges separated by a small distance *L* (see Figure 18-19).
The product of the magnitude of either charge and the vector **L** drawn from
the negative charge to the positive charge is called the *dipole moment* **p**:

$$\mathbf{p} = q\mathbf{L} \qquad\qquad\qquad 18\text{-}14$$

The dipole moment of a nonpolar atom or molecule in an external electric
field is called an *induced dipole moment*. It's direction is the same as that of
the electric field. If the electric field is uniform, there is no net force on the
dipole because the forces on the positive and negative charges are equal and
opposite. However, if the electric field is not uniform, there will be a net
external force acting on the dipole.

Figure 18-20 shows a nonpolar molecule in the external electric field of a
positive point charge *Q*. The induced dipole moment is parallel to **E** in the
radial direction from the point charge. Since the field is stronger at the
negative charge nearer the point charge, the net force on the dipole is
toward the point charge, and the dipole is attracted toward the point charge.

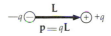

Figure 18-19
A dipole consists of equal and
opposite charges separated by some
distance *L*. The dipole moment
points from the negative charge to
the positive charge and has the
magnitude $p = qL$.

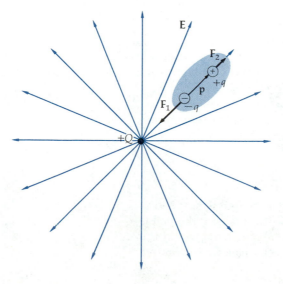

Figure 18-20
Nonpolar molecule in the electric
field of a positive point charge $+Q$.
The induced dipole moment **p** is
parallel to the field **E** of the point
charge. Since the point charge is
closer to the center of negative
charge than to the center of positive
charge, there is a net force of
attraction between the dipole and
the point charge.

If the point charge were negative, the induced dipole would be in the opposite direction, and the dipole would again be attracted to the point charge. The force produced by a nonuniform electric field on an electrically neutral charge is responsible for the familiar attraction of a charged comb for uncharged bits of paper. It is also responsible for the forces that hold an electrostatically charged balloon against a wall or ceiling. In this case, the charge on the balloon provides the nonuniform electric field that polarizes (that is, induces a dipole moment in) the molecules in the wall or ceiling and then attracts them.

In some molecules, the center of positive charge does not coincide with the center of negative charge. These *polar molecules* have a permanent electric dipole moment. When such a molecule is placed in a uniform electric field, there is no net force on it, but there is a torque, which tends to rotate the molecule so that the dipole lines up with the field (see Figure 18-21). In a nonuniform electric field, the molecule experiences a net force because the electric field has different magnitudes at the positive and negative charges. An example of a polar molecule is H_2O (see Figure 18-22). The dipole moment of the water molecule is what is mainly responsible for the energy absorption of food in a microwave oven. Microwaves, like all electromagnetic waves, have oscillating electric fields that can cause electric dipoles to vibrate. The vibration of the electric dipole moment of the water molecule in resonance with the oscillating electric field of the microwaves leads to the absorption of energy from the microwaves.

Polar molecules

Small bits of paper are polarized by the charged comb and then attracted to it and to each other.

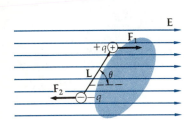

Figure 18-21
A dipole in a uniform electric field **E** experiences equal and opposite forces **F**$_1$ and **F**$_2$, which tend to rotate the dipole so that the dipole moment **p** lines up with the electric field.

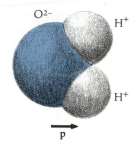

Figure 18-22
A water molecule can be thought of as an oxygen ion of charge -2 and two hydrogen ions of charge $+1$ each. This molecule has a permanent electric dipole moment in the direction shown.

Question

6. A small, nonconducting ball with no net electric charge is suspended from a thread. When a positive charge is brought near the ball, the ball is attracted toward the charge. How does this come about? How would the situation be different if the charge brought near the ball were negative instead of positive?

Benjamin Franklin (1706–1790)

I. Bernard Cohen
Harvard University

When Benjamin Franklin's contemporaries wanted to express their admiration for his scientific achievements, they could think only of comparing him to Newton. Joseph Priestley wrote that Franklin's book on electricity "bid fare to be handed down to posterity as expressive of the true philosophy of electricity; just as the Newtonian philosophy is of the true system of nature in general."

In order to appreciate Franklin's contribution, we must remember that electricity is a young branch of physics. Whereas we can trace an early history of statics and dynamics, heat and light, and even atomic theory back to the Greeks, electricity emerged as a proper subject of scientific study only in the days of Newton. How meager the information about electricity was in the early eighteenth century can be seen from the fact that the fundamental distinction between conductors and nonconductors had not yet been made, nor had it been discovered that there are two kinds of electric charge. When Franklin took up this new subject, he was almost forty, past the age we usually think of someone making great scientific discoveries. Largely self-educated, Franklin had already obtained a solid grounding in experimental physics through his study of Newton's *Opticks* and many of the primary textbooks of newtonian experimental science. When his printing and newspaper business was successful enough to allow him to retire from active participation, he eagerly seized upon the new subject and began to work at it intensively, together with a small group of coworkers.

Franklin's interest had first been sparked when he attended some public lectures on science given by a Dr. Adam Spencer, first in Boston (where Franklin was visiting his family) and then in Philadelphia (where Franklin sponsored Spencer's lectures). He purchased Spencer's apparatus in order to perform experiments himself. Soon thereafter, the Library Company of Philadelphia (of which Franklin was the principal founder) received a gift of electrical apparatus, with instructions for using it, from Peter Collinson, a London merchant.

Before long Franklin and his fellow experimenters realized that they had progressed in knowledge far beyond the literature that accompanied the apparatus. Franklin periodically sent reports of his work to Collinson and to other London correspondents, which eventually were assembled into a book entitled *Experiments and Observations on Electricity, Made at Philadelphia,* first published in England in 1751.

For his pioneering research in electricity, Franklin was elected a Fellow of the Royal Society of London, with the singular distinction of being forgiven the annual dues.

Benjamin Franklin.

Shortly thereafter, he was awarded the Society's Copley Gold Medal, the highest scientific honor then being awarded in England. His book on electricity was a spectacular success, going through five editions in English and being translated into French (three editions in two different translations), Italian, and German. In 1773, Franklin was elected a Foreign Associate of the French Academy of Sciences, an extraordinary honor since, according to the terms of the Academy's foundation, there could be only eight such foreign associates at any one time. No American was similarly honored again for another century.

Franklin's discoveries included not only important new experimental evidence but a new theory of electricity that made a science of the subject. This theory was based upon the fundamental postulate that all electrostatic phenomena (charging and discharging) result from the motion or transfer of a single electrical fluid. This hypothetical fluid was made up of "particles," or atoms, of electricity that repelled one

Essay: Benjamin Franklin (continued)

another but were attracted by the particles of "ordinary" matter. A charged body was one that had either lost or gained electrical fluid and was accordingly in a state that Franklin called "plus" (positive) or "minus" (negative). This postulate was closely associated with a major theoretical principle known today as the law of conservation of charge: whatever charge is lost by one body must be gained by one or more other bodies, and so negative and positive charges always appear simultaneously or are simultaneously canceled out in equal amounts. This law implied that electrical effects were not the results of the "creation" of some mysterious entity (as had often been supposed before Franklin) but merely followed from an alteration or redistribution of the amount of electrical fluid in a body. A variety of experiments confirmed the universality of this principle of the conservation of charge, which remains (along with the conservation of momentum) one of the most fundamental principles of physical science.

One of Franklin's most startling experiments was the analysis of the Leyden jar, the first "condenser" or capacitor. The Leyden jar is a glass jar with a metal-foil outer coating. Inside, the jar has another metal-foil coating or contains water or lead shot that is in contact with a wire passing through the cork of the jar (Figure 1). The jar is usually charged by bringing the wire into contact with an electrostatic generator or a bit of rubbed amber, glass, or sulfur while the outer coating is grounded. Franklin discovered that when a jar is charged and the two conductors are then separated from the glass, neither the inner nor the outer conductor will show any sign of being charged; but when they are placed once again on the two sides of the jar, the jar will produce all the familiar phenomena of being charged. Franklin said that the whole charge "resides" in the glass, but today we refer to this phenomenon under the name of "polarization of the dielectric." Franklin also showed that such a device does not depend upon the shape of the bottle, and he invented the parallel-plate capacitor.

In a variety of experiments, Franklin also showed the effects of grounding and insulation. He discovered that a grounded pointed conductor can actually "draw off" the charge of a nearby charged object and that, contrariwise, a charged pointed conductor, however well-insulated, will "throw off" the charge through its point. This led him to study the electrical nature of the lightning discharge. Franklin's most important experiment with lightning was not the familiar kite experiment but the experiment of the sentry box (Figure 2). Franklin proposed that a sentry box, with a long pointed conductor rising up through the roof, be erected on a high building. To determine whether the pointed conductor

Figure 1
An early Leyden jar from the Smithsonian Institution.

was charged, an experimenter would stand on an insulated stool inside the sentry box and bring up to the pointed conductor a grounded conductor set in an insulating handle. The objective was to draw a spark. If, as Franklin supposed, thunderclouds are electrically charged, then the pointed conductor would always become charged by induction when such clouds passed overhead, and a spark could be obtained. In practice, this experiment not only proved that clouds are electrically charged, and thus that lightning is only an ordinary electrical discharge from clouds on a large scale, but the rods also attracted a stroke of lightning and showed it to be an electrical phenomenon. Described in his book on electricity, this experiment was first performed in France. Franklin had been waiting for the completion of the spire of Christ Church in Philadelphia, where he hoped to erect a sentry box in order to perform the experiment. The kite was thought of as an alternative.

The lightning experiments brought fame to Franklin far and wide. Today the historical importance of these experiments is often misunderstood. In proving that lightning is an electrical discharge, Franklin showed that the electrical experiments performed in the laboratory are directly related to events in the natural world on a large scale. Thereafter, any general science of nature that did not include electricity would obviously be incomplete. Furthermore, the lightning experiments led Franklin to the invention of the lightning rod. For the first time in history, research in pure science led to a practical invention of major consequence. Bacon had indeed been correct in predicting that pure scientific knowledge would lead men to practical applications that would enable them to control their environment.

When Franklin began his research in electricity, the great French scientist Buffon (1707 – 1788) pointed out that electricity was not yet a science but only a collection of bizarre phenomena subject to no single law. After Franklin produced his theory of electricity, which explained and correlated the known phenomena and also predicted verifiable new ones, it was generally agreed that electricity had indeed become a science. In fact, electricity was the first new science (or branch of science) to arise since Newton.

It is often thought that Franklin was not really a pure scientist in the ordinary sense of the term, that he was, rather, a gadgeteer and inventor whose claim to science was aggrandized by his success as a statesman. In point of fact, the opposite is true. By 1776, when Franklin was sent to France as the American representative, he had already gained an international reputation for his scientific work, a factor of great importance in his diplomatic success. This was no unknown local patriot but one of the leading figures of the scientific world. We continually pay tribute to Franklin's scientific genius whenever we use the many words he introduced into the language of electricity: *plus* and *minus*, *positive* and *negative*, *electric battery*, and a host of others.

When Franklin retired from business, he hoped to devote the rest of his life to the peaceful pursuit of a career in science. The demands of his community and his country, however, all too soon drew him into a life of public service. Franklin's choice between his love of science and his duty to his fellow men was expressed by him as follows: "Had Newton been Pilot but of a single common Ship, the finest of his Discoveries would scarce have excused, or atoned for his abandoning the Helm one Hour in Time of Danger; how much less if she carried the Fate of the Commonwealth."

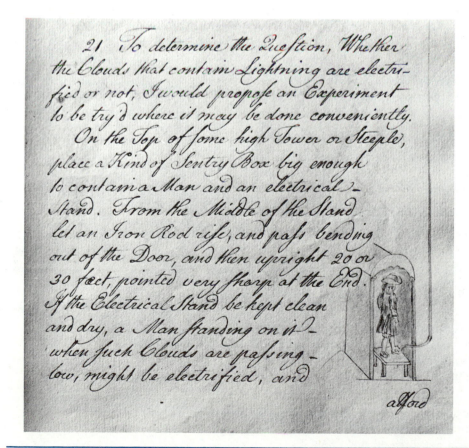

Figure 2
The sentry-box experiment, from a manuscript of Franklin's with drawings.

Summary

1. There are two kinds of electric charge, labeled positive and negative. Electric charge always occurs in integral multiples of the fundamental charge unit e. The charge of the electron is $-e$ and that of the proton is $+e$. Objects become charged by the transfer of electric charge from one object to another, usually in the form of electrons. Thus, charge is conserved. It is neither created nor destroyed; it is merely transferred.

2. The force exerted by one charge on another is proportional to the product of the charges and inversely proportional to the square of their separation, a result known as Coulomb's law:

$$F = k \frac{q_1 q_2}{r^2}$$

where the Coulomb constant is

$$k = 8.99 \times 10^9 \text{ N} \cdot \text{m}^2/\text{C}^2$$

3. The electric field at a point is defined as the force exerted by other charges on a test charge q_0 divided by q_0:

$$\mathbf{E} = \frac{\mathbf{F}}{q_0}$$

The electric field due to a point charge q at the origin is in the radial direction and is given by

$$E_r = \frac{kq}{r^2}$$

The electric field due to several charges is the vector sum of the fields due to the individual charges. A particle of charge q in an electric field experiences a force given by

$$\mathbf{F} = q\mathbf{E}$$

4. The electric field can be represented by lines of force that originate on positive charges and end on negative charges. The strength of the electric field is indicated by the density of the lines of force.

5. Electric flux $\Delta\phi$ through an element of area ΔA is the product of the component of the electric field perpendicular to the area times the area:

$$\Delta\phi = E_n \, \Delta A$$

The net flux through a closed surface equals $4\pi k$ times the net charge within the surface, a result known as Gauss' law:

$$\phi_{net} = \Sigma E_n \, \Delta A = 4\pi k Q_{inside}$$

6. An electric dipole is a system of two equal but opposite charges separated by a small distance. The dipole moment $\mathbf{p}$ is a vector that points in the direction from the negative charge to the positive charge with a magnitude equal to the charge times the separation:

$$\mathbf{p} = q\mathbf{L}$$

In the presence of a uniform electric field, the net force on the dipole is zero, but there is a torque that tends to align the dipole with the direction of the field. In a nonuniform field, there is a net force on the dipole.

7. Polar molecules such as H_2O have permanent dipole moments because the centers of positive and negative charge do not coincide. They behave like simple dipoles in an electric field. Nonpolar molecules do not have permanent dipole moments, but they acquire an induced dipole moment in the presence of an electric field.

Suggestions for Further Reading

Goldhaber, Alfred Scharff, and Michael Martin Nieto: "The Mass of the Photon," *Scientific American*, May 1976, p. 86.

Strange as it may seem, tests of Coulomb's law provide an upper limit on a possible mass for the photon. This article describes the history of such tests, which began before Coulomb started his investigations and continue today.

Kevles, Daniel J.: "Robert A Millikan," *Scientific American*, January 1979, p. 142.

The life and work of the second American scientist to win the Nobel prize in physics, in part for his beautiful experimental demonstration that all electrons carry the same charge and his measurement of that charge, are examined in this article.

Review

A. Objectives: After studying this chapter, you should:

1. Be able to state Coulomb's law and use it to find the force exerted by one point charge on another.

2. Know the value of the Coulomb constant in SI units.

3. Know the magnitude of the fundamental unit of electric charge e in coulombs.

4. Be able to use Coulomb's law to calculate the electric field due to a set of point charges.

5. Be able to draw the lines of force for simple charge distributions and obtain information about the direction and strength of an electric field from such a diagram.

6. Be able to describe the electric field produced by a spherically symmetric charged shell.

7. Be able to state Gauss' law.

8. Be able to state the difference between a polar and nonpolar molecule and describe the behavior of each in uniform and nonuniform electric fields.

9. Be able to explain why bits of paper are attracted to a comb and why a rubbed balloon will stick to the wall.

B. Define, explain, or otherwise identify:

Charge conservation, p. 425
Charge quantization, p. 426
Coulomb's law, p. 427
Electric field, p. 431
Electric flux, p. 440
Gauss' law, p. 441
Dipole moment, p. 443
Polar molecule, p. 444

C. True or false: If the statement is true, explain why. If it is false, give a counterexample.

1. The electric field of a point charge always points away from the charge.

2. The charge of the electron is the smallest unit of charge found.

3. Electric field lines never diverge from a point in space.

4. Electric field lines never cross at a point in space.

5. All molecules have electric dipole moments in the presence of an external electric field.

Exercises

Section 18-1   Electric Charge; Section 18-2   Coulomb's Law

1. Find the number of electrons in a charge of (*a*) 1 μC, (*b*) 1 nC, and (*c*) 1 pC.

2. A *faraday* (F) is a unit of charge equal to that of Avogadro's number of electron charges. Calculate the number of coulombs in a faraday. (Avogadro's number is 6.02×10^{23}.)

3. Two small spheres each have a charge of -0.02 μC. Find the magnitude of the electric force exerted by one sphere on the other (assuming them to be point charges) if their separation is (*a*) 2 cm, (*b*) 20 cm, and (*c*) 1 m.

4. In the hydrogen atom, the electron and proton are separated by $r = 0.053$ nm. (a) Calculate the force exerted by the proton on the electron. (b) Find the acceleration of the electron, whose mass is 9.11×10^{-31} kg.

5. A small charge $q_1 = 2$ nC is 3 cm from a large charge $q_2 = 4000$ nC. Find the magnitude of the force on (a) q_1 and (b) q_2.

6. Two charges exert a force of 20 N on each other. What is the force (a) if their separation is tripled and (b) if their separation is halved?

7. Two small neutral spheres are 15 cm apart. N electrons are transferred from one sphere to the other. Find the force exerted on either sphere if (a) $N = 2.0 \times 10^9$ and (b) $N = 5.0 \times 10^{12}$.

8. A charge of 3 nC is a distance r from a charge of 15 nC. The force on either charge is 1.58×10^{-9} N. Find the separation distance r.

9. Two protons in a helium nucleus are about 10^{-15} m apart. Calculate the electric force exerted by one proton on the other.

10. A charge $q_1 = 4.0$ μC is at the origin, and a second charge $q_2 = 6.0$ μC is on the x axis at $x = 3.0$ cm. (a) Find the force on charge q_2. (b) Find the force on q_1. (c) How would your answers differ if q_2 were -6.0 μC?

11. Three point charges are on the x axis; $q_1 = -6.0$ μC is at $x = -3.0$ m, $q_2 = 4.0$ μC is at the origin, and $q_3 = -6.0$ μC is at $x = 3.0$ m. Find the force on q_1.

12. Two equal charges of 3.0 μC are on the y axis, one at the origin and the other at $y = 6.0$ m. A third charge $q_3 = 2$ μC is on the x axis at $x = 8.0$ m. Find the force on q_3.

13. Two electrons are a distance r apart. Calculate the ratio of the electric force to the gravitational force exerted by one on the other.

Section 18-3   The Electric Field

14. When a test charge $q_0 = 3$ nC is placed at the origin, it experiences a force of 6.0×10^{-4} N in the positive y direction. (a) What is the electric field at the origin? (b) What would be the force on a charge $q = -4$ nC placed at the origin?

15. A charge of 4.0 μC is at the origin. Find the magnitude and direction of the electric field on the x axis at (a) $x = 6$ m, (b) $x = 10$ m, and (c) $x = -4$ m. (d) Sketch the function E_x versus x for both positive and negative values of x. (Remember that E_x is negative when E points in the negative x direction.)

16. Near the surface of the earth, there is an electric field of magnitude 100 N/C that points downward. What charge should be placed on a penny of mass 3 g so that the electric force balances the weight of the penny near the earth's surface? Should the charge be positive or negative?

17. A proton of mass 1.67×10^{-27} kg is in an electric field of 3000 N/C. What is the acceleration of the proton, assuming the electric force is the only force acting on it?

18. What is the electric field midway between two charges of $+18$ nC and -18 nC separated by 12 cm?

19. Two charges, each $+4$ μC, are on the x axis, one at the origin and the other at $x = 8$ m. Find the electric field on the x axis at (a) $x = 10$ m and at (b) $x = 2$ m. (c) At what point on the x axis is the electric field zero? (d) What is the direction of E at points on the x axis just to the right of the origin? Just to the left of the origin?

20. For the charge distribution of Exercise 19, find the electric field (a) on the y axis at $y = 6$ m and (b) at the point $x = 8$ m, $y = 8$ m.

21. Two equal positive charges of magnitude 6.0 nC are on the y axis at points $y_1 = +3$ cm and $y_2 = -3$ cm. (a) Find the magnitude and direction of the electric field at the point on the x axis at $x = 4$ cm. (b) What is the force exerted on a test charge $q_0 = 2.0$ nC placed at this point?

Section 18-4   Lines of Force

22. Two charges, $+q$ and $-3q$, are separated by a small distance. Draw the lines of force for this system.

23. Figure 18-23 shows the lines of force for a system of two point charges. (a) What are the relative magnitudes of the charges? (b) What are the signs of the charges? (c) In what regions of space is the electric field strong? In what regions is it weak?

Figure 18-23
Electric field lines for Exercise 23.

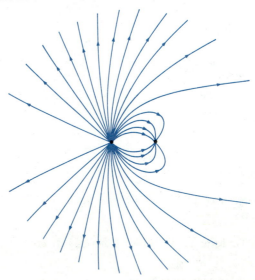

24. A spherical shell of radius r carries a uniform electric charge Q on its surface. The charge per unit area is called the charge density and denoted by σ. Show that the magnitude of the electric field at a point just outside the shell is given by $E_r = 4\pi k\sigma$.

25. A spherical shell of radius 12 cm carries a charge of 2 μC uniformly distributed on its surface. Find the magnitude of the electric field at the following distances from the center of the shell: (a) 5 cm, (b) 11.99 cm, (c) 12.01 cm, (d) 20 cm, and (e) 40 cm.

Section 18-5 Gauss' Law

26. What are the SI units of electric flux?

27. Use Gauss' law to derive Equations 18-8a and b for the electric field inside and outside a uniformly charged spherical shell of radius R. (*Hint:* Consider a spherical surface of radius $r < R$ to find the field inside the shell and one of radius $r > R$ to find the field outside the shell.)

28. Figure 18-24 shows a uniformly charged plane sheet. The charge per unit area on the sheet is σ. Far from the edges of the sheet, the electric field is uniform and points away from the sheet on both sides of the sheet. Apply Gauss' law to the closed surface in the shape of a pill box shown in the figure to show that the electric field near the sheet is given by

$$E_n = 2\pi k\sigma$$

Section 18-6 Electric Dipoles in Electric Fields

29. Two point charges lie on the x axis, $q_1 = 2.0$ pC at the origin and $q_2 = -2.0$ pC at $x = 3$ mm. Find the magnitude and direction of the dipole moment.

30. Since atomic charges are of the order of the fundamental electric charge e and their separation is of the order of nanometers (1 nm $= 10^{-9}$ m), a convenient unit for atomic dipole moments is $e \cdot$ nm. Find a conversion factor between $e \cdot$ nm and $C \cdot$ m.

31. An electron is 0.053 nm from a proton. Find the dipole moment in $e \cdot$ nm and $C \cdot$ m (see Exercise 30).

32. An electric dipole makes an angle θ with a uniform electric field **E**. Show that the magnitude of the torque acting on the dipole is

$$\tau = pE \sin \theta$$

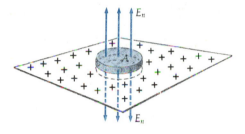

Figure 18-24
Uniformly charged plane sheet with a surface in the shape of a pill box for applying Gauss' law to calculate the electric field for Exercise 28.

Problems

1. A penny has a mass of 3 g. Each copper atom contains 29 electrons, and there are 6.02×10^{23} atoms in 64 g of copper. (a) Find the number of atoms, the number of electrons, and the total negative charge in coulombs in a penny of pure copper. (b) How long would it take for this much charge to flow through a wire at a rate of 1 ampere $= 1$ coulomb per second? (c) If 1 percent of the negative charge on one penny could be transferred to another penny, calculate the electric force exerted by one penny on the other when they are separated by 60 cm, assuming the pennies to be point charges. (*Note:* It is impossible to produce such a large charge on an object as small as a penny.)

2. Two equal positive charges q are on the y axis. One is at

$y = +a$ and the other is at $y = -a$. (a) Show that the electric field on the x axis is given by

$$E_x = \frac{2kqx}{(x^2 + a^2)^{3/2}}$$

(b) Show that when x is much larger than a, E_x is approximately $2kq/x^2$. Explain why you would expect this result even before calculating it.

3. An electric dipole consists of two charges, q and $-q$, separated by a distance $2a$. It lies along the x axis with its center at x_0. There is a nonuniform electric field in the x direction, $E_x = Ax$, where A is a constant. Show that the force acting on the dipole is in the x direction and given by $F_x = Ap$, where p is the magnitude of the dipole moment.

4. Two small spheres of mass m are suspended from a common point by threads of length L. When each sphere carries a charge q, each thread makes an angle θ with the vertical, as shown in Figure 18-25. Show that the charge q is given by

$$q = 2L \sin \theta \sqrt{(mg \tan \theta)/k}$$

where k is the Coulomb constant. Find q if $m = 10$ g, $L = 50$ cm, and $\theta = 10°$.

Figure 18-25
Problem 4.

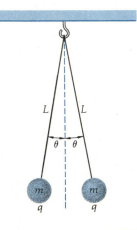

5. For the charge distribution in Example 18-8—an 8-nC charge at the origin and a 12-nC charge at $x = 4$ m—find the point on the x axis where the electric field is zero.

6. Two charges, q_1 and q_2, when combined give a total charge of 6 μC. When they are separated by 3 m, the force exerted by one charge on the other has the magnitude 8 mN. Find q_1 and q_2 if (a) both are positive, so that they repel each other; (b) one is positive and the other negative, so that they attract each other.

7. Four charges of equal magnitude are arranged at the corners of a square of side L as shown in Figure 18-26. (a) Find the magnitude and direction of the force exerted on the charge at the lower left corner by the other charges. (b) Show that the electric field at the midpoint of one of the sides of the square is directed along that side toward the negative charge and has a magnitude E given by

$$E = k \frac{8q}{L^2} \left(1 - \frac{\sqrt{5}}{25} \right)$$

Figure 18-26
Problem 7.

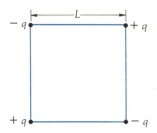

8. One model of the atomic nucleus considers it to be a uniformly charged ball with charge density of ρ coulombs per cubic metre. Let R be the radius of the ball. Use Gauss' law to show that the electric field at some distance r is given by

$$E_r = \frac{kQ}{R^3} r \qquad r < R$$

$$E_r = \frac{kQ}{r^2} \qquad r > R$$

where $Q = \frac{4}{3}\pi R^3 \rho$ is the total charge in the ball. [*Hint:* To find the electric field inside the charged ball, consider a spherical surface of radius $r < R$ and show that the total charge within this surface is $(r^3/R^3)(Q)$.]

CHAPTER NINETEEN

Electrostatics

Electrostatics is the study of electric fields produced by charges at rest and the effects of these fields on matter. The subject flourished in the days of Benjamin Franklin, but with the development of current electricity it became largely a historical curiosity. Recently, there has been a revival of interest in electrostatics with the invention of xerographic copying, medical x-ray imaging, and liquid crystal displays.

In this chapter, you will meet a new concept, potential, commonly called voltage. You will learn why static charge resides on the surface of conductors, why lightning rods have sharp points, and why Van de Graaff accelerators are spherical.

The knocking down of the six men was performed with two of my large Leyden jars not fully charged. I laid one end of my discharging rod upon the head of the first; he laid his hand on the head of the second; the second his hand on the head of the third, and so on to the last, who held in his hand the chain that was connected with the outside of the jars. When they were thus placed, I applied the other end of my rod to the prime-conductor, and they all dropt together. When they got up, they all declared they had not felt any stroke, and wondered how they came to fall.

BENJAMIN FRANKLIN

Long before we were able to produce and maintain electric currents, the study of electrostatics generated much interest. In eighteenth-century Leyden (in Holland), experimenters studying the effects of electric charges on people and animals got the idea of trying to store a large amount of charge in a bottle of water. One man held a jar of water in his hand while charge was conducted to the water by a chain from a static electric generator. When the man tried to lift the chain out of the water with his other hand, he was knocked unconscious. After many experiments, it was discovered that the hand holding the jar could be replaced by metal foil on the surface of the glass. A jar with gold foil inside and out was the first capacitor, which is a device for storing large electric charges. It was called a Leyden jar.

Benjamin Franklin realized that a capacitor did not have to be jar-shaped; foil-covered window glass, called Franklin panes, could be used instead. With several of these connected in parallel, he stored a large charge and tried to kill a turkey with it. Instead, he knocked himself out. "I tried to kill a turkey but nearly succeeded in killing a goose," he wrote. Later, Franklin proposed the famous experiment to find out whether lightning was electrical. Before he got around to doing the experiment, a French experimenter tried it and was killed. (Electrical experiments were dangerous in those days.) Not knowing this, Franklin tried it and, fortunately, survived.

Despite enormous advances in technology since those early experiments, our fascination with electrostatic phenomena continues. We are awed by lightning in a thunderstorm, and we are intrigued by clothes from the dryer sticking together and crackling when pulled apart or the small shocks we get after walking on a wool carpet. In this chapter, we will learn about electric potential, electrical energy, the electrostatic properties of conductors, and the storage of charge and energy in capacitors.

A lightning discharge is a spectacular electrostatic effect.

19-1 Electric Potential and Potential Difference

In our study of mechanics, particularly gravity, we found the concept of potential energy to be very useful. When we lift an object of mass m to a height h near the earth's surface, the work we do goes into potential energy mgh of the object. If the object is then dropped, the earth does work mgh, which decreases the potential energy of the object and increases its kinetic energy. The electric force between two charges has the same inverse square dependence as the gravitational force between two masses. Like the gravitational force, the electric force is conservative. If we exert a force on a body equal and opposite to the electric force so that the speed of the body is constant, the work we must do to move the body from one point to another equals the change in its electric potential energy (see Figure 19-1).

Consider a test charge q_0 in an electric field E_x produced by some system of charges. The electric force on the charge is q_0E_x. If we wish to move the charge (at constant speed), we must exert a force $F_x = -q_0E_x$ on it. If we give the charge a small displacement Δx, the work we do is

$$\Delta W = F_x \, \Delta x = -q_0E_x \, \Delta x \qquad\qquad 19\text{-}1$$

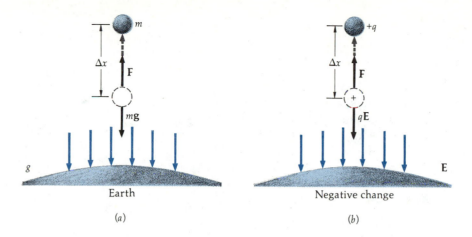

Figure 19-1
(a) The work done lifting a mass against the gravitational field increases the gravitational potential energy of the mass. (b) The work done moving a positive charge $+q$ against the electric field increases the electrical potential energy of the charge.

This work equals the increase in the potential energy of the test charge. The change in the potential-energy function ΔU is thus defined by

$$\Delta U = -q_0 E_x \, \Delta x \qquad\qquad 19\text{-}2$$

The potential-energy change is proportional to the test charge q_0. The potential-energy change per unit charge is called the *potential difference* ΔV.

$$\Delta V = \frac{\Delta U}{q_0} = -E_x \, \Delta x \qquad\qquad 19\text{-}3 \qquad \text{Potential difference defined}$$

The potential difference $V_b - V_a$ is the work per unit charge necessary to move a test charge at constant speed from point a to point b.

Equation 19-3 defines the change in the function V, which is called the *electric potential* (or sometimes just the *potential*).

Potential

Only *changes* in potential energy are important. We are free to choose the electric potential energy to be zero at any convenient point, just as we did for mechanical potential energy. (For example, in the expresson for the gravitational potential energy near the earth's surface, *mgh*, we could choose *h* to be zero at any convenient point, such as at the floor or at the top of a table. For two point masses, we chose the gravitational potential energy to be zero when their separation was infinite.) Since electric potential is potential energy per unit charge, it is only the change in potential that is important and we can choose $V = 0$ at any convenient point.

The SI unit for potential and potential difference is the joule per coulomb, which is called the volt (V):

$$1 \text{ V} = 1 \frac{\text{J}}{\text{C}} \qquad\qquad 19\text{-}4 \qquad \text{Volt defined}$$

The units for the electric field, newtons per coulomb, can also be written in terms of the volt. Using the fact that a newton-metre is a joule, we have

$$1 \frac{\text{N}}{\text{C}} = 1 \frac{\text{N} \cdot \text{m}}{\text{C} \cdot \text{m}} = 1 \frac{\text{J}}{\text{C} \cdot \text{m}}$$

But according to Equation 19-4, 1 J/C $=$ 1 V, so

$$1 \frac{N}{C} = 1 \frac{V}{m} \qquad\qquad\qquad 19\text{-}5$$

If we place a positive test charge q_0 in an electric field $\mathbf{E}$ and release the charge, it experiences a force in the direction of the field and will accelerate in the direction of $\mathbf{E}$ along a field line. The kinetic energy of the charge increases, and its potential energy decreases. The work done *by* the electric field, which increases the kinetic energy, *decreases* the potential energy. An alternate statement of the definition of potential difference is as follows:

The decrease in potential $-\Delta V$ equals the work per unit charge done by the electric field, $E_x \Delta x$.

We note that the charge moves toward a region of lower potential energy just as a massive body falls toward a region of lower gravitational potential energy. For a positive test charge, a region of lower potential energy is one of lower potential. Thus, *the electric field lines point in the direction of decreasing electric potential* (see Figure 19-2). For example, if we have a constant electric field of 10 N/C $=$ 10 V/m in the x direction, the potential decreases by 10 V in each metre in the x direction.

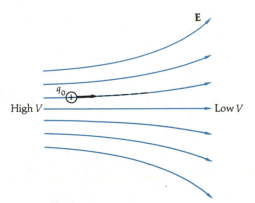

Figure 19-2
Electric field lines point in the direction of decreasing potential. When a positive test charge q_0 is placed in an electric field, it accelerates in the direction of the field. Its kinetic energy increases and its potential energy decreases.

Example 19-1   A proton of mass 1.67×10^{-27} kg and charge 1.6×10^{-19} C is placed in a uniform electric field $E_x = 5.0$ N/C $=$ 5.0 V/m and released from rest. After traveling 4 cm, how fast is it moving?

As the proton travels down an electric field line, its potential energy decreases and its kinetic energy increases by an equal amount. According to Equation 19-3, the change in the electric potential for $\Delta x = 4$ cm $=$ 0.04 m is

$$\Delta V = -E_x \Delta x = -(5 \text{ V/m})(0.04 \text{ m}) = -0.20 \text{ V}$$

The change in the potential energy of the proton is the product of its charge times its change in potential (also Equation 19-3):

$$\Delta U = q \Delta V = (1.6 \times 10^{-19} \text{ C})(-0.20 \text{ V}) = -3.2 \times 10^{-20} \text{ J}$$

By the law of conservation of energy, the loss in potential energy equals the gain in kinetic energy. Since the proton starts from rest, the gain in kinetic energy is just $\frac{1}{2}mv^2$, where v is its speed after traveling the 4 cm. We thus have

$$\Delta E_k + \Delta U = 0$$

$$\Delta E_k = -\Delta U = -(-3.2 \times 10^{-20} \text{ J})$$

$$\tfrac{1}{2}mv^2 = 3.2 \times 10^{-20} \text{ J}$$

$$v^2 = \frac{(2)(3.2 \times 10^{-20} \text{ J})}{1.67 \times 10^{-27} \text{ kg}} = 3.83 \times 10^7 \text{ J/kg}$$

$$v = \sqrt{3.83 \times 10^7 \text{ J/kg}} = 6.19 \times 10^3 \text{ m/s}$$

When the electric field is uniform (constant in space), the potential difference between two points can be found from Equation 19-3 simply by multiplying the field strength times the distance between the points. When the field varies in space, Equation 19-3 holds only for very small Δx. To find the potential difference between two points far apart in a nonconstant electric field, the appropriate average value of E must be used. An important example of a nonuniform electric field is that due to a single point charge q at the origin. The electric field of such a charge is radially outward (assuming q to be positive) and is given by

$$E_r = \frac{kq}{r^2} \qquad\qquad\qquad 19\text{-}6$$

If we move a very small distance Δr, the change in the electric potential is given by

$$\Delta V = -E_r\, \Delta r = -\frac{kq}{r^2}\, \Delta r$$

This equation holds only if Δr is very small because the electric field E_r varies with r. We studied this same situation in Chapter 9 when we discussed the change in gravitational potential energy in the gravitational field of a point mass at the origin. We pointed out there that when Δr is not very small, we have to use the proper average value of the force over the distance Δr, which can be obtained using calculus. If we move from r_1 to r_2 so that $\Delta r = r_2 - r_1$, the correct average value of $E_r = kq/r^2$ is

$$E_{r,\text{av}} = \frac{kq}{r_1 r_2}$$

This value is clearly between the original value kq/r_1^2 and the final value kq/r_2^2. Using this "average" value, we obtain for the potential difference between points r_1 and r_2

$$\Delta V = -\frac{kq}{r_1 r_2}\, \Delta r$$

If we use V_1 for the potential at r_1 and V_2 for that at r_2 and we use $r_2 - r_1$ for Δr, we obtain

$$\Delta V = V_2 - V_1 = -\frac{kq}{r_1 r_2}\,(r_2 - r_1)$$

or

$$V_2 - V_1 = \frac{kq}{r_2} - \frac{kq}{r_1} \qquad\qquad\qquad 19\text{-}7$$

In our study of gravity in Chapter 9, we found it convenient to choose the gravitational potential energy of two masses to be zero when the masses are very far apart, that is, when the separation is infinite. Similarly, we choose the electric potential function of a point charge at the origin to be zero at an infinite distance from the origin. We do this mathematically in Equation 19-7 by setting V_2 equal to zero when r_2 equals infinity. We then have

$$0 - V_1 = 0 - \frac{kq}{r_1}$$

since kq/r_2 is zero when $r_2 = \infty$. At a general distance r from the point charge q, the electric potential is

$$V(r) = \frac{kq}{r}$$ 19-8   Potential of a point charge

The potential is positive or negative depending on the sign of the charge q. Compare this result with Equation 9-24 for the gravitational potential energy of a mass m at a distance r from the earth. The negative sign in that equation arises because the gravitational force between two (positive) masses (of course, there are no negative masses) is attractive whereas the electric force between two positive charges is repulsive.

If we place a test charge q_0 at a distance r from the charge q at the origin, the potential energy of the system is the charge q_0 times the potential at that point due to the other charge q. (This follows from the fact that potential is just the potential energy per unit charge.)

$$U = q_0 V = \frac{kq\,q_0}{r}$$ 19-9

This potential energy is, of course, zero when the charges are infinitely separated because we choose the potential to be zero at $r = \infty$. If we begin with charge q at the origin and bring the test charge q_0 from an infinite distance away to a distance r (see Figure 19-3), the work we must do against the electric force equals the increase in potential energy, which is given by Equation 19-9.

The electric potential $V = kq/r$ is the work per unit charge that must be done to bring a positive test charge from infinity to the distance r from the point charge q.   Potential is work per unit charge

If we have two point charges q_1 and q_2, the potential at some point is the sum of the potentials due to each charge:

$$V = \frac{kq_1}{r_1} + \frac{kq_2}{r_2}$$ 19-10

where r_1 and r_2 are the distances from the point to the charges q_1 and q_2, respectively.

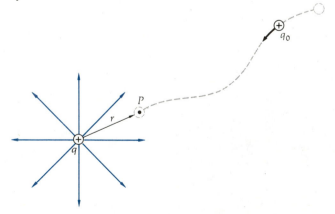

Figure 19-3
The work required to bring a test charge q_0 at constant speed from infinity to a point P that is a distance r from a charge q at the origin is kqq_0/r. The work per unit charge kq/r equals the electric potential at point P relative to zero potential at infinity.

Example 19-2   What is the electric potential at a distance $r = 0.529 \times 10^{-10}$ m from a proton? (This is the average distance between the proton and electron in the hydrogen atom.) What is the potential energy of the electron and the proton at this separation?

The charge of the proton is $q = 1.6 \times 10^{-19}$ C. Equation 19-8 gives

$$V = \frac{kq}{r} = \frac{(9 \times 10^9 \text{ N} \cdot \text{m}^2/\text{C}^2)(1.6 \times 10^{-19} \text{ C})}{0.529 \times 10^{-10} \text{ m}} = 27.2 \text{ N} \cdot \text{m/C}$$

$$= 27.2 \text{ J/C} = 27.2 \text{ V}$$

The charge of the electron is $-e = -1.6 \times 10^{-19}$ C. In SI units, the potential energy of the electron and proton separated by a distance of 0.529×10^{-10} m is

$$U = qV = (-1.6 \times 10^{-19} \text{ C})(27.2 \text{ V}) = -4.36 \times 10^{-18} \text{ J}$$

In atomic and nuclear physics, we often have elementary particles, such as electrons and protons, with charge of magnitude e moving through potential differences of several thousands or even millions of volts. Since energy has the dimensions of electric charge times electric potential, a convenient unit of energy is the product of the electron charge e times a volt. This unit is called an electronvolt and is written eV. The conversion between electronvolts and joules is obtained by expressing the fundamental electric charge in coulombs.

$$1 \text{ eV} = 1.6 \times 10^{-19} \text{ C} \cdot \text{V} = 1.6 \times 10^{-19} \text{ J} \qquad \text{19-11} \qquad \text{Electronvolt defined}$$

In Example 19-2, the potential energy of the electron and proton at the separation $r = 0.529 \times 10^{-10}$ m is -27.2 eV.

Example 19-3   Two equal positive point charges of magnitude $+5$ nC are on the x axis. One is at the origin and the other is at $x = 8$ cm, as shown in Figure 19-4. Find the potential at point P_1 on the x axis at $x = 4$ cm and at point P_2 on the y axis at $y = 6$ cm.

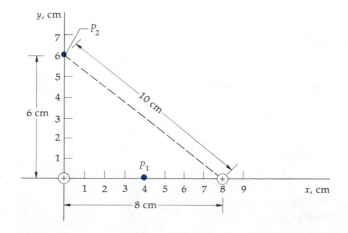

Figure 19-4
Two positive point charges on the x axis for Example 19-3. The potential is to be found at points P_1 and P_2.

Point P_1 is 4 cm from each charge. Using Equation 19-10, with $q_1 = q_2 = q = 5$ nC and $r_1 = r_2 = r = 0.04$ m, we find that the potential at P_1 is

$$V = \frac{kq_1}{r_1} + \frac{kq_2}{r_2} = \frac{2kq}{r}$$

$$= \frac{2(9 \times 10^9 \text{ N} \cdot \text{m}^2/\text{C}^2)(5 \times 10^{-9} \text{ C})}{0.04 \text{ m}} = 2250 \text{ V}$$

We note that the electric field is zero at this point midway between the charges but the potential is not. To bring a test charge from a long distance away to this point requires work because the electric field is zero only at the final position. Point P_2 is 6 cm from one charge and 10 cm from the other. The potential at P_2 is

$$V = \frac{(9 \times 10^9 \text{ N} \cdot \text{m}^2/\text{C}^2)(5 \times 10^{-9} \text{ C})}{0.06 \text{ m}}$$

$$+ \frac{(9 \times 10^9 \text{ N} \cdot \text{m}^2/\text{C}^2)(5 \times 10^{-9} \text{ C})}{0.10 \text{ m}}$$

$$= 750 \text{ V} + 450 \text{ V} = 1200 \text{ V}$$

Questions

1. Explain in your own words the distinction between electric potential and electric potential energy.

2. If a charge is moved a small distance in the direction of an electric field, does its electric *potential energy* increase or decrease? Does your answer depend on the sign of the charge?

3. If a charge is moved a small distance in the direction of an electric field, does its electric *potential* increase or decrease? Does your answer depend on the sign of the charge?

4. What direction can you move relative to an electric field so that the electric potential does not change?

5. A positive charge is released from rest in an electric field. Will it move toward a region of greater or smaller electric potential?

6. Is the electronvolt a larger or smaller energy unit than the joule?

19-2   Electric Conductors

The great difference in the electrical behavior of conductors and insulators was noted even before the discovery of electric conduction. Gilbert had classified materials according to their ability to be electrified. Objects that could be electrified he called *electrics;* those that could not (metals and some other materials) he called *nonelectrics.* After Gray discovered conduction, Du Fay showed that all materials can be electrified but that care must be

taken to insulate Gilbert's nonelectrics from the ground (or the experimenter) lest the charge be quickly conducted away. Using only physiological sensation for detection, Cavendish compared the conducting abilities of
many substances.*

> It appears from some experiments, of which I propose shortly to lay an
> account before this Society, that iron wire conducts about 400 million times
> better than rain or distilled water—that is, the electricity meets with no
> more resistance in passing through a piece of iron wire 400,000,000 inches
> long than through a column of water only one inch long. Sea water, or a
> solution of one part of salt in 30 of water, conducts 100 times, or a saturated
> solution of sea salt about 720 times, better than rainwater.

Because of the enormous variation in the ability of materials to conduct
electricity, it is possible and convenient to classify most materials as conductors or insulators (nonconductors). (Some materials, called semiconductors, do not fit easily into either of these two classifications.) The ability of a
material to conduct electricity is measured by its conductivity. The conductivity of a typical conductor is of the order of 10^{15} times that of a typical
insulator, whereas the conductivity within the group of conductors varies
only over several orders of magnitude.

Here we are concerned only with the behavior of conductors in electrostatic equilibrium, that is, when all electric charges are at rest. The property
of a conductor that is important in studying electrostatic fields is the availability of a charge that is free to move about inside the conductor. The source
of this free charge is electrons that are not bound to any atom. For example,
in a single atom of copper, 29 electrons are bound to the nucleus by electrostatic attraction of the positively charged nucleus. The outermost electrons
are more weakly bound than the innermost electrons because of their
greater distance from the positive nucleus and because of the repulsion of
the inner electrons. (This is called *screening*.) When many copper atoms
combine to form metallic copper, the electron binding of the individual
atom is charged by its interaction with neighboring atoms. One or more of
the outer electrons in each individual atom are no longer bound but are free
to move throughout the whole piece of metal, much as a gas molecule is free
to move about in a container. The number of free electrons depends on the
particular metal, but it averages about one per atom. The bonding of atoms
in an insulator is such that there are no free electrons that can roam about
away from their parent atoms.

In the presence of an external electric field, the free charge in a conductor
moves about the conductor until it is so distributed that it creates an electric
field that cancels the external field inside the conductor. Consider a free
charge q inside a conductor. If there were a field **E** inside the conductor, the
charge would not be in equilibrium. There would be a force $q\mathbf{E}$ on this
charge, and the charge would accelerate. Thus, electrostatic equilibrium is
impossible in a conductor unless the electric field is zero everywhere inside
the conductor.

---

* Henry Cavendish, *Philosophical Transactions of the Royal Society of London,* vol. 66, 1776, p.
196.

Figure 19-5 shows a conducting slab placed in an external electric field $\mathbf{E}_0$. The free electrons are initially distributed uniformly throughout the slab.

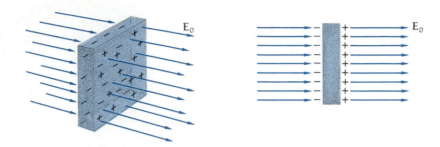

Figure 19-5
A conducting slab in an external electric field. Electrons are forced to the left (opposite to the field), leaving the left face with a negative charge and the right face with a positive charge. These charges produce an electric field inside the conductor that cancels the external field there so that there is no net electric field inside the conducting slab.

Since the slab is made up of neutral atoms, it is electrically neutral (assuming that no extra charge has been placed on it). If the external electric field is to the right, there will be a force on each electron $-e\mathbf{E}_0$ to the left because the electron has a negative charge. The free electrons accordingly accelerate to the left. At the surface of the conductor, atoms that have been left with an unbalanced positive charge exert forces on these electrons that balance the force due to the external field, so the electrons are bound to the conductor. (If the external field is very strong, the electrons can be stripped from the surface. In electronics, this is called *field emission.* We assume here that the external field is not strong enough to overcome the forces binding the electrons to the surface.) The result is a negative surface charge on the left side of the slab and a positive surface charge on the right side because of the removal of some of the free electrons from that side. Both these surface charges produce an electric field inside the slab that is opposite the external field. These two fields cancel everywhere inside the conductor, so there is no unbalanced force on the free electrons and electrostatic equilibrium results.

The behavior of the free charge in a conductor placed in an external electric field is the same no matter what the shape of the conductor may be. When an external field is applied, the free charge quickly distributes itself on the surface of the conductor until an equilibrium distribution is reached such that the net electric field is zero everywhere inside the conductor. The time it takes to reach equilibrium depends on the conductivity. For copper and other good conductors, it is less than 10^{-16} second. For all practical purposes, then, electrostatic equilibrium is reached instantaneously.

When there is a charge on the surface of a conductor, electric field lines leave the charge if it is positive or enter the charge if it is negative. We can calculate the electric field just outside a conductor in terms of the charge on the surface of the conductor from Gauss' law (Equation 18-12). We first note that the electric field at the surface of a conductor must be perpendicular to the surface. If there were a component of $\mathbf{E}$ parallel to the surface, the free charge on the conductor would move about until that component were zero. Let σ be the charge density (the charge per unit area) on the surface of a conductor. In general, σ will vary from point to point. We calculate the electric field E_n just outside the conductor by applying Gauss' law to the pillbox-shaped surface shown in Figure 19-6. Since $\mathbf{E}$ is parallel to the sides of this surface, the only possible flux through the surface is that through the

Lines of force for an oppositely charged cylinder and plate, shown by bits of fine thread suspended in oil. Note that the field lines are perpendicular to the conductors and that there are no lines inside the cylinder.

faces. Since the electric field inside the conductor is zero, there is flux only through the top face equal to $E_n A$, where A is the area of the face. The charge inside the pillbox is σA. Gauss' law then gives

$$\phi_{net} = 4\pi k Q_{inside}$$

$$E_n A = 4\pi k \sigma A$$

The field just outside the conductor is thus

$$E_n = 4\pi k\sigma \qquad\qquad 19\text{-}12$$

A simple and practical method of charging a conductor makes use of the free movement of charge in a conductor. In Figure 19-7a, two uncharged metal spheres are in contact. If we bring a positively charged rod near the spheres, free electrons on one flow to the other. The rod attracts the negatively charged electrons, so the sphere nearest the rod acquires electrons from the other, leaving the near sphere with a negative charge and the far sphere with an equal positive charge due to the lack of electrons. If the spheres are separated before the rod is removed, as in Figure 19-7b, they will have equal and opposite charges. A similar result is obtained, of course, with a negatively charged rod, which drives electrons from the nearest sphere to the other. In each case, the spheres are charged without being touched by the rod, and the charge on the rod is undisturbed. This is called *electrostatic induction* or *charging by induction*.

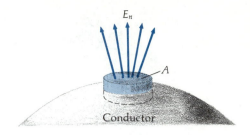

Figure 19-6
Pillbox-shaped surface for applying Gauss' law to find the electric field near the surface of a conductor. There is electric flux through only the face of the box that is outside the conductor.

Charging by induction

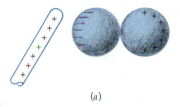

(a)   (b)

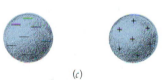

(c)

Figure 19-7
Charging by induction. (a) The two spherical conductors in contact become oppositely charged because the positively charged rod attracts electrons to the left sphere leaving the right sphere positively charged. (b) If the spheres are now separated with the rod in place, they retain equal and opposite charges. (c) When the rod is removed and the spheres are far apart, the spheres are uniformly charged with equal and opposite charges.

A convenient large conductor is the earth itself. For most purposes, we can consider the earth to be an infinitely large conductor. When a conductor is connected to the earth, it is said to be *grounded*. This is indicated schematically by a connecting wire and parallel horizontal lines, as illustrated in Figure 19-8b. We can use the earth to charge a single conductor by induction. In Figure 19-8a, a positively charged rod is brought near a neutral conductor, and the conductor becomes polarized as shown. Free electrons are attracted to the side near the positive rod, leaving the other side with a positive charge. If we ground the conductor while the charged rod is still

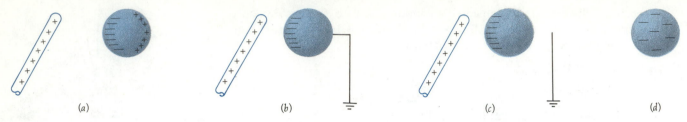

(a)   (b)   (c)   (d)

present (see Figure 19-8b), the conductor becomes charged oppositely to the rod because electrons from the earth travel along the connecting wire and neutralize the positive charge on the far side of the conductor. The connection to ground is broken before the rod is removed to complete the charging by induction (see Figure 19-8c). When the charged rod is removed, the sphere is left with a uniform negative charge, as shown in Figure 19-8d.

Questions

7. Distinguish between free charge in a conductor and net charge in a conductor.

8. An insulating rod is given a charge and is then used to charge a set of conductors by induction. What practical limit is there on the number of times the rod can be used before it must be recharged?

9. Can insulators as well as conductors be charged by induction?

Figure 19-8
(a) The free charge on the single conducting sphere is polarized by the charged rod such that it has a negative charge on the side closest to the rod and a positive charge on the other side. (b) When the conductor is grounded by connecting it by a wire to a very large conductor such as the earth, electrons from the ground neutralize the positive charge on the far side. The conductor is thus negatively charged. (c) The negative charge remains if the ground is broken before the rod is removed. (d) When the rod is removed, the sphere is uniformly negatively charged.

19-3   Equipotential Surfaces, Charge Sharing, and Dielectric Breakdown

We have said that, in electrostatic equilibrium, there can be no electric field inside a conductor. There is then no force on a test charge inside the conductor and no work is required to move the test charge about inside the conductor. The electric potential is thus the same throughout the conductor. If there is charge on the surface of the conductor, there will be electric field lines leaving or entering the charge. These lines are perpendicular to the surface of the conductor, as discussed in the previous section. Thus, if we move a test charge around on the surface of a conductor, we always move perpendicular to the electric field, so again we do no work. A conducting surface is an *equipotential surface*. (Since all parts of the conductor are at the same potential, a conductor is also an equipotential volume.) Figures 19-9 and 19-10 show equipotential surfaces near a spherical conductor and near a nonspherical conductor. Equipotential surfaces for a dipole and for two equal positive point charges are shown in Figures 19-11 and 19-12. Note that the electric field lines are always perpendicular to these surfaces. If we move a short distance Δs along a field line from one equipotential surface to another, the potential changes by

$$\Delta V = -E \, \Delta s \qquad\qquad 19\text{-}13$$

If E is large, neighboring equipotential surfaces with a fixed potential difference between them are closely spaced.

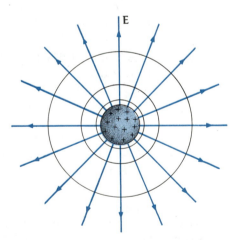

Figure 19-9
Equipotential surfaces and electric field lines outside a uniformly charged spherical conductor. The surfaces are spherical. The electric field lines are always perpendicular to an equipotential surface.

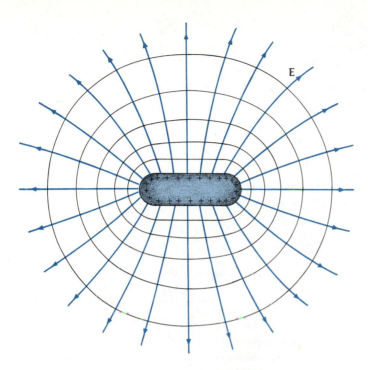

Figure 19-10
Equipotential surfaces and electric field lines
outside a nonspherical conductor.

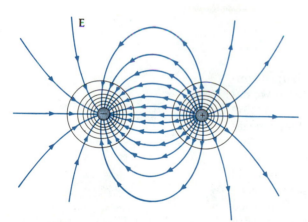

Figure 19-11
Equipotential surfaces and electric field lines
for an electric dipole.

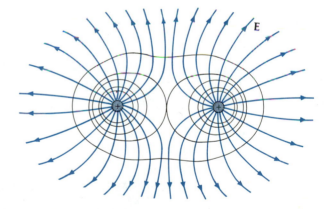

Figure 19-12
Equipotential surfaces and electric field lines
for two nearby equal positive charges.

   In general, two conductors that are separated in space will not be at the
same potential. The potential difference between the conductors depends
on their geometrical shapes, their separation, and the net charge on each.
When two conductors are brought into contact, the charge on the conductors distributes itself so that electrostatic equilibrium is established and the
electric field is zero inside both conductors. In this situation, the two conductors in contact may be considered a single conductor. In equilibrium,
each conductor has the same potential. The transfer of charge from one
conductor to another is called *charge sharing*.

Consider a spherical conductor carrying a charge $+Q$. The lines of force outside the conductor point radially outward, and the potential of the conductor relative to infinity is kQ/R. If we bring up a second, uncharged conductor, the potential and the field lines will change: Negative electrons on the uncharged conductor will be attracted to the positive charge Q, leaving the near side of the uncharged conductor with a negative charge and the far side with a positive charge (see Figure 19-13). This charge separation on the neutral conductor will affect the originally uniform charge distribution on the positive conductor. Although the detailed calculation of the charge distributions and potential in this case is quite complicated, we can see that some of the field lines leaving the positive conductor will end on the negative charge on the near side of the neutral conductor and that an equal number of lines will leave the far side of that conductor. Since electric field

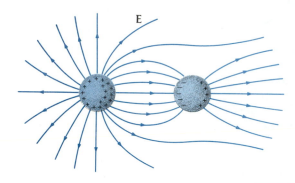

Figure 19-13
Electric field lines for a charged spherical conductor (left) near an uncharged spherical conductor (right).

lines point toward regions of lower potential, the positively charged conductor must be at a greater potential than the neutral conductor. If we put the two conductors in contact, positive charge will flow to the neutral conductor until both conductors are at the same potential. (Actually, negative electrons flow from the neutral conductor to the positive conductor. It is slightly more convenient, however, to think of this as a flow of positive charge in the opposite direction.) If the conductors are identical, they will share the original charge equally. If the conductors are now separated, each will carry a charge $\frac{1}{2}Q$, and both will be at the same potential. Coulomb used the method of charge sharing to produce various charges of known ratios in his experiment to find the force law between two small (point) charges.

In Figure 19-14, a small conductor carrying a positive charge q is inside the cavity of a second, larger conductor. In equilibrium, the electric field is zero inside the conducting material of both conductors. The lines of force that leave the positive charge q must end on the inner surface of the large conductor. This must occur no matter what the charge may be on the outside surface of the large conductor. Regardless of the charge on the large conductor, the small conductor in the cavity is at a greater potential because the lines of force go from this conductor to the larger conductor. If the conductors are now connected, say, with a fine conducting wire, *all* the charge originally on the smaller conductor will flow to the larger one. When the connection is broken, there is no charge on the small conductor and there

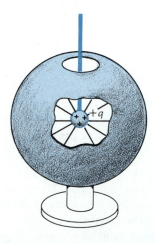

Figure 19-14
Small conductor carrying a positive charge inside a larger conductor.

are no field lines anywhere in the cavity. The positive charge transferred from the smaller conductor to the larger one resides completely on the outside surface of the larger conductor. If we put more positive charge on the small conductor in the cavity and again connect the conductors with a fine wire, all of the charge on the inner conductor will again flow to the outer conductor. This procedure can be repeated indefinitely. This method is used to produce large potentials in the Van de Graaff generator, where the charge is brought to the inner surface of a larger spherical conductor by a continuous charged belt (see Figures 19-15 and 19-16). Work must be done by the motor driving the belt to bring added charge to the high potential of the outer sphere. The greater the net charge on the outer conductor, the greater its potential. The maximum potential obtainable in this way is limited only by the fact that air molecules become ionized in very high electric fields and the air becomes a conductor. This phenomenon, which is called *dielectric breakdown*, occurs in air at electric field strengths of about $E_{max} \approx 3 \times 10^6$ V/m. The electric field strength for which dielectric breakdown occurs in a material is called the dielectric strength of that material. The dielectric strength of air is 3 MV/m. The resulting discharge through the conducting air is called *corona discharge*. The electric shocks you receive when you walk across a wool rug on a dry day and touch a metal door knob is a familiar example of corona discharge. (This occurs more often on dry days because moist air will conduct some of the charge away before you can accumulate enough charge by walking on the rug to reach a high potential.) Lightning is another familiar example of corona discharge.

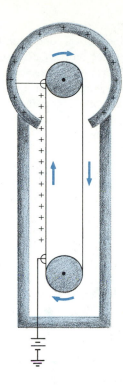

Figure 19-15
Schematic diagram of a Van de Graaff generator. Charge leaks off the pointed conductor near the bottom onto the belt. Near the top, the charge leaks off the belt onto pointed conductors attached to the large spherical conductor.

(a)

(b)

Figure 19-16
(a) Small demonstration Van de Graaff generator. (b) This girl has been charged to a very high potential by being in contact with a demonstration Van de Graaff generator and standing on an insulating block. Her hair has acquired a sufficient charge to show electrostatic repulsion. Some care must be taken to acquire the charge gradually to avoid a painful shock.

Figure 19-17
Nonspherical conductor. If a charge is placed on such a conductor, it will produce an electric field that is stronger near point A, where the radius of curvature is small, than near point B, where the radius of curvature is large.

Electric field lines near a nonspherical conductor and plate carrying equal and opposite charges. The lines are shown by small bits of thread suspended in oil. The electric field is strongest near points of small radius of curvature, such as at the ends of the plate and at the pointed left side of the conductor.

When a charge is placed on a conductor of nonspherical shape like that in Figure 19-17, the conductor will be an equipotential surface, but the charge density on the surface (the charge per unit area) and the electric field just outside the conductor will vary from point to point. Near a point where the radius of curvature is small (A in Figure 19-17), the charge density and electric field will be large, whereas near a point where the radius of curvature is large (B in the figure), they will be small. We can understand this qualitatively by considering the ends to be spheres of different radius. Let σ be the charge per unit area. The potential of a sphere of radius r is

$$V = \frac{kq}{r} \qquad\qquad 19\text{-}14$$

Since the area of a sphere is $4\pi r^2$, the charge on a sphere is related to the charge density by

$$q = 4\pi r^2 \sigma \qquad\qquad 19\text{-}15$$

Substituting this expression for q into Equation 19-14, we have

$$V = \frac{k4\pi r^2 \sigma}{r} = 4\pi kr\sigma$$

Solving for σ, we obtain

$$\sigma = \frac{V}{4\pi kr} \qquad\qquad 19\text{-}16$$

Since both "spheres" are at the same potential, the one with the smaller radius must have the greater charge density.

For an arbitrarily shaped conductor, the potential at which dielectric breakdown occurs depends on the smallest radius of curvature of any part of the conductor. If the conductor has sharp points of very small radius of curvature, dielectric breakdown will occur at relatively low potentials. In the Van de Graaff generator, the charge is transferred onto the belt by needle-shaped conductors placed very near the bottom of the belt. The charge is removed from the belt by a sharp-edged conductor near the top of the belt as shown in Figure 19-15. Lightning rods at the top of a tall building draw the charge off a nearby cloud before the potential of the cloud can build up to a very large value.

Corona discharge from overloaded power lines at General Electric test facilities in Pittsfield, Massachusetts. Because of the intense electric field near the wire, the air molecules are ionized and the air becomes conducting. The light is given off when the ions and electrons recombine.

Question

10. When you walk across a wool rug on a dry day and touch a friend, you draw a spark of about 2 mm. Estimate the potential difference between you and your friend before the spark.

19-4   Capacitance

A capacitor is a useful device for separating and storing charge and for storing energy. The first capacitor was the Leyden jar discussed in the introduction of this chapter. Capacitors have many uses. The flash attachment for a camera uses a capacitor to store the energy needed to provide a sudden flash of light. Capacitors are used to smooth out ripples in a direct current voltage supply (see Section 23-5) such as those used to power your calculator or radio when the batteries are low.

A typical capacitor, called a *parallel-plate capacitor,* consists of two large conducting plates insulated from each other and separated by a small distance. In practice, the plates may be thin metallic foils separated by a thin sheet of paper. This "paper" sandwich is then rolled up to save space. When the plates are connected to a charging device, for example, a battery, as shown in Figure 19-18, charge is transferred from one conductor to the other until the potential difference between the conductors due to their equal and opposite charges equals the potential difference between the battery terminals.

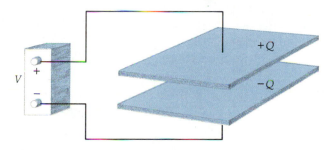

Figure 19-18
Capacitor consisting of two closely spaced parallel-plate conductors. When the conductors are connected to the terminals of a battery, the battery transfers charge from one conductor to the other until the potential difference between the conductors equals that between the battery terminals. The amount of charge transferred is proportional to the potential difference.

The amount of charge transferred depends on the potential difference and on the geometry of the capacitor, for example, on the area and separation of the plates in a parallel-plate capacitor. Let Q be the magnitude of the charge on either plate and V be the potential difference between the plates.* The ratio Q/V is called the *capacitance C.*

$$C = \frac{Q}{V}$$ 19-17   Capacitance defined

The capacitance C is a measure of the "capacity" for separating and storing charge for a given potential difference. The SI unit of capacitance is the coulomb per volt and is called the *farad* (F):

$$1\ \text{F} = 1\ \frac{\text{C}}{\text{V}}$$ 19-18

Since the farad is a rather large unit, submultiples such as the microfarad ($1\ \mu\text{F} = 10^{-6}\ \text{F}$) or the picofarad ($1\ \text{pF} = 10^{-12}\ \text{F}$) are often used.

---

* The use of V for the potential difference between the plates rather than ΔV is standard and simplifies many of the equations.

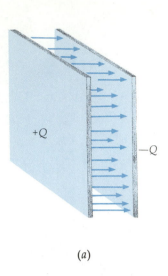

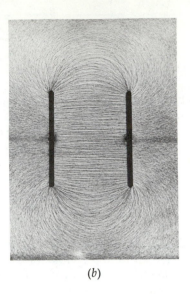

(a)   (b)

Figure 19-19
(a) The electric field lines between
the plates of a parallel-plate
capacitor are equally spaced,
indicating that the electric field
there is uniform. (b) Electric field
lines between plates of a parallel-
plate capacitor, shown by small bits
of thread suspended in oil. When
the plates are very close together,
the fringing of the field near the
edges can be neglected.

To calculate the capacitance of a parallel-plate capacitor, we first find the electric field between the plates when they carry charges $+Q$ and $-Q$ and then use our results to find the voltage difference between the plates. Figure 19-19 shows the electric field lines in the space between the plates of a parallel-plate capacitor. Except for the regions near the edges of the plates, the electric field is uniform. The electric field between the plates equals the electric field just outside either conductor. According to Equation 19-12, if the charge density on the surface of a conductor is σ, the electric field just outside the conductor is given by

$$E = 4\pi k\sigma$$

The charge density on a plate of area A carrying a charge Q is

$$\sigma = \frac{Q}{A}$$

Since the field between the plates of our capacitor is uniform, the potential difference between the plates equals the field times the plate separation d:

$$V = Ed \qquad\qquad\qquad\qquad\qquad\qquad 19\text{-}19$$

Writing $4\pi kQ/A$ for E, we obtain

$$V = Ed = \frac{4\pi Qd}{A} \qquad\qquad\qquad\qquad 19\text{-}20$$

The charge Q is thus related to the potential difference by

$$Q = \frac{A}{4\pi kd} V$$

Dividing by V, we obtain the capacitance of a parallel-plate capacitor:

$$C = \frac{A}{4\pi kd} \qquad\qquad\qquad\qquad\qquad 19\text{-}21 \quad \text{Parallel-plate capacitor}$$

The parallel-plate capacitor was originally called a condenser *and the property it exhibited was called* capacity. *This terminology is still used in connection with automobile ignition circuits. But since this device doesn't really condense anything, the original terms have been changed to* capacitor *and* capacitance, *respectively.*

The capacitance is proportional to the area of the plates and inversely proportional to the separation distance. We note in Equation 19-20 that the potential difference between the plates is proportional to the charge Q. Therefore, the ratio Q/V does not depend on either the charge Q or the potential difference V. In general, capacitance does not depend on either V or Q but on the size, shape, and geometrical arrangement of the conductors.

To avoid having to write the factor $4\pi k$ in the formula for capacitance and in many other formulas, a new constant ϵ_0, called the *permittivity of free space*, is often used. It is defined by

$$k = \frac{1}{4\pi\epsilon_0}$$

or

$$\epsilon_0 = \frac{1}{4\pi k} = 8.85 \times 10^{-12} \text{ C}^2/\text{N}\cdot\text{m}^2 \qquad 19\text{-}22$$

A variable parallel-plate capacitor.

Permittivity of free space

where we have used $k = 8.99 \times 10^9 \text{ N}\cdot\text{m}^2/\text{C}^2$. In terms of ϵ_0, the capacitance of a parallel-plate capacitor is given by

$$C = \epsilon_0 \frac{A}{d} \qquad 19\text{-}23$$

Parallel-plate capacitor

Since capacitance is in farads and A/d is in metres, we can see from Equation 19-23 that the SI unit for ϵ_0 can also be written as farads per metre.

$$\epsilon_0 = 8.85 \times 10^{-12} \text{ F/m} \qquad 19\text{-}24$$

A numerical calculation illustrates how large a unit of capacitance the farad is.

Example 19-4 A parallel-plate capacitor has square plates of side 10 cm separated by 1 mm. Calculate its capacitance.

Using Equation 19-23, we obtain for the capacitance

$$C = \epsilon_0 \frac{A}{d} = (8.85 \times 10^{-12} \text{ C}^2/\text{N}\cdot\text{m}^2) \frac{(0.1 \text{ m})^2}{0.001 \text{ m}}$$

$$= 8.85 \times 10^{-11} \text{ F} \approx 90 \text{ pF}$$

Thus, even though the plates are separated by only 1 mm, the capacitance is less than one ten-billionth of a farad.

Example 19-5 A 90-pF capacitor is connected to a 12-V battery and charged to 12 V. How many electrons are transferred from one plate to the other?

From the definition of capacitance (Equation 19-17), the charge transferred is

$$Q = CV = (90 \times 10^{-12} \text{ F})(12 \text{ V}) = 1.1 \times 10^{-9} \text{ C}$$

This is the magnitude of the charge on either plate. The number of electrons in a charge of 1.1×10^{-9} C is

$$N = \frac{Q}{e} = \frac{1.1 \times 10^{-9} \text{ C}}{1.6 \times 10^{-19} \text{ C/electron}} = 6.9 \times 10^9 \text{ electrons}$$

Your body contains millions of parallel-plate capacitors — the nerve cells whose cell membranes act as insulators separating sheets of positive and negative charge. Nerve impulses travel between your brain and the rest of your body by means of the charging and discharging of these microscopic capacitors. The potential difference on a typical nerve cell wall is 12 million volts per metre. In air, one quarter of this field would cause dielectric breakdown and would result in a spark. The typical capacitance of the nerve cell is about 10,000 $\mu F/m^2$, which is far greater than that of any capacitor yet produced by industrial engineers.

Dielectrics

A nonconducting material, such as glass, paper, or wood, is called a *dielectric*. When the space between the two conductors of a capacitor is occupied by a dielectric, the capacitance is increased by a factor K that is characteristic of the dielectric and is called the *dielectric constant*. This was discovered experimentally by Michael Faraday. The reason for this increase is that the electric field between the plates of a capacitor is weakened by the dielectric. Thus, for a given charge-on the plates, the potential difference is reduced and the ratio Q/V is increased. The dielectric weakens the electric field between the plates of a capacitor because the molecules in the dielectric become polarized by the original electric field. If the molecules are polar molecules, they will be partially aligned with the direction of the field; if they are nonpolar molecules, they will have induced dipole moments in the direction of the field. In either case, the net effect is the creation of an induced charge on the surface of the dielectric near the plates, as shown in Figure 19-20. This surface charge, which is bound to the dielectric, produces an electric field opposite the direction of the field due to the free charge on the conductors. Thus, the electric field between the plates is weakened, as is illustrated in Figure 19-21. If the original electric field without the dielectric is E_0, the field with the dielectric is

$$E' = \frac{E_0}{K} \qquad \qquad \text{19-25}$$

where K is the dielectric constant. The potential difference between the plates is then

$$V' = E'd = \frac{E_0 d}{K} = \frac{V_0}{K}$$

where $V_0 = E_0 d$ is the original potential difference without the dielectric. The new capacitance is

$$C = \frac{Q}{V'} = \frac{Q}{V_0/K} = K\frac{Q}{V_0}$$

or

$$C = kC_0 \qquad \qquad \text{19-26}$$

where $C_0 = Q/V_0$ is the original capacitance.

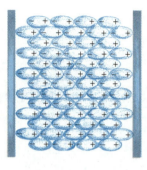

Dielectric constant

Figure 19-20
When a nonconducting material (a dielectric) is placed between the plates of a capacitor, the electric field of the capacitor polarizes the molecules of the dielectric. The result is a bound charge on the surface of the dielectric that produces its own electric field that opposes the external field. The electric field between the plates is thus weakened by the dielectric.

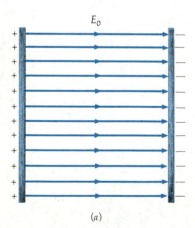
(a)

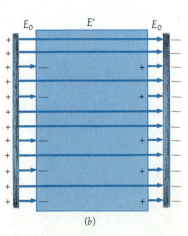
(b)

Figure 19-21
Electric field between the plates of a capacitor (a) with no dielectric and (b) with a dielectric. The surface charge on the dielectric weakens the field between the plates.

In the above calculation, we assumed that the charge on the plates of the capacitor did not change when the dielectric was inserted. This would be true if the capacitor was charged and then removed from the charging source (the battery) before the insertion of the dielectric. If the dielectric is inserted while the battery is still connected, the battery must supply more charge to maintain the original potential difference. The total charge on the plates is then $Q' = KQ_0$. In either case, the capacitance is increased by the factor K.

In addition to increasing the capacitance, a dielectric has two other functions in a capacitor. First, it provides a mechanical means of separating the two conductors, which must be very close together in order to obtain a large capacitance since the capacitance varies inversely with the separation. Second, the dielectric strength is increased because the dielectric strength of a dielectric is usually greater than that of air. We have already mentioned that the dielectric strength of air is 3 MV/m = 3 kV/mm. Fields with a magnitude greater than this cannot be maintained in air because of dielectric breakdown; that is, the air becomes ionized and conducts. Many materials have dielectric strengths greater than that of air, and so allow greater potential differences between the conducting plates of a capacitor.

Examples of the three dielectric functions can be seen in a parallel-plate capacitor made from two sheets of metal foil of large area (to increase the capacitance) separated by a sheet of paper. The paper increases the capacitance because of its polarization; that is, K is greater than 1. It also provides mechanical separation that allows the sheets to be very close together without being in electrical contact. Finally, the dielectric strength of paper is greater than that of air, so greater potential differences can be attained without breakdown. Table 19-1 lists the dielectric constants and dielectric strengths of some dielectrics.

Three functions of a dielectric

One important type of capacitor, the electrolytic capacitor, consists of a foil sheet coated with a semiliquid paste. The paste reacts chemically with the foil to form a thin oxide coating that acts as a dielectric. Thus, the foil serves as one plate and the paste as the other. Because the dielectric is extremely thin, the capacitance is quite high. If the dielectric strength is exceeded, the capacitor fails. Once the voltage is reduced, however, the paste and the foil again react chemically to "heal" the dielectric.

Table 19-1

Dielectric Constant and Strength of Various Materials

Material	Dielectric constant K	Dielectric strength, kV/mm
Air	1.00059	3
Bakelite	4.9	24
Glass (Pyrex)	5.6	14
Mica	5.4	10–100
Neoprene	6.9	12
Paper	3.7	16
Paraffin	2.1–2.5	10
Plexiglas	3.4	40
Polystyrene	2.55	24
Porcelain	7	5.7
Transformer oil	2.24	12
Water (20°C)	80	

Question

11. Two students are arguing about the effect of introducing a dielectric into the space between the plates of a capacitor. One says that the potential difference between the plates is decreased. The other says that the charge on the plates is increased. How would you settle the argument?

19-5   Combinations of Capacitors

Two or more capacitors are often used in combination. In electric circuits, a capacitor is indicated by the symbol ─┤├─ . Figure 19-22 shows two capacitors connected in parallel. The upper plates of the two capacitors are connected together by a conducting wire and are therefore at the same potential V_a. The lower plates are also connected together and are at a common potential V_b. The potential difference across both capacitors is $V = V_a - V_b$.

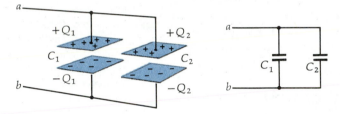

Figure 19-22
Two capacitors in parallel. The potential difference is the same across both capacitors.

It is clear that the effect of adding a second capacitor connected in this way is to increase the capacitance. Essentially, the area is increased, allowing more charge to be stored for the same potential difference. If the capacitances are C_1 and C_2, the charges Q_1 and Q_2 stored on the plates are given by

$$Q_1 = C_1 V \quad \text{and} \quad Q_2 = C_2 V$$

where V is the potential difference across either capacitor. The total charge stored is

$$Q = Q_1 + Q_2 = C_1 V + C_2 V = (C_1 + C_2)V$$

The effective capacitance of two capacitors in parallel is defined as the ratio of the total charge stored to the potential difference. The effective capacitance is thus

$$C_{\text{eff}} = \frac{Q}{V} = C_1 + C_2 \qquad \text{19-27}$$

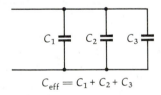

$$C_{\text{eff}} = C_1 + C_2 + C_3$$

Figure 19-23
Three capacitors in parallel. The effect of adding a parallel capacitor is to increase the effective capacitance.

Capacitors in parallel

The effective capacitance is that of a single capacitor that could replace the parallel combination and store the same amount of charge for a given potential difference. This reasoning can be extended to three or more capacitors connected in parallel, as in Figure 19-23. The effective capacitance equals the sum of the individual capacitances.

Two capacitors connected as shown in Figure 19-24 are said to be connected in series. When points a and b are connected to the terminals of a battery, there is a potential difference $V = V_a - V_b$ across the two capacitors, but the potential difference across one capacitor is not the same as that across the other. If a charge $+Q$ is placed on the upper plate of the first capacitor, there will be an equal negative charge $-Q$ induced on its lower plate. This charge comes from electrons drawn from the upper plate of the second capacitor. Thus there will be an equal charge $+Q$ on the upper plate of the second capacitor and $-Q$ on its lower plate. The potential difference across the upper capacitor is $V_a - V_c = Q/C_1$. Similarly, the potential difference across the second capacitor is $V_c - V_b = Q/C_2$. The potential difference across the two capacitors in series is the sum of these potential differences:

$$V = V_a - V_b = (V_a - V_c) + (V_c - V_b) = V_1 + V_2$$

where

$$V_1 = V_a - V_c = \frac{Q}{C_1}$$

is the potential difference across the first capacitor and

$$V_2 = V_c - V_b = \frac{Q}{C_2}$$

is the potential difference across the second capacitor. Thus,

$$V = \frac{Q}{C_1} + \frac{Q}{C_2} = Q \left(\frac{1}{C_1} + \frac{1}{C_2} \right) \qquad \text{19-28}$$

The effective capacitance of two capacitors in series is that of a single capacitor that could replace the two capacitors and give the potential difference V for the same charge Q. Thus

$$V = \frac{Q}{C_{\text{eff}}} \qquad \text{19-29}$$

Comparing Equations 19-28 and 19-29, we have

$$\frac{1}{C_{\text{eff}}} = \frac{1}{C_1} + \frac{1}{C_2}$$

This equation can be generalized to three or more capacitors connected in series:

$$\frac{1}{C_{\text{eff}}} = \frac{1}{C_1} + \frac{1}{C_2} + \frac{1}{C_3} + \cdots \qquad \text{19-30} \qquad \text{Capacitors in series}$$

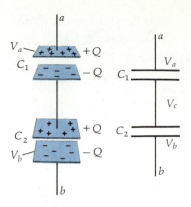

Figure 19-24
Two capacitors in series. The charge is the same on both capacitors. The potential difference across the series combination is the sum of the potential differences across each capacitor, $V_a - V_b = (V_a - V_c) + (V_c - V_b)$.

Example 19-6   Two capacitors have capacitances of 20 μF and 30 μF. Find the effective capacitance if the capacitors are connected (*a*) in parallel and (*b*) in series.

(*a*) Since the effective capacitance of capacitors in parallel is just the sum of the individual capacitances, the effective capacitance in parallel is

$$C_{\text{eff}} = 20 \ \mu\text{F} + 30 \ \mu\text{F} = 50 \ \mu\text{F}$$

(b) When the capacitors are connected in series, the effective capacitance is found from

$$\frac{1}{C_{eff}} = \frac{1}{20\ \mu F} + \frac{1}{30\ \mu F} = \frac{3}{60\ \mu F} + \frac{2}{60\ \mu F} = \frac{5}{60\ \mu F}$$

Thus,

$$C_{eff} = \frac{60\ \mu F}{5} = 12\ \mu F$$

Note that the effective capacitance of two capacitors in series is less than the capacitance of either capacitor.

Example 19-7   Find the effective capacitance of the network of three capacitors shown in Figure 19-25.

In this network, the 2-μF and 3-μF capacitors are in parallel and the parallel combination is in series with the 4-μF capacitor. The effective capacitance of the two capacitors in parallel is

$$C_{eff} = C_1 + C_2 = 2\ \mu F + 3\ \mu F = 5\ \mu F$$

If we replace the two capacitors in parallel with a single capacitor of capacitance 5 μF, we have a 5-μF capacitor in series with a 4-μF capacitor. The effective capacitance of the series combination is

$$\frac{1}{C_{eff}} = \frac{1}{C_1} + \frac{1}{C_2} = \frac{1}{5\ \mu F} + \frac{1}{4\ \mu F}$$

$$= \frac{4}{20\ \mu F} + \frac{5}{20\ \mu F} = \frac{9}{20\ \mu F}$$

The effective capacitance of the three-capacitor network is thus

$$C_{eff} = \frac{20\ \mu F}{9} = 2.22\ \mu F$$

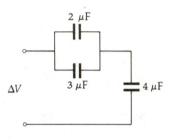

Figure 19-25
Capacitor network for Example 19-7.

Example 19-8   Two capacitors of capacitance 2 μF and 4 μF are connected in series across a 18-V battery as shown in Figure 19-26a. Find the charge on the capacitors and the potential difference across each.

In Figure 19-26b, we have replaced the two capacitors by one with the effective capacitance C_{eff} given by

$$\frac{1}{C_{eff}} = \frac{1}{C_1} + \frac{1}{C_2} = \frac{1}{2\ \mu F} + \frac{1}{4\ \mu F} = \frac{3}{4\ \mu F}$$

$$C_{eff} = \tfrac{4}{3}\ \mu F$$

The charge on either plate of the equivalent capacitor in Figure 19-26b is then

$$Q = C_{eff}V = (\tfrac{4}{3}\ \mu F)(18\ V) = 24\ \mu C$$

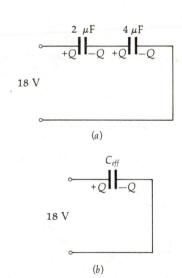

Figure 19-26
(a) Two capacitors in series and connected to a battery for Example 19-8. (b) An equivalent capacitor can replace the two capacitors in (a).

This is the charge that is on each plate of the two original capacitors. The potential difference across each capacitor is then

$$V_1 = \frac{Q}{C_1} = \frac{24 \ \mu C}{2 \ \mu F} = 12 \ V$$

and

$$V_2 = \frac{Q}{C_2} = \frac{24 \ \mu C}{4 \ \mu F} = 6 \ V$$

Note that the sum of these potential differences is 18 V, as required.

Example 19-9   The two capacitors in Example 19-8 are removed from the battery and are carefully disconnected from each other without disturbing the charge on the plates. They are then connected to each other with positive plate connected to positive plate and negative plate connected to negative plate, as shown in Figures 19-27a and b. Find the charge on each plate and the potential difference across the capacitors.

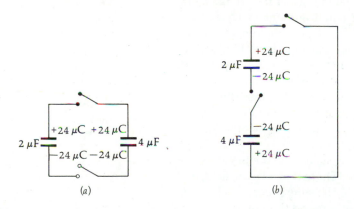

(a)   (b)

Figure 19-27
The two capacitors in Figure 19-26a just after they are disconnected from the battery but before they are connected to each other, positive plate to positive plate and negative plate to negative plate. (b) This is exactly the same situation as in (a) but it looks different. When two capacitors are connected together with nothing else in the network, they can be considered to be in parallel.

Let Q_1 be the final charge on the 2-μF capacitor and Q_2 the final charge on the 4-μF capacitor. Since the charge on each capacitor was originally 24 μC, the sum of the final charges must be 48 μC:

$$Q_1 + Q_2 = 48 \ \mu C$$

From Figure 19-27a we can see that the potential difference is the same across each capacitor. Thus,

$$V = \frac{Q_1}{2 \ \mu F} = \frac{Q_2}{4 \ \mu F}$$

From this equation, we can see that the charge Q_2 must be twice Q_1:

$$Q_2 = 2Q_1$$

Then,
$$Q_1 + Q_2 = Q_1 + 2Q_1 = 48 \ \mu\text{C}$$

$$3Q_1 = 48 \ \mu\text{C}$$

$$Q_1 = 16 \ \mu\text{C}$$

$$Q_2 = 2Q_1 = 32 \ \mu\text{C}$$

The potential difference is

$$V = \frac{Q_1}{2 \ \mu\text{F}} = \frac{16 \ \mu\text{C}}{2 \ \mu\text{F}} = 8 \text{ V}$$

19-6   Electrical Energy Storage

If a charge is moved in an electric field, work is done. Some or all of this work is often stored as electrostatic potential energy. A simple example of this is the charging of a capacitor. While a capacitor is being charged, a positive charge is transferred from the negatively charged conductor to the positively charged conductor. Since the positive conductor is at a greater potential than the negative conductor, the potential energy of the charge being transferred is increased. For example, if a small amount of charge q is transferred through a potential difference V, the potential energy of the charge is increased by the amount qV. (Remember that, by definition, potential difference is the potential-energy difference per unit charge.) Work must therefore be done to charge a capacitor. Some of this work is stored as electrostatic potential energy. If the capacitor is charged by connecting it to a battery, for example, we will see that only half of the work done by the battery is stored as electrostatic energy. The other half of the work goes into heat losses in the connecting wires.

Consider the charging of a parallel-plate capacitor. At the beginning of the charging process, neither plate is charged. There is no electric field, and both plates are at the same potential. After the charging process, a charge Q has been transferred from one plate to the other, and the potential difference is $V = Q/C$, where C is the capacitance. One might expect that the work needed to accomplish this would be just the charge Q times the potential energy per unit charge V, but only the last bit of charge must be moved through the full potential difference V. The potential difference between the plates increases from zero initially to its final value V. The average value of the potential difference during the charging process is just $\frac{1}{2}V$. The work needed to charge the capacitor from a zero initial charge to a final charge of Q is the final charge times the average potential difference, $W = QV_{\text{av}} = \frac{1}{2}QV$. The electrostatic energy stored in a charged capacitor is therefore

$$U = \tfrac{1}{2}QV \qquad\qquad\qquad \text{19-31} \qquad \text{Energy in a charged capacitor}$$

Since the potential difference V across the terminals of the battery remains constant, the work done by the battery during the charging process is QV, twice the energy stored in the capacitor. The other half of the work done by the battery goes into heating the wires connecting the battery to the capacitor. We can write Equation 19-31 in terms of the capacitance $C = Q/V$ by eliminating either Q or V.

$$U = \frac{1}{2} CV^2 = \frac{1}{2} \frac{Q^2}{C}$$
19-32   Energy in a charged capacitor

Equations 19-31 and 19-32 are general expressions for the energy stored in a capacitor.

Example 19-10   A 15-μF capacitor is charged to 60 V. How much energy is stored?

Since we are given the capacitance and voltage and not the charge on the capacitor, Equation 19-32 is more convenient to use than Equation 19-31.

$$U = \tfrac{1}{2}CV^2 = \tfrac{1}{2}(15 \times 10^{-6}\text{ F})(60\text{ V})^2 = 0.027\text{ J}$$

In the process of charging a capacitor, an electric field is produced between the plates. The work required to charge the capacitor can be thought of as the work required to create the electric field. That is, we can think of the energy stored in a capacitor as energy stored in the electric field. We can write Equation 19-32 for this energy in terms of the electric field. The electric field E between the plates of a parallel-plate capacitor is related to the potential difference V and the plate separation d by Equation 19-19:

$$E = \frac{V}{d}$$

Writing $C = \epsilon_0 A/d$ for the capacitance and $V = Ed$ for the potential difference, Equation 19-32 becomes

$$U = \frac{1}{2} CV^2 = \frac{1}{2} \frac{\epsilon_0 A}{d} (Ed)^2$$

or

$$U = \tfrac{1}{2}\epsilon_0 E^2 (Ad)$$
19-33

The quantity Ad is the volume of the space between the plates of the capacitor containing the electric field. The energy per unit volume is called the *energy density* η.

$$\eta = \frac{\text{energy}}{\text{volume}} = \frac{1}{2} \epsilon_0 E^2$$
19-34   Electric field energy density

The electrostatic field energy per unit volume is proportional to the square of the electric field. Although we have obtained Equation 19-34 by considering the special case of the electric field between the plates of a parallel-plate capacitor, the result is general. Whenever there is an electric field in space, the electrostatic energy per unit volume is given by Equation 19-34.

Questions

12. If the potential difference of a capacitor is doubled, by what factor does its stored electric energy change?

13. Half the charge is removed from a capacitor. What fraction of its stored energy is removed along with the charge?

Electrostatics and Xerography

Richard Zallen*
Virginia Polytechnic Institute and State University

There are many important and beneficial technological applications that could be included in a discussion of uses of electrostatic phenomena. For example, a powerful air-pollution preventer is the electrostatic precipitator, which years ago made life livable near cement mills and ore-processing plants and which is currently credited with extracting better than 99 percent of the ash and dust from the gases about to issue from chimneys of coal-burning power plants. The basic idea of this very effective antipollution technique is shown in Figure 1. The outer wall of a vertical metal duct is grounded, while a wire running down the center of the duct is kept at a very large negative voltage. In this concentric geometry, a very nonuniform electric field is set up, with lines of force directed radially inward toward the negative wire electrode.

* The author spent many years with Xerox Research Laboratories in Webster, N.Y.

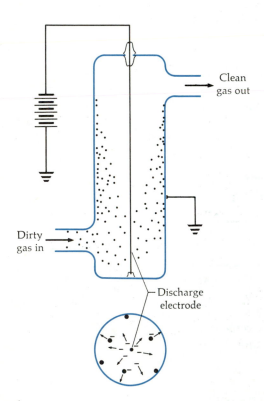

Figure 1
Schematic diagram of the use of a corona discharge in an electrostatic precipitator.

Close to the wire the field attains enormous values, large enough to produce an electrical breakdown of air, and the normal placid mixture of neutral gas molecules is replaced by a turmoil of free electrons and positive ions. The electrons from this corona discharge are driven outward from the wire by the electric field. Most of them quickly become attached to oxygen molecules to produce negative O_2^- ions, which are also accelerated outward. As this stream of ions passes across the hot waste gas rising in the duct, small particles carried by the gas become charged by capturing ions and are pulled by the field to the outer wall. If the noxious particles are solid, they are periodically shaken down off the duct into a hopper; if they are liquid, the residue simply runs down the wall and is collected below.

Besides electrostatic precipitation, other technological examples include electrocoating with spray paints and the electrostatic separation of granular mixtures used for the removal of rock particles from minerals, garlic seeds from wheat, even rodent excreta from rice. However, the application that is the main focus of this essay is xerography, the most widely used form of electrostatic imaging, or electrophotography. This is the most familiar use of electrostatics in terms of the number of people who have occasion to use plain-paper copying machines in offices, libraries, and schools, and it also provides a fine example of a process utilizing a sequence of distinct electrostatic events.

The xerographic process was invented in 1937 by Chester Carlson. The term xerography, literally "dry writing," was actually adopted a bit later to emphasize the distinction from wet chemical processes. Carlson's innovative concept did not find early acceptance, and a practical realization of his idea became available only after a small company (in a famous entrepreneurial success story) risked its future in its intensive efforts to develop the process.

Four of the main steps involved in xerography are illustrated in Figure 2. In the interest of clarity the process has been oversimplified, and several subtleties (as well as gaps in our understanding) have been suppressed. Electrostatic imaging takes place on a large thin plate of a photoconducting material supported by a grounded metal backing. A photoconductor is a solid that is a good insulator *in the dark* but becomes capable of conducting electric current when exposed to light. The unilluminated, insulating state is indicated by shading in Figure 2. In the dark, a uniform electrostatic charge is laid down on the surface of the photoconductor. This charging step (Figure 2*a*) is accomplished by means of a positive corona discharge surrounding a fine wire held at about +5000 V. This corona (a miniature version of, and opposite in sign to, the intense precipitator corona of Figure 1) is passed over the photoconductor surface, spraying positive ions onto it and charging it to a potential of the order of $+1000$ V. Since charge is free to flow within the grounded metal backing, an equal and opposite induced charge develops at the metal-photoconductor interface. In the dark the photoconductor contains no mobile charge, and the large potential difference persists across this dielectric layer, which is only 0.005 cm thick.

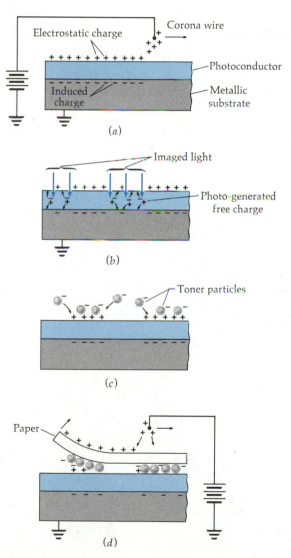

(a)

(b)

(c)

(d)

Figure 2
Steps in the xerographic process: (a) charging, (b) exposure, (c) development, and (d) transfer.

The photoconductor plate is next exposed to light in the form of an image reflected from the document being copied. What happens now is indicated in Figure 2b. Where light strikes the photoconductor, light quanta (photons) are absorbed, and pairs of mobile charges are created. Each photo-generated pair consists of a negative charge (an electron) and a positive charge (a hole; crudely, a missing electron). Photo-generation of this free charge depends not only on the photoconductor used and on the wavelength and intensity of the incident light but also on the electric field present. This large field (1000 V$/0.005$ cm $= 2 \times 10^5$ V/cm $= 2 \times 10^7$ V/m) helps to pull apart the mutually attracting electron-hole pairs so that they are free to move separately. The electrons then move under the influence of the field to the surface, where they neutralize positive charges, while the holes move to the photoconductor-substrate interface and neutralize negative charges there. Where intense light strikes the photoconductor, the charging step is totally undone; where weak light strikes it, the charge is partially reduced; and where no light strikes it, the original electrostatic charge remains on the surface. The critical task of converting an optical image into an electrostatic image, which is now recorded on the plate, has been completed. This latent image consists of an electrostatic potential distribution that replicates the light and dark pattern of the original document.

To develop the electrostatic image, fine negatively charged pigmented particles are brought into contact with the plate. These *toner particles* are attracted to positively charged surface regions, as shown in Figure 2c, and a visible image appears. The toner is then transferred (Figure 2d) to a sheet of paper that has been positively charged in order to attract the particles. Brief heating of the paper fuses the toner to it and produces a permanent photocopy ready for use.

Finally, to prepare the photoconductor plate for a repetition of the process, any toner particles remaining on its surface are mechanically cleaned off, and the residual electrostatic image is erased, that is, discharged, by flooding with light. The photoconductor is now ready for a new cycle, starting with the charging step. In high-speed duplicators, the photoconductor layer is often in the form of a moving continuous drum or belt, around the perimeter of which are located stations for performing the various functions of Figure 2. The speed of xerographic printing technology is presently on the order of a few copies per second.*

---

* For further information on electrostatics in xerography, consult J. H. Dessauer and H. E. Clark (eds.), *Xerography and Related Processes*, Focal Press, New York, 1965, and R. M. Schaffert, *Electrophotography*, rev. ed., Focal Press, New York, 1973. Other applications of electrostatics are discussed in A. D. Moore, *Scientific American*, March 1972.

Summary

1. The electric potential difference between two points is the work per unit charge we must do to move a test charge from one point to another in an electric field. It also equals the potential energy difference per unit charge. Since only differences in electric potential are important, we can choose the potential to be zero at any convenient point. The SI unit of potential and potential difference is the volt (V), which is defined as

$$1 \, V = 1 \, J/C$$

In terms of this unit, the units for the electric field can be expressed as

$$1 \, N/C = 1 \, V/m$$

2. A convenient unit of energy in atomic and nuclear physics is the electronvolt (eV), which is related to the joule by

$$1 \, eV = 1.6 \times 10^{-19} \, J$$

3. For a constant field in the x direction E_x the potential difference between points x_1 and x_2 is given by

$$\Delta V = -E_x \, \Delta x$$

where $\Delta x = x_1 - x_1$.

4. The electric potential due to a point charge q at the origin is given by

$$V = \frac{kq}{r}$$

where we have chosen the potential to be zero at an infinite distance from the charge.

5. Electric conductors are materials that have electrons that are not bound to any atom but are free to move about. A conductor cannot support an electrostatic field because the free electrons move until they set up a field that cancels the original field. Any excess charge in a conductor resides on the surface.

6. Just outside the surface of a conductor, the electric field is perpendicular to the surface and is related to the charge per unit area σ on the surface by

$$E_n = 4\pi k \sigma = \frac{\sigma}{\epsilon_0}$$

where ϵ_0 is the permittivity of free space:

$$\epsilon_0 = \frac{1}{4\pi k} = 8.85 \times 10^{-12} \, F/m$$

7. A capacitor is a device for storing charge and energy. It consists of two conductors carrying equal and opposite charges. The capacitance is the ratio of the magnitude of the charge Q on either conductor to the potential difference V between the conductors:

$$C = \frac{Q}{V}$$

Capacitance depends only on the geometrical arrangement of the conductors and not on the charge or the potential difference. The capacitance of a parallel-plate capacitor is proportional to the area of the plates and inversely proportional to the separation distance:

$$C = \epsilon_0 \frac{A}{d}$$

8. A nonconducting material is called a dielectric. When a dielectric is inserted between the plates of a capacitor, the molecules in the dielectric become polarized and the electric field is weakened, leading to an increase in the capacitance by the factor K, called the dielectric constant:

$$C' = KC$$

Dielectrics also provide physical separation of the plates of the capacitor and increase the charge that can be stored before dielectric breakdown.

9. The maximum electric field that can be supported in a dielectric before breakdown is called the dielectric strength. The dielectric strength of air is about 3 MV/m. When the electric field in air exceeds this value, the air molecules become ionized and the air becomes a conductor. The resulting spark is called corona discharge.

10. When two or more capacitors are connected in parallel, the effective capacitance of the combination is the sum of the individual capacitances:

$$C_{eff} = C_1 + C_2 + C_3 + \cdots \quad \text{(parallel)}$$

When two or more capacitors are connected in series, the effective capacitance is found by adding the reciprocals of the individual capacitances:

$$\frac{1}{C_{eff}} = \frac{1}{C_1} + \frac{1}{C_2} + \frac{1}{C_3} + \cdots \quad \text{(series)}$$

11. The electrostatic energy stored in a capacitor of charge Q, potential difference V, and capacitance C is

$$U = \frac{1}{2} QV = \frac{1}{2} CV^2 = \frac{Q^2}{2C}$$

This energy can be considered to be stored in the electric field. The energy per unit volume in an electric field E is given by

$$\eta = \frac{\text{energy}}{\text{volume}} = \frac{1}{2} \epsilon_0 E^2$$

Suggestions for Further Reading

de Santillana, Giorgio: "Alessandro Volta," *Scientific American,* January 1965, p. 82.

This article describes the argument between Luigi Galvani and Alessandro Volta over whether or not electricity was a living force and a key to the mystery of life or a phenomenon that could manifest itself without requiring the presence of a living being. Volta won the argument in 1800 when he announced his invention of what we would now call a battery.

Grundfest, Harry: "Electric Fishes," *Scientific American,* October 1960, p. 115.

Members of many families of fishes are able to produce appreciable voltages outside their bodies using specialized organs containing arrays of "electroplaque membranes" in series or in parallel.

McDonald, James E.: "The Earth's Electricity," *Scientific American,* April 1953, p. 32.

The earth and the ionosphere, separated by the atmosphere, can be thought of as the negative and positive elements of a huge spherical capacitor that would slowly lose its voltage difference due to the movement of atmospheric ions if not for the recharging effect of thunderstorms.

Moore, A. D.: "Electrostatics," *Scientific American*, March 1972, p. 46.

This article describes some modern uses of electrostatics, including precipitation of airborne industrial wastes, separation of granular solids such as minerals, efficient paint-spraying, and xerographic copying.

Rose, Peter H., and Andrew B. Wittkower: "Tandem Van de Graaff Accelerators," *Scientific American*, August 1970, p. 24.

These machines produce high-energy beams of charged particles for use in fundamental research by accelerating them between terminals maintained at potential differences of millions of volts.

Review

A. Objectives: After studying this chapter, you should:

1. Be able to give a definition of electric potential and discuss its relation to the electric field.

2. Be able to explain why there is no electrostatic field inside a conducting material.

3. Be able to describe the charging of a conductor by induction.

4. Be able to discuss the phenomena of dielectric breakdown and corona discharge.

5. Be able to give a definition of capacitance and calculate the effective capacitance of several capacitors in series or in parallel.

6. Be able to discuss the effect of a dielectric in a capacitor.

B. Define, explain, or otherwise identify:

Potential difference, p. 455
Electric potential, p. 455
Volt, p. 455
Electronvolt, p. 459
Charging by induction, p. 463
Equipotential surface, p. 464
Dielectric breakdown, p. 467
Dielectric strength, p. 467
Capacitor, p. 469
Dielectric, p. 472
Effective capacitance, p. 474

Capacitors in parallel, p. 474
Capacitors in series, p. 475
Electric field energy density, p. 479

C. True or false: If the statement is true, explain. If it is false, give a counterexample.

1. Electric field lines always point toward regions of lower potential.

2. The value of the electric potential can be chosen to be zero at any convenient point.

3. The electric field is always zero inside a conducting material.

4. In electrostatics, the electric field is perpendicular to the surface of a conductor.

5. In electrostatics, the surface of a conductor is an equipotential surface.

6. Dielectric breakdown occurs in air when the potential is 3×10^6 V.

7. The capacitance of a parallel-plate capacitor is proportional to the charge on its plates.

8. A dielectric inserted in a capacitor increases the capacitance.

9. The electrostatic energy per unit volume at some point is proportional to the square of the electric field at that point.

Exercises

Section 19-1   Electric Potential and Potential Difference

1. A uniform electric field of 3000 N/C is in the negative x direction. The potential is chosen to be zero at the origin. Find the potential at (a) $x = 2$ m, (b) $x = 4$ m, and (c) $x = -3$ m. (d) Make a sketch of the potential V as a function of x.

2. A uniform electric field of 1500 N/C is in the positive x direction. The potential is chosen to be zero at the origin.

Find the potential at (a) $x = 3$ m, (b) $x = 5$ m, (c) $x = 12$ m, and (d) $x = -3$ m.

3. A uniform electric field is along the z axis. The xy plane is at $V = 0$. As you move up from the xy plane, the potential increases by 15 V in each centimetre. Find the magnitude and direction of the electric field.

4. A positive test charge $q_0 = 4\ \mu C$ is released from rest at $x = -3$ m in the uniform electric field of Exercise 2. (*a*) What is its initial potential energy at $x = -3$ m, assuming its potential energy is zero at the origin? (*b*) What is its kinetic energy when it reaches the origin? (*c*) What is its kinetic energy when it reaches the point $x = +3$ m?

5. The electric potential has the value $V = 200$ V at $x = 2$ m and $V = 600$ V at $x = 10$ m. Find the magnitude and direction of the electric field, assuming it is uniform.

6. A uniform electric field of 2 kN/C is in the positive x direction. A point charge $q = 3\ \mu C$, initially at rest at the origin, is released. (*a*) By how much has the potential energy of the charge decreased when it reaches the point $x = 4$ m? (*b*) What is the kinetic energy of the charge when it reaches the point $x = 4$ m? (*c*) If the potential is chosen to be zero at the origin, what is its value at $x = 4$ m? If the potential is chosen to be zero at $x = 2$ m, what is its value at (*d*) $x = 4$ m and (*e*) the origin?

7. A point charge $q = 5\ \mu C$ is at the origin. Find the potential at (*a*) $r = 4$ m and at (*b*) $r = 8$ m, assuming it to be zero at infinity. (*c*) If a second point charge $q' = 2\ \mu C$ is placed at $x = 4$ m, what is its potential energy? (*d*) What is its potential energy at $r = 8$ m? (*e*) If this charge is released from rest at $r = 4$ m, what is its kinetic energy at $r = 8$ m? (Assume the original charge q remains fixed at the origin.)

8. A point charge $q = 12\ \mu C$ is at the origin. (*a*) How much work does it take to bring a second point charge $q' = 3\ \mu C$ from a large distance away to the point $r = 5$ m? (*b*) If this second charge is released from rest at $r = 5$ m, what is its kinetic energy when it is a great distance from the origin?

9. A point charge of $+2$ nC is at the origin and another one of $+2$ nC is on the x axis at $x = 4$ cm. Find the potential on the x axis at (*a*) $x = 2$ cm and at (*b*) $x = 6$ cm. (*c*) Find the potential on the y axis at $y = 3$ cm.

10. Work Exercise 9 for the case where the charge at the origin is -2 nC.

Section 19-2   Electric Conductors;
Section 19-3   Equipotential Surfaces, Charge Sharing, and Dielectric Breakdown

11. A penny has a mass of 3 g and is made of copper of molar mass 63.5 g/mol. Assume one free electron for each copper atom. (*a*) How many free electrons are there in a penny? (*b*) How much free charge (in coulombs) is there?

12. A solid spherical conductor has a radius of 16 cm and carries a net charge $Q = 8.0$ nC. Its center is at the origin. (*a*) Find the electric field everywhere in space. (*b*) Find the electric potential on the sphere, assuming $V = 0$ at infinity.

13. A second, identical spherical conductor is brought up to the charged sphere in Exercise 12 ($Q = 8.0$ nC). It is touched to the first sphere and is then removed to a great distance

away. Find (*a*) the charge on the first sphere and (*b*) the electric potential of that sphere, assuming $V = 0$ at infinity.

14. Find the maximum net charge that can be placed on a spherical conductor of radius 16 cm before dielectric breakdown of the air occurs.

15. Find the minimum radius of a spherical conductor that can hold 1 C of charge before dielectric breakdown of the air occurs.

16. A spherical conductor of radius R carries charge Q. The electric field just outside the conductor is

$$E_n = \frac{kQ}{R^2}$$

Show that this is the same as that given by Equation 19-12:

$$E_n = 4\pi k\sigma = \sigma/\epsilon_0$$

17. A conductor has a charge density of $1.2\ nC/m^2$. Find the magnitude of the electric field just outside the surface of the conductor.

18. Explain, giving each step, how a positively charged insulating rod can be used to give a metal sphere (*a*) a negative charge and (*b*) a positive charge. (*c*) Can the same rod be used to give one sphere a positive charge and another sphere a negative charge without recharging the rod? Explain how or why not.

19. Find the greatest charge density σ_{max} that can exist on a conductor before dielectric breakdown of the air occurs.

20. The potential at a distance $r > R$ from a uniformly charged sphere of radius R is kQ/r, where Q is the total charge on the sphere. Find the radius of the equipotential surface of potential (*a*) 200 V, (*b*) 220 V, and (*c*) 240 V for a sphere of radius 20 cm carrying a charge of 6 nC.

Section 19-4   Capacitance

21. The potential difference across the plates of a parallel-plate capacitor is 500 V when the charge on the plates is $40\ \mu C$. What is the capacitance?

22. A capacitor has a capacitance of $15\ \mu F$. (*a*) If it is charged to 60 V, what is the charge on the plates? (*b*) If the charge on the plates is $24\ \mu C$, what is the potential difference between the plates?

23. If a parallel-plate capacitor has a separation of 0.1 mm, what must its area be to have a capacitance of 1 F? If the plates are square, what is the length of their sides?

24. A parallel-plate capacitor has square plates of side 15 cm and separation 2.0 mm. (*a*) Find the capacitance. (*b*) Find the capacitance when a dielectric of $K = 3$ is inserted between the plates.

25. A parallel-plate capacitor has capacitance of $2.0\ \mu F$ and plate separation of 1.5 mm. (*a*) How much potential difference can be placed across the plates before dielectric breakdown of the air occurs? (b) What is the magnitude of the greatest charge that the capacitor can store before break-down?

26. A capacitor of capacitance 3.0 μF is charged by connecting it to a 12-V battery. It is then removed from the battery. (a) What is the charge on the plates? (b) A dielectric of constant $K = 4$ is inserted between the plates. What is the potential difference between the plates? (c) What is the charge on the plates? (d) What is the capacitance?

27. The capacitor of Exercise 26 is connected to the 12-V battery with the dielectric of constant $K = 4$ already inserted. What is the charge on the plates?

Section 19-5   Combinations of Capacitors

28. A 10.0-μF capacitor is connected in parallel with a 20.0-μF capacitor across a 6.0-V battery. (a) What is the effective capacitance of this combination? (b) Find the charge on each capacitor. (c) Find the potential difference across each capacitor.

29. A 10.0-μF capacitor is connected in series with a 20.0-μF capacitor across a 6.0-V battery. (a) What is the effective capacitance of this combination? (b) Find the charge on each capacitor. (c) Find the potential difference across each capacitor.

30. Three capacitors have capacitances of 2.0, 4.0, and 8.0 μF. Find the effective capacitance (a) if the capacitors are in parallel and (b) if they are in series.

31. A 3.0- and a 6.0-μF capacitor are connected in series, and the combination is connected in parallel with an 8.0-μF capacitor. What is the effective capacitance of this combination?

32. A 1.0-μF capacitor is connected in parallel with a 2.0-μF capacitor, and the combination is connected in series with a 6.0-μF capacitor. What is the effective capacitance of this combination?

33. (a) How many 1.0-μF capacitors would have to be connected in parallel to store 1 mC of charge with a potential difference of 10 V across each? (b) What would be the potential difference across the combination? (c) If these capacitors are connected in series and the potential difference across each is 10 V, find the charge on each and the potential difference across the combination.

Section 19-6   Electrical Energy Storage

34. A 3-μF capacitor is charged to 100 V. (a) How much energy is stored in the capacitor? (b) How much additional energy is required to charge the capacitor from 100 to 200 V?

35. A 10-μF capacitor is charged to $Q = 4 \mu C$. (a) How much energy is stored? (b) If half the charge is removed, how much energy remains?

36. (a) Find the energy stored in a 35-μF capacitor charged to $Q = 6 \mu C$. (b) Find the additional energy required to increase the charge from 6 μC to 12 μC.

37. A capacitor is charged to 400 V. The charge on the plates is 45 μC. How much energy is stored?

38. A parallel-plate capacitor is charged to 200 V. The plate separation is 2.5 mm. (a) What is the electric field between the plates? (b) Find the electrostatic energy density in the space between the plates. (c) If the plates are squares of side 15 cm, find the total energy stored in the field between the plates. (d) Calculate the capacitance of this capacitor from $C = \epsilon_0 A/d$, and use your result to calculate the total energy stored from $U = \frac{1}{2}CV^2$.

39. Repeat Exercise 38 for a parallel-plate capacitor that has plate separation of 0.5 mm and is charged to 300 V. (Assume the plates are 15 cm on a side.)

Problems

1. Protons are released from rest at an electric potential of 5 MeV obtained from a Van de Graaff accelerator and travel through a vacuum to a region at zero potential. (a) If this change in potential occurs over a distance of 2.0 m, find the electric field, assuming it to be uniform. (b) Find the speed of the 5-MeV protons. (The proton mass is 1.67×10^{-27} kg.)

2. The potential due to a charge Q uniformly distributed on a spherical conductor of radius R is kQ/r for $r \geq R$. (a) What is the potential in the region $r \leq R$? (b) A spherical conductor has a radius $r = 10$ cm and carries charge of 0.2 nC. Draw to scale the equipotential surfaces in potential steps of 2 V,

beginning at $r = R$. (That is, if V_0 is the potential at the surface of the sphere, sketch the equipotential surfaces for $V_0 - 2$ V, $V_0 - 4$ V, $V_0 - 6$ V, and so on.)

3. The membrane of an axon of a neuron cell is a thin cylindrical shell of radius $r = 10^{-5}$ m, length $L = 0.1$ m, and thickness $d = 10^{-8}$ m. The membrane has a positive charge on one side of it and a negative charge on the other and acts as a parallel-plate capacitor of area $A = 2\pi r L$ and separation d. Its dielectric constant is about $K = 3$. (a) Find the capacitance of the membrane. If the potential difference across the membrane is 70 mV, find (b) the charge on each side of the membrane, and (c) the electric field between the surfaces.

4. For the arrangement shown in Figure 19-28 find (*a*) the total effective capacitance between the terminals, (*b*) the charge stored on each capacitor, and (*c*) the total stored energy.

Figure 19-28
Problem 4.

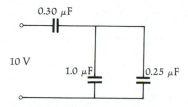

5. For the arrangement shown in Figure 19-29, find (*a*) the total effective capacitance between the terminals, (*b*) the charge stored on each capacitor, and (*c*) the total stored energy.

Figure 19-29
Problem 5.

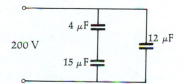

6. A conductor is in the shape of a spherical shell of inner radius R_1 and outer radius R_2. It carries a net charge Q. Prove that all of the charge must reside on the outer surface. (Do this by showing that if there is any charge q_1 on the inner surface, the electric field will not be zero inside the conducting material.)

7. Three identical capacitors are connected so that their maximum effective capacitance is $15\ \mu F$. Find the three other combinations possible using all three capacitors, and calculate the effective capacitance for each combination.

8. Find all the different possible effective capacitances that can be obtained using a 1.0-, a 2.0-, and a 4.0-μF capacitor in any combination that includes all three or any two capacitors.

9. A 20-pF capacitor is charged to 3.0 kV and is then removed and connected in parallel with an uncharged 50-pF capacitor. (*a*) What is the new charge on each capacitor? (*b*) Find the initial energy stored in the 20-pF capacitor and the final energy stored in the two capacitors. Is energy gained or lost in connecting the two capacitors?

10. Two capacitors of capacitance $C_1 = 4\ \mu F$ and $C_2 = 12\ \mu F$ are connected in series across a 12-V battery. They are then carefully disconnected without being discharged and connected together with positive side to positive side and negative side to negative side. (*a*) Find the charge on each capacitor after they are connected together and the potential difference across the combination. (*b*) Find the initial stored energy and the final stored energy.

11. Work Problem 10 for the same two capacitors initially connected in parallel across the 12-V battery and then disconnected and connected together in parallel with the positive side of one capacitor connected to the negative side of the other.

12. In Figure 19-30, $C_1 = 1\ \mu F$, $C_2 = 6\ \mu F$, and $C_3 = 3.5\ \mu F$. (*a*) Find the effective capacitance of this combination. (*b*) If the breakdown voltages of the individual capacitors are $V_1 = 100$ V, $V_2 = 50$ V, and $V_3 = 400$ V, what is the maximum voltage that can be placed across points *a* and *b*?

Figure 19-30
Problem 12.

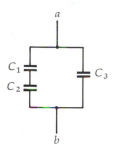

13. When uranium 235 captures a neutron, it splits into two nuclei, a process called nuclear fission. Assume that the two nuclei produced in the fission are equally charged with a charge of $+46e$ and that these nuclei are at rest just after the fission with a separation r equal to 1.3×10^{-14} m. (This is approximately equal to the diameter of such a nucleus.) (*a*) Using $U = kq_1q_2/r$ for the potential energy of these charges separated by r, calculate this energy in electronvolts. This is approximately the energy released per fission. (*b*) About how many fissions per second are needed to produce 1 megawatt of power in a reactor?

14. Radioactive polonium emits alpha particles with an energy of 5.30 MeV. Assume that just after the alpha particle is formed and escapes from the nucleus, the alpha particle with charge $+2e$ is a distance R from the center of the remaining nucleus, which has charge $+82e$. Calculate R by setting the electric potential energy of the two charges ($2e$ and $82e$) separated by R equal to 5.30 MeV. (This type of calculation provided one of the early estimates of the nuclear radius.)

CHAPTER TWENTY

Electric Current and Circuits

Franklin's electrical experiments had few practical applications, for they were limited to static electricity. In his day, there were no means for producing a steady electric current. Then in 1791, the year after Franklin's death, the biologist Luigi Galvani announced a discovery he had made while investigating the effects of atmospheric electricity on the muscular response of frogs. When a frog's tissues were in simultaneous contact with two different metals, muscular contractions were induced. This showed, Galvani claimed, that animal tissues generate electricity, which the metals then served to discharge. His contemporary Alessandro Volta disagreed. Volta held that the contractions were caused by ordinary physical electricity generated outside the animal by its contact with the two different metals. A dispute lasting several years resulted, with each scientist endeavoring to prove his view correct through further studies. Volta ultimately won out. More significantly, his work led him to the development of the first battery in 1800. With this source of electric current, modern electrical science was born.

Stretches, for leagues and leagues,
the Wire
A hidden path for a Child of Fire —
Over its silent spaces sent,
Swifter than Ariel ever went
Over continent to continent.

W. H. BURLEIGH, *The Rhyme of the Cable*

When we turn on a light, we connect the wire filament in the light bulb across a potential difference that drives electric charge through the wire, much like the pressure difference in a garden hose drives water through the hose. The rate of flow of electric charge past some point is called electric current. We usually think of electric currents in conducting wires, but any flow of charge constitutes a current. An example of a current that is not in a conducting wire is a beam of charged ions in a vacuum from an accelerator.

In this chapter, we will define electric current and relate it to the motion of charged particles. After a discussion of electrical resistance and Ohm's law, we will consider the energy aspects of electric currents. We will then discuss some rules for solving direct-current circuit problems involving batteries, resistors, and capacitors in various combinations. These circuits are called direct-current (dc) circuits because the direction of the current in any one part of the circuit does not vary. Circuits in which the current alternates in direction (ac circuits) will be discussed in Chapter 23.

20-1   Current and Motion of Charges

Electric current is defined as the rate of flow of electric charge. Figure 20-1 shows a segment of a current-carrying wire with charge carriers moving with some small average velocity. The current in the wire at point P is defined as the amount of charge flowing past point P per unit of time. If ΔQ is the charge that flows in time Δt, the current is

$$I = \frac{\Delta Q}{\Delta t} \qquad\qquad 20\text{-}1$$

Electric current defined

The SI unit of current, the coulomb per second, is called an ampere (A):

$$1\ \text{A} = 1\ \text{C/s} \qquad\qquad 20\text{-}2$$

The direction of electric current is taken to be the direction of flow of positive charge. In conducting wires, the current is due to the motion of electrons, which have negative charge. The motion of electrons is in the direction opposite to that of the current. A negative charge moving to the left is equivalent to a transport of positive charge to the right. This definition of the direction of the current is purely arbitrary. In a conducting wire, it is electrons that are free to move and that produce the flow of charge. By our definition, the current is in the direction opposite to the motion of the electrons. In an accelerator that produces a proton beam, the direction of motion of the positively charged protons is in the direction of the current. In electrolysis, the use of electrical energy to decompose a compound in solution, the current is produced by the motion of both the negative electrons and the positive ions resulting from the decomposition. In an applied electric field, these particles experience forces in opposite directions and therefore move in opposite directions. The movement of negative electrons in one direction and positive ions in the opposite direction both contribute to a current in the same direction. In nearly all applications, the motion of negative charges to the left is indistinguishable from the motion of positive ions to the right. We can always think of current as the motion of positive charges in the direction of the current and remember (if we need to) that in conducting wires, for example, the electrons are moving in the direction opposite to the current.

Electric current is related to the motion of the charged particles responsible for it. The motion of the free electrons in a conducting wire is quite complicated. When there is no electric field in the wire, these electrons move in random directions with relatively large speeds due to thermal excitation. Since the velocity vectors of the electrons are randomly oriented, the average velocity due to this thermal excitation is zero. When an electric field is applied, for example, by connecting the wire to a battery, which applies a potential difference across the wire, the free electrons experience a momentary acceleration due to the force $-e\mathbf{E}$, but they quickly collide with the fixed ions in the wire. The result is that the electrons have a small *drift velocity* opposite to the electric field superimposed on their large random thermal velocities. The behavior of the free electrons in a metal is somewhat similar to that of gas molecules in air. In still air, the gas molecules move with large instantaneous velocities between collisions, but the average velocity is zero. When there is a breeze, the air molecules have a small drift velocity in the

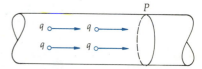

Figure 20-1
Segment of current-carrying wire. If the amount of charge that passes point P in time Δt is ΔQ, the current is $I = \Delta Q / \Delta t$.

direction of the breeze superimposed on their much larger instantaneous velocities. Similarly, when there is no current in a conductor, the electrons move about in random directions with very high speeds because of the thermal energy. When there is a current, the electrons have a small drift velocity (opposite to the direction of the current because of their negative charge) superimposed on their much larger instantaneous velocities (see Figures 20-2 and 20-3).

Figure 20-2
Motion of an electron in a wire. Superimposed on the random thermal motion is a slow drift velocity v_d in the direction of the force $q\mathbf{E}$. During a time Δt, the electron drifts a distance $v_d\,\Delta t$ while making many collisions with the fixed ions in the metal. In this figure, the drift velocity is greatly exaggerated.

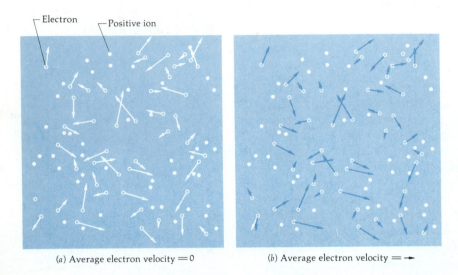

(a) Average electron velocity $=0$ (b) Average electron velocity $=$ →

Figure 20-3
(a) Stationary positive ions, and electrons with a random distribution of velocities. (b) A very small drift velocity has been added to the velocity of each electron as indicated for the electron in the lower left corner.

A calculation for a typical current in a copper wire shows that the drift velocity of the electrons is of the order of 0.1 mm/s. This is indeed a small velocity. The transport of a significant amount of charge in a wire is accomplished not by a few charges moving rapidly down the wire but by a very large number of charges slowly drifting down the wire.

20-2   Ohm's Law and Resistance

In our study of conductors in Chapter 19, we argued that the electric field inside a conductor must be zero in electrostatic equilibrium. If this were not so, the free charge inside the conductor would move about. We are now considering situations in which the free charge *does* move inside a conductor. That is, the conductor is *not* in electrostatic equilibrium. When there is a current in a conductor, there is an electric field in the conductor that produces the current. Since the force on a positive charge is in the direction of $\mathbf{E}$ and the current is in the direction of flow of positive charge, the current is in the direction of the electric field. Figure 20-4 shows a segment of a wire of length L and cross-sectional area A carrying a current I. Since there is an electric field in the wire, the potential at point a is greater than that at point b. The potential difference between points a and b is

$$V = V_a - V_b = EL \qquad\qquad 20\text{-}3$$

where E is the electric field.*

For most conducting materials, such as wire conductors, the current is proportional to the potential difference. This experimental result is known as *Ohm's law*. The constant of proportionality is written $1/R$ where R is called the *resistance*.

$$I = \frac{V}{R} \qquad\qquad 20\text{-}4$$

or

$$V = IR \qquad R \text{ constant} \qquad\qquad 20\text{-}5 \quad \text{Ohm's law}$$

The SI unit of resistance, the volt per ampere, is called an ohm (Ω).

$$1\ \Omega = 1\ \frac{V}{A} \qquad\qquad 20\text{-}6$$

The resistance of a conducting wire depends on the length of the wire, its cross-sectional area, the type of material, and the temperature, but for materials that obey Ohm's law, it does not depend on the current I. Such materials, which include most metals, are called *ohmic* materials. For non-ohmic materials, the current is not proportional to the potential difference.

The resistance between two points in any material is defined by

$$R = \frac{V}{I} \qquad\qquad 20\text{-}7 \quad \text{Resistance defined}$$

where V is the potential difference between the points. Figure 20-5 shows the potential difference V versus the current I for ohmic and nonohmic materials. For ohmic materials (the top curve), the relation is linear, so $R = V/I$ is the same at any point; for nonohmic materials (the bottom curve), $R = V/I$ is not constant but depends on the current I. Ohm's law is therefore not a fundamental law of nature like Newton's laws or the laws of thermodynamics but an empirical description of a property shared by many materials.

---

* Again we use V for the potential difference (which in this case is a potential decrease) to simplify the notation.

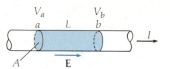

Figure 20-4
Segment of wire carrying current I. The potential difference is related to the electric field by $V_a - V_b = EL$.

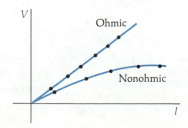

Figure 20-5
Plots of V versus I for ohmic and nonohmic materials. The resistance $R = V/I$ is independent of I for ohmic materials, as indicated by the constant slope of the line.

The resistance of a conducting wire is found to be proportional to the length of the wire and inversely proportional to its cross-sectional area:

$$R = \rho \frac{L}{A} \qquad\qquad 20\text{-}8$$

Resistivity defined

where ρ is a proportionality constant called the *resistivity* of the conducting material. The SI unit of resistivity is the ohm-metre ($\Omega \cdot m$).

Example 20-1   A nichrome wire ($\rho = 10^{-6}\ \Omega \cdot m$) has a radius of 0.65 mm. What length of wire is needed to obtain a resistance of 2.0 Ω?

The cross-sectional area of this wire is

$$A = \pi r^2 = (3.14)(6.5 \times 10^{-4}\ m)^2 = 1.33 \times 10^{-6}\ m^2$$

From Equation 20-8, we have

$$L = \frac{RA}{\rho} = \frac{(2\ \Omega)(1.33 \times 10^{-6}\ m^2)}{10^{-6}\ \Omega \cdot m} = 2.66\ m$$

We sometimes refer to a wire as a conductor, and at other times we call it a resistor, depending on which property we wish to emphasize. The reciprocal of the resistivity is called the *conductivity* σ*:

$$\sigma = \frac{1}{\rho} \qquad\qquad 20\text{-}9$$

Conductivity and resistivity

In terms of the conductivity, the resistance of a conducting wire is

$$R = \frac{L}{\sigma A} \qquad\qquad 20\text{-}10$$

Note that Equations 20-5 and 20-10 are the same as Equations 13-13 ($\Delta T = IR$) and 13-12 ($R = \Delta x/kA$) for thermal conduction and resistance, except that the temperature difference ΔT for thermal conduction is replaced by the potential difference V and the thermal conductivity k is replaced by the electrical conductivity σ. Ohm was, in fact, led to his law by the similarity between the conduction of electricity and the conduction of heat.

The resistivity (and conductivity) of any given metal depends on the temperature. Figure 20-6 shows the temperature dependence of the resistivity of copper. This graph is nearly a straight line. Except at very low temperatures, the resistivity varies nearly linearly with temperature. The resistivity is usually given in tables in terms of its value at 20°C, ρ_{20}, and the temperature coefficient of resistivity α, which is the slope of the ρ-versus-T curve. The resistivity at some other Celsius temperature t is then given by

$$\rho = \rho_{20}[1 + \alpha(t - 20\ C°)] \qquad\qquad 20\text{-}11$$

---

* The symbol σ used here for conductivity was used for surface charge density in the previous chapter. Care must be taken to distinguish between these two quantities. Usually, it is clear from the context whether σ means surface charge density or conductivity.

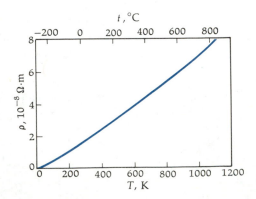

Figure 20-6
Plot of resistivity ρ versus temperature for copper.

(Since the Celsius and absolute temperature scales differ only in the choice of zero, the resistivity has the same slope whether plotted against t or T.) Table 20-1 gives the resistivity at 20°C and the temperature coefficient α for various materials. Note the tremendous difference between the values of ρ for conductors (the metals listed) and those for nonconductors.

Table 20-1
Resistivities and Temperature Coefficients

Material	Resistivity ρ at 20°C, $\Omega \cdot m$	Temperature coefficient α at 20°C, K^{-1}
Silver	1.6×10^{-8}	3.8×10^{-3}
Copper	1.7×10^{-8}	3.9×10^{-3}
Aluminum	2.8×10^{-8}	3.9×10^{-3}
Tungsten	5.5×10^{-8}	4.5×10^{-3}
Iron	10×10^{-8}	5.0×10^{-3}
Lead	22×10^{-8}	4.3×10^{-3}
Mercury	96×10^{-8}	0.9×10^{-3}
Nichrome	100×10^{-8}	0.4×10^{-3}
Carbon	3500×10^{-8}	-0.5×10^{-3}
Germanium	0.45	-4.8×10^{-2}
Silicon	640	-7.5×10^{-2}
Wood	$10^{8}-10^{14}$	
Glass	$10^{10}-10^{14}$	
Hard rubber	$10^{13}-10^{16}$	
Amber	5×10^{14}	
Sulfur	1×10^{15}	

Wires used to carry electric current are manufactured in standard sizes. The diameter of the circular cross section is indicated by a "gauge number;" higher numbers correspond to smaller diameters. For example, the diameter of a 10-gauge copper wire is 2.588 mm and that of a 14-gauge wire is 1.628 mm. Handbooks give the combination ρ/A in ohms per centimetre or ohms per foot for various wires.

Example 20-2   Calculate the resistance per unit length in ohms per metre for 14-gauge copper wire, which has a diameter $d = 1.63$ mm.

From Table 20-1, the resistivity of copper is

$$\rho = 1.7 \times 10^{-8} \ \Omega \cdot m$$

The cross-sectional area of 14-gauge wire is

$$A = \frac{\pi d^2}{4}$$

$$= \frac{\pi}{4}(0.00163 \text{ m})^2 = 2.09 \times 10^{-6} \text{ m}^2$$

Thus,

$$\frac{R}{L} = \frac{\rho}{A} = \frac{1.7 \times 10^{-8} \text{ }\Omega\cdot\text{m}}{2.09 \times 10^{-6} \text{ m}^2}$$

$$= 8.13 \times 10^{-3} \text{ }\Omega/\text{m}$$

This example shows that the copper connecting wires used in the laboratory have a very small resistance.

Georg Simon Ohm (1781–1854).

Example 20-3   What is the electric field in a 14-gauge copper wire that carries a current of 1 A?

According to Example 20-2, the resistance of a 1-m length of 14-gauge copper wire is $8.13 \times 10^{-3} \text{ }\Omega$. From Ohm's law, the voltage drop across 1 m of this wire is

$$V = IR = (1 \text{ A})(8.13 \times 10^{-3} \text{ }\Omega) = 8.13 \times 10^{-3} \text{ V}$$

and the electric field is

$$E = \frac{V}{L} = \frac{8.13 \times 10^{-3} \text{ V}}{1 \text{ m}} = 8.13 \times 10^{-3} \text{ V/m}$$

We note that the electric field in a conducting wire is very small.

Resistors for use in the laboratory are often made by winding a fine wire around an insulating tube to get a long wire in a short space. Carbon, which has a relatively high resistivity, is often used for resistors in electronic equipment. Such resistors are usually painted with colored stripes that provide a code for the value of the resistance.

Example 20-4   By what percentage does the resistance of a copper wire increase when the temperature increases from 20°C to 30°C?

From Equation 20-11, the fractional change in the resistivity is

$$\frac{\rho - \rho_{20}}{\rho_{20}} = \alpha(t - 20 \text{ C}^\circ)$$

The percentage is 100 percent times this fractional change. Using $\alpha = 3.9 \times 10^{-3} \text{ K}^{-1}$ from Table 20-1 and $(t - 20 \text{ C}^\circ) = 10°C = 10 \text{ K}$, we obtain

$$\frac{\rho - \rho_{20}}{\rho_{20}} = \alpha(t - 20 \text{ C}^\circ) = (3.9 \times 10^{-3} \text{ K}^{-1})(10 \text{ K}) = 3.9 \times 10^{-2}$$

The percentage change is thus 3.9 percent.

There are many metals for which the resistivity is zero below a certain temperature T_c, called the *critical temperature*. This phenomenon, called *superconductivity,* was discovered in 1911 by the Dutch physicist H. Kamerlingh Onnes. Figure 20-7 shows his plot of the resistance of mercury versus temperature. The critical temperature for mercury is 4.2 K. Critical tempera-

Figure 20-7
Plot by Kamerlingh Onnes of the resistance of mercury versus temperature showing a sudden decrease at the critical temperature $T_c = 4.2$ K.

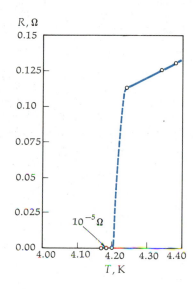

tures for other superconductors range from less than 0.1 K for hafnium and iridium to 9.2 K for niobium. Many alloys are also superconductors. The highest critical temperature yet found (23 K) is for the alloy Nb_3Ge. The search continues for higher temperature superconductors.

The conductivity of a superconductor cannot be defined since there can be a current even when the electric field in a superconductor is zero. Steady currents have been observed to persist with no apparent loss in superconducting rings for years despite the absence of an electric field. The phenomenon of superconductivity cannot be understood in terms of classical physics. Instead, quantum mechanics, developed in the twentieth century, is needed. We will discuss some of the ideas of quantum mechanics in the latter chapters of this book. The first successful theory of superconductivity was published by John Bardeen, Leon Cooper, and J. Robert Schrieffer in 1957 and is known as the BCS theory. These physicists were awarded the Nobel Prize in Physics (1972) for their accomplishment.

Questions

1. Wire a and wire b have the same electric resistance and are made of the same material. Wire a has twice the diameter of wire b. How do the lengths of the wires compare?

2. In our study of electrostatics, we concluded that there is no electric field within the material of a conductor. Why do we now find it possible to discuss electric fields inside conducting materials?

Demonstration of persistent currents. Oppositely directed superconducting current loops are set up in the ring and ball so that the magnetic force between the currents is repulsive. The ball floats above the ring, its weight balanced by the magnetic force of repulsion.

Electrical Conduction in Nerve Cells

Stephen C. Woods
University of Washington

An important example of many of the principles of current flow is the conduction of nerve impulses. To understand this phenomenon, we must first look at some of the special properties of the cell membrane. Every living cell creates and maintains an internal environment that is ideal for its functioning by means of the cell membrane. This membrane is an effective barrier that isolates the cell's interior from its external environment. It allows only certain needed materials to pass freely into and out of the cell. In particular, the cell membrane is relatively impermeable to the common ions found in the body, such as sodium (Na^+), chloride (Cl^-), calcium (Ca^{2+}), and potassium (K^+). Penetration by these ions normally occurs only when specific channels, or pores, in the membrane are opened, enabling those ions near the pores to pass into or out of the cell. Additionally, the cell membrane is able to transport, or "pump," specific ions through the membrane in one or the other direction. For example, Na^+ ions are continually pumped out of the cell and K^+ ions are continually pumped into the cell. As a result, the fluid inside the cell has a relatively high concentration of K^+ ions and a relatively low concentration of Na^+ ions, whereas the reverse is true outside the cell. The "pumps" in the cell membrane are able to counter effectively the slight leakage of ions that normally occurs and thus to maintain a relatively constant internal cellular environment.

A nerve cell, or *neuron*, has a receptive area, which includes the cell body and a variable number of nerve fibers, the *dendrites*, that conduct nerve impulses to the cell body. It also has a transmitting area, the *axon*, a nerve fiber that conducts nerve impulses away from the cell body (see Figure 1). The receptive area is where stimulatory chemicals released from other neurons *(neurotransmitters)* or those found in the blood and fluid outside the cells *(hormones)* cause changes in the permeability of the cell membrane to ions. These changes

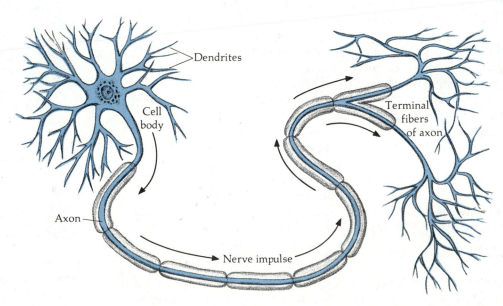

Figure 1

A simplified diagram of a neuron. The neuron receives information at the dendrites and/or cell body. When a nerve impulse is generated, it is conducted away from the cell body along the axon.

in permeability create a transient current flow through the membrane at that location. In simple terms, when a nerve cell membrane is stimulated by a neurotransmitter or hormone, it becomes momentarily leaky to ions. Each type of ion then undergoes a net flow in one or the other direction through the membrane due to the concentration difference, or gradient, that exists across the membrane. Because of this movement of ions, the stimulation of the cell membrane is associated with a current flow that is proportional to the number of ions moving through the membrane. When the stimulation is over, the pumps in the membrane rapidly restore the normal concentration gradients.

The membrane pumps, the normal relative permeability of the cell membrane to sodium and potassium ions, and the resultant concentration gradients all cause the establishment and maintenance of a small electric charge on each side of the membrane and thus a potential difference across the membrane. The cell membrane maintains slightly more positive and fewer negative ions on its outer surface than on its inner surface such that there is a slight negative charge on the inside of the cell relative to the outside (see Figure 2a). For most cells, the magnitude of this charge is from 50 to 90 millivolts. A cell membrane is therefore said to be polarized and is analogous to a small leaky battery with the negative terminal on the inside of the cell and the positive terminal on the outside of the cell.

Neurons, muscle cells, and a few other types of cells have electrically excitable cell membranes. When such cells are appropriately stimulated, Na^+ and K^+ (and other) ions flow through the cell membrane and there is a reduction of the normal potential (called the *resting potential*) across the membrane; that is, the membrane becomes depolarized at that location. Although this happens to some extent in all cells, electrically excitable cells respond to stimulation in a unique manner. Once the cell membrane is depolarized to a certain level (called the *threshold*), some of the specialized channels through the membrane are altered in a particular way. In neurons, there is a rapid increase in sodium perme-

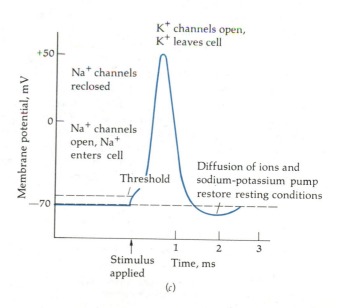

Figure 2
The development of an action potential. (a) In its resting state, the nerve cell membrane is polarized. (b) During an action potential, changes in the permeability of a portion of the membrane to certain ions result in a momentary reversal of polarity at that site. (c) A graph of the voltage change versus time for a site on a nerve cell membrane undergoing an action potential.

Essay: Electrical Conduction in Nerve Cells (continued)

ability, which allows Na$^+$ ions to penetrate the membrane more readily than usual. The result is a large influx of Na$^+$ ions due both to the concentration gradient (more Na$^+$ ions are normally outside the cell than inside it) and to the electrical gradient (the negative charge inside the cell attracts the positive sodium ions). This influx of Na$^+$ ions causes the inside of the cell to become less and less negatively charged. As more and more Na$^+$ ions rush in, the polarity on the inside of the cell actually becomes positive at the site of the stimulation (see Figure 2b). As the polarity across the cell membrane reverses, the Na$^+$ ion channels are reclosed, preventing further passage of Na$^+$ ions. In the meantime, channels in the membrane for K$^+$ ions have also opened, and K$^+$ ions flow out of the cell due both to the concentration gradient and to the electrical gradient as the inside of the cell becomes more positive. The K$^+$ ion channels are then also reclosed, and the pumps in the membrane soon restore the resting conditions at that location. Figure 2c depicts the change in voltage that results from these events. The entire process lasts only a few milliseconds. It is called a "spike" potential because of the shape of the graph or, more commonly, an *action potential.*

At the peak of the spike of an action potential, an unstable situation is created inside of the cell because a positively charged area is adjacent to areas that are negatively charged (see Figure 3) and a conducting medium (the ionic fluid bathing the cell contents) interconnects these areas. Likewise, on the outside of the cell a negatively charged area is adjacent to areas that are positively charged. As a result, current flows along the membrane away from the action potential. On the inside of the membrane, this causes a reduction of the positive charge at the site of the action potential and a concomitant reduction of the negative charge at the neighboring sites

(that is, the depolarization of the neighboring sections of the membrane).

In electrically excitable cells, the height of an action potential (that is, the change in voltage achieved during the spike) is sufficient to depolarize neighboring sites to the threshold. An action potential at any particular site on a cell membrane will therefore spread in all directions away from the point of origin. It cannot reverse itself and depolarize the original site because the membrane is refractory for a short interval after an action potential has occurred and cannot be restimulated. Neurons are normally stimulated by neurotransmitters at discrete points on the dendrites or the cell body. Action potentials therefore typically move along the axon, away from the cell body. Because neurons are able to restore their resting conditions rapidly after an action potential has occurred and because the refractory period is very short, it is possible for a second action potential to be generated soon after the first has occurred. Some neurons are capable of transmitting several hundred action potentials per second along their axons.

Axons are cylindrical and have very small diameters. An action potential travels along an axon as each site on the membrane depolarizes the next to the threshold. The overall movement along the axon is called a *nerve impulse.* The speed of a nerve impulse increases with the diameter of the axon and with the temperature. At the end of the axon, there are one or more terminals (sometimes there are hundreds) where the action potential causes the release of stored neurotransmitters, which then activate a neighboring cell or cells across a junction called a *synapse.*

Vertebrate animals have evolved a type of nerve cell insulation, called *Schwann cells,* that increases the speed and effi-

Direction of action potential

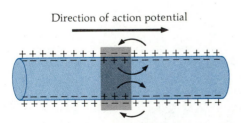

Figure 3

Conduction of a nerve impulse. The reversal of polarity of a nerve cell membrane at the site of an action potential leads to current flow along the membrane away from the original site of that action potential.

ciency with which action potentials are conducted. Schwann cells are very thin sheets that wrap tightly around axons in a series of concentric layers (see Figure 4a). They contain myelin, a fatty material that is resistant to electric current. Thus, they are analogous to the insulation used to cover electric wires. Because the Schwann cells are wrapped so closely around the axons, electric current cannot pass through the axon membrane except at the gaps between adjacent Schwann cells where no myelin exists. These gaps are called the *nodes of Ranvier*. Local currents created by action potentials moving along these axons therefore jump the distance between adjacent Schwann cells. This type of conduction is

therefore called *saltatory conduction* (see Figure 4b). Because it takes essentially the same amount of time for an action potential to jump from one node of Ranvier to the next as it does for it to move from one site to an adjacent site along a nonmyelinated axon, nerve conduction along a myelinated axon is much more rapid. Such conduction reaches speeds of up to 100 metres per second in large-diameter myelinated axons.

Invertebrates do not have myelinated nerve cells. The rapid conduction of nerve impulses can therefore be achieved only through a considerable increase in axon diameter. In neuronal circuits critical for rapid escape response to

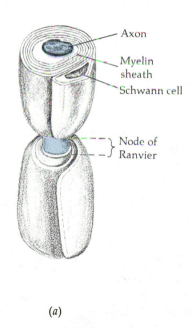

(a)

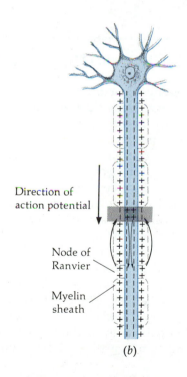

(b)

Figure 4

(a) A diagram of a myelinated neuron. As the Schwann cell grows, it wraps itself around the axon. The Schwann cell acts as an insulator because it contains the substance myelin which is resistant to electric current. Gaps, called nodes of Ranvier, occur between adjacent Schwann cells.

(b) In myelinated neurons, local currents created by action potentials jump from one gap between Schwann cells to the next in a rapid form of nerve impulse conduction called saltatory conduction.

Essay: Electrical Conduction in Nerve Cells (continued)

danger, some invertebrates have axons so large that they can be seen without the aid of a microscope. Most of the pioneering work in neurophysiology was done on these giant axons in such animals as the squid. The presence of myelin, with the associated decrease in required axon diameter, allows far more axons to be packed into a small space. For example, each human optic nerve contains over a million individual axons for carrying nerve impulses between the eye and the brain. The analogous nerve in insect eyes may have fewer than one hundred axons.

Problems with the formation or functioning of myelin have serious consequences. Muscular dystrophy is the disease caused by the loss of myelin in the motor nerves that pass to the muscles. Similarly, the nerve conduction problems associated with diabetes mellitus are thought to be due to abnormal metabolism in myelin-producing Schwann cells. Numerous human behavioral disorders are thought to be due to a deficiency or an excess of one or more neurotransmitters, including Parkinson's disease, depression, and schizophrenia.

20-3   Energy in Electric Circuits

When there is electric current in a conductor, electrical energy is continually converted into thermal energy in the conductor. For example, in a simple model of conduction, the electric field in the conductor accelerates the free electrons for a short time, giving them an increased kinetic energy; however, this extra energy is quickly transferred into thermal energy of the conductor by collisions between the electrons and the lattice ions of the conductor. Thus, though the electrons continually gain energy from the electric field, this energy is immediately transferred into thermal energy of the conductor and the electrons maintain a steady drift velocity.

When positive charge flows in a conductor, it flows from high potential to low potential in the direction of the electric field. (The negatively charged electrons, of course, move in the opposite direction.) The charge thus loses potential energy. The potential-energy loss does not appear as kinetic energy of the charge carriers, except momentarily, before it is transferred to the conducting material by collisions with the fixed ions. The potential-energy loss goes into thermal energy of the conductor. Consider the charge ΔQ, which passes point P_1 in Figure 20-8 during a time Δt. If the potential at that

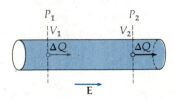

Figure 20-8
At point P_1 the charge ΔQ has potential energy $\Delta Q\, V_1$, and at P_2 it has potential energy $\Delta Q\, V_2$. The energy loss by the charge as it moves from one point to the other is $\Delta Q\, V$, where $V = V_1 - V_2$ is the potential drop. The rate of energy loss is the power loss IV.

point is V_1, the charge has potential energy $\Delta Q\, V_1$. During the time interval, the same amount of charge passes point P_2, where the potential is V_2. It has potential energy $\Delta Q\, V_2$, which is less than the original potential energy. The change in the potential energy of the charge is

$$\Delta U = \Delta Q\, (V_2 - V_1) = \Delta Q\, (-V)$$

where $V = V_1 - V_2$ is the potential *decrease* from point P_1 to point P_2. The energy lost by the charge passing through this segment of the conductor is then

$$-\Delta U = \Delta Q\, V$$

The rate of energy loss by the charge is

$$-\frac{\Delta U}{\Delta t} = \frac{\Delta Q}{\Delta t} V = IV$$

where $I = \Delta Q / \Delta t$ is the current. The energy loss per unit of time is the power put into the conductor, which we will call P:

$$P = IV \qquad\qquad\qquad\qquad\qquad 20\text{-}12 \quad \text{Power}$$

If I is in amperes and V in volts, the power is in watts.

This expression for electrical power can easily be remembered by recalling the definitions of V and I. The voltage drop is the potential-energy decrease per unit charge, and the current is the charge flowing per unit time. The product IV is thus the energy loss per unit time or the power put into the conductor. As we have seen, this power goes into heating the conductor.

Using the definition of resistance, $R = V/I$, we can write Equation 20-12 in other useful forms by eliminating either V or I:

$$P = (IR)I = I^2 R \qquad\qquad\qquad\qquad 20\text{-}13$$

or

$$P = \left(\frac{V}{R}\right)V = \frac{V^2}{R} \qquad\qquad\qquad 20\text{-}14$$

Equations 20-12, 20-13, and 20-14 all contain the same information. The choice of which to use depends on the particular problem. The energy put into a conductor is called *Joule heat*.

Example 20-5   A 12-Ω resistor carries a current of 3 A. Find the power dissipated in this resistor.

Since we are given the current and the resistance but not the potential drop, Equation 20-13 is the most convenient to use:

$$P = I^2 R = (3\ \text{A})^2 (12\ \Omega) = 108\ \text{W}$$

We could, of course, have found the potential drop across the resistor from $V = IR = (3\ \text{A})(12\ \Omega) = 36\ \text{V}$ and then used Equation 20-12 to find the power

$$P = IV = (3\ \text{A})(36\ \text{V}) = 108\ \text{W}$$

The first practical method of maintaining a steady electric current was the battery, invented by Alessandro Volta in 1800. The earliest battery was a series of electric cells containing a copper disk and a zinc disk separated by a moistened pasteboard. Volta is shown here demonstrating his electric cell to Napoleon.

To have a steady current in a conductor, we need to have a supply of electrical energy. A device that supplies electrical energy is called a *source of electromotive force* or simply a *source of emf.* It converts chemical, mechanical, or other forms of energy into electrical energy. Examples are a battery, which converts chemical energy into electrical energy, and a generator, which converts mechanical energy into electrical energy. A source of emf does work on the charge passing through it, raising the potential energy of the charge. The work per unit charge is called the *emf* $\mathcal{E}$ of the source. When a charge ΔQ flows through a source of emf, its potential energy is increased by the amount $\Delta Q\,\mathcal{E}$. The SI unit of emf is the volt, the same as the unit of potential difference. An example of a source of emf is an ideal battery, denoted by the symbol ⊣⊢. (We will discuss real batteries shortly.) The longer vertical line indicates the higher potential side. Ideally, a source of emf maintains a constant potential difference equal in magnitude to the emf $\mathcal{E}$ between its two sides independent of the rate of flow of charge, that is, independent of the current in the circuit.

Figure 20-9 shows a simple circuit consisting of a resistance R connected to an ideal battery. The resistance is indicated by the symbol —ᴧᴧᴧ—. The straight lines in the circuit diagram indicate connecting wires of negligible resistance. Since we are interested only in the potential differences between various points in the circuit, we can choose any point we wish to have zero potential. The source of emf maintains a constant potential difference equal to $\mathcal{E}$ between points a and b, with point a having the higher potential. There

emf

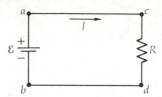

Figure 20-9
Simple circuit with an ideal source of emf, a resistance R, and a connecting wire that is assumed to be resistanceless.

is no potential difference between points a and c or between points d and b because the connecting wire is assumed to have negligible resistance. The potential difference between points c and d is therefore also equal to $\mathcal{E}$, and the current in the resistor is given by $I = \mathcal{E}/R$. The direction of the current in this circuit is clockwise, as indicated in the figure. Note that *in the source of emf, the charge flows from low potential to high potential.*[*]

When charge ΔQ flows through the source of emf, its potential energy is increased by the amount $\Delta Q\,\mathcal{E}$. It then flows through the conductor, where it loses this potential energy to heat. The rate at which energy is supplied by the source of emf is

$$P = \frac{\Delta W}{\Delta t} = \frac{\Delta Q\,\mathcal{E}}{\Delta t} = \mathcal{E}I \qquad\qquad 20\text{-}15$$

In this simple circuit, the power input by the source of emf equals that dissipated in the resistor. A source of emf can be thought of as a sort of charge pump. It pumps the charge from low to high electrical potential energy, like a water pump pumps water from low to high gravitational potential energy.

---

[*] As we will see later in this chapter, there are some situations, such as charging one battery with another, where the charge can flow in the opposite direction through a source of emf.

An unusual source of emf is the electric eel *(Electrophorus electricus)*, which lives in the Amazon and Orinoco rivers of South America. When attacking, the eel can develop up to 600 V between its head (positive) and its tail (negative).

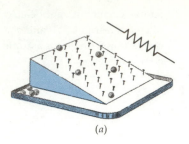

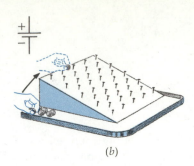

(a)   (b)

Figure 20-10 shows a mechanical analog to the simple electric circuit just discussed. Marbles of mass m roll along an inclined board with many nails in it. The marbles start at some height h above the bottom and are accelerated by the gravitational field between collisions with the nails. They transfer the kinetic energy they obtain between collisions to the nails during collisions. Because of the many collisions, the marbles move with a small drift velocity toward the ground. When they reach the bottom, a boy picks them up, lifts them to their original height h, and starts them again. The boy is the analog of the source of emf. He does work mgh on each marble. The work per mass is gh, which is analogous to the work per charge done by the source of emf. The energy source in this case is the internal chemical energy of the boy.

A real battery is more than merely a source of emf. The potential difference across the battery terminals, called the *terminal voltage*, is not simply equal to the emf of the battery. Consider the simple circuit of a battery and resistor. Figure 20-11 shows a typical dependence of the terminal voltage of the battery on the current in the battery. The terminal voltage decreases slightly as the current increases. As we have seen, an ideal source of emf maintains a constant potential difference between its sides, independent of the current. Because the decrease in terminal voltage is approximately linear, we can represent a real battery by a source of emf plus a small resistance r, called the *internal resistance* of the battery.

Figure 20-12 shows a simple circuit containing a battery, a resistor, and connecting wires. As before, we will ignore any resistance in the connecting wires. The circuit diagram for this circuit is shown in Figure 20-13. As charge

Figure 20-10
Mechanical analogy for resistance and emf. (a) As the marbles roll down the incline, their potential energy is converted into kinetic energy, which is quickly converted into heat because of collisions with the nails in the board. (b) A boy lifts the marbles up from low potential energy to high potential energy, converting his internal chemical energy into the potential energy of the marbles.

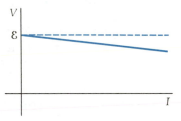

Figure 20-11
Terminal voltage V versus I for a real battery. The dashed line shows the open-circuit terminal voltage, which has the same magnitude as $\mathcal{E}$.

Figure 20-12
A simple circuit consisting of a battery, a resistor, and connecting wires.

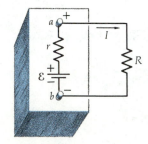

Figure 20-13
Circuit diagram representing the circuit with the battery and resistor in Figure 20-12. The battery can be represented by an ideal source of emf $\mathcal{E}$ and a small resistance r.

passes from point b to point a, its potential energy is first increased as it passes through the source of emf and is then decreased slightly as it passes through the internal resistance of the battery. (In the actual battery, these energy changes take place concurrently.) If the current in the circuit is I, the potential at point a is related to that at point b by

$$V_a = V_b + \mathcal{E} - Ir$$

The terminal voltage is thus

$$V_a - V_b = \mathcal{E} - Ir \qquad\qquad\qquad\qquad\qquad 20\text{-}16$$

The terminal voltage of the battery decreases linearly with current, as indicated in Figure 20-11. The potential drop across the resistor R is IR and is equal to the the terminal voltage:

$$IR = V_a - V_b = \mathcal{E} - Ir$$

Solving for the current I, we obtain

$$IR + Ir = \mathcal{E}$$

or

$$I = \frac{\mathcal{E}}{R + r} \qquad\qquad\qquad\qquad\qquad 20\text{-}17$$

Questions

3. What are several common kinds of sources of emf? What sort of energy is converted into electric energy in each?

4. In a simple electric circuit like that shown in Figure 20-13, the charge flows from positive voltage toward negative voltage outside the source of emf, but inside the source of emf the charge flows from negative to positive voltage. Explain how this is possible.

5. Figure 20-10 illustrates a mechanical analog to the simple electric circuit. Devise another analog in which the current is a flow of water instead of marbles.

6. A skier is towed up a hill and then skis down with constant speed because of friction. How is this analogous to a simple electric circuit?

20-4 Combinations of Resistors

Resistors in Series

Two or more resistors connected so that the same charge must flow through each are said to be connected in *series*. Resistors R_1 and R_2 in Figure 20-14a are examples of resistors in series. They must carry the same current I. We can often simplify the analysis of a circuit with resistors in series by replacing the resistors with a single equivalent resistance R_{eq} that gives the same total potential drop V when carrying the same current I (see Figure 20-14b).

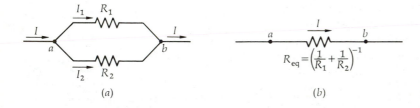

Figure 20-14
(a) Two resistors in series carry the
same current. (b) The resistors in (a)
can be replaced by a single
equivalent resistance $R_{eq} = R_1 + R_2$
that gives the same total potential
drop when carrying the same
current as in (a).

The potential drop across R_1 is IR_1 and that across R_2 is IR_2. The potential
drop across the two resistors is the sum of these:

$$V = IR_1 + IR_2 = I(R_1 + R_2)$$

Setting this potential drop equal to IR_{eq}, we obtain

$$R_{eq} = R_1 + R_2$$

Thus, the equivalent resistance for resistances in series is the sum of the
original resistances. When there are more than two resistances in series, the
equivalent resistance is

$$R_{eq} = R_1 + R_2 + R_3 + \cdots \qquad \text{20-18} \qquad \text{Series resistors}$$

Resistors in Parallel

Two resistors connected as shown in Figure 20-15a so that they have the
same potential difference across them are said to be connected in *parallel*.

Figure 20-15
(a) Two resistors are in parallel
when they are connected together
at both ends so that the potential
drop is the same across each. (b) The
two resistors in (a) can be replaced
by an equivalent resistance R_{eq}
related to R_1 and R_2 by $1/R_{eq} =$
$1/R_1 + 1/R_2$.

Note that both ends of both resistors are connected together by wires. Let I
be the current from point a to point b. At point a, the current splits into two
parts, I_1 through resistor R_1 and I_2 through R_2. The total current is the sum of
the individual currents:

$$I = I_1 + I_2 \qquad \text{20-19}$$

Let $V = V_a - V_b$ be the potential drop across either resistor. In terms of the
currents and resistances,

$$V = I_1 R_1 = I_2 R_2 \qquad \text{20-20}$$

The equivalent resistance of a combination of parallel resistors is defined as
that resistance R_{eq} for which the same total current I produces the same
potential drop V:

$$R_{eq} = \frac{V}{I}$$

Solving this equation for I and using $I = I_1 + I_2$, we have

$$I = \frac{V}{R_{eq}} = I_1 + I_2 \qquad\qquad 20\text{-}21$$

But according to Equation 20-20, $I_1 = V/R_1$ and $I_2 = V/R_2$. Equation 20-21 can therefore be written

$$I = \frac{V}{R_{eq}} = \frac{V}{R_1} + \frac{V}{R_2}$$

The equivalent resistance for two resistors in parallel is thus given by

$$\frac{1}{R_{eq}} = \frac{1}{R_1} + \frac{1}{R_2} \qquad\qquad 20\text{-}22$$

This result can be generalized to three resistors in parallel, as shown in Figure 20-16, or more:

$$\frac{1}{R_{eq}} = \frac{1}{R_1} + \frac{1}{R_2} + \frac{1}{R_3} + \cdots \qquad\qquad 20\text{-}23 \quad \text{Parallel resistors}$$

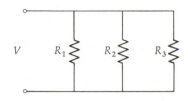

Figure 20-16
Three resistors in parallel.

Example 20-6   A 4-Ω resistor and a 6-Ω resistor are connected in parallel as shown in Figure 20-17, and a potential difference of 12 V is applied across the combination. Find the equivalent resistance, the total current, and the current through each resistor. Also find the power dissipated in each resistor.

We first calculate the equivalent resistance from Equation 20-22:

$$\frac{1}{R_{eq}} = \frac{1}{R_1} + \frac{1}{R_2} = \frac{1}{4\ \Omega} + \frac{1}{6\ \Omega} = \frac{3}{12\ \Omega} + \frac{2}{12\ \Omega} = \frac{5}{12\ \Omega}$$

or

$$R_{eq} = \frac{12\ \Omega}{5} = 2.4\ \Omega$$

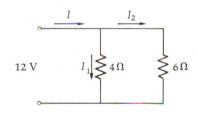

Figure 20-17
Two resistors in parallel across a potential difference of 12 V for Example 20-6.

The total current is then

$$I = \frac{V}{R_{eq}} = \frac{12\ \text{V}}{2.4\ \Omega} = 5\ \text{A}$$

We obtain the current in each resistor from the fact that the potential drop is 12 V across each resistor (Equation 20-20.) Calling the current through the 4-Ω resistor I_1 and that through the 6-Ω resistor I_2, we have

$$I_1(4\ \Omega) = V = 12\ \text{V}$$

$$I_1 = \frac{12\ \text{V}}{4\ \Omega} = 3.0\ \text{A}$$

and

$$I_2 = \frac{12\ \text{V}}{6\ \Omega} = 2.0\ \text{A}$$

The power dissipated in the 4-Ω resistor is

$$P = I_1^2 R_1 = (3.0\ \text{A})^2 (4\ \Omega) = 36\ \text{W}$$

The power dissipated in the 6-Ω resistor is

$$P = (2.0\ \text{A})^2(6\ \Omega) = 24\ \text{W}$$

This power comes from the source of emf, which maintains the 12-V potential difference across the combination of resistors. The power required to deliver 5.0 A at 12 V is

$$P = IV = (5.0\ \text{A})(12\ \text{V}) = 60\ \text{W}$$

which equals the total power dissipated in the two resistors.

We note from Example 20-6 that the equivalent resistance of two parallel resistances is less than either resistance by itself. This is a general result. Suppose we have a single resistor R_1 carrying current I_1 with a potential drop $V = I_1 R_1$. We now add a second resistor R_2 in parallel. If the potential drop is to remain the same, the second resistor must carry additional current $I_2 = V/R_2$ without affecting the original current I_1. The parallel combination thus carries more total current $I = I_1 + I_2$ for the same potential drop, so the ratio of the potential drop to the total current is less.

We also note from Example 20-6 that the ratio of the currents in the two parallel resistors equals the inverse ratio of the resistances. We could have obtained this in general from Equation 20-20:

$$\frac{I_1}{I_2} = \frac{R_2}{R_1} \qquad\qquad \textbf{20-24}$$

Example 20-7   For the circuit shown in Figure 20-18, find the equivalent resistance of the parallel combination of resistors, the current carried by each, and the total current through the source of emf.

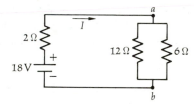

Figure 20-18
Circuit for Example 20-7. The 12-Ω and 6-Ω resistors are in parallel, and this parallel combination is in series with the 2-Ω resistor.

We first find the equivalent resistance of the 6- and 12-Ω resistors in parallel:

$$\frac{1}{R_{eq}} = \frac{1}{R_1} + \frac{1}{R_2} = \frac{1}{6\ \Omega} + \frac{1}{12\ \Omega} = \frac{3}{12\ \Omega} = \frac{1}{4\ \Omega}$$

$$R_{eq} = 4\ \Omega$$

Figure 20-19 shows the circuit with R_{eq} replacing the parallel combination. The 2-Ω resistance is in series with $R_{eq} = 4\ \Omega$. The equivalent resistance of this series combination is $R'_{eq} = R_{eq} + 2\ \Omega = 6\ \Omega$. The current in the circuit is therefore

$$I = \frac{\mathcal{E}}{R'_{eq}} = \frac{18\ \text{V}}{6\ \Omega} = 3\ \text{A}$$

This is the total current in the source of emf.

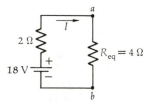

Figure 20-19
The circuit of Figure 20-18 is simplified by replacing the two parallel resistors with their equivalent resistance.

The potential drop from a to b across R_{eq} in Figure 20-19 is $V = IR_{eq} = (3\ A)(4\ \Omega) = 12\ V$. The current through the 6-Ω resistor is thus

$$I_1 = \frac{12\ V}{6\ \Omega} = 2\ A$$

and that through the 12-Ω resistor is

$$I_2 = \frac{12\ V}{12\ \Omega} = 1\ A$$

We note that the current through the 6-Ω resistance is twice that through the 12-Ω resistance, as expected.

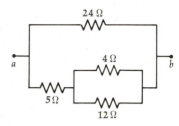

Figure 20-20
Resistor network for Example 20-8.

Example 20-8   Find the equivalent resistance between points a and b for the combination of resistors shown in Figure 20-20.

This combination of resistors may look complicated, but it is easily treated step by step. The only pair of resistors that are either in series or in parallel are the 4- and 12-Ω resistors, which are in parallel. We therefore first find the equivalent resistance of the 4- and 12-Ω resistors. From Equation 20-22, we obtain

$$\frac{1}{R_{eq}} = \frac{1}{4\ \Omega} + \frac{1}{12\ \Omega} = \frac{4}{12\ \Omega} = \frac{1}{3\ \Omega}$$

or

$$R_{eq} = 3\ \Omega$$

In Figure 20-21, we have replaced the 4- and 12-Ω resistors with their equivalent, a 3-Ω resistor. Since this 3-Ω resistor is in series with the 5-Ω resistor, the equivalent resistance of the bottom branch of this combination is 8 Ω. We are now left with an 8-Ω resistor in parallel with a 24-Ω resistor (see Figure 20-22). The equivalent resistance of these two parallel resistors is again found from Equation 20-22.

$$\frac{1}{R_{eq}} = \frac{1}{24\ \Omega} + \frac{1}{8\ \Omega} = \frac{4}{24\ \Omega} = \frac{1}{6\ \Omega}$$

$$R_{eq} = 6\ \Omega$$

The equivalent resistance between points a and b is thus 6 Ω.

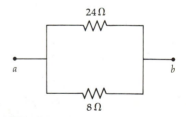

Figure 20-21
Simplification of the resistor network of Figure 20-20. The parallel 4- and 12-Ω resistors in that figure are replaced with their equivalent resistance of 3 Ω.

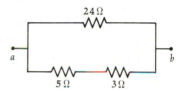

Figure 20-22
Further simplification of the resistor network of Figures 20-20 and 20-21. The 5- and 3-Ω resistors in series in Figure 20-21 are replaced by their equivalent resistance of 8 Ω. The network is now reduced to a 24- and an 8-Ω resistor in parallel.

Questions

7. For a given source of emf that remains constant, which will produce more heat when connected across it, a small resistance or a large resistance?

8. It is said that after the introduction of electric lighting, some people were careful to keep bulbs in all the sockets so that the electricity would not leak out. Why is this not necessary?

9. Occasionally, one sees a sign in a laboratory reading "DANGER 1,000,000 Ω." Is this serious?

20-5   Kirchhoff's Rules

The methods discussed in the previous section for replacing series and parallel combinations of resistors with their equivalent resistances are very useful for simplifying many combinations of resistors. However, they are not sufficient for the analysis of many simple circuits. Figure 20-23 is an example of a such a circuit. The two resistors R_1 and R_2 in this circuit look like they might be in parallel, but they are not. The potential drop is not the same across both resistors because of the presence of the emf $\mathcal{E}_2$ in series with R_2. Also, R_1 and R_2 are not in series because they do not carry the same current. There are two simple rules, called *Kirchhoff's rules,* that are applicable to any direct-current (dc) circuits. Kirchhoff's rules are as follows:

1. When any closed circuit loop is traversed, the sum of the potential increases must equal the sum of the potential decreases.

2. At any junction point in a circuit where the current can divide, the sum of the currents into the junction must equal the sum of the currents out of the junction.

Rule 1, called the loop rule, follows from the simple fact that in the steady state the potential difference between any two points is constant. As we move around a circuit loop, the potential may decrease or increase as we pass through a resistor or a battery, but when we have completely traversed the loop and arrive back at our starting point, *the net change in the potential must be zero.* This rule can also be related to the conservation of energy. If we have a charge q at some point where the potential is V, its potential energy is qV. As the charge traverses a loop in the circuit, it loses or gains energy as it passes through resistors and batteries, but when it arrives back at its starting point, its energy must again be qV.

Rule 2, called the junction rule, follows from the conservation of charge. This rule is needed for multiloop circuits containing points where the current can divide. In Figure 20-24, we have labeled the currents in the three wires I_1, I_2, and I_3. In a time interval Δt, charge $I_1 \Delta t$ flows into the junction from the left. In the same time interval, charges $I_2 \Delta t$ and $I_3 \Delta t$ flow out of the junction to the right. Since there is no way that charge can either originate or collect at this point, the conservation of charge implies rule 2, which for this case gives

$$I_1 = I_2 + I_3 \qquad\qquad 20\text{-}25$$

In this section we will illustrate the application of Kirchhoff's rules to various dc circuits. We begin with a simple circuit with one battery and one loop and then give examples of more complicated circuits containing more than one battery and more than one loop.

Figure 20-25 shows a simple circuit consisting of a battery of emf $\mathcal{E}$ with internal resistance r and an external resistance R across the terminals of the battery. The current I is in the direction shown. The connecting wires, indicated by straight lines in this diagram, are assumed to have negligible resistance. Any two points along such a line are therefore always at the same potential. If we start from point a and move around the circuit with the

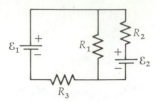

Figure 20-23
A simple circuit that cannot be analyzed by replacing series or parallel resistor combinations with their equivalent resistances. The potential drops across R_1 and R_2 are not equal because of the emf $\mathcal{E}_2$. Note that these resistors are not connected together at both ends and so are not in parallel. Also, there are no resistors in series.

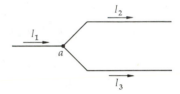

Figure 20-24
Illustration of Kirchhoff's junction rule. Since charge cannot originate or collect at point a, the current I_1 into point a must equal the sum $I_2 + I_3$ of the currents out of that point.

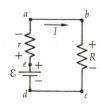

Figure 20-25
Simple circuit containing a battery of emf $\mathcal{E}$ and internal resistance r and an external resistance R. The $+$ and $-$ signs on the resistors are placed there to help us remember which side of the resistor is at higher potential.

current (clockwise in this case), we first encounter a potential drop IR from point b to point c. We have placed $+$ and $-$ signs at the ends of the resistor in the figure to remind us that the side nearest b is at the higher potential. We then encounter a potential increase equal to $\mathcal{E}$ from point d to point e across the emf of the battery and finally another potential drop Ir from point e to point a across the internal resistance of the battery. Setting the potential increases equal to the potential drops according to Kirchhoff's first rule gives

$$\mathcal{E} = IR + Ir \qquad\qquad\qquad 20\text{-}26$$

The current is then given by

$$I = \frac{\mathcal{E}}{r + R} \qquad\qquad\qquad 20\text{-}27$$

Gustav Robert Kirchhoff (1824–1887).

The power output $\mathcal{E}I$ of the source of emf goes partly into Joule heating of the external resistor and partly into Joule heating of the battery because of its internal resistance. We can see this by multiplying each term in Equation 20-26 by I, giving

$$\mathcal{E}I = I^2R + I^2r \qquad\qquad\qquad 20\text{-}28$$

The potential difference between the terminals of the battery, $V_a - V_d = \Delta V_{ad}$, is the terminal voltage, as discussed earlier. It is less than the emf of the battery because of the potential drop across the internal resistance. For this circuit, the terminal voltage is

$$\Delta V_{ad} = \mathcal{E} - Ir$$

Real batteries, such as a good car battery, usually have internal resistance of the order of a few hundredths of an ohm, so the terminal voltage is nearly equal to the emf unless the current is very large. One sign of a bad battery is an unusually high internal resistance. If you suspect that your car battery is bad, and you check the terminal voltage with a voltmeter (discussed in Section 20-7) that draws very little current, the voltmeter may read nearly 12 V. If you check the terminal voltage while drawing current from the battery, as when the car lights are on or while trying to start the car, the terminal voltage may drop considerably below 12 V.

Example 20-9   An 11-Ω resistance is connected across a battery of emf 6 V and internal resistance 1 Ω.* Find the current, the terminal voltage of the battery, the power delivered by the source of emf, and the power delivered to the external resistance.

From Equation 20-27, the current is

$$I = \frac{\mathcal{E}}{r + R} = \frac{6\text{ V}}{1\ \Omega + 11\ \Omega} = 0.5\text{ A}$$

The terminal voltage across the battery is

$$\Delta V_{ad} = \mathcal{E} - Ir = 6\text{ V} - (0.5\text{ A})(1\ \Omega) = 5.5\text{ V}$$

---

* We exaggerate the value of the internal resistance in this example to simplify calculations. In other examples, we may simply ignore the internal resistance.

The power delivered by the source of emf is

$$P = \mathcal{E}I = (6 \text{ V})(0.5 \text{ A}) = 3 \text{ W}$$

The power delivered to the external resistance is

$$P = I^2R = (0.5 \text{ A})^2(11 \text{ }\Omega) = 2.75 \text{ W}$$

This is dissipated as Joule heat. The other 0.25 W of power is dissipated as heat in the internal resistance of the battery.

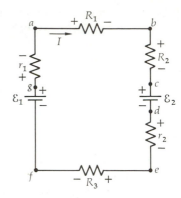

Figure 20-26
Circuit for Example 20-10.

Example 20-10   Find the current in the circuit shown in Figure 20-26.

In this circuit, we have two sources of emf and three external resistors. We cannot predict the direction of the current unless we know which emf is greater, but we do not have to know the direction of the current before solving the problem. We can assume any direction and solve the problem based on that assumption. If the assumption is incorrect, we will get a negative number for the current, indicating that the direction of the current is opposite to that assumed.

Let us assume I to be clockwise as indicated in the figure. We then apply Kirchhoff's first rule as we traverse the circuit in the assumed direction of the current, beginning at point a. For this choice of current direction, the high- and low-potential sides of the resistors are indicated by the $+$ and $-$ signs in the figure. The potential decreases and increases are given in Table 20-2. We encounter a potential drop traversing one source of emf (from c to d) and an increase traversing the other (from f to g). Kirchhoff's first rule gives

$$\mathcal{E}_1 = IR_1 + IR_2 + \mathcal{E}_2 + Ir_2 + IR_3 + Ir_1 \qquad \text{20-29}$$

Solving for the current I, we obtain

$$I = \frac{\mathcal{E}_1 - \mathcal{E}_2}{R_1 + R_2 + R_3 + r_1 + r_2} \qquad \text{20-30}$$

Note that if $\mathcal{E}_2$ is greater than $\mathcal{E}_1$, we will get a negative number for the current I, indicating that we have chosen the wrong direction for I. If $\mathcal{E}_2$ is greater than $\mathcal{E}_1$, the current is in the counterclockwise direction. On the other hand, if $\mathcal{E}_1$ is the greater emf, we get a positive number for I, indicating that the assumed direction is correct.

Table 20-2

Changes in Potential between Points Labeled in Circuit in Figure 20-26

$a \rightarrow b$	Drop IR_1
$b \rightarrow c$	Drop IR_2
$c \rightarrow d$	Drop $\mathcal{E}_2$
$d \rightarrow e$	Drop Ir_2
$e \rightarrow f$	Drop IR_3
$f \rightarrow g$	Increase $\mathcal{E}_1$
$g \rightarrow a$	Drop Ir_1

Let us assume for the example above that $\mathcal{E}_1$ is the greater emf and that the current is clockwise, as indicated in the figure. In battery 2, the charge moves from high potential to low potential. A charge ΔQ moving through battery 2 from point c to point d loses energy $\mathcal{E}_2 \Delta Q$. In this battery, electrical energy is converted into chemical energy and is stored. Thus, battery 2 is *charging*. We can account for the energy balance in this circuit by multiplying each term in Equation 20-29 by the current I. The term on the left, $\mathcal{E}_1 I$, is the power, the rate at which battery 1 puts energy into the circuit. This energy comes from the internal chemical energy of the battery. The first term on the right, I^2R_1, is the rate of production of Joule heat in resistor R_1. There are similar terms for each of the other resistors. The term $\mathcal{E}_2 I$ is the rate at which electric energy is converted into chemical energy in the second battery.

Example 20-11   The quantities in the circuit of Example 20-10 have the magnitudes $\mathcal{E}_1 = 12$ V, $\mathcal{E}_2 = 4$ V, $r_1 = r_2 = 1\ \Omega$, $R_1 = R_2 = 5\ \Omega$, and $R_3 = 4\ \Omega$, as indicated in Figure 20-27. Find the potentials at the points indicated in the figure, assuming that the potential at point f is zero, and discuss the energy balance in the circuit.

Since only potential differences are important, any point in a circuit can be chosen to have zero potential. The analysis of a problem is usually simplified if we choose one point to be at zero potential and then find the potentials of the other points relative to it. In this example we have chosen point f to be at zero potential, as indicated by the ground symbol (the three horizontal lines in the figure).

We first find the current in the circuit. From Equation 20-30, we have

$$I = \frac{12\ \text{V} - 4\ \text{V}}{5\ \Omega + 5\ \Omega + 4\ \Omega + 1\ \Omega + 1\ \Omega} = \frac{8\ \text{V}}{16\ \Omega} = 0.5\ \text{A}$$

We can now find the potentials at the indicated points relative to zero potential at point f. Since, by definition, the source of emf maintains a constant potential difference, $\mathcal{E}_1 = 12$ V, between point g and point f, the potential at point g is 12 V. The potential at point a is less than that at g by the potential drop $Ir_1 = 0.5(1) = 0.5$ V. Thus, the potential at point a is $12 - 0.5 = 11.5$ V. Similarly, the potential drops across the 5-Ω resistors R_1 and R_2 are each $IR_1 = 0.5(5) = 2.5$ V. The potential at point b is then $11.5 - 2.5 = 9$ V, and that at c is 6.5 V. The potential drop across $\mathcal{E}_2$ is 4 V. Thus, point d is at a potential of 2.5 V. Since the drop across the 1-Ω resistance r_2 is 0.5 V, the potential at e is 2 V. The potential drop across the 4-Ω resistance R_3 is $IR_3 = 2$ V. This gives zero for the potential at f, consistent with our original assumption. Figure 20-28 shows a sketch of the potential changes that occur as we traverse this circuit.

The power, the rate at which energy is delivered by the source of emf $\mathcal{E}_1$, is

$$P_{\mathcal{E}_1} = \mathcal{E}_1 I$$
$$= (12\ \text{V})(0.5\ \text{A}) = 6.0\ \text{W}$$

The power into the internal resistance of battery 1 is

$$P_{r_1} = I^2 r_1$$
$$= (0.5)^2(1) = 0.25\ \text{W}$$

Thus, the power delivered by the battery to the external circuit is $6 - 0.25 = 5.75$ W. This also equals $V_1 I$, where $V_1 = V_a - V_f = 11.5$ V is the terminal voltage of that battery. The total power into the external resistances in the circuit is

$$P_R = I^2 R_1 + I^2 R_2 + I^2 R_3 = I^2(R_1 + R_2 + R_3)$$
$$= 0.5^2(5\ \Omega + 5\ \Omega + 4\ \Omega) = 3.5\ \text{W}$$

The power into the battery being charged is $(V_c - V_e)I = (6.5 - 2)(0.50) = 2.25$ W. Part of this, $I^2 r_2 = 0.25$ W, is dissipated in the internal resistance, and part, $\mathcal{E}_2 I = 2$ W, is the rate at which energy is stored in that battery.

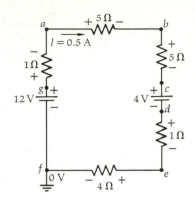

Figure 20-27
Same circuit as Figure 20-26 but with sample values for Example 20-11. The potential is chosen to be zero at point f. The three horizontal lines near that point indicate that the point is grounded.

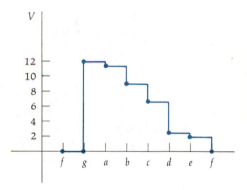

Figure 20-28
Changes in potential as the circuit of Figure 20-27 is traversed, beginning at point f, which is at zero potential.

Note that the terminal voltage of the battery that is being charged in Example 20-11 is $V_c - V_e = 4.5$ V, which is greater than the emf of the battery. Because of its internal resistance, a battery is not completely reversible. If the same 4-V battery were to deliver 0.5 A to an external circuit, its terminal voltage would be 3.5 V (again assuming a value of 1 Ω for its internal resistance). If the internal resistance is very small, the terminal voltage of the battery is nearly equal to its emf whether the battery is delivering current to an external circuit or is being charged. Some real batteries, such as the storage batteries used in automobiles, are nearly reversible and can easily be recharged. Other types of batteries are often not reversible. If you attempt to recharge them by driving current from their positive to their negative terminals, most, if not all, of the energy expended goes into heat rather than into the chemical energy of the battery.

We will now consider circuits containing more than one loop. In such circuits, we need to apply Kirchhoff's second rule at junction points where the current splits into two or more parts.

Example 20-12   Find the current in each part of the circuit shown in Figure 20-29.

Let I be the current through the 18-V battery in the direction shown. At point b, this current divides into currents I_1 and I_2. Until we know the solution for the currents, we cannot be sure of their direction. For example, we need to know whether point b or point c is at a higher potential in order to know the direction of current through the 6-Ω resistor. The current directions indicated on the figure are merely guesses. Applying the junction rule (rule 2) to point b, we obtain

$$I = I_1 + I_2$$

The junction rule applied to point c gives us the same information since the currents I_1 and I_2 join there to form the current I toward point d. There are three possible loops for applying rule 1, loops $abcd$, $befc$, and $abefcd$. We need only two more equations to determine the three unknown currents. Equations for any two of the loops will be sufficient. (The third loop will then give redundant information.) Traversing the loop $abcd$ in the clockwise direction, Kirchhoff's first rule gives

$$(12\ \Omega)I + (6\ \Omega)I_1 = 18\ \text{V}$$

Simplifying this by dividing each term by 6 Ω gives

$$2I + I_1 = 3\ \text{A} \qquad\qquad 20\text{-}31$$

Applying rule 1 to loop $befc$, we obtain another equation:

$$(1\ \Omega)I_2 + (1\ \Omega)I_2 = 12\ \text{V} + (6\ \Omega)I_1$$

Note that in moving from c to b, we encounter a voltage increase because the (assumed) current I_1 is in the opposite direction. Combining terms and substituting $I - I_1$ for I_2 in this equation, we obtain

$$2(I - I_1) = 12\ \text{A} + 6I_1$$
$$2I = 12\ \text{A} + 8I_1 \qquad\qquad 20\text{-}32$$

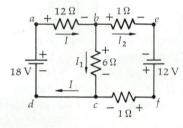

Figure 20-29
Circuit for Example 20-12. The direction of the current I_1 from b to c is not known before the problem is solved. The $+$ and $-$ signs on the 6-Ω resistor are for the assumed direction of I_1 from b to c.

Equations 20-31 and 20-32 can be solved for the unknown currents I and I_1. The results are

$$I = 2 \text{ A}$$

$$I_1 = -1 \text{ A}$$

Then,

$$I_2 = I - I_1 = 2 - (-1) = 3 \text{ A}$$

Our original guess about the direction of I_1 was incorrect. The current through the 6-Ω resistor is in the direction from point c to point b.

Example 20-13   Find the currents I, I_1, and I_2 for the circuit shown in Figure 20-30.

This circuit is the same as the one in Figure 20-23, with $\mathcal{E}_1 = 12$ V, $\mathcal{E}_2 = 5$ V, $R_1 = 4 \ \Omega$, $R_2 = 2 \ \Omega$, and $R_3 = 3 \ \Omega$. We choose directions for the currents as shown in the figure, and we obtain from the junction rule

$$I = I_1 + I_2$$

Using this to eliminate I and applying the loop rule to the outer loop, we get

$$12 \text{ V} = (2 \ \Omega)I_1 + 5 \text{ V} + (3 \ \Omega)(I_1 + I_2) \qquad \text{20-33}$$

Applying the loop rule to the first loop on the left gives

$$12 \text{ V} = (4 \ \Omega)I_2 + (3 \ \Omega)(I_1 + I_2) \qquad \text{20-34}$$

When we solve Equations 20-33 and 20-34 for the currents I_1 and I_2, we obtain $I_1 = 0.5$ A and $I_2 = 1.5$ A. The total current through the 12-V source of emf is then

$$I = 0.5 \text{ A} + 1.5 \text{ A} = 2.0 \text{ A}$$

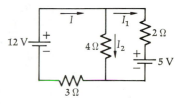

Figure 20-30
Circuit for Example 20-13. This is the same circuit as that in Figure 20-23 with sample values for the emfs and resistors.

Question

10. A weak car battery is to be connected (jumpered) to a good car battery to start the car. To which terminal of the second battery should the positive terminal of the first battery be connected? If they are connected in the opposite way, both batteries may explode in a shower of boiling battery acid. Why? (A typical car battery has an emf of 12 V and an internal resistance of 0.01 Ω. The cables may also have a total resistance of 0.01 Ω.)

20-6   RC Circuits

A circuit containing a resistor and a capacitor is called an *RC circuit*. In such a circuit, the current is not steady but varies with time. A practical example of an *RC* circuit is that in a flash attachment of a camera. Before a flash photograph is taken, a battery in the flash attachment charges the capacitor through a resistor. When this is accomplished, the flash is ready. When the picture is taken, the capacitor discharges through the flashbulb. The capacitor is then recharged by the battery, and a short time later it is ready for another flash. We will first look at the circuit for the discharge of a capacitor.

Figure 20-31 shows a capacitor with initial charges of $+Q_0$ on the upper plate and $-Q_0$ on the lower plate. It is connected to a resistor R and a switch S, which is open to prevent the charge from flowing through the resistor. The potential difference across the capacitor is initially $V_0 = Q_0/C$, where C is the capacitance. Since there is no current when the switch is open, there is no potential drop across the resistor. Thus, there is also a potential difference V_0 across the switch.

We close the switch at time $t = 0$. Since there is now a potential difference across the resistor, there must be a current in it. The initial current is

$$I_0 = \frac{V_0}{R} = \frac{Q_0}{RC} \qquad \text{20-35}$$

The current is due to the flow of charge from the positive plate to the negative plate through the resistor. After a time, the charge on the capacitor is reduced. Since the charge on the capacitor is *decreasing* and we are taking the current to be positive, the current equals the rate of decrease of the charge.

$$I = -\frac{\Delta Q}{\Delta t} \qquad \text{20-36}$$

We can write an equation for the charge and current as functions of time by applying Kirchhoff's first rule to the circuit after the switch is closed. Traversing the circuit in the direction of the current, we encounter a potential drop IR across the resistor and a potential increase Q/C across the capacitor. Kirchhoff's first rule gives

$$IR = \frac{Q}{C} \qquad \text{20-37}$$

where Q and I both depend on time and are related by Equation 20-36. Substituting $-\Delta Q/\Delta t$ for I in this equation, we have

$$-R\frac{\Delta Q}{\Delta t} = \frac{Q}{C}$$

or

$$\frac{\Delta Q}{\Delta t} = -\frac{1}{RC}Q \qquad \text{20-38}$$

Equation 20-38 states that the rate of change of the charge Q on the capacitor is proportional to the charge Q. Initially, when the charge is at its maximum Q_0, the current is large, but as the charge decreases, the current becomes smaller. We note from Equation 20-38 that the product RC must have dimensions of time. The quantity RC is called the *time constant τ* of the circuit:

$$\tau = RC \qquad \text{20-39} \qquad \text{Time constant}$$

The time constant is the time it takes for the charge and the current to decrease to $e^{-1} = 0.37$ of their initial values.

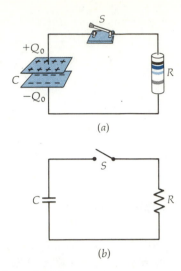

Figure 20-31
(a) A parallel-plate capacitor in series with a switch and a resistor R. (b) Circuit diagram for (a).

Figure 20-32 shows the charge on the capacitor as a function of time. The dashed line is the initial slope of the charge-versus-time function. If the charge continued to decrease at a constant rate equal to its initial rate, it would reach zero in a time equal to the time constant τ. The actual rate of decrease $(-\Delta Q/\Delta t)$ is not constant; instead, it also decreases with time. This is evident from Equation 20-38, which shows that the rate of decrease of the charge is proportional to the charge itself. After a time $t = \tau$, the charge is $0.37 Q_0$. After a time equal to several time constants, the charge on the capacitor is negligible. This type of decrease is very common in nature and is called *exponential decrease*. Exponential decrease occurs whenever the rate of decrease of a quantity is proportional to the quantity itself. (We encountered exponential decrease in Chapter 10 when we studied the decrease in air pressure with increasing altitude and again in Chapter 15 when we studied the decreases in the energy of a damped oscillator with time.)

Figure 20-33 shows the current as a function of time. The current, like the charge, decreases exponentially with time with the time constant $\tau = RC$.

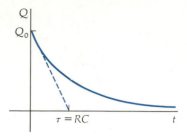

Figure 20-32
Charge on capacitor versus time for the circuit of Figure 20-31 when the switch is closed at time $t = 0$. The time constant $\tau = RC$ is the time for the charge to decrease to $0.37 Q_0$. After a time equal to two time constants, the charge is $(0.37)^2 Q_0$. This is an example of exponential decrease. The time constant is also the time that the capacitor would fully discharge if its discharge rate were constant, as indicated by the dashed line.

Figure 20-33
Current versus time for a discharging capacitor. The curve has the same shape as that in Figure 20-32.

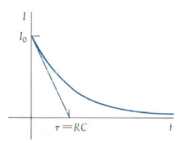

A circuit used for charging a capacitor is shown in Figure 20-34. Initially, the switch is open and the capacitor is uncharged. At time $t = 0$, we close the switch. Charge immediately begins to flow into the capacitor. If the charge on the capacitor at some time is Q and the current in the circuit is I, Kirchhoff's first rule gives

$$\mathcal{E} = V_R + V_C = IR + \frac{Q}{C} \qquad\qquad 20\text{-}40$$

In this circuit, the current equals the rate of increase of charge on the capacitor. We will discuss the behavior of this circuit qualitatively. Initially, there is no charge on the capacitor, so the voltage drop across the resistor must equal the emf. The initial current (just after the switch is thrown) is thus $\mathcal{E}/R$. As the capacitor charges, the voltage drop across the capacitor Q/C increases, and the current decreases. Eventually, the voltage drop across the capacitor equals the emf and the current is zero. The maximum charge on the capacitor is found from Equation 20-40 by setting $I = 0$. Then $Q_{max} = C\mathcal{E}$.

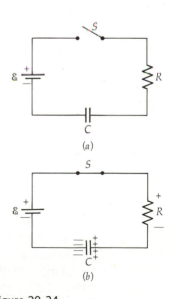

Figure 20-34
(a) Circuit for charging a capacitor to a potential difference $V = \mathcal{E}$. (b) After the switch is closed, there is a potential drop across the resistor and a charge on the capacitor.

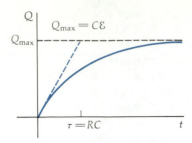

Figure 20-35
Charge versus time for the charging
circuit of Figure 20-34 when the
switch is closed at $t = 0$. After a
time of one time constant, $\tau = RC$,
the charge is $0.63C\mathcal{E}$, where $C\mathcal{E}$ is
the maximum charge. If the
charging rate were constant, the
capacitor would be fully charged
after time $t = \tau$.

Figure 20-35 shows the charge on the capacitor as a function of time. The time contant $\tau = RC$ is the time it takes for the charge to reach $0.63Q_{max}$. It is also the time it would take for the charge to reach its maximum value if the charging rate were constant and equal to the initial rate; that is, if the current were constant and equal to its initial value of $\mathcal{E}/R$. Figure 20-36 shows the current as a function of time. The current decreases exponentially, just as it did in the previous circuit for a discharging capacitor.

Example 20-14   A 6-V battery of negligible internal resistance is used to charge a 2-μF capacitor through a 100-Ω resistor. Find the initial current, the final charge, and the time constant of this circuit.

The initial current is

$$I_0 = \frac{\mathcal{E}}{R} = \frac{6\ V}{100\ \Omega} = 0.06\ A$$

The final charge is

$$Q_f = \mathcal{E}C = (6\ V)(2\ \mu F) = 12\ \mu C$$

The time constant for this circuit is

$$\tau = RC = (100\ \Omega)(2\ \mu F) = 200\ \mu s$$

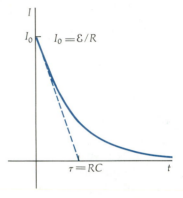

Figure 20-36
Current versus time for the charging
circuit of Figure 20-34. The current
I_0 is initially $\mathcal{E}/R$, and it decreases
exponentially with time.

20-7 Ammeters, Voltmeters, and Ohmmeters

We turn now to the consideration of the measurement of electrical quantities in dc circuits. Devices that measure current, potential difference, and resistance are called *ammeters, voltmeters,* and *ohmmeters,* respectively. You might use a voltmeter to measure the terminal voltage of your car battery or an ohmmeter to measure the resistance between two points in some electrical device at home (such as a toaster) when you suspect a short or possibly a broken wire. Some knowledge of the basic operation of these devices might prove very useful.

To measure the current through the resistor in the simple circuit shown in Figure 20-37, we place an ammeter in series with the resistor, as indicated in the figure. Since the ammeter has some resistance, the current in the circuit decreases slightly when the ammeter is inserted. Ideally, the ammeter should have a very small resistance so that only a small change will be introduced in the current to be measured.

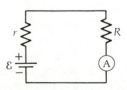

Figure 20-37
To measure the current in the
resistor R, an ammeter —Ⓐ— is
placed in series with the resistor.

Figure 20-38
To measure the voltage drop across a resistor, a voltmeter —Ⓥ— is placed in parallel with the resistor.

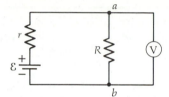

The potential difference across the resistor is measured by placing a voltmeter across the resistor in parallel with it, as shown in Figure 20-38. The voltmeter reduces the resistance between points a and b, thus increasing the total current in the circuit and changing the potential drop across the resistor. An ideal voltmeter has a very large resistance to minimize its effect on the circuit.

The principal component of an ammeter or a voltmeter is a *galvanometer*, a device that detects a small current that passes through it. A common type, the *d'Arsonval galvanometer*, consists of a coil of wire that is free to turn in a magnetic field, an indicator of some kind, and a scale. The galvanometer is designed so that the scale reading is proportional to the current in the galvanometer. Figure 20-39 shows a sketch of such a galvanometer. Many meters today have a digital readout rather than an indicator and a scale, but the basic operation of these newer voltmeters, ammeters, and ohmmeters is similar to that discussed here.

There are two properties of a galvanometer that are important for its use in the construction of an ammeter or voltmeter. These are the resistance of the galvanometer R_g and the current I_g needed to produce full-scale deflection. Typical values for a portable pivoted-coil laboratory galvanometer are $R_g = 20\ \Omega$ and $I_g = 0.5$ mA. The voltage drop across such a galvanometer for full-scale deflection is thus

$$\Delta V = I_g R_g = (20\ \Omega)(5.0 \times 10^{-4}\ \text{A}) = 10^{-2}\ \text{V}$$

To construct an ammeter from a galvanometer, we place a small resistor, called a *shunt resistor*, in parallel with the galvanometer. Since the shunt resistance is usually much smaller than the resistance of the galvanometer, most of the current is carried by the shunt so that the effective resistance of the ammeter is much smaller than that of the galvanometer. The effective resistance of the ammeter is, in fact, approximately equal to that of the shunt resistor. Resistors are added in series with a galvanometer to construct a voltmeter.

Figure 20-40 illustrates the construction of an ammeter and a voltmeter from a galvanometer. The resistance of the galvanometer R_g, shown separately in these schematic drawings, is actually part of the galvanometer; it is essentially the resistance of the galvanometer coil. The choice of appropriate resistors for the construction of an ammeter or voltmeter from a galvanometer is best illustrated by example.

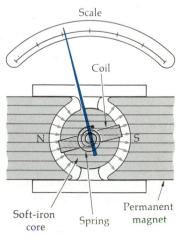

Scale

Coil

N   S

Soft-iron core   Spring   Permanent magnet

Figure 20-39
The d'Arsonval galvanometer. When the coil carries a current, the magnet exerts a torque on the coil, causing it to twist. The deflection of the scale is proportional to the current.

Ammeter

Voltmeter

Figure 20-40
(*a*) An ammeter consists of a galvanometer —Ⓖ— with resistance R_g and a small parallel resistance R_p, called a shunt resistance. (*b*) A voltmeter consists of a galvanometer —Ⓖ— and a large series resistance R_s. In these circuit diagrams, the galvanometer resistance is drawn next to the symbol for the galvanometer.

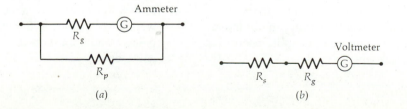

Ammeter
R_g
R_p
(*a*)

Voltmeter
R_s R_g
(*b*)

Example 20-15   Using a galvanometer with a resistance of 20 Ω, for which a current of 5×10^{-4} A through the galvanometer gives full-scale deflection, design an ammeter that will read full scale when the current is 5 A.

Since the total current through the ammeter must be 5 A when the current through the galvanometer is just 5×10^{-4} A, most of the current must go through the shunt resistor. Let R_p be the shunt resistance and I_p be the current through the shunt. Since the galvanometer and the shunt are in parallel, we have $I_g R_g = I_p R_p$ and $I_p + I_g = 5$ A or

$$I_p = 5 \text{ A} - I_g = 5 \text{ A} - 5 \times 10^{-4} \text{ A} \approx 5 \text{ A}$$

Thus, the value of the shunt resistor should be

$$R_p = \frac{I_g}{I_p} R_g = \frac{5 \times 10^{-4} \text{ A}}{5 \text{ A}} (20 \ \Omega) = 2 \times 10^{-3} \ \Omega$$

Since the resistance of the shunt is so much smaller than the resistance of the galvanometer, the effective resistance of the parallel combination is approximately equal to the shunt resistance.

Example 20-16   Using the same galvanometer as in Example 20-15, design a voltmeter that will read 10 V at full-scale deflection.

Let R_s be the value of a resistor in series with the galvanometer. We want to choose R_s so that a current of $I_g = 5 \times 10^{-4}$ A gives a potential drop of 10 V. Thus,

$$I_g(R_s + R_g) = 10 \text{ V}$$

$$R_s + R_g = \frac{10 \text{ V}}{5 \times 10^{-4} \text{ A}} = 2 \times 10^4 \ \Omega$$

$$R_s = (2 \times 10^4 \ \Omega) - R_g = (2 \times 10^4 \ \Omega) - 20 \ \Omega$$

$$= 19,980 \ \Omega \approx 20 \text{ k}\Omega$$

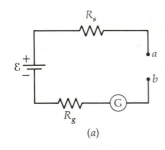

(a)

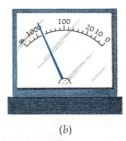

(b)

Figure 20-41 shows an ohmmeter used to measure resistance. It consists of a battery, a galvanometer, and a resistor. The resistance R_s is chosen so that there is a full-scale deflection when the terminals a and b are shorted (touched together). Thus, a full-scale deflection on the galvanometer is marked as zero resistance. When the terminals are connected across an unknown resistance R, the current is less than I_g and the galvanometer shows less than a full-scale deflection. The current in this case is

$$I = \frac{\mathscr{E}}{R + R_s + R_g} \qquad\qquad 20\text{-}41$$

Since the current depends on R, the scale can be calibrated in terms of the resistance measured, from zero resistance at full-scale deflection to infinite resistance at zero deflection. The calibration of the scale is far from linear and depends on the constancy of the emf of the battery. Such an ohmmeter is therefore not a high-precision instrument, but it can be quite useful for making quick, rough determinations of resistance.

Figure 20-41
(a) An ohmmeter consists of a battery in series with a galvanometer and a resistor R_s chosen so that the galvanometer reads a full-scale deflection when points a and b are shorted. (b) The galvanometer scale.

Some caution must be exercised in the use of an ohmmeter since it sends a current through the resistance to be measured. For example, consider an ohmmeter with a 1.5-V battery and a galvanometer similar to that in the previous examples. The series resistance needed is

$$I_g(R_s + R_g) = 1.5 \text{ V}$$

or

$$R_s = \frac{1.5}{5 \times 10^{-4}} - R_g = 3000 - 20 = 2980 \ \Omega$$

Suppose we were to use this ohmmeter to measure the resistance of a more sensitive laboratory galvanometer that gives a full-scale reading with a current of 10^{-5} A and has a resistance of about 20 Ω. When the terminals a and b are placed across this more sensitive galvanometer, the current will be just slightly less than 5×10^{-4} A because the total resistance is 3020 Ω, which is just slightly more than 3000 Ω. This current is about 50 times that needed to produce full-scale deflection. The result would probably be a puff of smoke, some popping sounds, one less sensitive galvanometer, and possibly some unkind words from the laboratory instructor.

Questions

11. Under what conditions might it be advantageous to use a galvanometer that is less sensitive, that is, requires a greater current I_g for full-scale deflection, than the one discussed in Examples 20-15 and 20-16?

12. When the series resistance R_s is properly chosen for a particular emf of an ohmmeter, any value of resistance from zero to infinity can be measured. Why are there different scales on a practical ohmmeter for different ranges of resistances to be measured?

13. A none-too-bright student decides to measure the internal resistance of his car battery with an ohmmeter borrowed from his physics laboratory. Why is this a bad idea?

Summary

1. Electric current is the flow of charge past a point. Its direction is chosen as that of the flow of positive charge. Electric current in conducting wires is the result of negatively charged electrons that are accelerated by an electric field in the wire. Since the electrons quickly collide with the conductor atoms, the result is a slow drift opposite the direction of the current. Typical drift velocities of electrons in wires are of the order of 1 mm/s.

2. The resistance of a wire segment is defined as the ratio of the voltage drop to the current. In ohmic materials, which include most metals, the resistance is independent of the current, an experimental result known as Ohm's law. For all materials, the current, resistance, and potential difference are related by

$$V = IR$$

3. The resistance of a wire is proportional to its length and inversely proportional to its cross-sectional area:

$$R = \rho \frac{L}{A}$$

where ρ is the resistivity of the material, which depends on temperature. The reciprocal of the resistivity is called the conductivity σ:

$$\sigma = \frac{1}{\rho}$$

4. The power supplied to a segment of a circuit equals the product of the current and the voltage drop in the segment:

$$P = IV$$

A device that supplies energy to a circuit is called a source of emf. The power supplied by a source of emf is the product of the emf $\mathcal{E}$ and the current:

$$P = \mathcal{E}I$$

Since the voltage drop across a resistor is IR, the power dissipated in a resistor is given by

$$P = IV = I^2R$$

A battery can be considered to be a source of emf in series with a small resistance called its internal resistance.

5. The equivalent resistance of a set of resistors connected in series equals the sum of the resistances:

$$R_{eq} = R_1 + R_2 + R_3 + \cdots \qquad \text{(series)}$$

For resistors connected in parallel, the reciprocal of the equivalent resistance equals the sum of the reciprocals of the resistances.

$$\frac{1}{R_{eq}} = \frac{1}{R_1} + \frac{1}{R_2} + \frac{1}{R_3} + \cdots \qquad \text{(parallel)}$$

6. Simple circuits can be solved using Kirchhoff's rules:

1. When any closed circuit loop is traversed, the sum of the potential increases must equal the sum of the potential decreases.

2. At any junction point in a circuit where the current can divide, the sum of the currents into the junction must equal the sum of the currents out of the junction.

7. When a capacitor is discharged though a resistor, the charge on the capacitor and the current both decrease exponentially with time. The time constant $\tau = RC$ is the time it takes for either to decrease to 0.37 times its original value. After a time equal to 2τ, they have decreased to $(0.37)^2$ times their original value, and so on.

8. A galvanometer is a device that detects a small current that passes through it. Its deflection is proportional to the current. An ammeter is a device for measuring current. It consists of a galvanometer plus a parallel resistor called a shunt resistor. To measure the current through a resistor, an ammeter is placed in series with it. The ammeter has a very small effective resistance so that it has little effect on the current to be measured. A voltmeter measures potential difference. It consists of a galvanometer plus a large series resistor. To measure the potential drop across a resistor, a voltmeter is placed in parallel with the resistor. The voltmeter has a very large effective resistance so that it has little effect on the potential drop to be measured. An ohmmeter is a device for measuring resistance. It consists of a galvanometer, a battery, and a resistor.

Suggestions for Further Reading

Allen, Philip B.: "Electrical Conductivity," *The Physics Teacher*, vol. 17, 1979, p. 362.

A brief but advanced treatment of the classical and quantum theories of electrical conduction.

Rosenfeld, L.: "Gustav Robert Kirchhoff," *The Dictionary of Scientific Biography*, vol. 7, Charles Coulston Gillespie (ed.), Charles Scribner's Sons, New York, 1973, p. 379.

This biography, contained in a sixteen-volume work, shows that Kirchhoff's famous rules for electrical circuits were only the first of a number of important discoveries, including an important result in electromagnetic theory arising from work on the flame spectra of salts undertaken with Bunsen.

Review

A. Objectives: After studying this chapter, you should:

1. Be able to define and discuss the concepts of electric current, drift velocity, resistance, and emf.

2. Be able to state Ohm's law and distinguish between it and the definition of resistance.

3. Be able to give the definition of resistivity and describe its temperature dependence.

4. Be able to discuss the simple model of a real battery in terms of an ideal emf and an internal resistance, and find the terminal voltage of a battery when it delivers a current I.

5. Be able to give the general relationship for potential difference, current, and power.

6. Be able to determine the equivalent resistances of resistors connected in series or in parallel.

7. Be able to state Kirchhoff's rules and use them to analyze various dc circuits.

8. Be able to sketch both the charge on a capacitor and the current as functions of time for charging and discharging the capacitor.

9. Be able to draw the circuit diagrams and calculate the proper series or shunt resistors needed to make an ammeter, a voltmeter, and an ohmmeter from a given galvanometer.

B. Define, explain, or otherwise identify:

Current, p. 489
Ampere, p. 489
Drift velocity, p. 489
Ohm's law, p. 491
Resistance, p. 491
Resistivity, p. 492
Conductivity, p. 492
Superconductivity, p. 495
Joule heat, p. 501
Source of emf, p. 502
emf, p. 502
Terminal voltage, p. 504
Internal resistance, p. 504
Series resistors, p. 505
Parallel resistors, p. 506
Kirchhoff's rules, p. 510
RC circuit, p. 515
Time constant, p. 516
Exponential decrease, p. 517
Ammeter, p. 518
Voltmeter, p. 518
Ohmmeter, p. 518
Galvanometer, p. 519

C. True or false: If the statement is true, explain. If it is false, give a counterexample.

1. Ohm's law is $V = IR$.

2. Electrons drift in the direction of the current.

3. A source of emf supplies power to an electrical circuit.

4. When the potential drops by V in a segment of a circuit, the power supplied to that segment is IV.

5. The equivalent resistance of two resistors in parallel is always less than that of either resistor.

6. The terminal voltage of a battery is always less than its emf.

7. The time constant of an RC circuit is the time needed to completely discharge the capacitor.

8. To measure the potential drop across a resistor, a voltmeter is placed in series with the resistor.

Exercises

Section 20-1   Current and Motion of Charges

1. A wire carries a current of 3.0 A. (*a*) How much charge flows past a point in the wire in 5.0 min? (*b*) If the current is due to the flow of electrons, how many electrons flow past a point in this time?

2. A charge $+q$ moves in a circle of radius r with speed v. (*a*) Express the frequency f with which the charge passes a point in terms of r and v. (*b*) Show that the average current is qf and express it in terms of v and r.

3. A charge of 20 C flows past a point in a wire in 2.0 min. Find the current in the wire.

Section 20-2   Ohm's Law and Resistance

4. A 10-m wire of resistance 0.4 Ω carries a current of 5 A. (*a*) What is the potential difference across the wire? (*b*) What is the magnitude of the electric field in the wire?

5. A potential difference of 80 V produces a current of 3 A in a certain resistor. (*a*) What is the resistance? (*b*) What is the current when the potential difference is 20 V?

6. A copper wire is 40 m long and has a radius of 0.814 mm. (*a*) Find its resistance. (*b*) Find the current when the potential difference across the wire is 0.5 V.

7. A piece of carbon is 6.0 cm long and has a square cross section 0.3 cm on a side. A potential difference of 6.5 V is maintained across its long dimension. (*a*) What is its resistance? (*b*) What is the current?

8. A tungsten rod is 50 cm long and has a square cross section 1.0 mm on a side. (*a*) What is its resistance at 20°C? (*b*) What is its resistance at 40°C?

9. The third (current-carrying) rail of a subway track is made of steel (resistivity $\rho = 2 \times 10^{-7}$ $\Omega \cdot$m) and has a cross-sectional area of about 55 cm². What is the resistance of 10 km of this track?

10. A copper wire and an iron wire have the same length and diameter and carry the same current. (*a*) Find the ratio of the potential drops across the wires. (*b*) In which wire is the electric field greatest?

11. What is the potential difference across one wire of a 30-m extension cord of 16-gauge copper wire (diameter 1.30 mm) carrying a current of 3.0 A?

12. At what temperature will the resistance of a copper wire be 10 percent greater than it is at 20°C?

13. How long is a 14-gauge copper wire (diameter 1.628 mm) that has a resistance of 3 Ω?

Section 20-3   Energy in Electric Circuits

14. What is the power dissipated in a 10.0-Ω resistor if the potential difference across it is 50 V?

15. Find the power dissipated in a resistor of resistance (*a*) 5 Ω and (*b*) 10 Ω connected across a potential difference of 110 V.

16. A 1-kW heater is designed to operate at 220 V. (*a*) What is its resistance and what current does it draw? (*b*) What is the power dissipated in this resistor if it operates at 110 V?

17. A 10.0-Ω resistor is rated capable of dissipating 5.0 W. (*a*) What maximum current can this resistor tolerate? (*b*) What voltage across the resistor will produce this current?

18. A battery has an emf of 12.0 V. How much work does it do in 5 s if it delivers 3 A?

19. If energy costs 8 cents per kilowatt-hour, (*a*) how much does it cost to operate an electric toaster for 4 min if the toaster has resistance 11.0 Ω and is connected across 110 V? (*b*) How much does it cost to operate a heater of resistance 8 Ω across 110 V for 8 hours?

20. A battery with 12-V emf has a terminal voltage of 11.4 V when it delivers 20 A to the starter of a car. What is the internal resistance of the battery?

21. (*a*) How much power is delivered by the source of emf of the battery in Exercise 20 when it delivers 20 A? How much of this power is delivered to the starter? (*b*) By how much does the chemical energy of the battery decrease when it delivers 20 A for 3 min in starting the car? (*c*) How much heat is developed in the battery when it delivers 20 A for 3 min?

22. A 12-V car battery has an internal resistance of 0.4 Ω. (*a*) What is the current if the battery is shorted momentarily? (*b*) What is the terminal voltage when the battery delivers 25 A?

Section 20-4   Combinations of Resistors

23. (*a*) Find the equivalent resistance between points a and b in Figure 20-42. (*b*) If the potential drop between a and b is 12 V, find the current in each resistor.

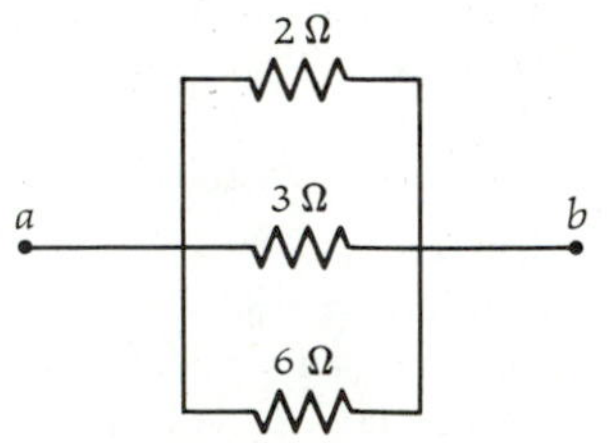

Figure 20-42
Exercise 23.

24. Repeat Exercise 23 for the resistors shown in Figure 20-43.

Figure 20-43
Exercise 24.

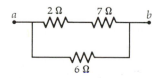

25. Repeat Exercise 23 for the resistors shown in Figure 20-44.

Figure 20-44
Exercises 25 and 27.

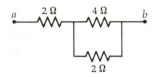

26. Repeat Exercise 23 for the resistors shown in Figure 20-45.

Figure 20-45
Exercises 26 and 28.

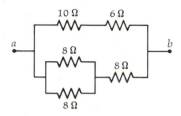

27. If the current in the 6-Ω resistor of Figure 20-44 is 3 A, (*a*) what is the potential drop between *a* and *b*? (*b*) What is the current in the 2-Ω resistor?

28. If the current in the 6-Ω resistor in Figure 20-45 is 2 A, (*a*) what is the potential drop between *a* and *b*? (*b*) What is the total current in the circuit?

29. Repeat Exercise 23 for the circuit shown in Figure 20-46.

Figure 20-46
Exercises 29 and 30.

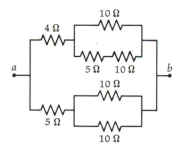

30. If the potential drop across the 4-Ω resistor in Figure 20-46 is 6 V, (*a*) what is the total current in the circuit? (*b*) What is the potential drop between *a* and *b*?

31. The battery in Figure 20-47 has negligible internal resistance. Find (*a*) the current in each resistor and (*b*) the power delivered by the battery.

Figure 20-47
Exercise 31.

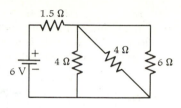

Section 20-5 Kirchhoff's Rules

32. A battery of emf 6 V and internal resistance 2.0 Ω is connected to a variable resistance R. Find the current and the power delivered by the battery when R is (*a*) 1.0 Ω, (*b*) 2.0 Ω, and (*c*) 3.0 Ω.

33. In Figure 20-48, the emf is 6 V and $R = 4\ \Omega$. The rate of Joule heating in R is 5.76 W. (*a*) What is the current in the circuit? (*b*) What is the value of *r*?

Figure 20-48
Exercise 33.

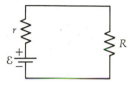

34. The batteries in Figure 20-49 have negligible internal resistance. Find (*a*) the current, (*b*) the power delivered or absorbed by each battery, and (*c*) the rate of Joule heat production in each resistor.

Figure 20-49
Exercise 34.

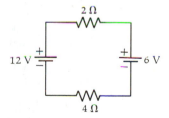

35. In the circuit shown in Figure 20-50, the batteries have negligible internal resistance and the ammeter has negligible resistance. (*a*) Find the current through the ammeter. (*b*) Find the energy delivered by the 12-V battery in 3 s. (*c*) Find the total heat produced in that time. (*d*) Account for the difference in your answers to parts (*b*) and (*c*).

Figure 20-50
Exercise 35.

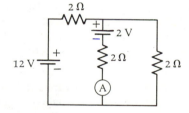

36. In the circuit shown in Figure 20-51, the batteries have negligible internal resistance. (*a*) Find the current in each resistor, (*b*) the potential difference between points *a* and *b*, and (*c*) the power supplied by each battery.

Figure 20-51
Exercise 36.

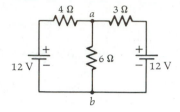

37. Repeat Exercise 36 for the circuit shown in Figure 20-52.

Figure 20-52
Exercise 37.

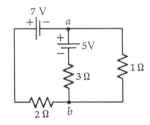

Section 20-6   RC Circuits

38. A 6-μF capacitor is initially charged to 100 V and is then connected across a 500-Ω resistor. (*a*) What is the initial charge on the capacitor? (*b*) What is the initial current just after the capacitor is connected? (*c*) What is the time constant of this circuit?

39. Repeat Exercise 38 for a 15-μF capacitor initially charged to 60 V and then connected across a 40-Ω resistor.

40. A 10-μF capacitor is initially charged to 1.2×10^{-4} C. It is then connected across a 6-Ω resistor. (*a*) What is the initial current when the capacitor is connected to the resistor? (*b*) What is the time constant? (*c*) How much charge is on the capacitor after a time equal to 1τ?

41. A 1.2-μF capacitor initially uncharged is connected in series with a 10-kΩ resistor and a 5.0-V battery of negligible internal resistance. (*a*) What is the initial current? (*b*) What is the charge on the capacitor after a very long time? (*c*) Sketch a graph of the charge on the capacitor as a function of time.

42. A 2-MΩ resistor is connected in series with a 1.5-μF capacitor and a 6.0 V battery of negligible internal resistance. The capacitor is initially uncharged. After a time $t = RC$, (*a*) what is the charge on the capacitor? (*b*) What is the current? (*c*) Find the values of the charge and current after a very long time.

Section 20-7   Ammeters, Voltmeters, and Ohmmeters

43. A galvanometer has a resistance of 100 Ω and requires 1 mA to give a full-scale deflection. (*a*) What resistance should be placed in parallel with the galvanometer to make an ammeter that reads 2 A at full-scale deflection? (*b*) What resistance should be placed in series to make a voltmeter that reads 5 V at full-scale deflection?

44. Sensitive galvanometers can detect currents as small as 1 pA. How many electrons per second does this current represent?

45. A sensitive galvanometer has a resistance of 80 Ω and requires 2.0 μA current to produce full-scale deflection. (*a*) Find the shunt resistance needed to construct an ammeter that reads 1.0 mA as a full-scale deflection. (*b*) What is the resistance of this ammeter? (*c*) Find the series resistance needed to construct a voltmeter that reads 3.0 V as full scale from this galvanometer.

46. A galvanometer of resistance 100 Ω gives full-scale deflection when its current is 1.0 mA. It is used with a 1.5-V battery of negligible internal resistance to make an ohmmeter. (*a*) What resistance R_s should be placed in series with the galvanometer? (*b*) What resistance R will give a half-scale deflection? (*c*) What resistance will give a deflection of one-tenth scale?

Problems

1. A 200-W heater is used to heat water in a cup. It consists of a single resistor R placed across 110 V. (*a*) Find R. (*b*) Assume that 90 percent of the energy goes into heating the water. How long does it take to heat 0.25 kg of water from 15 to 100°C? (*c*) How long does it take to boil this water away after it reaches 100°C?

2. A variable resistance R is connected across a potential difference V that remains constant. When R has the value R_1, the current is 6.0 A. When R is increased by 10 Ω from this value, the current drops to 2.0 A. Find R_1 and V.

3. A battery has emf $\mathscr{E}$ and internal resistance r. When a 5.0-Ω resistor is connected across the terminals, the current is 0.50 A. When this resistor is replaced by an 11.0-Ω resistor, the current is 0.25 A. Find the emf $\mathscr{E}$ and the internal resistance r.

4. A galvanometer of resistance 100 Ω gives a full-scale reading when its current is 0.1 mA. It is to be used in a multirange voltmeter as shown in Figure 20-53, where the connections refer to full-scale readings. Determine R_1, R_2, and R_3.

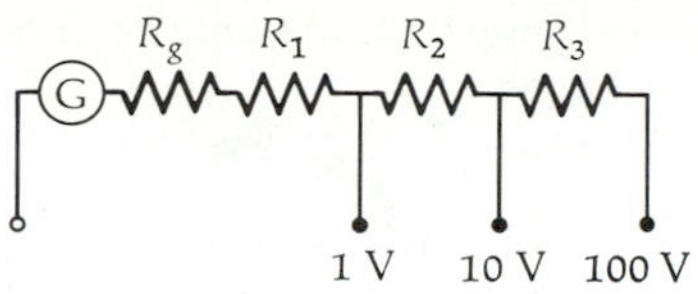

Figure 20-53
Problem 4.

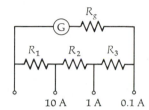

1 V   10 V   100 V

5. The galvanometer of Problem 4 is to be used in a multi-range ammeter with the full-scale readings indicated in Figure 20-54. Determine R_1, R_2, and R_3.

Figure 20-54
Problem 5.

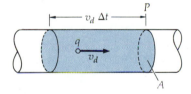

10 A   1 A   0.1 A

6. In this problem, you are to derive a relation between the current in a wire and the drift velocity of the charge carriers. Figure 20-55 shows a segment of wire with a typical charge carrier carrying charge q and moving with speed v_d. Convince yourself that in some time Δt, all the charge carriers in the shaded volume cross point P. If A is the cross-sectional area of the wire, and there are n charge carriers per unit volume, each with charge q, show that the current in the wire is given by

$$I = nqv_d A$$

Figure 20-55
Problem 6.

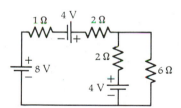

7. Copper has a density of 8.92 g/cm³ and a molar mass of 63.5 g/mol. (a) Use these data and the fact that there are 6.02×10^{23} atoms/mol to find the number of atoms per cubic centimetre in copper. This is also the number of free electrons per cubic centimetre in copper because there is one free electron per atom on the average. (b) A 14-gauge copper wire (radius 0.0814 cm) carries a current of 1 A. Calculate the drift velocity of the electrons from the equation given in Problem 6, using n from part (a) and $q = e$ for the charge of each carrier.

8. A galvanometer of resistance 80 Ω gives a full-scale deflection when its current is 1.2 mA. It is used to construct an ammeter whose full-scale reading is 10 A. (a) Find the shunt resistance needed. (b) If the shunt resistance consists of a

piece of 10-gauge copper wire (diameter 2.59 mm), what should its length be?

9. A sick car battery of emf 11.4 V and internal resistance 0.01 Ω is connected to a load of 1.0 Ω. To help the ailing battery, a second battery of emf 12.6 V and internal resistance 0.01 Ω is connected by jumper cables to the terminals of the first battery. Draw a circuit diagram for this situation and find the current in each part of the circuit. Find the power delivered by the second battery and discuss where this power goes, assuming that both emfs and internal resistances are constant.

10. For the ohmmeter of Exercise 46, indicate how the galvanometer scale should be calibrated by representing the scale on a straight line of some length L, where the end of the line at $x = L$ represents a full-scale reading of $R = 0$. Divide the line into 10 equal divisions and indicate the values of the resistance at each division.

11. A Van de Graaff generator belt carries a charge of 5 μC per square metre of surface. The belt is 0.5 m wide and moves at 20 m/s. (a) What current does it carry? (b) If this charge is raised to a potential of 100 kV, what is the minimum power of the motor needed to drive the belt?

12. A coil of nichrome wire is used as a heating element in a water boiler required to generate 8.0 g of steam per second from water at 100°C. The wire has a diameter of 1.80 mm and is connected to a 115-V power supply. Find the length of wire required.

13. In the circuit shown in Figure 20-56, find (a) the current in each resistor, (b) the power supplied by each emf source, and (c) the power dissipated in each resistor.

Figure 20-56
Problem 13.

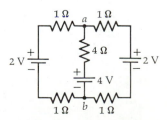

14. In the circuit shown in Figure 20-57, find the potential difference between points a and b.

Figure 20-57
Problem 14.

15. In the circuit shown in Figure 20-58, the battery has an internal resistance of 0.01 Ω. An ammeter of resistance 0.01 Ω is inserted at point a. (a) What is the reading of the ammeter? (b) By what percentage is the current changed because of the ammeter? (c) The ammeter is removed and a voltmeter of resistance 1 kΩ is connected from a to b. What is the reading of the voltmeter? (d) By what percentage is the voltage drop from a to b changed by the presence of the voltmeter?

Figure 20-58
Problem 15.

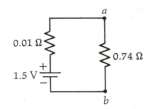

16. Figure 20-59 shows a circuit called a *Wheatstone bridge,* which is used for determining an unknown resistance R_x in terms of known resistances R_1, R_2, and R_0. The galvanometer is used as a null detector. When points a and b are at the same potential, no current flows through the galvanometer and the bridge is said to be balanced. In the slide-wire type bridge, resistances R_1 and R_2 are a wire one metre long. A pointer is moved along the wire to vary these two resistances (while keeping their sum constant) until the bridge is balanced. (a) Show that when the bridge is balanced,

$$R_1 I_1 = R_x I_2$$

and

$$R_2 I_1 = R_0 I_2$$

(b) Show that the unknown resistance is then given by

$$R_x = \frac{R_1}{R_2} R_0$$

In a slide-wire Wheatstone bridge, R_1 is proportional to the length from 0 to the pointer at a, and R_2 is proportional to the length from a to 100 cm (the end of the wire). If $R_0 = 200\ \Omega$, find R_x if the bridge balances at (c) the 18-cm mark and (d) the 60-cm mark.

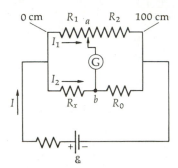

Figure 20-59
Wheatstone bridge circuit for Problem 16. In the slide-wire type, the resistances R_1 and R_2 consist of a wire 100 cm long.

CHAPTER TWENTY-ONE

The Magnetic Field

In this chapter, we investigate magnetism, a subject that still has something of a mysterious air about it. Partly this is due to our fascination with magnets. They are favorite playthings for many children, and most novelty shops carry a few magnetic "toys" for adults. Then again, who hasn't twisted a compass merely to watch the needle reorient itself to the north? But the mysteriousness of magnetism is also partly due to the fact that there is much about it we don't yet understand. Why, for instance, is the earth magnetized? Similarly, why did the direction of earth's magnetic field reverse itself in the distant past? More to the point, will it happen again?

For that stone not only attracts iron rings, but also imparts to them a similar power of attracting other rings; and sometimes you may see a number of pieces of iron and rings suspended from one another so as to form a chain: and all of them derive their power of suspension from the original stone. Now this is like the Muse, who first gives to men inspiration herself: and from these inspired persons a chain of other persons is suspended, who take the inspiration from them.

PLATO

As was mentioned in Chapter 18, it was known to the early Greeks more than 2000 years ago that lodestone (now called magnetite) attracts iron. There is written reference to the use of magnets for navigation as early as the twelfth century. Early experiments in magnetism were concerned chiefly with the behavior of permanent magnets. In 1269, an important discovery was made by Pierre de Maricourt when he laid a needle on a spherical natural magnet at various positions and marked the directions taken by the needle. The lines encircled the magnet just as the meridian lines encircle the earth, passing through two points at opposite ends of the sphere. Because of the analogy with the meridian lines of the earth, he called these points the *poles* of the magnet. It was noted by many early experimenters that every magnet of whatever shape has two poles, a north pole and a south pole, where the magnetic force exerted by the magnet is strongest. Like poles of two magnets repel each other, and unlike poles attract.

In 1600, William Gilbert discovered why a compass needle orients itself in definite directions: the earth itself is a permanent magnet. Since the north pole of a compass needle is attracted to a region near the north geographic pole of the earth, this region is actually a south magnetic pole. The attraction

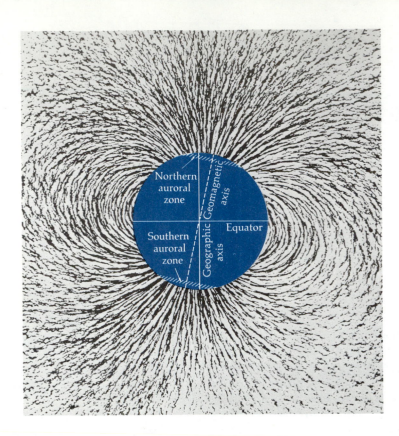

Magnetic field lines of the earth indicated by iron filings around a uniformly magnetized sphere.

and repulsion of magnetic poles was studied quantitatively by John Michell in 1750. Using a torsion balance, Michell showed that the attraction and repulsion of the poles of two magnets are of equal strength and vary inversely with the square of the distance between the poles. These results were confirmed a few years later by Coulomb. The force between two magnetic poles is similar to that between two electric charges, but there is an important difference: magnetic poles always occur in pairs. It is impossible to isolate a single magnetic pole. If a magnet is broken in half, equal and opposite poles appear at the break point so that there are two magnets, each with equal and opposite poles. Coulomb explained this result by assuming that magnetism is contained in each molecule of the magnet.

The connection between electricity and magnetism was not known until the nineteenth century, when Hans Christian Oersted discovered that an electric current affects the orientation of a compass needle. Subsequent experiments by André Marie Ampère and others showed that electric currents attract bits of iron and that parallel currents attract each other. Ampère proposed the theory that electric currents are the source of all magnetism. In ferromagnets, these currents were thought to be "molecular" current loops that are aligned when the material is magnetized. Ampère's model is the basis of the theory of magnetism today, though in many cases the "currents" in magnetic materials are related to the intrinsic electron spin, a quantum property of the electron that could not be envisioned in the nineteenth century. Further connections between magnetism and electricity were demonstrated by experiments of Michael Faraday and Joseph Henry, showing that a changing magnetic field produces an electric field, and by

Oersted's discovery of the connection between electricity and magnetism was made while delivering a physics lecture. It created immense excitement in the scientific world of his day. Indeed, when they heard about it, both Ampère and Faraday immediately stopped their investigations of other phenomena and turned to the study of electromagnetism.

James Clerk Maxwell's theory, showing that a changing electric field produces a magnetic field.

The basic magnetic interaction is one between two charges in motion. The force exerted by one charge in motion on another charge in motion (in addition to the electrostatic force) is called the magnetic force. As with electrostatic interaction, it is convenient to consider the magnetic force exerted by one moving charge on another to be transmitted by a third agent, the magnetic field. A moving charge produces a magnetic field, and that field in turn exerts a force on a second moving charge. Since a moving charge constitutes an electric current, magnetic interaction can also be thought of as an interaction between two currents.

21-1   Definition of the Magnetic Field

The existence of a magnetic field **B** at some point is easily verified. We simply place a compass needle at that point and see if it tends to align itself in a particular direction. If there are no magnets or current-carrying wires nearby, the needle will point in the direction of the magnetic field of the earth. If there are magnets or current-carrying wires nearby, the needle will point in the direction of the resultant magnetic field due to the earth and the magnets or currents. It has been observed experimentally that, when a charge q has velocity **v** in a magnetic field, there is a force on it that depends on q and on the magnitude and direction of the velocity. Let us assume that we know the direction of the magnetic field **B** at a point in space from a measurement with a compass. Experiments with various charges moving with various velocities at such a point give the following results for the magnetic force:

1. The force is proportional to the magnitude of the charge q.   Magnetic force on a moving charge

2. The force is proportional to the speed v.

3. The magnitude and direction of the force depend on the direction of the velocity **v** relative to the direction of the magnetic field **B**.

4. If the velocity of the particle is parallel or antiparallel to the magnetic field, the force is zero.

5. If the velocity is not parallel to **B**, there is a force that is perpendicular to the magnetic field and is also perpendicular to the velocity.

6. If the velocity makes an angle θ with **B**, the force is proportional to $\sin \theta$.

7. The force on a negative charge is in the direction opposite to that on a positive charge with the same velocity.

These experimental results can be summarized as follows. When a charge q moves with velocity **v** in a magnetic field **B**, there is a magnetic force **F** that is perpendicular to both **B** and **v** and has the magnitude

$$F = qvB \sin \theta \qquad\qquad 21\text{-}1$$

The direction of **F** is perpendicular to both **v** and **B** and is therefore perpendicular to the plane formed by these two vectors. We still need to specify the sense of **F**, that is, which of the two possible directions perpendicular to the plane of **v** and **B** it has. This is most easily done by the right-hand rule illustrated in Figure 21-1.

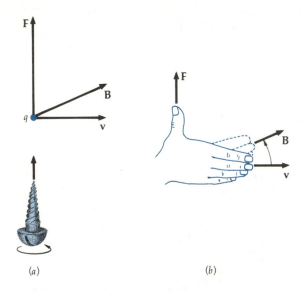

(a)   (b)

We imagine rotating the vector **v** into **B**. The direction of **F** is the direction in which a right-hand threaded screw would advance if turned in the same way. Alternatively, we point the fingers of the right hand along the direction of **v** so that they can be curled in the direction of **B**. The thumb then points in the direction of **F**. Figure 21-2 shows the force exerted on various moving charges when the magnetic-field vector **B** is in the vertical direction. For historical reasons, the magnetic field **B** is sometimes called the *magnetic-induction vector* or the *magnetic flux density*. We will usually refer to **B** as simply the magnetic field.

Equation 21-1 defines the magnetic field **B** in terms of the force exerted on a moving charge. The sources of magnetic fields will be discussed in Section 21-4. There are two other useful ways of writing Equation 21-1 for the magnitude of the magnetic force exerted on a moving charge. We note that the quantity $v \sin \theta$ is the component of the velocity $v_\perp$ perpendicular to **B**. Thus, Equation 21-1 can also be written

$$F = q v_\perp B \qquad\qquad 21\text{-}2$$

Alternatively, $B \sin \theta$ can be considered to be the component of the magnetic field $B_\perp$ perpendicular to the velocity **v**. Then Equation 21-1 is

$$F = q v B_\perp \qquad\qquad 21\text{-}3$$

The SI unit of magnetic field is the tesla (T). A charge of one coulomb moving with a velocity of one metre per second perpendicular to a magnetic field of one tesla experiences a force of one newton:

$$1\ \text{T} = 1\ \frac{\text{N/C}}{\text{m/s}} = \frac{\text{N}\cdot\text{s}}{\text{C}\cdot\text{m}} = 1\ \frac{\text{N}}{\text{A}\cdot\text{m}} \qquad\qquad 21\text{-}4$$

Figure 21-1
Right-hand rule for determining the direction of the force exerted on a charge moving in a magnetic field. (*a*) The force is perpendicular to both **v** and **B** and is in the direction of the advance of a right-hand threaded screw if it is turned in the direction that rotates **v** into **B**. (*b*) If the fingers of the right hand are in the direction of **v** such that they can be curled into **B**, the thumb points in the direction of **F**.

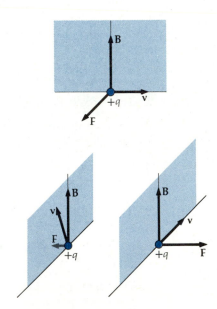

Figure 21-2
Direction of the magnetic force on a charged particle with velocity **v** at various orientations in a magnetic field **B**. The force is perpendicular to the plane of **v** and **B**.

This unit is rather large. A commonly used unit, derived from the cgs system, is the gauss (G), which is related to the tesla as follows:

$$1 \text{ T} = 10^4 \text{ G} \qquad\qquad 21\text{-}5$$

The magnitude of the magnetic field of the earth is of the order of 1 G. Since magnetic fields are often given in gauss, which is not an SI unit, it is important to remember to convert gauss to tesla when making calculations.

Example 21-1   The magnetic field of the earth has a magnitude of 0.6 G and is directed downward and northward, making an angle of about 70° with the horizontal. (These data are approximately correct for the central United States.) A proton moves horizontally in the northward direction with speed $v = 10^7$ m/s. Calculate the force on the proton.

Figure 21-3 shows the directions of the magnetic field **B** and the proton's velocity **v**. The angle between them is $\theta = 70°$. The magnetic force on a proton moving north is toward the west, as indicated in the figure. The magnitude of the force is

$$F = qvB \sin \theta$$

$$= (1.6 \times 10^{-19} \text{ C})(10^7 \text{ m/s})(0.6 \times 10^{-4} \text{ T})(0.94)$$

$$= 9.02 \times 10^{-17} \text{ N}$$

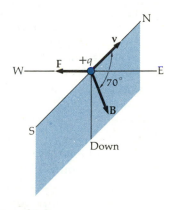

Figure 21-3
Example 21-1. Force on a proton moving north in the magnetic field of the earth, which makes an angle of 70° with the horizontal north direction. The force is directed toward the west.

When a wire carries a current in a magnetic field, there is a force on the wire that is the sum of the magnetic forces on the charged particles whose motions produce the current. Figure 21-4 shows a short segment of wire of cross-sectional area A and length ℓ carrying a current I. The current can be related to the number of charges n per unit volume in the wire, the charge q on each, and their drift velocity v_d. In some time Δt, all the charges shown in the figure will cross the boundary at the right if $\ell = v_d \Delta t$. The total charge that crosses this boundary is then the number of charges n per unit volume times the charge q times the volume $\ell A = v_d \Delta t A$:

$$\Delta Q = nq v_d \Delta t A$$

The current in the wire is thus

$$I = \frac{\Delta Q}{\Delta t} = nq v_d A \qquad\qquad 21\text{-}6$$

If there is a magnetic field **B** that makes an angle θ with the wire (and therefore with the velocity vector of each charge), there will be a force on each charge of magnitude $q v_d \sin \theta\, B$. Since there are $n\ell A$ charges in the wire segment of area A, the total force on the wire segment will be

$$F = q v_d B \sin \theta\; n\ell A$$

Using $I = nq v_d A$ from Equation 21-6, we have

$$\boxed{F = I\ell B \sin \theta} \qquad\qquad 21\text{-}7$$

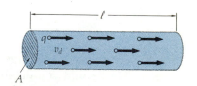

Figure 21-4
Wire segment of length ℓ carrying current I. In some time Δt, all the charges in the wire moving with speed v_d will cross the boundary at the right if $\ell = v_d \Delta t$. If the wire is in a magnetic field, there will be a force on each charge carrier resulting in a force on the wire.

This can also be written

$$F = I\ell B_\perp \qquad\qquad 21\text{-}8$$

where $B_\perp$ is the component of **B** perpendicular to the wire. For the current and magnetic field directions shown in Figure 21-5, the force on the wire is directed out of the paper as indicated.

Equations 21-7 and 21-8 assume that the wire is straight and that the magnetic field does not vary over the length of the wire. It is easily generalized for an arbitrarily shaped wire in any magnetic field. We merely choose a very small wire segment $\Delta\ell$ and write the force on this segment as ΔF:

$$\Delta F = IB \sin\theta\, \Delta\ell \qquad\qquad 21\text{-}9$$

Magnetic force on a current element

where B is the magnitude of the magnetic-field vector at the segment. The quantity $I\,\Delta\ell$ is called a *current element*. We find the total force on the wire by summing all the current elements using the appropriate field **B** at each element.

Equation 21-9 is equivalent to Equation 21-1. It defines the magnetic field **B** in terms of the force exerted on a current element. The source of a magnetic field is another current element or a moving charge, as will be discussed in Section 21-4.

Example 21-2   A wire carries a current of 3 A in a magnetic field of magnitude 0.02 T. What is the magnitude of the magnetic force exerted on an element of the wire 1 mm long that makes an angle of 30° with the magnetic field?

From Equation 21-9, we have

$$\Delta F = IB \sin\theta\, \Delta\ell = (3.0\ \text{A})(0.02\ \text{T})(\sin 30°)(0.001\ \text{m})$$

$$= 3 \times 10^{-5}\ \text{N}$$

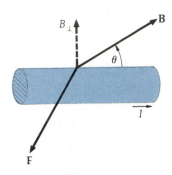

Figure 21-5
Magnetic force on a current-carrying wire in a magnetic field. The force **F** is perpendicular to both **B** and the current I, which are in the plane of the paper. Its magnitude is $I\ell B \sin\theta = I\ell B_\perp$.

Questions

1. Charge q moves with velocity **v** through a magnetic field **B**. At this instant, it experiences a magnetic force **F**. How would the force differ if the charge had the opposite sign? If the velocity had the opposite direction? If the magnetic field had the opposite direction?

2. For what angle between **B** and **v** is the magnetic force on q the greatest? The least?

3. A moving electric charge may experience both electric and magnetic forces. How could you determine whether a force that causes a charge to deviate from a straight path is an electric force or a magnetic force?

4. How can a charge move through a region of magnetic field without ever experiencing any magnetic force?

5. Show that the force on a current element is the same in direction and magnitude regardless of whether positive charges, negative charges, or a mixture of positive and negative charges create the current.

6. A current-carrying wire is in a magnetic field, but the wire does not experience any magnetic force. How is this possible?

21-2   Torques on Magnets and Current Loops

When a small permanent magnet, such as a compass needle, is placed in a magnetic field **B**, it tends to orient itself so that the north pole points in the direction of **B**. This effect also occurs with previously unmagnetized iron filings, which become magnetized in the presence of a magnetic field. Figure 21-6 shows a small magnet of length ℓ that makes an angle θ with a magnetic field **B**. There is a force $\mathbf{F_1}$ on the north pole in the direction of **B** and an equal but opposite force $\mathbf{F_2}$ on the south pole. These two forces produce no translational motion because they are equal and opposite, but they do produce a torque that tends to rotate the magnet so that it lines up with the field. The *pole strength of the magnet* q_m is defined as the ratio of the magnitude of the force **F** on the pole to the magnitude of the field **B***:

Figure 21-6
A small magnet in a uniform magnetic field experiences a torque that tends to rotate the magnet into the direction of the field.

$$q_m = \frac{F}{B}$$

Magnetic pole strength defined

Since a tesla is equal to a newton per ampere-metre, the SI unit of magnetic pole strength is the ampere-metre (A·m).

$$\frac{1\ \text{N}}{1\ \text{N/A}\cdot\text{m}} = 1\ \text{A}\cdot\text{m}$$

If we adopt the sign convention that the north pole is positive and the south pole negative, the force on a pole can be written as a vector equation:

$$\mathbf{F} = q_m\mathbf{B} \qquad \text{21-11}$$

The torque produced by equal and opposite forces is the same about any point. If we choose the south pole for computing the torque, the force is $F_1 = q_mB$ and the lever arm is $\ell \sin \theta$, so the torque is

$$\tau = q_mB\ell \sin \theta \qquad \text{21-12}$$

The *magnetic moment* **m** of a magnet is defined as

$$\mathbf{m} = q_m\boldsymbol{\ell} \qquad \text{21-13} \qquad \text{Magnetic moment}$$

where $\boldsymbol{\ell}$ is the vector length of the magnet pointing from the south pole to the north pole. The magnetic moment is analogous to the electric dipole moment $q\boldsymbol{\ell}$. The SI unit of magnetic moment is the ampere-metre squared (A·m²). In terms of the magnetic moment, the torque on a magnet is

$$\tau = mB \sin \theta \qquad \text{21-14} \qquad \text{Torque on a magnetic moment}$$

The sense of the torque is such that it tends to align the magnetic moment of the magnet in the direction of the external magnetic field **B**.

It is customary to indicate the direction of the magnetic field **B** by drawing lines that are parallel to **B** at each point in space and to indicate the magnitude of **B** by the density of the lines, just as for the electric field **E**. For a given magnetic field **B**, the lines can be found using a compass needle or iron filings since these small magnets align themselves in the direction of **B**.

Magnetic field lines

---

* The notation for magnetic pole strength q_m is used so that the magnetic equations resemble the corresponding equations for electric charges in electric fields. The subscript m reminds us that q_m denotes a magnetic pole and not an electric charge.

Figure 21-7 shows the magnetic field lines near a bar magnet. External to the magnet, the lines leave the north pole and enter the south pole.

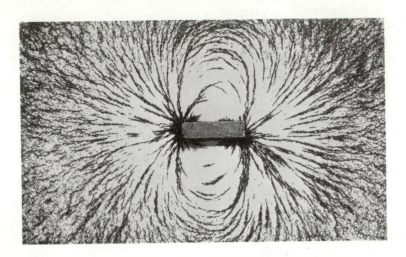

Figure 21-7
Magnetic field lines of a bar magnet indicated by the alignment of iron filings.

A small current loop placed in an external magnetic field acts very much like a small bar magnet. Figure 21-8 shows a rectangular wire loop carrying a current I and placed in an external magnetic field **B**. There is no magnetic force on either the top or bottom segments of the loop because the current in these segments is in the direction of the magnetic field. However, there are equal and opposite forces on the side segments that produce a torque on the current loop. The magnitude of these forces is just $F = IaB$ since the current is perpendicular to the magnetic field and the length of each wire segment is a. If we consider the center of the loop, each force produces a torque $Fb/2$, and the torques add. The resultant torque has the magnitude

$$\tau = Fb = IabB$$

The torque is proportional to the current I, the magnetic field **B**, and the area ab of the loop. The behavior of such a loop is most easily described by defining a magnetic moment for the loop whose magnitude equals the product of the current and the area of the loop and whose direction is perpendicular to the plane of the loop:

$$m = IA$$

Often a loop contains several turns of wire, each turn having the same area and carrying the same current. The total magnetic moment of such a loop is the product of the number of turns N and the magnetic moment of each turn. The total magnetic moment of a loop of N turns is thus

$$m = NIA \qquad \qquad 21\text{-}15$$

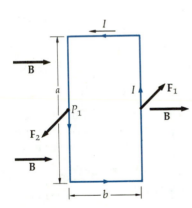

Figure 21-8
Forces exerted on a rectangular current loop in a uniform magnetic field **B** parallel to the plane of the loop. The forces produce a torque that tends to twist the loop.

Magnetic moment of a current loop

The sense of the magnetic moment is obtained from the right-hand rule illustrated in Figure 21-9. If the loop is grasped so that the fingers of the right hand point along the current, the thumb points in the direction of the magnetic moment **m**. With this definition, the torque on the loop is given by

$$\tau = mB \sin \theta$$

where θ is the angle between the magnetic moment vector **m** and the magnetic field **B**. This is the same as Equation 21-14 for the torque on a small magnet. As in that case, the torque tends to align the magnetic moment of the loop in the direction of the magnetic field **B**.

Although the torque is most easily calculated for a rectangular loop, the results apply to a loop of any shape. The galvanometer discussed in Chapter 20 makes use of the torque exerted by a magnetic field on a current loop. The pointer of the galvanometer (see Figure 21-10) is attached to a loop of several turns of wire that sits in a magnetic field produced by a permanent magnet. When the coil carries a current I, there is a torque on the loop proportional to I. The coil rotates until the magnetic torque is balanced by a torque due to a spring attached to the coil. The reading on the scale is thus proportional to the current through the coil.

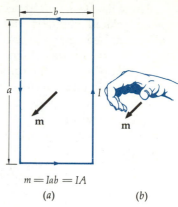

$$m = Iab = IA$$

(a)   (b)

Figure 21-9
(a) The magnetic moment **m** of a loop is perpendicular to the plane of the loop and has a magnitude IA, where A is the area of the loop. (b) Right-hand rule for determining the direction of **m**. When the fingers of the right hand curl around the loop in the direction of the current, the thumb points in the direction of **m**.

Figure 21-10
Galvanometer. When the coil carries a current, the magnet exerts a torque on the coil proportional to the current, causing the coil to twist. The coil is mounted on a spring that exerts a balancing torque. The deflection read on the scale is proportional to the current.

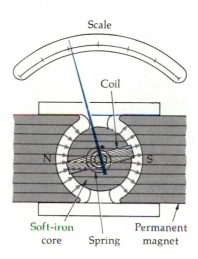

Scale

Coil

N   S

Soft-iron   Spring   Permanent
core   magnet

The fact that a current loop behaves like a small magnet is not accidental. The origin of the magnetic moment in a bar magnet is, in fact, microscopic current loops due to motion of electrons in the atoms of the magnet. This motion is partly due to the orbital motion of the electrons in their atoms and partly due to electron spin. (Although electron spin is a rather complicated quantum-mechanics concept, the simple qualitative picture of an electron as a charged ball spinning on its axis works quite well. A spinning ball of charge can be considered to be a set of current loops.) We can think of a narrow bar magnet as a line of current loops with their magnetic moments aligned, as is shown in Figure 21-11. Inside the magnet, the north and south poles cancel leaving a single north pole at one end and a single south pole at the other end. However, we can never break the magnet in such a way as to isolate a single magnetic pole.

Figure 21-11
A bar magnet can be thought of as a line of current loops, each of which is equivalent to a tiny magnet. Inside the bar magnet, the north and south poles of the tiny magnets cancel leaving one end with a north pole and the other with a south pole.

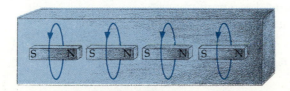

There has been much speculation throughout the years as to the existence of an isolated magnetic pole, and in recent years considerable experimental effort has been made to find such an object. Despite this effort, there seems to be no conclusive evidence that an isolated magnetic pole exists. At this time, it appears that the fundamental unit of magnetism is not the magnetic pole but is, instead, a current loop that behaves as a magnetic dipole. Since isolated magnetic poles do not exist, there are no regions in space where magnetic field lines begin or end. This is different from the situation for electric field lines, which begin on positive charges and end on negative charges. Magnetic field lines form closed loops, as illustrated in Figure 21-12, which shows the magnetic field lines both inside and outside a bar magnet.

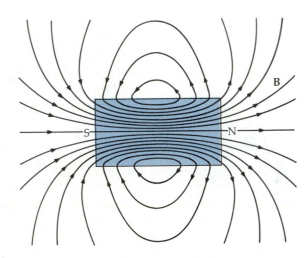

Figure 21-12
Magnetic field lines inside and outside a bar magnet. The lines have no beginning or end. Instead, they form closed loops.

Example 21-3   A circular loop of radius 2 cm has 10 turns of wire and carries a current of 3 A. The axis of the loop makes an angle of 30° with a magnetic field of 8000 G. Find the torque on the loop.

The magnitude of the magnetic moment of the loop is

$$m = NIA = (10)(3 \text{ A})\pi(0.02 \text{ m})^2 = 3.77 \times 10^{-2} \text{ A} \cdot \text{m}^2$$

The magnitude of the torque is then

$$\tau = mB \sin \theta = (3.77 \times 10^{-2} \text{ A} \cdot \text{m}^2)(0.8 \text{ T})(\sin 30°)$$

$$= 1.51 \times 10^{-2} \text{ N} \cdot \text{m}$$

where we have used the fact that 8000 G equals 0.8 T and 1 T = 1 N/A·m.

Questions

7. A current loop has its magnetic moment antiparallel to a uniform magnetic field. What is the torque on the loop? Is this equilibrium stable or unstable?

8. How are magnetic field lines similar to electric field lines? How are they different?

21-3   Motion of a Point Charge in a Magnetic Field

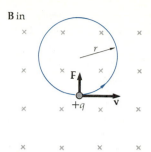

Figure 21-13
Charged particle moving in a plane perpendicular to a uniform magnetic field that is directed into the paper (as indicated by the crosses). The magnetic force is perpendicular to the velocity of the particle, causing it to move in a circular orbit.

An important characteristic of the magnetic force on a moving charged particle is that the force is always perpendicular to the velocity of the particle. The magnetic force therefore does no work on the particle, and the kinetic energy of the particle is unaffected by this force. The magnetic force changes the direction of the velocity but not its magnitude.

In the special case where the velocity of a particle is perpendicular to a uniform magnetic field, as shown in Figure 21-13, the particle moves in a circular orbit. The magnetic force provides the centripetal force necessary for circular motion. We can relate the radius of the circle r to the magnetic field $\mathbf{B}$ and the speed v of the particle by setting the resultant force equal to the mass m times the centripetal acceleration v^2/r in accordance with Newton's second law. The resultant force in this case is just qvB since $\mathbf{v}$ and $\mathbf{B}$ are perpendicular. Thus, Newton's second law gives

$$F = ma$$

$$qvB = \frac{mv^2}{r}$$

or

$$r = \frac{mv}{qB} \qquad \text{21-16}$$

The period of the circular motion is the time it takes the particle to travel once around the circle, a distance equal to the circumference of the circle, $2\pi r$. If T is the period and v the speed, we have

$$vT = 2\pi r$$

or

$$T = \frac{2\pi r}{v}$$

Substituting $r = mv/qB$ from Equation 21-16, we have for the period

$$T = \frac{2\pi(mv/qB)}{v} = \frac{2\pi m}{qB} \qquad \text{21-17}$$

The frequency for the circular motion is the reciprocal of the period:

$$f = \frac{1}{T} = \frac{qB}{2\pi m} \qquad \text{21-18}$$

Cyclotron frequency

Circular path of electrons moving in a magnetic field produced by two large coils. The electrons ionize the gas in the tube, causing it to give off a bluish glow that indicates the path of the beam.

Note that the period given by Equation 21-17 and the frequency given by Equation 21-18 do not depend on the radius of the orbit or the velocity of the particle. This period is called the *cyclotron period* and the frequency, the *cyclotron frequency*. (Two of the many interesting applications of the circular motion of charged particles in a uniform magnetic field, the mass spectrograph and the cyclotron, will be discussed later in this section.)

Example 21-4   A proton of mass $m = 1.67 \times 10^{-27}$ kg and charge $q = e = 1.6 \times 10^{-19}$ C moves perpendicularly to a magnetic field $B = 4000$ G. It moves in a circle of radius 21 cm. Find the period of the motion and the speed of the proton.

We don't need to know the radius of the circle to find the period. Using Equation 21-17 and converting the magnetic field to SI units (4000 G = 0.4 T), we have

$$T = \frac{2\pi m}{qB} = \frac{2\pi(1.67 \times 10^{-27}\text{ kg})}{(1.6 \times 10^{-19}\text{ C})(0.4\text{ T})} = 1.64 \times 10^{-7}\text{ s}$$

The speed v is related to the radius of the circle by Equation 21-16:

$$v = \frac{rqB}{m} = \frac{(0.21\text{ m})(1.6 \times 10^{-19}\text{ C})(0.4\text{ T})}{1.67 \times 10^{-27}\text{ m}} = 8.05 \times 10^6\text{ m/s}$$

In these calculations, we put each quantity in SI units so that the results will also be in SI units, namely, seconds for the period and metres per second for the speed. We note from Equation 21-16 that the radius of the circular motion is proportional to the speed. If we doubled the speed of the proton in this example, the radius would double, but the period and the frequency would remain unchanged. We can check our results by noting that the product of the speed v times the period T is the circumference of the circle $2\pi r$. Then

$$\frac{vT}{2\pi} = \frac{(8.05 \times 10^6\text{ m/s})(1.64 \times 10^{-7}\text{ s})}{2\pi}$$

$$= 0.21\text{ m} = 21\text{ cm}$$

Suppose a charged particle enters a region of uniform magnetic field with a velocity that is not perpendicular to **B**. We can resolve the velocity of the particle into components $v_\parallel$ parallel to **B** and $v_\perp$ perpendicular to **B**. The motion due to the perpendicular component is the same as that just discussed. The component of the velocity parallel to **B** is not affected by the magnetic field. It therefore remains constant. The path of the particle is thus a helix, as shown in Figure 21-14.

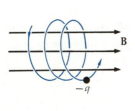

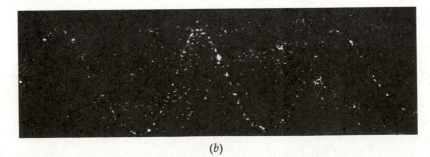

(a)   (b)

Figure 21-14

(a) When a charged particle has a velocity component parallel to a magnetic field as well as one perpendicular to it, the particle moves in a helical path around the field lines.

(b) Cloud-chamber photograph of the helical path of an electron moving in a magnetic field. The path of the electron is made visible by the condensation of water droplets in the cloud chamber.

The motion of charged particles in nonuniform magnetic fields is quite complicated. Figure 21-15 shows an interesting magnetic field configuration called a *magnetic bottle*. The field is weak at the center of the pattern and strong on both sides. (As in the case with the electric field, the strength of the magnetic field is indicated by the density of lines.) A detailed analysis of the motion of a charged particle in such a field shows that the particle will spiral around the field lines and become trapped, oscillating back and forth between points P_1 and P_2 in the "bottle."

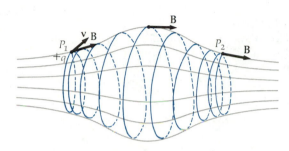

Figure 21-15
Magnetic bottle. When a charged particle moves in such a field, which is strong at both ends and weak in the middle, the particle becomes trapped and moves back and forth, spiraling around the field lines.

Such a magnetic field pattern is important for confining dense beams of charged particles, called *plasmas*, in nuclear fusion research. A similar phenomenon is the motion of protons and electrons along the magnetic field lines between the earth's magnetic poles. These particles oscillate back and forth in regions of space called the Van Allen belts as shown in Figure 21-16.

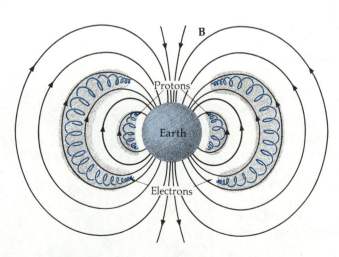

Figure 21-16
Van Allen belts. Protons (inner belts) and electrons (outer belts) are trapped in the earth's magnetic field and spiral around the field lines between the north and south poles.

Aurora borealis, sometimes called northern lights, is the result of ionization of oxygen and nitrogen in the upper atmosphere by protons that leak out of the Van Allen belts. The light is emitted when the ions and electrons recombine.

The Velocity Selector

The magnetic force on a charged particle moving in a uniform magnetic field can be balanced by an electrostatic force if the magnitudes and directions of the magnetic and electric fields are properly chosen. Since the electric force is in the direction of the electric field (for positive particles) and the magnetic force is perpendicular to the magnetic field, the electric and magnetic fields must be perpendicular to each other if the forces are to balance. Figure 21-17 shows a region of space between the plates of a capacitor where there is an electric field and a perpendicular magnetic field (which is produced by a magnet not shown). Such an arrangement of perpendicular fields is called *crossed fields*. Consider a particle of charge q entering this space from the left. If q is positive, the electric force of magnitude qE is down and the magnetic force of magnitude qvB is up. If the charge is negative, each of the forces is reversed. These two forces will balance if $qE = qvB$ or

$$v = \frac{E}{B} \qquad\qquad\qquad 21\text{-}19$$

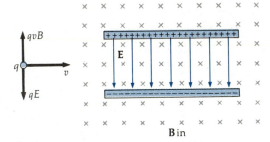

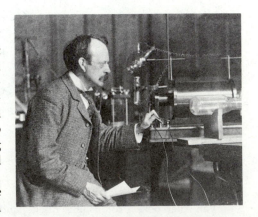

Figure 21-17
Crossed electric and magnetic fields. When a positive particle moves to the right, it experiences a downward electric force and an upward magnetic force. These forces balance if the speed of the particle is related to the field strengths by $vB = E$.

For given magnitudes of the electric and magnetic fields, the forces will balance only for particles with the speed given by Equation 21-19. Any particle, no matter what its mass or charge, will traverse the space unde-flected if its speed is that given by Equation 21-19. A particle with greater speed will be deflected in the direction of the magnetic force; and one with less speed will be deflected in the direction of the electric force. Such an arrangement of fields is called a *velocity selector*.

Thomson's Measurement of q/m for Electrons

An example of the use of a velocity selector is the famous experiment performed by J. J. Thomson in 1897 in which he showed that the rays in a cathode-ray tube can be deflected by electric and magnetic fields and there-fore consist of charged particles. By observing the deflection of these parti-cles with various combinations of electric and magnetic fields, Thomson showed that all the particles had the same charge-to-mass ratio q/m. He also showed that particles with this charge-to-mass ratio can be obtained using any material for the cathode, which means that these particles, now called electrons, are fundamental constituents of all matter.

Figure 21-18 shows the cathode-ray tube Thomson used. Electrons are emitted from a cathode C that is at a negative potential relative to the slits A and B. An electric field in the direction from A to C accelerates the electrons.

J. J. Thomson in his laboratory.

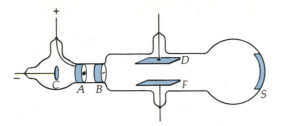

Figure 21-18
Thomson's tube for measuring q/m
for the particles of cathode rays
(electrons). Electrons from the
cathode C pass through the slits at
A and B and strike a phosphorescent
screen S. The beam can be deflected
by an electric field between plates D
and F or by a magnetic field (not
shown).

They pass through slits A and B into a field-free region and then encounter
an electric field that is perpendicular to their velocity between the plates D
and F. The acceleration produced by this electric field gives the electrons a
vertical component of velocity when they leave the region between the
plates. They strike the phosphorescent screen S at the far right side of the
tube at some displacement Δy from the point at which they would strike
were there no field between the plates D and F. When the electrons strike the
screen, it glows, indicating the location of the beam.

The deflection of the beam is proportional to the acceleration due to the
vertical electric field between the plates D and F. It is also proportional to the
time it takes the particles to reach the screen. If q is the charge of the electron
and m is its mass, the acceleration in an electric field $\mathbf{E}$ is qE/m. The time it
takes for the electrons to pass through the capacitor plates and reach the
screen is L/v, where L is the total distance traveled and v is their speed. Since
this distance is easily measured, if the speed is known, a measurement of the
deflection for a given electric field $\mathbf{E}$ yields a value for q/m. The speed v is
determined by introducing a magnetic field $\mathbf{B}$ between the plates perpendicular both to the electric field and to the initial velocity of the electrons. The
magnitude of the magnetic field is adjusted so that the beam is undeflected.
The speed is then found from Equation 21-19.

The Mass Spectrograph

The *mass spectrograph,* first developed by Francis William Aston in 1919 and
improved upon by Kenneth Bainbridge and others, was designed to measure the masses of isotopes. Measurement of the masses of isotopes plays an
important role in determining their existence and their abundance in nature.
For example, natural magnesium consists of 78.7 percent ^{24}Mg, 10.1 percent
^{25}Mg, and 11.2 percent ^{26}Mg. These isotopes have masses in the approximate ratio of $24:25:26$.

The mass spectrograph measures the mass-to-charge ratio of ions of
various isotopes by determining the velocity of the ions and then measuring
the radius of their circular orbit in a uniform magnetic field. Figure 21-19
shows a simple schematic drawing of a mass spectrograph. Ions from an ion
source are accelerated through a potential difference and enter a uniform
magnetic field produced by an electromagnet. A simple ion source can be
made by coating a wire filament with a solution containing the material
whose mass is to be measured, for example, natural magnesium. When the
filament is heated, ions such as ^{24}Mg$^+$, ^{25}Mg$^+$, and ^{26}Mg$^+$ spew off and are
accelerated by the potential difference ΔV. The kinetic energy of the ions
when they enter the magnetic field equals the loss in potential energy $q\,\Delta V$:

$$\tfrac{1}{2}mv^2 = q\,\Delta V \qquad\qquad 21\text{-}20$$

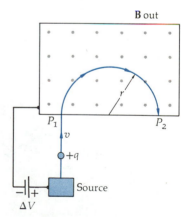

Figure 21-19
Schematic drawing of a mass
spectrograph. Ions from an ion
source are accelerated though a
potential difference ΔV and enter a
uniform magnetic field indicated by
the dots. They are bent into a
circular arc and emerge at P_2. The
radius of the circle varies inversely
with the mass of the ion.

The ions move in a semicircle of radius r given by Equation 21-16,

$$r = \frac{mv}{qB}$$

They strike the photographic film at point P_2, a distance $2r$ from the point where they entered the magnetic field. The speed v can be eliminated from Equations 21-16 and 21-20 to find m/q in terms of the known quantities V, B, and r. We first solve Equation 21-16 for v and square each term

$$v^2 = \frac{r^2 q^2 B^2}{m^2}$$

Substituting this expression for v^2 into Equation 21-20, we obtain

$$\frac{1}{2} m \left(\frac{r^2 q^2 B^2}{m^2} \right) = q \, \Delta V$$

Simplifying and solving for m/q, we obtain

$$\frac{m}{q} = \frac{B^2 r^2}{2 \, \Delta V} \qquad\qquad 21\text{-}21$$

In Aston's original spectrograph, mass differences could be measured to a precision of about 1 part in 10,000. The precision is improved by introducing a velocity selector between the ion source and the magnet, which makes it possible to limit the range of velocities of the incoming ions and to determine these velocities accurately.

The Cyclotron

The *cyclotron* was invented by E. O. Lawrence and M. S. Livingston in 1934 to accelerate particles such as protons or deuterons to high kinetic energies.

M. S. Livingston and E. O. Lawrence standing in front of their 27-in cyclotron in 1934. Lawrence won the Nobel Prize (1939) for the invention of the cyclotron.

(A deuteron is a nucleus of heavy hydrogen, ²H, consisting of a proton and neutron tightly bound together.) The high-energy particles are then used to bombard nuclei, causing nuclear reactions that are studied to obtain information about the nucleus. High-energy protons or deuterons are also used to produce radioactive materials for medical and other purposes. The operation of the cyclotron is based on the fact that the period of motion of a charged particle in a uniform magnetic field is independent of the velocity of the particle, as is shown by Equation 21-17:

$$T = \frac{2\pi m}{qB}$$

Figure 21-20 is a schematic drawing of a cyclotron. The particles move in two semicircular metal containers called *dees* (because of their shape). The region inside each metal dee is shielded from electric fields by the metal. The dees lie in a vacuum chamber that is in a uniform magnetic field provided by an electromagnet. (The region in which the particles move must be evacuated so that the particles will not lose energy and be scattered in collisions with air molecules.) The dees are maintained at a potential difference ΔV that varies sinusoidally with the period T, which is chosen so as to be equal to the cyclotron period given by Equation 21-17.

The charged particle is initially injected with a small velocity from an ion source S near the center of the magnetic field. It moves in a semicircle in one of the dees and arrives at the gap between the dees after a time $\frac{1}{2}T$, where T is the cyclotron period and is also the period of the alternating potential across the dees. Let us call the dee in which the particle traverses this first semicircle dee 1. The alternating potential is adjusted so that when the particle arrives at the gap between the dees, dee 1 is at a higher potential than dee 2. The particle is thus accelerated across the gap by the electric field across the gap and gains kinetic energy $q\,\Delta V$. Thus, the particle moves in a semicircle of larger radius in dee 2 and again arrives at the gap after a time $\frac{1}{2}T$. By this time, the potential between the dees has been reversed, and dee 2 is at the higher potential. Once more the particle is accelerated across the gap and gains additional kinetic energy $q\,\Delta V$. In each half revolution, the particle gains kinetic energy $q\,\Delta V$ and moves into a semicircular orbit of larger

Compact cyclotron that is used to produce short-lived positron-emitting radioactive nuclides for synthesis of radiotracers in the study of brain functions at the University of Michigan, Cyclotron/Positron Emission Tomography Facility.

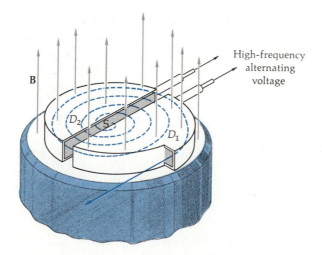

High-frequency alternating voltage

Figure 21-20
Schematic drawing of a cyclotron. The upper pole face of the magnet has been omitted. Charged particles, such as protons, from a source at the center are accelerated by the potential difference across the gap between the dees. When they arrive at the gap again, the potential difference has changed sign, so they are again accelerated across the gap and move in a larger circle. The potential difference across the gap alternates with the cyclotron frequency of the particle, which is independent of the radius of the circle.

radius until it eventually leaves the magnetic field. The energy of the particle leaving the cyclotron can be expressed in terms of the maximum radius using Equation 21-16:

$$r = \frac{mv}{qB}$$

$$v^2 = \frac{q^2 B^2 r^2}{m^2}$$

$$E_k = \frac{1}{2} mv^2 = \frac{1}{2}\left(\frac{q^2 B^2}{m}\right) r^2 \qquad\qquad \text{21-22}$$

In a typical cyclotron, a particle may make 50 to 100 revolutions and exit with an energy of several hundred megaelectronvolts (MeV).

Questions

9. How can you determine by observing the path of a particle whether the particle is deflected by a magnetic field or an electric field?

10. A beam of positively charged particles passes undeflected from left to right through a velocity selector in which the electric field is up. The beam is then reversed so that it travels from right to left. Will the beam be deflected in the velocity selector? If so, in which direction?

21-4   Sources of the Magnetic Field

We now turn to a consideration of the origins of the magnetic field **B**. The earliest known sources of magnetism were permanent magnets. In 1820, Hans Christian Oersted discovered that a compass needle is deflected by an electric current, thus showing that electric currents produce magnetic fields. Subsequently, many scientists investigated the properties of the magnetism associated with electric currents. One month after Oersted announced his discovery, Jean Baptiste Biot and Felix Savart announced the results of their measurements of the force on a magnetic pole near a long, current-carrying wire and analyzed these results in terms of the magnetic field produced by each element of the current. André Marie Ampère extended these experiments and showed that the current elements themselves experience a force in the presence of a magnetic field. In particular, he showed that two currents exert forces on each other.

From their investigations of the force on a magnetic pole exerted by a long wire carrying a current, Biot and Savart proposed an expression relating the magnetic field **B** at a point in space to an element of the current that produces it. Let $I\,\Delta\boldsymbol{\ell}$ be an element of current (see Figure 21-21). The magnetic field $\Delta\mathbf{B}$ produced by this element at a field point P a distance r away has the magnitude

$$\Delta B = \frac{\mu_0}{4\pi} \frac{I\,\Delta\ell\,\sin\theta}{r^2} \qquad\qquad \text{21-23} \qquad \text{Biot-Savart law}$$

where θ is the angle between $\Delta\boldsymbol{\ell}$ and the vector **r** from the current element to

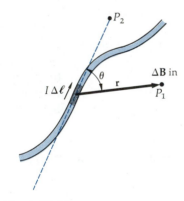

Figure 21-21
The current element $I\,\Delta\boldsymbol{\ell}$ produces a magnetic field at point P_1 that is perpendicular to both $\boldsymbol{\ell}$ and **r**. It produces no magnetic field at point P_2, which is directed along the line of $\boldsymbol{\ell}$.

the field point and μ_0 is a constant called the *permeability of free space*. It has the value

$$\mu_0 = 4\pi \times 10^{-7} \text{ T·m/A} = 4\pi \times 10^{-7} \text{ N/A}^2 \qquad \text{21-24}$$

(The units of μ_0 are such that B is in teslas when I is in amperes and the distances $\Delta\ell$ and r are in metres. The unit N/A^2 comes from the fact that $1 \text{ T} = 1 \text{ N/A·m}$.) The direction of the magnetic field is perpendicular to the current element $I\,\Delta\ell$ and to $\mathbf{r}$, as shown in Figure 21-21. $\mathbf{B}$ is directed into the paper in the figure.

Equation 21-23, known as the *Biot-Savart law,* was also deduced by Ampère. The Biot-Savart law is analogous to the expression for the electric field of a point charge, $E = kq/r^2$, from Coulomb's law. The source of the magnetic field is the current element $I\,\Delta\ell$, just as the charge q is the source of the electrostatic field. Also, the magnetic field decreases as the square of the distance from the current element, like the electric field decreases with distance from a point charge. But the directional aspects of these fields are quite different. Whereas the electrostatic field points in the radial direction $\mathbf{r}$ from the point charge to the field point (assuming the charge is positive), the magnetic field is perpendicular both to the radial direction and to the direction of the current element $I\,\Delta\ell$. At a point along the line of the current element, point P_2 in Figure 21-21, for example, the magnetic field is zero because θ is zero for the vector $\mathbf{r}$ to that point.

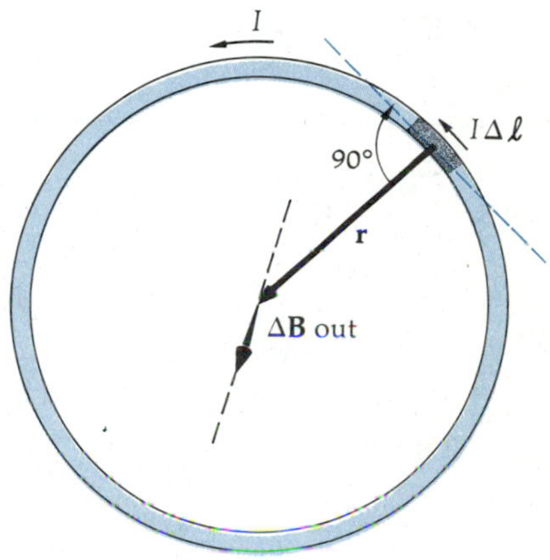

The magnetic field due to the total current in a circuit can be found by using the Biot-Savart law to find the field due to each current element and then summing all the current elements in the circuit. This calculation is very difficult for all but the most simple circuit geometries. One calculation that we can do is to find the magnetic field at the center of a circular loop. Figure 21-22 shows a current element $I\,\Delta\ell$ and the radius vector $\mathbf{r}$ to the center of the loop. The magnetic field at the center due to this element is directed along the axis, and its magnitude is given by

$$\Delta B = \frac{\mu_0}{4\pi} \frac{I\,\Delta\ell \sin\theta}{r^2}$$

For each such current element, the angle θ is $90°$, so $\sin\theta = 1$. The magnetic field due to the entire current is found by summing all the current elements. This amounts to summing $\Delta\ell$ around the complete loop, which results in a total length of $2\pi r$, the circumference of the loop. The magnetic field due to the entire loop is thus

$$B = \frac{\mu_0}{4\pi} \frac{I2\pi r}{r^2} = \frac{\mu_0}{2} \frac{I}{r} \qquad \text{21-25} \qquad \textbf{B at center of a loop}$$

The calculation of the magnetic field of a current loop at points other than the center is much more difficult and will not be discussed here.

An important result that can be obtained from the Biot-Savart law is the magnetic field due to a current in a long, straight wire (see Figure 21-23). Here, we will merely state this result, which was found experimentally by Biot and Savart in 1820. At a distance r from the wire, the magnetic field has the magnitude

$$B = \frac{\mu_0}{2\pi} \frac{I}{r} \qquad \text{21-26} \qquad \textbf{B due to a very long, straight wire}$$

Figure 21-22
Current element for calculating the magnetic field at the center of a circular loop. Each element produces a magnetic field directed out of the paper. Since the distance from the center of the loop to each element is the same, summing these fields merely requires summing all the current elements over the circumference of the loop.

The direction of **B** is tangent to concentric circles that encircle the wire, as shown in Figure 21-23. This direction can be remembered by the right-hand rule also shown in the figure.

Example 21-5   Find the magnetic field at a distance of 20 cm from a long, straight wire carrying a current of 5 A.

From Equation 21-26, we have

$$B = \frac{\mu_0}{2\pi} \frac{I}{r} = \frac{4\pi \times 10^{-7} \text{ T·m/A}}{2\pi} \frac{5 \text{ A}}{0.2 \text{ m}} = 5.00 \times 10^{-6} \text{ T}$$

We note from this example that the magnetic field near a wire carrying a current of ordinary magnitude is very small. In this case, it is only about 10 percent of the magnetic field due to the earth.

Figure 21-23
Right-hand rule for determining the direction of the magnetic field due to a current in a long wire. The field lines encircle the wire in the direction of the fingers of the right hand when the thumb is in the direction of the current.

We can use the expression for the magnetic field due to a current in a long, straight wire and Equation 21-7 for the force exerted by a magnetic field on a current to find the force exerted by one long, straight current on another. Figure 21-24 shows two long, parallel wires carrying currents in the same direction. We consider the force on a segment $\Delta\ell_2$ carrying current I_2, as shown in the figure. The magnetic field $\mathbf{B}_1$ at this segment (and everywhere on current 2) due to current I_1 is directed into the paper. The magnetic force on current element $I_2 \Delta\ell_2$ is directed toward current 1. The parallel currents thus attract each other. If one of the currents is reversed, the force will be reversed. Thus, two antiparallel currents will repel each other. The attraction or repulsion of parallel or antiparallel currents was discovered experimentally by Ampère one week after he heard of Oersted's discovery of the effect of a current on a compass needle.

Since the magnetic field at segment $I_2 \Delta\ell_2$ is perpendicular to the current segment, the magnitude of the force is just $I_2 \Delta\ell_2 B_1$. Assuming the currents are close together, we can use our result for the magnetic field of a long, current-carrying wire. The force on the segment of wire 2 is therefore

$$F_2 = \frac{\mu_0}{2\pi} \frac{I_1 I_2 \Delta\ell_2}{r}$$

where r is the distance between the currents. The force per unit length is

$$\frac{F_2}{\Delta\ell_2} = \frac{\mu_0}{2\pi} \frac{I_1 I_2}{r} \qquad\qquad 21\text{-}27$$

In Chapter 18, we deferred the definition of the coulomb as a unit of charge, mentioning that it is defined in terms of the ampere. The ampere is defined as follows:

If two very long parallel wires one metre apart carry equal currents, the current in each is defined to be one ampere if the force per unit length on each wire is 2×10^{-7} N/m.

This definition allows the unit of current (and therefore of electric charge) to be determined by a mechanical experiment. When this is done, of course,

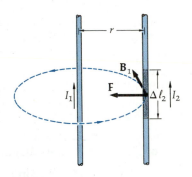

Figure 21-24
Two long, straight wires carrying parallel currents. The magnetic field $\mathbf{B}_1$ due to current I_1 is directed into the paper at the current I_2. The force **F** on current I_2 is thus toward current I_1. There is an equal and opposite force exerted by current I_2 on I_1. The currents thus attract each other.

Ampere defined

the currents are chosen to be much closer together than 1 m so that the wires need not be so long and the force is large enough to be measured accurately. Figure 21-25 shows a *current balance* that can be used to calibrate an ammeter from the fundamental definition of the ampere. The upper conductor is free to rotate about the knife edges and is balanced so that the wires are a small distance apart. The conductors are wired in series so that they carry the same current but in opposite directions. The wires thus repel rather than attract each other. The force of repulsion can be measured by placing weights on the upper conductor until it balances again at the original separation. This definition of the ampere makes the permeability of free space μ_0 exactly equal to $4\pi \times 10^{-7}$.

Figure 21-25
Current balance used in an elementary physics laboratory to calibrate an ammeter.

The constant k that occurs in Coulomb's law for the electric force between two charges has the value

$$k = \frac{1}{4\pi\epsilon_0} = 9 \times 10^9 \ \text{N} \cdot \text{m}^2/\text{C}^2$$

and the constant $\mu_0/4\pi$ that occurs in the Biot-Savart law for the magnetic force between two current elements has the value

$$\frac{\mu_0}{4\pi} = 10^{-7} \ \text{N/A}^2$$

The ratio of these two constants is

$$\frac{k}{\mu_0/4\pi} = \frac{1/4\pi\epsilon_0}{\mu_0/4\pi} = \frac{1}{\mu_0\epsilon_0}$$

$$= \frac{9 \times 10^9 \ \text{N} \cdot \text{m}^2/\text{C}^2}{10^{-7} \ \text{N/A}^2} = 9 \times 10^{16} \ \text{m}^2/\text{s}^2$$

where we have used the fact that $1 \ \text{A} = 1 \ \text{C/s}$. This ratio equals the square of the speed of light. In 1860, the English physicist James Clerk Maxwell demonstrated that the laws of electricity and magnetism imply that an

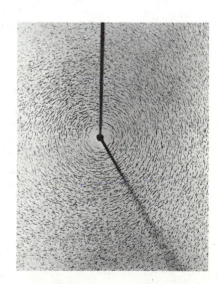

Magnetic field lines due to a current in a long wire indicated by iron filings.

accelerated charge radiates energy in the form of waves that travel with the speed

$$c = \frac{1}{\sqrt{\epsilon_0 \mu_0}} = 3 \times 10^8 \text{ m/s} \qquad \text{21-28}$$

Since this speed is the same as that of light, Maxwell correctly speculated that light is an electromagnetic wave produced by the acceleration of atomic charges.

André Marie Ampère (1775–1836).

21-5   Ampere's Law

We have noted that the lines of the magnetic field **B** for a long, straight, current-carrying wire encircle the wire. These lines are quite different from the lines of the electric field, which begin and end on electric charges. As we have seen, magnetic field lines form closed curves. In Chapter 18, we found an important equation called Gauss' law relating the normal component of the electric field summed over a closed surface to the net charge inside the surface. There is an analogous equation for the magnetic field called *Ampere's law* that relates the tangential components of **B** summed around a closed curve to the current that passes through the curve. In mathematical form, Ampere's law is

$$\Sigma B_t \, \Delta \ell = \mu_0 I_{\text{total}} \qquad \text{21-29}$$

Ampere's law

where B_t is the component of **B** tangent to an element of length $\Delta \ell$ of a closed curve and I_{total} is the total current that penetrates the area bounded by the curve. Ampere's law can be used to obtain an expression for the magnetic field in situations where there is a high degree of symmetry. The simplest application of Ampere's law is in finding the magnetic field of a long, straight, current-carrying wire. In Figure 21-26, we have drawn a circle around a long wire with its center at the wire. We assume that we are far from the ends of the wire. Then, by symmetry, the magnetic field will be tangent to this circle and will have the same magnitude B at any point on the circle. Ampere's law then gives

$$\Sigma B_t \, \Delta \ell = B \Sigma \, \Delta \ell = \mu_0 I$$

where we have taken B out of the sum because it has the same value everywhere along the circle. The sum of $\Delta \ell$ around the circle is just the circumference of the circle, which is $2\pi r$. We thus obtain

$$B(2\pi r) = \mu_0 I$$

$$B = \frac{\mu_0}{2\pi} \frac{I}{r}$$

which is Equation 21-26. We will use Ampere's law in the next section to obtain an expression for the magnetic field inside a long, tightly wound solenoid.

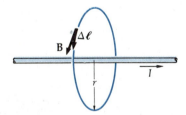

Figure 21-26
The magnetic field at a circle around a long, straight, current-carrying wire is constant and tangent to the circle. The sum of $B_t \, \Delta \ell$ around this circle is $B(2\pi r)$. Ampere's law then gives $B(2\pi r) = \mu_0 I$ or $B = \mu_0 I / 2\pi r$.

21-6   Current Loops, Solenoids, and Magnets

We saw in Section 21-2 that a small current loop experiences a torque in an external magnetic field and behaves like a small magnet with a magnetic moment of magnitude IA, where I is the current and A is the area of the loop. Figure 21-27 shows the magnetic field lines for the magnetic field *produced* by a small current loop. The lines of **B** are identical to those of a small magnet. The magnitude of the magnetic field at any point can, in principle, be calculated from the Biot-Savart law. We found **B** at the center of the loop in Section 21-4. For a general point, that is, a point some place other than at the center of the loop, this calculation is very difficult.

Since the magnetic field due to a single circular current loop is the same as that due to a small magnet, the magnetic field due to many closely spaced loops is the same as that due to a line of small magnets, which, as discussed in Section 21-2, is equivalent to a bar magnet. A solenoid consists of a wire wound into a helix as shown in Figure 21-28. If the turns are tightly wound, as is usually the case, the solenoid can be considered to be a series of closely spaced current loops. Figure 21-29 shows the magnetic field lines of a tightly wound solenoid. The magnetic field of the solenoid is the same as that of bar magnet of the same shape, as can be seen by comparing this figure with Figure 21-12, which shows the magnetic field lines of a bar magnet. We can see from the magnetic field lines in Figure 21-29 that the magnetic field is relatively uniform everywhere inside the coil except near the ends and that outside the coil the magnetic field is much weaker. We can use Ampere's law

$$\Sigma B_t \, \Delta \ell = \mu_0 I_{\text{total}}$$

to find an expression for the magnetic field inside the solenoid, assuming that it is uniform there and zero outside. For our closed curve, we choose the

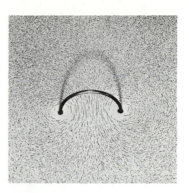

Figure 21-27
Magnetic field lines of a circular loop indicated by iron filings.

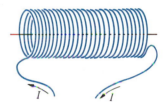

Figure 21-28
A tightly wound solenoid can be considered to be a line of circular loops carrying the same current. It produces a uniform magnetic field inside it.

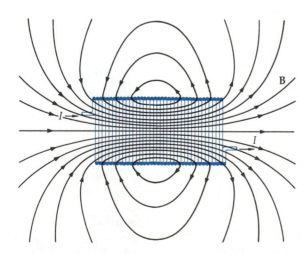

(a)

(b)

Figure 21-29
(*a*) Magnetic field lines of a solenoid. The lines are identical to those of a permanent magnet of the same shape (see Figure 21-12).
(*b*) Magnetic field lines of a solenoid indicated by iron filings.

rectangle of sides a and b shown in Figure 21-30. The current that penetrates this curve is the current I in each turn times the number of turns in the length b. If the solenoid has N turns in a total length ℓ, the number of turns in the length b will be $(N/\ell)b$. The only contribution to the sum of $B\,\Delta\ell$ for this curve is along the long side of the rectangle inside the solenoid, which gives Bb. Ampere's law thus gives

$$\Sigma B_t\,\Delta\ell = Bb = \mu_0(N/\ell)bI$$

or

$$B = \mu_0(N/\ell)I \qquad\qquad 21\text{-}30$$

Figure 21-31 shows a plot of **B** versus distance from the center of the solenoid. The field right at either end is half of that at the center. Because of the uniformity of this field, a solenoid is very useful for producing a uniform magnetic field in some region of space. A solenoid is somewhat analogous to a parallel-plate capacitor, which produces a uniform electric field between the plates.

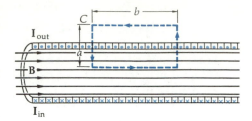

Figure 21-30
The magnetic field inside a solenoid can be calculated by applying Ampere's law to the curve C shown, assuming that **B** is uniform inside the solenoid and zero outside.

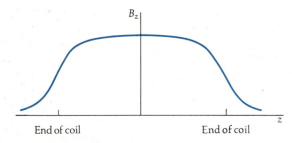

Figure 21-31
Plot of the magnetic field inside a solenoid versus z on the axis of the solenoid. The field is nearly constant inside the solenoid except near the ends.

We can understand why the field of a bar magnet is the same as that of a solenoid from a simple model first suggested by Ampere. According to Ampere's model, all magnetic fields are due to currents of some kind. In permanent magnets or other magnetized materials, these currents are due to the intrinsic motions of the electrons of the atoms of the material. Although these motions are very complicated, for this model we need only assume that the motions are equivalent to closed-circuit loops (see Figure 21-32). The magnetic moments of all the current loops shown in this figure are in the direction parallel to the axis. If the material of the magnet is homogeneous, the net current at any point inside the magnet is zero because of the cancellation of neighboring current loops. However, since there is no cancellation on the surface of the material, the result of these current loops is equivalent to a current on the surface of the material called an *amperian current*. This surface current is similar to the real conduction current in a tightly wound solenoid. The magnetic field due to the surface current is the same as that due to an equivalent "surface" current in a solenoid.

Figure 21-32
A model of electron current loops inside a cylindrical magnet when the magnetic moments of all the current loops are parallel to the axis of the cylinder. The net current at any point inside the magnet is zero due to cancellation of neighboring current loops. The result is a surface current similar to that of a solenoid.

21-7 Magnetism in Matter

In studying electric fields in matter, we found that the electric field is affected by the presence of electric dipoles. For polar molecules, which have a permanent electric dipole moment, the dipoles are aligned by the external electric field, whereas for nonpolar molecules, electric dipoles are induced by the external electric field. In both cases, the dipoles are aligned parallel to the external electric field, and the alignment tends to weaken the external field.

Similar but more complicated effects occur with magnetism. Atoms have magnetic moments due to the motions of their electrons. In addition, each electron has an intrinsic magnetic moment associated with its spin. The net magnetic moment of an atom depends on the arrangement of the electrons in the atom. Unlike the situation with electric dipoles, the alignment of magnetic dipoles parallel to an external magnetic field tends to *increase* the field. We can see this difference by comparing the electric field lines for an electric dipole with the magnetic field lines for a magnetic dipole. Figure 21-33*a* shows an electric dipole. The small current loop shown in Figure 21-33*b* is an example of a magnetic dipole. Far from the dipoles, the electric field lines and magnetic field lines are identical. Between the charges of the electric dipole, however, the electric field lines are opposite the direction of the dipole moment. Conversely, inside the current loop, the magnetic field lines are parallel to the magnetic dipole moment. Thus, inside an electrically polarized material, the dipoles create an electric field *antiparallel* to their dipole-moment vectors, whereas in a magnetically polarized material, the dipoles create a magnetic field *parallel* to the magnetic-dipole-moment vectors.

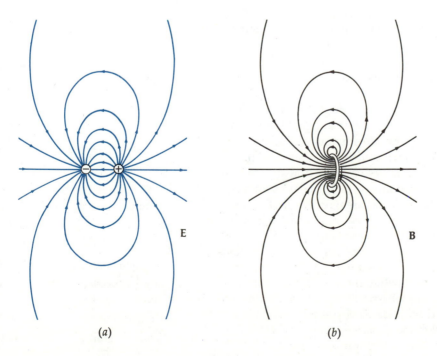

(a) (b)

Figure 21-33
(*a*) Electric field lines of an electric dipole. (*b*) Magnetic field lines of a magnetic dipole. Far from the dipoles, the field patterns are identical. In the region between the charges in (*a*), the electric field lines are opposite to the dipole moment, whereas inside the loop in (*b*), the magnetic field lines are parallel to the dipole moment.

Materials fall into three categories—*paramagnetic, diamagnetic,* and *ferromagnetic*—according to their behavior in an external magnetic field. Paramagnetic and ferromagnetic materials have molecules with permanent magnetic dipole moments.

In paramagnetic materials, the magnetic dipole moments do not interact strongly with each other and are normally randomly oriented. In the presence of an external magnetic field, the dipoles are partially aligned in the direction of the field, thereby increasing the field. However, at ordinary temperatures and in ordinary external fields, only a very small fraction of the molecules are aligned because thermal motion tends to randomize their orientation. The increase in the total magnetic field is therefore very small.

Ferromagnetism is much more complicated. Because of strong interactions between neighboring magnetic dipoles, a high degree of alignment can be achieved even in weak external magnetic fields, thereby causing a very large increase in the total field. Even when there is no external magnetic field, ferromagnetic materials may have magnetic dipoles aligned, as is the case for permanent magnets.

Diamagnetism is the result of an induced magnetic moment opposite in direction to the external field. The induced dipoles thus weaken the resultant magnetic field. This effect, which is independent of temperature, occurs in all materials, but it is very small and is often masked by the paramagnetic or ferromagnetic effects if the individual molecules have permanent magnetic dipole moments.

Suppose we place some material in an external magnetic field B_0. The resultant magnetic field inside the material will be B_0 plus some contribution due to the magnetization of the material. This contribution is usually proportional to B_0. The resultant magnetic field can then be written

$$B = B_0 + \chi_m B_0 \qquad\qquad 21\text{-}31$$

where χ_m is called the *magnetic susceptibility.* For paramagnetic materials, χ_m is a small positive number that depends on temperature. For diamagnetic materials, it is a small negative constant independent of temperature. Table 21-1 lists the magnetic susceptibility of various paramagnetic and diamagnetic materials. We can see that the magnetic susceptibility for these materials is of the order of 10^{-5}.

For ferromagnetic materials, χ_m is very large, often greater than 1000. However, Equation 21-31 is not very useful for ferromagnetic materials since χ_m depends on B_0. It also depends on the previous state of magnetization of the material. Ferromagnetism occurs in pure iron, cobalt, and nickel and in alloys composed of combinations of these metals. It also occurs in gadolinium, dysprosium, and a few compounds. In these substances, a small external magnetic field can produce a very large degree of alignment of the magnetic dipole moments of the atoms. In some cases, this alignment can persist even when there is no external magnetic field. This occurs because the magnetic dipole moments of the atoms of these substances exert strong forces on their neighbors; thus, over a small region of space, the moments are aligned with each other even when there is no external field. The dipole forces for these substances can be predicted by quantum mechanics, but they cannot be explained with classical physics. At temperatures above a critical temperature, called the *Curie temperature,* these forces disappear and ferromagnetic materials become paramagnetic.

Three types of magnetic materials

Table 21-1

Magnetic Susceptibility of Various Materials at 20°C

Material	χ_m
Aluminum	2.3×10^{-5}
Bismuth	-1.66×10^{-5}
Copper	-0.98×10^{-5}
Diamond	-2.2×10^{-5}
Gold	-3.6×10^{-5}
Magnesium	1.2×10^{-5}
Mercury	-3.2×10^{-5}
Silver	-2.6×10^{-5}
Sodium	-0.24×10^{-5}
Titanium	7.06×10^{-5}
Tungsten	6.8×10^{-5}
Hydrogen (1 atm)	-9.9×10^{-9}
Carbon dioxide (1 atm)	-2.3×10^{-9}
Nitrogen (1 atm)	-5.0×10^{-9}
Oxygen (1 atm)	2090×10^{-9}

The region of space in which the magnetic dipole moments are aligned is called a *domain*. The size of a domain is usually microscopic. Within the domain, all the magnetic moments are aligned, but the direction of alignment varies from domain to domain, so the net magnetic moment of a macroscopic piece of material is zero in the normal state. Figure 21-34 illustrates this situation. When an external magnetic field is applied, the boundaries of the domains may shift so that the domains aligned parallel to the field are enlarged, or the alignment within a domain may change. In either case, the result is a net magnetic moment in the direction of the applied field. Since the degree of alignment is great even for a small external field, the magnetic field produced in the material by the dipoles is often much greater than the external field. Ferromagnetism has many applications: permanent magnets, magnetic tape, and magnetic memory in computers, to name a few.

Domain

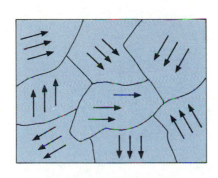

(a)

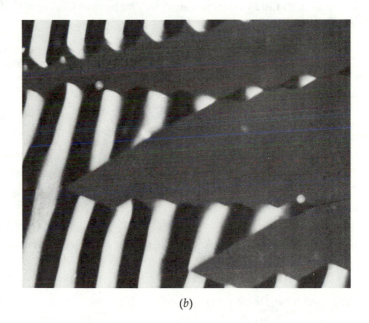

(b)

Figure 21-34
(*a*) Illustration of ferromagnetic domains. Within a domain, the magnetic dipoles are aligned, but the direction varies from domain to domain such that the net magnetic moment is zero. (*b*) Photomicrograph showing magnetic domains in a thin single crystal of yttrium iron garnet magnified 330 times. The photograph was made with transmitted light through the crystal between crossed Polaroids. The different directions of magnetization of the domains give rise to different colors in the transmitted light. The magnetization is pointing down into the paper in the light striped regions and out of the paper in the dark striped regions. In the large gray areas, the magnetization lies in the plane of the paper.

Questions

11. Why are some values of the magnetic susceptibility in Table 21-1 positive whereas others are negative?

12. Faraday discovered that a piece of bismuth is repelled by either pole of a magnet. Explain how this is possible.

Summary

1. Moving charges interact with each other via the magnetic force. Since electric currents consist of moving charges, they exert magnetic forces on each other. This force is described by saying that one moving charge or current creates a magnetic field that, in turn, exerts a force on the other moving charge or current. All magnetic fields are caused ultimately by charges in motion. The magnetic fields of permanent magnets are the result of current loops due to the orbital motion or intrinsic spin of the electrons of the atoms of the material.

2. When a charge q moves with velocity **v** in a magnetic field **B**, it experiences a force that is perpendicular to both **v** and **B** and has the magnitude

$$F = qvB \sin \theta$$

where θ is the angle between **v** and **B**. The SI unit of magnetic field is the tesla (T):

$$1 \text{ T} = 1 \text{ N/A} \cdot \text{m}$$

The force exerted by a magnetic field on a wire of length ℓ carrying a current I is perpendicular to the wire and to the magnetic field **B** and has the magnitude

$$F = I\ell B \sin \theta$$

where θ is the angle between the magnetic field and the wire.

3. A magnetic field exerts a torque on a magnet or current loop that tends to align the magnetic moment of the magnet or current loop with the external field. The magnitude of this torque is

$$\tau = mB \sin \theta$$

where θ is the angle between **B** and the magnetic moment **m**. The magnetic moment of a current loop equals the product of the current and the area of the loop:

$$m = IA$$

It is directed along the axis of the loop in the sense given by the right-hand rule.

4. A particle of mass m and charge q moving with speed v in a plane perpendicular to a magnetic field **B** moves in a circle of radius r given by

$$r = \frac{mv}{qB}$$

The period and frequency of this circular motion are independent of the radius of the circle and the speed of the particle. The period, called the cyclotron period, is given by

$$T = \frac{2\pi m}{qB}$$

The frequency of this motion, called the cyclotron frequency, is given by

$$f = \frac{1}{T} = \frac{qB}{2\pi m}$$

5. The magnetic field $\Delta \mathbf{B}$ at a distance r from a current element $I\ \Delta\ell$ is given by the Biot-Savart law

$$\Delta B = \frac{\mu_0}{4\pi}\frac{I\ \Delta\ell\ \sin\theta}{r^2}$$

where θ is the angle between the current element and the radius vector to the field point. The constant μ_0 is called the permeability of free space and has the value

$$\mu_0 = 4\pi \times 10^{-7}\ \text{N/A}^2$$

The magnetic field is perpendicular to both the current element and the radius vector.

6. Ampere's law relates the sum of the tangential components of the magnetic field around a closed curve to the total current passing through the area bounded by the curve:

$$\Sigma B_t\ \Delta\ell = \mu_0 I_{\text{total}}$$

Ampere's law can be used to derive expressions for the magnetic field of a long, straight, current-carrying wire, for the magnetic field inside a solenoid, and for the magnetic fields in other situations where there is a high degree of symmetry.

7. The magnetic field a distance r from a long, straight wire carrying a current I is

$$B = \frac{\mu_0}{2\pi}\frac{I}{r}$$

The ampere is defined such that two long wires, each carrying a current of one ampere and separated by one metre, exert a force of exactly 2×10^{-7} N/m on each other.

8. The magnetic field produced by a small current loop of area A is exactly the same as that of a small magnet of magnetic moment IA. The magnetic field of a solenoid is the same as that of a bar magnet of the same shape. Inside a solenoid and far from the ends, the magnetic field is uniform and has the magnitude

$$B = \mu_0(N/\ell)I$$

where N is the number of turns of wire of the solenoid and ℓ is its length.

9. All materials can be classified as paramagnetic, ferromagnetic, or diamagnetic. Paramagnetic materials have atoms with permanent magnetic moments that have random directions in the absence of an external magnetic field. In an external magnetic field, some of these dipoles are aligned, producing a small increase in the total magnetic field. The degree of alignment is small except in very strong fields and at very low temperatures. At ordinary temperatures, thermal motion tends to maintain the random directions of the magnetic moments. Diamagnetic materials are those in which the magnetic moments of the atoms cancel, leaving zero magnetic moment in the absence of an external field. In an external field, a small magnetic moment is induced that tends to weaken the field. This effect is independent of temperature. Ferromagnetic materials have atoms with magnetic moments that are aligned in small regions called magnetic domains. When unmagnetized, the direction of alignment in one domain is independent of that in another, so no net magnetic field is produced. When magnetized, the domains of a ferromagnetic material are aligned, producing a very strong contribution to the magnetic field. This alignment can persist even when the external field is removed, thus leading to permanent magnetism.

Suggestions for Further Reading

Becker, Joseph J.: "Permanent Magnets," *Scientific American,* December 1970, p. 92.

Magnets made from the new alloys described in this article can be made many times stronger than those made from the conventional metals.

Carrigan, Charles R., and David Gubbins: "The Source of the Earth's Magnetic Field," *Scientific American,* February 1979, p. 118.

The earth may act as a huge dynamo in which electric currents in the molten and flowing metallic core maintain themselves by producing the magnetic field that deflects compass needles on the surface.

Nier, Alfred O. C.: "The Mass Spectrometer," *Scientific American,* March 1953, p. 68.

This device, which has made possible great advances in chemistry and other sciences, allows the determination of the composition of a substance based on the principle of the deflection of a beam of charged particles in a magnetic field.

Van Allen, James A.: "Interplanetary Particles and Fields," *Scientific American,* September 1975, p. 160.

This article, written by the man for whom the earth's Van Allen Radiation Belts are named, describes the deflection of the "solar wind" of charged particles in the earth's magnetic field.

Review

A. Objectives: After studying this chapter, you should:

1. Be able to calculate the magnetic force on a current element and on a moving charge in a given magnetic field.

2. Be able to calculate the magnetic moment of a current loop and the torque exerted on a current loop in a magnetic field.

3. Be able to discuss the Thomson experiment for measuring q/m for electrons.

4. Be able to describe a velocity selector, a mass spectrograph, and a cyclotron.

5. Be able to discuss the experimental definition of the ampere.

6. Be able to sketch the magnetic field lines of a loop, a solenoid, and a bar magnet.

B. Define, explain, or otherwise identify:

Magnetic field, pp. 531–532
Tesla, p. 532
Gauss, p. 533
Current element, p. 534
Pole strength, p. 535
Magnetic moment, p. 535
Cyclotron period, p. 539
Cyclotron frequency, p. 539
Crossed fields, p. 542
Velocity selector, p. 542
Mass spectrograph, pp. 543–544

Cyclotron, pp. 544–546
Biot-Savart law, pp. 546–547
Permeability of free space, p. 547
Current balance, p. 549
Ampere's law, p. 550
Solenoid, p. 551
Paramagnetism, p. 554
Diamagnetism, p. 554
Ferromagnetism, p. 554

C. True or false: If the statement is true, explain. If it is false, give a counterexample.

1. The magnetic force on a moving particle is always perpendicular to the velocity of the particle.

2. The torque on a magnet tends to align the magnetic moment in the direction of the magnetic field.

3. The period of a particle moving in a circle in a magnetic field is proportional to the radius of the circle.

4. Magnetic field lines never diverge from a point in space.

5. The magnetic field due to a current element is parallel to the current element.

6. The magnetic field due to a current in a long wire varies inversely with the square of the distance from the wire.

7. Paramagnetism occurs in materials whose molecules have permanent magnetic dipole moments.

8. Ferromagnetism occurs in all metals.

Exercises

Section 21-1   Definition of the Magnetic Field

1. Find the force on a proton moving with velocity 4 Mm/s in the positive x direction in a magnetic field of 2.0 T in the positive z direction.

2. A charge of -2.0 nC moves with speed 3.0×10^6 m/s in the negative x direction. Find the force on the charge if the magnetic field is (a) 0.6 T in the positive y direction, (b) 0.4 T in the positive z direction, and (c) 1.3 T in the positive x direction.

3. A particle of charge $+3.0$ nC moves in the positive x direction with speed 4.0×10^6 m/s. It experiences a force of 2.4×10^{-2} N in the positive z direction. Find whatever information you can about the magnetic field in this exercise.

4. A uniform magnetic field of magnitude 1.5 T is in the positive z direction. Find the force on a particle of charge $+2.5$ nC if its velocity is (a) 400 km/s in the positive y direction, (b) 800 km/s in the positive z direction, (c) 200 km/s in the negative z direction, and (d) 400 km/s in the yz plane upward along a line making an angle of 30° with the z axis.

5. An electron moves with velocity 5 Mm/s in the xy plane at an angle of 30° to the x axis and 60° to the y axis. A magnetic field of 1.5 T is in the positive y direction. Find the force on the electron.

6. A straight wire segment 2 m long makes an angle of 60° with a uniform magnetic field of 4000 G. Find the magnitude of the force on the wire if it carries a current of 2.5 A.

7. A straight wire 20 cm long carrying a current of 3.0 A is in a uniform magnetic field of 0.8 T. The wire makes an angle of 37° with the direction of **B**. What is the magnitude of the force on the wire?

8. A long wire parallel to the x axis carries a current of 14 A in the direction of increasing x. There is a uniform magnetic field of magnitude 8000 G in the positive y direction. Find the force per unit length on the wire.

Section 21-2   Torques on Magnets and Current Loops

9. A small magnet is placed in a magnetic field of 0.15 T. The maximum torque experienced by the magnet is 0.20 N·m. (a) What is the magnetic moment of the magnet? (b) If the length of the magnet is 4.0 cm, what is its pole strength?

10. A small circular coil of 20 turns of wire lies in a uniform magnetic field of 0.5 T such that the normal to the plane of the coil makes an angle of 60° with the direction of **B**. The radius of the coil is 4 cm, and it carries a current of 3.0 A. (a) What is the magnitude of the magnetic moment of the coil? (b) What torque is exerted on the coil?

11. A rectangular, 50-turn coil has sides 6.0 and 8.0 cm long and carries a current of 2.0 A. It is oriented as shown in Figure 21-35 and is pivoted about the z axis. The side in the

xy plane makes an angle of 63° with the x axis as shown. (a) Find the magnitude of the magnetic moment of the coil, and indicate its direction on a diagram. (b) What angle does the magnetic moment of the coil make with the x axis? (c) Find the torque exerted on the coil if there is a uniform magnetic field of 15,000 G in the positive x direction.

Figure 21-35
Exercise 11.

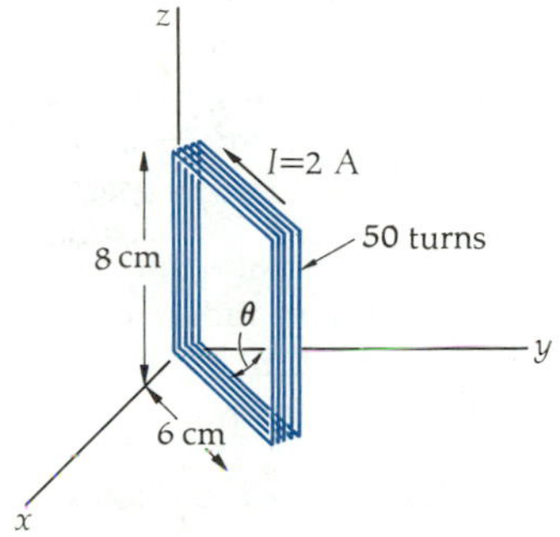

12. What is the maximum torque on a 500-turn circular coil of radius 0.5 cm that carries a current of 2.0 mA and resides in a uniform magnetic field of 0.3 T?

13. A small magnet of length 5 cm is placed at an angle of 45° to the direction of a uniform magnetic field of 400 G. The observed torque on the magnet is 0.12 N·m. (a) Find the magnetic moment of the magnet. (b) Find the pole strength.

Section 21-3   Motion of a Point Charge in a Magnetic Field

14. A proton moves in a circular orbit of radius 80 cm perpendicular to a uniform magnetic field of magnitude 0.5 T. (a) What is the period for this motion? (b) Find the speed of the proton. (c) Find the kinetic energy of the proton.

15. An electron of kinetic energy 25 keV moves in a circular orbit in a magnetic field of 0.2 T. (a) Find the radius of the orbit. (b) Find the period of the motion.

16. A deuteron, which consists of a neutron and a proton, has charge $+e$ and mass approximately twice that of a proton. An alpha particle (two neutrons plus two protons) has a charge $+2e$ and mass approximately four times that of a proton. Show that the cyclotron frequency is the same for deuterons and for alpha particles and that this frequency is half that for a proton in the same magnetic field.

17. An alpha particle of charge $+2e$ and mass 6.65×10^{-27} kg moves in a circular path of radius 0.5 m in a magnetic field of 1.4 T. Find the (a) period, (b) the speed, and (c) the kinetic energy of the alpha particle.

18. A beam of protons moves along the x axis in the positive x direction with speed 10 km/s through a region of crossed fields. (a) If there is a magnetic field of magnitude 0.8 T in the positive y direction, find the magnitude and direction of the electric field such that the beam is undeflected. (b) Would electrons of the same speed be deflected by these fields? If so, in what direction?

19. A velocity selector has a magnetic field of magnitude 0.2 T perpendicular to an electric field of magnitude 0.4 MV/m. (*a*) What must the speed of a particle be to pass through undeflected? What energy must (*b*) protons and (*c*) electrons have to pass through undeflected?

20. A $^{24}Mg^+$ ion has mass 3.99×10^{-26} kg. It is accelerated through a potential difference of 2000 V and is bent in a magnetic field of 0.05 T in a mass spectrograph. (*a*) Find the radius of curvature of the orbit for this ion. (*b*) The mass of ^{26}Mg is 26/24 times that of ^{24}Mg. Find the difference in radius for these two ions in this spectrograph.

21. A cyclotron for accelerating protons has a magnetic field of 1.5 T and a maximum radius of 0.5 m. (*a*) What is the cyclotron frequency? (*b*) Find the maximum kinetic energy of the protons when they emerge.

Section 21-4   Sources of the Magnetic Field

22. A small current element $I \, \Delta\ell$, with $I = 2$ A and $\Delta\ell = 2$ mm, points in the z direction and is centered at the origin. Find the magnetic field $\Delta\mathbf{B}$ at the following points: (*a*) on the x axis at $x = 3$ m, (*b*) on the x axis at $x = -6$ m, (*c*) on the y axis at $y = 3$ m, and (*d*) on the z axis at $z = -4$ m.

23. Find the magnetic field due to the current element in Exercise 22 at the point $x = 3$ m, $z = 4$ m.

24. A long, straight wire carries a current of 60 A. Find the magnitude of the magnetic field at distances (*a*) 10 cm, (*b*) 50 cm, and (*c*) 2 m from the wire.

Exercises 25 to 29 refer to Figure 21-36, which shows two long, straight wires in the xy plane parallel to the x axis. One wire is at y = +6 cm and the other is at y = -6 cm. The current in each wire is 20 A.

Figure 21-36
Two long, straight, parallel wires
for Exercises 25 to 29.

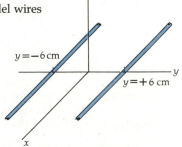

25. If the currents in Figure 21-36 are in the negative x direction, find **B** on the y axis at (*a*) $y = -3$ cm, (*b*) $y = 0$, (*c*) $y = +3$ cm, and (*d*) $y = +9$ cm.

26. Find **B** at the points on the y axis of Exercise 25 with the current in the wire at $y = -6$ cm in the negative x direction but the current in the wire at $y = +6$ cm in the positive x direction.

27. Find **B** on the z axis at $z = 8$ cm if (*a*) the currents are parallel as in Exercise 25 and (*b*) the currents are antiparallel as in Exercise 26.

28. Find the magnitude of the force per unit length exerted by one wire on the other.

29. The two wires carry currents I of equal magnitude. They repel each other with a force per unit length of 4.0 nN/m. (*a*) Are the currents parallel or antiparallel? (*b*) Find I.

30. In a student experiment using a current balance, the upper wire of length 30 cm is pivoted such that with no current it balances at 2 mm above a fixed parallel wire also 30 cm long. When the wires carry equal and opposite currents I, the upper wire again balances at its original position when a mass of 2.4 g is placed on it. What is the current I?

31. Three long, parallel, straight wires pass through the corners of an equilateral triangle of side 10.0 cm as shown in Figure 21-37, where a dot means that the current is directed out of the paper and a cross means that it is directed into the paper. If each current is 15.0 A, find (*a*) the force per unit length on the upper wire and (*b*) the magnetic field **B** at the upper wire due to the two lower wires. (*Hint:* It is easier to find the force per unit length directly from Equation 21-27 first and then use your result to find **B** than it is to find **B** first and then use it to find the force.)

Figure 21-37
Three long, straight, parallel wires
for Exercise 31.

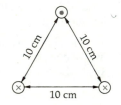

32. Work Exercise 31 with the current in the lower right corner of Figure 21-37 reversed.

Section 21-5   Ampere's Law

There are no exercises for this section.

Section 21-6   Current Loops, Solenoids, and Magnets

33. Find the magnetic field at the center of a circular loop of radius 15 cm carrying a current of 6 A.

34. Find the magnetic field at point P in Figure 21-38.

Figure 21-38
Exercise 34.

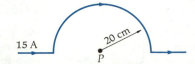

35. A single-turn circular loop of radius 10.0 cm is to produce a field at its center that will just cancel the earth's field at the equator, which is 0.7 G directed north. Find the current in the wire and make a sketch showing the orientation of the loop and current.

36. In Figure 21-39, find the magnetic field at point P, which is at the common center of the two semicircular arcs.

Figure 21-39
Exercise 36.

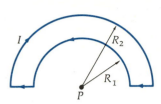

37. A solenoid has length 25 cm, radius 1.2 cm, and 400 turns. It carries 3 A of current. Find the magnetic field at its center.

38. Find the magnetic field at the center of a solenoid of length 30 cm, radius 2.2 cm, and 800 turns if it carries a current of 2.5 A.

Section 21-7   Magnetism in Matter

39. Which of the substances listed in Table 21-1 are paramagnetic and which are diamagnetic?

Problems

1. A 10.0-cm length of wire has mass 5.0 g and is connected to a source of emf by light, flexible leads. A magnetic field $B = 0.5$ T is horizontal and perpendicular to the wire. Find the current necessary to float the wire, that is, the current for which the magnetic force balances the weight of the wire.

2. Particles of charge q and mass m are accelerated through a potential difference ΔV and enter a region of uniform magnetic field B perpendicular to their velocity. If r is the radius of their circular orbit, show that $q/m = 2\,\Delta V/r^2 B^2$.

3. A mass spectrograph uses a velocity selector consisting of parallel plates separated by 2.00 mm and having a potential difference of 160 V. The magnetic field between the plates is 0.42 T. The magnetic field in the mass spectrograph is 1.2 T. Find (a) the speed of the ions entering the mass spectrograph and (b) the separation distance of the peaks on the photographic film for $^{238}\mathrm{U}^+$ (mass 3.95×10^{-25} kg) and $^{235}\mathrm{U}^+$ (mass 3.90×10^{-25} kg).

4. A certain cyclotron has a magnetic field of 2.0 T and is designed to accelerate protons to 20 MeV. (a) What is the cyclotron frequency? (b) What must the minimum radius of the magnet be to achieve the 20-MeV emergence energy? (c) If the alternating potential difference applied to the dees has a maximum value of 50 kV, how many orbits must the protons make before they will emerge with 20 MeV of energy?

5. Three very long, parallel wires are at the corners of a square, as shown in Figure 21-40. Find the magnetic field B at the unoccupied corner of the square when (a) all the currents are directed into the paper, (b) I_1 and I_3 are directed in and I_2 is out, and (c) I_1 and I_2 are in and I_3 is out.

Figure 21-40
Three long parallel wires for Problems 5 and 8.

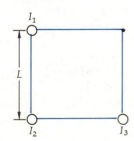

6. Show that the radius of the orbit in a cyclotron is proportional to the square root of the number of orbits completed.

7. Four long, straight, parallel wires each carry current I. In a plane perpendicular to the wires, the wires are at the corners of a square of side a. Find the force per unit length on one of the wires if (a) all the currents are in the same direction and (b) the currents in the wires at adjacent corners are oppositely directed.

8. Find the magnetic field at the center of the square of Figure 21-40 if (a) all three currents are directed in, (b) I_1 is in and the other two currents are out, (c) I_2 is in and the other two currents are out, and (d) I_3 is in and the other two currents are out.

9. A very long, straight wire carries a current of 20.0 A. An electron is 1.0 cm from the center of the wire and is moving with speed 5.0×10^6 m/s. Find the force on the electron when it moves (a) directly away from the wire, (b) parallel to the wire in the direction of the current, and (c) perpendicular to the wire and tangent to a circle around the wire.

10. A relatively inexpensive ammeter, called a *tangent galvanometer*, can be made using the earth's magnetic field. A plane circular coil of N turns and radius R is oriented such that the field B_c it produces in the center of the coil is either east or west. A compass is placed at the center of the coil. When there is a current I, the compass points in the direction of the resultant magnetic field B at an angle θ to the north. Show that the current I is related to θ and to the horizontal component of the earth's field B_e by

$$I = \frac{2RB_e}{\mu_0 N}\tan\theta$$

11. A particle of positive charge q moves in a helical path around a long, straight wire carrying current I. Its velocity has components $v_\parallel$ parallel to the current and v_t tangent to a circle around the current. Find the relationship between $v_\parallel$ and v_t such that the magnetic force provides the necessary centripetal force.

CHAPTER TWENTY-TWO

Magnetic Induction

Around the magnet Faraday

Was sure that Volta's lightnings
play:

But how to draw them from the
wire?

He took a lesson from the heart:

'Tis when we meet, 'tis when we
part

Breaks forth the celestial fire.

As Written by a Contemporary of Faraday

In the last chapter you learned that a current in a wire creates a magnetic field. Is the reverse true? That is, can a magnetic field induce a current in a wire? Michael Faraday in England and Joseph Henry in the United States carried out similar experiments simultaneously, unknown to each other, and both announced in the same year their success in producing induced currents, but in a surprising way. A magnetic field can induce a current only when the field is changing. Their discovery made possible the generation of alternating current which is used today in our homes, offices, and factories.

The fact that electric currents are induced by changing magnetic fields, discovered independently in the early 1830s by Faraday and Henry, can be easily demonstrated. The ends of a coil are attached to a galvanometer and a strong magnet is moved toward or away from the coil (see Figure 22-1). The momentary deflection of the galvanometer *during* the movement of the magnet indicates that there is an induced electric current in the coil–galvanometer circuit. In another demonstration, a circuit containing a large electromagnet made of many turns of wire wrapped around an iron core is broken by a knife switch. As the circuit is broken, a spark jumps the gap across the switch. The changing magnetic field caused by breaking the circuit induces a large emf across the switch. This emf causes dielectric breakdown of the air, indicated by the spark.

The results of these and many other experiments can be expressed by a single relation known as Faraday's law, which relates the induced emf in a circuit to the change in magnetic flux through the circuit. Faraday's law is the basis for the generation of alternating current (ac). We will study Faraday's law and some of its consequences in this chapter. In Chapter 23, we will study some simple ac circuits.

Figure 22-1
Demonstration of induced emf. When the magnet is moved toward or away from the coil, an emf is induced in the coil, as shown by the deflection of the galvanometer. No deflection is observed when the magnet is stationary.

22-1   Magnetic Flux and Faraday's Law

Magnetic flux is related to the number of magnetic field lines that pass through a given area. It is the magnetic analog of electric flux, which we defined in Section 18-5. In Figure 22-2, the magnetic field is perpendicular to the area bound by a simple circuit consisting of one turn of wire. In this case, the *magnetic flux* is defined as the product of the magnetic field **B** and the area A bound by the circuit. In the more general case when the magnetic field is not perpendicular to the area, as illustrated in Figure 22-3, the magnetic flux ϕ_m is defined as

$$\phi_m = B_\perp A = BA \cos \theta$$

where $B_\perp = B \cos \theta$ is the component of **B** perpendicular to the area and θ is the angle between **B** and line perpendicular to the area. For a coil of N turns, the magnetic flux through the coil is

$$\phi_m = NB_\perp A = NBA \cos \theta \qquad\qquad \text{22-1} \quad \text{Magnetic flux}$$

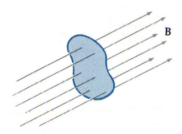

Figure 22-2
When the magnetic field **B** is perpendicular to the area enclosed by a loop, the magnetic flux through the loop is BA.

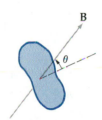

Figure 22-3
When the magnetic field **B** makes an angle θ with the normal to the area of a loop, the flux through the loop is $(B \cos \theta)A$.

The units of magnetic flux are those of magnetic field times area or tesla-metre squared, which is called a weber (Wb):

$$1 \text{ Wb} = 1 \text{ T} \cdot \text{m}^2 \qquad\qquad \text{22-2}$$

Example 22-1   A uniform magnetic field of magnitude 2000 G makes an angle of 30° with the axis of a circular loop of 300 turns and radius 4 cm. Find the magnetic flux through the coil.

Using $1 \text{ G} = 10^{-4} \text{ T}$, this magnetic field in SI units is 0.2 T. The area of the coil is

$$A = \pi r^2 = (3.14)(0.04 \text{ m})^2 = 0.00502 \text{ m}^2$$

The flux through the coil is then

$$\phi_m = NBA \cos 30° = (300)(0.2 \text{ T})(0.00502 \text{ m}^2)(0.866)$$

$$= 0.26 \text{ Wb}$$

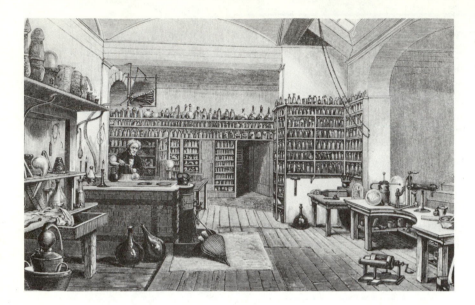

An engraving of Michael Faraday (1791–1867) in his laboratory in the basement of the Royal Institution in London, where he carried out his research on electricity and magnetism.

Experiments of Faraday, Henry, and others show that if the magnetic flux is changed by any means, an emf $\mathcal{E}$ is induced in the circuit equal in magnitude to the rate of change of the flux.

$$\mathcal{E} = (-)\frac{\Delta\phi_m}{\Delta t} \qquad\qquad 22\text{-}3$$

Faraday's law

This result is known as *Faraday's law*. (The reason for the negative sign in this equation will be discussed shortly.) The emf is usually detected by observing a current in the circuit, but it is present even if the circuit is open so that no current exists.

The magnetic flux through a circuit can be changed in many different ways. For example, the current producing the magnetic field may be increased or decreased, permanent magnets may be moved toward or away from the circuit, the circuit itself may be moved toward or away from the source of the flux, or the orientation of the circuit or the area of the circuit in a fixed magnetic field may be changed. In every case, an emf is induced in the circuit that is equal in magnitude to the rate of change of the magnetic flux.

Example 22-2   Show that a weber per second is a volt.

As we saw in Chapter 21, the unit of magnetic field, the tesla (T), is related to other SI units by Equation 21-4:

$$1\,\text{T} = 1\,\frac{\text{N}}{\text{A}\cdot\text{m}}$$

(We can remember this relation from the fact that the force on a current element of length ℓ is given by $F = BI\ell$. Thus $1\,\text{N} = 1\,\text{T}\cdot\text{A}\cdot\text{m}$.) The unit of the rate of change of magnetic flux is then

$$1\,\frac{\text{Wb}}{\text{s}} = 1\,\frac{\text{T}\cdot\text{m}^2}{\text{s}} = 1\,\frac{\text{N}\cdot\text{m}}{\text{A}\cdot\text{s}}$$

Faraday and his wife in the 1850s.

But a newton-metre is a joule, and an ampere-second is a coulomb. There-fore,

$$1\,\frac{Wb}{s} = 1\,\frac{J}{C} = 1\,V$$

The negative sign in Faraday's law has to do with the direction of the induced emf. In practice, we can find the direction of the induced emf and the induced current from a general statement known as *Lenz's law*:

The induced emf and the induced current are in such a direction as to tend to oppose the change that produces them.

Lenz's law

In our statement of Lenz's law, we have not specified just what kind of change causes the induced emf. This was purposely left vague to allow for a variety of interpretations. A few illustrations will clarify this point.

Figure 22-4 shows a bar magnet moving toward a loop that has a resist-ance R. Since the magnetic field from the bar magnet is directed to the right, out of the north pole of the magnet, the movement of the magnet toward the loop increases the magnetic flux through the loop due to the magnet. (The closer the magnet is to the loop, the stronger is the magnetic field due to the magnet at the loop.) The change in magnetic flux through the loop due to the movement of the magnet induces a current in the loop in the direction shown. This induced current produces a magnetic field of its own. The flux of this field is opposite to that of the magnet, so it tends to decrease the total magnetic flux through the loop.

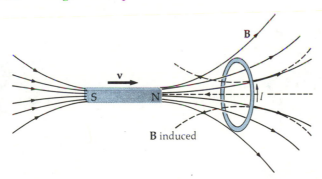

Figure 22-4
When the bar magnet is moving toward the loop, the emf induced in the loop produces a current in the direction shown. The magnetic flux due to the induced current in the loop opposes the increase in flux through the loop due to the motion of the magnet.

If the magnet is moved away from the loop (which decreases the magnetic flux through the loop due to the magnet), the current induced in the loop is in the opposite direction from that shown in Figure 22-4. The magnetic field of the loop is now to the right, which tends to increase the total flux through the loop. That is, the increasing flux from the loop directed to the right opposes the decreasing flux from the magnet directed to the left.

When the magnet is moving toward the loop, as in Figure 22-4, the current it induces in the loop causes the loop to act like a small magnet with the induced magnetic moment shown in Figure 22-5. The north pole of the loop is to the left and the south pole is to the right. Since opposite poles attract and like poles repel, the induced magnetic moment of the loop exerts a force on the bar magnet to the left that opposes the motion of the magnet toward the loop. We thus see that we can use Lenz's law in terms of forces rather than flux. If the bar magnet is moved toward the loop, the induced current produces a magnetic moment to oppose this change.

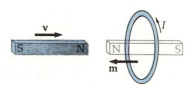

Figure 22-5
The induced magnetic moment of the loop (indicated by the outlined magnet) due to the induced current is such as to oppose the motion of the bar magnet. Here the bar magnet is moving toward the loop, so the induced magnetic moment repels the bar magnet.

We note that Lenz's law is required by the law of conservation of energy. If the current in the loop in Figure 22-5 were opposite the direction shown, the induced magnetic moment of the loop would attract the magnet when it was moving toward the loop and cause it to accelerate toward the loop. If we began with the magnet a great distance from the loop and gave it a very slight push toward the loop, the force on the magnet due to the induced current would be toward the loop, increasing the velocity of the magnet. As the speed of the magnet increased, the rate of change of the flux would increase, which would increase the induced current. This would further increase the force on the magnet. Hence, the kinetic energy of the magnet and the Joule heat produced in the loop (I^2R) would both increase with no source of energy. This would violate the law of conservation of energy.

In Figure 22-6, the bar magnet is at rest, and the loop is moved away from it. The induced current and magnetic moment are shown in the figure. In this case, the magnetic moment of the loop attracts the bar magnet, thus opposing the motion of the loop as required by Lenz's law.

Figure 22-7 shows two circuits close enough together so that flux in one produces flux in the other. When the current in circuit 1 is changing, there is a changing flux through circuit 2. Suppose the switch in circuit 1 is initially open, with no current in either circuit (see Figure 22-7a). When the switch is closed (see Figure 22-7b), the current in circuit 1 does not reach its steady value of $\mathcal{E}_1/R_1$ instantaneously but takes some time to change from zero to this value. During this time, while the current is increasing, the flux through circuit 2 is changing, which causes an induced current in that circuit in the direction shown. When the current in circuit 1 reaches its steady value, the flux is no longer changing, so there is no induced current in circuit 2.

Figure 22-6
When the loop is moving away from the stationary bar magnet, the induced magnetic moment in the loop attracts the bar magnet, again opposing the relative motion.

Figure 22-7
(a) Two adjacent circuits. (b) Just after the switch is closed, I_1 is increasing in the direction shown. The changing flux through circuit 2 induces current I_2. The flux due to I_2 opposes the increase in flux due to I_1. (c) When the switch is opened again, I_1 decreases and **B** decreases. The induced current I_2 then tends to maintain the flux through circuit 2, opposing the change in flux.

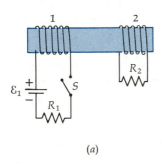

(a)

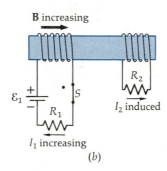

(b)

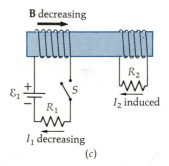

(c)

When the switch in circuit 1 is opened again (see Figure 22-7c) and the current is decreasing to zero, an induced current in circuit 2 in the opposite direction appears momentarily. It is important to understand that there is an induced emf only *while the flux is changing.* The emf does not depend on the magnitude of the flux, only on its rate of change. If there is a large, steady flux through a circuit, there is no induced emf.

For our next example, we consider a single, isolated circuit (see Figure 22-8). When there is a current in the circuit, there is a magnetic flux through the coil due to this current. When the current is changing, the flux through the coil is changing, causing an induced emf in the circuit. This *self-induced* emf, called a *back emf,* opposes the change in the current. It is because of this back emf that the current in a circuit cannot jump instantaneously from zero to some finite value or from some value to zero. Joseph Henry first noticed this effect when he was experimenting with a circuit consisting of many

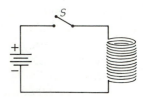

Figure 22-8
The many turns of wire in the coil give a large flux for a given current in the circuit. When the current changes, a large emf that opposes the change is induced in the coil.

turns of a wire like the one in Figure 22-8. This arrangement gives a large flux through the circuit for even a small current. When Henry tried to break the circuit, he noticed a spark across the switch. This spark was due to the large induced emf that occurs in the circuit when the current varies rapidly, such as during the opening of the switch. In this case, the induced emf tries to maintain the original current. The large induced emf produces a large voltage drop across the switch as it is opened. The electric field between the poles of the switch is large enough to tear the electrons from the air molecules, causing dielectric breakdown. When the molecules in the air are ionized, the air conducts electric current in the form of a spark.

Question

1. Figure 22-9*a* shows a square loop in a uniform magnetic field that is directed into the paper. Indicate the direction of the current induced in the loop as it is twisted into the position shown in Figure 26-9*b*.

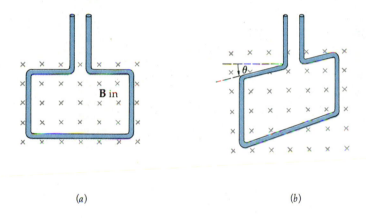

(*a*)   (*b*)

Figure 22-9
(*a*) A coil whose plane is perpendicular to a magnetic field **B**. (*b*) When the coil is twisted, the flux through the coil is changed and an emf is induced in the coil.

22-2 Motional emf

Figure 22-10 shows a conducting rod sliding along conducting rails that are connected by a resistor. A uniform magnetic field is directed into the paper. Since the area of the circuit increases as the rod moves to the right, the magnetic flux through the circuit is increasing. An emf is therefore induced in the circuit. Let ℓ be the separation of the rails and x be the distance from the left end to the rod at some time. The area enclosed by the circuit is then ℓx, and the magnetic flux through the circuit at this time is

$$\phi_m = BA = B\ell x$$

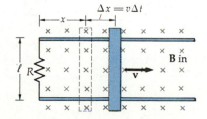

Figure 22-10
A conducting rod sliding on conducting rails in a magnetic field. As the rod moves to the right, the area of the circuit increases and the magnetic flux through the circuit directed into the paper increases. An emf is induced in the circuit, which produces a counterclockwise current whose flux is directed out of the paper to oppose the change.

When the rod moves through a distance Δx, the area enclosed by the circuit changes by $\Delta A = \ell\, \Delta x$ and the flux changes by $\Delta \phi = B\ell\, \Delta x$. The rate of change of the flux is

$$\frac{\Delta \phi_m}{\Delta t} = B\ell\, \frac{\Delta x}{\Delta t} = B\ell v$$

where $v = \Delta x / \Delta t$ is the speed of the rod. The magnitude of the emf induced in this circuit is therefore

$$\mathcal{E} = \frac{\Delta \phi_m}{\Delta t} = B\ell v \qquad\qquad 22\text{-}4 \qquad \text{Motional emf}$$

(We will often drop the negative sign from Faraday's law when we are just interested in the magnitude of the emf. We can always find the direction of the emf from Lenz's law.) The direction of the emf in this case is such as to produce a current in the counterclockwise sense. The flux produced by this induced current is directed out of the paper, opposing the increase in flux due to the motion of the rod. Because of the induced current in the rod, which is upward, there is a magnetic force on the rod of magnitude $I\ell B$. The direction of this force, obtained from the right-hand rule, is to the left, opposing the motion of the rod. If the rod is given some initial velocity **v** to the right and is then released, the force due to the induced current slows the rod down until it stops. Thus, to maintain the motion of the rod, an external force to the right must be exerted on it.

Suppose we exert an external force of magnitude $F = I\ell B$ on the rod to the right to keep the rod moving with constant speed v. The power input of this force is the force times the speed.

$$P = Fv = I\ell Bv$$

This power goes into Joule heat in the resistor. Setting the power equal to $I^2 R$, we obtain

$$I\ell Bv = I^2 R$$

or

$$B\ell v = IR$$

The induced emf, $B\ell v$, equals IR, the potential drop across the resistor, consistent with Kirchhoff's first rule (Section 20-5).

The emf in this case is called *motional emf*. It is induced in a conducting rod or wire moving in a magnetic field even if there is no complete circuit and therefore no steady current. Figure 22-11 shows an electron in a conducting rod that is moving through a uniform magnetic field directed into the paper. There is a magnetic force on the electron of magnitude qvB directed downward. Because of this magnetic force, free electrons in the rod move downward, producing a net negative charge at the bottom of the rod and leaving a net positive charge at the top. The free electrons continue to move downward until the electric field produced by these charges exerts a force of magnitude qE on the electrons that balances the magnetic force qvB. In equilibrium, the electric field in the rod is thus

$$E = vB$$

The potential difference across the rod equals the magnitude of the induced emf

$$V = E\ell = vB\ell$$

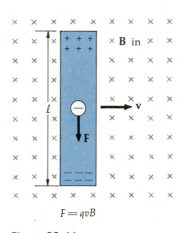

$$F = qvB$$

Figure 22-11
Magnetic force **F** on an electron in a conducting rod that is moving through a magnetic field. Negative charges move to the bottom of the rod, leaving the top of the rod positive. These charges produce an electric field of magnitude $E = vB$. The potential at the top of the rod is greater than that at the bottom by $E\ell = vB\ell$.

22-3   Eddy Currents

In the examples we have discussed thus far, the currents produced by a changing magnetic flux were set up in definite circuits. Often, though, a changing flux sets up circulating currents, called *eddy currents,* in a piece of bulk metal such as the core of a transformer. (We will discuss transformers in the next chapter.) The heat produced by eddy currents constitutes a power loss in the transformer.

Consider a conducting slab between the pole faces of an electromagnet (see Figure 22-12). If the magnetic field **B** between the pole faces is changing with time (as it will if the current in the magnet windings is alternating current), the flux through any closed loop in the slab will be changing. For example, the flux through the curve C indicated in the figure is the product of the magnetic field B and the area enclosed by the curve. If B varies, the flux will vary and there will be an induced emf around the curve C. Since path C is in a conductor, there will be a current equal to the emf divided by the resistance of the path. In this figure, we have indicated just one of the many closed paths that will contain currents if the magnetic field between the pole faces varies.

Eddy currents are usually unwanted because the heat produced is a power loss; in addition, this heat must be dissipated. The power loss can be reduced by increasing the resistance of the possible paths for the eddy currents, as illustrated in Figure 22-13. Here the conducting slab is laminated; that is, it is made up of thin strips of metal glued together. Because of the insulating glue between the strips, the eddy currents are essentially confined to the strips. The large eddy-current loops are broken up, and the power loss is greatly reduced.

The existence of eddy currents can be demonstrated by pulling a copper or aluminum sheet between the poles of a strong permanent magnet (see Figure 22-14). Part of the area enclosed by curve C in the figure is in the magnetic field, and part is outside the field. As the sheet is pulled to the right, the magnetic flux through the curve decreases (assuming that the flux directed into the paper is positive). According to Faraday's law and Lenz's law, a clockwise current will be induced around the curve. Since the current is directed upward in the region between the pole faces, the magnetic field exerts a force on the current to the left, opposing the motion of the sheet. You can feel this force on the sheet if you try to pull a conducting sheet quickly through a strong magnetic field. If the sheet has cuts in it, as in Figure 22-15, the eddy currents are broken up, and so the force is greatly reduced.

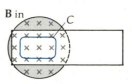

Figure 22-12
Eddy currents in a metal conducting slab. When the magnetic field is changing, an emf is induced in any closed path in the metal slab, such as the curve C shown. The induced emf causes a current in the closed path.

Figure 22-13
Eddy currents in a metal slab can be reduced by constructing the slab from thin metal strips. The resistance of the path indicated by C is now large because of the insulating glue between the strips.

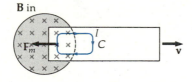

Figure 22-14
Demonstration of eddy currents. When the metal slab is pulled to the right, there is a magnetic force to the left on the induced current opposing the motion.

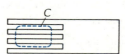

Figure 22-15
If the metal slab has cuts as shown, the eddy currents are greatly reduced because the conducting paths are broken up.

Eddy currents are not always undesired, however. They can, for example, be used to damp unwanted oscillations in sensitive mechanical balance scales used to weigh small masses. Such scales have a tendency to oscillate back and forth around their equilibrium reading many times. They are therefore usually designed so that a small piece of metal moves between the poles of a magnet as the scales oscillate. The resulting eddy currents damp the oscillations so that equilibrium is reached quickly.

Questions

2. A bar magnet is dropped inside a long vertical pipe. The pipe is evacuated so that there is no air resistance, but the falling magnet still reaches a terminal velocity. Explain why.

3. A sheet of metal fixed to the end of a pivoted rod will swing like a pendulum about the pivot. If the sheet is made to swing through the gap between two poles of a magnet, the oscillation will rapidly damp out. Why?

22-4 Inductance

The magnetic flux through a circuit can be related to the current in that circuit and the currents in other nearby circuits. (We will assume that there are no permanent magnets around.) Consider a coil carrying a current I. The current produces a magnetic field that could, in principle, be calculated from the Biot-Savart law. Since the magnetic field at every point in the neighborhood of the coil is proportional to I, the magnetic flux through the coil is also proportional to I.

$$\phi_m = LI \qquad\qquad\qquad 22\text{-}5$$

where L is a constant called the *self-inductance* of the coil. It depends on the geometric shape of the coil, just as capacitance depends on the geometric arrangement of the conducting plates of a capacitor. The SI unit of inductance is the henry (H). From Equation 22-5, we can see that the unit of inductance equals that of flux divided by that of current. Since flux has units of webers and current, amperes,

    Self-inductance

$$1\ \text{H} = 1\ \text{Wb/A} = 1\ \text{T}\cdot\text{m}^2/\text{A}$$

In principle, the calculation of self-inductance is straightforward for any given coil or circuit. In actual practice, the calculation is very difficult. There is one case, however, that of a tightly wound solenoid, for which the self-inductance can be easily calculated. The magnetic field inside a tightly wound solenoid carrying a current I is given by Equation 21-30:

$$B = \mu_0(N/\ell)I$$

where ℓ is the length and N is the number of turns. If the solenoid has a cross-sectional area A, the magnetic flux through the N turns is

$$\phi_m = NBA = \mu_0 N^2(A/\ell)I \qquad\qquad 22\text{-}6$$

As expected, the flux is proportional to the current I. The proportionality constant is the self-inductance L:

$$L = \frac{\phi_m}{I} = \mu_0 N^2 A / \ell \qquad\qquad 22\text{-}7$$

Self-inductance of a solenoid

The self-inductance of a solenoid is proportional to the square of the number of turns and to the area A. We note from Equation 22-7 that the SI units of μ_0 can be expressed as henrys per metre,

$$\mu_0 = 4\pi \times 10^{-7} \text{ H/m}$$

Example 22-3 Find the self-inductance of a solenoid of length 10 cm, area 5 cm², and 100 turns.

We can calculate the self-inductance from Equation 22-7 in henrys if we put all the quantities in SI units. Using $\ell = 0.1$ m, $A = 5 \times 10^{-4}$ m², $N = 100$, and $\mu_0 = 4\pi \times 10^{-7}$ H/m, we obtain

$$L = \mu_0 N^2 A / \ell$$

$$= (4\pi \times 10^{-7} \text{ H/m})(100)^2 (5 \times 10^{-4} \text{ m}^2)/(0.1 \text{ m}) = 2\pi \times 10^{-5} \text{ H}$$

When the current in a circuit is changing, the magnetic flux due to the current is also changing, and there is an emf induced in the circuit. Since the self-inductance of a particular circuit is constant, the change in flux is related to the change in current by

$$\Delta \phi_m = \Delta(LI) = L(\Delta I)$$

According to Faraday's law, we have

$$\mathcal{E} = -\frac{\Delta \phi_m}{\Delta t} = -L\frac{\Delta I}{\Delta t} \qquad\qquad 22\text{-}8$$

Thus the self-induced emf is proportional to the rate of change of the current.

Example 22-4 At what rate must the current in the solenoid of Example 22-3 change to induce an emf of 20 V?

Since we are interested only in magnitudes, we can omit the negative sign from Equation 22-8:

$$\mathcal{E} = L\frac{\Delta I}{\Delta t} = 20 \text{ V}$$

Then,

$$\frac{\Delta I}{\Delta t} = \frac{\mathcal{E}}{L} = \frac{20 \text{ V}}{2\pi \times 10^{-5} \text{ H}} = 3.18 \times 10^5 \text{ A/s}$$

The emf generated in a solenoid because of a changing current, as in Example 22-4, is in the direction to oppose the emf that supplies the current. It is therefore a back emf.

When two or more circuits are close to each other, as in Figure 22-16, the magnetic flux through one circuit depends not only on the current in that circuit but also on the current in the nearby circuits. Let I_1 be the current in the circuit on the left in Figure 22-16 and I_2 be that in the circuit on the right. The magnetic field at some point P consists of a part due to I_1 and a part due to I_2. These fields are proportional to the currents that produce them. We can therefore write the flux through circuit 2, ϕ_{m2}, as the sum of two parts, one proportional to current I_1 and the other to current I_2:

$$\phi_{m2} = L_2 I_2 + M_{12} I_1 \qquad 22\text{-}9a$$

where L_2 is the self-inductance of circuit 2 and M_{12} is called the *mutual inductance* of the two circuits. The mutual inductance depends on the geometrical arrangement of both circuits. In particular, we can see that if the circuits are far apart, the flux through circuit 2 due to the current I_1 will be small and the mutual inductance will be small. An equation similar to Equation 22-9a can be written for the flux through circuit 1:

$$\phi_{m1} = L_1 I_1 + M_{21} I_2 \qquad 22\text{-}9b$$

where L_1 is the self-inductance of circuit 1.

Figure 22-17 shows a long, narrow, tightly wound solenoid inside another solenoid of equal length but larger radius. For this situation, we can actually calculate the mutual inductance of the two solenoids. Let ℓ be the length of both solenoids, and let the inner solenoid have N_1 turns and radius r_1 and the outer solenoid have N_2 turns and radius r_2. We first calculate the mutual inductance M_{12} by assuming that the inner solenoid carries a current I_1 and finding the flux ϕ_{m2} due to this current through the outer solenoid. The magnetic field due to the current in the inner solenoid has the magnitude

$$B_1 = \mu_0 (N_1 / \ell) I_1 \qquad 22\text{-}10$$

and is constant in the space within the solenoid. Outside the inner solenoid, the magnetic field is zero. The flux through the outer solenoid due to this magnetic field is therefore

$$\phi_{m2} = N_2 B_1 (\pi r_1^2) = \mu_0 N_2 (N_1 / \ell)(\pi r_1^2) I_1$$

Note that the area used to compute the flux through the outer solenoid is not the area of that solenoid (πr_2^2) but is instead πr_1^2 because there is no magnetic field outside the inner solenoid. The mutual inductance is thus

$$M_{12} = \frac{\phi_{m2}}{I_1} = \mu_0 N_2 (N_1 / \ell) \pi r_1^2 \qquad 22\text{-}11$$

We will now calculate M_{21} by finding the flux through the inner solenoid due to a current I_2 in the outer solenoid. When the outer solenoid carries a current I_2, there is a uniform magnetic field inside that solenoid given by Equation 22-10 with I_2 replacing I_1 and N_2 replacing N_1:

$$B_2 = \mu_0 (N_2 / \ell) I_2$$

The magnetic flux through the inner solenoid is then

$$\phi_{m1} = N_1 B_2 (\pi r_1^2) = \mu_0 N_1 (N_2 / \ell)(\pi r_1^2) I_2$$

The area used here is also πr_1^2 because that is the cross-sectional area of the

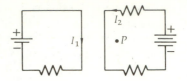

Figure 22-16
Two adjacent circuits. The magnetic field at point P is partly due to current I_1 and partly due to I_2. The flux through either circuit is the sum of two terms, one proportional to I_1 and the other to I_2.

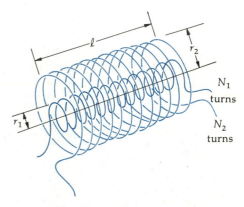

Figure 22-17
A long, narrow solenoid inside a second solenoid of the same length. A current in either solenoid produces magnetic flux in the other.

inner solenoid, and the magnetic field is uniform everywhere inside that solenoid. The mutual inductance is then

$$M_{21} = \frac{\phi_{m1}}{I_2} = \mu_0 N_1 (N_2/\ell)\pi r_1^2 \qquad \text{22-12}$$

We note that Equations 22-11 and 22-12 are the same. That is, $M_{21} = M_{12}$. It can be shown that this is a general result. We will therefore drop the subscripts for mutual inductance and simply write M.

Question

4. How would the self-inductance of a solenoid be changed if the same length of wire were wound onto a cylinder of the same diameter but twice the length? If twice as much wire were wound onto the same cylinder?

22-5   *LR* Circuits and Magnetic Energy Density

In Section 19-6, we showed that it takes work to charge a capacitor and that a charged capacitor stores energy given by

$$U = \frac{1}{2} QV = \frac{1}{2} CV^2 = \frac{1}{2}\frac{Q^2}{C}$$

where Q is the charge on either plate, V is the potential difference between the plates, and C is the capacitance. We showed that this energy can be considered to be stored in the electric field E between the plates. We also stated that, in general, when there is an electric field E in space, the electric energy per unit volume is

$$\eta = \tfrac{1}{2}\epsilon_0 E^2$$

There is a similar expression for the energy in a magnetic field.

A coil or solenoid, which usually has a large self-inductance, is called an *inductor*. The symbol for an inductor is ⟳ . An inductor is a device for producing a uniform magnetic field; it is analogous to the capacitor, which produces a uniform electric field. It takes work to produce a current in an inductor.

Figure 22-18 shows an inductance L and resistance R in series with a battery of emf $\mathcal{E}_0$ and a switch. Such a circuit is called an *LR* circuit. We assume that the resistance R includes the resistance of the inductor coil and that the inductance of the rest of the circuit is negligible compared with that of the inductor. The switch is initially open, so there is no current in the circuit. Just after the switch is closed, the current is still zero, but is changing at a rate $\Delta i/\Delta t$. There is a back emf in the inductor of magnitude $L\,\Delta i/\Delta t$, which equals the emf of the battery $\mathcal{E}_0$. In the circuit diagram, we have placed $+$ and $-$ signs on the inductor to indicate the direction of the emf when the current is increasing. A short time after the switch is closed, there is a current i in the circuit. There is then a potential drop iR across the

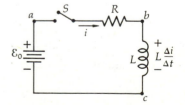

Figure 22-18
A typical *LR* circuit. Just after the switch is closed, the current begins to increase in the circuit and a back emf of magnitude $L(\Delta i/\Delta t)$ is generated in the inductor. The potential drop across the resistor iR plus the potential drop across the inductor equals the emf of the battery.

resistor, and the back emf of the inductor is less than its initial value. Kirchhoff's rule 1 applied to this circuit gives

$$\mathcal{E}_0 = iR + L\frac{\Delta i}{\Delta t} \qquad \text{22-13}$$

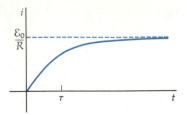

Figure 22-19
Current versus time in an *LR* circuit. After a time $\tau = L/R$, the current is 63 percent of its maximum value $\mathcal{E}_0/R$.

As the current increases, the potential drop iR increases, and the back emf, which is proportional to the rate of change of the current, decreases. Figure 22-19 shows a graph of the current in this circuit as a function of time. This graph is similar to that for the charge on a capacitor when the capacitor is charged in an *RC* circuit (see Figure 20-35). After a long time, the current reaches its maximum value

$$I = \frac{\mathcal{E}_0}{R} \qquad \text{22-14}$$

The time τ in Figure 22-19 is the *time constant* for this circuit, which is given by

$$\tau = \frac{L}{R} \qquad \text{22-15}$$

Time constant in a *LR* circuit

If the rate of increase of the current were constant and equal to its original rate, the current would reach its maximum value after a time $t = \tau$. However, the rate of increase of the current is not constant but decreases with time. Thus, after a time $t = \tau$, the current is $0.63I$. We can see that, if the inductance L is large, it takes a long time for the current to build up to its maximum value.

If we multiply each term in Equation 22-13 by the current i, we obtain

$$\mathcal{E}_0 i = i^2R + Li\frac{\Delta i}{\Delta t} \qquad \text{22-16}$$

The first term in Equation 22-16, $\mathcal{E}_0 i$, is the power input of the battery. The second term, i^2R, is the power dissipated as heat in the resistance in the circuit. The term $Li\,\Delta i/\Delta t$ is the power input to the inductance. The energy put into the inductance in the time interval Δt is

$$\Delta U = P\,\Delta t = Li\,\Delta i$$

The energy increase is the product of Li and the increase in current Δi. The total energy put into the inductor in increasing the current from $i = 0$ to its final value $i = I$ is the product of the average value of Li, which is $\frac{1}{2}LI$, and the final current I:

$$U = \tfrac{1}{2}LI^2 \qquad \text{22-17}$$

Energy in an inductor

This result is analogous to that for the energy for charging a capacitor,

$$U = \frac{1}{2}\frac{Q^2}{C}$$

In the process of producing a current in an inductor, a magnetic field is created in the space within the inductor coil. The work done in producing a current in an inductor can be thought of as the work required to create a magnetic field. That is, we can think of the energy stored in an inductor as energy stored in the magnetic field of the inductor. The magnetic field in a solenoid is related to the number of turns N, the length of the solenoid ℓ, and the current I by

$$B = \mu_0(N/\ell)I$$

Writing $I = B\ell/\mu_0 N$ for the current and $L = \mu_0 N^2 A/\ell$ for the inductance, Equation 22-17 becomes

$$U = \tfrac{1}{2}(\mu_0 N^2 A/\ell)(B\ell/\mu_0 N)^2 = \frac{B^2}{2\mu_0}(A\ell)$$

The quantity $A\ell$ is the volume of the space in the solenoid containing the magnetic field. The energy per unit volume is the *magnetic energy density* η_m.

$$\eta_m = \frac{B^2}{2\mu_0}$$

22-18   Magnetic energy density

Although we have obtained Equation 22-18 by considering the special case of the magnetic field in a solenoid, the result is general. Whenever there is a magnetic field in space, the magnetic energy per unit volume is given by Equation 22-18.

22-6   Generators and Motors

Most electrical energy used today is produced by electrical generators in the form of alternating current (ac). A simple generator of alternating current is a coil that is rotating in a uniform magnetic field, as illustrated in Figure 22-20a. The ends of the coil are connected to rings that rotate with the loop.

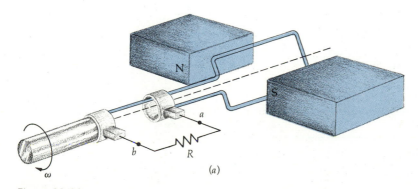
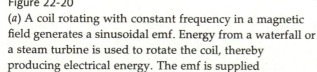
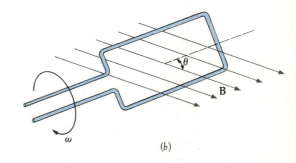

(a)

(b)

Figure 22-20

(a) A coil rotating with constant frequency in a magnetic field generates a sinusoidal emf. Energy from a waterfall or a steam turbine is used to rotate the coil, thereby producing electrical energy. The emf is supplied to an external circuit by the brushes in contact with the rings. (b) At this instant, the normal to the area of the coil makes an angle θ with the magnetic field and the flux is $BA \sin \theta$.

Electrical contact is made with the loop by stationary metal brushes in contact with the rings. When the line perpendicular to the area of the loop makes an angle θ with a uniform magnetic field **B**, as shown in Figure 22-20b, the magnetic flux through the coil is

$$\phi_m = NBA \cos \theta \qquad\qquad\qquad 22\text{-}19$$

where N is the number of turns and A is the cross-sectional area of the coil. When the coil is mechanically rotated, the flux through the coil will change and, according to Faraday's law, an emf will be induced in the coil. Let f be the frequency of rotation of the coil. During a time of one period $T = 1/f$, the angle θ changes by 2π radians. The angle thus increases with time according to

$$\theta = \frac{2\pi}{T} t = 2\pi ft = \omega t \qquad\qquad 22\text{-}20$$

where $\omega = 2\pi f$ is the angular frequency of rotation. Substituting this expression for θ into Equation 22-19, we obtain for the flux

$$\phi_m = NBA \cos 2\pi ft = NBA \cos \omega t \qquad\qquad 22\text{-}21$$

The emf in the coil will then be

$$\mathcal{E} = -\frac{\Delta\phi_m}{\Delta t} \qquad\qquad\qquad 22\text{-}22$$

Except for the minus sign in Equation 22-22, Equations 22-21 and 22-22 are of the same form as the equations for the position and velocity of a simple harmonic oscillator (Chapter 15):

$$x = A \cos 2\pi ft = A \cos \omega t$$

and

$$v_x = \frac{\Delta x}{\Delta t}$$

We showed in Section 15-2 that v_x is given by Equation 15-14:

$$v_x = -(2\pi f)A \sin 2\pi ft = -\omega A \sin \omega t$$

Using this analogy, we can obtain the rate of change of the flux, $\Delta\phi_m/\Delta t$:

$$\frac{\Delta\phi_m}{\Delta t} = -NBA(2\pi f) \sin 2\pi ft = -NBA\omega \sin \omega t$$

The emf in the coil is thus

$$\mathcal{E} = +NBA\omega \sin \omega t \qquad\qquad 22\text{-}23$$

The maximum value of the emf occurs when $\sin \omega t = 1$:

$$\mathcal{E}_{max} = NBA\omega \qquad\qquad\qquad 22\text{-}24$$

Equation 22-23 can therefore be written

$$\mathcal{E} = \mathcal{E}_{max} \sin \omega t \qquad\qquad\qquad 22\text{-}25$$

Thus, we can produce a sinusoidal emf in a coil by rotating it with constant frequency in a magnetic field. In this source of emf, mechanical energy is converted into electric energy. The mechanical energy usually comes from a waterfall or a steam engine. In circuit diagrams, an ac generator is represented by the symbol ⊝. Although practical generators are considerably more complicated than the one just discussed, they work on the principle that there is an alternating emf in a coil rotating in a magnetic field, and they are designed so that the emf produced is sinusoidal.

The same coil in a magnetic field that can be used to generate an alternating emf can also be used as an ac motor. Instead of mechanically rotating the coil to generate an emf, we apply an alternating current to the coil from another ac generator, as illustrated in Figure 22-21. We learned in Chapter 21 that a current loop in a magnetic field experiences a torque that tends to rotate the loop so that its magnetic moment points in the direction of **B** and the plane of the loop is perpendicular to **B**. If direct current were supplied to the loop in Figure 22-21, the torque on the loop would change directions when the loop rotated past its equilibrium position, which occurs when the plane of the loop is vertical in the figure. The loop would therefore oscillate about its equilibrium position and eventually come to rest there with its plane vertical. However, if the direction of the current is reversed just as the loop passes the vertical position, the torque does not change direction but continues to rotate the loop in the same direction. As the loop rotates in the magnetic field, a back emf is generated that tends to counter the emf that supplies the current. When the motor is first turned on, there is no back emf and the current is very large, being limited only by the resistance in the circuit. As the motor begins to rotate, the back emf increases and the current decreases. If the load on the motor is increased greatly, the frequency of rotation decreases, thereby decreasing the back emf and increasing the current.

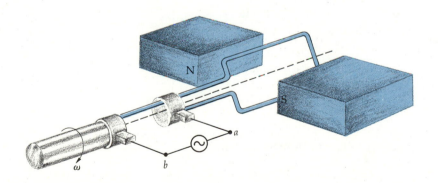

Figure 22-21
When alternating current is supplied to the coil of Figure 22-20, the coil becomes a motor. As the coil rotates, a back emf is generated, limiting the current.

Questions

5. Does the sinusoidal nature of the emf depend on the size or shape of the coil?

6. How could such a coil be used to generate a nonsinusoidal emf?

7. When a generator delivers electric energy to a circuit, where does the energy come from?

8. A motor sometimes burns out when its load is suddenly increased. Why?

Summary

1. Magnetic flux in a loop is the product of the component of the magnetic field perpendicular to the plane of the loop and the area of the loop. For a coil of N turns and cross-sectional area A in a magnetic field **B** that makes an angle θ with the normal to the area, the flux is

$$\phi_m = NBA \cos \theta$$

2. When the flux changes, there is an emf induced in the loop given by Faraday's law:

$$\mathcal{E} = -\frac{\Delta \phi_m}{\Delta t}$$

The direction of the emf is given by Lenz's law: The induced emf and the induced current are in such a direction as to oppose the change that produces them.

3. The emf induced in a conducting wire or rod of length ℓ moving with velocity **v** perpendicular to a magnetic field **B** is called motional emf. Its magnitude is

$$\mathcal{E} = B\ell v$$

4. Circulating currents set up in bulk metal by a changing flux are called eddy currents. Such currents are usually unwanted because they produce heat and power loss in transformers or other electrical devices.

5. The flux through a coil or a circuit is related to the current in the circuit by

$$\phi_{m1} = L_1 I_1$$

where L is the self-inductance of the circuit, which depends on the geometrical arrangement of the circuit. If there is another circuit nearby carrying a current I_2, there is additional flux

$$\phi_{m12} = M I_2$$

where M is the mutual inductance, which depends on the geometrical arrangement of both circuits. The self-inductance of a tightly wound solenoid of length ℓ and cross-sectional area A with N turns is given by

$$L = \mu_0 N^2 A / \ell$$

6. When the current in an inductor is changing, the emf induced in the inductor is given by

$$\mathcal{E} = -L \frac{\Delta I}{\Delta t}$$

7. In an LR circuit consisting of a resistance, an inductance, a switch, and a battery in series, the current does not reach its maximum value instantaneously but takes some time to build up. The time for the current to reach 63 percent of its maximum value is the time constant τ, which is given by

$$\tau = \frac{L}{R}$$

8. The energy stored in an inductor carrying a current I is

$$U = \tfrac{1}{2}LI^2$$

This energy can be considered to be stored in the magnetic field inside the inductor. In general, the energy per unit volume in a magnetic field **B** is given by

$$\eta_m = \frac{B^2}{2\mu_0}$$

9. A coil rotating with frequency f in a magnetic field generates an alternating emf given by

$$\mathcal{E} = \mathcal{E}_{max} \sin 2\pi f t = \mathcal{E}_{max} \sin \omega t$$

where $\omega = 2\pi f$ is the angular frequency of rotation. If alternating current is supplied to a coil in a magnetic field, the coil becomes a motor.

Suggestions for Further Reading

Kondo, Herbert: "Michael Faraday," *Scientific American,* October 1953, p. 90.

This article describes Faraday's series of experiments, in which he discovered electromagnetic induction and made possible the electric motor and dynamo, and his revolutionary concept of the electromagnetic field.

Review

A. Objectives: After studying this chapter, you should:

1. Be able to state Faraday's law and use it to find the emf induced by a changing magnetic flux.

2. Be able to state Lenz's law and use it to find the direction of the induced current in various applications of Faraday's law.

3. Be able to discuss eddy currents.

4. Be able to sketch a graph of current versus time in an *LR* circuit.

5. Be able to describe how simple ac generators and motors work.

B. Define, explain, or otherwise identify:

Magnetic flux, p. 563
Faraday's law, p. 564
Lenz's law, p. 565
Motional emf, p. 568
Eddy currents, p. 569
Self-inductance, p. 570
Mutual inductance, p. 572
Inductor, p. 573
LR circuit, p. 573
Magnetic energy density, p. 575

C. True or false: If the statement is true, explain why. If it is false, give a counterexample.

1. The induced emf in a circuit is proportional to the magnetic flux through the circuit.

2. There can be an induced emf at an instant when the magnetic flux through a circuit is zero.

3. Lenz's law is related to the conservation of energy.

4. The inductance of a solenoid is proportional to the rate of change of the current in it.

5. The magnetic energy density at some point in space is proportional to the square of the magnetic field at that point.

6. When a motor slows down, it draws more current because its back emf decreases.

Exercises

Section 22-1   Magnetic Flux and Faraday's Law

1. A circular coil of radius 3.0 cm has 6 turns. A magnetic field $B = 5000$ G is perpendicular to the coil. (*a*) Find the magnetic flux through the coil. (*b*) Find the magnetic flux through the coil if the coil is turned so that the normal to the coil makes an angle of $20°$ with the magnetic field.

2. A magnetic field of 1.2 T is perpendicular to a square coil of 14 turns. The length of each side of the coil is 5 cm. (*a*) Find the magnetic flux through the coil. (*b*) Find the magnetic flux if the magnetic field makes an angle of $60°$ with the normal to the plane of the coil.

3. The magnetic field in Exercise 1 is steadily reduced to zero in 1.2 s. Find the emf induced in the coil when (*a*) the magnetic field is perpendicular to the coil and (*b*) the magnetic field makes an angle of $20°$ with the normal to the coil.

4. A circular coil of radius 3 cm has its plane perpendicular to a magnetic field of 400 G. (*a*) What is the magnetic flux through the coil if it has 75 turns? (*b*) How many turns should the coil have so that the flux is 0.015 Wb?

5. The magnetic field in Exercise 4 is steadily reduced to zero in 0.8 s. What is the magnitude of the emf induced in the coil of part (*b*)?

6. The two loops in Figure 22-22 have their planes parallel to each other. As viewed from *A* toward *B*, there is a counter-clockwise current in *A*. Give the direction of the current in loop *B*, and state whether the loops attract or repel each other if the current in loop *A* is (*a*) increasing and (*b*) decreasing.

Figure 22-22

Exercise 6.

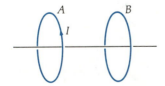

7. A 100-turn coil has radius 4.0 cm and resistance 25 Ω. At what rate must a perpendicular magnetic field change to produce a current of 4.0 A in the coil?

8. Give the direction of the induced current in the circuit on the right in Figure 22-23 when the resistance in the circuit on the left is suddenly (*a*) increased and (*b*) decreased.

Figure 22-23

Exercise 8.

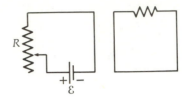

9. A solenoid of length 25 cm and radius 0.8 cm with 400 turns is in an external magnetic field of 600 G that makes an angle of $50°$ with the axis of the solenoid. (*a*) Find the magnetic flux through the solenoid. (*b*) Find the magnitude of the emf induced in the solenoid if the external magnetic field is reduced to zero in 1.4 s.

10. Find the flux through the solenoid of Exercise 9 if there is no external magnetic field but the solenoid carries a current of 3.5 A.

11. A 100-turn circular coil has a diameter of 2.0 cm and a resistance of 50 Ω. The plane of the coil is perpendicular to a uniform magnetic field of magnitude 1.3 T. The field is suddenly reversed in direction. If the reversal takes place in 0.1 s, find the emf and current induced in the coil. (*Hint:* When the field is reversed, the flux change is $2\phi_m$, where ϕ_m is the original flux.)

Section 22-2   Motional emf

12. A rod 25 cm long moves at 8 m/s in a plane perpendicular to a magnetic field of 600 G. Its velocity is perpendicular to the length of the rod. Find (*a*) the magnitude of the magnetic force on an electron in the rod, (*b*) the magnitude of the electric field *E* in the rod, and (*c*) the potential difference *V* between the ends of the rod.

13. Find the speed at which the rod in Exercise 12 would have to move in the same magnetic field to make the potential difference between its ends 6 V.

14. In Figure 22-10 (p. 567), let $B = 0.8$ T, $v = 12.0$ m/s, $\ell = 25$ cm, and $R = 2.5$ Ω. Find (*a*) the induced emf in the circuit, (*b*) the current in the circuit, and (*c*) the force *F* needed to move the rod with constant velocity, assuming negligible friction. (*d*) With the force *F* found in part (*c*), use $P = Fv$ to calculate the power input of the force and compare your result with the rate of heat production, I^2R.

15. Work Exercise 14 for $B = 1.5$ T, $v = 4$ m/s, $\ell = 45$ cm, and $R = 1.4$ Ω.

Section 22-3   Eddy Currents

There are no exercises for this section.

Section 22-4   Inductance

16. A coil with self-inductance of 8.0 H carries a current of 3 A that is changing at a rate of 200 A/s. Find (*a*) the magnetic flux through the coil and (*b*) the induced emf in the coil.

17. A solenoid has a length of 25 cm, a radius of 1.2 cm, and 400 turns. It carries a current of 3 A. Find (*a*) B on the axis at the center; (*b*) the magnetic flux through the solenoid, assuming B to be uniform; (*c*) the self-inductance of the solenoid; and (*d*) the induced emf in the solenoid when the current changes at 150 A/s.

18. Work parts (*a*), (*b*), and (*c*) of Exercise 17 for a solenoid of 500 turns, radius 0.8 cm, and length 15 cm carrying a current of 4.5 A. (*d*) If the current is reduced to zero at a steady rate in 0.2 s, find the induced emf in the solenoid.

19. A circular coil of N_1 turns and cross-sectional area A_1 encircles a long, tightly wound solenoid of N_2 turns, area A_2, and length ℓ_2, as shown in Figure 22-24. The solenoid carries a current I_2. (*a*) Show that the flux through the circular coil due to the current in the solenoid is given by

$$\phi_{m1} = \mu_0 N_1 (N_2 / \ell_2) A_2 I_2$$

(*b*) Explain why the flux depends on the area A_2 of the solenoid and not on that of the coil. (*c*) Find the mutual inductance of the solenoid and the coil.

Figure 22-24
Exercise 19.

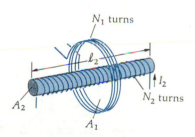

N_1 turns
ℓ_2
I_2
N_2 turns
A_2
A_1

20. A very small circular coil of N_1 turns and cross-sectional area A_1 is completely inside a long, tightly wound solenoid with its plane perpendicular to the axis of the solenoid. The solenoid has N_2 turns, area A_2, and length ℓ_2 and carries current I_2. Find (*a*) the flux through the small coil and (*b*) the mutual inductance of the coil and the solenoid.

21. The outer solenoid of Figure 22-17 (p. 572) has 600 turns and a radius of 4 cm. The inner solenoid has 200 turns and a

radius of 1.8 cm. Both are 30 cm long. Find the mutual inductance of the two solenoids.

Section 22-5   *LR* Circuits and Magnetic Energy Density

22. What is the time constant of a circuit of resistance 200 Ω and inductance 4.3 mH?

23. A solenoid has an inductance of 32 mH. What resistance should be placed in series with the solenoid to make an *LR* circuit with a time constant of 2.50 μs?

24. A coil with a self-inductance of 2.5 H and a resistance of 12.0 Ω is connected across a 24-V battery of negligible internal resistance. (*a*) What is the final current? (*b*) How much energy is stored in the inductor when the final current is attained?

25. Find the energy in a volume of space of 1 L that contains a magnetic field of magnitude $B = 2000$ G.

26. A solenoid of 2000 turns, area 4 cm², and length 30 cm carries a current of 4.0 A. (*a*) Calculate the magnetic energy stored in the solenoid from $U = \frac{1}{2} L I^2$. (*b*) Divide your answer in part (*a*) by the volume of the solenoid to find the magnetic energy per unit volume in the solenoid. (*c*) Find B in the solenoid. (*d*) Compute the magnetic energy density from $\eta_m = B^2 / 2\mu_0$ and compare your answer with the answer to part (*b*).

Section 22-6   Generators and Motors

27. A 200-turn coil has an area of 4 cm². It rotates in a magnetic field of 0.5 T. (*a*) What should its frequency be to generate a maximum emf of 10 V? (*b*) If the coil rotates at 60 Hz, what is $\mathcal{E}_{max}$?

28. In what magnetic field should the coil of Exercise 27 rotate to generate a maximum emf of 10 V at 60 Hz?

29. A rectangular coil with sides 2 cm by 1.5 cm has 300 turns and rotates in a magnetic field of 4000 G. (*a*) What is the maximum emf generated when the coil rotates at 60 Hz? (*b*) What should its frequency be to generate a maximum emf of 110 V?

30. The coil of Exercise 29 rotates at 60 Hz in a magnetic field B. What should the value of B be so that the emf generated has a maximum value of 24 V?

Problems

1. (*a*) Find the magnetic energy, the electric energy, and the total energy in a 25-L volume of space containing an electric field of 10^4 V/m and a magnetic field of 2 T. (*b*) In a plane electromagnetic wave such as a light wave, the magnitudes

of the electric and magnetic fields are related by $E = cB$, where $c = 1/\sqrt{\epsilon_0 \mu_0}$ is the speed of light. Show that in this case, the electric and magnetic energy densities are equal.

2. A conducting rod of length ℓ rotates at constant angular velocity ω about one end in a plane perpendicular to a uniform magnetic field **B**, as shown in Figure 22-25. (*a*) Find the magnetic force on a charge a distance r from the pivot point and show that the electric field there is given by $E = B\omega r$. (*b*) Use $V = E_{av}\ell$ to find the potential difference between the ends of the rod, where E_{av} is the average electric field in the rod. (Since the electric field varies linearly from 0 to $B\omega\ell$, the average value is half the maximum value.) (*c*) Draw any radial line in the plane from which to measure $\theta = \omega t$. Show that the area of the pie-shaped region between the reference line and the rod is $A = \frac{1}{2}\ell^2\theta$. Compute the magnetic flux through this area and show that $\mathcal{E} = \frac{1}{2}B\omega\ell^2$ follows from Faraday's law applied to this area.

Figure 22-25
Problem 2.

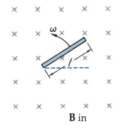

B in

3. A rectangular loop 10 by 5.0 cm with a resistance of 1.5 Ω is pulled through a region of uniform magnetic field $B = 0.85$ T (see Figure 22-26) with constant speed $v = 2.0$ cm/s. The front end of the loop enters the region of the magnetic field at time $t = 0$. (*a*) Find and graph the magnetic flux through the loop as a function of time. (*b*) Find and graph the induced emf and the current in the loop as functions of time. Neglect any self-inductance in the loop, and extend your graphs from $t = 0$ to $t = 18$ s.

Figure 22-26
Problem 3.

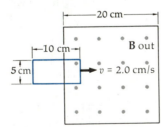

4. A certain coil has a self-inductance L and a resistance R. When the current in the coil is 5.0 A and is increasing at 10.0 A/s, the potential difference across the coil is 140 V. When the current is 5.0 A and is decreasing at 10.0 A/s, the potential difference is 60 V. Find L and R. (*Hint:* You may replace the coil with an inductance L in series with a resistance R. The potential differences given are then across the combination.)

5. In Figure 22-27, a conducting rod of mass m rests on conducting rails perpendicular to a uniform magnetic field **B**. The resistance of the rod and rails is negligible, and there is no friction between them. With the rod at rest on the rails, the switch is closed at time $t = 0$, and the battery of emf $\mathcal{E}_0$ supplies current to the circuit. (*a*) Find the initial current in the circuit and the initial magnetic force on the rod before it starts to move. (*b*) When the rod is moving with speed v, there is a back emf in the circuit that tends to counter the emf of the battery. Show that the current i is then obtained from

$$\mathcal{E}_0 - B\ell v = iR$$

(*c*) Find the force on the rod when it moves with speed v, and write Newton's second law for the rod. (*d*) By setting $\Delta v/\Delta t = 0$ in your equation for part (*c*), show that the rod eventually moves with a terminal speed that is given by $v_t = \mathcal{E}_0/\ell B$.

Figure 22-27
Problem 5.

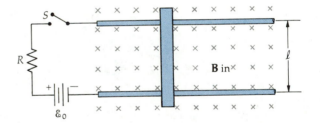

CHAPTER TWENTY-THREE

Alternating Current Circuits

Back and forth the charges flow
As motors turn and heaters glow

PAUL ALLEN

In this final chapter on electricity and magnetism, we consider alternating current, the form of electricity that we use as our power source and in our telephone, radio, and television circuits among others. You will again study resonance, first encountered in the study of vibrations and waves. Everything you learned about resonance in vibrating bodies will be applicable to resonance in ac circuits, a phenomenon that enables us to tune our radios and television sets to particular stations.

More than 99 percent of the electrical energy used today is produced by electrical generators in the form of alternating current. The current alternates in direction 60 times each second. Such a current is easily produced by magnetic induction in an ac generator, as discussed in Chapter 22.

Toward the end of the nineteenth century, there was a heated debate as to whether direct or alternating current should be used to deliver electrical energy to consumers in the United States. Thomas Edison lobbied for the use of direct current, whereas Nikola Tesla and George Westinghouse were proponents of the use of alternating current. In 1893, alternating current was chosen to light the World's Columbian Exposition in Chicago, and a contract was awarded to Westinghouse to deliver alternating current generated at Niagara Falls to American homes and factories.

Alternating current has a great advantage in that it allows electrical energy to be transported over long distances at very high voltages and low currents to reduce energy losses in the form of Joule heat. It can then be transformed, with almost no energy loss, to lower and safer voltages and correspondingly higher currents for everyday use. The transformer that accomplishes this change in voltage and current works on the basis of magnetic induction. An alternating current in a coil sets up an alternating magnetic field that induces an alternating current in a nearby coil.

In this chapter, we will first study simple alternating current circuits containing resistors, inductors, and capacitors. We will then learn how transformers work, and finally, we will learn how diodes are used to convert alternating current to direct current.

23-1   Alternating Current in a Resistor

Figure 23-1 shows a simple ac circuit with a generator and a resistor. The current in the resistor is

$$I = \frac{\mathcal{E}}{R} = \frac{\mathcal{E}_{max}}{R} \sin \omega t = I_{max} \sin \omega t \qquad \text{23-1}$$

where the maximum current is

$$I_{max} = \frac{\mathcal{E}_{max}}{R} \qquad \text{23-2}$$

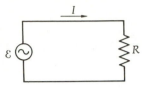

Figure 23-1
An ac generator in series with a resistor R.

The current is negative as often as it is positive and has an average value of zero.

The instantaneous power P dissipated in the resistor is

$$P = I^2 R = (I_{max} \sin \omega t)^2 R = I_{max}^2 R \sin^2 \omega t$$

Figure 23-2 shows the power as a function of time. It varies from zero to its maximum value of $I_{max}^2 R$, as shown. We are usually interested in the average power over one or more cycles. The function $\sin^2 \omega t$ varies from 0 to 1 and has an average value of $\frac{1}{2}$.* The average power dissipated in the resistor is thus

$$P_{av} = (I^2 R)_{av} = \frac{1}{2} I_{max}^2 R \qquad \text{23-3}$$

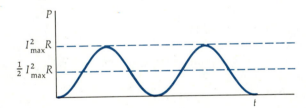

Figure 23-2
Plot of the power in the resistor versus time. The power P varies from zero to a maximum value of $I_{max}^2 R$. The average power is half the maximum power.

Most ac ammeters and voltmeters measure root-mean-square (rms) values of current and voltage rather than maximum or peak values. The rms value of the current, I_{rms}, is defined by

$$\boxed{I_{rms} = \sqrt{(I^2)_{av}}} \qquad \text{23-4}$$ rms current defined

The average value of I^2 is

$$(I^2)_{av} = (I_{max} \sin \omega t)^2_{av} = \frac{1}{2} I_{max}^2$$

The rms value of the current is thus

$$\boxed{I_{rms} = \frac{1}{\sqrt{2}} I_{max}} \qquad \text{23-5}$$

The rms value of any quantity that varies sinusoidally equals the maximum

* This can be seen from the identity $\sin^2 \omega t + \cos^2 \omega t = 1$. A plot of $\cos^2 \omega t$ looks the same as a plot of $\sin^2 \omega t$ except that it is shifted by 90°. Both have the same average value over one or more periods, and since their sum is 1, the average value of each must be $\frac{1}{2}$.

value of that quantity divided by $\sqrt{2}$. In terms of the rms current, the average power dissipated in the resistor is

$$P_{av} = (I^2R)_{av} = I_{rms}^2 R \qquad\qquad 23\text{-}6$$

The average power delivered by the generator is, of course, equal to that dissipated in the resistor for this simple circuit:

$$P_{av} = (\mathcal{E}I)_{av} = \mathcal{E}_{max}I_{max} (\sin^2 \omega t)_{av}$$

or

$$P_{av} = \tfrac{1}{2}\mathcal{E}_{max}I_{max} = \mathcal{E}_{rms}I_{rms} \qquad\qquad 23\text{-}7$$

where $\mathcal{E}_{rms} = \mathcal{E}_{max}/\sqrt{2}$. The rms current is related to the rms emf in the same way as the maximum current is related to the maximum emf (Equation 23-2):

$$I_{rms} = \frac{\mathcal{E}_{rms}}{R} \qquad\qquad 23\text{-}8$$

We can see from Equations 23-6 and 23-7 why meters are calibrated to read rms values. If we use rms values for current and emf, we can calculate the power input and heat generated using the same equations as we did for direct current.

The power company supplies ac power to your house of frequency 60 Hz and voltage 110 V rms. (For some special high-power appliances, such as an electric clothes dryer or oven, separate lines carrying power at 220 V are often supplied. For a given power requirement, only half as much current is required at 220 V as at 110 V, but 220 V is much more dangerous than 110 V. A shock at 220 V is much more likely to be fatal than one at 110 V.) If you plug in a 1600-W heater, it will draw a current of

$$I_{rms} = \frac{P_{av}}{\mathcal{E}_{rms}} = \frac{1600 \text{ W}}{110 \text{ V}} = 14.5 \text{ A}$$

The voltage across the outlets is maintained at 110 V, independent of what is plugged in. Thus, all appliances plugged into the outlets are essentially in parallel. If you plug a 500-W toaster into another outlet, it will draw a current of (500 W/110 V) = 4.5 A. Most household wiring is rated at 20 or 30 A. A current greater than this will overheat the wiring and create a fire hazard. Each circuit is therefore equipped with a circuit breaker (or a fuse in older houses). For a 20-A circuit, the circuit breaker trips (or the fuse blows), opening the circuit, when the current exceeds 20 A. The maximum power than can be handled by a circuit with a 20-A circuit breaker is

$$P_{av} = \mathcal{E}_{rms}I_{rms} = (110 \text{ V})(20 \text{ A}) = 2.2 \text{ kW}$$

Since most modern houses require considerably more than 2.2 kW of power, many circuits are supplied, each with its own circuit breaker and each supplying several outlets.

Questions

1. What is the average current in the resistor in Figure 23-1?

2. Is the instantaneous power in the resistor in Figure 23-1 ever negative?

23-2 Alternating Current in Inductors and Capacitors

The behavior of alternating current in inductors and capacitors is very different from that of direct current. For example, when a capacitor is in series in a dc circuit, the current stops completely when the capacitor becomes fully charged. When the current alternates, however, charge continually flows onto or off the plates of a capacitor. We will see shortly that, if the frequency of the alternating current is great, a capacitor hardly impedes the current at all. Similarly, an inductor coil usually has a very small resistance and therefore has little effect on direct current. But when the current is changing in an inductor, a back emf is generated that is proportional to the rate of change of the current. The greater the frequency of the alternating current in an inductor, the greater the rate of change of current and, therefore, the greater the back emf. Thus, an inductor has just the opposite effect on alternating current as a capacitor. At very low frequencies, an inductor hardly impedes the current at all, but at high frequencies, it impedes the current flow greatly because of its back emf.

Figure 23-3 shows an inductor coil across the terminals of an ac generator. When the current increases in the inductor, a back emf of magnitude $L\,\Delta I/\Delta t$ is generated due to the changing flux. Usually, the potential drop across the inductor due to this back emf is much greater than the IR drop due to the resistance of the coil. Hence, we can neglect any resistance in the coil. Setting the emf of the generator equal to the magnitude of the potential drop across the inductor, we obtain

$$\mathcal{E} = L\frac{\Delta I}{\Delta t} = \mathcal{E}_{max}\sin\omega t \qquad\qquad 23\text{-}9$$

The current that satisfies this equation can be obtained by once again using the analogy with the equations for the harmonic oscillator, as we did in Chapter 22. In Chapter 15, we found that the position x, the velocity v, and the acceleration a obeyed equations of the form

$$x = A\cos\omega t \qquad\qquad 23\text{-}10$$

$$v = \frac{\Delta x}{\Delta t} = -\omega A\sin\omega t = -v_0\sin\omega t \qquad\qquad 23\text{-}11$$

and

$$a = \frac{\Delta v}{\Delta t} = -\omega v_0\cos\omega t = -\omega^2 A\cos\omega t \qquad\qquad 23\text{-}12$$

We note that in Equation 23-9, $\Delta I/\Delta t$ is proportional to $\sin\omega t$, like $\Delta x/\Delta t$ is in Equation 23-11. Then, by analogy with Equation 23-10, the current I is given by

$$I = -\frac{\mathcal{E}_{max}}{\omega L}\cos\omega t = -I_{max}\cos\omega t \qquad\qquad 23\text{-}13$$

where

$$I_{max} = \frac{\mathcal{E}_{max}}{\omega L} \qquad\qquad 23\text{-}14$$

Figure 23-3
An ac generator in series with an inductor L.

The current in this circuit is not in phase with the generator voltage (which is also the potential drop across the inductor). Using the trigonometric identity

$$\cos \omega t = -\sin (\omega t - 90°)$$

we have

$$I = I_{max} \sin (\omega t - 90°) \qquad\qquad \text{23-15}$$

The current and voltage across the inductor are plotted in Figure 23-4. From this figure, we can see that the maximum value of the voltage occurs 90° or one-fourth period before the corresponding maximum value of the current. The voltage across an inductor is therefore said to *lead* the current by 90° or one-fourth period.

The relation between the maximum (or rms) current and the maximum (or rms) voltage can be written in a form similar to Equation 23-8 for a resistor:

$$I_{max} = \frac{\mathscr{E}_{max}}{\omega L} = \frac{\mathscr{E}_{max}}{X_L} \qquad\qquad \text{23-16}a$$

or

$$I_{rms} = \frac{\mathscr{E}_{rms}}{X_L} \qquad\qquad \text{23-16}b$$

where

$$X_L = \omega L = 2\pi f L \qquad\qquad \text{23-17} \qquad \textcolor{blue}{\text{Inductive reactance}}$$

The quantity X_L is called the *inductive reactance*. Like resistance, inductive reactance has units of ohms. As seen from Equation 23-16a, the larger the reactance for a given emf, the smaller the maximum value of the current. Unlike resistance, the inductive reactance depends on the frequency of the current—the greater the frequency, the greater the reactance.

The instantaneous power input by the generator is

$$P = \mathscr{E}I = (\mathscr{E}_{max} \sin \omega t)(I_{max} \cos \omega t)$$

$$= \mathscr{E}_{max} I_{max} \sin \omega t \cos \omega t$$

The average power input by the generator is zero. We can see this by using

$$\sin \omega t \cos \omega t = \tfrac{1}{2} \cos 2\omega t$$

This term oscillates twice during each cycle and is negative as often as it is positive. The result that the average power input into an inductor is zero is, of course, true only if the resistance of the inductor can be neglected. There is no energy dissipated in an inductor like there is in a resistor.

Figure 23-5 shows a capacitor connected across the terminals of a generator. Since the potential drop across the capacitor equals Q/C, we have

$$\mathscr{E} = \frac{Q}{C} = \mathscr{E}_{max} \sin \omega t$$

or

$$Q = \mathscr{E}_{max} C \sin \omega t$$

The current, $\Delta Q/\Delta t$, is obtained by analogy with Equations 23-11 and 23-12 for simple harmonic motion. Q is proportional to $\sin \omega t$ like v is in

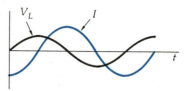

Figure 23-4
Current I and voltage V_L across the inductor versus time. The maximum voltage occurs one-fourth period before the maximum current. The voltage thus leads the current by one-fourth period or 90°.

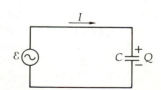

Figure 23-5
An ac generator in series with a capacitor C.

Equation 23-11. We then obtain $\Delta Q/\Delta t$ by analogy with Equation 23-12 for $\Delta v/\Delta t$:

$$I = \frac{\Delta Q}{\Delta t} = \omega C \mathscr{E}_{max} \cos \omega t$$

$$= I_{max} \sin (\omega t + 90°) \qquad\qquad \textbf{23-18}$$

where we have written I_{max} for $\omega C \mathscr{E}_{max}$ and have used the trigonometric identity $\cos \omega t = \sin (\omega t + 90°)$. Once again, the current is not in phase with the generator voltage. From Figure 23-6, we can see that the maximum value of the voltage occurs 90° or one-fourth period after the maximum value of the current. The voltage across the capacitor is therefore said to *lag* the current by 90° or one-fourth period.

Again, the relation between the maximum (or rms) current and the maximum (or rms) voltage can be written in a form similar to Ohm's law:

$$I_{max} = \omega C \mathscr{E}_{max} = \frac{\mathscr{E}_{max}}{1/\omega C} = \frac{\mathscr{E}_{max}}{X_C} \qquad\qquad \textbf{23-19a}$$

and similarly

$$I_{rms} = \frac{\mathscr{E}_{rms}}{X_C} \qquad\qquad \textbf{23-19b}$$

where

$$X_C = \frac{1}{\omega C} = \frac{1}{2\pi f C} \qquad\qquad \textbf{23-20}$$

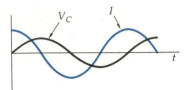

Figure 23-6
Current I and voltage V_C across the capacitor versus time. The maximum voltage occurs one-fourth period after the maximum current. The voltage lags the current by one-fourth period or 90°.

Capacitive reactance

The quantity X_C is called the *capacitive reactance* of the circuit. Like resistance and inductive reactance, capacitive reactance has units of ohms, and like inductive reactance, capacitive reactance depends on the frequency of the current. In this case, the greater the frequency, the smaller the reactance. As it was for an ac generator and an inductor, the average power input of the generator into the capacitor is zero. This is because the emf is proportional to $\sin \omega t$ and the current is proportional to $\cos \omega t$, and $(\sin \omega t \cos \omega t)_{av} = 0$. Thus, like inductors, capacitors dissipate no energy.

Since charge certainly cannot pass through the space between the plates of a capacitor, it may seem strange that there is a continuing alternating current in the circuit of Figure 23-5. Remember, though, that when we place an uncharged capacitor across the terminals of a dc voltage source, such as a battery, there is a momentary current that decreases exponentially with time until the plates are charged to the same potential as the battery. Consider an initially uncharged capacitor in series with a source of emf with the upper plate connected to the positive terminal. Initially, positive charge flows to the upper plate and away from the lower plate. (Of course, it is really the negative electrons that flow in the opposite direction.) The effect is the same as if the charge had actually flowed *through* the space between the plates. If the source of emf is an ac generator, the potential difference changes sign after one-half period. Then, before the capacitor can become fully charged, the upper plate is connected to the negative terminal and the charge begins to flow in the opposite direction. The greater the frequency of the ac emf, the shorter the connection time for either polarity and, hence, the less the capacitor impedes the flow of charge.

Example 23-1   A 40-mH inductor is placed across an ac generator that has a maximum emf of 120 V. Find the reactance and the maximum current when the frequency is 60 Hz and when it is 2000 Hz.

The inductive reactances at these frequencies are

$$X_{L1} = 2\pi f_1 L$$

$$= 2\pi(60 \text{ Hz})(40 \times 10^{-3} \text{ H}) = 15.1 \ \Omega$$

and

$$X_{L2} = 2\pi f_2 L$$

$$= 2\pi(2000 \text{ Hz})(40 \times 10^{-3} \text{ H}) = 504 \ \Omega$$

The maximum values of the currents for these frequencies are

$$I_{1,\text{max}} = \frac{\mathscr{E}_{\text{max}}}{X_{L1}} = \frac{120 \text{ V}}{15.1 \ \Omega} = 7.95 \text{ A}$$

and

$$I_{2,\text{max}} = \frac{120 \text{ V}}{504 \ \Omega} = 0.238 \text{ A}$$

Example 23-2   A 20-μF capacitor is placed across a generator that has a maximum emf of 100 V. Find the reactance and the maximum current when the frequency is 60 Hz and when it is 5000 Hz.

The capacitive reactances at these frequencies are

$$X_{C1} = \frac{1}{2\pi f_1 C}$$

$$= [2\pi(60 \text{ Hz})(20 \times 10^{-6})]^{-1} = 133 \ \Omega$$

and

$$X_{C2} = \frac{1}{2\pi f_2 C}$$

$$= [2\pi(5000 \text{ Hz})(20 \times 10^{-6})]^{-1} = 1.59 \ \Omega$$

The maximum currents are then

$$I_{1,\text{max}} = \frac{\mathscr{E}_{\text{max}}}{X_{C1}} = \frac{100 \text{ V}}{133 \ \Omega} = 0.754 \text{ A}$$

and

$$I_{2,\text{max}} = \frac{100 \text{ V}}{1.59 \ \Omega} = 62.8 \text{ A}$$

Question

3. In a circuit with a generator and an inductor, are there any times when the inductor absorbs power from the generator? Are there any times when the inductor supplies power to the generator?

23-3   An *LCR* Circuit with a Generator

An important circuit that exhibits many of the features of most ac circuits is a series *LCR* circuit with a generator, as shown in Figure 23-7. The mathematical analysis of this circuit is complicated, but the analysis and results are identical to those for the damped driven oscillator discussed in Chapter 15. After the switch is closed, the current in the circuit consists of two parts: a transient current that depends on the initial conditions (for example, the phase of the generator when the switch is closed and the initial charge on the capacitor) and a steady-state current that is independent of the initial conditions. The transient current decreases exponentially with time and is eventually negligible compared with the steady-state current. We will ignore the transient current and concentrate on the steady-state current. We will state without proof some of the algebraic results of the analysis of this circuit and then discuss these results qualitively.

We assume that the emf of the generator varies with time as

$$\mathcal{E} = \mathcal{E}_{max} \sin \omega t \qquad \text{23-21}$$

where $\omega = 2\pi f$ is the angular frequency, as usual. The steady-state current is then given by

$$I = I_{max} \sin (\omega t - \phi) \qquad \text{23-22}$$

where

$$\tan \phi = \frac{X_L - X_C}{R} \qquad \text{23-23}$$

and

$$I_{max} = \frac{\mathcal{E}_{max}}{\sqrt{R^2 + (X_L - X_C)^2}} = \frac{\mathcal{E}_{max}}{Z} \qquad \text{23-24}$$

where

$$Z = \sqrt{R^2 + (X_L - X_C)^2} \qquad \text{23-25} \qquad \text{Impedance}$$

The quantity Z is called the *impedance*. Combining these results and converting to the rms values $I_{rms} = I_{max}/\sqrt{2}$ and $\mathcal{E}_{rms} = \mathcal{E}_{max}/\sqrt{2}$, we have

$$I_{rms} = \frac{\mathcal{E}_{rms}}{Z} \sin (\omega t - \phi) \qquad \text{23-26}$$

In the previous sections, we showed that the voltage across a resistance is in phase with the current, whereas the voltage across an inductor leads the current by 90° and the voltage across a capacitor lags the current by 90°. These phase relations are illustrated in Figure 23-8, where we have plotted the voltage across the resistance V_R as the magnitude of a vector that lies along the *x* axis. This voltage is in phase with the current. The voltage across the inductor V_L is the magnitude of a vector along the *y* axis. This voltage leads the current by 90°. Similarly, the voltage across the capacitor V_C is the magnitude of a vector in the negative *y* direction, which lags the current by 90°. The emf is the magnitude of the sum of these vectors, which leads the current by the phase angle ϕ, as shown. (Alternatively, we could say that the current lags the emf by ϕ, as shown by Equation 23-26.) Figure 23-9 shows a

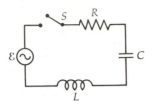

Figure 23-7
A series *LRC* circuit with an ac generator.

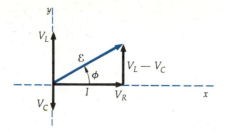

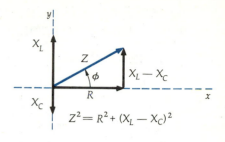

$$Z^2 = R^2 + (X_L - X_C)^2$$

Figure 23-8
Phase relations among voltages in a series *LRC* circuit. The voltage across the resistor V_R is in phase with the current and is plotted along the *x* axis. The voltage across the inductor V_L leads the current by 90°, so it is plotted along the *y* axis. The voltage across the capacitor V_C lags the current by 90°, so it is plotted along the negative *y* axis. The sum of the vectors representing these voltages gives a vector at an angle ϕ with the current representing the applied emf. For the case shown here, V_L is greater than V_C and the emf leads the current by ϕ.

Figure 23-9
Vector model for relating capacitive and inductive reactance, resistance, impedance, and phase angle in an *LRC* circuit.

simple vector diagram for the resistance R, the inductive reactance X_L, and the capacitive reactance X_C. The magnitude of the vector sum is the impedance Z.

Although these results appear to be complicated, we can use them to learn some simple and important features of the behavior of this circuit. Since both the inductive reactance, $X_L = \omega L$, and the capacitive reactance, $X_C = 1/\omega C$, depend on the frequency f of the applied emf, the impedance Z and the rms current I_{rms} also depend on f. As we increase f, the inductive reactance increases and the capacitive reactance decreases. When X_L and X_C are equal, the impedance Z has its minimum value, $Z_{\text{min}} = R$, and I_{rms} has its greatest value. The value of f for which X_L and X_C are equal is obtained from

$$X_L = X_C$$

$$2\pi fL = \frac{1}{2\pi fC}$$

or

$$f = \frac{1}{2\pi\sqrt{LC}} = f_0 \qquad\qquad 23\text{-}27 \qquad \text{Resonance frequency}$$

The quantity f_0 is called the natural frequency or resonance frequency of the circuit. The impedance is minimum and the rms current is maximum when the frequency of the emf equals the natural frequency $f_0 = 1/2\pi\sqrt{LC}$. At this frequency, the circuit is said to be at *resonance*. This resonance condition in a driven *LCR* circuit is exactly the same as that in a driven simple harmonic oscillator. We note from Equation 23-23 that when $X_L = X_C$, the phase angle ϕ is zero. Hence, at resonance the current is in phase with the applied emf.

We pointed out in the previous section that neither an inductor nor a capacitor dissipates energy. The average power delivered to a series *LRC* circuit equals the average power supplied to the resistor:

$$P_{\text{av}} = I_{\text{rms}}^2 R = I_{\text{rms}} V_{R,\text{rms}}$$

From Figure 23-8, we see that $V_R = \mathcal{E} \cos \phi$. The average power supplied to the circuit can thus be written

$$P_{av} = \mathcal{E}_{rms} I_{rms} \cos \phi \qquad \text{23-28}$$

The quantity $\cos \phi$ is called the *power factor*. At resonance, ϕ is zero, and the power factor is 1. Figure 23-10 shows the average power supplied by the generator to the circuit as a function of generator frequency for two different values of resistance R. These curves, called resonance curves, are the same as the power-versus-frequency curves for the driven damped oscillator (see Section 15-5). The average power is maximum when the generator frequency equals the resonance frequency. When the resistance is small, the resonance curve is narrow; when it is large, the curve is broad. These curves can be characterized by the *width* at half-maximum, Δf, shown in the figure. This width is the frequency difference between the two points on the curve where the power is half its maximum value. When the width is small compared with the resonance frequency, the resonance is sharp, that is, the resonance curve is narrow. The quality of the resonance is described by the Q *factor*, which is defined as

$$Q = \frac{2\pi f_0 L}{R} \qquad \text{23-29}$$

When the resonance is reasonably narrow (that is, for Q factors greater than about 2 or 3), the Q factor can be approximated by

$$Q = \frac{f_0}{\Delta f} \qquad \text{23-30}$$

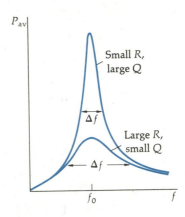

Figure 23-10
Plot of average power versus frequency for a series LRC circuit. The power is maximum when the frequency of the generator f equals the natural frequency of the circuit $f_0 = 1/2\pi\sqrt{LC}$. If the resistance is small, the Q factor is large and the resonance is sharp.

A common application of series resonance circuits is a radio receiver, where the resonance frequency of the circuit is varied by varying the capacitance. Resonance occurs when the natural frequency of the circuit equals the frequency of the radio waves picked up at the antenna. At resonance, there is a relatively large current in the antenna circuit. If the Q value of the circuit is sufficiently high, currents due to other stations off resonance will be negligible compared with those due to the station to which the circuit is tuned.

Example 23-3   A series LCR circuit with $L = 2$ H, $C = 2 \ \mu F$, and $R = 20 \ \Omega$ is driven by a generator of maximum emf 100 V and variable frequency. Find the resonance frequency f_0, the phase ϕ, and the maximum current I_{max} when the generator frequency is $f = 60$ Hz.

The resonance frequency is

$$f_0 = \frac{1}{2\pi\sqrt{LC}} = \frac{1}{2\pi\sqrt{(2 \text{ H})(2 \times 10^{-6} \text{ F})}} = 79.6 \text{ Hz}$$

When the generator frequency is 60 Hz, it is well below the resonance frequency. The capacitive and inductive reactances at 60 Hz are

$$X_C = \frac{1}{\omega C} = \frac{1}{(2\pi)(60 \text{ Hz})(2 \times 10^{-6} \text{ F})} = 1326 \ \Omega$$

and

$$X_L = \omega L = (2\pi)(60 \text{ Hz})(2 \text{ H}) = 754 \ \Omega$$

The total reactance is $X_L - X_C = 754\ \Omega - 1326\ \Omega = -572\ \Omega$. This is of much greater magnitude than the resistance, a result that always holds far from resonance. The total impedance is

$$Z = \sqrt{R^2 + (X_L - X_C)^2} = \sqrt{(20\ \Omega)^2 + (-572\ \Omega)^2} \approx 572\ \Omega$$

since $(20)^2$ is negligible compared with $(572)^2$. The maximum current is then

$$I_{max} = \frac{\mathcal{E}_{max}}{Z} = \frac{100\ \text{V}}{572\ \Omega} = 0.175\ \text{A}$$

This is small compared with I_{max} at resonance, which is $(100\ \text{V})/(20\ \Omega) = 5\ \text{A}$. The phase angle ϕ is given by

$$\tan \phi = \frac{X_L - X_C}{R} = \frac{-572\ \Omega}{20\ \Omega} = -28.6$$

$$\phi = -88°$$

From Equation 23-22 and Figure 23-8, we can see that a negative phase angle means that the generator voltage lags the current.

Example 23-4   Find the average power delivered by the emf in Example 23-3 at 60 Hz.

Since we are given the maximum emf and have calculated the maximum current in Example 23-3, it is convenient to write the average power in terms of these quantities:

$$P_{av} = \mathcal{E}_{rms} I_{rms} \cos \phi = \tfrac{1}{2} \mathcal{E}_{max} I_{max} \cos \phi$$

$$= \tfrac{1}{2}(100\ \text{V})(0.175\ \text{A})(\cos -88°) = 0.306\ \text{W}$$

As we have said, this power goes into Joule heat in the resistor. We could also have calculated the average power from

$$P_{av} = I_{rms}^2 R = \tfrac{1}{2} I_{max}^2 R = \tfrac{1}{2}(0.175\ \text{A})^2 (20\ \Omega) = 0.306\ \text{W}$$

Example 23-5   Find the resonance width and Q value of the circuit of Example 23-3.

In that example, we found the resonance frequency to be $f_0 = 79.6$ Hz. The Q value is then

$$Q = \frac{2\pi f_0 L}{R}$$

$$= \frac{2\pi(79.6\ \text{Hz})(2\ \text{H})}{20} = 50$$

and the width of the resonance is

$$\Delta f = \frac{f_0}{Q}$$

$$= \frac{79.6\ \text{Hz}}{50} = 1.6\ \text{Hz}$$

This is a very sharp resonance. The width is only 1.6 Hz at the resonance frequency of 79.6 Hz.

Example 23-6   Find the maximum voltage across the inductor, the capacitor, and the resistor at resonance for the circuit of Examples 23-3, 23-4, and 23-5.

At resonance, the impedance is just the resistance $R = 20\ \Omega$. Since the maximum emf is 100 V, the maximum current is

$$I_{max} = \frac{\mathcal{E}_{max}}{Z} = \frac{100\ \text{V}}{20\ \Omega} = 5\ \text{A}$$

The maximum voltage across the resistor is

$$V_R = I_{max}R = (5\ \text{A})(20\ \Omega) = 100\ \text{V}$$

The maximum voltage across the inductor is

$$V_L = I_{max}X_L$$

and that across the capacitor is

$$V_C = I_{max}X_C$$

The resonance frequency found in Example 23-3 was $f_0 = 79.6$ Hz. The inductive and capacitive reactances at resonance are

$$X_L = \omega_0 L = (2\pi)(79.6\ \text{Hz})(2\ \text{H}) = 1000\ \Omega$$

and

$$X_C = \frac{1}{\omega_0 C} = \frac{1}{(2\pi)(79.6\ \text{Hz})(2 \times 10^{-6}\ \text{F})} = 1000\ \Omega$$

The inductive and capacitive reactances are equal, as expected, since we found the resonance frequency by setting them to be equal. The maximum voltage across the inductor is then

$$V_L = I_{max}X_L = (5\ \text{A})(1000\ \Omega) = 5000\ \text{V}$$

and that across the capacitor is

$$V_C = I_{max}X_C = (5\ \text{A})(1000\ \Omega) = 5000\ \text{V}$$

Figure 23-11 shows the vector diagram for these voltages. Note that although the maximum voltage across the resistor is a relatively safe 100 V, equal to the maximum emf of the generator, the maximum voltage across the inductor and the capacitor is a dangerously high 5000 V. These voltages are out of phase by 180°. At resonance, the voltage across the inductor at any instant is the negative of that across the capacitor so that they always sum to zero, leaving the voltage across the resistor equal to the emf in the circuit.

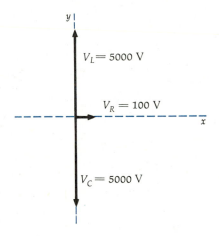

Figure 23-11
Voltages in Example 23-6. The voltages across the inductor and capacitor in a series *LRC* circuit are always 180° out of phase. At resonance, they are equal in magnitude, so they sum to zero, leaving the sum across all three elements equal to just V_R. In this example, the maximum voltage across the resistor is just 100 V, whereas the maximum drop across the inductor and across the capacitor is 5000 V.

Questions

4. Does the power factor depend on frequency?

5. Are there any disadvantages in a radio circuit having an extremely large value of Q?

23-4   The Transformer

A *transformer* is a device for changing ac voltage and current without an appreciable loss of power. Its operation is based on the fact that an alternating current in one circuit will induce an alternating emf in a nearby circuit because of the mutual inductance of the two circuits. Figure 23-12 shows a simple transformer consisting of two coils of wire around a common iron core. The coil carrying the input power is called the *primary,* and the other coil is called the *secondary.* Either coil of a transformer can be the primary or the secondary. The function of the iron core is to increase the magnetic field for a given current and to guide it so that nearly all the magnetic flux through one coil goes through the other coil. The iron core is laminated to reduce eddy-current losses. Other power losses arise because of the small resistance in both coils. We will neglect these losses and consider an ideal transformer of 100 percent efficiency, for which all of the power supplied to the primary coil appears in the secondary coil. Actual transformers are often 98 to 99 percent efficient.

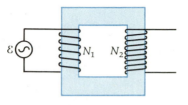

Figure 23-12
Transformer with N_1 turns in the primary and N_2 turns in the secondary.

Consider an ac emf $\mathcal{E}$ across the primary of a transformer which has N_1 turns. Assume the secondary coil of the transformer has N_2 turns and is open. Because of the iron core, there is a large flux through each coil even with a very small magnetizing current I_m in the primary circuit. We can neglect the resistance of the coil compared with the inductive reactance. The primary is then a simple circuit consisting of an ac generator and a pure inductance, as discussed in Section 23-2. The current and voltage in the primary are out of phase by 90°, and the average power in the primary circuit is zero. The induced voltage in the primary coil is equal in magnitude to the applied emf. If $\Delta\phi_{\text{turn}}$ is the magnetic flux in one turn of the primary coil, the induced voltage in the primary coil is given by Faraday's law:

$$V_1 = -N_1 \frac{\Delta\phi_{\text{turn}}}{\Delta t} \qquad\qquad 23\text{-}31$$

Assuming there is no flux leakage out of the iron core, the flux through each turn is the same for both coils. Thus the total flux through the secondary coil is $N_2\phi_{\text{turn}}$, and the voltage across the secondary coil is

$$V_2 = -N_2 \frac{\Delta\phi_{\text{turn}}}{\Delta t} \qquad\qquad 23\text{-}32$$

Comparing these equations, we see that

$$V_2 = \frac{N_2}{N_1} V_1 \qquad\qquad 23\text{-}33$$

If N_2 is greater than N_1, the voltage in the secondary coil is greater than that in the primary coil, and the transformer is called a *step-up transformer.* Since the power in either circuit is VI, when the voltage is increased, the current is correspondingly decreased. If N_2 is less than N_1, the voltage in the secondary coil is less than that in the primary coil, and the transformer is called a *step-down transformer.* In this case the current is increased.

Example 23-7   A doorbell requires 0.4 A at 6 V. It is connected to a transformer whose primary contains 2000 turns and is connected to 110-V household current. How many turns should there be in the secondary? What is the current in the primary?

Since the input voltage is 110 V and the output is 6 V, the turns ratio from Equation 23-33 is

$$\frac{N_2}{N_1} = \frac{V_2}{V_1} = \frac{6\ V}{110\ V}$$

The number of turns in the secondary is thus

$$N_2 = \frac{6}{110}\ 2000\ \text{turns} = 109\ \text{turns}$$

Since we are assuming 100 percent efficiency in power transmission, the input and output currents are related by

$$V_2 I_2 = V_1 I_1$$

The current in the primary is thus

$$I_1 = \frac{V_2}{V_1} I_2 = \frac{6}{110}\ (0.4\ \text{A}) = 0.02\ \text{A}$$

Suburban power substation where transformers step down the voltage from the high-voltage transmission lines.

An important use of transformers is in the transport of electrical power. To minimize the I^2R heat loss in transmission lines, it is economical to use a high voltage and a low current. On the other hand, safety and other considerations, such as, insulation, make it necessary to use power at lower voltage and higher current to run motors and other electrical appliances. Suppose, for example, that each person in a city with a population of 50,000 uses 1.1 kW of electric power. (The per capita consumption in the United States is actually somewhat greater than this.) The current required for each person at 110 V is

$$I = \frac{1100\ W}{110\ V} = 10\ \text{A}$$

The total current for 50,000 people would then be 500,000 A. The transport of such a current from a power-plant generator to a city many kilometres away would require wires of enormous size (actually, the wires would probably have to be large copper cylinders), and the I^2R power loss would still be substantial. Rather than transmit the power at 110 V, step-up transformers are used at the power plant to step up the voltage to some very large value, such as 600,000 V. The current needed is thus reduced to

$$I = \frac{110\ V}{600,000\ V}\ (500,000\ \text{A}) = 91.7\ \text{A}$$

To reduce the voltage to a safer level for transport within the city, power stations are located just outside the city to step down the voltage to a safer value, such as 10,000 V. Transformers in boxes attached to the power poles outside your house again step down the voltage to 110 V (or 220 V) for distribution to your house. It is because of the ease of stepping its voltage up or down with transformers that alternating current rather than direct current is in common use.

Power box with transformer for stepping down voltage of electrical power for distribution to homes.

Example 23-8   A transmission line has a resistance of 0.02 Ω/km. Calculate the I^2R power loss if 200 kW of power is transmitted from a power generator to a city 10 km away at (*a*) 220 V and (*b*) 4.4 kV.

(*a*) The total resistance of 10 km of wire is

$$R = (0.02\ \Omega/\text{km})(10\ \text{km}) = 0.20\ \Omega$$

The current required to transmit 200 kW at 220 V is

$$I = \frac{200{,}000\ \text{W}}{220\ \text{V}} = 909\ \text{A}$$

The power loss is then

$$I^2R = (909\ \text{A})^2(0.20\ \Omega) = 165\ \text{kW}$$

Thus, more than 80 percent of the power is wasted as heat loss.

(*b*) If the transmission voltage is 4.4 kV, the current is

$$I = \frac{200\ \text{kW}}{4.4\ \text{kV}} = 45.5\ \text{A}$$

The power loss is then

$$I^2R = (45\ \text{A})^2(0.20\ \Omega) = 414\ \text{W}$$

The energy loss is considerably less.

23-5   Rectification and Amplification

Although alternating current is readily available in every home, direct current is often needed to power such devices as portable radios or calculators. These devices often come with batteries and with ac–dc converters to save the batteries when ac power is available. The converters contain a transformer for stepping down the voltage from 110 V to whatever voltage is needed (typically 9 V) and a circuit for converting alternating current to direct current, a process called *rectification*. The principal element in a rectifier circuit is a *diode*. The first of these were vacuum tubes containing two main elements, a cathode that emitted electrons and a plate that collected them, from which the name diode came. The important feature of a diode is that it conducts current in one direction and not the other. The symbol for a diode as a circuit element is ➤⊢. The arrow indicates the direction of the current that can pass through the diode. Most diodes in use today are semiconductor devices, which will be discussed in Chapter 31.

Figure 23-13 shows a vacuum-tube diode, the first of which was developed by Sir John Fleming in 1904. When the cathode is heated (by a heating element in a separate circuit), it emits electrons in a process called *thermionic emission* (discovered by Thomas Edison in 1883). If the plate is at a higher potential than the cathode, it attracts the electrons and the tube conducts a

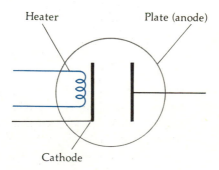

Figure 23-13
Vacuum-tube diode. When the cathode is heated, it emits electrons. The electrons are drawn to the plate, which is at a higher potential than the cathode.

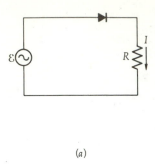

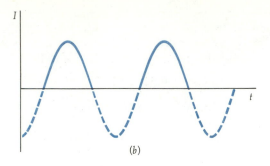

(a)

(b)

Figure 23-14

(a) Simple circuit containing an alternating current source, a diode, and a resistor. (b) Current versus time in the resistor in the circuit in (a). The negative current indicated by the dashed line does not get through the diode.

current. This current is called the plate current. If the plate is at a lower potential than the cathode, the electrons are repelled and there is no current through the tube.

Figure 23-14a shows a simple circuit containing an ac generator, a diode, and a resistor. The current in the resistor is shown in Figure 23-14b. The diode is said to be a *half-wave rectifier* because there is current in the resistor for only half of the cycle of the ac generator. Figure 23-15 shows a circuit that is a *full-wave rectifier*. Two diodes are connected to the terminals a and b of a transformer. The output plates of the diodes are connected together and to a resistor. The other side of the resistor is connected to the midpoint c of the transformer. When point a is at a higher potential than point c, diode 1 conducts current I_1 to the resistor. One-half cycle later, point b is at a higher potential than point c and diode 2 conducts current I_2 to the resistor.

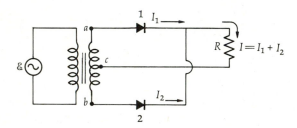

Figure 23-15

Full-wave rectifying circuit. When the potential of point a is greater than that of point c, current I_1 passes through diode 1. One-half cycle later, the potential of point b is greater than that of point c, and current I_2 passes through diode 2.

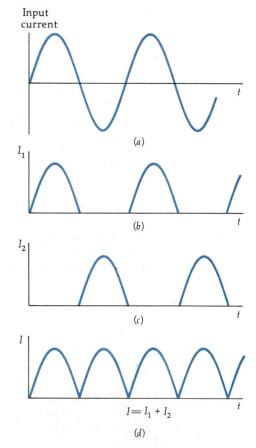

Input current

(a)

I_1

(b)

I_2

(c)

$I = I_1 + I_2$

(d)

Figure 23-16

(a) Input current to the circuit shown in Figure 23-15. (b) Current I_1 through diode 1. (c) Current I_2 through diode 2. (d) Total current $I = I_1 + I_2$ through the resistor in Figure 23-15.

The current $I = I_1 + I_2$ through the resistor is shown in Figure 23-16d. In Figure 23-17a, a capacitor has been added in parallel with the resistor. When the current in the resistor starts to increase, some of the charge flows onto the capacitor plates, which tends to reduce the current in the resistor. When this current starts to decrease, charge flows off the capacitor plates through

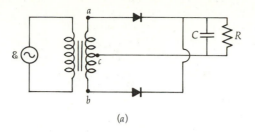

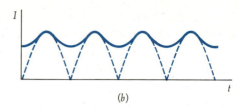

(a)

(b)

Figure 23-17
(a) Full-wave rectifying circuit of Figure 23-15 with a capacitor in parallel with the resistor. The capacitor tends to smooth out some of the ripples in the rectified current. (b) Current through the resistor in Figure 23-17a.

the resistor. The effect of the capacitor is to smooth out the ripples in the current, as shown in Figure 23-17b.

In 1907 Lee de Forest discovered that the plate current could be greatly modified by small voltage changes on a third electrode inserted between the cathode and the plate. A vacuum-tube *triode* is shown in Figure 23-18. The third electrode is a fine wire mesh called the *grid*. As in the case of the diode, the cathode emits electrons that are collected by the plate, which is at a higher potential (typically 100 to 200 V) than the cathode. Because the grid is closer to the plate than the cathode is, the potential of the grid relative to the cathode has a large effect on the plate current. When the grid is at the same potential, the plate current is essentially unaffected by the grid, but when the grid is negative relative to the cathode, the electrons emitted by the cathode are repelled by the grid, and the plate current is greatly diminished. Similarly, when the grid is positive relative to the cathode, the plate current is increased. Figure 23-19 shows how a triode can be used as an amplifier. The input signal is a small sinusoidal voltage applied between the grid and the cathode. The output signal is the voltage across the resistance R. The output signal is considerably larger than the input signal because small voltage changes on the grid produce large changes in the plate current. Today, vacuum tube triodes have been largely replaced by transistors, which are discussed in Chapter 31.

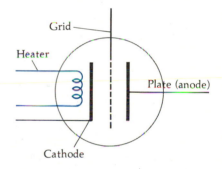

Figure 23-18
Vacuum-tube triode. The grid placed near the cathode controls the plate current. When the grid is negative relative to the cathode, it repels the electrons emitted by the cathode and diminishes the plate current. When the grid is positive relative to the cathode, it attracts the electrons emitted by the cathode and increases the plate current.

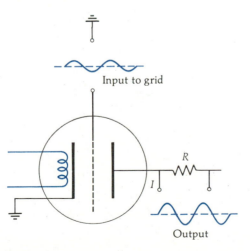

Figure 23-19
Amplification from a triode. A small sinusoidal signal applied to the grid results in a large sinusoidal signal across the output resistance R.

Question

6. Explain why the rms current in the output of a half-wave rectifier circuit is half that of a full-wave rectifier circuit.

Summary

1. The root-mean-square (rms) value of alternating current, I_{rms}, is defined by

$$I_{rms} = \sqrt{(I^2)_{av}}$$

It is related to the maximum current by

$$I_{rms} = \frac{I_{max}}{\sqrt{2}}$$

The average power dissipated in a resistor carrying a sinusoidal current is

$$P_{av} = \tfrac{1}{2}\mathcal{E}_{max}I_{max} = \mathcal{E}_{rms}I_{rms} = I_{rms}^2 R$$

2. In a circuit consisting of an ac emf and an inductor, the voltage across the inductor leads the current by $90°$. The rms or maximum current is related to the rms or maximum emf by

$$I = \frac{\mathcal{E}}{X_L}$$

where

$$X_L = 2\pi f L$$

is the inductive reactance of the inductor. The average power dissipated in this circuit is zero.

3. In a circuit consisting of an ac emf and a capacitor, the voltage across the capacitor lags the current by $90°$. The rms or maximum current is related to the rms or maximum emf by

$$I = \frac{\mathcal{E}}{X_C}$$

where

$$X_C = \frac{1}{2\pi f C}$$

is the capacitive reactance. The average power dissipated in this circuit is also zero. Like resistance, inductive and capacitive reactances have units of ohms.

4. The current in a series LRC circuit driven by an ac emf is given by

$$I = \frac{\mathcal{E}_{max}}{Z} \sin(\omega t - \phi)$$

where the impedance Z is

$$Z = \sqrt{R^2 + (X_L - X_C)^2}$$

and the phase angle ϕ is found from

$$\tan \phi = \frac{X_L - X_C}{R}$$

The average power input to such a circuit depends on the frequency and is given by

$$P_{av} = \mathcal{E}_{rms} I_{rms} \cos \phi$$

where $\cos \phi$ is called the power factor. The average power is maximum at the resonance frequency, which is given by

$$f_0 = \frac{1}{2\pi\sqrt{LC}}$$

At the resonance frequency the phase angle ϕ is zero, the power factor is 1, the inductive and capacitive reactances are equal, and the impedance equals the resistance R.

5. The quality of the resonance is described by the Q factor, which is defined by

$$Q = \frac{2\pi f_0 L}{R}$$

When the resonance is reasonably narrow, the Q factor can be approximated by

$$Q = \frac{f_0}{\Delta f}$$

where Δf is the width of the curve.

6. A transformer is a device for changing ac voltage and current without an appreciable loss of power. For a transformer with N_1 turns in the primary and N_2 turns in the secondary, the voltage across the secondary coil is related to that across the primary coil by

$$V_2 = \frac{N_2}{N_1} V_2$$

The transformer is called a step-up transformer if N_2 is greater than N_1 so that the output voltage is greater than the input voltage. If N_2 is less than N_1, it is a step-down transformer.

7. A diode that conducts current in one direction only can be used to convert alternating current to direct current, a process called rectification.

8. Small changes in the voltage of the grid in a triode produce large changes in the plate current, an effect that can be used to amplify ac signals.

Suggestions for Further Reading

Barthold, L. O., and H. G. Pfeiffer: "High-Voltage Power Transmission," *Scientific American*, May 1964, p. 38.

Practical considerations in the transmission of electrical power are discussed, including how high the voltage should be for greatest efficiency and whether dc transmission could be more economical than ac. The answers to such questions are complex and may need to include a consideration of weather patterns along the transmission line.

Rieder, Werner: "Circuit Breakers," *Scientific American*, January 1971, p. 76.

The principles of operation of circuit breakers in high-voltage transmission lines and why the formation of a plasma arc between the separating contacts is actually a desirable thing are discussed in this article.

Shiers, George: "The Induction Coil," *Scientific American*, May 1971, p. 80.

This device, which makes use of electromagnetic induction to produce high voltages, made possible a number of discoveries, including the existence of X rays and electrons and the possibility of wireless electromagnetic communication.

Review

A. Objectives: After studying this chapter, you should:

1. Be able to state the definition of rms current and relate it to the maximum current in an ac circuit.

2. Be able to state the definition of capacitive reactance, inductive reactance, and impedance.

3. Be able to give the phase relations between the voltage across a resistor, an inductor, or a capacitor and the current.

4. Be able to state the resonance condition for an *LRC* circuit with generator and sketch a graph of the power versus frequency for both high-*Q* and low-*Q* circuits.

5. Be able to describe a step-up and a step-down transformer.

6. Be able to describe how a diode can be used to convert alternating current to direct current.

B. Define, explain, or otherwise identify:

rms current, p. 584
Inductive reactance, p. 587
Capacitive reactance, p. 588
Impedance, p. 590
Resonance in *LRC* circuit, p. 591

Power factor, p. 592
Resonance width, p. 592
Q factor, p. 592
Transformer, p. 595
Step-up transformer, p. 595
Step-down transformer, p. 595
Diode, p. 597
Thermionic emission, p. 597
Half-wave rectifier, p. 598
Full-wave rectifier, p. 598
Triode, p. 599

C. True or false: If the statement is true, explain why. If it is false, give a counterexample.

1. Alternating current in a resistor produces no power because the current is negative as often as it is positive.

2. At very high frequencies, a capacitor acts like a short circuit.

3. An *LCR* circuit with a high *Q* factor has a narrow resonance curve.

4. At resonance, the impedance of an *LCR* circuit equals the resistance R.

Exercises

Section 23-1   Alternating Current in a Resistor

1. (*a*) Show that the rms emf of a generator is related to the maximum emf by $\mathcal{E}_{rms} = 0.707\mathcal{E}_{max}$. (*b*) The rms value of standard house voltage is 110 V. What is the maximum voltage?

2. A 100-W light bulb is plugged into a standard 110-V rms outlet. Find (*a*) I_{rms}, (*b*) I_{max}, and (*c*) the maximum power.

3. A 3-Ω resistor is placed in series with a 12.0-V (maximum) generator of frequency 60 Hz. Find (*a*) I_{max}, (*b*) I_{rms}, (*c*) the maximum power into the resistor, and (*d*) the rms power.

4. A 5.0-kW electric clothes dryer runs on 220 V rms. Find (*a*) I_{rms} and (*b*) I_{max}. (*c*) Find these same quantities for a dryer of the same power that operates at 110 V rms.

Section 23-2   Alternating Current in Inductors and Capacitors

5. Find the reactance of a 1.6-mH inductor at (*a*) 60 Hz, (*b*) 600 Hz, and (*c*) 6 Hz.

6. An inductor has a reactance of 120 Ω at 80 Hz. (*a*) What is its inductance? (*b*) What is the reactance at 160 Hz?

7. Find the reactance of a 1.4-nF capacitor at (*a*) 60 Hz, (*b*) 600 Hz, and (*c*) 6 Hz.

8. Find the reactance of a 15.0-μF capacitor at (*a*) 20 Hz, (*b*) 200 Hz, and (*c*) 2 MHz.

9. At what frequency would the reactance of a 10.0-μF capacitor equal that of a 1.0-mH inductor?

10. Sketch a graph of X_C versus frequency f for $C = 100 \ \mu F$.

11. An emf of 12.0 V rms and frequency 20 Hz is applied to a 2.5-mH inductor. Find (a) I_{rms} and (b) I_{max}.

12. An emf of 110 V rms and frequency 60 Hz is applied to a 20-μF capacitor. Find (a) I_{rms} and (b) I_{max}.

13. At what frequency is the reactance of a 10-μF capacitor equal to (a) 1 Ω, (b) 100 Ω, and (c) 0.01 Ω?

14. At what frequency is the reactance of a 2.5-mH inductor equal to (a) 1 Ω, (b) 100 Ω, and (c) 0.1 Ω?

Section 23-3 An *LCR* Circuit with a Generator

15. A series *LCR* circuit with $L = 10$ mH, $C = 2 \ \mu F$, and $R = 5 \ \Omega$ is driven by a generator of maximum emf 100 V and variable frequency. Find (a) the resonance frequency f_0 and (b) I_{rms} at resonance. When the generator frequency is 1250 Hz, find (c) X_C, X_L, and the impedance Z; (d) I_{rms}; and (e) the phase angle ϕ.

16. If the generator frequency is 1 kHz in the circuit of Exercise 15, find (a) X_C, X_L, and Z; (b) I_{rms}; and (c) the phase angle ϕ.

17. For the circuit of Exercise 15, find the power factor and the average power delivered to the circuit when the generator frequency is (a) 1250 Hz and (b) 1000 Hz.

18. Find the maximum voltages V_R, V_L, and V_C across the resistor, inductor, and capacitor for the circuit in Exercise 15 (a) at resonance, (b) when the generator frequency is 1250 Hz, and (c) when the generator frequency is 1000 Hz. In each case, make a vector plot showing the phase relations between these quantities and calculate the vector sum.

19. A series *LCR* circuit in a radio receiver is tuned by a variable capacitor so that it can resonate at frequencies from 500 kHz to 1600 kHz. If $L = 1.0 \ \mu H$, find the range of capacitances necessary to cover this range of frequencies.

20. FM radio stations have carrier frequencies that are separated by 0.20 MHz. When the radio is tuned to a station, such as 100.1 MHz, the resonance width of the receiver circuit should be much smaller than 0.20 MHz so that adjacent stations are not received. If $f_0 = 100.1$ MHz and $\Delta f = 0.05$ MHz, what is the Q factor of this circuit?

21. Find the Q factor and the resonance width for the circuit of Exercise 15.

Section 23-4 The Transformer

22. A transformer has 400 turns in the primary and 8 turns in the secondary. (a) Is this a step-up or a step-down transformer? (b) If the primary is connected across 110 V rms, what is the open-circuit voltage across the secondary? (c) If the primary current is 0.1 A, what is the secondary current?

23. The primary of a step-down transformer has 250 turns and is connected to a 110-V rms line. The secondary is to supply 20 A at 9 V rms. Find (a) the current in the primary and (b) the number of turns in the secondary.

24. A transformer has 500 turns in its primary, which is connected to 110 V rms. Its secondary coil is tapped at three places to give outputs of 2.5, 7.5, and 9 V rms. How many turns are needed for each part of the secondary coil?

Section 23-5 Rectification and Amplification

25. The maximum output current in a half-wave rectifier circuit is 3.5 A. (a) Find the rms current. (b) Find the rms current if the maximum output current is 3.5 A in a full-wave rectifier circuit.

26. Sketch the current versus time if a capacitor is added in parallel with the resistance in Figure 23-14.

Problems

1. Show that the maximum power delivered by a generator to a capacitor is

$$P_{max} = \mathcal{E}_{rms} I_{rms}$$

What is the average power?

2. Show that in a series circuit with resistance and capacitance but no inductance, the power factor is given by

$$\cos \phi = \frac{2\pi f RC}{\sqrt{1 + (2\pi f RC)^2}}$$

Sketch the power factor versus frequency f.

3. A coil can be considered to be a resistance and an inductance in series. Assume that $R = 120 \ \Omega$ and $L = 0.4$ H. The coil is connected across a 120-V, 60-Hz line. Find (a) the power factor, (b) the rms current, and (c) the average power supplied.

4. A coil with resistance and inductance is connected to a 120-V, 60-Hz line. The average power supplied to the coil is 60 W, and the rms current is 1.5 A. Find (a) the power factor, (b) the resistance of the coil, and (c) the inductance of the coil. (d) Does the voltage lead or lag the current? Find the phase angle ϕ.

5. In a certain *LRC* circuit, the capacitive reactance is 16 Ω and the inductive reactance is 4 Ω. The resonance frequency is 1.60 kHz. Find (*a*) *L* and *C*. If $R = 5\ \Omega$ and $\mathcal{E}_{max} = 26$ V, find (*b*) *Q* and (*c*) the maximum current.

6. Show that Equation 23-24 can be written

$$I_{max} = \frac{f\mathcal{E}_{max}}{\sqrt{4\pi^2 L^2 (f^2 - f_0^2)^2 + R^2 f^2}}$$

7. When an *LRC* series circuit is connected to a 110-V rms, 60-Hz line, the rms current is 11.0 A and the emf lags the current by 45°. (*a*) Find the average power supplied to the circuit. (*b*) Find the resistance. (*c*) If the inductance is 0.05 H, find the capacitance. (*d*) What inductance or capacitance would you add to make the power factor 1?

8. Sketch the impedance versus frequency for a series *LRC* circuit.

9. A certain electrical device draws 10 A rms and has an average power of 720 W when connected to a 120-V, 60-Hz power line. (*a*) What is the impedance of this device? (*b*) What series combination of resistance and reactance is equivalent to this device? (*c*) If the emf lags the current, is the reactance inductive or capacitive?

10. (*a*) Show that Equation 23-23 can be written

$$\tan \phi = \frac{2\pi L(f^2 - f_0^2)}{fR}$$

Find the approximate phase angle ϕ at (*b*) very high frequencies and (*c*) very low frequencies.

11. A method for measuring inductance is to connect the inductor in series with a variable-frequency signal generator, an ammeter, a known capacitance, and a known resistance. The frequency of the generator is varied while the emf is held constant until the current is maximum. If $C = 10\ \mu$F, $\mathcal{E}_{max} = 10$ V, $R = 100\ \Omega$, and *I* is maximum at 800 Hz, what is *L*? What is the maximum current?

CHAPTER TWENTY-FOUR

Light

For the rest of my life
I want to reflect
on what light is.

ALBERT EINSTEIN, 1916

Light is essential to most of the living forms we know of. It is electromagnetic radiation that we can see, a narrow frequency range of less than an octave; yet, it brings us much of the beauty of the world and is our chief contact with our environment in every other way. Its importance to us is literally incalculable, but only in the last four generations has its nature been understood. We are lucky to be the beneficiaries of the discoveries in optics, from those of the ancient Arabs to the latest innovations in lasers and holograms.

Light has intrigued humankind for centuries. The question of whether light is a beam of particles or some kind of wave motion caused heated debate throughout much of the history of science. We now know that in its transmission, light exhibits wave characteristics, whereas in its interaction with matter, it exhibits particle characteristics. (This wave-particle duality is not confined to light. As we will see in Chapter 30, the transmission of electrons and other "particles" is a wave phenomenon whereas their interaction with matter is a particle phenomenon.) In this chapter and the next two, we will be mainly concerned with the wave nature of light.

To gain historical perspective, we will first look briefly at some early theories and experiments that will help us to understand light as a wave motion. We will then discuss the basic phenomena of reflection, refraction, and polarization. These phenomena can all be adequately understood by using rays to describe the straight-line propagation of light and neglecting interference and diffraction effects. As discussed in Chapter 17, this ray approximation is valid for the propagation of any wave motion if the wavelength is small compared with any apertures or obstacles encountered. This is often the situation in optics because the wavelengths of visible light range from about 400 nm (1 nm $= 10^{-9}$ m) for violet light to 700 nm for red light.

605

24-1   Waves or Particles

The controversy over the nature of light is one of the most interesting in the history of science. Early theories considered light to be something that emanated *from* the eye. In later theories, it was realized that light must come from the objects seen and that light entering the eye caused the sensation of vision. The most influential proponent of the *particle theory* of light was Newton. Using it, he was able to explain the laws of reflection and refraction. As we will see shortly, his derivation of the law of refraction depends on the assumption that light travels faster in water or glass than in air, an assumption later shown to be false.

When light strikes a boundary between two transparent media, for example, air and water or air and glass, part of the light energy is reflected and part is transmitted. The reflected ray makes an angle with the normal to the surface (a line perpendicular to the surface) equal to that of the incident ray (see Figure 24-1). The transmitted ray is bent toward the normal if the light is moving from a less dense to a more dense medium, such as from air to water or glass, as shown in Figure 24-1. It is bent away from the normal if the light is leaving a more dense medium, such as water or glass, and is entering a less dense one, such as air. This bending of the transmitted ray is called *refraction*.

The law of reflection of light from a plane boundary is easily explained by the particle theory. Figure 24-2 shows a particle bouncing off a hard plane surface. If there is no friction, the component of the particle momentum parallel to the surface is not changed by the collision, but the component perpendicular to the wall is reversed. Thus, the angle of reflection equals the angle of incidence, as is observed for light rays reflected from a plane mirror.

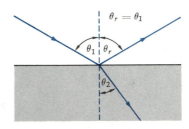

Figure 24-1
When a light ray strikes the boundary between two different media, for example, air and glass, part of the energy is reflected and part is transmitted. The angle of reflection θ_r equals the angle of incidence θ_1. The angle of refraction θ_2 is less than the angle of incidence for air-to-glass refraction.

Figure 24-2
Reflection of a particle. If the surface is frictionless, the angle of reflection θ_r equals the angle of incidence θ_1.

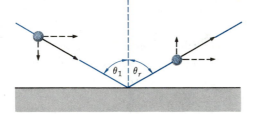

Figure 24-3 illustrates Newton's explanation of refraction at an air–water surface. Newton assumed that light particles are strongly attracted to water. Therefore, when they approach the surface, they receive a momentary impulse that increases the component of momentum perpendicular to the surface. Thus, the direction of the momentum of the light particles changes, and the light beam is bent toward the normal to the surface, which is in agreement with observation. An important condition of this theory is that the speed of light must be greater in water than in air to account for the bending of the beam toward the normal when light enters water from air and away from the normal when it enters air from water. Experimental determination of the speed of light in water did not come until about 1850, nearly 200 years later, when Foucault showed that light travels more slowly in water. This result showed that Newton's derivation of the refraction of

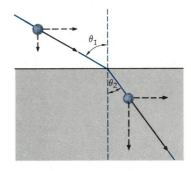

Figure 24-3
Refraction of a light particle from air into water. According to Newton's particle theory, the light particle is attracted by the water, which increases the component of its momentum normal to the surface. The direction of motion is therefore bent toward the normal to the surface. This theory requires that light travel faster in water than in air.

light as a particle phenomenon was incorrect. We will use the wave theory of light to derive the laws of reflection and refraction in Sections 24-4 and 24-5.

The chief proponents of the wave theory of light propagation were Christian Huygens and Robert Hooke. Using his wavelet construction (Section 17-2), Huygens was able to explain reflection and refraction by assuming that light travels more slowly in glass or water than in air. Newton saw the virtues of the wave theory of light, particularly as it explained the colors formed by thin films, which he had studied extensively. However, he rejected the wave theory because of the observed straight-line propagation of light:

> To me the fundamental supposition itself seems impossible, namely, that the waves or vibrations of any fluid can, like the rays of light, be propagated in straight lines, without a continual and very extravagant spreading and bending every way into the quiescent medium, where they are terminated by it. I mistake if there be not both experiment and demonstration to the contrary.*

Because of Newton's great reputation and authority, this reluctant rejection of the wave theory of light, based on lack of evidence of diffraction, was strictly adhered to by Newton's followers. Even after evidence of diffraction was available, they sought to explain it as the scattering of light particles from the edges of slits. Newton's particle theory of light was accepted for more than a century.

In 1801, Thomas Young revived the wave theory of light. He was one of the first to introduce the idea of interference as a wave phenomenon in both light and sound. Figure 24-4 shows his drawing of the combination of waves from two sources. His observation of interference with light was a clear demonstration of the wave nature of light. Young's work went unnoticed by the scientific community for more than a decade, however.

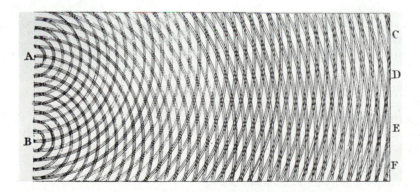

Perhaps the greatest advance in the general acceptance of the wave theory of light was due to the work of the French physicist Augustin Fresnel (1788–1827), who performed extensive experiments on interference and diffraction and put the wave theory on a mathematical basis. He showed, for example, that the observed rectilinear propagation of light is a result of

Portrait of Thomas Young, British physicist, physician, and egyptologist (1773–1829).

Figure 24-4
Thomas Young's drawing of a two-source interference pattern for light.

---

* Newton, *Opticks*, 1704. Reprint, Dover, New York, 1952

the very short wavelengths of visible light. An interesting triumph of the wave theory was an experiment suggested to Fresnel by S. Poisson, who sought to discredit the wave theory. Poisson noted that if an opaque disk is illuminated by light from a source on its axis, the Fresnel wave theory predicts that light waves bending around the edge of the disk should meet and interfere constructively on the axis, producing a bright spot in the center of the shadow of the disk. Poisson considered this to be a ridiculous contradiction of fact, but Fresnel immediately demonstrated experimentally that such a spot does in fact exist. This demonstration convinced many doubters that the wave theory of light is valid (see Figure 24-5).

In 1850, Jean Foucault measured the speed of light in water and showed that it is less there than in air, thus ruling out Newton's particle theory. In 1860, James Clerk Maxwell published his mathematical theory of electromagnetism, which predicted the existence of electromagnetic waves propagated with a speed calculated from the laws of electricity and magnetism to be 3×10^8 m/s, the same as the speed of light. Maxwell correctly suggested that this result was not accidental but instead indicated that light is an electromagnetic wave. Maxwell's theory was confirmed in 1887 by Hertz, who used a tuned electrical circuit to generate electromagnetic waves and a similar circuit to detect them. In the latter half of the nineteenth century, Kirchhoff and others applied Maxwell's theory to explain the interference and diffraction of light and other electromagnetic waves and put Huygens' empirical methods of construction of wavefronts on a firm mathematical basis.

This discussion of the wave-particle controversy would not be complete if we did not mention the discovery in the twentieth century that the wave theory, although generally correct in describing the propagation of light (and other electromagnetic waves), fails to account for all the properties of light, particularly those involving the interaction of light with matter. One of the ironies of the history of science is that in his famous 1887 experiment that confirmed Maxwell's wave theory, Hertz also discovered the *photoelectric effect* (which will be discussed in detail in Chapter 29). This effect can be explained only by a particle model of light, as Einstein showed in 1905. Thus a particle model of light was re-introduced.

Complete understanding of the dual nature of light did not come until the 1920s, when experiments by C. J. Davisson and L. Germer and by G. P. Thompson showed that electrons (and other, similar "particles") also have a dual nature and exhibit the wave properties of interference and diffraction as well as their well-known particle properties. The behavior of fundamental quantities such as light, electrons, and other subatomic particles is correctly described by the theory of quantum mechanics worked out by E. Schrödinger, W. Heisenberg, P. A. M. Dirac, and others. Although the quantum theory differs from both classical wave theory and classical particle theory, in some circumstances it resembles one and in other circumstances, the other. For example, the propagation of these fundamental quantities can always be described as a wave that exhibits the usual wave effects of interference and diffraction. On the other hand, exchanges of energy between these fundamental quantities is described in terms of particle mechanics. As discussed in Chapter 17, when the wavelength of any wave is very small compared with the various obstacles and apertures it encounters, the ray approximation is valid and the propagation is in straight

Nineteenth-century engraving of Augustin Jean Fresnel (1788–1827).

Figure 24-5
Diffraction of light by an opaque disk showing the bright spot in the center of the shadow.

lines. The wave nature of electrons and other, similar particles is not easily observed because of their extremely short wavelengths. (This is the same difficulty that prevented Newton from observing the wave nature of light.) The particle nature of light evident in its energy exchanges also is often not noticed because the number of light particles, which are called *photons*, in a   Photons
light beam is so enormous and the energy of each individual photon is so small. This difficulty is similar to that of observing the particle nature of gas molecules that exert pressure on the walls of a container.

Question

1. Why is the diffraction of sound so much more noticeable than the diffraction of light?

24-2 Electromagnetic Waves

Electromagnetic waves include light, radio waves, x-rays, gamma rays, microwaves, and others. They are transverse waves involving the propagation of electric and magnetic fields through space with speed $c = 3 \times 10^8$ m/s (in a vacuum). The various types of electromagnetic waves differ only in wavelength and frequency, which is related to wavelength in the usual way:

$$f = \frac{c}{\lambda} \qquad\qquad\qquad 24\text{-}1$$

Table 24-1 gives the electromagnetic spectrum and the names usually associated with the various frequency and wavelength ranges. These ranges often are not well defined and sometimes overlap. For example, electromagnetic waves with wavelengths of about 0.1 nm are usually called x-rays, but if they originate from nuclear radioactivity they are called gamma rays.

The human eye is sensitive to electromagnetic radiation of wavelengths from about 400 to 700 nm, the range called *visible light*. The shortest wavelengths in the visible spectrum correspond to blue or violet light and the longest to red light, with all the colors of the rainbow falling between these extremes. Electromagnetic waves with wavelengths slightly less than those of visible light are called ultraviolet rays. Those with wavelengths slightly greater than those of visible light are called infrared. There are no limits on the wavelengths of electromagnetic radiation. All frequencies and wavelengths are theoretically possible.

The differences in the wavelengths of the various kinds of electromagnetic waves are very important. As we know, the behavior of waves depends strongly on the relative sizes of the wavelength and the physical objects or apertures the waves encounter. Since the wavelengths of light are in the rather narrow range from about 400 to 700 nm, they are much smaller than most obstacles, so the ray approximation is often valid. Wavelength and frequency are also important in determining the kinds of interactions between electromagnetic waves and matter. X-rays, for example, which have very short wavelengths and high frequencies, easily penetrate many

Table 24-1

Electromagnetic Spectrum

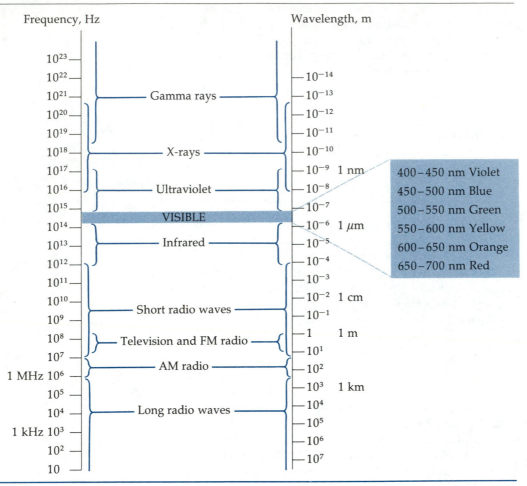

materials that are opaque to lower-frequency light waves because of absorption. Microwaves have wavelengths of the order of a few centimetres and frequencies that are close to the natural resonance frequencies of vibration of water molecules in solids and liquids. Microwaves are therefore readily absorbed by water molecules in foods, which is the mechanism responsible for heating and cooking in microwave ovens.

Electromagnetic waves are produced by the oscillation of electric charges. The frequency of oscillation determines the frequency and therefore the wavelength of the waves. Light waves of frequencies of the order of 10^{14} Hz originate from the motion of atomic charges. Radio waves of frequencies of the order of a kilohertz for AM radio waves and a megahertz for FM radio waves are produced by macroscopic electric currents oscillating in a radio antenna.

Figure 24-6 is a schematic drawing of an *electric-dipole radio antenna,* which consists of two conducting rods along a line fed by an alternating-current generator. At the time $t = 0$, the ends of the rods are charged, and there is an electric field near the rods parallel to the rods, as shown in Figure 24-6a. There is also a magnetic field (not shown) encircling the rods due to the current in the rods. The magnetic field is perpendicular to the page. These fields move outward from the rods with the speed of light. After one fourth period, at $t = \frac{1}{4}T$, the rods are uncharged, and the electric field near the rods is zero, as shown in Figure 24-6b. At $t = \frac{1}{2}T$, the rods are again charged, but the charges are opposite to those at $t = 0$, as shown in Figure 24-6c.

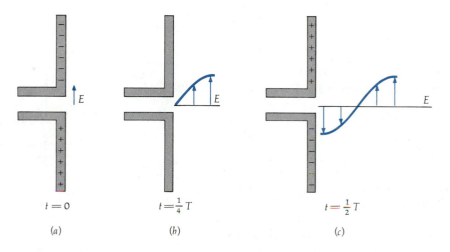

$t = 0$ $t = \frac{1}{4}T$ $t = \frac{1}{2}T$

(a)   (b)   (c)

Figure 24-6
An electric-dipole antenna. Alternating current is supplied to the antenna by a generator not shown. The electric field due to the charges in the antenna propagates outward at the speed of light. There is also a magnetic field (not shown) perpendicular to the page due to the current in the antenna. (a) At $t = 0$, the electric charges produce the electric field E shown. (b) At $t = \frac{1}{4}T$, the antenna is uncharged. The field E produced at $t = 0$ has moved outward. (c) The field due to the antenna charges at $t = \frac{1}{2}T$ is now downward.

Figure 24-7 shows the electric and magnetic field vectors in the electromagnetic wave far from the antenna. The electric and magnetic fields oscillate in phase with simple harmonic motion, and they are perpendicular to each other and to the direction of propagation of the wave. The wave is thus a transverse wave.

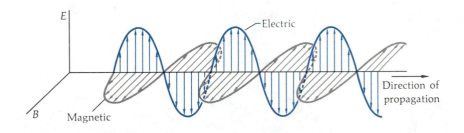

E

Electric

Direction of propagation

B Magnetic

Figure 24-7
Electric and magnetic field vectors in an electromagnetic wave. The vectors are perpendicular to each other and to the direction of propagation.

As discussed in previous chapters, there is electric energy associated with an electric field in space and magnetic energy associated with a magnetic field in space. The energy per unit volume, or energy density, associated with an electric field E is

$$\eta_e = \tfrac{1}{2}\epsilon_0 E^2$$

and that associated with a magnetic field B is

$$\eta_m = \frac{B^2}{2\mu_0}$$

In an electromagnetic wave, the magnitude of the electric and magnetic fields are related by

$$E = cB \qquad\qquad 24\text{-}2$$

where c is the speed of light and other electromagnetic waves, and the electric and magnetic energy densities are equal.

$$\eta_e = \eta_m$$

The total energy density associated with an electromagnetic wave is then equal to twice that of either field:

$$\eta = \epsilon_0 E^2 = \frac{B^2}{\mu_0}$$

In Chapter 16, we showed that the intensity of a wave (the average incident power per unit area) is the product of the average energy density and the wave speed. The intensity of an electromagnetic wave is therefore

$$I = \eta c = c\epsilon_0 (E^2)_{av} = \frac{c(B^2)_{av}}{\mu_0} \qquad\qquad 24\text{-}3$$

We note that the intensity of a light wave is proportional to E^2 (or to B^2), which is consistent with the general result that the intensity of a harmonic wave is proportional to the square of its amplitude.

Electromagnetic waves of radio or television frequencies can be detected by a dipole antenna (see Figure 24-8) placed parallel to the electric field so that the field induces an alternating current in the antenna. They can also be detected by a loop antenna (see Figure 24-9) placed perpendicular to the magnetic field so that the changing magnetic flux through the loop induces an alternating current in the loop. Electromagnetic waves of frequencies in the visible light range can be detected by the eye or by photographic film, both of which are mainly sensitive to the electric field. Because of this, we will, for simplicity, consider the electric field E to be the wave amplitude analogous to the displacement of a string in a string wave or to the variation in density or pressure in a sound wave. In terms of the root-mean-square value of the electric field E_{rms}, the intensity is

$$I = \eta c = c\epsilon_0 E^2_{rms} \qquad\qquad 24\text{-}4$$

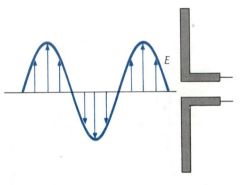

Figure 24-8
An electric-dipole antenna for detecting electromagnetic radiation. The alternating electric field of the radiation produces an alternating current in the antenna.

Figure 24-9
A loop antenna for detecting electromagnetic radiation. The alternating magnetic flux through the loop due to the magnetic field of the radiation induces an alternating current in the loop.

The radiation from an electric-dipole antenna, such as that in Figure 24-6, is called electric-dipole radiation. Many electromagnetic waves including light exhibit the characteristics of electric-dipole radiation. A feature of electric-dipole radiation is that the intensity of electromagnetic waves radiated by a dipole antenna is zero along the axis of the antenna and is

Electric-dipole radiation

maximum in the directions perpendicular to the axis. If the dipole is in the z direction with its center at the origin, the intensity is zero along the z axis and is maximum in the xy plane. In the direction of a line making an angle θ with the z axis, as shown in Figure 24-10, the intensity is proportional to $\sin^2 \theta$.

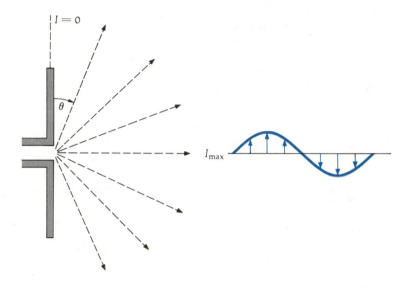

Figure 24-10
The intensity of electromagnetic radiation from an electric-dipole antenna is maximum in the directions perpendicular to the antenna and zero in the directions along the antenna.

Questions

2. Which waves have greater frequencies, light waves or x-rays?

3. Is the frequency of ultraviolet radiation greater or less than that of infrared radiation?

4. What kind of waves have wavelengths of the order of a few metres?

24-3   The Speed of Light

The first effort to measure the speed of propagation of light was made by Galileo. He and a partner stood on hilltops about a kilometre apart, each with a lantern and a shutter to cover it. Galileo proposed to measure the time it took for light to traverse twice the distance between the experimenters. First one would uncover his lantern; when the other saw the light, he would uncover his. The time between the first experimenter's uncovering his lantern and his seeing the light from the other lantern would be the time it took for the light to travel back and forth between the experimenters. Though this method is sound in principle, the speed of light is so great that the time interval that had to be measured was much smaller than fluctuations in human response time, so Galileo was unable to obtain any value for the speed of light.

The first indication of the true magnitude of the speed of light came from astronomical observations of the period of Io, one of the moons of Jupiter.

"But enough about the world situation, the economy, and all that. Let's talk about the speed of light."

This period is determined by measuring the time between eclipses (when Io disappears behind Jupiter). The eclipse period is about 42.5 hours, but measurements made when the earth is moving away from Jupiter, as from point A to C in Figure 24-11, give a greater time for this period than do measurements made when the earth is moving toward Jupiter, as from point C to A in the figure. Since these measurements differ by only about 15 seconds from the average value, the discrepancies were difficult to measure accurately. In 1675, the astronomer Ole Roemer attributed these discrepancies to the fact that the speed of light is not infinite. During the 42.5 hours between eclipses of Io, the distance between the earth and Jupiter changed, making the path of the light longer or shorter. Roemer devised the following method for measuring the cumulative effect of these discrepancies. We neglect the motion of Jupiter since it is much less than that of the earth.

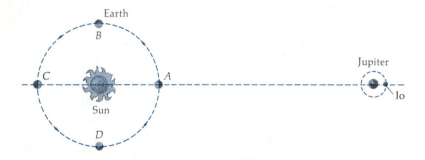

Figure 24-11
Figure 24-11
Roemer's method for measuring the speed of light. The time between eclipses of Jupiter's moon Io appears to be greater when the earth is moving from A to C than when it is moving from C to A. The difference is due to the time it takes light to travel the distance traveled by the earth along the line of sight during one period of Io. (The distance traveled by Jupiter in 1 earth year is negligible.)

When the earth is at point A, the distance between the earth and Jupiter is changing negligibly. The period of Io's eclipse is measured, providing the time between the beginning of successive eclipses. Based on this measurement the number of eclipses in six months is computed, and the time is predicted when an eclipse should begin a half-year later when the earth is at point C. At C, the observed beginning of the eclipse is about 16 minutes later than predicted. This, then, is the time it takes light to travel a distance equal to the diameter of the earth's orbit.

Example 24-1   The diameter of the earth's orbit is 3.00×10^{11} m. If light takes 16 min to travel this distance, what is the speed of light?

The number of seconds in 16 min is $(16 \text{ min}) \times (60 \text{ s/min}) = 960$ s. The speed of light is thus

$$c = \frac{\Delta x}{\Delta t} = \frac{3.00 \times 10^{11} \text{ m}}{960 \text{ s}} = 3.12 \times 10^8 \text{ m/s}$$

The first nonastronomical measurement of the speed of light was made by the French physicist A. H. L. Fizeau in 1849. On a hill in Paris, Fizeau placed a light source and a system of lenses that reflected the light from a half-silvered mirror and focused it on a gap in a toothed wheel. On a distant hill (about 8.63 km away), he placed a mirror to reflect the light back to the first hill to be viewed by an observer there. A schematic representation of Fizeau's set up is shown in Figure 24-12. The toothed wheel could be rotated and the speed of rotation could be varied. At low speeds of rotation, little light was visible because the path of the reflected light was obstructed by the

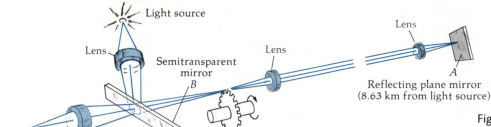

Figure 24-12
Fizeau's method of measuring the speed of light. Light from the source is reflected by mirror B and passes through a gap in the toothed wheel to mirror A. The speed of light is determined by measuring the angular speed of the wheel that will permit the reflected light to pass through the next gap in the toothed wheel so that an image of the source is observed.

teeth of the rotating wheel. The speed of rotation was then increased. The light suddenly became visible when the rotation speed was such that the reflected light passed through the next gap in the wheel.

Example 24-2   In Fizeau's experiment, the wheel had 720 teeth and the light was bright when the wheel rotated at 25.2 revolutions per second. If the distance from the wheel to the distant mirror was 8.63 km, what was Fizeau's value for the speed of light?

The total distance traveled by the light from the gap in the wheel to the mirror and back was 2×8.63 km $= 17.26$ km. The reflected light passed through the next gap in the wheel, so during this time, the wheel made $1/720$ rev. Since the wheel made 25.2 rev in 1 s, the time to make $1/720$ rev was

$$\Delta t = \frac{1 \text{ s}}{25.2 \text{ rev}} \left(\frac{1}{720} \text{ rev} \right) = 5.51 \times 10^{-5} \text{ s}$$

The value for the speed of light from this experiment is then

$$c = \frac{\Delta x}{\Delta t} = \frac{8.63 \text{ km}}{5.51 \times 10^{-5} \text{ s}}$$

$$= 3.13 \times 10^5 \text{ km/s} = 3.13 \times 10^8 \text{ m/s}$$

Fizeau's method was improved by Foucault, who replaced the toothed wheel with an eight-sided rotating mirror, as illustrated in Figure 24-13.

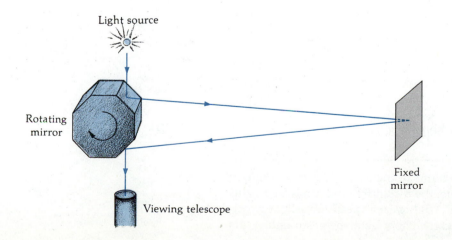

Figure 24-13
Simplified drawing of Foucault's method of measuring the speed of light. Essentially, Fizeau's rotating toothed wheel is replaced by a rotating octagonal mirror. Light is seen when the mirror makes an eighth of a revolution during the time it takes for the light to travel to the fixed mirror and back so that the next face on the mirror is in the proper position to reflect the light into the telescope.

Light strikes one face of the mirror and is reflected from a distant fixed mirror to another face and then to an observing telescope. (The actual experimental arrangement is somewhat more complicated than the simple diagram of Figure 24-13.) When the mirror rotates through one-eighth of a turn (or $n/8$ turns, where n is an integer) another face of the mirror is in the right position for the reflected light to enter the telescope. Around 1850, Foucault measured the speed of light in air and in water and showed that it is less in water thus disproving Newton's particle theory of refraction. Using essentially the same method as Foucault, the American physicist A. A. Michelson made precise measurements of the speed of light from 1880 to 1930.

Another method of determining the speed of light involves the measurement of electrical constants rather than the speed of light directly. According to Maxwell's electromagnetic theory, the speed of light is given by

$$c = \frac{1}{\sqrt{\epsilon_0 \mu_0}}$$

The constant ϵ_0 can be measured by measuring the capacitance of a parallel-plate capacitor. The constant μ_0 is defined in terms of the definition of the ampere, which in turn determines the coulomb.

The various methods for measuring the speed of light all are in general agreement. Today, the speed of light is defined to be exactly

$$c = 299,792,457 \text{ m/s}$$

The standard unit of length, the metre, is now defined in terms of this speed. A measurement of the speed of light is therefore now a measurement of the size of the metre, which is the distance light travels in $1/299,792,457$ s. The value 3×10^8 m/s for the speed of light is accurate enough for nearly all calculations. The speed of radio waves and of all electromagnetic waves (in a vacuum) is the same as the speed of light.

Example 24-3   Space travelers on the moon use electromagnetic waves to communicate with the space control center on earth. What is the time delay for their signal to reach the earth, which is 3.84×10^8 m away?

Using 3×10^8 m/s for the speed of light and other electromagnetic waves, we find the time for a signal to travel from the moon to the earth to be

$$\Delta t = \frac{\Delta x}{c} = \frac{3.84 \times 10^8 \text{ m}}{3 \times 10^8 \text{ m/s}} = 1.28 \text{ s}$$

The time delay is thus 1.28 s each way.

Example 24-4   The sun is 1.50×10^{11} m from the earth. How long does it take for the sun's light to reach the earth?

$$\Delta t = \frac{\Delta x}{c} = \frac{1.50 \times 10^{11} \text{ m}}{3 \times 10^8 \text{ m/s}} = 500 \text{ s} = 8.33 \text{ min}$$

Large distances are often given in terms of the time that it takes for light to travel those distances. For example, the distance to the sun is 8.33 light-minutes, which is written 8.33 c-min. A light year is the distance light travels in one year. We can easily find a conversion factor between light-years and metres. The number of seconds in one year is

$$1 \text{ y} = 1 \text{ y} \times \frac{365.24 \text{ d}}{1 \text{ y}} \times \frac{24 \text{ h}}{1 \text{ d}} \times \frac{3600 \text{ s}}{1 \text{ h}} = 3.156 \times 10^7 \text{ s}$$

The number of metres in one light-year is therefore

$$1 \text{ } c\text{-year} = (2.998 \times 10^8 \text{ m/s})(3.156 \times 10^7 \text{ s}) = 9.46 \times 10^{15} \text{ m}$$

Questions

5. Estimate the time required for light to make the round trip in Galileo's experiment to determine c.

6. Fizeau noticed that when the toothed wheel in his experiment rotated slowly, some faint light was seen, but no light was seen when the wheel rotated at 12.6 revolutions per second. Explain.

24-4 Reflection

When waves of any type strike a plane barrier, new waves are generated that move away from the barrier. This phenomenon is called *reflection*. Reflection occurs at a boundary between two different media, such as an air–glass surface (in which case part of the incident energy is reflected and part is transmitted). Figure 24-14 shows a light ray striking a smooth air–glass surface. The angle θ_1 between the incident ray and the normal is called the *angle of incidence*, and the plane formed by these two lines is called the *plane of incidence*. The reflected ray lies in the plane of incidence and makes an angle θ_r with the normal, which is called the *angle of reflection* and is equal to the angle of incidence, as is shown in the figure. This result is known as the *law of reflection*. The law of reflection holds for any type of wave. Figure 24-15 shows the wavefronts of ultrasonic waves in water reflected from a steel plate.

The fraction of light energy reflected at a boundary such as an air–glass surface depends in a complicated way on the angle of incidence and on the relative speed of light in the first medium (air) and in the second medium (glass). The speed of light in a medium, such as glass, water, or air, is characterized by the *index of refraction n*, which is defined as the ratio of the speed of light in a vacuum c to the speed of light in the medium v:

$$n = \frac{c}{v} \qquad\qquad\qquad 24\text{-}5$$

For the special case of normal incidence, the reflected intensity is given by

$$I = \left(\frac{n_2 - n_1}{n_2 + n_1}\right)^2 I_0 \qquad \text{(normal incidence)} \qquad 24\text{-}6$$

where I_0 is the incident intensity. For a typical case of reflection from an air–glass surface for which $n_1 = 1$ and $n_2 = 1.5$, Equation 24-6 gives $I = I_0/25$. That is, only about 4 percent of the energy is reflected; the rest is transmitted.

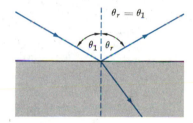

Figure 24-14
The angle of reflection θ_r equals the angle of incidence θ_1.

Figure 24-15
Ultrasonic plane waves in water reflecting from a steel plate.

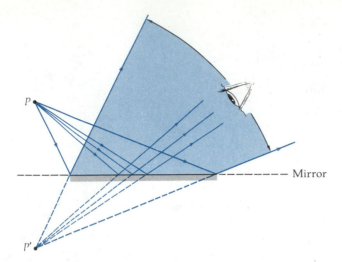

Figure 24-16
Rays from source P reflected by a
mirror into the eye appear to come
from the image point P' behind the
mirror. The image can be seen by
an eye located anywhere in the
shaded region.

Figure 24-16 shows a narrow bundle of light rays originating from a point source P reflected from a smooth surface. After reflection, the rays diverge exactly as if they came from a point P' behind the surface. The point P' is called the image of point P. When these rays enter the eye, they cannot be distinguished from rays originating from a source at P' with no reflecting surface present. (We will study the formation of images by reflecting and refracting surfaces in the next chapter.)

Reflection from a smooth surface is called *specular reflection*. It differs from *diffuse reflection*, which is illustrated in Figure 24-17. Here, because the surface is rough, rays enter the eye from many different points on the surface and there is no reflected image. The reflection from the page of this book is diffuse reflection. Ground glass is sometimes used in a picture frame to give diffuse reflection and cut out the glare from the light used to illuminate the picture. Also it is diffuse reflection that allows you to see the road when driving at night because some of the light from your headlights reflects back toward you.

The law of reflection can be derived from Huygens' principle (Section 17-2), according to which each point on a given wavefront can be considered to be a point source of secondary waves. Figure 24-18 shows a plane wavefront AA' striking a barrier at point A. As can be seen from the figure, the angle ϕ_1 between the wavefront and the barrier is the same as the angle of incidence θ_1, which is the angle between the perpendicular to the barrier

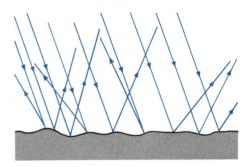

Figure 24-17
Diffuse reflection from a rough
surface.

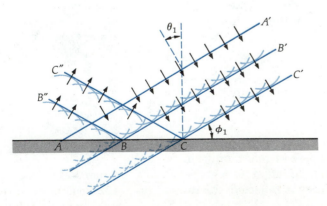

Figure 24-18
Plane wave reflected at a plane
barrier. The angle θ_1 between the
incident ray and the normal to the
barrier is the angle of incidence. It is
equal to the angle ϕ_1 between the
incident wavefront and the barrier.

and the rays that are perpendicular to the wavefronts. The position of the wavefront after a time t is found by constructing wavelets of radius vt with their centers on the wavefront AA'. Wavelets that do not strike the barrier form the portion of the new wavefront BB'. Wavelets that do strike the barrier are reflected and form the portion of the new wavefront BB''. By a similar construction, the wavefront $C''CC'$ is obtained from the Huygens' wavelets originating on the wavefront $B''BB'$.

Figure 24-19 is an enlargement of a portion of Figure 24-18 that shows the part of the original wavefront AP that strikes the barrier in time t. In this time, the wavelet from point P reaches the barrier at point B, and the wavelet from point A reaches point B''. The reflected wave BB' makes an angle ϕ_r with the barrier that is equal to the angle of reflection θ_r between the reflected ray and the normal to the barrier. The triangles ABP and BAB'' are both right triangles with a common side AB and equal sides $AB'' = BP = vt$. Hence, these triangles are congruent, and the angles ϕ_1 and ϕ_r are equal, implying that the angle of reflection θ_r equals the angle of incidence θ_1.

Question

7. How does a thin layer of water on the road affect your seeing the light reflected off the road from your own headlights? How does it affect the reflected light you see from the headlights of an oncoming car?

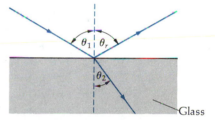

Figure 24-19
Geometry of Huygens' construction for the calculation of the law of reflection. The wavefront AP originally intersects the barrier at point A. After a time t, the Huygens' wavelet from P strikes the barrier at point B, and the one from A reaches point B''. Since the triangles ABP and BAB'' are congruent, the angles θ_1 and θ_r are equal.

24-5   Refraction

When a beam of light strikes a boundary surface separating two different media, such as an air–glass surface, part of the light energy is reflected and part enters the second medium. Figure 24-20 shows light striking a smooth air–glass surface. The ray that enters the glass is called the *refracted ray*, and the angle θ_2 is called the *angle of refraction*. The angle of refraction is less than the angle of incidence, as is shown in the figure; that is, the refracted ray is bent toward the normal. If, however, the light beam originates in the glass and is refracted into the air, the angle of refraction is greater than the angle

Figure 24-20
Incident, reflected, and refracted rays for light striking an air–glass surface. The angle of refraction θ_2 is less than the angle of incidence θ_1.

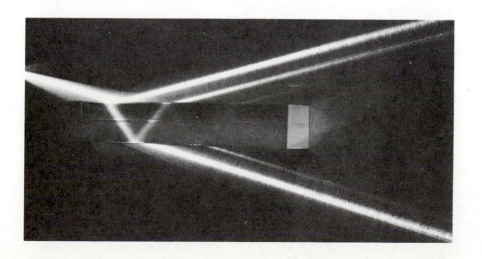

Beam of light incident on glass slab. Part of the beam is reflected and part refracted into the glass. The refracted beam is partially reflected and partially refracted at the bottom glass-air surface.

of incidence and the refracted ray is bent away from the normal, as shown in Figure 24-21.

The angle of refraction θ_2 depends on the ratio of the speeds of the waves in the two media as well as on the angle of incidence. It is related to the index of refraction of the original medium n_1, that of the second medium n_2, and the angle of incidence θ_1 by

$$n_1 \sin \theta_1 = n_2 \sin \theta_2 \qquad\qquad 24\text{-}7$$

This result was discovered experimentally in 1621 by W. Snell and is known as *Snell's law* or the *law of refraction*. If we substitute c/v for n in Equation 24-7, the law of refraction can be written

$$\frac{\sin \theta_1}{v_1} = \frac{\sin \theta_2}{v_2} \qquad\qquad 24\text{-}8$$

Equation 24-8 holds for the refraction of any kind of wave incident on a boundary surface separating two media. Figure 24-22 shows the refraction of plane waves in a ripple tank at a boundary at which the wave speed changes because the depth of the water changes.

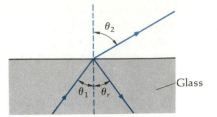

Figure 24-21
Refraction from a more dense medium to a less dense medium. Here the angle of refraction is greater than the angle of incidence, and the light ray is bent away from the normal.

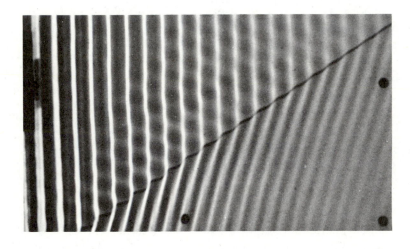

Figure 24-22
Refraction of plane waves in a ripple tank at a boundary at which the wave speed changes because the depth of water changes. Note that reflection also occurs at the boundary.

Example 24-5   Light traveling in air enters water with an angle of incidence of 45°. If the index of refraction of water is 1.33, what is the angle of refraction?

Taking $n = 1$ for air and using Equation 24-7, we obtain

$$(1.00) \sin 45° = (1.33) \sin \theta_2$$

$$\sin \theta_2 = \frac{(1.00) \sin 45°}{1.33} = \frac{(1.00)(0.707)}{1.33} = 0.53$$

The angle whose sine is 0.53 is 32°.

We can derive Snell's law using Huygens' principle as we did to derive the law of reflection. Figure 24-23 shows a plane wave incident on an air–glass surface. We apply Huygens' construction to find the wavefront of the transmitted wave. Line AP indicates a portion of the wavefront in medium 1 that strikes the glass surface at an angle of incidence θ_1. In time t, the wavelet from P travels the distance $v_1 t$ and reaches the point B on the line AB separating the two media while the wavelet from point A travels a shorter distance $v_2 t$ into the second medium. The new wavefront BB' is not parallel to the original wavefront because the speeds v_1 and v_2 are different. From the triangle APB,

$$\sin \phi_1 = \frac{v_1 t}{AB}$$

or

$$AB = \frac{v_1 t}{\sin \phi_1} = \frac{v_1 t}{\sin \theta_1}$$

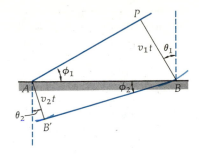

Figure 24-23
Application of Huygens' principle to the refraction of plane waves at the surface separating a medium in which the wave speed is v_1 from a medium in which the wave speed v_2 is less than v_1. The angle of refraction in this case is less than the angle of incidence.

using the fact that the angle ϕ_1 equals the angle of incidence θ_1. Similarly, from triangle $AB'B$,

$$\sin \phi_2 = \frac{v_2 t}{AB}$$

or

$$AB = \frac{v_2 t}{\sin \phi_2} = \frac{v_2 t}{\sin \theta_2}$$

where $\theta_2 = \phi_2$ is the angle of refraction. Equating the two values for AB we obtain

$$\frac{\sin \theta_1}{v_1} = \frac{\sin \theta_2}{v_2}$$

This is the same as Equation 24-8. Substituting $v_1 = c/n_1$ and $v_2 = c/n_2$ in this equation and multiplying by c, we obtain

$$n_1 \sin \theta_1 = n_2 \sin \theta_2$$

which is the same as Equation 24-7.

Figure 24-24 shows a point source in glass with rays striking the glass–air surface at various angles. All these rays are bent away from the normal. As the angle of incidence is increased, the angle of refraction increases until a critical angle of incidence θ_c is reached for which the angle of refraction is 90°. For incident angles greater than this critical angle, there is no refracted ray: that is, no waves enter the second medium, which is air in this case. All the energy is reflected. This phenomenon is called *total internal reflection*. The critical angle can be found in terms of the indexes of refraction of the two media by solving Equation 24-7 for $\sin \theta_1$ and setting θ_2 equal to 90°:

$$\sin \theta_1 = \left(\frac{n_2}{n_1}\right) \sin \theta_2$$

Setting $\theta_2 = 90°$, we obtain

$$\sin \theta_c = \frac{n_2}{n_1} \qquad\qquad 24\text{-}9$$

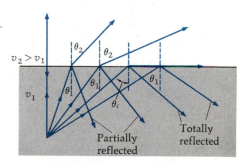

Figure 24-24
Total internal reflection. For refraction from glass or water to air, the refracted ray is bent away from the normal. As the angle of incidence is increased, the angle of refraction is increased until, at a critical angle of incidence θ_c, the angle of refraction is 90°. For angles of incidence greater than the critical angle, there is no refracted ray. All the energy is reflected.

We note that total internal reflection occurs only when the light is originally in the medium with the higher index of refraction. When light is incident on glass or water from air, the refracted ray is bent toward the normal. Even at angles of incidence near 90°, there is a refracted ray at an angle of refraction less than 90°.

Example 24-6   A particular glass has an index of refraction of $n = 1.50$. What is the critical angle for total internal reflection for light leaving this glass and entering air, which has $n = 1.00$?

We can obtain the critical angle from Equation 24-9:

$$\sin \theta_c = \frac{1.00}{1.50} = 0.667$$

The angle whose sine is 0.667 is 42°.

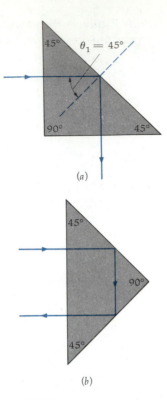

(a)

(b)

Figure 24-25a shows light incident normally on the face of a 45-45-90° glass prism. If the index of refraction of the prism is 1.5, the critical angle for total internal reflection is 42°, as found in the preceding example. Since the angle of incidence of the ray on the glass–air surface is 45°, the light will be totally reflected and will exit perpendicular to the other face of the prism, as shown in the figure. In Figure 24-25b, the light is incident perpendicular to the hypotenuse of the prism, and it is totally reflected twice such that it emerges backward at an angle of 180° to its original direction. Prisms are used to change the direction of a light ray. In binoculars, four prisms are used to reinvert the image that is inverted by the binocular lens. Total internal reflection occurs in diamonds, so all the light that enters is eventually reflected back out, giving the diamond its sparkle.

An interesting application of total internal reflection is the transmission of a beam of light down a light pipe, a long, narrow transparent glass fiber (see Figure 24-26). If the beam begins approximately parallel to the axis of the fiber, it will always strike the walls of the fiber at angles greater than the critical angle, and no light energy will be lost through the walls of the fiber. A bundle of fibers can be used for imaging, as illustrated in Figure 24-27. Fiber optics has many applications in medicine. For example, a tiny bundle of fibers can be used as a probe to examine various internal organs without surgery.

Figure 24-25
(a) Light entering through one of the short sides of a 45-45-90° glass prism is totally reflected in the prism and emerges through the other short side at 90° to the direction of the incident light. (b) Light entering through the long side of the prism is totally reflected twice and emerges in a direction opposite to that of the incident light.

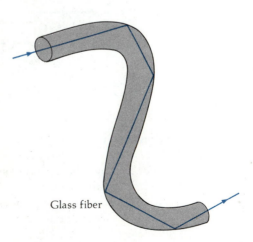

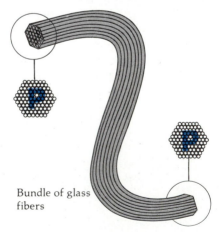

Glass fiber

Bundle of glass fibers

Figure 24-26
A light pipe. Light inside the pipe is always incident at an angle greater than the critical angle, so no light escapes the pipe by refraction.

Figure 24-27
A bundle of glass fibers. Light from the object is transported by the fibers to form an image of the object at the other end.

Because of total internal reflection, light is transmitted along a glass fiber.

When the index of refraction of a medium changes gradually, the refraction is continuous, leading to a gradual bending of the light. An interesting example of this is the formation of a mirage. On a hot day, there is often a layer of very hot air near the ground. This air is warmer than the air just above it and is therefore less dense. The speed of light is slightly greater in this less dense layer than in the layer above, and a light beam passing through it is therefore bent. Figure 24-28a shows light from a tree when all the air is at the same temperature. The wavefronts are spherical and the rays are straight lines. In Figure 24-28b, the air near the ground is warmer, resulting in a greater speed of light there. The portions of the wavefronts near the ground travel faster and get ahead of the higher portions, resulting in nonspherical wavefronts and a curving of the rays. The ray shown heading toward the ground in Figure 24-28a is bent back upwards in Figure 24-28b. A viewer thus sees an inverted image of the tree and thinks that the light has been reflected from the ground. The viewer attributes this reflection to a water surface near the tree. When driving on a very hot day, you may have noticed apparent wet spots on the highway that disappear when you get to them. This is due to the refraction of light from the hot air layer near the pavement.

An optical fiber consists of a core with a high index of refraction coated by a material with a lower index of refraction. Fiber optics is beginning to revolutionize telephone communications. Hundreds of messages can be transmitted simultaneously via a light beam through a single fiber.

Figure 24-28
A mirage. (a) When the air is at a uniform temperature, the wavefronts of the light from the tree are spherical. (b) When the air near the ground is warmer, the wavefronts are not spherical and the light from the tree is continuously refracted into a curved path. Because an inverted image of the tree is seen, the viewer thinks there is a body of water in front of the tree.

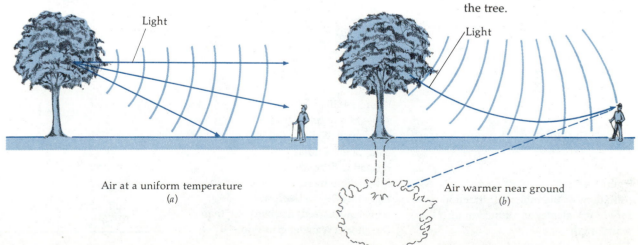

Light

Air at a uniform temperature
(a)

Light

Air warmer near ground
(b)

An unusual mirage produced by heat from the wall of a building. Light rays from Jay that are originally headed toward the wall are refracted by the warm air near the wall and appear to be coming from the image behind the wall.

Table 24-2 lists indexes of refraction for various transparent materials. These values are for sodium light of wavelength 589 nm. The index of refraction of a material has a slight dependence on wavelength. Figure 24-29 shows this dependence for several materials. We see that for these materials, the index of refraction decreases slightly as the wavelength increases. This dependence of the index of refraction on wavelength (and therefore on frequency) is called *dispersion*. When a beam of light is incident at some angle on a glass surface, the angle of refraction for shorter wavelengths (toward the violet end of the visible spectrum) is slightly larger than that for the longer wavelengths (toward the red end of the spectrum). Violet light is therefore bent more than red light, and a beam of white light is spread out or dispersed into its component colors or wavelengths (see Figure 24-30). The formation of a rainbow is a familiar example of the dispersion of sunlight by refraction in water droplets.

Table 24-2

Index of Refraction for Yellow Sodium Light ($\lambda = 589$ nm)

Substance	Index of refraction
Solids	
Ice (H_2O)	1.309
Fluorite (CaF_2)	1.434
Rock salt (NaCl)	1.544
Quartz (SiO_2)	1.544
Zircon ($ZrO_2 \cdot SiO_2$)	1.923
Diamond (C)	2.417
Glasses (typical values)	
Crown	1.52
Light flint	1.58
Medium flint	1.62
Dense flint	1.66
Liquids at 20°C	
Methyl alcohol (CH_3OH)	1.329
Water (H_2O)	1.333
Ethyl alcohol (C_2H_5OH)	1.36
Carbon tetrachloride (CCl_4)	1.460
Turpentine	1.472
Glycerine	1.473
Benzene	1.501
Carbon disulfide (CS_2)	1.628

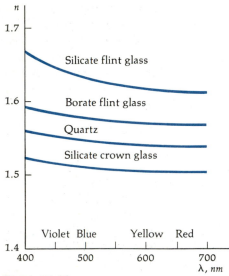

Figure 24-29
Variations of the index of refraction for various glasses as a function of wavelength.

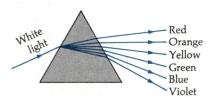

Figure 24-30
A beam of white light incident on a glass prism is dispersed into its component colors. The index of refraction decreases as the wavelength increases, as shown in Figure 24-29, so the longer wavelengths (red) are bent less than the shorter wavelengths (violet).

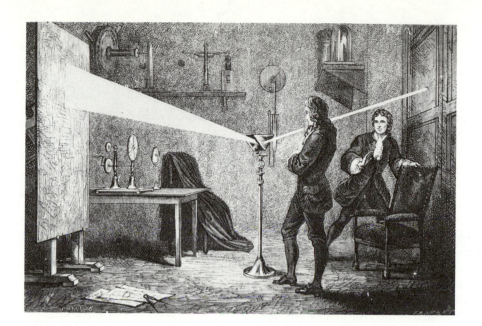

Newton demonstrating the dispersion of sunlight with a glass prism.

24-6   Polarization

In any transverse wave, the vibration is perpendicular to the direction of propagation of the wave. For example, for waves moving down a string, the elements of the string move in a plane perpendicular to the string. Similarly, in a light wave traveling in the z direction, the electric field is in a plane parallel to the xy plane. (The magnetic field in a light wave is also perpendicular to the z direction.) If the vibration in a transverse wave remains parallel to a fixed line in space, the wave is said to be *linearly polarized*.

We can visualize polarization most easily by considering mechanical waves in a string. If one end is moved up and down, the resulting waves in the string are linearly polarized, and each element of the string vibrates in the vertical direction. Similarly, if the string is moved along a horizontal line (perpendicular to the string), the displacements of the string are polarized in the horizontal direction. If the end of the string is moved with constant speed in a circle, the resulting wave is said to be *circularly polarized;* in this case, each element of the string moves in a circle. Unpolarized waves can be produced by moving the end of the string vertically and horizontally in a random way. Then, assuming the string to be in the z direction, the vibrations will have both x and y components that vary in a random way.

Most waves produced by a single source are polarized. For example, string waves produced by the regular vibration of one end of a string or electromagnetic waves produced by a single atom or by a single antenna are polarized. Waves produced by many sources are usually unpolarized. A typical light source, for example, contains millions of atoms acting independently. The electric field for such a wave can be resolved into x and y components that vary randomly because there is no correlation between the individual atoms producing the light.

There are four phenomena that produce polarized light from unpolarized light. They are as follows:

Methods of polarization

1. Absorption

2. Reflection

3. Scattering

4. Birefringence (also called double refraction)

Polarization by Absorption

A common method of polarization is the absorption of light in a sheet of polarizing film. (The most common type of polarizing film is a commercial material called Polaroid, invented by E. H. Land in 1938.) This material contains long-chain hydrocarbon molecules that are aligned when the sheet is stretched in one direction during the manufacturing process. These chains become conducting at optical frequencies when the sheet is dipped in a solution containing iodine. When light is incident on the sheet with its electric field vectors parallel to the chains, electric currents are set up along the chains and the light energy is absorbed. If the electric field vectors are perpendicular to the chain, the light is transmitted. The direction perpendicular to the chains is called the *transmission axis.* We will make the simplifying assumptions that all the light is transmitted when the electric field vectors are parallel to the transmission axis and that all is absorbed when they are perpendicular to the transmission axis.

Consider a light beam in the z direction incident on a polarizing film that has its transmission axis in the y direction. On the average, half of the incident light has its electric field in the y direction and half in the x direction. Thus, half the intensity is transmitted, and the transmitted light is linearly polarized with its electric field in the y direction.

Suppose we have a second piece of polarizing material whose transmission axis makes an angle θ with that of the first, as shown in Figure 24-31. If **E** is the electric field between the films, its component along the direction of the transmission axis of the second piece of polarizing material is $E \cos \theta$. Since the intensity of light is proportional to E^2, the intensity of the light transmitted by both films is given by

$$I = I_0 \cos^2 \theta \qquad\qquad 24\text{-}10 \quad \text{Malus' law}$$

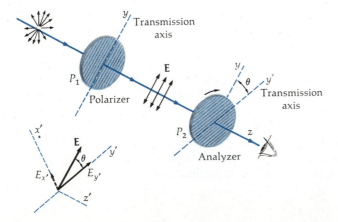

Figure 24-31
Two polarizing films with their transmission directions making an angle θ with each other. Only the component $E \cos \theta$ is transmitted through the second film. If the intensity between the films is I_0, then that transmitted by the second film is $I_0 \cos^2 \theta$.

where I_0 is the intensity incident on the second film. I_0 is, of course, half the intensity incident on the first film. When two polarizing elements are placed in succession in a beam of light as described here, the first is called the *polarizer* and the second, the *analyzer.* If the polarizer and analyzer are crossed, that is, if they have their axes perpendicular to each other, no light gets through if the absorption is complete. Equation 24-10 is known as *Malus' law,* after its discoverer E. L. Malus (1775 – 1812). It applies to any two polarizing elements whose transmission directions make an angle θ with each other.

Crossed polarizers from two pairs of sunglasses block out all of the light.

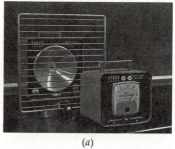
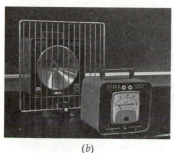

(a)   (b)

Polarization of microwaves. The electric field of the microwaves is vertical, parallel to the vertical dipole radiator. (a) When the metal wires are horizontal (transmission axis vertical), the waves are transmitted, as indicated by the high reading on the detector. (b) When the wires are vertical (transmission axis horizontal), the waves are absorbed, as indicated by the low reading on the detector.

A mechanical analogy is sometimes useful in helping us to visualize polarization. If we vibrate one end of a long rope back and forth and up and down in a random way, we produce unpolarized mechanical waves in the rope. If the rope passes through a long vertical slot, such as between the slats in a picket fence (see Figure 24-32), the horizontal vibrations are absorbed and only the vertical vibrations are transmitted. Thus, on the other side of the slot, we have waves polarized in the vertical direction.

Example 24-7   Unpolarized light of intensity 3.0 W/m² is incident on two polarizing films whose transmission axes make an angle of 60°. What is the intensity of light transmitted by both films?

The intensity of the light that is transmitted by the first film is half the incident intensity or 1.5 W/m². Calling this intensity I_0, that transmitted by the second film is

$$I = I_0 \cos^2 \theta = (1.5 \text{ W/m}^2) \cos^2 60°$$

$$= (1.5 \text{ W/m}^2)(0.500)^2 = 0.375 \text{ W/m}^2$$

Figure 24-32
Unpolarized waves in a rope are polarized when they pass through a vertical slot because the horizontal vibrations are not transmitted.

Polarization by Reflection

When unpolarized light is reflected from a plane surface boundary between two transparent media, such as air and glass or air and water, the reflected light is partially polarized. The degree of polarization depends on the angle of incidence and the indexes of refraction of the two media. At one particular angle of incidence, called the polarizing angle θ_p, the reflected light is completely polarized. The polarizing angle is related to the indexes of refraction of the two media by

$$\tan \theta_p = \frac{n_2}{n_1}$$ 24-11   Brewster's law

This result was discovered experimentally by Sir David Brewster in 1812 and is known as *Brewster's law*. The transmitted light is only slightly polarized because only a small fraction of the incident light is removed from the beam by reflection.

When light is incident at the polarizing angle, the reflected and refracted rays are perpendicular to each other (see Figure 24-33). (This can be derived by combining Brewster's law and Snell's law; see Problem 6.) The electric field of the incident light can be resolved into components parallel to and perpendicular to the plane of incidence. The reflected light is completely polarized with its electric field perpendicular to the plane of incidence.

Because of the polarization of reflected light, sunglasses made of polarizing material can be very effective in cutting out glare. If light is reflected from a horizontal surface, such as a lake or snow on the ground, the plane of incidence will be vertical and the electric field of the reflected light will be predominantly horizontal. Polarized sunglasses with their transmission axis vertical will then reduce the glare by absorbing much of the reflected light. If you have polarized sunglasses, you can demonstrate this by looking at reflected light and rotating the glasses 90°, at which point much more of the reflected light is transmitted.

Polarization by Scattering

The phenomenon of absorption and reradiation is called *scattering*. A familiar example of light scattering is that from clusters of air molecules (due to random fluctuations in the density of air), which tend to scatter short wavelengths more than long wavelengths, thereby giving the sky its blue color. Scattering can be demonstrated by adding a small amount of powdered milk to a container of water. The milk particles absorb light and reradiate it. Laser beams can be made visible by introducing chalk or smoke particles into the air to scatter the light to your eye.

We can understand polarization by scattering if we think of an absorbing molecule as an electric-dipole antenna that radiates waves with maximum intensity in the directions perpendicular to the antenna. The electric field vectors will be parallel to the antenna and the intensity will be zero in the direction along the axis of the antenna. Figure 24-34 shows a beam of unpolarized light traveling along the z axis and striking a scattering center at the origin. The electric field in the light beam has components in both the x and y directions perpendicular to the direction of motion of the light beam. These fields set up oscillations in the scattering center in both the x and y directions, but there is no oscillation in the z direction. The oscillations of the scattering center in the x direction produce light along the y axis but not along the x axis, which is along the line of oscillation. The light radiated along the y axis is thus polarized in the x direction. Similarly, the light radiated along the x axis is polarized in the y direction. This is easily demonstrated by examining the scattered light with a piece of polarizing film.

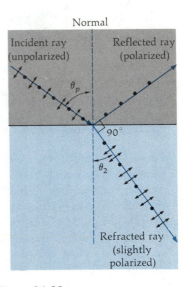

Figure 24-33
Polarization by reflection. The incident wave is unpolarized and has electric-field components both parallel to the plane of incidence (arrows) and perpendicular to this plane (dots). For incidence at the polarizing angle, the reflected wave is completely polarized, with its electric field perpendicular to the plane of incidence. Waves reflected from a horizontal surface will be polarized in the horizontal plane and will not be transmitted by polarized sunglasses with a vertical transmission axis.

Polarization by Birefringence

Birefringence, or double refraction, is a complicated phenomenon that occurs in calcite and other noncubic crystals and in some stressed plastics such as cellophane. In most materials, the speed of light is the same in all directions. Such materials are *isotropic*. Because of their atomic structure, birefringent materials are *anisotropic*. The speed of light and other optical properties depend on the direction of propagation of light through the material. When a light ray is incident on such materials, it may be separated into two rays called the *ordinary ray* and the *extraordinary ray*. These rays are polarized in mutually perpendicular directions, and they travel with different speeds. Depending on the relative orientation of the material and the incident light, the rays may also travel in different directions. The ordinary ray obeys Snell's law of refraction, but the extraordinary ray does not.

There is one particular direction in a birefringent material in which both rays propagate with the same speed. This direction is called the *optic axis* of the material. Nothing unusual happens when light travels along the optic axis. (The optic axis is actually a direction rather than a line or axis in the material.) However, when light is incident at an angle to the optic axis, as shown in Figure 24-35, the rays travel in different directions and emerge separated in space. If the crystal is rotated, the extraordinary ray (labeled the *e* ray in the figure) rotates in space.

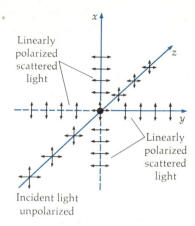

Figure 24-34
Polarization by scattering. Unpolarized light propagating in the *z* direction is incident on a scattering center at the origin. The light scattered in the *x* direction is polarized in the *y* direction and that scattered in the *y* direction is polarized in the *x* direction.

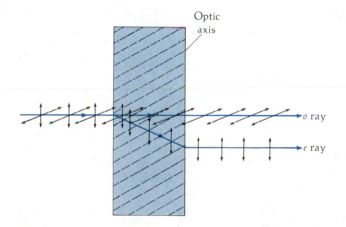

Figure 24-35
A narrow beam of light incident on a birefringent crystal such as calcite is split into two beams called the ordinary ray (*o*) and the extraordinary ray (*e*), which have mutually perpendicular polarizations. If the crystal is rotated, the extraordinary ray rotates in space.

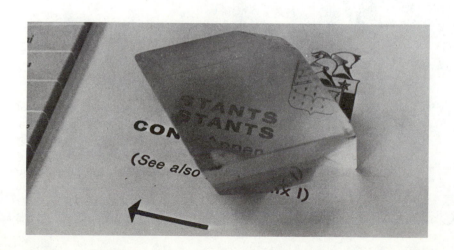

Double image produced by birefringent calcite crystal. (*From Richard T. Weidner and Robert L. Sells*, Elementary Classical Physics, *vol. 2, 2d ed., fig. 36.17, p. 743. Copyright © 1973 by Allyn and Bacon, Inc. Used with permission.*)

If light is incident on a birefringent material perpendicular to the crystal face and perpendicular to the optic axis, the two rays travel in the same direction but at different speeds. If the incident light is polarized, as in Figure 24-36, the direction of polarization of the emerging light is rotated relative to that of the incident light. The amount of rotation depends on the thickness of the material and on the wavelength of the light. Suppose we have two pieces of polarizing film with their transmission axes perpendicular to each other (crossed polarizing films). Ordinarily, no light is transmitted through both films because the polarization direction of light transmitted by the first film is perpendicular to the axis of the second film. However, if we place a birefringent material between the crossed films, the direction of polarization is rotated and some light gets through both films. Interesting and beautiful patterns are produced because light of certain colors (depending on the thickness of the material) is transmitted due to this rotation.

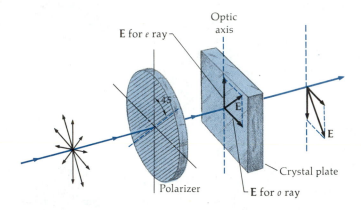

Figure 24-36
Polarized light emerging from the polarizer is incident on a birefringent crystal such that the electric field vector makes a 45° angle with the optic axis that is perpendicular to the light beam. The ordinary (*o*) and extraordinary (*e*) rays travel in the same direction but at different speeds. When the light emerges, its polarization direction has been rotated.

Various glasses and plastics become birefringent when they are under stress. These stress patterns can be observed when the materials are placed between crossed polarizing films (see Figure 24-37).

Figure 24-37
When under stress, the piece of plastic becomes birefringent and rotates the plane of polarization of light. When viewed between crossed polarizing films, the stress pattern is visible. A plastic model can thus be used to determine stress patterns in an object of similar shape.

The New Age of Exploration: Voyagers to the Outer Solar System

Donald Goldsmith

In 1977, two automated spacecraft, Voyager 1 and Voyager 2, were launched from the Kennedy Space Center in Florida, beginning the most successful voyages of exploration in human history. To find anything comparable, we must look back to the great expeditions of the 1570s and 1580s by explorers such as Sir Francis Drake and Sir Walter Raleigh that opened the "new world" of the Americas to colonization.

To make their epic journeys, the Voyagers have relied on two key sources of acceleration: the thrust of their launch rockets and the gravitational forces of the very planets they were sent to explore. Since it takes considerable rocket power to escape Earth's gravitational force, we may think of gravity as "hostile" to spaceflight. But a planet's gravitational force can *help* a spacecraft travel deep into space, not during the initial launch but later, by changing its trajectory and accelerating it without the need for a new supply of rocket fuel.

When a spacecraft coasting through the solar system approaches a planet, say Jupiter, two types of trajectories are possible. The spacecraft can fall *into* the planet, or it can fall *around* the planet, the way a comet "falls" around the sun. When a spacecraft falls *around* a planet, the planet's gravitational force pulls the spacecraft toward the planet, accelerating it in the process. Because of the precise direction and speed of the spacecraft's approach, however, it curves around the planet and back out into space—along a different trajectory than it would have had the planet not been there. Astronavigators call this use of a planet's gravity to accelerate and redirect a spacecraft the "planetary slingshot."

Voyager 1 and Voyager 2 were launched from earth in the plane of the planets' orbits around the sun at a time when the sun's four giant planets—Jupiter, Saturn, Uranus, and Neptune—were all approximately "lined up" on the same side of the sun, an event that occurs only once every few centuries. The Voyager spacecraft could thus use each planetary encounter to extend their journeys. The Voyagers first travelled about a billion kilometres to reach Jupiter, 5.2 times further than the earth from the sun, in 1979. Jupiter's "slingshot" boost helped propel them outward nearly twice as far, to Saturn, 9.5 times further than the earth from the sun. Voyager 1 arrived at Saturn in November 1980 and Voyager 2, in August 1981. Voyager 1 then headed into deep space without passing close to any more planets, but Voyager 2, carefully directed from earth, used Saturn's gravitational force to redirect its trajectory towards Uranus, the sun's seventh planet, three billion kilometres from the Earth. By the time of Voyager 2's encounter with Uranus in January 1986, the spacecraft had traveled so far that the radio messages carrying photographs and other information took nearly three hours to reach the earth. The Voyagers could never have carried enough fuel to make the sweeping turns required to visit one planet after another. But thanks to the careful work of NASA's scientists and engineers, who used the small fuel supply aboard the spacecraft to line up each

Montage of the rings and satellites of Saturn prepared from images taken by Voyager 1 in November 1980.

Essay: The New Age of Exploration (continued)

planetary encounter perfectly, and thanks to the planets' gravitational "slingshots," the Voyagers could make a "grand tour" of the giant planets, a trip that will not be repeated in our lifetimes.

What did the Voyagers discover? Nothing more than a host of new satellite worlds never before seen by human beings, plus new information about the giant planets that shed light on their origins, composition, and atmospheric structure. Jupiter has four large satellites—Io, Europa, Ganymede, and Callisto—named after mythological lovers of its namesake, the king of gods. From the Voyagers, we know that these four satellites, two of them close in size to our moon and two of them much larger, differ markedly from each other.

Io has perhaps the most amazing exterior of any satellite in the solar system, a cratered, orange-red surface mottled with whitish patches. But Io's craters are not the result of impact, as are the craters on the moon. Instead, Io's craters are volcanic. Indeed, it is the only place in the solar system besides the earth where active volcanoes are known to exist. Sulfur-rich material spewn from these volcanoes continually renews the crust of Io, so the surface we see has been there for "only" the past million years or so.

Europa is the smallest, most highly reflecting, and least pockmarked of Jupiter's large moons. Its surface is scored with hundreds of intersecting lines, many of them thousands of kilometres in length. The scientists who interpreted the Voyager photographs concluded that Europa likely has a crust of ice, water frozen forever at more than a hundred degrees below zero on the Celsius scale. But beneath Europa's icy crust, it is possible that a global ocean exists, protected from freezing by the outside crust and warmed by Europa's internal heat. If this is so, this European ocean is one of the most likely sites in the solar system for extraterrestrial life.

Ganymede, the largest and most massive satellite in the solar system, larger even than the planet Mercury, resembles our moon in its cratered surface. One huge basin, nearly 3000 kilometres across, may be the remnant of the impact of a huge object that pierced the soft crust of Ganymede billions of years ago. The surface is also scored by vast networks of thousand-kilometre long, parallel grooves, some of them with offsets along their length that testify to the plate-tectonic motions of the satellite's crust, similar to those which move the plates of the earth's crust.

Callisto, the outermost and least dense of the four large Jovian satellites, has the most heavily cratered surface in the solar system, a surface so pocked and pitted that there ap-

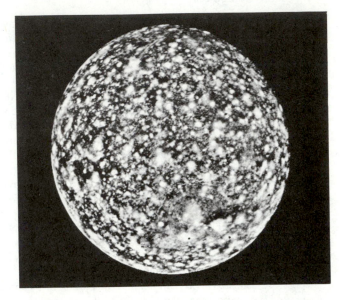

Photograph of Callisto taken by Voyager 2 from about 1.1 million kilometres.

pears to be no room for a single crater more. But, mysteriously, this moon's surface is distinguished for being flat. Unlike the crater rims on the moon, the ridges of the craters on Callisto rise no more than a hundred metres or so above the surface. This suggests that most of the craters on Callisto were formed before the moon's surface had become rigid; that is, they date from close to the time of the formation of the solar system, 4.6 billion years ago.

And what about Jupiter itself? Voyagers 1 and 2 photographed the planet, hour after hour, day after day, producing a mammoth movie of the giant planet, full of fascinating new information. Jupiter rotates more rapidly than any other planet, once every 10 hours, and has more variegated colors, mostly subtle hues of red and brown, than any of the other giant planets. Jupiter is entirely gaseous, or nearly so. A relatively small solid core may exist beneath the thousands of kilometres of "atmosphere," mostly hydrogen and helium, that comprises the bulk of the planet. The colors that distinguish the planet may arise either from inorganic compounds, such as those of sulfur and phosphorus, or from organic (carbon-containing) materials. More detailed observations are needed to distinguish between these two possibilities. Jupiter is slowly contracting, giving it an internal source of heat, like a star in formation. But Jupiter has too little mass to

Photograph of Jupiter showing the great red spot taken by Voyager 2.

become a star. We may imagine the sun's largest planet to be a star that failed, a would-be solar companion, with a thousandth the mass of the sun or about 318 times the mass of the earth.

Saturn, the second-largest and most beautiful of the planets, is famous for its rings. Saturn has a fantastic broad and flat set of rings, spread out over 50,000 kilometres. We now know them to consist of millions upon millions of solid particles, ranging in size from great boulders to microscopic bits of dust, that orbit the planet in the same plane and the same direction. Saturn's rings are full of amazing variations, with changing, spokelike patterns that are the result of subtle gravitational interactions among the various particles that compose them. The Voyagers photographed these rings in exquisite detail, providing "ring theorists" with grist for their calculations—calculations that also reveal new facts about the motions of the stars in the disk of the Milky Way galaxy, a gravitationally interacting system of billions of particles that is analogous to Saturn's rings but on a scale a hundred million times larger.

Saturn has a single large satellite—Titan, which is almost as large as Ganymede—and a host of smaller satellites, none of them more than 1500 kilometers in diameter. Before the Voyager encounters, nine moons of Saturn had been found

during the three centuries of telescopic observation. The Voyager spacecraft discovered an additional dozen smaller moons. Titan is unique among satellites. It is the only one with a thick atmosphere, which completely hides the surface of the satellite from view. Titan's atmosphere consists mostly of nitrogen—just as earth's does. In addition, Voyager found that it contains hydrocarbon compounds, such as ethane, propane, hydrogen cyanide, cyanogen, and cyanoacetylene. Titan thus offers us what the planetary expert Tobias Owen has called "an immense natural laboratory free of experimental bias," a place where we could see whether roughly earthlike conditions, though at much lower temperatures than earth's, might have led to the formation of complex organic molecules—or even life itself.

Far beyond Saturn, the "smaller" giant planets, Uranus and Neptune, orbit the sun in slow majesty: Uranus once in 84 years, Neptune once in 185. Before Voyager 2 reached Uranus, we knew only that it and Neptune are near twins in size (about four earth diameters) and mass (about 15 earth masses). Since Uranus, like Jupiter and Saturn, has a completely gaseous outer layer, our views of it are limited to swirls of muted color. But the five larger moons of Uranus—Miranda, Ariel, Umbriel, Titania, and Oberon—provided yet another set of images of startling diversity. In addition, Voyager 2 discovered ten small moons orbiting Uranus.

At the time of the encounter with Uranus, Voyager 2 had been on its journey for nearly nine years. Though a bit stiff in its motions, it was still functioning better than anyone expected. The Voyager scientists, ever inventive, figured out how to program the spacecraft's scans of Uranus and its moons so that the images sent to earth were even *better* than those from the Jupiter and Saturn encounters. It verified that the outer layers of Uranus consist mainly of hydrogen and helium, at a temperature of only about −218°C. The spacecraft also obtained new images of Uranus' rings. These were first detected by telescope in 1977. Like some of Saturn's rings, the rings of Uranus are kept in well-defined orbits by small "shepherd" moons.

Some 12 years after launch, Voyager 2 will fly by Neptune and its satellites in late August of 1989. The Neptune observations, which will include photographs of Neptune's large satellite Triton (almost as large as Ganymede and Titan) will be the culminating event in a career that already boasts the record for planets observed at close range (three), satellites photographed in detail (nineteen), and satellites discovered (nineteen, also counting those discovered by Voyager 1). From Neptune, which is almost twice as far from the earth as Uranus, Voyager 2's signals will require six hours to reach us.

Essay: The New Age of Exploration (continued)

These signals are generated by a tiny nuclear power plant that emits no more energy than a household light bulb but has already provided the power for more than ten thousand separate images and a host of other scientific observations.

The success of Voyager 1 and Voyager 2, which has exceeded even the boldest hopes of the scientists and engineers responsible for the Voyager missions, underscores the usefulness of automated spacecraft in the exploration of the solar system. The automated Voyagers survived years of travel through the unimaginable cold of space and were ready to function once they reached their destinations. As the Space Shuttle tragedy in January 1986 demonstrated, human fragility makes the exploration of the solar system *in person* a dangerous business, fraught with potential disaster. When Ferdinand Magellan or Francis Drake circumnavigated the globe during the sixteenth century, it was simply expected that some members of the crew would perish on the journey. (In the case of Magellan's expedition, this included Magellan himself.) Today, we place a higher worth on human life and cannot easily justify losing a certain number of brave astronauts to achieve relatively limited space-exploration capabilities. For this reason, automated spacecraft seem the best means of exploring our cosmic environment, at least until we gain greater confidence in our launch vehicles and determine a greater need for human beings in space. We should not forget that whenever an automated probe such as the Voyager spacecraft visits other planets and their moons, *we* are there, too.

Thanks to the Voyager spacecraft—and the people who developed them and made them work—we know the outer solar system as never before. During the 1990s, we hope that we will investigate the outer planets in still more detail, building on the Voyager results to achieve ever-increasing knowledge. The Voyagers have shown the way to explore the outer planets—and they're not done yet.

Summary

1. Light consists of electromagnetic waves with wavelengths in the range from about 400 nm (violet) to 700 nm (red). Electromagnetic waves with wavelengths shorter than those of visible light include ultraviolet radiation, x-rays, and gamma rays. Electromagnetic waves with wavelengths longer than those of visible light include infrared radiation, microwaves, radio waves, and television waves.

2. Electromagnetic waves are transverse waves with oscillating electric and magnetic fields that are perpendicular to each other and to the direction of propagation. They are produced when electric charges accelerate, and they travel through a vacuum with a speed of about 3×10^8 m/s. The intensity of an electromagnetic wave is proportional to E^2.

$$I = \frac{\text{Power}}{\text{Area}} = c\epsilon_0 E_{\text{rms}}^2$$

3. Oscillating charges in an electric-dipole antenna radiate electromagnetic waves with an intensity that is maximum in directions perpendicular to the antenna and zero in the direction along the axis of the antenna. In the direction perpendicular to the antenna and far away from it, the electric field of the electromagnetic wave is parallel to the antenna.

4. When light is incident on a surface separating two media in which the speed of light differs, part of the light energy is transmitted and part is reflected. The angle of reflection equals the angle of incidence. The angle of refraction depends on the angle of incidence and on the indexes of refraction of the two media and is given by Snell's law:

$$n_1 \sin \theta_1 = n_2 \sin \theta_2$$

where the index of refraction of a medium n is the ratio of the speed of light in a vacuum c to that in the medium v:

$$n = \frac{c}{v}$$

5. When light is traveling in a medium with an index of refraction n_1 and is incident on the boundary of a second medium with a lower index of refraction $n_2 < n_1$, the light is totally reflected if the angle of incidence is greater than the critical angle θ_c given by

$$\sin \theta_c = \frac{n_2}{n_1}$$

6. The speed of light in a medium and, therefore, the index of refraction of that medium depend slightly on the wavelength of light, a phenomenon known as dispersion. Because of dispersion, a beam of white light incident on a refracting prism is dispersed into its component colors.

7. The four phenomena that produce polarized light from unpolarized light are

1. Absorption

2. Reflection

3. Scattering

4. Birefringence

When two polarizers have their transmission axes at an angle θ, the intensity transmitted by the second polarizer is reduced by the factor $\cos^2 \theta$, a result known as Malus' law. If I_0 is the intensity of light between the polarizers, the intensity transmitted by the second polarizer is

$$I = I_0 \cos^2 \theta$$

Suggestions for Further Reading

Boyle, W. S.: "Light-Wave Communications," *Scientific American*, August 1977, p. 40.

The physics and technology of a telephone system that transmits signals via light pulses carried along optical fibers are discussed in this article.

"Light," *Scientific American* special issue, September 1968.

How light interacts with both living and nonliving matter, how images are formed, vision, and laser light are some of the topics covered in this issue.

Walker, Jearl: "The Amateur Scientist: Studying Polarized Light with Quarter-Wave and Half-Wave Plates of One's Own Making," *Scientific American*, December 1977, p. 172.

Walker, Jearl: "The Amateur Scientist: More about Polarizers and How to Use Them, Particularly for Studying Polarized Sky Light," *Scientific American*, January 1978, p. 132.

These two articles are instructive even if one doesn't choose to repeat the experiments.

Wehner, Rüdiger: "Polarized-Light Navigation by Insects," *Scientific American*, July 1976, p. 106.

This article describes how ants and bees are able to use the natural polarization of sky light as an aid to navigation.

Review

A. Objectives: After studying this chapter, you should:

1. Be able to state the range of wavelengths in the visible spectrum.

2. Be able to state the value of the speed of light in a vacuum.

3. Be able to state the law of reflection and Snell's law of refraction.

4. Be able to derive an expression relating the critical angle for total internal reflection to the index of refraction of a substance.

5. Be able to list the four means of producing polarized light from unpolarized light.

6. Be able to state Malus' law and use it in problems involving the transmission of polarized light through a polarizer.

7. Be able to state Brewster's law for polarization by reflection.

B. Define, explain, or otherwise identify:

Photon, p. 609

Electromagnetic wave, p. 609
Electric-dipole antenna, p. 610
Intensity, p. 612
Plane of incidence, p. 617
Law of reflection, p. 617
Index of refraction, p. 617
Snell's law, p. 620
Malus' law, p. 626
Brewster's law, p. 628
Scattering, p. 628
Birefringence, p. 629

C. True or false: If the statement is true, explain why. If it is false, give a counterexample.

1. Light and radio waves travel with the same speed through a vacuum.

2. An electric-dipole antenna radiates energy uniformly in all directions.

3. The angle of refraction of light is always less than the angle of incidence.

4. Longitudinal waves cannot be polarized.

Exercises

Section 24-1   Waves or Particles;
Section 24-2   Electromagnetic Waves;
Section 24-3   The Speed of Light

1. What is the frequency of a 3-cm microwave?

2. What is the frequency of an x-ray of wavelength 0.1 nm?

3. Find the wavelength for (a) a typical AM radio wave of frequency 1000 kHz and (b) a typical FM radio wave of frequency 1000 MHz.

4. An electromagnetic wave has intensity 100 W/m². Find (a) the average energy density, (b) E_{rms}, and (c) B_{rms}.

5. The rms value of the electric field in an electromagnetic wave is $E_{rms} = 400$ V/m. Find (a) B_{rms}, (b) the average energy density, and (c) the intensity.

6. Show that the units of $E = cB$ are consistent; that is, when B is in teslas and c is in metres per second, the units of cB are volts per metre or newtons per coulomb.

7. The root-mean-square value of the magnetic field in an electromagnetic wave is $B_{rms} = 0.245\ \mu$T. Find (a) E_{rms}, (b) the average energy density, and (c) the intensity.

8. A laser beam has a diameter of 1.0 mm and average power of 1.5 mW. Find (a) the intensity of the beam in W/m², (b) E_{rms}, and (c) B_{rms}.

9. The intensity of sunlight striking the upper atmosphere is 1.4 kW/m². Find (a) E_{rms} and (b) B_{rms} due to this sunlight at the upper atmosphere.

10. The spiral galaxy in the Andromeda constellation is about 2×10^{19} km away from us. Find this distance in light years.

11. On a rocket sent to Mars to take pictures, the camera is triggered by radio waves. What is the time delay between sending and receiving the signal from the earth to Mars, a distance of 9.7×10^{10} m?

12. The distance from a point on the surface of the earth to one on the surface of the moon is measured by aiming a laser light beam at a reflector on the surface of the moon and measuring the time required for the round trip. The uncertainty in the measured distance, Δx, is related to the uncertainty in the time, Δt, by $\Delta x = c \Delta t$. If time intervals can be measured to ± 1.0 ns, find the uncertainty in the distance in metres.

Section 24-4   Reflection

13. Calculate the fraction of light energy reflected from an air–water surface at normal incidence ($n = 1.33$ for water).

14. Light is incident normally on a slab of glass with index of refraction $n = 1.5$. Reflection occurs at both surfaces of the slab. What percentage of the incident light energy is transmitted by the slab?

Section 24-5   Refraction

15. Light in air is incident at an angle of $30°$ on the surface of a slab of rock salt. The angle of refraction is $18.9°$. (a) What is the index of refraction of rock salt? (b) What would be the angle of refraction for an incident angle of $60°$?

16. Light in air is incident at an angle of $45°$ on a piece of dense flint glass. The angle of refraction is $25.2°$. Find (a) the index of refraction of the flint glass and (b) the angle of refraction if the incident angle is $60°$.

17. The index of refraction of water is 1.33. Find the angle of refraction of a beam of light in air hitting the water surface at an angle of incidence of (a) $20°$, (b) $30°$, (c) $45°$, and (d) $60°$, and indicate the refracted rays on a diagram.

18. What is the critical angle for total internal reflection for light traveling from water ($n = 1.33$) to air ($n = 1.00$)?

19. Repeat Exercise 17 for a beam of light initially in water and incident on a water–air surface at the same angles.

20. A slab of glass with index of refraction 1.50 is submerged in water with index of refraction 1.33. Light in the water is incident on the glass. Find the angle of refraction if the angle of incidence is (a) $60°$, (b) $45°$, and (c) $30°$.

21. A glass surface ($n = 1.50$) has a layer of water ($n = 1.33$) on it. Light in the glass is incident on the glass–water surface. Find the critical angle for total internal reflection.

22. Repeat Exercise 20 for a beam of light initially in the glass incident on the glass–water surface at the same angles.

23. In Figure 24-38, light is initially in a medium (such as air) with index of refraction n_1. It is incident at angle θ_1 on the surface of a liquid (such as water) with index of refraction n_2. The light passes through the layer of water and enters glass with index of refraction n_3. If θ_3 is the angle of refraction in the glass, show that $n_1 \sin \theta_1 = n_3 \sin \theta_3$. That is, show that the second medium can be neglected when finding the angle of refraction in the third medium.

Figure 24-38
Light in a medium with index of refraction n_1 incident on a liquid with index n_2 and then entering a third medium with index n_3 for Exercise 23.

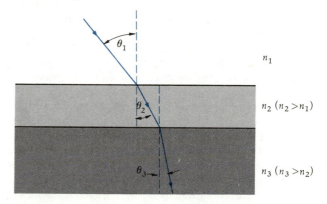

24. Find the speed of light in water ($n = 1.33$) and in glass ($n = 1.50$).

25. A beam of red light of wavelength 700 nm enters water. What is the wavelength in water?

26. The critical angle for total internal reflection in diamond is about $24°$. What is the index of refraction of diamond?

27. The index of refraction for silicate flint glass is 1.66 for light of wavelength 400 nm and 1.61 for light of wavelength 700 nm. Find the angles of refraction for light of these wavelengths incident on this glass at an angle of $45°$.

Section 24-6   Polarization

28. Two polarizing sheets have their transmission directions crossed so that no light gets through. A third sheet is inserted between the two such that its transmission direction makes an angle θ with that of the first sheet. Unpolarized light of intensity I_0 is incident on the first sheet. Find the intensity of light transmitted through all three sheets if (a) $\theta = 45°$ and (b) $\theta = 30°$.

29. Two polarizing sheets are crossed as in Exercise 28. Two additional sheets are inserted between the original two such that the transmission direction of each sheet makes an angle of $30°$ with that of the sheet next to it. Find the intensity of light transmitted by all four sheets if unpolarized light of intensity I_0 is incident on the first one.

30. The polarizing angle for reflection for a certain substance is $60°$. (a) What is the angle of refraction of light incident at $60°$? (b) What is the index of refraction of this substance?

31. The critical angle for total internal reflection for a substance is $45°$. What is the polarizing angle for reflection for this substance?

32. What is the polarizing angle for (a) water with $n = 1.33$ and (b) glass with $n = 1.5$?

Problems

1. Two affluent students decide to improve on Galileo's attempt to measure the speed of light. One student goes to London and calls the other in New York on the telephone. The telephone signals are transmitted by reflecting electromagnetic waves from a satellite that is 37.9 Mm above the earth's surface. Neglecting the distance between London and New York, the distance traveled by the waves is twice the altitude of the satellite. One student claps his hands; when the other student hears the sound over the phone, she claps her hands. The first student measures the time between his clap and his hearing the second one. Calculate this time lapse. Do you think this experiment would be successful? What improvements for measuring the time interval would you suggest? (Time delays in the electronic circuits that are greater than those due to the light traveling to the satellite and back make this experiment not feasible.)

2. This problem is a refraction analogy. A band is marching down a football field with a constant speed v_1. About midfield, the band comes to a section of muddy ground, which has a sharp boundary making an angle of 30° with the 50-yd line, as shown in Figure 24-39. In the mud, the marchers move with speed $v_2 = \frac{1}{2}v_1$. Diagram how each line of marchers is bent as it encounters the muddy section of the field such that the band is eventually marching in a different direction. Indicate the original direction by a ray and the final direction by a second ray, and find the angles between the rays and the line perpendicular to the boundary. Is their direction of motion bent toward the perpendicular to the boundary or away from it?

Figure 24-39
Band marching down a football field for Problem 2.

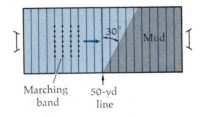

Marching band   50-yd line

3. (a) Find the intensity of light at a distance of 6 m from a 500-W light bulb, assuming that half the electrical energy goes into light energy. (Assume that the light bulb is a point source and use Equation 16-18.) (b) Find E_{rms} and B_{rms} at this distance.

4. Michelson used a rotating octagonal mirror on Mount Wilson and a fixed mirror on Mount San Antonio, 35 km away, to measure the speed of light. Find the first frequency of rotation of the mirror that gives a bright image in the telescope in Figure 24-13.

5. Investigate the effect on the critical angle for total reflection of a thin film of water on a glass surface for rays originating in the glass. (a) What is the critical angle for total internal reflection at the glass–water surface? (Take $n = 1.5$ for glass and 1.33 for water.) (b) Is there any range of incident angles that are greater than θ_c for glass-to-air refraction and for which light rays will leave the glass *and* the water and pass into the air?

6. Use Brewster's law, Snell's law, and the trigonometric identity $\tan \theta = (\sin \theta)/(\cos \theta)$ to show that the reflected and refracted rays are perpendicular when light is incident at the polarizing angle. (*Hint:* Use a diagram to first show that these rays will be perpendicular if $\theta_r + \theta_2 = 90°$, where θ_r is the angle of reflection and θ_2 is the angle of refraction.)

7. Light is incident normally on one face of a glass prism with index of refraction n (see Figure 24-40). The light is totally reflected at the right side. (a) What is the minimum value n can have? (b) When this prism is immersed in a liquid whose index of refraction is 1.15, there is still total reflection, but in water whose index is 1.33, there no longer is total reflection. Use this information to establish limits for possible values of n.

Figure 24-40
Problem 7.

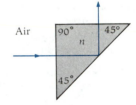

8. When light is transmitted through a glass slab, some of the energy is reflected at each face. Show that the transmitted intensity through a glass slab with index of refraction n for normally incident light is given by

$$I_T = I_0 \left[\frac{4n}{(n+1)^2} \right]^2$$

9. Use the result of Problem 8 to find the ratio of the transmitted intensity to the incident intensity for three slabs of glass ($n = 1.5$) for light of normal incidence.

10. Show that a light ray transmitted through a glass slab emerges parallel to the incident ray but displaced from it. For an incident angle of 60°, index of refraction 1.5, and slab thickness of 10 cm, find the displacement of the transmitted ray measured along the glass surface from where the incident ray would be if there were no slab.

Geometric Optics

Now we see through a glass, darkly. . . .

I Corinthians, 13:12

Drops of water were the first lenses and pools the first mirrors. Have you ever seen a tiny upside-down landscape in a dewdrop? (If you want to try this, a magnifying glass helps.) Lenses and mirrors are familiar to everyone, but there are many questions that one might ask about them. Why, if a mirror interchanges left and right, does it not also interchange up and down? Why are some images in a curved makeup mirror right-side up and some upside down? Are the images on the retinas of our eyes upside down? What causes rainbows? Why do our legs look shortened when we stand up to our waist in water? There are many more such questions that you should be able to answer after you have studied this chapter.

The wavelength of light is very small compared with most obstacles and apertures encountered. Because of this, diffraction (the bending of waves around corners) is often negligible and the ray approximation, in which waves are considered to propagate in straight lines, is usually valid. Geometric optics is the study of those phenomena for which the ray approximation is valid. (Physical optics, which will be treated in Chapter 27, is the study of those phenomena for which the wavelength of light is not small compared with the dimensions of objects and apertures encountered, and thus, diffraction is important.) In this chapter, we will apply the laws of reflection and refraction to study the formation of images by mirrors and lenses.

The lake serves as a plane mirror producing virtual images of the trees. The images are inverted and the same size as the trees.

25-1   Plane Mirrors

Figure 25-1 shows a narrow bundle of light rays from a point object P reflected from a plane mirror. After reflection, the rays diverge exactly as if they came from a point P' behind the plane of the mirror. The point P' is called the *image* of the object P. When these rays enter the eye, they cannot be distinguished from rays originating from a point object at P' with no mirror present. The image is called a *virtual image* because the light does not actually emanate from the image but only appears to.

Virtual image

Geometric construction using the law of reflection shows that the image point P' lies on the line through the object P perpendicular to the plane of the mirror and at a distance behind the plane equal to that from the plane to the object. The image can be seen by an eye anywhere in the region indicated in the figure in which a straight line from the image to the eye passes through the mirror. We note that the object need not be directly in front of the mirror. An image can be seen as long as the object is not behind the plane of the mirror.

Figure 25-2 shows the image you see if you hold the palm of your right hand up to a plane mirror. The image is the same size as your hand, but it is not the same as what would be seen by someone facing you or what you see when you look at the palm of your right hand. The image of a right hand in the mirror is a left hand. An object and its image in a plane mirror have left and right interchanged. The image formed by a plane mirror is therefore called a *perverted image*. Figure 25-3 shows the image of a simple rectangular coordinate system with its x and y axes parallel to the plane of the mirror. The image of the arrow along the x or y axes is parallel to the object arrow. But the image of the arrow along the z axis is directed oppositely to the object arrow along the z axis.

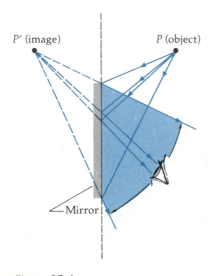

Figure 25-1
Image formed by a plane mirror. The rays from object point P that strike the mirror and enter the eye appear to come from the image point P' behind the mirror. The image can be seen by the eye anywhere in the shaded region.

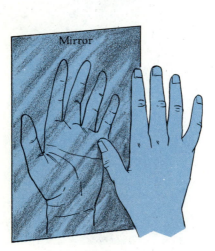

Figure 25-2
The image of a right hand in a plane mirror is a left hand. Such an image is called a perverted image.

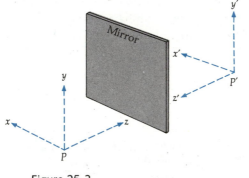

Figure 25-3
Image of a rectangular coordinate system in a plane mirror. The arrows along the x and y axes, which are parallel to the plane of the mirror, are in the same direction in the image as in the object. The direction of the arrow along the z axis is reversed in the image.

Figure 25-4
Images formed by two plane mirrors. P_1' is the image of the object P in mirror 1, and P_2' is the image of the object in mirror 2. Point $P_{1,2}''$ is the image of P_1' in mirror 2 seen when light rays from the object reflect first from mirror 1 and then from mirror 2. The image P_2' does not have an image in mirror 1 because it is behind the plane of that mirror.

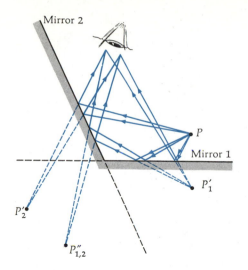

© 1987 by Sidney Harris.

"Most mirrors reverse left and right. This one reverses top and bottom."

Figure 25-4 illustrates the formation of multiple images by two plane mirrors making an angle with each other. We frequently see this phenomenon in clothing stores that provide adjacent mirrors. Light reflected from mirror 1 strikes mirror 2 just as if it came from the image point P_1'. The image P_1' is called the *object point* for mirror 2. Its image is at point $P_{1,2}''$. This image will be formed whenever the image point P_1' is in front of the plane of mirror 2. The image at point P_2' is due to rays from the object that reflect directly from mirror 2. Since P_2' is behind the plane of mirror 1, it cannot serve as an object point for a further image in mirror 1. The number of multiple images formed by two mirrors depends on the angle between the mirrors and the position of the object.

Figure 25-5 shows two mirrors at right angles to each other. Rays from the object to the eye that strike mirror 1 and then mirror 2 are shown in Figure 25-5a. In this case, the image point $P_{1,2}''$ is the same as that for rays that strike mirror 2 first and then mirror 1 as can be seen by Figure 25-5b. If you stand in front of two vertical mirrors that are perpendicular to each other, as in the corner of a room, you see an image of yourself that *is* the same as others see when facing you. Since each mirror perverts the image, the image formed by reflection by the two mirrors is not perverted.

Figure 25-5
Two plane mirrors at right angles to each other. (a) Rays that strike mirror 1 first and then mirror 2. The image of P_1' in mirror 2 is $P_{1,2}''$. (b) Rays that strike mirror 2 first and then mirror 1. The image of P_2' in mirror 1 is $P_{1,2}''$, which coincides with $P_{1,2}'$ for perpendicular mirrors.

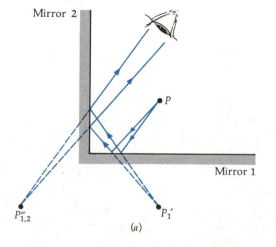

(a)

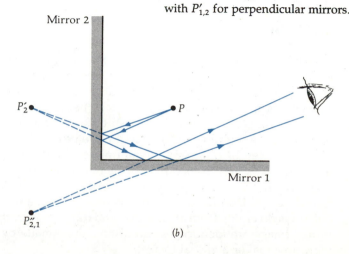

(b)

Figure 25-6
A ray striking one of two perpendicular plane mirrors is reflected from the second mirror in the direction opposite the original direction for any angle of incidence.

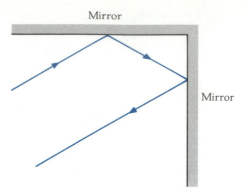

This girl standing at the edge of a large plane mirror at the Exploritorium in San Francisco appears to be floating in air.

Figure 25-6 illustrates the fact that a horizontal ray reflected from two perpendicular vertical mirrors is exactly reversed in direction no matter what angle the ray makes with the mirrors. If three mirrors are placed perpendicular to each other like the inside corner of a box, any ray incident on any of the mirrors from any direction is exactly reversed. A set of mirrors of this type (actually a set of internally reflecting prisms) was placed on the moon facing the earth. A laser beam from earth directed at the corner is reflected back to the same place on the earth. Such a beam has been used to measure the distance to the mirrors to within a few centimetres by measuring the time for the light to reach the mirrors and return.

Questions

1. What is the minimum length of a mirror in which a standing person can see her entire reflection? Draw a ray diagram.

2. A coordinate system like the one in Figure 25-3 is painted such that each axis is a different color. A photograph is taken of the coordinate system and another is taken of its image in a plane mirror. Is it possible to tell by looking at the photographs that one is of a mirror image rather than both being photographs of the real coordinate system taken from different angles?

25-2   Spherical Mirrors

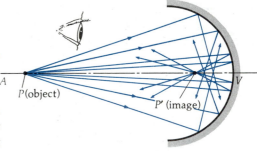

Figure 25-7 shows a bundle of rays from a point object P on the axis of a concave spherical mirror reflecting from the mirror and converging at point P'. The rays then diverge from this point just as if there were an object at that point. This image is called a *real image* because the light actually does emanate from the image point. It can be seen by an eye at the left of the image looking into the mirror. It could also be observed on a ground-glass viewing screen or photographic film placed at the image point. A virtual image, such as that formed by a plane mirror as discussed in the previous section, cannot be observed on a screen at the image point because there is no light there. Despite this distinction between real and virtual images, the light rays diverging from a real image and those appearing to diverge from a virtual image are identical, so no distinction is made by the eye between viewing a real or a virtual image.

Figure 25-7
Rays from a point object P on the axis AV of a concave spherical mirror form an image at P', which is sharp if the rays strike the mirror near the axis. Nonparaxial rays striking the mirror at points far from V are not reflected through the image point P' and therefore blur the image.

From Figure 25-7 we see that only rays that strike the mirror at points near the axis AV are reflected through the image point. Such rays are called *paraxial rays*. Because other, nonparaxial rays converge to different points near the image point, the image appears blurred, an effect called *spherical aberration*. The image can be sharpened by blocking off all but the central part of the mirror so that nonparaxial rays do not strike it. Although the image is then sharper, its brightness is reduced because less light is reflected and focused at the image point.

Spherical aberration

Figure 25-8 shows a ray from an object point P reflecting off the mirror and passing through the image point P'. Point C is the center of curvature of the mirror. The incident and reflected rays make equal angles with the radial line CA, which is perpendicular to the surface of the mirror. The image distance s' from the vertex of the mirror V to P' can be related to the object distance s from the vertex V to point P and to the radius of curvature r by elementary geometry. The result is

$$\frac{1}{s} + \frac{1}{s'} = \frac{2}{r}$$

25-1

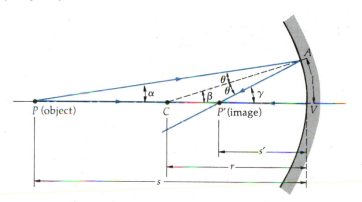

Figure 25-8
Geometry for calculating the image distance s' from the object distance s and the radius of curvature r. The angle β is an exterior angle to the triangle PAC and is therefore equal to $\alpha + \theta$. Similarly, from triangle PAP', $\gamma = \alpha + 2\theta$. Eliminating θ from these equations gives $2\beta = \gamma + \alpha$. Equation 25-1 follows from the small-angle approximations $\alpha \approx l/s$, $\beta \approx l/r$, and $\gamma \approx l/s'$.

The derivation of this equation assumes that angles made by the incident and reflected rays with the axis are small. This is equivalent to assuming that the rays are paraxial.

When the object distance is many times greater than the radius of curvature of the mirror, the term $1/s$ in Equation 25-1 is much smaller than $2/r$ and can be neglected. The image distance is then $s' = \frac{1}{2}r$. This distance is called the *focal length f* of the mirror, and the image point is called the *focal point F*.

$$f = \tfrac{1}{2}r$$

25-2

In terms of the focal length f, the mirror equation is

$$\frac{1}{s} + \frac{1}{s'} = \frac{1}{f}$$

25-3

The focal point F is the point at which parallel rays incident on the mirror are focused, as illustrated in Figure 25-9a. (Again, only paraxial rays are focused at a single point.) When an object is very far from the mirror, the wavefronts are approximately planes, as shown in Figure 25-9b, and the rays are parallel. In Figure 25-9b, note how the edges of the wavefront hit the concave mirror surface before the central portion near the axis, resulting in a spherical wavefront upon reflection.

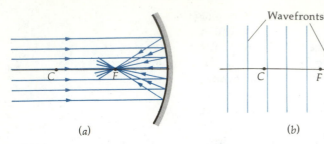

(a)   (b)

Parallel rays focused by a concave
mirror.

Figure 25-9
(a) Parallel rays strike a concave mirror and are reflected
through the focal point F, which is midway between the
center of curvature C and the mirror, at a distance $r/2$.
(b) The incoming wavefronts are plane waves; upon
reflection, they become spherical waves that converge at
the focal point.

Figure 25-10 shows the wavefronts and rays for plane waves striking a
convex mirror. In this case, the central part of the wavefront strikes the
mirror first and the reflected wave appears to come from the focal point
behind the mirror.

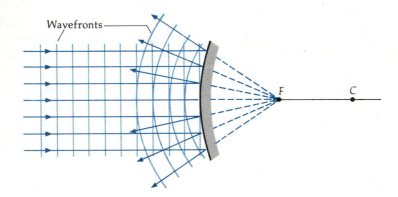

Figure 25-10
Reflection of plane waves from a
convex mirror. The reflected
wavefronts are spherical, just as if
they emanate from the focal point F
behind the mirror. The rays,
perpendicular to the wavefronts,
diverge from F.

Figure 25-11 shows rays from a point source at the focal point that strike a
concave mirror and are reflected parallel to the axis. This illustrates a property of waves called *reversibility*. If we reverse the direction of a reflected
ray, the law of reflection assures us that its reflection will be along the
original incoming ray but in the opposite direction. (Reversibility also holds
for refracted rays, which are discussed in later sections.) If we have a real
image of an object formed by a reflecting (or refracting) surface, we can
place a source object at the image point and a new image will be formed at
the position of the original object.

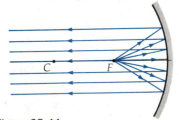

Figure 25-11
Illustration of reversibility. Rays
diverging from a point source at the
focal point of a concave mirror are
reflected from the mirror as parallel
rays. The rays are the same as those
in Figure 25-9a but in the reverse
direction.

Example 25-1   An object is 12 cm from a concave mirror of radius of curvature 6 cm. Find the focal length of the mirror and the image distance.

From Equation 25-2, the focal length is

$$f = \tfrac{1}{2}r = \tfrac{1}{2}(6 \text{ cm}) = 3 \text{ cm}$$

From Equation 25-3

$$\frac{1}{s} + \frac{1}{s'} = \frac{1}{f}$$

$$\frac{1}{12 \text{ cm}} + \frac{1}{s'} = \frac{1}{3 \text{ cm}}$$

$$\frac{1}{s'} = \frac{4}{12 \text{ cm}} - \frac{1}{12 \text{ cm}} = \frac{3}{12 \text{ cm}}$$

$$s' = 4 \text{ cm}$$

A useful method for locating images is the geometric construction of a *ray diagram*. This is illustrated in Figure 25-12, where the object is a human figure that is perpendicular to the axis a distance s from the mirror. By a judicious choice of rays from the head of the figure, we can quickly locate the image. Three principle rays are used in constructing a ray diagram:

1. The *parallel ray* drawn parallel to the axis. This ray is reflected through the focal point, as shown in the figure.   **Principle rays for mirrors**

2. The *focal ray* drawn through the focal point. This ray is reflected parallel to the axis.

3. The *radial ray* drawn through the center of curvature. This ray strikes the mirror perpendicular to its surface and is thus reflected back on itself, as shown in the figure.

The intersection of any two of these rays locates the image point of the head. The third ray provides a useful check.

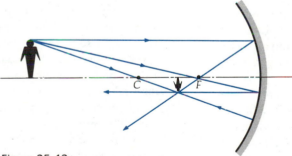

Figure 25-12
Ray diagram for the location of the image by geometric construction.

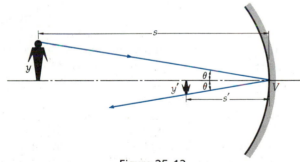

Figure 25-13
Geometry for finding the magnification of a spherical mirror. From the upper triangle, $\tan \theta = y/s$; and from the lower triangle, $\tan \theta = -y'/s$, where the negative sign is introduced because the image is inverted, so y' is negative. The magnification is then $m = y'/y = -s'/s$.

We see from the figure that the image is inverted and is not the same size as the object. The lateral magnification of the image is defined as the ratio of the image size to the object size. In Figure 25-13, we have drawn a ray from the top of the object to the center of the mirror; this ray makes an angle θ with the axis. The reflected ray to the top of the image makes an equal angle with the axis. Comparison of the triangle formed by the incident ray, the axis, and the object with the triangle formed by the reflected ray, the axis, and the image in this figure shows that the lateral magnification $m = y'/y$ equals the (negative) ratio of the distances $-s'/s$.

When the object is between the mirror and its focal point, the rays reflected from the mirror do not converge but appear to diverge from a point behind the mirror, as illustrated in Figure 25-14. In this case, the image is virtual and erect. ("Erect" merely means that the image is not inverted relative to the object.) For an object between the mirror and the focal point, s is less than $\frac{1}{2}r$. Then the image distance s' calculated from Equation 25-1 turns out to be negative. We can apply Equations 25-1, 25-2, and 25-3 to this case and to convex mirrors if we adopt a convenient sign convention. Whether the mirror is convex or concave, real images can be formed only in front of the mirror, that is, on the same side of the mirror as the object. Virtual images are formed behind the mirror, where there are no actual light rays. Our sign convention is as follows:

Sign convention for reflection

s + if object is in front of the mirror (real object)
     − if object is behind the mirror (virtual object)*

s' + if image is in front of the mirror (real image)
     − if image is behind the mirror (virtual image)

r, f + if center of curvature is in front of the mirror (concave mirror)
     − if center of curvature is behind the mirror (convex mirror)

With this sign convention, Equations 25-1, 25-2, and 25-3 can be used for all situations with any type of mirror. The *lateral magnification m* of the image is then given by

$$m = \frac{y'}{y} = -\frac{s'}{s} \qquad\qquad 25\text{-}4$$

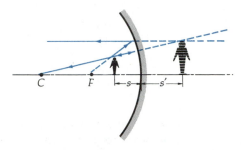

Figure 25-14
A virtual image formed by a concave mirror. Here the image is located by the radial ray, which is reflected back on itself, and the focal ray, which is reflected parallel to the axis. These two rays appear to diverge from a point behind the mirror found by extending the rays. A third ray (not shown) could be drawn from the object parallel to the axis. It would be reflected through the focal point, and its extension would intersect the other two rays at the image point.

A negative magnification, which occurs when both s and s' are positive, indicates that the image is inverted.

For plane mirrors, discussed in the previous section, the radius of curvature is infinite. The focal length given by Equation 25-2 is thus infinite also. Equation 25-3 then gives $s' = -s$, indicating that the image is behind the mirror at a distance equal to the object distance. The magnification given by Equation 25-4 is then $+1$, indicating that the image is erect and the same size as the object.

Although these equations coupled with this sign convention are relatively easy to use, practical work in optics often requires knowing only whether the image is real or virtual and its approximate location. This knowledge is usually most easily obtained by constructing a ray diagram. It is usually a good idea, though, to use both the graphical method and the algebraic method to locate an image so that one method serves as a check on the results of the other.

There are three important properties of images. They can be

     Real or virtual

     Upright or inverted

     Enlarged or diminished

Types of images

* You may wonder how an object can be behind the mirror. This occurs when there is a lens in front of the mirror such that the rays to the image of the lens are intercepted by the mirror. The image of the lens is then never formed, but the distance to the unformed image behind the mirror is taken as the object distance for the mirror, and the object is called a virtual object.

Figure 25-15 shows a ray diagram for an object in front of a convex mirror. The central ray heading toward the center of curvature C is perpendicular to the mirror and is reflected back on itself. The parallel ray is reflected as if it came from the focal point F behind the mirror. The focal ray (not shown) would be drawn towards the focal point and would be reflected parallel to the axis. We see from the figure that the image is behind the mirror and is therefore virtual. It is also erect and smaller than the object.

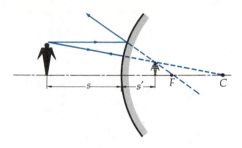

Figure 25-15
Ray diagram for a convex mirror. The parallel ray is reflected as if it came from the focal point behind the mirror, and the radial ray is reflected back on itself. These rays appear to diverge from a point behind the mirror. A third ray (not shown) could be drawn toward the focal point. It would be reflected parallel to the axis, and its extension would intersect the other two rays at the image point.

Example 25-2   An object 2 cm high is 10 cm from a convex mirror with a radius of curvature of 10 cm. Locate the image and find its height.

Since the center of curvature of a convex mirror is behind the mirror, the radius and focal length are negative:

$$f = \tfrac{1}{2}r = \tfrac{1}{2}(-10 \text{ cm}) = -5 \text{ cm}$$

We find the image distance from Equation 25-3:

$$\frac{1}{s} + \frac{1}{s'} = \frac{1}{f}$$

$$\frac{1}{10 \text{ cm}} + \frac{1}{s'} = -\frac{1}{5 \text{ cm}}$$

$$\frac{1}{s'} = -\frac{2}{10 \text{ cm}} - \frac{1}{10 \text{ cm}} = -\frac{3}{10 \text{ cm}}$$

$$s' = -3.33 \text{ cm}$$

The image distance is negative, indicating a virtual image behind the mirror. The magnification is

$$m = -\frac{s'}{s} = -\frac{-3.33 \text{ cm}}{10 \text{ cm}} = +0.333$$

The image is erect and one-third the size of the object. Since the object size is 2 cm, the image size is 2/3 cm. The ray diagram for this example is similar to that in Figure 25-15.

Questions

3. Under what conditions will a concave mirror produce an erect image? A virtual image? An image smaller than the object? An image larger than the object?

4. Answer Question 3 for a convex mirror.

5. A convex mirror is sometimes used on a car or truck as a rear-view mirror because it gives a wide-angle view. Below the mirror is written "Warning: Objects are closer than they appear." Yet according to a ray diagram, such as that in Figure 25-15, the image distance for distant objects is much smaller than the object distance. Why then do the images appear more distant?

Convex mirror resting on paper with equally spaced parallel stripes. The image of each point in front of the mirror is virtual and behind the mirror. Note the reduction in size and distortion in the shape of the image and the large number of stripes that are seen.

25-3   Images Formed by Refraction

Figure 25-16 illustrates the formation of an image by refraction from a spherical surface separating two media with indexes of refraction n_1 and n_2. In this figure, n_2 is greater than n_1, so the waves travel more slowly in the second medium. Again, only paraxial rays converge to one point. An equa-

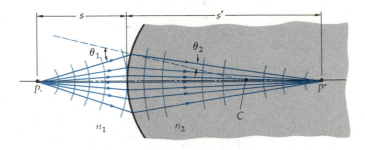

tion relating the image distance to the object distance, the radius of curvature, and the indexes of refraction can be derived by applying Snell's law of refraction to these rays and using small angle approximations in the geometry. We omit the derivation and merely state the result:

$$\frac{n_1}{s} + \frac{n_2}{s'} = \frac{n_2 - n_1}{r} \qquad\qquad 25\text{-}5$$

We use a sign convention for this equation similar to those for mirrors. We note that for refraction, real images are formed in back of the surface, which we will call the transmission side, whereas virtual images occur in front of the surface, that is, on the incident side. The sign convention for refraction is as follows:

Sign convention for refraction

s + real object; for objects in front of the surface (incident side)
   − virtual object; for objects in back of the surface (transmission side)

s' + real image; for images in back of the surface (transmission side)
   − virtual image; for images in front of the surface (incident side)

r + if center of curvature is on the transmission side
   − if center of curvature is on the incident side

If we compare this sign convention with that for mirrors, we note that s' is positive and the image is real when the image is on the side of the surface traversed by the reflected or refracted light. For reflection this side is in front of the mirror, whereas for refraction it is behind the refracting surface. Similarly r is positive when the center of curvature is on the side traversed by the reflected or refracted light.

We can obtain an expression for the magnification at a refracting surface by considering Figure 25-17, which shows a ray from the top of an object to the top of the image. The ray is bent toward the normal as it crosses the surface, so θ_2 is less than θ_1. These angles are related by Snell's law

$$n_1 \sin \theta_1 = n_2 \sin \theta_2$$

Figure 25-16
Image formed by refraction at a spherical surface between two media where the waves move slower in the second medium.

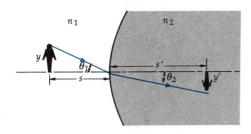

Figure 25-17
The lateral magnification for an image formed by refraction at a single spherical surface can be obtained from $\tan \theta_1 = y/s$, $\tan \theta_2 = -y'/s'$ and from Snell's law $n_1 \sin \theta_1 = n_2 \sin \theta_2$. When the angles are small, $\sin \theta \approx \tan \theta$. Then $n_1 y/s \approx n_2(-y'/s')$ and the lateral magnification is $m = y'/y = -n_1 s'/n_2 s$.

The object and image sizes are related to the angles by

$$\tan \theta_1 = \frac{y}{s} \qquad \tan \theta_2 = \frac{-y'}{s'}$$

where the minus sign arises because y' is negative. Since we are considering only paraxial rays for which the angles are small, we may assume that the sine of the angle is approximately equal to its tangent. Using this approximation, Snell's law becomes

$$n_1 \left(\frac{y}{s} \right) = n_2 \left(\frac{-y'}{s'} \right)$$

The magnification is therefore

$$m = \frac{y'}{y} = -\frac{n_1 s'}{n_2 s} \qquad\qquad 25\text{-}6$$

Example 25-3   A fish is in a spherical bowl of water of index of refraction 1.33. The radius of the bowl is 15 cm. The fish looks through the bowl and sees a cat sitting on the table with its nose 10 cm from the bowl. Where is the image of the cat's nose, and what is its magnification?

We neglect any effect of the thin glass wall of the bowl. The object distance between the cat and the bowl is 10 cm. The indexes of refraction are $n_1 = 1$ and $n_2 = 1.33$. The radius is $+15$ cm. We obtain the image distance from Equation 25-5:

$$\frac{n_1}{s} + \frac{n_2}{s'} = \frac{n_2 - n_1}{r}$$

$$\frac{1.00}{10 \text{ cm}} + \frac{1.33}{s'} = \frac{1.33 - 1.00}{15 \text{ cm}}$$

Solving for s', we obtain

$$s' = -17.1 \text{ cm}$$

The negative image distance means that the image is virtual and in front of the refracting surface on the same side as the object, as shown in Figure 25-18. The magnification of the image is

$$m = -\frac{n_1 s'}{n_2 s} = -\frac{-17.1 \text{ cm}}{1.33(10 \text{ cm})} = 1.29$$

Thus the cat appears to be farther away but larger.

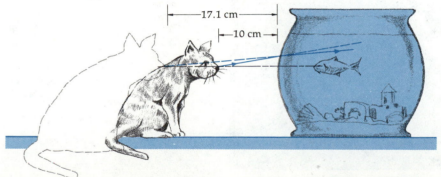

Figure 25-18
Fish looking at cat for Example 25-3. Because of refraction at the spherical surface, the cat appears farther away but slightly larger.

We can use Equation 25-5 to find the apparent depth of an underwater object when it is viewed from directly overhead. For this case, the surface is a plane surface, the radius of curvature is infinite, and the image and object distances are related by

$$\frac{n_1}{s} + \frac{n_2}{s'} = 0$$

$$s' = -\frac{n_2}{n_1} s \qquad\qquad 25\text{-}7$$

The negative sign indicates that the image is virtual and on the same side of the refracting surface as the object, as shown in the ray diagram in Figure 25-19. The magnification is

$$m = -\frac{n_1 s'}{n_2 s} = +1$$

Example 25-4   Find the apparent depth of a fish resting 1 m below the surface of water, which has an index of refraction $n = 4/3$.

Using $n_1 = 4/3$ and $n_2 = 1$ in Equation 25-7, we obtain

$$s' = -\frac{1}{(4/3)} (1\text{ m}) = -\frac{3}{4} (1\text{ m}) = -0.75\text{ m}$$

The apparent depth is three-fourths the actual depth, so the fish appears to be 75 cm below the surface. Note that this result holds only when the object is viewed from directly overhead so that the rays are paraxial.

Question

6. If a fish under water is viewed from a point not directly overhead, is its apparent depth greater or less than three-fourths its actual depth? (Draw a ray diagram to help you answer this question.)

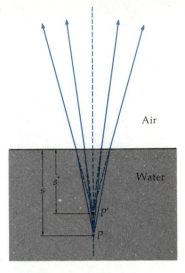

Figure 25-19
Ray diagram for image of an object in water viewed from directly overhead. The image is at a depth that is less than the depth of the object. The *apparent depth* equals the real depth divided by the index of refraction of water.

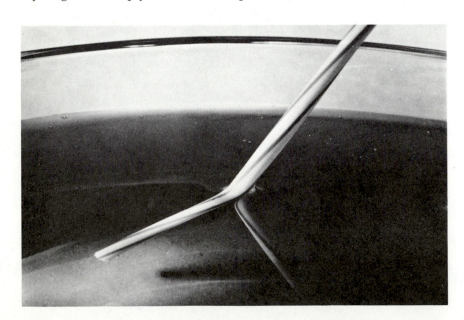

Because of refraction, the apparent depth of the submerged portion of the straw is less than the real depth. Consequently, the straw appears to be bent.

25-4   Thin Lenses

The most important application of Equation 25-5 is in finding the position of the image formed by a lens. This is done by considering the refraction at each surface separately to derive an equation relating the image distance to the object distance, the radius of curvature of each surface of the lens, and the index of refraction of the lens.

We consider a very thin lens of index of refraction n with air on both sides. Let the radii of curvature of the surfaces of the lens be r_1 and r_2. If an object is at a distance s from the first surface (and therefore from the lens), the distance s_1' of the image due to refraction at the first surface can be found by substituting into Equation 25-5:

$$\frac{1}{s} + \frac{n}{s_1'} = \frac{n-1}{r_1} \qquad\qquad 25\text{-}8$$

This image is not formed because the light is refracted again at the second surface. Figure 25-20 shows the case when the image distance s_1' for the first surface is negative, indicating a virtual image to the left of the surface. Rays in the glass refracted from the first surface diverge as if they came from the image point. They strike the second surface at the same angles as if there were an object at this image point. The image of the first surface therefore becomes the object for the second surface. Since the lens is of negligible thickness, the object distance is equal in magnitude to s_1', but since object distances in front of the surface are positive whereas s_1' is negative there, the object distance for the second surface is $s_2 = -s_1'$. (If s_1' is positive, the rays

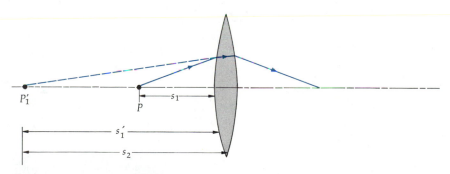

are converging as they strike the second surface. The object for the second surface is then to the right of the surface. This is called a virtual object. Again, $s_2 = -s_1'$.) We now write Equation 25-5 for the second surface, with $n_1 = n$, $n_2 = 1$, and $s = -s_1'$. The image distance for the second surface is the final image distance s' for the lens.

$$\frac{n}{-s_1'} + \frac{1}{s'} = \frac{1-n}{r_2} \qquad\qquad 25\text{-}9$$

We can eliminate the image distance for the first surface s_1' from Equations 25-8 and 25-9 by adding these equations. We thereby obtain

$$\frac{1}{s} + \frac{1}{s'} = (n-1)\left(\frac{1}{r_1} - \frac{1}{r_2}\right) \qquad\qquad 25\text{-}10$$

Figure 25-20
Refraction occurs at both surfaces of a lens. Here, the refraction at the first surface leads to a virtual image at P_1'. The rays strike the second surface as if they came from P_1'. Since image distances are negative when the image is on the incident side of the surface whereas object distances are positive for objects there, $s_2 = -s_1'$ is the object distance for the second surface of the lens.

Equation 25-10 relates the image distance s' to the object distance s and to the properties of the thin lens, that is, r_1, r_2, and the index of refraction n. As with mirrors, the focal length of a thin lens is defined as the image distance when the object distance is very large. Setting s equal to infinity and writing f for the image distance s', we obtain

$$\frac{1}{f} = (n - 1)\left(\frac{1}{r_1} - \frac{1}{r_2}\right)$$

25-11   Lens-maker's equation

Equation 25-11 is called the *lens-maker's equation* because it gives the focal length of a thin lens in terms of the properties of the lens. Substituting $1/f$ for the right side of Equation 25-10 we obtain

$$\frac{1}{s} + \frac{1}{s'} = \frac{1}{f}$$

25-12   Thin-lens equation

Equation 25-12 is the same as that for a mirror, except that the sign convention for lenses is somewhat different. For lenses, the image distance s' is positive when the image is on the transmission side of the lens, that is, on the side opposite the side on which the light is incident. The sign convention for r in Equation 25-11 is the same as for refraction at a single surface. The radius is positive if the center of curvature is on the transmission side of the lens and negative if it is on the incident side.

Figure 25-21a shows the wavefronts of plane waves incident on a double convex lens. The central parts of the wavefronts strike the lens first. Since the wave speed in the lens is less than that in air (assuming $n > 1$), the central parts of the wavefronts lag behind the outer parts, resulting in spherical waves that converge at the focal point F'. The rays for this situation are shown in Figure 25-21b. Such a lens is called a *converging lens*. Since its focal length, as calculated from Equation 25-11, is positive, it is also called a *positive lens*. Any lens that is thicker in the middle than at the edges is a

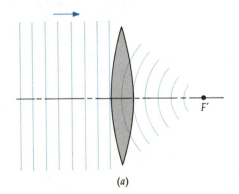

(a)

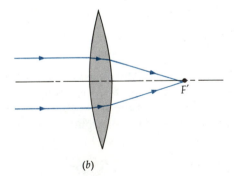

(b)

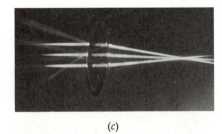

(c)

Figure 25-21
(a) Wavefronts for plane waves striking a converging lens. The central part of the wavefronts are retarded more by the lens than the outer parts, resulting in spherical

waves that converge at the focal point F'. (b) Rays for plane waves striking a converging lens. The rays are bent at each surface and converge at the focal point.

(c) In the photograph, reflected rays from each surface of the lens can also be seen.

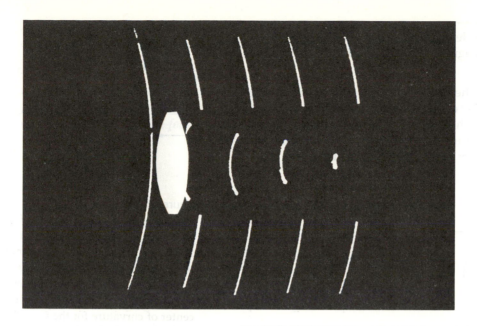

Photograph of the wavefronts of light passing through a lens. The photograph uses a technique called *light-in-flight-recording* which uses a pulsed laser to make a hologram of the wavefronts.

converging lens (provided that the index of refraction of the lens is greater than that of the surrounding medium). Figures 25-22a and 25-22b show wavefronts and rays for plane waves incident on a double concave lens. In this case, the outer parts of the wavefronts lag behind the central parts, resulting in outgoing spherical waves diverging from a point F' on the incident side of the lens. The focal length of this lens is negative. A lens (with an index of refraction greater than that of the surrounding medium) that is thinner in the middle than at the edges is a *diverging* or *negative lens.*

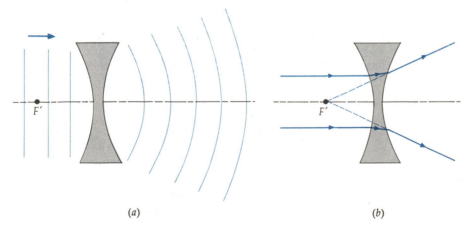

(a)   (b)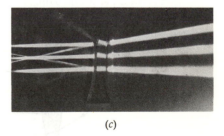

(c)

Figure 25-22
(a) Wavefronts for plane waves striking a diverging lens. Here the outer parts of the wavefronts are retarded more than the central part, resulting in spherical waves that

move out as if they came from the focal point F' in front of the lens. (b) Rays for plane waves striking the same diverging lens. The rays are bent outward and diverge as if they

came from the focal point F'. (c) As in the photograph in Figure 25-21, reflected rays from each surface of the lens can also be seen.

In laboratory experiments involving lenses, it is usually much easier simply to measure the focal length rather than calculate it from the radii of curvature of the surfaces.

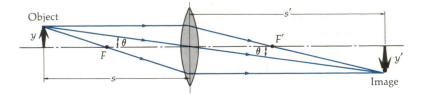

As with images formed by mirrors, it is convenient to locate the images of lenses by graphical methods. Figure 25-28 illustrates the graphical method for a converging lens. As with mirrors, we use three principle rays:

1. The *parallel ray* drawn parallel to the axis. This ray is bent through the second focal point of the lens.

2. The *central ray* drawn through the center of the lens (vertex). This ray is undeflected. (The faces of the lens are parallel at this point, so the ray emerges in the same direction but displaced slightly, as in transmission through a glass slab. Since the lens is thin, the displacement is negligible.)

3. The *focal ray* through the first focal point. This ray emerges parallel to the axis.

Principle rays for lenses

These three rays converge to the image point, as indicated in the figure. In this case the image is real and inverted. From Figure 25-28, we have tan $\theta = y/s = -y'/s'$. The lateral magnification m is then

$$m = \frac{y'}{y} = -\frac{s'}{s} \qquad\qquad 25\text{-}13$$

This expression is the same as that for mirrors. Again, a negative magnification indicates that the image is inverted.

The principle rays for a negative lens are the same as those for a positive lens, except the parallel ray diverges from the lens as if it came from the second focal point, and the focal ray is drawn toward the first focal point and is refracted parallel to the axis. The ray diagram for a diverging lens is shown in Figure 25-29.

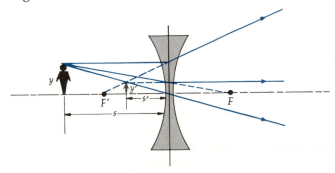

Figure 25-28
Ray diagram for a thin converging lens. For simplicity, we assume that all the bending takes place at a line. The ray through the center is undeflected because the lens surfaces there are nearly parallel and are close together.

Figure 25-29
Ray diagram for a thin diverging lens. The parallel ray is bent away from the axis as if it came from the second focal point F'. The ray toward the first focal point F emerges parallel to the axis. The central ray is undeflected. The three rays appear to diverge from the image point.

Example 25-7   An object 1.2 cm high is placed 6 cm from the double convex lens of Example 25-5. Locate the image, state whether it is real or virtual, and find its size.

The focal length *f* for this lens was found in Example 25-5 to be 12 cm. Figure 25-30 shows a ray diagram for an object placed 6 cm in front of a positive lens of focal length 12 cm. The parallel ray is bent through the second focal point and the central ray is undeflected. These rays are diverging on the transmission side of the lens. The image is located by extending the rays back until they meet. These two rays are sufficient to locate the image. (As a check, we could draw the third ray, the focal ray, along the line from the first focal point *F* on the incident side of the lens. This ray would then leave the lens parallel to the axis.) We can see immediately from this drawing that the image is virtual, erect, and enlarged. It is on the same side of the lens as the object and is about twice as far away from the lens. Since it is quite easy to make an error in the calculation of the image distance from Equation 25-12, it is always a good idea to check your result with a ray diagram.

The image distance is found algebraically from Equation 25-12.

$$\frac{1}{s} + \frac{1}{s'} = \frac{1}{f}$$

$$\frac{1}{6 \text{ cm}} + \frac{1}{s'} = \frac{1}{12 \text{ cm}}$$

$$s' = -12 \text{ cm}$$

The image distance is negative, indicating that the image is virtual and on the incident side of the lens. The magnification is

$$m = -\frac{s'}{s} = -\frac{-12 \text{ cm}}{6 \text{ cm}} = +2$$

The image is thus twice as large as the object and is erect. Since the size of the object is given as 1.2 cm, the size of the image is 2.4 cm.

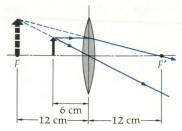

Figure 25-30
Ray diagram for Example 25-7. When the object is between the first focal point and the converging lens, the image is virtual and erect.

If we have two or more thin lenses, we can find the final image produced by the system by finding the image distance for the first lens and using it along with the distance between lenses to find the object distance for the second lens. That is, we consider each image, whether it is real or virtual and whether it is formed or not, as the object for the next lens.

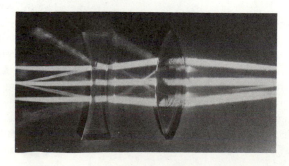

Parallel light rays incident on a diverging lens followed by a converging lens. The net effect of such a lens system can be converging rays, diverging rays, or parallel rays, depending on the lens separation and the relative size of the focal lengths of the two lenses. Here the net effect is to produce converging rays from parallel rays.

Example 25-8   A second lens of focal length $+6$ cm is placed 12 cm to the right of the lens in Example 25-7. Locate the final image.

Figure 25-31 shows a ray diagram for this example. The rays used to locate the image of the first lens are not necessarily convenient rays for the second lens. If they are not, we merely draw two additional rays from the first image that are principle rays for the second lens; for example, a ray from the image parallel to the axis and one from the image through the first focal point of the second lens or one through the vertex of the second lens. In this case, two of the principle rays for the first lens are also principle rays for the second lens. The parallel ray for the first lens turns out to be the central ray for the second lens. Also, the focal ray for the first lens emerges parallel to the axis and is therefore refracted through the focal point of the second lens, as shown in the figure. We can see that the final image is real, inverted, and just outside the second focal point of the second lens. We locate its position algebraically by noting that the virtual image of the first lens is 12 cm to the left of that lens, so it is 24 cm to the left of the second lens. Using $s_2 = 24$ cm and $f = 6$ cm, we have

$$\frac{1}{s_2} + \frac{1}{s_2'} = \frac{1}{f}$$

$$\frac{1}{24} + \frac{1}{s_2'} = \frac{1}{6}$$

giving

$$s_2' = 8 \text{ cm}$$

Example 25-9   Two positive lenses of focal length 10 cm each are 15 cm apart. Find the final image of an object 15 cm from one of the lenses.

In the ray diagram of Figure 25-32, the image of the first lens would be 30 cm from that lens if the second lens were not there. We calculate this using $s_1 = 15$ cm and $f_1 = 10$ cm in the thin-lens equation:

$$\frac{1}{s_1} + \frac{1}{s_1'} = \frac{1}{f}$$

$$\frac{1}{15} + \frac{1}{s_1'} = \frac{1}{10}$$

Solving for s_1', we obtain

$$s_1' = 30 \text{ cm}$$

This image is not formed because the light rays strike the second lens before they reach the image position. We can locate the final image graphically by choosing rays that are heading toward the unformed image when they strike the second lens. These rays need not be the principle rays for the first lens. Any ray that leaves the object and strikes the first lens is directed toward the image of the first lens. We choose a ray that leaves the first lens parallel to the axis and one that goes through the center of the second lens. We can see that the final image is between the second lens and its focal point. The final image can be located algebraically by using the first image as

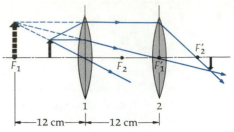

Figure 25-31
Ray diagram for Example 25-8. The image of the first lens acts as the object for the second lens. The final image is located by drawing two rays from the first image through the second lens. In this case, one of the original rays used to locate the first image happens to be the central ray for the second lens. A second ray parallel to the axis from the first image locates the final image.

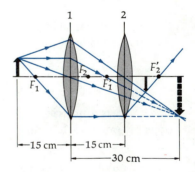

Figure 25-32
Ray diagram for Example 25-9. The image of the first lens is to the right of the second lens. This image isn't formed because the rays are refracted by the second lens before they get to the first image. Nevertheless, this unformed image acts as a virtual object for the second lens. The final image is found by drawing rays toward the first image as shown. One ray through the center of the second lens and one parallel to the axis as it strikes the second lens are used.

the object for the second lens. Since this unformed image is on the transmission side of the second lens, it is a *virtual object*. Since it is 15 cm from the second lens, the object distance is $s_2 = -15$ cm. Then,   **Virtual object**

$$\frac{1}{s_2} + \frac{1}{s_2'} = \frac{1}{f}$$

$$\frac{1}{-15} + \frac{1}{s_2'} = \frac{1}{f_2} = \frac{1}{10}$$

Solving for s_2', we obtain

$$s_2' = 6 \text{ cm}$$

Example 25-10   Two thin lenses of focal lengths f_1 and f_2 are placed together. Show that the equivalent focal length of the combination f is given by

$$\frac{1}{f} = \frac{1}{f_1} + \frac{1}{f_2} \qquad\qquad\qquad 25\text{-}14$$

Let s be the object distance for the first lens (and, therefore, for the lens combination) and s_1' be the image distance. Applying the thin-lens equation, we have

$$\frac{1}{s} + \frac{1}{s_1'} = \frac{1}{f_1}$$

Since the lenses are together, the object distance for the second lens is the negative of the image distance for the first lens, $s_2 = -s_1'$. Calling the final image distance s', we have for the second lens

$$\frac{1}{-s_1'} + \frac{1}{s'} = \frac{1}{f_2}$$

Adding these two equations to eliminate s_1', we obtain

$$\frac{1}{s} + \frac{1}{s'} = \frac{1}{f_1} + \frac{1}{f_2} = \frac{1}{f}$$

where f is given by Equation 25-14.

Example 25-10 gives us the important result that when two lens are placed in contact (or very close together), the reciprocals of the focal lengths add. The reciprocal of the focal length of a lens is called the *power* of the lens. When the focal length f is expressed in metres, the power P is given in reciprocal metres called *diopters* (D):

$$P = \frac{1}{f} \qquad \text{diopters} \qquad\qquad 25\text{-}15 \qquad \text{Power of a lens}$$

The power of a lens measures its ability to focus parallel light at a short distance from the lens. The shorter the focal length, the greater the power. For example, a lens with a focal length of 25 cm $= 0.25$ m has a power of 4.0 diopters. A lens of focal length 10 cm $= 0.10$ m has a power of 10 diopters. Since the focal length of a diverging lens is negative, its power is also negative.

Example 25-11   A lens has a power of -2.5 D. What is its focal length?

Solving Equation 25-15 for the focal length, we obtain

$$f = \frac{1}{P} = \frac{1}{-2.5 \text{ D}} = -0.40 \text{ m} = -40 \text{ cm}$$

where we have used the fact that a diopter is the same as a reciprocal metre, $1 \text{ D} = 1 \text{ m}^{-1}$.

In terms of the power of a lens, the results of Example 25-10 can be simply stated. When two lenses are in contact, the power of the combination equals the sum of the powers of the lenses

$$P = P_1 + P_2 \qquad\qquad\qquad\qquad \text{25-16} \quad \text{Power of two lenses in contact}$$

Questions

7. Under what conditions will the focal length of a thin lens be positive? Negative?

8. The focal length of a lens is different for different colors of light. Why?

25-5   Aberrations

When all the rays from a point object are not focused at a single image point, the resulting blurring of the image is called *aberration*. Figure 25-33 shows rays from a point source on the axis traversing a thin lens with spherical surfaces. Rays that strike the lens far from the axis are bent more than those near the axis, with the result that all the rays are not focused at a single point. Instead, the image appears as a circle. The *circle of least confusion* is at point C, where the diameter of this circle is minimum. This type of aberration is called *spherical aberration*. It is similar to the spherical aberration of mirrors discussed in Section 25-2.

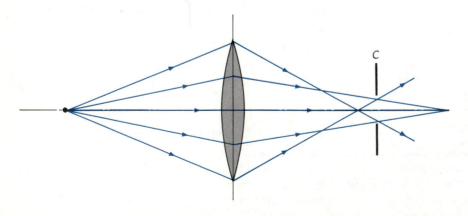

Figure 25-33
Spherical aberration. Rays from a point object on the axis are not focused at a point. The image is thus a circle about the axis rather than a point. At point C, the circle has its least diameter. This is called the circle of least confusion. Spherical aberration can be reduced by blocking off the outer parts of the lens, which reduces the diameter of the circle of least confusion. This also reduces the amount of light reaching the image.

There are several other types of aberrations that are similar to but more complicated than spherical aberrations. *Coma* (for the comet-shaped image) and *astigmatism* occur for objects off axis. *Distortion* is due to the fact that the magnification depends on the distance of the object point from the axis, and for an extended object this distance varies. We will not discuss these aberrations further except to point out that they do not arise from any defect in a lens (or mirror) but instead result from the laws of refraction (and reflection for mirrors) applied to spherical surfaces. They are not evident in our simple equations because we have used small-angle approximations in the derivation of these equations.

Some aberrations can be eliminated or partially avoided for by using nonspherical surfaces for mirrors or lenses, but nonspherical surfaces are usually much more difficult and costly to produce than spherical surfaces. One example of a nonspherical reflecting surface is the parabolic mirror illustrated in Figure 25-34. Rays that are parallel to the axis of a parabolic surface are reflected and focused at a common point no matter how far they are from the axis. Parabolic reflecting surfaces are important in large astronomical telescopes, where a large reflecting surface is needed to gather as much light as possible to make the image as intense as possible. A parabolic surface can also be used in a searchlight to produce a parallel beam of light from a small source placed at the focus of the surface.

An important aberration of lenses not found in mirrors is *chromatic aberration*, which is due to variations in the index of refraction with wavelength. From Equation 25-11, we can see that the focal length of a lens depends on its index of refraction, so it is different for different wavelengths. Since n is generally slightly greater for violet light than for red light (see Figure 24-29), the focal length for violet light will be shorter than that for red light. Since chromatic aberration does not occur for mirrors, many large telescopes use mirrors rather than lenses. Chromatic and other aberrations can be partially corrected for by using combinations of lenses instead of a single lens. For example, a positive lens and a negative lens of greater focal length can be used together to produce a converging lens system that has much less chromatic aberration than a single lens of the same focal length.

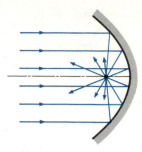

Figure 25-34
A parabolic mirror focuses all rays parallel to the axis to a single point with no spherical aberration. Parabolic surfaces are more costly to produce than spherical surfaces.

Chromatic aberration

Summary

1. The image formed by a spherical mirror of radius of curvature r is at a distance s' from the mirror, which is related to the object distance s by

$$\frac{1}{s} + \frac{1}{s'} = \frac{2}{r} = \frac{1}{f}$$

where the focal length $f = r/2$ is the distance to the point where parallel rays are focused. The focal length f is the image distance for objects that are very far from the mirror.

2. The mirror equation can be applied to any spherical mirror using the following sign convention. The object distance s is positive if the object is in front of the mirror and negative if it is behind the mirror. Similarly, the image distance s' is positive if the image is in front of the mirror and negative if it is behind the mirror. The radius of curvature r is positive for concave mirrors, for which the center of curvature is in front of the mirror, and negative for convex mirrors, for which the center of curvature is behind the mirror. When s' is positive, the image is real, meaning that light rays actually diverge from the image point. Real images can be seen on a screen or a piece of ground glass placed at the image point. When s' is negative, the image is virtual, meaning that no light actually diverges from the image point. The eye cannot distinguish a real image from a virtual image. Virtual images occur on the side of the mirror where there is no light.

3. The lateral magnification of the image is given by

$$m = \frac{y'}{y} = -\frac{s'}{s}$$

where y is the object size and y' the image size. A negative magnification means that the image is inverted.

4. For a plane mirror, r and f are infinite, $s' = -s$, and the image is virtual, erect, and the same size as the object.

5. Images for spherical mirrors can be conveniently located by a ray diagram using three principle rays: the parallel ray parallel to the axis, the focal ray through the focal point, and the radial ray through the center of curvature of the mirror. The point from which these rays diverge or appear to diverge is the image point.

6. The image distance s' for refraction at a single spherical surface of radius r is related to the object distance s and the radius of curvature of the surface by

$$\frac{n_1}{s} + \frac{n_2}{s'} = \frac{n_2 - n_1}{r}$$

where n_1 is the index of refraction of the medium on the incident side of the surface and n_2 is the index of refraction on the transmission side. The magnification of the image due to refraction at a single surface is

$$m = -\frac{n_1 s'}{n_2 s}$$

7. The image distance for a thin lens is related to the object distance by

$$\frac{1}{s} + \frac{1}{s'} = \frac{1}{f}$$

where the focal length is related to the radii of curvature of the lens surfaces and to the index of refraction by the lens-maker's equation

$$\frac{1}{f} = (n - 1)\left(\frac{1}{r_1} - \frac{1}{r_2}\right)$$

The equation for the lateral magnification of an image formed by a thin lens is the same as that for the lateral magnification of an image formed by a mirror.

8. The sign convention used for refraction at a single surface, the lens-maker's equation, and the thin-lens equation is as follows: s is positive if the object is on the incident side of the lens or surface and negative if it is on the transmission side; s' is positive if the image is on the transmission side of the lens or surface and negative if it is on the incident side. Similarly, the radius of a spherical surface is positive if the center of curvature is on the transmission side of the surface and negative if it is on the incident side.

9. A positive or converging lens is one that is thicker at the middle than at the edges. Parallel light incident on a positive lens is focused at the second focal point, which is on the transmission side of the lens. A negative or

diverging lens is one that is thicker at the edges than at the middle. Parallel light incident on a negative lens emerges as if it originated from the second focal point, which is on the incident side of the lens.

10. Images for thin lenses can also be conveniently located by a ray diagram using three principle rays: the parallel ray parallel to the axis, the focal ray through the second focal point, and the central ray through the center of the lens.

11. The power of a lens equals the reciprocal of the focal length. When the focal length is in metres, the power is in diopters (D):

$$P = \frac{1}{f}$$

$$1 \text{ D} = 1 \text{ m}^{-1}$$

12. The fact that the image of a single object point is sometimes blurred is known as aberration. Spherical aberration results from the fact that a spherical surface focuses only paraxial rays (those nearly parallel to the axis) at a single point. Nonparaxial rays are focused at nearby points, depending on the angle made with the axis. Spherical aberration can be reduced by reducing the size of the spherical surface, which also unfortunately reduces the amount of light focused. Chromatic aberration, which occurs for lenses but not mirrors, results from variations in the index of refraction with wavelength.

Suggestions for Further Reading

Walker, Jearl: "The Amateur Scientist: The Kaleidoscope Now Comes Equipped with Flashing Diodes and Focusing Lenses," *Scientific American*, December 1985, p. 134.

Multiple reflections created by arrangements of different numbers of plane mirrors are investigated.

Walker, Jearl: "The Amateur Scientist: What Is a Fish's View of a Fisherman and the Fly He Has Cast on the Water? *Scientific American*, March 1984, p. 138.

Refraction causes the world above water to appear to a fish the way things appear to us when viewed through a "fish-eye" lens.

Review

A. Objectives: After studying this chapter, you should:

1. Be able to draw simple ray diagrams for mirrors and lenses to locate images and determine whether they are real or virtual, erect or inverted, and enlarged or reduced.

2. Be able to determine algebraically the location of the image formed by a mirror, single spherical refracting surface, or a thin lens, and calculate the magnification of the image.

3. Be able to use the lens-maker's equation to determine the focal length of a lens from the radii of curvature of the surfaces.

4. Be able to discuss the various aberrations that occur with mirrors and lenses.

B. Define, explain, or otherwise identify:

Virtual image, p. 640
Real image, p. 642
Paraxial ray, p. 643
Focal length, p. 643
Focal point, p. 643
Lateral magnification, p. 646
Apparent depth, p. 650
Converging lens, p. 652
Positive lens, p. 652
Diverging lens, p. 653
Negative lens, p. 653
Virtual object, p. 659
Power of a lens, p. 659

Diopter, p. 659
Spherical aberration, p. 660
Chromatic aberration, p. 661

C. True or false: If the statement is true, explain why. If it is false, give a counterexample.

1. Virtual and real images look the same to the eye.

2. A virtual image cannot be displayed on a screen.

3. Aberrations occur only for real images.

4. A negative image distance implies that the image is virtual.

5. All rays parallel to the axis of a spherical mirror are reflected through a single point.

6. A diverging lens cannot form a real image from a real object.

7. The image distance for a positive lens is always positive.

8. Chromatic aberration does not occur with mirrors.

Exercises

Section 25-1   Plane Mirrors

1. The image of the point object shown in Figure 25-35 is viewed by an eye as shown. Draw a bundle of rays from the object that reflect from the mirror and enter the eye. For this object position and mirror, indicate the region of space in which the eye can be placed to see the image.

Figure 25-35   Eye   P•
Exercise 1.

───────────── Mirror

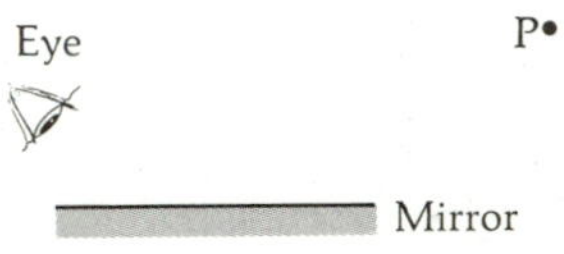

2. When two plane mirrors are parallel, such as on opposite walls in a barber shop, multiple images arise because each image in one mirror serves as an object for the other mirror. A point object is placed between parallel mirrors separated by 30 cm. The object is 10 cm in front of the left mirror and 20 cm in front of the right mirror. (*a*) Find the distance from the left mirror to the first four images in that mirror. (*b*) Find the distance from the right mirror to the first four images in that mirror.

3. Two plane mirrors make an angle of 90°. Show by considering various object positions that there are three images for any position of the object. Draw appropriate bundles of rays from the object to the eye for viewing each image.

4. Two plane mirrors make an angle of 60° with each other. Consider a point object on the bisector of the angle between the mirrors. Show on a sketch the location of all the images formed.

5. Repeat Exercise 4 for two mirrors at an angle of 120°.

Section 25-2   Spherical Mirrors

6. A concave spherical mirror has a radius of curvature of 40 cm. Draw ray diagrams to locate the image (if one is formed) for an object at a distance of (*a*) 100 cm, (*b*) 40 cm, (*c*) 20 cm, and (*d*) 10 cm from the mirror. For each case, state whether the image is real or virtual; erect or inverted; and enlarged, reduced, or the same size as the object.

7. Use the mirror equation to locate and describe the images for the object distances and mirror of Exercise 6.

8. Repeat Exercise 6 with a convex mirror of the same radius of curvature.

9. Repeat Exercise 7 with a convex mirror of the same radius of curvature.

10. Show that a convex mirror cannot form a real image of a real object no matter where the object is placed by showing that s' is always negative for positive s.

11. Convex mirrors are used in stores to give a wide angle of surveillance for a reasonable mirror size. The mirror shown in Figure 25-36 allows a clerk 5 m away to survey the entire store. It has a radius of curvature of 1.0 m. (*a*) If a customer is 10 m from the mirror, how far from the mirror surface is his or her image? (*b*) Is the image in front of or behind the mirror? (*c*) If the customer is 2 m tall, how tall is his or her image?

Figure 25-36
Exercise 11.

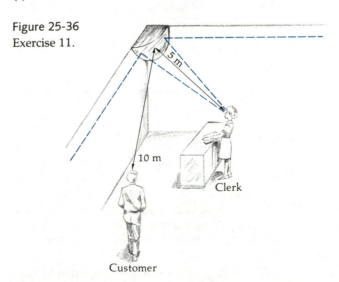

12. Show that a concave mirror forms a virtual image of a real object when s is less than f and a real image when s is greater than f. (See Exercise 10.)

13. A certain telescope uses a concave spherical mirror of radius 8 m. Find the location and diameter of the image of the moon formed by this mirror. The moon has a diameter of 3.5×10^6 m and is 3.8×10^8 m from the mirror.

Section 25-3   Images Formed by Refraction

14. A sheet of paper with writing on it is protected by a thick glass plate having an index of refraction of 1.5. If the plate is 2 cm thick, at what distance beneath the top of the plate does the writing appear when viewed from directly overhead?

15. A very long glass rod has one end ground to a convex hemispherical surface of radius 5 cm. Its index of refraction is 1.5. (a) A point object is on the axis in air a distance 20 cm from the surface. Find the image and state whether it is real or virtual. Repeat (a) for (b) an object 5 cm from the surface and (c) an object very far from the surface. Using your results, draw a ray diagram for each case.

16. At what distance from the rod of Exercise 15 should an object be placed so that the light rays in the rod are parallel to the axis? Draw a ray diagram for this situation.

17. Repeat Exercise 15 for a glass rod with a concave hemispherical surface of radius −5 cm.

18. Repeat Exercise 15 for the glass rod and the objects immersed in water of index of refraction 1.33.

19. Repeat Exercise 15 for a glass rod with a concave hemispherical surface ($r = -5$ cm) and the objects immersed in water of index of refraction 1.33.

20. A fish is 10 cm from the front surface of a fish bowl of radius 20 cm. (a) Where does the fish appear to be to someone in air viewing it from in front of the bowl? (b) Where does the fish appear to be when it is 30 cm from the front surface of the bowl?

21. A scuba diver wears a diving mask with a face plate that bulges outward with a radius of curvature of 0.5 m. There is thus a convex spherical surface between the water and the air in the mask. A fish is 2.5 m from the diver's mask. Where does the fish appear to be? What is the magnification of the image of the fish?

Section 25-4   Thin Lenses

22. The following thin lenses are made of glass with an index of refraction of 1.6. Make a sketch of the lens, and find the focal length in air for each: (a) double convex, $r_1 = 10$ cm and $r_2 = -21$ cm; (b) plano-convex, $r_1 = \infty$ and $r_2 = -10$ cm; (c) double concave, $r_1 = -10$ cm and $r_2 = +10$ cm; and (d) plano-concave, $r_1 = \infty$ and $r_2 = +20$ cm.

23. Glass with an index of refraction of 1.5 is used to make a thin lens that has radii of equal magnitude. Find the radii of curvature and make a sketch of the lens if the focal length in air is (a) +5 cm and (b) −5 cm.

24. The following thin lenses are made of glass of index of refraction 1.6. Make a sketch of the lens and find the focal length in air for each: (a) $r_1 = 20$ cm, $r_2 = 10$ cm; (b) $r_1 = 10$ cm, $r_2 = 20$ cm; and (c) $r_1 = -10$ cm, $r_2 = -20$ cm.

25. For the following object distances and focal lengths of thin lenses in air, find the image distance and the magnification, and state whether the image is real or virtual and erect or inverted: (a) $s = 40$ cm, $f = 20$ cm; (b) $s = 10$ cm, $f = 20$ cm; (c) $s = 40$ cm, $f = -30$ cm; and (d) $s = 10$ cm, $f = -30$ cm.

26. An object 3.0 cm high is placed 20 cm from a thin lens of power 20 D. Carefully draw a ray diagram to find the position and size of the image and check your result using the thin-lens equation.

27. Repeat Exercise 26 for an object 1.0 cm high placed 10 cm from a thin lens of power 20 D.

28. Repeat Exercise 26 for an object 1.0 cm high placed 10 cm from a thin lens whose power is −20 D.

29. Show that a diverging lens can never form a real image from a real object. (Hint: Show that s' is always negative.)

30. Show that a converging lens forms a real image from a real object when s is greater than f. (Hint: Show that s' is then positive.)

Section 25-5   Aberrations

31. A double convex lens of radii $r_1 = +10$ cm and $r_2 = -10$ cm is made from glass with an index of refraction of 1.53 for violet light and 1.47 for red light. Find the focal lengths of this lens for red and violet light.

Problems

1. A thin converging lens of focal length 10 cm is used to obtain an image that is twice as large as a small object. Find the object and image distances if (a) the image is erect and (b) the image is inverted. Draw a ray diagram for each case.

2. A concave mirror has a radius of curvature of 6.0 cm. Draw rays parallel to the axis at heights 0.5, 1.0, 2.0, and 4.0 cm, and find the points at which the reflected rays cross the axis. (Use a compass to draw the mirror and a protractor to find the angle of reflection for each ray.) What is the spread Δx of the points where these rays cross the axis? By what percentage could this spread be reduced if the edge of the mirror were blocked off so that parallel rays of heights greater than 2.0 cm could not strike the mirror?

3. Two converging lenses, each of focal length 10 cm, are separated by 35 cm. An object is 20 cm to the left of the first lens. Find the position of the final image using both ray diagrams and the lens equation. Is the image real or virtual and erect or inverted? What is the overall lateral magnification?

4. A point object is on the axis of the concave mirror of Problem 2. Carefully construct a ray diagram showing the rays from the object that make angles of 5°, 10°, 30°, and 60° with the axis, strike the mirror, and are reflected back across the axis. (Use a compass to draw the mirror and a protractor to measure the angles needed to find the reflected rays.) What is the spread along the axis of the image for these rays?

5. Work Problem 3 when the second lens is a diverging lens with a focal length of −15 cm.

6. What is meant by a negative object distance? How can it occur? Find the image distance and magnification and state whether the image is virtual or real and erect or inverted for a thin lens in air when (*a*) $s = -20$ cm, $f = +20$ cm and (*b*) $s = -10$ cm, $f = -30$ cm. Draw a ray diagram for each of these cases.

7. An object is 15 cm in front of a positive lens of focal length 15 cm. A second positive lens of focal length 15 cm is 20 cm from the first lens. Find the final image and draw a ray diagram.

8. You wish to see an image of your face for putting on makeup or shaving. If you want the image to be upright, virtual, and magnified 1.5 times when your face is 30 cm from the mirror, what kind of mirror should you use, convex or concave, and what should its focal length be?

9. A glass rod 96 cm long with an index of refraction of 1.6 has its ends ground to convex spherical shapes of radii 8 cm and 16 cm. A point object is in air on the axis 20 cm from the 8-cm radius end. (*a*) Find the image distance due to refraction at the first surface. (*b*) Find the final image due to refraction at both surfaces. (*c*) Is the final image real or virtual?

10. Repeat Problem 9 for a point object in air 20 cm from the 16-cm radius end.

11. A small object is 15 cm from a thin positive lens of focal length 10 cm. To the right of the lens is a plane mirror that crosses the axis at the second focal point of the lens and is tilted so that the reflected rays do not go back through the lens (see Figure 25-37). Sketch a ray diagram showing the final image. Is this image real or virtual?

12. Find the final image for the situation in Problem 11 if the mirror is not tilted. Assume that the image is viewed by an eye to the left of the object looking through the lens into the mirror.

13. When a bright light source is placed 30 cm in front of a lens, there is an erect image 7.5 cm from the lens. There is also a faint inverted image 6 cm in front of the lens due to reflection from the front surface of the lens. When the lens is turned around, this weaker inverted image is 10 cm in front of the lens. Find the index of refraction of the lens.

14. A horizontal concave mirror with radius of curvature 50 cm holds a layer of water of index of refraction 1.33 and maximum depth 1 cm. At what height above the mirror must an object be placed so that its image is at the same position as the object?

15. In the seventeenth century, Anton von Leeuwenhoek, the first great microscopist, used simple spherical lenses, first made of water droplets and then of glass, for his first instruments. He made staggering discoveries with these simple lenses. Consider a glass sphere of radius 2.0 mm with an index of refraction of 1.50. Find the focal length of this lens. (*Hint:* Use the equation for refraction at a single spherical surface to find the image distance for an infinite object distance for the first surface. Then use this image point as the object point for the second surface.)

16. Find the focal length of a *thick* double convex lens with index of refraction 1.5, thickness 4 cm, and radii +20 cm and −20 cm.

17. A convenient form of the thin-lens equation used by Newton measures the object and image distance from the focal points. Show that if $x = s - f$ and $x' = s' - f$, the thin-lens equation can be written $xx' = f^2$ and the lateral magnification is given by $m = -x'/f = -f/x$. Indicate x and x' on a sketch of a lens.

18. A glass ball of radius 10 cm has an index of refraction of 1.5. The back half of the ball is silvered so that it acts as a concave mirror (see Figure 25-38). Find the position of the final image seen by an eye to the left of the object and the ball for an object at (*a*) 30 cm and (*b*) 20 cm to the left of the front surface of the ball.

Figure 25-38
Problem 18.

Figure 25-37
Problem 11.

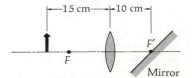

CHAPTER TWENTY-SIX

Optical Instruments

But optics sharp it needs, I ween
To see what is not to be seen

JOHN TRUMBULL

The invention of the microscope gave scientists access to that part of our environment too tiny to be seen with the naked eye. Similarly, astronomy was largely limited to the observation and prediction of the positions of stars and planets until Galileo turned his version of the newly invented telescope on the night sky. The development of the camera allowed formerly transient observations to be saved for unhurried study and contemplation. In short, much of the progress of science has gone hand in hand with the invention of optical instruments and further developments in their designs.

In this chapter, we use what we have learned about mirrors and lenses to examine the workings of various optical instruments, such as the camera, the simple magnifier, the microscope, and the telescope. The most important optical instrument is the eye, which we will study first. Though many optical instruments used today are quite complicated, the basic principles behind their workings are often quite simple.

Table salt magnified 50 times by an electron microscope.

667

26-1   The Eye

The optical system of prime importance is the eye, shown in Figure 26-1.

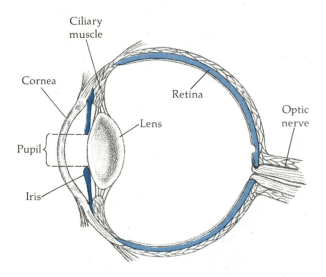

Ciliary
muscle

Cornea

Retina

Optic
nerve

Lens

Pupil

Iris

Figure 26-1
Cutaway view of the human eye.
The amount of light entering the
eye is controlled by the iris, which
regulates the size of the pupil. The
lens thickness is controlled by the
ciliary muscle. The cornea and lens
together focus the image on the
retina, which contains about 125
million receptors called rods and
cones.

Light enters the eye through a variable aperture, the pupil, and is focused by the cornea–lens system on the retina, a film of nerve fibers covering the back surface of the eyeball. The retina contains tiny sensing structures, called *rods* and *cones*, that receive the image and transmit the information along to the optic nerve to the brain. The shape of the lens can be altered slightly by the action of the ciliary muscle. When the eye is focused on an object far away, the muscle is relaxed and the cornea–lens system has its maximum focal length of about 2.5 cm, the distance from the cornea to the retina. When the object is brought closer to the eye, the ciliary muscle increases the curvature of the lens slightly, thereby decreasing its focal length so that the image is again focused on the retina. This process is called *accommodation.* If the object is too close to the eye, the lens cannot focus the light on the retina and the image is blurred. The closest point for which the lens can focus an image on the retina for extended periods of time without undue strain is called the *near point.* For this reason, the distance to the near point is called the *distance of most distinct vision.* The distance from the eye to the near point varies greatly from one person to another and changes with

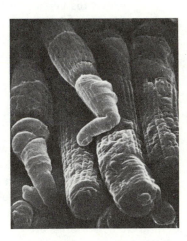

Rods (cylindrical structures
extending into lower foreground)
and cones (upper middle and left) as
seen through a scanning electron
microscope, magnified 1600 times.

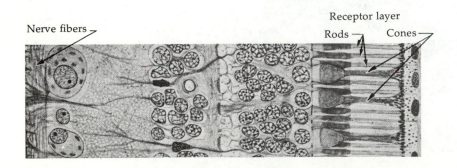

Nerve fibers

Receptor layer

Rods   Cones

Section of the retina of the human
eye, magnified 500 times.

age. At the age of 10 years, the near point may be as close as 7 cm, whereas at 60 years, it may recede to 200 cm because of loss of flexibility of the lens. The standard value used for the near point is 25 cm.

If the eye underconverges, resulting in images being focused behind the retina, the person is said to be farsighted. A farsighted person can see distant objects where little convergence is required but has trouble seeing close objects. Farsightedness is corrected with a converging lens (see Figure 26-2). On the other hand, the eye of a nearsighted person overconverges and focuses light from distant objects in front of the retina. A nearsighted person can see nearby objects, for which the widely diverging incident rays can be focused on the retina, but has trouble seeing distant objects. Nearsightedness is corrected with a diverging lens (see Figure 26-3).

Farsightedness

Nearsightedness

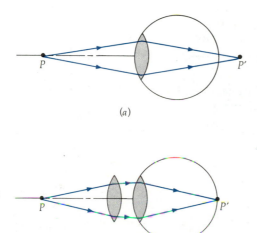

(a)

(b)

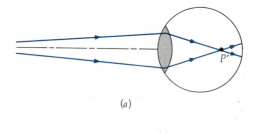

(a)

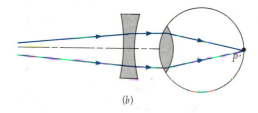

(b)

Figure 26-2
(a) The lens of a farsighted eye focuses rays from a nearby object P to a point P' behind the retina. (b) A converging lens corrects this defect by bringing the image onto the retina.

Figure 26-3
(a) The lens of a nearsighted eye focuses rays from a distant object to a point in front of the retina. (b) A diverging lens corrects this defect.

Another common defect of vision is astigmatism, which is caused by the cornea being not quite spherical but having a different curvature in one plane than in another, resulting in a blurring of the image of a point object into a short line. Astigmatism is corrected by spectacles using lenses of cylindrical rather than spherical shape.

Example 26-1   By how much must the focal length of the cornea–lens system of the eye change when the object is moved from infinity to the near point at 25 cm? Assume that the distance from the cornea to the retina is 2.5 cm.

When the object is at infinity, the rays from the object are parallel and are focused by the eye on the retina, giving a focal length for the cornea–lens system of 2.5 cm. When the object is at 25 cm, the focal length f must be such that the image distance is 2.5 cm. Using $s = 25$ cm for the object distance and $s' = 2.5$ cm for the image distance in the thin-lens equation (Equation 25-12), we have

$$\frac{1}{25 \text{ cm}} + \frac{1}{2.5 \text{ cm}} = \frac{1}{f}$$

$$\frac{1}{f} = \frac{1}{25 \text{ cm}} + \frac{10}{25 \text{ cm}} = \frac{11}{25 \text{ cm}}$$

$$f = \frac{25 \text{ cm}}{11} = 2.27 \text{ cm}$$

The focal length must therefore decrease by 0.23 cm. In terms of the power of the cornea–lens system, when the focal length is 2.5 cm = 0.025 m for distant objects, the power is $P = 1/f = 40$ D. When the focal length is 2.27 cm, the power is 44 D.

The eye is a wonderful optical instrument. By accommodation, it can focus on objects anywhere from the near point of about 25 cm to infinity!

The apparent size of an object is determined by the size of the image on the retina. From Figures 26-4a and b, we can see that the size of the image on the retina is greater when the object is close than when it is far away, even though the actual size of the object does not change. A convenient measure of the size of the image on the retina is the angle θ subtended by the object at the eye, as shown in Figure 26-4. From Figure 26-4c, we can see that the angle θ is related to the image size y' by

$$\theta = \frac{y'}{2.5 \text{ cm}} \qquad\qquad\qquad 26\text{-}1$$

The image size is therefore directly proportional to the angle subtended by the object. From Figure 26-4a or b, we can see that the angle θ is related to

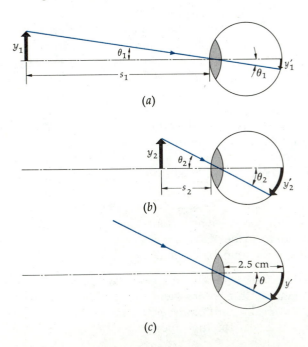

(a)

(b)

(c)

Figure 26-4
(a) A distant object of height y_1 looks small because the image on the retina is small. (b) When the same object is closer, it looks larger because the image on the retina is larger. The size of the image on the retina is proportional to the angle θ subtended by the object, which is in turn inversely proportional to the object distance. (c) The angle subtended is $\theta = y'/(2.5 \text{ cm})$.

the object size y and object distance s:

$$\tan \theta = \frac{y}{s}$$

For small angles, we can use the approximation $\tan \theta \approx \theta$ and write

$$\theta \approx \frac{y}{s} \qquad\qquad 26\text{-}2$$

Combining Equations 26-1 and 26-2, we have

$$y' = (2.5 \text{ cm})\theta \approx (2.5 \text{ cm})\frac{y}{s} \qquad\qquad 26\text{-}3$$

Thus, the size of the image on the retina is proportional to the size of the object and inversely proportional to the distance between the object and the eye.

Example 26-2   Assume that the near point of your eye is 75 cm. What power reading glasses should be used to bring your near point to 25 cm?

If your near point is 75 cm, you are farsighted. To read a book, you must hold the book at least 75 cm from your eye so that you can focus on the print. The image of the print on your retina is then very small. A converging lens, as is used in reading glasses, allows the book to be brought closer to the eye so that the image of the print is larger. When the book is 25 cm from your eye, we want the image of the converging lens to be 75 cm from your eye. We recall that a converging lens forms a virtual, erect image when the object is between the lens and the focal point. We therefore expect the focal length of the lens to be greater than 25 cm.

   Figure 26-5 shows a diagram of an object 25 cm from a converging lens that produces a virtual, erect image at $s' = -75$ cm. Using the thin-lens equation with $s = 25$ cm and $s' = -75$ cm, we obtain

$$\frac{1}{25 \text{ cm}} + \frac{1}{-75 \text{ cm}} = \frac{1}{f}$$

$$\frac{1}{f} = \frac{2}{75 \text{ cm}} = \frac{1}{0.375 \text{ m}} = 2.67 \text{ D}$$

The power is 2.67 D and the focal length is

$$f = \frac{1}{2.67 \text{ D}} = 0.375 \text{ m} = 37.5 \text{ cm}$$

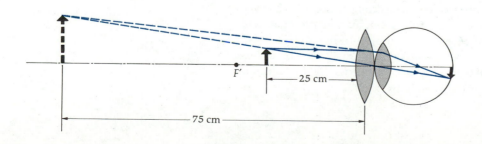

Figure 26-5
Ray diagram for Example 26-2. When the object is placed just inside the first focal point of a converging lens, the image is virtual, erect, enlarged, and far from the lens. In this example, the image distance is chosen to be -75 cm, the near point of a farsighted eye, and the object distance is chosen to be 25 cm. The focal length of the lens for these choices is then calculated from the thin-lens equation.

There is an alternate way to solve this problem that uses the result from Section 25-4 that the power of two lenses in contact is the sum of the powers of the individual lenses. Without reading glasses, when the object is at 75 cm $= 0.75$ m from the eye, the image is at 2.5 cm $= 0.025$ m, the distance from the eye lens to the retina. The thin-lens equation then gives for the focal length of the eye

$$\frac{1}{0.75 \text{ m}} + \frac{1}{0.025 \text{ m}} = \frac{1}{f_e}$$

The power of the eye lens is then

$$P_e = \frac{1}{f_e} = 1.33 + 40.00 = 41.33 \text{ D}$$

When the reading-glass lens is used, the image distance of the combination should be 0.025 m when the object is at a distance of 25 cm. If f_c is the focal length of the combination and $P_c = 1/f_c$ is the power of the combination, we have

$$\frac{1}{0.25 \text{ m}} + \frac{1}{0.025 \text{ m}} = \frac{1}{f_c} = P_c$$

or

$$P_c = 4.00 + 40 = 44.0 \text{ D}$$

The power of the combination equals the sum of the power of the eye lens and that of the reading-glass lens:

$$P_c = P_e + P_g$$

The power of the reading-glass lens is thus

$$P_g = P_c - P_e = 44.0 - 41.33 = 2.67 \text{ D}$$

This is in agreement with our first calculation.

Question

1. Glasses of power -2 D are prescribed for a certain person. Is that person nearsighted or farsighted?

26-2   The Simple Magnifier

We saw in Example 26-2 that the apparent size of an object can be increased by using a converging lens to allow the object to be brought closer to the eye, thus increasing the size of the image on the retina. Such a converging lens is called a *simple magnifier*. In Figure 26-6a a small object of height y is at the near point of the eye, at distance x_{np}. As discussed previously, the size of the image on the retina is proportional to the angle θ_0 subtended by the object at the eye. In this case θ_0 is given approximately by

$$\theta_0 = \frac{y}{x_{np}}$$

Figure 26-6

(a) An object at the near point subtends an angle θ_0 at the eye. (b) When the object is at the focal point of the converging lens, the rays emerge from the lens parallel and enter the eye as if they came from an object a very large distance away. The image can thus be viewed at infinity by the relaxed eye. When f is less than the near point, the converging lens allows the object to be brought closer to the eye, increasing the angle subtended by the object to θ and thereby increasing the size of the image on the retina.

In Figure 26-6b, a converging lens of focal length f, which is smaller than x_{np}, is placed in front of the eye, and the object is placed at the focal point of the lens. The rays emerge from the lens parallel, indicating that the image is at an infinite distance in front of the lens. The parallel rays are focused by the relaxed eye on the retina. Assuming the lens is close to the eye, the angle subtended by the object is now approximately

$$\theta = \frac{y}{f}$$

The ratio θ/θ_0 is called the *angular magnification* or *magnifying power M* of the lens:

$$M = \frac{\theta}{\theta_0} = \frac{x_{np}}{f} \qquad\qquad 26\text{-}4 \quad \textbf{Angular magnification}$$

Example 26-3   A person with a near point of 25 cm uses a 40-D lens as a simple magnifier. What magnification is obtained?

The focal length of a 40-D lens is

$$f = \frac{1}{P} = \frac{1}{40 \text{ m}^{-1}}$$

$$= 0.025 \text{ m} = 2.5 \text{ cm}$$

Using $x_{np} = 25$ cm and $f = 2.5$ cm in Equation 26-4, we obtain for the magnification

$$M = \frac{25 \text{ cm}}{2.5 \text{ cm}} = 10$$

An object looks 10 times larger because it can be placed at 2.5 cm rather than at 25 cm from the eye thus increasing the image on the retina tenfold, while using the additional converging power of the magnifier to focus the very divergent rays from the close object.

The magnification of a simple magnifier can be increased slightly by moving the object closer to the magnifier. When the object is inside the focal point of the magnifier, the image is virtual and erect as in Figure 26-5. As the object is moved toward the magnifier, the image moves closer to the eye, and the angle subtended increases slightly. The largest usable magnification occurs when the image is at the near point of the eye, as it was in Example 26-2. Calculation shows that the magnification in this case is just 1 greater than it is with the image at infinity (see Problem 26-1). For example, if the near point distance is 25 cm, a lens with a focal length of 2.5 cm would give a magnification of 11 with the image at 25 cm rather than a magnification of 10 with the image at infinity. In practice, this gain in magnification is not worth the additional eye strain incurred in viewing the image at the near point rather than at infinity with a relaxed eye, so we will use Equation 26-4 for the magnification of a simple magnifier.

Simple magnifiers are used as eyepieces in compound microscopes and telescopes to view the image formed by another lens or lens system. To correct for aberrations, combinations of lenses with a resulting short positive focal length may be used in place of a single lens, but the principle of the simple magnifier is the same.

26-3   The Camera

A camera consists of a positive lens, a variable aperture called the iris or f-stop, a shutter than can be opened for a short time that can be varied, a light-tight box, and a film (see Figure 26-7). Unlike the eye, which has a lens of variable focal length, the focal length of a camera lens is fixed. Typically, the focal length for a normal lens in a 35-mm camera is 50 mm. (The 35-mm refers to the width of the film). Focusing is accomplished by varying the distance from the lens to the film by moving the lens closer or farther from the film.

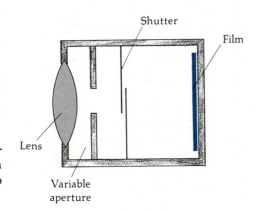

Figure 26-7
Schematic diagram of a camera. The positive lens focuses the light onto the film. The variable aperture limits the amount of light entering the camera and the area of the lens used. The shutter speed can be varied to vary the exposure time.

Example 26-4   The focal length of a camera lens is 50 mm. By how much must the lens be moved to change from focusing on an object far away to one 2 m away?

When the object is far away, the image of the lens is at the focal length, so the film should be 50 mm from the lens. When the object is 2 m away, the image distance is

$$\frac{1}{2 \text{ m}} + \frac{1}{s'} = \frac{1}{f} = \frac{1}{50 \text{ mm}}$$

$$\frac{1}{s'} = \frac{1}{50 \text{ mm}} - \frac{1}{2000 \text{ mm}} = \frac{40}{2000 \text{ mm}} - \frac{1}{2000 \text{ mm}}$$

$$s' = \frac{2000 \text{ mm}}{39} = 51.3 \text{ mm}$$

The lens must therefore be moved 1.3 mm farther away from the film.

The amount of light that strikes the film is varied by varying the time that the shutter is opened and by varying the size of the aperture. For a given film type, there is an optimum amount of light that will give a good picture of proper contrast. Too little light results in a dark picture. Too much light results in a washed out picture with too little contrast. The amount of light needed for proper contrast is related to the film's "speed," which is rated by an ASA number. The higher the ASA number, the faster the film, meaning that less light is needed. A film with a high ASA number, such as ASA 400 or ASA 1000, is good for pictures taken indoors where there is little available light. To make high-speed films, some sacrifice in picture quality (sharpness in the image or true color reproduction) is usually required, so for photographs taken outdoors where plenty of light is available, it is usually preferable to use lower-speed film, such as ASA 100 or ASA 64. Lower-speed film may also be needed in bright-light situations if the shutter speed variability of the camera is limited. Shutter speeds on good cameras can often be varied from exposures of several seconds for low-light photography to 1/1000 of a second for stop-action photography. For a hand-held camera, exposure times of more than about 1/60 of a second will result in blurring of the image because of camera motion.

The size of the aperture is limited by the size of the lens, which is in turn limited by the various lens aberrations discussed in Section 25-5. Although we will treat the camera lens as a single positive lens, all optical systems in good cameras use combinations of lenses to reduce chromatic, spherical, and other aberrations. The aperture size is given by the $f/number$, which is the ratio of the focal length to the diameter of the aperture:

$$f/\text{number} = \frac{f}{D} \qquad\qquad\qquad\qquad \text{26-5} \quad \textit{f/number}$$

The maximum aperture is the $f/number$ of the lens. For example, an $f/2.8$ lens of focal length 50 mm has a usable diameter given by

$$D = \frac{f}{f/\text{number}} = \frac{50 \text{ mm}}{2.8} = 17.9 \text{ mm}$$

Large-diameter lenses (small $f/numbers$) are costly to make because of the expense of correcting for aberrations. Aperture settings on a camera are usually marked $f/22$, $f/16$, $f/11$, $f/8$, $f/5.6$, $f/4$, $f/2.8$, $f/2.0$, $f/1.4$, $f/1.0$, and on down to the lowest $f/number$ corresponding to the largest usable diameter of the lens. Note that each successive setting corresponds to an aperture diameter $\sqrt{2} = 1.4$ times that of the previous one. Since the amount of light that enters is proportional to the area of the lens, which is in turn proportional to the square of the diameter, stopping down the aperture by one $f/stop$, for example, from $f/2.0$ to $f/1.4$, increases the area and therefore the amount of light entering in a given time by a factor of 2.

When the camera is focused for a certain distance, objects at nearby distances are also in relatively good focus. The distance over which objects are in relatively good focus is called the *depth of field*. The depth of field depends on the size of the lens aperture. The depth of field is larger when the lens aperture is small, that is, when the lens is stopped down (high $f/number$). A large depth of field is not always desirable. For example, you might want to have a person in a photograph in sharp focus with the background slightly blurred. To do this, you would use a large aperture (low f/number) and increase the shutter speed to reduce the amount of light.

Example 26-5 The instructions for a certain film say to set the aperture at f/11 and the shutter speed at 1/250 s for pictures on bright, sunny days. What should the shutter speed be if the aperture is set at f/5.6?

Since $11^2 = 121$ and $5.6^2 = 31.4$, the area of the aperture at f/5.6 is 121/31.4 = 3.85 $\approx$ 4 times as great as called for. Thus, to get the same light, you should use a shutter speed 4 times faster or 1/1000 s.

    The focal length of 50 mm for a 35-mm camera is chosen to obtain a field of view approximately the same as that for ordinary vision, which is about 45°. To increase the field of view, a wide-angle lens of smaller focal length, for example 24 mm, is used. When the object distance is much greater than the focal length, which is usually the case for a camera, the image distance s' is approximately equal to f. Since the lateral magnification of a lens is $m = -s'/s$, the size of the image on the film is approximately proportional to the focal length. A wide-angle lens thus gives a smaller image on the film than does a normal lens for a given object size. A telephoto lens has a large focal length to increase the size of the image on the film and thus make the object seem closer. A telephoto lens with a focal length of 200 mm would give a magnification approximately 4 times that of an ordinary lens of focal length 50 mm.

Photograph of the University of Michigan football stadium using a wide-angle lens.

Questions

2. What are the advantages of having a camera with a fast shutter?

3. When might you want to use a small aperture rather than a large aperture?

4. Why is an f/1.0 lens more expensive than an f/2.8 lens?

26-4   The Compound Microscope

The compound microscope (see Figure 26-8) is used to look at very small objects at short distances. In its simplest form, it consists of two converging lenses. The lens nearest the object, called the *objective,* forms a real image of the object. This image is enlarged and inverted. The lens nearest the eye, called the *eyepiece* or *ocular,* is used as a simple magnifier for viewing the

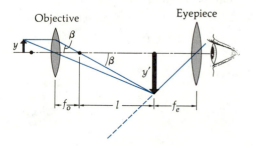

Figure 26-8
Schematic diagram of a compound microscope consisting of two positive lenses, the objective of focal length f_o and the eyepiece or ocular of focal length f_e. The real image of the object formed by the objective is viewed by the eyepiece, which acts as a simple magnifier. The final image is at infinity.

image formed by the objective. The eyepiece is placed so that the image of the objective falls at the first focal point of the eyepiece. The light thus emerges from the eyepiece as a parallel beam as if coming from a point a great distance in front of the lens. (This is commonly called viewing the image at infinity.)

As discussed in Section 26-2, the function of a simple magnifier, the eyepiece in this case, is to allow the object (which is the image of the objective) to be brought closer to the eye than the near point. Since a simple magnifier produces a virtual image that is erect, the final image produced by the two lenses is inverted. The distance between the second focal point of the objective and the first focal point of the eyepiece is called the *tube length l* and is usually fixed. The object is placed just outside the first focal point of the objective so that an enlarged image is formed at the first focal point of the eyepiece, a distance $l + f_o$ from the objective, where f_o is the focal length of the objective. From Figure 26-8, $\tan \beta = y/f_o = -y'/l$. The lateral magnification of the objective is therefore

$$m_o = \frac{y'}{y} = -\frac{l}{f_o} \tag{26-6}$$

The angular magnification of the eyepiece is

$$M_e = \frac{x_{np}}{f_e}$$

The first microscopic investigations were performed by Anton von Leeuwenhoek during the seventeenth century. He used for his first instruments simple spherical lenses, first of water droplets and then of glass.

where x_{np} is the near point of the viewer and f_e is the focal length of the eyepiece. As discussed in Section 26-2, a slightly greater angular magnification can be obtained by placing the object (the image of the objective) at a point just inside the first focal point of the eyepiece so that the final image is at the near point. The slight gain in angular magnification of the eyepiece is usually not worth the eye strain in viewing the image at the near point rather than viewing it at infinity with a relaxed eye. The overall magnification of the compound microscope is the product of the lateral magnification of the objective and the angular magnification of the eyepiece.

$$m = m_o M_e = -\left(\frac{l}{f_o}\right)\frac{x_{np}}{f_e}$$

26-7   Microscope magnification

Example 26-6   A microscope has an objective lens of focal length 1.2 cm and an eyepiece of focal length 2.0 cm. They are separated by 20 cm. Find the overall magnification if the near point of the viewer is 25 cm. Where should the object be placed if the final image is to be viewed at infinity?

The distance between the second focal point of the objective and the first focal point of the eyepiece is 20 cm $-$ 2 cm $-$ 1.2 cm $=$ 16.8 cm. The overall magnification is given by Equation 26-7 with $l = 16.8$ cm, $f_o = 1.2$ cm, $f_e = 2.0$ cm, and $x_{np} = 25$ cm:

$$m = -\left(\frac{16.8\ \text{cm}}{1.2\ \text{cm}}\right)\frac{25\ \text{cm}}{2\ \text{cm}} = -175$$

We can calculate the object distance between the original object and the objective from the thin-lens equation. From Figure 26-8, we can see that the image distance is

$$s' = f_o + l = 1.2\ \text{cm} + 16.8\ \text{cm} = 18\ \text{cm}$$

The object distance is then found from

$$\frac{1}{s} + \frac{1}{s'} = \frac{1}{f}$$

$$\frac{1}{s} + \frac{1}{18\ \text{cm}} = \frac{1}{1.2\ \text{cm}}$$

Solving for s, we obtain $s = 1.286$ cm. The object should therefore be placed at 1.286 cm from the objective or 0.086 cm outside its first focal point.

26-5 The Telescope

A telescope is used to view objects that are far away and often large. Its purpose is to bring the object closer, that is, to increase the angle subtended so that the object appears larger. The astronomical telescope, illustrated schematically in Figure 26-9, consists of two positive lenses, an objective lens that forms a real, inverted image and an eyepiece or ocular that is used as a simple magnifier to view that image. Since the object is very far away, the image of the objective lies at the focal point of the objective, and the image distance equals the focal length f_o. Since the object distance is much larger than the focal length of the objective, the image of the objective is much smaller than the object. For example, if we are looking at the moon, the image of the moon formed by the objective is much smaller than the moon itself. The purpose of the objective is not to magnify the object, but to

Seventeenth-century telescope used by Galileo to study the heavens. With it, he discovered, among other things, the mountains on the moon, sunspots, Saturn's rings, and the bands of Jupiter. One of his most important discoveries was the existence of the moons of Jupiter. He viewed Jupiter and its moons as a model of the solar system. This, in turn, convinced him that Copernicus' claim that the planets revolved around the sun was correct.

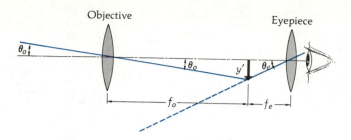

Objective   Eyepiece

Figure 26-9
Schematic diagram of an
astronomical telescope. The
objective forms a real image of a
distant object near its second focal
point, which coincides with the first
focal point of the eyepiece. The
eyepiece serves as a simple
magnifier for viewing the image.

produce an image that is much closer than the object to be viewed by the
eyepiece. Since this image is at the second focal point of the objective and at
the first focal point of the eyepiece , the objective and eyepiece must be
separated by the sum of their focal lengths, $f_o + f_e$, where f_e is the focal
length of the eyepiece.

We find the angular magnification of the telescope by comparing the
angle θ_o subtended by the object when it is viewed directly by the unaided
eye, to the angle θ_e subtended by the final image as it is viewed through the
eyepiece. The angular magnification of the telescope is θ_o / θ_e. The angle θ_o is
the same as that subtended by the object at the objective, as shown in Figure
26-9. From this figure, we can see that

$$\tan \theta_o = -\frac{y'}{f_o} \approx \theta_o$$

where we have used the small-angle approximation $\tan \theta \approx \theta$ and have
introduced a negative sign to make θ_o positive when y' is negative. The angle
θ_e in the figure is that subtended by the final image.

$$\tan \theta_e = \frac{y'}{f_e} \approx \theta_e$$

Since y' is negative, θ_e is negative, indicating that the image is inverted. The
angular magnification of the telescope is then

$$M = \frac{\theta_e}{\theta_o} = -\frac{f_o}{f_e}$$

26-8   **Telescope magnification**

A large angular magnification is thus obtained with an objective of large
focal length and an eyepiece of short focal length.

Example 26-7   The world's largest refracting telescope is at the Yerkes Observatory of the University of Chicago at Williams Bay, Wisconsin. The objective has a diameter of 102 cm and a focal length of 19.5 m. The focal length
of the eyepiece is 10 cm. What is the magnification of this telescope?

From Equation 26-8, we have

$$M = -\frac{f_o}{f_e} = -\frac{195 \text{ m}}{0.10 \text{ m}} = -195$$

The main consideration with an astronomical telescope is not its magnification but its light-gathering power, which depends on the size of the
objective. The larger the objective, the brighter the image. Very large lenses

Refracting telescope at Kitt Peak
Observatory in Tucson, Arizona.

without aberrations are difficult to produce. In addition, there are mechanical problems in supporting very large, heavy lenses since they must be supported by their edges. A reflecting telescope uses a concave mirror in place of a lens for its objective. This has several advantages. For one, a mirror does not exhibit chromatic aberration. In addition, mechanical support is much simpler since the mirror can be supported over its entire back surface and is much lighter in weight than a lens of equivalent optical quality.

One problem with the reflecting telescope is that the image of the objective mirror is viewed in the region of the incoming rays (see Figure 26-10). In very large reflecting telescopes, such as the 200-in (5.1 m) diameter telescope at Mt. Palomar, California, the viewer sits in an observer's cage near the focal point of the mirror. To obstruct as little light as possible, the cage is very small, resulting in highly cramped quarters and little space for sophisticated instruments such as spectrographs. With smaller telescopes, the fraction of light obstructed by such an arrangement is too great. One scheme for reducing the amount of light obstructed is to use a second small mirror to reflect the rays through a hole cut in the center of the objective, as illustrated in Figure 26-11. This has the further advantage of making the viewing area more accessible and providing more room for auxiliary instrumentation.

The fact that the final image is inverted in a simple telescope is not a disadvantage for viewing astronomical objects such as stars and planets, but it is for viewing terrestrial objects. Binoculars use two 45-45-90° prisms in each side to provide a second inversion of the image so that the final image is upright. Figure 26-12 shows a 45-45-90° prism with its hypotenuse horizontal and its reflecting sides vertical. Light entering through the long face is reflected twice and emerges back through the long face in the opposite direction. Horizontal images are inverted but vertical ones are not. A second prism with its hypotenuse vertical redirects the light back into its original direction and inverts vertical images without changing horizontal ones. The multiple reflections of the prisms also increase the path length of the light so that a relatively long focal length for the objective can be used in a relatively short space.

Objective mirror

Figure 26-10
A reflecting telescope uses a mirror for its objective. Because the viewer compartment blocks off some of the incoming light, this arrangement is used only in telescopes with very large objective mirrors.

The 200-in reflecting telescope at Mt. Palomar, California. A researcher is shown sitting in a small compartment near the focal point.

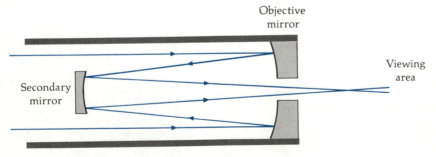

Objective mirror

Viewing area

Secondary mirror

Figure 26-11
Reflecting telescope with a secondary mirror to redirect the light through a small hole in the objective mirror. This arrangement has a further advantage over that of the reflecting telescope in Figure 26-10 in that there is more room for instrumentation in the viewing region.

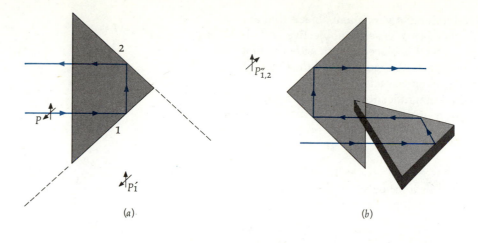

(a)   (b)

Figure 26-12
(*a*) Top view of a 45-45-90° prism
with its hypotenuse horizontal and
the reflecting faces 1 and 2 vertical.
The image due to reflection from
surfaces 1 and 2 is inverted in the
horizontal direction but not in the
vertical direction. (*b*) If the light
enters a second, identical prism
oriented with its hypotenuse
vertical, the image will be inverted
in the vertical direction. After
passing through both prisms, the
light emerges in its original direction
with the image inverted in both the
vertical and horizontal directions.
Two such prisms are inserted in
each side of binoculars to reinvert
the image that was originally
inverted by the binocular objective.
The prisms also allow for a long
focal length for the objective in a
short space.

Summary

1. One of the most important optical systems is the eye. The cornea – lens
system of the eye focuses light on the retina, where it is sensed by the rods
and cones, which send information along the optic nerve to the brain. The
focal length of the cornea – lens system when it is relaxed is about 2.5 cm,
the distance from the lens to the retina. When objects are brought near the
eye, the lens changes to decrease the overall focal length so that the image is
again focused on the retina. The closest distance for which the image can be
focused on the retina without undue strain is called the near point; it is
typically about 25 cm, but it varies with age and from person to person. The
apparent size of an object depends on the size of the image on the retina,
which in turn depends on the distance from the object to the eye. The closer
the object, the larger the image on the retina and, therefore, the larger the
apparent size of the object.

2. A simple magnifier consists of a lens with a positive focal length that is
smaller than the near point. The angular magnification of a simple magnifier is the ratio of the near-point distance to the focal length of the lens:

$$M = \frac{x_{np}}{f}$$

3. A camera consists of a lens, a variable aperture, a shutter, a light-tight
box, and a film. Since the focal length of the lens is fixed, focusing is
accomplished by moving the lens toward or away from the film. The
f/number of the aperture is the ratio of the focal length to the diameter of
the lens

$$f/\text{number} = \frac{f}{D}$$

The focal length of a typical lens in a 35-mm camera is about 50 mm. A
telephoto lens has a larger focal length, giving a larger image on the film, but
a narrower field of view. A wide-angle lens has a smaller focal length, giving
a smaller image on the film, but a wider field of view.

4. The compound microscope is used to look at very small objects that are nearby. In its simplest form, it consists of two lenses, an objective and an eyepiece or ocular. The object to be viewed is placed just outside the first focal point of the objective, which forms an enlarged image of the object at the first focal point of the eyepiece. The eyepiece acts as a simple magnifier for viewing the final image. The overall magnification of the microscope is the product of the lateral magnification of the objective and the angular magnification of the eyepiece:

$$m = m_o M_e = -\left(\frac{l}{f_o}\right)\frac{x_{np}}{f_e}$$

where l is the tube length, which is the distance between the second focal point of the objective and the first focal point of the eyepiece.

5. The telescope is used to view objects that are far away. The objective of the telescope forms a real image that is much smaller than the object but much closer. The eyepiece is then used as a simple magnifier for viewing the image. A reflecting telescope uses a mirror for its objective. The magnification of a telescope equals the (negative) ratio of the focal length of the objective to the focal length of the eyepiece

$$M = -\frac{f_o}{f_e}$$

The most important feature of an astronomical telescope is its light gathering power, which is proportional to the area of the objective.

Suggestions for Further Reading

Everhart, Thomas E., and Thomas L. Hayes; "The Scanning Electron Microscope," *Scientific American*, January 1972, p. 54.

This article describes how the interaction between a beam of high-energy electrons and matter is used by the scanning electron microscope to create an image of three-dimensional appearance.

Land, Michael F.: "Animal Eyes with Mirror Optics," *Scientific American*, December 1978, p. 126.

Several sea creatures, described in this article, have been found to use reflection rather than refraction to form images of their surroundings.

Price, William H.: "The Photographic Lens," *Scientific American*, August 1976, p. 72.

A history of lens design, including a description of the defects and aberrations one wishes to minimize in a lens, and a discussion of modern computer-aided design are presented in this article.

Review

A. Objectives: After studying this chapter, you should:

1. Be able to discuss how the eye works.

2. Be able to show with a simple diagram why an object appears larger when it is brought closer to the eye.

3. Be able to describe how a simple magnifier works and calculate its angular magnification.

4. Be able to discuss how a camera works.

5. Be able to describe with diagrams and equations how a microscope and a telescope work.

B. Define, explain, or otherwise identify:

Accommodation, p. 668
Near point, p. 668
Farsightedness, p. 669
Nearsightedness, p. 669
Angular magnification, p. 673

f/number, p. 675
Objective, p. 677
Eyepiece, p. 677
Ocular, p. 677
Tube length, p. 677

C. True or false: If the statement is true, explain why. If it is false, give a counterexample.

1. The lens of the eye always forms a real image.

2. A simple magnifier should have a short focal length.

3. A simple magnifier forms a virtual image.

4. The lens of a camera always forms a real image.

5. The area of a camera aperture is proportional to the *f*/number.

6. The focal length of a telephoto lens is greater than that of a wide-angle lens.

7. The image formed by the objective of a microscope is inverted and larger than the object.

8. The image formed by the objective of a telescope is inverted and larger than the object.

9. The final image of a telescope is virtual.

10. A reflecting telescope uses a mirror for its eyepiece.

Exercises

Section 26-1   The Eye

In the following exercises, take the distance from the cornea–lens system of the eye to the retina to be 2.5 cm.

1. Suppose the eye were designed like a camera with a lens of fixed focal length of 2.5 cm that could move toward or away from the retina. Approximately how far would the lens have to move to focus the image of an object that is 25 cm from the eye? (*Hint:* Find the distance from the retina to the image behind it for an object at 25 cm.)

2. Find the change in the focal length of the eye when an object originally at 3 m is brought to 30 cm from the eye.

3. A nearsighted person needs to read from a computer screen that is 45 cm from her eye. Her near point is at 80 cm. (*a*) Find the focal length of the lenses in reading glasses that will produce an image of the screen at 80 cm from her eye. (*b*) What is the power of the lenses?

4. Find (*a*) the focal length and (*b*) the power of a lens that will produce an image at 80 cm from the eye of a book that is 30 cm from the eye.

5. Since the index of refraction of the lens of the eye is not very different from that of the surrounding material, most of the refraction takes place at the cornea, where *n* changes abruptly from 1.0 in air to about 1.4. Using the formula for refraction at a single spherical surface (Equation 25-5), find the radius of a cornea that will image light from a distant object on the retina. Do you expect your result to be larger or smaller than the actual radius of the cornea?

6. If two point objects close together are to be seen as two distinct objects (see Figure 26-13), the images must fall on the retina on two different cones that are not adjacent. That is, there must be an unactivated cone between them. The separation of the cones is about 1 μm. (*a*) What is the smallest angle the two points can subtend? (*b*) How close together can two points be if they are 20 m from the eye?

Figure 26-13

Exercise 6. The two points will look like two separate points only if the retinal images are on two different, nonadjacent cones.

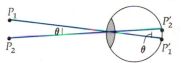

Section 26-2   The Simple Magnifier

7. A person with a near point distance of 30 cm uses a simple magnifier of power 20 D. What is the magnification obtained if the final image is at infinity?

8. A person with a near point distance of 25 cm wishes to obtain a magnification of 5 times with a simple magnifier. What should be the focal length of the lens used?

9. A lens of focal length 6 cm is used as a simple magnifier with the image at infinity by one person whose near point is 25 cm and by another whose near point is 40 cm. What is the effective magnifying power of the lens for each person? Compare the size of the image on the retina when each looks at the same object with the magnifier.

Section 26-3   The Camera

10. What is the diameter of an *f*/1.4 lens if its focal length is 50 mm?

11. A lens has a usable diameter of 2.5 cm. What is its *f*/number if its focal length is 50 mm?

12. A telephoto lens has a focal length of 200 mm. By how much must it move to change from focusing on an object at infinity to one at a distance of 30 m?

13. A wide-angle lens has a focal length of 28 mm. By how much must it move to change from focusing on an object at infinity to one at a distance of 5 m?

14. Light conditions on a certain day for a certain film call for an aperture stop of $f/8$ at $1/250$ s. (a) If you wish to take a picture of a hummingbird at $1/1000$ s, what f/number aperture should you use? (b) If you set the aperture at $f/22$ to maximize the depth of field for another picture, what shutter speed should you use?

Section 26-4   The Compound Microscope

15. A microscope objective has a focal length of 0.5 cm. It forms an image at 16 cm from its second focal point. What is the magnification if the focal length of the eyepiece is 3 cm?

16. A microscope has an objective of focal length 16 mm and an eyepiece that gives a magnification of 5 times for a person whose near point is 25 cm. The tube length is 18 cm. (a) What is the lateral magnification of the objective? (b) What is the overall magnification?

17. A crude symmetric hand-held microscope consists of two converging 20-D lenses fastened in the ends of a tube 30 cm long. (a) What is the "tube length" of this microscope? (b) What is the lateral magnification of the objective?

(c) What is the overall magnification of the microscope? (d) How far from the objective should the object be placed?

18. Repeat Exercise 17 for the same two lenses separated by 40 cm.

Section 26-5   The Telescope

19. A simple telescope has an objective with a focal length of 100 cm and an ocular of focal length 5 cm. It is used to look at the moon, which subtends an angle of about 0.009 rad. (a) What is the diameter of the image formed by the objective? (b) What angle is subtended by the final image at infinity? (c) What is the overall magnification of the telescope?

20. The objective lens of the refracting telescope at the Yerkes Observatory has a focal length of 19.5 m. What is the diameter of the image of the moon formed by the objective. (The angle subtended by the moon is 0.009 rad.)

21. The 200-in reflecting telescope at Mt. Palomar has a mirror (diameter 200 in = 5.1 m) of focal length 1.68 m. (a) By what factor is the light-gathering power increased over the 40-in (1.016 m) diameter refracting lens of the Yerkes Observatory telescope? (b) If the focal length of the eyepiece is 1.25 cm, what is the overall magnification of this telescope?

Problems

1. (a) Show that if the final image of a simple magnifier is to be at the near point of the eye rather than at infinity, the magnification is given by

$$M = \frac{x_{np}}{f} + 1$$

(b) Find the magnification of a 20-D lens for a person with a near point of 30 cm if the final image is at the near point. Draw a ray diagram for this situation.

2. Show that when the image of a simple magnifier is viewed at the near point, the lateral and angular magnification of the magnifier are equal.

3. A 35-mm camera has a picture size of 24 mm by 36 mm. It is used to take a picture of a person 175 cm tall such that the image just fills the height (24 mm) of the film. How far should the person stand from the camera if the focal length of the lens is 50 mm?

4. An astronomical telescope has a magnification of 7. The two lenses are 32 cm apart. Find the focal length of each lens.

5. The near point of a certain person is 80 cm. Reading glasses are prescribed so that he can read a book at 25 cm from his eyes. The glasses are 2 cm from the eyes. What diopter lens should be used in the glasses?

6. A 35-mm camera with interchangeable lenses is used to take a picture of a hawk that has a wingspan of 2 m. The hawk is 30 m away. What would be the ideal focal length for the lens so that the image of the wings just fills the width of the film, which is 36 mm?

7. A galilean telescope uses a converging lens as its objective but a diverging lens as its eyepiece. The image formed by the objective is behind the eyepiece at its focal point, so the final image is virtual, erect, and at infinity. (a) Show that the angular magnification is $M = -f_o/f_e$, where f_o is the focal length of the objective and f_e is that of the eyepiece (which is negative). Draw a ray diagram to show that the final image is indeed virtual, erect, and at infinity.

8. A galilean telescope is designed such that the final image is at the near point, which is 25 cm (rather than at infinity). The focal length of the objective is 100 cm and that of the eyepiece is -5 cm. (a) If the object distance is 30 m, where is the image of the objective? (b) What is the object distance for the eyepiece so that the final image is at the near point? (c) How far apart are the lenses? (d) If the object height is 1.5 m, what is the height of the final image? What is the angular magnification? (See Problem 7.)

CHAPTER TWENTY-SEVEN

Physical Optics: Interference and Diffraction

Whenever two portions of the same light arrive at the eye by different routes, either exactly or very nearly in the same direction, the light becomes most intense where the difference of the routes is a multiple of a certain length, and least intense in the intermediate state for the interfering portions; and this length is different for light of different colours.

THOMAS YOUNG (1802)

The beautiful colors in a soap film and in butterflies' wings are the result of interference. The bending of light around corners is known as diffraction. It is diffraction that limits the ability of optical instruments to resolve the details of a distant or very small source. Interference and diffraction are consequences of the wave nature of light. Their study is called physical optics, the subject of this chapter.

Interference and diffraction are the important phenomena that distinguish waves from particles. Interference is the combining by superposition of two or more waves that meet at one point in space. Diffraction, the bending of waves around corners, occurs when a portion of the wavefront is cut off by a barrier or obstacle. The pattern of the resulting wave can be calculated by treating each point on the original wavefront as a point source according to Huygens' principle and calculating the interference pattern resulting from these sources.

In Chapter 17, we discussed the interference of sound waves from two point sources and gave a brief, qualitative account of the diffraction of sound. Since the analytical treatment of interference and diffraction is the same for all waves, whether they are sound waves, waves in strings, water waves, or electromagnetic waves, you should review Chapter 17 before you begin this chapter.

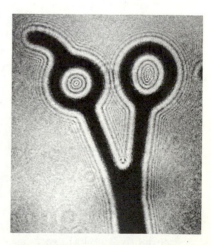

Diffraction pattern of light incident on a pair of scissors.

27-1   Phase Difference and Coherence

When two harmonic waves of the same frequency and wavelength but differing in phase combine, the resultant wave is a harmonic wave whose amplitude depends on the phase difference. If the phase difference is 0 or an integer times 360°, the waves are in phase and the interference is constructive. The resultant amplitude equals the sum of the individual amplitudes, and the intensity (which is proportional to the square of the amplitude) is maximum. If the phase difference is 180°, the waves are out of phase and the interference is destructive. The resultant amplitude is then the difference between the individual amplitudes, and the intensity is minimum. If the amplitudes are equal, the maximum intensity is 4 times that of either source and the minimum intensity is 0.

A common cause of phase difference between two waves is a difference in the path length traveled by each of the waves. A path difference of one wavelength produces a phase difference of 360°, which is equivalent to no phase difference at all. A path difference of one-half wavelength produces a 180° phase difference. In general, a path difference of Δx contributes a phase difference δ given by

$$\delta = \frac{\Delta x}{\lambda} \, 360° \qquad\qquad 27\text{-}1$$

Another cause of phase difference is the 180° phase change a wave undergoes upon reflection from a boundary surface when the wave speed is greater in the first (incident) medium than in the second. This phase change is analogous to the inversion of a pulse on a string when it reflects from a point where the density suddenly increases, as when a light string is attached to a heavy string or rope, as was illustrated in Figure 16-2. The inversion of the reflected pulse is equivalent to a phase change of 180° for a sinusoidal wave, which can be thought of as a series of pulses. Thus, when light traveling in air strikes a boundary surface of a medium in which light travels more slowly, such as glass or water, there is a 180° phase change in the reflected light.

When the pulse on a string is reflected from a point where the density suddenly decreases (such as a point where a heavy string is attached to a light string), the pulse is not inverted, as was also illustrated in Figure 16-2. Similarly, when light is originally traveling in glass or water, there is no phase change when the light is reflected from a glass–air or water–air surface. These results can be summarized by stating that when a wave is reflected from a boundary surface between two media in which the wave speed differs, there is a phase change of 180° if the wave speed is greater in the original medium and no phase change if the wave speed is less in the original medium.

As was mentioned in Chapter 17, interference of waves from two sources is not observed unless the sources are coherent, that is, unless the phase difference between the waves from the sources is constant over time. Because a light beam is usually the result of millions of atoms radiating independently, two light sources are usually not coherent. The phase difference between the waves from such sources fluctuates randomly many times per second. Coherence in optics is usually achieved by splitting a light beam from a single source into two or more beams, which can then be combined to produce an interference pattern. This splitting can be caused by the reflection from two nearby surfaces of a thin film (Section 27-2); simultaneous reflection and transmission from a half-silvered mirror, as in the Michelson interferometer (Section 27-3); or diffraction of the beam from two small openings or slits in an opaque barrier (Section 27-4). Coherent sources can also be obtained by using a single point source and its image in a plane mirror for the two sources, an arrangement called Lloyd's mirror. Lasers are also sources of coherent light.

27-2   Interference in Thin Films

Undoubtedly, you have noticed the colored bands in the film of a soap bubble or in an oil film on a water-covered street. These bands are due to the interference of light reflected from top and bottom surfaces of the film. The different colors arise because the thickness of the film varies, causing constructive interference for different wavelengths at different points. When a thin film is viewed with monochromatic light, such as the yellow light from a sodium lamp, light and dark bands called *fringes* are observed. We will now see how to calculate the spacing of these fringes.

Consider viewing, at small angles with the normal, a thin film of water of uniform thickness (see Figure 27-1). Part of the light is reflected from the upper air–water surface. Since light travels more slowly in water than in air, there is a 180° phase change in this reflection. Part of the light enters the film and is partially reflected by the bottom water–air surface. There is no phase change in this reflection. If the light is nearly perpendicular to the surfaces, both the ray reflected from the top surface and the one reflected from the bottom surface can enter the eye at point P in the figure. The path difference between these two rays is $2t$, where t is the thickness of the film.

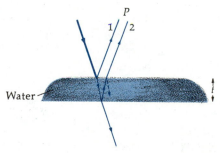

This path difference produces a phase difference of $360°(2t/\lambda')$, where λ' is the wavelength of the light in the film. It is related to the wavelength λ_0 in air by

$$\lambda' = \frac{v}{f} = \frac{c/n}{f} = \frac{\lambda_0}{n}$$

27-2

where n is the index of refraction. The phase difference between the two rays is thus 180° plus that due to the path difference. Destructive interference occurs when the path difference $2t$ is zero or a whole number of

Figure 27-1
Light rays reflected from the top and bottom surfaces of a thin film are coherent. If the light is incident nearly normally, the two reflected rays will be very close to each other and will produce interference. A phase change of 180° is introduced upon reflection from the boundary where the index of refraction increases. If the path difference $2t$ is an integral number of wavelengths, the interference will be destructive because one of the rays undergoes a 180° phase change upon reflection whereas the other does not.

wavelengths λ' (in the film). Constructive interference occurs if the path difference is an odd number of half wavelengths. We can express these conditions mathematically. When there is one phase change of 180° due to reflection, the conditions for interference are

$$\frac{2t}{\lambda'} = m \qquad m = 0, 1, 2, 3, \ldots \quad \text{(destructive)} \qquad \text{27-3a}$$

$$\frac{2t}{\lambda'} = m + \tfrac{1}{2} \qquad m = 0, 1, 2, 3, \ldots \quad \text{(constructive)} \qquad \text{27-3b}$$

Interference with one 180° phase change

When a thin water film lies on a glass surface (see Figure 27-2), the ray that reflects from the lower water–glass surface also undergoes a 180° phase change because the index of refraction of glass (about 1.50) is greater than that of water (about 1.33). Thus, both rays shown in the figure undergo a 180° phase change upon reflection. The phase difference between these rays is due solely to the path difference and is given by $\delta = 360°(2t/\lambda')$. Thus when there are two 180° phase changes upon reflection (or if there is no phase change) the conditions for interference are

$$\frac{2t}{\lambda'} = m \qquad m = 0, 1, 2, 3, \ldots \quad \text{(constructive)} \qquad \text{27-4a}$$

$$\frac{2t}{\lambda'} = m + \tfrac{1}{2} \qquad m = 0, 1, 2, 3, \ldots \quad \text{(destructive)} \qquad \text{27-4b}$$

Interference with no phase change or two 180° phase changes

Figure 27-2
Interference of light reflected from a thin film of water resting on a glass surface. In this case, both rays undergo a change in phase of 180° upon reflection.

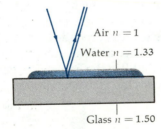

Air $n = 1$

Water $n = 1.33$

Glass $n = 1.50$

When a thin film of varying thickness is viewed with monochromatic light, alternating bright and dark lines called *fringes* are observed. The distance between a bright fringe and a dark fringe is that distance for which the film thickness changes such that the path difference $2t$ is $\lambda'/2$. Figure 27-3 illustrates the interference pattern observed when light is reflected from an air film between a spherical glass surface and a plane glass surface in contact. These circular interference fringes are known as *Newton's rings*. Typical rays from reflection at the top and bottom of the air film are shown in Figure 27-4. Near the point of contact, where the path difference between the ray reflected from the upper glass–air surface and the ray reflected from the lower air–glass surface is essentially zero, or at least small compared with the wavelength of light, the interference is perfectly destructive because of the 180° phase shift of the ray reflected from the lower air–glass surface. This region is therefore dark. The first bright fringe occurs at a

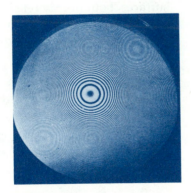

Figure 27-3
Newton's rings observed with light reflected from a thin film of air between a spherical glass surface and a plane glass surface. At the center, the thickness of the air film is negligible and the interference is destructive because of the phase change of one of the rays.

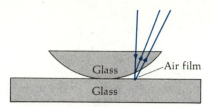

radius such that the path difference is $\lambda/2$, which contributes a phase difference of 180°. This adds to the phase shift due to reflection and produces a total phase difference of 360°, which is equivalent to zero phase difference. The second dark region occurs at a radius such that the path difference is λ, and so on.

Figure 27-4
Glass surfaces for the observation of Newton's rings shown in Figure 27-3. The thin film in this case is the film of air between the glass surfaces.

Example 27-1   A wedge-shaped film of air is made by placing a small slip of paper between the edges of two flat pieces of glass, as shown in Figure 27-5. Light of wavelength 500 nm is incident normally on the glass plates, and interference fringes are observed by reflection. If the angle made by the plates is 3×10^{-4} rad, how many interference fringes per unit length are observed?

Because of the 180° phase change of the ray reflected from the bottom plate, the first fringe near the point of contact (where the path difference is zero) will be dark. Let x be the horizontal distance to the mth dark fringe, where the plate separation is t, as shown in the figure. Since the angle θ is very small, it is given approximately by

$$\theta = \frac{t}{x}$$

Using Equation 27-3a for m, we have

$$m = \frac{2t}{\lambda'} = \frac{2t}{\lambda}$$

since the film is an air film. Substituting $t = x\theta$ gives

$$m = \frac{2x\theta}{\lambda}$$

or

$$\frac{m}{x} = \frac{2\theta}{\lambda} = \frac{2(3 \times 10^{-4})}{5 \times 10^{-5} \text{ cm}} = 12 \text{ cm}^{-1}$$

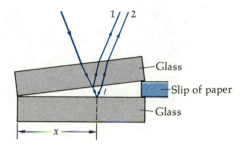

Figure 27-5
Light incident normally on a wedge of air between two glass plates. The path difference $2t$ is proportional to x. When viewed from above, alternate bright and dark bands or fringes are seen.

where we have used $\lambda = 5 \times 10^{-7}$ m $= 5 \times 10^{-5}$ cm. We therefore observe 12 dark fringes per centimetre. In practice, the number of fringes per centimetre, which is easy to count, can be used to determine the angle.

We note that if the angle of the wedge is increased, the fringes become more closely spaced. The distance along the glass between adjacent dark (or adjacent bright) fringes is the distance that gives an additional path difference equal to the wavelength of light in the film. If the angle of the wedge is increased, this distance is decreased.

Figure 27-6*a* shows the interference fringes produced by a wedge-shaped air film between the two flat glass plates, as in Example 27-1. The straightness of the fringes indicates the flatness of the glass plates. Such plates are said to be *optically flat.* A similar air wedge formed by two ordinary glass plates yields the irregular fringe pattern shown in Figure 27-6*b*, indicating that these plates are not optically flat.

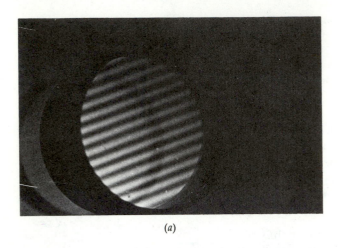

(a)

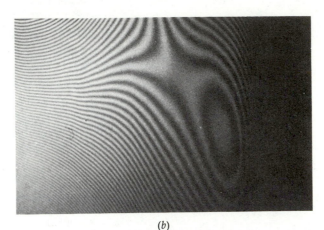

(b)

Figure 27-6
(*a*) Straight-line fringes from a wedge of air, as in Figure 27-5. The straightness of the fringes indicates the optical flatness of the glass plates. (*b*) Fringes from a wedge of air between glass plates that are not optically flat.

Nonreflecting coatings for lenses are made by covering the lens with a thin film of material that has an index of refraction of about 1.22, which is between that of glass and that of air, so that the intensities of light reflected from the top and bottom surface are approximately equal. The thickness of the film is chosen to be $\lambda'/4$, where $\lambda' = \lambda/n$ and λ is in the middle of the visible spectrum. There is a phase change at each reflection and therefore a net phase change of 180° due to the path difference of $\lambda'/2$.

Questions

1. Why must a film used to observe interference colors be thin?

2. If the angle of a wedge like that in Example 27-1 is too large, fringes are not observed. Why?

3. The spacing of Newton's rings decreases rapidly as the diameter of the rings increases. Explain qualitatively why this happens.

27-3   The Michelson Interferometer

An *interferometer* is a device that uses interference fringes to make precise measurements. Figure 27-7 is a schematic diagram of a Michelson interferometer. Light from a broad source strikes plate A, a beam-splitter that is partially silvered. The light is partially reflected and partially transmitted by this plate. The reflected beam travels to mirror M_2 and is reflected back toward the eye at O. The transmitted beam travels through a compensating plate B of the same thickness as A to mirror M_1 and is reflected back to the eye at O. The purpose of the compensating plate is to make both beams pass through the same thickness of glass. Mirror M_1 is fixed, but mirror M_2 can be moved back and forth with a fine and accurately calibrated screw adjustment. The two beams combine at O and form an interference pattern.

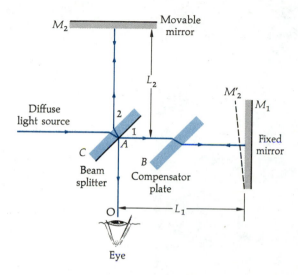

Figure 27-7
Michelson interferometer. The dashed line M_2' is the image of mirror M_2 in the mirror A. The interference fringes observed are those of a small wedge formed between M_1 and M_2'. As M_2 (and therefore M_2') is moved, the fringes move past the field of view.

The interference pattern is most easily understood by considering the light sources to be mirror M_1 and the image of mirror M_2 produced by the mirror in the beam splitter. This image is labeled M_2' in the diagram. If the mirrors M_1 and M_2 are exactly perpendicular to each other and equidistant from the beam splitter, the image M_2' will coincide with M_1. If they are not, M_2' will be slightly displaced and will make a small angle with M_1, as shown in the diagram. The interference pattern at O will then be that of a thin wedge-shaped film of air between M_1 and M_2', as discussed in Example 27-1.

If mirror M_2 is now moved, the fringe pattern will shift. Suppose, for example, that the mirror M_2 is moved back $\frac{1}{4}\lambda$. The image M_2' will then move an additional distance $\frac{1}{4}\lambda$ away from M_1, increasing the thickness of the wedge by $\frac{1}{4}\lambda$ at each point. This will introduce an additional path difference of $\frac{1}{2}\lambda$ everywhere in the wedge, and the fringe pattern will move over by one-half fringe; that is, a previously dark fringe will now be a bright fringe, etc. As mirror M_2 is moved, the displacement of the fringe pattern is observed. If the distance the mirror is moved is known, the wavelength of the light can be determined. Michelson used such an interferometer to measure the wavelength of a spectral line of light emitted by krypton 86 in terms of the standard metre bar. This measurement was then used to redefine the standard metre in terms of this wavelength. (The definition of the standard metre has since been changed. It is now defined in terms of the speed of light.)

Another use for the Michelson interferometer is to measure the index of refraction of air (or of some other gas). One of the beams from plate A is enclosed in a container that can be evacuated. The wavelength in air λ' is related to that in a vacuum by $\lambda' = \lambda/n$, where n is the index of refraction of air (about 1.0003). When the container is evacuated, the wavelength in-

A student-type Michelson interferometer. The fringes are produced on a ground-glass screen by light from a laser.

creases so that there are fewer waves in this distance, causing a shift in the fringe pattern. By measuring the shift, the index of refraction can be determined (see Exercise 11).

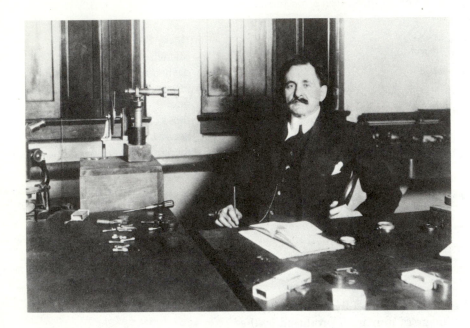

Albert A. Michelson in his laboratory.

Michelson also used his interferometer in a famous experiment with Edward W. Morley in 1887 in an attempt to measure the difference between the speed of light relative to the earth in the direction of motion of the earth and in a perpendicular direction. This experiment will be described in some detail in the next chapter.

The interferometer is used extensively in industry for the fine calibration of gauges and the precise measurement of the thickness of such materials as thin films.

Example 27-2   A film of index of refraction 1.33 and thickness 12 μm is inserted in one arm of a Michelson interferometer, which is illuminated with light of wavelength 589 nm in air. By how many fringes is the pattern shifted?

The number of waves in the thin film, N_f, is the thickness t divided by the wavelength $\lambda' = \lambda/n$, where n is the index of refraction.

$$N_f = \frac{t}{\lambda'} = \frac{nt}{\lambda} = \frac{(1.33)(12 \times 10^{-6}\ \text{m})}{(589 \times 10^{-9}\ \text{m})} = 27.1 \text{ waves}$$

The number of waves originally in this space when it was occupied by air, N_a, is

$$N_a = \frac{t}{\lambda} = \frac{12 \times 10^{-6}\ \text{m}}{589 \times 10^{-9}\ \text{m}} = 20.4 \text{ waves}$$

There are thus $27.1 - 20.4 = 6.7$ more waves in one arm of the interferometer, so the pattern will shift by 6.7 fringes.

27-4   The Two-Slit Interference Pattern

Interference patterns of light from two or more sources can be observed only if the sources are coherent, that is, are in phase or have a phase difference that is constant over time. We have mentioned that the randomness of the emission of light by atoms means that two different light sources are generally incoherent. The interference in thin films discussed previously can be observed because the two beams come from the same light source but are separated by reflection.

In the famous experiment devised by Thomas Young in 1801 by which he demonstrated the wave nature of light, two coherent light sources are produced by illuminating two parallel slits with a single source. We assume here that the width of each slit is small compared with the wavelength. (We will treat the general case in a later section.) We saw in Chapter 17 when we studied the diffraction of sound waves that when a wave encounters a barrier with a very small opening, the opening acts as a point source of waves. In this case, each slit acts as a line source (which is equivalent to a point source in two dimensions). The interference pattern consisting of alternate dark and light bands is observed on a screen far from the slits. (Or a converging lens is placed between the slits and the screen so that parallel light from the slits is focused on the screen.) This pattern is similar to that of two sound sources discussed in Section 17-1.

At very large distances from the slits, the lines from the two slits to some point P on the screen are approximately parallel, and the path difference is approximately $d \sin \theta$, where d is the separation of the slits as shown in Figure 27-8. We thus have interference maxima at an angle given by

$$d \sin \theta = m\lambda \qquad\qquad 27\text{-}5$$

Two-slit interference maxima

and interference minima at

$$d \sin \theta = (m + \tfrac{1}{2})\lambda \qquad\qquad 27\text{-}6$$

where $m = 0, 1, 2, 3, \ldots$.

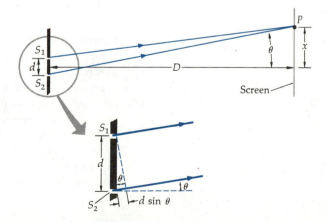

Figure 27-8
Two slits act as coherent sources of light for the observation of interference in Young's experiment. Since the screen is very far away compared with the slit separation, the rays from the slits to a point on the screen are approximately parallel. The path difference between the two rays is $d \sin \theta$.

The distance y_m measured along the screen from the central point to the mth bright fringe is related to the angle θ and the distance L to the screen by

$$\tan \theta = \frac{y_m}{L}$$

For small θ, we have

$$\sin \theta \approx \tan \theta = \frac{y_m}{L}$$

Substituting this into Equation 27-5, we have

$$d\,\frac{y_m}{L} = m\lambda$$

Thus for small angles (which is nearly always the case), the distance measured along the screen to the mth bright fringe is given by

$$y_m = m\,\frac{\lambda L}{d} \qquad\qquad 27\text{-}7$$

Distance to mth interference maximum

We note from this result that the fringes are equally spaced on the screen, with the distance between two successive bright fringes given by

$$\Delta y = \frac{\lambda L}{d}$$

Figure 27-9a shows the intensity pattern as seen on a screen. A graph of intensity as a function of $\sin \theta$ is shown in Figure 27-9b. For small θ, this is equivalent to a graph of intensity versus y, since $y \approx L \sin \theta$. The intensity I_0 is that from each slit separately. The dashed line shows the average intensity $2I_0$, which is the result of averaging the intensity over many interference maxima and minima. This is the intensity that would arise from the two sources if they acted independently without interference. It is the intensity we would observe if the sources were incoherent, because then there would be an additional phase difference between them that would fluctuate randomly so that only the average intensity could be observed.

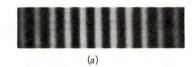

(a)

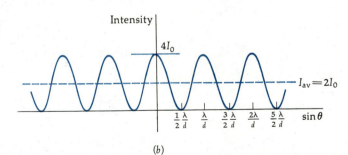

(b)

Figure 27-9
(a) Intensity pattern for the interference observed on a screen far away from the two slits of Figure 27-8. (b) Plot of intensity versus $\sin \theta$. The maximum intensity is $4I_0$, where I_0 is the intensity of the light from each slit separately. The average intensity (dashed line) is $2I_0$. Since for small θ, $\sin \theta \approx y/L$, this is also a plot of intensity versus the distance y measured along the screen.

Example 27-3 Two narrow slits separated by 1.5 mm are illuminated by sodium light of wavelength 589 nm. Interference fringes are observed on a screen 3 m away. Find the spacing of the fringes on the screen.

The distance y_m measured along the screen to the mth bright fringe is given by Equation 27-7, with $L = 3$ m, $d = 1.5$ mm, and $\lambda = 589$ nm. The spacing of the fringes is this distance divided by the number of fringes, or y_m / m. Solving Equation 27-7 for y_m / m and substituting in the given values, we obtain

$$\frac{y_m}{m} = \lambda \frac{L}{d} = \frac{(589 \times 10^{-9} \text{ m})(3 \text{ m})}{(0.0015 \text{ m})}$$

$$= 1.18 \times 10^{-3} \text{ m} = 1.18 \text{ mm}$$

The fringes are thus 1.18 mm apart.

Figure 27-10 shows another method of producing the two-source interference pattern, an arrangement known as *Lloyd's mirror*. A single slit is placed at a distance $\frac{1}{2}d$ above the plane of a mirror. Light striking the screen directly from the source interferes with that reflected from the mirror. The reflected light can be considered to come from the virtual image of the slit formed by the mirror. Because of the 180° change in phase on reflection at the mirror, the interference pattern is that of two coherent line sources that differ in phase by 180°. The pattern is the same as that shown in Figure 27-9 for two slits, except that the maxima and minima are interchanged. The central fringe just above the mirror at a point equidistant from the sources is dark. Constructive interference occurs at points for which the path difference is a half wavelength or any odd number of half wavelengths. At these points, the 180° phase difference due to the path difference combines with the 180° phase difference of the sources to produce constructive interference.

Lloyd's mirror

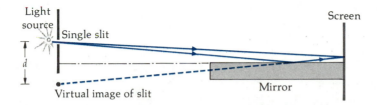

Light source
Single slit
d
Virtual image of slit
Screen
Mirror

Figure 27-10
Lloyd's mirror for observations of double-slit interference with light. The two sources (the slit and its image) are coherent and 180° out of phase. The central interference band at points equidistant from the sources is dark.

If we have three or more sources that are equally spaced and in phase with each other, the intensity pattern on a screen far away is similar to that due to two sources, but there are important differences. The position on the screen of the intensity maxima is the same no matter how many sources we have (assuming that they are equally spaced and in phase), but these maxima have much greater intensity and are much sharper if there are many sources.

Figure 27-11 compares the intensity pattern on a screen far away from four equally spaced narrow slits with that due to two slits. Again, I_0 is the intensity of the light from each slit separately. We can see that there are some very small submaxima between the main maxima and that the main maxima are sharper and more intense than those due to just two slits. As the number of slits is increased, the main maxima become more intense and sharper, and the intensity of the submaxima becomes negligible.

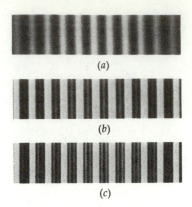

(a)

(b)

(c)

Interference patterns for (a) two, (b) three, and (c) four coherent sources. In (b) there is a secondary maximum between each pair of principal maxima, and in (c) there are two such secondary maxima.

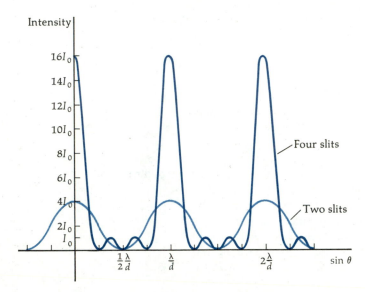

Figure 27-11
Comparison of intensity pattern for four slits with that for two slits when the space between each pair of adjacent slits is d. The maxima for four slits are much stronger and much narower than those for two slits.

Question

4. When destructive interference occurs, what happens to the energy in the light waves?

27-5   Diffraction Pattern of a Single Slit

In our discussion of the interference patterns produced by two or more slits, we assumed that the widths of the slits were much less than the wavelength of the light. This enabled us to consider the slits to be line sources of cylindrical waves (which in two-dimensional diagrams would be point sources of circular waves). We could therefore assume that the intensity due to the light from one slit alone, I_0, was the same at any point P on the screen independent of the angle θ made between the ray to point P and the normal line between the slit and the screen. When the slit width is not small compared with the wavelength of light, the intensity on a screen far away is not independent of the angle but decreases as the angle increases.

Let us consider a slit of width a. Figure 27-12 shows the intensity pattern on a screen far away from the slit as a function of $\sin \theta$. We can see that the intensity is maximum in the forward direction ($\sin \theta = 0$) and that it decreases to zero at an angle that depends on the slit width a and the wavelength λ.

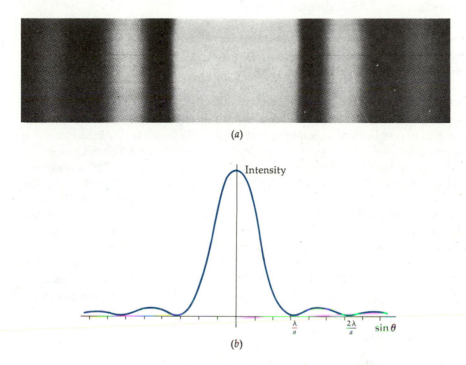

(a)

(b)

Figure 27-12
(a) Diffraction pattern of a single slit as observed on a screen far away. (b) Plot of intensity versus $\sin \theta$ for the pattern in (a).

Most of the light intensity is concentrated in the broad *central maximum*, and there are minor submaximum bands on either side of the central maximum. The first zeros in the intensity occur at an angle given by

$$\sin \theta = \lambda/a \qquad\qquad 27\text{-}8$$

We note that for a given wavelength λ, the width of the central maximum varies inversely with the width of the slit. That is, if we increase the slit width a, the angle θ at which the intensity first becomes zero decreases, giving a narrower central maximum. Conversely, if we decrease the slit width, the angle of the first zero increases, giving a wider central maximum. When a is less than the wavelength, there is no zero in the intensity pattern, and the slit acts as a line source, radiating light energy essentially equally in all directions.

We can write Equation 27-8 slightly differently. Multiplying both sides by a, we obtain

$$a \sin \theta = \lambda \qquad\qquad 27\text{-}9$$

The quantity $a \sin \theta$ is the path difference between a light ray leaving the top of the slit and one leaving the bottom of the slit. We see that the first diffraction minimum occurs when these two rays are in phase, that is, when their path difference equals one wavelength. We can understand this result by considering each point on a wavefront to be a point source of light in accordance with Huygens' principle. In Figure 27-13, we have placed a line of dots on the wavefront at the slit to represent these point sources schematically. Suppose, for example, that we have 100 such dots and we look at an angle θ satisfying $a \sin \theta = \lambda$. At this angle, the waves from the top and bottom of the slit are in phase. Let us consider the slit to be divided into two regions, with the souces 1 through 50 in the upper region and sources 51 through 100 in the lower region. When the path difference between the top and bottom of the slit equals one wavelength, the path difference between source 1 (the first source in the upper region) and source 51 (the first source in the lower region) is one-half wavelength. The waves from these two sources will be out of phase by 180° and will thus cancel. Similarly, waves from the second source in each region (source 2 and source 52) will cancel. We can continue this argument and see that waves from each pair of sources separated by $a/2$ will cancel. Thus there will be no light energy at this angle.

We can extend this argument to the second and third minima in the diffraction pattern of Figure 27-12. At an angle such that $a \sin \theta = 2\lambda$, we can divide the slit into four regions, two for the top half and two for the bottom half. Using the same argument as before, the light from the top half is zero because of cancellation of pairs of sources, and, similarly, that from the bottom half is zero. The general expression for the zeros in the diffraction patterns of a single slit is thus

$$a \sin \theta = m\lambda \qquad m = 1, 2, 3, \ldots \qquad \text{27-10}$$

Usually, we are just interested in the first occurrence of a minimum in the light intensity because nearly all of the light energy is contained in the central maximum.

In Figure 27-14, the distance y from the center of the central maximum to the first diffraction minimum is related to the angle θ and the distance L from the slit to the screen by

$$\tan \theta = \frac{y}{L}$$

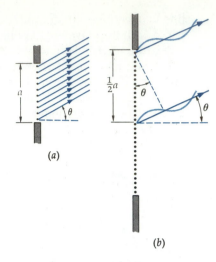

(a)

(b)

Figure 27-13
Diagram for the calculation of the single-slit diffraction pattern. (a) The slit of width a is assumed to contain a large number of point sources in phase. The interference pattern of these sources produces the single-slit diffraction pattern. (b) At an angle θ such that the path difference between the source at the top and the one in the middle is one-half wavelength, the waves from these two sources cancel, as shown in the enlargement.

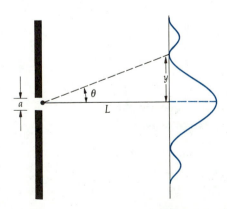

Figure 27-14
The distance y measured along the screen from the center of the central maximum to the first diffraction minimum is related to the angle θ by $\tan \theta = y/L$, where L is the distance from the slit to the screen. Since the angle is very small, $\tan \theta$, $\sin \theta$, and θ are all aproximately equal. Thus, $a \sin \theta = \lambda$ gives $y = L\lambda/a$.

Since this angle is very small, $\tan \theta \approx \sin \theta$. Then, according to Equation 27-8, we have

$$\sin \theta = \frac{\lambda}{a} \approx \frac{y}{L}$$

or

$$y = \frac{L\lambda}{a} \qquad\qquad\qquad\qquad 27\text{-}11$$

Example 27-4 In a lecture demonstration of diffraction, a laser beam of wavelength 700 nm passes through a vertical slit 0.2 mm wide and hits a screen 6 m away. Find the horizontal length of the central diffraction maximum on the screen, that is, the distance between the first minimum on the left and the first minimum on the right of the central maximum.

The width of the central diffraction maximum is $2y$ in Figure 27-14. From Equation 27-11, we have

$$y = \frac{L\lambda}{a} = \frac{(6 \text{ m})(700 \times 10^{-9} \text{ m})}{(0.0002 \text{ m})} = 2.1 \times 10^{-2} \text{ m} = 2.1 \text{ cm}$$

The width of the central maximum is thus $2y = 4.2$ cm.

The intensity pattern of Figure 27-12 is called a *Fraunhofer diffraction pattern* of a single slit. It is the pattern that is observed when the screen is far away from the slit and the slit width is not greater than a few wavelengths of the light. Under these conditions, the shape of the pattern can be derived mathematically using Huygens' principle. (The derivation is somewhat difficult and will not be considered here.)

Fraunhofer diffraction

When the diffraction pattern is observed near the opening, it is called a *Fresnel diffraction pattern*. This pattern is much more difficult to calculate. Figure 27-15 shows how the diffraction pattern of a single slit changes as we move toward the slit.* Note that, except for the peaks at the sides shown in Figure 27-15d, the pattern we see when we are close to the slit approximates the shape of the slit. That is, the screen is approximately uniformly illuminated in a region the size of the slit and is dark outside that region.

Fresnel diffraction

When there are two or more slits, the intensity pattern on a screen far away is a combination of the single-slit diffraction pattern and the multiple-

Figure 27-15
Diffraction patterns for a single slit at various screen distances. As you move closer to the slit, the Fraunhofer pattern (a) gradually merges into the Fresnel pattern (d) observed near the slit.

* See Richard E. Haskell, "A Simple Experiment on Fresnel Diffraction," *American Journal of Physics,* vol. 38, 1970, p. 1039.

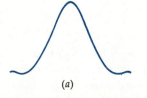

(a)

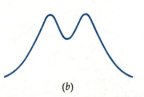

(b)

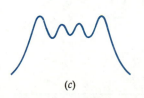

(c)

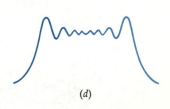

(d)

slit interference pattern we have studied. Figure 27-16 shows the Fraunhofer diffraction–interference pattern on a screen far from two slits whose separation d is 10 times the width a of each slit. The pattern is the same as the two-slit interference pattern except that it is modulated by the Fraunhofer single-slit diffraction pattern; that is, the intensity of the light from each slit separately, I_0, is now not constant but decreases with angle, as shown in Figure 27-12.

Question

5. As the width of the slit producing a single-slit diffraction pattern is slowly and steadily reduced, how will the diffraction pattern change?

Figure 27-16
Interference–diffraction pattern for two slits with separation d equal to 10 times their width a. The tenth interference maxima on either side of the central interference maximum is missing because it falls at the first diffraction minimum.

27-6   Diffraction and Resolution

Diffraction occurs whenever a portion of a wavefront is limited by an obstacle or aperture of some kind. The intensity of light at any point in space can be computed using Huygens' principle by taking each point on the wavefront to be a point source and computing the resulting interference pattern. Fraunhofer patterns can be observed at great distances from the obstacle or aperture where the rays reaching any point are approximately parallel, or they can be observed using a lens to focus parallel rays on a viewing screen placed at the focal point of the lens. Fresnel patterns are observed at points close to the obstacle or aperture. Diffraction of light is often difficult to observe because the wavelength is so small or because the light intensity is not great enough. Except for the Fraunhofer pattern produced by a long, narrow slit, diffraction patterns are usually difficult to calculate.

When waves are incident on a barrier having two slits with size and separation of the order of the wavelength, both diffraction and interference can be observed. Here plane ultrasonic waves in water are incident on slits of width 2λ and separation 8λ.

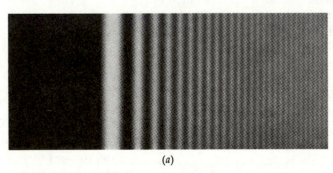

(a)

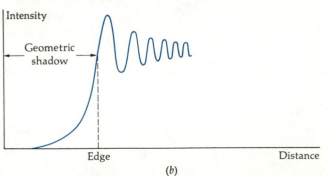

(b)

Figure 27-17
(a) Fresnel diffraction pattern for a straight edge. (b) Intensity versus distance measured along a line perpendicular to the edge.

Figure 27-17*a* shows the Fresnel diffraction pattern of a straight edge illuminated by light from a point source. A graph of the intensity versus distance measured along a line perpendicular to the edge is shown in Figure 27-17*b*. The light intensity does not fall abruptly to zero in the geometric shadow, but it decreases rapidly and is negligible within a few wavelengths of the edge. The Fresnel diffraction pattern of a rectangular opening is shown in Figure 27-18. These patterns cannot be seen with broad light sources like an ordinary light bulb because the dark fringes of the pattern produced by light from one point on the source overlap the bright fringes of the pattern produced by light from another point.

Figure 27-19 shows the Fraunhofer diffraction pattern for a circular aperture, which has important applications for the resolution of many optical instruments.

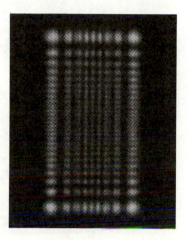

Figure 27-19
Fraunhofer diffraction pattern for a circular opening.

Figure 27-18
Fresnel diffraction pattern for a rectangular opening.

The angle θ subtended by the first diffraction minimum is related to the wavelength and the diameter of the opening D by

$$\sin \theta = 1.22 \frac{\lambda}{D} \qquad \text{27-12}$$

Equation 27-12 is similar to Equation 27-8 except for the factor 1.22. This factor arises from the mathematical analysis, which is similar to that for a single slit but more complicated because of the circular geometry. In many applications, the angle θ is small, so $\sin \theta$ can be replaced by θ. The first diffraction minimum is then at an angle θ given by

$$\theta \approx 1.22 \frac{\lambda}{D} \qquad \text{27-13}$$

Figure 27-20 shows two point sources that subtend an angle α at a circular aperture far from the sources. The Fraunhofer diffraction patterns are also sketched in this figure. If α is much greater than $1.22\lambda/D$, the two sources will be seen as two sources. However, as α is decreased, the overlap of the

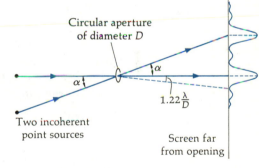

Figure 27-20
Two distant sources that subtend an angle α. If α is much greater than $1.22\lambda/D$, where λ is the wavelength of the light and D is the diameter of the aperture, the diffraction patterns have little overlap and the sources are easily seen as two sources. If α is not much greater than $1.22\lambda/D$, the overlap of the diffraction patterns makes it difficult to distinguish two sources from one.

diffraction patterns increases, and it becomes difficult to distinguish the two sources from one source. At the critical angular separation α_c given by

$$\alpha_c = 1.22 \frac{\lambda}{D} \qquad \text{27-14}$$

Critical angular separation for the resolution of two sources

the first minimum of the diffraction pattern of one source falls at the central maximum of the other source. The objects are then said to be "just resolved" by the *Rayleigh criterion for resolution.* Figure 27-21 shows the diffraction patterns for two sources for α greater than the critical angle for resolution and for α just equal to the critical angle for resolution.

Equation 27-14 has many applications. The *resolving power* of an optical instrument, such as a microscope or telescope, refers to the ability of the instrument to resolve two objects that are close together. We can see from Equation 27-14 that resolving power can be increased either by increasing the diameter D of the lens (or mirror) or by decreasing the wavelength λ of the light. Astronomical telescopes use as large an objective lens or mirror as possible to increase their resolution as well as their light-gathering power. In a microscope, a transparent oil with an index of refraction of about 1.55 is sometimes used under the objective to decrease the wavelength of the light ($\lambda' = \lambda/n$). The wavelength can be reduced further by using ultraviolet light and photographic film; however, ordinary glass is opaque to ultraviolet light, so the lenses in an ultraviolet microscope must be made from quartz or fluorite. In Chapter 30, we will see that electrons exhibit the wave properties of interference and diffraction just as light does. The wavelengths of electrons vary inversely with the square root of their kinetic energy and can be made as small as desired. Microscopes of very high resolution that use electrons rather than light are called electron microscopes.

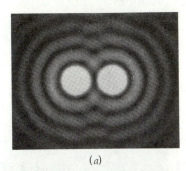

(a)

(b)

Figure 27-21
Diffraction patterns for a circular aperture and two incoherent point sources for (a) α much greater than 1.22 λ/D, and (b) the limit of resolution $\alpha = 1.22\ \lambda/D$.

Example 27-5   What is the minimum angular separation of two point objects that will allow them to be just resolved by the eye? How far apart must they be if they are 100 m away? Assume that the diameter of the pupil of the eye is 5 mm and that the wavelength of the light is 600 nm.

Using Equation 27-14 with $D = 5$ mm and $\lambda = 600$ nm, we can obtain the minimum angular separation:

$$\alpha_c = 1.22 \frac{\lambda}{D} = 1.22 \left(\frac{6 \times 10^{-7}\ \text{m}}{5 \times 10^{-3}\ \text{m}} \right) = 1.46 \times 10^{-4}\ \text{rad}$$

If the objects are separated by a distance y and are 100 m away, they will be just barely resolved if $\tan \theta = y/(100\ \text{m})$. Then

$$y = (100\ \text{m}) \tan \alpha_c \approx (100\ \text{m})\alpha_c$$

$$= 1.46 \times 10^{-2}\ \text{m} = 1.46\ \text{cm}$$

where we have used the small-angle approximation $\alpha_c \approx \tan \alpha_c$.

It is interesting to compare the limitation on resolution of the eye due to diffraction, as shown in the preceding example, with that due to the separa-

The maximum resolution that can be achieved with a modern light microscope is approximately equal to the average wavelength of visible light. This is 2000 to 3000 times greater than the diameter of small molecules. Thus, a light microscope can never resolve the image of individual molecules.

tion of the receptors (cones) on the retina. To be seen as two distinct objects, the images of the objects must fall on the retina on two nonadjacent cones (see Exercise 6 in Chapter 26). Assuming the retina is 2.5 cm from the eye lens, the distance y on the retina corresponding to an angular separation of 1.5×10^{-4} rad is found from

$$\theta = 1.5 \times 10^{-4} = \frac{y}{2.5 \text{ cm}}$$

or

$$y \approx 4 \times 10^{-4} \text{ cm} = 4 \times 10^{-6} \text{ m} = 4 \text{ } \mu m$$

The actual separation of the cones on the retina is about 1 μm in the fovea centralis where these color-sensitive cells are the most tightly packed. Outside this region, they are about 3 to 5 μm apart.

The Soviet Union has built a telescope with a 6-metre-diameter mirror as its objective. It can resolve two stars that are separated by only 9×10^{-8} radians. This is equivalent to taking a readable picture of a page of a telephone book from a distance of 16 kilometres.

27-7   Diffraction Gratings

A useful tool for the analysis of light is the *diffraction grating,* which consists of a large number of equally spaced lines or slits. Such a grating can be made by cutting parallel, equally spaced grooves on a glass or metal plate with a precision ruling machine. In a reflection grating, light is reflected from the ridges between the lines. A phonograph record exhibits some of the properties of a reflection grating. In a transmission grating, the light passes through the clear gaps between the rulings. Inexpensive plastic replica gratings with 10,000 or more slits per centimetre are not uncommon. The spacing of the slits for a grating with 10,000 slits per centimetre is $d = (1 \text{ cm})/10{,}000 = 10^{-4}$ cm.

Consider a plane light wave incident normally on a transmission grating (see Figure 27-22) and assume that the width of each slit is very small so that each slit produces a wide diffracted beam. Then, the interference pattern produced on a screen a large distance from the grating is just that due to a large number of equally spaced line sources. The interference maxima are at angles θ given by

$$d \sin \theta = m\lambda \qquad\qquad\qquad 27\text{-}15$$

The position of an interference maximum does not depend on the number of sources, but the more sources there are, the sharper and more intense the maximum will be, as was illustrated in Figure 27-11.

Figure 27-23 shows a typical student spectroscope that uses a diffraction grating to analyze light from a source such as a tube containing atoms of a gas, for example, helium or sodium vapor. The atoms are excited through bombardment by electrons accelerated by the high voltage across the tube. The light emitted by such a source does not consist of a continuous spectrum but contains only certain wavelengths that are characteristic of the atoms in the source. Light from the source passes through a narrow slit and is made parallel by a lens. The parallel light from the lens is incident on the grating.

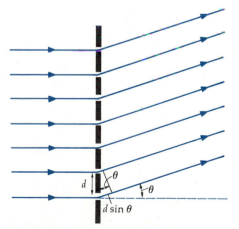

Figure 27-22
Light incident normally on a diffraction grating. At angle θ, the path difference between adjacent slits is $d \sin \theta$.

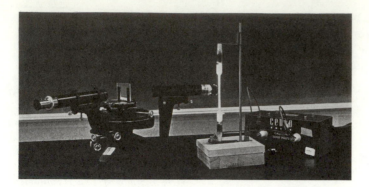

Figure 27-23
Typical student spectroscope. Light
from a slit near the source on the
right is made parallel by a lens and
falls on a grating. The diffracted
light is viewed with a telescope on
the left at an angle that can be
accurately measured. The
wavelength of the light is then
found from Equation 27-15.

Instead of falling on a screen a large distance away, the parallel light from
the grating is focused by a telescope and viewed by the eye. The telescope is
mounted on a rotating platform, which is calibrated so that the angle θ can
be measured. In the forward direction ($\theta = 0$), the central maximum for all
wavelengths is seen. If light of a particular wavelength λ is emitted by the
source, the first interference maximum is seen at the angle θ given by Equation 27-15 with $m = 1$. Each wavelength emitted by the source produces a
separate image of the slit in the spectroscope, which is called a *spectral line*.
The set of lines corresponding to $m = 1$ is called the *first-order spectrum*. The
second-order spectrum corresponds to $m = 2$ for each wavelength. Higher
orders may be seen as long as the angle θ given by Equation 27-15 is less
than 90°. Depending on the wavelengths and the grating spacing, the
orders may be mixed; that is, the third-order line corresponding to one
wavelength may occur before the second-order line corresponding to another wavelength. By measuring the angle θ and knowing the spacing of the
slits on the grating, the wavelengths emitted by the source can be measured.

Example 27-6   Sodium light is incident on a diffraction grating of 10,000 lines
per centimetre. At what angles will the two yellow lines of wavelengths
589.00 nm and 589.59 nm be seen in the first order?

Using $m = 1$, $d = 10^{-4}$ cm $= 10^{-6}$ m, and $\lambda = 589 \times 10^{-9}$ m in Equation
27-15, we obtain

$$\sin \theta = \frac{\lambda}{d} = \frac{589 \times 10^{-9} \text{ m}}{10^{-6} \text{ m}} = 0.589$$

$$\theta = 36.09°$$

For $\lambda = 589.59$ nm, a similar calculation gives $\sin \theta = 0.58959$, resulting in
$\theta = 36.13°$.

An important feature of a spectroscope is its ability to measure light of two
nearly equal wavelengths, λ_1 and λ_2. For example, the two prominent yellow lines in the spectrum of sodium have wavelengths of 589.00 and
589.59 nm, which can be seen as two separate spectral lines if their interference maxima do not overlap. According to the Rayleigh criterion of resolution, these wavelengths are resolved if the angular separation of their inter-

Spectrometer used in the late
nineteenth century by Gustav
Kirchhoff. In this drawing, a prism
replaces the diffraction grating.

ference maxima is greater than the angular separation between one interference maximum and the first interference minimum on either side of it. The *resolving power* of a diffraction grating is defined as $\lambda/|\Delta\lambda|$, where $|\Delta\lambda|$ is the difference between two nearby wavelengths, each approximately equal to λ. The resolving power is proportional to the number of slits illuminated because the more slits, the sharper the interference maxima. The resolving power can be shown to be

$$R = \frac{\lambda}{|\Delta\lambda|} = mN \qquad \text{27-16}$$

Resolving power of a grating

where N is the number of slits and m is the order number. We can see from Equation 27-16 that to resolve the two yellow lines in the sodium spectrum, the resolving power must be

$$R = \frac{589.00 \text{ nm}}{589.59 - 589.00 \text{ nm}} = 998$$

Thus, to resolve the two yellow sodium lines in the first order ($m = 1$), we need a grating containing about 1000 slits in the area illuminated by the light.

Summary

1. Two light rays interfere constructively if their phase difference is zero or an integer times $360°$. They interfere destructively if their phase difference is $180°$ or an odd integer times $180°$. A common source of phase difference is a path difference. A path difference Δx introduces a phase difference δ given by

$$\delta = \frac{\Delta x}{\lambda} 360°$$

A phase difference of $180°$ is introduced when a wave is reflected from a boundary between two media for which the wave speed is greater in the original medium, as between air and glass.

2. The Michelson interferometer uses interference fringes to measure the wavelength of light or to measure small differences in the index of refraction of different media, such as that between air and a vacuum.

3. When two narrow slits are separated by a small distance d, the path difference at an angle θ to a screen a large distance away is $d \sin \theta$. When this path difference is an integer times the wavelength, the interference is constructive; when it is an integer plus one-half times the wavelength, the interference is destructive:

$$d \sin \theta = m\lambda \qquad \text{(constructive)}$$
$$d \sin \theta = (m + \tfrac{1}{2})\lambda \qquad \text{(destructive)}$$

If the intensity of the light from each slit separately is I_0, the intensity at constructive interference points is $4I_0$ and that at destructive interference

A very large array of radio telescopes is used to detect and locate radio signals from distant galaxies. The signals from the telescopes add constructively when Equation 27-15 is satisfied, where d is the separation of the telescopes.

points is 0. When there are many equally spaced slits, the interference maxima occur at the same points as for two slits, but the maxima are much more intense and much narrower.

4. When light is incident on a single slit of width a, the intensity pattern on a screen far away shows a broad maximum that decreases to zero at an angle θ given by

$$a \sin \theta = \lambda$$

The width of the central diffraction maximum is thus inversely proportional to the width of the slit.

5. Diffraction patterns are produced whenever a portion of the wavefront is limited, as by a sharp edge, a small obstacle, or a barrier with a small opening. When the diffraction pattern is observed far away from the obstacle or opening, it is called a Fraunhofer diffraction pattern. When it is observed close to the obstacle or opening, it is called a Fresnel diffraction pattern.

6. When light from two point sources that are close together passes through an aperture, the diffraction patterns of the sources may overlap. If the overlap is too great, the two sources cannot be resolved as two separate sources. When the central diffraction maximum of one source falls at the diffraction minimum of the other source, the two sources are said to be just resolved according to the Rayleigh criterion for resolution. The critical angular separation for resolution by the Rayleigh criterion is given by

$$\alpha_c = 1.22 \frac{\lambda}{D}$$

where D is the diameter of the aperture.

7. A diffraction grating consists of a large number of closely spaced slits. It is used to measure the wavelength of light emitted by a source. The position of the interference maximum for a grating is given by

$$d \sin \theta = m\lambda$$

where m is the order number. Two nearly equal wavelengths can be resolved by a diffraction grating if

$$\frac{\lambda}{|\Delta\lambda|} = mN$$

where λ is the average wavelength, $|\Delta\lambda|$ is the difference in wavelengths, m is the order number, and N is the number of slits illuminated.

Suggestions for Further Reading

Baumeister, Philip, and Gerald Pincus: "Optical Interference Coatings," *Scientific American*, December 1970, p. 58.

Color television cameras, lasers, projector bulbs, and lenses of various types employ thin films to reflect or transmit light of certain wavelengths.

Feinberg, Gerald: "Light," *Scientific American*, September 1968, p. 50.

This article presents an introduction to our present understanding of light as a phenomenon of both wavelike and particlelike properties, as manifested by diffraction, two-slit interference, the photoelectric effect, and blackbody radiation.

Walker, Jearl: "The Amateur Scientist: A Ball Bearing Aids in the Study of Light and Also Serves as a Lens," *Scientific American*, November 1984, p. 186.

This is a report of an unusual investigation into the properties of the diffraction pattern of a ball bearing placed in the beam of a laser.

Review

A. Objectives: After studying this chapter, you should:

1. Be able to work problems involving interference in thin films.

2. Be able to describe the Michelson interferometer.

3. Be able to describe both interference and diffraction and discuss how they differ from each other.

4. Be able to sketch the two-slit interference intensity pattern and calculate the positions of the interference maxima and minima.

5. Be able to sketch the single-slit diffraction pattern and calculate the position of the first diffraction minimum.

6. Be able to state the Rayleigh criterion for resolution and use it to investigate the conditions for resolution of two nearby objects.

7. Be able to discuss the use of diffraction gratings.

B. Define, explain, or otherwise identify:

Newton's rings, p. 688
Interferometer, p. 690

Lloyd's mirror, p. 695
Fraunhofer diffraction pattern, p. 699
Fresnel diffraction pattern, p. 699
Rayleigh's criterion for resolution, p. 702
Diffraction grating, p. 703

C. True or false: If the statement is true, explain why. If it is false, give a counterexample.

1. When waves interfere destructively, the energy is converted into heat energy.

2. Interference is observed only for waves from coherent sources.

3. In the Fraunhofer diffraction pattern for a single slit, the narower the slit, the wider the central maximum of the diffraction pattern.

4. A circular aperture can produce both a Fraunhofer and a Fresnel diffraction pattern.

5. The ability to resolve two point sources depends on the wavelength of light.

Exercises

Section 27-1 Phase Difference and Coherence; Section 27-2 Interference in Thin Films

1. (*a*) What minimum path difference is needed to introduce a phase shift of 180° in light of wavelength 600 nm? (*b*) What phase shift will that path difference introduce in light of wavelength 800 nm?

2. Which of the following pairs of light sources are coherent? (*a*) Two candles. (*b*) One candle and its image in a plane mirror. (*c*) Two pinholes uniformly illuminated by the same source. (*d*) Two headlights of a car. (*e*) Two images of a candle due to reflection from the front and back surfaces of a glass window pane.

3. Light of wavelength 500 nm is incident normally on a film of water 10^{-4} cm thick. The index of refraction of water is 1.33. (*a*) What is the wavelength of the light in the water? (*b*) How many waves are contained in the distance $2t$, where t is the film thickness? (*c*) What is the phase difference between the wave reflected from the top of the film and the one reflected from the bottom after it has traveled this distance?

4. A loop of wire is dipped in soapy water and held so that the soap film is vertical. Viewed by reflection with white light, the top of the film appears black. Explain why. Below the black region are colored bands. Is the first band red or violet? Describe the appearance of the film when viewed by transmitted light. (*Hint:* The thickness of such a film varies from top to bottom with the thinnest part at the top.)

5. A wedge-shaped film of air is made by placing a small slip of paper between the edges of two flat pieces of glass. Light of wavelength 600 nm is incident normally on the glass plates, and interference bands are observed by reflection. (*a*) Is the first band near the point of contact of the plates dark or bright? Why? (*b*) If there are 4 dark bands per centimetre, what is the angle of the wedge?

6. A thin layer of transparent material with index of refraction 1.22 is used as a nonreflective coating on the surface of glass with index of refraction 1.50. What should the thickness be for the film to be nonreflecting for light of wavelength 700 nm (in a vacuum)?

7. The diameter of fine wires can be accurately measured by interference patterns. Two optically flat pieces of glass of length L are arranged with the wire as shown in Figure 27-24. The set up is illuminated by monochromatic light, and the resulting interference fringes are detected. Suppose $L = 25$ cm and yellow sodium light of wavelength 589 nm is used for illumination. If 23 bright fringes are seen along this 25-cm distance, what are the limits on the diameter of the wire? (*Hint:* Fringe 23 might not be right at the end, but you do not see 24 fringes.)

Figure 27-24
Exercise 7.

8. A thin layer of oil ($n = 1.45$) floats on the surface of water ($n = 1.33$). Find the minimum thickness of the oil film for constructive interference of the reflection of red light of wavelength 700 nm in air.

9. Photographic slides are sometimes mounted between glass plates. When the slide is held under a lamp, interference fringes are often seen due to an air space of variable thickness between the glass and the film. Find the thickness of the thinnest air film that gives constructive interference for yellow light of wavelength 600 nm.

Section 27-3   The Michelson Interferometer

10. A thin film of index of refraction $n = 1.4$ for light of wavelength 600 nm is inserted in one arm of a Michelson interferometer. (*a*) If a fringe shift of 14 fringes occurs, what is the thickness of this film? (*b*) When the illuminating light is changed to 400 nm, the fringe shift when this film is inserted becomes 18 fringes. What is the index of refraction of this film for light of wavelength 400 nm?

11. A hollow cell of length 5 cm with glass windows is inserted into one arm of a Michelson interferometer. The air is pumped out of the cell and the mirrors are adjusted to give a bright fringe at the center. As the air is gradually let back into the cell, there is a shift of 49.6 fringes when light of wavelength 589.29 nm is used. (*a*) How many waves are there in the 5.0-cm long cell when it is evacuated? (*b*) How many waves are there in the cell when it contains air? (*c*) What is the index of refraction of air as determined by this experiment?

Section 27-4   The Two-Slit Interference Pattern

12. Two slits separated by 1 mm are illuminated with light of wavelength 600 nm. Interference fringes are observed on a screen 5 m away. Find the spacing of the fringes on the screen.

13. A long, narrow, horizontal slit lies 0.6 mm above a plane mirror, which is in the horizontal plane. The interference pattern produced by the slit and its image is viewed on a screen a distance 4 m from the slit. The wavelength of the light is 700 nm. (*a*) Find the distance from the mirror to the first bright fringe. (*b*) How many dark fringes per centimetre are seen on the screen?

14. When light of wavelength 589 nm is used and the screen is 3 m from two slits, there are 28 bright fringes per centimetre on the screen. What is the slit separation?

15. In a lecture demonstration, laser light is used to illuminate two slits separated by 0.5 mm, and the interference pattern is observed on a screen 5 m away. The distance on the screen to the thirty-seventh bright fringe is 25.7 cm. What is the wavelength of the light used?

16. Light is incident at an angle ϕ with the normal to a plane containing two slits of separation d (see Figure 27-25). Show that the interference maxima are located at angles θ given by $\sin \theta + \sin \phi = m(\lambda/d)$.

Figure 27-25
Light incident on two slits at an angle ϕ with the normal for Exercise 16.

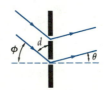

17. Light from a helium–neon laser of wavelength 633 nm is incident normally on a plane containing two slits. The interference pattern is observed on a screen 12.0 m away. The first interference maximum is 82 cm from the central maximum on the screen. (*a*) Find the separation of the slits. (*b*) How many interference maxima can be observed?

Section 27-5   Diffraction Pattern of a Single Slit

18. Equation 27-5, $d \sin \theta = m\lambda$, and Equation 27-10, $a \sin \theta = m\lambda$, are sometimes confused. For each equation, define the symbols and explain the equation's application.

19. Light of wavelength 600 nm is incident on a long, narrow slit. Find the angle of the first diffraction minimum if the width of the slit is (*a*) 0.01 mm, (*b*) 0.1 mm, and (*c*) 1 mm.

20. The single-slit diffraction pattern of light is observed on a screen a distance L from the slit. Find the width of the central maximum if $L = 4$ m, $\lambda = 600$ nm, and (*a*) $a = 0.1$ mm, (*b*) $a = 0.01$ mm, and (*c*) $a = 0.001$ mm.

21. In a lecture demonstration of diffraction, a laser beam of wavelength 700 nm passes through a vertical slit 0.3 mm wide and hits a screen 8 m away. Find the horizontal width of the central diffraction maximum on the screen.

22. Plane microwaves are incident on a long, narrow slit of width 5 cm. The first diffraction minimum is observed at $\theta = 37°$. What is the wavelength of the microwaves?

Section 27-6   Diffraction and Resolution

23. Light of wavelength 600 nm is incident on a pinhole of diameter 0.1 mm. (*a*) What is the angle between the central maximum and the first diffraction minimum for a Fraunhofer diffraction pattern? (*b*) What is the distance between the central maximum and the first diffraction minimum on a screen 4 m away?

24. Two light sources of wavelength 600 nm are 10 m away from a pinhole of diameter 0.1 mm. How far apart must the sources be for their diffraction patterns to be just resolved according to the Rayleigh criterion?

25. (*a*) How far apart must two objects be on the moon to be resolved by the eye? Take the diameter of the pupil of the eye to be 5.00 mm, $\lambda = 600$ nm, and $L = 380,000$ km for the distance to the moon. (*b*) How far apart must the objects on the moon be to be resolved by a telescope that has a 5.00-m diameter mirror?

26. Two sources of wavelength 700 nm are separated by a horizontal distance x. They are 8 m from a vertical slit of width 0.2 mm. What is the least value of x that permits the diffraction pattern of the sources to be resolved by the Rayleigh criterion?

27. What minimum aperture (in millimetres) is required for perfect opera glasses (binoculars) if an observer is to be able to distinguish the soprano's individual eyelashes (separated by 0.5 mm) at 25 m? Assume the effective wavelength of the light to be 550 nm.

28. The headlights on a small car are separated by 112 cm. At what maximum distance could you resolve them if the diameter of your pupil is 5.0 mm and the effective wavelength of the light is 550 nm?

29. You are told not to shoot until you see the whites of their eyes. If the eyes are separated by 6.5 cm, and the diameter of your pupil is 5.0 mm, at what distance can you resolve the two eyes using light of wavelength 550 nm?

Section 27-7   Diffraction Gratings

30. A diffraction grating with 2000 slits per centimetre is used to measure the wavelengths emitted by hydrogen gas. At what angles θ in the first-order spectrum would you expect to find the two violet lines of wavelength 434 and 410 nm?

31. With the grating used in Exercise 30, two other lines in the hydrogen spectrum are found in first order at angles $\theta_1 = 9.72 \times 10^{-2}$ rad and $\theta_2 = 1.32 \times 10^{-1}$ rad. Find the wavelengths of these lines.

32. Repeat Exercise 30 for a grating with 15,000 slits per centimetre.

33. A grating of 10,000 slits per centimetre is used to analyze the spectrum of mercury. (*a*) Find the angular deviation in first order of the two lines of wavelength 579.0 and 577.0 nm. (*b*) How wide must the beam be on the grating for these lines to be resolved?

34. What is the longest wavelength that can be observed in fifth order using a grating with 4000 slits per centimetre?

Problems

1. Laser light falls normally on three evenly spaced, very narrow slits. The angle of the first interference maximum is 0.60° from the normal. Find the angle to the first interference maximum if (*a*) one of the side slits is covered and the other two are open and (*b*) the center slit is covered and the other two are open.

2. Two slits are separated by a distance d and their interference pattern is observed on a screen a large distance L away. (*a*) Calculate the spacing of the interference maxima for light of wavelength 500 nm with $L = 1$ m and $d = 1$ cm. Would you expect to observe the interference of light on the screen for this situation? If you want the interference maxima to be separated by 1 mm, (*b*) how far away should you place the screen for a 1-cm slit separation and (*c*) how close together should you place the slits for a screen distance of 1 m?

3. A thin film of index of refraction 1.4 is surrounded by air. It is illuminated normally with white light and is viewed by reflection. Analysis of the resulting reflected light shows that the wavelengths 360, 450, and 600 nm are the only missing wavelengths near the visible portion of the spectrum. That is, for these wavelengths, there is destructive interference. (*a*) What is the thickness of the film? (*b*) What visible wavelengths would be especially bright in the reflected interference pattern?

4. A mica sheet 1.20 μm thick is suspended in air. In reflected light, there are gaps in the visible spectrum at 421, 474, 542, and 632 nm. Find the index of refraction of this sheet.

5. A two-slit Fraunhofer diffraction–interference pattern is observed with light of wavelength 500 nm. The slits have a separation of 0.1 mm and a width a. Find the slit width a if the fifth interference maximum is at the same angle as the first diffraction minimum. Sketch the pattern on a screen far from the slits. How many interference maxima are contained in the central diffraction maximum?

6. A two-slit Fraunhofer diffraction–interference pattern is observed with light of wavelength 700 nm. The slits have width 0.01 mm and are separated by 0.2 mm. How many bright fringes will be seen in the central diffraction maximum? (See Problem 5.)

7. Suppose that the central *diffraction* maximum for a Fraunhofer diffraction–interference pattern from two slits contains 17 interference maxima for some wavelength of light. How many interference maxima would you expect in the first *secondary* diffraction maximum? Make a sketch of the pattern on the screen.

8. The ceiling of your lecture hall is probably covered with acoustical tile that has small holes separated by about 6.0 mm. (*a*) Using $\lambda = 500$ nm, how far could you be from this tile and still resolve these holes? Take 5.0 mm for the diameter of your pupil. (*b*) Could you "see" these holes better with red or violet light?

9. The telescope on Mt. Palomar has a diameter of about 5.0 m. Assuming "ideal" sky conditions, the resolution would be diffraction-limited. Suppose a double star were 4 light-years away. What would the separation of these stars have to be for their images to be resolved by the telescope? (Take 500 nm for the wavelength of light.)

10. For a ruby laser of wavelength 694 nm, the aperture of the laser determines the diameter of the light beam emitted. At great distances, the size of the light beam is determined mainly by diffraction. If the aperture has a diameter of 2.00 cm, find the approximate diameter of the light beam at the moon a distance 0.38 Gm away.

CHAPTER TWENTY-EIGHT

Relativity

The most famous person in all of science may well be Albert Einstein. Millions of people know both his name and the name of one of his major contributions to physics, the theory of relativity, though most of them have no idea what it is all about. Somewhere along the line, the fiction developed that only a few people could understand relativity. As you study this chapter, you will discover that the first part of Einstein's theory — special relativity — is actually fairly easy to grasp. At the same time, you will undoubtedly see why the theory astounded the scientific world when Einstein first presented it. In fact, you may even be astonished yourself, for what you will learn will cause you to reexamine the basic concepts of time and space.

There was a young lady named Bright,
Whose speed was far faster than light;
She set out one day
In a relative way,
And returned home the previous night.

ARTHUR BULLER, *Punch* **(December 19, 1923)**

Near the end of the nineteenth century and the beginning of the twentieth, many thought that all the important laws of physics had been discovered and that there was little left for physicists to do other than work out some of the details. Newton's laws of motion and gravity seemed to describe all known motion on earth as well as that of the planets and other heavenly bodies, whereas Maxwell's equations of electricity and magnetism seemed to give a complete description of electromagnetic phenomena. Even as evidence of the microscopic world of molecules and atoms began to accumulate, it was assumed that these new phenomena could be adequately described by the theories of Newton and Maxwell. However, the discovery of radioactivity by Becquerel in 1896, the theoretical papers of Planck in 1897 and Einstein in 1905, and the work of Rutherford, Millikan, Bohr, de Broglie, Schrödinger, Heisenberg, and others in the early twentieth century led to two completely new theories: relativity and quantum mechanics. These theories revolutionized the world of science and became the foundations for new technologies that have changed the face of civilization.

In this chapter we will study relativity. The theory of relativity consists of two rather different theories, the special theory and the general theory. The special theory, developed by Einstein and others in 1905, concerns the comparison of measurements made in different inertial reference frames moving with constant velocity relative to one another. Its consequences, which can be derived with a minimum of mathematics, are applicable in a wide variety of situations encountered in physics and engineering. On the other hand, the general theory, developed by Einstein and others around 1916, is concerned with accelerated reference frames and gravity. A thorough understanding of the general theory requires sophisticated mathematics, and the applications of this theory are chiefly in the area of gravitation. It is of great importance in cosmology, but it is rarely encountered in other areas of physics or in engineering. We will therefore concentrate on the special theory (often referred to as *special relativity*) and discuss the general theory only briefly in the last section of this chapter.

28-1   Newtonian Relativity

Newton's first law does not distinguish between a particle at rest and one moving with constant velocity. If no external forces are acting, the particle will remain in its initial state—either at rest or moving with its initial velocity. Consider a particle at rest relative to you with no forces acting on it. According to Newton's first law, it will remain at rest. Now consider the same particle from the point of view of a second observer moving with constant velocity relative to you. From this observer's "frame of reference," both you and the particle are moving with constant velocity. Newton's first law holds also for him. (Note that if the second observer were accelerating relative to you, he would see the particle accelerating relative to him, with no external forces acting on it. Thus Newton's first law would not hold for the second observer.) How might we determine whether you and the particle are at rest and the second observer is moving with constant velocity or the second observer is at rest and you and the particle are moving?

Let us consider some simple experiments. Suppose we have a train moving along a straight, flat track with a constant velocity v. (We assume that there are no bumps or shakes in the motion.) Let us chose a coordinate system xyz, with the x axis along the track, as shown in Figure 28-1. It doesn't matter where we choose the origin along the track. For different choices, the position (relative to the origin) of the train and its contents will differ, but their velocities will be the same. A set of coordinate systems at rest relative to each other is called a *reference frame*. We will call the reference frame at rest relative to the track frame S. We now consider doing various mechanics experiments in the closed box car. For these, we choose a coordinate system at rest relative to the box car. This coordinate system is in reference frame S', which is moving to the right with speed v relative to frame S. We note that a ball at rest in the train remains at rest. If we drop a ball, it falls straight down with acceleration g due to gravity. (The ball, of course, when viewed in frame S moves in a parabolic path because it has an initial velocity v to the right.) No mechanics experiment that we can do— measuring the period of a pendulum or a body on a spring, observing the collisions between two bodies, etc. — will tell us whether the train is moving

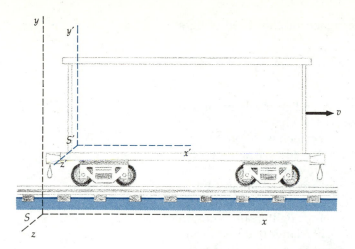

Figure 28-1
Box car moving with constant velocity along a straight track. The reference frame S' is at rest relative to the car and is moving with speed v relative to S, which is at rest relative to the track. It is impossible to tell by doing mechanics experiments inside the car whether the car is moving to the right with speed v or the track is moving to the left with speed v.

and the track is at rest or the track is moving and the train is at rest. Newton's laws hold for reference frame S' as well as for reference frame S.

A reference frame in which Newton's laws hold is called an *inertial reference frame. All reference frames moving at constant velocity relative to an inertial reference frame are also inertial reference frames.* If we have two inertial reference frames moving with constant velocity relative to each other, such as S and S', there is no mechanics experiment that can tell us which is at rest and which is moving or even if they are both moving. This result is known as the principle of newtonian relativity (also called galilean relativity):

Absolute motion cannot be detected.

Principle of newtonian relativity

This principle was well known by Galileo, Newton, and others in the seventeenth century. By the late nineteenth century, however, this view had changed. It was then generally thought that newtonian relativity was not valid and that absolute motion could be detected in principle by a measurement of the speed of light.

From our study of wave motion, we know that all mechanical waves require a medium for their propagation. The speed of such waves depends only on the properties of the medium. The speed of sound waves, for example, depends on the temperature of air or on the bulk modulus of a solid or liquid. This speed is relative to still air. Motion relative to still air can indeed be detected. It was therefore natural to expect that some kind of medium supports the propagation of light and other electromagnetic waves. This medium was called the *ether.* As proposed, the ether had to have unusual properties. It had to have great rigidity to support waves of such high velocity. (Recall that the velocity of waves in a string depends on the tension of the string and that of longitudinal sound waves in a solid depends on the bulk modulus of the solid.) Yet the ether could introduce no drag force on the planets, as their motion is fully accounted for by the law of gravitation. It was suspected that the ether was at rest relative to the distant stars, but this was considered to be an open question. It was therefore of considerable interest to determine the velocity of the earth relative to the ether. Experiments to do this were undertaken by Albert Michelson, first in 1881 and then again with Edward Morley in 1887 with greater precision. The idea was to measure the speed of light relative to the earth.

According to Maxwell's theory of electromagnetism, the speed of light and other electromagnetic waves is

$$c = \frac{1}{\sqrt{\epsilon_0 \mu_0}} = 3 \times 10^8 \text{ m/s}$$

where ϵ_0 and μ_0 are, respectively, the permittivity and permeability of free space. There is nothing in this equation that tells us in what reference frame the speed of light will have this value, but the expectation was that this was the speed of light relative to its natural medium, the ether. It was thought that a measurement of the speed of light relative to some reference frame moving though the ether would yield a result greater or less than c by an amount that depended on the speed of the frame relative to the ether and the direction of motion relative to the direction of the light beam. Thus, in 1881 Michelson set out to measure the speed of light relative to the earth and from this measurement to determine the velocity of the earth relative to the ether.

28-2   The Michelson–Morley Experiment

The object of the Michelson–Morley experiment was to determine the velocity of the earth relative to the ether by measuring the speed of light relative to the earth. In the usual measurements of the speed of light (Section 24-3), the time for a light pulse to travel to and from a mirror is determined. Figure 28-2 shows a light source and mirror a distance L apart. If we assume that both are moving with speed v through the ether, classical theory predicts that the light will travel toward the mirror with speed $c - v$ and back with speed $c + v$ (both speeds relative to the mirror and the light source). The time for the total trip will be

$$t_1 = \frac{L}{c - v} + \frac{L}{c + v} = \frac{2cL}{c^2 - v^2} \qquad \text{28-1}$$

$$= \frac{2L}{c}\left(1 - \frac{v^2}{c^2}\right)^{-1}$$

We can see that this differs from the time $2L/c$ by the term $(1 - v^2/c^2)^{-1}$, which is very nearly equal to 1 if v is much less than c. We can simplify this expression for small v/c by using the binomial expansion

$$(1 + x)^n = 1 + nx + n(n - 1)\frac{x^2}{2} + \cdots$$

$$\approx 1 + nx \qquad \text{28-2}$$

when x is much less than 1. Using $n = -1$ and $x = -v^2/c^2$, Equation 28-1 becomes

$$t_1 \approx \frac{2L}{c}\left(1 + \frac{v^2}{c^2}\right) \qquad \text{28-3}$$

The orbital speed of the earth about the sun is approximately 3×10^4 m/s. If

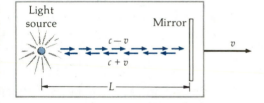

Figure 28-2
Light source and mirror moving with speed v relative to the "ether." According to classical theory, the speed of light relative to the source and mirror is $c - v$ toward the mirror and $c + v$ away from the mirror.

we take this for an estimate of v, we have $v = 3 \times 10^4$ m/s $= 10^{-4}\,c$ and $v^2/c^2 = 10^{-8}$. Thus, the correction for the earth's motion is small indeed.

Michelson realized that though the effect is too small to be measured directly, it should be possible to determine v^2/c^2 by a difference measurement. For this measurement, he used the Michelson interferometer, which we discussed in Section 27-3. In this experiment one beam of light moves along the direction of the earth's motion and another moves perpendicular to that direction. The difference in the round-trip times of these beams depends on the speed of the earth and can be determined by an interference measurement. Let us assume that the interferometer (see Figure 27-7) is oriented so that the beam transmitted by the beam splitter is in the direction of the assumed motion of the earth. Equation 28-3 then gives the classical result for the round-trip time t_1 for the transmitted beam. The reflected beam travels with some velocity $\mathbf{u}$ (relative to the earth) perpendicular to the earth's velocity. Relative to the ether, it travels with velocity $\mathbf{c}$, as shown in Figure 28-3. The velocity $\mathbf{u}$ (according to classical theory) is then the vector difference $\mathbf{u} = \mathbf{c} - \mathbf{v}$, as shown in the figure. The magnitude of $\mathbf{u}$ is $\sqrt{c^2 - v^2}$, so the round-trip time t_2 for this beam is

$$t_2 = \frac{2L}{\sqrt{c^2 - v^2}} = \frac{2L}{c}\left(1 - \frac{v^2}{c^2}\right)^{-1/2} \qquad \text{28-4}$$

Again using the binomial expansion, we obtain

$$t_2 \approx \frac{2L}{c}\left(1 + \frac{1}{2}\frac{v^2}{c^2}\right) \qquad \text{28-5}$$

This expression is slightly different from that for t_1 given in Equation 28-3. The difference in these two times is

$$\Delta t = t_1 - t_2 \approx \frac{L}{c}\frac{v^2}{c^2} \qquad \text{28-6}$$

This time difference is to be detected by observing the interference of the two beams of light. Because of the difficulty of making the two paths of equal length to the precision required, the interference pattern of the two beams is observed and the whole apparatus is rotated 90°. The rotation produces a time difference given by Equation 28-6 for each beam. The total time difference of $2\,\Delta t$ is equivalent to a path difference of $2c\,\Delta t$. The interference fringes observed in the first orientation should thus shift by a number of fringes ΔN given by

$$\Delta N = \frac{2c\,\Delta t}{\lambda} = \frac{2L}{\lambda}\frac{v^2}{c^2} \qquad \text{28-7}$$

where λ is the wavelength of the light.

In Michelson's first attempt in 1881, L was about 1.2 m and λ was 590 nm. For $v^2/c^2 = 10^{-8}$, ΔN was expected to be 0.04 fringe. Even though the experimental uncertainties were estimated to be about this same magnitude, when no shift was observed, Michelson reported this as evidence that the earth did not move relative to the ether. In 1887, when he repeated the experiment with Edward W. Morley, he used an improved system for rotating the apparatus without introducing a fringe shift because of mechanical strains, and he increased the effective path length L to about 11 m by a series

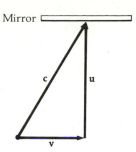

Figure 28-3
Light beam reflected from the beam splitter in a Michelson interferometer. The interferometer is moving to the right with speed v. The light beam moves perpendicular to the interferometer with speed u and reflects from the mirror. The velocity of light is $\mathbf{c}$ in the frame in which the interferometer is moving. Relative to the earth, in which the interferometer is fixed, the velocity of light is $\mathbf{u} = \mathbf{c} - \mathbf{v}$. The speed of light relative to the earth is then $u = (c^2 - v^2)^{1/2} = c(1 - v^2/c^2)^{1/2}$ according to classical theory.

of multiple reflections. Figure 28-4 shows the configuration of the Michelson–Morley apparatus. For this attempt, ΔN was expected to be 0.4 fringe, about 20 to 40 times the minimum possible to observe. Once again, no shift was observed. The experiment has since been repeated under various conditions by a number of people, and no shift has ever been found.

In 1905, at the age of 26, Albert Einstein published a paper on the electrodynamics of moving bodies.* In this paper, he postulated that absolute motion cannot be detected by any experiment. (We will discuss the Einstein postulates in detail in the next section.) The null result of the Michelson–Morley experiment is therefore expected: we can consider the whole apparatus and the earth to be at rest. Thus, no fringe shift is expected when the apparatus is rotated 90° since all directions are equivalent. Einstein did not set out to explain the Michelson–Morley experiment. His theory arose from his considerations of the theory of electricity and magnetism and the unusual property of electromagnetic waves that they propagate in a vacuum. In this first paper, which contains the complete theory of special relativity, he made only a passing reference to the Michelson–Morley experiment, and in later years he could not recall whether he was aware of the details of this experiment before he published his theory.

* *Annalen der Physik,* vol. 17, 1905, p. 841. For a translation from the original German, see W. Perrett and G. B. Jeffery (trans.), *The Principle of Relativity: A Collection of Original Memoirs on the Special and General Theory of Relativity,* by H. A. Lorentz, A. Einstein, H. Minkowski, and W. Weyl, Dover, New York, 1923.

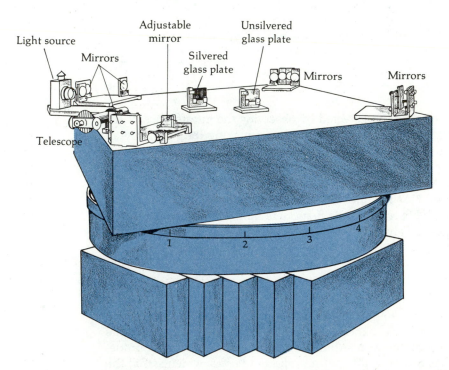

Figure 28-4
Drawing of Michelson and Morley's apparatus for their 1887 experiment. The optical parts were mounted on a sandstone slab 5 ft square, which was floated in mercury, thereby reducing the strains and vibrations that had affected the earlier experiments. Observations could be made in all directions by rotating the apparatus in the horizontal plane.

28-3   Einstein's Postulates and Their Consequences

The theory of special relativity can be derived from two postulates proposed by Einstein in his original paper in 1905. Simply stated, these postulates are

Postulate 1. Absolute, uniform motion cannot be detected.

Postulate 2. The speed of light is independent of the motion of the source.

Einstein's postulates

Postulate 1 is merely an extension of the newtonian principle of relativity to include all types of physical measurements (not just those that are mechanical). Postulate 2 describes a common property of all waves. For example, the speed of sound waves does not depend on the motion of a sound source. When a car is approaching you sounding its horn, the frequency you hear increases according to the doppler effect we studied in Section 16-4, but the speed of the waves traveling through the air does not depend on the speed of the car. The speed depends only on the properties of the air, such as its temperature.

Although each postulate seems quite reasonable, many of the implications of the two together are quite surprising and contradict what is often called common sense. For example, an immediate consequence of these postulates is that

Every observer measures the same value for the speed of light independent of the relative motion of the source and observer.

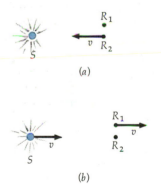

(a)

(b)

Consider a light source S and two observers, R_1 at rest relative to S and R_2 moving toward S with speed v, as shown in Figure 28-5a. The speed of light measured by R_1 is $c = 3 \times 10^8$ m/s. What is the speed measured by R_2? The answer is *not* $c + v$. By postulate 1, Figure 28-5a is equivalent to Figure 28-5b, in which R_2 is pictured at rest and the source S and R_1 are moving with speed v. That is, since absolute motion has no meaning, it is not possible to say which is really moving and which is at rest. By postulate 2, the speed of light from a moving source is independent of the motion of the source. Thus, looking at Figure 28-5b, we see that R_2 measures the speed of light to be c, just as R_1 does.

This result—that all observers measure the same value for the speed of light—contradicts our intuitive ideas about relative velocities. If a car moves at 50 km/h away from an observer and another car moves at 80 km/h in the same direction, the velocity of the second car relative to the first car is 30 km/h. This result is easily measured and conforms to our intuition. However, according to Einstein's postulates, if a light beam is moving in the direction of the cars, observers in both cars will measure the same speed for the light beam. Our intuitive ideas about the combination of velocities are approximations that hold only when the speeds are very small compared with the speed of light. Even in an airplane moving with the speed of sound, it is not possible to measure the speed of light accurately enough to distinguish the difference between the result c and $c + v$, where v is the speed of the plane. In order to make such a distinction, we must either move with a

Figure 28-5
(a) A stationary light source S and an observer R_1 with a second observer R_2 moving toward the source with speed v. (b) In the reference frame in which observer R_2 is at rest, the light source S and observer R_1 move to the right with speed v. If absolute motion cannot be detected, both views are equivalent. Since the speed of light does not depend on the motion of the source, observer R_2 measures the same value for that speed as observer R_1.

very great velocity (much greater than that of sound) or make extremely accurate measurements, as in the Michelson–Morley experiment.

Einstein recognized that his postulates had important consequences for measuring time intervals and space intervals as well as relative velocities. He showed that the size of time and space intervals between two events depends on the reference frame in which the events are observed. The changes in such measurements from one reference frame to another are called *time dilation* and *length contraction.* Instead of developing the general formalism of relativity, known as the *Lorentz transformation,* we will derive some of the famous relativistic effects directly from the Einstein postulates by considering some simple special cases of measurements of time and space intervals. Other important results we will discuss without derivation.

In these discussions, we will be comparing measurements made by observers who are moving relative to each other. We will use a rectangular coordinate system xyz with origin O, called the S reference frame, and another system $x'y'z'$ with origin O', called the S' reference frame, that is moving with constant velocity relative to the S frame. For simplicity, we will consider the S' frame to be moving with speed v along the x (or x') axis relative to S. In each frame, we assume that there are as many observers as needed, equipped with clocks, metresticks, and so forth, that are identical when compared at rest (see Figure 28-6).

Time Dilation

The results of measurements of the time and space intervals between events do not depend on the kind of apparatus used for the measurements or on the events. Convenient events for understanding Einstein's postulates are those that produce light flashes. We will consider the time interval between two light flashes as measured by observers in two reference frames moving relative to each other. Figure 28-7a shows an observer A' a distance D from a mirror. A' and the mirror are in a spaceship that is at rest in some frame, which we will call S'. A' explodes a flash gun and measures the time interval $\Delta t'$ between the original flash and the return flash from the mirror. Since light travels with speed c, this time is

$$\Delta t' = \frac{2D}{c} \qquad\qquad 28\text{-}8$$

We now consider these same two events, the original flash of light and the returning flash, as observed in reference frame S, where observer A' and the

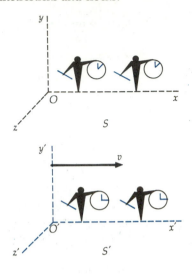

Figure 28-6
Coordinate reference frames S and S' moving with relative speed v. In each frame are observers with metresticks and clocks.

Figure 28-7
(a) A' and the mirror are in a spaceship at rest in frame S'. The time for the light pulse to reach the mirror and return is measured by A' to be $2D/c$. (b) In frame S, the spaceship is moving to the right with speed v. If the speed of light is the same in both frames, the time for the light to reach the mirror and return is longer than $2D/c$ in S because the distance traveled is greater than $2D$. (c) Right triangle for computing the time Δt in frame S.

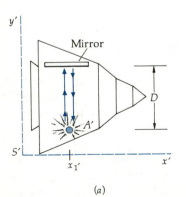

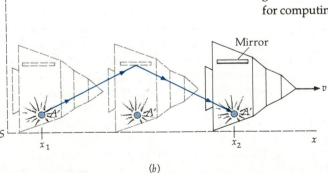

(a)

(b)

(c)

mirror are moving to the right with speed v, as shown in Figure 28-7b. The events happen at two different places, x_1 and x_2, in frame S. During the time interval Δt (as measured in S) between the original flash and the return flash, observer A' and the spaceship have moved a horizontal distance $v\,\Delta t$. In Figure 28-7b, we see that the path traveled by the light is longer in S than in S'. However, by Einstein's postulates, light travels with the same speed c in frame S as it does in frame S'. Since it travels farther in S at the same speed, it takes longer in S to reach the mirror and return. The time interval in S is thus longer than it is in S'. We can easily calculate Δt in terms of $\Delta t'$. From the triangle in Figure 28-7c we have

$$\left(\frac{c\,\Delta t}{2}\right)^2 = D^2 + \left(\frac{v\,\Delta t}{2}\right)^2$$

or

$$\Delta t = \frac{2D}{\sqrt{c^2 - v^2}} = \frac{2D}{c}\,\frac{1}{\sqrt{1 - v^2/c^2}}$$

Using $\Delta t' = 2D/c$, we have

$$\Delta t = \frac{\Delta t'}{\sqrt{1 - v^2/c^2}} = \gamma\,\Delta t' \qquad\qquad 28\text{-}9 \qquad \text{Time dilation}$$

where

$$\gamma = \frac{1}{\sqrt{1 - v^2/c^2}} \geq 1 \qquad\qquad 28\text{-}10$$

Observers in S would say that *the clock held by A' runs slow*, since A' claims a shorter time interval for these events.

A' measures the times of the light flash and return at the same point in S', whereas in S these events happen at two different places. A single clock can be used in S' to measure the time interval, but in S, two synchronized clocks are needed, one at x_1 and one at x_2. The time between events that happen at the *same place* in a reference frame (as with A' in S' in this case) is called *proper time*. The time interval measured in any other reference frame is always longer than the proper time. This expansion is called *time dilation.*

Example 28-1   Astronauts in a spaceship traveling at $v = 0.6c$ past the earth sign off from space control saying they are going to take a nap for one hour and then call back. How long does their nap last as measured on earth?

Since the astronauts go to sleep and wake up at the same place in their reference frame, the time interval for their nap of one hour as measured by them is proper time. In the earth's reference frame, they move a considerable distance between these two events. The time interval measured in the earth's frame (using two clocks) is longer by the factor γ. For this example with $v = 0.6c$, we have

$$1 - \frac{v^2}{c^2} = 1 - (0.6)^2 = 0.64$$

Then γ is

$$\gamma = \frac{1}{\sqrt{1 - v^2/c^2}} = \frac{1}{\sqrt{0.64}} = \frac{1}{0.8} = 1.25$$

The nap thus lasts for 1.25 hours as measured on earth.

Length Contraction

Time dilation is closely related to another phenomenon, *length contraction*. The length of an object measured in the reference frame in which the object is at rest is called its *proper length*. In a reference frame in which the object is moving, the measured length is shorter than its proper length. Suppose that x_1 and x_2 in Figure 28-7 are at the ends of a measuring rod of length $L_0 = x_2 - x_1$, measured in frame S, in which the rod is at rest. In frame S, A' is moving with speed v, so the distance traveled in time Δt is $v \, \Delta t$. Since A' moves from point x_1 to point x_2 in this time, this distance is $L_0 = x_2 - x_1 = v \, \Delta t$. However, in frame S' (see Figure 28-8), the measuring rod moves with speed v and takes a shorter time $\Delta t'$ to move past observer A'. (Note that observers in both frames always agree on the relative speed v.) The length of the rod as measured by A' is $L' = v \, \Delta t'$. Since the time interval $\Delta t'$ is less than Δt, the length L' is less than L_0. These lengths are related by

$$L' = v \, \Delta t' = \frac{v \, \Delta t}{\gamma} = \frac{L_0}{\gamma} = \sqrt{1 - v^2/c^2} \, L_0 \qquad \text{28-11} \qquad \text{Length contraction}$$

Thus, the length of the rod is shorter when it is measured in a frame in which it is moving. Before Einstein's paper was published, Lorentz and FitzGerald tried to explain the null result of the Michelson–Morley experiment by assuming that distances in the direction of motion contracted by the amount given in Equation 28-11. This contraction is now known as the Lorentz–FitzGerald contraction.

Example 28-2   How fast must a metrestick travel to measure the same length as a yardstick?

The relation between a metre and a yard is

$$1 \text{ m} = 1.094 \text{ yd}$$

From Equation 28-11 with $L = 1$ yd and $L_0 = 1.094$ yd, we have

$$\gamma = \frac{L_0}{L} = 1.094 = \frac{1}{\sqrt{1 - v^2/c^2}}$$

$$\sqrt{1 - v^2/c^2} = \frac{1}{1.094} = 0.914$$

$$1 - \frac{v^2}{c^2} = (0.914)^2 = 0.835$$

$$\frac{v^2}{c^2} = 1 - 0.835 = 0.165$$

$$v = 0.406c$$

An interesting example of the observation of these phenomena is afforded by the appearance of muons as secondary radiation from cosmic rays. Muons decay according to the statistical law of radioactivity (Chapter 32):

$$N(t) = N_0 e^{-t/\tau} \qquad \text{28-12}$$

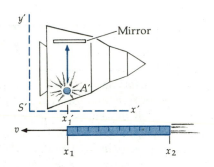

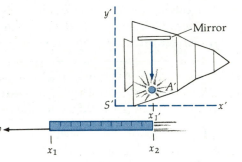

Figure 28-8
Measuring the length of a moving object. In frame S, the rod is at rest and the spaceship moves from one end of the rod to the other in time Δt, so the length of the rod is $L_0 = v \, \Delta t$. In frame S', the rod moves past the spaceship in time $\Delta t'$, which is shorter than Δt. The length of the rod in S' is $L' = v \, \Delta t' = \sqrt{1 - v^2/c^2} \, L_0$.

where N_0 is the number of muons at time $t = 0$, $N(t)$ is the number at time t, and τ is the mean lifetime, which is about $2\ \mu s$ for muons at rest. Since muons are created (from the decay of pions) high in the atmosphere, usually several thousand metres above sea level, few muons should reach sea level. A typical muon moving with speed $0.998c$ would travel only about 600 m in $2\ \mu s$. However, the lifetime of the muon measured in the earth's reference frame is increased by the factor $1/\sqrt{1 - v^2/c^2}$, which is 15 for this particular speed. The mean lifetime measured in the earth's reference frame is therefore $30\ \mu s$, and a muon with speed $0.998c$ travels about 9000 m in this time. From the muon's point of view, it lives only $2\ \mu s$, but the atmosphere is rushing past it with a speed of $0.998c$. The distance of 9000 m in the earth's frame is thus contracted to only 600 m, as indicated in Figure 28-9.

It is easy to distinguish experimentally between the classical and relativistic predictions of the observation of muons at sea level. Suppose that we observe 10^8 muons with a muon detector at an altitude of 9000 m in some time interval. How many would we expect to observe at sea level in the same time interval? According to the nonrelativistic prediction, the time taken for these muons to travel 9000 m is $(9000\ \text{m})/0.998c \approx 30\ \mu s$, which is 15 lifetimes. Putting $N_0 = 10^8$ and $t = 15\tau$ into Equation 28-12, we obtain

$$N = 10^8 e^{-15} = 30.6$$

We would thus expect all but about 31 of the original 100 million muons to decay before reaching sea level.

According to the relativistic prediction, the earth must travel only the contracted distance of 600 m in the rest frame of the muon. This takes only $2\ \mu s = 1\tau$. Thus the number expected at sea level is

$$N = 10^8 e^{-1} = 3.68 \times 10^7$$

Thus relativity predicts that we would observe 36.8 million muons in the same time interval. Experiments of this type have confirmed the relativistic predictions.

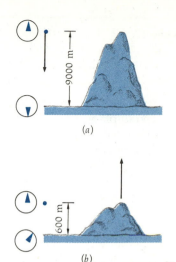

Figure 28-9
Although muons are created high above the earth and their mean lifetime is only about $2\ \mu s$ when at rest, many appear at the earth's surface. (a) In the earth's reference frame, a typical muon moving at $0.998c$ has a mean lifetime of $30\ \mu s$ and travels 9000 m in this time. (b) In the reference frame of the muon, the distance traveled by the earth is only 600 m in the muon's lifetime of $2\ \mu s$.

The Velocity Transformation

A third result that can be derived from Einstein's postulates but which we will merely state is the relativistically correct expression for combining velocities. We consider an object moving with velocity components u'_x, u'_y, and u'_z as measured in frame S', which is moving along the x or x' axis with speed v relative to frame S. (These components can be either positive or negative.) The velocity components of the object measured in frame S are given by

$$u_x = \frac{u'_x + v}{1 + vu'_x/c^2} \qquad\qquad 28\text{-}13a$$

$$u_y = \frac{u'_y}{\gamma(1 + vu'_x/c^2)} \qquad\qquad 28\text{-}13b$$

$$u_z = \frac{u'_z}{\gamma(1 + vu'_x/c^2)} \qquad\qquad 28\text{-}13c$$

Velocity transformation

These differ from the classical and intuitive result, $u'_x = u_x + v$, $u'_y = u_y$, and $u'_z = u_z$, because the denominators in Equations 28-13 are not equal to 1. When v and u'_x are small compared with the speed of light c, $\gamma \approx 1$, $vu'_x/c^2 \ll 1$, and the relativistic and classical expressions are the same.

Example 28-3   A supersonic plane moves with speed 1000 m/s (about three times the speed of sound) along the x axis relative to you. Another plane moves along the x axis at speed 500 m/s relative to the first plane. How fast is the second plane moving relative to you?

According to the classical formula for combining velocities, the speed of the second plane relative to you is $1000 + 500 = 1500$ m/s. If we assume that you are fixed in the S reference frame and the first plane is at rest in S' moving at $v = 1000$ m/s, the second plane has velocity $u'_x = 500$ m/s in S'. The correction term is then

$$\frac{vu'_x}{c^2} = \frac{(1000)(500)}{(3 \times 10^8)^2} = \frac{5 \times 10^5}{9 \times 10^{16}} \approx 5 \times 10^{-12}$$

This correction term is so small that the classical and relativistic results are essentially the same.

Example 28-4   Work Example 28-3 if the first plane moves with speed $v = 0.8c$ relative to you and the second plane moves with the same speed $0.8c$ relative to the first plane.

In this case, the correction term is

$$\frac{vu'_x}{c^2} = \frac{(0.8c)(0.8c)}{c^2} = 0.64$$

The speed of the second plane in frame S is then

$$u_x = \frac{0.8c + 0.8c}{1 + 0.64} = 0.98c$$

This is quite different from the classically expected result of $0.8c + 0.8c = 1.6c$. It can be shown, in fact, from Equations 28-13 that if the speed of an object is less than c in one frame, it is less than c in all other frames moving relative to that frame with speed less than c. We will see in Section 28-7 that it takes an infinite amount of energy to accelerate a particle to the speed of light. The speed of light c is thus an upper, unattainable limit for the speed of a (massive) particle. (Massless particles, such as photons, always move at the speed of light.)

Example 28-5   A light beam moves along the x axis in frame S' with speed c. What is its speed measured in frame S?

Putting $u'_x = c$ in Equation 28-13a, we obtain

$$u_x = \frac{c + v}{1 + vc/c^2} = \frac{c(1 + v/c)}{1 + v/c} = c$$

The speed of the light beam measured in frame S is also c, as required by the Einstein postulates.

Question

1. You are standing on a corner and a friend is driving past in an automobile. Both of you note the times when the car passes two different intersections and determine from your watch readings the time that elapses between the two events. Which of you has determined the proper time interval?

28-4   Clock Synchronization and Simultaneity

At first glance, time dilation and length contraction seem contradictory not only to our intuition but also to our ideas of self-consistency. If each reference frame can be considered to be at rest with the other moving, the clocks in the "other" frame should run slow. How can there be any self-consistency if each observer sees the clocks of the other run slow? The answer to this puzzle lies in the problem of clock synchronization and in the concept of simultaneity. We note that the time intervals Δt and $\Delta t'$ considered in Section 28-3 were measured in quite different ways. The events in frame S' happened at the same place, so the times could be measured on a single clock. But in frame S, the two events happened at different places, so the time of each event was measured on a different clock and the interval found by subtraction. This procedure requires that the clocks be synchronized. We will show in this section that

Two clocks synchronized in one reference frame are not synchronized in any other frame moving relative to the first frame.

A corollary to this result is that

Two events that are simultaneous in one reference frame are not simultaneous in another frame moving relative to the first.

(This is true unless the events and clocks are in the same plane and are perpendicular to the relative motion.) Comprehension of these facts usually resolves all relativity paradoxes. Unfortunately, the intuitive (and incorrect) belief that simultaneity is an absolute relation is difficult to abandon.

Suppose we have two clocks at rest at points A and B at a distance L apart in frame S. How can we synchronize these two clocks? If an observer at A looks at the clock at B and sets her clock to read the same time, the clocks will not be synchronized because of the time L/c it takes light to travel from one clock to another. To synchronize the clocks, the observer at A must set her clock ahead by the time L/c. Then she will see that the clock at B reads a time that is L/c behind the time on this clock, but she will calculate that the clocks are synchronized when she allows for the time L/c for the light to reach her. All observers except those midway between the clocks will see the clocks reading different times, but they will also compute that the clocks are synchronized when they correct for the time it takes the light to reach them. An equivalent method for the synchronization of two clocks would be for a third observer at a point C midway between the clocks to send a light signal and for the observers at A and B to set their clocks to some prearranged time when they receive the signal.

We now examine the question of simultaneity. Suppose A and B agree to explode bombs at t_0 (having previously synchronized their clocks). Observer C will see the light from the two explosions at the same time, and since he is equidistant from A and B, he will conclude that the explosions were simultaneous. Other observers in S will see the light from A or B first, depending on their locations, but after correcting for the time the light takes to reach them, they also will conclude that the explosions were simultaneous. *We will thus define two events in a reference frame to be simultaneous if the light signals from the events reach an observer halfway between the events at the same time.*

To show that two events that are simultaneous in frame S are not simultaneous in another frame S' moving relative to S, we will use an example introduced by Einstein. A train is moving with speed v past the station platform. We will consider the train to be at rest in S' and the platform to be at rest in S. We have observers A', B', and C' at the front, back, and middle of the train. We now suppose that the train and the platform are struck by lightning at the front and back of the train and that the lightning bolts are simultaneous in the frame of the platform (S) (see Figure 28-10). That is, an observer C halfway between the positions A and B, where the lightning strikes, observes the two flashes at the same time. It is convenient to suppose that the lightning scorches the train and platform so that the events can be easily located in each reference frame. Since C' is in the middle of the train, halfway between the places on the train that are scorched, the events can be simultaneous in S' only if C' sees the flashes at the same time. However, C' sees the flash from the front of the train before he sees the flash from the back.

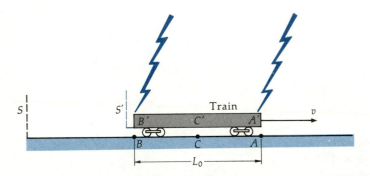

Figure 28-10
Simultaneous lightning bolts strike the ends of a train traveling with speed v in frame S attached to the platform. The light from these simultaneous events reaches observer C midway between the events at the same time. The distance between the bolts is L_0.

We can understand this by considering the motion of C' as seen in frame S (see Figure 28-11). When the light from the front flash reaches C', he has moved some distance toward it, so the flash from the back has not yet reached him, as indicated in the figure. He must therefore conclude that the events were not simultaneous. The front of the train was struck before the back. As we have discussed previously, all observers in S' on the train will agree with C' when they have corrected for the time it takes the light to reach them.

It is instructive to examine these events in reference frame S' attached to the train. Let L_0 be the distance between the scorch marks on the platform as measured in frame S. L_0 is the length of the train measured in frame S. This distance is smaller than the proper length of the train L'_T because of length contraction. Figure 28-12 shows the situation in frame S', in which the train is at rest and the platform is moving. In this frame, the distance between the burns on the platform is contracted. When the lightning bolt strikes the front of the train at A', the front of the train is at point A, and the back of the train has not yet reached point B. Later, when the lightning bolt strikes the back of the train at B', the back has reached point B on the platform.

In reference frame S, the lightning bolts strike A and B simultaneously. Suppose there are clocks at A and B, synchronized in frame S. From the point of view of frame S' attached to the train, the clocks and the platform are

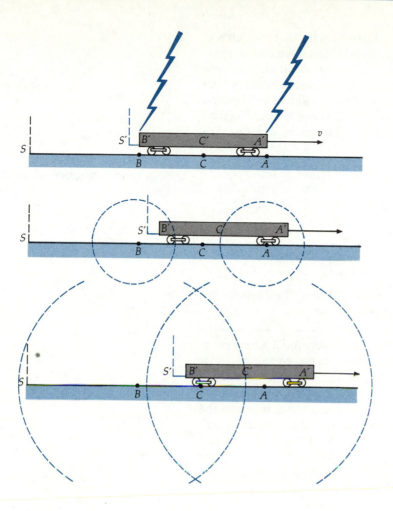

Figure 28-11
The light from the lightning bolt at the front of the train reaches observer C' in the middle of the train before that from the bolt at the back of the train. Since C' is midway between the events (which occur at the front and rear of the train), these events are not simultaneous for him.

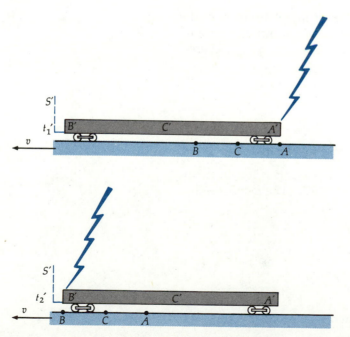

Figure 28-12
The lightning bolts of Figure 28-10 as seen in the reference frame of the train. In this frame, the distance between A and B on the platform is less than L_0 and the train is longer than L_0. The first lightning bolt occurs at the front of the train when A' and A are coincident. The second bolt occurs at the rear of the train when B' and B are coincident.

moving past the train. Lightning first strikes the front of the train, which is at point A, and some time later, lightning strikes the back of the train, which is now at point B. The moving clocks are thus not synchronized as seen from frame S'. The clock at B is slow; that is, it reads a time too small. Another way of saying this is that the clock at A leads the clock at B as seen in S'. The time discrepancy of two clocks synchronized in frame S as seen in frame S' can be worked out from our formulas for time dilation and length contraction. We will merely state the result:

If two clocks are synchronized in the frame in which they are at rest, they will be out of synchronization in another frame. In the frame in which they are moving, the "chasing clock" leads (shows a later time) by an amount

Lack of synchronization of moving clocks

$$\Delta t_s = \frac{L_0 v}{c^2}$$

where L_0 is the proper distance between the clocks.

A numerical example should help to clarify time dilation, clock synchronization, and the internal consistency of these results.

Example 28-6   An observer in a spaceship has a flash gun and a mirror (as in our time dilation example in Figure 28-7). The distance to the mirror is 15 light-minutes, written 15 $c \cdot$min, and the spaceship travels with speed $v = 0.8c$. The spaceship travels past a very long space platform that has two synchronized clocks, one at the point where the spaceship is when the observer explodes her flash gun and another at the point where the spaceship is when the light returns from the mirror. Find the time intervals between the events (exploding the flash gun and receiving the flash from the mirror) in the frame of the ship and in that of the platform, the distance traveled by the ship, and the amount by which the clocks are out of synchronization as viewed from the ship.

We will call the frame of the ship S' and that of the platform S. In the spaceship, the light travels from the gun to the mirror and back, a total distance of 30 $c \cdot$min. The time for light to travel 30 $c \cdot$min is, of course, 30 min.

$$\Delta t' = \frac{2D}{c} = \frac{30 \ c \cdot \text{min}}{c} = 30 \text{ min}$$

During this time, the ship travels

$$\Delta x' = v \, \Delta t' = (0.8c)(30 \text{ min}) = 24 \ c \cdot \text{min}$$

In frame S on the platform, the time between the events is longer by the factor γ. Since $v/c = 0.8$, $1 - v^2/c^2 = 1 - 0.64 = 0.36$. The factor γ is thus

$$\gamma = \frac{1}{\sqrt{1 - v^2/c^2}} = \frac{1}{\sqrt{0.36}} = \frac{1}{0.6} = \frac{5}{3}$$

The time between the events in S is therefore

$$\Delta t = \gamma \, \Delta t' = \frac{5}{3} (30 \text{ min}) = 50 \text{ min}$$

During this time, the spaceship travels a distance in S given by

$$\Delta x = v \, \Delta t = (0.8c)(50 \text{ min}) = 40 \ c \cdot \text{min}$$

From the point of view of observers on the platform in frame S, the space-ship's clocks are running slow because the clock on the ship records a time of only 30 min between the events whereas the time measured on the platform is 50 min. Figure 28-13 shows the situation viewed from the spaceship in S'. The platform is traveling past the ship with speed $0.8c$. There is a clock at point x_1 that coincides with the ship when the gun is exploded and another at point x_2 that coincides with the ship when the flash is received from the mirror. The separation in S of these clocks is $L_0 = \Delta x = 40\ c \cdot min$. We assume that the clock at x_1 reads noon at the time of the light flash. The clocks at x_1 and x_2 are synchronized in S but not in S'. In S', the clock at x_2, which is chasing the one at x_1, leads by

$$\frac{L_0 v}{c^2} = \frac{(40\ c \cdot min)(0.8c)}{c^2} = 32\ min$$

When the spaceship coincides with x_2, the clock there reads 50 min past noon. The time between events is therefore 50 min in S. Note that according to observers in S', this clock ticks off $50 - 32 = 18$ min for a trip that takes 30 min in S'. Thus this clock runs slow by the factor $30/18 = 5/3$.

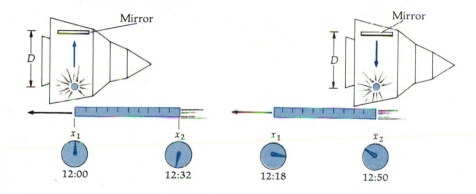

Figure 28-13

Example 28-6. During the time $\Delta t' = 30$ min it takes for the platform to pass the spaceship, the clocks on the platform run slow and tick off $(30\ min)/\gamma = 18$ min. But the clocks are unsynchronized with the chasing clock leading by $L_0 v/c^2$, which for this case is 32 min. The time for the spaceship to pass as measured on the platform is therefore $32\ min + 18\ min = 50\ min$.

Each observer sees clocks in the other frame run slow. According to observers in S, who measure 50 min for the time interval, the time interval in S' is too small (30 min) because the single clock in S' runs too slow by the factor $5/3$. According to the observers in S', the observers in S measure a time that is too *long*, despite the fact that their clocks run too slow, because they are out of synchronization. The clocks tick off only 18 min, but the second one leads the first by 32 min, so the time interval is 50 min.

Questions

2. Two observers are in relative motion. In what circumstances can they agree on the simultaneity of two different events?

3. If event A occurs before event B in some frame, might it be possible for there to be a reference frame in which event B occurs before event A?

4. Two events are simultaneous in a frame in which they also occur at the same point in space. Are they simultaneous in other reference frames?

Albert Einstein (1879 – 1955)

Gerald Holton
Harvard University

A friend visiting Albert Einstein in his modest walkup apartment in Bern, Switzerland, found him seated at a kitchen table, "pipe in his mouth, his left hand moving a children's carriage back and forth, in his right hand a shabby stub of a pencil." The year was 1905, and Einstein was writing his great paper on the special theory of relativity. Some of Einstein's personality traits can already be glimpsed here — the simplicity of his personal life and his ability to lift himself out of his surroundings by concentrating on the work before him.*

The paper of 1905 was the culmination of ideas that had started to preoccupy Einstein 10 years earlier, when he was about 16. At that age, he later wrote in his autobiography,† the following paradox occurred to him. As one knows from ample experience, galilean relativity holds in mechanics; if you throw a ball forward in a moving carriage, observing the motion of the ball cannot tell you how fast you and the carriage are moving. But when it comes to optics, it seems to be different. For example, if I move along "a beam of light with the velocity c [velocity of light in a vacuum], I should observe such a beam of light as a spatially oscillatory electromagnetic field." Looking back along the beam a distance of one whole wavelength, one should see that the local magnitudes of the electric and magnetic field vectors increase point by point from, say, zero to full strength, and then decrease again to zero, one wavelength away. Seeing such a curious field in free space would tell me that I am going at the speed of light with respect to absolute space, or the ether.

Einstein suspected that the imagined result of this *Gedankenexperiment* (thought experiment) must somehow be in error. In any case, he said later, "from the very beginning it appeared to me intuitively clear that, judged from the standpoint of such an observer, everything would have to happen according to the same laws as for an observer who, relative to the earth, was at rest."

Einstein solved the apparent paradox in 1905 by showing that the expectation of what one would see in pursuing a light beam is false, being grounded in a wrong idea that "unrecognizedly was anchored in the unconscious" of all scientists of the time — namely the absolute character of time and of simultaneity. Instead, Einstein showed that a sound view of how physical nature operates can be gained by boldly postulating two principles.

---

* See Einstein's essay "Motive of Research," pp. 224 – 227 in Sonja Bargmann (trans. and ed.), *Ideas and Opinions*, Crown, New York, 1954; look also at the other essays in this collection.

† Albert Einstein, *Autobiographical Notes*, in P. A. Schilpp (ed.), *Albert Einstein: Philosopher-Scientist*, Harper, New York, 1959.

Albert Einstein in 1905, at the time of his greatest productivity.

The first principle of relativity for inertial systems generalizes galilean relativity to encompass not only mechanics but also optics and electromagnetism; the second principle says that light in a vacuum will always be found to move with the same speed c, regardless of the state of motion of the emitting body. The solutions to many of the paradoxes and experimentally puzzling observations at the time were derivable from these two principles.

At least as important to Einstein was the fact that this approach hugely simplified our view of nature in two ways: it broke down barriers between hitherto entirely separate notions, and it cleansed physics of unnecessary conceptions that had produced pseudo problems. Now electromagnetism was on the same footing as mechanics, instead of allowing "privileged" systems. Time and space were found to have interpenetrating meaning — what really existed was a space-time continuum. Electric fields and magnetic fields were at bottom the same reality perceived in different experimental conditions. Mass and energy were equivalent. Even the boldly intuitive approach to scientific discovery and the rigorously rational method were joined into one powerful approach in his paper. Moreover, the notion of the ether was at last declared to be "superfluous." It had long been embedded in physics but had required the assumption of ever more puzzling properties. While he was the first to give up the idea of an ether, to the end of his life Einstein was dedicated to the theme of the *field* (or, in general, the continuum) as the basic conceptual tool for the fundamental explanation of phenomena.

The preoccupation with the continuum, with explaining mysterious orderliness, with finding pleasing simplicity—these were conceptions in physics that characterized Einstein's work from the beginning. But as with other highly creative persons, the power of his work was derived not merely from good physical ideas. Instead it came from a fusion of his scientific interests and his characteristics as a human being, a synthesis of his life-style and his perception of the laws of nature. Much of what was most daring or novel in his great work in physics was present in Einstein's ways of everyday thinking and behaving, even as a child or a young student. Take, for example, the ability to come back, again and again, for years, to a difficult puzzle. In early life, this trait showed up as what may have seemed mere obstinacy. He was commonly reported to have been withdrawn as a child, preferring to play by himself, erecting complicated constructions. Before he was 10, with infinite patience he was making fantastic card houses that had as many as 14 floors. He was unable or unwilling to talk until the age of 3. In school, he was not an exceptional student, preferring to follow his own thoughts. Later, he stuck to his ideas with the same persistence when the experiments of others seemed to disconfirm his theories (in time, those experiments usually turned out to be wrong). And during the last 30 years of his life, he persisted in his skepticism concerning the fundamental explanatory power of quantum mechanics and in his dedication to the problems of field theory, unlike most physicists of the time.

With this single-mindedness and concentration on his own revolutionary ideas went his deep suspicion of established authority, in science no less than in daily life. The same young student who quietly challenged the established ideas in physics also abhorred the current political, religious, and social conventions. From the age of 12 years on, he confessed later, he had "suspicion against every kind of authority." One result was that his teachers were quite delighted to see him drop out of high school at the age of $15\frac{1}{2}$, when he no longer could stand the regimented and militaristic way of life in his native Germany and in his school. Moving to the freer, more democratic atmosphere of Switzerland, he found at last a school to his liking, and there had a glorious year before entering the Polytechnic Institute of Zurich. It was there, too, that he had his first ideas on relativity.

Related to this trait of uncompromisingly sticking to his own identity was his search for what is really *necessary*. In his early work in physics, he said, the thought of having to explain electric and magnetic field effects as "two fundamentally different cases was for me unbearable." The highest aim was nothing less than finding the most economical, simple principles, the barest bones of nature's frame, cleansed of everything that is ad hoc, redundant, unsymmetrical. "What really interests me," he once said, "is whether God had any choice in the creation of the world." Nature does not like anything that is unnecessary. Nor did Einstein, in his personal life—in his clothing, in his manner of speech and writing, in his behavior, from his preference for the classical music of Bach and Mozart down to his preference for using the same bar of soap for washing and shaving instead of complicating life unnecessarily by facing two kinds of soap every morning.

The preference for an egalitarian democracy characterized Einstein's political life just as it did his physics. The man who declared every inertial system to be created equal before all the laws of physics was also, from youth on, fiercely opposed to every antidemocratic and narrowly nationalistic political or social system.

Albert Einstein at 53.

Einstein kept an abiding interest in philosophy, which penetrated his work; his ideas, in turn, influenced the development of modern philosophy itself. Here, too, he was cutting across unnecessary barriers. This is not surprising. Genius discovers itself not in splendid solutions to little puzzles but in the struggle with deep and perhaps eternal problems at the point where science and philosophy join.*

* For further discussion on Einstein's relativity theory, its genesis and influence, see Gerald Holton, *Thematic Origins of Scientific Thought, Kepler to Einstein*, chaps. 5–10, Harvard University Press, Cambridge, Mass., 1973. Among more recent biographies, see Banesh Hoffmann, with the collaboration of Helen Dukas, *Albert Einstein: Creator and Rebel*, Viking, New York, 1972; Jeremy Bernstein, *Einstein*, Viking, New York, 1973; or the best of the older biographies: Philipp Frank, *Einstein: His Life and Times*, Knopf, New York, 1947. For a selection of Einstein's correspondence, see *Albert Einstein, the Human Side: New Glimpses from His Archives*, edited by Helen Dukas and Banesh Hoffmann, Princeton University Press, Princeton, New Jersey, 1979.

28-5   The Twin Paradox

Homer and Ulysses are identical twins. Ulysses travels at high speed to a planet beyond the solar system and returns while Homer remains at home. When they are together again, which twin is older, or are they the same age? The correct answer is that Homer, the twin who stays at home, is older. This problem, with variations, has been the subject of spirited debate for decades, though there are very few who disagree with the answer.* The problem is a paradox because of the seemingly symmetric roles played by the twins with the asymmetric result in their aging. The paradox is resolved when the asymmetry of the twins' roles is noted. The relativistic result conflicts with common sense based on our strong but incorrect belief in absolute simultaneity. We will consider a particular case with some numerical magnitude that, though impractical, make the calculations easy.

Let planet P and Homer on earth be fixed in reference frame S a distance L_0 apart, as illustrated in Figure 28-14. We neglect the motion of the earth. Reference frames S' and S'' are moving with speed v toward and away from the planet, respectively. Ulysses quickly accelerates to speed v, then coasts in S' until he reaches the planet. When he stops, he is momentarily at rest in S. To return, he quickly accelerates to speed v toward earth and coasts in S'' until he reaches earth, where he stops. We can assume that the acceleration times are negligible compared with the coasting times. We use the following values for illustration: $L_0 = 8$ light-years and $v = 0.8c$. Then $\sqrt{1 - v^2/c^2} = 3/5$ and $\gamma = 5/3$.

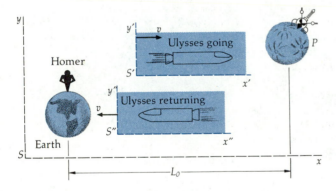

Figure 28-14
The twin paradox. The earth and a distant planet P are fixed in frame S. Ulysses coasts in frame S' to the planet and then coasts back in frame S''. His twin Homer stays on earth. When Ulysses returns, he is younger than his twin. The roles played by the twins are not symmetric. Homer remains in one inertial reference frame, but Ulysses must accelerate if he is to return home.

It is easy to analyze the problem from Homer's point of view on earth. According to Homer's clock, Ulysses coasts in S' for a time $L_0/v = 10$ y and in S'' for an equal time. Thus, Homer is 20 y older when Ulysses returns. The time interval in S' between Ulysses' leaving earth and arriving at the planet is shorter because it is proper time. The time it takes to reach the planet by Ulysses' clock is

$$\Delta t' = \frac{\Delta t}{\gamma} = \frac{10 \text{ y}}{5/3} = 6 \text{ y}$$

---

* A collection of some important papers concerning this paradox can be found in *Special Relativity Theory, Selected Reprints,* American Association of Physics Teachers, New York, 1963.

Since the same time is required for the return trip, Ulysses will have recorded 12 y for the round trip and will be 8 y younger than Homer.

From Ulysses' point of view, the calculation of this trip time is not difficult. The distance from the earth to the planet is contracted and is only

$$L = \frac{L_0}{\gamma} = \frac{8 \text{ light-years}}{5/3} = 4.8 \text{ light-years}$$

At $v = 0.8c$, it takes only 6 y each way. The real difficulty in this problem is for Ulysses to understand why his twin ages 20 y during his absence. If we consider Ulysses at rest and Homer moving away, Homer's clock should run slow and measure only $(3/5)(6) = 3.6$ y. Then why shouldn't Homer age only 7.2 y during the round trip? This, of course, is the paradox. The difficulty with the analysis from the point of view of Ulysses is that he does not remain in an inertial frame. What happens while Ulysses is stopping and starting? To investigate this problem in detail, we would need to treat accelerated reference frames, a subject dealt with in the study of general relativity but beyond the scope of this book. However, we can get some insight into the problem by considering the lack of synchronization of moving clocks.

Suppose that there is a clock on the planet P synchronized in S with Homer's clock on earth. In reference frame S', these clocks are unsynchronized by the amount $L_0 v/c^2$. For our example, this is 6.4 y. Thus, when Ulysses is coasting in S' near the planet, the clock on the planet leads that on the earth by 6.4 y. After Ulysses stops, he is in the frame S, in which these two clocks are synchronized. Thus, in the negligible time (according to Ulysses) it takes him to stop, the clock at earth must gain 6.4 y. Accordingly, his twin on earth ages 6.4 y. This 6.4 y plus the 3.6 y that Homer aged during the coasting makes him 10 y older by the time Ulysses is stopped in frame S. When Ulysses is in frame S'' coasting home, the clock on earth leads that on the planet by 6.4 y, and it will run another 3.6 y before he arrives home. We do not need to know the detailed behavior of the clocks during the acceleration in order to know the cumulative effect. The special theory of relativity is enough to show us that if the clocks on earth and on the planet are synchronized in S, the clock on earth lags that on P by $L_0/c^2 = 6.4$ y when viewed in S', and the clock on earth leads that on P by this amount when viewed in S''. The difficulty in understanding the analysis of Ulysses lies in the difficulty of giving up the idea of absolute simultaneity.

The predictions of the special theory of relativity concerning the twin paradox have been tested many times using small particles that can be accelerated to such large speeds that γ is appreciably greater than 1. Unstable particles can be accelerated and trapped in circular orbits in a magnetic field, for example, and their lifetimes can be compared with those of identical particles at rest. In all such experiments, the accelerated particles live longer on the average than those at rest, as predicted. These predictions are also confirmed by the results of an experiment using high-precision atomic clocks flown around the world in commercial airplanes, but the analysis of this experiment is complicated by the necessity of including gravitational effects treated in the general theory of relativity.*

---

* The details of these tests can be found in J. C. Hafele and Richard E. Keating, "Around-the-World Atomic Clocks: Predicted Relativistic Time Gains" and "Around-the-World Atomic Clocks: Observed Relativistic Time Gains," *Science*, July 14, 1972, p. 166.

28-6   Relativistic Momentum

We have seen in previous sections that Einstein's postulates require important modifications in our ideas of simultaneity and in our measurements of time and length. Perhaps more importantly, they also lead to modifications in our concepts of mass, momentum, and energy. In classical mechanics, the momentum of a particle is defined as the product of its mass and its velocity, $\mathbf{p} = m\mathbf{u}$, where $\mathbf{u}$ is the velocity. If there is no unbalanced force acting on the particle, its momentum remains constant. When there is an unbalanced force acting, Newton's second law states that the force equals the mass times the acceleration. Since the acceleration is the rate of change of the velocity and the mass is constant in classical mechanics, Newton's second law can be written

$$\mathbf{F} = \frac{m\,\Delta \mathbf{v}}{\Delta t} = \frac{\Delta(m\mathbf{v})}{\Delta t} = \frac{\Delta \mathbf{p}}{\Delta t}$$

In an isolated system of particles, with no unbalanced force acting on the system, the total momentum of the system remains constant.

In this section, we will show from a simple thought experiment that the classical expression for momentum, $\mathbf{p} = m\mathbf{u}$, is just an approximation. That is, this quantity is not conserved in an isolated system. We consider two observers: observer A in reference frame S and observer B in frame S'. Each observer has a ball of mass m. The two balls are identical when compared at rest. Each observer throws her ball vertically with speed u_0 such that it travels a distance L, makes an elastic collision with the other ball, and returns. Figure 28-15 shows how the collision looks in each reference frame. Classically, each ball has vertical momentum of magnitude mu_0. Since the vertical components of momentum are equal and opposite, the total vertical component of momentum is zero before the collision. The collision merely reverses the momentum of each ball, so the vertical momentum is zero after the collision.

Relativistically, however, the vertical components of momentum of the two balls as seen by either observer are not equal and opposite. Thus, when they are reversed by the collision, momentum is not conserved. Consider the time interval between throwing and catching the ball as observed by B. Since these events occur at the same place in space (see Figure 28-15b), this interval is proper time for B. (It equals $2L/u_0$.) However, from A's point of view, B is moving, so B throws the ball at one point in space and catches it at another (see Figure 28-15a). The time interval between these events is thus longer by the factor γ. The vertical component of velocity of B's ball as seen by A is u_0/γ, and the vertical components of momentum of the two balls are not equal and opposite as seen by observer A. Since the balls are reversed by the collision, momentum is not conserved. Of course the same result is observed by B. In the classical limit, when v is much less than c, γ is approximately 1, and the momentum of the system is conserved as seen by either observer.

The reason for defining momentum to be $\Sigma\, m\mathbf{u}$ in classical mechanics was that this quantity is conserved when there are no external forces, as in collisions. We now see this quantity is conserved only approximately. We

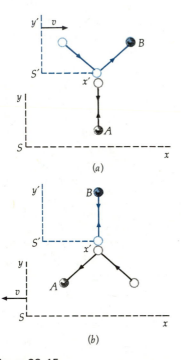

Figure 28-15
(a) Elastic collision of two identical balls as seen in frame S. The vertical component of the velocity of ball B is u_0/γ in S if it is u_0 in S'. (b) The same collision as seen in S'. In this frame, ball A has a vertical component of velocity u_0/γ.

will define the *relativistic momentum* **p** of a particle to have the following properties:

1. In collisions, **p** is conserved.

2. As u/c approaches zero, **p** approaches $m\mathbf{u}$.

We will state without proof that the quantity meeting these conditions is

$$\mathbf{p} = \frac{m\mathbf{u}}{\sqrt{1 - u^2/c^2}} \qquad \text{28-14}$$

Relativistic momentum

We will take this equation as the definition of the relativistic momentum of a particle. Because of the similarity between the factor $1/\sqrt{1 - u^2/c^2}$ and the factor γ in the equations for time dilation and length contraction, Equation 28-14 is often written

$$\mathbf{p} = \gamma m\mathbf{u} \qquad \text{with } \gamma = \frac{1}{\sqrt{1 - u^2/c^2}} \qquad \text{28-15}$$

The use of the symbol γ for two different quantities can cause some confusion. The notation is standard, however, and it simplifies many of the equations in relativity.

One interpretation of Equations 28-14 and 28-15 is that the mass of an object increases with speed from m_0 at rest to $m = \gamma m_0$ when it is moving at speed u. The quantity γm_0 is sometimes called the *relativistic mass* of the particle. Although this makes the expression mu for relativistic momentum look like the nonrelativistic expression, the use of relativistic mass often leads to mistakes. For example, the expression $\frac{1}{2}mu^2$ is not the correct relativistic expression for kinetic energy. We will avoid using relativistic mass in this book. The symbol m will always refer to the mass of a body at rest, sometimes called the rest mass.

28-7 Relativistic Energy

In classical mechanics, the work done by an unbalanced force acting on a particle equals the change in the kinetic energy of the particle. In relativistic mechanics, we define the unbalanced force acting on a particle as the rate of change of the relativistic momentum. The work done by such a force can then be calculated and set equal to the change in kinetic energy. We will give the result of this calculation without proof. The kinetic energy of a particle moving with speed u is given by

$$E_k = \gamma mc^2 - mc^2 \qquad \text{28-16}$$

where γ is given by

$$\gamma = \frac{1}{\sqrt{1 - u^2/c^2}}$$

(It is important to remember that γ is always greater than 1 and approaches 1 when u is much less than c.) The expression for kinetic energy consists of two terms. One, γmc^2, depends on the speed of the particle (through the factor γ), and the other, mc^2, is independent of the speed. The quantity mc^2 is called the *rest energy* of the particle. The total energy E is then defined as the sum of the kinetic energy and the rest energy:

Rest energy

$$E = E_k + mc^2 = \gamma mc^2 \qquad \text{28-17}$$

Total relativistic energy

Thus, the work done by an unbalanced force increases the energy from the rest energy mc^2 to γmc^2.

The expression for kinetic enegy given by Equation 28-16 doesn't look much like the classical expression $\frac{1}{2}mu^2$. However, when u is much less than c, we can approximate γ using the binomial expansion (Equation 28-2):

$$\gamma = (1 - u^2/c^2)^{-1/2} \approx 1 + \frac{1}{2}\frac{u^2}{c^2} \qquad \text{28-18}$$

Using this result, when u is much less than c, the expression for relativistic kinetic energy becomes

$$E_k = mc^2(\gamma - 1) \approx mc^2\left(1 + \frac{1}{2}\frac{u^2}{c^2} - 1\right) = \frac{1}{2}mu^2$$

Thus, at low speeds the relativistic expression is the same as the classical expression. The identification of the term mc^2 as rest energy is not merely a convenience. The conversion of rest energy to kinetic energy with a corresponding loss in rest mass is a common occurrence in radioactive decay and in nuclear reactions, including nuclear fission and nuclear fusion. We will give some examples of this in this section.

We note from Equation 28-18 that as the speed u approaches the speed of light c, the energy of the particle becomes very large because γ becomes very large. At $u = c$, $\gamma \to \infty$ and the energy becomes infinite. For u greater than c, γ is the square root of a negative number and is therefore imaginary. A simple interpretation of the result that it takes an infinite amount of energy to accelerate a particle to the speed of light is that no particle (that has rest mass) can travel as fast as or faster than the speed of light c. As we mentioned in Example 28-4, if the speed of a particle is less than c in one reference frame, it is less than c in all other reference frames moving relative to that frame at speeds less than c.

In practical applications, the momentum or energy of a particle is often known rather than the speed. Equation 28-14 for the relativistic momentum and Equation 28-17 for the relativistic energy can be combined to eliminate the speed u (see Problem 2). The result is

$$E^2 = p^2c^2 + (mc^2)^2 \qquad \text{28-19}$$

If the energy of a particle is much greater than its rest energy mc^2, the second term on the right of Equation 28-19 can be neglected, giving the useful approximation

$$E \approx pc \qquad \text{for} \qquad E \gg mc^2 \qquad \text{28-20}$$

Equation 28-20 is an exact relation between energy and momentum for particles with no rest mass such as photons.

Example 28-7   An electron with a rest energy of 0.511 MeV moves with speed $u = 0.8c$. Find its total energy, kinetic energy, and momentum.

We first calculate γ:

$$\gamma = \frac{1}{\sqrt{1 - u^2/c^2}} = \frac{1}{\sqrt{1 - 0.64}} = \frac{5}{3} = 1.67$$

The total energy is then

$$E = \gamma mc^2 = 1.67(0.511 \text{ MeV}) = 0.8533 \text{ MeV}$$

The kinetic energy is the total energy minus the rest energy:

$$E_k = E - mc^2 = 0.853 \text{ MeV} - 0.511 \text{ MeV} = 0.342 \text{ MeV}$$

The magnitude of the momentum is

$$p = \gamma mu = \gamma m(0.8c) = \frac{(0.8)\gamma mc^2}{c}$$

$$= \frac{(0.8)(1.67)(0.511 \text{ MeV})}{c} = 0.683 \text{ MeV}/c$$

The unit MeV/c is a convenient momentum unit.

Some numerical examples from atomic and nuclear physics will illustrate changes in rest mass and rest energy. Energies in atomic and nuclear physics are usually expressed in units of electron volts (eV) or mega electron volts (MeV).

$$1 \text{ eV} = 1.6 \times 10^{-19} \text{ J}$$

A convenient unit for the masses of atomic particles is eV/c^2 or MeV/c^2, which is just the rest energy of the particle divided by c^2.

The rest energies of some elementary particles and light nuclei are given in Table 28-1.

Table 28-1

Rest Energies of Some Elementary Particles and Light Nuclei

Particle	Symbol	Rest energy, MeV
Photon	γ	0
Electron (positron)	e or e^- (e^+)	0.5110
Muon	$\mu^\pm$	105.7
Pi meson	π^0	135
	$\pi^\pm$	139.6
Proton	p	938.280
Neutron	n	939.573
Deuteron	^{2}H or d	1875.628
Triton	^{3}H or t	2808.944
Alpha	^{4}He or α	3727.409

Example 28-8   A deuteron consists of a proton and neutron bound together. It is the nucleus of the deuterium atom, which is an isotope of hydrogen called heavy hydrogen and written ^2H. How much energy is required to separate the proton from the neutron in the deuteron?

From Table 28-1, we can see that the rest energy of the deuteron is 1875.63 MeV, the rest energy of the proton is 938.28 MeV, and that of the neutron is 939.57 MeV. The sum of the rest energies of the proton and neutron is 938.28 MeV + 939.57 MeV = 1877.85 MeV. This is greater than the rest energy of the deuteron by 1877.85 − 1875.63 = 2.22 MeV. This energy is called the *binding energy* of the deuteron. To break up a deuteron into a proton plus a neutron, 2.22 MeV of energy must be added to the deuteron. This can be done by bombarding the deuteron with energetic particles or with electromagnetic radiation of energy of at least 2.22 MeV.

When a deuteron is formed by the combination of a neutron and a proton, energy must be released. When neutrons from a reactor are incident on protons, some neutrons are captured to form deuterons. In the capture process, 2.22 MeV of energy is released, usually in the form of electromagnetic radiation.

Example 28-8 illustrates an important property of atoms and nuclei. Any stable composite particle, such as a deuteron or a helium nucleus (2 neutrons plus 2 protons), that is made up of other particles has a rest energy that is less than the sum of the rest energies of its parts. The difference is the binding energy of the composite particle. Binding energies of electrons in atoms and molecules are of the order of a few electron volts, which leads to a negligible difference in mass between the composite particle and its parts. Binding energies of nuclei are of the order of several MeV, which leads to a noticeable mass difference. Some very heavy nuclei, such as radium, are radioactive and decay into a lighter nucleus plus an alpha particle. (An alpha particle is the nucleus of the ^4He atom.) In this case, the original nucleus has a rest energy greater than that of the decay particles. The excess energy appears as kinetic energy of the decay particles. We will study this in more detail in Chapter 32.

Example 28-9   In a typical nuclear fusion reaction, a tritium nucleus (^3H) and a deuterium nucleus (^2H) fuse together to form a helium nucleus (^4He) plus a neutron. How much energy is released in this fusion reaction?

The reaction is written

$$^2H + {}^3H \rightarrow {}^4He + n$$

From Table 28-1, we see that the rest energy of the deuterium plus tritium nuclei is 1875.628 MeV + 2808.944 MeV = 4684.572 MeV. The rest energy of the helium nucleus plus the neutron is 3728.409 + 939.573 = 4666.982 MeV. This is less than that of deuterium plus tritium by 4684.572 − 4666.982 = 17.59 MeV. The energy released in this reaction is 17.59 MeV. This and other fusion reactions occur in the sun and are responsible for the energy supplied to the earth. As the sun gives off energy, its mass continually decreases.

Example 28-10   A hydrogen atom consisting of a proton and an electron has a binding energy of 13.6 eV. By what percentage is the mass of the proton plus the electron greater than that of the hydrogen atom?

The rest energy of a proton plus that of an electron is 938.28 MeV + 0.511 MeV = 938.791 MeV. The sum of the masses of these two particles is 938.791 MeV/c^2. The mass of the hydrogen atom is less than this by 13.6 eV/c^2. The percentage difference is

$$\frac{13.6 \text{ eV}/c^2}{938.791 \times 10^6 \text{ eV}/c^2} = 1.45 \times 10^{-8} = 1.45 \times 10^{-6}\%$$

This mass difference is so small as to be hardly measurable.

Example 28-11   If the rest energy of one gram of dirt could be converted completely into electrical energy, for how long would it light a 100-W light bulb?

Since one watt is one joule per second, the energy needed to light a 100-W bulb for a time t is

$$E = Pt = (100 \text{ J/s})(t) = 100t \text{ J/s}$$

The rest energy of one gram of dirt in SI units is

$$E = mc^2 = (10^{-3} \text{ kg})(3 \times 10^8 \text{m/s})^2 = 9 \times 10^{13} \text{ J}$$

Note that this is an enormous amount of energy. Setting this equal to $100t$ J/s and solving for t, we obtain

$$100t \text{ J/s} = 9 \times 10^{13} \text{ J}$$

$$t = 9 \times 10^{11} \text{ s}$$

The number of seconds in one year is 3.16×10^7. Thus if the rest energy of one gram of dirt could be converted completely into electrical energy, it could light a 100-W bulb for a time of

$$t = (9 \times 10^{11} \text{ s})\left(\frac{1 \text{ y}}{3.16 \times 10^7 \text{ s}}\right) = 2.85 \times 10^4 \text{ y}$$

$$= 28,500 \text{ years}$$

28-8   General Relativity

The generalization of relativity theory to noninertial reference frames by Einstein in 1916 is known as the general theory of relativity. It is much more difficult mathematically than the special theory of relativity, and there are fewer situations in which it can be tested. Nevertheless, its importance calls for a brief qualitative discussion.

The basis of the general theory of relativity is the principle of equivalence:

A homogeneous gravitational field is completely equivalent to a uniformly accelerated reference frame.   Principle of equivalence

This principle arises in newtonian mechanics because of the apparent identity of gravitational and inertial mass. In a uniform gravitational field, all
objects fall with the same acceleration **g** independent of their mass because
the gravitational force is proportional to the (gravitational) mass whereas
the acceleration varies inversely with the (inertial) mass. Consider a compartment in space far from any matter and undergoing uniform acceleration
a, as shown in Figure 28-16a. If objects are dropped in the compartment,
they fall to the "floor" with acceleration $\mathbf{g} = -\mathbf{a}$. If people stand on a spring
scale it will read their "weight" of magnitude ma. No mechanics experiment
can be performed *inside* the compartment that will distinguish whether the
compartment is actually accelerating in space or is at rest (or moving with
uniform velocity) in the presence of a uniform gravitational field $\mathbf{g} = -\mathbf{a}$.

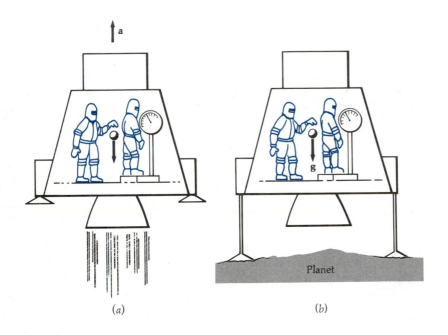

(a)   (b)

Figure 28-16
The results of mechanics
experiments in a uniformly
accelerated reference frame (*a*)
cannot be distinguished from those
in a uniform gravitational field (*b*) if
the acceleration **a** and the
gravitational field **g** have the same
magnitude.

   Einstein assumed that the principle of equivalence applied to all physics
and not only to mechanics. In effect, he assumed that there is no experiment
of any kind that can distinguish uniformly accelerated motion from the
presence of a gravitational field. We will look qualitatively at a few of the
consequences of this assumption.
   The first consequence of the principle of equivalence we will discuss, the
deflection of a light beam in a gravitational field, was one of the first to be
tested experimentally. Figure 28-17 shows a beam of light entering a compartment that is accelerating. Successive positions of the compartment are
shown at equal time intervals. Because the compartment is accelerating, the
distance it moves in each time interval increases with time. The path of the
beam of light as observed from inside the compartment is therefore a parabola. But according to the equivalence principle, there is no way to distinguish between an accelerating compartment and one with uniform velocity
in a uniform gravitational field. We conclude, therefore, that a beam of light,

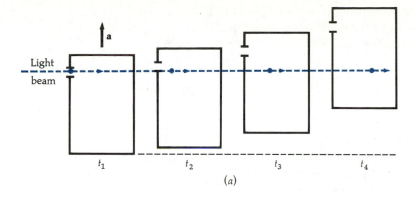

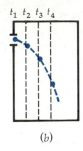

(b)

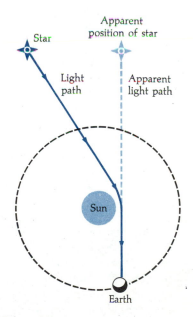

(a)

Figure 28-17
(a) Light beam moving in a straight line through a compartment that is undergoing uniform acceleration. The position of the beam is shown at equally spaced times t_1, t_2, t_3, and t_4. (b) In the reference frame of the compartment, the light travels in a parabolic path, as a ball would if it were projected horizontally. The vertical displacements are greatly exaggerated in both (a) and (b) for emphasis.

like massive objects, will accelerate in a gravitational field. For example, near the surface of the earth, light will fall with acceleration 9.81 m/s^2. This is difficult to observe because of the enormous speed of light. For example, in a distance of 3000 km, which takes about 0.01 s to traverse, a beam of light should fall about 0.5 mm. Einstein pointed out that the deflection of a light beam in a gravitational field might be observed when light from a distant star passes close to the sun, as illustrated in Figure 28-18. Because of the brightness of the sun, such a star cannot ordinarily be seen. Such a deflection was first observed in 1919 during an eclipse of the sun. This well-publicized observation brought instant worldwide fame to Einstein.

A second prediction from Einstein's theory of general relativity (which we will not discuss in detail) is the excess precession of the perihelion of the orbit of Mercury of about $0.01°$ per century. This effect had been known and unexplained for some time, so in a sense, explaining it represented an immediate success of the theory. There is, however, some difficulty in comparing the prediction of general relativity with experimental results because of other effects, for example, the perturbations due to other planets and the nonspherical shape of the sun.

A third prediction of general relativity concerns the change in time intervals and frequencies of light in a gravitational field. In Chapter 9, we found the gravitational potential energy between two masses, M and m, a distance r apart to be

$$U = -\frac{GMm}{r}$$

where G is the universal gravitational constant, and the zero of potential energy has been chosen to be when the separation of the masses is infinite. The potential energy per unit mass near a mass M is called the gravitational potential ϕ:

$$\phi = -\frac{GM}{r} \qquad\qquad 28\text{-}21$$

According to the general theory of relativity, clocks run more slowly in regions of low gravitational potential. (Since the gravitational potential is negative, as can be seen from Equation 28-21, low gravitational potential occurs near the mass where the *magnitude* of the potential is large.) If Δt_1 is a

Figure 28-18
Deflection (greatly exaggerated) of a beam of light due to the gravitational attraction of the sun.

time interval measured by a clock where the gravitational potential is ϕ_1 and Δt_2 is the same interval measured by a clock where the gravitational potential is ϕ_2, general relativity predicts that the fractional difference between these times will be approximately

$$\frac{\Delta t_2 - \Delta t_1}{\Delta t} = \frac{1}{c^2}(\phi_2 - \phi_1) \qquad 28\text{-}22$$

(Since this shift is usually very small, it does not matter by which interval we divide the left side of the equation.) A clock in a region of low gravitational potential will therefore run slower than one in a region of high potential. Since a vibrating atom can be considered to be a clock, the frequency of vibration in a region of low potential, such as near the sun, will be lower than that of the same atom on earth. This shift toward a lower frequency and therefore a longer wavelength is called the *gravitational red shift*.

Gravitational red shift

   As our final example of the predictions of general relativity, we mention *black holes*, which were first predicted by Oppenheimer and Snyder in 1939. According to the general theory of relativity, if the density of an object such as a star is great enough, the gravitational attraction will be so great that once inside a critical radius, nothing can escape, not even light or other electromagnetic radiation. (The effect of a black hole on objects outside the critical radius is the same as that of any other mass.) A remarkable property of such an object is that nothing that happens inside it can be communicated to the outside. As sometimes occurs in physics, a simple but incorrect calculation gives the correct results for the relation between the mass and the critical radius of a black hole. In newtonian mechanics, the speed needed for a particle to escape from the surface of a planet or star is found by requiring the kinetic energy $\frac{1}{2}mv^2$ to be equal in magnitude to the potential energy $-GMm/r$ so that the total energy is zero. The resulting escape speed is

$$v_e = \sqrt{\frac{2GM}{r}}$$

(If M and r are replaced by the mass and radius of the earth, respectively, this equation becomes identical to Equation 9-22 for escape from the gravitational field of the earth.) If we set the escape speed equal to the speed of light and solve for the radius, we obtain the critical radius R_G, called the *Schwarzschild radius*:

$$R_G = \frac{2GM}{c^2} \qquad 28\text{-}23$$

For an object of mass equal to that of our sun to be a black hole, its radius must be about 3 km. Since no radiation is emitted from a black hole and its radius is expected to be small, the detection of such an object is not easy. The best chance of detection would occur if a black hole were a companion to a normal star in a binary star system. The black hole would affect a number of properties of its companion. Measurements of the doppler shift of the light from the normal star, for example, might allow a computation of the mass of the unseen companion to determine whether it is great enough to be a black hole. At present, there are several excellent candidates — one in the constellation Cygnus, one in the Small Magellanic Cloud, and perhaps one in our own galaxy — but the evidence is not conclusive.

Summary

1. The special theory of relativity is based on two postulates of Albert Einstein:

>Postulate 1. Absolute, uniform motion cannot be detected.

>Postulate 2. The speed of light is independent of the motion of the source.

An important consequence of these postulates can be stated as follows: Every observer measures the same value for the speed of light independent of the relative motion of the sources and observer. All of the results of special relativity can be derived from these postulates.

2. The Michelson–Morley experiment was an attempt to measure the absolute velocity of the earth by comparing the speed of light in the direction of motion of the earth with that in a direction perpendicular to that motion. Their null result for the difference in these speeds is consistent with Einstein's postulates.

3. The time interval measured between two events that occur at the same point in space in some reference frame is called *proper time.* In another reference frame, in which the events occur at different places, the time interval between the events is longer by the factor γ, where

$$\gamma = \frac{1}{\sqrt{1 - v^2/c^2}}$$

This result is known as time dilation. A related phenomenon is length contraction. The length of an object measured in a frame in which it is at rest is called the object's proper length L_0. When measured in another reference frame, the length of the object is L_0/γ.

4. Two events that are simultaneous in one reference frame are not simultaneous in another frame moving relative to the first. If two clocks are synchronized in the frame in which they are at rest, they will be out of synchronization in another frame. In the frame in which they are moving, the "chasing clock" leads by an amount $\Delta t_s = L_0 v/c^2$, where L_0 is the proper distance between the clocks.

5. The relativistic momentum of a particle is related to its mass and velocity by

$$\mathbf{p} = \frac{m\mathbf{u}}{\sqrt{1 - u^2/c^2}} = \gamma m \mathbf{u}$$

6. The kinetic energy of a particle is given by

$$E_k = \gamma mc^2 - mc^2$$

where mc^2 is the rest energy. The total energy is

$$E = E_k + mc^2 = \gamma mc^2$$

The total energy is related to the momentum by

$$E^2 = p^2c^2 + (mc^2)^2$$

For particles with energy much greater than their rest energy, a useful approximation is

$$E \approx pc$$

for $E \gg mc^2$. This is an exact equation for massless particles such as photons.

7. The total mass of a bound system of particles, such as nuclei or atoms, is less than the sum of the masses of the particles making up the system. The mass difference times c^2 equals the binding energy of the system. The binding energy is the energy that must be added to break up the system into its parts. Binding energies of electrons in atoms are of the order of eV or keV, leading to a negligible mass difference. Binding energies of nuclei are of the order of MeV, so that mass difference is noticeable.

8. The basis of the general theory of relativity is the principle of equivalence: A homogeneous gravitational field is completely equivalent to a uniformly accelerated reference frame. Important consequences of general relativity include the bending of light in a gravitational field; the prediction of the precession of the perihelion of the orbit of Mercury; the gravitational red shift; and, probably, the existence of black holes.

Suggestions for Further Reading

Bondi, Hermann: *Relativity and Common Sense: A New Approach to Einstein*, Doubleday, Garden City, New York, 1964.

This book uses familiar phenomena to help show how logical and easy to understand the ideas of special relativity really are.

Chaffee, Frederic H., Jr.: "The Discovery of a Gravitational Lens," *Scientific American*, November 1980, p. 70.

General relativity predicts that light should be deflected by concentrations of matter. This article describes how an elliptical galaxy can act as a giant lens in space.

Gamow, George: "Gravity," *Scientific American*, March 1961, p. 94.

Einstein's general theory of relativity is explained in an entertaining and nonmathematical fashion in this article.

MacKeown, P.K.: "Gravity Is Geometry," *The Physics Teacher*, vol. 22, 1984, p. 557.

This article is an excellent brief exposition of the ideas of general relativity.

Shankland, R.S.: "The Michelson–Morley Experiment," *Scientific American*, November 1964, p. 107.

This article sets the experiment in its historical context and considers its influence on the development of the theory of relativity.

Review

A. Objectives: After studying this chapter, you should:

1. Be able to discuss the results and significance of the Michelson–Morley experiment.

2. Be able to state the Einstein postulates of special relativity.

3. Be able to define proper time and proper length and state the equations for time dilation and length contraction.

4. Be able to discuss the lack of synchronization of clocks in moving reference frames.

5. Be able to discuss the twin paradox.

6. Be able to state the definition of relativistic momentum and the equations relating the kinetic energy and total energy of a particle to its speed.

7. Be able to discuss the relation between mass and energy in special relativity and compute the binding energy of various systems from the known rest masses of their constituents.

8. Be able to state the principle of equivalence and discuss three predictions derived from it.

B. Define, explain, or otherwise identify:

Principle of newtonian relativity, p. 713
Michelson–Morley experiment, pp. 714–716
Einstein postulates, pp. 717
Proper time, p. 719
Time dilation, p. 719
Length contraction, p. 720
Proper length, p. 720
Synchronized clocks, p. 723
Simultaneity, p. 724
Twin paradox, p. 730
Relativistic momentum, p. 733
Rest energy, p. 734
Binding energy, p. 736
Principle of equivalence, p. 737
Gravitational red shift, p. 740
Black hole, p. 740
Schwarzschild radius, p. 740

C. True or false: If the statement is true, explain why. If it is false, give a counterexample.

1. The speed of light is the same in all reference frames.

2. Proper time is the shortest time interval between two events.

3. Absolute motion can be determined by means of length contraction.

4. The light-year is a unit of distance.

5. Simultaneous events must occur at the same place.

6. If two events are not simultaneous in one frame, they cannot be simultaneous in any other frame.

7. Rest mass can sometimes be converted into energy.

8. If two particles are tightly bound together by strong attractive forces, the mass of the system is less than the sum of the masses of the individual particles when separated.

Exercises

Section 28-1   Newtonian Relativity

There are no exercises for this section.

Section 28-2   The Michelson–Morley Experiment

1. In one series of measurements of the speed of light, Michelson used a path length L of 27.4 km (22 mi). (*a*) What is the time needed for light to travel the round-trip distance of $2L$? (*b*) What is the classical correction term in seconds in Equation 28-1, assuming the earth's speed is $v = 10^{-4}c$? (*c*) From about 1600 measurements, Michelson quoted the result for the speed of light as $299{,}796 \pm 4$ km/s. Is this experiment accurate enough to be sensitive to the correction term in Equation 28-1?

2. An airplane flies with speed u relative to still air from point A to point B and returns. Compare the time required for the round trip when the wind blows from A to B with speed v with that when the wind blows perpendicularly to the line AB with speed v.

Section 28-3   Einstein's Postulates and Their Consequences

3. The proper mean life of pions is 2.6×10^{-8} s. If a beam of these particles has speed of $0.9c$, (*a*) what would their mean life be as measured in the laboratory? (*b*) How far would they travel on the average before they decay? (*c*) What would your answer be to part (*b*) if you neglected time dilation?

4. (*a*) In the reference frame of the pion in Exercise 3, how far does the laboratory travel in a typical lifetime of 2.6×10^{-8} s? (*b*) What is this distance in the laboratory frame?

5. The proper mean life of a muon is 2 μs. Muons in a beam are traveling at $0.999c$. (*a*) What is their mean lifetime as measured in the laboratory? (*b*) How far do they travel on the average before they decay?

6. (*a*) In the reference frame of the muon in Exercise 5, how far does the laboratory travel in a typical lifetime of 2 μs? (*b*) What is this distance in the laboratory frame?

7. A spaceship of proper length 100 m passes you at a high speed. You measure the length of the spaceship to be 95 m. What was the speed of the spaceship?

8. A spaceship departs from earth for the star Alpha Centauri, which is 4 light-years away. The spaceship travels at $0.9c$. How long does it take to get there (*a*) as measured on earth and (*b*) as measured by a passenger on the spaceship?

9. A spaceship travels to a star 75 light-years away at a speed of 2×10^8 m/s. How long does it take to get there (*a*) as measured on earth and (*b*) as measured by a passenger on the spaceship?

10. The mean lifetime of a pion traveling at high speed is measured to be 8.0×10^{-8} s. Its lifetime when measured at rest is 2.6×10^{-8} s. How fast is the pion traveling?

11. How fast must a muon travel so that its mean lifetime is 50 μs if its mean lifetime at rest is 2 μs?

12. A metrestick moves with speed $v = 0.6c$ relative to you in the direction parallel to the stick. (*a*) Find the length of the stick as measured by you. (*b*) How long does it take for the stick to pass you?

13. How fast must a metrestick travel relative to you in the direction parallel to the stick so that its length as measured by you is 50 cm?

14. Use the binomial expansion (Equation 28-2) to derive the following results for v much less than c and use them when applicable in the following exercises and problems.

(a) $\gamma \approx 1 + \dfrac{1}{2}\dfrac{v^2}{c^2}$

(b) $\dfrac{1}{\gamma} \approx 1 - \dfrac{1}{2}\dfrac{v^2}{c^2}$

(c) $\gamma - 1 \approx 1 - \dfrac{1}{\gamma} \approx \dfrac{1}{2}\dfrac{v^2}{c^2}$

15. Supersonic jets achieve maximum speeds of about $3 \times 10^{-6}c$. (a) By what percentage would you see such a jet contracted in length? (b) During a time of $1\ \mathrm{y} = 3.16 \times 10^7\ \mathrm{s}$ on your clock, how much time would elapse on the pilot's clock? How many minutes are lost by the pilot's clock in 1 y of your time?

16. How great must the relative speed of two observers be for their time-interval measurements to differ by 1 percent? (See Exercise 14.)

17. Two spaceships are approaching each other. (a) If the speed of each is $0.9c$ relative to the earth, what is the speed of one relative to the other? (b) If the speed of each relative to the earth is 30,000 m/s (about 100 times the speed of sound), what is the speed of one relative to the other?

18. A spaceship is moving east at $0.99c$ relative to the earth. A second spaceship is moving west at $0.99c$ relative to the earth. What is the speed of one spaceship relative to the other?

Section 28-4   Clock Synchronization and Simultaneity

Exercises 19 to 23 refer to the following situation: An observer in S' lays out a distance $L' = 100\ c \cdot min$ between points A' and B' and places a flashbulb at the midpoint C'. She arranges for the bulb to flash and for clocks at A' and B' to be started at 0 when the light from the flash reaches the clocks (see Figure 28-19). Frame S' is moving to the right with speed $0.6c$ relative to an observer C in frame S who is at the midpoint between A' and B' when the bulb flashes and sets her clock to zero at that time.

Figure 28-19
Exercises 19 to 23.

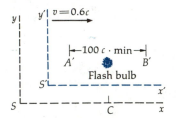

19. What is the separation distance between clocks A' and B' according to the observer in S?

20. As the light pulse from the flashbulb travels toward A' with speed c, A' travels toward C with speed $0.6c$. Show that the clock in S reads 25 min when the flash reaches A'. (*Hint:* In time t, the light travels a distance ct and A' travels $0.6ct$. The sum of these distances must equal the distance between A' and the flashbulb as seen in S.)

21. Show that the clock in S reads 100 min when the light flash reaches B', which is traveling away from C with speed $0.6c$. (See hint in Exercise 20.)

22. The time interval between reception of the flashes as calculated in Exercises 20 and 21 is 75 min according to the observer in S. How much time does she expect to have elapsed on the clock at A' during this 75 min?

23. The time interval calculated in Exercise 22 is the amount that the clock at A' leads that at B' according to observers in S. Compare this result with $L_0 v/c^2$.

Section 28-5   The Twin Paradox

24. A friend of yours who is the same age as you travels at $0.999c$ to a star 15 light-years away. She spends 10 years on one of the star's planets and then returns at $0.999c$. How long is she away (a) as measured by you and (b) as measured by her?

Section 28-6   Relativistic Momentum;
Section 28-7   Relativistic Energy

25. How much rest mass must be converted into energy (a) to produce 1 J and (b) to keep a 100-W light bulb lighted for 10 years?

26. If you can sell energy for 10 cents per kilowatt-hour, how much would you get for the rest energy in one gram of dirt?

27. Find the ratio of the total energy to the rest energy of a particle of mass m moving with speed (a) $0.1c$, (b) $0.5c$, (c) $0.8c$, and (d) $0.99c$.

28. An electron with rest energy 0.511 MeV moves with speed $v = 0.2c$. Find its total energy, kinetic energy, and momentum.

29. A muon has a rest energy of 105.7 MeV. Calculate its mass in kilograms.

30. A proton of rest energy 938 MeV has a total energy of 1200 MeV. (a) What is its speed? (b) What is its momentum?

31. The total energy of a particle is twice its rest energy. (a) What is γ? (b) Find u/c for the particle. (c) Show that its momentum is given by $p = \sqrt{3}\ mc$, where m is its mass. (*Hint:* Use Equation 28-19 to find the momentum.)

32. For the fusion reaction of Example 28-9, calculate the number of reactions per second necessary to generate 1 kW of power.

33. Use Table 28-1 to find how much energy is needed to remove one neutron from ^{4}He, leaving ^{3}H $+ n$.

34. A free neutron decays into a proton plus an electron:

$$n \rightarrow p + e$$

Use Table 28-1 to calculate the energy released in this reaction.

35. How much energy would be required to accelerate a particle of mass m from rest to a speed of (a) $0.5c$, (b) $0.9c$, and (c) $0.99c$? Express your answers as multiples of the rest energy.

36. If the kinetic energy of a particle equals its rest energy, what error is made by using $p = mu$ for its momentum?

37. In another nuclear fusion reaction, two ^{2}H nuclei are combined to produce ^{4}He. (a) How much energy is released in this reaction? (b) How many such reactions must take place per second to produce 1 W of power?

Section 28-8   General Relativity

There are no exercises for this section.

Problems

1. A friend of yours who is the same age as you travels to the star Alpha Centuari, 4 light-years away, and returns immediately. He claims that the entire trip took just 6 years. How fast did he travel?

2. Use Equations 28-14 and 28-17 to derive the equation $E^2 = p^2 c^2 + m^2 c^4$.

3. If a plane flies at a speed of 2000 km/h, how long must it fly before its clock loses 1 s because of time dilation?

4. Two spaceships, each 100 m long when measured at rest, travel toward each other with speeds of $0.8c$ relative to earth. (a) How long is each ship as measured by someone on earth? (b) How fast is each ship traveling as measured by the other? (c) How long is one ship when measured by the other?

5. (a) Show that the speed u of a particle of mass m and total energy E is given by

$$\frac{u}{c} = \left[1 - \frac{(mc^2)^2}{E^2} \right]^{1/2}$$

and that when E is much greater than mc^2, this can be approximated by

$$\frac{u}{c} = 1 - \frac{(mc^2)^2}{2E^2}$$

Find the speed of an electron of kinetic energy (b) 0.51 MeV and (c) 10 MeV.

6. A light beam moves along the y' axis with speed c in frame S', which is moving to the right with speed v relative to frame S. (a) Find u_x and u_y, the x and y components of the velocity of the light beam in frame S. (b) Show that the magnitude of the velocity of the light beam in S is c.

7. An electron of rest energy 0.511 MeV has a total energy of 5 MeV. (a) Find its momentum in units of MeV/c from Equation 28-19. (b) Find the ratio of its speed u to the speed of light.

8. The rest energy of a proton is about 938 MeV. If its kinetic energy is also 938 MeV, find (a) its momentum and (b) its speed.

CHAPTER TWENTY-NINE

The Origins of Quantum Theory

Scientific research consists in seeing what everyone else has seen, but thinking what no one else has thought.

A. SZENT-GYORGYI

Now that you have some acquaintance with Einstein's theory of relativity, you are ready to learn more about the discoveries that rocked the scientific world during the first part of the twentieth century. The most significant of these had to do with the gradual unfolding of quantum theory, which now dominates much of physics, chemistry, and biology. Quantum theory had its origins in Planck's theory of blackbody radiation, Einstein's explanation of the photoelectric effect, the behavior of x-rays, and Bohr's model of the hydrogen atom, all of which are discussed in this chapter.

In Chapter 28, we saw that Newton's laws must be modified when they are applied to objects that move at speeds comparable to the speed of light. In the last 20 years of the nineteenth century and the first 30 years of the twentieth century, many startling discoveries, both experimental and theoretical, demonstrated that the laws of classical physics also break down when they are applied to microscopic systems, such as the particles within an atom. This failure is as drastic as the failure of newtonian mechanics at high speeds. The interior of the atom can be described only in terms of *quantum theory* (sometimes called *quantum mechanics*), which requires the modification of some of our fundamental ideas about the relationships between physical theory and the physical world. As with special relativity, quantum theory reduces to classical physics when it is applied to macroscopic (large-scale) systems, that is, objects in our familiar, everyday world.

The development of quantum theory was very different from that of the theory of relativity. Special relativity was presented as essentially a complete theory in 1905 (and general relativity in 1916) by a single scientist, Albert Einstein. Quantum theory, on the other hand, was developed over a long period of time by many different people. Many of the discoveries seemed unrelated to each other, and it wasn't until the late 1920s that any consistent theory emerged. This theory, known as quantum theory, or quantum mechanics, is now the basis of all our understanding of the microscopic world. It is extremely successful, yet there is still controversy over many of its philosophical interpretations. (Table 29-1 lists the approximate dates of some of the important experiments performed and theories proposed between 1881 and 1932.)

Table 29-1

Approximate Dates of Some Important Experiments and Theories, 1881–1932

1881	Michelson obtains null result for absolute velocity of earth
1884	Balmer finds empirical formula for spectral lines of hydrogen
1887	Hertz produces electromagnetic waves, verifying Maxwell's theory and accidently discovering photoelectric effect
1887	Michelson repeats his experiment with Morley, again obtaining null result
1895	Roentgen discovers x-rays
1896	Becquerel discovers nuclear radioactivity
1897	J. J. Thomson measures e/m for cathode rays, showing that electrons are fundamental constituents of atoms
1900	Planck explains blackbody radiation using energy quantization involving new constant h
1900	Lenard investigates photoelectric effect and finds energy of electrons independent of light intensity
1905	Einstein proposes special theory of relativity
1905	Einstein explains photoelectric effect by suggesting quantization of radiation
1907	Einstein applies energy quantization to explain temperature dependence of heat capacities of solids
1908	Rydberg and Ritz generalize Balmer's formula to fit spectra of many elements
1909	Millikan's oil-drop experiment shows quantization of electric charge
1911	Rutherford proposes nuclear model of atom based on alpha-particle scattering experiments of Geiger and Marsden
1912	Friedrich and Knipping and von Laue demonstrate diffraction of x-rays by crystals showing that x-rays are waves and that crystals are regular arrays
1913	Bohr proposes model of hydrogen atom
1914	Moseley analyzes x-ray spectra using Bohr model to explain periodic table in terms of atomic number
1914	Franck and Hertz demonstrate atomic energy quantization
1915	Duane and Hunt show that the short-wavelength limit of x-rays is determined from quantum theory
1916	Wilson and Sommerfeld propose rules for quantization of periodic systems
1916	Millikan verifies Einstein's photoelectric equation
1923	Compton explains x-ray scattering by electrons as collision of photon and electron and verifies results experimentally
1924	De Broglie proposes electron waves of wavelength h/p
1925	Schrödinger develops mathematics of electron wave mechanics
1925	Heisenberg invents matrix mechanics
1925	Pauli states exclusion principle
1927	Heisenberg formulates uncertainty principle
1927	Davisson and Germer observe electron wave diffraction by single crystal
1927	G. P. Thomson observes electron wave diffraction in metal foil
1928	Gamow and Condon and Gurney apply quantum mechanics to explain alpha-decay lifetimes
1928	Dirac develops relativistic quantum mechanics and predicts existence of positron
1932	Chadwick discovers neutron
1932	Anderson discovers positron

The origins of quantum theory were, strangely enough, not in the discovery of radioactivity or x-rays or atomic spectra, but in thermodynamics. In his study of the radiation spectrum of a blackbody, Max Planck found that he could reconcile theory and experiment if he assumed that radiant energy was emitted and absorbed not continuously but in discrete lumps or quanta. It was Einstein who first recognized that this "quantization" of radiant energy was not just a calculational device, but a general property of radiation. Niels Bohr then applied Einstein's ideas of energy quantization to the energy of an atom and proposed a model of the hydrogen atom that was spectacularly successful in calculations of the wavelengths of the radiation emitted by hydrogen. In this chapter, we will look qualitatively at the origins of the idea of energy quantization.

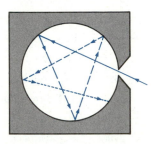

Figure 29-1
Cavity approximating an ideal blackbody. Radiation entering the cavity has little chance of leaving before it is completely absorbed.

29-1 Blackbody Radiation

One of the most puzzling phenomena studied near the end of the nineteenth century was the spectral distribution of blackbody radiation.* A blackbody is an ideal system that absorbs all the radiation incident on it. It can be approximated by a cavity with a very small opening, as illustrated in Figure 29-1. The characteristics of the radiation in such a cavity of a body depend only on the temperature of the walls. At ordinary temperatures (below about 600°C), the thermal radiation emitted by a blackbody is not visible because the energy is concentrated in the infrared region of the electromagnetic spectrum. As the body is heated, the amount of energy radiation increases (according to the Stefan–Boltzmann law, Equation 13-20), and the energy concentration moves to shorter wavelengths. Between about 600 and 700°C, there is enough energy in the visible spectrum for the body to glow a dull red. At higher temperatures, it becomes bright red or even "white hot."

Figure 29-2 shows the power radiated by a blackbody as a function of wavelength for several different temperatures. These curves are known as *spectral distribution curves*. The quantity P in this figure is the power radiated per unit area per unit wavelength. It is a function of both the wavelength λ and the temperature T and is called the *spectral distribution function*. This function, $P(\lambda, T)$, has a maximum at a wavelength λ_m that varies inversely with the temperature according to the Wien displacement law (Equation 13-24), as discussed in Section 13-5:

$$\lambda_m = \frac{2.898 \text{ mm} \cdot \text{K}}{T}$$

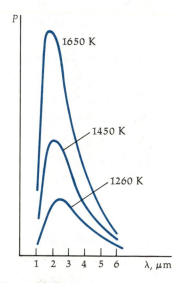

Figure 29-2
Spectral distribution curves for the radiation from a blackbody for three different temperatures.

The spectral distribution function $P(\lambda, T)$ can be calculated from classical thermodynamics in a straightforward way, and the result can be compared with the experimental curves of Figure 29-2. The result of this classical calculation, known as the *Rayleigh-Jeans law*, is

$$P(\lambda, T) = 2\pi ckT\lambda^{-4} \qquad\qquad 29\text{-}1$$

Rayleigh-Jeans law

* Blackbodies were discussed briefly in Section 13-5. Some of that discussion is repeated here.

where k is Boltzmann's constant. This result agrees with experimental results in the region of long wavelengths, but it disagrees violently at short wavelengths. As λ approaches zero, the experimentally determined $P(\lambda, T)$ also approaches zero, but the calculated function becomes infinite because it is proportional to λ^{-4}. Thus, according to the classical calculation, blackbodies radiate an infinite amount of energy concentrated in the very short wavelengths. This result was known as the *ultraviolet catastrophe*.

In 1900, the German physicist Max Planck announced that by making a strange modification in the classical calculation he could derive a function $P(\lambda, T)$ that agreed with the experimental data at all wavelengths. Planck first found an empirical function that fit the data and then searched for a way to modify the usual calculation. He found that he could "derive" a function that fit the experimental results if he made the unusual assumption that the energy was not continuously emitted or absorbed by the blackbody but was instead emitted or absorbed in discrete packets or quanta. The size of an energy quantum was proportional to the frequency of the radiation:

$$E = hf \qquad\qquad\qquad\qquad 29\text{-}2$$

where h is the proportionality constant now known as *Planck's constant*. The value of h was determined by Planck by fitting his function to the data. The accepted value of this constant is now

$$h = 6.626 \times 10^{-34} \text{ J} \cdot \text{s} = 4.136 \times 10^{-15} \text{ eV} \cdot \text{s} \qquad 29\text{-}3 \qquad \text{Planck's constant}$$

Planck's result is shown in Figure 29-3 along with experimental data and the Rayleigh-Jeans law.

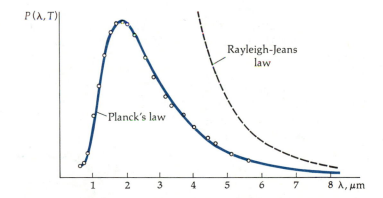

$P(\lambda, T)$

Rayleigh-Jeans law

Planck's law

1 2 3 4 5 6 7 8 $\lambda, \mu m$

Figure 29-3
Spectral distribution curve for blackbody radiation at $T = 1600$ K. The classical theoretical calculation gives the Rayleigh-Jeans law, which agrees with experimental results at very large wavelengths but disagrees with them violently at short wavelengths. Experimental data are indicated by the small circles.

Although Planck tried to fit the constant h into the framework of classical physics, he was unable to do so. The fundamental importance of his assumption of energy quantization implied by Equation 29-2 was not generally appreciated until Einstein applied similar ideas to explain the photoelectric effect and suggested that quantization is a fundamental property of electromagnetic radiation.

29-2   The Photoelectric Effect

In 1905, Einstein applied Planck's idea of energy quantization to explain the photoelectric effect. His paper appeared in the same journal that contained the special theory of relativity. It was this paper that marked the beginning of quantum theory and for which Einstein received the Nobel Prize in physics. Planck looked at energy quantization in his blackbody radiation theory as a calculational device. Einstein, however, made the bold suggestion that energy quantization is a fundamental property of electromagnetic energy. Later, the ideas of energy quantization were applied to atomic energies by Niels Bohr to give the first explanation of atomic spectra.

As mentioned in Chapter 24, the photoelectric effect was discovered by Hertz in 1887 in his experiment to generate and detect electromagnetic waves. In that experiment, Hertz used a spark gap in a tuned circuit to generate waves and another similar circuit to detect them. He noticed that the passage of sparks in the receiving gap of his apparatus was facilitated by light from the generating gap. In 1900, P. Lenard investigated this effect and found that light falling on a metal surface ejects electrons and that the energies of these electrons do *not* depend on the intensity of the light. Figure 29-4 shows a schematic diagram of the basic apparatus Lenard used. When light is incident on a clean metal surface, the cathode C, electrons are emitted. If some of these electrons strike the anode A, there is a current in the external circuit. The number of emitted electrons that reach the anode can be increased or decreased by making the anode positive or negative with respect to the cathode. Let V be the increase in potential from the cathode to anode. Figure 29-5 shows the current versus V for two values of the intensity of light incident on the cathode. When V is positive, the electrons are attracted to the anode. At sufficiently large V, all the emitted electrons reach the anode and the current is at its maximum value. A further increase in V does not affect the current. Lenard observed that the maximum current is proportional to the light intensity, an expected result since doubling the energy per unit time incident on the cathode should double the number of electrons emitted. When V is negative, the electrons are repelled from the anode. Only electrons with initial kinetic energy greater than the potential energy $|eV|$ can then reach the anode. From Figure 29-5 we see that if V is less than $-V_0$, no electrons reach the anode. The potential V_0 is called the *stopping potential.* It is related to the maximum kinetic energy of the emitted electrons by

$$(\tfrac{1}{2}mv^2)_{\text{max}} = eV_0$$

The experimental result that V_0 is independent of the intensity of the incident light was surprising. Apparently, increasing the rate of energy falling on the cathode does not increase the maximum kinetic energy of the electrons emitted.

In 1905, Einstein demonstrated that this result can be understood if light energy is not distributed continuously in space but is quantized in small bundles called *photons.* The energy of each photon is hf, where f is the frequency and h is Planck's constant. An electron emitted from a metal surface exposed to light receives its energy from a single photon. When the intensity of the light of a given frequency is increased, more photons fall on the surface in a unit time, but the energy absorbed by each electron is

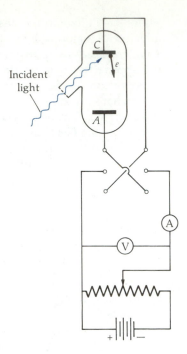

Figure 29-4
Schematic drawing of the photoelectric-effect apparatus used by Lenard. When light strikes the cathode C, electrons are emitted. The number of electrons that reach the anode A is measured by the current in the ammeter. The anode can be made positive or negative with respect to the cathode to attract or repel the electrons.

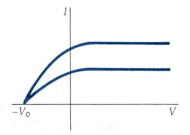

Figure 29-5
Photoelectric current I versus voltage V for two values of light intensity. There is no current when V is less than $-V_0$. The current observed for large V is proportional to the intensity of the incident light.

unchanged. If ϕ is the energy necessary to remove an electron from the surface of a metal, the maximum kinetic energy of the electrons emitted will be

$$(\tfrac{1}{2}mv^2)_{\text{max}} = eV_0 = hf - \phi \qquad\qquad 29\text{-}4$$

Einstein's photoelectric equation

The quantity ϕ, called the *work function,* is a characteristic of the metal. Some electrons will have kinetic energy less than $hf - \phi$ because of energy loss from traveling through the metal. Equation 29-4 is known as *Einstein's photoelectric equation.* From Equation 29-4, we can see that the slope of V_0 versus f should equal h/e.

Einstein's photoelectric equation was a bold prediction, for at the time it was made, there was no evidence that Planck's constant had any applicability outside of blackbody radiation, and there were no experimental data on the stopping potential V_0 as a function of frequency. The experimental verification of the equation proved to be quite difficult. Careful experiments by R. C. Millikan, reported first in 1914 and then in more detail in 1916, showed that Einstein's equation was correct and that measurements of h agreed with the value found by Planck. Figure 29-6 shows a plot of Millikan's data.

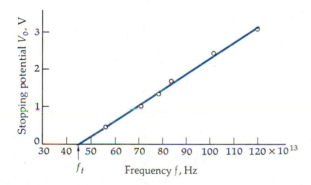

Figure 29-6
Millikan's data for the stopping potential versus frequency for the photoelectric effect. The data fall on a straight line that has slope h/e, as predicted by Einstein a decade before the experiment was performed.

The threshold frequency f_t and the corresponding threshold wavelength $\lambda_t = c/f_t$ are related to the work function ϕ by setting V_0 in Equation 29-4 equal to zero. Then,

$$\phi = hf_t = \frac{hc}{\lambda_t} \qquad\qquad 29\text{-}5$$

Threshold frequency and wavelength

Photons of frequency less than f_t (and therefore of wavelengths greater than λ_t) do not have enough energy to eject an electron from the metal. Work functions for metals are typically a few electronvolts. Since wavelengths are usually given in nanometres and energies in electronvolts, it is useful to have the value of hc in electronvolt–nanometres:

$$hc = (4.14 \times 10^{-15} \text{ eV}\cdot\text{s})(3 \times 10^8 \text{ m/s})$$

$$= 1.24 \times 10^{-6} \text{ eV}\cdot\text{m}$$

or

$$hc = 1240 \text{ eV}\cdot\text{nm} \qquad\qquad 29\text{-}6$$

Example 29-1   Calculate the photon energy for light of wavelength 400 nm (violet) and 800 nm (red). (These are approximately the extreme wavelengths in the visible spectrum.)

Using Equations 29-2 and 29-6, we have for $\lambda = 400$ nm

$$E = hf = \frac{hc}{\lambda} = \frac{1240 \text{ eV} \cdot \text{nm}}{400 \text{ nm}} = 3.1 \text{ eV}$$

For $\lambda = 800$ nm, the photon energy is half that for $\lambda = 400$ nm or 1.55 eV. We can see from these calculations that visible light contains photons with energies from about 1.5 to 3.0 eV.

Example 29-2   The intensity of sunlight at the earth's surface is approximately 1400 W/m². Assuming the average photon energy is 2 eV (corresponding to a wavelength of about 600 nm), calculate the number of photons that strike an area of 1 cm² in one second.

Since 1 watt is 1 joule per second, the energy striking the earth's surface in one second is 1400 J/m². The energy per second per square centimetre is then

$$\frac{1400 \text{ J}}{\text{m}^2} \times \frac{1 \text{ m}^2}{(100 \text{ cm})^2} = 0.14 \text{ J/cm}^2$$

If N is the number of 2-eV photons that have a total energy of 0.14 J, we have

$$N(2 \text{ eV}) = 0.14 \text{ J}$$

$$N = \frac{0.14 \text{ J}}{2 \text{ eV}} \times \frac{1 \text{ eV}}{1.6 \times 10^{-19} \text{ J}} = 4.38 \times 10^{17} \text{ photons}$$

This is an enormous number of photons. In most everyday situations, the number of photons is so great that a few more or less makes no difference. That is, quantization is not noticed.

Example 29-3   The threshold wavelength for potassium is 564 nm. What is the work function for potassium? What is the stopping potential when light of wavelength 400 nm is used?

Using Equation 29-5,

$$\phi = hf_t = \frac{hc}{\lambda_t} = \frac{1240 \text{ eV} \cdot \text{nm}}{564 \text{ nm}} = 2.20 \text{ eV}$$

The energy of a photon of wavelength 400 nm was calculated in Example 29-1 to be 3.1 eV. The maximum kinetic energy of the emitted electrons is then

$$(\tfrac{1}{2}mv^2)_{\text{max}} = hf - \phi = 3.10 \text{ eV} - 2.20 \text{ eV} = 0.90 \text{ eV}$$

The stopping potential is therefore 0.90 V.

Another interesting feature of the photoelectric effect is the absence of any lag between the time the light is turned on and the time the electrons appear. In the classical theory, the time can be calculated for enough energy of a given intensity (power per unit area) to fall on the area of an atom to eject an electron. However, even when the intensity is so small that such a calculation gives a time lag of hours, essentially no time lag is observed. The explanation of this result is simple. When the intensity is low, the number of photons hitting the metal per unit time is very small, but each photon has enough energy to eject an electron. There is therefore a good chance that one photon will be absorbed immediately. (The classical calculation gives the correct *average* number of electrons ejected per unit time.)

The photoelectric effect is used in photocells, which have many practical applications. For example, when a beam of light falls on a photocell, a photoelectric current is produced by the photocell. When the beam is interrupted, the current ceases. This stopping of the current can be used to trigger an alarm or open a door, for example. Very large arrays of photocells can be used to generate electrical power from available sunlight.

29-3   X-Rays

X-rays were discovered in 1895 by W. Roentgen when he was working with a cathode-ray tube. He found that "rays" from the tube could pass through materials that were opaque to light and activate a fluorescent screen or expose photographic film. These rays originated from a point where the electrons in the tube hit a target within the tube or hit the glass tube itself. He investigated this phenomenon extensively and found that all materials were transparent to these rays to some degree and that the transparency decreased with increasing density. This fact led to the medical use of x-rays within months after Roentgen's first paper. Roentgen was the first recipient of the Nobel Prize in 1901.

Roentgen was unable to deflect these rays with a magnetic field, as would be expected if they were charged particles; nor was he able to observe diffraction or interference, as would be expected if they were waves. He therefore gave the rays the somewhat mysterious name of *x-rays*. Since classical electromagnetic theory predicts that electric charges will radiate electromagnetic waves when they are accelerated (or decelerated), it was natural to assume that x-rays are electromagnetic waves produced when electrons decelerate as they are stopped by a target. In 1899, H. Haga and C. H. Wind observed a slight broadening of an x-ray beam after it passed through a slit a few thousandths of a millimetre wide. Assuming this to be due to diffraction, they estimated the wavelength of x-rays to be about 0.1 nm. In 1912, M. Laue suggested that, since the wavelengths of x-rays were of the same order of magnitude as the spacing of atoms in a crystal, the regular array of atoms in a crystal might act as a three-dimensional grating for the diffraction of x-rays. Acting on his suggestion, W. Friedrich and P. Knipping allowed a collimated beam of x-rays to pass through a crystal, behind which was a photographic plate (see Figure 29-7a). In addition to the central beam, they observed a regular array of spots as shown in Figure 29-7b. From an analysis of the positions of the spots, they were able to calculate that their x-ray beam contained wavelengths ranging from about 0.01 to 0.05 nm. This important experiment confirmed two important assumptions: (1) x-rays are electromagnetic radiation and (2) atoms in crystals are arranged in a regular array.

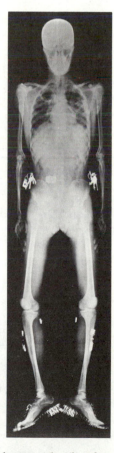

X-ray photograph taken by Roentgen. Note the nails in the shoes and the keys in the pockets of the man's jacket.

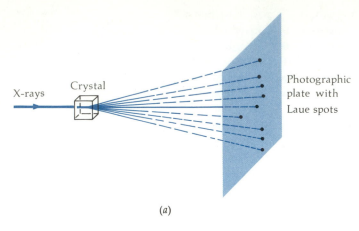

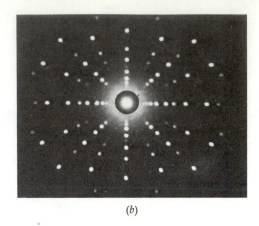

(a)   (b)

A simple and convenient way of analyzing the diffraction of x-rays by crystals was proposed by William L. Bragg in 1912. He considered the interference of x-rays due to scattering from various sets of parallel planes of atoms, now called *Bragg planes*. Two sets of Bragg planes for NaCl, which has a simple crystal structure called face centered cubic, are illustrated in Figure 29-8. Figure 29-9 shows x-rays incident on a Bragg plane at an angle ϕ. Waves scattered from two successive atoms within a plane will be in phase and will therefore interfere constructively independent of the wavelength if the scattering angle equals the incident angle. (This condition is the same as for reflection.) Waves scattered at equal angles from atoms in two different planes will be in phase if the difference in path length is an integral number of wavelengths. From Figure 29-9, we see that the difference in path length is $2d \sin \theta$, where θ is the angle between the incident beam and the Bragg plane and d is the separation between successive Bragg planes. The condition for constructive interference is thus

Figure 29-7
(a) Schematic sketch of the Laue experiment. The crystal acts as a three-dimensional grating that diffracts the x-ray beam and produces a regular array of spots, called a *Laue pattern*, on a photographic plate. (b) Modern Laue-type x-ray diffraction pattern from a niobium diboride crystal and 20-kV molybdenum x-rays.

$$2d \sin \theta = m\lambda \qquad\qquad 29\text{-}7$$

Bragg's law

where m is an integer.

Equation 29-7 is called *Bragg's law*. At angles satisfying his equation, waves will be strongly scattered because the waves scattered from many

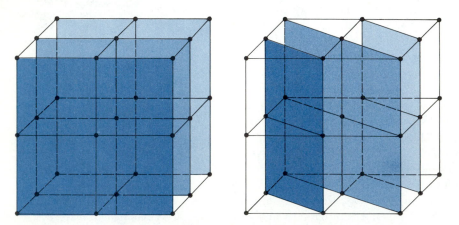

Figure 29-8
A crystal of NaCl showing two sets of Bragg planes.

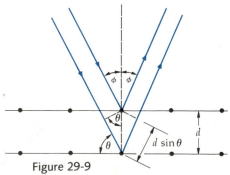

Figure 29-9
Bragg scattering from two successive planes. The waves from the two atoms shown have a path difference of $2d \sin \theta$. They will be in phase if the Bragg condition, $2d \sin \theta = m\lambda$, is met.

atoms interfere constructively. Figure 29-10 shows the main features of a
crystal spectrometer first built by William H. Bragg, father of W. L. Bragg.
X-rays with wavelengths satisfying the Bragg condition are scattered at an
angle θ equal to the incident angle. A measurement of the scattered intensity
versus angle gives the distribution of wavelengths in the incident x-ray
beam if the spacing d is known. Conversely, the scattering of x-rays of
known wavelength from crystals can be used to obtain information about
the structure of the crystals. W. H. Bragg and W. L. Bragg were awarded the
Nobel Prize in 1915 for their contribution to crystal analysis.

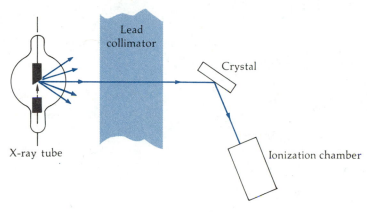

Figure 29-10
Schematic diagram of a Bragg
crystal spectrometer. A collimated
x-ray beam is incident on a crystal
and is scattered into an ionization
chamber that serves as a detector.
The crystal and the ionization
chamber can be rotated so as to
vary the angle of incidence while
keeping the scattering angle equal to
the angle of incidence. Since the
wavelength of the scattered x-rays
is related to the incident and
scattering angle by the Bragg
condition, the x-ray spectrum of the
source can be determined by
counting the number of scattered
x-rays for each angle of incidence.

Figure 29-11 shows a plot of the intensity versus wavelength for the
spectrum emitted from a typical x-ray tube when a target (molybdenum in
this case) in the x-ray tube is bombarded with electrons. The spectrum
consists of a series of sharp lines called the *characteristic spectrum* superimposed on a continuous spectrum called a *bremsstrahlung spectrum* (from the
German for "braking radiation"). The line spectrum is characteristic of the
target material and varies from element to element. We will discuss the line
spectrum in Chapter 31. The bremsstrahlung, or continuous, spectrum is

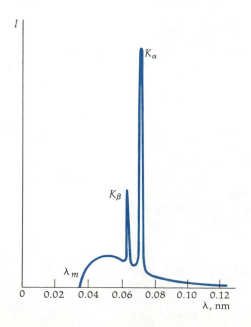

Figure 29-11
X-ray spectrum of molybdenum.
The sharp peaks labeled K_α and K_β
are characteristic of the target
element. The cutoff wavelength λ_m
is independent of the target
element; it is related to the voltage
of the x-ray tube by $\lambda_m = hc/eV_0$.

produced by the rapid deceleration of the bombarding electrons when they crash into the target. If the voltage across the x-ray tube is V_0, the maximum kinetic energy of the electrons is eV_0 when they hit the target. Usually, several photons are emitted as an electron slows down. However, sometimes just one photon is emitted with the maximum energy of eV_0. Since the wavelength of a photon varies inversely with the energy ($\lambda = hc/hf = hc/E$), the minimum wavelength emitted in the bremsstrahlung spectrum corresponds to a photon with maximum energy eV_0. The minimum wavelength is called the cutoff wavelength and is labeled λ_m in the figure. The cutoff wavelength is related to the x-ray tube voltage by

$$\lambda_m = \frac{hc}{E} = \frac{hc}{eV_0} \qquad\qquad 29\text{-}8 \quad \text{Cutoff wavelength}$$

Example 29-4   What is the minimum wavelength of the x-rays emitted by a television picture tube that is at a voltage of 2000 V?

The maximum kinetic energy of the electrons is 2000 eV, so this will be the maximum energy of the photons in the x-ray spectrum. The wavelength of a photon of this energy is the cutoff wavelength, which from Equation 29-8 is

$$\lambda_m = \frac{hc}{E} = \frac{1240 \text{ eV} \cdot \text{nm}}{2000 \text{ eV}} = 0.62 \text{ nm}$$

29-4   Compton Scattering

Further evidence of the correctness of the photon concept was furnished by Arthur H. Compton, who measured the scattering of x-rays by free electrons. According to classical theory, when an electromagnetic wave of frequency f_1 is incident on material containing charges, the charges will oscillate with this frequency and reradiate electromagnetic waves of the same frequency. Compton pointed out that if the scattering process were considered to be a collision between a photon and an electron, the electron would absorb energy due to recoil and the scattered photon would have less energy and therefore a lower frequency than the incident photon.

Figure 29-12 shows the geometry of a collision between a photon of wavelength λ_1 and an electron at rest. According to classical theory, the energy and momentum of an electromagnetic wave are related by

$$E = pc$$

This result is also consistent with the relativistic expression relating the energy and momentum of a particle (Equation 28-19),

$$E^2 = p^2c^2 + (mc^2)^2,$$

if the mass of the photon is assumed to be zero. Using this result and $E = hc/\lambda$, the momentum of a photon is related to its wavelength by

$$p = \frac{h}{\lambda} \qquad\qquad 29\text{-}9$$

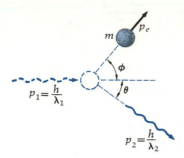

Figure 29-12
Compton scattering of an x-ray by
an electron. The scattered photon
has less energy and therefore
greater wavelength than the
incident photon because of the
recoil energy of the electron. The
change in wavelength is found
from the conservation of energy
and momentum.

By treating the scattering as a relativistic-mechanics problem using the
conservation of energy and momentum, Compton related the scattering
angle θ to the incident and scattered wavelengths, λ_1 and λ_2, respectively.
Compton's result is

$$\lambda_2 - \lambda_1 = \frac{h}{mc}(1 - \cos\theta) \qquad \text{29-10} \qquad \text{Compton scattering}$$

The change in wavelengths is independent of the original wavelength. The
quantity h/mc depends only on the mass of the electron. It has dimensions
of length and is called the *Compton wavelength*. Its value is

$$\lambda_C = \frac{h}{mc} = \frac{hc}{mc^2} \qquad \text{29-11} \qquad \text{Compton wavelength}$$

$$= \frac{1240 \text{ eV} \cdot \text{nm}}{5.11 \times 10^5 \text{ eV}} = 2.43 \times 10^{-12} \text{ m} = 2.43 \text{ pm}$$

where 1 pm (picometre) $= 10^{-12}$ m. Since λ_C is small, the wavelength shift
$\lambda_1 - \lambda_2$ is difficult to observe unless λ_1 is so small that the fractional change
$(\lambda_2 - \lambda_1)/\lambda_1$ is appreciable. Compton used x-rays of wavelength 71.1 pm $=$
0.0711 nm. The energy of a photon of this wavelength is $E = hc/\lambda = (1240$
eV $\cdot$ nm$)/(0.0711$ nm$) = 17.4$ keV. Since this is much greater than the binding energy of the valence electrons in carbon, these electrons can be considered to be essentially free. Compton's experimental results for $\lambda_2 - \lambda_1$ as a
function of scattering angle θ agreed with Equation 29-10, thereby confirm-
ing the correctness of the photon concept.

Example 29-5   Calculate the percentage change in wavelength observed in
Compton scattering of 20-keV photons at $\theta = 60°$.

The change in wavelength at $\theta = 60°$ is given by Equation 29-10:

$$\lambda_2 - \lambda_1 = \lambda_C (1 - \cos\theta) = 2.43 \text{ pm} (1 - \cos 60°) = 1.22 \text{ pm}$$

The wavelength of the incident 20-keV photons is

$$\lambda_1 = \frac{1240 \text{ eV} \cdot \text{nm}}{20,000 \text{ eV}} = 0.062 \text{ nm} = 62 \text{ pm}$$

The percentage change in wavelength is thus

$$\frac{\Delta\lambda}{\lambda_1} = \frac{1.22 \text{ pm}}{62 \text{ pm}} \times 100\% = 1.97\%$$

29-5   Quantization of Atomic Energies: The Bohr Model

The most famous application of energy quantization to microscopic systems was that of Niels Bohr. In 1913, Bohr proposed a model of the hydrogen atom that had spectacular success in calculations of the wavelengths of lines in the known hydrogen spectrum and in predicting new lines (later found experimentally) in the infrared and ultraviolet spectra.

Near the turn of the century, many data were collected on the emission of light by the atoms of a gas when excited by an electric discharge. Viewed through a spectroscope with a narrow-slit aperture, this light appears as a discrete set of lines of different colors or wavelengths; the spacing and intensities of the lines are characteristic of the element. It was possible to determine the wavelengths of these lines accurately, and much effort went into finding regularities in the spectra. In 1884, a Swiss schoolteacher, Johann Balmer, found that the wavelengths of some of the lines in the spectrum of hydrogen can be represented by the formula

$$\lambda = (364.6 \text{ nm}) \frac{n^2}{n^2 - 4}$$ 29-12   Balmer formula

where n is an integer that takes on the values $n = 3, 4, 5, \ldots$. Figure 29-13 shows the set of spectral lines of hydrogen, now known as the *Balmer series*, whose wavelengths are given by Equation 29-12.

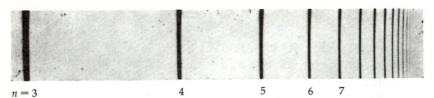

$n = 3$ 4   5   6   7

Figure 29-13
Balmer series for hydrogen. The wavelengths of these lines are given by Equation 29-12 for different values of the integer n.

Balmer suggested that his formula might be a special case of a more general expression that would be applicable to the spectra of other elements. Such an expression, found by Johannes R. Rydberg and Walter Ritz, gives the reciprocal wavelength as

$$\frac{1}{\lambda} = R\left(\frac{1}{n_2^2} - \frac{1}{n_1^2}\right) \qquad n_1 > n_2$$ 29-13   Rydberg-Ritz formula

where R, called the *Rydberg constant* or simply the *Rydberg*, is the same for all spectral series of the same element and varies only slightly in a regular way from element to element. For very massive elements, R approaches the value

$$R_\infty = 10.97373 \ \mu m^{-1}$$ 29-14

If we take the reciprocal of Equation 29-12 for the Balmer series, we obtain

$$\frac{1}{\lambda} = (1/364.6 \text{ nm}) \frac{n^2 - 4}{n^2} = (1/364.6 \text{ nm})\left(\frac{1}{1} - \frac{4}{n^2}\right)$$

$$= (4/364.6 \text{ nm})\left(\frac{1}{4} - \frac{1}{n^2}\right) = (10.97 \ \mu m^{-1})\left(\frac{1}{2^2} - \frac{1}{n^2}\right)$$

We can see, then, that the Balmer formula is a special case of the Rydberg-Ritz formula (Equation 29-13) with $n_2 = 2$. The Rydberg-Ritz formula and various modifications of it were very successful in predicting other spectra. For example, other lines for hydrogen outside the visible spectrum were predicted and found. Setting $n_2 = 1$ in Equation 29-13 leads to a series in the ultraviolet region called the *Lyman series* whereas setting $n_2 = 3$ leads to the *Paschen series* in the infrared region.

Many attempts were made to construct a model of the atom that would yield these formulas for its radiation spectrum. The most popular model, developed by J. J. Thomson, considered electrons to be embedded in some kind of fluid that contained most of the mass of the atom and had enough positive charge to make the atom electrically neutral. Thomson's model, called the "plum pudding" model, is illustrated in Figure 29-14. Since classical electromagnetic theory predicted that a charge oscillating with frequency f would radiate light of that frequency, Thomson searched for configurations of electrons that were stable and had normal modes of vibration of frequencies equal to those of the spectrum of the atom. A difficulty with this model and all others was that electric forces alone cannot produce stable equilibrium. Thomson was unable to find a configuration that predicted the observed frequencies for any atom.

The Thomson model was essentially ruled out by a set of experiments by H. W. Geiger and E. Marsden under the supervision of E. Rutherford in about 1911, in which alpha particles from radioactive radium were scattered by atoms in a gold foil. Rutherford showed that the number of alpha particles scattered at large angles could not be accounted for by an atom in which the positive charge was distributed throughout the atomic volume (known to be about 0.1 nm in diameter). Their findings required that the positive charge and most of the mass of the atom be concentrated in a very small region, now called the nucleus, with a diameter of the order of 10^{-6} nm = 1 fm. (Before the establishment of the SI unit system, the femtometre, 1 fm = 10^{-15} m, was called the *fermi* after the Italian physicist Enrico Fermi.)

Niels Bohr, who was working in the Rutherford laboratory at the time, proposed a model of the hydrogen atom that combined the work of Planck, Einstein, and Rutherford and successfully predicted the observed spectra. Bohr assumed that the electron in the hydrogen atom moved under the influence of the Coulomb attraction between it and the positive nucleus according to classical mechanics, which predicts circular or elliptical orbits with the force center at one focus, as in the motion of the planets around the sun. For simplicity, he chose a circular orbit, as shown in Figure 29-15. Although mechanical stability is achieved because the Coulomb attractive force provides the centripetal force necessary for the electron to remain in orbit, such an atom is unstable electrically according to classical theory because the electron must accelerate when moving in a circle and must therefore radiate electromagnetic energy of a frequency equal to that of its motion. According to classical electromagnetic theory, such an atom would quickly collapse, the electron spiraling into the nucleus as it radiates away its energy.

Bohr "solved" this difficulty by modifying the laws of electromagnetism by *postulating* that the electron could move in certain nonradiating orbits. He called these stable orbits *stationary states*. The atom radiates only when

Figure 29-14
J. J. Thomson's plum pudding model of the atom. In this model, the negative electrons are embedded in a fluid of positive charge. For a given configuration of electrons in such a system, the resonance frequencies of oscillations of the electrons can be calculated.

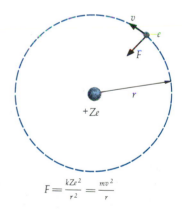

$$F = \frac{kZe^2}{r^2} = \frac{mv^2}{r}$$

Figure 29-15
Electron of charge $-e$ traveling in a circular orbit of radius r around the nuclear charge $+Ze$. The attractive electrical force kZe^2/r^2 provides the centripetal force that holds the electron in its orbit.

Bohr's first postulate: Stationary states

the electron somehow makes a transition from one stationary state to another. The frequency of radiation is not the frequency of the electron's motion in either stable orbit. Instead, it is related to the energies of the orbits by

$$f = \frac{E_i - E_f}{h} \qquad\qquad 29\text{-}15$$

where h is Planck's constant and E_i and E_f are the total energies in the initial and final orbits. This assumption, which is equivalent to that of energy conservation with the emission of a photon, is a key one in the Bohr theory because it deviates from the classical theory, which requires the frequency of radiation to be that of the motion of the charged particle.

If the nuclear charge is $+Ze$ and the electron charge is $-e$, the potential energy at a distance r is (see Equation 19-9)

$$U = -\frac{kZe^2}{r}$$

where k is the Coulomb constant. (For hydrogen, $Z = 1$, but it is convenient not to specify Z at this time, so that the results can be applied to other atoms.) The total energy of the electron moving in a circular orbit with speed v is then

$$E = \tfrac{1}{2}mv^2 + U = \tfrac{1}{2}mv^2 - \frac{kZe^2}{r}$$

The kinetic energy can be obtained as a function of r by using Newton's law $F = ma$. Setting the Coulomb attractive force equal to the mass times the centripetal acceleration, we have

$$\frac{kZe^3}{r^2} = m\,\frac{v^2}{r}$$

or

$$\tfrac{1}{2}mv^2 = \frac{1}{2}\frac{kZe^2}{r} \qquad\qquad 29\text{-}16$$

Niels Bohr explaining a point in front of the blackboard (1956).

For circular orbits, the kinetic energy equals half the magnitude of the potential energy, a result that holds for circular motion in any inverse-square-law force field. The total energy is then

$$E = \frac{1}{2}\frac{kZe^2}{r} - \frac{kZe^2}{r} = -\frac{1}{2}\frac{kZe^2}{r} \qquad\qquad 29\text{-}17$$

Using Equation 29-15 for the frequency of radiation when the electron changes from an initial orbit of radius r_i to a final orbit of radius r_f, we obtain

$$f = \frac{E_i - E_f}{h} = \frac{1}{2}\frac{kZe^2}{h}\left(\frac{1}{r_f} - \frac{1}{r_i}\right) \qquad\qquad 29\text{-}18$$

To obtain the frequency $f = c/\lambda = cR(1/n_2^2 - 1/n_1^2)$ from the Rydberg-Ritz formula, it is evident that the radii of the stable orbits must be proportional to the squares of integers. Bohr searched for a quantum condition for the radii of the stable orbits that would yield this result. After much trial and error, he found that he could obtain the correct results if he postulated that the angular momentum of the electron in a stable orbit equals an integer times Planck's constant divided by 2π. Since the angular momentum of a circular orbit is just mvr, this postulate for the quantum condition is

$$mvr = \frac{nh}{2\pi} \qquad \text{29-19}$$

Bohr's third postulate: Quantized angular momentum

We can determine r by eliminating v between Equations 29-16 and 29-19. Solving Equation 29-19 for v and squaring, we obtain

$$v^2 = n^2 \frac{(h/2\pi)^2}{m^2 r^2}$$

But from Equation 29-16, we have

$$v^2 = \frac{kZe^2}{mr}$$

Eliminating v^2 and solving for r, we obtain

$$r = n^2 \frac{(h/2\pi)^2}{mkZe^2} = n^2 \frac{a_0}{Z} \qquad \text{29-20}$$

Radii of nonradiating orbits

where

$$a_0 = \frac{(h/2\pi)^2}{mke^2} \approx 0.0529 \text{ nm} \qquad \text{29-21}$$

is called the *first Bohr radius*. Combining Equations 29-20 and 29-18, we get

$$f = Z^2 \frac{2m\pi^2 k^2 e^4}{h^3} \left(\frac{1}{n_2^2} - \frac{1}{n_1^2} \right) \qquad \text{29-22}$$

When we compare this for $Z = 1$ with the empirical Rydberg-Ritz formula (Equation 29-13), we have for the Rydberg constant

$$R = \frac{2m\pi^2 k^2 e^4}{ch^3} \qquad \text{29-23}$$

Using the values of m, e, and h known in 1913, Bohr calculated R and found his result to agree (within the limits of the uncertainties of the constants) with the value obtained from spectroscopy. Figure 29-16 illustrates the Bohr model of the hydrogen atom.

The possible values of the energy of the hydrogen atom predicted by the Bohr model are given by Equation 29-17, with r given by Equation 29-20:

$$E_n = -\frac{2\pi^2 k^2 e^4 m}{h^2} \frac{Z^2}{n^2} = -Z^2 \frac{E_0}{n^2} \qquad \text{29-24}$$

Allowed energies

where

$$E_0 = \frac{2\pi^2 k^2 e^4 m}{h^2} \approx 13.6 \text{ eV} \qquad \text{29-25}$$

It is convenient to represent these energies in an *energy-level diagram*, as in

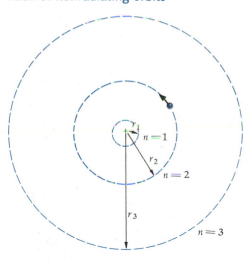

Figure 29-16
Stable orbits in the Bohr model of the hydrogen atom. The radii of the stable orbits are given by $r_n = n^2(a_0/Z)$ where n is an integer and a_0 is the smallest radius.

Figure 29-17. The lowest energy level is called the ground state. The energy of the hydrogen atom in the ground state is -13.6 eV. The binding energy of the hydrogen atom is 13.6 eV. To remove the electron from the atom, a process called *ionization*, 13.6 eV must be added. The measured ionization energy of the hydrogen atom is 13.6 eV. Various series of transitions are indicated in Figure 29-17 by vertical arrows between the energy levels. The frequency of the light emitted in one of these transitions is the energy difference divided by h, according to Equation 29-15. At the time of Bohr's paper (1913), the Balmer series, corresponding to $n_2 = 2, n_1 = 3, 4, 5, \ldots$, and the Paschen series, corresponding to $n_2 = 3, n_1 = 4, 5, 6, \ldots$, were known. In 1916, T. Lyman found the series corresponding to $n_2 = 1$, and in 1922 and 1924, F. Brackett and H. A. Pfund, respectively, found the series corresponding to $n_2 = 4$ and $n_2 = 5$. As can be determined by computing the wavelengths of these series, only the Balmer series lies in the visible portion of the electromagnetic spectrum.

In our derivations, we have assumed the electron to revolve around a stationary nucleus. This is equivalent to assuming the nucleus to have infinite mass. Since the mass of the hydrogen nucleus is not infinite but is actually about 2000 times that of the electron, a correction must be made for the motion of the nucleus. This correction leads to a very slight dependence of the Rydberg constant, as given in Equation 29-23, on the nuclear mass, which is in precise agreement with the observed variations in it.

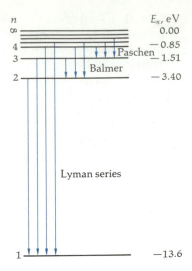

Figure 29-17
Energy-level diagram for hydrogen showing a few transitions in the Lyman, Balmer, and Paschen series. The energies of the levels are given by Equation 29-24.

Example 29-6   Find the energy and the wavelength of the line with the longest wavelength in the Lyman series.

From Figure 29-17, we can see that the Lyman series corresponds to transitions ending at the ground state of energy $E_1 = -13.6$ eV. Since λ varies inversely with energy, the longest-wavelength transition is the lowest-energy transition, which is from the first excited state $n = 2$ to the ground state. The first excited state has energy $E_2 = (-13.6 \text{ eV})/4 = -3.40$ eV. Since this is 10.2 eV above the ground-state energy, the energy of the photon emitted is 10.2 eV. The wavelength of this photon is

$$\lambda = \frac{hc}{E} = \frac{1240 \text{ eV} \cdot \text{nm}}{10.2 \text{ eV}} = 121.6 \text{ nm}$$

This photon is outside the visible spectrum and in the ultraviolet. Since all the other lines in the Lyman series have even greater energies and shorter wavelengths, the Lyman series is completely in the ultraviolet region.

Questions

1. If an electron moves to a larger orbit, does its total energy increase or decrease? Does its kinetic energy increase or decrease?

2. How does the spacing of adjacent energy levels change as n increases?

3. What is the energy of the shortest-wavelength photon emitted by the hydrogen atom?

X-Rays and Medical Diagnosis

John R. Cameron
Department of Medical Physics, University of Wisconsin – Madison

When Roentgen discovered x-rays in the fall of 1895 in Würzburg, Germany, it didn't take him long to recognize the great medical potential of a ray that will pass through the body and cast shadows of bones. Today, over half of the people in the United States have one or more x-rays each year with the total x-rays taken a year probably over a half-billion. Before learning about the physics of medical x-rays, you might be interested in knowing if they are a health hazard. X-rays do pose a slight risk, but it is no greater than many risks we accept on a daily basis without any thought, such as riding in a car (see Table I). This is not to suggest that you should be unconcerned about getting an x-ray. However, if your doctor or dentist needs an x-ray to make a diagnosis, the small risk is usually well worth the information it can produce.

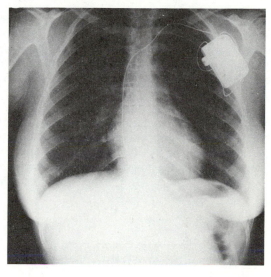

A chest x-ray showing an implanted pacemaker.

Table 1

Risks in Individual Actions*

Individual action	Minutes of life expectancy lost
Smoking a cigarette	10
Eating a calorie-rich dessert	50
Drinking a nondiet soft drink	15
Drinking a diet soft drink	0.15
Crossing a street	0.4
Not fastening seat belt	0.1/mile
Chest x-ray exposure	45
Flying coast to coast	100
Driving coast to coast	1000
Routinely driving a small car	7000

* Adapted from B. L. Cohen and I-Sing Lee, "A Catalog of Risks," *Health Physics 36*, June 1979, p. 721.

X-rays are a type of electromagnetic radiation, as is light. Unlike light, their speed does not change when they pass through matter; that is, their index of refraction is essentially 1.00 in matter. Thus, they cannot be focused and so the images they produce are shadows. In fact, the original name given to an x-ray image was "skiagraph," which is derived from the Greek for "shadow picture."

When x-ray photons enter the body, some are absorbed by the photoelectric effect, especially those of lower energy. The probability of absorption increases with increasing atomic number of the elements that comprise the materials they pass through. This is why bones, with their high calcium ($Z = 20$) content, show up so effectively in an x-ray compared to soft tissues, which consist mostly of hydrogen ($Z = 1$), carbon ($Z = 6$), and oxygen ($Z = 8$).

Many of the higher-energy photons are scattered by collisions with electrons—Compton scattering. The probability of Compton scattering is independent of the atomic number. In Compton scattering, the incident photon gives part of its energy to an electron and the remainder is emitted as a photon of lower energy. This scattered photon can, in turn, be absorbed through the photoelectric effect, suffer another Compton scattering, or leave the patient. The thicker the body part being x-rayed, the greater the amount of scatter. In an x-ray of the spine of an adult, less than 1 percent of the original x-ray photons pass through the body undeflected to produce the final image on the x-ray film.

X-ray photons are easy to produce. Every television set produces x-rays when the electrons strike the screen. These x-rays have low energy and are absorbed in the glass wall of the TV tube. The basic components of an x-ray unit are

Essay: X-Rays and Medical Diagnosis (continued)

shown in Figure 1. A hot filament produces electrons that are accelerated toward and strike the tungsten target or anode, which is at a high positive voltage, typically between 30 and 140 thousand volts (30 to 140 kV). The current of the electron beam striking the target is usually from 0.1 to 1.0 ampere. That means that the power into the target (*VI*) is 3 to 140 kilowatts. Over 99 percent of this energy is converted to heat; less than 1 percent goes into the production of x-rays. With high accelerating potentials and high currents, as are used when x-raying the abdomen of a large patient, there is danger of melting the tungsten target if the exposure time is

too long. Modern x-ray units use rotating targets to distribute the heat over a larger surface area. They also have protective circuits to block an exposure if the time setting will exceed a safe heat limit for the x-ray target. The x-ray photons emitted by the target have energies from just above the ultraviolet band to a maximum determined by the accelerating potential; for example, for an 80 kV accelerating potential, the maximum photon energy is 80 keV. The low-energy photons are absorbed in the wall of the x-ray tube, and so the lowest-energy photons leaving the tube are typically 15 keV. An x-ray beam consists of photons of many different energies just as white light consists of many colors. The beam emerging from an x-ray tube contains more low-energy photons that are more easily absorbed than the more penetrating high-energy photons. To reduce the risk due to absorption of too many low-energy photons in the first few centimetres of body tissue, an aluminum filter is used in medical x-ray units. It absorbs the low-energy x-ray photons much the way a blue-light filter absorbs the lower-energy red-light photons.

To produce an x-ray image, the radiographer consults a "technique chart" to select the optimum accelerating potential, tube current, and exposure time. The distance between the x-ray tube and the film is usually standardized to 1 m, except for chest x-rays, which are typically 1.8 m. The x-ray beam is directed at the area of interest, and the beam size is adjusted by moving lead sheets in the collimator attached to the tube. A light beam from the collimator shows where the x-ray beam will strike. The image is produced on a special x-ray film with a light sensitive coating (emulsion) on both sides in a light-tight holder, called a cassette. Inside the cassette, in close contact with each side of the x-ray film, is an intensifying screen, a sheet of stiff cardboard covered with a fluorescent coating. The film is not very efficient at absorbing x-rays. The intensifying screens absorb most of the x-ray photons, and their fluorescent light exposes the x-ray film. Thus, an x-ray image is really a second-hand picture since it is exposed by light from the intensifying screens. This technique greatly reduces the radiation exposure to the patient. If the image is made by the x-ray film alone, as in a dental x-ray, the exposure has to be about 10 to 20 times greater.

After the x-ray is taken, the cassette is taken to the darkroom, where the film is placed in an automatic film processor that produces a dry, finished film in only 90 seconds. The film, a photographic negative, is placed on a viewbox to be reviewed by the radiographer or radiologist to see if it is satisfactory. The dark areas on the film represent a large number of x-ray photons, whereas the clear (white) areas represent very few photons. For example, in an x-ray of the

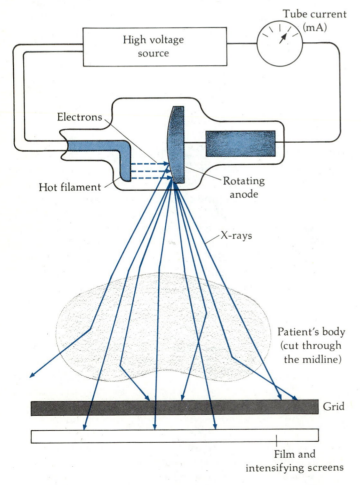

Figure 1
Schematic diagram of an x-ray unit.

A fluoroscope used in shoe stores from about 1930 to 1950 before the health hazards of x-rays were understood.

chest, the lungs will be dark because the low density of the lungs allows many photons to pass through, whereas in the center (the mediastinum), the film will be relatively clear because of the much greater amount of tissue in the path of the beam. If the overall film is too dark or too light, its diagnostic usefulness is reduced, so the x ray is often retaken. Many x-ray units have a phototimer that terminates the exposure automatically, like a modern camera. Because of the large variability of humans' body shapes and thicknesses, this device does not always work well.

You may think we can forget about the scattered x-ray photons that leave the patient. Unfortunately, many of these photons strike the cassette and produce an overall blurring of the image. The effect of these scattered photons is to reduce the information carried by the nonscattered x-ray photons. This effect is analogous to taking a snapshot with a dirty lens. The light scattered by the dirt obscures the detail one would see in a photo taken with a clean lens. The number of scattered x-ray photons reaching the cassette can be greatly reduced by placing an antiscatter grid (usually just called a "grid") between the patient and the cassette. The grid is made of many narow strips of thin lead with thicker plastic strips separating them. The lead strips are typically eight times as high as the spacing between them and appropriately slanted so that x-rays from the x-ray source can pass easily between them whereas x-rays that have been scattered are likely to strike the lead strips and be absorbed. It is important to use grids for x-raying thick tissues, such as the abdomen of adults, where the number of scattered photons leaving the patient far exceeds the nonscattered photons that carry useful information. Since the grid makes it necessary to give a greater exposure to the patient because of absorption in the grid, it is usually undesirable to use one for x-rays of infants and small children where scatter is not a problem.

In addition to scattered radiation, there are several other factors that can affect the quality of the x-ray image. One of these is the size of the focal spot. The focal spot is the area on the target from which the x-rays originate. Its size is determined by the dimensions of the electron beam incident on the target. The maximum dimension of the focal spot is typically a few millimetres. The smaller the focal spot, the greater the detail of the x-ray shadows in the image. The principle is the same as that for casting animal shadows on the wall with your fingers; the smaller the diameter of the light source, the sharper the shadows. At the same time, all of the energy in an x-ray tube is going into a small area of the target, which will overheat if the power is too large. Therefore, if one needs a lot of x-ray photons to penetrate a thick body part, it is better to use a large focal spot. Most x-ray tubes allow a choice between a "large" and a "small" focal spot.

Many other factors besides scattered radiation and the size of the focal spot affect the diagnostic quality of the x-ray image. These include the accuracy with which the accelerating potential, the current, and the exposure time are chosen. For example, an error in the accelerating potential can reduce the visibility of a critical component, such as small calcific specks in an x-ray of the breast. The most common source of problems is in the functioning of the automatic film processor. A quality control program is necessary to detect the problems before they become serious.

Early x-ray images were good at showing bones, but soft tissues, such as blood vessels or the digestive system, could not be distinguished from their surroundings. The use of contrast media was developed to solve this problem. For example, to make an angiogram (an x-ray of the arteries), an iodine compound is injected into an artery. A series of x-rays are taken immediately after the injection. Because of the high atomic number of iodine ($Z = 53$), the absorption of x-rays by the arteries is increased, thus casting shadows of them on the x-ray film. To study different parts of the digestive system, a barium ($Z = 56$) compound is used as a contrast medium. It is swallowed for x-rays of the stomach and upper gastrointestinal (GI) tract and is administered by enema for x-rays of the lower GI tract.

Essay: X-Rays and Medical Diagnosis (continued)

Computer technology is playing an increasing role in radiology; an example is *computerized axial tomography* (CAT). In computerized tomography, a narrow x-ray beam scans across a slice of the body while the amount of the beam transmitted through each small area of the slice is measured by a row of many detectors and recorded in a computer. Such measurements are continually made and recorded as the x-ray tube rotates through 180 or 360 degrees so that the body is measured from many different angles in a procedure that may take as little as two seconds. A computer program then calculates the density of different parts of the body to give a three-dimensional image with much more information than can be obtained from a conventional x-ray and from which any two-dimensional slice desired can be displayed. Another example is *digital subtraction angiography* (DSA). In this technology, a live x-ray image of the body is viewed with a television camera. The information from the TV signal is digitized and stored both before and after a contrast medium is injected into the bloodstream. The image which results from subtracting the information before from that after shows the arteries without the confusing shadows due to bone and air. With DSA, less contrast medium is needed, reducing the risk of a serious reaction by the patient to the contrast medium. In addition, the DSA can often be done as an outpatient procedure, reducing hospitalization costs.

New technologies for creating images of the body are replacing x-rays in some cases, such as using echoes from

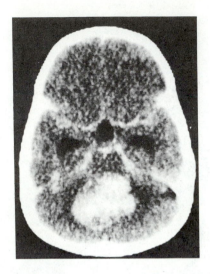

A CAT scan of a brain. The bright spot near the bottom of the photograph reveals a tumor.

pulses of ultrasound or measuring the change in orientation of hydrogen nuclei in a strong magnetic field, *magnetic resonance imaging* (MRI). Neither ultrasound nor MRI has any known risks. Nevertheless, conventional x-ray images will continue to be used because of their simplicity and their relatively low cost.

Summary

1. The energy in electromagnetic radiation is not continuous but comes in quanta with energy given by

$$E = hf = \frac{hc}{\lambda}$$

where f is the frequency, λ is the wavelength, and h is Planck's constant, which has the value

$$h = 6.626 \times 10^{-34} \text{ J} \cdot \text{s} = 4.136 \times 10^{-15} \text{ eV} \cdot \text{s}$$

The quantity hc occurs often in calculations and has the value

$$hc = 1240 \text{ eV} \cdot \text{nm}$$

The quantum nature of light is exhibited in the photoelectric effect, in which a photon is absorbed by an atom with the emission of an electron; and in

Compton scattering, in which a photon collides with a free electron and emerges with reduced energy and therefore a greater wavelength.

2. X-rays are emitted when electrons are decelerated by crashing into a target in an x-ray tube. An x-ray spectrum consists of a series of sharp lines called the characteristic spectrum superimposed on the continuous bremsstrahlung spectrum. The minimum wavelength in the bremsstrahlung spectrum, λ_m, corresponds to the maximum-energy photon emitted, which equals the maximum kinetic energy of the electrons eV_0, where V_0 is the voltage of the x-ray tube. The minimum wavelength is then given by

$$\lambda_m = \frac{hc}{eV_0}$$

3. The wavelengths of x-rays are typically a few nanometres, which is also the approximate spacing of atoms in a crystal. Diffraction maxima are observed in x-rays scattered from a crystal at angles satisfying the Bragg condition.

$$2d \sin \theta = m\lambda$$

where θ is the angle between the incident beam and a plane of atoms called a Bragg plane, and d is the separation distance between successive Bragg planes.

4. In order to derive the Balmer formula for the spectrum of the hydrogen atom, Bohr proposed the following postulates:

Postulate 1. The electron in the hydrogen atom can move only in certain nonradiating circular orbits called stationary states.

Postulate 2. The atom radiates a photon when the electron makes a transition from one stationary state to another. The frequency of the photon is given by

$$f = \frac{E_i - E_f}{h}$$

where E_i and E_f are the initial and final energies.

Postulate 3. The radius of a stationary state orbit is determined by classical physics plus the quantum condition that the angular momentum of the electron must equal an integer times Planck's constant divided by 2π:

$$mvr = nh/2\pi$$

These postulates lead to allowed energy levels in the hydrogen atom given by

$$E_n = -\frac{2\pi^2 k^2 e^4 m}{h^2} \frac{Z^2}{n^2} = -Z^2 \frac{E_0}{n^2}$$

where n is an integer and

$$E_0 = 2\pi^2 k^2 e^4 m/h^2 \approx 13.6 \text{ eV}$$

The radii of the stationary states are given by

$$r_n = n^2 \frac{a_0}{Z}$$

where a_0 is the first Bohr radius given by

$$a_0 = \frac{(h/2\pi)^2}{mke^2} \approx 0.0529 \text{ nm}$$

Suggestions for Further Reading

Moran, Paul R., R. Jerome Nickles, and James A. Zagzebski: "The Physics of Medical Imaging," *Physics Today*, vol. 36, no. 7, 1983, p. 36.

This article briefly describes such new medical imaging techniques as digital subtraction angiography, computed tomography (CAT), nmr imaging (MRI), positron-emission tomography (PET), and ultrasound imaging.

Ter-Pogossian, Michel M., Marcus E. Raichle, and Burton E. Sobol: "Positron-Emission Tomography," *Scientific American*, October 1980, p. 170.

Images of the brain, heart, or other organs can be produced and metabolic processes monitored with this new technique, which relies on the decay of short-lived isotopes introduced into the body in various ways.

Wheeler, John Archibald: "Niels Bohr, the Man," *Physics Today*, vol. 38, no. 10, 1985, p. 66.

Bohr's very personal approach to science is recounted by a former collaborator, who is himself a highly respected physicist. This article appears as part of a special issue commemorating the centennial of Bohr's birth.

Review

A. Objectives: After studying this chapter, you should:

1. Be able to sketch the spectral distribution curves for blackbody radiation and the curve predicted by the Rayleigh-Jeans law.

2. Be able to discuss the photoelectric effect and state the Einstein equation describing it.

3. Be able to discuss how the photon concept explains all the features of the photoelectric effect and the Compton scattering of x-rays.

4. Be able to sketch a typical x-ray spectrum and relate the minimum wavelength of the spectrum to the voltage of the x-ray tube.

5. Be able to state the Bohr postulates and describe the Bohr model of the hydrogen atom.

6. Be able to draw an energy-level diagram for hydrogen, to indicate on it the transitions involving the emission of a photon, and to use it to calculate the wavelengths of the emitted photons.

B. Define, explain, or otherwise identify:

Blackbody radiation, p. 748
Ultraviolet catastrope, p. 749
Photoelectric effect, p. 750
Work function, p. 751
Bragg plane, p. 754

Bragg's law, p. 754
Characteristic spectrum, p. 755
Bremsstrahlung spectrum, p. 755
Compton wavelength, p. 757
Balmer series, p. 758
Rydberg, p. 758
Stationary states, p. 760
Energy-level diagram, p. 761

C. True or false: If the statement is true, explain why. If it is false, give a counterexample.

1. The spectral distribution of radiation in a blackbody depends only on the temperature of the body.

2. In the photoelectric effect, the maximum current is proportional to the intensity of the incident light.

3. The work function of a metal depends on the frequency of the incident light.

4. The maximum kinetic energy of electrons emitted in the photoelectric effect varies linearly with the frequency of the incident light.

5. The energy of a photon is proportional to its frequency.

6. One of Bohr's assumptions was that atoms never radiate light.

7. In the Bohr model, the energy of a hydrogen atom is quantized.

Exercises

Section 29-1   Blackbody Radiation

1. Find the photon energy in joules and in electronvolts for an electromagnetic wave in the FM radio band of frequency 100 MHz.

2. Repeat Exercise 1 for an electromagnetic wave in the AM radio band of frequency 800 kHz.

3. What is the frequency of a photon of energy (*a*) 1 eV, (*b*) 10 keV, and (*c*) 1 MeV?

Section 29-2   The Photoelectric Effect

4. Find the photon energy for light of wavelength (*a*) 500 nm, (*b*) 600 nm, and (*c*) 700 nm.

5. Find the wavelength of a photon whose energy is (*a*) 0.5 eV, (*b*) 1.0 eV, (*c*) 1.5 eV, (*d*) 2.0 eV, and (*e*) 10 eV.

6. Find the photon energy if the wavelength is (*a*) 0.1 nm (about 1 atomic diameter) and (*b*) 1 fm (1 fm $= 10^{-15}$ m, about 1 nuclear diameter).

7. The work function for tungsten is 4.58 eV. (*a*) Find the threshold frequency and wavelength for the photoelectric effect. Find the stopping potential if the wavelength of the incident light is (*b*) 200 nm and (*c*) 250 nm.

8. When light of wavelength 300 nm is incident on potassium, the emitted electrons have a maximum kinetic energy of 2.03 eV. (*a*) What is the energy of the incident photon? (*b*) What is the work function for potassium? (*c*) What is the stopping potential if the incident light has a wavelength of 400 nm? (*d*) What is the threshold wavelength for the photoelectric effect with potassium?

9. The threshold wavelength for the photoelectric effect for silver is 262 nm. (*a*) Find the work function for silver. (*b*) Find the stopping potential if the incident radiation has a wavelength of 200 nm.

10. The work function for cesium is 1.9 eV. (*a*) Find the threshold frequency and wavelength for the photoelectric effect. Find the stopping potential if the wavelength of the incident light is (*b*) 300 nm and (*c*) 400 nm.

11. A light beam of wavelength 400 nm has an intensity of 100 W/m². (*a*) What is the energy of each photon in the beam? (*b*) How much energy strikes an area of 1 cm² perpendicular to the beam in 1 s? (*c*) How many photons strike this area in 1 s?

Section 29-3   X-Rays

12. An x-ray tube operates at a potential of 40 kV. What is the minimum wavelength of the continuous x-ray spectrum from this tube?

13. The minimum wavelength in the continuous x-ray spectrum from a television tube is 0.124 nm. What is the voltage of the tube?

14. What is the minimum wavelength of the continuous x-ray spectrum from a television tube operating at 1500 V?

Section 29-4   Compton Scattering

15. Find the shift in wavelength of photons scattered at $\theta = 60°$.

16. When photons are scattered by electrons in carbon, the shift in wavelength is 0.33 pm. Find the scattering angle.

17. Find the momentum of a photon in eV/*c* and in kg·m/s if the wavelength is (*a*) 400 nm, (*b*) 2 nm, (*c*) 0.1 nm, and (*d*) 3 cm.

18. The wavelength of Compton-scattered photons is measured at $\theta = 90°$. If $\Delta\lambda/\lambda$ is to be 1 percent, what should the wavelength of the incident photons be?

19. Compton used photons of wavelength 0.0711 nm. (*a*) What is the energy of these photons? (*b*) What is the wavelength of the photon scattered at $\theta = 180°$? (*c*) What is the energy of the photon scattered at this angle?

20. For the photons used by Compton (see Exercise 19), find the momentum of the incident photon and that of the photon scattered at 180° and use momentum conservation to find the momentum of the recoil electron in this experiment.

Section 29-5   Quantization of Atomic Energies: The Bohr Model

21. Use the known values of the constants in Equation 29-21 to show that a_0 is approximately 0.0529 nm.

22. The wavelength of the longest wavelength of the Lyman series was calculated in Example 29-6. Find the wavelengths for the transitions (*a*) $n_1 = 3$ to $n_2 = 1$ and (*b*) $n_1 = 4$ to $n_2 = 1$. (*c*) Find the shortest wavelength in the Lyman series.

23. Find the photon energy for the three longest wavelengths in the Balmer series and calculate the wavelengths.

24. (*a*) Find the photon energy and wavelength for the series limit (the shortest wavelength) in the Paschen series ($n_2 = 3$). (*b*) Calculate the wavelengths for the three longest wavelengths in this series and indicate their positions on a horizontal linear scale.

25. Repeat Exercise 24 for the Brackett series ($n_2 = 4$).

Problems

1. Data for stopping potential versus wavelength for the photoelectric effect using sodium are

λ, nm	200	300	400	500	600
V_0, V	4.20	2.06	1.05	0.41	0.03

Plot these data so as to obtain a straight line. From your plot, find (a) the work function, (b) the threshold frequency, and (c) the ratio h/e.

2. The diameter of the pupil of the eye is about 5 mm. (It can vary from about 1 mm to about 8 mm.) Find the intensity of light of wavelength 600 nm such that 1 photon per second passes through the pupil.

3. A light bulb radiates 100 W uniformly in all directions. The intensity at some distance r is then given by

$$I = \frac{100 \text{ W}}{4\pi r^2}$$

(a) Find the intensity at a distance of 1.5 m. (b) If the wavelength of the light is 600 nm, find the number of photons per second that strike a 1-cm^2 area oriented so that its normal is along the line to the bulb.

4. In blackbody radiation, the power per unit area radiated with wavelengths between λ and $\lambda + \Delta\lambda$ is given by $P(\lambda, T) \Delta\lambda$, where $P(\lambda, T)$ is the quantity shown experimentally in Figure 29-3 and found classically by the Rayleigh-Jeans law (Equation 29-1). (a) Show from the Wien displacement law, that at $T = 300$, $P(\lambda, T)$ is maximum at $\lambda_m = 9.66 \ \mu\text{m}$. Use Equation 29-1 with $T = 300$ K to calculate the power per unit area in $\Delta\lambda = 1$ nm expected from the Rayleigh-Jeans law for (b) $\lambda = 9.66 \ \mu\text{m}$, (c) $\lambda = 4.83 \ \mu\text{m}$, and (d) $\lambda = 0.966 \ \mu\text{m}$, and compare your results with Figure 29-3.

5. This problem is one of estimating the time lag (expected classically but not observed) in the photoelectric effect. Let the intensity of the incident radiation be 0.01 W/m^2. (a) Assuming the area of the atom is 0.01 nm^2, find the energy per second falling on an atom. (b) If the work function is 2 eV, how long would it take classically for this much energy to fall on one atom?

6. The kinetic energy of rotation of a diatomic molecule can be written

$$E_k = \frac{L^2}{2I}$$

where L is its angular momentum and I is the moment of inertia. (a) Assuming that the angular momentum is quantized as in the Bohr model of the hydrogen atom, show that the energy is given by

$$E_n = n^2 E_1$$

where $E_1 = h^2/8\pi^2 I$. (b) Draw an energy-level diagram for such a molecule. (c) Estimate E_1 for the hydrogen molecule, assuming the separation of the atoms to be $r = 0.1$ nm and considering rotation about an axis through the center of mass and perpendicular to the line joining the atoms. (The moment of inertia is then $I = m(r/2)^2 + m(r/2)^2$.) Express your answer in electronvolts.

CHAPTER THIRTY

Electron Waves and Quantum Theory

The idea that energy is not continuous but comes in lumps or quanta launched a revolution in science. Understanding the behavior of electrons and atoms requires a completely new kind of theory, called wave mechanics or quantum theory. In this theory, electrons which we once thought to be particles have wave properties and exhibit interference and diffraction. Newtonian mechanics in which the paths and interactions of particles can be predicted from initial conditions is replaced by quantum theory in which only probabilities can be calculated. If you find these ideas difficult to understand, don't worry, you are in the company of many of the greatest minds of science. Despite its conceptual difficulties, quantum theory has been extremely fruitful. It is the basis for our understanding of the modern world, from the inner workings of the cell in biology to the radiation spectrum of distant galaxies in cosmology.

The most beautiful thing we can experience is the mysterious. It is the source of all true art and science.

I shall never believe that God plays dice with the world.

ALBERT EINSTEIN

In the previous chapter, we saw that light, which was thought to be a wave phenomenon, has particle properties. In this chapter we will see that electrons, which were thought to be particles, also have wave properties. The first suggestion of the possible wave properties of electrons came from a French student, L. de Broglie, in his dissertation in 1924. His reasoning was based on the symmetry of nature. Since light is known to have both wave and particle properties, perhaps matter—especially electrons—may also have both wave and particle characteristics. This suggestion was highly speculative since there was no evidence at that time for any wave aspects of electrons. De Broglie showed that Bohr's quantum condition for the electron in the hydrogen atom could be understood as a standing-wave condition for electron waves.

In 1927, electron diffraction was observed accidentally by C. J. Davisson and L. H. Germer at the Bell Telephone Laboratories. From the diffraction pattern they observed, the wavelength of the electrons could be calculated, and it was found to agree with that predicted by de Broglie. Since then, diffraction of electrons and other "particles," such as neutrons and helium atoms, has been observed in many experiments.

De Broglie's ideas were developed into a detailed mathematical theory called *wave mechanics* by Erwin Schrödinger in 1928. In this theory, the electron is described by a wave function that obeys a wave equation that is somewhat similar to the classical wave equations for sound and light waves. The frequency and wavelength of electron waves are related to the energy and momentum of the electron, just as the frequency and wavelength of light waves are related to the energy and momentum of photons. Diffraction and interference of these waves, as observed by Davisson and Germer and others, is a natural consequence of their propagation. The quantization of energy in atoms, molecules, and other microscopic systems results from standing-wave patterns of electron waves in these systems.

In this chapter, we will look at some of the evidence for electron waves and at how Schrödinger's wave mechanics theory leads to the idea of the quantization of energy in atoms and molecules.

30-1   The de Broglie Equations

De Broglie's suggestion that electrons have wave properties was based on the analogy of electrons with light and the fact that light has both wave and particle properties. For the frequency and wavelength of electron waves, de Broglie chose the equations

$$f = \frac{E}{h} \tag{30-1}$$

de Broglie equations

$$\lambda = \frac{h}{p} \tag{30-2}$$

where p is the momentum and E is the energy of the electron. Equation 30-1 is the same as the Planck–Einstein equation for the energy of a photon (Equation 29-2). Equation 30-2 also holds for photons, as can be seen from $\lambda = c/f = hc/hf = hc/E = h/(E/c) = h/p$, since the momentum of a photon is related to its energy by $p = E/c$. De Broglie pointed out that with these relations, the Bohr postulate of quantized angular momentum (Equation 29-19) is equivalent to a standing-wave condition.

$$mvr = \frac{nh}{2\pi}$$

Substituting h/λ for the momentum mv gives

$$\frac{h}{\lambda} r = \frac{nh}{2\pi}$$

or

$$n\lambda = 2\pi r = C \tag{30-3}$$

where C is the circumference of the Bohr orbit. Thus, Bohr's quantum condition is equivalent to saying that an integral number of electron waves must fit into the circumference of the circular orbit, as shown in Figure 30-1.

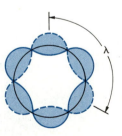

Figure 30-1
Standing waves around the circumference of a circle.

The idea of explaining the discrete energy stages of matter by standing waves seemed promising. In classical wave theory, standing waves lead to a quantization of frequency. For example, for standing waves in a string of length L fixed at both ends (see Figure 30-2), the standing-wave condition is

$$n\frac{\lambda}{2} = L$$

For waves traveling with speed v, the frequency of such standing waves in a string is then given by

$$f = \frac{v}{\lambda} = n\frac{v}{2L}$$

If energy is associated with the frequency of a standing wave, as in Equation 30-1, standing waves imply quantized energies. These ideas were developed in 1930 by Erwin Schrödinger, who found a wave equation for electron waves and solved the standing-wave problem for the hydrogen atom, the simple harmonic oscillator, and other systems of interest. He found that the allowed frequencies combined with the de Broglie relation $E = hf$ led to the same set of energy levels for the hydrogen atom as found by Bohr (Equation 29-24). Schrödinger's *wave mechanics* therefore gave a general method for finding the quantized energy levels for a given system.

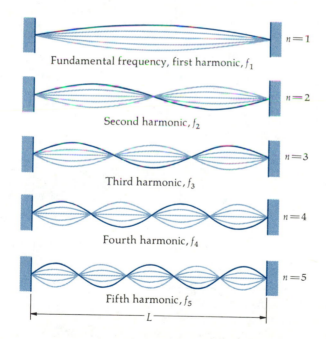

Fundamental frequency, first harmonic, f_1 $n = 1$

Second harmonic, f_2 $n = 2$

Third harmonic, f_3 $n = 3$

Fourth harmonic, f_4 $n = 4$

Fifth harmonic, f_5 $n = 5$

L

Figure 30-2
Standing waves in a string fixed at both ends. The frequencies of these waves are quantized and take on the values $f_n = nf_1$, where f_1 is the fundamental frequency.

De Broglie's equations are thought to apply to all matter. However, for macroscopic objects, the wavelengths calculated from Equation 30-2 are so small that it is impossible to observe the usual wave properties of interference and diffraction. Even a particle as small as 1 μg is much too massive for any wave characteristics to be noticed, as we will see in the following example.

Example 30-1 Find the de Broglie wavelength of a particle of mass 10^{-6} g moving with speed 10^{-6} m/s.

From Equation 30-2, we have

$$\lambda = \frac{h}{p} = \frac{h}{mv} = \frac{6.63 \times 10^{-34}\ \text{J} \cdot \text{s}}{(10^{-9}\ \text{kg})(10^{-6}\ \text{m/s})}$$

$$= 6.63 \times 10^{-19}\ \text{m}$$

Since this wavelength is much smaller than any possible apertures or obstacles (the diameter of the nucleus of an atom is about 10^{-15} m, roughly 10,000 times this wavelength), diffraction and interference of such waves cannot be observed. As we have discussed, the propagation of waves of very small wavelength is indistinguishable from the propagation of particles. Note that we have chosen an extremely small value for the momentum in Example 30-1. Any macroscopic particle with greater momentum would have an even smaller de Broglie wavelength. We therefore do not need to worry about the wave properties of such macroscopic objects as baseballs or billiard balls. However, the situation is different for low-energy electrons. Consider an electron of kinetic energy E_k. Assuming the electron to be nonrelativistic, its momentum is found from

$$E = \frac{p^2}{2m}$$

or

$$p = \sqrt{2mE_k}$$

Its wavelength is then

$$\lambda = \frac{h}{p} = \frac{h}{\sqrt{2mE_k}} = \frac{hc}{\sqrt{2mc^2E_k}}$$

Using $hc = 1240$ eV $\cdot$ nm and $mc^2 = 0.511$ MeV, we obtain

$$\lambda = \frac{1240\ \text{eV} \cdot \text{nm}}{\sqrt{2(0.511 \times 10^6\ \text{eV})E_k}}$$

or

$$\lambda = \frac{1.226}{\sqrt{E_k}}\ \text{nm} \qquad \text{where } E_k \text{ is in electronvolts} \qquad \text{30-4} \qquad \text{Wavelength of electron}$$

Example 30-2 The kinetic energy of the electron in the ground (lowest energy) state of the hydrogen atom is 13.6 eV. (Its potential energy is -27.2 eV and its total energy is -13.6 eV, leading to a binding energy of 13.6 eV.) Find the de Broglie wavelength for this electron.

Substituting $E_k = 13.6$ eV into Equation 30-4, we have

$$\lambda = \frac{1.226}{\sqrt{13.6}}\ \text{nm} = 0.332\ \text{nm} = 2\pi(0.0529\ \text{nm})$$

This is the circumference of the first Bohr orbit in the hydrogen atom.

From Equation 30-4, we can see that electrons with energies of the order of tens of electronvolts have de Broglie wavelengths of the order of nanometres. This is the order of magnitude of the size of the typical atom and of the spacing of atoms in a crystal. Thus, when electrons with energies of the order of 10 eV are incident on a crystal, they are scattered in much the same way as are x-rays of the same wavelength. Interference maxima are seen at angles obeying the Bragg condition (Equation 29-7), and diffraction patterns similar to those obtained with x-rays are observed.

30-2   Electron Diffraction

In our study of light, we found that the important properties that distinguish waves from particles are interference and diffraction. In interference, waves from two sources combine to give cancellation (destructive interference) when the waves are out of phase by 180° or enhancement (constructive interference) when the waves are in phase. Diffraction is the bending of waves around obstacles; it is observed easily for sound waves of long wavelengths and with much more difficulty for light waves, which have very short wavelengths compared with ordinary sized obstacles and apertures.

   The crucial test for the existence of the wave properties of electrons is the observation of diffraction and interference of electron waves. This was first accomplished accidentally in 1927 by C. J. Davisson and L. H. Germer, who were studying electron scattering from a nickel target at the Bell Telephone Laboratories. After heating the target to remove an oxide coating that had accumulated during an accidental break in the vacuum system, Davisson and Germer found that the intensity of the scattered electrons as a function of the scattering angle showed maxima and minima. Their target had crystallized, and by accident they had observed electron diffraction. They then prepared a target consisting of a single crystal of nickel and investigated this phenomenon extensively. Figure 30-3 illustrates their experiment. Electrons from an electron gun are directed at a crystal and are then detected at some angle ϕ, which can be varied. Figure 30-4 shows a typical pattern observed. There is a strong scattering maximum at an angle of 50°. This angle can be related to the Bragg angle for the scattering of waves from the crystal. Using the known spacing of the atoms in their crystal, Davisson and Germer calculated the wavelength that could produce such a maximum and found that it agreed with the de Broglie equation (Equation 30-2) for the electron energy they were using. By varying the energy of the incident electrons, they could vary the electron wavelengths and produce maxima and minima at different locations in the diffraction patterns. In all cases, the measured wavelengths agreed with de Broglie's hypothesis.

   In the same year, G. P. Thomson (son of J. J. Thomson) also observed electron diffraction in the transmission of electrons through thin metal foils. A metal foil consists of tiny crystals, randomly oriented. If a crystal in the foil is oriented at an angle θ with the incident beam, where θ satisfies the Bragg condition, the crystal will strongly scatter waves at an equal angle θ; thus, there will be a scattered beam making an angle 2θ with the original beam, and the diffraction pattern will consist of concentric circles. Since Thom-

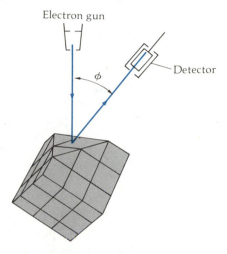

Figure 30-3
The Davisson–Germer experiment. Electrons from the electron gun are incident on a crystal and are scattered into a detector at some angle ϕ, which can be varied.

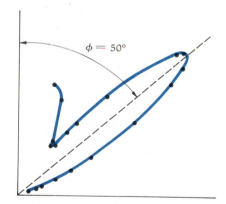

Figure 30-4
Polar plot of the scattered intensity versus angle for 54-eV electrons. (The intensity at each angle is indicated by the distance of the point from the origin.) There is a maximum intensity at $\phi = 50°$, as predicted for Bragg scattering of waves having wavelength $\lambda = h/p$.

(a)

(b)

(c)

Figure 30-5
Diffraction pattern produced by (a) x-rays and (b) electrons incident on an aluminum foil target and (c) by neutrons incident on a polycrystalline copper target. Note the similarity in the patterns produced.

son's experiment, diffraction has been observed for neutrons, protons, and other particles of small mass. Figure 30-5a to c shows the diffraction patterns of x-rays, electrons, and neutrons of similar wavelength after transmission through thin foils. Figure 30-6 shows a diffraction pattern produced by electrons incident on two narrow slits. This experiment is equivalent to Young's famous double-slit diffraction–interference experiment with light. The pattern is identical to that observed with photons of the same wavelength.

The Electron Microscope

Shortly after the wave properties of the electron were demonstrated, it was suggested that electrons rather than light might be used to "see" small objects. Today the *electron microscope* is an important research tool. Figure 30-7 is a schematic drawing of an electron microscope. The electron beam is made parallel and is focused by specially designed magnets that serve as lenses. The energy of the electrons is typically 100 keV, resulting in a wavelength of about 0.004 nm. The target specimen must be very thin so that the transmitted beam will not be slowed down or scattered too much. The final image is projected onto a fluorescent screen or photographic film.

Figure 30-6
A two-slit electron diffraction–interference pattern. This pattern is the same as that usually obtained with photons.

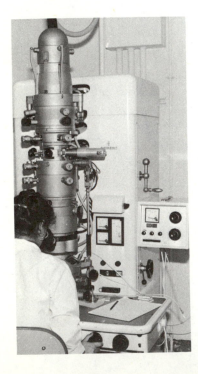

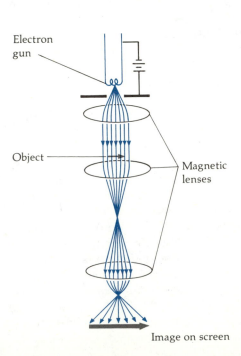

Electron
gun

Object

Magnetic
lenses

Image on screen

Figure 30-7
Electron microscope. Electrons from a heated filament (the electron gun) are accelerated by a large voltage difference. The electron beam is made parallel by a magnetic focusing lens. The electrons strike a thin target and are then focused by a second magnetic lens that is equivalent to the objective lens in an ordinary microscope. The third magnetic lens takes the place of the eyepiece in a microscope. It projects the electron beam onto a fluorescent screen for the viewing of the image.

Various distortions resulting from focusing problems with the magnetic lenses limit the resolution to a few tenths of a nanometre, which is about a thousand times better than can be achieved with visible light.

30-3   Wave–Particle Duality

We have seen that light, which we ordinarily think of as a wave, exhibits particle properties when it interacts with matter, as in the photoelectric effect or in Compton scattering. At the same time, electrons, which we usually think of as particles, exhibit the wave properties of interference and diffraction. All carriers of momentum and energy—electrons, atoms, light, sound, and so forth—have both particle and wave characteristics. It might be tempting to say that an electron, for example, is both a wave and a particle, but the meaning of such a statement is not clear. In classical physics, the concepts of waves and particles are mutually exclusive. A *classical particle* behaves like a piece of shot; it can be localized and scattered, it exchanges energy suddenly in lumps, and it obeys the laws of conservation of energy and momentum in collisions. It does *not* exhibit interference or diffraction. A *classical wave,* on the other hand, behaves like a water wave; it exhibits diffraction and interference, and its energy is spread out continuously in space and time. Nothing can be a classical particle and a classical wave at the same time.

After Thomas Young observed the two-slit interference pattern with light in 1801, it had been thought that light was a classical wave. Similarly, after J. J. Thomson's experiment in 1897 in which he deflected electrons in electric and magnetic fields, it had been thought that electrons were classical particles. We now see that these concepts of classical waves and particles do not adequately describe the complete behavior of any phenomenon.

Everything propagates like a classical wave and exchanges energy like a classical particle.

Often, the classical-particle and classical-wave concepts give the same results. When the wavelength is very small, the propagation of a classical wave cannot be distinguished from that of a classical particle. For waves of very small wavelengths, diffraction effects are negligible, so the waves travel in straight lines. Similarly, interference is not seen for waves of very small wavelength because the interference fringes are too closely spaced to be observed. It then makes no difference which concept we use. We can think of light as a wave propagating along rays, as in geometric optics, or as a beam of photon particles. Similarly, we can think of an electron as a wave propagating in straight lines along rays or, more commonly, as a particle.

We can also use either the wave or particle concept to describe exchanges of energy if we have a large number of particles and we are interested only in average values of energy and momentum exchanges. For example, we need the particle concept of light to explain the existence of a threshold for the photoelectric effect and the dependence of the maximum electron energy on the frequency of the light. But if we are interested only in the total photoelectric current (above the threshold), the wave theory of light correctly predicts that this current is proportional to the intensity of the light.

Electron micrograph of grain weevil emerging from barley, magnified 25 times.

© 1987 Sidney Harris.

"If this is correct, then everything we thought was a wave is really a particle and everything we thought was a particle is really a wave."

30-4 The Uncertainty Principle

In classical mechanics, the path taken by a particle can be determined if the forces acting on the particle are given and the initial position and velocity of the particle are known. Although there are always experimental uncertainties in any measurement of the initial position and velocity, it is assumed in classical mechanics that such uncertainties can be made as small as desired. However, because of the wave–particle duality of both radiation and matter, we now understand that it is impossible *in principle* to measure both the position and the velocity of a particle simultaneously with infinite precision. This result is known as the *uncertainty principle,* first stated by Werner Heisenberg in 1927. For one-dimensional motion, it is expressed in terms of the momentum p and the position x of a particle. Let Δx and Δp be the uncertainties in the position and momentum, respectively. According to the uncertainty principle, the product of Δx and Δp can never be less than $h/4\pi$:

$$\Delta x \, \Delta p \geq h/4\pi \qquad\qquad 30\text{-}5$$

Uncertainty principle

Usually, the uncertainty product is much greater than $h/4\pi$. The equality holds only if the measurements of both x and p are ideal.

We can get a qualitative understanding of the uncertainty principle by considering the measurement of the position and momentum of a particle. If we know the mass of the particle, we can determine its momentum by measuring its position at two nearby times and computing its velocity. A common way to measure the position of an object is to look at it with light. When we do this, we scatter light from the object and determine the position by the direction in which the light is scattered. If we use light of wavelength λ, we can measure the position only to an uncertainty of the order of λ because of diffraction effects. To reduce the uncertainty in position, we therefore use light of very short wavelength, perhaps even x-rays.

In principle, there is no limit to the accuracy of such a position measurement because there is no limit to how small a wavelength λ we can use. However, since all electromagnetic radiation carries momentum, the scattering of the radiation by the particle will deflect the radiation and change its original momentum in an uncontrollable way. By momentum conservation, the momentum of the particle is also changed in an uncontrollable way. According to classical wave theory, this effect on the momentum of the particle could be reduced by reducing the intensity of the radiation. However, the energy and momentum of the radiation are quantized; each photon has momentum h/λ. When the wavelength of the radiation is small, the momentum of each photon will be large and the momentum measurement will have a large uncertainty. This uncertainty cannot be eliminated by reducing the intensity of the light; such a reduction merely reduces the number of photons in the beam. To "see" the particle, we must scatter at least one photon. Therefore, the uncertainty in the momentum measurement of the particle will be large if λ is small, and the uncertainty in the position measurement of the particle will be large if λ is large. A detailed analysis shows that the product of these uncertainties will always be at least of the order of Planck's constant h. Of course we could always "look at" the particles by scattering electrons instead of photons, but we would still have the same difficulty. If we use low-momentum electrons to reduce the uncertainty in the momentum measurement, we will have a large uncertainty in the position measurement because of the diffraction of the electrons. The relation between the wavelength and momentum, $\lambda = h/p$, is the same for electrons as for photons.

One consequence of the uncertainty principle is that when a particle is confined in some region of space, it cannot have zero kinetic energy. The minimum energy of a particle required by the uncertainty principle is called its *zero-point energy*. Suppose, for example, that a particle is confined to some region of space of length L. Its uncertainty in position is then no greater than L. Consequently, we can see from Equation 30-5 that its uncertainty in momentum Δp is

Zero-point energy

$$\Delta p \geq \frac{h}{4\pi L} \qquad\qquad 30\text{-}6$$

The kinetic energy of the particle is

$$E_k = \tfrac{1}{2}mv^2 = \frac{p^2}{2m} \qquad\qquad 30\text{-}7$$

The magnitude of the momentum p must be at least as large as its uncertainty Δp. Then

$$E_k = \frac{p^2}{2m} \geq \frac{h^2}{32\pi^2 mL^2} \qquad\qquad 30\text{-}8$$

The smaller the region of space L, the greater the minimum kinetic energy.

Example 30-3 A marble of mass 25 g is in a box of length 10 cm. Find its minimum uncertainty in momentum, its speed v, and its minimum kinetic energy assuming that $p = \Delta p$.

Two significant figures will be accurate enough for this problem. From Equation 30-5, with $\Delta x = 10$ cm, we have

$$(\Delta p)_{min} = \frac{h}{4\pi \,\Delta x} = \frac{6.6 \times 10^{-34} \text{ J} \cdot \text{s}}{(4\pi)(0.1 \text{ m})}$$

$$= 5.3 \times 10^{-34} \text{ kg} \cdot \text{m/s}$$

The speed corresponding to a momentum of this magnitude is

$$v = p/m = \frac{5.3 \times 10^{-34} \text{ kg} \cdot \text{m/s}}{0.025 \text{ kg}} = 2.1 \times 10^{-32} \text{ m/s}$$

We would be quite safe in saying that the marble is at rest. The minimum kinetic energy is

$$E_{k,min} = \frac{(\Delta p)^2_{min}}{2m} = \frac{(5.3 \times 10^{-34} \text{ kg} \cdot \text{m/s})^2}{0.050 \text{ kg}} = 5.6 \times 10^{-66} \text{ J}$$

Because Planck's constant is so small, the uncertainty relation of Equation 30-5 is not significant for macroscopic systems.

Example 30-4 Work Example 30-3 for an electron confined to a region of space of length $L = 0.1$ nm. This distance is of the order of the diameter of an atom.

In this case, the minimum uncertainty in momentum is

$$(\Delta p)_{min} = \frac{h}{4\pi \, \Delta x} = \frac{6.6 \times 10^{-34} \, \text{J} \cdot \text{s}}{(4\pi)(10^{-10} \, \text{m})} = 5.3 \times 10^{-25} \, \text{kg} \cdot \text{m/s}$$

The speed of an electron with momentum of this magnitude is

$$v = \frac{p}{m} = \frac{5.3 \times 10^{-25} \, \text{kg} \cdot \text{m/s}}{9.1 \times 10^{-31} \, \text{kg}} = 5.8 \times 10^5 \, \text{m/s}$$

We note that this is a significant speed. The minimum kinetic energy is

$$E_{k,min} = \frac{(\Delta p)^2_{min}}{2m} = \frac{(5.3 \times 10^{-25} \, \text{kg} \cdot \text{m/s})^2}{2(9.1 \times 10^{-31} \, \text{kg})} = 1.5 \times 10^{-19} \, \text{J}$$

This is approximately 1 eV, which is about the order of magnitude of the kinetic energy of an electron in an atom.

30-5   The Electron Wave Function

In our study of classical waves, such as waves in a string, sound waves, or light waves, we found that the energy density (the energy per unit volume in a wave) is proportional to the square of the amplitude of the wave. The intensity, which equals the energy density times the wave speed (Equation 16-19 in Section 16-5), is also proportional to the square of the wave amplitude. For waves in a string, the wave amplitude is the displacement of the string. For sound waves in air, the amplitude is the displacement of the air molecules from their equilibrium positions or, alternatively, the pressure variations due to the sound wave. For light and other electromagnetic waves, the amplitude is the electric field E associated with the wave.

The amplitude of electron waves is called the *wave function* and is designated by the greek letter ψ (psi). The wave function for a particular problem is found by solving the Schrödinger wave equation for that problem, just as the amplitude E for light waves is found by solving the classical wave equation for light.

Wave function

When Schrödinger first published his wave equation for electrons, it was not clear to him or to anyone else just what the wave function represented. We can get a hint as to how to interpret ψ by considering the quantization of light waves.

Since the energy per unit volume in a light wave is proportional to E^2 and the energy is quantized in units of hf for each photon, we expect that the number of photons in a unit volume is proportional to E^2. Let us consider Young's famous double-slit experiment (see Figure 30-8). The pattern observed on the screen is determined by the interference of the waves from the slits. At a point on the screen where the wave from one slit is 180° out of phase with that from the other, the resultant electric field E is zero. There is no light energy at that point and the point is dark. At points where the waves from the slits are in phase, E is maximum and the points are bright. If we reduce the intensity of the light incident on the slits, we can still observe the interference pattern if we replace the screen by a photographic film and wait a sufficient time to expose the film.

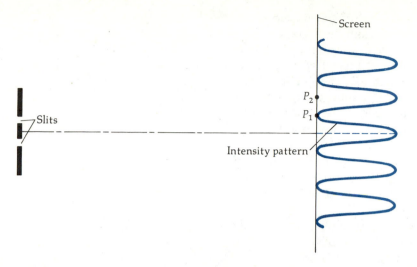

Figure 30-8
Young's double-slit experiment.
Many photons go to point P_2 but no
photons go to P_1. This experiment
can be done with light intensity
that is so low that only one photon
at a time arrives at the two slits and
at the screen. The intensity pattern
must then be interpreted as a
measure of the probability of a
photon arriving at a particular point.

The interaction of light with photographic film is a quantum phenomenon. If we expose the film for a very short time with a low-intensity light
source, we do not see merely a weaker version of the high-intensity pattern.
Instead, we see "dots" on the film caused by the interactions of the individual photons (see Figure 30-9). At points where the waves from the slits
interfere destructively, there are no dots; that is, no photons arrive at these
points. At points where the waves interfere constructively, there are many

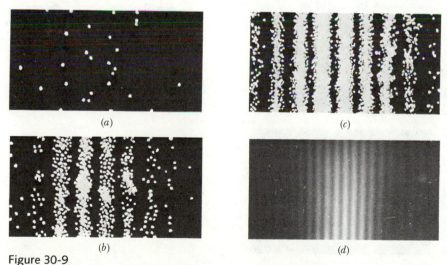

Figure 30-9
Growth of the two-slit interference pattern. The drawing
(a) shows the expected pattern after the film has been
exposed to about 28 photons or electrons. The drawings
(b) and (c) show the expected patterns for 1000 and 10,000
photons or electrons, respectively. Note that there are no
dots in the regions of the interference minima. The
photograph (d) is an actual two-slit electron interference
pattern in which the film was exposed to millions of
electrons. The pattern is identical to that usually obtained
with photons.

dots, indicating many photons. When the exposure time is very short and the light source is weak, random fluctuations from the average predictions of the wave theory are clearly evident. If the exposure time is long or the light source is strong so that many photons interact with the film, the fluctuations average out and the quantum nature of the light is not noticed. The interference pattern depends on the total number of photons interacting with the film and not on the rate of interaction. Even when the intensity is so low that only one photon at a time hits the film, the wave theory predicts the correct average pattern.

For low intensities, we therefore interpret E^2 to be proportional to the probability of detecting a photon in a unit volume.

At points on the film or screen where E^2 is zero, no photons are observed, whereas at points where E^2 is large, they are most likely to be observed.

As we have seen, we obtain the same double-slit pattern if we use electrons instead of photons. Figure 30-10 shows the interference pattern with just a few electrons and with many electrons. In the wave theory of electrons, the motion of a single electron is described by the wave function ψ.

The quantity ψ^2 is proportional to the probability of finding the electron in a unit volume of space.

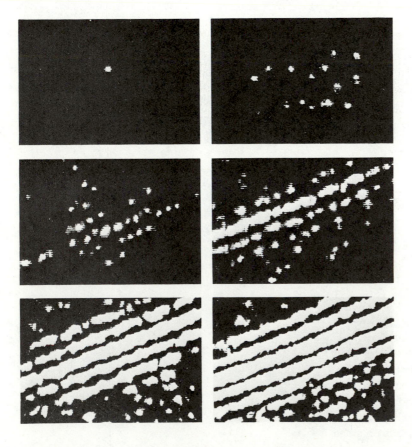

Figure 30-10
Actual electron interference patterns filmed from a television monitor at increasing electron-beam densities.

30-6   A Particle in a Box

To illustrate some of the ideas of quantum theory, we consider a classic, though somewhat artificial problem of a particle, such as an electron, confined to a one-dimensional box of length L. Classically, the particle bounces back and forth between the walls of the box, which we assume are at $x = 0$ and $x = L$. The particle is equally likely to be found anywhere in the box, and its energy and momentum can take on any values. According to quantum theory, the particle is described by a wave function ψ. Since the particle cannot be outside the box, the wave function must be zero at $x = 0$ and $x = L$ and everywhere outside the box. This "boundary condition" on the wave function leads to the standing-wave condition

$$n\frac{\lambda}{2} = L \qquad\qquad\qquad 30\text{-}9$$

where n is an integer, $n = 1, 2, 3, \ldots$. This is the same standing-wave condition as for waves in a string of length L that is fixed at both ends. Assuming zero potential energy, the total energy is the kinetic energy, which is given by

$$E = \tfrac{1}{2}mv^2 = \frac{(mv)^2}{2m} = \frac{p^2}{2m} \qquad\qquad 30\text{-}10$$

where p is the momentum of the particle. Using the de Broglie relation (Equation 30-2) $\lambda = h/p$, the kinetic energy can be written in terms of the wavelength:

$$E = \frac{p^2}{2m} = \frac{h^2}{2m\lambda^2}$$

Substituting $\lambda = 2L/n$ from the standing-wave condition of Equation 30-9, we obtain for the kinetic energy

$$E = n^2\left(\frac{h^2}{8mL^2}\right) \qquad\qquad 30\text{-}11$$

Energy levels for a particle in a box

The energy is thus quantized. The lowest energy state, the ground state, occurs for $n = 1$. The energy of the ground state is

$$E_1 = \frac{h^2}{8mL^2} \qquad\qquad\qquad 30\text{-}12$$

Note that the lowest energy is not zero. According to quantum theory, the particle cannot remain at rest in the box. This result, which, as we saw in Section 30-4, is also a consequence of the uncertainty principle, is a general feature of quantum theory. When a particle is confined to some region of space, it has a minimum energy, the *zero-point energy*. The smaller the region of space, the greater the zero-point energy, as is indicated by the fact that E_1 varies as $1/L^2$ in Equation 30-12. The energies of the other states are given by

$$E_n = n^2 E_1 \qquad\qquad\qquad 30\text{-}13$$

The number n is called a quantum number. It arises from the boundary condition on the wave function that the wave function be zero at $x = 0$ and $x = L$. (In three-dimensional problems, three quantum numbers arise, each associated with a boundary condition in each dimension.)

Figure 30-11 shows plots of ψ^2 for the ground state $n = 1$, the first excited state $n = 2$, the second excited state $n = 3$, and for the state $n = 10$. In the ground state, the particle is most likely to be found near the center of the box, as indicated by the maximum ψ^2 at $x = L/2$. In the first excited state, the particle is never found exactly in the center of the box because ψ^2 is zero at $x = L/2$. For very large n, the maxima and minima of ψ^2 are very close together, as is illustrated for $n = 10$. The average value of ψ^2 is indicated in this figure by the dashed line. For very large n, the maxima and minima are so closely spaced that ψ^2 cannot be distinguished from its average value.

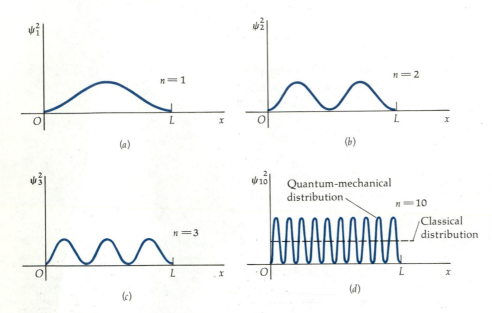

(a)

(b)

(c)

(d)

Figure 30-11
Graphs of ψ^2 versus x for a particle in a box of length L for (a) the ground state, $n = 1$; (b) the first excited state, $n = 2$; (c) the second excited state, $n = 3$; and (d) the state $n = 10$. Compare these with the standing wave patterns for a string fixed at both ends (Figure 30-2).

The fact that $(\psi^2)_{av}$ is constant across the whole box means that the particle is equally likely to be found anywhere in the box, the same as the classical result. This is an example of Bohr's correspondence principle:

In the region of very large quantum numbers, classical calculation and quantum calculation must yield the same results.

Bohr's correspondence principle

The region of very large quantum numbers is also the region of very large energies. It can be shown (see Problem 3) that for large energies, the percentage change in energy between adjacent quantum states is very small, so energy quantization is not important.

We are so accustomed to thinking of the electron as a particle that we tend to think of an electron in a box as a particle bouncing back and forth between the walls. But the probability distributions shown in Figure 30-11 are stationary; that is, they do not depend on time. (The analogous patterns for the electron in the hydrogen atom, which are discussed in the next chapter, are the quantum theoretical counterparts of the stationary orbits of the Bohr model.) An alternative picture of the electron is as a *cloud of charge* with a charge density proportional to ψ^2. The graphs in Figure 30-11 can then be thought of as plots of the charge density versus x for the various states. In the ground state, $n = 1$, the maximum charge density of the electron clouds is in the middle of the box, but the cloud is spread out over most of the box, as is indicated in Figure 30-11a. In the first excited state, $n = 2$, the charge density of the electron cloud has two maxima, as indicated in Figure 30-11b. For very large n, there are many closely spaced maxima and minima in the charge density, resulting in an average charge density that is approximately uniform throughout the box. This electron-cloud picture is very useful for understanding the structure of atoms and molecules, which will be discussed in the next chapter.

Example 30-5 Find the energy in the ground state of an electron confined to a one-dimensional box of length $L = 0.1$ nm. (This box is roughly the size of an atom.)

The energy in the ground state is given by Equation 30-12. Multiplying the numerator and denominator by c^2, we obtain an expression in terms of hc and the rest energy mc^2:

$$E_1 = \frac{(hc)^2}{8mc^2L^2} \qquad\qquad 30\text{-}14$$

Substituting $hc = 1240$ eV·nm and $mc^2 = 0.511$ MeV, we obtain

$$E_1 = \frac{(1240 \text{ eV·nm})^2}{8(5.11 \times 10^5 \text{ eV})(0.1 \text{ nm})^2} = 37.6 \text{ eV}$$

This is greater than the minimum energy of about 1 eV that we found from the uncertainty principle in Example 30-4. It is of the same order of magnitude as the kinetic energy of the electron in the ground state of the hydrogen atom, which is 13.6 eV. In that case, the wavelength of the electron equals the circumference of a circle of radius 0.0529 nm, or about 0.33 nm, whereas for the electron in a one-dimensional box of length 0.1 nm, the wavelength in the ground state is $2L = 0.2$ nm.

Summary

1. The wave nature of electrons was first suggested by de Broglie, who postulated the equations

$$f = \frac{E}{h}$$

and

$$\lambda = \frac{h}{p}$$

for the frequency and wavelength of electron waves. With these postulates, the Bohr postulate of quantized angular momentum can be understood as a standing-wave condition. The wave nature of electrons was observed experimentally first by Davisson and Germer and later by G. P. Thomson, who measured the diffraction and interference of electrons.

2. Light, electrons, neutrons, and all carriers of momentum and energy exhibit both wave and particle properties. Everything propagates like a classical wave exhibiting diffraction and interference but exchanges energy in discrete lumps like a classical particle. Because the wavelength of macroscopic objects is so small, diffraction and interference are not observed. Also, when a macroscopic amount of energy is exchanged, so many quanta are involved that the particle nature of the energy is not evident.

3. The wave–particle duality of nature leads to the uncertainty principle, which states that the product of the uncertainty in a measurement of position and the uncertainty in a measurement of momentum must be greater than $h/4\pi$, where h is Planck's constant:

$$\Delta x\, \Delta p \geq h/4\pi$$

An important consequence of the uncertainty principle is that a particle confined in space has a minimum energy called the zero-point energy.

4. The state of a particle such as an electron is described by its wave function ψ, which is the solution of the Schrödinger wave equation. The square of the wave function, ψ^2, measures the probability of finding the particle in some region of space.

5. A useful picture of an electron is a cloud of charge with charge density proportional to ψ^2.

6. The wavelength for a particle confined to a one-dimensional box of length L obeys the standing-wave condition

$$n\frac{\lambda}{2} = L$$

This results in the energy of the particle being quantized to the values

$$E_n = n^2 E_1$$

where E_1 is the ground-state energy given by

$$E_1 = \frac{h^2}{8mL^2} = \frac{(hc)^2}{8mc^2L^2}$$

The number n is called a quantum number.

7. When the quantum numbers of a system are very large, quantum calculations and classical calculations agree, a result known as Bohr's correspondence principle.

Suggestions for Further Reading

See the article by Gerald Feinberg in the Suggestions for Further Reading for Chapter 27 for a discussion of the wave–particle duality of light.

Everhart, Thomas E., and Thomas L. Hayes; "The Scanning Electron Microscope," *Scientific American*, January 1972, p. 54.

This article describes how the interaction between a beam of high-energy electrons and matter is used by the scanning electron microscope to create an image of three-dimensional appearance.

Review

A. Objectives: After studying this chapter, you should:

1. Be able to state the de Broglie relations for the frequency and wavelength of electron waves and use them and the standing-wave condition to derive the Bohr postulate for the quantization of angular momentum in the hydrogen atom.

2. Be able to discuss the experimental evidence for the existence of electron waves.

3. Be able to discuss wave–particle duality.

4. Be able to discuss the uncertainty principle.

B. Define, explain, or otherwise identify:

De Broglie wavelength, p. 772
Classical particle, p. 777
Classical wave, p. 777
Wave–particle duality, p. 777
Uncertainty principle, p. 778

Electron wave function, p. 780
Zero-point energy, p. 783
Bohr's correspondence principle, p. 784
Electron charge cloud, p. 785

C. True or false: If the statement is true, explain why. If it is false, give a counterexample.

1. The de Broglie wavelength of an electron varies inversely with the electron's momentum.

2. Electrons can be diffracted.

3. Neutrons can be diffracted.

4. An electron microscope is used to look at electrons.

5. The measurement of the position of a particle interferes with the measurement of its momentum.

6. A particle confined in a region of space cannot have zero kinetic energy.

Exercises

Section 30-1   The de Broglie Equations

1. Use Equation 30-4 to calculate the de Broglie wavelength for an electron of kinetic energy (a) 1 eV, (b) 100 eV, (c) 1 keV, and (d) 10 keV.

2. An electron is moving at $v = 2.5 \times 10^5$ m/s. Find its de Broglie wavelength.

3. An electron has a wavelength of 200 nm. Find (a) its momentum and (b) its kinetic energy.

4. Through what potential must an electron be accelerated so that its de Broglie wavelength is (a) 5 nm and (b) 0.01 nm?

5. A thermal neutron in a reactor has kinetic energy of about 0.02 eV. Calculate the de Broglie wavelength of this neutron from

$$\lambda = \frac{hc}{\sqrt{2mc^2E_k}}$$

where $mc^2 = 940$ MeV is the rest energy of the neutron.

6. Find the de Broglie wavelength of a proton ($mc^2 = 938$ MeV) that has a kinetic energy of 1 MeV (see Exercise 5).

7. A proton is moving at $v = 0.002c$, where c is the speed of light. Find its de Broglie wavelength.

8. What is the kinetic energy of a proton whose de Broglie wavelength is (a) 1 nm and (b) 10 fm?

9. Find the de Broglie wavelength of a baseball of mass 0.145 kg moving at 30 m/s.

Section 30-2   Electron Diffraction

10. The energy of the electron beam in Davisson and Germer's experiment was 54 eV. Calculate the wavelength for these electrons.

11. The distance between Li^+ and Cl^- ions in a LiCl crystal is 0.257 nm. Find the energy of electrons that have wavelengths equal to this spacing.

12. An electron microscope uses electrons of energy 50 keV. Find the wavelength of these electrons.

Section 30-3   Wave–Particle Duality

There are no exercises for this section.

Section 30-4   The Uncertainty Principle

13. A particle of mass 10^{-6} g moves with speed 1 cm/s. If its speed is uncertain by 0.01 percent, what is the minimum uncertainty in its position?

14. The uncertainty in the position of an electron in an atom cannot be greater than the diameter of the atom. (a) Calculate the minimum uncertainty in momentum associated with an uncertainty in position of an electron of 0.1 nm. (b) If an electron has momentum p equal in magnitude to the uncertainty Δp found in (a), what is its kinetic energy? Express your answer in electronvolts.

15. An electron has kinetic energy of 25 eV. If its momentum is uncertain by 10 percent, what is the minimum uncertainty in its position?

Section 30-5   The Electron Wave Function;
Section 30-6   A Particle in a Box

16. (*a*) Draw an energy-level diagram to scale for an electron in a one-dimensional box of length 0.1 nm. Include the states from $n = 1$ to $n = 5$. Calculate the wavelength of the electromagnetic radiation emitted when the electron makes a transition from (*b*) $n = 2$ to $n = 1$, (*c*) $n = 3$ to $n = 2$, and (*d*) $n = 5$ to $n = 1$.

17. (*a*) Find the energy of the ground state and first two excited states of a proton in a one-dimensional box of length $L = 10^{-14}$ m $= 10$ fm. (These are of the order of magnitude of nuclear energies.) Calculate the wavelength of the electromagnetic radiation emitted when the proton makes a transition from (*b*) $n = 2$ to $n = 1$, (*c*) $n = 3$ to $n = 2$, and (*d*) $n = 3$ to $n = 1$.

18. A golf ball of mass 46 g is in a one-dimensional box of length 10 cm. Find its minimum kinetic energy. Discuss the possibility of ever observing this minimum kinetic energy.

19. A proton is in a one-dimensional box of length 0.2 nm (about the diameter of a H_2 molecule. (*a*) Find the energies of the ground state and first two excited states. Calculate the wavelength of the electromagnetic radiation emitted when the proton makes a transition from (*b*) $n = 2$ to $n = 1$, (*c*) $n = 3$ to $n = 2$, and (*d*) $n = 3$ to $n = 1$.

20. A small particle of mass 1 μg is confined to a one-dimensional box of length 1 cm. (*a*) Find the minimum energy E_1 for a particle of this mass in this box. (*b*) If the particle moves with a speed of about 1 mm/s, calculate its kinetic energy and find the approximate value of the quantum number n.

Problems

1. When the kinetic energy of an electron is much greater than its rest energy, the relativistic approximation $E_k \approx pc$ is good. (*a*) Show that in this case, photons and electrons of the same energy have the same wavelength. (*b*) Find the de Broglie wavelength of an electron of energy 100 MeV.

2. Sketch ψ^2 as a function of x for the states $n = 4$ and $n = 5$ for a particle in a one-dimensional box.

3. For the electron in the one-dimensional box of Example 30-5, find the percentage change in energy between the states $n_1 = 1000$ and $n_2 = 1001$, and comment on how this result is related to Bohr's correspondence principle.

CHAPTER THIRTY-ONE

Atoms, Molecules, and Solids

In this chapter you will see how the ideas of quantum theory explain the structure and spectra of atoms and molecules, molecular bonding, and the electrical properties of solids. You will also learn some of the physics behind the operation of such devices as lasers, semiconductor junction diodes, and transistors, which are the bases for many of the technological advances in our modern world.

According to convention there is a sweet and a bitter, a hot and a cold, and according to convention there is order. In truth there are atoms and a void.

DEMOCRITUS (c. 400 B.C.)

Slightly more than 100 different elements have been discovered, 92 of which are found in nature. Each is characterized by an atom that contains a number of protons, an equal number of electrons, and a number of neutrons. The number of protons, Z, is called the *atomic number*. The lightest atom, hydrogen (H), has $Z = 1$; the next lightest, helium (He), has $Z = 2$; the next lightest, lithium (Li), has $Z = 3$; and so forth. The structure of an atom is somewhat like that of our solar system. Nearly all the mass of the atom is in a tiny nucleus, which contains the protons and neutrons. The nuclear radius is typically about 1 to 10 fm (1 fm $= 10^{-15}$ m). The distance between the nucleus and the electrons is about 0.1 nm $= 100,000$ fm. This distance determines the "size" of the atom.

The chemical and physical properties of an element are determined by the number and arrangement of the electrons in the atom. Because each proton has a positive charge $+e$, the nucleus has a total positive charge $+Ze$. The electrons are negatively charged $(-e)$, and so they are attracted to the nucleus and repelled by each other. The electrons are arranged in shells. The first shell has up to two electrons. The second shell, about four times farther out than the first, can contain up to eight electrons. The third shell, about nine times farther out than the first, can also contain up to eight electrons. This shell structure accounts for the periodic nature of the properties of the elements as shown in the periodic table (Table 31-1). Elements with a single electron in an outer shell (hydrogen, lithium, sodium, etc.) or those with a single vacancy in an outer shell (fluorine, chlorine, bromine, etc.) are very active chemically and readily combine to form molecules. Those with com-

Table 31-1 Periodic Table of the Elements

The atomic-mass values listed are based on $^{12}_{6}C = 12$ u exactly. For artificially produced elements, the approximate atomic mass of the most stable isotope is given in brackets.

Period	Series	Group									
		I	II	III	IV	V	VI	VII	VIII		0
1	1	1 H 1.00797									2 He 4.003
2	2	3 Li 6.942	4 Be 9.012	5 B 10.81	6 C 12.011	7 N 14.007	8 O 15.9994	9 F 19.00			10 Ne 20.183
3	3	11 Na 22.990	12 Mg 24.31	13 Al 26.98	14 Si 28.09	15 P 30.974	16 S 32.064	17 Cl 35.453			18 Ar 39.948
4	4	19 K 39.102	20 Ca 40.08	21 Sc 44.96	22 Ti 47.90	23 V 50.94	24 Cr 52.00	25 Mn 54.94	26 Fe 55.85 27 Co 58.93 28 Ni 58.71		
	5	29 Cu 63.54	30 Zn 65.37	31 Ga 69.72	32 Ge 72.59	33 As 74.92	34 Se 78.96	35 Br 70.909			36 Kr 83.80
5	6	37 Rb 85.47	38 Sr 87.62	39 Y 88.905	40 Zr 91.22	41 Nb 92.91	42 Mo 95.94	43 Tc [98]	44 Ru 101.1 45 Rh 102.905 46 Pd 106.4		
	7	47 Ag 107.870	48 Cd 112.40	49 In 114.82	50 Sn 118.69	51 Sb 121.75	52 Te 127.60	53 I 126.90			54 Xe 131.30
6	8	55 Cs 132.905	56 Ba 137.34	57-71 Lanthanoid series*	72 Hf 178.49	73 Ta 180.95	74 W 183.85	75 Re 186.2	76 Os 190.2 77 Ir 192.2 78 Pt 195.09		
	9	79 Au 196.97	80 Hg 200.59	81 Tl 204.37	82 Pb 207.19	83 Bi 208.98	84 Po [210]	85 At [210]			86 Rn [222]
7	10	87 Fr [223]	88 Ra [226]	89-103 Actinoid series**							

* Lanthanoid Series (rare earths)	57 La 138.91	58 Ce 140.12	59 Pr 140.91	60 Nd 144.24	61 Pm [147]	62 Sm 150.35	63 Eu 152.0	64 Gd 157.25	65 Tb 158.92	66 Dy 162.50	67 Ho 164.93	68 Er 167.26	69 Tm 168.93	70 Yb 173.04	71 Lu 174.97
** Actinoid Series	89 Ac [227]	90 Th 232.04	91 Pa [231]	92 U 238.03	93 Np [237]	94 Pu [242]	95 Am [243]	96 Cm [247]	97 Bk [247]	98 Cf [251]	99 Es [254]	100 Fm [253]	101 Md [256]	102 No [254]	103 Lw [257]

pletely filled outer shells (helium, neon, argon, etc.) are more or less chemically inert. The calculation of the electron configurations of the atoms and the resulting chemical properties was one of the great triumphs of quantum mechanics in the 1920s.

Since electrons and protons have equal but opposite charges and there are an equal number of electrons and protons in an atom, atoms are electrically neutral. Atoms that lose or gain one or more electrons are then electrically charged and are called *ions*. Atoms with just 1 electron in an outer shell (such as sodium, which has 11 electrons) tend to lose it readily and become positive ions; those lacking just one electron for a complete outer shell tend to gain an electron and become negative ions (for example, chlorine). Atoms bond together to form molecules, such as H_2O, or solids. This bonding involves only the outer electrons. In a molecule or solid, the separation of the atomic nuclei is about one atomic diameter, that is, of the order of 0.1 nm.

In this chapter, we begin by discussing the quantum-mechanical model of the hydrogen atom. We will then apply our qualitative knowledge of quantum theory to various topics in atomic, molecular, and solid state physics, including atomic structure, the periodic table, atomic and molecular spectra, molecular bonding, lasers, the band structure of solids, semiconductor junctions, and transistors. Because of the complexity of these subjects, much of our treatment will be descriptive.

31-1   Quantum Theory of the Hydrogen Atom

Despite its spectacular successes, the Bohr model of the hydrogen atom had many shortcomings. There was no justification for the postulates of stationary states or for the quantization of angular momentum other than the fact that these postulates led to energy levels that agreed with spectroscopic data. Furthermore, the Bohr model gave no information about the intensities of the spectral lines, and attempts to apply it to more complicated atoms had little success. The quantum mechanical theory resolved these difficulties. The stationary states of the Bohr model correspond to the standing-wave solutions of the Schrödinger wave equation. As we have mentioned, energy quantization is a direct consequence of the frequency quantization that results from standing waves and the de Broglie relation, $E = hf$. The quantized energies resulting from the standing-wave solutions of the Schrödinger equation agree with those obtained from the Bohr model and with experiment. The quantization of angular momentum that had to be postulated in the Bohr model is predicted by the quantum theory.

In the quantum theory, the electron is described by its *wave function ψ*. The square of the electron wave function ψ^2 is proportional to the probability of finding the electron in some region of space. Boundary conditions on the wave function lead to the quantization of the wavelengths and frequencies and thereby to the quantization of the electron energy. In our example of a particle in a one-dimensional box in Section 30-6, we saw that the boundary condition on the wave function—namely, that it is zero at the ends of the box—led to the standing-wave condition, resulting in the quantum number n that appears in the expression for energy quantization. We mentioned that in three-dimensional problems, three quantum numbers arise. Since the hydrogen atom is, of course, three-dimensional, the solution of the Schrödinger wave equation leads to three quantum numbers. These are labeled n, ℓ, and m. The possible values of these quantum numbers are

$$n = 1, 2, 3, \ldots \qquad\qquad\qquad\qquad \text{31-1}$$

$$\ell = 0, 1, 2, \ldots, n-1 \qquad\qquad\qquad \text{31-2}$$

$$m = -\ell, -\ell+1, -\ell+2, \ldots, +\ell \qquad \text{31-3}$$

That is, n can be any integer; ℓ can be 0 or any integer up to $n-1$; and m can have $2\ell + 1$ possible values, ranging from $-\ell$ to $+\ell$ in integral steps.

The number n is called the *principal quantum number*. It is associated with the dependence of the wave function on distance r and therefore with the probability of finding the electron at various distances from the nucleus. Figure 31-1 shows the probability of finding the electron at a distance r as a function of r for $n = 1$ and for $n = 2$ (and $\ell = 1^*$). For $n = 1$, the most likely distance between the electron and the nucleus is a_0; for $n = 2$ and $\ell = 1$, it is $4a_0$. These are the Bohr radii for the first and second Bohr orbits (Equation 29-20). For $n = 3$ and $\ell = 2$, the most likely distance between the electron and the nucleus is $9a_0$, the radius of the third Bohr orbit.

The quantum numbers ℓ and m are associated with the angular momentum of the electron and with the angular dependence of the electron wave function. The quantum number ℓ is called the *orbital quantum number*. The *orbital angular momentum* of the electron L is related to ℓ by

$$L = \sqrt{\ell(\ell+1)}\, h/2\pi = \sqrt{\ell(\ell+1)}\,\hbar \qquad \text{31-4}$$

where $\hbar = h/2\pi$ is Planck's constant divided by 2π. (The constant $\hbar$, read "h bar," is used because the combination $h/2\pi$ occurs so often in equations.)

The quantum number m is called the *magnetic quantum number*. It is related to the component of angular momentum in some direction. Ordinarily, all directions are equivalent, but one particular direction can be specified by placing the atom in a magnetic field. If the z direction is chosen for the magnetic field, the z component of the angular momentum of the electron is given by

$$L_z = m\,\frac{h}{2\pi} = m\hbar \qquad\qquad\qquad \text{31-5}$$

If we measure the angular momentum of the electron in units of $\hbar$, we can see that the angular momentum is quantized to the values $\sqrt{\ell(\ell+1)}$ units, and its component in any direction can have only the $2\ell + 1$ values ranging from $-\ell$ to $+\ell$ units. Figure 31-2 shows a vector-model diagram illustrating the possible orientations of the angular momentum vector. It is worth remembering that for a given value of ℓ, there are $2\ell + 1$ possible values of m, ranging from $+\ell$ to $-\ell$ in integral steps.

---

* The probability distribution has a slight ℓ dependence. The correspondence with the Bohr model is closest for the maximum value of ℓ, which is $n - 1$.

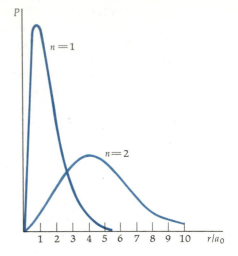

Figure 31-1
Probability distribution P for the electron of the hydrogen atom in the ground state, $n = 1$, and in the excited state, $n = 2$ and $\ell = 1$. For $n = 1$, the electron is most likely to be found near $r = a_0$, the radius of the first Bohr orbit; for $n = 2$ and $\ell = 1$, it is most likely to be found near $r = 4a_0$, the radius of the second Bohr orbit.

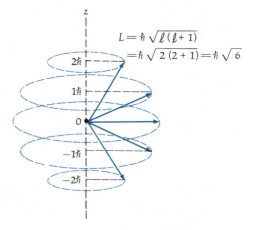

Figure 31-2
Vector model illustrating the possible values of the z component of the angular momentum vector for the case $\ell = 2$.

Example 31-1 If an atom has angular momentum characterized by the quantum number $\ell = 2$, what are the possible values of L_z, and what is the smallest possible angle between **L** and the z axis?

The possible values of L_z are $m\hbar$, where the values of m are $-2, -1, 0, +1$, and $+2$. The magnitude of L is $L = \sqrt{\ell(\ell+1)}\,\hbar = \sqrt{6}\,\hbar$. From Figure 31-2, the angle between **L** and the z axis is given by

$$\cos\theta = \frac{L_z}{L} = \frac{m\hbar}{\sqrt{\ell(\ell+1)}\,\hbar} = \frac{m}{\sqrt{\ell(\ell+1)}}$$

The smallest angle occurs when $m = +\ell$ or $-\ell$, which for $\ell = 2$ gives $\cos\theta = 2/\sqrt{6} = 0.816$ or $\theta = 35.3°$.

We note the somewhat strange result that the angular momentum vector cannot lie along the z axis. This is related to an uncertainty relation for angular momentum that implies that no two components of angular momentum can be precisely known except when $\ell = 0$.

The quantum numbers are related to the spatial dependence of the wave function. Figure 31-3 depicts the electron as a cloud of charge density $e\psi^2$ for $n = 2$, $\ell = 0$, and $m = 0$ (Figure 31-3a); $n = 2$, $\ell = 1$, and $m = 0$ (Figure 31-3b); and $n = 2$, $\ell = 1$, and $m = +1$ or -1 (Figure 31-3c). For $\ell = 0$, the electron cloud is spherically symmetric, whereas for other values of ℓ, it is not.

The energies of the hydrogen atom that result from the solution of the Schrödinger equation are given by

$$E_n = -Z^2 \frac{E_0}{n^2} \qquad\qquad 31\text{-}6$$

Energy levels in hydrogen

where $E_0 = 13.6$ eV, as in the Bohr model. These energies are the same as in the Bohr model. That they depend only on the principal quantum number n is a special result for the hydrogen atom. For more complicated atoms having several electrons, the interaction of the electrons leads to a dependence of the energy on the orbital quantum number. In general, the lower the ℓ value, the lower the energy for these atoms. The energy for any atom

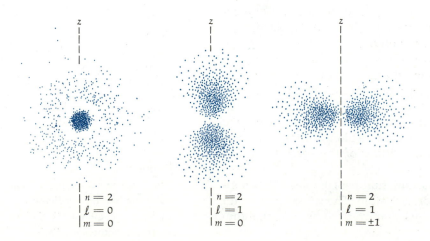

Figure 31-3
Graphical representation of ψ^2 for the electron in the $n = 2$ states in hydrogen. For $\ell = 0$, ψ^2 is spherically symmetric. For $\ell = 1$ and $m = 0$, ψ^2 is proportional to $\cos^2\theta$, and for $\ell = 2$ and $m = +1$ or -1, ψ^2 is proportional to $\sin^2\theta$.

does not depend on the magnetic quantum number m unless the atom is in a magnetic field. The splitting of the energy levels of an atom in a magnetic field for different values of m was discovered by P. Zeeman and is known as the *Zeeman effect*.

In the lowest energy state, the ground state, the principal quantum number n has the value 1, ℓ is 0, and m is 0. The energy is -13.6 eV, the same as in the Bohr model, but the angular momentum has a value of zero rather than the $1\hbar$ as in the Bohr model. The electron can be found anywhere in the atom, with its most likely position being near the radius of the Bohr orbit. In the first excited state, $n = 2$ and the energy is $-13.6/4 = -3.4$ eV, the same as in the Bohr model. The orbital quantum number ℓ can be 1 or 0. If $\ell = 1$, the magnetic quantum number can be -1, 0, or $+1$. All of these possible states have the same energy. Thus, the energy-level diagram for the hydrogen atom given by the quantum theory is the same as that given by the Bohr model (see Figure 29-17). The wavelength of the spectral lines emitted by hydrogen (and by other atoms) are related to the energy levels by the Bohr formula

$$hf = \frac{hc}{\lambda} = E_i - E_f \qquad\qquad\qquad 31\text{-}7$$

where E_i and E_f are the energies of the initial and final states.

When a spectral line of hydrogen is viewed under high resolution, it is found to consist of two closely spaced lines. (The spectral lines for other atoms may consist of more than two closely spaced lines.) These lines are called the *fine structure* of the spectral lines. To explain this fine structure and to clear up a major difficulty with the quantum-mechanical explanation of the periodic table (which will be discussed in the next section), it was suggested by W. Pauli in 1925 that the electron has an additional quantum number that can take on just two values. At first, Pauli thought that this quantum number was associated with the time dimension in a relativistic theory, but this idea did not prove fruitful. In the same year, S. Goudsmit and G. Uhlenbeck, graduate students at Leyden, suggested that the fourth quantum number was the z component of an intrinsic angular momentum of the electron called *electron spin*. In the particle picture of the electron, it is visualized as a spinning ball that orbits the nucleus (see Figure 31-4), much as the earth rotates about its axis as it revolves around the sun. The idea of electron spin has proved fruitful. The spin of the electron is described by a quantum number s which always equals $\frac{1}{2}$. The fourth quantum number needed to understand fine structure and the periodic table is labeled m_s. It can take on two values, $+\frac{1}{2}$ and $-\frac{1}{2}$, corresponding to the z components of intrinsic electron spin of $+\frac{1}{2}\hbar$ and $-\frac{1}{2}\hbar$.

One consequence of electron spin is that the electron possesses an intrinsic magnetic moment—a result to be expected since a spinning charge is equivalent to a current loop. There is also a magnetic moment associated with the orbital angular momentum of the electron, as there would be if the electron moved in a circular orbit. It is the interaction of these magnetic moments that causes the fine-structure splitting of the spectral lines of hydrogen and other atoms.

The addition of the electron-spin quantum number completes the quantum-mechanical description of the hydrogen atom. The wave function for the electron in the hydrogen atom is thus characterized by four quantum numbers: n, ℓ, m, and m_s.

Fine structure

Figure 31-4
The electron can be pictured as a spinning ball that orbits the nucleus, somewhat like the spinning earth orbits the sun.

Question

1. The Bohr theory and the quantum theory of the hydrogen atom give the same results for the energy levels of the electron. Discuss the advantages and disadvantages of each model.

31-2   The Periodic Table

In applying the quantum-mechanical theory to atoms with more than one electron, the state of each electron is described by the quantum numbers n, ℓ, m, and m_s. The energy of the electron is determined mainly by the principal quantum number n (which is related to the radial dependence of the wave function) and by the orbital angular momentum quantum number ℓ. Generally, the lower the values of n and ℓ, the lower the energy. The dependence of the energy on ℓ is due to the interaction of the electrons in the atom with each other. In hydrogen, of course, there is only one electron and the energy is independent of ℓ. The specification of n and ℓ for each electron in an atom is called the *electron configuration.* A letter rather than a numerical value is customarily used for the specification of ℓ. The code is

	s	p	d	f	g	h
ℓ value	0	1	2	3	4	5

(These code letters are remnants of spectroscopists' descriptions of various spectral lines as *sharp, principal, diffuse,* and *fundamental.* For values of ℓ greater than 3, the letters follow alphabetically; thus, g is used for $\ell = 4$, and so forth.) The n values are sometimes referred to as shells, which have another letter code: $n = 1$ is called the K shell; $n = 2$, the L shell; and so on.

An important principle that governs the electron configurations of atoms is the *Pauli exclusion principle:*

No two electrons in an atom can be in the same quantum state; that is, they cannot have the same set of values for the quantum numbers n, ℓ, m, and m_s.

Pauli exclusion principle

We can understand much of the structure of the periodic table using the exclusion principle and the restrictions on the quantum numbers mentioned in the previous section, namely, n is an integer, ℓ is an integer that ranges from 0 to $n - 1$, m can have $2\ell + 1$ values from $-\ell$ to ℓ in integral steps, and m_s can be either $+\frac{1}{2}$ or $-\frac{1}{2}$. We have already discussed the lightest element hydrogen, which has just one electron. In the ground (lowest energy) state, the electron of hydrogen has $n = 1$, $\ell = 0$, $m = 0$, and m_s equal to either $+\frac{1}{2}$ or $-\frac{1}{2}$. We call this a 1s electron. The 1 signifies $n = 1$, and the s signifies $\ell = 0$.

Helium ($Z = 2$)

The next element is helium ($Z = 2$), which has two electrons. In the ground state, these are both in the K shell, with $n = 1$, $\ell = 0$, and $m = 0$; one of the electrons has $m_s = +\frac{1}{2}$ and the other, $m_s = -\frac{1}{2}$. The electron configuration for helium is writen 1s^2. The 1 signifies $n = 1$, the s signifies $\ell = 0$, and the 2 signifies that there are two electrons in this state. Since ℓ can only be 0 for $n = 1$, these two electrons fill the K ($n = 1$) shell. The energy required to remove an electron from an atom is called the *ionization energy*. For helium, this is 24.6 eV, which is relatively large. Helium is therefore basically inert.

Lithium ($Z = 3$)

The next element, lithium (Li), has three electrons. Since the K shell is completely filled with two electrons, the third electron must go into the $n = 2$ or L shell. This outer electron is much farther from the nucleus than are the two inner, $n = 1$ electrons. It is most likely to be found at the radius of the second Bohr orbit, which is four times the radius of the first Bohr orbit. (The radii of the Bohr orbits, as given by Equation 29-20, are proportional to n^2.) The outer electron is thus partly screened from the nuclear charge by the two inner electrons. Recall that the electric field outside a spherically symmetric charge density is the same as if all the charge were at the center of the sphere. If the outer electron were completely outside of the charge cloud of the two inner electrons, the electric field it would see would be that of a single charge $+e$ at the origin due to the nuclear charge of $+3e$ and the inner electron cloud charge of $-2e$. However, this outer electron does not have a well defined orbit; instead, it is itself a charge cloud that penetrates the charge cloud of the inner electrons to some extent. Because of this penetration, the effective nuclear charge, $Z'e$, is somewhat greater than $+1e$. The energy of the outer electron at a distance r from a point charge $+Z'e$ is

$$E = -\frac{1}{2}\frac{kZ'e^2}{r} \qquad \text{31-8}$$

(This is Equation 29-17 with the nuclear charge $+Ze$ replaced by $+Z'e$). Thus, the greater the penetration of the inner electron cloud, the larger the effective nuclear charge $Z'e$ and the lower the energy. Because the penetration is greater for lower ℓ values, the energy of the outer electron of lithium is lower for the s state ($\ell = 0$) than for the p state ($\ell = 1$). The configuration of lithium is therefore $1s^2 2s$. The ionization energy of lithium is only 5.39 eV. Because this outer electron is so loosely bound to the atom, lithium is very active chemically. It behaves like a "one electron atom," and its spectrum is similar to that of hydrogen.

Beryllium ($Z = 4$)

The fourth electron has the least energy in the $2s$ state. There can be two electrons with $n = 2$, $\ell = 0$, and $m = 0$ because of the two possible values for the spin quantum number m_s. The configuration of beryllium is therefore $1s^2 2s^2$.

Boron to Neon ($Z = 5$ to $Z = 10$)

Since the $2s$ subshell is filled, the fifth electron must go into the $2p$ subshell, that is, $n = 2$ and $\ell = 1$. Since there are three possible values of m ($+1$, 0, and -1) and two values of m_s for each, there can be six electrons in this subshell. The electron configuration for boron is $1s^2 2s^2 2p$. The electron configurations for the elements carbon ($Z = 6$) to neon ($Z = 10$) differ from that for boron only in the number of electrons in the $2p$ subshell. The ionization energy increases slightly with Z for these elements, reaching the value of 21.6 eV for the last element in the group, neon. Neon has the maximum number of electrons allowed in the $n = 2$ shell. Its electron configuration is $1s^2 2s^2 2p^6$. Because of its very high ionization energy, neon, like helium, is basically chemically inert. The element just before neon, flourine, has a "hole" in this shell; that is, it has room for one more electron. It readily combines with elements such as lithium that have one outer electron. Lithium, for instance, will donate its single outer electron to the fluorine atom to make an F$^-$ ion and a Li$^+$ ion. These ions then bond together to form lithium flouride.

Sodium to Argon (Z = 11 to Z = 18)

The eleventh electron must go into the $n = 3$ shell. Since this electron is very far from the nucleus and from the inner electrons, it is weakly bound in the sodium atom. The ionization energy of sodium is only 5.14 eV. Sodium therefore combines readily with atoms such as flourine. With $n = 3$, the value of ℓ can be 0, 1, or 2. Because of the lowering of the energy due to the penetration of the electron shield formed by the other 10 electrons (similar to that discussed for lithium), the 3s state is lower than the 3p or 3d states. This energy difference in the subshells of the same n value becomes greater as the number of electrons increases. The configuration of sodium is $1s^2 2s^2 2p^6 3s^1$. As we move to elements of higher Z, the 3s subshell and then the 3p subshell begin to fill. These two subshells can accommodate $2 + 6 = 8$ electrons. The configuration of argon ($Z = 18$) is $1s^2 2s^2 2p^6 3s^2 3p^6$. One might expect the nineteenth electron to go into the third subshell (the d shell with $\ell = 2$), but the penetration effect is now so strong that the energy is lower in the 4s shell than in the 3d shell. There is thus another large energy difference between the eighteenth and nineteenth electrons, and so argon, with 18 electrons and thus a full 3p subshell, is basically stable and inert.

Atoms with Z > 18

The nineteenth electron in potassium ($Z = 19$) and the twentieth electron in calcium ($Z = 20$) go into the 4s rather than the 3d subshell. The electron configurations of the next 10 elements, scandium ($Z = 21$) through zinc ($Z = 30$), differ only in the number of electrons in the 3d subshell, except for chromium ($Z = 24$) and copper ($Z = 29$), each of which has only one 4s electron. These 10 elements are called *transition elements*. Because their chemical properties are mainly due to their 4s electrons, they are quite similar chemically.

Figure 31-5 shows a plot of the ionization energy versus Z for $Z = 1$ to $Z = 60$. The peaks in ionization energy at $Z = 2, 10, 18, 36$, and 54 mark the closing of a shell or subshell. Table 31-2 gives the electron configurations of all the elements.

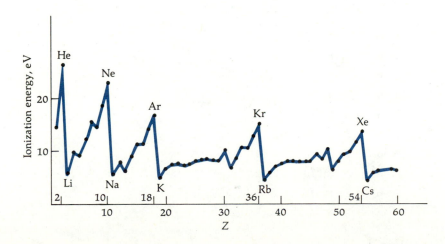

Figure 31-5
Ionization energy versus Z for $Z = 1$ to $Z = 60$. This energy is the binding energy of the last electron in the atom. The binding energy increases with Z until a shell is closed at $Z = 2, 10, 18, 36$, and 54. Atoms, such as sodium ($Z = 11$), with a closed shell plus one outer electron have very low binding energies because the outer electron is very far from the nucleus and is shielded by the inner electrons.

Table 31-2

Electron Configurations* of the Atoms in Their Ground States.

Z	Element		K 1 s	L 2 s	p	M 3 s	p	d	N 4 s	p	d	f	O 5 s	p	d	f	P 6 s	p	d	Q 7 s
1	H	hydrogen	1																	
2	He	helium	2																	
3	Li	lithium	2	1																
4	Be	beryllium	2	2																
5	B	boron	2	2	1															
6	C	carbon	2	2	2															
7	N	nitrogen	2	2	3															
8	O	oxygen	2	2	4															
9	F	fluorine	2	2	5															
10	Ne	neon	2	2	6															
11	Na	sodium	2	2	6	1														
12	Mg	magnesium	2	2	6	2														
13	Al	aluminum	2	2	6	2	1													
14	Si	silicon	2	2	6	2	2													
15	P	phosphorus	2	2	6	2	3													
16	S	sulfur	2	2	6	2	4													
17	Cl	chlorine	2	2	6	2	5													
18	Ar	argon	2	2	6	2	6													
19	K	potassium	2	2	6	2	6	.	1											
20	Ca	calcium	2	2	6	2	6	.	2											
21	Sc	scandium	2	2	6	2	6	1	2											
22	Ti	titanium	2	2	6	2	6	2	2											
23	V	vanadium	2	2	6	2	6	3	2											
24	Cr	chromium	2	2	6	2	6	5	1											
25	Mn	manganese	2	2	6	2	6	5	2											
26	Fe	iron	2	2	6	2	6	6	2											
27	Co	cobalt	2	2	6	2	6	7	2											
28	Ni	nickel	2	2	6	2	6	8	2											
29	Cu	copper	2	2	6	2	6	10	1											
30	Zn	zinc	2	2	6	2	6	10	2											
31	Ga	gallium	2	2	6	2	6	10	2	1										
32	Ge	germanium	2	2	6	2	6	10	2	2										
33	As	arsenic	2	2	6	2	6	10	2	3										

Table 31-2 *(continued)*

Z	Element		K 1 s	L 2 s	p	M 3 s	p	d	N 4 s	p	d	f	O 5 s	p	d	f	P 6 s	p	d	Q 7 s
34	Se	selenium	2	2	6	2	6	10	2	4										
35	Br	bromine	2	2	6	2	6	10	2	5										
36	Kr	krypton	2	2	6	2	6	10	2	6										
37	Rb	rubidium	2	2	6	2	6	10	2	6	.	.	1							
38	Sr	strontium	2	2	6	2	6	10	2	6	.	.	2							
39	Y	yttrium	2	2	6	2	6	10	2	6	1	.	2							
40	Zr	zirconium	2	2	6	2	6	10	2	6	2	.	2							
41	Nb	niobium	2	2	6	2	6	10	2	6	4	.	1							
42	Mo	molybdenum	2	2	6	2	6	10	2	6	5	.	1							
43	Tc	technetium	2	2	6	2	6	10	2	6	6	.	1							
44	Ru	ruthenium	2	2	6	2	6	10	2	6	7	.	1							
45	Rh	rhodium	2	2	6	2	6	10	2	6	8	.	1							
46	Pd	palladium	2	2	6	2	6	10	2	6	10	.	.							
47	Ag	silver	2	2	6	2	6	10	2	6	10	.	1							
48	Cd	cadmium	2	2	6	2	6	10	2	6	10	.	2							
49	In	indium	2	2	6	2	6	10	2	6	10	.	2	1						
50	Sn	tin	2	2	6	2	6	10	2	6	10	.	2	2						
51	Sb	antimony	2	2	6	2	6	10	2	6	10	.	2	3						
52	Te	tellurium	2	2	6	2	6	10	2	6	10	.	2	4						
53	I	iodine	2	2	6	2	6	10	2	6	10	.	2	5						
54	Xe	xenon	2	2	6	2	6	10	2	6	10	.	2	6						
55	Cs	cesium	2	2	6	2	6	10	2	6	10	.	2	6	.	.	1		
56	Ba	barium	2	2	6	2	6	10	2	6	10	.	2	6	.	.	2		
57	La	lanthanum	2	2	6	2	6	10	2	6	10	.	2	6	1	.	2		
58	Ce	cerium	2	2	6	2	6	10	2	6	10	1	2	6	1	.	2		
59	Pr	praseodymium	2	2	6	2	6	10	2	6	10	3	2	6	.	.	2		
60	Nd	neodymium	2	2	6	2	6	10	2	6	10	4	2	6	.	.	2		
61	Pm	promethium	2	2	6	2	6	10	2	6	10	5	2	6	.	.	2		
62	Sm	samarium	2	2	6	2	6	10	2	6	10	6	2	6	.	.	2		
63	Eu	europium	2	2	6	2	6	10	2	6	10	7	2	6	.	.	2		
64	Gd	gadolinium	2	2	6	2	6	10	2	6	10	7	2	6	1	.	2		
65	Tb	terbium	2	2	6	2	6	10	2	6	10	9	2	6	.	.	2		
66	Dy	dysprosium	2	2	6	2	6	10	2	6	10	10	2	6	.	.	2		
67	Ho	holmium	2	2	6	2	6	10	2	6	10	11	2	6	.	.	2		
68	Er	erbium	2	2	6	2	6	10	2	6	10	12	2	6	.	.	2		

Table 31-2 (continued)

			K	L		M			N				O				P			Q
		n:	1	2		3			4				5				6			7
Z	Element	l:	s	s	p	s	p	d	s	p	d	f	s	p	d	f	s	p	d	s
69	Tm	thulium	2	2	6	2	6	10	2	6	10	13	2	6	.	.	2			
70	Yb	ytterbium	2	2	6	2	6	10	2	6	10	14	2	6	.	.	2			
71	Lu	lutetium	2	2	6	2	6	10	2	6	10	14	2	6	1	.	2			
72	Hf	hafnium	2	2	6	2	6	10	2	6	10	14	2	6	2	.	2			
73	Ta	tantalum	2	2	6	2	6	10	2	6	10	14	2	6	3	.	2			
74	W	tungsten (wolfram)	2	2	6	2	6	10	2	6	10	14	2	6	4	.	2			
75	Re	rhenium	2	2	6	2	6	10	2	6	10	14	2	6	5	.	2			
76	Os	osmium	2	2	6	2	6	10	2	6	10	14	2	6	6	.	2			
77	Ir	iridium	2	2	6	2	6	10	2	6	10	14	2	6	7	.	2			
78	Pt	platinum	2	2	6	2	6	10	2	6	10	14	2	6	9	.	1			
79	Au	gold	2	2	6	2	6	10	2	6	10	14	2	6	10	.	1			
80	Hg	mercury	2	2	6	2	6	10	2	6	10	14	2	6	10	.	2			
81	Tl	thallium	2	2	6	2	6	10	2	6	10	14	2	6	10	.	2	1		
82	Pb	lead	2	2	6	2	6	10	2	6	10	14	2	6	10	.	2	2		
83	Bi	bismuth	2	2	6	2	6	10	2	6	10	14	2	6	10	.	2	3		
84	Po	polonium	2	2	6	2	6	10	2	6	10	14	2	6	10	.	2	4		
85	At	astatine	2	2	6	2	6	10	2	6	10	14	2	6	10	.	2	5		
86	Rn	radon	2	2	6	2	6	10	2	6	10	14	2	6	10	.	2	6		
87	Fr	francium	2	2	6	2	6	10	2	6	10	14	2	6	10	.	2	6	.	1
88	Ra	radium	2	2	6	2	6	10	2	6	10	14	2	6	10	.	2	6	.	2
89	Ac	actinium	2	2	6	2	6	10	2	6	10	14	2	6	10	.	2	6	1	2
90	Th	thorium	2	2	6	2	6	10	2	6	10	14	2	6	10	.	2	6	2	2
91	Pa	protactinium	2	2	6	2	6	10	2	6	10	14	2	6	10	1	2	6	2	2
92	U	uranium	2	2	6	2	6	10	2	6	10	14	2	6	10	3	2	6	1	2
93	Np	neptunium	2	2	6	2	6	10	2	6	10	14	2	6	10	4	2	6	1	2
94	Pu	plutonium	2	2	6	2	6	10	2	6	10	14	2	6	10	6	2	6	.	2
95	Am	americium	2	2	6	2	6	10	2	6	10	14	2	6	10	7	2	6	.	2
96	Cm	curium	2	2	6	2	6	10	2	6	10	14	2	6	10	7	2	6	1	2
97	Bk	berkelium	2	2	6	2	6	10	2	6	10	14	2	6	10	8	2	6	1	2
98	Cf	californium	2	2	6	2	6	10	2	6	10	14	2	6	10	10	2	6	.	2
99	Es	einsteinium	2	2	6	2	6	10	2	6	10	14	2	6	10	11	2	6	.	2
100	Fm	fermium	2	2	6	2	6	10	2	6	10	14	2	6	10	12	2	6	.	2
101	Md	mendelevium	2	2	6	2	6	10	2	6	10	14	2	6	10	13	2	6	.	2
102	No	nobelium	2	2	6	2	6	10	2	6	10	14	2	6	10	14	2	6	.	2
103	Lw	lawrencium	2	2	6	2	6	10	2	6	10	14	2	6	10	14	2	6	1	2

* For some of the rare-earth elements (Z = 57 to 71) and the heavy elements (Z > 89), the configurations are not firmly established.

Questions

2. Why is the energy of the 3s state considerably lower than that of the 3p state for sodium, whereas in hydrogen, these states have essentialy the same energy?

3. Discuss the evidence from the periodic table of the need for a fourth quantum number. How would the properties of helium differ if there were only three quantum numbers, n, ℓ, and m?

31-3   Atomic Spectra

When an atom is in an excited state (that is, when one or more of its electrons is in an energy state above the ground state), the electrons make transitions to lower energy states and, in doing so, emit electromagnetic radiation. The frequency of the electromagnetic radiation emitted is related to the initial and final energy states of the electron by the Bohr formula (Equation 29-15), $f = (E_i - E_f)/h$, where E_i and E_f are the initial and final energies and h is Planck's constant. The wavelength of the radiation is, of course, related to the frequency by $\lambda = c/f$. An atom can be excited to a higher energy state by bombarding it with a beam of electrons in a spectral tube with a high voltage across it. Since the excited energy states of an atom are discrete rather than continuous, only certain wavelengths are emitted. The wavelengths of the emitted radiation determine the spectral lines that constitute the emission spectrum of the atom.

To understand atomic spectra, we need to understand the excited states of an atom. The situation for an atom with many electrons is, in general, much more complicated than that of hydrogen with its one electron. An excited state of an atom may involve a change in the state of any one of its electrons or even two or more of its electrons. Fortunately, in most cases, an excited state of an atom involves the excitation of just one of its electrons. The energies of excitation of the outer, valence electrons of an atom are of the order of a few electron volts. Transitions involving these electrons result in photons in or near the visible or *optical spectrum.* (Recall that the energies of visible photons range from about 1.5 to 3 eV). Excitation energies can often be calculated using a simple model in which the atom is pictured as a single electron plus a stable core consisting of the nucleus and the other, inner electrons (see Figure 31-6). This model works particularly well for the alkali metals: lithium, sodium, potassium, rubidium, and cesium. These elements are in the first column of the periodic table. The optical spectra of these elements are similar to that of hydrogen. The optical spectra for atoms that have two outer electrons, such as helium, beryllium, and magnesium, are considerably more complex because of the interaction of the two outer electrons.

The energy needed to excite an inner core electron, for example, an electron in the $n = 1$ state (K shell), is much greater than that needed for an outer, valence electron. Such an electron cannot be excited to any of the filled states (for example, the $n = 2$ states in sodium) because of the Pauli exclusion principle. The energy required to excite an inner core electron to an unoccupied state is typically of the order of several keV. An inner core electron can be excited by the bombardment of the atom by a high-energy

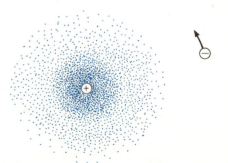

Figure 31-6
An atom can be thought of as a stable core consisting of a positive nucleus and a negative cloud of inner core electrons plus one or more outer electrons far away from the core. Optical spectra result from the excitation of only the outer electrons.

electron beam in, for example, an x-ray tube. If an electron is knocked out of the $n = 1$ (K) shell, there is a vacancy left in this shell. This vacancy can be filled by an electron from the L shell or a higher shell that makes a transition to the K shell. The photons emitted by electrons making such transitions have energies of the order of 1 keV and constitute the *characteristic x-ray spectrum* shown in Figure 29-11. The K_α line arises from transitions from the $n = 2$ (L) shell to the $n = 1$ (K) shell. The K_β line arises from transitions from the $n = 3$ (M) shell to the $n = 1$ shell. These and other lines arising from transitions that end at the $n = 1$ shell make up the K series of the characteristic x-ray spectrum of a target element. Similarly, a second series, the L series, is produced by transitions from higher energy states to a vacated place in the $n = 2$ (L) shell.

Characteristic x-ray spectrum

We can use the Bohr theory to calculate the approximate frequencies of the characteristic x-ray spectrum. According to the Bohr model, the energy of a single electron in a state n is given by (Equations 29-24 and 29-25)

$$E_n = -Z^2 \frac{13.6 \text{ eV}}{n^2}$$

Since for any atom other than hydrogen there are two electrons in the K shell, the effective charge seen by one of the electrons is less than Ze because of the shielding of the nucleus by the other electron. Assuming the effective charge is $(Z - 1)e$, the energy of an electron in the K shell is given by this equation when $n = 1$ and Z is replaced by $(Z - 1)$.

$$E_1 = -(Z - 1)^2(13.6 \text{ eV})$$

The energy of an electron in the state n (assuming the same effective charge) is given by

$$E_n = -(Z - 1)^2 \frac{13.6 \text{ eV}}{n^2}$$

When an electron from state n (when is greater than 1) drops into the vacated state in the $n = 1$ shell, a photon of energy $E_n - E_1$ is emitted. The wavelength of this photon is

$$\lambda = \frac{hc}{E_n - E_1} = \frac{hc}{(Z - 1)^2(13.6 \text{ eV})(1 - 1/n^2)} \qquad 31\text{-}9$$

In 1913, the English physicist H. Moseley measured the wavelengths of the characteristic x-ray spectra for about 40 elements. From his data, he was able to determine the atomic number Z for each element.

Example 31-2   Calculate the wavelength of the K_α x-ray line for molybdenum ($Z = 42$) and compare it with the value $\lambda = 0.0721$ nm measured by Moseley.

The K_α-line corresponds to a transition from $n = 2$ to $n = 1$. The wavelength is given by Equation 31-9, with $Z = 42$ and $n = 2$:

$$\lambda = \frac{hc}{(41)^2(13.6 \text{ eV})(1 - \frac{1}{4})}$$

$$= \frac{1240 \text{ eV} \cdot \text{nm}}{(41)^2(13.6 \text{ eV})(3/4)} = 0.0723 \text{ nm}$$

This result is in good agreement with the measured value.

Questions

4. Would you expect the optical spectrum of potassium to be like that of hydrogen or that of helium?

5. Would you expect the optical spectrum of beryllium to be like that of hydrogen or that of helium?

31-4 Molecular Bonding

We rarely see single atoms in nature. Instead, atoms normally bond together to form molecules or solids. Molecules may exist as separate entities, as in gaseous oxygen (O_2) or nitrogen (N_2), or they may bond together to form liquids or solids. Two principal types of bonds are the ionic bond, found in sodium chloride and most other salts, and the covalent bond, which is, for instance, responsible for the bonding of two oxygen atoms to form O_2 or two nitrogen atoms to form N_2, molecules found in air. Other types of bonds that are important in the bonding of liquids and solids are van der Waals bonds, hydrogen bonds, and metallic bonds. In many cases, bonding is a mixture of these types.

The Ionic Bond

The simplest type of bond is the *ionic bond,* which is found in most salts. Consider sodium chloride (NaCl) as an example. The sodium atom has one $3s$ electron outside a stable core. It takes just 5.1 eV to remove this electron from sodium (see Figure 31-5). The ionization energy for other alkali metals is also low. The removal of one electron from Na leaves a positive ion with a spherically symmetric, closed-shell core. Chlorine, on the other hand, is only one electron short of having a closed core. The energy released by the acquisition of one electron is called the *electron affinity,* which in the case of Cl is 3.8 eV. The acquisition of one electron by chlorine leaves a negative chlorine ion with a spherically symmetric, closed-shell core. Thus, the formation of a Na^+ ion and a Cl^- ion by the donation of one electron of Na to Cl requires only $5.1 - 3.8 = 1.3$ eV. The electrostatic potential energy of the two ions a distance r apart is $-ke^2/r$. When the separation of the ions is less than about 1.1 nm, the negative potential energy of attraction is of greater magnitude than the 1.3 eV of energy needed to create the ions. The result is a net attraction of the ions to form NaCl. A solid crystal of sodium chloride has an alternating arrangement of sodium and chlorine ions.

Since the electrostatic attraction increases as the ions get closer, it would seem that equilibrium could not exist. However, when the separation between ions is very small, there is a strong repulsion that is quantum mechanical in nature and is related to the exclusion principle. This "exclusion-principle repulsion" is responsible for the repulsion of the atoms in all molecules (except H_2) no matter what the bonding mechanism is. (In H_2, the repulsion is due simply to the two positively charged protons.) We can understand this exclusion-principle repulsion qualitatively as follows.

When the ions are very far apart, the wave function for a core electron of one of the ions does not overlap that of any electron in the other ion. We can distinguish the electrons by the ion to which they belong. However, when the ions are close, the core-electron wave functions begin to overlap, and some of the electrons must go into higher-energy quantum states because of the exclusion principle. This is not a sudden process. The energy states of the electrons are gradually changed as the ions are brought together. A sketch of the potential energy of the Na^+ and Cl^- ions versus separation is shown in Figure 31-7. The energy is lowest at an equilibrium separation of about 0.24 nm. At smaller separations, the energy rises steeply as a result of the exclusion principle. The energy required to separate the ions and form Na and Cl atoms is called the *dissociation energy*, which is about 4.23 eV.

Dissociation energy

The separation distance of 0.24 nm is for gaseous diatomic NaCl, which can be obtained by the evaporation of solid NaCl. Normally, NaCl exists in a cubic crystal structure, with Na^+ and Cl^- at alternate corners of a cube. The separation of the ions in a crystal is somewhat larger, about 0.28 nm. Because of the presence of neighboring ions of opposite charge, the Coulomb energy per ion pair is lower when the ions are in a crystal.

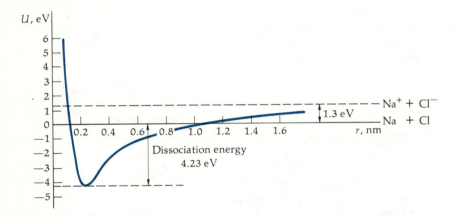

Figure 31-7
Potential energy for Na^+ and Cl^- ions as a function of separation distance. The energy at infinite separation was chosen to be 1.3 eV, corresponding to the energy needed to form the ions from neutral atoms. The minimum energy in this curve is at the equilibrium separation for the ions in the molecule.

The Covalent Bond

A completely different mechanism, the *covalent bond*, is responsible for the bonding of such molecules as H_2, N_2, and CO. If, for example, we calculate the energy needed to form the ions H^+ and H^- by the transfer of an electron from one atom to the other and then add this energy to the electrostatic energy, we find that there is no separation distance for which the total energy is negative. Thus, an ionic bond cannot be formed. Instead, the attraction of two hydrogen atoms is entirely a quantum-mechanical effect.

The decrease in energy when two hydrogen atoms approach each other is due to the sharing of the two electrons by both atoms and is intimately connected with the symmetry properties of the electron wave functions. A symmetry requirement related to the Pauli exclusion principle states that if the spins of the two electrons in H_2 are parallel, the wave function ψ must be antisymmetric in space, whereas if the spins are antiparallel ($m_s = +\frac{1}{2}$ for one electron and $-\frac{1}{2}$ for the other), the wave function ψ must be symmetric in space. A symmetric wave function ψ_S and an antisymmetric wave function ψ_A are shown in Figure 31-8*a* and *b* for two different separations of two hydrogen atoms. The squares of the wave functions in Figure 31-8*b* are shown in Figure 31-8*c*. The important feature of these plots is that the probability distribution ψ^2 in the region between the protons is large for the

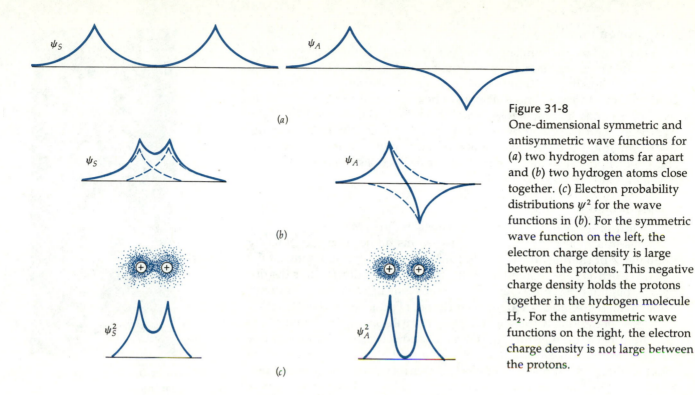

Figure 31-8
One-dimensional symmetric and antisymmetric wave functions for (a) two hydrogen atoms far apart and (b) two hydrogen atoms close together. (c) Electron probability distributions ψ^2 for the wave functions in (b). For the symmetric wave function on the left, the electron charge density is large between the protons. This negative charge density holds the protons together in the hydrogen molecule H_2. For the antisymmetric wave functions on the right, the electron charge density is not large between the protons.

symmetric case and small for the antisymmetric case. Thus, when the spins of the two electrons are antiparallel, the electrons are often found in the region between the protons and the protons are bound together by the negatively charged electrons between them. The negatively charged electron cloud representing these electrons is concentrated in the space between the protons, as shown in the left part of Figure 31-8c. Conversely, when the electron spins are parallel, the electrons spend little time between the protons, so the atoms do not bind together to form a molecule. In this case, the electron cloud is not concentrated in the space between the protons, as shown in the right part of Figure 31-8c.

The van der Waals Bond

Any two separated molecules will be attracted to one another by complex electrostatic forces called *van der Waals forces*. So will atoms that do not form ionic or covalent bonds. The van der Waals forces arise from the interaction of the instantaneous electric dipole moments of the molecules or atoms. Even nonpolar molecules, which have zero electric dipole moments on the average, have instantaneous dipole moments that interact with the instantaneous dipole moments of nearby molecules in such a way as to produce attraction.

The bonds formed because of the van der Waals forces are much weaker than those already discussed. At high enough temperatures, these forces are not strong enough to overcome the ordinary thermal agitation of atoms or molecules, but at sufficiently low temperatures thermal agitation becomes negligible and because of van der Waals forces all substances will condense into a liquid.

The Hydrogen Bond

Another bonding mechanism of great importance is the *hydrogen bond,* which often holds different groups of molecules together and is responsible for the cross-linking that allows giant biological molecules and polymers to hold their fixed shapes. The well-known helical structure of DNA is due to hydrogen bonds linking across turns of the helix (see Figure 31-9). The hydrogen bond is formed by the sharing of a proton (the nucleus of the hydrogen atom) between two atoms, frequently two oxygen atoms. This sharing of a proton is similar to the sharing of electrons responsible for the covalent bond already discussed.

The Metallic Bond

The nature of the bonding of atoms in a metal is different from the bonding of atoms in a molecule. In a metal, two atoms do not bond together by exchanging or sharing an electron to form a molecule. Instead, some of the valence electrons are shared by many atoms. The bonding is thus distributed throughout the entire metal rather than being between just two atoms. A metal can therefore be thought of as a lattice of positive ions held together by a "gas" of essentially free electrons that roam throughout the solid. In the quantum-mechanical picture, these free electrons form a cloud of negative charge density between the positively charged lattice ions that holds the ions together. The number of free electrons varies from metal to metal but is of the order of one per atom. Since there are about 10^{28} atoms per cubic metre in a solid, the number of electrons free to move about in a metal is of this same order.

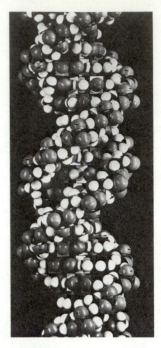

Figure 31-9
The DNA molecule.

31-5 Molecular Spectra

As is the case with atoms, electromagnetic radiation is often emitted when a molecule makes a transition from an excited state to a state of lower energy. Conversely, an atom or molecule can absorb radiation and make a transition from a lower state to a higher state. The study of molecular emission and absorption spectra thus gives information about the excited energy states of molecules.

The energy of a molecule can be conveniently separated into three parts: energy due to the excitation of its electrons, energy due to the vibration of the molecule, and energy due to the rotation of the molecule. Fortunately, the magnitudes of these energies are sufficiently different that they can be treated separately. The energies of electronic excitations of a molecule are of the order of magnitude of 1 eV, the same as for the excitation of atoms. The energies of vibration and rotation are much smaller than this. In this section, we will consider only diatomic molecules for simplicity. We will begin by considering rotation.

Figure 31-10 shows a simple schematic model of a diatomic molecule consisting of mass m_1 and m_2 separated by a distance r and rotating about its center of mass. Classically, the kinetic energy of rotation (see Section 8-3) is

$$E = \tfrac{1}{2}I\omega^2 \qquad\qquad 31\text{-}10$$

where I is the moment of inertia and ω is the angular frequency of rotation.

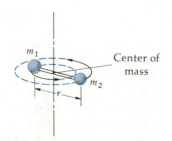

Figure 31-10
Diatomic molecule rotating about an axis through its center of mass.

If we write this equation in terms of the angular momentum $L = I\omega$, we have

$$E = \frac{(I\omega)^2}{2I} = \frac{L^2}{2I}$$ 31-11

The solution of the Schrödinger equation for rotation leads to quantization of the angular momentum with values given by

$$L^2 = \ell(\ell + 1)\hbar^2$$ 31-12

where ℓ can be $0, 1, 2, \ldots$. This is the same quantum condition on angular momentum that holds for the hydrogen atom and for all other atoms (Equation 31-4). Note, however, that in Equation 31-4, L refers to the angular momentum of the electron moving about the nucleus, whereas here L refers to the angular momentum of the entire molecule rotating about its center of mass. The energy levels of a rotating molecule are therefore given by

$$E = \frac{\ell(\ell + 1)\hbar^2}{2I} = \ell(\ell + 1)E_{0r}$$ 31-13

where

$$E_{0r} = \frac{\hbar^2}{2I}$$ 31-14

The energy E_{0r} is characteristic of a particular molecule and can be calculated from its moment of inertia. Conversely, a measurement of the rotational energy of a molecule from its rotational spectrum can be used to determine the moment of inertia of the molecule, and from that, the separation of the atoms in the molecule.

Example 31-3   Estimate the rotational energies of an O_2 molecule, assuming that the mass of the oxygen atom is about 16 times that of a proton and the separation of the atoms is 10^{-10} m.

If r is the separation of the atoms in the oxygen molecule, the center of mass is a distance $r/2$ from each atom and the moment of inertia about the center of mass is

$$I = m(r/2)^2 + m(r/2)^2 = \tfrac{1}{2}mr^2$$

where m is the mass of the oxygen atom. Substituting $mr^2/2$ for the moment of inertia in Equation 31-14, we obtain

$$E_{0r} = \frac{\hbar^2}{mr^2} = \frac{(6.6 \times 10^{-34} \text{ J} \cdot \text{s})^2}{4\pi^2(16)(1.67 \times 10^{-27} \text{ kg})(10^{-10} \text{ m})^2} \approx 4 \times 10^{-23} \text{ J}$$

Converting this to electron volts, we obtain

$$E_{0r} \approx (4 \times 10^{-23} \text{ J}) \frac{1 \text{ eV}}{1.6 \times 10^{-19} \text{ J}} = 2.5 \times 10^{-4} \text{ eV}$$

We see from this example that the rotational energy levels are several orders of magnitude smaller than those due to electronic excitation. Transitions between pure rotational energy levels yield photons in the far infrared region of the electromagnetic spectrum.

The vibrational energies are about 10 to 100 times those of rotation, but they are still much smaller than the energies associated with electronic transitions. The problem of the quantization of energy in a simple harmonic oscillator was one of the first problems solved by Schrödinger in his paper proposing his wave equation. Solution of the Schrödinger equation for a simple harmonic oscillator gives

$$E = (n + \tfrac{1}{2})hf \qquad n = 0, 1, 2, 3, \ldots .^*$$ 31-15

An interesting feature of this result is that the energy levels are equally spaced at intervals of hf, where f is the frequency of the oscillator. A *selection rule* for transitions between vibrational states requires that n can only change by ± 1, so the energy of a photon emitted by such a transition is hf and the frequency is f, the same as the frequency of vibration. (There is a similar selection rule that ℓ must change by ± 1 for transitions between rotational states.) A typical measured frequency for a transition between vibrational states is of the order of 10^{12} Hz, which gives a vibrational energy of

$$E = hf = (4.14 \times 10^{-15} \text{ eV} \cdot \text{s})(10^{12} \text{ s}^{-1}) = 4.14 \times 10^{-3} \text{ eV}$$

Vibrational energies are typically of this order of magnitude.

Figure 31-11 is a schematic sketch of some electronic, vibrational, and rotational energy levels of a molecule. The levels are labeled by the quantum numbers n for vibration and ℓ for rotation. The lower vibrational levels are evenly spaced, with $\Delta E = hf$. For higher vibrational levels, the approximation that the vibration is simple harmonic is not valid and the levels are not quite evenly spaced. The spacing of the rotational levels increases with increasing ℓ. Since the energies of vibration or rotation are so much smaller than those of the excitation of an atomic electron, molecular vibration and rotation show up in optical spectra as a fine splitting of the lines. When this

---

* Note that the n used here is not the same n as the principal quantum number for electronic energy levels in atoms, such as in Equation 31-6.

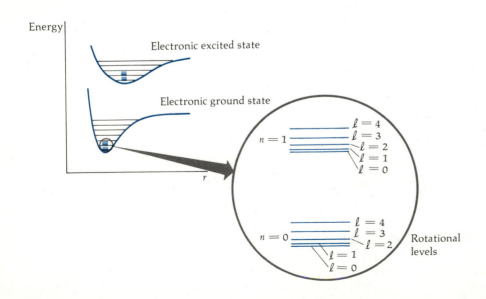

Figure 31-11
Electronic, vibrational, and rotational energy levels of a diatomic molecule. The rotational levels are shown in an enlargement of the $n = 0$ and $n = 1$ vibrational levels.

structure is not resolved, the spectrum appears as bands, as shown in Figure 31-12*a*. The components of the bands are due to transitions between the vibrational levels of two electronic states, as indicated in the diagram. Close inspection of these bands, as shown in the enlargement of this figure (Figure 31-12*b*), reveals that the bands have a fine structure due to the rotational energy levels.

   Much of molecular spectroscopy is done by infrared absorption techniques in which only the vibrational and rotational energy levels are excited. Figure 31-13 shows the absorption spectrum of hydrogen chloride (HCl). The moment of inertia of HCl can be calculated from the spacing of the peaks in the figure. The frequency at the center of the large gap between the peaks is the frequency of vibration of the molecule. The double peak structure results from the fact that chlorine occurs naturally in two isotopes, ^{35}Cl and ^{37}Cl.

Figure 31-12
Part of the emission spectrum of N_2. (*a*) The components of the bands are due to transitions between the vibrational levels of two electronic states, as indicated in the energy-level diagram below the photograph. (*b*) An enlargement of part of (*a*) shows that the apparent lines in (*a*) are in fact bands with structure caused by the rotational energy levels.

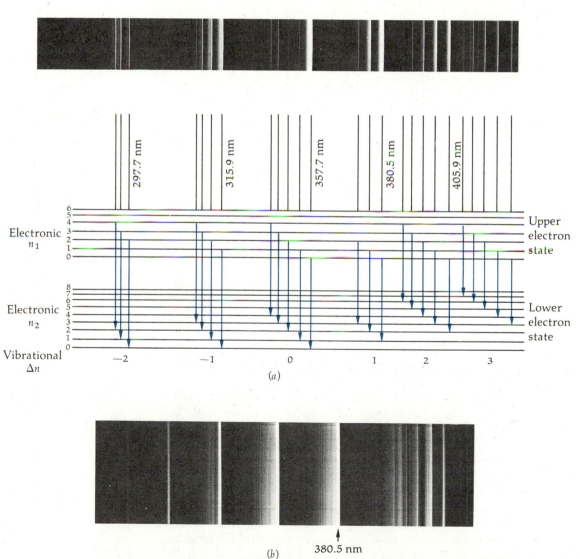

(*a*)

(*b*)   380.5 nm

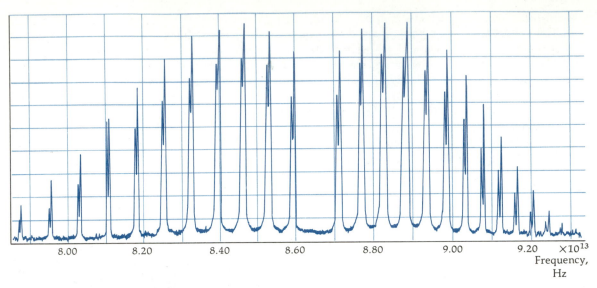

8.00   8.20   8.40   8.60   8.80   9.00   9.20   ×10¹³

Frequency, Hz

Figure 31-13
Absorption spectrum of the diatomic molecule HCl. The equally spaced peaks are due to transitions between rotational levels from ℓ to $\ell + 1$ or to $\ell - 1$. The frequency at the center of the large gap is the vibrational frequency. The double-peak structure results from the two isotopes of chlorine, ^{35}Cl (abundance, 75.5 percent) and ^{37}Cl (abundance, 24.5 percent).

31-6   Absorption, Scattering, and Stimulated Emission

Information about the energy levels of an atom or molecule is usually obtained from the radiation emitted when the atom or molecule makes a transition from an excited state to a state of lower energy. As mentioned previously, we can also obtain information about such energy levels from the absorption spectrum. When atoms or molecules are irradiated with a continuous spectrum of radiation, the transmitted radiation shows dark lines corresponding to the absorption of light at discrete wavelengths. Absorption spectra of atoms were the first line spectra observed. In 1817, Fraunhofer labeled the most prominent absorption lines in the spectrum of sunlight. It is for this reason that the two intense yellow lines in the spectrum of sodium are called the *Fraunhofer D lines.* Since at normal temperatures atoms and molecules are in their ground states or in low-lying excited states, absorption spectra are usually simpler than emission spectra. For example, only those lines corresponding to the Lyman emission series are seen in the absorption spectrum of atomic hydrogen because nearly all the atoms are normally in their ground states.

In addition to absorption, several other interesting phenomena occur when a photon is incident on an atom or molecule. These are illustrated in Figure 31-14. In Figure 31-14*a*, the photon is absorbed and the system makes a transition to an excited state. Later, the system makes a transition to a lower state or back to the ground state with the emission of a photon. This two-step process is called *resonance absorption.* The emitted photon is not correlated with the incident photon.

Figure 31-14*b* illustrates *elastic* or *Rayleigh scattering,* which is a one-step process in which the incident and emitted or scattered photon are correlated. The incident and scattered photons are also correlated in the *inelastic scattering* process shown in Figure 31-14*c*. Such scattering of light from

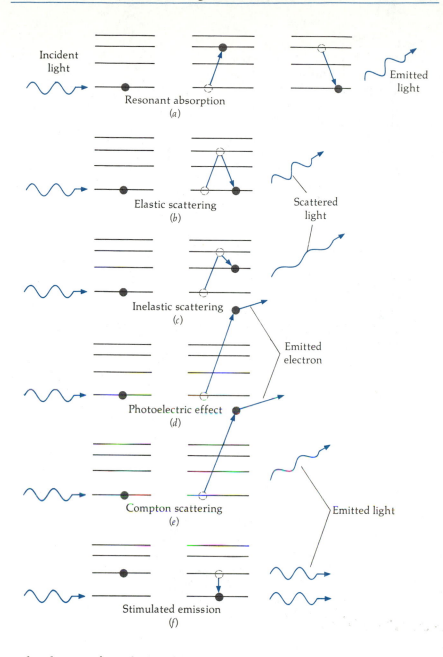

Incident
light

Resonant absorption
(a)

Emitted
light

Elastic scattering
(b)

Scattered
light

Inelastic scattering
(c)

Photoelectric effect
(d)

Emitted
electron

Compton scattering
(e)

Emitted light

Stimulated emission
(f)

Figure 31-14
Descriptions of photon interactions with an atom. In (a) the photon is absorbed and the atom is left in an excited state. The atom emits a photon as it decays to a lower state. This is a two-step process, and the emitted photon is not correlated with the incident photon. Rayleigh scattering (b) and Raman scattering (c) differ from (a) in that they are single-step processes and the incident and emitted photons are correlated. The photoelectric effect is illustrated in (d) and Compton scattering in (e). In (f), the atom is initially in an excited state and is stimulated to make a transition to a lower state by an incident photon of just the right energy. The emitted and incident photons have the same energy and are coherent.

molecules was first observed by the Indian physicist C. V. Raman and is known as *Raman scattering*.

Figure 31-14d illustrates the photoelectric effect, in which the absorption of the photon ionizes the atom or molecule with the emission of an electron. Figure 31-14e illustrates the Compton scattering that occurs if the incident photon energy is much greater than the ionization energy. (Note that in Compton scattering, Figure 31-14e, a photon is emitted, whereas in the photoelectric effect, Figure 31-14d, the photon is absorbed and none is emitted.)

Figure 31-14f illustrates *stimulated emission,* which we will discuss more fully here because of its important applications. This process occurs if the atom or molecule is initially in an excited state and the energy of the incident photon is equal to $E_2 - E_1$, where E_2 is the excited energy of the atom or molecule and E_1 is the energy of a lower state or the ground state. In this case, the oscillating electromagnetic field associated with the incident photon stimulates the excited atom or molecule, which then emits a photon in the same direction as the incident photon and in phase with it.

Stimulated emission is important because the resulting light is coherent; that is, the phase of the light emitted from one atom is related to that from every other atom. In the more usual case of spontaneous emission (Figure 31-14a), the phase of the light from one atom is unrelated to that from another atom and the resultant light is incoherent. In stimulated emission, the light from all the atoms is also in phase. Important applications of stimulated emission are the *maser* (*m*icrowave *a*mplification by *s*timulated *e*mission of *r*adiation) and the *laser* (*l*ight *a*mplification by *s*timulated *e*mission of *r*adiation). We will describe only the basic ideas of these important devices.

Laser

The laser and maser do not differ in theory but only in the frequency range of the electromagnetic spectrum in which they operate. The frequency is determined by the energy levels involved. Consider a system of atoms that have a ground state energy E_1 and an excited state energy E_2. If these atoms are irradiated by photons of energy $E_2 - E_1$, those atoms in the ground state can absorb a photon and make a transition to state E_2, whereas those atoms already in the excited state may be stimulated to decay back to the ground state. The relative probabilities of stimulated emission and absorption were first worked out by Einstein, who showed them to be equal. At normal temperatures, nearly all the atoms will ordinarily be initially in the ground state, so absorption will be the main effect. In order to produce more stimulated-emission transitions than absorption transitions, we must arrange to have more atoms in the excited state than in the ground state. This is called *population inversion*. It can be achieved if the excited state is relatively stable, that is, if it lasts a long time before spontaneous transitions occur. Such a state is called a *metastable state*. Population inversion is usually obtained by a process called *optical pumping*.

Figure 31-15 shows the energy levels of chromium, which are important for the operation of a ruby laser. Ruby is a transparent crystal of Al_2O_3 containing a small amount (about 0.05 percent) of chromium. The energy level labeled E_2 in the figure is a metastable state about 1.79 eV above the ground state. Level E_3 is about 2.25 eV above the ground state. An intense auxiliary radiation of energy $E_3 - E_1 = 2.25$ eV is provided to cause many of the atoms to make a transition from the ground state E_1 to level E_3 by absorption. This radiation has a wavelength of 550 nm and is green. These atoms quickly decay by spontaneous emission to the metastable energy level E_2. (The level labeled E_3 in the figure is usually a band of levels, so the incident radiation need not be monochromatic.) If the incident radiation is intense enough, more atoms will be transferred to state E_2 than remain in the ground state, with the result that the populations of these two states are inverted. When some atoms in state E_2 decay to the ground state by spontaneous emission, they emit photons of energy 1.79 eV and wavelength 694.3 nm. Some of these photons stimulate other excited atoms to emit photons of the same energy and wavelength. In the ruby laser, both ends of

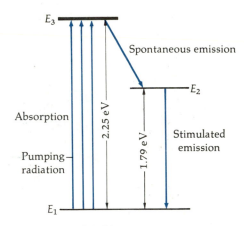

Figure 31-15
Energy levels in a ruby laser. To obtain amplification, the population of level E_2 must be greater than that of the ground state E_1. This is accomplished by an intense auxiliary radiation that excites atoms from the ground state to E_3, a band of levels from which they decay to level E_2 by spontaneous emission.

the crystal are silvered, one end such that it is totally reflecting and the other such that it is partially reflecting (99 percent or more) so that some of the beam is transmitted. If the ends are parallel, an intense beam of coherent light emerges from the partially reflecting end. Figure 31-16 illustrates the build up of the beam inside the laser. Photons traveling parallel to the axis of the crystal strike the silvered ends. All are reflected from the back face and most from the front face, with a few escaping through the partially-silvered front face. During each pass, the photons stimulate more and more atoms, causing the photon beam to build up and an intense beam to be emitted.

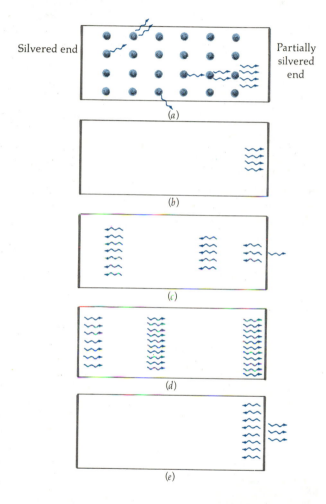

(a)

(b)

(c)

(d)

(e)

Figure 31-16
Build up of photon beam in a laser. In (a), some atoms spontaneously emit photons, some of which stimulate other atoms to emit photons parallel to the axis of the crystal. In (b), four photons strike the partially silvered right face of the laser. In (c), one photon has been transmitted and the others reflected. As the reflected photons traverse the laser crystal, they stimulate other atoms to emit photons and the beam builds up. By the time the beam reaches the right face again in (d) there are many photons. Some of these are transmitted (e) and the rest are reflected.

Silvered end

Partially silvered end

Population inversion is achieved somewhat differently in a helium-neon (He–Ne) laser, which consists of about 15 percent helium gas and 85 percent neon gas. Figure 31-17 shows the energy levels of helium and neon that are important for the laser operation. Helium has an excited state E_{H2} that is 20.61 eV above the ground state E_{H1}. The helium atoms are excited to state E_{H2} by an electric discharge. Neon has an excited state E_{N3} that is 20.66 eV above its ground state. This is just 0.05 eV above the first excited state of helium. The neon atoms are excited to this state by collisions with excited

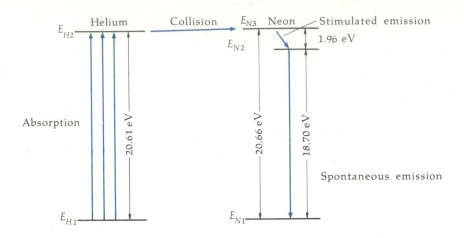

Figure 31-17
Energy levels of helium and neon important for the He−Ne laser. The helium atoms are excited by electrical discharge to an energy state 20.61 eV above the ground state. They collide with neon atoms and excite some of them to an energy state 20.66 eV above the ground state. Population inversion is thus achieved between this level and one 1.96 eV below it. Photons of energy 1.96 eV stimulate atoms in the upper state to emit other photons of energy 1.96 eV.

helium atoms, whose kinetic energy provides the extra 0.05 eV energy needed to excite the neon atoms. There is another excited state of neon, E_{N2}, that is 18.70 eV above the ground state and 1.96 eV below the state E_{N3}. Since this state is normally unoccupied, population inversion is obtained immediately between states E_{N3} and E_{N2}, and stimulated emission can occur between these two states. Atoms in state E_{N2} after stimulated emission then decay by spontaneous emission to the ground state. We note that there are four energy states involved in this laser, whereas the ruby laser involved only three levels. In a three-level laser, population inversion is difficult to obtain because more than half of the atoms in the ground state must be excited. In the four-level laser, population inversion is easily obtained because the final state after stimulated emission is not the ground state but an excited state that is normally unpopulated.

Because the laser beam is coherent, it emerges as a very narrow, intense beam. The angular spread of the beam is essentially limited only by diffraction effects. Its narrowness and precise direction make it useful as a medical tool. It is also used by surveyors for alignment over large distances. New uses for lasers are continually being found.

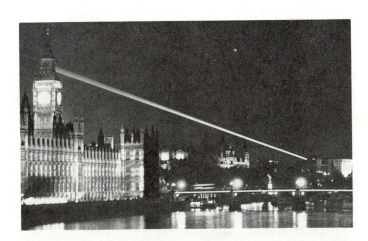

A laser beam transmitted from the New London Theater in Drury Lane, Covent Garden crosses the Thames to highlight Big Ben.

Lasers

John R. Cameron
Department of Medical Physics,
University of Wisconsin – Madison

The theoretical basis for the laser was originally described by Einstein in 1917. The first actual design for a laser was proposed by Townes and Schawlow in 1958, and the first laser to operate at visible wavelengths was reported by Maiman in 1960. Since then, many others have been involved in the evolution of lasers. At the time the laser was invented, there were predictions of a brilliant future for it in medicine. The laser still has a brilliant future in medicine—it's just been a slow starter.

Laser light is very different from a beam of ordinary light. First of all, it is made up of photons of a single wavelength; that is, it is monochromatic light. Second, the photons in the beam are all in phase with each other. This condition of being in phase is called *coherence* and the laser is referred to as a *coherent light source.* Because the laser beam is coherent, it emerges as a very narrow, intense beam. Its narrowness and precise direction have led to a number of unique medical uses of laser light. One of the most common is as a pointer. A series of radiation treatments for cancer may last a month, and it is important to have the radiation beam strike the same tissues each day. For this reason, many devices for radiation therapy have several rigidly mounted lasers whose narrow beams are used for repositioning the patient each day using marks on the skin.

Other common uses of lasers are in ophthalmology—the treatment of diseases of the eye. One of the most important of these is in the treatment of diabetic retinopathy, a disease which produces additional blood vessels in the retina that hinder vision and can lead to blindness if not treated. To stop these blood vessels from spreading, they are burned with a laser pulse. The laser beam is focused on a small area of the retina at the back of the eye, where the light-detection cells are located. When the laser pulse strikes the retina, most of it is absorbed within a few cell depths, causing this small volume of tissue to heat up rapidly. This is similar to using a lens to concentrate sunlight on a small area to start a fire. The resulting high temperature destroys the cells.

Another use of lasers in ophthalmology is to reattach a torn retina to the tissue layer behind it. When a retina tears, say, due to a blow to the head, there is often some fluid near the retina that may enter through the tear, causing the retina to separate or "detach" from the tissues behind it. This detachment usually starts at the periphery of the retina, affecting vision at the edge of the visual field. It may not be noticed at first, but if the detachment progresses into the main viewing area, double vision and severe distortion in vision may result. With proper medical care, the fluid behind the retina will return to its original location, but it is then desirable to

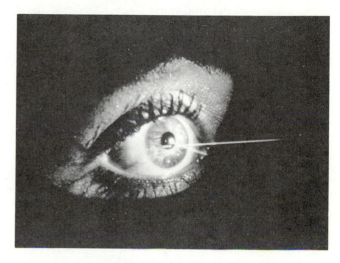

A beam from a helium-neon laser uses the doppler effect to monitor the flow of blood in vessels in the eye.

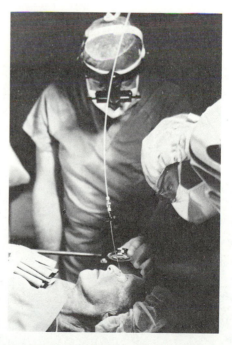

Eye surgery using a laser beam.

Essay: Lasers (continued)

reattach the retina so that the detachment will not recur. A laser pulse can be used to destroy small spots of tissue in the retina and its backing. The resulting scar tissue anchors the retina to the tissue layer behind it. The laser "spot welds" destroy the vision in those small areas, but this is usually not noticed since these areas are in the region of peripheral vision.

One of the original predictions for the laser was that it would be used in surgery as a bloodless scalpel that would seal the blood vessels as they were cut. This prediction has been slow in being fulfilled. However, progress is being made. For laser surgery, a 25-watt continuous-wave (CW), carbon-dioxide laser is often used. The carbon-dioxide laser is almost ideal for this purpose because its infrared wavelength is rapidly absorbed by the water in the tissues.

Another use of a laser that is primarily cosmetic is for the removal of tatoos. There are other techniques for this, but they are often not very effective and are usually quite painful. The laser removal technique takes advantage of the fact that the cells containing the tatoo ink absorb the pulsed laser energy, which causes them to vaporize. The same principle is useful in reducing the dark tissues of some birthmarks. It is obviously necessary to choose a laser wavelength that is strongly absorbed in the dark tissues.*

There are many nonmedical uses of lasers. Intense laser beams are used in industry to cut and weld various materials. Distances are accurately measured by reflecting a laser pulse from a mirror and measuring the transit time for the pulse to travel to the mirror and back. The distance to the moon has been measured to within about 15 cm using a corner mirror placed on the moon for that purpose. Three-dimensional photographs called holograms are made by splitting a laser beam into two beams, a reference beam and an object beam. The object beam reflects from the object to be photographed and the interference pattern between it and the reference beam is recorded on photographic film. When the film is illuminated with a laser, a three-dimensional replica of the object is produced. Another, nonmedical use of lasers is in fusion research. An intense laser pulse is focused on tiny pellets of deuterium-tritium in a combustion chamber. The pellets are heated to temperatures of the order of 10^8 K in a very short time causing the deuterium and tritium to fuse and release energy. As research continues, new uses are being found for this powerful tool.

*Many other medical uses of lasers are described in J. S. Carruth and A. McKenzie, *Medical Lasers,* Adam Hilger, 1986.

Corner reflecting mirror placed on the moon by the Apollo 14 astronauts.

31-7   Band Theory of Solids

One of the most interesting properties of solids is the enormous variation in electrical resistivity and conductivity between conductors and insulators. The resistivity of a typical conductor, such as copper, is of the order of $10^{-8}\ \Omega\cdot m$, whereas for a typical insulator, such as wood or glass, it is greater than $10^{8}\ \Omega\cdot m$ (see Table 20-1). We have explained this difference by saying that there are electrons in a conductor that are free to roam about the entire solid, whereas all of the electrons in an insulator are bound to their individual atoms. In this section, we will examine the difference between conductors and insulators in terms of the allowed energy levels of the atoms in a solid.

We have seen that the allowed energy levels in an isolated atom are often far apart. For example, in hydrogen, the energy for $n = 1$ is -13.6 eV and that for $n = 2$ is $-13.6/4 = -3.4$ eV. Let us consider two identical atoms that are far apart and focus our attention on one particular level, such as the $n = 2$ in hydrogen. There is a level of the same energy for each atom. If the atoms are brought close together, the energy of this level changes because of the influence of the other atom. The two previously identical energy levels each split into two levels of slightly different energies for the two-atom system. Similarly, if we have N identical atoms, a particular energy level of the isolated atom splits into N different, nearly equal energy levels when the atoms are close together. In a macroscopic solid, N is very large—of the order of Avogadro's number. A single atomic level identified with one isolated atom splits into a *band* of a very large number of levels when there are a large number of atoms close together in a solid. Because the number of levels in the band is so large, they are spaced almost continuously within the band. There is a separate band of levels for each particular energy level of the isolated atom. The energy bands corresponding to these individual levels in a solid may be widely separated in energy, or they may be close together or they may even overlap, depending on the kind of atom and the type of bonding in the solid.

Figure 31-18 shows three possible kinds of band structures for a solid. The band structure for copper, a conductor, is shown in Figure 31-18a. The lower bands (not shown) are filled with the inner electrons of the atoms. According to the Pauli exclusion principle, no more electrons can occupy levels in these bands. The uppermost band containing electrons is only about half full. In the normal state, at low temperatures, the lower half of this band is filled and the upper half is empty. At higher temperatures, a few of the electrons are in the higher energy states in this band because of thermal excitation, but there are still many unfilled energy states above the filled ones.

When an electric field is established in a conductor, the electrons in the partially filled band are accelerated, which means that their energy is increased. This is consistent with the Pauli exclusion principle because there are many empty energy states just above those occupied by electrons in this band. These electrons are the conduction electrons, and the band is called the *conduction band*.

Figure 31-18b shows the band structure for a typical insulator. At $T = 0$ K, the highest energy band that contains electrons is completely full. The next energy band, the conduction band, which contains empty energy

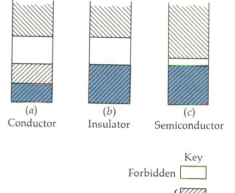

(a)	(b)	(c)
Conductor	Insulator	Semiconductor

Key
Forbidden
Allowed, empty
Allowed, occupied

Figure 31-18
Three possible band structures for a solid. (a) In a conductor, the highest occupied band is only partially full, so there are nearby energy states available for the excitation of electrons. (b) In an insulator, the highest occupied band, the valence band, is completely full and there is a large gap between it and the next allowed band, the conduction band. (c) In a semiconductor, the energy gap between the filled valence band and the empty conduction band is small enough that at normal temperatures there are some electrons in the conduction band and some holes in the valence band.

states, is separated from the last filled band by an energy gap called a *forbidden energy band.* At $T = 0$, the conduction band is empty. At ordinary temperatures or even very high temperatures, only a few electrons can be excited to states in the conduction band. Most electrons cannot be because the energy gap is large compared with the energy an electron might obtain by thermal excitation, which, on the average, is of the order of kT, where k is Boltzmann's constant (Chapter 11). For example, at $T = 300$ K, kT is only 0.025 eV whereas the energy gap in an insulator is typically 10 eV or greater. When an electric field is established in the solid, electrons cannot be accelerated because there are no empty energy states at nearby energies. We describe this by saying that there are no free electrons. The small conductivity that is observed is due to the very few electrons that are thermally excited into the upper, nearly empty conduction band.

In some materials, the energy gap between the top filled band and the empty conduction band is very small, as shown in Figure 31-18c. At ordinary temperatures, there is an appreciable number of electrons in the conduction band due to thermal excitation. Such a material is called a *semiconductor.* In the presence of an electric field, the electrons in the conduction band can be accelerated because there are empty states nearby. Also, for each electron in the conduction band, there is a vacancy, or hole, in the nearly filled band (this band is called the *valence band*). In the presence of an electric field, electrons in the valence band can be excited to a vacant energy level. This contributes to the electric current and is most easily described as the motion of a hole in the direction of the field and opposite to the motion of the electrons. The hole thus acts like a positive charge. (An analogy is a line of cars with a space the size of one car; as the cars move forward to fill the space, the space moves backward, in the direction opposite the motion of the cars.)

An interesting feature of a semiconductor is that as the temperature increases, the conductivity increases (and the resistivity decreases), which is contrary to the usual behavior of conductors. The reason is that as the temperature is increased, the number of free electrons is increased because more electrons are excited into the conduction band. The number of holes in the valence band also increases, of course. In semiconductors, the effect of the increase in the number of charge carriers, both electrons and holes, outweighs the effect of the increase in resistivity due to the increased scattering of the electrons resulting from thermal vibration of the lattice atoms.

The semiconductors we have been describing are called *intrinsic semiconductors.* A typical intrinsic semiconductor is silicon, which has four valence electrons in the $n = 3$ shell. In a solid crystal of silicon, each atom forms a covalent bond with four neighboring atoms by sharing one of its valence electrons with each neighbor, as illustrated schematically in Figure 31-19. The energy gap between the nearly filled valence band and the nearly empty conduction band in silicon is about 1.1 eV.

Most semiconductor devices, such as the semiconductor diode and the transistor, make use of *impurity semiconductors,* which result from the controlled addition of certain impurities to intrinsic semiconductors. This process is called *doping.* Figure 31-20a is a schematic illustration of the effect of adding a small amount of arsenic to silicon. Arsenic has five electrons in its valence shell rather than the four of silicon. Four of these electrons take part

Semiconductor

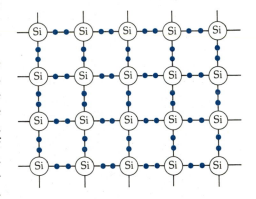

Figure 31-19
A two-dimensional representation of solid silicon. Each atom forms a covalent bond with four neighbors sharing one of its four valence electrons with each neighbor.

Impurity semiconductor

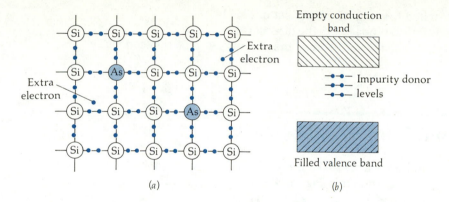

(a)   (b)

Figure 31-20

Figure 31-20
(a) Two-dimensional representation
of silicon doped with arsenic.
Because arsenic has five valence
electrons, there is an extra, unbound
electron that is easily excited to the
conduction band, thus contributing
to electrical conduction. (b) Band
structure of an n-type
semiconductor, such as silicon
doped with arsenic. The impurity
atoms provide filled energy levels
just below the conduction band,
which donate electrons to the
conduction band.

in the covalent bonds with the four neighboring silicon atoms, and the fifth
electron is very loosely bound to the atom, as illustrated. This extra electron
occupies an energy level that is just slightly below the conduction band in
the solid, and it is easily excited into the conduction band, where it can
contribute to electrical conduction.

The effect on the band structure of a silicon crystal of doping it with
arsenic is shown in Figure 31-20b. The levels just below the conduction
band are due to the extra electrons of the arsenic atoms. These levels are
called *donor* levels because they donate electrons to the conduction band
without leaving holes in the valence band. Such a semiconductor is called an
n-type semiconductor because the major charge carriers are *negative* electrons. The conductivity of such a doped semiconductor can be controlled by
the amount of impurity added. The addition of just one part per million
causes a significant change in the conductivity.

n-type semiconductor

Another type of impurity semiconductor can be made by replacing a
silicon atom with a gallium atom, which has three electrons in its valence
shell (see Figure 31-21a). The galium atom accepts electrons from the valence band to complete its four covalent bonds, thus creating a hole in the
valence band. The effect on the band structure of doping with gallium is
shown in Figure 31-21b. The empty levels just above the valence band are
due to the extra holes from the gallium atoms. These levels are called
acceptor levels because they accept electrons from the filled valence band

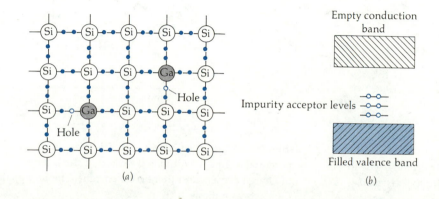

(a)   (b)

Figure 31-21
(a) Two-dimensional representation
of silicon doped with gallium.
Because gallium has only three
valence electrons, there is a hole in
one of its bonds. As electrons move
into the hole, the hole moves about
contributing to the conduction of
electrical current. (b) Band structure
of a p-type semiconductor, such as
silicon doped with gallium. The
impurity atoms provide empty
energy levels just above the filled
valence band, which accept
electrons from the valence band.

when these electrons are thermally excited to a higher energy state. This creates a hole in the valence band that is free to propagate in the direction of an electric field. This is called a *p-type semiconductor* because the charge carriers are *positive* holes.

p-type semiconductor

31-8   Semiconductor Junctions and Devices

Semiconductor devices such as diodes and transistors make use of *n*-type and *p*-type semiconductors joined together as shown in Figure 31-22. In practice, the two types of semiconductors are often a single silicon crystal that is doped with donor impurities on one side and acceptor impurities on the other. The region in which the semiconductor changes from a *p*-type to an *n*-type is called a *junction*.

Suppose we join an *n*-type and a *p*-type semiconductor together. Initially each side is uncharged, but the unequal concentrations of electrons and holes result in the diffusion of electrons across the junction from the *n* side to the *p* side and of holes, from the *p* side to the *n* side. The result of this diffusion is a net transport of positive charge from the *p* side to the *n* side. This creates a double layer of charge similar to that on a parallel-plate capacitor. There is then a potential difference across the junction that tends to inhibit further diffusion. In equilibrium, the *n* side, with its net positive charge, will be at some potential V greater than the *p* side, which has a net negative charge. In the junction region, there will be very few charge carriers of either type, so the junction region is one of high resistance.

The semiconductor with a *pn* junction that we have described can be used as a simple diode rectifier. In Figure 31-23, we have applied an external potential difference across the junction by connecting a battery and resistor to the semiconductor. When the positive terminal of the battery is connected to the *p* side of the junction, as shown in Figure 31-23*a*, the junction is said to be *forward biased*. The effect of forward biasing is to lower the potential across the junction. The diffusion of electrons and holes is thereby increased as they attempt to reestablish equilibrium, resulting in a current in the

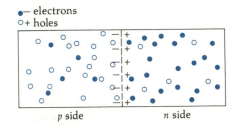

Figure 31-22

A *pn* junction. Because of the difference in concentrations, holes diffuse from the *p* side to the *n* side, and electrons diffuse from the *n* side to the *p* side. As a result, there is a double layer of charge at the junction, with the *p* side being negative and the *n* side being positive.

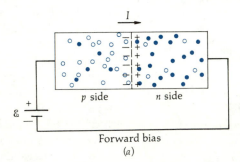

Forward bias
(a)

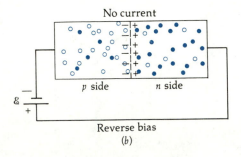

Reverse bias
(b)

Figure 31-23

(a) Forward-biased *pn* junction. The applied potential difference aids in the diffusion of holes from the *p* side to the *n* side and of electrons from the *n* side to the *p* side, resulting in a current *I*.

(b) Reverse-biased *pn* junction. In this case, the applied potential difference inhibits diffusion of the majority of charge carriers and there is no current. The *pn* junction thus acts as a diode.

circuit. If we connect the positive terminal of the battery to the *n* side of the junction, as shown in Figure 31-23*b*, the junction is said to be *reverse biased*, and the battery tends to increase the potential difference across the junction, thereby further inhibiting diffusion. Figure 31-24 shows the current versus voltage for a typical semiconductor junction. Essentially, the junction conducts only in one direction, the same as the vacuum-tube diode discussed in Section 23-5. Junction diodes have replaced vacuum diodes in nearly all applications except those for which a very high current is required.

Another use for the *pn* junction is the solar cell, which is illustrated schematically in Figure 31-25. When a photon of energy greater than 1.1 eV strikes the cell, it can excite an electron from the valence band of silicon into the conduction band, leaving a hole in the valence band. Suppose the photon strikes the *p*-type region, as shown in the figure. This region is already rich in holes. Some of the electrons created by the photons will recombine with holes, but some will migrate to the junction, where they will be accelerated into the *n*-type region by the electric field between the double layer of charge. This creates an excess of positive charge in the *n*-type region and an excess of negative charge in the *p*-type region. The result is a potential difference between the two regions, which in practice is about 0.6 V. If a load resistance is connected across the two regions, a charge flows through the resistor. Some of the incident light energy is thus converted into electrical energy. The current in the resistor is proportional to the number of incident photons, which is in turn proportional to the intensity of the incident light.

There are many other applications of *pn* junctions. Particle detectors called *surface-barrier detectors* consist of a *pn* junction with a large reverse bias such that ordinarily there is no current. When a high-energy particle, such as an electron, passes through the semiconductor, it loses energy and creates many electron–hole pairs. The resulting current pulse signals the arrival of the particle. Light-emitting diodes (LEDs) are *pn* junctions with a large forward bias, that produces large excess concentrations of electrons on the *p* side and holes on the *n* side of the junction. Under these conditions, the diode emits light due to electron–hole recombination. This is essentially the reverse of the process in a solar cell, where electron–hole pairs are created by the absorption of light. LEDs are commonly found in digital watch and calculator displays.

Transistors

The transistor, invented in 1948 by John Bardeen, Walter H. Brattain, and William Shockley has revolutionized the electronics industry and our everyday world. A simple junction transistor consists of three distinct semiconductor regions called the *emitter*, the *base*, and the *collector*. The base is a very thin slice of one type of semiconductor sandwiched between two semiconductors of the opposite type. The emitter semiconductor is much more heavily doped than either the base or the collector. In an *npn* transistor, the emitter and collector are *n*-type semiconductors and the base is a *p*-type semiconductor, whereas in a *pnp* transistor, the base is a *n*-type semiconductor and the emitter and collector are *p*-type semiconductors. The emitter, base, and collector behave somewhat similarly to the cathode, grid, and plate in a vacuum-tube triode (Section 23-5), except that in a *pnp* transistor it is holes that are emitted rather than electrons.

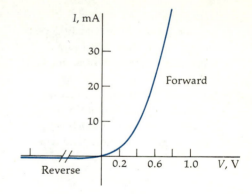

Figure 31-24
Current versus voltage across a junction diode. The current is essentially zero when the junction is reverse biased.

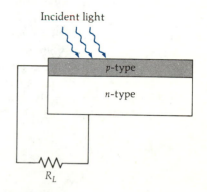

Figure 31-25
A *pn* junction as a solar cell. When light strikes the *p*-type semiconductor, electron-hole pairs are created, resulting in a current through the load resistance R_L.

Figure 31-26
A *pnp* transistor. The heavily doped emitter emits holes that pass through the narrow base to the collector. (*b*) Symbol for a *pnp* transistor in a circuit. The arrow points in the direction of the conventional current, which is the same as that of the holes emitted.

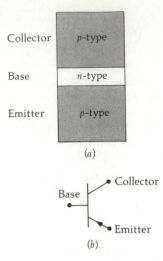

(*a*)

Base ●──── ● Collector

── ● Emitter

(*b*)

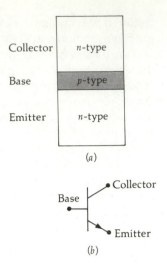

(*a*)

Base ●──── ● Collector

── ● Emitter

(*b*)

Figure 31-27
An *npn* transistor. The heavily doped emitter emits electrons that pass through the narrow base to the collector. (*b*) Symbol for an *npn* transistor. The arrow points in the direction of the conventional current, which is opposite the direction of the electrons emitted.

Figures 31-26 and 31-27 show, respectively, a *pnp* transistor and an *npn* transistor and the symbols used to describe each transistor in circuit diagrams. We can see that a transistor consists of two *pn* junctions. We will describe the operation of a *pnp* transistor here. The operation of an *npn* transistor is similar.

In normal operation, the emitter–base junction is forward biased, and the base–collector junction is reverse biased. The heavily doped *p*-type emitter emits holes that flow across the emitter–base junction into the base. Because the base is very thin, most of the holes flow across the base into the collector. This flow constitutes a current I_C from the emitter to the collector. However, some of the holes recombine in the base, producing a positive charge that inhibits the further flow of current. To prevent this, some of the holes that do not reach the collector are drawn off the base as a base current I_B in a circuit connected to the base. The base current I_B is usually only a small fraction of the collector current I_C. However, a small change in the base current produces a large change in the collector current. This is similar to the operation of a vacuum-tube triode discussed in Section 23-5, where a small change in the potential of the grid produces a large change in the plate current. Figure 31-28 shows a simple amplifier circuit containing a *pnp* transistor. The bias voltages are supplied by the batteries V_{EB} and V_{EC}. The input signal in the base-current circuit produces a small change in the base current that, in turn, produces a large change in the collector current. The output voltage across the load resistance R_L is then much larger than the voltage of the input signal.

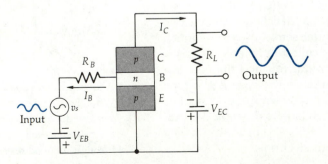

Figure 31-28
A *pnp* transistor used as an amplifier. A small change in the base current I_B results in a large change in the collector current I_C. A small signal v_s in the base circuit thus results in a large signal across the load resistor R_L in the collector circuit.

Electronics: Vacuum Tube to Solid State

Larry C. Burton
Virginia Polytechnic Institute & State University

The electronics industry originated early in this century with the development of the vacuum tube. Many widely-used commercial, consumer, and military electronic devices, such as radio and television transmitters and receivers and computers, were initially developed using vacuum tubes. During the past 35 years, however, vacuum-tube circuits have been almost totally replaced by circuits based on the transistor. This more recent type of electronics is called *solid state electronics* or, due to the extremely small size of many of the components, *microelectronics*.

Because of the historical importance of vacuum tubes and the fact that they are still used—the cathode ray tubes in television sets and oscilloscopes, for instance—we will briefly discuss their operation before going on to transistors and integrated circuits. The major physical phenomenon by which the vacuum tube operates is the thermionic emission of electrons from a hot filament (see Section 23-5). In a vacuum tube, the cathode—a tungsten filament—is heated electrically to a temperature above 1000°C. The more energetic electrons "boil" off the cathode and are collected by a metal anode maintained at a positive voltage. Additional electrodes placed between the cathode and the anode are used to control the electron flow resulting in amplification, rectification, and other applications.

By about 1950, vacuum-tube circuits formed the backbone of the electronics industry, including the fledgling computer industry being developed by IBM, General Electric, Burroughs, and other companies. Even though vacuum-tube circuits performed admirably and gave us an unprecedented electronics technology, they had several severe drawbacks. For one thing, vacuum tubes were quite large. Early computers, for instance, required literally rooms of space just to house their vacuum tubes. Their size also made vacuum tubes relatively slow because of the time it took electrons to traverse them in response to applied signals. In addition, vacuum tubes and their supports were quite heavy, which limited electronic applications in situations where size and weight were factors, such as in airplanes or on ships. Also, because the tube filaments had to be heated, vacuum tubes consumed large amounts of power, which made them expensive to operate. Finally, the filaments would eventually burn out just like the filaments in light bulbs, so the tubes would have to be replaced. This both increased costs and reduced reliability.

Spurred in part by the desire to surmount the inherent problems of vacuum tubes, two physicists at the Bell Laboratories, J. Bardeen and W. Brattain, invented the transistor in 1948. (The name *transistor* is a shortened form of the term "transfer-resistor.") Subsequently, W. Shockley added much to the understanding of how the transistor operated and devised ingenious schemes for improving it. In 1956, these three men received the Nobel Prize in Physics for the invention and development of the transistor.

The early experimental transistors did not work very well. Still, their potential was obvious, and transistor technology developed rapidly. The first transistor was made by butting two wires against a piece of the semiconductor germanium. It soon became obvious, however, that silicon was a more effective semiconductor for use in transistors. It could be made quite pure, and then, by adding the correct type of impurities to it, *p* and *n* regions could be formed. This allowed the development of the silicon *pn* junction, which became the fundamental building block of many solid state circuits.

One of the earliest computers, the ENIAC, in 1946.

Essay: Electronics: Vacuum Tube to Solid State (continued)

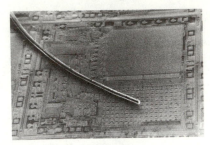

An electron micrograph of a human
hair on an integrated circuit
computer chip, magnified 58 times.

Initially, many different processes were used to make transistors. Then, in 1958, two breakthroughs occurred that, when they were combined, revolutionized electronics and paved the way for solid state electronics. At Fairchild, the planar process for making silicon pn junctions was developed. At Texas Instruments, a process for incorporating all of the components of a circuit on a single silicon chip was devised. Such a circuit is called a "monolithic integrated circuit," though it is more commonly referred to simply as an *integrated circuit* or *IC*.

The importance of the IC can best be seen by comparing it with a vacuum-tube circuit that performs the same function. The IC wins on all counts—cost, power requirements, size, weight, speed, and reliability. These features have greatly expanded the applications of microelectronics in a variety of commercial and military areas, the automotive and aerospace fields to name but two. As a specific example, consider the hand-held calculator based on silicon ICs. It has greater speed and calculating power than did early vacuum-tube computers thousands of times larger.

The major steps in the fabrication of a silicon chip integrated circuit are illustrated in Figure 1. The process used is called the *planar process* because all of the required components of the circuit are formed at or near the surface plane of the silicon. Due to the minute size of the circuit components in an IC, this process must be very precise. In addition, it must be carried out under sterile manufacturing conditions, as a few specks of dust on a wafer can degrade the performance of the fabricated ICs.

To begin the planar process, a slice of silicon several centimetres in diameter and roughly 1 millimetre thick is cut from a large, cylindrical, silicon "ingot" that has been doped with n- or p-type impurities. This slice of silicon, also called a *silicon wafer*, will be fabricated into several hundred identical silicon chips, each with an area of about 0.4 square centimetres. In our example, we will assume that we are working with an n-type silicon wafer, and we will focus on two of the many pn junctions fabricated on a given chip.

The silicon wafer is polished so that it is optically flat. It is then placed in a furnace at an elevated temperature (900 to 1000°C), which causes a layer of silicon dioxide several tenths of a micrometre thick to form on the surface by the reaction $Si + O_2 \rightarrow SiO_2$. A cross-sectional diagram of the silicon wafer with its SiO_2 layer is shown in Figure 1a. The SiO_2 layer is one of the reasons silicon is the dominant semiconductor used in microelectronics. It is a good electrical insulator, so it can be used to separate electrically active regions on the chip and as a dielectric for capacitors on the chip. In addition, "windows" can be etched in it through which certain impurities can be deposited to form pn junctions.

Next, a layer of photosensitive material, called *photoresist*, is applied (see Figure 1b). A glass "mask" with opaque regions to define the circuit elements is placed over the wafer, and it is then exposed briefly to a collimated light beam (see Figure 1c), which causes those areas of the photosensitive material not masked to polymerize. (In the actual fabrication process of a sophisticated microprocessor chip, several masks, each with a different pattern array, are used.)

The wafer is then placed in a developer. The exposed photoresist is resistant to the developer, so it is not affected by it, but the unexposed areas are developed away, forming windows in the photoresist (see Figure 1d). The wafer is then immersed in a hydrofluoric acid (HF) solution, and those regions of the SiO_2 exposed through the windows in the photoresist are etched away down to the silicon surface (see Figure 1e). Another chemical is then used to strip off the remaining photoresist.

The wafer is again placed in a furnace, where a p-type dopant, usually boron, is diffused into the n-type silicon through the windows in the SiO_2 to form pn junctions in those regions (see Figure 1f). At the same time, another layer of SiO_2 forms in the exposed regions. As before, windows are made in the SiO_2 layer photolithographically. These *contact windows*, so named because they provide access between the pn junctions and the contact metal that is to be added, are situated above the pn junctions. They are smaller than the original windows, however, to prevent the contact metal from shorting out across the surface of the pn junction. The wafer is then covered with the contact metal, usually aluminum (see Figure 1g), and the photolithographic process is used once again to form the contact pattern (see Figure 1h).

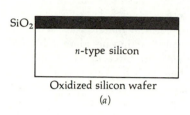

SiO₂

n-type silicon

Oxidized silicon wafer
(a)

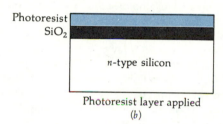

Photoresist
SiO₂

n-type silicon

Photoresist layer applied
(b)

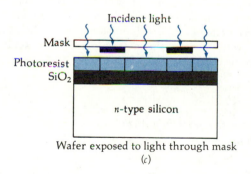

Incident light

Mask

Photoresist
SiO₂

n-type silicon

Wafer exposed to light through mask
(c)

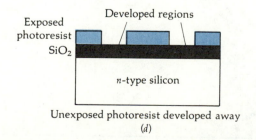

Exposed photoresist

Developed regions

SiO₂

n-type silicon

Unexposed photoresist developed away
(d)

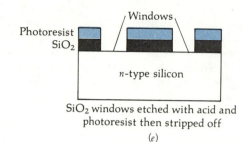

Windows

Photoresist
SiO₂

n-type silicon

SiO₂ windows etched with acid and photoresist then stripped off
(e)

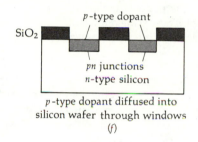

p-type dopant

SiO₂

pn junctions
n-type silicon

p-type dopant diffused into silicon wafer through windows
(f)

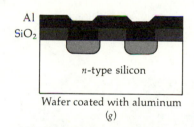

Al
SiO₂

n-type silicon

Wafer coated with aluminum
(g)

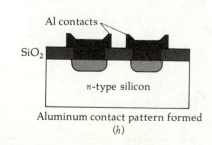

Al contacts

SiO₂

n-type silicon

Aluminum contact pattern formed
(h)

Figure 1
Fabrication of silicon *pn* junctions.

Essay: Electronics: Vacuum Tube to Solid State (continued)

The wafer is then scribed into a checkerboard pattern with a diamond-tipped instrument and is broken into chips, each of which contains an IC. The chips are mounted in "packages," which are little boxes made of materials—usually combinations of metal, glass, and plastic—that do not react with silicon, and leads are attached with a precision bonding tool. In the final step of the fabrication process, the packages are sealed in an inert ambient material, such as helium or argon, to insure that they contain no moisture or oxygen that might subsequently react with the silicon.

Another class of semiconductors that has received much attention in recent years are the III–V compounds formed by elements from groups III and V of the periodic table. The major one is gallium arsenide (GaAs). These semiconductors are superior to silicon in some respects and inferior in others for microelectronic applications.

One advantage of transistors made from GaAs and its III–V relatives is that when a large forward bias is applied across a GaAs *pn* junction, photons are emitted. (This does not happen with silicon.) These photons can be modulated by an electric signal, which means that they can be made to carry information. Indeed, because of the high frequencies of photons, they can carry more information than can microwaves. Two of the devices that make use of this property of GaAs transistors are the light emitting diode (LED) and the junction laser, both of which have applications in the growing field of optical communications. For instance, the information-packed photon (light) signal generated by a GaAs transistor can be transmitted by optical fibers, which can carry an enormous number of separate signals, as noted in Section 24-5.

A second advantage has to do with the fact that the energy gap of GaAs is greater than that of silicon (see Section 31-7). Devices made of GaAs are therefore less sensitive to temperature, which means that they are more stable and thus more reliable. Lastly, electrons travel faster in GaAs than in silicon, which means that GaAs ICs can operate at higher speeds. These two advantages are exploited in such military applications as missile guidance systems.

The major disadvantage of GaAs and its III–V relatives is cost. Gallium, for instance, is a rare metal and is therefore expensive. Furthermore, the III–V compounds are relatively difficult to form and process due to the volatility of the group V elements. Arsenic, for example, starts to vaporize out of GaAs at the comparatively low temperature of 450°C. Finally, the III–V compounds do not form stable oxides. (Recall the importance of the SiO_2 layer in the fabrication of silicon transistors.) These properties of the III–V compounds mean that a different fabrication process from that used to make silicon transistors must be used to make III–V transistors. The process used is still quite expensive, though progress in reducing its cost is being made.

Nevertheless, GaAs transistors have a definite niche in solid state electronics in those areas where they perform better than silicon transistors. This niche is growing, but it presently comprises only a small fraction of the transistor market. The inherent advantages of silicon plus the nearly 40 years of intense development of silicon transistors make it likely that silicon will continue to be the most prominent semiconductor used in the microelectronics industry for the foreseeable future.

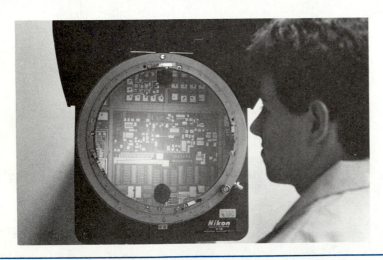

To check the manufacturing of this integrated circuit, this worker uses a device which greatly magnifies the circuit.

Summary

1. In quantum theory, the hydrogen atom is described by a wave function, the square of which gives the probability of finding the electron in a given region of space. The wave function is characterized by four quantum numbers:

$$n = 1, 2, 3, \ldots$$
$$\ell = 0, 1, \ldots, n-1$$
$$m = -\ell, -\ell+1, \ldots, +\ell;$$

and

$$m_s = +\tfrac{1}{2} \text{ or } -\tfrac{1}{2}$$

The energy depends only on the principal quantum number n and is the same as in the Bohr model. In the ground state, $n = 1$, $\ell = 0$, and $m = 0$, and the probability distribution is spherically symmetric, with the electron most likely to be found near the first Bohr radius. It is convenient to think of the electron as a charged cloud with a charge density proportional to the probability distribution.

2. In multielectron atoms, the energy of the electron is determined mainly by the principal quantum number n (which is related to the radial dependence of the wave function) and by the orbital angular momentum quantum number ℓ. Generally, the lower the values of n and ℓ, the lower the energy. The specification of n and ℓ for each electron in an atom is called the *electron configuration*. In the specification of ℓ, a letter rather than the numerical value is customarily used. The code is

	s	p	d	f	g	h
ℓ value	0	1	2	3	4	5

3. An important principle that governs the electron configurations of atoms is the Pauli exclusion principle: No two electrons in an atom can be in the same quantum state: that is, they can not have the same set of values for the quantum numbers n, ℓ, m, and m_s. Using the exclusion principle, we can understand much of the structure of the periodic table of the elements.

4. Atomic spectra consist of optical spectra and x-ray spectra. Optical spectra can be understood in terms of transitions between energy levels of a single outer electron moving in the field of the nucleus and core electrons of the atom. Characteristic x-ray spectra result from the excitation of inner core electrons and the subsequent filling of the vacancies by other electrons in the atom.

5. Bonding mechanisms for atoms and molecules include ionic, covalent, van der Waals, hydrogen, and metallic. Ionic bonds result when an electron is transferred from one atom to another, leaving positive and negative ions that bond together. The covalent bond is a quantum effect that arises from the sharing of one or more electrons by identical or similar atoms. Van der Waals bonds are weak bonds resulting from the attraction of instantaneous dipole moments of atoms or molecules. The hydrogen bond results from the sharing of proton (a hydrogen atom) by other atoms. The metallic bond results from free electrons in a metal that form a cloud of negative charge that holds the positive lattice ions together.

6. Optical spectra of molecules have a band structure due to transitions between vibrational and rotational levels. Information about the structure and bonding of a molecule can be found from its rotational and vibrational absorption spectra. The energies of rotation of a molecule with moment of inertia I are given by

$$E = \ell(\ell + 1)E_{0r}$$

where $\ell = 0, 1, 2, 3, \ \ldots\ $, and

$$E_{0r} = \frac{\hbar^2}{2I}$$

7. Stimulated emission occurs if an atom is initially in an excited state and if a photon of energy equal to the excitation energy is incident on the atom. The oscillating electromagnetic field of the incident photon stimulates the excited atom to emit another photon in the same direction and in phase with the incident photon. The operation of a laser depends on the possibility of population inversion, in which there are more atoms in an excited state than in the ground state or a lower state. A laser produces an intense, coherent, narrow beam of photons.

8. The allowed energy states for electrons in solids form bands of nearly continuous energy levels. In a conductor, the upermost band containing electrons is only partially full, so there are many available states for excited electrons. In an insulator, the uppermost band containing electrons, the valence band, is completely full and there is a large energy gap between it and the next allowed band, the conduction band. In a semiconductor, the energy gap between the filled valence band and the empty conduction band is small, so at ordinary temperatures, a few electrons are thermally excited into the conduction band.

9. The conductivity of a semiconductor can be greatly increased by doping. In an n-type semiconductor, the doping adds electrons just below the conduction band. In a p-type semiconductor, holes are added just above the valence band. A junction of an n-type and a p-type semiconductor has many applications, such as diodes, solar cells, and light-emitting diodes.

10. A transistor consists of a very thin semiconductor of one type sandwiched between two semiconductors of the opposite type. Transistors have many applications, such as in amplifiers, because a small variation in the base current results in a large variation in the collector current.

Suggestions for Further Reading

Leith, Emmett N., and Juris Upatnicks: "Photography by Laser," *Scientific American*, June 1965, p. 24.

The interference of coherent light produced by a laser is employed in wavefront reconstruction photography, more commonly known as holography.

Nassau, Kurt: "The Causes of Color," *Scientific American*, October 1980, p. 124.

In this article, the ways in which matter can appear colored, including electronic transitions in atoms and molecules, transitions between molecular orbitals, transitions in materials having energy bands, scattering, and interference are described.

Pauling, Linus, and Roger Hayward: *The Architecture of Molecules*, W. H. Freeman and Company, San Francisco and London, 1964.

This is a picture book illustrating the relative positions of the various atoms in some common molecules and solids, with commentary by one of the men most responsible for our modern understanding of the chemical bond.

Ronn, Avigdor M.: "Laser chemistry," *Scientific American* May 1979, p. 114.

This article explains how laser light of the correct frequency can efficiently induce the electronic or molecular transitions required for a desired chemical reaction to occur.

Schawlow, Arthur L.: "Laser Light," *Scientific American*, September 1968, p. 120.

How lasers work and how laser light differs from ordinary light are discussed in this article.

Schewe, Phillip F.: "Lasers," *The Physics Teacher*, vol. 19, no. 8, 1981, p. 534.

This is an excellent and comprehensive exposition of the principles of laser operation, the types of lasers in use today, and applications of laser light.

Seaborg, Glenn T., and Justin L. Bloom: The Synthetic Elements: IV" *Scientific American*, April 1969, p. 56.

The synthetic elements produced in laboratories, nuclear reactors, or nuclear explosions have extended the known periodic table beyond element number 100 and may extend it much further according to this article.

Vali, Victor: "Measuring Earth Strains by Laser," *Scientific American*, December 1969, p. 88.

One important modern use for laser interferometers is to measure small changes in the compression of the earth's crust, both steady and sudden, as in earthquakes.

Walker, Jearl: "The Amateur Scientist: The Spectra of Streetlights Illuminate Basic Principles of Quantum Mechanics," *Scientific American*, January 1984, p. 138.

A straightforward account of the development and application of quantum mechanics to the explanation of atomic spectra, illustrated with novel photographs.

Zare, Richard N.: "Laser Separation of Isotopes," *Scientific American*, February 1977, p. 86.

This article explains how laser irradiation of atoms or molecules in a beam can be useful in separating the isotopes of an element.

Review

A. Objectives: After studying this chapter, you should:

1. Be able to compare the Schrödinger and Bohr models of the hydrogen atom.

2. Be able to discuss the shell structure of atoms and the periodic table.

3. Be able to compare the optical and x-ray spectra of an atom.

4. Be able to list the ways atoms can bond together and describe each briefly.

5. Be able to describe the general features of the energy-level diagram for a diatomic molecule and discuss the vibrational and rotational spectra.

6. Be able to discuss why only certain lines in the emission spectrum of a molecule are seen in the absorption spectrum.

7. Be able to state the meaning of the terms *population inversion, optical pumping,* and *metastable state.*

8. Be able to describe the operation of a ruby laser and a helium–neon laser.

9. Be able to discuss qualitatively the band theory of solids.

10. Be able to discuss the characteristics of *np* junctions and list some of their applications.

B. Define, explain, or otherwise identify:

Principal quantum number, p. 792
Orbital quantum number, p. 792
Magnetic quantum number, p. 792
Electron spin, p. 794
Electron configuration, p. 795
Pauli exclusion principle, p. 795

K shell, p. 795
Optical spectra, p. 801
Characteristic x-ray spectrum, p. 802
Ionic bond, p. 803
Covalent bond, p. 804
Van der Waals bond, p. 805
Hydrogen bond, p. 806
Metallic bond, p. 806
Band spectra, p. 809
Absorption spectra, p. 809 or 810
Stimulated emission, p. 811–812
Maser, p. 812
Laser, p. 812
Population inversion, p. 812
Metastable state, p. 812
Conduction band, p. 817
Valence band, p. 818
Intrinsic semiconductor, p. 818
n-type semiconductor, p. 819
p-type semiconductor, p. 820
pn junction, p. 820

C. True or false: If the statement is true, explain why. If it is false, give a counterexample.

1. No two electrons can be in the same quantum state.

2. Atoms with one electron outside a closed shell have small ionization energies and are chemically active.

3. Visible light results from transitions involving only the outermost electrons in an atom.

4. Characteristic x-rays result from transitions made by inner electrons.

5. Ionic bonds are formed by atoms whose outer electron shells are filled.

6. The energies of vibration and rotation of a molecule are usually much greater than the energy of electronic excitation.

7. Semiconductors conduct current in one direction only.

Exercises

Section 31-1   Quantum Theory of the Hydrogen Atom

1. For $\ell = 1$, find (*a*) the magnitude of the angular momentum *L* and (*b*) the possible values of *m*. (*c*) Draw to scale a vector diagram showing possible orientations of **L** with the *x* axis.

2. Work Exercise 1 for $\ell = 3$.

3. If $n = 3$, (*a*) what are the possible ℓ values? (*b*) For each ℓ value in (*a*), list the possible *m* values. (*c*) Using the fact that there are two quantum states for each value of ℓ and *m* because of electron spin, find the total number of electron states with $n = 3$.

4. Find the total number of electron states with (*a*) $n = 2$ and (*b*) $n = 4$.

5. The moment of inertia of a phonograph record is about 10^{-3} kg·m^2. Find the angular momentum $L = I\omega$ when it rotates at $\omega/2\pi = 33.3$ rev/min and find the approximate value of the quantum number ℓ.

Section 31-2   The Periodic Table

6. Write the electron configuration of (*a*) carbon and (*b*) oxygen.

7. Write the electron configuration of (*a*) aluminum and (*b*) chromium.

8. What element has the electron configuration $1s^2 2s^2 2p^6 3s^2 3p^2$?

9. What element has the electron configuration $1s^2 2s^2 2p^6 3s^2 3p^6 4s^2$?

10. The properties of iron ($Z = 26$) and cobalt ($Z = 27$), which have adjacent atomic numbers, are similar, whereas the properties of neon ($Z = 10$) and sodium ($Z = 11$), which also have adjacent atomic numbers, are very different. Explain why this is so.

11. In Figure 31-5, there are small dips in the ionization potential curve at $Z = 31$ (gallium) and $Z = 49$ (indium) that are not labeled in the figure. Explain these dips using the electron configurations of these atoms given in Table 31-2.

12. If the outer electron in sodium did not penetrate the inner core, its energy would be -13.6 eV$/3^2 = -1.51$ eV. Because it does penetrate, it sees a higher effective *Z* and its energy is lower. Use the measured ionization energy of 5.14 eV for sodium to calculate Z_{eff} seen by this electron.

13. Separate the following six elements—potassium, calcium, titanium, chromium, manganese, and copper—into two groups of three each such that those in each group have similar properties.

Section 31-3   Atomic Spectra

14. The optical spectra of atoms with two electrons in the same outer shell are similar, but they are quite different from the spectra of atoms with just one outer electron because of the interaction of the two electrons. Separate the following elements into two groups such that those in each group have similar spectra—lithium, beryllium, sodium, magnesium, potassium, calcium, chromium, nickel, cesium, and barium.

15. Use Equation 31-9 to calculate the next two longest wavelengths in the *K* series after the K_α line of molybdenum.

16. Find the wavelength of the K_α line in magnesium ($Z = 12$).

17. Find the wavelength of the K_α line in copper ($Z = 29$).

18. The wavelength of the K_α x-ray line for an element is measured to be 0.0794 nm. Calculate Z from this result and give the name of the element.

19. The wavelength of the K_α x-ray line for an element is measured to be 0.3368 nm. Calculate Z from this result and give the name of the element.

Section 31-4 Molecular Bonding

20. What kind of bonding mechanism would you expect for (a) the HCl molecule, (b) the O_2 molecule, and (c) Cu atoms in a solid?

21. What kind of bonding mechanism would you expect for (a) the N_2 molecule, (b) the KF molecule, and (c) Ag atoms in a solid?

Section 31-5 Molecular Spectra

22. Explain why the moment of inertia of a diatomic molecule increases slightly with increasing angular momentum.

23. The characteristic rotational energy E_{0r} for the rotation of the N_2 molecule is 2.48×10^{-4} eV. From this, find the separation distance of the N atoms in N_2.

24. The separation of the O atoms in O_2 is actually slightly greater than the 0.1 nm used in Example 31-3, and the characteristic energy of rotation E_{0r} is 1.78×10^{-4} eV rather than the result obtained in that example. Use this value to calculate the separation distance of the O atoms in O_2.

Section 31-6 Absorption, Scattering, and Stimulated Emission

There are no exercises for this section.

Section 31-7 Band Theory of Solids;
Section 31-8 Semiconductor Junctions and Devices

25. State whether an n-type or p-type semiconductor is obtained if silicon is doped with (a) aluminum or (b) phosphorus. (Hint: See Table 31-2 for the electron configurations of the elements.)

26. State whether an n-type or p-type semiconductor is obtained if germanium is doped with (a) boron, (b) gallium, or (c) arsenic. (See hint in Exercise 25.)

27. State whether an n-type or p-type semiconductor is obtained if silicon is doped with (a) indium or (b) antimony. (See hint in Exercise 25.)

28. The energy gap between the valence and conduction bands in silicon is 1.1 eV. What is the maximum wavelength of a photon that can excite an electron from the valence band to the conduction band?

29. Work Exercise 28 for germanium, for which the energy gap is 0.74 eV.

30. Work Exercise 28 for diamond, for which the energy gap is 7.0 eV.

31. The donor energy levels in a n-type semiconductor are 0.01 eV below the conduction band. Find the temperature such that $kT = 0.01$ eV.

32. A photon of wavelength 3.35 μm has just enough energy to raise an electron from the valence band to the conduction band in a lead sulfide crystal. (a) Find the energy gap between these bands in lead sulfide. (b) Find the temperature T such that kT equals this energy gap.

Problems

1. The equilibrium separation of the K^+ and Cl^- ions in KCl is about 0.267 nm. (a) Calculate the potential energy of attraction of the ions, assuming them to be point charges at this separation. (b) The ionization energy of potassium is 4.34 eV and the electron affinity of Cl is 3.82 eV. Find the dissociation energy, neglecting any energy of repulsion (see Figure 31-7). (The measured dissociation energy is 4.43 eV, which differs slightly from your answer in (b) because of a small potential energy associated with the repulsion of the ions.)

2. The equilibrium separation of the K^+ and F^- ions in KF is about 0.217 nm. (a) Calculate the potential energy of attraction of the ions, assuming them to be point charges at this separation. (b) The ionization energy of potassium is 4.34 eV and the electron affinity of F is 4.07 eV. Find the dissociation energy, neglecting any energy of repulsion (see Figure 31-7).

3. The L shell in an atom has $n = 2$ and $\ell = 1$. For atoms with eight or more electrons, there are two electrons in the K shell and six in the L shell. An electron in the L shell is thus shielded from the nuclear charge by the two electrons in the K shell and is partially shielded by the five other electrons in the L shell. There is also some shielding because of the penetration of the wave functions of the electrons in the outer shells. The frequencies for x-rays in the L series involve transitions from some level n to the level $n_2 = 2$. Equation 31-9 is a good approximation for the photon energy for the L-series x-rays if ($Z - 1$) in that equation is replaced by ($Z - 7.4$). Use this result to calculate the shortest wavelength in the L series for molybdenum ($Z = 42$) and for zinc ($Z = 30$).

4. The energy of vibration and rotation of a diatomic molecule can be written

$$E = E_v + \ell(\ell + 1)E_{0r}$$

where E_v is the energy of vibration, which is related to the frequency of vibration f_v by $E_v = (n + \frac{1}{2})hf_v$, and E_{0r} is a rotational energy related to the moment of inertia of a molecule by $E_{0r} = \hbar^2/2I$. (a) Draw an energy-level diagram for the lowest two vibrational states ($n = 0$ and $n = 1$), including the rotational states corresponding to $\ell = 0, 1, 2, 3,$ and 4. On your diagram, indicate absorption transitions ($n = 0$ to $n = 1$) obeying the selection rule $\Delta\ell = +1$ or -1. (b) Show that the energies of these transitions are given by

$$E_{\ell \to \ell+1} = hf_v - 2(\ell + 1)E_{0r} \qquad \ell = 0, 1, 2, \ldots$$

and

$$E_{\ell \to \ell-1} = hf_v + 2\ell E_{0r} \qquad \ell = 1, 2, \ldots$$

(In the second equation, ℓ begins at $\ell = 1$, because for $\ell \to \ell - 1$, ℓ cannot be 0.) (c) Show from these results that the absorption spectrum contains frequencies equally spaced by $2E_{0r}/h$ except for a gap of $4E_{0r}/h$ at the vibrational frequency.

5. The central frequency for the absorption band of HCl shown in Figure 31-13 is at $f_v = 8.66 \times 10^{13}$ Hz, and the absorption peaks are separated by about $\Delta f = 6 \times 10^{11}$ Hz. Use this information and the results of Problem 4 to find (a) the lowest (zero-point) vibrational energy for HCl and (b) the moment of inertia of HCl.

6. The L_α x-ray line arising from the transition $n_1 = 3$ to $n_2 = 2$ for a certain element has a wavelength of 0.3617 nm. What is the element? (See Problem 3.)

CHAPTER THIRTY-TWO

Nuclear Physics

I am become death, the destroyer of worlds.

J. ROBERT OPPENHEIMER, quoting from the Bhaghavad Gita upon watching the first nuclear explosion in New Mexico, July 1945

The atomic nucleus has a radius of only one hundred thousandth that of an atom, yet it has a complex structure and interesting properties. The study of nuclear physics began with the discovery of radioactivity long before the nuclear atom was proposed by Rutherford. With the discovery of fission in 1939 and its devastating use in World War II, nuclear physics became of prime importance to everyone. The development of nuclear fusion provides us with hope for nearly unlimited energy and fear of a nuclear holocaust. In this chapter, you will learn about the structure and properties of atomic nuclei, about the various kinds of radioactivity, and about nuclear reactions, including fission and fusion.

The first information about the atomic nucleus came from the discovery of radioactivity by A. H. Becquerel in 1896. The rays emitted by radioactive nuclei were studied by many physicists in the early decades of the twentieth century. They were first classified by E. Rutherford as alpha (α), beta (β), and gamma (γ) rays, according to their ability to penetrate matter and ionize air: α rays penetrate the least and produce the most ionization and γ rays penetrate the most and produce the least ionization. It was later found that α rays are helium nuclei, β rays are electrons (β^-) or positrons (β^+) (a positron is an antielectron, a particle identical to the electron except that its charge is positive), and γ rays are high-energy photons (that is, electromagnetic radiation of very short wavelength). The α-particle scattering experiments of H. W. Geiger and E. Marsden in 1911 and the successes of the Bohr model of the atom led to the modern picture of an atom as consisting of a tiny, massive nucleus with a radius of the order of 1 to 10 fm (1 fm $= 10^{-15}$ m) surrounded by a cloud of electrons at a relatively great distance, of the order of 0.1 nm $= 100,000$ fm, from the nucleus.

In 1919, Rutherford bombarded nitrogen with α particles and observed scintillations on a zinc sulfide screen due to protons. This was the first observation of artificial nuclear disintegration. Such experiments were extended to many other elements in the next few years.

In 1932, the neutron was discovered by J. Chadwick, the positron was discovered by C. Anderson, and the first nuclear reaction using artificially accelerated particles was observed by J. D. Cockcroft and E. T. S. Walton. It is therefore quite reasonable to mark this year as the beginning of modern nuclear physics. With the discovery of the neutron, it became possible to understand some of the properties of nuclear structure, and the advent of nuclear accelerators made many experimental studies possible without the severe limitations on particle type and energy imposed by naturally occurring radioactive sources.

In this chapter, we will first discuss some of the general properties of the atomic nucleus and the important features of radioactivity. We will then look at some nuclear reactions, including the important reactions of fission and fusion. Finally, we will look at the interactions of nuclear particles with matter, a subject important for understanding the detection of nuclear particles, the shielding of reactors, and the effects of radiation on the human body. Our discussions will be descriptive and phenomenological, with the aim of presenting general information rather than a theoretical understanding of nuclear physics.

The Cockcroft–Walton accelerator. Walton is sitting in the foreground. J. D. Cockcroft and E. T. S. Walton produced the first transmutation of nuclei with artificially accelerated particles in 1932, for which they received the Nobel prize in 1951.

32-1   Properties of Nuclei

The nucleus of an atom contains just two kinds of particles: protons and neutrons. (The hydrogen nucleus contains a single proton.) These particles have approximately the same mass (the neutron is about 0.2 percent more massive). The proton has a charge of $+e$ and the neutron is uncharged. The number of protons Z is the atomic number of the atom, which also equals the number of electrons in the atom. The number of neutrons N is approximately equal to Z for light nuclei and is slightly greater than Z for heavier nuclei. The total number of nucleons $A = N + Z$ is called the *mass number* of the nucleus. (The term *nucleon* refers to either a neutron or a proton.)

A particular nuclear species is called a *nuclide.* Two or more nuclides with the same atomic number Z but different N and A numbers are called *isotopes.* A particular nuclide is designated by its atomic symbol (H for hydrogen, He for helium, and so forth) with the mass number A as a pre-superscript. The lightest element, hydrogen, has three isotopes: ordinary hydrogen ^{1}H, whose nucleus is just a single proton; deuterium ^{2}H, whose nucleus contains one proton and one neutron; and tritium ^{3}H, whose nucleus contains one proton and two neutrons. Although the mass of the deuterium atom is about twice that of the hydrogen atom and that of the tritium atom is three times that of hydrogen, these three isotopes have the same chemical properties because they each have one electron. On the average, there are about three stable isotopes for each element (some have only one; others have five or six), so there are about 300 stable nuclei for the approximately 100 different elements. The most common isotope of the second lightest atom, helium, is ^{4}He. The nucleus of the ^{4}He atom is also known as an α particle. Another isotope of helium is ^{3}He.

Inside the nucleus, the nucleons exert strong attractive forces on their nearby neighbors. This force, called the *strong nuclear force* or the *hadronic*

Isotopes

Hadronic force

force, is much stronger than the electric force of repulsion between the protons and is very much stronger than the gravitational forces between the nucleons. (Gravity is so weak it can always be neglected in nuclear physics.) The hadronic force between two neutrons is roughly the same as that between two protons or between a neutron and proton. Two protons, of course, also exert a repulsive electric force on each other due to their charges, which tends to weaken their attraction somewhat. The hadronic force decreases rapidly with distance; when two nucleons are more than a few femtometres apart, the force is negligible.

Nuclear Size and Shape

The size and shape of the nucleus can be determined by bombarding it with high-energy particles and observing the scattering (as in the experiments of Geiger and Marsden) or, in some cases, from measurements of radioactivity. The results depend somewhat on the kind of experiment. For example, a scattering experiment using electrons measures the charge distribution of the nucleus (see Figure 32-1), whereas one using neutrons determines the region of influence of the hadronic force. Despite these differences, a wide variety of experiments suggest that most nuclei are approximately spherical, with radii given approximately by

$$R = R_0 A^{1/3}$$

32-1   Nuclear radius

where R_0 is about 1.5 fm. (Some nuclei in the rare-earth region of the periodic table are ellipsoidal, the major and minor axes differing by about 20 percent or less.) The fact that the radius of a spherical nucleus is proportional to $A^{1/3}$ implies that the volume of the nucleus is proportional to the mass number A. Since the mass of the nucleus is also approximately proportional to A, the density is approximately the same for all nuclei. The fact that a liquid drop has a constant density independent of its size has led to the analogy of the nucleus with a liquid drop, a model that has proved quite successful for understanding nuclear behavior, especially in the fission of heavy nuclei.

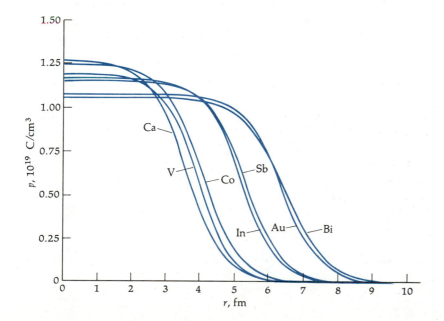

Figure 32-1
Charge density ρ versus distance r for several nuclei as determined by electron-scattering experiments.

N and Z Numbers

One characteristic of the nuclear force is that for light nuclei, greatest stability is achieved with approximately equal numbers of protons and neutrons, $N \approx Z$. For heavier nuclei, the electric repulsion among the protons leads to greater stability when there are more neutrons than protons. We can see this by looking at the N and Z numbers for the most abundant isotopes of various elements, for example, $^{16}_{8}O$, $^{40}_{20}Ca$, $^{56}_{26}Fe$, $^{207}_{82}Pb$, and $^{238}_{92}U$. (We have written the atomic number Z here as a pre-subscript for emphasis. Ordinarily, this is not needed because the atomic number is implied by the chemical symbol.)

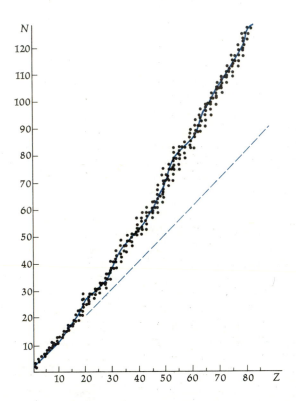

Figure 32-2
Plot of neutron number N versus proton number Z for the stable nuclides. The dashed line is $N = Z$.

Figure 32-2 shows a plot of N versus Z for the known stable nuclei. The dashed line $N = Z$ is followed for small N and Z. We can understand this tendency for N and Z to be equal by considering the total energy of A particles in a one-dimensional box (see Section 30-6). Figure 32-3 shows the energy levels for eight neutrons and for four neutrons and four protons. Like electrons, neutrons and protons each have spin quantum number $s = \frac{1}{2}$ and obey the Pauli exclusion principle. Because of the exclusion principle, only two identical particles (one with spin $+\frac{1}{2}$ and one with spin $-\frac{1}{2}$) can be in the same energy level. Since protons and neutrons are not identical, we can put two each in a level, as shown in Figure 32-3b. This gives a total energy that is less for four protons and four neutrons than for eight neutrons (or eight protons). When the Coulomb energy of repulsion is included, the result changes somewhat. This potential energy is proportional to Z^2. At large A values (and therefore large Z values), the energy is increased less by adding two neutrons than by adding one neutron and one proton because of the electrostatic repulsion.

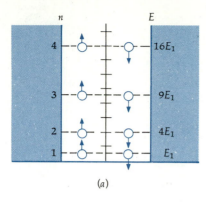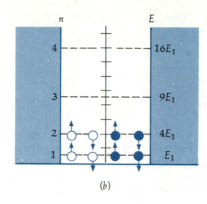

Figure 32-3
(a) Eight neutrons in a one-dimensional box. In accordance with the exclusion principle, only two neutrons (with opposite spins) can be in a given energy level. (b) Four neutrons and four protons in a one-dimensional box. Because protons and neutrons are not identical particles, two of each can be in each energy level. The total energy is much less for this case than for (a).

Mass and Binding Energy

In our study of relativistic energy (Section 28-7), we pointed out that the mass of a nucleus is not equal to the sum of the masses of the nucleons that make up the nucleus. When two or more nucleons fuse together to form a nucleus, the total mass decreases and energy is given off. Conversely, to break up a nucleus into its parts, energy must be put into the system to increase the rest mass. The energy involved is c^2 times the change in mass, where c is the speed of light in a vacuum. The difference between the rest energy of the parts of a nucleus and the rest energy of the nucleus is the total *binding energy* of the nucleus.

A unit of mass convenient for discussing atomic and nuclear masses is the *unified mass unit* (u), which is defined as one-twelfth the mass of the neutral carbon atom, which consists of the ^{12}C nucleus plus six electrons. Since 1 mol of carbon contains Avogadro's number, $N_A = 6.022 \times 10^{23}$, of atoms and has a mass of 12 g, the relation between the unified mass unit and the gram is

Unified mass unit

$$1 \, u = \frac{1 \, g}{N_A} = \frac{1 \, g}{6.022 \times 10^{23}}$$

$$= 1.6606 \times 10^{-24} \, g = 1.6606 \times 10^{-27} \, kg$$

In rough calculations, we can write

$$1 \, u = 1.66 \times 10^{-24} \, g = 1.66 \times 10^{-27} \, kg \qquad 32\text{-}2$$

The rest energy of a unified mass unit is

$$(1 \, u)c^2 = 931.5 \, MeV \qquad 32\text{-}3$$

Consider ^4He, for example, which consists of two protons and two neutrons. The mass of an atom can be accurately measured in a mass spectrometer. The mass of the ^4He atom is 4.00263 u. This includes the masses of the two electrons in the atom. The mass of the ^1H atom is 1.007825 u, and that of the neutron is 1.008665 u. The sum of the masses of two ^1H atoms plus two neutrons is $2(1.007825 \, u) + 2(1.008665 \, u) = 4.03298 \, u$, which is greater than the mass of the ^4He atom by 0.030377 u. Note that by using the masses of two ^1H atoms rather than two protons, the masses of the electrons in the atoms cancel out. We do this because it is atomic masses that are measured directly and listed in mass tables.

We can find the binding energy of the ^{4}He nucleus from this mass difference of 0.030377 u by using the energy conversion factor $1\ uc^2 = 931.5$ MeV from Equation 32-3:

$$(0.030377\ u)c^2 = 0.030377\ uc^2 \times \frac{931.5\ \text{MeV}}{1\ uc^2} = 28.20\ \text{MeV}$$

The total binding energy of ^{4}He is thus 28.2 MeV. In general, the binding energy E_b of the nucleus of an atom of atomic mass M_A containing Z protons and N neutrons is found by calculating the difference between the mass of the parts and the mass of the nucleus and then multiplying by c^2:

$$E_b = (ZM_H + Nm_n - M_A)c^2 \qquad \text{32-4}$$

Binding energy

where M_H is the mass of the ^{1}H atom and m_n is that of the neutron. The atomic masses of the neutron and some selected isotopes are listed in Table 32-1.

Table 32-1

Atomic Masses of the Neutron and Selected Isotopes

Element	Symbol	Z	Atomic mass, u
Neutron	n	0	1.008 665
Hydrogen	^{1}H	1	1.007 825
	^{2}H	1	2.014 102
	^{3}H	1	3.016 050
Helium	^{3}He	2	3.016 030
	^{4}He	2	4.002 603
Lithium	^{6}Li	3	6.015 125
Boron	^{10}B	5	10.012 939
Carbon	^{12}C	6	12.000 000
	^{14}C	6	14.003 242
Oxygen	^{16}O	8	15.994 915
Sodium	^{23}Na	11	22.989 771
Potassium	^{39}K	19	38.963 710
Iron	^{56}Fe	26	55.939 395
Copper	^{63}Cu	29	62.929 592
Silver	^{107}Ag	47	106.905 094
Gold	^{197}Au	79	196.966 541
Lead	^{208}Pb	82	207.976 650
Polonium	^{212}Po	84	211.989 629
Radon	^{222}Rn	86	222.017 531
Radium	^{226}Ra	88	226.025 360
Uranium	^{238}U	92	238.048 608
Plutonium	^{242}Pu	94	242.058 725

Example 32-1   Find the binding energy of the last neutron in ^{4}He.

From Table 32-1, the rest mass of ^{4}He is 4.00260 u and that of ^{3}He is 3.01603 u. The rest mass of ^{3}He plus that of the neutron is 3.01603 + 1.00866 = 4.02469 u. This is greater than the rest mass of ^{4}He by 4.02469 u − 4.00260 u = 0.02209 u. The binding energy of the last neutron is thus

$$(\Delta m)c^2 = (0.02209\ \text{u})c^2 \times \frac{931.5\ \text{MeV}}{1\ \text{u}c^2} = 20.58\ \text{MeV}$$

Once the atomic mass has been determined, the binding energy can be computed from Equation 32-4. It is found that the total binding energy of a nucleus is approximately proportional to the total number of nucleons A in the nucleus. Figure 32-4 shows the binding energy per nucleon E_b/A versus A. The mean value is about 8.3 MeV. The flatness of this curve shows that E_b is approximately proportional to A. This indicates that there is saturation of nuclear forces in the nucleus. That is, each nucleon bonds to only a certain number of other nucleons, independent of the total number of nucleons in the nucleus. If, for example, there were no saturation and each nucleon bonded to every other nucleon, there would be $A - 1$ bonds for each nucleon and a total of $A(A - 1)$ bonds altogether. The total binding energy, which is a measure of the energy needed to break all these bonds, would then be proportional to $A(A - 1)$ and E_b/A would not be approximately constant. Figure 32-4 indicates that, instead, there are a fixed number of bonds per nucleon, which implies that each nucleon is attracted only to its nearest neighbors. Such a situation also leads to a constant nuclear density, which is consistent with actual measurements of nuclear radii.

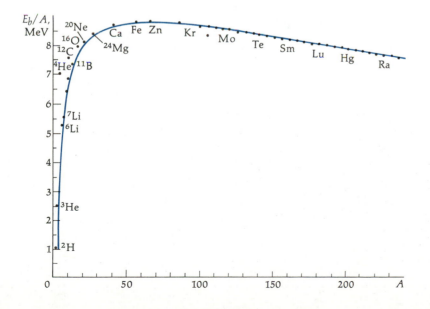

Figure 32-4
The binding energy per particle versus mass number A. For nuclei with A greater than about 50, the curve is approximately flat, indicating that the total binding energy is approximately proportional to A.

Questions

1. How does the strong nuclear force differ from the electromagnetic force?

2. What property of the strong nuclear force is indicated by the fact that all nuclei have about the same density?

3. The mass of ^{12}C, which contains six protons and six neutrons, is exactly 12.000 u by the definition of the unified mass unit. Why isn't the mass of ^{16}O, which contains eight protons and eight neutrons, exactly 16.000 u?

32-2   Radioactivity

Nuclei that are not stable are radioactive; that is, they decay to another nucleus by the emission of radiation. The term radiation includes the emission of particles, such as electrons, neutrons, or alpha particles, as well as electromagnetic radiation. There are three kinds of radioactivity, alpha (α) decay, beta (β) decay, and gamma (γ) decay.

In 1900, Rutherford discovered that the rate of emission of radioactive particles from a substance is not constant over time but decreases exponentially. *This exponential time dependence is characteristic of all radioactivity and indicates that the decay is a statistical process.* Because each nucleus is well shielded from others by the atomic electrons, pressure and temperature changes have little or no effect on nuclear properties.

Let N be the number of radioactive nuclei at some time t. For a statistical decay, in which the decay of any individual nucleus is a random event, we expect the number of nuclei that decay in some time interval Δt to be proportional to N and to Δt. Because of these decays, the number N will decrease. The change in N, ΔN, is given by

$$\Delta N = -\lambda N\, \Delta t \qquad\qquad 32\text{-}5$$

where λ is a proportionality constant called the *decay constant*. The rate of change of N, $\Delta N/\Delta t$, is proportional to N. This is characteristic of exponential decay. The solution of Equation 32-5 for the number N is

$$N = N_0 e^{-\lambda t} \qquad\qquad 32\text{-}6$$

where N_0 is the number of nuclei at $t = 0$ and $e = 2.718$, the base of natural logarithms. The number of radioactive decays per second is called the decay rate R:

$$R = -\frac{\Delta N}{\Delta t} = \lambda N = \lambda N_0 e^{-\lambda t} \qquad\qquad 32\text{-}7$$

or

$$R = R_0 e^{-\lambda t} \qquad\qquad 32\text{-}8 \quad \text{Decay rate}$$

where

$$R_0 = \lambda N_0 \qquad\qquad 32\text{-}9$$

It is the decrease in the rate of decay R that is determined experimentally. The reciprocal of the decay constant is called the mean lifetime τ:

$$\tau = \frac{1}{\lambda} \qquad\qquad 32\text{-}10$$

The mean lifetime is analogous to the time constant in the exponential decrease of the charge on a capacitor in an RC circuit discussed in Chapter 20. After a time equal to the mean life, the number of radioactive nuclei and the decay rate have each decreased to 37 percent of their original values. The *half-life* $t_{1/2}$ is defined as the time it takes for the number of nuclei and the decay rate to decrease by half. It is related to the mean lifetime by

$$t_{1/2} = 0.693\tau = \frac{0.693}{\lambda} \qquad\qquad 32\text{-}11 \quad \text{Half-life}$$

Figure 32-5 shows N versus t. If we multiply the numbers on the N axis by λ, this figure becomes a graph of R versus t. After each time interval of one half-life, the number of nuclei left and the decay rate have decreased to half of their previous values. For example, if the decay rate is R_0 initially, it will be $\frac{1}{2}R_0$ after one half-life and $(\frac{1}{2})(\frac{1}{2})R_0$ after two half-lives. After n half-lives, the decay rate will be

$$R = \left(\frac{1}{2}\right)^n R_0 \qquad\qquad 32\text{-}12$$

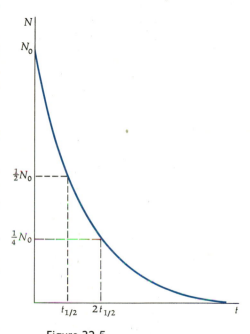

Figure 32-5
Exponential radioactive-decay law. After each half-life $t_{1/2}$, the number of nuclei remaining has decreased by one-half. The decay rate $R = \lambda N$ has the same time dependence.

The half-lives of various radioactive nuclei vary. For alpha decay, they range from a fraction of a second to millions of years. For beta decay, they range up to hours or days. For gamma decay, half-lives are usually less than a microsecond.

Example 32-2   A radioactive source has a half-life of 1 min. At time $t = 0$, it is placed near a detector, and the counting rate is observed to be 2000 counts/s. Find the counting rate at times $t = 1, 2, 3,$ and 10 min.

Since the half-life is 1 min, the counting rate will be half as great at $t = 1$ min as at $t = 0$, so $R_1 = 1000$ counts/s at 1 min, $R_2 = 500$ counts/s at $t = 2$ min, and $R_3 = 250$ counts/s at 3 min. At $t = 10$ min $= 10t_{1/2}$, the rate will be $R_{10} = (\frac{1}{2})^{10}$ (2000) = 1.95 counts/s.

Example 32-3   If the detection efficiency in Example 32-2 is 20 percent, how many radioactive nuclei are there at time $t = 0$? At time $t = 1$ min? How many nuclei decay in the first minute?

The detection efficiency depends on the distance from the source to the detector and the chance that a radioactive decay particle entering the detector will produce a count. If the source is smaller than the detector and is placed very close, about half the emitted particles will enter the detector. If the counting rate at $t = 0$ is 2000 counts/s and the efficiency is 20 percent, the decay rate at $t = 0$ must be 10,000 s^{-1}. The number of radioactive nuclei at $t = 0$ can be found from Equation 32-9:

$$R_0 = \lambda N_0$$

The decay constant λ is related to the half-life by Equation 32-11:

$$\lambda = \frac{0.693}{t_{1/2}} = \frac{0.693}{1 \text{ min}}$$

The number of nuclei at $t = 0$ is then

$$N_0 = \frac{R_0}{\lambda} = \frac{10,000 \text{ s}^{-1}}{0.693 \text{ min}^{-1}} \times \frac{60 \text{ s}}{1 \text{ min}} = 8.66 \times 10^5$$

At time $t = 1 \text{ min} = t_{1/2}$, there are half as many nuclei as at $t = 0$, so $N_1 = \frac{1}{2}(8.66 \times 10^5) = 4.33 \times 10^5$. The number of nuclei that decay in the first minute is therefore 4.33×10^5.

The SI unit of radioactive decay is the becquerel (Bq), which is defined as one decay per second:

$$1 \text{ Bq} = 1 \text{ decay/s} \qquad\qquad 32\text{-}13 \qquad \text{Becquerel}$$

A historical unit that applies to all types of radioactivity is the curie (Ci), which is defined as

$$1 \text{ Ci} = 3.7 \times 10^{10} \text{ decays/s} = 3.7 \times 10^{10} \text{ Bq} \qquad 32\text{-}14 \qquad \text{Curie}$$

The curie is the amount of radiation emitted by 1 g of radium. Since this is a very large unit, the millicurie (mCi) or microcurie (μCi) is often used.

Alpha Decay

All very heavy nuclei ($Z > 83$) are theoretically unstable to α decay. That is, for heavy nuclei, the mass of the original radioactive nucleus is greater than the sum of the masses of the decay products, which consist of an α particle and what is called the daughter nucleus. An example of α decay is the decay of Th ($Z = 90$) into Ra ($Z = 88$) plus an α particle. This is written

$$^{232}\text{Th} \rightarrow \, ^{228}\text{Ra} + \, ^4\text{He} \qquad\qquad 32\text{-}15$$

The mass of the ^{232}Th atom is 243.038124 u. The mass of the daughter atom ^{228}Ra is 228.031139 u. Adding this to the mass of ^{4}He, we get 232.03374 for the total mass on the right side of the reaction. This is less than that of ^{232}Th by 0.00438 u, which, when multiplied by 931.50 MeV/c^2, gives 4.08 MeV/c^2 for the excess rest mass of ^{232}Th over that of the decay products. Therefore, ^{232}Th is theoretically unstable to α decay, and this decay does in fact occur in nature with the emission of an α particle of energy 4.08 MeV. (The energy is actually somewhat less than 4.08 MeV because some of the decay energy is shared by the recoiling ^{228}Ra nucleus.)

In general, when a nucleus emits an α particle, both N and Z decrease by 2 and A decreases by 4. The daughter of a radioactive nucleus is often itself radioactive and decays by either α or β decay or both. If the original decaying nucleus has a mass number A that is 4 times an integer, the daughter nucleus and all those in the chain will also have mass numbers equal to 4 times an integer. Similarly, if the mass number of the original nucleus is $4n + 1$, where n is an integer, all the nuclei in the decay chain will have mass numbers given by $4n + 1$, with n decreasing by one at each decay. We see, therefore, that there are four possible α-decay chains, depending on whether A equals $4n$, $4n + 1$, $4n + 2$, or $4n + 3$, where n is an integer. All but one of these decay chains are found in nature. The $4n + 1$ series is not because its longest-lived member (other than the stable end product ^{209}Bi) is ^{237}Np, which has a half-life of only 2×10^6 y. As this is much less than the age of the earth, the series has essentially disappeared.

Figure 32-6 illustrates the thorium series, which has $A = 4n$ and begins with an α decay from ^{232}Th to ^{228}Ra. The daughter nuclide of an α decay is on the left or neutron-rich side of the stability curve (the dashed line in the figure), so it often decays by β^- decay, in which one neutron changes to a proton by emitting an electron. In Figure 32-6, ^{228}Ra decays by β decay to ^{228}Ac, which in turn decays to ^{228}Th. There are then four α decays to ^{212}Pb, which decays by β^- to ^{212}Bi. There is a branch point at ^{212}Bi, which decays either by α decay to ^{208}Tl or by β^- decay to ^{212}Po. The branches meet at the stable lead isotope ^{208}Pb.

Beta Decay

Beta decay occurs for nuclei that have too many or too few neutrons for stability, as indicated on the N-versus-Z curve of Figure 32-2. The energy released in β decay can also be determined by computing the rest mass of the original nucleus and that of the decay products. In β decay, A remains the same while Z either increases by 1 (β^- decay) or decreases by 1 (β^+ decay).

The simplest example of β decay is that of the free neutron, which decays into a proton plus an electron with a half-life of about 10.8 min. The energy of decay is 0.79 MeV, which is the difference between the rest energy of the neutron (939.57 MeV) and that of the proton plus electron ($938.28 + 0.511$ MeV). More generally, in β^- decay, a nucleus of mass number A and atomic number Z decays into one with mass number A and atomic number $Z' = Z + 1$ with the emission of an electron. If the decay energy is shared by the daughter nucleus and the emitted electron, the energy of the electron is uniquely determined by the conservation of energy and momentum. Experimentally, however, the energies of the electrons emitted in decay are observed to vary from zero to the maximum energy available. A typical energy spectrum is shown in Figure 32-7.

To explain the apparent nonconservation of energy in β decay, W. Pauli in 1930 suggested that a third particle, which he called the *neutrino*, is also emitted. The mass of the neutrino was assumed to be zero because the maximum energy of the electrons emitted is equal to the total energy available for the decay. In 1948, measurements of the momentum of the emitted electron and that of the recoiling nucleus showed that the neutrino was also needed to conserve linear momentum in beta decay. The neutrino was observed experimentally in 1957. Like the electron, the neutrino also has an antiparticle written $\bar{\nu}_e$. It is the antineutrino that is emitted in the decay of the neutron, which is written

$$n \rightarrow p + \beta^- + \bar{\nu}_e \qquad \text{32-16}$$

In β^+ decay, a proton changes into a neutron with the emission of a positron (and a neutrino). A free proton cannot decay by positron emission because of the conservation of energy (the rest mass of the neutron plus positron is greater than that of the proton), but because of binding-energy effects, a proton inside a nucleus can. A typical decay is

$$^{13}_{7}\text{N} \rightarrow {}^{13}_{6}\text{C} + \beta^+ + \nu_e \qquad \text{32-17}$$

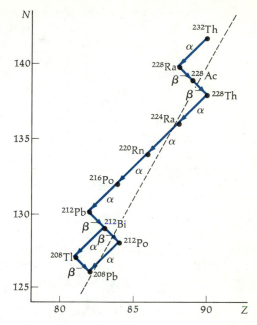

Figure 32-6
The thorium $(4n)$ α-decay series.

Neutrino

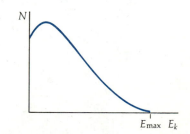

Figure 32-7
Number of electrons emitted in β decay versus kinetic energy. The fact that all the electrons do not have the same energy, E_{max}, suggests that there is another particle emitted that shares the energy available for decay.

The electrons or positrons emitted in β decay do not exist inside the nucleus. They are created in the process of decay, just as photons are created when an atom makes a transition from a higher to a lower energy state.

An important example of β decay is that of ^{14}C, which is used in radioactive dating:

Carbon dating

$$^{14}C \rightarrow {}^{14}N + \beta^- + \bar{\nu}_e \qquad\qquad 32\text{-}18$$

The half-life for this decay is 5730 y. The radioactive isotope ^{14}C is produced in the upper atmosphere from nuclear reactions caused by cosmic rays. The chemical behavior is the same for carbon atoms with ^{14}C nuclei as for those with ordinary ^{12}C nuclei. For example, atoms with these nuclei combine with oxygen to form CO_2 molecules. Since living organisms continually exchange CO_2 with the atmosphere, the ratio of ^{14}C to ^{12}C in a living organism is the same as the equilibrium ratio in the atmosphere, which is about 1.3×10^{-12}. When an organism dies, it no longer absorbs ^{14}C from the atmosphere, so the $^{14}C/^{12}C$ ratio in the organism continually decreases due to the radioactive decay of ^{14}C. A measurement of the decay rate per gram of carbon makes it possible to calculate when the organism died. The number of ^{14}C decays per minute per gram of carbon in a living organism can be calculated from the known half-life and the number of ^{14}C nuclei in a gram of carbon. The result is that there are about 15 decays per minute per gram of carbon in a living organism. Using this number and the measured number of decays per minute per gram of carbon in a nonliving sample of bone or wood or some other object containing carbon, we can determine the age of the sample. For example, if the measured rate were 7.5 decays per minute per gram, the sample would be one half-life, that is, 5730 years old.

Example 32-4   A bone containing 200 g of carbon has a β-decay rate of 400 decays/min. How old is the bone?

We first obtain a rough estimate. If the bone were from a living organism, we would expect the decay rate to be $(15 \text{ decays/min} \cdot g)(200 \text{ g}) = 3000 \text{ decays/min}$. Since $400/3000$ is roughly $\frac{1}{8}$ (actually $1/7.5$), the sample must be about 3 half-lives old, or about $3(5730)$ y. To find the age more accurately, we note that after n half-lives, the decay rate decreases by a factor of $(\frac{1}{2})^n$. We can find n from

$$\left(\frac{1}{2}\right)^n = \frac{400}{3000}$$

or

$$2^n = \frac{3000}{400} = 7.5$$

We solve for n by taking the logarithm of each side.

$$n \ln 2 = \ln 7.5$$

$$n = \frac{\ln 7.5}{\ln 2} = 2.91$$

The age is therefore $t = n t_{1/2} = 2.91(5730 \text{ y}) = 16,700 \text{ y}$.

Gamma Decay

In γ decay, a nucleus in an excited state decays to a lower-energy state by the emission of a photon. This is the nuclear counterpart of spontaneous emission by atoms and molecules. It usually happens very quickly and is observed only because it usually follows either α or β decay. In γ decay, the radioactive nucleus remains the same nucleus as it decays from an excited state to the ground state or an excited state of lower energy than that of the original state. Since the spacing of the nuclear energy levels is of the order of millions of electronvolts (compared with electronvolts in atoms), the wavelengths of the emitted photons are of the order of

$$\lambda = \frac{hc}{E} \approx \frac{1240 \text{ eV} \cdot \text{nm}}{1 \text{ MeV}} = 0.00124 \text{ nm}$$

Emission of γ rays usually follows β or α decay. For example, if a radioactive parent nucleus decays by β decay to an excited state of the daughter nucleus, the daughter nucleus often decays down to its ground state by γ emission. The mean life for γ decay is often very short. Direct measurements of mean lives down to about 10^{-11} s are possible. Measurements of lifetimes smaller than 10^{-11} s are difficult but can sometimes be accomplished by indirect methods.

Questions

4. Why do extreme changes in the temperature and pressure of a radioactive sample have little or no effect on the radioactivity?

5. Why is the decay series $A = 4n + 1$ not found in nature?

6. A decay by α emission is often followed by β decay. When this occurs, it is by β^- and not β^+ decay. Why?

7. The half-life of ^{14}C is much less than the age of the universe, yet ^{14}C is found in nature. Why?

8. What effect would a long-term variation in cosmic-ray activity have on the accuracy of ^{14}C dating?

32-3   Nuclear Reactions

Most information about nuclei is obtained by bombarding them with various particles and observing the results. Although the first experiments were limited by the need to use radiation from naturally occurring sources, they produced many important discoveries. In 1932, J. D. Cockcroft and E. T. S. Walton succeeded in producing the reaction

$$p + {}^{7}\text{Li} \rightarrow {}^{8}\text{Be} \rightarrow {}^{4}\text{He} + {}^{4}\text{He}$$

using artificially accelerated protons. At about the same time, the Van de Graaff electrostatic generator was built (by R. Van de Graaff in 1931) as was the first cyclotron (by E. O. Lawrence and M. S. Livingston in 1932). Since then, an enormous technology has been developed for accelerating and detecting particles, and many nuclear reactions have been studied.

Three-mile-long linear accelerator at Stanford University, used for accelerating electrons to very high energies.

When a particle is incident on a nucleus, several different things can happen. The particle may be scattered elastically or inelastically (in which case the nucleus is left in an excited state and decays by emitting photons or other particles), or the original particle may be absorbed and another particle or particles emitted.

When energy is released by a nuclear reaction, the reaction is said to be *exothermic*. The amount of energy released is called the Q *value* of the reaction. In an exothermic reaction, the total mass of the incoming particles is greater than that of the outgoing particles. The Q value equals c^2 times this mass difference. If the total mass of the incoming particles is less than that of the outgoing particles, energy is required for the reaction to take place and the reaction is said to be *endothermic*. The Q value of an endothermic reaction is also equal to c^2 times the initial mass minus the final mass, but in this case Q is negative. An endothermic reaction cannot take place below a certain threshold energy. The threshold energy is usually somewhat greater than $|Q|$ because the outgoing particles must have some kinetic energy to conserve momentum.

A measure of the effective size of a nucleus for a particular reaction is the *cross section* σ. If I is the number of incident particles per unit time per unit area (the incident intensity) and R is the number of reactions per unit time per nucleus, the cross section is

$$\sigma = \frac{R}{I} \qquad\qquad 32\text{-}19$$

The cross section σ has the dimensions of area. Since nuclear cross sections are of the order of the square of the nuclear radius, a convenient unit for a nuclear cross section is the *barn*, defined as

$$1 \text{ barn} = 10^{-28} \text{ m}^2 \qquad\qquad 32\text{-}20$$

The cross section for a particular reaction is a function of energy. For an endothermic reaction, it is zero for energies below the threshold.

Example 32-5   The mass of ^{7}Li is 7.016004 u. Find the Q value of the reaction

$$p + {}^7\text{Li} \rightarrow {}^4\text{He} + {}^4\text{He}$$

and state whether the reaction is exothermic or endothermic.

Using 1.007825 u for the mass of ^{1}H and 4.002603 for the mass of ^{4}He from Table 32-1, the total mass of the incoming particles is

$$m_i = 1.007825 \text{ u} + 7.016004 \text{ u} = 8.023829 \text{ u}$$

and the mass of the outgoing particles is

$$m_o = 2(4.002603 \text{ u}) = 8.005206 \text{ u}$$

Since the initial mass is greater than the final mass by $\Delta m = m_i - m_o = 8.023829 - 8.005206 = 0.018623$ u, mass is converted into energy and the reaction is exothermic. The Q value is positive and given by

$$Q = (\Delta m)c^2 = (0.018623 \text{ u})c^2(931.5 \text{ MeV}/uc^2) = 17.3 \text{ MeV}$$

Q value of reaction

Reactions with Neutrons

Nuclear reactions involving neutrons are important for understanding nuclear reactors. The most likely reaction with a nucleus for a neutron of energy of more than about 1 MeV is scattering. However, even if the scattering is elastic, the neutron loses some energy to the nucleus because the nucleus recoils. If a neutron is scattered many times in a material, its energy decreases until it is of the order of the energy of thermal motion kT, where k is Boltzmann's constant and T is the absolute temperature. (At ordinary room temperatures, kT is about 0.025 eV.) The neutron is then equally likely to gain or lose energy from a nucleus when it is elastically scattered. A neutron with energy of the order of kT is called a *thermal neutron.*

At low energies, a neutron is likely to be captured, with the emission of a γ ray from the excited nucleus. Figure 32-8 shows the neutron-capture cross section for silver as a function of energy. The large peak in this curve is called a *resonance.* Except for the resonance, the cross section varies fairly smoothly with energy, decreasing with increasing energy roughly as $1/v$, where v is the speed of the neutron. We can understand this energy dependence as follows. Consider a neutron moving with speed v near a nucleus of diameter $2R$. The time it takes the neutron to pass the nucleus is $2R/v$. Thus, the capture cross section is proportional to the time spent by the neutron in the vicinity of the nucleus. The dashed line in Figure 32-8 indicates this $1/v$ dependence. At the maximum of the resonance, the value of the cross section is very large ($\sigma > 5000$ barns) compared with the value of about 10 barns just past the resonance. Many elements show similar resonances in the neutron-capture cross section. For example, the maximum cross section for ^{113}Cd is about 57,000 barns. This material is thus very useful for shielding against low-energy neutrons.

An important nuclear reaction that involves neutrons is fission. Very heavy nuclei (atomic numbers greater than 100) are subject to spontaneous fission. They break apart into two nuclei even if left to themselves with no outside disturbance. We can understand this by considering the analogy of a charged liquid drop. If the drop is not too large, surface tension can overcome the repulsive forces of the charges and hold the drop together. There is, however, a certain maximum size beyond which the drop will be unstable and will spontaneously break apart. Spontaneous fission puts an upper limit on the size of a nucleus and therefore on the number of elements that are possible. Some heavy nuclei, uranium and polonium in particular, can be induced to fission by the capture of a neutron. We will study fission and another important nuclear reaction, fusion, in the next section.

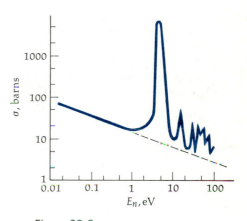

Figure 32-8
Cross section for neutron capture for silver versus energy. The straight line indicates $1/v$ dependence, which is proportional to the time spent by the neutron near the silver nucleus. Superimposed on this dependence are a large resonance and several smaller ones. At the resonance energy, the capture cross section is very large.

Questions

9. What is meant by the cross section for a nuclear reaction?

10. Why is the neutron-capture cross section (excluding resonances) proportional to $1/v$?

11. What is meant by the Q value of a reaction?

12. Why isn't there an element with $Z = 130$?

32-4   Fission, Fusion, and Nuclear Reactors

Two nuclear reactions, *fission* and *fusion*, are of particular importance. In the fission of ^{235}U, for example, the uranium nucleus is excited by the capture of a neutron and splits into two nuclei, each with about half the total mass, and emits several neutrons. The Coulomb force of repulsion drives the fission fragments apart, with the energy eventually showing up as thermal energy. In fusion, two light nuclei, such as those of deuterium and tritium (^2H and ^3H), fuse together to form a heavier nucleus (in this case, He plus a neutron). A typical fusion reaction is

$$^2H + {}^3H \rightarrow {}^4He + n + 17.6 \text{ MeV} \qquad\qquad 32\text{-}21$$

A plot of the mass difference per nucleon, $(M - Zm_p - Nm_n)/A$, versus A in units of MeV/c^2 is shown in Figure 32-9. This is just the negative of the binding-energy curve of Figure 32-4. From this figure, we can see that the rest mass per particle for both very heavy ($A \approx 200$) and very light ($A \approx 20$) nuclides is more than that for nuclides of intermediate mass. Thus, in both fission and fusion, the total rest mass decreases and energy is released.

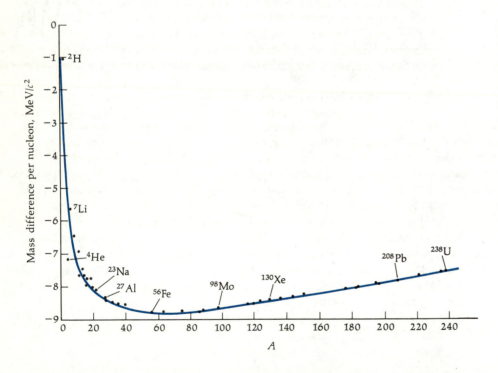

Figure 32-9
Plot of mass difference per particle, $(M - Zm_p - Nm_n)/A$, versus A in units of MeV/c^2. The rest mass per particle is less for intermediate-mass nuclei than for either very light or very heavy nuclei.

Consider, for example, the fission of a nucleus of mass number $A = 200$ into two nuclei of mass number $A = 100$. Since the rest energy for $A = 200$ is about 1 MeV per nucleon greater than that for $A = 100$, about 200 MeV per nucleus is released in such a fission. This is a large amount of energy. In the chemical reaction of combustion, for example, only about 4 eV of energy is released per molecule of oxygen burned. The energy released in fusion depends on the particular reaction. For the $^2H + {}^3H$ reaction, 17.6 MeV is released. Although this is less than the total energy released in fission, it is a greater amount of energy per unit mass. For example, the energy released per nucleon in this fusion reaction is 17.6 MeV/(5 nucleons) = 3.52 MeV/nucleon. This is about 3.5 times as great as the 1 MeV/nucleon released in fission.

Example 32-6   Calculate the total energy in kilowatt-hours released in the fission of 1 g of ^{235}U, assuming that 200 MeV is released per fission.

Since 1 mol of ^{235}U has a mass of 235 g and contains $N_A = 6.02 \times 10^{23}$ nuclei, the number of ^{235}U nuclei in 1 g is

$$N = \frac{6.02 \times 10^{23} \text{ nuclei/mol}}{235 \text{ g/mol}} = 2.56 \times 10^{21} \text{ nuclei/g}$$

The energy released per gram is then

$$\frac{200 \text{ MeV}}{\text{nucleus}} \times \frac{2.56 \times 10^{21} \text{ nuclei}}{\text{g}} \times \frac{1.6 \times 10^{-19} \text{ J}}{\text{eV}} \times \frac{1 \text{ h}}{3600 \text{ s}}$$

$$\times \frac{1 \text{ kW}}{1000 \text{ J/s}} = 2.28 \times 10^4 \text{ kW·h}$$

The application of both fission and fusion to the development of nuclear weapons has had a profound effect on our lives during the past 40 years. The peaceful application of these reactions to the development of our energy resources may have an even greater effect on the future. In this section, we will look at some of the features of fission and fusion that are important for their application in reactors to generate power.

The fission of uranium was discovered in 1939 by Hahn and Strassmann, who found by careful chemical analysis that medium-mass elements, such as barium and lanthanum, were produced in the bombardment of uranium with neutrons. The discovery that several neutrons are emitted in the fission process led to speculation concerning the possibility of using these neutrons to cause further fissions, thereby producing a chain reaction.

Uranium occurs naturally as 99.3 percent ^{238}U and 0.7 percent ^{235}U. When ^{235}U captures a neutron, the nucleus emits γ rays as it deexcites to the ground state about 15 percent of the time and fissions about 85 percent of the time. The fission process is analogous to the oscillation of a liquid drop. If the oscillations are violent enough, the drop splits in two, as shown in Figure 32-10. Using a liquid-drop model, Bohr and Wheeler calculated the critical energy E_c needed by the nucleus ^{236}U to fission. (The ^{236}U nucleus is formed momentarily by the capture of a neutron by ^{235}U.) For this nucleus, the critical energy is 5.3 MeV, which is less than the 6.4 MeV of excitation energy produced when ^{235}U captures a neutron. The addition of a neutron to ^{235}U therefore produces an excited state of the ^{236}U nucleus with more than enough energy to break apart. On the other hand, the critical energy for fission of the ^{239}U nucleus is 5.9 MeV, while the capture of a neutron by ^{238}U nucleus produces an excitation energy of only 5.2 MeV. Thus, when a neutron is captured by ^{238}U to form ^{239}U, the excitation energy is not great enough for fission. In this case, the excited ^{239}U nucleus deexcites by γ emission.

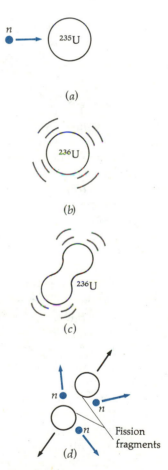

(a)

(b)

(c)

(d)

Figure 32-10
Schematic description of nuclear fission. (a) The absorption of a neutron by ^{235}U leads to (b) ^{236}U in an excited state. In (c), the oscillation of ^{236}U has become unstable. (d) The nucleus splits apart, emitting several neutrons that can produce fission in other nuclei.

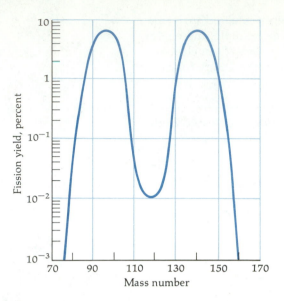

Figure 32-11
Distribution of fission fragments of
^{235}U. The splitting into two
fragments of unequal mass is more
likely than the splitting into
equal-mass fragments.

A fissioning nucleus can break into two medium-mass fragments in many different ways (see Figure 32-11). Depending on the particular reaction, 1, 2, or 3 neutrons may be emitted. The average number of neutrons emitted in the fission of ^{235}U is about 2.5. In order to sustain a chain reaction, one of these neutrons (on the average) must be captured by another ^{235}U nucleus, causing it to fission. The *reproduction constant k* of a reactor is defined as the average number of neutrons from each fission that cause a further fission. The maximum possible value of k is 2.5, but it is less than this for two important reasons: (1) neutrons may escape from the region containing fissionable nuclei, and (2) neutrons may be captured by other, nonfissioning nuclei in the reactor. If k is exactly 1, the reaction will be self-sustaining. If it is less than 1, the reaction will die out. If k is significantly greater than 1, the reaction rate will increase rapidly and "run away." In the design of nuclear bombs, such a runaway reaction is desired; *in power reactors, the value of k must be kept very nearly equal to 1.*

Since the neutrons emitted in fission have energies of the order of 1 MeV, whereas the cross section for capture leading to fission in ^{235}U is largest at thermal energies, the chain reaction can only be sustained if the neutrons are slowed down before they escape from the reactor. At high energies (1 to 2 MeV), neutrons lose energy rapidly by inelastic scattering from ^{238}U. Such collisions are quite likely since natural uranium contains 99.3 percent ^{238}U. Once the neutron energy is below the excitation energies of the reactor elements (about 1 MeV), the main process of energy loss is by elastic scattering, in which a fast neutron collides with a nucleus at rest and transfers some of its kinetic energy to that nucleus. However, energy transfers by elastic scattering are efficient only if the masses of the two bodies are comparable. A neutron will not transfer much energy in an elastic collision with a heavy uranium nucleus. Such a collision is like one between a cue ball and a bowling ball. The cue ball will be deflected by the much more massive bowling ball, but very little kinetic energy will be transferred to the bowling ball. A *moderator* consisting of material that contains light nuclei, such as water or carbon, is therefore placed around the fissionable material in the core of the reactor to slow down the neutrons. Neutrons are slowed down by elastic collisions with the nuclei of the moderator until the neutrons are in thermal equilibrium with the moderator. Because of the relatively large neutron-capture cross section of the hydrogen in water, reactors using ordinary water as a moderator cannot easily achieve $k \approx 1$ unless they use enriched uranium, in which the ^{235}U content has been increased from 0.7 percent to between 1 and 4 percent. Natural uranium can be used if heavy water (D_2O) replaces ordinary (light) water (H_2O) as the moderator. Although heavy water is expensive, most Canadian reactors use it for a moderator to avoid the cost of constructing uranium-enrichment facilities.

The ability to control the reproduction factor k is important if a power reactor is to operate with any degree of safety. There are both natural negative-feedback mechanisms and mechanical methods of control. If k is greater than 1 and the reaction rate increases, the temperature of the reactor increases. If water is used as a moderator, its density decreases with increasing temperature and it becomes a less effective moderator. A second important method is the use of control rods made of a material such as cadmium that has a very large neutron-capture cross section. When a reactor is started up, the control rods are inserted, so that k is less than 1. As the rods are gradually withdrawn from the reactor, the amount of neutron capture decreases and k increases to 1. If k becomes greater than 1, the rods are inserted again.

Control of the reaction rate of a nuclear reactor with mechanical control rods is possible only because some of the neutrons emitted in the fission process are *delayed*. The time needed to slow down a neutron from 1 or 2 MeV to thermal energy is only of the order of a millisecond. If all the neutrons emitted in fission were prompt neutrons, that is, if they were emitted immediately in the fission process, mechanical control would not be possible because the reactor would run away before the rods could be inserted. However, about 0.65 percent of the neutrons emitted are delayed by an average time of about 14 s. These neutrons are emitted not in the fission process itself but in the decay of the fission fragments. The effect of the delayed neutrons can be seen in the following examples.

Example 32-7   If the average time between fission generations (the time for a neutron emitted in one fission to cause another) is 1 ms = 0.001 s and the reproduction factor is 1.001, how long would it take for the reaction rate to double?

If $k = 1.001$, the rate after N generations is 1.001^N. Setting this rate equal to 2 and solving for N, we obtain

$$(1.001)^N = 2$$

$$N \ln (1.001) = \ln 2$$

$$N = \frac{\ln 2}{\ln (1.001)} = 693 \approx 700$$

It thus takes about 700 generations for the reaction rate to double. The time for 700 generations is then 700(0.001 s) = 0.70 s. This is not enough time for the mechanical control by insertion of control rods.

Example 32-8 Assuming that 0.65 percent of the neutrons emitted are delayed by 14 s, find the average generation time and the doubling time if $k = 1.001$.

Since 99.35 percent of the generations are 0.001 s and 0.65 percent are 14 s, the average time is

$$t_{av} = (0.9935)(0.001 \text{ s}) + (0.0065)(14 \text{ s}) = 0.092 \text{ s}$$

Note that these few delayed neutrons increase the generation time by nearly a hundredfold. The time for 700 generations is then

$$700(0.092 \text{ s}) = 64.4 \text{ s}$$

This is plenty of time for the mechanical insertion of control rods.

Because of the small fraction of ^{235}U in natural uranium and the limited capacity of enrichment facilities, reactors using the fission of ^{235}U cannot meet our energy needs for very long. Two other possibilities hold promise for the future: breeder reactors and controlled nuclear-fusion reactors.

The *breeder reactor* makes use of the fact that when the relatively plentiful but nonfissionable ^{238}U nucleus captures a neutron, it decays by β decay (half-life, 20 min) to ^{239}Np, which in turn decays by β decay (half-life, 2 days) to the fissionable nuclide ^{239}Pu. Since plutonium fissions with fast neutrons, no moderator is needed. A reactor with a mixture of ^{238}U and ^{239}Pu will breed as much or more fuel than it uses if one or more of the neutrons emitted in the fission of ^{239}Pu is captured by ^{238}U. Practical studies indicate that a typical breeder reactor can be expected to double its fuel supply in 7 to 10 years.

There are several safety problems inherent with breeder reactors. Since fast neutrons are used, the time between generations is essentially determined by the fraction of delayed neutrons, which is only 0.3 percent for the fission of ^{239}Pu. Mechanical control is therefore much more difficult. Also the material used as a heat exchanger (usually liquid sodium) normally absorbs some neutrons. When the temperature of the reactor increases, the density of the heat exchanger decreases and it absorbs fewer neutrons, leading to positive feedback. There is also the general safety problem of the large-scale production of plutonium, which is extremely poisonous and is

Breeder reactor

© 1987 Sidney Harris.

the material used in nuclear bombs. In addition, there is the problem of the storage of the radioactive waste products produced in any reactor. For example, a single 1-GW ordinary reactor produces about 3 million Ci of radioactive ^{90}Sr (half-life 28.8 y) in 1 y, enough to contaminate Lake Michigan above the legal limit if it were mixed uniformly in the lake. Despite elaborate storage methods for this and other long-lived radioactive waste products, the safety of the large-scale production of these wastes over the long term is always open to question.

The production of power from the fusion of light nuclei holds great promise because of the relative abundance of the fuel and the lack of some of the dangers presented in fission reactors. Unfortunately, technology has not yet been developed sufficiently for the use of this plentiful energy source. We will consider the $^{2}H + {}^{3}H$ reaction (Equation 32-21); other reactions present similar problems.

Because of the Coulomb repulsion between the ^{2}H and ^{3}H nuclei, kinetic energies of the order of 10 keV or greater are needed to get the nuclei close enough for the attractive nuclear forces to become effective and cause fusion. Such energies can be obtained in an accelerator, but since the scattering of one nucleus by the other is much more probable than fusion, the bombardment of one nucleus by another in an accelerator requires the input of more energy than is recovered. To obtain energy from fusion, the particles must be heated to a temperature great enough for the fusion reaction to occur as the result of random thermal collisions. The temperature corresponding to $kT = 10$ keV is of the order of 10^{8} K. (This is roughly the temperature in the interiors of stars, where such reactions actually take place.) At such temperatures, a gas consists of positive ions and negative electrons; such a gas is called a *plasma*. One of the problems arising in the attempt to produce controlled fusion reactions is that of confining such a plasma long enough for the reactions to take place. The energy required to heat a plasma is proportional to the density of the ions n, whereas the collision rate is proportional to the square of the density n^2. If τ is the confinement time, the output energy is proportional to $n^2\tau$. If the output energy is to exceed the input energy, we must have

$$C_1 n^2 \tau > C_2 n$$

where C_1 and C_2 are constants. In 1957, the British physicist J. D. Lawson evaluated these constants from estimates of the efficiencies of various hypothetical fusion reactors and derived the following relation between density and confinement time, which is known as *Lawson's criterion:*

$$n\tau > 10^{14} \text{ s·particles/cm}^3 \qquad \text{32-22}$$

Lawson's criterion

Two schemes for achieving Lawson's criterion are currently under investigation. (At present, the product of density and confinement time attainable by either scheme is about an order of magnitude too small.) In one scheme, a magnetic field is used to confine a hot plasma (see Section 21-3). In a second scheme, called *inertial confinement,* a pellet of solid deuterium and tritium is bombarded from all sides by intense pulsed laser beams of energy of the order of 10^4 J for about 10^{-8} s. Computer simulation studies indicate that the pellet should be compressed to about 10^4 times its normal density and heated to a temperature greater than 10^8 K. This should produce about 10^6 J of fusion energy in 10^{-11} s, which is so brief that confinement is achieved by inertia alone.

The Tokamak Fusion Test Reactor at the Plasma Physics Laboratory in Princeton uses a hollow donut-shaped vessel with magnetic fields to confine a hot plasma for the production of fusion energy. Plasma temperatures of 2×10^8 K have been achieved in this device.

Questions

13. Why is a moderator needed in an ordinary nuclear-fission reactor?

14. What happens to the neutrons produced in fission that do not produce another fission?

15. What is the advantage of a breeder reactor over an ordinary one? What are the disadvantages?

16. Why does fusion occur spontaneously in the sun but not on earth?

32-5   The Interaction of Particles with Matter

In this section, we will discuss briefly the main interactions of charged particles, neutrons, and photons with matter. Understanding these interactions is important for the study of nuclear detectors, shielding, and the effects of radiation on living organisms. We will not attempt to give a detailed theory but will instead indicate the principal factors involved in stopping or attenuating a beam of particles.

Charged Particles

When a charged particle traverses matter, it loses energy mainly by collisions with electrons, which often leads to the ionization of the atoms in the matter. The particle thus leaves a trail of ionized atoms in its path. If the particle energy is large compared with the ionization energies of the atoms, the energy loss in each encounter with an electron will be only a small fraction of the particle energy. (A heavy particle cannot lose a large fraction of its energy to a free electron because of the conservation of momentum. For example, when a billiard ball collides with a marble, only a very small fraction of the energy of the billiard ball can be lost.) Since the number of electrons in matter is so large, we can treat the problem as that of a continuous loss of energy. After a fairly well-defined distance, called the *range,* the particle has lost all its kinetic energy and stops. Near the end of the range, the continuous picture of energy loss is not valid because individual encounters are important. The variation of the path length for monoenergetic incident particles is called *straggling.* For electrons, this can be quite important, but for heavy particles of several MeV or more, the path lengths vary by only a few percent or less.

Figure 32-12 shows the rate of energy loss per unit path length $-\Delta E_k/\Delta x$ versus the kinetic energy of the ionizing particle. We can see from this figure that the rate of energy loss is maximum at low energies and that the rate for high energies is approximately independent of energy. Particles with kinetic energy greater than their rest energy mc^2 are called *minimum ionizing particles.* Their energy loss per unit path length is approximately constant, and the range is roughly proportional to the energy. Figure 32-13 shows the range-versus-energy curve for protons in air.

Since the energy loss of a charged particle is due to interactions with the electrons in the material, the greater the number of electrons, the greater the

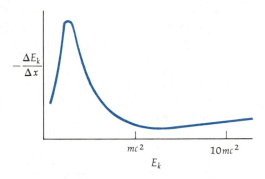

Figure 32-12
Energy loss per unit path length $-\Delta E_k/\Delta x$ versus kinetic energy for a charged particle. For particles with kinetic energy greater than their rest energy, the energy loss per unit path length is approximately constant. The distance traveled by such a particle is then approximately proportional to its energy.

rate of energy loss. The energy loss rate $-\Delta E_k/\Delta x$ is approximately proportional to the density of the material. For example, the range of a 6-MeV proton is about 40 cm in air. In water, which is about 800 times more dense, the range is only 0.5 mm.

If the energy of the charged particle is large compared with its rest energy, the particle also radiates some energy away in the form of Bremsstrahlung.

The fact that the rate of energy loss for heavy charged particles is very great at very low energies (as can be seen from the low energy peak in Figure 32-12) has important applications in nuclear radiation therapy. Figure 32-14 shows the energy loss versus penetration distance of charged particles in water. Most of the energy is deposited near the end of the range. The peak in this curve is called the *Bragg peak.* A beam of heavy charged particles can be used to destroy cancer cells at a given depth in the body without destroying healthy cells if the energy is carefully chosen so that most of the energy loss occurs at the proper depth.

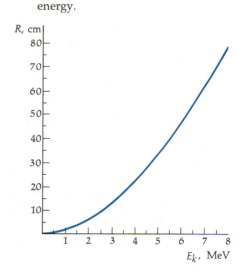

Figure 32-13
Range versus energy for protons in dry air. Except at low energies, the range is approximately linear with energy.

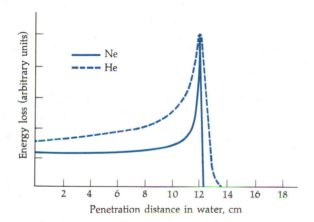

Figure 32-14
Energy loss of helium ions and neon ions in water versus depth of penetration. Most of the energy loss occurs near the end of the path in the Bragg peak. The heavier the ion, the narrower the peak.

Neutrons

Since neutrons are uncharged, they do not interact with electrons in matter. Neutrons are removed from a beam by nuclear scattering or by capture. For energies that are large compared with thermal energies (kT), the most important processes are elastic and inelastic scattering. If we have a collimated neutron beam of intensity I, any scattering or absorption will remove neutrons from the beam. This is much different from the case of a charged particle, which suffers many collisions that decrease the energy of the particle but do not remove it from the beam until its energy is essentially zero. A neutron, on the other hand, is removed from the beam at its first collision. The chance of a neutron being removed from a beam in a given path distance is proportional to the number of neutrons in the beam and to the path distance. This leads to an exponential decrease in the neutron intensity with penetration. That is, after a certain characteristic distance, half the neutrons in a beam will have been removed. Then after an equal distance, half of the remaining neutrons will have been removed, and so on. Thus, there is no well-defined range.

Photons

The intensity of a photon beam, like that of a neutron beam, decreases exponentially with distance through an absorbing material. The important processes that remove photons from a beam are the photoelectric effect, Compton scattering, and pair production. The total cross section for absorption is the sum of the partial cross sections for these three processes: σ_{pe}, σ_{cs}, and σ_{pp}. The cross section for the photoelectric effect dominates at very low energy but decreases rapidly with increasing energy. If the photon energy is large compared with the binding energy of the electrons, the electrons can be considered to be free and Compton scattering is the principal mechanism for the removal of photons from the beam. If the photon energy is greater than $2m_e c^2 = 1.02$ MeV, the photon can disappear with the creation of an electron–positron pair, a process called *pair production*. The cross section for pair production increases rapidly with the photon energy and is the dominant term in σ at high energies. Pair production cannot occur in free space. Conservation of momentum requires that a nucleus nearby absorb the momentum by recoil. The cross section for pair production is proportional to Z^2 of the absorbing matter. The three partial cross sections σ_{pe}, σ_{cs}, and σ_{pp} are shown with the total cross section as functions of energy in Figure 32-15.

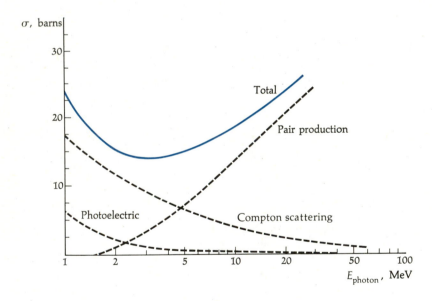

Figure 32-15
Photon interaction cross sections versus energy for lead. The cross sections for the photoelectric effect, Compton scattering, and pair production are shown by the dashed curves. The total cross section is the sum of these.

Dosage

The biological effects of radiation are principally due to the ionization produced. Even a small amount of ionization can be damaging to sensitive living cells. It can seriously disrupt their functions or even kill them. Three different quantities are used to measure these effects: exposure, absorbed dose, and dose equivalent. The *roentgen* (R) measures exposure in terms of the ionization produced in air. The roentgen has been largely replaced by the *rad* (radiation *a*bsorbed *d*ose), which measures the absorbed dose. One   Rad

rad is the amount of radiation that deposits 10^{-2} J/kg of energy in any material. The SI unit joules per kilogram is now called a *gray* (Gy):

$$1 \text{ rad} = 10^{-2} \text{ Gy} \qquad \qquad 32\text{-}23$$

If tissue is placed at a point which is exposed to 1 R, it will absorb about 1 rad.

The amount of biological damage depends not only on the energy absorbed, which is equivalent to a dependence on the number of ions formed, but also on the density of the ions. If the ions are closely spaced, as in ionization by α particles, the biological effect is enhanced. The unit *rem* (*roentgen equivalent in man*) is the dose equivalent that has the same biological effect as 1 rad of β or γ radiation:

$$1 \text{ rem} = 1 \text{ rad} \times QF \qquad \qquad 32\text{-}24 \qquad \text{Rem}$$

where QF (quality factor) is a factor that has been tabulated for α particles, protons, and neutrons of various energies. For example, QF is 1 for γ and β rays, about 4 or 5 for slow neutrons, about 10 for fast neutrons, and between 10 and 20 for α particles of energies from 5 to 10 MeV. The SI unit of dose equivalent is the seivert (Sv), defined as the gray times the quality factor:

$$1 \text{ Sv} = 1 \text{ Gy} \times QF \qquad \qquad 32\text{-}25$$

Since 1 Gy = 100 rad,

$$1 \text{ Sv} = 100 \text{ rem} \qquad \qquad 32\text{-}26$$

Table 32-2 compares the various radiation units we have discussed.

Table 32-2
Radiation and Dose Units*

Quantity	Customary unit		SI unit		Conversion
	Name	Symbol	Name	Symbol	
Energy	electronvolt	eV	joule	J	1 MeV = 1.602×10^{-13} J
Exposure	roentgen	R	coulomb per kilogram	C/kg	1 R = 2.58×10^{-4} C/kg
Absorbed dose	rad	rad or rd	gray	Gy = J/kg	1 rad = 10^{-2} J/kg = 10^{-2} Gy
Dose equivalent	rem	rem	seivert	Sv	1 rem = 10^{-2} Sv
Activity	curie	Ci	becquerel	Bq = 1/s	1 Ci = 3.7×10^{10} decays/s $= 3.7 \times 10^{10}$ Bq

* Adapted from S. C. Bushong, *The Physics Teacher*, vol. 15, no. 3, p. 135, 1977.

Our knowledge of the effects of large radiation doses comes mainly from studies of the victims of atomic bomb explosions. Doses under 25 rem over the entire body seem to have no immediate effects. Doses over 100 rem damage the blood-forming tissues, and those over 500 rem usually lead to death in a short time.

The long-time effects of sublethal doses acquired over a period of time are more difficult to measure. The chances of dying of cancer are doubled by a dose somewhere between 100 and 500 rem. Not much is known about the effects of very low-level doses. It is possible that there is some threshold dose below which the damage done is repaired so that there is no resulting increase in the chance of cancer. But it is also possible (and generally believed) that there is no such threshold and that the cancer-causing effects of radiation are proportional to the dose even at low levels. Some typical human radiation exposures are listed in Table 32-3. The largest source of artificial exposure is currently medical diagnostic x-rays. The dose received in a medical x-ray varies enormously, depending on the type of machine used, the sensitivity of the film, and so forth. Using the best procedures, a chest x-ray can be limited to 6 millirems, but the average dose for a chest x-ray is around 30 millirems. It should be clear that, given our lack of knowledge about the risks, we should limit our exposure to radiation as much as possible. Table 32-4 lists some maximum dose recommendations set by the National Council on Radiation Protection.

Table 32-3

Sources and Average Intensities of Human Radiation Exposure*

Source	mrad/y
Natural	
Cosmic rays	45
Terrestrial, external exposure	60
Terrestrial, internally deposited radionuclides	25
	130
Artificial	
Diagnostic x-rays	70
Weapons testing	<1
Power generation	<1
Occupational	≤1
	70
Total	200

* From S. C. Bushong, *The Physics Teacher,* vol. 15, no. 3, p. 135, 1977.

Table 32-4

Recommended Dose Limits*

	Maximum permissible dose equivalent for occupational exposure
Combined whole-body occupational exposure	
Prospective annual limit	5 rems in any one year
Retrospective annual limit	10–15 rems in any one year
Long-term accumulation to age N years	$(N - 18) \times 5$ rems
Skin	15 rems in any one year
Hands	75 rems in any one year (25 per quarter)
Forearms	30 rems in any one year (10 per quarter)
Other organs, tissues, and organ systems	15 rems in any one year (5 per quarter)
Fertile women (with respect to fetus)	0.5 rem in gestation period
Dose limits for nonoccupationally exposed	
Population average	0.17 rem in any one year
An individual in the population	0.5 rem in any one year
Students	0.1 rem in any one year

* Adapted from *NCRP Rep.* 39, 1971, as given in S. C. Bushong, *The Physics Teacher,* vol. 15, no. 3, p. 135, 1977.

Summary

1. Nuclei have N neutrons, Z protons, and mass number $A = N + Z$. Two or more nuclei having the same Z but different N and A are called isotopes. There are about three stable isotopes for each stable atom, on the average.

2. For low-mass nuclei, N and Z are approximately equal, whereas for heavy nuclei, N is greater than Z.

3. Many nuclei are approximately spherical in shape, with a volume proportional to A, implying that nuclear density is independent of A. The radius of a nucleus is given approximately by

$$R = R_0 A^{1/3}$$

where R_0 is about 1.5 fm.

4. The mass of a stable nucleus is less than the sum of the masses of the nucleons. The mass difference times c^2 equals the binding energy of the nucleus. The binding energy is approximately proportional to the mass number A.

5. Unstable nuclei are radioactive and decay by emitting alpha (α) particles (^{4}He nuclei), beta (β) particles (electrons or positrons), or gamma (γ) rays (photons). All radioactivity is statistical in nature and follows the exponential decay law:

$$N = N_0 e^{-\lambda t}$$

where λ is the decay constant. The rate of decay is given by

$$R = \lambda N = R_0 e^{-\lambda t}$$

The time for the number of nuclei or the decay rate to decrease by half is called the half-life:

$$t_{1/2} = \frac{0.693}{\lambda}$$

Half-lives of alpha decay range from a fraction of a second to millions of years. For beta decay, they range up to hours or days, and for gamma decay half-lives are usually less than a microsecond. The number of decays per second of one gram of radium is the curie, which is 3.7×10^{10} decays/s $= 3.7 \times 10^{10}$ Bq.

6. The cross section σ is a measure of the effective size of a nucleus for a particular nuclear reaction. Cross sections are measured in barns, where 1 barn $= 10^{-28}$ m^2. An important nuclear reaction is neutron capture, in which a neutron is captured by a nucleus with the emission of a photon. The neutron capture cross section exhibits strong resonances superimposed on a gradual $1/v$ dependence.

7. Fission occurs when some heavy elements, such as ^{235}U or ^{239}Pu, capture a neutron and split apart into two medium-mass nuclei. The two nuclei then fly apart because of electrostatic repulsion, releasing a large amount of energy. A chain reaction is possible because several neutrons are emitted in the fission. A chain reaction can be sustained if, on the average, one of the emitted neutrons is slowed down by scattering in the reactor and is then captured by another fissionable nucleus. Very heavy nuclei ($Z > 100$) are subject to spontaneous fission.

8. A large amount of energy is also released when two light nuclei, such as ^{2}H and ^{3}H, fuse together. Fusion takes place spontaneously inside the sun and other stars, where the temperature is great enough (about 10^8 K) for thermal motion to bring the charged hydrogen ions close enough to fuse. Although controlled fusion in the laboratory holds great promise as a future energy source, practical difficulties have thus far prevented its development.

9. Charged particles lose energy nearly continuously when traversing matter because they interact with the electrons in the matter. They have a fairly well-defined range in matter, which is roughly proportional to the energy of the particle and varies inversely with the density of the matter. Neutrons and photons do not have a well-defined range. Instead, the intensity of a neutron or photon beam decreases exponentially with distance. Neutrons are removed from a beam by scattering or by capture. Photons are absorbed by the photoelectric effect at very low energies and by pair production at high energies. At intermediate energies, they undergo Compton scattering from electrons in the material.

10. Radiation dosage is measured in rads, where one rad is the amount of radiation that deposits 0.01 J/kg of energy in any material. A rem is the dose that has the same biological effect as 1 rad of beta or gamma radiation. It equals the rad multiplied by the quality factor (QF), which is 1 for gamma and beta rays, about 4 or 5 for slow neutrons, about 10 for fast neutrons, and between 10 and 20 for alpha particles of energies of the order of 5 to 10 MeV.

Suggestions for Further Reading

Bulletin of the Atomic Scientists, published monthly except for July and August by the Educational Foundation for Nuclear Science, Chicago, Illinois.

Founded in 1945 by Western scientists concerned about the consequences of our newfound ability to induce nuclear fission, this magazine numbers among its sponsors many eminent physicists including Albert Einstein. Its articles deal with current issues relating mainly to nuclear weapons and the policies of countries possessing them but also to nuclear reactors.

Cerny, Joseph, and Arthur M. Poskanzer: "Exotic Light Nuclei," *Scientific American,* June 1978, p. 60.

Many isotopes of most of the light elements can be produced, but some decay so quickly that there is only just sufficient time to detect them before they are gone.

Engelman, Donald M., and Peter B. Moore: "Neutron-Scattering Studies of the Ribosome," *Scientific American,* October 1976, p. 44.

This article describes how a beam of neutrons can be used to obtain information about the relative positions of large molecules in biological structures such as the ribosome.

Furth, Harold P.: "Progress Toward a Tokamak Fusion Reactor," *Scientific American,* August 1979, p. 50.

Nuckolls, John H.: "The Feasibility of Inertial-Confinement Fusion," *Physics Today,* vol. 35, no. 9, 1982, p. 24.

These articles describe two possible methods of achieving the fusion of light nuclei under controlled conditions suitable for power generation.

Lewis, Harold W.: "The Safety of Fission Reactors," *Scientific American*, March 1980, p. 53.

Levi, Barbara G.: "Radionuclide Releases from Severe Accidents at Nuclear Power Plants," *Physics Today*, vol. 38, no. 5, 1985, p. 67.

Both of these articles deal with risk assessment studies of American fission reactor technology undertaken after the Three Mile Island reactor accident.

Perrow, Charles: *Normal Accidents: Living with High-Risk Technologies*, Basic Books, New York, 1984.

This book is also the result of a study undertaken after the Three Mile Island reactor accident. It provides an alternative view to that of risk assessment in making decisions about such complex, high-risk systems as fission reactors.

Schramm, David N.: "The Age of the Elements," *Scientific American*, January 1974, p. 69.

The age of the universe and structures in it can be determined from abundance ratios of various radioactive elements and their daughter nuclei, based on certain assumptions about the processes by which they were formed.

Schroeer, Dietrich: *Science, Technology, and the Nuclear Arms Race*, John Wiley and Sons, New York, 1984.

This book describes in a clear and understandable way the principles of operation of modern fission and fusion bombs. The technology of bomb delivery systems, nuclear deterrence, alternatives to deterrence, and arms control and disarmament are discussed with reference to the specific technologies involved.

Review

A. Objectives: After studying this chapter, you should:

1. Be able to give the order of magnitude of the radius of an atom and of a nucleus.

2. Be able to sketch the N-versus-Z curve for stable nuclei.

3. Be able to sketch the binding energy per nucleon versus A and discuss the significance of this curve for fission and fusion.

4. Know the exponential law of radioactive decay and be able to work problems using it.

5. Be able to describe the nuclear-fission chain reaction and discuss the advantages and disadvantages of fission reactors.

6. Be able to state Lawson's criterion for nuclear-fusion reactors.

7. Be able to discuss the chief mechanisms of energy loss of particles in matter and explain why some particles have well-defined ranges and others do not.

8. Be able to discuss the radiation dosage units rad and rem.

B. Define, explain, or otherwise identify:

Mass number, p. 834
Isotope, p. 834
Hadronic force, p. 834
Binding energy, p. 837
Decay constant, p. 840
Half-life, p. 841
Curie, p. 842
α decay, pp. 842–843
β decay, pp. 843–844
Neutrino, p. 843
Carbon dating, p. 844
γ decay, p. 845
Q value of reaction, p. 846
Cross section, p. 846
Thermal neutron, p. 847
Fission, p. 848
Fusion, p. 848
Moderator, p. 850
Delayed neutron, p. 851
Breeder reactor, p. 852
Plasma, p. 853
Lawson's criterion, p. 853
Inertial confinement, p. 853
Range, p. 854
Bragg peak, p. 855
Pair production, p. 856
Rad, p. 856
Rem, p. 857

C. True or false: If the statement is true, explain why. If it is false, give a counterexample.

1. The atomic nucleus contains protons, neutrons, and electrons.

2. The mass of ^{2}H is less than the mass of a proton plus a neutron.

3. After two half-lives, all the radioactive nuclei in a given sample will have decayed.

4. Exothermic reactions have no threshold energy.

5. In a breeder reactor, fuel can be produced as fast as it is consumed.

6. The rad is a measure of the energy deposited per unit mass in matter.

Exercises

Section 32-1   Properties of Nuclei

1. Give symbols for one other possible isotope of (a) ^{14}N, (b) ^{56}Fe, and (c) ^{118}Sn.

2. Calculate the binding energy and the binding energy per nucleon from the masses given in Table 32-1 for (a) ^{12}C, (b) ^{56}Fe, and (c) ^{238}U.

3. Repeat Exercise 2 for (a) ^{6}Li, (b) ^{39}K, and (c) ^{208}Pb.

4. Use Equation 32-1 to compute the radii of the following nuclei: (a) ^{16}O, (b) ^{56}Fe, and (c) ^{197}Au.

5. Find the energy needed to remove a neutron from ^{4}He.

Section 32-2   Radioactivity

6. The counting rate from a radioactive source is 4000 counts/s at time $t = 0$. After 10 s, the counting rate is 1000 counts/s. (a) What is the half-life? (b) What is the counting rate after 20 s?

7. A certain source gives 2000 counts/s at time $t = 0$. Its half-life is 2 min. (a) What is the counting rate after 4 min? (b) After 6 min? (c) After 8 min?

8. The counting rate from a radioactive source is 6400 counts/s. The half-life of the source is 10 s. Make a plot of the counting rate as a function of time for times up to 1 min. What is the decay constant for this source?

9. The counting rate from a radioactive source is 8000 counts/s at time $t = 0$, and 10 min later the rate is 1000 counts/s. (a) What is the half-life? (b) What is the decay constant? (c) What is the counting rate after 20 min?

10. The half-life of radium is 1620 y. Calculate the number of disintegrations per second of 1 g of radium, and show that the disintegration rate is approximately 1 Ci.

11. A radioactive silver foil ($t_{1/2} = 2.4$ min) is placed near a Geiger counter and 1000 counts/s are observed at time $t = 0$. (a) What is the counting rate at $t = 2.4$ min and at $t = 4.8$ min. (b) If the counting efficiency is 20 percent, how many radioactive nuclei are there at time $t = 0$? At time $t = 2.4$ min? (c) At what time will the counting rate be about 30 counts/s?

12. The stable isotope of sodium is ^{23}Na. What kind of radioactivity would you expect of (a) ^{22}Na and (b) ^{24}Na?

13. Use Table 32-1 to calculate the energy in MeV for the alpha decay of (a) ^{226}Ra and (b) ^{242}Pu.

14. A sample of wood contains 10 g of cabon and shows a ^{14}C decay rate of 100 counts/min. What is the age of the sample?

15. A bone claimed to be 10,000 years old contans 15 g of carbon. What should the decay rate of ^{14}C be for this bone?

Section 32-3   Nuclear Reactions

16. Use Table 32-1 to find the Q value for the reactions (a) ^{1}H + ^{3}H → ^{3}He + n + Q and (b) ^{2}H + ^{2}H → ^{3}He + n + Q.

17. Find the Q value for the reactions (a) ^{2}H + ^{2}H → ^{3}H + ^{1}H + Q, (b) ^{2}H + ^{3}He → ^{4}He + ^{1}H + Q, and (c) ^{6}Li + n → ^{3}H + ^{4}He + Q.

Section 32-4   Fission, Fusion, and Nuclear Reactors

18. Explain why water is more effective than lead in slowing down fast neutrons.

19. Assuming an average energy of 200 MeV per fission, calculate the number of fissions per second needed for a 500-MW reactor.

20. If the reproduction factor in a reactor is $k = 1.1$, find the number of generations needed for the power level to (a) double, (b) increase by a factor of 10, and (c) increase by a factor of 100. Find the time needed in each case (d) if there are no delayed neutrons, so the time between generations is 1 ms; and (e) if there are delayed neutrons that make the average time between generations 100 ms.

Section 32-5   The Interaction of Particles with Matter

21. The range of 4-MeV alpha particles in air ($\rho = 1.29$ mg/cm³) is 2.5 cm. Assuming the range to be inversely proportional to the density of matter, find the range of 4-MeV alpha particles in (a) water and (b) lead ($\rho = 11.2$ g/cm³).

22. The range of 6-MeV protons in air is approximately 45 cm. Find the approximate range of 6-MeV protons in (a) water and (b) lead (see Exercise 21).

23. A neutron beam has its intensity reduced by a factor of 2 in 3 cm of iron. How great a thickness of iron is needed to reduce the intensity by a factor of (a) 8 and (b) 128?

24. A 1.0-cm-thick piece of lead shielding reduces the intensity of a beam of 15-MeV gamma rays by a factor of 2. (a) By how much will 5 cm of lead reduce the intensity of this beam? (b) Approximately what thickness is needed to reduce the intensity by a factor of 1000?

Problems

1. The counting rate from a radioactive source is measured every minute. The resulting counts per second are 1000, 820, 673, 552, 453, 371, 305, 250, ... Plot the counting rate versus time, and use your graph to find the half-life of the source.

2. Show that $ke^2 = 1.44$ MeV·fm, where k is the Coulomb constant and e is the electronic charge.

3. The electrostatic potential energy of two charges q_1 and q_2, separated by a distance r, is $U = kq_1q_2/r$, where k is the Coulomb constant. (a) Use Equation 32-1 to calculate the radii of ^{2}H and ^{3}H. (b) Find the electrostatic potential energy when these two nuclei are just touching; that is, when their centers are separated by the sum of their radii.

4. (a) Calculate the radii of $^{141}_{56}$Ba and $^{92}_{36}$Kr from Equation 32-1. (b) Assume that after fission of ^{235}U into ^{141}Ba and ^{92}Kr, the two nuclei are momentarily separated by a distance r that is equal to the sum of the radii found in (a), and calculate the electrostatic potential energy for these two nuclei at this separation. (See Problem 3.) Compare your result with the measured fission energy of 175 MeV.

5. Energy is generated in the sun and other stars by fusion. One of the fusion cycles, the proton–proton cycle, consists of the following reactions:

$$^1H + {}^1H \rightarrow {}^2H + \beta^+ + \nu$$
$$^1H + {}^2H \rightarrow {}^3He + \gamma$$

followed by

$$^1H + {}^3He \rightarrow {}^4He + \beta^+ + \nu$$

(a) Show that the net effect of these reactions is

$$4\,{}^1H \rightarrow {}^4He + 2\beta^+ + 2\nu + \gamma$$

(b) Show that the rest-mass energy of 25.7 MeV is released in this cycle (not counting the energy of 1.02 MeV released when each positron meets an electron and the two annihilate). (c) The sun radiates energy at the rate of about 4×10^{26} W. Assuming this is due to the conversion of four protons into helium plus γ rays and neutrinos, which releases 26.7 MeV, what is the rate of proton consumption in the sun? How long will the sun last if it continues to radiate at its present level? (Assume that protons constitute a mass of 10^{30} kg, which is about half the total mass of the sun.)

CHAPTER THIRTY-THREE

Elementary Particles

The study of elementary particles is one of the most fascinating and exciting fields of science. The idea that all matter is composed of a small number of elementary particles has appealed to mankind for centuries, but only in recent years has theory and experiment come together to give us a picture of these building blocks of nature. There are many questions yet to be answered, but many think that we are finally on the right path to a fundamental understanding of the nature of matter.

Who sees with equal eye, as god of
  all,
A hero perish or a sparrow fall,
Atoms or systems into ruin hurl'd
And now a bubble burst, and now
  a world.

ALEXANDER POPE

Three quarks for Muster Mark

JAMES JOYCE, *Finnegans Wake*

In Dalton's atomic theory of matter (1808), the atom was considered to be the smallest indivisible constituent of matter, that is, the elementary particle. But with the discovery of the electron by Thomson (1897), the development of the Bohr–Rutherford theory of the nuclear atom (1913), and the discovery of the neutron (1932), it became clear that atoms and even nuclei have considerable structure. For a time, the structure of matter and its interactions could be described with just four "elementary" particles: the proton, neutron, electron, and photon. However, the positron, or "antielectron," was discovered in 1932, and shortly thereafter, the muon, the pion, and many other particles were predicted and discovered. Since the 1950s, enormous sums of money have been spent constructing particle accelerators of greater and greater energy in hopes of finding particles predicted by various theories. At present, we know of several hundred particles that at one time or another have been considered to be elementary, and research teams at the giant accelerator laboratories around the world are searching for and are finding new particles. Some of these have such short lifetimes (of the order of 10^{-23} s) that they can be detected only indirectly. Many are observed only in nuclear reactions in high-energy accelerators. In addition to the usual particle properties of mass, charge, and spin, new properties

"Particles, particles, particles."

have been found and given such whimsical names as strangeness, charm, color, truth, and bottomness.

In this chapter, we will first look at the various ways of classifying the multitude of particles that have been found. We will then describe the current theory of elementary particles, in which all matter in nature — from electrons or hydrogen atoms to galaxies, from the exotic particles produced in the giant accelerator laboratories to an ordinary grain of sand — is considered to be constructed from just two families of elementary particles, leptons and quarks. These fundamental particles are subject to four fundamental forces or interactions — gravitational, electromagnetic, weak, and hadronic — that ultimately describe all of the forces in nature, from ordinary friction to those involved in supernova explosions. There are thought to be just six leptons, which include the most common particle, the electron, as well as the muon, a new particle called the tauon, and three neutrinos. There are also thought to be just six quarks. A single, isolated quark has never been observed, and according to current thinking, it may be impossible ever to isolate a single quark. Nonetheless, the quark is thought to be the fundamental particle from which the proton, the neutron, the pion, and all the various particles that interact via the strong or hadronic force are constructed.

33-1   Spin and Antiparticles

There are various ways to classify particles. One important characteristic of a particle is its intrinsic spin angular momentum. We mentioned in Section 31-1 that the electron has a quantum number m_s that corresponds to the z component of its intrinsic spin characterized by the quantum number $s = \frac{1}{2}$. Protons, neutrons, neutrinos, and various other particles also have intrinsic spin characterized by the quantum number $s = \frac{1}{2}$. Such particles are called "spin-$\frac{1}{2}$" particles. Particles that have spin $\frac{1}{2}$ (or $\frac{3}{2}$, $\frac{5}{2}$, and so forth) obey the Pauli exclusion principle. There are other particles, such as pions, which we will discuss below, that have zero spin or integral spin ($s = 0, 1, 2, \ldots$). Particles with zero or integral spin do not obey the Pauli exclusion principle. Any number of these particles can be in the same quantum state.

Spin-$\frac{1}{2}$ particles are described by the Dirac equation, an extension of the Schrödinger equation to include special relativity. A feature of Dirac's theory, which was proposed in 1927, is the prediction of the existence of antiparticles. In special relativity, the energy of a particle is related to the mass and momentum of the particle by $E = \pm\sqrt{p^2c^2 + m^2c^4}$. We usually choose the positive sign and ignore the negative-energy solution with a physical argument. However, the Dirac equation requires the existence of wave functions that correspond to these negative-energy states. Dirac got around this difficulty by postulating that all the negative energy states were filled and would therefore not be observable. Only holes in this "infinite sea" of negative-energy states would be observed. This interpretation received little attention until the positron was discovered in 1932 by Carl Anderson (see Figure 33-1). Antiparticles are never created alone but always in particle–antiparticle pairs. Figure 33-2 illustrates the creation of an electron–positron pair by a photon of energy greater than $2m_ec^2$, where m_e is

Figure 33-1
The first cloud-chamber photograph of a positron track. The particle came in from the bottom and followed a circular path in the magnetic field (directed into the page). It was slowed down in traversing the 6-mm lead plate in the middle of the photograph. The direction of motion of the particle is known because of the greater curvature of the track above the plate.

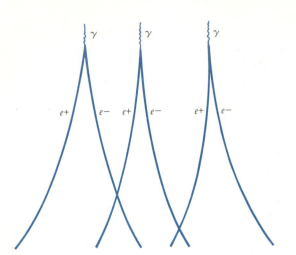

the mass of the electron. The creation of a proton–antiproton pair requires
energy of at least $2m_p c^2 = 1896$ MeV $= 1.896$ GeV, which was not available
in laboratories before the development of high-energy accelerators in the
1950s. The antiproton ($\bar{p}$) was discovered in 1955 by E. Segre and O. Chamberlain (see Figure 33-3) using a beam of protons in the Bevatron at Berkeley
to produce the reaction

$$p + p \rightarrow p + p + p + \bar{p} \qquad\qquad 33\text{-}1$$

Although the positron is stable, it has only a short-term existence in our
universe because of the large supply of electrons in matter. The fate of a
positron is annihilation according to the reaction*

$$e^+ + e^- \rightarrow \gamma + \gamma \qquad\qquad 33\text{-}2 \qquad \text{Positron annilhilation}$$

* The symbols e^+ or β^+ are used interchangeably for positrons. Similarly, electrons are denoted
by e^- or β^-.

Figure 33-3
Bubble-chamber tracks showing the
creation of a proton–antiproton
pair in the collision of an incident
25-GeV proton with a stationary
proton in liquid hydrogen.

The probability of this reaction is large only if the positron is at rest or nearly at rest. Two photons are needed to conserve linear momentum. Antiprotons annihilate with protons, also producing two gamma rays, in a similar reaction.

The fact that we call electrons *particles* and positrons *antiparticles* does not imply that positrons are less fundamental than electrons; it merely reflects the nature of our part of the universe. If our matter were made up of negative protons and positive electrons, then particles such as positive protons and negative electrons would suffer quick annihilation and would be called antiparticles.

33-2   Hadrons and Leptons

All the different forces observed in nature can be understood in terms of four basic interactions that occur among elementary particles. In order of decreasing strength, these are (1) the strong nuclear or hadronic interaction, (2) the electromagnetic interaction, (3) the weak (nuclear) interaction, and (4) the gravitational interaction. Molecular forces and most of the everyday forces that we observe between macroscopic objects (for example, friction, contact forces, and forces exerted by springs and strings) are complex manifestations of the electromagnetic interaction. Although gravity plays an important role in our life, it is so weak compared with the other basic interactions that it essentially plays no role in the interactions between elementary particles. The *weak interaction* describes the interaction between electrons or positrons and nucleons (neutrons and protons). The weak interaction is responsible for beta decay, which was discussed in Chapter 32. The *hadronic interaction* describes forces between nucleons. It is this interaction that holds nuclei together.

The four basic interactions

The four basic interactions provide a convenient structure for the classification of particles. Some particles participate in all four interactions, whereas others participate in only three, two, or one. In general, as we move from the strongest interaction, the hadronic, to the weakest, gravity, we add to the list of particles that participate.

Particles that interact via the hadronic interaction are called *hadrons*. There are two kinds of hadrons, *baryons* (from the Greek *bary*, meaning "heavy"), which have spin $\frac{1}{2}$ (or $\frac{3}{2}$, $\frac{5}{2}$, . . .), and *mesons*, which have zero or integral spin. Baryons, which include nucleons, are the most massive of the elementary particles. Mesons have intermediate masses between the mass of the electron and the mass of the proton. Their existence was predicted in 1935 by the Japanese physicist H. Yukawa in a theory of nuclear forces that involved the exchange of a particle whose mass is related to the range of the nuclear (hadronic) force. Yukawa estimated the mass of his particle to be about 100 times the mass of the electron. A particle called the muon of approximately this mass was discovered in 1937 and was thought to be Yukawa's particle, but it was later shown that the muon did not interact via the hadronic interaction. The π meson or pion was discovered about 10 years later and was shown to have the properties described by Yukawa.

Baryons and mesons

Radioactive decay can occur by any of the basic interactions. Particles that decay via the hadronic interaction have very short lifetimes of the order of

10^{-23} s, which is about the time it takes light to travel a distance equal to the diameter of a nucleus. On the other hand, the lifetimes of particles that decay via the weak interaction are much longer, of the order of 10^{-10} s.

Hadrons are rather complicated entities with complex structures. If we use the term "elementary particle" to mean a point particle without structure that is not constructed from some more elementary entity, hadrons do not fill the bill. Today, it is believed that all hadrons are composed of more fundamental entities called *quarks*, which are truly elementary particles. We will discuss quarks later. Table 33-1 lists some properties of the hadrons that are stable against decay via the hadronic interaction.

Particles that participate in the weak interaction but not in the hadronic interaction are called *leptons*. These include electrons, muons, and neutrinos, which are all less massive than the lightest hadron. The term "lep-

Leptons

Table 33-1

Hadrons Stable Against Hadronic Decay

Name	Symbol	Mass, MeV/c^2	Spin, $\hbar$	Charge, e	Antiparticle	Mean lifetime, s	Typical decay products*
Baryons							
Nucleon	p^+ (proton)†	938.3	$\frac{1}{2}$	$+1$	$\bar{p}$	infinite	
	n (neutron)	939.6	$\frac{1}{2}$	0	$\bar{n}$	930	$p^+ + e^- + \bar{\nu}_e$
Lambda	Λ^0	1116	$\frac{1}{2}$	0		2.5×10^{-10}	$p^+ + \pi^-$
Sigma	Σ^+	1189	$\frac{1}{2}$	$+1$	$\bar{\Sigma}^-$	0.8×10^{-10}	$n + \pi^+$
	Σ^0	1193	$\frac{1}{2}$	0	$\bar{\Sigma}^0$	10^{-20}	$\Lambda^0 + \gamma$
	Σ^-	1197	$\frac{1}{2}$	-1	$\bar{\Sigma}^+$	1.7×10^{-10}	$n + \pi^-$
Xi	Ξ^0	1315	$\frac{1}{2}$	0	$\bar{\Xi}^0$	3.0×10^{-10}	$\Lambda^0 + \pi^0$
	Ξ^-	1321	$\frac{1}{2}$	-1	$\bar{\Xi}^+$	1.7×10^{-10}	$\Lambda^0 + \pi^-$
Omega	Ω^-	1672	$\frac{3}{2}$	-1	$\bar{\Omega}^+$	1.3×10^{-10}	$\Xi^0 + \pi^-$
Mesons							
Pion	π^+	139.6	0	$+1$	π^-	2.6×10^{-8}	$\mu^+ + \nu_\mu$
	π^0	135	0	0	π^0	0.8×10^{-16}	$\gamma + \gamma$
	π^-	139.6	0	-1	π^+	2.6×10^{-8}	$\mu^- + \bar{\nu}_\mu$
Kaon	K^+	493.7	0	$+1$	K^-	1.24×10^{-8}	$\pi^+ + \pi^0$
	K^0	497.7	0	0	$\bar{K}^0$	0.88×10^{-10} and	$\pi^+ + \pi^-$
						5.2×10^{-8} ‡	$\pi^+ + e^- + \bar{\nu}_e$
Eta	η^0	549	0	0		2×10^{-19}	$\gamma + \gamma$

* Other decay modes also occur for most particles.

† The symbol p without the $+$ sign is often used.

‡ The K^0 has two distinct lifetimes, sometimes referred to as K^0_{short} and K^0_{long}. All other particles have a unique lifetime.

ton" meaning "light particle" was chosen to reflect the relatively small mass of these particles. However, the most recently discovered lepton, the *tauon*, found by Perl in 1975, has a mass of about 1780 MeV$/c^2$, nearly twice that of the proton (938 MeV$/c^2$), so we now have a "heavy lepton." As far as we know, leptons are point particles with no structure and can truly be considered to be elementary in the sense that they are not composed of other particles.

It is currently thought that there are six leptons, each of which has an antiparticle. They are the electron, the muon, and the tauon, and a distinct neutrino associated with each of these three particles. (The neutrino associated with the tauon has not yet been observed experimentally.) The masses of these particles are quite different. The mass of the electron is 0.511 MeV$/c^2$, the mass of the muon is 105 MeV$/c^2$, and that of the tauon is 1780 MeV$/c^2$. The neutrinos are thought to be massless, but there is considerable debate as to the possibility that they have a very small but nonzero mass, perhaps of the order of a few eV$/c^2$. Experiments designed to detect neutrinos from the sun have found a much smaller number than expected, which could be explained if the mass of the neutrino were not zero. In addition, a mass as small as 40 eV$/c^2$ would have great cosmological significance. The answer to the question of whether the universe will continue to expand indefinitely or will reach a maximum size and then begin to contract depends on the total mass in the universe, which could be quite different if the rest mass of the neutrino is merely small rather than zero.

33-3   The Conservation Laws

One of the maxims of nature is "anything that can happen does." If a conceivable decay or reaction does not occur, there must be a reason. The reason is usually expressed in terms of a conservation law. The conservation of energy rules out any decay in which the total rest mass of the decay products is greater than the initial rest mass of the particle before decay. Similarly, the conservation of linear momentum requires that when an electron and positron at rest annihilate, two photons must be emitted. A third conservation law that restricts the possible particle decays and reactions is that of electric charge. The net electric charge before a decay or reaction must equal the net charge after the decay or reaction.

There are two other conservation laws that are important in elementary particle reactions and decays: the conservation of baryon number and the conservation of lepton number. Consider the possible decay $p^+ \rightarrow \pi^0 + e^+$. This decay would conserve charge, energy, angular momentum, and linear momentum, but it does not occur because it does not conserve either leptons or baryons. The conservation of leptons and baryons implies that whenever a lepton or baryon particle is created, an antiparticle of the same type is also created. We describe this further by assigning a lepton number $L = +1$ to all leptons, $L = -1$ to all antileptons, and $L = 0$ to all other particles. Similarly, a baryon number $B = +1$ is assigned to all baryons, $B = -1$ to antibaryons, and $B = 0$ to all other particles. The baryon and lepton numbers cannot change in a reaction or decay. The conservation of baryons along with the

Lepton number and baryon number

conservation of energy implies that the least massive baryon, the proton, must be stable.

The conservation of lepton number implies that the neutrino emitted in the beta decay of the free neutron is an antineutrino:

$$n \rightarrow p^+ + e^- + \bar{\nu}_e \qquad\qquad 33\text{-}3$$

The fact that neutrinos and antineutrinos are different is illustrated by an experiment sensitive to the detection of the reaction $^{37}\text{Cl} + \bar{\nu}_e \rightarrow {}^{37}\text{Ar} + e^-$ using an intense antineutrino beam from the decay of reactor neutrons. This reaction, which does not conserve lepton number, is not observed. However, the reaction $p^+ + \bar{\nu}_e \rightarrow n + e^+$ is observed.

Not only are neutrinos and antineutrinos distinct particles, but the neutrinos associated with electrons are distinct from the neutrinos associated with muons. It is also believed that the recently discovered "heavy lepton," the tauon, has a neutrino associated with it. Electron-like leptons (e and ν_e), muon-like leptons (μ and ν_μ), and presumably tauon-like leptons (τ and ν_τ) are each separately conserved. This is easily handled by assigning separate lepton numbers L_e, L_μ, and L_τ to all particles. L_e, for instance, is $+1$ for e and ν_e, -1 for their antiparticles, and zero for all other particles.

There are some conservation laws that are not universal but apply only to certain kinds of interactions. In particular, there are quantities that are conserved in hadronic decays and reactions but not in weak decays or reactions. A particularly important quantity is *strangeness*, which was introduced by M. Gell-Mann and K. Nishijima in 1952 to explain the strange behavior of the heavy baryons and mesons. Consider the reaction for the production of lamda particles and kaons.

Strangeness

$$p^+ + \pi^- \rightarrow \Lambda^0 + \text{K}^0 \qquad\qquad 33\text{-}4$$

The cross section for this reaction is large, as would be expected since it takes place via the hadronic interaction. However, the decay times for both the Λ^0 and K^0 are of the order of 10^{-10} s, which is characteristic of the weak interaction, rather than 10^{-23} s, as would be expected for hadronic decay. Other particles showing similar behavior are called *strange particles*. These particles are always produced in pairs, never singly, even when all other conservation laws are met. This behavior is described by assigning a new property called strangeness to these particles. The strangeness S of the ordinary hadrons, the nucleons and pions, was arbitrarily chosen to be zero. The strangeness of the K^0 particle was arbitrarily chosen to be $+1$. Thus, the strangeness of the Λ^0 particle must be -1 so that strangeness is conserved in the reaction of Equation 33-4. The strangeness of other particles could then be assigned by looking at their various reactions and decays. In hadronic interactions, strangeness is conserved. In weak interactions, strangeness can change by ± 1.

The values of strangeness assigned in this way turn out to be related to the mass and charge of the particle. Figure 33-4 shows the mass of the hadronically stable baryons and mesons versus electric charge. We see from this figure that these particles cluster in groups, or "multiplets," of 1, 2, or 3 particles of approximately equal mass. The lightest baryons are the nucleons, a "doublet" consisting of a proton and a neutron. The next lightest is the lamda particle which occurs as a single particle Λ^0. The next lightest is the sigma triplet, Σ^-, Σ^0, and Σ^+. The strangeness of a multiplet is related to

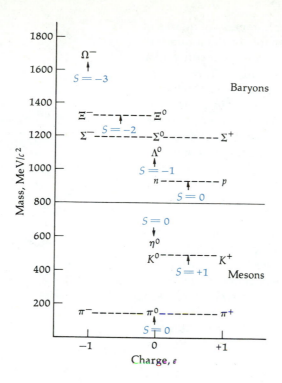

Figure 33-4
Strangeness S of hadrons shown in a plot of mass versus charge. The strangeness of a baryon charge multiplet equals the number of places the center of charge of the multiplet is displaced from that of the nucleon doublet. For mesons, the strangeness is the number of places the center of charge is displaced from that of the pion triplet.

its center of charge as shown. The nucleon doublet has center of charge at $Q = +\frac{1}{2}e$ and has strangeness 0. The Λ^0 and Σ particles have center of charge 0 and strangeness -1. The Ξ^- and Ξ^0 has center of charge $-\frac{1}{2}e$ and strangeness -2. The Ω^- has center of charge $-1e$ and strangeness -3.

33-4   The Quark Model

We have noted that leptons seem to be truly elementary particles in that they do not break down into smaller entities and they seem to have no measureable size or structure. Hadrons, on the other hand, are complex particles with size and structure that do decay into other hadrons. In addition, at the present time, there are only six known leptons, whereas there are many more hadrons. Table 33-1 includes only hadrons that are hadronically stable. Hundreds of other hadrons have been discovered and their properties, such as charge, spin, mass, strangeness, and decay schemes, have been measured.

The most important advance in our understanding of elementary particles is the quark model, proposed by M. Gell-Mann and G. Zweig in 1963. In this model, all hadrons are thought to consist of combinations of two or three truly elementary particles called *quarks*. (The name quark was chosen by Gell-Mann from a quotation from *Finnegans Wake* by James Joyce.) In the original model, quarks came in three types, called *flavors*, labeled u, d, and s

(for *up*, *down*, and *sideways* or *strange*). An unusual property of quarks is that they carry fractional electronic charges. The charge of the *u* quark is $+\frac{2}{3}e$ and that of the *d* and *s* quarks is $-\frac{1}{3}e$. Each quark has spin $\frac{1}{2}\hbar$ and a baryon number of $\frac{1}{3}$. The strangeness of the *u* and *d* quarks is 0 and that of the *s* quark is -1. Each quark has an antiquark with opposite electric charge, baryon number, and strangeness. These properties are listed in Table 33-2. Baryons consist of three quarks or three antiquarks, whereas mesons consist of a quark and an antiquark, giving them a 0 baryon number, as required. The proton consists of the combination *uud* and the neutron, *udd*. Baryons with a strangeness of -1 contain one *s* quark. All the particles listed in Table 33-1 can be constructed from these three quarks. The great strength of the quark model is that all the allowed combinations of three quarks or quark–antiquark pairs result in known hadrons.

In 1967, a fourth quark was proposed to explain some descrepancies between calculations based on the quark model and experimental determinations of certain decay rates. The fourth quark, labeled *c*, contains a new property called *charm* that, like strangeness, is conserved in strong interactions but changes by ± 1 in weak interactions. In 1975, a new heavy meson originally called the ψ/J particle and now known as the ψ particle was discovered simultaneously by a group led by S. Ting at the Brookhaven National Laboratory and a group led by B. Richter at the Stanford Linear Accelerator (SLAC) that has the properties expected of a $c\bar{c}$ combination. Since then, other mesons corresponding to various combinations, such as $c\bar{d}$ or $\bar{c}d$, have been discovered, as have baryons containing the charmed quark. More recently, two more quarks have been proposed. They are labeled *t* and *b* (for *top* and *bottom* or, as some prefer, *truth* and *beauty*). Some very heavy

Charm

Table 33-2

Properties of Quarks and Antiquarks

Flavor	Spin	Charge	Baryon number	Strangeness	Charm	Topness	Bottomness
			Quarks				
u (up)	$\frac{1}{2}\hbar$	$+\frac{2}{3}e$	$+\frac{1}{3}$	0	0	0	0
d (down)	$\frac{1}{2}\hbar$	$-\frac{1}{3}e$	$+\frac{1}{3}$	0	0	0	0
s (strange)	$\frac{1}{2}\hbar$	$-\frac{1}{3}e$	$+\frac{1}{3}$	-1	0	0	0
c (charmed)	$\frac{1}{2}\hbar$	$+\frac{2}{3}e$	$+\frac{1}{3}$	0	$+1$	0	0
t (top)	$\frac{1}{2}\hbar$	$+\frac{2}{3}e$	$+\frac{1}{3}$	0	0	$+1$	0
b (bottom)	$\frac{1}{2}\hbar$	$-\frac{1}{3}e$	$+\frac{1}{3}$	0	0	0	$+1$
			Antiquarks				
$\bar{u}$	$\frac{1}{2}\hbar$	$-\frac{2}{3}e$	$-\frac{1}{3}$	0	0	0	0
$\bar{d}$	$\frac{1}{2}\hbar$	$+\frac{1}{3}e$	$-\frac{1}{3}$	0	0	0	0
$\bar{s}$	$\frac{1}{2}\hbar$	$+\frac{1}{3}e$	$-\frac{1}{3}$	$+1$	0	0	0
$\bar{c}$	$\frac{1}{2}\hbar$	$-\frac{2}{3}e$	$-\frac{1}{3}$	0	-1	0	0
$\bar{t}$	$\frac{1}{2}\hbar$	$-\frac{2}{3}e$	$-\frac{1}{3}$	0	0	-1	0
$\bar{b}$	$\frac{1}{2}\hbar$	$+\frac{1}{3}e$	$-\frac{1}{3}$	0	0	0	-1

mesons corresponding to these quarks have already been found and others are being sought.

At present, the six quarks and six leptons (and their antiparticles) are thought to be the fundamental, elementary particles of which all matter is composed. Table 33-3 lists these particles along with some of their properties. In this table, the masses of the neutrinos are upper limits, and those of the quarks are "best guesses." Despite considerable experimental effort, no isolated quark has ever been observed. It is now believed that it is impossible to obtain an isolated quark. This would be true, for example, if the force between two quarks increases with separation distance (like the force of a spring does) rather than decreasing with distance (as do other fundamental forces, such as the electric force between two charges, the gravitational force between two masses, and the hadronic force between two hadrons).

Table 33-3

Fundamental Particles and Their Approximate Rest Energies (mc^2)

Light		Medium	Heavy	Charge
Quarks	u (360 MeV)	c (1500 MeV)	t (≈ 100 GeV)	$+\frac{2}{3}e$
	d (360 MeV)	s (540 MeV)	b (5 GeV)	$-\frac{1}{3}e$
Lepton	e^- (511 keV)	μ^- (107 MeV)	τ^- (1784 MeV)	$-1e$
	ν_e (<30 eV)	ν_μ (<0.5 MeV)	ν_τ (<250 MeV)	0

33-5   Field Particles and Unified Theories

In addition to the six fundamental leptons and six fundamental quarks, there are other particles called *field particles* that are associated with the forces exerted by one elementary particle on another. In classical electrodynamics, to avoid action at a distance, we describe the force between two electric charges as being carried by the electromagnetic field. In an alternative description, called *quantum electrodynamics*, the electromagnetic field is described in terms of photons or *field quanta* that are emitted by one charged particle and absorbed by the other. Because these photons are not observed directly, they are called *virtual photons*. The ability to emit or absorb a virtual photon is proportional to the electric charge of the particle, which comes in two varieties, + and −. The field of a single charged particle is described by virtual photons that are continuously being emitted and reabsorbed by the particle. If we put energy into the system by accelerating the charge, some of these virtual photons can be "shaken off" to become real, observable photons. The photon is said to mediate the electromagnetic interaction.

Each of the four basic interactions can be described in this way. The field quantum associated with the gravitational interaction, called the *graviton*, has not yet been observed. The gravitational "charge" analogous to electric charge is mass.   **Graviton**

The weak interaction is thought to be mediated by three field quanta called *vector bosons*: $W^\pm$ and Z^0. These particles were predicted by S. Glashow, A. Salam, and S. Weinberg in a theory called the *electroweak theory*.

In this theory, the electromagnetic and weak interactions are considered to be two different manifestations of a more fundamental electroweak interaction. These particles were first observed in 1983 by a group of over 100 scientists led by C. Rubbia using the high-energy accelerator at CERN, Geneva, Switzerland. The mass of the $W^\pm$ particle measured in this experiment, about 82 GeV/c^2, is in excellent agreement with that predicted by the electroweak theory.

The field quanta associated with the hadronic force between quarks are called *gluons*. Isolated gluons have not been observed experimentally. The "charge" responsible for the hadronic interactions comes in three varieties. In analogy with the three primary colors, these are labeled red, green, and blue, and the hadronic charge is called the *color charge*. The field theory for hadronic interactions analogous to quantum electrodynamics for electromagnetic interactions is called *quantum chromodynamics*.

Gluon

With the success of the electroweak theory, attempts have recently been made to combine the hadronic, electromagnetic, and weak interactions in a *grand unification theory* known as GUT. In one version of this theory, leptons and quarks are considered to be two aspects of a single class of particles. According to this theory, under certain conditions, a quark could change into a lepton and vice versa. Then the conservation of leptons and of baryons would be violated. One of the exciting predictions of this theory is that the proton is not stable but merely has a very long lifetime of the order of 10^{31} y. Such a long lifetime makes proton decay difficult to observe.

It was Einstein's dream to be able to describe all the forces in nature with one unified theory. Whether this will ever be accomplished is an open question. There is considerable experimental effort underway to observe the decay of the proton and to test other predictions of the grand unification theory. At the same time, much theoretical work is being done to refine these theories and construct others to further our understanding of our universe.

"Sure I'm depressed. First I find that the dollar is unstable,
then I find that the family is unstable,
and now I find that the proton is unstable."

The Origin of the Universe

Donald Goldsmith
President, Interstellar Media

For as long as human beings have realized that there was a universe, many have wondered how it came into being. Has it always been around? Has matter in the universe always been "clumped" as we find it today? Or did it once fill space smoothly and homogeneously?

The Science of Cosmology

Physics basically describes how particles in the universe move and interact, and it has a branch, *cosmology*, that attempts to describe the universe itself, its origin, and its ultimate fate. The Greek philosopher Parmenides, who lived during the sixth century B.C., wrote that all things are either existence or nonexistence, Ent or Nonent, and that the diversity of the forms we see conceals the fact that all forms of existence are varieties of a single essence, the Ent. Half a century later, another great philosopher, Democritus, built on Parmenides' idea when he wrote that all that exists is atoms and the void, that is, particles and the space between them. As insightful as they were, Parmenides and Democritus would both be amazed at the richness of nature that may have been hidden in the "void," the space between particles. The big news from particle physics today, so far as cosmology is concerned, is that space deserves more attention than it has received. In the early universe, space turns out to be where all the particles came from. To understand what this means requires a mental excursion—a trip back through fifteen billion years to the initial moments of the universe.

The Expanding Universe

Until 1929, no one had taken seriously the notion that the universe might be expanding. In that year, Edwin Hubble, the world's expert on estimating the distances to galaxies, put his estimates of those distances together with other astronomers' measurements of galaxies' motions along our line of sight—towards us or away from us—as determined by the doppler effect. Hubble found a nearly straight-line relationship. Except for a few of the closest galaxies, *all* galaxies are receding from our Milky Way galaxy. Furthermore, the more distant galaxies are receding with velocities that are directly proportional to their *distances* from us; that is, the farther away a galaxy is, the faster it is receding.

Does this mean that we are at the center of this expanding universe? No. *Any* observer anywhere in the universe should see what we see—galaxies receding at recession velocities that are proportional to the galaxies' distances from *that* observer.

A model of this apparently paradoxical concept of the universe is the surface of a balloon with dots painted on it to

represent galaxies. If you blow up the balloon, *each* of the dots moves away from all the other dots, and each dot "sees" the others receding with speeds proportional to their distances from *that* dot. In the real universe, *all* of space somehow resembles the *surface* of the balloon. Thus, the galaxies are not moving through "still" space. Rather, *space itself is expanding*, like the surface of the expanding balloon.

The Big Bang

The fact that the universe is expanding leads naturally to the questions: What was the universe like in the past? And where will it end up in the future? These two questions have interrelated answers. The problem is that we don't know these answers, only that different possible answers exist.

An expanding universe means (by definition) that galaxies are moving apart from one another everywhere. This being so, there are two basic possibilities for the past and future of the universe. The first is that new matter is being created, and so as galaxies move farther apart, the new matter coming into existence keeps the average density of matter in the universe constant. The second is that new matter is not being created, and so as galaxies move farther apart, the number of

Essay: The Origin of the Universe (continued)

galaxies in a given volume, say a million cubic light-years, diminishes and the average density of matter in the universe steadily decreases.

The first possibility is called the *steady-state model* of the universe. Initially, it may seem too strange for serious consideration. Why should we expect new matter to appear continuously as the universe expands? On reflection, however, the steady-state model is a reasonable hypothesis. After all, the matter in the universe *was* created somehow at some time in the past. And if matter was created long ago, it can hardly be completely impossible that matter is being created now.

The second possibility, that new matter is not being continuously created, leads to the *big-bang model* of the universe. If the average density of matter in the universe is steadily decreasing, then the density used to be greater. If we run a mental "movie history" of the universe backwards, *all* of the universe—all of space and all of the matter in it—moves closer together, leaving nothing "outside." If we keep running the movie backward, we approach a time—about fifteen billion years ago—when the cosmos had near-infinite density. This is the time of the big bang, the moment when everything in the universe sprang into being. We still can't say what made the big bang, but we have plenty to say about the incredible moments immediately afterwards. Before heading for those moments, however, we ought to pause along the way to admire a relic from the *relatively* early history of the universe—cosmic microwave background radiation—which seems to prove the big-bang theory correct.

Cosmic Microwave Background Radiation

Slightly more than two decades ago, scientists near Princeton, New Jersey, first detected the whisper of a sea of radiation that continually bathes all the matter in the universe. But where did this radiation come from?

Immediately after the big bang the universe was entirely different from what we see now. It consisted of tremendously dense matter at correspondingly enormous temperatures. During the first half hour after the big bang, particles collided with one another so violently that various types of nuclei were formed and destroyed. The expansion of the universe reduced the fury of these collisions, leaving behind a mix of nuclei (mostly hydrogen and helium) at temperatures of millions of degrees. Matter at such tremendously high density and temperatures quite naturally produced copious amounts of blackbody radiation consisting of high-energy photons. This radiation filled all of space and kept all the matter in a state of high agitation, breaking apart any agglomerations that might momentarily have formed. As a result of this agitation, no stars, planets, rocks, dust, or even atoms existed in any meaningful sense. If an atom did form momentarily, its electron or electrons were immediately knocked loose by yet another collision with a high-energy photon.

As the universe kept on expanding, the energies (and the frequencies) of these photons decreased. One way to see why this occurred is to imagine an observer anywhere in the universe. As time went on, the photons bombarding this observer would arrive from progressively more distant regions of space. Because of the doppler effect, the photons would arrive with progressively lower energies. After enough time had gone by, almost none of the photons had enough energy to break apart an atom upon colliding with it. This period, when the sea of photons filling the universe ceased to interact significantly with matter, is known as the *era of decoupling*. From then on, atoms could form and persist. As a result, the universe changed from a homogenous, photon-frothy foam into a place where particles, attracted to one another by the gravitational force, could stick together to form highly-localized clumps of matter much denser than the average density of the universe.

The era of decoupling occurred about a million years after the big bang. Since that time, matter and radiation in the universe have basically gone their separate ways, though in the same space. The photons still fill the universe as background radiation, most of them now having the energy of microwaves.

The existence of the cosmic microwave background radiation was predicted in 1956 by George Gamow, and it was first detected by Arno Penzias and Robert Wilson nine years later. It provides strong evidence that a big bang actually did occur. Alternative explanations for the background radiation have been proposed, but the generally accepted view is that it consists of photons that have not interacted significantly with matter since the era of decoupling, fifteen billion years ago.

Will the Universe Expand Forever?

The discovery that the universe has been expanding for the last fifteen billion years or so leads naturally to one key question. Will this expansion go on for eternity, or will it one day cease, to be replaced by a contraction of the universe, leading perhaps to another big bang in a form of cosmic recycling? The intriguing answer to this question, according to Einstein's theory of general relativity, depends on the "average density" of matter in the universe. We don't know this answer yet, but we hope to find out fairly soon.

We can think of the big bang as having somehow given a tremendous "kick" to all the pieces of the universe that sent them flying apart from one another. The future of the universe depends on the "struggle" between the tendency of matter to remain in motion at a constant velocity (unless acted upon by an unbalanced external force) and the gravitational attraction of every piece of matter for every other piece of matter in the universe. If the average density of matter in the universe is sufficiently high, gravitation will "win," and the universe will eventually contract. If it is not, the universe will expand through eternity. The critical density of matter in the universe that separates the two possibilities can be calculated from Einstein's theory. It is now approximately 10^{-30} grams per cubic centimetre. Small though this value may be, it separates two entirely different futures for the universe.

How does the actual value of the average density of matter in the universe compare with the critical value calculated from general relativity theory? The answer is that they are remarkably close, but we can't yet determine whether the actual value is larger or smaller than the critical value. Estimates based on observations of the matter that *shines* in stars or galaxies yield an average density that is about one-fiftieth of the critical density. But other estimates that take into account "dark," invisible matter as well as luminous matter are about twenty times higher, which is much closer to the critical value. This implies that most of the matter in the universe has an unknown form—perhaps neutrinos or perhaps even more exotic types of particles.

We now know that the average density of *all* matter appears to be at least one-fifth of the critical density and may equal it. In view of the fact that the average density might easily have turned out to be, say, one million times the critical density or one millionth of it, the most remarkable thing about the estimated actual density is *how close* it is to the critical density. Modern theories of the universe now imply that this closeness may not be a coincidence but may instead be the result of what happened during the earliest history of the universe.

Pushing Back the Frontiers of our Knowledge of Cosmic Time

Not long ago, cosmologists hesitated to say much about the universe at times earlier than a few seconds after the big bang. But advances in particle physics, particularly in the branch called "high-energy physics," have spurred new speculations about the first moments of the universe. They have done so by generating new theories of how matter behaves at densities and temperatures so enormous that we cannot begin to approximate them here on earth. Even

though these theories have yet to be verified, their implications for cosmology are staggering.

Grand Unification Theories of Forces

Ever since physicists began to understand how elementary particles interact, they have recognized four fundamental forces in nature: gravitational, electromagnetic, strong (hadronic), and weak. For the five decades since these four forces were first categorized, physicists have dreamed of a unified theory that would show how the four forces actually represent different aspects of a *single* force. In fact, Albert Einstein spent the last thirty years of his career seeking a theory that would unify the gravitational and electromagnetic forces. Although Einstein did not succeed, his quest has been continued by modern particle physicists.

Part of the goal of developing a unified theory was achieved during the 1970s, when scientists in Geneva, Switzerland, found evidence supporting the theory of Steven Weinberg, Abdus Salam, and Sheldon Glashow that the electromagnetic and weak forces are aspects of a single force, the "electroweak" force. At the relatively low energies of interaction that exist in our familiar world, these two aspects appear to be so different that we are justified in calling them different. But at the immensely high energies that are just barely attainable in modern particle accelerators, the two aspects are clearly a single force. At still higher energies, we can anticipate the unification of the strong force with the electroweak force in a single force described by a *grand unification theory*. The ultimate unification, which would incorporate the gravitational force as well, has been named the *super grand unification theory*.

Physicists have now proposed an entire range of grand unification theories or *GUTs*, the most popular of which was advanced by Howard Georgi and Sheldon Glashow in 1974. The GUTs have tremendous implications for cosmology because of what they predict about the time when energies in the universe were so high that the electromagnetic, strong, and weak forces were unified. According to the GUTs, during the first 10^{-32} seconds following the big bang, the universe was far different from the way it is now, not simply because it was so hot and dense, but because there was energy hidden in space itself.

The Phase Transition from the False Vacuum

According to the GUTs, the energy locked within space turned into particles as the universe underwent a *phase transition* about 10^{-32} seconds after the big bang. A phase transition typically describes the passage of matter from one state

Essay: The Origin of the Universe (continued)

to another as, for example, when water changes from a liquid to a solid or vice versa. Liquid water and ice are made of the same type of molecules, but these states have quite different properties. If you knew nothing about water, you might be amazed to discover that as you cool the liquid, it gradually becomes a solid crystalline substance as its temperature goes below 273.16 K. If you cool purified water carefully, you can manage to cool it well below 273.16 K, down to about 255 K or so, before it freezes. Once freezing begins, however, the supercooled water freezes with amazing rapidity, so that within a few seconds all of the liquid becomes ice.

The freezing of supercooled water provides an analogy to a phase transition that occurred in the early universe. Between 10^{-34} and 10^{-32} seconds after the big bang, space in the universe was a false vacuum, a cosmic analog to supercooled water. This false vacuum contained energy that would have been invisible to the eye but was capable of turning into the mass energy of matter. As the universe expanded, the energy contained within each cubic centimetre of the false vacuum did not change for the first 10^{-32} seconds. The expansion produced greater amounts of the false vacuum however, and hence the total amount of energy increased. This energy drove the expansion at such fantastic speeds that all distances doubled about every 10^{-34} seconds! The result, according to the most recent speculation, was the inflation of the universe.

The New Inflationary Theory of the Universe

During the brief interval from 10^{-34} to 10^{-32} seconds after the big bang, tiny parts of the universe doubled in size and then doubled and doubled and doubled again. Each tiny part, which may have begun these doublings with a size far smaller than that of an atomic nucleus, went through perhaps two hundred doublings and grew to a size 10^{60} times larger than its original size. This means that what today forms the visible universe, which has a radius of 10^{10} light-years, grew from a region initially less than 10^{-30} centimetres across!

At 10^{-32} seconds after the big bang, the universe cooled to temperatures below 10^{27} K, and a phase transition occurred that turned the false vacuum into the true vacuum of actual, empty space and converted the energy of the false vacuum into the mass energy of particles. Thus, this phase transition brought into being all the particles of the universe, which eventually formed the stars, the galaxies, the planets, and ourselves. Furthermore, it marked the end of the era of inflation because the energy hidden in the false vacuum that drove the inflation no longer existed. Since that time, the universe has expanded in accordance with the well-accepted big-bang model we have already described.

What Difference Does the Inflationary Theory Make?

The new inflationary model of the universe may appear to be one of those "scientific" theories that turn out not to be scientific at all because there is no way of checking them against reality. After all, the first 10^{-32} seconds of the universe are gone forever. Isn't the new inflationary theory of the universe therefore simply an untestable attempt to explain how the universe got to be as it is?

Fortunately, it is not. The new inflationary theory allows definite predictions about what we should observe around us. One of these predictions explains the coincidence we noted earlier of the average density of matter in the universe being roughly equal to the critical density that separates the ever-expanding from the eventually-contracting model of the universe. The new theory predicts, as a theoretically testable aspect of its coherence, that the average density of matter in the universe should *exactly* equal the critical density. If this is so, it means that the universe will eventually stop expanding, but only after an *infinite* amount of time. Thus, in any finite amount of time, the universe will keep on expanding, but at a slower and slower rate.

The success of the new inflationary theory of the universe in explaining why the universe is as we see it is not limited to its explanation for the average density of matter. The theory also explains the evenness of the cosmic microwave background radiation on the grounds that the inflation smoothed the universe into near perfect homogeneity. Finally (at least for our purposes), the model provides a useful starting point for calculations of how the matter that emerged from the phase transition evenly spread through space might have gone on to become highly clumped into galaxies and galaxy clusters.

If the GUTs that underlie the inflationary model of the universe prove correct, then the era of inflation may be judged to have actually occurred. In this case, we have solved some of the riddles of the universe (for example, why the average density of matter is close to the critical density) while introducing some others (for example, what might be happening in other domains that grew from subatomic size during the inflationary era). But this is the way science progresses. Clearing up one mystery quite often introduces others. This may dishearten the timid, but it encourages each new generation of scientists-to-be.

Summary

1. Particles and their antiparticles have identical masses but opposite values in their other properties, such as charge, lepton number, or baryon number. Particle–antiparticle pairs can be produced in various nuclear reactions if the energy available is greater than $2mc^2$, where m is the particle mass.

2. There are four fundamental forces or interactions: hadronic, electromagnetic, weak, and gravitational. The hadronic force, also called the strong nuclear force, is the force exerted among nucleons and is responsible for keeping the nucleus together. The weak force is associated with beta decay. In the electroweak theory, the electromagnetic and weak interactions are considered to be manifestations of a more general interaction called the electroweak interaction.

3. Hadrons are particles that interact via the hardonic interaction. There are two types, baryons and mesons. Baryons, which include the neutron and proton, have spin quantum number $\frac{1}{2}$ and are generally the most massive of particles. Mesons, which include pions and kaons, have zero or integral spin quantum numbers. Hadrons have size and structure and are therefore not the basic building blocks of matter.

4. Leptons are particles that interact via the weak interaction but not the hadronic interaction. There are six leptons, the electron e^- and its neutrino v_e, the muon μ^- and its neutrino v_μ, and the tauon τ^- and its neutrino v_τ. Leptons are true point particles with no structure.

5. Some quantities, such as energy, momentum, electric charge, angular momentum, baryon number, and lepton number, are strictly conserved in all reactions and decays. Others, such as strangeness and charm, are conserved in reactions and decays that proceed via the hadronic interactions but not in those that proceed via the weak interaction.

6. All hadrons are thought to be composed of quarks, which have fractional electric charge and baryon number $\frac{1}{3}$. Quarks come in six flavors, up (u), down (d), strange (s), charmed (c), top (t), and bottom (b). Baryons consist of three quarks whereas mesons consist of a quark and an antiquark. The properties of every hadron that has been observed experimentally can be understood in terms of the quark model. All possible hadrons that contain only the u, d, or s quarks have been observed. Several very massive hadrons predicted to contain one or more of the charmed, top, or bottom quarks have been found and others are being sought in high-energy accelerator experiments.

7. In quantum field theory, each interaction is mediated by the exchange of one or more field particles. The field particle associated with the electromagnetic interaction is the photon; that associated with the gravitational interaction is the graviton, which has not yet been observed. The field particles associated with the weak interaction are the vector bosons: $W^\pm$ and Z^0. The field particle associated with the hadronic interaction between quarks are called gluons. Gluons have not yet been observed experimentally.

Suggestions for Further Reading

Bloom, Elliott D., and Gary J. Feldman: "Quarkonium," *Scientific American,* May 1982, p. 66.

How quark-antiquark pairs have been observed and are being investigated in order to learn more about the color force binding them together is discussed in this article.

Jackson, J. David, Maury Tigner, and Stanley Wojcicki: "The Superconducting Supercollider," *Scientific American,* March 1986, p. 66.

This article explains why a giant, new colliding-beam particle accelerator 52 mi in circumference is needed to test possible extensions of the electroweak theory and new theories such as supersymmetry.

LoSecco, J. M., Frederick Reines, and Daniel Sinclair: "The Search for Proton Decay," *Scientific American,* June 1985, p. 54.

Experimenters who have taken part in the design and operation of the largest of the water Cerenkov proton-decay detectors describe the history of the search and the contribution of their experiment to it.

Quigg, Chris: "Elementary Particles and Forces," *Scientific American,* April 1985, p. 84.

This article provides an overview of leptons, hadrons, and quarks; the fundamental interactions; the unified theory of the electromagnetic and weak interactions; and possible further unification encompassing the strong force. New experimental directions such as the search for proton decay and the Superconducting Supercollider are also discussed.

Sutton, Christine: *The Particle Connection: The Most Exciting Scientific Chase Since DNA and the Double Helix,* Simon and Schuster, New York, 1984.

An excellent 175-page account of the search for the particles which mediate the electroweak interaction: W^+, W^-, and Z^0. The book requires as background only an introductory physics course.

Weinberg, Steven: "The Decay of the Proton," *Scientific American,* June 1981, p. 64.

One of the three men who shared the 1979 Nobel prize in physics for the theory unifying the electromagnetic and weak interactions describes why attempts at unification of these two forces with the strong force lead to predictions of an extremely small but finite rate of decay for the proton.

Review

A. Objectives: After studying this chapter, you should:

1. Be able to list the four basic interactions and name some of the particles that participate in each interaction.

2. Be able to name the particle or particles that mediate each of the basic interactions.

3. Be able to discuss the difference between hadrons and leptons and between baryons and mesons.

4. Be able to discuss the quark model of hadrons.

B. Define, explain, or otherwise identify:

Positron, p. 865
Antiproton, p. 866
Hadronic interaction, p. 867
Weak interaction, p. 867
Hadron, p. 867
Baryon, p. 867
Meson, p. 867
Lepton, p. 868
Muon, pp. 868–869
Tauon, p. 869
Lepton number, p. 869
Baryon number, p. 869

Strangeness, p. 870
Quark, p. 871
Charm, p. 872
ψ/J particle, p. 872
Virtual photon, p. 873
Graviton, p. 873
W particle, p. 873
Gluon, p. 874
Color charge, p. 874

C. True or false: If the statement is true, explain why. If it is false, give a counterexample.

1. All baryons are hadrons.

2. All hadrons are baryons.

3. Mesons are spin-$\frac{1}{2}$ particles.

4. The decay time for weak interactions is typically much longer than that for hadronic interactions.

5. The electron interacts with the proton by the hadronic interaction.

6. Strangeness is not conserved in weak interactions.

7. Neutrons have no charm.

Exercises

Section 33-1 Spin and Antiparticles

1. A proton and antiproton at rest annihilate according to the reaction $p^+ + \bar{p} \rightarrow \gamma + \gamma$. (a) Why must the energies of the two γ rays be equal? (b) Find the energy of each gamma ray. (c) Find the wavelength of each gamma ray.

2. Find the minimum energy of the photon needed for the following pair-production reactions: (a) $\gamma \rightarrow \pi^+ + \pi^-$, (b) $\gamma \rightarrow p + \bar{p}$, and (c) $\gamma \rightarrow \mu^- + \mu^+$.

Section 33-2 Hadrons and Leptons

There are no exercises for this section.

Section 33-3 The Conservation Laws

3. Indicate which of the decays or reactions that follow violate one or more of the conservation laws, and name the law or laws in each case: (a) $p^+ \rightarrow n + e^+ + \bar{\nu}_e$; (b) $n \rightarrow p^+ + \pi^-$; (c) $e^+ + e^- \rightarrow \gamma$; (d) $p^+ + \bar{p} \rightarrow \gamma + \gamma$; and (e) $\nu_e + p^+ \rightarrow n + e^+$.

4. Compute the change in strangeness in each reaction that follows and state whether the reaction can proceed via the hadronic interaction or the weak interaction. (*Hint:* Use Figure 33-4 to determine the strangeness of the hadrons listed.) (a) $\Omega^- \rightarrow \Xi^0 + \pi^-$; (b) $\Xi^0 \rightarrow p^+ + \pi^- + \pi^0$; (c) $\Lambda^0 \rightarrow p^+ + \pi^-$.

Section 33-4 The Quark Model

5. Find the baryon number, charge, and strangeness for the following quark combinations and identify the corresponding hadron: (a) *uud*, (b) *udd*, (c) *uus*, (d) *dds*, (e) *uss*, and (f) *dss*. (*Hint:* Use Figure 33-4 to find the strangeness of hadrons.)

6. Repeat Exercise 5 for the following quark combinations: (a) $u\bar{d}$, (b) $\bar{u}d$, and (c) $u\bar{s}$.

7. The Δ^{++} particle is a baryon that decays via the hadronic interaction. Its strangeness, charm, topness, and bottomness are all zero. What combination of quarks gives a particle with these properties?

Section 33-5 Field Particles and Unified Theories

There are no exercises for this section.

Appendix A   Review of Mathematics

In this appendix, we will review some of the basic results of algebra, geometry, and trigonometry. In many cases, we will merely state results without proof. Table A-1 lists some mathematical symbols and abbreviations.

Table A-1
Mathematical Symbols and Abbreviations

$=$	is equal to	Δx	change in x
$\neq$	is not equal to	$|x|$	absolute value of x
$\approx$	is approximately equal to	$n!$	$n(n-1)(n-2)\cdots 1$
$\sim$	is of the order of	Σ	sum
$\propto$	is proportional to	lim	limit
$>$	is greater than	$\Delta t \to 0$	Δt approaches zero
$\gg$	is much greater than		
$<$	is less than		
$\ll$	is much less than		

Equations

The following operations can be performed on mathematical equations to facilitate their solution:

1. The same quantity can be added or subtracted from each side of the equation.

2. Each side of the equation can be multiplied or divided by the same quantity.

3. Each side of the equation can be raised to the same power; for example, each side can be squared or cubed, and the square (or any other) root of each side can be taken.

It is important to understand that the preceding rules apply to each *side* of the equation and not to each *term* in the equation.

Example A-1   Solve the following equation for x:

$$(x - 3)^2 + 7 = 23$$

We first subtract 7 from each side of the equation to obtain $(x - 3)^2 = 16$. We then take the square root of each side to obtain $\pm(x - 3) = \pm 4$. We have included the plus-or-minus signs because either $(+4)^2 = 16$ or $(-4)^2 = 16$. We do not need to write $\pm$ on both sides of the equation as all the possibilities are included in $x - 3 = \pm 4$. We can now solve for x by adding 3 to each side. There are two solutions: $x = 4 + 3 = 7$ and $x = -4 + 3 = -1$. These values can be checked by substituting them into the original equation.

Example A-2   Solve the following equation for x:

$$\frac{1}{x} + \frac{1}{4} = \frac{1}{3}$$

This type of equation occurs both in geometric optics and in analyses of electric circuits. Although it is easy to solve, errors are often made. We solve it by first subtracting $\frac{1}{4}$ from each side to obtain

$$\frac{1}{x} = \frac{1}{3} - \frac{1}{4} = \frac{4}{12} - \frac{3}{12} = \frac{1}{12}$$

We then multiply each side by $12x$ to obtain $x = 12$. Note that this is equivalent to taking the reciprocal of each side of the equation. A typical mistake in handling this type of equation is to take the reciprocal of each term first to obtain $x + 4 = 3$. This operation is not allowed; it changes the relative values of each side of the equation and leads to incorrect results.

Direct and Inverse Proportion

The relationships of direct proportion and inverse proportion are so important in physics that they deserve special consideration. Often much algebraic manipulation can be avoided through a simple knowledge of these relationships. Suppose, for example, that you work for 5 days at a certain pay rate and earn $200. How much would you earn at the same pay rate if you worked 8 days? In this problem, the money earned is *directly proportional* to the time worked. We can write an equation relating the money earned M to the time worked t using a constant of proportionality R.

$$M = Rt$$

The constant of proportionality in this case is the pay rate. We can express R in dollars per day. Since $200 was earned in 5 d, the value of R is $200/(5\text{ d}) = \$40/\text{d}$. In 8 d the amount earned is then

$$M = (\$40/\text{d})(8\text{ d}) = \$320$$

However, we do not have to find the rate explicitly to work the problem. Since the amount earned in 8 d is $\frac{8}{5}$ times that earned in 5 d, this amount is $M = (\frac{8}{5})(\$200) = \$320$.

We can use the same type of example to illustrate inverse proportion. If you get a 25 percent raise, how long would you need to work to earn $200?

"It's comforting in one's later years to know that the old truths still apply."

Here we consider R to be a variable and we wish to solve for t.

$$t = \frac{M}{R}$$

In this equation, the time t is *inversely proportional* to the rate R. Thus, if the new rate is $5/4$ times the old rate, the new time will be $4/5$ times the old time, or 4 d.

There are some situations in which one quantity varies as the square or some other power of another quantity where the ideas of proportion are still very useful. Suppose, for example, that a 10-in diameter pizza costs \$8.50. How much would you expect a 12-in diameter pizza to cost? We expect the cost of a pizza to be approximately proportional to the amount of its contents, which is proportional to the area of the pizza. Since the area is in turn proportional to the square of the diameter, the cost should be proportional to the square of the diameter. If we increase the diameter by a factor of $12/10$, the area increases by a factor of $(12/10)^2 = 1.44$, so we should expect the cost to be $(1.44)(\$8.50) = \$12.24$.

Example A-3 The intensity of light from a point source varies inversely with the square of the distance from the source. If the intensity is 3.20 W/m² at 5 m from the source, what is it at 6 m from the same source?

The equation expressing the fact that the intensity varies inversely with the square of the distance can be written

$$I = \frac{C}{r^2}$$

where C is some constant. Then, if $I_1 = 3.20$ W/m² at $r_1 = 5$ m and I_2 is the unknown intensity at $r_2 = 6$ m, we have

$$\frac{I_2}{I_1} = \frac{C/r_2^2}{C/r_1^2} = \frac{r_1^2}{r_2^2} = \left(\frac{5}{6}\right)^2 = 0.694$$

The intensity at 6 m from the source is thus $I_2 = 0.694(3.20 \text{ W/m}^2) = 2.22$ W/m².

Linear Equations

An equation in which the variable or variables occur only to the first power is said to be linear because a graph of the equation is a straight line. A linear equation relating y and x can always be put in the standard form

$$y = mx + b \qquad\qquad\qquad\qquad \text{A-1}$$

where m and b are constants that may be either positive or negative. Figure A-1 shows a graph of the values of x and y that satisfy Equation A-1. The constant b, called the *intercept*, is the value of y at $x = 0$. The constant m is the slope of the line, which equals the ratio of the change in y to the corresponding change in x. In the figure, we have indicated two points on the line, x_1, y_1 and x_2, y_2, and the changes $\Delta x = x_2 - x_1$ and $\Delta y = y_2 - y_1$. The slope m is then

$$m = \frac{y_2 - y_1}{x_2 - x_1} = \frac{\Delta y}{\Delta x}$$

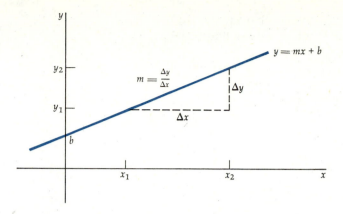

Figure A-1
Graph of the linear equation $y = mx + b$, where the constant b is the intercept and $m = \Delta y / \Delta x$ is the slope.

If x and y are both unknown, there is no unique solution for their values. Any pair of values x_1, y_1 on the line in Figure A-1 will satisfy the equation. If we have two equations, each with the same two unknowns x and y, the equations can be solved simultaneously for the unknowns.

Example A-4   Find the values of x and y that satisfy

$$3x - 2y = 8 \qquad\qquad\qquad \text{A-2}$$

and

$$y - x = 2 \qquad\qquad\qquad \text{A-3}$$

Figure A-2 shows a graph of each of these equations. At the point where the lines intersect, the values of x and y satisfy both equations.

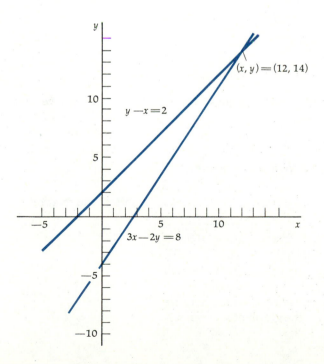

Figure A-2
Graph of Equations A-2 and A-3. At the point where the lines intersect, the values x and y satisfy both equations.

We can solve two simultaneous equations by first solving either equation for one variable in terms of the other and then substituting the result into the other equation. From Equation A-3, we have

$$y = x + 2$$

Substituting this for y in Equation A-2, we obtain

$$3x - 2(x + 2) = 8$$

$$3x - 2x - 4 = 8$$

$$x = 12$$

Then,

$$y = x + 2 = 14$$

An alternative method that is sometimes easier is to multiply one equation by a constant such that one of the unknown terms is eliminated when the equations are added or subtracted. If we multiply Equation A-3 by 2, we can add the resulting equation to Equation A-2 and eliminate y.

$$3x - 2y = 8$$

$$2y - 2x = 4$$

Adding, we obtain $3x - 2x = 12$ or $x = 12$, as before.

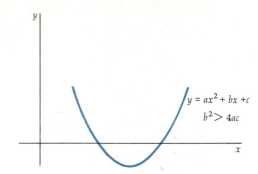

Figure A-3
Graph of y versus x when $y = ax^2 + bx + c$ for the case $b^2 > 4ac$. The two x values for which $y = 0$ satisfy the quadratic equation (Equation A-4).

Factoring

Equations can often be simplified by factoring. Three important examples are

1. Common factor: $2ax + 3ay = a(2x + 3y)$

2. Perfect square: $x^2 \pm 2xy + y^2 = (x \pm y)^2$

3. Difference of squares: $x^2 - y^2 = (x + y)(x - y)$

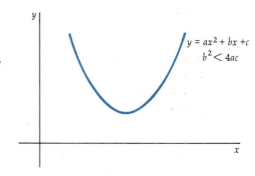

Figure A-4
Graph of y versus x when $y = ax^2 + bx + c$ for the case $b^2 < 4ac$. In this case there are no (real) x values for which $y = 0$.

The Quadratic Formula

An equation that contains a variable to the second power is called a quadratic equation. The standard form for a quadratic equation is

$$ax^2 + bx + c = 0 \qquad\qquad \text{A-4}$$

where a, b, and c are constants. The general solution of this equation is

$$x = -\frac{b}{2a} \pm \frac{1}{2a}\sqrt{b^2 - 4ac} \qquad\qquad \text{A-5}$$

Roots of the quadratic equation

When b^2 is greater than $4ac$, there are two solutions corresponding to the $+$ and $-$ signs. Figure A-3 shows a graph of $y = ax^2 + bx + c$ versus x. The curve, called a *parabola*, crosses the x axis twice. The values of x for which $y = 0$ are the solutions to Equation A-4. When $b^2 < 4ac$, the graph of y versus x does not intersect the x axis, as is shown in Figure A-4, and there are no real solutions to Equation A-4. When $b^2 = 4ac$, the graph of y versus x is tangent to the x axis at the point $x = -b/2a$.

Exponents

The notation x^n stands for the quantity obtained by multiplying x times itself n times. For example, $x^2 = x \cdot x$, and $x^3 = x \cdot x \cdot x$. The quantity n is called the *power* or the *exponent* of x. When two powers of x are multiplied, the exponents add:

$$(x^m)(x^n) = x^{m+n} \qquad\qquad\qquad \text{A-6}$$

This can be readily seen from an example:

$$x^2 x^3 = (x \cdot x)(x \cdot x \cdot x) = x^5$$

Any number raised to the 0 power is defined to be 1.

$$x^0 = 1 \qquad\qquad\qquad \text{A-7}$$

Then

$$x^n x^{-n} = x^0 = 1$$

$$x^{-n} = \frac{1}{x^n} \qquad\qquad\qquad \text{A-8}$$

When two powers are divided, the exponents subtract:

$$\frac{x^n}{x^m} = x^n x^{-m} = x^{n-m} \qquad\qquad\qquad \text{A-9}$$

Using these rules, we have

$$x^{1/2} \cdot x^{1/2} = x$$

so

$$x^{1/2} = \sqrt{x}$$

When a power is raised to another power, the exponents multiply:

$$(x^n)^m = x^{nm} \qquad\qquad\qquad \text{A-10}$$

Logarithms

When y is related to x by

$$y = a^x$$

the number x is said to be the logarithm of y to the base a and is written

$$x = \log_a y$$

If $y_1 = a^n$ and $y_2 = a^m$, then

$$y_1 y_2 = a^n a^m = a^{n+m}$$

and

$$\log_a y_1 y_2 = n + m = \log_a y_1 + \log_a y_2 \qquad\qquad\qquad \text{A-11}$$

It follows then that

$$\log_a y^n = n \log_a y \qquad\qquad\qquad \text{A-12}$$

Since $a^1 = a$ and $a^0 = 1$,

$$\log_a a = 1 \qquad\qquad\qquad \text{A-13}$$

and

$$\log_a 1 = 0 \qquad\qquad\qquad \text{A-14}$$

There are two bases in common use: base 10, called *common logarithms*, and base e ($e = 2.718$ . . .), called *natural logarithms*. When no base is specified, the base is understood to be 10. Thus, log 100 = $\log_{10}$ 100 = 2 since $100 = 10^2$.

The symbol ln is used for natural logarithms. Thus,

$$y = \ln x \qquad\qquad\qquad A\text{-}15$$

implies

$$x = e^y \qquad\qquad\qquad A\text{-}16$$

Logarithms can be changed from one base to another using

$$\ln x = (\ln 10)\log_{10} x = (2.30)\log x \qquad\qquad A\text{-}17$$

The Exponential Function

When the rate of change of a quantity is proportional to the quantity itself, the quantity increases or decreases exponentially. An example of exponential *decrease* is nuclear decay. If N is the number of radioactive nuclei at some time, then the change ΔN in some very small time interval Δt will be proportional to N and to Δt:

$$\Delta N = -\lambda N \, \Delta t$$

where λ, the constant of proportionality, is called the decay rate. The minus sign arises because a nuclear decay reduces the number of radioactive nuclei. The rate of change of N is given by

$$\frac{\Delta N}{\Delta t} = -\lambda N$$

Then N satisfies the equation

$$N = N_0 e^{-\lambda t} \qquad\qquad\qquad A\text{-}18$$

where N_0 is the number of nuclei at time $t = 0$. Figure A-5 shows N versus t.

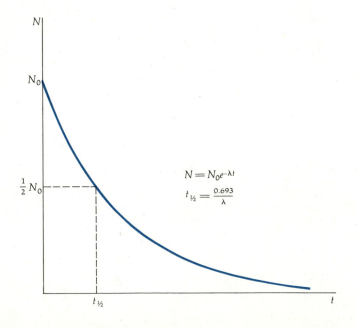

$$N = N_0 e^{-\lambda t}$$
$$t_{1/2} = \frac{0.693}{\lambda}$$

Figure A-5
Graph of N versus t when N decreases exponentially. The time $t_{1/2}$ is the time for N to decrease by one-half.

Exponential decay is characterized by a half-life $t_{1/2}$ that is related to the decay rate by

$$t_{1/2} = \frac{\ln 2}{\lambda} = \frac{0.693}{\lambda}$$

The half-life is the time it takes for N to decrease to half its original value. After each time interval of $t_{1/2}$, N decreases to half its value at the beginning of the interval.

A good example of exponential *increase* is population growth. If the number of people is N, the change in N after a small time interval Δt is given by

$$\Delta N = +\lambda N\, \Delta t$$

where λ is a constant that characterizes the rate of increase. Then N satisfies

$$N = N_0 e^{\lambda t} \qquad \text{A-19}$$

A graph of this function is shown in Figure A-6. Exponential increase is characterized by a doubling time T_2 given by

$$T_2 = \frac{\ln 2}{\lambda} = \frac{0.693}{\lambda} \qquad \text{A-20}$$

If the rate of increase λ is expressed as a percentage, $r = \lambda/100\%$, the doubling time is

$$T_2 = \frac{69.3}{r} \qquad \text{A-21}$$

For example, if the population increases by 2 percent per year, the population will double every $69.3/2 \approx 35$ years. Table A-2 lists some useful relations for exponential and logarithmic functions.

Geometry

The ratio of the circumference of a circle to its diameter is a natural number π, which has the approximate value

$$\pi = 3.141592 \ldots \qquad \text{A-22}$$

The circumference of a circle C is then related to its diameter d and its radius r by

$$C = \pi d = 2\pi r \qquad \text{(circumference of circle)} \qquad \text{A-23}$$

The area of a circle is

$$A = \pi r^2 \qquad \text{(area of circle)} \qquad \text{A-24}$$

The area of a parallelogram is the base times the height (see Figure A-7) and that of a triangle is one-half the base times the height (see Figure A-8). A sphere of radius r (see Figure A-9) has a surface area given by

$$A = 4\pi r^2 \qquad \text{(spherical surface area)} \qquad \text{A-25}$$

and a volume

$$V = \tfrac{4}{3}\pi r^3 \qquad \text{(spherical volume)} \qquad \text{A-26}$$

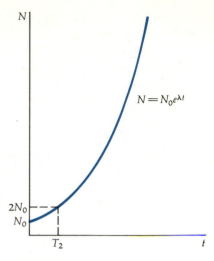

Figure A-6
Graph of N versus t when N increases exponentially. The time T_2 is the time for N to double.

Table A-2
Exponential and Logarithmic Functions

$e = 2.71828 \qquad e^0 = 1$

If $y = e^x$, then $x = \ln y$.

$e^{\ln x} = x$

$e^x e^y = e^{(x+y)}$

$(e^x)^y = e^{xy} = (e^y)^x$

$\ln e = 1 \qquad \ln 1 = 0$

$\ln xy = \ln x + \ln y$

$\ln \dfrac{x}{y} = \ln x - \ln y$

$\ln e^x = x \qquad \ln a^x = x \ln a$

$\ln x = (\ln 10) \log x = 2.3026 \log x$

$\log x = \log e \ln x = 0.43429 \ln x$

$e^x = 1 + x + \dfrac{x^2}{2!} + \dfrac{x^3}{3!} + \cdots$

$\ln (1 + x) = x - \dfrac{x^2}{2} + \dfrac{x^3}{3} - \dfrac{x^4}{4} + \cdots$

Figure A-7
Area of a parallelogram.

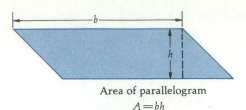

Area of parallelogram
$A = bh$

Figure A-8
Area of a triangle.

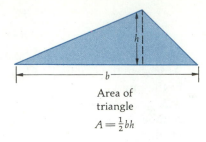

Area of
triangle
$A = \frac{1}{2}bh$

Figure A-9
Surface area and volume of a sphere.

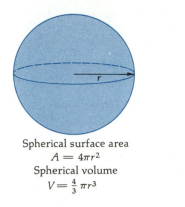

Spherical surface area
$A = 4\pi r^2$
Spherical volume
$V = \frac{4}{3}\pi r^3$

A cylinder of radius r and length L (see Figure A-10) has surface area (not including the end faces) of

$$A = 2\pi rL \qquad \text{(cylindrical surface)} \qquad \text{A-27}$$

and a volume

$$V = \pi r^2 L \qquad \text{(cylindrical volume)} \qquad \text{A-28}$$

Trigonometry

The angle between two intersecting straight lines is measured as follows. A circle is drawn with its center at the intersection of the lines, and the circular arc is divided into 360 parts called *degrees*. The number of degrees in the arc between the lines is the measure of the angle between the lines. For scientific work, a more useful measure of an angle is the radian (rad), which is defined as the length of the circular arc between the lines divided by the radius of the circle (see Figure A-11). If s is the arc length and r is the radius of the circle, the angle θ, measured in radians, is defined as

$$\theta = \frac{s}{r} \qquad \text{A-29}$$

Since the angle measured in radians is the ratio of two lengths, it is dimensionless. The radian is therefore a dimensionless unit. In dimensional analyses, it can be ignored. We can relate these two measures of angle by noting that a complete circle contains 360°; since the circumference is $2\pi r$, its radian measure is $2\pi r/r = 2\pi$ rad. The conversion relation is therefore

$$360° = 2\pi \text{ rad}$$

or

$$1 \text{ rad} = \frac{360°}{2\pi} = 57.3° \qquad \text{A-30}$$

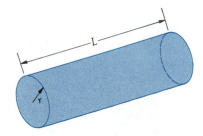

Cylindrical surface
$A = 2\pi rL$
Cylindrical volume
$V = \pi r^2 L$

Figure A-10
Surface area (not including ends) and volume of a cylinder.

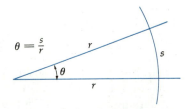

Figure A-11
The angle θ is defined to be the ratio s/r, where s is the arc length intercepted on a circle of radius r.

Figure A-12 shows some useful relations for angles.

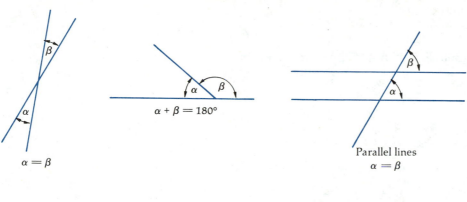

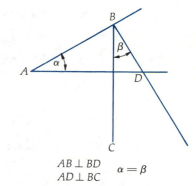

$\alpha = \beta$

$\alpha + \beta = 180°$

Parallel lines
$\alpha = \beta$

$AB \perp BD$
$AD \perp BC$ $\alpha = \beta$

Figure A-12
Some useful relations for angles.

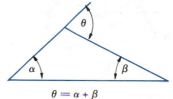

$\alpha + \beta + \gamma = 180°$

$\theta = \alpha + \beta$

Figure A-13 shows a right triangle formed by drawing the line BC perpendicular to AC. The lengths of the sides are labeled a, b, and c. The trigonometric functions $\sin \theta$, $\cos \theta$, and $\tan \theta$ are defined for an acute angle θ by

$$\sin \theta = \frac{a}{c} = \frac{\text{opposite side}}{\text{hypotenuse}} \qquad \text{A-31}$$

$$\cos \theta = \frac{b}{c} = \frac{\text{adjacent side}}{\text{hypotenuse}} \qquad \text{A-32}$$

$$\tan \theta = \frac{a}{b} = \frac{\text{opposite side}}{\text{adjacent side}} = \frac{\sin \theta}{\cos \theta} \qquad \text{A-33}$$

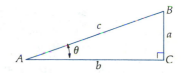

Figure A-13
A right triangle with sides a and b and hypotenuse c.

Three other trigonometric functions are defined as the inverse of these functions:

$$\sec \theta = \frac{c}{b} = \frac{1}{\cos \theta} \qquad \text{A-34}$$

$$\csc \theta = \frac{c}{a} = \frac{1}{\sin \theta} \qquad \text{A-35}$$

$$\cot \theta = \frac{b}{a} = \frac{1}{\tan \theta} = \frac{\cos \theta}{\sin \theta} \qquad \text{A-36}$$

The pythagorean theorem gives some useful identities:

$$a^2 + b^2 = c^2 \qquad \text{A-37}$$

If we divide each term in this equation by c^2, we obtain

$$\frac{a^2}{c^2} + \frac{b^2}{c^2} = 1 \qquad\qquad\qquad\qquad\text{A-38}$$

or, from the definitions of $\sin \theta$ and $\cos \theta$,

$$\sin^2 \theta + \cos^2 \theta = 1 \qquad\qquad\qquad\qquad\text{A-39}$$

Similarly, we can divide each term in Equation A-37 by a^2 or b^2 and obtain

$$1 + \cot^2 \theta = \csc^2 \theta \qquad\qquad\qquad\qquad\text{A-40}$$

and

$$1 + \tan^2 \theta = \sec^2 \theta \qquad\qquad\qquad\qquad\text{A-41}$$

These and other identities that are sometimes useful are listed in Table A-3.

Table A-3
Trigonometric Formulas

$$\sin^2 \theta + \cos^2 \theta = 1 \qquad \sec^2 \theta - \tan^2 \theta = 1 \qquad \csc^2 \theta - \cot^2 \theta = 1$$

$$\sin 2\theta = 2 \sin \theta \cos \theta$$

$$\cos 2\theta = \cos^2 \theta - \sin^2 \theta = 2 \cos^2 \theta - 1 = 1 - 2 \sin^2 \theta$$

$$\tan 2\theta = \frac{2 \tan \theta}{1 - \tan^2 \theta}$$

$$\sin \frac{1}{2}\theta = \sqrt{\frac{1 - \cos \theta}{2}} \qquad \cos \frac{1}{2}\theta = \sqrt{\frac{1 + \cos \theta}{2}} \qquad \tan \frac{1}{2}\theta = \sqrt{\frac{1 - \cos \theta}{1 + \cos \theta}}$$

$$\sin (A \pm B) = \sin A \cos B \pm \cos A \sin B$$

$$\cos (A \pm B) = \cos A \cos B \mp \sin A \sin B$$

$$\tan (A \pm B) = \frac{\tan A \pm \tan B}{1 \mp \tan A \tan B}$$

$$\sin A \pm \sin B = 2 \sin [\tfrac{1}{2}(A \pm B)] \cos [\tfrac{1}{2}(A \mp B)]$$

$$\cos A + \cos B = 2 \cos [\tfrac{1}{2}(A + B)] \cos [\tfrac{1}{2}(A - B)]$$

$$\cos A - \cos B = 2 \sin [\tfrac{1}{2}(A + B)] \sin [\tfrac{1}{2}(B - A)]$$

$$\tan A \pm \tan B = \frac{\sin (A \pm B)}{\cos A \cos B}$$

Example A-5   Use the isosceles right triangle shown in Figure A-14 to find the sine, cosine, and tangent of 45°.

It is clear from the figure that the two acute angles of this triangle are equal. Since the sum of the three angles in a triangle must equal 180° and the right angle is 90°, each acute angle must be 45°. If we multiply each side of any triangle by a common factor, we obtain a similar triangle with the same angles as before. Since the trigonometric functions involve the ratios of only two sides of a triangle, we can choose any convenient length for one side. Let the equal sides of this triangle have a length of 1 unit. The length of the hypotenuse can then be found from the pythagorean theorem (Equation A-37):

$$c = \sqrt{a^2 + b^2} = \sqrt{1^2 + 1^2} = \sqrt{2} \text{ units}$$

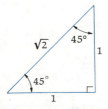

Figure A-14
An isosceles right triangle for Example A-5.

The trigonometric functions for the angle 45° are then given by Equations A-31, 32, and 33.

$$\sin 45° = \frac{1}{\sqrt{2}} = 0.707 \qquad \cos 45° = \frac{1}{\sqrt{2}} = 0.707 \qquad \tan 45° = \frac{1}{1} = 1.00$$

Example A-6 The sine of 30° is exactly $\frac{1}{2}$. Find the ratios of the sides of a 30°–60° right triangle.

This common triangle is shown in Figure A-15. We choose a length of 1 unit for the side opposite the 30° angle. The hypotenuse is then obtained from the fact that $\sin 30° = 0.5$:

$$c = \frac{a}{\sin 30°} = \frac{1}{0.5} = 2 \text{ units}$$

The length of the side opposite the 60° is found from the pythagorean theorem:

$$b = \sqrt{c^2 - a^2} = \sqrt{2^2 - 1} = \sqrt{3} \text{ units}$$

From these results, we can obtain the other trigonometric functions for the angles 30° and 60°:

$$\cos 30° = \frac{b}{c} = \frac{\sqrt{3}}{2} = 0.866$$

$$\tan 30° = \frac{a}{b} = \frac{1}{\sqrt{3}} = 0.577$$

$$\sin 60° = \frac{b}{c} = \cos 30° = \frac{\sqrt{3}}{2} = 0.866$$

$$\cos 60° = \frac{a}{c} = \sin 30° = \frac{1}{2} = 0.500$$

$$\tan 60° = \frac{b}{a} = \frac{\sqrt{3}}{1} = 1.732$$

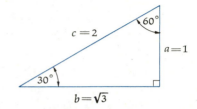

Figure A-15
A 30°–60° right triangle for Example A-6.

For small angles, the length a is nearly equal to the arc length s, as can be seen in Figure A-16. The angle $\theta = s/c$ is therefore nearly equal to $\sin \theta = a/c$:

$$\sin \theta \approx \theta \qquad (\text{small } \theta) \tag{A-42}$$

Similarly, the lengths c and b are nearly equal, so $\tan \theta = a/b$ is nearly equal to both θ and $\sin \theta$ for small θ:

$$\boxed{\tan \theta \approx \sin \theta \approx \theta \qquad (\text{small } \theta) \tag{A-43}}$$

Since $\cos \theta = b/c$ and these lengths are nearly equal for small θ, we have

$$\cos \theta \approx 1 \qquad (\text{small } \theta) \tag{A-44}$$

Equations A-42 and A-43 hold only if θ is measured in radians.

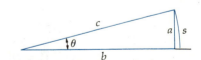

Figure A-16
For small angles, $\sin \theta = a/c$, $\tan \theta = a/b$, and the angle $\theta = s/c$ are all approximately equal.

Example A-7 By how much do sin θ, tan θ, and θ differ when $\theta = 15°$?

This angle in radians is

$$\theta = 15° \frac{2\pi \text{ rad}}{360°} = 0.262 \text{ rad}$$

From a calculator or a table of trigonometric functions,

$$\sin 15° = 0.259 \qquad \tan 15° = 0.268$$

Thus sin θ and θ (in radians) differ by 0.003 or about 1 percent, and tan θ and θ differ by 0.006 or about 2 percent. For smaller angles, the approximation $\theta \approx \sin \theta \approx \tan \theta$ is even more accurate.

Example A-7 shows that, if accuracy of a few percent is needed, small angle approximations can be used only for angles of about 15° or less. Figure A-17 shows graphs of θ, sin θ, and tan θ versus θ for small θ.

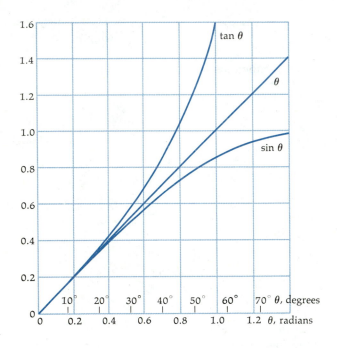

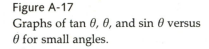

Figure A-17
Graphs of tan θ, θ, and sin θ versus θ for small angles.

Figure A-18 shows an obtuse angle with its vertex at the origin and one side along the x axis. The trigonometric functions for a general angle are defined by

$$\sin \theta = \frac{y}{c} \qquad\qquad\qquad\qquad\qquad \text{A-45}$$

$$\cos \theta = \frac{x}{c} \qquad\qquad\qquad\qquad\qquad \text{A-46}$$

$$\tan \theta = \frac{y}{x} \qquad\qquad\qquad\qquad\qquad \text{A-47}$$

Figure A-19 shows plots of these functions versus θ. All trigonometric functions have a period of 2π. That is, when an angle changes by 2π rad or

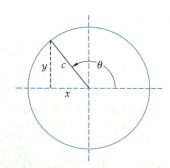

Figure A-18
Diagram for defining trigonometric functions for an obtuse angle.

360°, then the original value of the function is obtained again. Thus, $\sin(\theta + 2\pi) = \sin\theta$ and so forth. Some other useful relations that we will have occasion to use are

$$\sin(\pi - \theta) = \sin\theta \qquad\qquad\qquad A\text{-}48$$

$$\cos(\pi - \theta) = -\cos\theta \qquad\qquad\qquad A\text{-}49$$

$$\sin(\pi/2 - \theta) = \cos\theta \qquad\qquad\qquad A\text{-}50$$

$$\cos(\pi/2 - \theta) = \sin\theta \qquad\qquad\qquad A\text{-}51$$

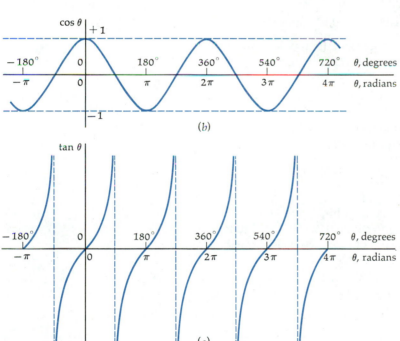

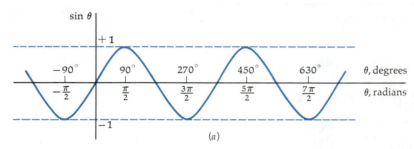

Figure A-19
The trigonometric functions sin θ, cos θ, and tan θ versus θ.

The Binomial Expansion

The binomial theorem is very useful for making approximations. One form of this theorem is

$$(1 + x)^n = 1 + nx + \frac{n(n-1)}{2}x^2 + \frac{n(n-1)(n-2)}{(3)(2)}x^3$$

$$+ \frac{n(n-1)(n-2)(n-3)}{(4)(3)(2)}x^4 + \cdots \qquad A\text{-}52$$

If n is a positive integer, there are just $n + 1$ terms in this series. If n is a real number other than a positive integer, there are an infinite number of terms. The series is valid for any value of n if x^2 is less than 1. It is also valid for $x^2 = 1$ if n is positive. The series is particularly useful if $|x|$ is much less than 1. Then each term in Equation A-52 is much smaller than the previous term, and we can drop all but the first two or three terms in the equation. If $|x|$ is much less than 1, we have

$$(1 + x)^n \approx 1 + nx \qquad |x| \ll 1 \qquad\qquad\qquad \text{A-53}$$

Binomial approximation

Example A-8   Use Equation A-53 to find an approximate value for the square root of 101.

First state the problem to give an expression of the form $(1 + x)^n$ with x much less than 1:

$$(101)^{1/2} = (100 + 1)^{1/2} = (100)^{1/2}(1 + 0.01)^{1/2} = 10(1 + 0.01)^{1/2}$$

Now we can use Equation A-53 with $n = \frac{1}{2}$ and $x = 0.01$:

$$(1 + 0.01)^{1/2} \approx 1 + \tfrac{1}{2}(0.01) = 1.005$$

$$(101)^{1/2} = 10(1 + 0.01)^{1/2} \approx 10(1.005) = 10.05$$

We can get an idea of the accuracy of this approximation by looking at the first term that has been neglected in Equation A-52. This term is

$$\frac{n(n - 1)}{2} x^2$$

Since x in this case is 1 percent of 1, the quantity x^2 is 0.01 percent of 1. By dropping this term, we are making an error of the order of

$$\frac{n(n - 1)}{2} x^2 = -\tfrac{1}{8}(0.01)^2 \approx -0.001\%$$

We therefore expect our answer to be correct to within about 0.001 percent. The value of $(101)^{1/2}$ to eight significant figures is 10.049876. The difference between this value and our approximation is 0.000124, which is about 0.001 percent of 10.05.

Example A-9   Using $\sin\theta \approx \theta$ and $\cos\theta = (1 - \sin^2\theta)^{1/2}$, find an approximation for $\cos\theta$ for small θ.

We have

$$\cos\theta = (1 - \sin^2\theta)^{1/2} \approx (1 - \theta^2)^{1/2} \approx 1 + \tfrac{1}{2}(-\theta^2) = 1 - \tfrac{1}{2}\theta^2$$

This approximation is often useful as a correction to the more drastic approximation $\cos\theta \approx 1$ given by Equation A-44.

Appendix B   SI Units

Basic Units

Length	The *metre* (m) is the distance traveled by light in a vacuum in 1/299,792,458 s
Time	The *second* (s) is the duration of 9,192,631,770 periods of the radiation corresponding to the transition between the two hyperfine levels of the ground state of the ^{133}Cs atom
Mass	The *kilogram* (kg) is the mass of the international standard body preserved at Sèvres, France
Current	The *ampere* (A) is that current in two very long parallel wires 1 m apart that gives rise to a magnetic force per unit length of 2×10^{-7} N/m
Temperature	The *kelvin* (K) is 1/273.16 of the thermodynamic temperature of the triple point of water
Luminous intensity	The *candela* (cd) is the luminous intensity, in the perpendicular direction, of a surface of area 1/600,000 m^2 of a blackbody at the temperature of freezing platinum at pressure of 1 atm

---

Derived Units

Force	newton (N)	$1\,\text{N} = 1\,\text{kg} \cdot \text{m/s}^2$
Work, energy	joule (J)	$1\,\text{J} = 1\,\text{N} \cdot \text{m}$
Power	watt (W)	$1\,\text{W} = 1\,\text{J/s}$
Frequency	hertz (Hz)	$1\,\text{Hz} = \text{s}^{-1}$
Charge	coulomb (C)	$1\,\text{C} = 1\,\text{A} \cdot \text{s}$
Potential	volt (V)	$1\,\text{V} = 1\,\text{J/C}$
Resistance	ohm (Ω)	$1\,\Omega = 1\,\text{V/A}$
Capacitance	farad (F)	$1\,\text{F} = 1\,\text{C/V}$
Magnetic induction	tesla (T)	$1\,\text{T} = 1\,\text{N/A} \cdot \text{m}$
Magnetic flux	weber (Wb)	$1\,\text{Wb} = 1\,\text{T} \cdot \text{m}^2$
Inductance	henry (H)	$1\,\text{H} = 1\,\text{J/A}^2$

Appendix C   Numerical Data

For additional data, see the front and back endpapers and the following tables in the text:

Table 1-1	Prefixes for powers of 10, p. 3
Table 1-2	Order of magnitude of some masses, p. 9
Table 1-3	Order of magnitude of some time intervals, p. 9
Table 1-4	Order of magnitude of some lengths, p. 9
Table 8-1	Moments of inertia of uniform bodies of various shapes, p. 179
Table 8-2	Comparison of linear motion and rotational motion, p. 184
Table 10-1	Densities of selected substances, p. 223
Table 10-2	Young's modulus Y and strengths of various materials, p. 224
Table 10-3	Approximate values of the shear modulus M_s and the bulk modulus B of various materials, p. 226
Table 10-4	Coefficients of viscosity of various fluids, p. 245
Table 12-1	Specific heat and molar heat capacity for various solids and liquids at 20°C, p. 273
Table 12-2	Molar heat capacities in $J/mol \cdot K$ of various gases at 25°C, p. 284
Table 13-1	Approximate values of thermal-expansion coefficients, p. 293
Table 13-2	Normal melting point (MP), latent heat of fusion L_f, normal boiling point (BP), and latent heat of vaporization L_v for various substances at 1 atm, p. 297
Table 13-3	Vapor pressure of water versus temperature, p. 301
Table 13-4	Critical temperatures T_c for various substances, p. 302
Table 13-5	Thermal conductivities k for various materials, p. 305
Table 13-6	R factors $\Delta x/k$ and $1/k$ for various building materials, p. 307
Table 18-1	Some electric fields in nature, p. 432
Table 19-1	Dielectric constant and strength of various materials, p. 473
Table 20-1	Resistivities and temperature coefficients, p. 493
Table 21-1	Magnetic susceptibility of various materials at 20°C, p. 554
Table 24-1	Electromagnetic spectrum, p. 610
Table 24-2	Index of refraction for yellow sodium light ($\lambda = 589$ nm), p. 624
Table 28-1	Rest energies of some elementary particles and light nuclei, p. 735
Table 31-1	Periodic table of the elements, p. 790
Table 31-2	Electron configurations of the atoms in their ground states, p. 798
Table 32-1	Atomic masses of the neutron and selected isotopes, p. 838
Table 32-2	Radiation and dose units, p. 857
Table 32-3	Sources and average intensities of human radiation exposure, p. 858
Table 32-4	Recommended dose limits, p. 858
Table 33-1	Hadrons stable against hadronic decay, p. 868
Table 33-2	Properties of quarks and antiquarks, p. 872
Table 33-3	Fundamental particles and their approximate rest energies (mc^2), p. 873

Terrestrial Data

Acceleration of gravity g	9.80665 m/s^2
Standard value	32.1740 ft/s^2
At sea level, at equator*	9.7804 m/s^2
At sea level, at poles*	9.8322 m/s^2
Mass of earth M_E	$5.98 \times 10^{24} \text{ kg}$
Radius of earth R_E, mean	$6.37 \times 10^6 \text{ m}$
	3960 mi
Escape speed $\sqrt{2R_E g}$	$1.12 \times 10^4 \text{ m/s}$
	6.95 mi/s
Solar constant†	1.35 kW/m^2
Standard temperature and pressure (STP):	
Temperature	273.15 K
Pressure	101.325 kPa
	1.00 atm
Molar mass of air	28.97 g/mol
Density of air (STP), ρ_{air}	1.293 kg/m^3
Speed of sound (STP)	331 m/s
Heat of fusion of H_2O (0°C, 1 atm)	333.5 kJ/kg
Heat of vaporization of H_2O (100°C, 1 atm)	2.257 MJ/kg

* Measured relative to the earth's surface.
† Average power incident normally on 1 m² outside the earth's atmosphere at the mean distance from the earth to the sun.

Astronomical Data

Earth	
Distance to moon*	$3.844 \times 10^8 \text{ m}$
	$2.389 \times 10^5 \text{ mi}$
Distance to sun, mean*	$1.496 \times 10^{11} \text{ m}$
	$9.30 \times 10^7 \text{ mi}$
	1.00 AU
Orbital speed, mean	$2.98 \times 10^4 \text{ m/s}$
Moon	
Mass	$7.35 \times 10^{22} \text{ kg}$
Radius	$1.738 \times 10^6 \text{ m}$
Period	27.32 d
Acceleration of gravity at surface	1.62 m/s^2
Sun	
Mass	$1.99 \times 10^{30} \text{ kg}$
Radius	$6.96 \times 10^8 \text{ m}$

* Center to center.

Physical Constants

Gravitational constant	G	6.672×10^{-11} N·m^2/kg^2
Speed of light	c	2.997925×10^8 m/s
Electron charge	e	1.60219×10^{-19} C
Avogadro's number	N_A	6.0220×10^{23} particles/mol
Gas constant	R	8.314 J/mol·K
		1.9872 cal/mol·K
		8.206×10^{-2} L·atm/mol·K
Boltzmann's constant	$k = R/N_A$	1.3807×10^{-23} J/K
		8.617×10^{-5} eV/K
Unified mass unit	$u = 1/N_A$ g	1.6606×10^{-24} g
Coulomb constant	$k = 1/4\pi\epsilon_0$	8.98755×10^9 N·m^2/C^2
Permittivity of free space	ϵ_0	8.85419×10^{-12} C^2/N·m^2
Permeability of free space	μ_0	$4\pi \times 10^{-7}$ N/A^2
		1.256637×10^{-6} N/A^2
Planck's constant	h	6.6262×10^{-34} J·s
		4.1357×10^{-15} eV·s
	$\hbar = h/2\pi$	1.05459×10^{-34} J·s
		6.5822×10^{-16} eV·s
Mass of electron	m_e	9.1095×10^{-31} kg
		511.0 keV/c^2
Mass of proton	m_p	1.67265×10^{-27} kg
		938.28 MeV/c^2
Mass of neutron	m_n	1.67495×10^{-27} kg
		939.57 MeV/c^2

Appendix D   Conversion Factors

Conversion factors are written as equations for simplicity; relations marked with an asterisk are exact.

Length

1 km = 0.6215 mi

1 mi = 1.609 km

1 m = 1.0936 yd = 3.281 ft = 39.37 in

*1 in = 2.54 cm

*1 ft = 12 in = 30.48 cm

*1 yd = 3 ft = 91.44 cm

1 light-year = 1 $c \cdot y$ = 9.461 $\times$ 10^{15} m

*1 Å = 0.1 nm

Area

*1 m^2 = 10^4 cm^2

1 km^2 = 0.3861 mi^2 = 247.1 acres

*1 in^2 = 6.4516 cm^2

1 ft^2 = 9.29 $\times$ 10^{-2} m^2

1 m^2 = 10.76 ft^2

*1 acre = 43,560 ft^2

1 mi^2 = 640 acres = 2.590 km^2

Volume

*1 m^3 = 10^6 cm^3

*1 L = 1000 cm^3 = 10^{-3} m^3

1 gal = 3.786 L

1 gal = 4 qt = 8 pt = 128 oz = 231 in^3

1 in^3 = 16.39 cm^3

1 ft^3 = 1728 in^3 = 28.32 L = 2.832 $\times$ 10^4 cm^3

Time

*1 h = 60 min = 3.6 ks

*1 d = 24 h = 1440 min = 86.4 ks

1 y = 365.24 d = 31.56 Ms

Speed

1 km/h = 0.2778 m/s = 0.6215 mi/h

1 mi/h = 0.4470 m/s = 1.609 km/h

1 mi/h = 1.467 ft/s

Angle and Angular Speed

*π rad = 180°

1 rad = 57.30°

1° = 1.745 $\times$ 10^{-2} rad

1 rev/min = 0.1047 rad/s

1 rad/s = 9.549 rev/min

Mass

*1 kg = 1000 g

*1 tonne = 1000 kg = 1 Mg

1 u = 1.6606 $\times$ 10^{-27} kg

1 kg = 6.022 $\times$ 10^{23} u

1 slug = 14.59 kg

1 kg = 6.852 $\times$ 10^{-2} slug

1 u = 931.50 MeV/c^2

Density

*1 g/cm^3 = 1000 kg/m^3 = 1 kg/L

(1 g/cm^3)g = 62.4 lb/ft^3

Force

1 N = 0.2248 lb = 10^5 dyn

1 lb = 4.4482 N

(1 kg)g = 2.2046 lb

Pressure

*1 Pa = 1 N/m^2

*1 atm = 101.325 kPa = 1.01325 bars

1 atm = 14.7 lb/in^2 = 760 mmHg

= 29.9 inHg = 33.8 ftH$_2$O

1 lb/in^2 = 6.895 kPa

1 torr = 1 mmHg = 133.32 Pa

1 bar = 100 kPa

Energy

*1 kW $\cdot$ h = 3.6 MJ

*1 cal = 4.1840 J

1 ft $\cdot$ lb = 1.356 J = 1.286 $\times$ 10^{-3} Btu

*1 L $\cdot$ atm = 101.325 J

1 L $\cdot$ atm = 24.217 cal

1 Btu = 778 ft $\cdot$ lb = 252 cal = 1054.35 J

1 eV = 1.602 $\times$ 10^{-19} J

1 u $\cdot$ c^2 = 931.50 MeV

*1 erg = 10^{-7} J

Power

1 horsepower = 550 ft $\cdot$ lb/s = 745.7 W

1 Btu/min = 17.58 W

1 W = 1.341 $\times$ 10^{-3} horsepower

= 0.7376 ft $\cdot$ lb/s

Magnetic Induction

*1 G = 10^{-4} T

*1 T = 10^4 G

Thermal Conductivity

1 W/m $\cdot$ K = 6.938 Btu $\cdot$ in/h $\cdot$ ft^2 $\cdot$ F°

1 Btu $\cdot$ in/h $\cdot$ ft^2 $\cdot$ F° = 0.1441 W/m $\cdot$ K

Appendix E

Trigonometric Tables

Angle θ Degree	Angle θ Radian	sin θ	cos θ	tan θ	Angle θ Degree	Angle θ Radian	sin θ	cos θ	tan θ
0	0.0000	0.0000	1.0000	0.0000					
1	0.0175	0.0175	0.9998	0.0175	46	0.8029	0.7193	0.6947	1.0355
2	0.0349	0.0349	0.9994	0.0349	47	0.8203	0.7314	0.6820	1.0724
3	0.0524	0.0523	0.9986	0.0524	48	0.8378	0.7431	0.6691	1.1106
4	0.0698	0.0698	0.9976	0.0699	49	0.8552	0.7547	0.6561	1.1504
5	0.0873	0.0872	0.9962	0.0875	50	0.8727	0.7660	0.6428	1.1918
6	0.1047	0.1045	0.9945	0.1051	51	0.8901	0.7771	0.6293	1.2349
7	0.1222	0.1219	0.9925	0.1228	52	0.9076	0.7880	0.6157	1.2799
8	0.1396	0.1392	0.9903	0.1405	53	0.9250	0.7986	0.6018	1.3270
9	0.1571	0.1564	0.9877	0.1584	54	0.9425	0.8090	0.5878	1.3764
10	0.1745	0.1736	0.9848	0.1763	55	0.9599	0.8192	0.5736	1.4281
11	0.1920	0.1908	0.9816	0.1944	56	0.9774	0.8290	0.5592	1.4826
12	0.2094	0.2079	0.9781	0.2126	57	0.9948	0.8387	0.5446	1.5399
13	0.2269	0.2250	0.9744	0.2309	58	1.0123	0.8480	0.5299	1.6003
14	0.2443	0.2419	0.9703	0.2493	59	1.0297	0.8572	0.5150	1.6643
15	0.2618	0.2588	0.9659	0.2679	60	1.0472	0.8660	0.5000	1.7321
16	0.2793	0.2756	0.9613	0.2867	61	1.0647	0.8746	0.4848	1.8040
17	0.2967	0.2924	0.9563	0.3057	62	1.0821	0.8829	0.4695	1.8807
18	0.3142	0.3090	0.9511	0.3249	63	1.0996	0.8910	0.4540	1.9626
19	0.3316	0.3256	0.9455	0.3443	64	1.1170	0.8988	0.4384	2.0503
20	0.3491	0.3420	0.9397	0.3640	65	1.1345	0.9063	0.4226	2.1445
21	0.3665	0.3584	0.9336	0.3839	66	1.1519	0.9135	0.4067	2.2460
22	0.3840	0.3746	0.9272	0.4040	67	1.1694	0.9205	0.3907	2.3559
23	0.4014	0.3907	0.9205	0.4245	68	1.1868	0.9272	0.3746	2.4751
24	0.4189	0.4067	0.9135	0.4452	69	1.2043	0.9336	0.3584	2.6051
25	0.4363	0.4226	0.9063	0.4663	70	1.2217	0.9397	0.3420	2.7475
26	0.4538	0.4384	0.8988	0.4877	71	1.2392	0.9455	0.3256	2.9042
27	0.4712	0.4540	0.8910	0.5095	72	1.2566	0.9511	0.3090	3.0777
28	0.4887	0.4695	0.8829	0.5317	73	1.2741	0.9563	0.2924	3.2709
29	0.5061	0.4848	0.8746	0.5543	74	1.2915	0.9613	0.2756	3.4874
30	0.5236	0.5000	0.8660	0.5774	75	1.3090	0.9659	0.2588	3.7321
31	0.5411	0.5150	0.8572	0.6009	76	1.3265	0.9703	0.2419	4.0108
32	0.5585	0.5299	0.8480	0.6249	77	1.3439	0.9744	0.2250	4.3315
33	0.5760	0.5446	0.8387	0.6494	78	1.3614	0.9781	0.2079	4.7046
34	0.5934	0.5592	0.8290	0.6745	79	1.3788	0.9816	0.1908	5.1446
35	0.6109	0.5736	0.8192	0.7002	80	1.3963	0.9848	0.1736	5.6713
36	0.6283	0.5878	0.8090	0.7265	81	1.4137	0.9877	0.1564	6.314
37	0.6458	0.6018	0.7986	0.7536	82	1.4312	0.9903	0.1392	7.115
38	0.6632	0.6157	0.7880	0.7813	83	1.4486	0.9925	0.1219	8.144
39	0.6807	0.6293	0.7771	0.8098	84	1.4661	0.9945	0.1045	9.514
40	0.6981	0.6428	0.7660	0.8391	85	1.4835	0.9962	0.0872	11.430
41	0.7156	0.6561	0.7547	0.8693	86	1.5010	0.9976	0.0698	14.301
42	0.7330	0.6691	0.7431	0.9004	87	1.5184	0.9986	0.0523	19.081
43	0.7505	0.6820	0.7314	0.9325	88	1.5359	0.9994	0.0349	28.636
44	0.7679	0.6947	0.7193	0.9657	89	1.5533	0.9998	0.0175	57.290
45	0.7854	0.7071	0.7071	1.0000	90	1.5708	1.0000	0.0000	∞

Appendix F

Periodic Table of the Elements

The atomic-mass values listed are based on $^{12}_{6}C = 12$ u exactly. For artificially produced elements, the approximate atomic mass of the most stable isotope is given in brackets.

Period	Series	Group I		II		III		IV		V		VI		VII		VIII			0
1	1	1 **H** 1.00797																	2 **He** 4.003
2	2	3 **Li** 6.942		4 **Be** 9.012		5 **B** 10.81		6 **C** 12.011		7 **N** 14.007		8 **O** 15.9994		9 **F** 19.00					10 **Ne** 20.183
3	3	11 **Na** 22.990		12 **Mg** 24.31		13 **Al** 26.98		14 **Si** 28.09		15 **P** 30.974		16 **S** 32.064		17 **Cl** 35.453					18 **Ar** 39.948
4	4	19 **K** 39.102		20 **Ca** 40.08		21 **Sc** 44.96		22 **Ti** 47.90		23 **V** 50.94		24 **Cr** 52.00		25 **Mn** 54.94		26 **Fe** 55.85	27 **Co** 58.93	28 **Ni** 58.71	
	5		29 **Cu** 63.54		30 **Zn** 65.37		31 **Ga** 69.72		32 **Ge** 72.59		33 **As** 74.92		34 **Se** 78.96		35 **Br** 70.909				36 **Kr** 83.80
5	6	37 **Rb** 85.47		38 **Sr** 87.62		39 **Y** 88.905		40 **Zr** 91.22		41 **Nb** 92.91		42 **Mo** 95.94		43 **Tc** [98]		44 **Ru** 101.1	45 **Rh** 102.905	46 **Pd** 106.4	
	7		47 **Ag** 107.870		48 **Cd** 112.40		49 **In** 114.82		50 **Sn** 118.69		51 **Sb** 121.75		52 **Te** 127.60		53 **I** 126.90				54 **Xe** 131.30
6	8	55 **Cs** 132.905		56 **Ba** 137.34		57-71 Lanthanoid series*		72 **Hf** 178.49		73 **Ta** 180.95		74 **W** 183.85		75 **Re** 186.2		76 **Os** 190.2	77 **Ir** 192.2	78 **Pt** 195.09	
	9		79 **Au** 196.97		80 **Hg** 200.59		81 **Tl** 204.37		82 **Pb** 207.19		83 **Bi** 208.98		84 **Po** [210]		85 **At** [210]				86 **Rn** [222]
7	10	87 **Fr** [223]		88 **Ra** [226]		89-103 Actinoid series**													

* Lanthanoid Series (rare earths)	57 **La** 138.91	58 **Ce** 140.12	59 **Pr** 140.91	60 **Nd** 144.24	61 **Pm** [147]	62 **Sm** 150.35	63 **Eu** 152.0	64 **Gd** 157.25	65 **Tb** 158.92	66 **Dy** 162.50	67 **Ho** 164.93	68 **Er** 167.26	69 **Tm** 168.93	70 **Yb** 173.04	71 **Lu** 174.97
** Actinoid Series	89 **Ac** [227]	90 **Th** 232.04	91 **Pa** [231]	92 **U** 238.03	93 **Np** [237]	94 **Pu** [242]	95 **Am** [243]	96 **Cm** [247]	97 **Bk** [247]	98 **Cf** [251]	99 **Es** [254]	100 **Fm** [253]	101 **Md** [256]	102 **No** [254]	103 **Lw** [257]

Answers to Odd-Numbered Exercises and Problems

These answers are calculated using $g = 9.81$ m/s^2 unless otherwise specified in the exercise or problem. The results are usually rounded to three significant figures. Differences in the last figure can easily result from differences in rounding the input data and are not important.

CHAPTER 1

True or False

1. True **2.** False; for example, $x = vt$, where the speed v and time t have different dimensions **3.** True

Exercises

1. (a) 1 MW (b) 2 mg.(c) 3 μm (d) 30 ks

3. (a) 1 picaboo (b) 1 gigalow (c) 1 microphone (d) 1 attoboy (e) 1 megaphone (f) 1 nanogoat (g) 1 terabull

5. (a) C_1, length; C_2, length/time (b) C_1, length/(time)^2 (c) C_1, length/(time)^2 (d) C_1, length; C_2, (time)$^{-1}$ (e) C_1, length/time; C_2, (time)$^{-1}$

7. (a) 4×10^7 m (b) 6.37×10^6 m (c) 4×10^7 m $= 2.48 \times 10^4$ mi; 6.37×10^6 m $= 3.96 \times 10^3$ mi

9. (a) Unitless (b) s (c) m

11. (a) 31.56 Ms (b) 31.7 y (c) 1.91×10^{16} y

13. (a) 30,000 (b) 0.0062 (c) 0.000004 (d) 217,000,000,000,000,000,000,000,000,000,000

15. (a) 1.14×10^5 (b) 2.25×10^{-8} (c) 8.27×10^3 (d) 6.27×10^2

CHAPTER 2

True or False

1. False; this equation holds only for constant acceleration **2.** False; a ball at the top of its flight is momentarily at rest but has acceleration g **3.** False; for example, motion with constant velocity **4.** True **5.** True; definition of average velocity

Exercises

1. (a) 3.3 m/s (b) 7.45 mi/h

3. (a) 53.8 mi/h (b) 46.1 mi/h (c) -50 mi/h, relative to the direction of (b)

5. (a) 0 (b) 0.33 m/s (c) -2 m/s (d) $+1$ m/s

7. (a) 1.82 h (b) 4.85 h (c) 945 km/h (d) 621 km/h

9. (a) 8.3 min (b) 1.3 s (c) 9.45×10^{12} km $= 5.87 \times 10^{12}$ mi

11. (a) Velocity and speed at t_2 less than at t_1 (b) Both equal (c) Velocity greater but speed less at t_2 (d) Velocity less but speed greater at t_2

13. 2 m/s; 2.7 m/s; 3.2 m/s; 4.0 m/s; $v = 4.2$ m/s from measurement of slope at 0.75 s

15. For $t_2 = 3$ s, $\Delta t = 1.0$ s, $\Delta x = 25$ m, $v_{av} = 25$ m/s; for $t_2 = 2.5$ s, $\Delta t = 0.5$ s, $\Delta x = 11.25$ m, $v_{av} = 22.5$ m/s; for $t_2 = 2.2$ s, $\Delta t = 0.2$ s, $\Delta x = 4.2$ m, $v_{av} = 21.0$ m/s; for $t_2 = 2.1$ s, $\Delta t = 0.1$ s, $\Delta x = 2.05$ m, $v_{av} = 20.5$ m/s; for $t_2 = 2.01$ s, $\Delta t = 0.01$ s, $\Delta x = 0.2005$ m, $v_{av} = 20.05$ m/s; for $t_2 = 2.001$ s, $\Delta t = 0.001$ s, $\Delta x = 0.020005$ m, $v_{av} = 20.005$ m/s; $v = 20$ m/s

17. -2 m/s^2

19. (a) At $t = 1$ s, $v = 55$ mi/h; at $t = 2$ s, $v = 65$ mi/h

21. (a) 80 m/s (b) 40 m/s (c) 400 m

23. (a) 25.1 s (b) 3.08 km

25. 300 km/s^2

27. 12.5 y

29. 5.3 km

31. 301 m

33. (a) 1.41 Mm/s (b) 14.1 ns (c) 0.707 ns

35. (a) 105 cm/s (b) 0.381 s (c) -78.7 cm/s^2

37. (a) 0.759 g (b) 3.75 s (c) 81.3 m

Problems

1. (a) 790 km/s (b) 31.6 Mm/s (c) 2.0×10^{10} y

3. (a) -6.31 m/s (b) 5.42 m/s (c) 587 m/s^2 up

5. (a) 312 m/s^2 (b) 2.08×10^3 m/s^2

7. 66.7 km/h

9. (b) Tortoise by 2.4 km or 10 min (c) 125 min

11. (b) $t = 20$ s (c) 160 km/h

13. (b) 65 km/h; 5.54 s

15. (a) Yes (c) 8 m/s (d) 4 m/s (e) 80 m

CHAPTER 3

True or False

1. True **2.** False **3.** False; for example, circular motion **4.** True **5.** False; for example, equal and opposite vectors **6.** True **7.** True

Exercises

1. (*b*) 18.4 m at 22° N of E

3. (*a*) 2 km at 0°; 1 km at 135°; and 3 km at 90° (*b*) 0 north, 2 km east; 0.707 km north, −0.707 km east; 3 km north, 0 east (*c*) 3.93 km at 70.8° N of E (*d*) Not necessarily (*e*) Not necessarily

5. $R_x = -5$ m; $R_y = +5$ m

7. (*a*) $A = 5.83$ m; $\theta = 31°$ (*b*) $B = 12.2$ m; $\theta = -35°$ (*c*) $C = 3.6$ m; $\theta = 236°$

9. (*a*) $A = 8.06$ at 204°; $B = 3.61$ at −33.7° $|A + B| = 9.06$ at 264° (*b*) $A = 4.12$ at −76°; $B = 6.32$ at 71.6°; $|A + B| = 3.61$ at 33.7°

11. (*a*) $A_x = A_y = 1.41$ m; $B_x = 1.73$ m, $B_y = -1$ m (*b*) $R_x = 3.14$ m; $R_y = 0.41$ m; $R = 3.17$ m at 7.44°, where $R = A + B$ (*c*) $R_x = -0.32$ m; $R_y = 2.41$ m; $R = 2.43$ m at 97.6°, where $R = A - B$

13. 14.7 km/h at 16.3° S of E

15. (*a*) $AB, 90°; BC, 45°; CD, 0°; DE, -45°; EF, -90°$; all with the *x* axis (*b*) *AB*, along the +*y* axis; *BC*, toward the center of the circular arc; *DE*, toward the center of the circular arc (*c*) Magnitude of acceleration along *DE* is greater than that along *BC*

17. (*a*) 13.1° W of N (*b*) 300 km/h

19. (*a*) 0.553 s (*b*) 136 m

21. (*a*) 2.3 km (*b*) 43.3 s (*c*) 9.18 km

23. (*a*) 25 m/s horizontally (*b*) 9.81 m/s² downward

25. (*a*) Increases by a factor of 4 (*b*) Decreases by a factor of 2

27. (*a*) 3.84 m/s (*b*) 24.4 rev/min

29. 55.1 km/h = 34.2 mi/h

Problems

1. (*a*) 14.1 m at 45° at 15 s; 20 m at 90° at 30 s; 14.1 m at 135° at 45 s; 0 at 60 s (*b*) The magnitude is 14.1 m for each interval; the directions are 45° for the first interval, 135° for the second interval, 225° for the third interval, and 315° for the fourth interval (*c*) Perpendicular to the first and of the same magnitude (*d*) Equal in magnitude and opposite in direction

3. (*a*) 42.7 m/s at 38.7° to the *x* axis (*b*) 3.48 m/s² at 210° to the *x* axis

5. (*a*) 0.314 m/s (*b*) 5 m at 180° at $t = 50$ s; 5 m at 90° at $t = 25$ s; 5 m at 36° at $t = 10$ s; 5 m at 0° at $t = 0$ (*c*) 0.2 m/s; 0.283 m/s; 0.095 m/s (*d*) $v_{av} = 0.309$ m/s at 107°; $v = 0.314$ m/s at 90°

7. $R = 408$ m

9. (*a*) 8.14 m/s (*b*) 23.1 m/s

11. (*a*) 11.0 m (*b*) 10 m/s horizontal; 14.7 m/s vertical (*c*) 17.8 m/s (*d*) 55.8°

13. 3.18 m/s towards the ball

15. 14.2 m/s = 51 km/h

17. 34.6 m/s

CHAPTER 4

True or False

1. True **2.** False **3.** False; the *acceleration* is in the direction of the unbalanced force **4.** True **5.** False **6.** False

Exercises

1. 3 kg

3. (*a*) 8 m/s² (*b*) 0.5 (*c*) 8/3 m/s²

5. 12 m/s²

7. 10 kg

9. (*a*) −300 N (*b*) 500 N

11. (*a*) 79.5 kg (*b*) 7.95 × 10⁴ g

13. (*a*) 49,000 dyn (*b*) 0.49 N

17. (*a*) 1.67 m/s² (*b*) 1.67 m/s²; 1.67 N; contact force exerted by the 2-kg body (*c*) 3.33 N

19. (*a*) 16.7 kN/m (*b*) 5.87 cm

21. (*a*) 327 N/m (*b*) 8.18 m/s² up the plane

23. (*a*) $T = 424$ N; $F_n = 245$ N (*b*) $T = mg \sin \theta$; 0 at 0°; mg at 90°

25. (*a*) 6 m/s (*b*) 9 m

27. (*a*) −9 m/s² (*b*) −8 m/s² (*c*) +10 m/s²

29. 3.53 m/s²

31. (*a*) 600 N (*b*) 480 N (*c*) 480 N

33. 2.47 s

Problems

1. (*a*) 17.5 μm/s² (*b*) 5.63 min (*c*) 0.59 cm/s

3. (*b*) 7.2 N (*c*) 9.3 cm

5. 500 N, assuming $m = 80$ kg and $\Delta x = 50$ m

7. (*a*) 80 N (*b*) 600 N; 680 N (*c*) 6.8 m/s²

9. (*a*) 492 N; 3.71 m/s (*b*) 21.0 m/s

11. 80 kg; 785 N; 2.19 m/s²

13. 478 N

15. (*a*) 397 N (*b*) 367 N

17. (*a*) 8.85 m/s (*b*) 15.3 m/s downwards, which is $\sqrt{3}$ times the speed upwards

CHAPTER 5

True or False

1. False 2. False; the resultant torque must also be zero
3. True 4. True 5. False; only if **g** is constant over the body
6. True

Exercises

1. 21.1°

3. (a) 5.88 m/s² (b) 76.5 m (c) 91.8 m

5. 0.229

7. (a) Pulling upwards reduces the normal force and therefore reduces the friction (b) 519 N for 30° downward; 252 N for 30° upward; 294 N for 0°

9. 1.4 m from pivot

11. 318 N

13. 101 N

15. 26.6°

17. (a) 9.42 m/s (b) 14.8 m/s² (c) 7.40 N

19. 15.3 m/s

21. (a) 4.90 kN (b) 89.8 m

23. (a) 5.88 N (b) 0.941 kg/m

Problems

1. $w_1 = 1.5$ N; $w_2 = 7$ N; $w_3 = 3.5$ N

3. (a) 16.3 m/s² (b) 19.6 N (c) No

5. (a) 400 N (b) 625 N

7. (a) 42.4 N along the strut (b) 57 N at 52° to the horizontal

11. 23.6 rev/min

13. B can go anywhere from the far end to 1.83 m from A

15. (a) 17.6 N (b) 1.96 m/s²; 7.84 m/s² (c) 1.47 m/s²; 1.47 m/s²; 2.94 N on each

17. (a) 21.8 kg/m (b) 78.5 kN

19. 44.1 N to the left for the upper hinge; 44.1 N to the right for the lower hinge

21. $mg\sqrt{2hr - h^2}/(r - h)$

CHAPTER 6

True or False

1. False 2. True 3. True 4. True 5. False; it is a unit of energy
6. False 7. False 8. True

Exercises

1. (a) 490 J (b) −490 J (c) 0

3. (a) 296 J (b) −196 J

5. (a) 7.2 kJ (b) 1.8 kJ

7. 266 J

9. (a) 320 J; −196 J (b) 124 J

11. (a) 63.2 N (b) 190 J (c) −118 J (d) 72 J

13. (a) The normal force does no work; the weight does 102 J; the force of friction does −11.8 J (b) 90.2 J (c) 5.48 m/s

15. (a) 9 J (b) 12 J (c) 4.58 m/s

17. (a) 3.92 kJ (b) 3.92 kJ (c) Some work is done to accelerate his body upwards and decelerate it downwards during each step, as is required for walking on a level surface.

19. (a) 392 J (b) 4.9 m; 9.81 m/s (c) 296 J; 96 J; 392 J (d) 392 J; 19.8 m/s

21. 20.4 m

23. (a) 1.22 J; −1.22 J (b) 1.22 J; 1.22 J

25. (a) 7.67 m/s (b) −58.8 J (c) 0.333

27. 1.40 m/s

29. 3.62 m/s

31. (a) 3.4 (b) 14.7 N

33. (a) 78.5 W (b) 157 J

35. (a) 17.6 N; 0.197 (b) 1.73 W (c) −5.19 J

37. (a) 2.20 MJ (b) 0.150 lb

39. 1.00 kW

41. (a) 15.0 kJ (b) 3.58 kcal (c) 100 m/s

Problems

1. (a) 2.7 J (b) 6.08 J (c) 2.01 m/s (d) 3.6 J (e) 2.2 m/s

3. 2.62 W

5. (a) 1.76 MJ (b) 238 kW

7. (a) 2.60 MJ (b) 2.08 GJ (c) 1.46 GJ (d) 238 kW

9. 46.6 cm

11. (a) $-(3/8)mv_0^2$ (b) $3v_0^2/(16\pi r g)$ (c) $\frac{1}{3}$ rev

13. 6.0 m

15. (a) 5.10 m (b) 10.2 m

CHAPTER 7

True or False

1. True 2. False; this occurs only in the center-of-mass reference frame 3. True 4. True 5. True 6. True

Exercises

1. (a) 6 kg·m/s (b) 264 kg·m/s (c) 160 Mg·km/h

3. (a) 10.7 N·s (b) 13.4 kN

5. -35.0 kN·s

7. (a) 3.76 N·s (b) 2.89 kN

9. (a) 90 kg·m/s (b) 90 kg·m/s (c) 4.5 m/s

11. 5 m/s

13. Between them at 4.36 m from the man

15. $y_{cm} = 2.81$ cm

17. (a) 20 m/s (b) 800 kJ

19. 1.08 m/s to the right

21. 17.0 cm/s at $61.9°$ to the horizontal

23. 4.11 m

25. 0.837

Problems

1. (a) 0.6 m/s² (b) 960 N

3. 4.8 m/s for the 2-kg body; 0.8 m/s for the 3-kg body

5. 12.6 cm/s

7. 15.0 km/s

9. Assuming 4 m for the length of the car, $\Delta t = 0.16$ s and $F_{av} = 312$ kN

11. (a) 10.8 m/s (b) 867 N·s upward (c) 0.185 s (d) 469 kN upward

13. (a) $v_n = -(\frac{11}{13})v$; $v_C = (\frac{2}{13})v$ (b) 0.284

15. 4.99 m/s for the 1-kg body; 9.99 m/s for the 1-g body

17. 211 m/s; 467 J

19. (a) 3.5 m/s (b) 10.5 J (c) 0.75

21. 39.8 m/s

CHAPTER 8

True or False

1. False **2.** True **3.** True **4.** True **5.** False **6.** False; the *rate of change* of the angular momentum is zero **7.** False; it is true for a rigid body

Exercises

1. (a) 0.188 rad/s (b) 0.358 rev

3. (a) 0.524 rad (b) 1.05 rad (c) 3.14 rad (d) 12.6 rad

5. (a) 3.49 rad/s (b) 52.3 cm/s

7. (a) 15 rad/s (b) 6 m/s (c) 45 rad

9. (a) -0.0387 rad/s² (b) 1.74 rad/s (c) 25 rev

11. 0.624 s

13. (a) 0.56 kg·m² (b) 0.373 N·m

15. 36 kg·m²

17. (a) 3.5 N·m (b) 59.5 rad/s² (c) 298 rad/s

19. (a) 2.16 N·m (b) 2.88 kg·m² (c) 75 rad/s² (d) 375 rad/s

21. (a) 0.4 m/s and 0.24 J for the 3-kg body; 0.8 m/s and 0.32 J for the 1-kg body (b) 1.12 J

23. (a) 6.06×10^2 J (b) 3.86×10^{-2} kg·m²/s

25. 27.1 W

27. (a) 1.66×10^{-3} J (b) 5.30×10^{-5} kg·m²/s

29. 2.27 rev/s

31. 600 rev/min

33. 1008 J

35. (a) 28.6% (b) 33.3% (c) 50%

39. (a) Right (b) Down (to conserve angular momentum in each case)

Problems

1. (a) π rad/s² (b) $3\pi/2$ rad/s

3. (a) 3.53×10^4 J (b) -4.73 N·m (c) 1.20×10^5 rev

5. 2.73 m/s

7. 5.42 rad/s

9. $a = 59mg/(5m + 2M)$; $T = 2Mmg/(5m + 2M)$

11. 2.7R

13. $a_{cm} = 5g/7$; $T = 2Mg/7$

15. If **L** is horizontal and the car turns to the left, the front end of the car tends to move upward; if **L** is vertically upward and the car goes up a hill, the car tends to veer to the right.

CHAPTER 9

True or False

1. False **2.** True **3.** True

Exercises

1. 17.94 AU

3. 29.5 y

5. 83.7 y

7. 248 y

9. 2.45 m/s²

11. 8.67×10^{-7} N

13. (a) 8.79×10^{25} kg $= 14.7M_E$ (b) 1.16×10^6 s

15. 4.43×10^7 m $= 6.64R_E$

17. 6.02×10^{24} kg

19. 35.5 km/s

21. (a) -6.25×10^9 J (b) -3.12×10^9 J; 7.91 km/s

Problems

1. 2.01×10^{30} kg

3. 51.2 km/s

5. 6.95 km/s

7. (a) 6.27×10^7 J (b) 17.4 kW·h (c) $139.40

CHAPTER 10
True or False

1. False 2. True 3. False; it can float because of surface tension 4. True

Exercises

1. 0.676 kg

3. (a) 7.97 kg/L (b) Iron

5. 1.034 kg/L

7. 13,621 kg/m³

9. 0.975 mm

11. 20,000 kPa

13. 5.01°

15. (a) 10.3 m (b) 75.8 cm

17. 64.8 kN; it does not collapse because there is an equal force on the bottom of the table

19. 230 N

21. 39 kN, which is about 4.4 ton; to get the door open the occupant should roll down the window to let the water in to partially equalize the pressure inside and outside

23. 15.8 cm

25. 25.5 cm

27. 4.36 N

29. (a) 11.1 kg/L (b) Lead

31. 184 m³

33. (a) 69.1° (b) 5.62×10^{-6} kg

35. 1.49 m

37. (a) 65 m/s (b) 2213 kPa

39. $2\sqrt{h(H-h)}$

41. 1.31×10^5 N

43. 1.427 mm

Problems

1. ± 0.441 Pa

3. 3.88 kg

5. Upper scale, 12.3 N; lower scale, 36.7 N

7. 5.13 cm

9. 241 kPa

11. (a) 14.65 g (b) 1.0325 g/cm³

13. (b) 1.12 mm

CHAPTER 11
True or False

1. False; the degree size is also different 2. True 3. False 4. True

Exercises

1. 10.4°F to 19.4°F

3. (a) 90.29 K (b) -297.15°F

5. 37°C

7. 56.7°C and -62.2°C

9. (a) 10^7°C (b) 1.8×10^7°F

11. (a) 21.97 mmHg (b) 2.176 K

13. 3.22×10^8

15. (a) 1.8×10^{-2} kg (b) 55.5 mol

17. (a) 24.62 L (b) 0.5 atm

19. (a) 183 K (b) 244 K $= -29$°C (c) 1.43 atm

21. 1.35 kN

23. 498 km/s; 2.07×10^{-16} J

25. 435 m/s

27. 8.84 K

Problems

1. 1.66×10^{-27} kg

3. (a) 1.16 kg/m³ (b) 1.29 kg/m³

5. (a) 231 kPa (b) 201 kPa

7. (b) 507 m/s

9. (a) 6.07×10^{-21} J (b) 1.91 km/s (c) 7.31 kJ

11. 1.85 km/s for H_2, 0.463 km/s for O_2, and 0.394 km/s for CO_2; O_2 and CO_2 could be in Mars' atmosphere, but not H_2 because its rms speed is greater than one sixth the escape speed

CHAPTER 12
True or False

1. False 2. False 3. False 4. True 5. True

Exercises

1. 837 kJ

3. 61.8 kJ

5. 5×10^5 kJ

7. 20.8°C

9. 1.70 kJ/kg·K

11. +315 J

13. 0.023 C°

15. The temperature of the lead increases by 1.53 C° to 21.53°C

17. (a) 0.117 C° (b) 1.73 C°

19. −750 J

21. 1.39 kJ

23. 0.713 J/g·K

25. (a) 3.5 mol (b) $C_v = 43.6$ J/K; $C_p = 72.7$ J/K (c) $C_v = 72.7$ J/K; $C_p = 101.8$ J/K

27. 190.2×10^{-3} kg/mol; osmium

Problems

1. 2.44 min

3. 5.5 L·atm by counting squares; 5.54 L·atm from Equation 12-15

5. (a) 405 J (b) 861 J

7. (a) 507 J (b) 963 J

9. $Q_{AB} = 506$ J,　$\Delta U_{AB} = 303$ J;　$Q_{BC} = \Delta U_{BC} = -303$ J; $Q_{CD} = -253$ J, $\Delta U_{CD} = -152$ J; $Q_{DA} = \Delta U_{DA} = 152$ J

11. (a) $T_A = 243.7$ K; $T_B = 731.1$ K; $T_C = 365.6$ K; $T_D = 121.9$ K (b) $Q_{AB} = 10.13$ kJ; $Q_{BC} = -4.56$ kJ; $Q_{CD} = -5.06$ kJ; $Q_{DA} = +1.52$ kJ (c) $W_{AB} = 5.05$ kJ; $W_{BC} = 0$; $W_{CD} = -2.03$ kJ; $W_{DA} = 0$ (d) $U_A = 3.04$ kJ; $U_B = 9.12$ kJ; $U_C = 4.56$ kJ; $U_D = 1.52$ kJ (e) $W_{net} = 2.02$ kJ; $Q_{net} = 2.03$ kJ, which equals W_{net} within round-off errors

CHAPTER 13

True or False
1. False **2.** True **3.** True **4.** True **5.** True **6.** True

Exercises
1. 30.0264 cm

3. 7.7 cm

5. 0.7803 L

7. 12.25 kJ

9. 10.9°C

11. 48.8 g

13. 354 m/s

15. Argon, helium, hydrogen, neon, nitric oxide, and oxygen

17. 52.6%

19. (a) 15.9 K/W (b) 6.30 W (c) 50 K/m (d) 87.5°C

21. (a) 0.0831 K/W for Cu; 0.1406 K/W for Al (b) 0.2237 K/W (c) 358 W (d) 70.3°C

23. 49.7°C

25. 0.026 W/m·K

27. 9.47×10^{-3} mm

Problems
1. 3.70 cm³

3. 93.5 cm²

5. 99.8 g

7. 145°C

9. 34.7 cm

11. 496.4 g

13. (a) 28% (b) 17.5°C (c) 18°C

15. (a) 2.99°C (b) 188 g (c) No

17. 5.77×10^3 K

19. 101.27°C

21. (a) 6.75 W (b) 27.4 h

23. 626°C

25. 96.6 g/h

27. 134 kN

CHAPTER 14

True or False
1. False **2.** False **3.** False **4.** False **5.** False **6.** True **7.** False **8.** True

Exercises
1. (a) 500 J (b) 400 J

3. (a) 69.2 J (b) 24.2 J

5. (a) 20 J (b) 66.7 J absorbed; 46.7 J rejected

7. (a) 5 kJ (b) 30 kJ

9. (a) 1.67 (b) 37.5%

13. −130°C

15. (a) 140 J (b) 70% (c) 704°C

17. (a) 33.3% (b) 40 J (c) 2

19. (a) 16.7% (b) 77.9%

21. (a) 20% (b) 60% (c) 33.3 J

23. (a) 23.3% (b) 3.29 kJ

25. (*a*) 4.22 (*b*) 3.83 kW (*c*) 5.67 kW

27. (*a*) 2.89 kJ (*b*) 2.02 kJ

29. (*a*) -1.35 J/K (*b*) $+1.67$ J/K (*c*) $+0.42$ J/K (*d*) 125 J

31. 6.05 kJ/K

33. 1.56 kJ/K·d

Problems

1. 1.98 kJ/K

3. (*b*) 2.49 kJ (*c*) 162 kJ (*d*) 13.7 kJ (*e*) 15.4%

5. (*a*) 3.59 kJ (*b*) 80.2°C

9. (*a*) 46.7 kJ (*b*) 66.7 kJ

11. (*a*) 33.3% (*b*) 33.3 J; 66.7 J (*c*) 2

13. (*a*) 11.5 J/K (*b*) 0 (*c*) 11.5 J/K (*d*) 11.5 J/K

15. 13.1%

CHAPTER 15

True or False

1. False **2.** True **3.** True **4.** False **5.** True **6.** True **7.** True **8.** True **9.** True

Exercises

1. (*a*) 7.96 Hz (*b*) 0.126 s (*c*) 10 cm (*d*) 5 m/s (*e*) 250 m/s² (*f*) 0.0314 s; 0

3. (*a*) 653 N/m (*b*) 2.88 Hz (*c*) 0.348 s

5. 6.2 cm

7. 0.566 Hz

9. (*a*) 3 kN/m (*b*) 3.90 Hz (*c*) 0.256 s

11. (*a*) 2 Hz (*b*) 0.5 s (*c*) 5 cm (*d*) $\frac{1}{8}$ s; in the negative x direction

13. (*a*) 0.628 m/s (*b*) 7.90 m/s²

15. (*a*) 0.318 Hz (*b*) 3.14 s (*c*) $x = (0.4$ m) cos 2t, where t is in seconds

17. (*a*) 0.314 m/s (*b*) 2.09 rad/s (*c*) $x = (15$ cm) cos 2.09t, where t is in seconds

19. 1.92 J

21. (*a*) 3 cm (*b*) 0.779 m/s

23. 1.38 kN/m

25. 12.2 s

27. 24.8 cm

29. 6.2 cm

31. (*a*) 1.01 Hz (*b*) 2.01 Hz (*c*) 0.352 Hz

33. (*a*) 314 (*b*) 0.955 Hz

Problems

1. (*b*) 2.93 cm; 7.07 cm; 7.07 cm; 2.93 cm

3. (*a*) 1.57 m/s (*b*) 1.23 m/s² (*c*) 882 N; 686 N

5. (*a*) $x = (10$ cm) cos $(\pi/2)t$ (*b*) $v = -(15.7$ cm/s) sin $(\pi/2)t$ (*c*) $a = -(24.7$ cm/s²) cos $(\pi/2)t$, where t is in seconds

7. (*a*) 1.51 kg (*b*) 0.82 cm (*c*) $x = (2.5$ cm) cos 11πt, $v = -(0.864$ m/s) sin 11πt, and $a = -(29.9$ m/s²) cos 11πt, where t is in seconds

9. (*a*) 70 J (*b*) 22 J (*c*) 7.33 W

11. (*a*) No (*b*) 55.8 cm

13. 0.236 rad

15. (*a*) 52.3 kN/m (*c*) 2.00×10^{11} N/m²

CHAPTER 16

True or False

1. True **2.** False **3.** False **4.** False **5.** False; it has 1000 times the intensity **6.** False

Exercises

1. 251 m/s

3. (*a*) 265 m/s (*b*) 15 g

5. (*a*) 400 N (*b*) 327 m/s

7. 0.252 s

9. (*a*) 354 m/s (*b*) 319 m/s

11. 1.02 km/s

13. 5.09 km/s

15. 2.70×10^{10} N/m²

17. (*a*) 0.773 m (*b*) 3.27 m

19. 10^{10} Hz

21. 1.0×10^5 Hz

23. (*a*) 15 m; 20 pies/min (*b*) 13.5 m; 22.2 pies/min (*c*) 15 m; 22.0 pies/min

25. (*a*) 2.1 m (*b*) 162 Hz

27. 153 Hz

29. (*a*) 436 Hz (*b*) 370 Hz

31. (*a*) 20 dB (*b*) 100 dB

33. 99%

35. 8×10^{-9} W

Problems

1. The speed of sound is about 0.34 km/s, so it takes about 3 s to reach you. The accuracy depends on the accuracy of measuring time by counting. The correction for the finite speed of light is not important because it is so large compared to the speed of sound.

3. (*a*) 2.27 ms; 440 Hz (*b*) 346 m/s (*c*) 0.787 m

5. (*a*) 1.615 m; 211 Hz (*b*) 221 Hz

7. (*a*) 39.8 MHz (*b*) 39.6 MHz

9. 88 dB

11. 57 dB

CHAPTER 17

True or False

1. False 2. False 3. True 4. True 5. True 6. True 7. True 8. True 9. False

Exercises

1. (*a*) 85 Hz and 255 Hz (*b*) 170 Hz and 340 Hz

3. (*a*) 425 Hz and 1275 Hz (*b*) 850 Hz and 1700 Hz

5. (*a*) 0 (*b*) $4I_0$ (*c*) $2I_0$

7. (*a*) 113 Hz (*b*) 11.3 kHz (*c*) 567 Hz; 56.7 kHz

9. 4 Hz

11. (*a*) 504 Hz or 496 Hz (*b*) 496 Hz

13. (*a*) 2 m; 25 Hz (*b*) 6 m; 8.33 Hz

15. (*a*) 125 Hz (*b*) 3.75 Hz (*c*) 6.25 Hz

17. (*a*) 4.25 m (*b*) 8.5 m

Problems

1. 3.3 cm; 6.2 cm; 7.6 cm; 10 cm

3. (*a*) 0 (*b*) 66 dB (*c*) 63 dB

5. (*a*) $n = 8$ and $n + 1 = 9$ (*b*) 2.16 m

7. 22.0 m/s $= 79.0$ km/h $= 49.1$ mi/h

9. 345 m/s; 1.25 cm

11. (*a*) 85 Hz (*b*) 34 m/s (*c*) 28.9 N

CHAPTER 18

True or False

1. False; it points towards a negative charge 2. True 3. False; they diverge from a positive point charge 4. True 5. True

Exercises

1. (*a*) 6.25×10^{12} electrons (*b*) 6.25×10^{9} electrons (*c*) 6.25×10^{6} electrons

3. (*a*) 9×10^{-3} N (*b*) 9×10^{-5} N (*c*) 3.6×10^{-6} N

5. (*a*) 0.08 N (*b*) 0.08 N

7. (*a*) 4.10×10^{-8} N (*b*) 0.256 N

9. 230 N

11. 0.015 N toward q_2

13. 4.16×10^{42}

15. (*a*) 1.0 kN/C in the $+x$ direction (*b*) 360 N/C in the $+x$ direction (*c*) 2.25 kN/C in the $-x$ direction

17. 2.87×10^{11} m/s²

19. (*a*) 9.36 kN/C in the $+x$ direction (*b*) 8 kN/C in the $+x$ direction (*c*) $x = 4$ m (*d*) $+x$ just to the right of the origin and $-x$ just to the left

21. (*a*) 34.6 kN/C along the $+x$ axis (*b*) 6.91×10^{-5} N along the $+x$ axis

23. (*a*) $q_1/q_2 = 4$ (*b*) q_1 is positive and q_2 is negative (*c*) Strong between the charges and at points near the charges; weak far from the charges and at points to the right of q_2

25. (*a*) 0 (*b*) 0 (*c*) 1.25 MN/C (*d*) 450 kN/C (*e*) 112 kN/C

29. 6×10^{-15} C·m in the $-x$ direction

31. 5.3×10^{-2} e·nm $= 8.48 \times 10^{-30}$ C·m

Problems

1. (*a*) 2.82×10^{23} atoms; 8.18×10^{23} electrons; 1.31×10^{5} C (*b*) 36.3 h (*c*) 4.29×10^{16} N

5. $x = 1.8$ m

7. $(kq^2/L^2)(\sqrt{2} - \frac{1}{2}) = 0.914kq^2/L^2$ along the diagonal away from point 2

CHAPTER 19

True or False

1. True 2. True 3. False; it is zero in electrostatics but not when there are currents 4. True 5. True 6. False; breakdown depends on the electric field, not the potential 7. False 8. True 9. True

Exercises

1. (*a*) 6 kV (*b*) 12 kV (*c*) -9 kV

3. 1.5 kV/m in the $-z$ direction

5. 50 V/m in the $-x$ direction

7. (*a*) 11.25 kV (*b*) 5.625 kV (*c*) 2.25×10^{-2} J (*d*) 1.125×10^{-2} J (*e*) 1.125×10^{-2} J

9. (*a*) 1.8 kV (*b*) 1.2 kV (*c*) 960 V

11. (*a*) 1.84×10^{22} electrons (*b*) 4.55×10^{3} C

13. (*a*) 4 nC (*b*) 225 V

15. 54.8 m

17. 136 N/C

19. 26.55 μC/m²

21. 0.08 μF

23. 1.13×10^7 m^2; 3.36 km

25. (*a*) 4.5 kV (*b*) 9 mC

27. 144 μC

29. (*a*) 6.67 μF (*b*) 40 μC (*c*) 4 V across the 10-μF capacitor; 2 V across the 20-μF capacitor

31. 10 μF

33. (*a*) 100 (*b*) 10 V (*c*) 10 μC; 1 kV

35. (*a*) 0.8 μJ (*b*) 0.2 μJ

37. 9.00 mJ

39. (*a*) 600 kV/m (*b*) 1.593 J/m^3 (*c*) 17.9 μJ (*d*) 398 μF; 17.9 μJ

Problems

1. (*a*) 2.5 MV/m (*b*) 3.01×10^7 m/s

3. (*a*) 16.7 nF (*b*) 1.17 nC (*c*) 7 MV/m

5. (*a*) 15.2 μF (*b*) 1400 μC on the 12-μF capacitor and 630 μC on each of the other two capacitors (*c*) 0.303 J

7. 1.67 μF for the three in series; 7.50 μF for two in series with the combination in parallel with the third; 3.33 μF for two in parallel with the combination in series with the third

9. (*a*) 17.1 nC on the 20-pF capacitor, 42.9 nC on the 50-pF capacitor (*b*) 90 μJ; 25.7 μJ; lost

11. (*a*) 24 μC on the 4-μF capacitor; 72 μC on the 12-μF capacitor; 6 V (*b*) 1.15 mJ; 0.288 mJ

13. (*a*) 234 MeV (*b*) 2.67×10^{16} fissions/s

CHAPTER 20

True or False

1. False; this is the definition of resistance; Ohm's law requires the additional statement that R is constant **2.** False **3.** True **4.** True **5.** True **6.** False; it is greater when the battery is being charged **7.** False **8.** False; it is placed in parallel

Exercises

1. (*a*) 900 C (*b*) 5.62×10^{21} electrons

3. 0.167 A

5. (*a*) 26.7 Ω (*b*) 0.75 A

7. (*a*) 0.233 Ω (*b*) 27.9 A

9. 0.364 Ω

11. 1.15 V

13. 367 m

15. (*a*) 2.42 kW (*b*) 1.21 kW

17. (*a*) 0.707 A (*b*) 7.07 V

19. (*a*) 0.59 cents (*b*) 96.8 cents

21. (*a*) 240 W; 228 W (*b*) 43.2 kJ (*c*) 2.16 kJ

23. (*a*) 1 Ω (*b*) $I_2 = 6$ A, $I_3 = 4$ A, and $I_6 = 2$ A, where I_2 is the current through the 2-Ω resistor, etc.

25. (*a*) 3.6 Ω (*b*) $I_6 = 2.0$ A; $I_2 = I_7 = 1.33$ A

27. (*a*) 18 V (*b*) 2.0 A

29. (*a*) 5 Ω (*b*) 1.2 A through the 4-Ω resistor and the single 5-Ω resistor in lower branch; 0.72 A through the 10-Ω resistor at the top of the upper branch; 0.48 A through the 5-Ω and 10-Ω resistors in series; 0.6 A through each of the parallel 10-Ω resistors in the lower branch

31. (*a*) 2.0 A through the 1.5-Ω resistor; 0.75 A through each of the 4-Ω resistors; 0.5 A through the 6-Ω resistor (*b*) 12 W

33. (*a*) 1.2 A (*b*) 1.0 Ω

35. (*a*) 1.33 A (*b*) 132 J (*c*) 124 J (*d*) 8 J goes into charging the 2-V battery

37. (*a*) $I_2 = 3.0$ A toward *b*; $I_3 = 2.0$ A from *b* to *a*; $I_1 = 1.0$ A toward *a* (*b*) -1.0 V (*c*) 21 W by the 7-V battery; 10 W by the 5-V battery

39. (*a*) 900 μC (*b*) 1.5 A (*c*) 0.60 ms

41. (*a*) 0.5 mA (*b*) 0

43. (*a*) 50.0 mΩ (*b*) 4.9 kΩ

45. (*a*) 0.160 Ω (*b*) 0.1597 Ω (*c*) 1.5 MΩ

Problems

1. (*a*) 60.5 Ω (*b*) 8.22 min (*c*) 52.2 min

3. 3 V; 1.0 Ω

5. 1.0 mΩ; 9.0 mΩ; 90.1 mΩ

7. (*a*) 8.46×10^{22} atoms/cm^3 (*b*) 3.55×10^{-3} cm/s

9. 831.6 W; 43.56 W goes into heating its own internal resistance, 29.16 W goes into heating the internal resistance of the sick battery, 144 W goes into heating the load resistance, and 615.6 W goes into charging the sick battery

11. (*a*) 50 μA (*b*) 5.0 W

13. (*a*) 2.0 A in the 1-Ω resistor and the nearby 2-Ω resistor towards *a*; 1.0 A in the other 2-Ω resistor from *a* to *b*; 1.0 A in the 6-Ω resistor towards *b* (*b*) 16.0 W by the 8-V battery; 8.0 W by the 4-V battery at the top; -4.0 W by the 4-V battery near *b* (*c*) 4.0 W in the 1-Ω resistor; 8.0 W in the upper 2-Ω resistor; 2.0 W in the other 2-Ω resistor; 6.0 W in the 6-Ω resistor

15. (*a*) 1.974 A (*b*) 1.3% (*c*) 1.47998 V (*d*) 0.001%

CHAPTER 21

True or False

1. True 2. True 3. False; it is independent of the radius
4. True 5. False 6. False 7. True 8. False

Exercises

1. 1.28 pN in the $-y$ direction

3. $B_y = 2.0$ T, $B_z = 0$, B_x is unknown

5. 1.04×10^{-12} N in the $-z$ direction

7. 0.289 N

9. (a) 1.33 A·m² (b) 33.3 A·m

11. (a) 0.48 A·m² (b) 27° with the x axis (c) 0.327 N·m

13. (a) 4.24 A·m² (b) 84.8 A·m

15. (a) 2.7 nm (b) 0.18 ns

17. (a) 93.3 ns (b) 3.37×10^7 m/s (c) 3.77×10^{-12} J = 23.6 MeV

19. (a) 2×10^6 m/s (b) 3.34×10^{-15} J = 20.9 keV (c) 1.82×10^{-18} J = 11.4 eV

21. (a) 22.9 MHz (b) 4.31×10^{-12} J = 26.9 MeV

23. 9.6×10^{-12} T in the y direction

25. (a) 8.89×10^{-5} T in the $-z$ direction (b) 0 (c) 8.89×10^{-5} T in the $+z$ direction (d) 1.6×10^{-4} T in the $-z$ direction

27. (a) 6.4×10^{-5} T in the y direction (b) 4.8×10^{-5} T in the $-z$ direction

29. (a) Antiparallel (b) 49.0 mA

31. (a) 4.5×10^{-4} N/m upward (b) 5.20×10^{-5} T to the right

33. 2.51×10^{-5} T

35. 11.1 A

37. 6.03×10^{-3} T

39. Bismuth, copper, diamond, gold, mercury, silver, sodium, hydrogen, carbon dioxide, and nitrogen have negative susceptibilities and are therefore diamagnetic. Aluminum, magnesium, titanium, tungsten, and oxygen have positive susceptibilities and are therefore paramagnetic.

Problems

1. 0.98 A

3. (a) 190.5 km/s (b) 9.92 mm

5. (a) $B_x = (\mu_0/4\pi L)(I_2 + 2I_3)$; $B_y = -(\mu_0/4\pi L)(I_2 + 2I_1)$ (b) $B_x = (\mu_0/4\pi L)(2I_3 - I_2)$; $B_y = (\mu_0/4\pi L)(I_2 - 2I_1)$ (c) $B_x = (\mu_0/4\pi L)(I_2 - 2I_3)$; $B_y = -(\mu_0/4\pi L)(2I_1 + I_2)$

7. (a) $(\mu_0 I^2/4\pi a)(3\sqrt{2})$ along the diagonal toward the center (b) $-(\mu_0 I^2 \sqrt{2})/(4\pi a)$ along the diagonal away from the center

9. (a) 3.2×10^{-16} N opposite to the current (b) 3.2×10^{-16} N away from the wire (c) 0

11. $v_\parallel = [(2\pi m)/(q\mu_0 I)]v_t^2$

CHAPTER 22

True or False

1. False; it is proportional to the rate of change of the flux
2. True 3. True 4. False 5. True 6. True

Exercises

1. (a) 8.48 mWb (b) 7.97 mWb

3. (a) 7.07 mV (b) 6.64 mV

5. 18.75 mV

7. 199 T/s

9. (a) 3.10 mWb (b) 2.21 mV

11. 816 mV; 16.3 mA

13. 400 m/s

15. (a) 2.7 V (b) 1.93 A (c) 1.30 N (d) 5.21 W

17. (a) 6.03 mT (b) 1.09 mWb (c) 0.364 mH (d) 54.7 mV

19. (b) B due to the solenoid is zero outside the solenoid (c) $\mu_0 N_1 N_2 A_2/\ell_2$

21. 0.512 mH

23. 12.8 kΩ

25. 15.9 J

27. (a) 39.8 Hz (b) 15.1 V

29. (a) 13.6 V (b) 485 Hz

Problems

1. $U_m = 39.8$ kJ; $U_e = 11.1$ μJ; $U_{total} = 39.8$ kJ

3. (a) $\phi_m = (8.5 \times 10^{-4}$ Wb/s$)t$ (b) 850 μV; 567 μA

5. (a) $I_0 = \mathcal{E}_0/R$; $F = \mathcal{E}_0 \ell B/R$ (c) $F = (B\ell/R)(\mathcal{E}_0 - B\ell v) = m\Delta v/\Delta t$

CHAPTER 23

True or False

1. False 2. True 3. True 4. True

Exercises

1. (b) 156 V

3. (a) 4.0 A (b) 2.83 A (c) 48.0 W (d) 24.0 W

5. (*a*) 0.603 Ω (*b*) 6.03 Ω (*c*) 6.03 $\times$ 10^{-2} Ω

7. (*a*) 1.89 MΩ (*b*) 0.189 MΩ (*c*) 18.9 MΩ

9. 1.59 kHz

11. (*a*) 38.2 A (*b*) 54.0 A

13. (*a*) 15.9 kHz (*b*) 159 Hz (*c*) 1.59 MHz

15. (*a*) 1.12 kHz (*b*) 14.1 A (*c*) X_C = 63.7 Ω; X_L = 78.5 Ω; Z = 15.6 Ω (*d*) 4.53 A (*e*) 71.3°

17. (*a*) 0.320; 103 W (*b*) 0.286; 81.6 W

19. 9.95 nF to 0.101 μF

21. 14.1; 79.4 Hz

23. (*a*) 1.64 A (*b*) 20 turns

25. (*a*) 1.24 A (*b*) 2.47 A

Problems
1. P_{av} = 0

3. (*a*) 0.622 (*b*) 0.622 A (*c*) 46.4 W

5. (*a*) 0.794 mH; 12.4 μF (*b*) 1.60 (*c*) 2.0 A

7. (*a*) 856 W (*b*) 7.07 Ω (*c*) 102 μF (*d*) Add 0.019 H inductance

9. (*a*) 12 Ω (*b*) R = 7.2 Ω; X = 9.6 Ω (*c*) Capacitive

11. L = 3.96 mH; I_{max} = 0.10 A

CHAPTER 24
True or False
1. True **2.** False **3.** False **4.** True

Exercises
1. 1 $\times$ 10^{10} Hz

3. (*a*) 300 m (*b*) 0.3 m

5. (*a*) 1.33 μT (*b*) 1.42 μJ/m^3 (*c*) 425 W/m^2

7. (*a*) 73.5 V/m (*b*) 47.8 nJ/m^3 (*c*) 14.3 W/m^2

9. (*a*) 726 V/m (*b*) 2.42 μT

11. 10.8 min

13. 2.0%

15. (*a*) 1.54 (*b*) 34.2°

17. (*a*) 14.9° (*b*) 22.1° (*c*) 32.1° (*d*) 40.6°

19. (*a*) 27.1° (*b*) 41.7° (*c*) 70.1° (*d*) No refraction because the incident angle is greater than the critical angle

21. 62.5°

25. 526 nm

27. 25.2° for λ = 400 nm; 26.0° for λ = 700 nm

29. 0.211 I_0

31. 54.7°

Problems
1. 0.505 s

3. (*a*) 0.553 W/m^2 (*b*) E_{rms} = 14.4 V/m; B_{rms} = 48.1 nT

5. (*a*) 62.5° (*b*) No

7. (*a*) $n \geq$ 1.41 (*b*) 1.63 $\leq n \leq$ 1.88

9. 0.783

CHAPTER 25
True or False
1. True **2.** True **3.** False **4.** True **5.** False; only paraxial rays are focused **6.** True **7.** False **8.** True

Exercises
7. (*a*) s' = 25 cm, m = $-$0.25; real, inverted, reduced (*b*) s' = 40 cm, m = $-$1; real, inverted, the same size (*c*) s' = ∞; no image is formed (*d*) s' = $-$20 cm, m = +2; virtual, erect, enlarged

9. (*a*) s' = $-$16.7 cm, m = +0.167; virtual, erect, reduced (*b*) s' = $-$13.3 cm, m = +0.333; virtual, erect, reduced (*c*) s' = $-$10.0 cm, m = +0.500; virtual, erect, reduced (*d*) s' = $-$6.67 cm, m = +0.667; virtual, erect, reduced

11. (*a*) 0.476 m (*b*) Behind the mirror (*c*) 9.52 cm

13. s' = 4.0 m; y' = 3.68 cm

15. (*a*) s' = 30 cm; real (*b*) s' = $-$15 cm; virtual (*c*) s' = 15 cm; real

17. (*a*) s' = $-$10.0 cm; virtual (*b*) s' = $-$5.0 cm; virtual (*c*) s' = $-$15.0 cm; virtual

19. (*a*) s' = $-$14.9 cm; virtual (*b*) s' = $-$5.0 cm; virtual (*c*) s' = $-$44.1 cm; virtual

21. 0.839 m from the diver's mask; m = +0.446

23. (*a*) r_1 = 5.0 cm; r_2 = $-$5.0 cm (*b*) r_1 = $-$5.0 cm; r_2 = +5.0 cm

25. (*a*) s' = 40.0 cm, m = $-$1; real, inverted (*b*) s' = $-$20 cm, m = +2; virtual, erect (*c*) s' = $-$17.1 cm, m = +0.429; virtual, erect (*d*) s' = $-$7.5 cm, m = +0.75; virtual, erect

27. s' = 10.0 cm; y' = $-$1.0 cm

31. 9.434 cm for blue; 10.638 cm for red

Problems
1. s = 5.0 cm; s' = $-$10.0 cm (*b*) s = 15 cm; s' = 30 cm

3. 30 cm to the right of the second lens; real and erect; m = +2

5. 7.5 cm to the left of the second lens; virtual and inverted; $m = -0.5$

7. 15 cm to the right of the second lens (at its focal point)

9. (a) $s' = 64$ cm (b) $s'_2 = -80$ cm (c) Virtual

11. Real

13. 1.6

15. 1.00 mm

CHAPTER 26

True or False
1. True **2.** True **3.** True **4.** True **5.** False; it varies inversely with the square of the f/number **6.** True **7.** True **8.** False; it is smaller **9.** True **10.** False; it uses a mirror for its objective

Exercises
1. 2.3 mm

3. (a) 103 cm (b) 0.972 diopter

5. 7.14 mm; smaller

7. 6

9. 4.17 for the person with the 25-cm near point, 6.67 for the person with the 40-cm near point; the size of the image on the retina is the same for each person

11. $f/2$

13. 0.16 mm toward the object

15. -267

17. (a) 20 cm (b) -4 (c) -20 (d) 6.25 cm

19. (a) 0.90 cm (b) 0.180 rad (c) -20

21. (a) 25 (b) -134

Problems
1. (b) 7

3. 3.70 m

5. 3.07 diopter

CHAPTER 27

True or False
1. False; it is redistributed in space **2.** True **3.** True **4.** True **5.** True

Exercises
1. (a) 300 nm (b) 135°

3. (a) 376 nm (b) 5.32 (c) 295°

5. (a) Dark; the ray reflected from the lower surface will have a phase change of 180° (b) 1.2×10^{-4} rad

7. 6626 nm $\leq d \leq$ 6921 nm

9. 150 nm

11. (a) 84,848 waves (b) 84,873 waves (c) 1.00029

13. (a) 1.17 mm (b) 4.29 fringes/cm

15. 695 nm

17. (a) 9.26×10^{-4} cm (b) 29

19. (a) 0.06 rad (b) 6×10^{-3} rad (c) 6×10^{-4} rad

21. 3.73 cm

23. (a) 7.32×10^{-3} rad (b) 2.93 cm

25. (a) 55.6 km (b) 55.6 m

27. 33.6 mm

29. 0.484 km

31. 486 nm and 660 nm

33. (a) 0.14° (b) 0.298 mm

Problems
1. (a) 0.6° (b) 0.3°

3. (a) 643 nm (b) 720 nm; 514 nm; 400 nm

5. 0.02 mm; 11

7. 8

9. 4.62×10^6 km

CHAPTER 28

True or False
1. True **2.** True **3.** False **4.** True **5.** False **6.** False **7.** True **8.** True

Exercises
1. (a) 0.183 μs (b) 1.83×10^{-12} s (c) No

3. (a) 5.96×10^{-8} s (b) 16.1 m (c) 7.02 m

5. (a) 44.7 μs (b) 13.4 km

7. $0.312c = 9.37 \times 10^7$ m/s

9. (a) 112.5 y (b) 83.8 y

11. $0.9992c$

13. $0.866c$

15. (a) 4.5×10^{-10} % (b) 1 y $- 1.42 \times 10^{-4}$ s $\approx$ 1 y; 2.37×10^{-6} min

17. (a) $0.994c$ (b) 60,000 m/s $- 6 \times 10^{-4}$ m/s

19. 80 $c \cdot$ min

23. $L_0 v/c^2 = 60$ min

25. (a) 1.11×10^{-17} kg (b) 0.351 μg

27. (a) 1.005 (b) 1.155 (c) 1.667 (d) 7.089

29. 1.88×10^{-28} kg

31. (a) 2 (b) 0.866

33. 21.108 MeV

35. (a) $0.155E_0$ (b) $1.294E_0$ (c) $6.089E_0$

37. (a) 23.847 MeV (b) 2.62×10^{11} fusions/s

Problems

1. $0.8c$

3. 1.85×10^4 y

5. (b) $0.866c$ (c) $0.9988c$

7. (a) 4.974 MeV/c (b) 0.9948

CHAPTER 29

True or False

1. True **2.** True **3.** False **4.** True **5.** True **6.** False **7.** True

Exercises

1. 6.63×10^{-26} J $= 4.14 \times 10^{-7}$ eV

3. (a) 2.42×10^{14} Hz (b) 2.42×10^{18} Hz (c) 2.42×10^{20} Hz

5. (a) 2480 nm (b) 1240 nm (c) 828 nm (d) 621 nm (e) 124 nm

7. (a) 1.11×10^{15} Hz; 271 nm (b) 1.63 V (c) 0.39 V

9. (a) 4.74 eV (b) 1.47 V

11. (a) 4.97×10^{-19} J (b) 0.01 J (c) 2.01×10^{16} photons/s

13. 10 kV

15. 1.215 pm

17. (a) 3.11 eV/$c = 1.66 \times 10^{-27}$ kg·m/s (b) 621 eV/$c = 3.32 \times 10^{-25}$ kg·m/s (c) 12.4 keV/$c = 6.63 \times 10^{-24}$ kg·m/s (d) 4.14×10^{-5} eV/$c = 2.21 \times 10^{-32}$ kg·m/s

19. (a) 17.5 keV (b) 76.0 pm (c) 16.3 keV

23. 1.89 eV, 656 nm; 2.55 eV, 486 nm; 2.86 eV, 434 nm

25. (a) 0.85 eV, 1459 nm (b) 4052 nm; 2627 nm; 2168 nm

Problems

1. (a) 2.07 eV (b) 5.00×10^{14} Hz (c) 4.15×10^{-14} V·s

3. (a) 3.53 W/m² (b) 1.07×10^{15}

5. (a) 10^{-22} W (b) 3200 s

CHAPTER 30

True or False

1. True **2.** True **3.** True **4.** False **5.** True **6.** True

Exercises

1. (a) 1.23 nm (b) 0.123 nm (c) 0.0388 nm (d) 0.0123 nm

3. (a) 6.2 eV/c (b) 3.76×10^{-5} eV

5. 0.203 nm

7. 660 fm

9. 1.52×10^{-34} m

11. 22.8 eV

13. 5.27×10^{-20} m

15. 0.195 nm

17. (a) 2.05 MeV; 8.20 MeV; 18.45 MeV (b) 202 fm (c) 121 fm (d) 75.6 fm

19. (a) 4.13×10^{-3} eV; 2.02×10^{-2} eV; 4.61×10^{-2} eV (b) 80.8 μm (c) 48.5 μm (d) 30.3 μm

Problems

1. (b) 12.4 fm

3. 0.002%

CHAPTER 31

True or False

1. True **2.** True **3.** True **4.** True **5.** False **6.** False **7.** False

Exercises

1. (a) 1.414 $\hbar$ (b) $-1, 0, +1$

3. (a) 0, 1, 2 (b) 0 for $\ell = 0$; $-1, 0, 1$ for $\ell = 1$; $-2, -1, 0, 1, 2$ for $\ell = 2$ (c) 18

5. 3.49×10^{-3} kg·m²/s; 3.31×10^{31}

7. (a) $1s^2 2s^2 2p^6 3s^2 3p^1$ (b) $1s^2 2s^2 2p^6 3s^2 3p^6 3d^5 4s^1$

9. Calcium

11. Gallium has one outer electron in the $4p$ state that is shielded by the closed $n = 3$ shell and the closed $4s$ subshell. Indium has one electron in the $5p$ shell that is similarly shielded by the closed $4d$ and $5s$ subshells.

13. Potassium, chromium, and copper each has one outer $4s$ electron. Calcium, titanium, and manganese each has two outer $4s$ electrons.

15. 0.0611 nm; 0.058 nm

17. 0.155 nm

19. Z = 20; calcium

21. (a) Covalent (b) Ionic (c) Metallic

23. 0.110 nm

25. (a) p-type (b) n-type

27. (a) p-type (b) n-type

29. 1676 nm

31. 116 K

Problems

1. (*a*) −5.39 eV (*b*) 4.87 eV

3. 0.549 nm; 1.29 nm

5. (*a*) 0.18 eV (*b*) 1.94×10^{-49} kg·m²

CHAPTER 32

True or False

1. False; there are no electrons in the nucleus **2.** True **3.** False **4.** True **5.** True **6.** True

Exercises

1. (*a*) ^{15}N (*b*) ^{57}Fe (*c*) ^{118}Sn

3. (*a*) 32.0 MeV; 5.33 MeV (*b*) 333.7 MeV; 8.56 MeV (*c*) 1636 MeV; 7.87 MeV

5. 20.6 MeV

7. (*a*) 500 counts/s (*b*) 250 counts/s (*c*) 125 counts/s

9. (*a*) 200 s (*b*) 0.00347/s (*c*) 125 counts/s

11. (*a*) 500 counts/s; 250 counts/s (*b*) 1.04×10^6 nuclei; 0.52×10^6 nuclei (*c*) 728 s

13. (*a*) 4.87 MeV (*b*) 7.00 MeV

15. 67 decays/s

17. (*a*) +4.03 MeV (*b*) +18.35 MeV (*c*) +4.78 MeV

19. 1.56×10^{19} fissions/s

21. (*a*) 0.00323 cm (*b*) 0.000288 cm

23. (*a*) 9 cm (*b*) 21 cm

Problems

1. 3.50 s

3. (*a*) 1.89 fm; 2.16 fm (*b*) 0.356 MeV

5. (*c*) 3.75×10^{38} protons/s; 5.0×10^{10} y

CHAPTER 33

True or False

1. True **2.** False; some are mesons **3.** False **4.** True **5.** False **6.** True **7.** True

Exercises

1. (*a*) Two photons are needed to conserve linear momentum, which is zero before annihilation and must therefore be zero afterwards (*b*) 938 MeV (*c*) 1.33 fm

3. (*a*) Energy (*b*) Energy (*c*) Linear momentum (*d*) No violation (*e*) Lepton (energy can be conserved if the neutrino and proton have sufficient kinetic energy)

5. (*a*) $B = 1, Q = 1e, S = 0$; proton, p (*b*) $B = 1, Q = 0, S = 0$; neutron, n (*c*) $B = 1, Q = +1e, S = -1$; Σ^+ (*d*) $B = 1, Q = -1e, S = -1$; Σ^- (*e*) $B = 1, Q = 0, S = -2$; Ξ^0 (*f*) $B = 1, Q = -1e, S = -2$; Ξ^-

7. *uuu*

Illustration Credits

Figure 8-24   © Russ Kinne/Photo Researchers.

CHAPTER 9

p. 198   © Photo by G. D. Hackett, New York.

p. 199   The Granger Collection.

p. 205   N.A.S.A.

p. 207   Courtesy Deutsches Museum, Munich.

p. 208   The Bettmann Archive.

p. 209   (bottom) N.A.S.A.

p. 215   National Optical Astronomy Observatories.

CHAPTER 10

p. 221   © Peter Menzel 1984/Stock, Boston.

p. 226   Courtesy Michael Feld, Spectroscopy Laboratory, MIT, Cambridge, Massachusetts.

p. 227   Courtesy Deutsches Museum, Munich.

p. 230   © National Geographic Society.

p. 233   The Bettmann Archive.

p. 235   Culver Pictures.

p. 236   © Russ Kinne/Photo Researchers.

p. 237   © Tom Edwards/Earth Scenes.

p. 238   © Lawrence Pringle/Photo Researchers.

Figure 10-15   © Richard Megna/Fundamental Photographs.

CHAPTER 11

Figure 11-1   © Richard Megna/Fundamental Photographs.

p. 255   Alinari/Art Resource, New York.

p. 258   The Bettmann Archive.

p. 259   Movie Still Archives.

CHAPTER 12

Figure 12-1   (b) Trustees of the Science Museum, London.

CHAPTER 13

p. 292   Courtesy Battelle Memorial Institute.

p. 293   Wide World Photos.

p. 295   © John Maher 1984/Stock, Boston.

p. 296   © Jeffry W. Myers/Stock, Boston.

Figure 13-2   F. London, *Superfluids,* Dover, New York, 1964. Reprinted by permission of the publisher.

p. 299   © Leonard Lessin/Peter Arnold.

p. 303   (left) © Jack Spratt 1981/The Image Works; (right) © Jerry Howard/Stock, Boston.

p. 308   Courtesy Albert A. Bartlett.

p. 313   © Dr. R. P. Clark & M. Goff/Science Photo Library, Photo Researchers.

Essay Figure 2   © Peter Menzel 1982/Stock, Boston.

Essay Figure 3   Courtesy Laurent Hodges.

CHAPTER 14

p. 325   The Granger Collection.

p. 330   Culver Pictures.

p. 331   The Granger Collection.

Essay Figure 2   Courtesy The Marley Company, Mission, Kansas.

CHAPTER 15

p. 357   N.A.S.A.

p. 363   © Mark Antman/The Image Works.

p. 367   © George Bellerose/Stock, Boston.

CHAPTER 16

Figure 16-5   © Berenice Abbott 1973/Photo Researchers.

p. 383   Courtesy Nell Ubbelohde and William Stryk.

Figure 16-16   (a) Courtesy Education Development Center, Inc., Newton, Massachusetts.

Figure 16-18   (b) Courtesy Education Development Center, Inc., Newton, Massachusetts.

Essay Figure 3   Dr. Harold E. Edgerton, MIT, Cambridge, Massachusetts.

Essay Figure 4   Courtesy Air France.

CHAPTER 17

Figure 17-2   (a) © Berenice Abbott 1973/Photo Researchers.

Figure 17-8   Courtesy Education Development Center, Inc., Newton, Massachusetts.

Figure 17-10   Courtesy Education Development Center, Inc., Newton, Massachusetts.

p. 409   © Eric Kroll/Taurus Photos.

p. 411   Wide World Photos.

p. 416   © Bohdan Hrynewych/Stock, Boston.

Figure 17-22   Redrawn from Charles A. Culver, *Musical Acoustics,* 4th ed., McGraw-Hill, New York, 1956, p. 103. Reprinted by permission of the publisher.

Figure 17-23   Redrawn from Charles A. Culver, *Musical Acoustics,* 4th ed., McGraw-Hill, New York, 1956, p. 104. Reprinted by permission of the publisher.

CHAPTER 18

p. 424   Courtesy Deutsches Museum, Munich.

p. 426   Courtesy Smithsonian Institution.

p. 428   The Granger Collection.

Figure 18-9   (b) Courtesy Harold M. Waage.

Figure 18-10   (b) Courtesy Harold M. Waage.

Figure 18-11   (b) Courtesy Harold M. Waage.

p. 444   © 1962 Fundamental Photographs.

p. 445   The Bettmann Archive.

Essay Figure 1   Courtesy Smithsonian Institution.

Essay Figure 2   Photograph by Charles Eames.

CHAPTER 19

p. 454   Photography of Kennan Ward.

p. 462   Courtesy Harold M. Waage.

Figure 19-16   (a) Courtesy Larry Langrill; (b) © Karen R. Preuss 1984/Taurus Photos.

p. 468   (top) Courtesy Harold M. Waage; (bottom) William Vandivert.

Figure 19-19   (b) Courtesy Harold M. Waage.

p. 471   Courtesy Larry Langrill.

CHAPTER 20

Figure 20-3   Adapted from E. M. Purcell, *Electricity and Magnetism,* Berkeley Physics Course, McGraw-Hill, New York, 1965, p. 123. Reprinted by permission of the publisher and the Education Development Center, Inc, Newton, Massachusetts.

Figure 20-6   D. Halliday and R. Resnick, *Fundamentals of Physics,* Wiley and Sons, New York, 1974, p. 512. Reprinted by permission of the publisher.

p. 494   The Bettmann Archive.

Figure 20-7   C. Kittel, *Introduction to Solid State Physics,* 3d ed., Wiley and Sons, New York, 1966. Reprinted by permission of the publisher.

p. 495   Courtesy Cryogenic Technology, Inc.

p. 502   Brown Brothers.

p. 503   © Tom McHugh 1972/Photo Researchers.

Figure 20-12 Courtesy Larry Langrill.

p. 511 The Bettmann Archive.

CHAPTER 21

p. 530 Fred Weiss.

Figure 21-7 Fred Weiss.

p. 539 Courtesy Larry Langrill.

Figure 21-14 (b) Courtesy Carl E. Nielsen.

p. 541 © Charles J. Ott/NAS, Photo Researchers.

p. 542 Cavendish Laboratory, University of Cambridge.

Figure 21-28 J. J. Thompson, *Philosophical Magazine,* ser. 5, vol. 44, 1897.

p. 544 Courtesy of University of Michigan, photo by Roger Lininger.

Figure 21-25 Courtesy Larry Langrill.

p. 549 Courtesy Education Development Center, Inc., Newton, Massachusetts.

p. 550 Brown Brothers.

Figure 21-27 Courtesy Avco Research Laboratory.

Figure 21-29 (b) Fred Weiss.

Figure 21-34 (b) William Vandivert.

CHAPTER 22

Figure 22-1 Courtesy Larry Langrill.

p. 564 (top) Courtesy Neils Bohr Library, American Institute of Physics; (bottom) The Bettmann Archive.

CHAPTER 23

p. 596 (top) © Daniel S. Brody/Stock, Boston; (bottom) © Yoav/Phototake.

CHAPTER 24

p. 607 Trustees of the Science Museum, London.

Figure 24-4 Courtesy Niels Bohr Library, American Institute of Physics.

p. 608 The Bettmann Archive.

Figure 24-5 Courtesy Michel Cagnet.

Figure 24-12 Adapted from Bernard Jaffe, *Michelson and the Speed of Light,* Doubleday, New York, 1960. Reprinted by permission of the publisher.

Figure 24-15 Courtesy Battelle-Northwest Photography.

p. 619 © Fundamental Photographs.

Figure 24-22 Courtesy Education Development Center, Inc., Newton, Massachusetts.

p. 623 © Vaughan Fleming/Science Photo Library, Photo Researchers.

p. 624 Courtesy Jay Bolemon.

p. 625 The Bettmann Archive.

p. 627 (top right) © Richard Megna 1985/Fundamental Photographs; (center left), (center right) Courtesy Larry Langrill.

Figure 24-37 © Fundamental Photographs.

p. 631 N.A.S.A.

p. 632 N.A.S.A.

p. 633 N.A.S.A.

CHAPTER 25

p. 639 © Harold Hoffman 1981/Photo Researchers.

p. 642 Steve Takatsuno.

p. 644 © Richard Megna 1982/Fundamental Photographs.

p. 647 © Fundamental Photographs 1968.

p. 650 © Fundamental Photographs.

Figure 25-21 (c) Courtesy Bausch & Lomb.

p. 653 (top) Courtesy Nils Abramson.

Figure 25-22 (c) Courtesy Bausch & Lomb.

p. 657 Courtesy Bausch & Lomb.

CHAPTER 26

p. 667 © David Scharf 1977/Peter Arnold.

p. 668 (center) Courtesy E. R. Lewis; (bottom) F. W. Sears and M. W. Zemansky, *University Physics,* Addison-Wesley, Reading, Massachusetts, 1970, p. 582. Reprinted with permission.

p. 676 © 1976 Michael Hayman/Photo Researchers.

p. 678 The Bettmann Archive.

p. 679 © Baron Wolman/Woodfin Camp & Associates.

p. 680 California Institute of Technology.

CHAPTER 27

p. 685 © Richard Megna 1985/Fundamental Photographs.

Figure 27-3 Courtesy Bausch & Lomb.

Figure 27-6 Courtesy T. A. Wiggins.

p. 691 Courtesy L. Velinsky.

p. 692 Elmer-Taylor/Courtesy Niels Bohr Library, American Institute of Physics.

Figure 27-9 (a) Courtesy Michel Cagnet.

p. 696 Courtesy Michel Cagnet.

Figure 27-12 Courtesy Michel Cagnet.

Figure 27-15 Courtesy Richard E. Haskell, Oakland University.

Figure 27-16 Courtesy Michel Cagnet.

p. 700 Courtesy Battelle-Northwest Photography.

Figure 27-17 (a) Courtesy Michel Cagnet.

Figure 27-18 Courtesy Michel Cagnet.

Figure 27-19 Courtesy Michel Cagnet.

Figure 27-21 Courtesy Michel Cagnet.

Figure 27-23 Courtesy Larry Langrill.

p. 704 The Bettmann Archive.

p. 705 © Joe Munroe 1975/Photo Researchers.

CHAPTER 28

Figure 28-4 R. S. Shankland, "The Michelson-Morley Experiment," *Scientific American,* November 1964. All rights reserved.

p. 728 Lotte Jacobi.

p. 729 Courtesy California Institute of Technology.

CHAPTER 29

Figure 29-3 Adapted from F. K. Richtmyer et al., *Introduction to Modern Physics,* 5th ed., McGraw-Hill, New York, 1955. Reprinted by permission of the publisher.

Figure 29-6 R. C. Millikan, *Physical Review,* vol. 7, p. 362, 1916.

p. 753 Courtesy Deutsches Museum, Munich.

Figure 29-7 (b) Courtesy General Electric Company.

Figure 29-13 From G. Herzberg, *Annalen de Physik,* vol. 84, p. 565, 1927.

p. 760 Courtesy Niels Bohr Library, American Institute of Physics, Margrethe Bohr Collection.

p. 763 © Biophoto Associates/Photo Researchers.

Essay Figure 1 Adapted from J. R. Cameron and J. G. Skofronick, *Medical Physics,* Wiley, New York, 1978. © 1978 John Wiley & Sons, Inc. Reprinted by permission of John Wiley & Sons, Inc.

p. 765 Courtesy New York Department of Health.

p. 766 © Martin M. Rotker/Taurus Photos.

CHAPTER 30

Figure 30-5 (a), (b) Courtesy Education Development Center, Inc., Newton, Massachusetts; (c) C. G. Shull.

Figure 30-6   Claus Jönsson.

Figure 30-7   *(a)* © Martin M. Rotker/ Taurus Photos.

p. 777   © Professor G. F. Leedale/Biophoto Associates, Photo Researchers.

Figure 30-9   *(a), (b), (c)* E. R. Huggins; *(d)* Claus Jönsson.

Figure 30-10   From G. F. Missiroli and G. Pozzi, *American Journal of Physics,* vol. 44, no. 3, 1976, p. 306.

CHAPTER 31

Figure 31-3   Courtesy Paul Doherty, Oakland University.

Figure 31-12   Courtesy J. A. Marquisee.

Figure 31-13   Courtesy T. Faulkner and T. Nestrick, Oakland University.

p. 814   UPI/Bettmann Newsphotos.

p. 815   *(top)* © Alexander Tsiaras/Photo Researchers; *(bottom)* © Chuck O'Rear/ Woodfin Camp & Associates.

p. 816   N.A.S.A.

p. 823   UPI/Bettmann Archive.

p. 824   © David Scharf/Peter Arnold.

p. 826   © Ellis Herwig/Stock, Boston.

CHAPTER 32

p. 834   Cavendish Laboratory, University of Cambridge.

Figure 32-1   R. Hofstedter, *Annual Review of Nuclear Science,* vol. 7, 1957, p. 231. © 1957 Annual Reviews Inc. Reprinted by permission.

Figure 32-4   R. Leighton, *Principles of Modern Physics,* McGraw-Hill, New York, 1959. Reprinted by permission of the publisher.

p. 845   © Georg Gerster/Rapho, Photo Researchers.

Figure 32-8   R. Evans, *The Atomic Nucleus,* McGraw-Hill, New York, 1955. Reprinted by permission of the publisher.

p. 853   Courtesy Princeton University Plasma Physics Laboratory.

Figure 32-15   C. Davisson and R. Evans, *Reviews of Modern Physics,* vol. 24, 1952, p. 79. Reprinted by permission.

CHAPTER 33

Figure 33-1   Courtesy Carl Anderson.

Figure 33-2   Courtesy Lawrence Berkeley Laboratory, University of California, Berkeley.

Figure 33-3   Courtesy Richard Ehrlich.

p. 875   Inside cover illustration from Franklyn M. Branley, Mark R. Chartrand III, and Helmut K. Wimmer, *Astronomy,* Thomas Y. Crowell, New York, 1975. © 1975 Helmut K. Wimmer. Reprinted by permission of Harper & Row, Publishers, Inc.

Index

Aberration
  astigmatism, 660
  chromatic, 661, 680
  coma, 660
  distortion, 660
  of lenses, 660–661
  spherical, 643, 660
Absolute motion, 387, 713, 717
Absolute pressure, and gauge pressure, 230
Absolute temperature scale, 256–259
Absorption spectrum, of HCL, 809–810
Absorption, resonance, 810–811
ac circuits, 583–600
ac generator, 575
Acceleration
  angular, 175
  average, 22
  centripetal, 50, 51, 100, 175
  constant, 24–29
    formulas for, 25, 27, 29
  of gravity, 25, 63, 363
  instantaneous, 23
  in one dimension, 22–24
Acceptor levels, 820
Accommodation, 668
Action and reaction forces, 66–67
Action at a distance, 70
Action potential, 497
Adhesive force, 237
Adiabatic process, 327
Affinity, electron, 803
Alpha decay, 840, 842–843
Alpha rays, 833
Alternating current
  circuits, 583–600
  in inductors and capacitors, 586–589
  in resistors, 584–585
  rms, 584
Ammeter, 518–519
Ampère, André Marie, 546
Ampere (A), unit of current, 3, 489, 548

Ampere's law, 550
Ampère's model of magnetism, 552
Amperian current, 552
Amplification, with triode, 599
Amplifier, 599
  transistor, 822
Amplitude, in simple harmonic motion, 354, 360
Anderson, Carl D., 747, 824
Angular acceleration, 175
Angular displacement, 174
Angular frequency, of simple harmonic motion, 358
Angular magnification, 673
Angular momentum, 182
  conservation of, 182
  and Kepler's second law, 203
  quantization of, 761
  spin, 794
  vector model, 792
Angular velocity, 175
Anisotropic material, 629
Annihilation, positron, 866
Antielectron, 833, 865
Antineutrino, 870
Antinode, 409
Antiparticle, and spin, 865–867
Antiproton, 866
Aperture, in camera, 674
Apparent depth, 650
Apparent size, 670
Archimedes' principle, 232
  and buoyancy, 232–236
Argon, 797
Aristotle, 16
ASA number, 675
Astigmatism, 660, 669
Aston, Francis William, 543
Atmosphere (atm), unit of pressure, 231
Atomic spectra, 801–803
Atoms, molecules, and solids, 789–828

Atwood's machine, 131
Aurora borealis, 541
Avogadro's number, 260, 837
Axon, 496

Back emf, 566
Bainbridge, Kenneth, 543
Balance, torsion, 207, 426
Balmer, Johann, 747, 758
Balmer formula, 758
Band
  conduction, 817
  valence, 818
Band theory of solids, 817–820
Bar magnet, 536, 537, 538
Bar, unit of pressure, 231
Bardeen, John, 495, 821, 823
Barn, unit of cross section, 846
Baryon, 867
Baryon number, 869
  conservation of, 869
  of quarks, 872
Base, of transistor, 821
Bassoon, 415
Battery, electric, 502, 504
  emf, 502
  internal resistance, 504
  terminal voltage, 504
BCS theory of superconductivity, 495
Beat frequency, 407
Beats, 407–408
Beauty (see Bottom quark)
Becquerel, Antoine Henri 747, 833
Becquerel (Bq), unit of radioactivity, 842
Bernoulli's equation, 241
Beryllium, 796
Beta decay, 840, 843–844
Beta rays, 833
Bevatron, 866

Big bang theory, 875
Binding energy
    of deuteron, 736
    of hydrogen atom, 737
    nuclear, 838
    of nuclei, 837–840, 839
Binoculars, 622, 680–681
Biot, Jean Baptiste, 546
Biot-Savart law, 546–547
Birefringence, 629
Black hole, 740
Blackbody, 311
Blackbody radiation, 311, 748–749
    Planck's theory, 749
    Rayleigh-Jeans law, 749
    spectral distribution, 748
    Wien's displacement law, 311, 748
Bohr, Niels, 747, 748, 760, 849, 864
Bohr model, 758, 791
    energy levels, 761–762
    first postulate, 759
    radius, 761
    second postulate, 760
    third postulate, 761
Bohr postulates, 759–761
Bohr radius, 761
Bohr's correspondence principle, 784
Boiling point, table 297
Boltzmann, Ludwig, 310
Boltzmann's constant, 260, 847
Bonding
    covalent, 803, 804–805
    hydrogen, 803, 806
    ionic, 803–804
    metallic, 803, 806
    molecular, 803–806
    nuclear, 839
    van der Waals, 803, 805
Boron, 796
Boson, 873
Bottom quark, 872
Bottomness, of quarks, 872
Boundary condition, 783
Boyle's law, 259
Brackett, F., 762
Brackett series, 762
Bragg, W. H., 755
Bragg, William L., 754, 755
Bragg condition, 775 (see also Bragg's law)
Bragg peak, 855
Bragg plane, 754
Bragg scattering, 754
Bragg's law, 754
Brahe, Tycho, 199
Brattain, Walter H., 821, 823
Breeder reactor, 852
Bremsstrahlung spectrum, 755
Brewster, Sir David, 628
Brewster's law, 628
British thermal unit (Btu), unit of energy, 138, 272
Bubble chamber, 866
Bulk modulus, 226, 379
    table, 226
Buoyancy, and Archimedes' principle, 232–236
Buoyant force, 232

Cadmium, 851
Callisto, moon of Jupiter, 632
Calorie (cal), unit of energy, 138, 272
Camera, 674–676
Capacitance, 469–474
    effective, 474, 475
    of parallel-plate capacitor, 470–471
Capacitive reactance, 588
Capacitors, 469
    alternating current in, 587
    in parallel, 474
    parallel-plate, 469, 471
    in series, 475
Capillarity, 238
Capillary action, 238
Carbon dating, 844
Carbon 14, 844
Carnot, Nicolas Leonard Sadi, 331
Carnot cycle, 333
Carnot engine, 331
Carnot theorem, 331
CAT scan, 766
Cavendish experiment, 207
Cavendish, Henry, 207, 461
Celsius temperature scale, 254
Center of gravity, 95 (see also Center of mass)
Center of mass, 154 (see also Center of gravity)
    motion of, 155
    reference frame, 158
Centigrade temperature scale, 254
Centimetre (cm), unit of length, 3
Centripetal acceleration, 50, 51, 100, 175
Centripetal force, 100
cgs system, 3
Chadwick, Sir James, 747, 834
Chamberlain, Owen, 866
Characteristic x-ray spectrum, 755, 802
Charge density
    on a conductor, 468
    of nuclei, 835
Charge, color, 874
Charge, electric, 424
    on a conductor, 468
    conservation of, 425
    motion of, 489–490
    quantization, 426
    of quarks, 872
    sharing, 464–465
Charging by induction, 463
Charm, 872
Charmed quark, 872
Chromatic aberration, 661, 680
Ciliary muscle, of eye, 668
Circle of least confusion, 680
Circuits
    alternating current, 583–600
    amplifier, 822
    direct current, 488–523
    LR, 573–575
    RC, 515–518
Circular motion, 100–104
    period of, 101
    and simple harmonic motion, 358–360
Circularly polarized light, 625
Clarinet, 416
Classical particle, 777

Classical wave, 777
Clausius, Rudolf Julius Emanuel, 330
Clausius statement of the second law of thermodynamics, 329
Clock synchronization, and simultaneity, 723–727
Cloud, electron, 784, 793, 801
Cockcroft, J. D., 845
Cockcroft-Walton accelerator, 834
Coefficient of absorption, 311
Coefficient of performance (COP), 329
    Carnot, 338
Coefficient of restitution, 160
Coefficient of surface tension, 236
Coefficient of thermal expansion, 293
    table, 293
Coefficient of viscosity, 244
    table, 245
Coherence, 404, 686–687
Coherent sources, 404
Cohesive force, 237
Collector, of transistor, 821
Collisions, 156–164
    elastic, 156, 159–160
    perfectly inelastic, 157, 158–159
    in three dimensions, 163–164
Color charge, 874
Coma, 660
Comets (essay), 214–216
Components of vectors, 39
Compound microscope, 677–678
Compressibility, 226
Compressive strength, table, 224
Compton, Arthur H., 747, 756
Compton scattering, 756–757, 856
Compton wavelength, 757
Computerized axial tomography (CAT), 766
Concave mirror, 642, 644
Condon, E., 747
Conduction
    electrical, 490–500
        in nerve cells (essay), 496–500
    thermal, 304–310
Conduction band, 817
Conductivity
    electrical, 492
    thermal, table, 305
Conductor, 817
Conductors, electrical, 460–464
Cones, of retina, 668
Conservation of angular momentum, 182
Conservation of baryon number, 869
Conservation of electric charge, 425, 510
Conservation of energy, 125–132
Conservation laws, 869
Conservation of lepton number, 869
Conservation of momentum, 150–153
Constant acceleration, 24–29
Constraint, 71
Constructive interference, 377, 402
Continuity equation, 239
Control rods, 851
Converging lens, 652
Conversion factor, 5
Convex mirror, 647
Cooper, Leon, 495
Copernicus, 199

Cornea, 668
Cornet, 416
Corona discharge, 467, 468
Correspondence principle, 784
Cosmology, 875
Coulomb, Charles Augustin de, 207, 426, 428
Coulomb (C), unit of charge, 428
Coulomb constant, 428
Coulomb's law, 427–431, 547
Covalent bond, 803, 804–805
Critical angle, for total internal reflection, 621
Critical density, of universe, 877
Critical point, 301
Critical temperature, 301
    table, 302
Cross section, 846
    for neutron capture, 847
    for photon interactions, 856
Crossed fields, 542
Crossed polarizers, 555, 627
Crystal spectrometer, 755
Curie (Ci), unit of radioactivity, 842
Curie temperature, 554
Current (*see also* Circuits)
    alternating, in inductors and capacitors, 586–589
    alternating, in resistors, 584–585
    direct, 488
    eddy, 569–570
    electric, 489
    heat, 304–305
    rms, 584
    transient, 590
Current balance, 549
Current element, 534
Cyclotron, 544–546, 845
Cyclotron frequency and period, 539
Cygnus, 740

Dalton, 864
Damped oscillations, 364–365
Davisson, Clinton J., 608, 747, 771, 772
dc circuits, 488–523
de Broglie, Louis Victor, 711, 772
de Broglie equations, 772–775
de Forest, Lee, 599
Decay chain, 842–843
Decay constant, 840
Decay rate, 840
Decibel (dB), unit of intensity level, 390
    table, 391
Decimal system, 3
Degree of freedom, 285
Delayed neutrons, 851
Dendrites, 496
Density, 221–223
    critical, of universe, 877
    definition, 222
    electric charge, of nuclei, 835
    energy, 389, 479
    variations in sound waves, 381
    weight density, 222
Depth of field, 675
Destructive interference, 377, 403
Deuteron, 736
Dew point, 303

Diamagnetism, 554
Dielectric, 472–474
Dielectric breakdown, 467, 468
Dielectric constant, 472
    table, 473
Dielectric strength, table, 473
Diffraction, 404–406, 639, 685–706
    of electrons, 775–777
    Fraunhofer and Fresnel, 699
    grating, 703–705
    and interference, 685–706
    of mechanical waves, 404–406
    by opaque disk, 608
    pattern
        of circular opening, 702
        electron, 776
        neutron, 776
        of rectangular opening, 701
        of single slit, 697
        of straight edge, 700
        x-ray, 776
    and resolution, 700–703
    due to single slit, 696–700
    of sound, 406
Diffuse reflection, 618
Diffusion, 820
Digital subtraction angiography (DSA), 766
Dimensions, 6
Dinosaurs, extinction of, 215–216
Diode, 597
    semiconductor, 820
    vacuum-tube, 597
Diopter (D), unit of lens power, 659
Dipole, electric, 442–444
Dipole moment, 443
    electric, 443
    magnetic, 535
    magnetic, of current loop, 536
Dipole radiation, 612
Dirac, Paul A. M., 608, 747
Dirac equation, 865
Disorder, and entropy, 340–343
Dispersion, 624
Displacement, 17
    angular, 174
    in one dimension, 17–20
Dissociation energy, 804
Distortion, of lens, 660
Diverging lens, 653
Domain, magnetic, 555
Donor levels, 819
Doping, 818
Doppler effect, 384–388
Dosage, 856–858
Dose limits, table, 858
Double refraction (*see* Birefringence)
Double-slit, interference pattern, 693–696
Down quark, 872
Drift velocity, 489
Driven oscillator, 364–367
Du Fay, Charles François, 425
Dulong-Petit law, 273, 286
Dyne (dyn), unit of force, 65

Eddy currents, 569–570
Edison, Thomas, 583

Efficiency
    Carnot, 334
    of heat engine, 328
    of power plants, table, 337
    second-law, 335
Einstein, Albert, 388, 711, 716, 728–729 (essay), 746, 747, 874
    postulates of special relativity, 717
    principle of equivalence, 208
Einstein's photoelectric equation, 751
Einstein's postulates, 717
Elastic collisions, 156, 159–160
Elastic scattering, of photons, 810–811
Electric charge, 424 (*see also* Charge)
Electric conduction, 490–500 (*see also* Conduction)
Electric current and circuits, 488–523 (*see also* Current; Circuits)
Electric dipole, 442–444
    antenna, 610
    moment, 443
    radiation, 612
Electric field, 431–435
    of an electric dipole, 438
    inside a conductor, 461–462
    in a light wave, 611
    outside a conductor, 463
    of a point charge, 433
    of a spherical shell, 439
    table, 432
Electric fields and forces, 423–449
Electric flux, 440
Electric generator, 575–577
Electric motor, 577
Electric potential, 453–460 (*see also* Potential)
Electric potential energy, 455
Electrical conduction, in nerve cells (essay), 496–500
Electrical conductors, 460–464
Electrical energy storage, 478–480
Electrolysis, 489
Electromagnetic interaction, 867
Electromagnetic waves, 388, 609–613
    of electric dipole, 438
    energy density, 612
    intensity, 612
    spectrum, 610
Electromotive force (emf), 502
Electron, 426
    charge, 428
    charge-to-mass ratio, 542
    diffraction, 775–777
    spin, 794
    wave function, 974
    wavelength, 774
    waves, and quantum theory, 771–782
Electron affinity, 803
Electron cloud, 784, 793, 801
Electron configuration, 795
    of the elements, table, 798–800
Electron diffraction, 775–777
Electron microscope, 667, 702, 776–777
Electron spin, 794
Electron wave function, 780–782
Electron wavelength, 774
Electron waves, and quantum theory, 771–786

Electronic charge (e), 428
Electronic excitation, of molecule, 806
Electronics (essay), 823–826
Electronvolt (eV), unit of energy, 459
Electrostatic energy, 478–479
Electrostatics, 453–483
Electrostatics, and xerography (essay), 480–481
Electroweak theory, 873
Elementary particles, 864–879
    rest masses, table, 735
Elements, transition, 797
Ellipse, 200
emf, 502
    back, 566
    induced, 562, 564
    motional, 567–568
    rms, 585
Emission spectrum, of nitrogen, 809
Emissivity, 310
Emitter, of transistor, 821
Endothermic reaction, 846
Energy (see also Energy levels)
    availability of, 324
    binding
        of deuteron, 736
        of hydrogen atom, 737
    in a capacitor, 478–479
    conservation of, 125–132
    in electric circuits, 500–505
    in an electric field, 479
    electric potential, 455
    equipartition of, 264
    gravitational potential, 124, 210, 213
    in an inductor, 574
    and intensity in waves, 389–392
    internal, 276
    kinetic, 115
    loss, of changed particles, 854
    mechanical, 126
    molecular, 264
    nuclear, 136
    potential, 123–124
        electrical, 455
        gravitational, 124, 210, 213
        of spring, 124
    relativistic, 733–737
    resources (essay), 136–137
    rest, 724
    table, 735
    rotational kinetic, 181
    in simple harmonic motion, 360
    solar, 137
    thermal, 138–139
    work-energy theorem, 116, 127
    zero-point, 779
Energy bands, 809, 817–820
Energy density, 389, 479, 611
    in electric fields, 479
    in electromagnetic waves, 612
    in magnetic fields, 575
    in a wave, 389
Energy levels
    in helium-neon laser, 814
    in hydrogen, 793
    molecular, 808
    for particle in a box, 783
    in ruby laser, 812

Energy loss, of charged particles, 854
Energy-level diagram, 761–762
Engine
    Carnot, 331
    internal combustion, 327
    reversible, 331
    steam, 325
English horn, 415
Entropy
    and disorder, 340–343
    and probability, 344–345
    of universe, 341
Equation of continuity, 239
Equation of state, 261
Equilibrium
    and balance, 98–100
    conditions for, 94
    of extended body, 92–98
    neutral, 99
    stability of, 98–100
    stable, 98
    thermal, 254
    unstable, 99
Equipartition theorem, 264, 285
Equipotential surface, 464–468
Equivalence principle, 737, 208
Escape speed, 212, 265
Eta particle, 868
Ether, 713
Europa, moon of Jupiter, 632
Exclusion principle, 795
Exothermic reaction, 846
Expanding universe, 875
Expansion, thermal, 292–296
Expansivity, 293
Exponent, 7
Exponential decay, 840, 841
Exponential decrease, 231, 517
Extraordinary ray, 629
Eye, 668–672
Eyepiece, 677, 679

Fahrenheit temperature scale, 254
Farad (F), unit of capacitance, 469
Faraday, Michael, 424, 562, 564
Faraday's law, 564
Farsightedness, 669
Femtometre (fm), unit of length, 759
Fermi (fm), unit of length, 759
Ferromagnetism, 554–555
Fiber optics, 623
Field
    electric, 431 (see also Electric field)
    gravitational, 63
    magnetic, 531
Field emission, 462
Field lines
    electric, 436–440
    of electric and magnetic dipoles, 553
    magnetic, 535
    magnetic, of bar magnet, 537
Field particle, 873
Field quanta, 873
Fine structure, 794
First law of thermodynamics, 275–279, 277
Fission, 847

and fusion, 848–854
    spontaneous, 847
Fizeau, Armand Hippolyte Louis, 614–615
Flavor, 871, 872
Fleming, Sir John, 597
Fluids, 227–246
    Archimedes' principle, 232
    Bernoulli's equation, 241
    bulk modulus, 226
    buoyancy, 232–236
    coefficient of viscosity, 244
        table, 245
    compressibility, 236
    continuity equation, 239
    Pascal's principle, 228
    Poiseuille's law, 245
    pressure, 227–231
    Reynolds number, 245
    surface tension and capillarity, 236–238
    Torricelli's law, 242
    turbulent flow, 239, 245
    Venturi effect, 242
    viscous flow, 244–246
    volume flow rate, 239
Flux
    electric, 440
    magnetic, 563
Flux density, magnetic, 532
f/number, 675
Focal length
    of lens, 652
    of mirror, 643
Focal point, 643
Foot (ft), unit of length, 4
Foot-pound (ft·lb), unit of work and energy, 114
Force
    action-at-a-distance, 70
    adhesive and cohesive, 237
    basic interactions, 198
    buoyant, 232
    centripetal, 100
    cohesive, 237
    conservative and nonconservative, 126
    constraint, 71
    contact, 70
    drag, 104–105
    electric, 427
    frictional, 88–92
    gravitational, 63, 202
    hadronic, 834
    magnetic, 531
        on a current element, 534
        on a moving charge, 531
    and mass, 60–63
    molecular, 803–806
    normal, 70
    nuclear, 834, 867
    of spring, 68
    of string, 68
    of support, 69
    tension, 70
    units, 64–66
    weak nuclear, 867
    weight, 63–64
Forced oscillator (see Driven oscillator)
Foucault, Jean, 608, 615
Fourier analysis, 416

Franck, J., 747
Franklin, Benjamin, 425, 445–447 (essay), 453
Fraunhofer D lines, 810
Fraunhofer diffraction, 699
Free expansion, 344
Free-body diagram, 70
Frequency
    angular, 358
    beat, 407
    cyclotron, 539
    de Broglie, for electrons, 772
    of mass on a spring, 355
    and period of harmonic waves, 381
    resonance, 366
    resonance, for waves, 410, 412
    in simple harmonic motion, 354
    threshold, for photoelectric effect, 751
Fresnel, Augustin, 607, 608
Fresnel diffraction, 699
Friction, 69, 88–92
    coefficient of, 88–89
    kinetic, 88
    static, 88
Friedrich, W., 753
Fringes, interference, 687
Fundamental frequency, of standing waves, 409
Fundamental mode, of standing waves, 409
Fundamental particles, table, 873
Fundamental unit of charge (e), 428
Fusion
    and fission, 848–854
    nuclear, 736
Fusion reactor, 853

Galilean relativity (see Newtonian relativity)
Galilei, Galileo, 15, 678, 713
Galvanometer, 519, 537
Gamma decay, 840, 845–846
Gamma rays, 609, 833
Gamow, George, 747, 876
Ganymede, moon of Jupiter, 632
Gas constant, 260
Gauge number, 493
Gauge pressure, 230
Gauss (G), unit of magnetic field, 533
Gauss' law, 440–442
Geiger, H. W., 759, 833
Gell-Mann, M., 870, 871
General relativity, 712, 737–740
Generator, and motor, 575–577
Germer, Lester H., 608, 771, 747, 772
Gerogi, Howard, 877
Gilbert, William, 424, 529
Glashow, Sheldon, 877
Gluon, 874
Goddard, Robert, 167
Goudsmit, Samuel, 794
Gram (g), unit of mass, 65
Grand unification theory (GUT), 874, 877
Grating, diffraction, 703–705
Gravitation (see Gravity)
Gravitational attraction, of light, 738
Gravitational field, 63
Gravitational interaction, 867

Gravitational mass, and inertial mass, 208–209
Gravitational potential energy, 124, 210, 213
Gravitational red shift, 740
Graviton, 873
Gravity, 198–218
    acceleration of, 25, 63, 363
    gravitational constant, 202
    Newton's law of, 202
    specific, 222
Gray, Stephen, 424
Gray (Gy), unit of radiation absorbed, 857
Grid, in triode, 599
Ground, electrical, 463
Ground state, 762
Gurney, R., 747
Gyroscope
    motion of, 188–190
    nutation of, 190
    pression of, 190

Hadronic force, 834
Hadronic interaction, 867
Hadrons, 867
    table, 868
Haga, H., 753
Hahn, O., 849
Half-life, 841
Harmonic (first, second, etc.), 409
Harmonic analysis, and synthesis, 416–418
Harmonic series, 411
Harmonic synthesis, 417
Hearing threshold, 392
Heat, 270–288
    capacity, 271–275
        and equipartition theorem, 282–287
        of gases, table, 284
        molar, 272
        table, 284
        of solids and liquids, table, 273
        and specific heat, 271–275
    conduction, 304
    convection, 304
    current, 304–305
    engine, 325–329
    and the first law of thermodynamics, 270
    latent, table, 297
    pump, 338–339
    radiation, 310
    transfer, 304–313
Heavy water, 851
Heisenberg, Werner, 608, 711, 747, 778
Heisenberg's uncertainty principle, 778
Helium, 795–796
Helium-neon laser, 813
Henry, Joseph, 562
Henry (H), unit of inductance, 570
Hertz, Heinrich, 608, 747, 750
Hertz (Hz), unit of frequency, 354
Hooke, Robert, 76, 607
Hooke's law, 68, 353
Horsepower (hp), unit of power, 135
Hubble, Edwin, 875
Hubble expansion, 33, 875
Hubble's law, 33, 875
Humidity, 302–303

Huygens, Christian, 405, 607
Huygens' construction, 406
Huygens' principle, 618, 620
Hydraulic lift, 229
Hydrogen bond, 803, 806
Hydrostatic paradox, 229

Ice point, 254
Ideal gas, 259–262
    Carnot cycle, 333
    heat capacity, 283–284
    internal energy, 283
    law, 261
Image
    formed by reflection, 640–647
    formed by refraction, 648–650
    perverted, 640
    real, 642
    types, 646
    virtual, 640
Impedance, 590
Impulse, 148
Incoherent sources, 404
Index of refraction, 617
    table, 624
Inductance, 570–573
    mutual, 572
    self-, 570
    of solenoid, 571
Induction, charging by, 463
Inductive reactance, 587
Inductor, 573
    alternating current in, 587
    energy in, 574
Inelastic scattering
    of neutrons, 855
    of photons, 810–811
Inertia
    law of, 60
    moment of, 178
        table 179
Inertial confinement, 853
Inertial mass, 208
Inertial reference frame, 713
Inflationary theory of universe, 878
Infrared absorption, 809
Infrared light, 609, 610
Insulator, 817
Integrated circuit (IC), 824
Intensity
    electromagnetic waves, 612
    of reflected light, 617
    single-slit pattern, 697
    of sound, table, 391
    two-slit pattern, 694
    of waves, 389
    of waves from two sources, 403, 404
Intensity level, 390
    table, 391
Interaction, of particles with matter, 854–858
Interactions, basic, 867
Interference, 377, 401–404, 685–706
    beats, 407–408
    destructive and constructive, 377
    and diffraction, 404–406, 685–706
    and diffraction of electrons, 775

fringes, 687
Lloyd's mirror, 695
Newton's rings, 77
pattern
  and diffraction, 700
  electron, 776, 781, 782
  of multiple slits, 696
  two-slit, 693–696
and standing waves, 409
in thin films, 687–690
of wave pulses, 377
of waves from two point sources,
  401–404
Interferometer, Michelson, 690–692, 715
Internal combustion engine, 327
Internal resistance, 504
International system of units (SI), 2
Io, moon of Jupiter, 632
Ionic bond, 803
Ionization, by radiation, 856
Ionization energy, 796
  versus atomic number, 797
Irreversible process, 325, 331–332
  and entropy, 341
Isothermal expansion, 281
Isotopes, 834
Isotropic material, 629

Jet propulsion, 165–167
Joule, James P., 271, 275–276
Joule (J), unit of work and energy, 114
Junction, semiconductor, 820–822
Jupiter, 632–633

K shell, 795
Kamerlingh Omnes, Heike, 485
Kaon, 868
Kelvin, 3
Kelvin-Planck statement of the second law
  of thermodynamics, 328
Kepler, Johannes, 199
Kepler's laws, 200
  second law, 203
  third law, 201, 204
Kilogram (kg), unit of mass, 2, 65
Kinematics, 16
Kinetic energy, 115
  rotational, 181
Kirchhoff, Gustav Robert, 511, 608
Kirchhoff's rules, 510–515
Kitt Peak Observatory, 679
Knipping, P., 753

L shell, 795
Lambda particle, 868
Lambda point, 298
Land, E. H., 626
Laser, 812–814, 815–816 (essay)
  fusion, 853
  helium-neon, 813
  ruby, 812
Latent heat
  and phase change, 296–299

of fusion, table, 297
of vaporization, table, 297
Lateral magnification, 646
  by refraction, 648–649
  of a lens, 656
Laue, M., 753
Laue pattern, 754
Law of interaction, 66
Law of reflection, 617
Lawrence, E. O., 544, 845
Lawson, J. D., 853
Lawson's criterion, for fusion, 853
LCR circuit, with generator, 590–594
Lenard, P., 747, 750
Length contraction, 720–721
Lens
  combinations, 657–660
  converging, 652
  diverging, 653
  double concave, 655
  double convex, 654
  equation, 652
  of eye, 668, 669
  focal length of, 652
  negative, 653
  positive, 652
  power of, 659
  telephoto, 676
  thin, 651–660
  wide-angle, 676
Lens-maker's equation, 652
Lenz's law, 565
Lepton, 865, 868
Lepton number, 869
  conservation of, 869
Lever, 134
Lever arm, 93
Leyden jar, 453
Light, 605–634
  diffraction, 696–706
  infrared, 609, 610
  interference, 687–696
  polarization, 625–630
  reflection, 617–619
  refraction, 619–624
  speed of, 2, 550
  ultraviolet, 609, 610
  visible, 609
Light amplification by stimulated emission
  of radiation (see Laser)
Light pipe, 622
Light-emitting diode (LED), 821
Light-in-flight-recording, 653
Light-year (c·y), unit of distance, 31
Limiting process, 20
Linear accelerator, 845
Linearly polarized light, 625
Lines of force, 436–440
  rules for drawing, 437
Line spectrum
  optical, 704
  x-ray, 755
Liquid drop, and nucleus, 835, 849
Liquid helium I and II, 298
Lithium, 796
Litre (L), unit of volume, 222
Livingston, M. S., 544, 845
Lloyd's mirror, 695

Longitudinal wave, 376
Lorentz-FitzGerald contraction, 720
Lorentz transformation, 718
LR circuit, 573–575
  time constant, 574
Luminous intensity, 3
Lyman, T., 762
Lyman series, 762

Machine, Atwood's, 131
Machine, simple, 132–134
  inclined plane, 132–133
  lever, 134
  mechanical advantage of, 132, 133
  wheel and axle, 134
Magnet
  bar, 536
  dipole moment, 535
  magnetic field of, 536
  pole strength, 535
  torque on, 535
Magnetic bottle, 541
Magnetic domain, 555
Magnetic energy density, 575
Magnetic field, 531
  of a circular loop, 547, 551
  of a current element, 546
  in a light wave, 611
  of a long straight wire, 547, 549
  of a solenoid, 551–552
  sources of, 546–550
Magnetic flux, 563
  density, 532
Magnetic force
  on a current element, 534
  on a moving charge, 531
Magnetic induction, 532
Magnetic moment, 535
  of current loop, 536
Magnetic pole, 529
  strength, 535
Magnetic quantum number, 792
Magnetic resonance imaging (MRI), 766
Magnetic susceptibility, 554
  table, 554
Magnetism in matter, 553–555
Magnification
  angular, 673
  lateral, 646
  of a lens, 656
  of a microscope, 678
  of a mirror, 646
  by refraction, 648–649
  of a telescope, 679
Magnifier, simple, 672
Magnifying power, 673
Malus, E. L., 627
Malus' law, of polarization, 626
Maricourt, Pierre de, 529
Marsden, Ernest, 759, 833
Maser, 812
Mass, 2
  atomic, table, 838
  and binding energy of nuclei, 837–840
  gravitational and inertial, 208–209
Mass number, 834

Mass spectrograph, 543
Maxwell, James C., 426, 531, 549, 608, 616, 711
Mean lifetime, 841
Mechanical advantage, 132
    ideal, 133
Mechanical energy, 126 (see also Energy)
Mechanical equivalence of heat, 276
Melting point, table, 297
Mercury, precession of perihelion, 739
Meson, 867
Metabolic rate, 138–139
Metallic bond, 803, 806
Metastable state, 812
Metre (m), unit of length, 2
Michell, John, 207, 530
Michelson interferometer, 690–692, 715
Michelson, Albert, 616, 691, 692, 713, 747
Michelson-Morely experiment, 714
Microscope
    compound, 677–678
    electron, 667, 702, 776–777
Microwave amplification by stimulated
            emission of radiation (see Maser)
Microwave background, 876
Millikan, Robert, 426, 711, 747, 751
Minimum ionizing particle, 854
Mirage, 623
Mirror
    concave, 642, 644
    convex, 647
    corner reflecting, 816
    equation, 643
    Lloyd's, 695
    parabolic, 661
    plane, 640–642
    ray diagram for, 645
    spherical, 642–647
Moderator, 850
Modulus
    bulk, 226, 379
    shear, 226
    Young's, 224
Moiré pattern, 408
Molar heat capacity, 272
Molar mass, 260
Mole (mol), unit of quantity of matter, 3
Molecular bonding, 803–906
Molecular interpretation of temperature,
            263–265
Molecular spectra, 806–810
Molecular speed, 265
Molecular weight (see Molar mass)
Molecule
    nonpolar, 442
    polar, 444
Moment of inertia, 178
Moment
    electric dipole, 443
    magnetic, 535
        of current loop, 535
Momentum, 148
    angular, 182
    conservation of, 150–153
    relativistic, 732–733
    of system, 154
Moog, Robert, 418
Moons, of Jupiter, 632

Morley, Edward, 713
Mosely, H., 747, 802
Motion
    absolute, 713, 717
    with constant acceleration, 24–29
    in one dimension, 15–29
    of point charge in magnetic field, 539
    ordered and disordered, 340
Motional emf, 567–568
Motor, 577
Mt. Palomar Observatory, 680
Muon, 720, 867, 868
Musical instruments, and standing waves,
            415
Mutual inductance, 572

n-type semiconductor, 819
Near point, 668
Nearsightedness, 669
Negative lens, 653
Neon, 796
Neptune, 633
Neurons, 496
Neutrino, 843, 868–869
Neutron, 834
    delayed, 851
    interaction with matter, 855
    reactions with, 847
    thermal, 847
Newton, Isaac, 58, 76–78 (essay), 606, 625,
            711, 713
Newton's law of gravity, 202
Newton's law of motion, 59
    applications to problem solving, 69–75
    first law, law of inertia, 60
    second law, 62
        for rotation, 182
    third law, law of interaction, 66
        and jet propulsion, 165
Newton's rings, 77, 688
Newton (N), unit of force, 60
Newtonian relativity, 712–714
    principle of, 713
Nishijima, K., 870
Node, 409
Nonpolar molecule, 442
Normal boiling point, table, 297
Normal force, 70
Normal melting point, table, 297
Northern lights (see Aurora borealis)
Notation, scientific, 6
Nuclear force, 834
Nuclear fusion, 736
Nuclear radius, 835
Nuclear reactions, 845–847
    cross section, 846
    endothermic, 846
    exothermic, 846
    fission, 848–853
    fusion, 853
    with neutrons, 847
    Q value, 846
    threshold, 846
Nuclei
    size and shape, 835
    properties of, 834–840

Nucleon, 834, 868
Nuclide, 834
Nutation, of gyroscope, 190

Object point, 641
Objective,
    of microscope, 677
    of telescope, 678
Oboe, 415
Ocular, 677
Oersted, Hans Christian, 530, 546
Ohm, George, 494
Ohm (Ω), unit of resistance, 491
Ohm's law, 490–495
Ohmic material, 491
Ohmmeter, 520
Omega particle, 868
Oort Cloud, 214
Optic nerve, 668
Optical instruments, 667–682
Optical pumping, 812
Optical spectrum, 801
Optically flat, 690
Optics
    geometric, 639–663
    physical, 685–706
Orbital quantum number, 792
Order of magnitude, 9
    of lengths, table, 9
    of masses, table, 9
    of time intervals, table, 9
Ordinary ray, 629
Organ pipe, 414–416
Oscillations, 352–368
    damped, 364
    driven, 365
    of masses on a spring, 353–357
    of pendulums, 362–364
Otto cycle, 327
Overtone, 411

p-type semiconductor, 820
Pair production, 856
Parabolic mirror, 661
Paradox, twin, 730–731
Parallel capacitors, 474
Parallel resistors, 507
Parallel-plate capacitor, 469, 471
Paramagnetism, 554
Paraxial rays, 643
Particle, 16
    in a box, 783–785
    classical, 777
    elementary, 864–879
    field, 873
Particles and waves, 606
Particle theory of light, 606
Particles, system of, 16
Pascal, Blaise, 228
Pascal (Pa), unit of pressure, 227
Pascal's principle, 228
Paschen series, 762
Pauli, Wolfgang, 747, 794
Pauli exclusion principle, 795, 836

Pendulum,
    ballistic, 160–161
    frequency and period of, 362, 363
    simple, 128, 362
Penzias, Arno, 876
Performance, coefficient of (COP), 329
Period
    of circular motion, 51
    cyclotron, 539
    and frequency of harmonic waves, 381
    of mass on a spring, 355
Periodic table of the elements, 790, 795–801
Perl, M., 869
Permeability of free space, 547, 549
Permittivity of free space, 471
Perverted image, 640
Pfund, H. A., 762
Pfund series, 762
Phase, 587, 588
Phase constant, 401
Phase diagram, 301
Phase difference, 401
    and coherence, 686–687
    due to path difference, 686
    between waves, 686
Phase transition, of universe, 877
Phase change and latent heat, 296–299
Phase relations in LCR circuit, 591
Photoelectric effect, 750–753, 856
    Einstein's equation, 751
    stopping potential, 750
    threshold frequency and wavelength, 751
    work function, 751
Photon, 609, 722, 750
    interactions, 810–814
    interactions with matter, 856
    virtual, 873
Physical optics, 685–706
Pi meson, 867
Pion, 868
Pipe, light, 622
Planck, Max, 312, 711, 748
Planck's constant, 749
Plane mirrors, 640–642
Plane waves, 382
Plasma, 541, 853
pn junction, 820
    fabrication of, 825
Poise, unit of viscosity, 244–245
Poiseuille's law, 245
Poisson, S., 608
Poisson spot, 608
Polar molecule, 444
Polarization, 625–630
    by absorption, 626–627
    by birefringence, 629
    circular, 625
    of light, 625–630
    linear, 625
    methods of, 626
    of microwaves, 627
    by reflection, 627–628
    by scattering, 628
Polaroid, 626
Pole, magnetic, 529
Population inversion, 812
Positive lens, 652
Positron, 833, 843, 865

Positron annihilation, 866
Postulates, Einstein's, 717
Potential,
    action, 497
    electric, 453–460, 456
        of a point charge, 458
    stopping, 750
Potential difference, 455
    and electric current, 491
Potential energy, 123–125
    electric, 455
    gravitational, 124, 210, 213
    of spring, 124
Pound (lb), unit of force, 4, 65
Power, 135
    in conductor, 501
    delivered by ac generator, 585
    in driven oscillator, 366
    in electric circuits, 501
    in LCR circuit, 592
    of lens, 659
    of lenses in contact, 660
    magnifying, 673
    radiated by a blackbody, 748
    rotational, 182
Power factor, 592
Power plants and thermal pollution (essay),
        336–337
Power reactor, 850
Precession
    of gyroscope, 190
    of perihelion of Mercury, 739
Prefixes, table, 3
Pressure
    in a fluid, 227–231
    gauge and absolute, 230
    units of, 227, 231
    vapor, 300
        of water, table, 301
    variations in sound waves, 381
Priestley, Joseph, 426
Primary, of transformer, 595
Principal quantum number, 792
Principle of equivalence, 208, 737
Principle of newtonian relativity, 713
Principle of superposition, 377
Principle of uncertainty, 778
Principle rays, for lenses, 656
Prism, 622
Probability
    distribution, for electron, 784, 793
    and entropy, 344–345
    interpretation of wave function, 782
Problem solving, 10
    general methods, 71
Proton, stability of, 874
Ptolemy's model of the universe, 199
PV diagram, 280
    for Carnot cycle, 333

Q factor
    for damped oscillations, 365, 366
    in LCR circuit, 592
    for radiation absorption, 857
Q value, of reaction, 846

Quality factor (QF), 857 (see also Q factor)
    in LCR circuit, 592
Quantization
    of angular momentum, 761
    of atomic energies, 761, 762, 773, 783
    of charge, 426
Quantum chromodynamics, 874
Quantum electrodynamics, 873
Quantum mechanics (see Quantum theory)
Quantum number, 783, 792
    magnetic, 792
    orbital, 792
    principal, 792
Quantum theory, 746, 771–786
    dates of important experiments and
        theories, table, 747
    of hydrogen atom, 791–795
    origins, 746–767
Quark model, 426, 871–873
Quark, 865, 868

R factor, 306
    table, 307
RC circuits, 515–518
Rad, unit of radiation absorbed, 856–857
Radiation
    and dose units, table, 857
    electric dipole, 612
    exposure, table, 858
    thermal, 310–313
Radioactivity, 840–845
    alpha decay, 842–843
    beta decay, 843–844
    decay rate, 840
    exponential time dependence, 840
    gamma decay, 845
    half-life, 841
Radius
    of nucleus, 835
    Schwarzschild, 740
Rainbow, 624
Raman scattering, 811
Range
    of charged particles in matter, 854
        versus energy, 855
    of projectile, 47
Ray approximation, 406, 639
Ray diagram
    for lenses, 656
    for mirrors, 645
Rayleigh criterion, for resolution, 702
Rayleigh scattering, 810–811
Rayleigh-Jeans law, 749
Rays, 382
    paraxial, 643
    principle, 645
Reactance
    capacitive, 588
    inductive, 587
Reaction
    endothermic, 846
    exothermic, 846
    with neutrons, 847
    threshold, 846
Reactor
    breeder, 136, 852

fusion, 136
fission and fusion, 848–854
Real image, 642
Rectification and amplification, 597–600
Rectifier
full-wave, 598
half-wave, 598
Red shift, gravitational, 740
Reference frame, 713
center-of-mass, 158
inertial, 713
Reflection
diffuse, 618
Huygens' principle, 618–619
law of, 617
of light, 616–619
of a particle, 606
polarization by, 627–628
specular, 618
total internal, 621
Refraction, 619–624
formation of images, 648–650
Huygens' principle, 620–621
index of, 617
table, 624
mirage, 623
of particle, 606
Snell's law, 620
Refrigerator, 329
coefficient of performance (COP), 329
Relative humidity, 303
Relativistic energy, 733–737
Relativistic momentum, 732–733
Relativity, 388, 711–742
galilean (see newtonian relativity)
general, 712, 737–740, 865
general theory of, 208
newtonian, 712–714
special theory, 712
Rem, unit of equivalent dose, 857
Reproduction constant, 850
Resistance
electric, 491, 492
internal, 504
shunt, 519
thermal, 306
Resistivity, electrical, table, 493
Resistors
electrical
in parallel, 507
in series, 505–506
shunt, 519
thermal
in parallel, 309
in series, 306
Resolution
and diffraction, 700–793
of diffraction grating, 705
Rayleigh criterion, 702
Resolving power, of diffraction grating, 705
Resonance
absorption, 810–811
condition
for string fixed at both ends, 410
for string fixed at one end, 412
of driven oscillator, 366
frequency, 366, 410, 412
in LCR circuit, 591–592

in neutron capture, 847
Rest energy, 724
of fundamental particles, table, 873
of neutrino, 869
table, 735
Rest mass, 733
of neutrino, 869
Restitution, coefficient of, 160
Retina, 668
Reversibility
conditions for, 331–332
and entropy, 341
of waves, 644
Reversible engine, 331
Reynolds number, 245
Ritz, Walter, 758
rms current, 584
rms emf, 585
Rocket motion, 165–167
equation for, 166
Rods
control, 851
of retina, 668
Roentgen, Wilhelm, 747, 753
Roentgen (R), unit of exposure, 856
Rolling bodies, 186–188
Rolling condition, 186
Root-mean-square current (see rms current)
Rotation, 173–192
of molecule, 806
Rotational motion
table of equations, 184
kinetic energy of, 181
Rubbia, C., 874
Ruby laser, 812
Rutherford, Ernest, 711, 747, 759, 833, 864
Rydberg, Johannes, 747, 758
Rydberg constant, 758, 761
Rydberg-Ritz formula, 758

Salam, Abdus, 877
Satellite motion, 50, 51
Saturn, 631
Savart, Felix, 546
Saxophone, 415
Scattering
Bragg, 754
Compton, 756–757
elastic, of photons, 810–811
inelastic, of photons, 810–811
of neutrons, 850–851
and polarization, 628
Raman, 811
Rayleigh, 810–811
Schrieffer, J. Robert, 495
Schroedinger, Erwin, 608, 711
Schroedinger's equation, 791, 865
Schwarzschild radius, 740
Scientific notation, 6
Second (s), unit of time, 2
Second law of thermodynamics, 324
Clausius statement, 329
in terms of entropy, 341
Kelvin-Planck statement, 328
Secondary, of transformer, 595

Second-law efficiency, 335
of heat pump, 339
Segre, Emil, 865
Seivert (Sv), unit of dose equivalent, 857
Selection rule, 808
Self-inductance, 570
Semiconductor, 817, 818
doped, 818
impurity, 818
intrinsic, 818
junctions, 820–822
n-type, 819
p-type, 820
Series capacitors, 475
Series resistors, 505–506
Shear modulus, 226
table, 226
Shear stress, 225
Shell, atomic, 795
Shock waves, 388, 393–394
Shockley, William, 821, 823
Shunt resistor, 519
Shutter, in camera, 674
SI (international system of units), 2
Sigma particles, 868
Sign convention
for lenses, 652
for mirrors, 646
for refraction, 648
Significant figures, 8
Silicon chip, 824
Silicon wafer, 824
Simple harmonic motion, 353–357
and circular motion, 358–360
conditions for, 354
displacement, velocity, and acceleration, 354
Simple machine (see Machine)
Simple magnifier, 672
Simultaneity, 724
and clock synchronization, 723–727
Single-slit diffraction, 696–700
Slingshot, planetary, 631
Slope of line, 19, 21
Small Magellanic Cloud, 740
Snell's law, of refraction, 620, 628
Sodium, 797
wavelengths, 704
Solar cell, 821
Solar energy, 314–317 (essay)
forms of, table, 314
Solenoid
inductance of, 570
magnetic field of, 551–552
Solids, band theory, 817–820
Solid-state electronics, 823
Sommerfeld, Arnold, 747
Sonic booms, 393–394 (essay)
Sound, 373–394
beats, 407–408
diffraction of, 406
doppler effect, 384–388
energy and intensity, 389–392
harmonic analysis and synthesis, 416–418
intensity level, 390
table, 391
shock wave, 393

speed of, 379
standing waves, 513–416
synthesis, 417–418
tone quality, 416
ultrasound, 383, 406
Special relativity, 712, 865
Specific gravity, 222
Specific heat, 271
Specific heat, of helium, graph, 198
Spectra
atomic, 801–803
molecular, 806–810
Spectral line, 704
Spectrograph
Bragg, 755
mass, 543
Spectrometer, optical, 704
Spectrum
absorption, of HCL, 809–810
beta decay, 843
bremsstrahlung, 755
characteristic x-ray, 755
emission, of nitrogen, 809
optical, 801
x-ray, 755
Specular reflection, 618
Speed, 16, 17–20 (see also Velocity)
escape, 212, 265
of film, 675
of harmonic waves 381
of light, 2, 550, 613–617, 616
of molecules, 265
of sound waves, 379
terminal, 105
of wave pulses, 374
of waves on a string, 378
Spherical aberration, 643
Spin
and antiparticles, 865–867
electron, 794
of gyroscope, 190
of quarks, 872
Spin-$\frac{1}{2}$ particles, 865
Spontaneous emission, 812, 845
Spontaneous fission, 847
Spring, 68
oscillation of mass on a, 353–357
Square wave, 417
Standing waves, 409–416
electron, 772, 783
fundamental, 409
harmonic series, 411
nodes and antinodes, 409
resonance condition, 410
sound, 413–416
on a string, 409–413, 773
Standing-wave pattern
around circumference of circle, 772
in a closed or open pipe, 415
for a string fixed at both ends, 409
for a string fixed at one end, 412
Static friction, 88
Steam engine, 325
Steam point, 254
Stefan, Josef, 310
Stefan-Boltzmann law, 310
Stimulated emission, 811
Stopping distance, 27

Stopping potential, 750
Straggling, 854
Strain, 224
shear, 226
Strange particles, 870
Strange quark, 872
Strangeness, 870
of hadrons, 871
Strassmann, Otto, 849
Stress, 222
compressive, 225
patterns, 630
shear, 225
and strain, 223–226
Strings, 69
Strong nuclear interaction (see Hadronic interaction)
Sublimation, 301
Super grand unification theory, 877
Superconductivity, 495
Superfluid, 298
Superposition, 377
Surface tension, 236
coefficient of, 236
Surface-barrier detector, 821
Susceptibility, magnetic, table, 554
Synchronization of clocks, and simultaneity, 723–727
Synthesis, and harmonic analysis, 416–418
Synthesizer, 418
System of particles, 16
Système International (SI), 2

Tangential acceleration, 52, 175
Tauon, 869
Telephoto lens, 676
Telescope, 678
radio, 705
reflecting, 680
refracting, 679
Temperature, 253–266
absolute, 256–259
Celsius, 254
centigrade, 254
critical, 300
table, 301
Curie, 554
Fahrenheit, 254
gradient, 304
Kelvin scale (see absolute temperature)
molecular interpretation of, 263–265
scales, 254–259
triple point, 301
Temperature coefficient of resistivity, table, 493
Temperature gradient, 305
Tensile strength, table, 224
Tension, 70
Terminal speed, 105
Terminal voltage, 504
Tesla, Nikola, 583
Tesla (T), unit of magnetic field, 532
Thales of Miletus, 424
Thermal conduction, 305
Thermal conductivity, table, 305
Thermal equilibrium, 254

Thermal expansion, 292–296
coefficient of, table, 293
Thermal neutron, 847
Thermal pollution, and power plants (essay), 336–337
Thermal radiation, 310–313
Thermal resistance, 306
R factors, table, 307
Thermionic emission, 597
Thermograph, 313
Thermometer, 254
gas, 257
Thin lenses, 651–660
equation, 652
Thomson, G. P., 747, 775
Thomson, J. J., 426, 542, 747, 759, 864
Thomson's plum pudding model, 759
Threshold
for hearing, 391, 392
for nuclear reactions, 846
pain, for hearing, 391
for photoelectric effect, 751
Time constant
in LR circuit, 574
in RC circuit, 516
Time dilation, 718–719
Titan, moon of Saturn, 633
Tokamak fusion reactor, 853
Tone quality, 416
Top quark, 872
Topness, of quarks, 872
Torque, 92–93, 177
on magnetic moment, 535
and moment of inertia, 177–180
Torr, unit of pressure, 231
Torricelli's law, 242
Torsion balance, 207, 426, 530
Total internal reflection, 621
Transformer, 595–597
step-down, 595
step-up, 595
Transient current, 590
Transistor, 599, 821
Transition elements, 797
Translation and rotation, 186–188
Transverse waves, 376
Triode, 599
Triple point, 301
of water, 258
Truth (see Top quark)
Turbulence, 239
and Reynolds number, 245
Twin paradox, 730–731
Two-slit diffraction-interference pattern, 693–696
electron, 776, 781, 782

Uhlenbeck, G., 794
Ultrasonic waves, 383, 406
Ultraviolet catastrophe, 749
Ultraviolet light, 609, 610
Uncertainty principle, 778
Unified mass unit (u), unit of mass, 837
Unified theory, 873, 874
Units, 2
conversion of, 4–5

cgs, 3
SI, 2, 3
U.S. customary system, 4
Universal gas constant, 260
Universal gravitational constant, 202
Up quark, 872
U.S. customary system of units, 4
Uranium, fission of, 849
Uranium enrichment, 851
Uranus, 633

Vacuum-tube diode, 597
Valence band, 818
Van Allen belts, 541
Van de Graaff generator, 467, 845
Van der Waals bond, 803, 805
Van der Waals equation, 300
Vapor pressure, 300
    of water, table, 301
Vector, 36
    acceleration, 43
    addition, 37
        by components, 38–41
        by parallelogram method, 38
    displacement, 37, 41
    subtraction, 39
    velocity, 41
Vector boson, 873
Vector model
    for angular momentum, 792
    for impedance, 591
Velocity, 21
    angular, 175
    average, 17
    of center of mass, 154
    drift, 489
    instantaneous, 20–22
    in one dimension, 17–20
    transformation, 721–722
Velocity selector, 542
Venturi effect, 242
Vibration, of molecule, 806
Virtual image, 640
Virtual object, 646, 658, 659
Virtual photon, 873
Viscosity
    coefficient of, 244
        table, 245

Volt (V), unit of electric potential, 455
Volta, Alessandro, 502
Voltage, terminal, 504
Voltmeter, 519
Voyager (essay), 631–634

W particle, 873
Walton, E. T. S., 845
Watt (W), unit of power, 135
Wave function, electron, 780–782, 803
Wave pulses, 374–378
    speed of, 374
Waves
    classical, 777
    diffraction, 404, 696–705
    dispersion, 624
    doppler effect, 384–388
    electromagnetic, 388, 609–612
    electron, 771–786
    energy, 389
    harmonic, 380–383
    harmonic analysis and synthesis,
        416–418
    intensity, 389, 613
    intensity level of sound waves, 390–392
        table, 391
    interference, 401–404, 687–696
    longitudinal, 373
    mechanical, 373–394
    and particles, 606–609
    plane, 382
    radio and television, 610
    shock, 388, 393–394
    sound, 373–394
    square, 417
    standing, 409–416
    transverse, 376
    ultrasonic, 383, 406
    water, 377
Waveform (for clarinet, cornet, and tuning
        fork), 416
Wavefront, 382
Wavelength, 381
    Compton, 757
    cutoff, for x-rays, 756
    de Broglie, for electrons, 772
    of electron, 774
    threshold, for photoelectric effect, 751

Weak interaction, 867
Weak nuclear force, 867
Weber (Wb), unit of magnetic flux, 563
Weight, 63
    apparent weight, 64, 75
    weightlessness, 64, 75
Weight density, 222
Weinberg, Steven, 877
Westinghouse, George, 583
Wheeler, 849
Wide-angle lens, 676
Wien's displacement law, 311, 748
Wilson, Robert, 747, 876
Wind, C. H., 753
Work
    by a constant force, 114
    and energy, 114–119
    internal, 118
    lost, 342
    and potential energy, 455
    and PV diagram for a gas, 279–282
    by a variable force, 119–123
Work function, 751
Work-energy theorem, 116, 127

Xerography, and electrostatics (essay),
        480–481
Xi particles, 868
X-ray spectrum, characteristic, 802
X-rays, 609, 753–756
    and medical diagnosis (essay), 763–766

Yard (yd), unit of length, 4
Yerkes Observatory, 679
Young, Thomas, 607, 693
Young's double-slit experiment, 693, 781
Young's modulus, 224
    table, 224
Yukawa, H., 867

Z particle, 873
Zeeman, P., 794
Zeeman effect, 794
Zero-point energy, 779, 783
Zweig, G., 871

$1 \text{ m} = 39.37 \text{ in} = 3.281 \text{ ft} = 1.094 \text{ yd}$

$1 \text{ m} = 10^{15} \text{ fm} = 10^{10} \text{ Å} = 10^{9} \text{ nm}$

$1 \text{ km} = 0.6215 \text{ mi}$

$1 \text{ mi} = 5280 \text{ ft} = 1.609 \text{ km}$

$1 \text{ in} = 2.540 \text{ cm}$

$1 \text{ L} = 10^{3} \text{ cm}^{3} = 10^{-3} \text{ m}^{3} = 1.057 \text{ qt}$

$1 \text{ y} = 365.24 \text{ d} = 3.156 \times 10^{7} \text{ s}$

$1 \text{ km/h} = 0.278 \text{ m/s} = 0.6215 \text{ mi/h}$

$1 \text{ ft/s} = 0.3048 \text{ m/s} = 0.6818 \text{ mi/h}$

$1 \text{ rev} = 2\pi \text{ rad} = 360°$

$1 \text{ rad} = 57.30°$

$1 \text{ rev/min} = 0.1047 \text{ rad/s}$

$1 \text{ slug} = 14.59 \text{ kg}$

$1 \text{ tonne} = 10^{3} \text{ kg} = 1 \text{ Mg}$

$1 \text{ atm} = 101.3 \text{ kPa} = 1.013 \text{ bar} = 76.00 \text{ cmHg} = 14.70 \text{ lb/in}^{2}$

$1 \text{ N} = 10^{5} \text{ dyn} = 0.2248 \text{ lb}$

$1 \text{ lb} = 4.448 \text{ N}$

$1 \text{ Pa} \cdot \text{s} = 10 \text{ poise}$

$1 \text{ J} = 10^{7} \text{ erg} = 0.7373 \text{ ft} \cdot \text{lb} = 9.869 \times 10^{-3} \text{ L} \cdot \text{atm}$

$1 \text{ kW} \cdot \text{h} = 3.6 \text{ MJ}$

$1 \text{ cal} = 4.184 \text{ J} = 4.129 \times 10^{-2} \text{ L} \cdot \text{atm}$

$1 \text{ L} \cdot \text{atm} = 101.3 \text{ J} = 24.22 \text{ cal}$

$1 \text{ eV} = 1.602 \times 10^{-19} \text{ J}$

$1 \text{ Btu} = 778 \text{ ft} \cdot \text{lb} = 252 \text{ cal} = 1054 \text{ J}$

$1 \text{ horsepower} = 550 \text{ ft} \cdot \text{lb/s} = 746 \text{ W}$

$1 \text{ W/m} \cdot \text{K} = 6.938 \text{ Btu} \cdot \text{in/h} \cdot \text{ft}^{2} \cdot \text{F}°$

$1 \text{ T} = 10^{4} \text{ G}$

$1 \text{ kg weighs about } 2.205 \text{ lb}$